C0-ASH-610

KIRK-OTHMER

ENCYCLOPEDIA OF CHEMICAL TECHNOLOGY

THIRD EDITION

VOLUME 21

SILVER AND SILVER ALLOYS
TO
SULFOLANES AND SULFONES

A WILEY-INTERSCIENCE PUBLICATION
John Wiley & Sons
NEW YORK • CHICHESTER • BRISBANE • TORONTO • SINGAPORE

Library of Congress Cataloging in Publication Data:

Main entry under title:
Encyclopedia of chemical technology.

At head of title: Kirk-Othmer.
"A Wiley-Interscience publication."
Includes bibliographies.
1. Chemistry, Technical—Dictionaries. I. Kirk, Raymond
Eller, 1890–1957. II. Othmer, Donald Frederick, 1904—
III. Grayson, Martin. IV. Eckroth, David. V. Title:
Kirk-Othmer encyclopedia of chemical technology.

TP9.E685 1978 660′.03 77-15820
ISBN 0-471-02074-5

Printed in the United States of America

CONTENTS

EDITORIAL STAFF FOR VOLUME 21

CONTRIBUTORS TO VOLUME 21

Thomas A. Augurt, *Propper Manufacturing Co., Inc., Long Island City, New York,* Sterilization techniques
Kim Badenhop, *B. W. Dyer & Co., New York, New York,* Sugar economics under Sugar
Robert S. Bailey, *Lilly Industrial Coatings, Inc., Indianapolis, Indiana,* Stains, industrial
S. M. Balaban, *Monsanto Co., St. Louis, Missouri,* Sorbic acid
Elmer B. Bell, *E. I. du Pont de Nemours & Co., Inc., Wilmington, Delaware,* Sulfamic acid and sulfamates
G. N. Bollenback, *The Sugar Foundation, Washington, D. C.,* Special sugars under Sugar
D. E. Brownlee, *University of Washington, Seattle, Washington,* Space chemistry
Thomas F. Canning, *Kerr-McGee Chemical Corporation, Trona, California,* Sodium sulfates under Sodium compounds
C. E. Capes, *National Research Council of Canada, Ottawa, Canada,* Size enlargement
Frank G. Carpenter, *United States Department of Agriculture, New Orleans, Louisiana,* Sugar analysis; Cane sugar both under Sugar
C. M. Cooper, *Michigan State University, East Lansing, Michigan,* Solvent recovery
James R. Daniel, *Purdue University, West Lafayette, Indiana,* Starch

V. G. DiFate, *Monsanto Co., St. Louis, Missouri,* Sorbic acid
Frederick Disque, *Alpha Metals, Inc., Jersey City, New Jersey,* Solder and brazing alloys
Charles Drum, *PPG Industries, Inc., New Martinsville, West Virginia,* Sodium sulfides under Sodium compounds
James Fair, *University of Texas, Austin, Texas,* Sprays
Russell E. Farris, *Sandoz Colors & Chemicals, Inc., Martin, South Carolina,* Stilbene dyes
B. P. Faulkner, *Allis Chalmers Corporation, Milwaukee, Wisconsin,* Size reduction
C. Hagopian, *The Badger Co., Inc., Cambridge, Massachusetts,* Styrene
Dan Halacy, *Solar Energy Research Institute, Golden, Colorado,* Solar energy
John F. Heiss, *Diamond Crystal Salt Co., St. Clair, Michigan,* Sodium chloride under Sodium compounds, sodium halides
C. S. Helling, *Agricultural Research Service, USDA, Beltsville, Maryland,* Soil chemistry of pesticides
C. S. Hickey, *Monsanto Co., St. Louis, Missouri,* Sorbic acid
Otakar Jonas, *Westinghouse Electric Co., Philadelphia, Pennsylvania,* Steam
P. C. Kearney, *Agricultural Research Service, USDA, Beltsville, Maryland,* Soil chemistry of pesticides
C. S. Keller, *Monsanto Co., St. Louis, Missouri,* Sorbic acid
Robert J. King, *United States Steel Corporation, Monroeville, Pennsylvania,* Steel
P. Koch, *The Badger Co., Inc., Cambridge, Massachusetts,* Styrene
John Kraljic, *Allied Corporation, Sovay, New Jersey,* Sodium nitrite under Sodium compounds
Eugene J. Kuhajek, *Morton Salt, Woodstock, Illinois,* Sodium chloride under Sodium compounds, sodium halides
Martin Laborde, *CODELCO, Santiago, Chile,* Sodium nitrate under Sodium compounds
Charles H. Lemke, *University of Delaware, Newark, Delaware,* Sodium and sodium alloys
George R. Lenz, *Searle Laboratories, Chicago, Illinois,* Steroids
P. J. Lewis, *The Badger Co., Inc., Cambridge, Massachusetts,* Styrene
Merle Lindstrom, *Phillips Research Center, Bartlesville, Oklahoma,* Sulfolanes and sulfones
Haines B. Lockhart, Jr., *Eastman Kodak Company, Rochester, New York,* Silver compounds
Samuel Maya, *Beecham Products, Parsippany, New Jersey,* Sodium nitrate under Sodium compounds
R. A. McGinnis, *Consultant, San Rafael, California,* Beet sugar under Sugar
P. H. Merrell, *Mallinckrodt, Inc., St. Louis, Missouri,* Sodium iodide under Sodium compounds, sodium halides
Clyde Orr, *Micrometrics Instrument Corp., Norcross, Georgia,* Size measurement of particles
Frederick S. Osmer, *Lever Brothers Co., Inc., Edgewater, New Jersey,* Soap
K. J. Parker, *Tate & Lyle, Ltd., Reading, Berks, United Kingdom,* Sugar derivatives under Sugar
Clyde F. Parrish, *Indiana State University, Terre Haute, Indiana,* Solvents, industrial
E. M. Peters, *Mallinckrodt, Inc., St. Louis, Missouri,* Sodium iodide under Sodium compounds, sodium halides
A. E. Platt, *The Dow Chemical Co., Midland, Michigan,* Styrene plastics

J. R. Plimmer, *Agricultural Research Service, USDA, Beltsville, Maryland,* Soil chemistry of pesticides

H. W. Rimmer, *Allis Chalmers Corporation, Oak Creek, Wisconsin,* Size reduction

Robert E. Sequeira, *Amstar Corporation, Woodland, California,* Properties of sucrose under Sugar

George Sistare, *Consultant, Fairfield, Connecticut,* Silver and silver alloys; Solders and brazing alloys

V. A. Stenger, *Dow Chemical U.S.A., Midland, Michigan,* Sodium bromide under Sodium compounds, sodium halides

T. C. Wallace, *The Dow Chemical Co., Midland, Michigan,* Styrene plastics

William J. Welstead, Jr., *A. H. Robins Company, Inc., Richmond, Virginia,* Stimulants

Leonard A. Wenzel, *Lehigh University, Bethlehem, Pennsylvania,* Simultaneous heat and mass transfer

Roy L. Whistler, *Purdue University, West Lafayette, Indiana,* Starch

Ralph Williams, *Phillips Research Center, Bartlesville, Oklahoma,* Sulfolanes and sulfones

Theodore J. Williams, *Purdue University, West Lafayette, Indiana,* Simulation and process design

Leon O. Windstrom, *Consultant, East Aurora, New York,* Succinic acid and succinic anhydride

Walter J. Wolf, *U.S. Department of Agriculture, Peoria, Illinois,* Soybeans and other oilseed proteins

Andrew F. Zeller, *FMC Corporation, Philadelphia, Pennsylvania,* Strontium and strontium compounds

NOTE ON CHEMICAL ABSTRACTS SERVICE REGISTRY NUMBERS AND NOMENCLATURE

Chemical Abstracts Service (CAS) Registry Numbers are unique numerical identifiers assigned to substances recorded in the CAS Registry System. They appear in brackets in the *Chemical Abstracts* (CA) substance and formula indexes following the names of compounds. A single compound may have many synonyms in the chemical literature. A simple compound like phenethylamine can be named β-phenylethylamine or, as in *Chemical Abstracts*, benzeneethanamine. The usefulness of the *Encyclopedia* depends on accessibility through the most common correct name of a substance. Because of this diversity in nomenclature careful attention has been given the problem in order to assist the reader as much as possible, especially in locating the systematic CA index name by means of the Registry Number. For this purpose, the reader may refer to the CAS Registry Handbook-Number Section which lists in numerical order the Registry Number with the *Chemical Abstracts* index name and the molecular formula; eg, **458-88-8,** Piperidine, 2-propyl-, (*S*)-, $C_8H_{17}N$; in the *Encyclopedia* this compound would be found under its common name, coniine [*458-88-8*]. The Registry Number is a valuable link for the reader in retrieving additional published information on substances and also as a point of access for such on-line data bases as Chemline, Medline, and Toxline.

In all cases, the CAS Registry Numbers have been given for title compounds in articles and for all compounds in the index. All specific substances indexed in *Chemical Abstracts* since 1965 are included in the CAS Registry System as are a large number of substances derived from a variety of reference works. The CAS Registry System identifies a substance on the basis of an unambiguous computer-language description of its molecular structure including stereochemical detail. The Registry Number is a machine-checkable number (like a Social Security number) assigned in sequential order to each substance as it enters the registry system. The value of the number lies in the fact that it is a concise and unique means of substance identification, which is independent of, and therefore bridges, many systems of chemical nomenclature. For polymers, one Registry Number is used for the entire family; eg, polyoxyethylene (20) sorbitan monolaurate has the same number as all of its polyoxyethylene homologues.

Registry numbers for each substance will be provided in the third edition cumulative index and appear as well in the annual indexes (eg, Alkaloids shows the Registry Number of all alkaloids (title compounds) in a table in the article as well, but the intermediates have their Registry Numbers shown only in the index). Articles such as Analytical methods, Batteries and electric cells, Chemurgy, Distillation, Economic evaluation, and Fluid mechanics have no Registry Numbers in the text.

Cross-references are inserted in the index for many common names and for some systematic names. Trademark names appear in the index. Names that are incorrect, misleading or ambiguous are avoided. Formulas are given very frequently in the text to help in identifying compounds. The spelling and form used, even for industrial names, follow American chemical usage, but not always the usage of *Chemical Abstracts* (eg, *coniine* is used instead of *(S)-2-propylpiperidine*, *aniline* instead of *benzenamine*, and *acrylic acid* instead of *2-propenoic acid*).

There are variations in representation of rings in different disciplines. The dye industry does not designate aromaticity or double bonds in rings. All double bonds and aromaticity are shown in the *Encyclopedia* as a matter of course. For example, tetralin has an aromatic ring and a saturated ring and its structure appears in the

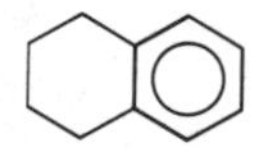

Encyclopedia with its common name, Registry Number enclosed in brackets, and parenthetical CA index name, ie, tetralin, [*119-64-2*] (1,2,3,4-tetrahydronaphthalene). With names and structural formulas, and especially with CAS Registry Numbers the aim is to help the reader have a concise means of substance identification.

CONVERSION FACTORS, ABBREVIATIONS, AND UNIT SYMBOLS

SI Units (Adopted 1960)

A new system of measurement, the International System of Units (abbreviated SI), is being implemented throughout the world. This system is a modernized version of the MKSA (meter, kilogram, second, ampere) system, and its details are published and controlled by an international treaty organization (The International Bureau of Weights and Measures) (1).

SI units are divided into three classes:

BASE UNITS

length	meter[†] (m)
mass[‡]	kilogram (kg)
time	second (s)
electric current	ampere (A)
thermodynamic temperature[§]	kelvin (K)
amount of substance	mole (mol)
luminous intensity	candela (cd)

[†] The spellings "metre" and "litre" are preferred by ASTM; however "-er" are used in the Encyclopedia.

[‡] "Weight" is the commonly used term for "mass."

[§] Wide use is made of "Celsius temperature" (t) defined by

$$t = T - T_0$$

where T is the thermodynamic temperature, expressed in kelvins, and $T_0 = 273.15$ K by definition. A temperature interval may be expressed in degrees Celsius as well as in kelvins.

SUPPLEMENTARY UNITS

plane angle	radian (rad)
solid angle	steradian (sr)

DERIVED UNITS AND OTHER ACCEPTABLE UNITS

These units are formed by combining base units, supplementary units, and other derived units (2–4). Those derived units having special names and symbols are marked with an asterisk in the list below:

Quantity	*Unit*	*Symbol*	*Acceptable equivalent*
*absorbed dose	gray	Gy	J/kg
acceleration	meter per second squared	m/s^2	
*activity (of ionizing radiation source)	becquerel	Bq	1/s
area	square kilometer	km^2	
	square hectometer	hm^2	ha (hectare)
	square meter	m^2	
*capacitance	farad	F	C/V
concentration (of amount of substance)	mole per cubic meter	mol/m^3	
*conductance	siemens	S	A/V
current density	ampere per square meter	A/m^2	
density, mass density	kilogram per cubic meter	kg/m^3	g/L; mg/cm^3
dipole moment (quantity)	coulomb meter	C·m	
*electric charge, quantity of electricity	coulomb	C	A·s
electric charge density	coulomb per cubic meter	C/m^3	
electric field strength	volt per meter	V/m	
electric flux density	coulomb per square meter	C/m^2	
*electric potential, potential difference, electromotive force	volt	V	W/A
*electric resistance	ohm	Ω	V/A
*energy, work, quantity of heat	megajoule	MJ	
	kilojoule	kJ	
	joule	J	N·m
	electron volt†	eV†	
	kilowatt-hour†	kW·h†	

† This non-SI unit is recognized by the CIPM as having to be retained because of practical importance or use in specialized fields (1).

Quantity	*Unit*	*Symbol*	*Acceptable equivalent*
energy density	joule per cubic meter	J/m^3	
*force	kilonewton	kN	
	newton	N	$kg{\cdot}m/s^2$
*frequency	megahertz	MHz	
	hertz	Hz	1/s
heat capacity, entropy	joule per kelvin	J/K	
heat capacity (specific), specific entropy	joule per kilogram kelvin	J/(kg·K)	
heat transfer coefficient	watt per square meter kelvin	$W/(m^2{\cdot}K)$	
*illuminance	lux	lx	lm/m^2
*inductance	henry	H	Wb/A
linear density	kilogram per meter	kg/m	
luminance	candela per square meter	cd/m^2	
*luminous flux	lumen	lm	cd·sr
magnetic field strength	ampere per meter	A/m	
*magnetic flux	weber	Wb	V·s
*magnetic flux density	tesla	T	Wb/m^2
molar energy	joule per mole	J/mol	
molar entropy, molar heat capacity	joule per mole kelvin	J/(mol·K)	
moment of force, torque	newton meter	N·m	
momentum	kilogram meter per second	kg·m/s	
permeability	henry per meter	H/m	
permittivity	farad per meter	F/m	
*power, heat flow rate, radiant flux	kilowatt	kW	
	watt	W	J/s
power density, heat flux density, irradiance	watt per square meter	W/m^2	
*pressure, stress	megapascal	MPa	
	kilopascal	kPa	
	pascal	Pa	N/m^2
sound level	decibel	dB	
specific energy	joule per kilogram	J/kg	
specific volume	cubic meter per kilogram	m^3/kg	
surface tension	newton per meter	N/m	
thermal conductivity	watt per meter kelvin	W/(m·K)	
velocity	meter per second	m/s	
	kilometer per hour	km/h	
viscosity, dynamic	pascal second	Pa·s	
	millipascal second	mPa·s	
viscosity, kinematic	square meter per second	m^2/s	

Quantity	*Unit*	*Symbol*	*Acceptable equivalent*
	square millimeter per second	mm^2/s	
volume	cubic meter	m^3	
	cubic decimeter	dm^3	L(liter) (5)
	cubic centimeter	cm^3	mL
wave number	1 per meter	m^{-1}	
	1 per centimeter	cm^{-1}	

In addition, there are 16 prefixes used to indicate order of magnitude, as follows:

Multiplication factor	*Prefix*	*Symbol*	*Note*
10^{18}	exa	E	
10^{15}	peta	P	
10^{12}	tera	T	
10^{9}	giga	G	
10^{6}	mega	M	
10^{3}	kilo	k	
10^{2}	hecto	h[a]	[a] Although hecto, deka, deci, and centi are SI prefixes, their use should be avoided except for SI unit-multiples for area and volume and nontechnical use of centimeter, as for body and clothing measurement.
10	deka	da[a]	
10^{-1}	deci	d[a]	
10^{-2}	centi	c[a]	
10^{-3}	milli	m	
10^{-6}	micro	μ	
10^{-9}	nano	n	
10^{-12}	pico	p	
10^{-15}	femto	f	
10^{-18}	atto	a	

For a complete description of SI and its use the reader is referred to ASTM E 380 (4) and the article Units and Conversion Factors which will appear in a later volume of the *Encyclopedia.*

A representative list of conversion factors from non-SI to SI units is presented herewith. Factors are given to four significant figures. Exact relationships are followed by a dagger. A more complete list is given in ASTM E 380-79(4) and ANSI Z210.1-1976 (6).

Conversion Factors to SI Units

To convert from	*To*	*Multiply by*
acre	square meter (m^2)	4.047×10^3
angstrom	meter (m)	1.0×10^{-10}†
are	square meter (m^2)	1.0×10^{2}†
astronomical unit	meter (m)	1.496×10^{11}
atmosphere	pascal (Pa)	1.013×10^{5}
bar	pascal (Pa)	1.0×10^{5}†
barn	square meter (m^2)	1.0×10^{-28}†

† Exact.

To convert from	*To*	*Multiply by*
barrel (42 U.S. liquid gallons)	cubic meter (m^3)	0.1590
Bohr magneton (μ_β)	J/T	9.274×10^{-24}
Btu (International Table)	joule (J)	1.055×10^{3}
Btu (mean)	joule (J)	1.056×10^{3}
Btu (thermochemical)	joule (J)	1.054×10^{3}
bushel	cubic meter (m^3)	3.524×10^{-2}
calorie (International Table)	joule (J)	4.187
calorie (mean)	joule (J)	4.190
calorie (thermochemical)	joule (J)	4.184†
centipoise	pascal second (Pa·s)	1.0×10^{-3}†
centistoke	square millimeter per second (mm^2/s)	1.0†
cfm (cubic foot per minute)	cubic meter per second (m^3/s)	4.72×10^{-4}
cubic inch	cubic meter (m^3)	1.639×10^{-5}
cubic foot	cubic meter (m^3)	2.832×10^{-2}
cubic yard	cubic meter (m^3)	0.7646
curie	becquerel (Bq)	3.70×10^{10}†
debye	coulomb·meter (C·m)	3.336×10^{-30}
degree (angle)	radian (rad)	1.745×10^{-2}
denier (international)	kilogram per meter (kg/m)	1.111×10^{-7}
	tex‡	0.1111
dram (apothecaries')	kilogram (kg)	3.888×10^{-3}
dram (avoirdupois)	kilogram (kg)	1.772×10^{-3}
dram (U.S. fluid)	cubic meter (m^3)	3.697×10^{-6}
dyne	newton (N)	1.0×10^{-5}†
dyne/cm	newton per meter (N/m)	1.0×10^{-3}†
electron volt	joule (J)	1.602×10^{-19}
erg	joule (J)	1.0×10^{-7}†
fathom	meter (m)	1.829
fluid ounce (U.S.)	cubic meter (m^3)	2.957×10^{-5}
foot	meter (m)	0.3048†
footcandle	lux (lx)	10.76
furlong	meter (m)	2.012×10^{-2}
gal	meter per second squared (m/s^2)	1.0×10^{-2}†
gallon (U.S. dry)	cubic meter (m^3)	4.405×10^{-3}
gallon (U.S. liquid)	cubic meter (m^3)	3.785×10^{-3}
gallon per minute (gpm)	cubic meter per second (m^3/s)	6.308×10^{-5}
	cubic meter per hour (m^3/h)	0.2271
gauss	tesla (T)	1.0×10^{-4}
gilbert	ampere (A)	0.7958
gill (U.S.)	cubic meter (m^3)	1.183×10^{-4}
grad	radian	1.571×10^{-2}
grain	kilogram (kg)	6.480×10^{-5}
gram force per denier	newton per tex (N/tex)	8.826×10^{-2}

† Exact.

‡ See footnote on p. xiv.

To convert from	*To*	*Multiply by*
hectare	square meter (m^2)	1.0×10^{4}†
horsepower (550 ft·lbf/s)	watt (W)	7.457×10^{2}
horsepower (boiler)	watt (W)	9.810×10^{3}
horsepower (electric)	watt (W)	7.46×10^{2}†
hundredweight (long)	kilogram (kg)	50.80
hundredweight (short)	kilogram (kg)	45.36
inch	meter (m)	2.54×10^{-2}†
inch of mercury (32°F)	pascal (Pa)	3.386×10^{3}
inch of water (39.2°F)	pascal (Pa)	2.491×10^{2}
kilogram force	newton (N)	9.807
kilowatt hour	megajoule (MJ)	3.6†
kip	newton (N)	4.48×10^{3}
knot (international)	meter per second (m/s)	0.5144
lambert	candela per square meter (cd/m^2)	3.183×10^{3}
league (British nautical)	meter (m)	5.559×10^{3}
league (statute)	meter (m)	4.828×10^{3}
light year	meter (m)	9.461×10^{15}
liter (for fluids only)	cubic meter (m^3)	1.0×10^{-3}†
maxwell	weber (Wb)	1.0×10^{-8}†
micron	meter (m)	1.0×10^{-6}†
mil	meter (m)	2.54×10^{-5}†
mile (statute)	meter (m)	1.609×10^{3}
mile (U.S. nautical)	meter (m)	1.852×10^{3}†
mile per hour	meter per second (m/s)	0.4470
millibar	pascal (Pa)	1.0×10^{2}
millimeter of mercury (0°C)	pascal (Pa)	1.333×10^{2}†
minute (angular)	radian	2.909×10^{-4}
myriagram	kilogram (kg)	10
myriameter	kilometer (km)	10
oersted	ampere per meter (A/m)	79.58
ounce (avoirdupois)	kilogram (kg)	2.835×10^{-2}
ounce (troy)	kilogram (kg)	3.110×10^{-2}
ounce (U.S. fluid)	cubic meter (m^3)	2.957×10^{-5}
ounce-force	newton (N)	0.2780
peck (U.S.)	cubic meter (m^3)	8.810×10^{-3}
pennyweight	kilogram (kg)	1.555×10^{-3}
pint (U.S. dry)	cubic meter (m^3)	5.506×10^{-4}
pint (U.S. liquid)	cubic meter (m^3)	4.732×10^{-4}
poise (absolute viscosity)	pascal second (Pa·s)	0.10†
pound (avoirdupois)	kilogram (kg)	0.4536
pound (troy)	kilogram (kg)	0.3732
poundal	newton (N)	0.1383
pound-force	newton (N)	4.448
pound per square inch (psi)	pascal (Pa)	6.895×10^{3}
quart (U.S. dry)	cubic meter (m^3)	1.101×10^{-3}

† Exact.

To convert from	*To*	*Multiply by*
quart (U.S. liquid)	cubic meter (m^3)	9.464×10^{-4}
quintal	kilogram (kg)	$1.0 \times 10^{2\dagger}$
rad	gray (Gy)	$1.0 \times 10^{-2\dagger}$
rod	meter (m)	5.029
roentgen	coulomb per kilogram (C/kg)	2.58×10^{-4}
second (angle)	radian (rad)	4.848×10^{-6}
section	square meter (m^2)	2.590×10^{6}
slug	kilogram (kg)	14.59
spherical candle power	lumen (lm)	12.57
square inch	square meter (m^2)	6.452×10^{-4}
square foot	square meter (m^2)	9.290×10^{-2}
square mile	square meter (m^2)	2.590×10^{6}
square yard	square meter (m^2)	0.8361
stere	cubic meter (m^3)	$1.0^{\dagger}$
stokes (kinematic viscosity)	square meter per second (m^2/s)	$1.0 \times 10^{-4\dagger}$
tex	kilogram per meter (kg/m)	$1.0 \times 10^{-6\dagger}$
ton (long, 2240 pounds)	kilogram (kg)	1.016×10^{3}
ton (metric)	kilogram (kg)	$1.0 \times 10^{3\dagger}$
ton (short, 2000 pounds)	kilogram (kg)	9.072×10^{2}
torr	pascal (Pa)	1.333×10^{2}
unit pole	weber (Wb)	1.257×10^{-7}
yard	meter (m)	$0.9144^{\dagger}$

Abbreviations and Unit Symbols

Following is a list of commonly used abbreviations and unit symbols appropriate for use in the *Encyclopedia.* In general they agree with those listed in *American National Standard Abbreviations for Use on Drawings and in Text (ANSI Y1.1)* (6) and *American National Standard Letter Symbols for Units in Science and Technology (ANSI Y10)* (6). Also included is a list of acronyms for a number of private and government organizations as well as common industrial solvents, polymers, and other chemicals.

Rules for Writing Unit Symbols (4):

1. Unit symbols should be printed in upright letters (roman) regardless of the type style used in the surrounding text.

2. Unit symbols are unaltered in the plural.

3. Unit symbols are not followed by a period except when used as the end of a sentence.

4. Letter unit symbols are generally written in lower-case (eg, cd for candela) unless the unit name has been derived from a proper name, in which case the first letter of the symbol is capitalized (W,Pa). Prefix and unit symbols retain their prescribed form regardless of the surrounding typography.

5. In the complete expression for a quantity, a space should be left between the numerical value and the unit symbol. For example, write 2.37 lm, *not* 2.37lm, and 35 mm, *not* 35mm. When the quantity is used in an adjectival sense, a hyphen is often used, for example, 35-mm film. *Exception:* No space is left between the numerical value and the symbols for degree, minute, and second of plane angle, and degree Celsius.

6. No space is used between the prefix and unit symbols (eg, kg).

7. Symbols, not abbreviations, should be used for units. For example, use "A," not "amp," for ampere.

8. When multiplying unit symbols, use a raised dot:

N·m for newton meter

In the case of W·h, the dot may be omitted, thus:

Wh

An exception to this practice is made for computer printouts, automatic typewriter work, etc, where the raised dot is not possible, and a dot on the line may be used.

9. When dividing unit symbols use one of the following forms:

$$\mathrm{m/s} \ \mathit{or} \ \mathrm{m{\cdot}s^{-1}} \ \mathit{or} \ \frac{\mathrm{m}}{\mathrm{s}}$$

In no case should more than one slash be used in the same expression unless parentheses are inserted to avoid ambiguity. For example, write:

$$\mathrm{J/(mol{\cdot}K)} \ \mathit{or} \ \mathrm{J{\cdot}mol^{-1} \cdot K^{-1}} \ \mathit{or} \ \mathrm{(J/mol)/K}$$

but *not*

J/mol/K

10. Do not mix symbols and unit names in the same expression. Write:

joules per kilogram *or* J/kg *or* $\mathrm{J{\cdot}kg^{-1}}$

but *not*

joules/kilogram *nor* joules/kg *nor* $\mathrm{joules{\cdot}kg^{-1}}$

ABBREVIATIONS AND UNITS

A	ampere
A	anion (eg, H*A*); mass number
a	atto (prefix for 10^{-18})
AATCC	American Association of Textile Chemists and Colorists
ABS	acrylonitrile–butadiene–styrene
abs	absolute
ac	alternating current, *n.*
a-c	alternating current, *adj.*
ac-	alicyclic
acac	acetylacetonate
ACGIH	American Conference of Governmental Industrial Hygienists
ACS	American Chemical Society
AGA	American Gas Association
Ah	ampere hour
AIChE	American Institute of Chemical Engineers
AIME	American Institute of Mining, Metallurgical, and Petroleum Engineers
AIP	American Institute of Physics
AISI	American Iron and Steel Institute
alc	alcohol(ic)
Alk	alkyl
alk	alkaline (not alkali)
amt	amount
amu	atomic mass unit
ANSI	American National Standards Institute
AO	atomic orbital
AOAC	Association of Official Analytical Chemists
AOCS	American Oil Chemist's Society
APHA	American Public Health Association

API American Petroleum Institute
aq aqueous
Ar aryl
ar- aromatic
as- asymmetric(al)
ASHRAE American Society of Heating, Refrigerating, and Air Conditioning Engineers
ASM American Society for Metals
ASME American Society of Mechanical Engineers
ASTM American Society for Testing and Materials
at no. atomic number
at wt atomic weight
av(g) average
AWS American Welding Society
b bonding orbital
bbl barrel
bcc body-centered cubic
BCT body-centered tetragonal
Bé Baumé
BET Brunauer-Emmett-Teller (adsorption equation)
bid twice daily
Boc *t*-butyloxycarbonyl
BOD biochemical (biological) oxygen demand
bp boiling point
Bq becquerel
C coulomb
°C degree Celsius
C- denoting attachment to carbon
c centi (prefix for 10^{-2})
c critical
ca circa (approximately)
cd candela; current density; circular dichroism
CFR Code of Federal Regulations
cgs centimeter–gram–second
CI Color Index
cis- isomer in which substituted groups are on same side of double bond between C atoms
cl carload
cm centimeter
CMA See MCA
cmil circular mil
cmpd compound
CNS central nervous system
CoA coenzyme A
COC Cleveland open cup
COD chemical oxygen demand
coml commercial(ly)
cp chemically pure
cph close-packed hexagonal
CPSC Consumer Product Safety Commission
cryst crystalline
cub cubic
D Debye
D- denoting configurational relationship
d differential operator
d- dextro-, dextrorotatory
da deka (prefix for 10^{1})
dB decibel
dc direct current, *n.*
d-c direct current, *adj.*
dec decompose
detd determined
detn determination
Di didymium, a mixture of all lanthanons
dia diameter
dil dilute
DIN Deutsche Industrie Normen
dl-; DL- racemic
DMA dimethylacetamide
DMF dimethylformamide
DMG dimethyl glyoxime
DMSO dimethyl sulfoxide
DOD Department of Defense
DOE Department of Energy
DOT Department of Transportation
DP degree of polymerization
dp dew point
DPH diamond pyramid hardness
dstl(d) distill(ed)
dta differential thermal analysis
(*E*)- entgegen; opposed
ϵ dielectric constant (unitless number)
e electron

ECU	electrochemical unit
ed.	edited, edition, editor
ED	effective dose
EDTA	ethylenediaminetetraacetic acid
emf	electromotive force
emu	electromagnetic unit
en	ethylene diamine
eng	engineering
EPA	Environmental Protection Agency
epr	electron paramagnetic resonance
eq.	equation
esp	especially
esr	electron-spin resonance
est(d)	estimate(d)
estn	estimation
esu	electrostatic unit
exp	experiment, experimental
ext(d)	extract(ed)
F	farad (capacitance)
F	faraday (96,487 C)
f	femto (prefix for 10^{-15})
FAO	Food and Agriculture Organization (United Nations)
fcc	face-centered cubic
FDA	Food and Drug Administration
FEA	Federal Energy Administration
fob	free on board
fp	freezing point
FPC	Federal Power Commission
FRB	Federal Reserve Board
frz	freezing
G	giga (prefix for 10^9)
G	gravitational constant = 6.67×10^{11} N·m^2/kg^2
g	gram
(g)	gas, only as in $H_2O(g)$
g	gravitational acceleration
gem-	geminal
glc	gas-liquid chromatography
g-mol wt; gmw	gram-molecular weight
GNP	gross national product
gpc	gel-permeation chromatography
GRAS	Generally Recognized as Safe
grd	ground
Gy	gray
H	henry
h	hour; hecto (prefix for 10^2)
ha	hectare
HB	Brinell hardness number
Hb	hemoglobin
hcp	hexagonal close-packed
hex	hexagonal
HK	Knoop hardness number
HRC	Rockwell hardness (C scale)
HV	Vickers hardness number
hyd	hydrated, hydrous
hyg	hygroscopic
Hz	hertz
^{i}(eg, Pr^i)	iso (eg, isopropyl)
i-	inactive (eg, *i*-methionine)
IACS	International Annealed Copper Standard
ibp	initial boiling point
IC	inhibitory concentration
ICC	Interstate Commerce Commission
ICT	International Critical Table
ID	inside diameter; infective dose
ip	intraperitoneal
IPS	iron pipe size
IPTS	International Practical Temperature Scale (NBS)
ir	infrared
IRLG	Interagency Regulatory Liaison Group
ISO	International Organization for Standardization
IU	International Unit
IUPAC	International Union of Pure and Applied Chemistry
IV	iodine value
iv	intravenous
J	joule
K	kelvin
k	kilo (prefix for 10^3)
kg	kilogram
L	denoting configurational relationship
L	liter (for fluids only)(5)
l-	*levo*-, levorotatory
(l)	liquid, only as in $NH_3(l)$

LC_{50}	conc lethal to 50% of the animals tested
LCAO	linear combination of atomic orbitals
LCD	liquid crystal display
lcl	less than carload lots
LD_{50}	dose lethal to 50% of the animals tested
LED	light-emitting diode
liq	liquid
lm	lumen
ln	logarithm (natural)
LNG	liquefied natural gas
log	logarithm (common)
LPG	liquefied petroleum gas
ltl	less than truckload lots
lx	lux
M	mega (prefix for 10^6); metal (as in *MA*)
M	molar; actual mass
$\overline{M}_w$	weight-average mol wt
$\overline{M}_n$	number-average mol wt
m	meter; milli (prefix for 10^{-3})
m	molal
m-	meta
max	maximum
MCA	Chemical Manufacturers' Association (was Manufacturing Chemists Association)
MEK	methyl ethyl ketone
meq	milliequivalent
mfd	manufactured
mfg	manufacturing
mfr	manufacturer
MIBC	methyl isobutyl carbinol
MIBK	methyl isobutyl ketone
MIC	minimum inhibiting concentration
min	minute; minimum
mL	milliliter
MLD	minimum lethal dose
MO	molecular orbital
mo	month
mol	mole
mol wt	molecular weight
mp	melting point
MR	molar refraction
ms	mass spectrum
mxt	mixture
μ	micro (prefix for 10^{-6})
N	newton (force)
N	normal (concentration); neutron number
N-	denoting attachment to nitrogen
n (as n_D^{20})	index of refraction (for 20°C and sodium light)
^{n} (as Bu^n), *n*-	normal (straight-chain structure)
n	neutron
n	nano (prefix for 10^9)
na	not available
NAS	National Academy of Sciences
NASA	National Aeronautics and Space Administration
nat	natural
NBS	National Bureau of Standards
neg	negative
NF	*National Formulary*
NIH	National Institutes of Health
NIOSH	National Institute of Occupational Safety and Health
nmr	nuclear magnetic resonance
NND	New and Nonofficial Drugs (AMA)
no.	number
NOI-(BN)	not otherwise indexed (by name)
NOS	not otherwise specified
nqr	nuclear quadruple resonance
NRC	Nuclear Regulatory Commission; National Research Council
NRI	New Ring Index
NSF	National Science Foundation
NTA	nitrilotriacetic acid
NTP	normal temperature and pressure (25°C and 101.3 kPa or 1 atm)
NTSB	National Transportation Safety Board
O-	denoting attachment to oxygen

o-	ortho
OD	outside diameter
OPEC	Organization of Petroleum Exporting Countries
o-phen	*o*-phenanthridine
OSHA	Occupational Safety and Health Administration
owf	on weight of fiber
Ω	ohm
P	peta (prefix for 10^{15})
p	pico (prefix for 10^{-12})
p-	para
p	proton
p.	page
Pa	pascal (pressure)
pd	potential difference
pH	negative logarithm of the effective hydrogen ion concentration
phr	parts per hundred of resin (rubber)
p-i-n	positive-intrinsic-negative
pmr	proton magnetic resonance
p-n	positive-negative
po	per os (oral)
POP	polyoxypropylene
pos	positive
pp.	pages
ppb	parts per billion (10^9)
ppm	parts per million (10^6)
ppmv	parts per million by volume
ppmwt	parts per million by weight
PPO	poly(phenyl oxide)
ppt(d)	precipitate(d)
pptn	precipitation
Pr (no.)	foreign prototype (number)
pt	point; part
PVC	poly(vinyl chloride)
pwd	powder
py	pyridine
qv	quod vide (which see)
R	univalent hydrocarbon radical
(*R*)-	rectus (clockwise configuration)
r	precision of data
rad	radian; radius
rds	rate determining step
ref.	reference
rf	radio frequency, *n.*
r-f	radio frequency, *adj.*
rh	relative humidity
RI	Ring Index
rms	root-mean square
rpm	rotations per minute
rps	revolutions per second
RT	room temperature
^{s} (eg, Bu^s); *sec*-	secondary (eg, secondary butyl)
S	siemens
(*S*)-	sinister (counterclockwise configuration)
S-	denoting attachment to sulfur
s-	symmetric(al)
s	second
(s)	solid, only as in $H_2O(s)$
SAE	Society of Automotive Engineers
SAN	styrene–acrylonitrile
sat(d)	saturate(d)
satn	saturation
SBS	styrene–butadiene–styrene
sc	subcutaneous
SCF	self-consistent field; standard cubic feet
Sch	Schultz number
SFs	Saybolt Furol seconds
SI	Le Système International d'Unités (International System of Units)
sl sol	slightly soluble
sol	soluble
soln	solution
soly	solubility
sp	specific; species
sp gr	specific gravity
sr	steradian
std	standard
STP	standard temperature and pressure (0°C and 101.3 kPa)
sub	sublime(s)
SUs	Saybolt Universal seconds
Sv	silvert (dose equivalent, J/kg)
syn	synthetic

t (eg, Bu^t), t-, *tert*-	tertiary (eg, tertiary butyl)
T	tera (prefix for 10^{12}); tesla (magnetic flux density)
t	metric ton (tonne); temperature
TAPPI	Technical Association of the Pulp and Paper Industry
TCC	Tagliabue closed cup
tex	tex (linear density)
T_g	glass-transition temperature
tga	thermogravimetric analysis
THF	tetrahydrofuran
tlc	thin layer chromatography
TLV	threshold limit value
trans-	isomer in which substituted groups are on opposite sides of double bond between C atoms
TSCA	Toxic Substance Control Act
TWA	time-weighted average
Twad	Twaddell
UL	Underwriters' Laboratory
USDA	United States Department of Agriculture
USP	*United States Pharmacopeia*
uv	ultraviolet
V	volt (emf)
var	variable
vic-	vicinal
vol	volume (not volatile)
vs	versus
v sol	very soluble
W	watt
Wb	Weber
Wh	watt hour
WHO	World Health Organization (United Nations)
wk	week
yr	year
(*Z*)-	zusammen; together; atomic number

Non-SI (Unacceptable and Obsolete) Units		*Use*
Å	angstrom	nm
at	atmosphere, technical	Pa
atm	atmosphere, standard	Pa
b	barn	cm^2
bar†	bar	Pa
bbl	barrel	m^3
bhp	brake horsepower	·W
Btu	British thermal unit	J
bu	bushel	m^3; L
cal	calorie	J
cfm	cubic foot per minute	m^3/s
Ci	curie	Bq
cSt	centistokes	mm^2/s
c/s	cycle per second	Hz
cu	cubic	exponential form
D	debye	C·m
den	denier	tex
dr	dram	kg
dyn	dyne	N
dyn/cm	dyne per centimeter	mN/m
erg	erg	J
eu	entropy unit	J/K
°F	degree Fahrenheit	°C; K
fc	footcandle	lx
fl	footlambert	lx
fl oz	fluid ounce	m^3; L
ft	foot	m
ft·lbf	foot pound-force	J

† Do not use bar (10^5Pa) or millibar (10^2Pa) because they are not SI units, and are accepted internationally only for a limited time in special fields because of existing usage.

Non-SI (Unacceptable and Obsolete) Units		*Use*
gf den	gram-force per denier	N/tex
G	gauss	T
Gal	gal	m/s^2
gal	gallon	m^3; L
Gb	gilbert	A
gpm	gallon per minute	(m^3/s); (m^3/h)
gr	grain	kg
hp	horsepower	W
ihp	indicated horsepower	W
in.	inch	m
in. Hg	inch of mercury	Pa
in. H_2O	inch of water	Pa
in.-lbf	inch pound-force	J
kcal	kilogram-calorie	J
kgf	kilogram-force	N
kilo	for kilogram	kg
L	lambert	lx
lb	pound	kg
lbf	pound-force	N
mho	mho	S
mi	mile	m
MM	million	M
mm Hg	millimeter of mercury	Pa
$m\mu$	millimicron	nm
mph	miles per hour	km/h
μ	micron	μm
Oe	oersted	A/m
oz	ounce	kg
ozf	ounce-force	N
η	poise	Pa·s
P	poise	Pa·s
ph	phot	lx
psi	pounds-force per square inch	Pa
psia	pounds-force per square inch absolute	Pa
psig	pounds-force per square inch gauge	Pa
qt	quart	m^3; L
°R	degree Rankine	K
rd	rad	Gy
sb	stilb	lx
SCF	standard cubic foot	m^3
sq	square	exponential form
thm	therm	J
yd	yard	m

BIBLIOGRAPHY

1. The International Bureau of Weights and Measures, BIPM (Parc de Saint-Cloud, France) is described on page 22 of Ref. 4. This bureau operates under the exclusive supervision of the International Committee of Weights and Measures (CIPM).
2. *Metric Editorial Guide* (*ANMC-78-1*) 4th ed., American National Metric Council, 1625 Massachusetts Ave. N.W., Washington, D.C. 20036, 1978.
3. *SI Units and Recommendations for the Use of Their Multiples and of Certain Other Units* (*ISO 1000-1981*), American National Standards Institute, 1430 Broadway, New York, N. Y. 10018, 1981.
4. Based on *ASTM E 380-82* (*Standard for Metric Practice*), American Society for Testing and Materials, 1916 Race Street, Philadelphia, Pa. 19103, 1982.
5. *Fed. Regist.*, Dec. 10, 1976 (41 FR 36414).
6. For ANSI address, see Ref. 3.

R. P. LUKENS
American Society for Testing and Materials

S *continued*

SILVER AND SILVER ALLOYS

Silver [*7440-22-4*] is the whitest element and has the highest electrical and thermal conductivity of all the metals. Silver, along with copper and gold its colorful nearest neighbors in the periodic table, have been used throughout recorded history. Silver was once used primarily for decorative articles, jewelry, tableware, and coinage. Recently coinage has become a relatively insignificant use and the aesthetic uses for the metal have been surpassed by industrial and technical uses. From 1976–1979 for example, the ratio of aesthetic to industrial and technical uses was about 1:3. The latter include photographic materials, electrical products, brazing alloys, catalysts, etc.

Properties

The properties of silver are summarized in Table 1. Its optical reflectance and neutron-absorption characteristics are illustrated in Figures 1 and 2, respectively. The primary valence is 1, although divalent silver oxide and other higher valent compounds exist (see Silver compounds). Two stable isotopes, ie, ^{107}Ag and ^{109}Ag, exist with an abundance ratio of 1.075:1. Twenty-five radioactive isotopes exist with atomic weights of 102–117 and half-lives of seconds–253 d (see Radioisotopes).

The chemical nobility of silver is implied by its position in the electrochemical series. It is much more noble than copper, somewhat less noble than palladium and platinum, and much less noble than gold. This indicates corrosion resistance, but it tends to inhibit the formation of protective films, such as those that protect aluminum and stainless steel.

Pure silver, much like gold, can be rolled into foil, drawn into fine wire and beaten to leaf. Extensive working can be done between annealing and, with high purity silver,

Table 1. Properties of Silver

Property	Value	Reference
melting point, °C	961.9[a]	1
recrystallization temperature, °C[b]	20–200	1
structure, at 25°C, nm	a = 0.408621	1
atomic radius, nm	0.144	1
ionic radius, nm	0.126	2
boiling point, °C	2163	1
density, g/cm^3		
at 20°C, hard drawn,	10.43	3
annealed	10.49	1
at 960.5–1300°C (liquid)	9.30–9.00	1
thermal conductivity, W/(m·k)[c]		
at 20°C	428	
at 450°C	356	1
mean specific heat, J/(kg·K)[d]		
solid:		
at 25°C	235	
at 127°C	239	
at 527°C	282	
at 961°C	297	1
liquid:		
at 961–2227°C	310	1
electrical resistivity, R		
at 0°C, μΩ·cm	1.59	1
temperature coefficient (0–100°C)	0.0041	1
$R(t)$:R(0°C) at −272°C	0.0068	
at −253°C	ca 0.1	
at 78°C	0.684	1
at 100°C	1.41	
at 200°C	1.83	
at 400°C	2.71	
at 800°C	4.62	
at 960.5°C	5.14	1
elastic properties		
elastic modulus, GPa[e]	71.0	1
Poisson's ratio	0.39 (hard drawn)	1
	0.37 (annealed)	1
latent heat of fusion, J/g[d]	104.2	1
latent heat of vaporization, kJ/g[d]	2.636	1
molal heat capacity (0–800°C), J/(g·°C)	$Cp = 0.2318 + 0.06031 \times 10^{-4} t - 0.67848 \times 10^{-8} t^2$	3
vapor pressure (T = K), Pa[e]		
solid: log p	$-14020/T + 11.012$	1
liquid: log p	$-13350/T + 10.486$	1
temp[f], K	*Pa*	
1304	1.013×10^2	
1510	1.013×10^3	
1783	1.013×10^4	
2163	1.013×10^5	1
coefficient of thermal expansion, μm/mK		
$L_T = L_o (1 + 19.494 \times 10^{-6} T - 1.0379 \times 10^{-9} T^2 + 2.375 \times 10^{-12} T^3)$ T, °C true from 0–900°C		

Table 1 (*continued*)

Property	Value	Reference
−190–0°C	17.0	
0–100°C	19.68	
0–500°C	20.61	1
volume change on melting, %	5	1
viscosity (liquid), mPa·s (= cP)		
at 1043°C	3.97	3
surface tension (liquid), mN/m (= dyn/cm)		
at 995°C	923	1
thermal electromotive force against platinum, mV (cold junction at 0°C, hot junction at 100°C)	0.074	1
electronic work function, eV		
thermionic	3.09–4.31	3
photoelectric	3.67–4.81	3
mechanical properties, HV	considerable spread in values of tensile strength and hardness of high purity silver	
	100 HV on electrodeposited silver, which has higher electrical resistivity than wrought silver	
average tensile strength, 125 MPa[e]		
5-mm wire annealed at 600°C		
hardness	27	
air-annealed at 650°C	27	
hydrogen-annealed at 650°C	25	

[a] A partial pressure of 20 kPa (2.9 psi) oxygen results in a freezing point of ca 950°C.
[b] Recrystallization temperature depends on the purity and the amount of cold work.
[c] To convert W/(m·k) to (Btu·in.)/(h·ft^2·°F), divide by 0.1441.
[d] To convert J to cal, divide by 4.148.
[e] To convert Pa to psi, divide by 6895; to convert MPa to psi, multiply by 145.
[f] At higher temperatures, oxidizing gases have higher volatization rates.

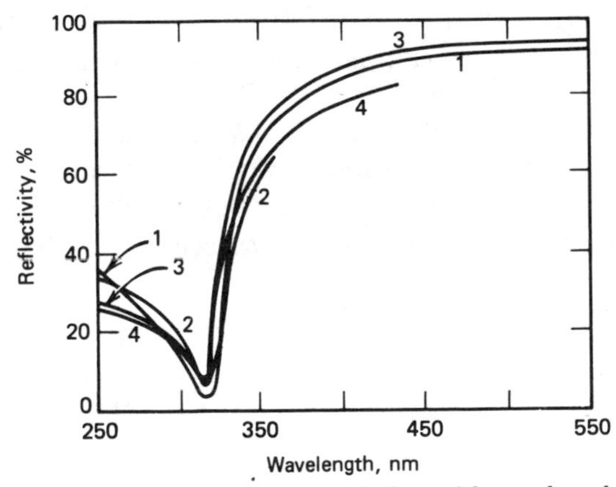

Figure 1. Change of reflectivity of silver with wavelength (4).

the heat generated by cold working may lead to the recrystallization of the metal during cold working.

Other properties of silver and its alloys are described in ref. 3.

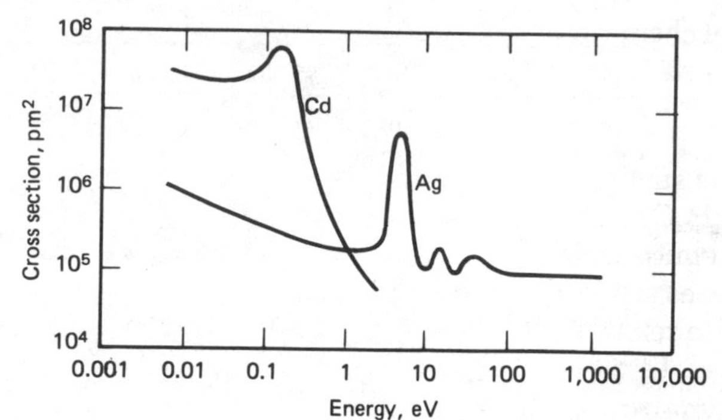

Figure 2. Total neutron cross sections of silver and cadmium (5). To convert pm^2 to barn, multiply by 10^{-4}.

Oxygen and Hydrogen. Molten silver dissolves ca 10 times its own volume of oxygen (ca 0.32 wt %) just above its melting point and ejects it violently just before it solidifies. Solid silver also dissolves oxygen (see Table 2). At ca 810°C, silver also dissolves an appreciable amount of hydrogen. Silver is seldom melted and cast in air because of the spitting that takes place before it solidifies. Moreover, the alternate exposure of gas-free 99.99 wt % Ag to oxygen and hydrogen or to hydrogen and oxygen at ca 810°C blisters and embrittles the metal.

With respect to the two electrical contact alloys Ag–CdO and Ag–MgO–NiO, the solubility of oxygen in silver at elevated temperatures enables them to be internally oxidized in the solid state.

Corrosive Reagents. Some substances that corrode silver are chlorine, mercury, and sulfer; chromic, nitric, and sulfuric acids; alkali metal cyanides; hydrogen peroxide; hydrogen sulfide; ferric sulfate; and permanganates (1,3). Silver is dissolved by alkali metal cyanide solutions as the anode or in the presence of oxygen to form $NaAg(CN)_2$ or $KAg(CN)_2$.

$$Ag + 2\,NaCN \rightarrow NaAg(CN)_2 + Na^+ + e$$

$$2\,Ag + 4\,NaCN + H_2O + \tfrac{1}{2}\,O_2 \rightarrow 2\,NaAg(CN)_2 + 2\,NaOH$$

These reactions are of commercial importance in electroplating (qv) and have been used to extract silver from its ores.

At elevated temperatures, silver and the halogens combine quickly and quantitatively with liberation of heat. Thus molten silver chloride is prepared for use in batteries by the direct reaction of chlorine gas on silver at above 445°C, which is the melting point of AgCl (see Batteries and electric cells).

Table 2. Solubility of Oxygen in Silver[a]

Oxygen, at. %	Temperature, °C
0.041	900
0.026	800
0.014	700
0.0072	600
0.0030	500
0.00094	400
0.00020	300
0.000022	200

[a] Ref. 6.

The principal chemical reaction of silver in the production of the chemical compounds of silver is its reaction with nitric acid:

$$4\ Ag + 6\ HNO_3 \rightarrow 4\ AgNO_3 + NO + NO_2 + 3\ H_2O$$

Nitric acid attacks silver at all concentrations, but the reaction usually employs ca 50:50 hot, HNO_3:H_2O. The silver nitrate solution produced is the starting material for most other chemical compounds of silver.

Sulfur tarnishes silver, coating it with a light-colored to black film that is sufficiently protective to retard its progressive formation. Rarely, if ever, is this responsible for the loss of silver. In the early stages, the tarnish does not seriously affect the conductivity of a silver electrical contact. Hydrogen sulfide in the atmosphere, which is chiefly the product of the burning of sulfur-containing fossil fuels, causes most of the tarnish, but sulfur dioxide contributes significantly. Increased concentrations of sulfur, moisture in the air, or the addition of copper to silver, eg, in sterling silver, coin silver, etc, accelerate the tarnishing process. A comprehensive list of all the corrosive materials is given in ref. 7.

Occurrence

There are 55 silver minerals including Ag; argentite [*1332-04-3*], Ag_2S; cerargyrite [*14358-96-4*], AgCl; polybasite; [*53810-31-4*], $Ag_{16}Sb_2S_{11}$; proustite [*15152-58-4*], Ag_3AsS_3; pyrargyrite [*15123-77-0*], Ag_3SbS_3; stephanite [*1302-12-1*], Ag_5SbS_4; tetrahedrite, $Cu_3(AsSb)S_3$; and the tellurides, stromerite [*12249-47-7*], and pearcite (8). These are most commonly associated with lead, but they also are associated with copper, zinc, and gold. Ores in which silver is the main component are associated with igneous rocks of intermediate felsic composition. Examples include the gold–silver ore Comstock lode in Virginia City, Nevada, where almost 31,100 t were produced, and the present cobalt–silver ores in the Cobalt District of Canada.

The majority of world silver production, however, has been from cavity fillings or replacement-type deposits where it is associated as a by-product constituent with mainly copper, lead, and/or zinc, or gold. The Kidd Creek Mine at Timmins, Ontario Canada is one of the largest zinc–silver–lead–copper mines and is a leading silver producer.

The Coeur d'Alene mining district of Idaho is the main silver-producing area in the U.S. Most of the ore is tetrahedrite with varying amounts of silver in isomorphic form. It occurs in veins along fault fissures and shear zones of precambrian quartzites and argillites. The mineralized zones range from a few centimeters–2.4 m wide, and the mineralized zones of randomly disseminated tetrahedrite are up to 9.1 m wide.

Mining and Processing

Silver is mined by open-pit methods and with subsurface shafts and drifts (8). The method used varies from one ore body to another, depending on the steepness of the terrain, availability of transportation, reserves, ore body or vein shapes, depth of mining, character of host rock, and economic factors peculiar to the individual mine. At the Kidd Creek Mine in Ontario, the zinc–silver–lead–copper ore is scooped up by shovels with capacities of 4.6 m^3, transferred to 45-metric ton capacity dump trucks, and shipped by rail for milling and concentration. At the Sunshine Mine in Coeur d'Alene, the tetrahedrite silver ore is extracted through one main vertical shaft and several winzes and development drifts along the veins. Mining is principally by the cut-and-fill stoping method; explosives break the vein and host rock material into pieces for transport by mine car and electric locomotive to the main shaft. There it is hoisted to the surface for milling. Hazardous rock bursts are minimized by the use of sandfill, rock bolts, and timber shoring.

Extraction. Silver ornaments and utensils of ca 3000–2500 BC attest to the antiquity of silver processing techniques. Prehistoric man probably found and washed silver from its ores and may have discovered that silver could be reduced from silver chloride over a charcoal fire. By 1000 BC, processing techniques had advanced to the stage where argentiferous lead ores in Anatolia (Turkey) were concentrated by washing and were extracted by reductive smelting and cupellation of the resulting lead bullion (3,8).

By the end of the 15th century, slag, mint sweepings, and scrap were amalgamated and, from the 16th century to the last decades of the 19th century, amalgamation was one of the principal methods used in extracting silver from its ores. Prior to its amalgamation, the crushed ore was usually chloride-roasted (Patio process, Mexico, 1557), but in some instances the ground ore was mulled with NaCl, H_2O, and Hg in a copper cauldron (Cazo process, Bolivia, 1590) or NaCl, $CuSO_4$, H_2SO_4, and Hg in a cast-iron pot (Washoe process, Nevada, 1861). The amalgam was washed and cleaned, the excess Hg was removed, and then the remainder was distilled from cast-iron pots. The residues were fluxed, melted, and cast into doré bullion containing minor amounts of impurities.

Several leaching processes were developed in the 19th century. These included chloride roasting followed by leaching in hot brine and precipitation of the silver on copper, leaching with sodium thiosulfate followed by precipitation of Ag_2S with Na_2S (Patera process, 1858), or leaching with $FeCl_3$, precipitation of AgI with ZnI_2, and reduction of the AgI with zinc (Clandot process, late 19th century).

Cyanidation had by 1907 largely superseded amalgamation and leaching. Multicrushed and rod-, ball-, or tube-milled finely ground ores were cyanided in alkaline cyanide solutions as ground ore or as classified ore subsequent to grinding. Careful control of the unsaturated and saturated solutions and the residues was required to prevent cyanide losses, and ores containing complex sulfides or elements such as Sb, Cu, graphite, or Zn had to be roasted prior to cyanidation and froth-floated, or water- or acid-leached.

Amalgamation, leaching, and cyanidation were adapted to the extraction of silver from fairly high grade ores. As these ores become less plentiful and as processes were rapidly developed for treating low grade ores, amalgamation, leaching, and cyanidation were almost discontinued. In the mid-20th century, amalgamation was only being used on placer and lode gold ores with cyanidation following amalgamation on lode gold ores to remove Cu, Zn, Sb, etc. In the 1970s, however, improvements in the cyanidation of low grade gold–silver ores, etc and the development of the activated-carbon stripping of the pregnant solutions led to the use of cyanidation for such materials (9–10).

Concentration of Low Grade Ores. By 1900 increasing tonnages of low grade, argentiferous, base metal ores, often containing gold as well as silver, were concentrated by gravitational methods, and in 1911 in the U.S., gravitation began to be replaced by flotation (qv). The latter improved the concentration as well as crushing, grinding, classification, thickening, filtering, and flotation of these ores. Results of 1963 concentration processes are listed in Table 3.

Table 4. Estimated Silver Supplies in the Noncommunist World, Metric Tons [a]

Sources	1971	1974	1977	1980
New production				
Western Hemisphere				
Mexico	1,141.4	1,244.1	1,461.8	1,601.8
Canada	1,430.7	1,309.4	1,331.2	1,035.7
Peru	1,194.3	1,244.1	1,122.8	1,259.6
U.S.	1,293.9	1,051.2	1,188.1	964.2
other countries	503.8	528.7	637.6	637.6
Total	*5,564.1*	*5,377.5*	*5,741.5*	*5,498.9*
Eastern Hemisphere				
Australia	674.9	684.2	852.2	855.3
other countries	1,440.0	1,449.4	1,723.1	1,576.9
Total	*2,114.9*	*2,133.6*	*2,575.3*	*2,432.2*
Total (new production)	*7,679.0*	*7,511.1*	*8,316.8*	*7,931.1*
Other sources of supply				
U.S. Treasury	77.7	31.1	12.4	3.1
stocks of foreign governments	155.5	622.0	155.5	161.7
demonetized coin	622.0	1,088.6	715.3	1,710.6
Indian stocks	995.3	1,306.3	1,262.8	1,297.0
salvage and other miscellaneous sources	880.2	2,189.6	2,929.9	3,779.0
liquidation of (additions to) private bullion stocks	1,399.6	1,555.1	821.1	(3,813.12)
Total (sources of supply)	*4,130.3*	*6,792.7*	*5,897.0*	*3,138.3*
Grand total	*11,809.3*	*14,303.8*	*14,213.8*	*11,069.4*

[a] To convert metric ton to 10^6 troy oz, divide by 31.1.

of ores and their mining and processing. From 1971–1979, however, new silver accounted for 51–65% of the total silver production. The other sources, in particular coinage, Indian stocks, and hoarded stocks, still provide a fairly large reservoir of silver.

The world's industrial consumption of silver has risen dramatically in a little more than one hundred years. From 1873, the consumption of silver rose to 900 metric tons in 1916, 1,200 t in 1929, 5,000 t in 1945, 10,100 t in 1963–1969, and 12,900 t in 1971–1979. Table 5 presents estimated data on consumption in the noncommunist world from 1971–1980.

Production and consumption are described in ref. 3.

Market Prices (New York)

New York market prices of silver from 1971–1980 are given in Table 6. Economic aspects of silver and its alloys are reviewed in refs. 3 and 8. In the summer of 1982, the price was $157–279/kg ($5–9/troy oz) (12). However this may be a reasonable inflated price.

Standards and Specifications

Commercial fine silver contains a minimum of 999.0 ppt silver as measured by the Gay-Lussac-Volhard method. This was the standard prior to 1964, when ASTM issued a tentative specification for 99.90 wt % Ag, which was modified in 1966 to include a higher grade (see Table 7).

Table 5. Estimated Silver Consumption in the Noncommunist World, t[a]

Uses	1971	1974	1977	1980
Industrial				
U.S.	4,015.4	5,505.3	4,777.4	3,723.0
Canada	186.6	267.4	273.7	211.5
Mexico	158.6	202.1	171.0	108.8
FRG	1,863.0	1,710.6	1,850.6	905.1
Italy	948.6	933.1	1,051.2	762.0
UK	777.5	1,060.6	1,001.5	637.6
France	485.2	653.1	640.7	628.2
Belgium			500.7	488.3
Japan	1,446.3	1,794.6	1,965.7	1,919.0
India	497.6	466.5	547.4	497.6
other countries	550.5	684.2	706.0	699.8
Total	*10,929.3*	*13,277.5*	*13,485.9*	*10,580.9*
Coinage				
Mexico			130.6	158.6
Canada	6.2	314.1	9.3	9.3
U.S.	77.7	31.1	12.4	3.1
France	12.4	111.9	214.6	
Austria	99.5	177.2	93.3	133.7
FRG	597.1	236.3	80.8	
other countries	87.0	155.5	186.6	186.6
Total	*879.2*	*1026.1*	*727.6*	*491.3*
Grand total consumption	*11,809.2*	*14,304.6*	*14,213.5*	*11,072.2*

[a] To convert to 10^6 troy oz, divide by 31.1.

Table 6. New York Market Prices for Silver, $/kg[a,b]

Year	Low	High	Av
1971	41.40	56.20	49.80
1972	44.60	65.90	54.30
1973	63.00	105.40	82.30
1974	105.10	215.40	151.40
1975	125.70	168.10	142.10
1976	122.80	163.90	139.80
1977	138.20	159.40	148.50
1978	155.20	202.50	173.60
1979	191.60	900.20	356.50
1980	347.20	1543.20	663.20
1981	255.60	528.90	338.30
1982	156.90	278.50	241.20

[a] Ref. 11.
[b] To convert $/kg to $/troy oz, multiply by 0.0311.

Table 7. Specifications for Silver, wt %

Method	Ag, min	Cu, max	Pb, max	Fe, max	Bi, max
Gay-Lussac-Volhard	99.90	0.08	0.025	0.002	0.001
ASTM	99.95	0.04	0.015	0.001	0.001

99.99 wt % silver is sometimes required and it may include the following: 0.01 wt % Cu; 0.001 wt % Pb, Fe, or Pd; and 0.0005 wt % Bi, Se, and/or Te.

See also ref. 3.

Assaying and Analysis

The analytical methods used in silver metallurgy are described in refs. 3, 13–14.

Fire assaying is used on low grade, voluminous materials, eg, ores, concentrates, and low grade sweeps and bullion. The sample is fused with fluxes, litharge, and reducing agents at ca 1040°C and the reduced lead picks up the reduced metals. The base metals are removed by oxidizing the fluxed impure lead in a scorifying dish. Then the purified lead is reoxidized in a porous cupeling cup, which absorbs the lead oxide and leaves a precious metal bead behind which may contain Ag, Au, and Pt. The latter can be analyzed spectrographically or by atomic absorption. Fire assaying is an extremely sensitive analytical method, but it requires furnaces and balances and is more time-consuming than any other method.

The gravimetric chemical method is used for silver alloys containing no interfering base metals and not more than 90 wt % Ag. The sample is dissolved in nitric acid and the silver precipitates as AgCl and is weighed.

Gay-Lussac-Volhard titration is used for silver alloys containing 90 wt % or more Ag. The procedures are described in ref. 3.

If a sample is completely volatilized in an electrical arc, the elements in the sample emit characteristic spectra, by which they can be identified. From 30–50 elements can be readily identified at quite low levels by optical emission spectrography and selected elements can be quantitatively analyzed. Most of the precious metals, however, can not be identified below 28 g/t in ores and concentrates.

If a sample is exposed to a high intensity x-ray beam, each element present in the sample emits characteristic x rays that can be used to identify it, and the intensity of the secondary radiation from the sample is directly proportional to the concentration of each element present. The elements can be identified or the radiation from a particular element can be measured and compared with those of known standards to determine the element's concentration. X-ray fluorescence analysis is simple, rapid, highly reliable, and quite accurate, ie, to 56–85 g/t for most noble metals; but each type of sample, ie, ore, concentrate, alloy, etc, must be individually calibrated.

A free atom absorbs light of the same wavelength that it normally emits. Therefore, an element can be analyzed by aspirating a suitable solution of the sample containing the element into a flame hot enough to atomize the element. Its absorption characteristics are then compared to known standards. Atomic absorption is fast, accurate, and economical as it involves no expensive equipment for concentrations below 2–3 wt %. It cannot be used for ores or concentrates, but if the precious metals in their solutions can be concentrated by organic extraction or fire assaying, it can be accurate to 0.28 g/t Au.

Health and Safety Factors

A problem that may be encountered in the melting, handling, and working of silver and its alloys in massive form is argyria (15). This results from the absorption of chemical or colloidal silver into the body tissues and its subsequent precipitation as an irreversible, permanent slate gray, blue gray, or sometimes purple pigmentation of the skin. This is apparently completely harmless, but it is disfiguring. Thus, excessive ingestion or inhalation of chemical or colloidal silver or its injection or application to body cavities and mucous membranes is to be avoided. Precautions are advisable in the handling and use of silver pharmaceuticals and against exposure to silver dusts and other wastes including condensed vapors and spray from melting operations. Ventilation of melting pots and buffing wheels, use of masks if the ventilation is inadequate, and simple hygiene, ie, daily bathing, washing before eating, etc, are helpful. The melting of silver alloys containing cadmium is hazardous (see Cadmium).

Uses

Table 8 presents U.S. Bureau of Mines data on the U.S. industrial consumption of silver from 1971–1980 (10). The data illustrate the increasing use of silver through 1979 in photographic materials and electrical and electronic products and its declining use in aesthetic products subsequent to 1976.

Photographic Materials. Conventional photographic emulsions contain fcc AgBr and AgCl and may contain up to 10 wt % hexagonal AgI. Their primary sensitivity to light is amplified up to 10^{11} times when the emulsions are developed (see Photography). A high degree of purity is required of the silver halides, the gelatin, and any other ingredients. Products have been developed that can be used to record images in black and white or color, negative or positive after exposure to x-rays, gamma rays, ultraviolet

Table 8. Estimated U.S. Industrial Consumption of Silver by Use, t[a]

Use	1971	1974	1977	1980
photographic materials	1122.8	1542.7	1670.2	1533.4
electrical and electronic products				
contacts and conductors	870.8	973.5	973.5	805.5
batteries	174.1	130.6	180.4	199.0
Total	*1044.9*	*1104.1*	*1153.9*	*1004.5*
aesthetic products				
sterling ware	622.0	690.4	519.4	251.9
jewelry	105.7	161.7	251.9	155.5
electroplated ware	339.0	410.5	211.5	130.6
coins, medallions, and commemorative objects	248.8	693.6	130.6	130.6
Total	*1315.5*	*1956.2*	*1113.4*	*668.6*
brazing alloys and solders	376.3	451.0	385.6	276.8
catalysts	52.8	227.0	276.8	124.4
mirrors	34.2	121.3	65.3	18.6
dental and medical supplies	46.6	74.6	68.4	59.0
bearings	12.4	12.4	15.5	6.2
miscellaneous	9.3	15.5	27.9	31.1
Grand total	*4014.8*	*5504.8*	*4777.0*	*3722.6*

[a] To convert t to 10^6 troy oz, divide by 31.1.

or infrared light, or visible light intensities from levels barely visible to the eye to those of intense arc lights. The images have packing densities of up to 10^8 bits or images per square centimeter and they do not fade on prolonged exposure to light as do most images formed from dyes and pigments.

Uses are described in ref. 3.

Electrical Contacts. Silver's high electrical and thermal conductivity and superior oxidation resistance make it a significant element with respect to electrical contacts. Where the contact or switching demands are minimal, silver, alone or silver alloys, eg, AgCd [*12002-62-9*] (15–25 wt % Cd), AgCu [*12249-45-5*] (7.5–28 wt % Cu), AgPd [*60495-83-2*] (3–10 wt % Pd), AgAu [*11126-80-0*] (10 wt % Au), and AgPt [*31189-85-8*] (3 wt % Pt) are used. If the demands are greater, silver and the semirefractory elements are used. Silver graphite [*7782-42-5*] (0.25–10 wt % C) and AgNi (10–40 wt % Ni) are sintered products, whereas Ag–MgO–NiO (0.41 wt % MgO and 0.25 wt % NiO) and Ag–CdO (2.5–15 wt % CdO) are internally oxidized; the former at or near finish and the latter during processing. The 10 and 15 wt % CdO contacts are unique and popular. If the demands are very severe, silver and the refractory elements are used. These include AgMo (50–60 wt % Mo), AgW (50–70 wt % W), and AgWC (10–60 wt % WC), which are all sintered (see Electrical connectors).

Electronic Products. Silver paints are of two types: the air-drying paints are used on wood, paper, plastics, etc, and those that must be fired are used on ceramics, mica, glass, and other refractory materials. Silver paints are used for producing electrically conducting surfaces on nonconductors, solder bonding to nonconductors and nonmetallic surfaces, and decorating porcelain or glass.

Batteries. Silver batteries include silver oxide–potassium hydroxide–zinc secondary batteries and silver oxide–potassium hydroxide–zinc, and silver chloride–seawater–magnesium primary batteries. The secondary, ie, rechargeable, 100-cycle batteries deliver up to 318 kJ/kg [88 (W·h)/kg] with a life of only 1.5 yr, whereas conventional Pb–acid, NiFe and NiCd batteries deliver only 79–111 kJ/kg [22–31 (W·h)/kg] with a life of 7–11 yr. An output of 88 (W·h)/kg is useful in aircraft, portable military radios, missiles, satellites, sounding balloons, submarines, and torpedoes, but commercial users prefer longer life batteries.

Silver oxide–potassium hydroxide–zinc primary batteries have been used as watch batteries and as remotely activated batteries. In the latter, stored electrolytes are injected into the batteries just before they are to be used. They are used in torpedoes and rockets. They have a shelf life of 5 yr at less than 40°C.

The silver chloride–seawater–magnesium primary batteries have been used in torpedoes. Seawater is injected into the battery to start the battery. A 227-kg battery develops 500 A at 250 V and discharges in 6–20 min.

Brazing Alloys and Solders. The silver–copper equilibrium (Fig. 3) is basic to an understanding of the common silver–copper alloys, ie, sterling, coin, 80 wt % Ag–20 wt % Cu, eutectic, ie, 7.5, 10, 20, and 28 wt % Cu, but especially with respect to low to medium-high temperature brazing alloys. With respect to modern brazing alloys, silver is present in five of the seven copper–phosphorus alloys, all of the seventeen silver–copper alloys, and four of the six silver–copper vacuum-tube alloys (see Solders and brazing alloys) (15).

Catalysts. Silver catalysts are used most frequently in oxidation reactions, eg, the production of formaldehyde from methanol and air by means of silver screens or crystallites containing 99.95 wt % min Ag and no zinc or any element of the fifth and

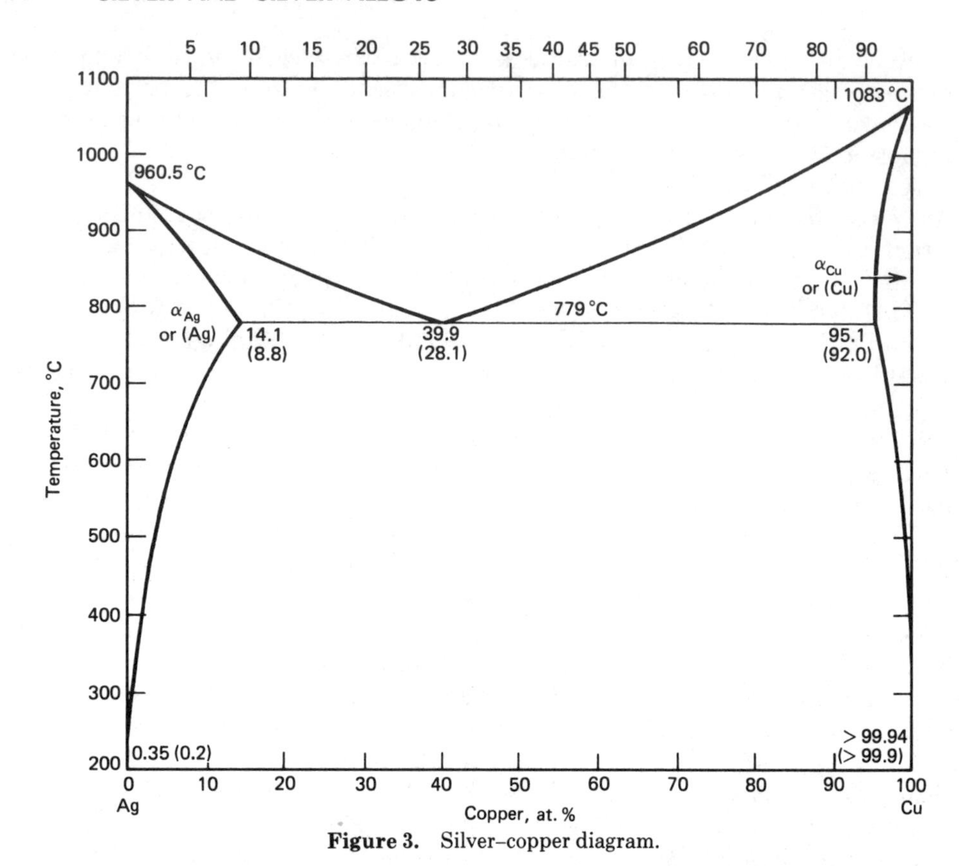

Figure 3. Silver–copper diagram.

sixth groups; the production of ethylene oxide from ethylene and air in a tubular reaction chamber filled with silver; and the production of glyoxal from the oxidation of ethylene glycol in an air–nitrogen mixture over finely dispersed silver or silver oxide or a blend of the two (see Catalysis).

Mirrors. Mirrors are silver-coated by reducing $AgNO_3$ solutions with reducing solutions on glass, that is pretreated with stannous chloride. The reducing solutions include sugar, rochelle salts, or formaldehyde.

Dental Amalgam. Dental amalgam is a universal product with which most people are familiar (see Dental materials). The comminuted product that is supplied to the dentist is basically the intermetallic epsilon silver–tin phase, which contains 67–70 wt % Ag, 25.3–27.7 wt % Sn, 0–5.2 wt % Cu, and 0–1.2 wt % Zn. The dentist mixes 5 parts of this with up to 8 parts of Hg, much of which he expresses in packing the cavity. The restoration for a filling contains ca 52 wt % Hg, 32 wt % Ag, 12.5 wt % Sn, 2 wt % Cu, and 0.5 wt % Zn.

Antiseptic. Storing water in silver vessels, or less than 10 ppm $AgNO_3$ in aqueous solution destroys the bacteria in the water. Silver nitrate is also used as an antiseptic but it is quite caustic. Colloidal solutions of AgI (Neo-Silvol) and strong-to-mild silver proteins, eg, Protargol (strong) and Argyrol (mild), are effective and are much less caustic (see Disinfectants and antiseptics).

Bearings. Lead-coated, silver-plated SAE 1010 or 1015 steel bearings proved to be the most satisfactory aircraft engine bearings during World War II. They had twice the load-carrying capacity of the available bearing materials, high thermal conductivity and corrosion resistance, good fatigue characteristics and, when lead-coated, satisfactory embedding and antiscoring characteristics. Their present use is limited by their cost and availability, but they are used on the wrist-pin bushings on diesel locomotives, the critical master rod and other bearings on reciprocating aircraft engines, and the rolling element or antifriction-bearing cages in mainshaft gas-turbine engine bearings (see Bearing materials).

BIBLIOGRAPHY

"Silver and Silver Alloys" in *ECT* 1st ed., Vol. 12, pp. 426–438, by E. H. Konrad and F. A. Meier (Analysis), The American Platinum Works; "Silver and Silver Alloys" in *ECT* 2nd ed., Vol. 18, pp. 279–294, by C. D. Coxe, Handy & Harman.

1. S. C. Carapella, Jr., and D. A. Corrigan in D. Benjamin, ed., *Metals Handbook: Pure Metals*, 9th ed., Vol. 2, American Society for Metals, Metals Park, Ohio, 1979, pp. 794–796.
2. V. N. Goldschmidt, *Geochemistry*, Oxford University Press, Oxford, England, 1954.
3. A. Butts and C. D. Coxe, *Silver—Economics, Metallurgy and Use*, D. Van Nostrand Co., Inc., Princeton, N.J., 1967.
4. W. W. Coblentz and R. Stair, *J. Res. Nat. Bur. Stand.* **2,** 243 (1929).
5. E. Bleuler and G. J. Goldsmith, *Experimental Nucleonics*, Holt, Rinehart and Winston, Inc., New York, 1952.
6. F. A. Shunk, *Constitution of Binary Alloys*, 2nd Suppl. (Hansen), McGraw-Hill Book Co., Inc., New York, 1969.
7. L. Addicks, *Silver in Industry*, Rheinhold Publishing Corp., New York, 1940, pp. 357–400.
8. H. J. Drake, *Silver Mineral Commodity Profile MCP-14*, U.S. Bureau of Mines, Washington, D.C., Sept. 1978.
9. S. J. Hussey, H. B. Salisbury, and G. M. Potter, *Carbon-in-Pulp Gold Adsorption From Cyanide Leach Slurries*, RI 8368, U.S. Bureau of Mines, Washington, D.C., 1979.
10. H. J. Heinen, G. E. McClelland, and R. E. Lindstrom, *Enhancing Percolation Rates in Heap Leaching of Gold–Silver Ores*, RI 8388, U.S. Bureau of Mines, Washington, D.C., 1979.
11. *The Silver Market in 1970, 1975, 1980, and 1982*, Handy & Harman, New York, 1971, 1976, 1981, 1982.
12. *Chem. Week*, 13 (June 30, 1982).
13. H. H. Heady and K. G. Broadhead, *Assaying Ores, Concentrates and Bullion*, Information Circular 8714r (Revision of I.C. 7695), U.S. Bureau of Mines, Washington, D.C.
14. W. W. White and P. J. Murphy, *J. Am. Chem. Soc.* **51**(11), 1864 (1979).
15. W. R. Hill and D. H. Pillsbury, *Argyria, The Industrial Pharmacology of Silver*, Williams and Wilkins, Co., Baltimore, Md., 1934.
16. M. Hansen and K. Anderko, *Constitution of Binary Alloys*, 2nd ed., McGraw-Hill Book Co., Inc., New York, 1958.

General References

I. C. Smith and B. L. Carson, *Trace Metals in the Environment*, Vol. 2, Ann Arbor Science Publishers, Inc., Ann Arbor, Mich., 1977.
Market Statistics, The Silver Institute, Washington, D.C.

GEORGE H. SISTARE
Consultant

SILVER COMPOUNDS

Silver is a white, lustrous metal, slightly less malleable and ductile than gold. It has high thermal and electrical conductivities. Most silver compounds are made from silver nitrate, which is prepared from silver metal.

Some silver metal is found in nature, frequently alloyed with other metals such as copper, lead, or gold; however, naturally occurring silver compounds are the primary sources of silver. The most abundant naturally occurring silver compound is silver sulfide, Ag_2S, known as argentite. It is found alone or in solid solution with iron, copper, and lead sulfides. Other naturally occurring silver compounds are silver sulfoantimonite [*15983-65-0*], Ag_3SbS_3 (pyragyrite), silver arsenite [*15122-57-3*], Ag_3AsS_3 (proustite), silver selenide [*1302-09-6*], and silver telluride [*12653-91-7*]. Silver chloride (cerargyrite) and silver iodide (iodagyrite) have been found in substantial quantities in the western United States.

The average concentration of silver in the earth's crust has been estimated to be 0.1 mg/kg (1). An examination of 130 sources of natural waters in the United States between 1962 and 1967 detected silver in 6.6% of the samples in concentrations from 0.1–38 μg/L, with a mean concentration of 2.6 μg/L (2).

Silver belongs to group IB of the periodic table, and silver metal has a $4d^{10}5s^1$ outer electronic configuration. Silver has been shown to have three possible positive ionic states, but only silver(I) is stable in solution. The silver(II) ion, a powerful oxidizing agent, is a transient species in aqueous solution. The two silver(II) compounds that have been well-defined are the oxide and the fluoride. Silver(III) exists only when stabilized through complex formation.

Only three simple silver salts, ie, the fluoride, nitrate, and perchlorate, are soluble to the extent of at least one mole per liter. Silver acetate, chlorate, nitrite, and sulfate are considered moderately soluble. All other silver salts are, at most, sparingly soluble; the sulfide is one of the most insoluble salts known. Silver(I) ion forms stable complexes with excess ammonia, cyanide, thiosulfate, and the halides; complex formation often results in the solubilization of otherwise insoluble salts. Silver bromide and iodide are colored, even though their respective ions are colorless. This is considered evidence for the partially covalent nature of these salts.

Silver compounds are available from commercial suppliers and are often very expensive. Reagent grades of silver(I) carbonate, chloride, cyanate, diethyldithiocarbamate, iodate, nitrate, oxide, phosphate, and sulfate are available. Standardized solutions of silver nitrate are also available for analytical uses. Purified grades of silver(I) acetate, bromide, cyanide, and iodide can be purchased; silver nitrate is also made as a USP XX grade for medicinal uses (3).

Since many silver compounds are unstable to light, they are shipped in brown glass or opaque plastic bottles. Silver compounds that are oxidants, eg, silver nitrate and iodate, must be so identified according to DOT regulations. Toxic silver compounds, eg, silver cyanate and cyanide, must carry a poison label.

Silver(I) Compounds

The solubilities and solubility product constants, K_{sp}, for many of the silver compounds described below, are given in Table 1.

Silver Acetate. Silver acetate, H_3CCO_2Ag, is prepared from silver nitrate and acetate ion. Colorless silver acetate crystals and solutions made from this salt are unstable to light.

Silver Azide. Silver azide, AgN_3, is prepared by treating a solution of silver nitrate with hydrazine or hydrazoic acid. It is shock-sensitive and decomposes violently when heated.

Silver Acetylide. Silver acetylide [*7659-31-6*], or silver carbide, Ag_2C_2, is prepared by bubbling acetylene through an ammoniacal solution of silver nitrate. Silver acetylide is sensitive to detonation on contact.

Silver Carbonate. Silver carbonate, Ag_2CO_3, is produced by the addition of an alkaline carbonate solution to a concentrated solution of silver nitrate. The pH and temperature of the reaction must be carefully controlled to prevent the formation of silver oxide. A suspension of Ag_2CO_3 is slightly basic because of the extensive hydrolysis of the ions present. Heating solid Ag_2CO_3 to 218°C gives Ag_2O and CO_2.

Table 1. Solubility and Solubility Product (K_{sp}) of Some Silver(I) Compounds

Silver compound	CAS Registry Number	Solubility, g/L[a] H_2O at 25°C[a]	K_{sp}
acetate	[*563-63-3*]	1.1×10	
azide	[*13863-88-2*]		2.9×10^{-9} [b]
carbonate	[*534-16-7*]	3.3×10^{-2}	8.2×10^{-12} [d]
chromate	[*7784-01-2*]	3.6×10^{-2}	1.9×10^{-12} [b]
cyanide	[*506-64-9*]	2.3×10^{-5}	1.6×10^{-14} [d]
chloride	[*7783-90-6*]	1.9×10^{-3}	1.8×10^{-10} [d]
bromide	[*7785-23-1*]	1.3×10^{-4}	3.3×10^{-13} [d]
iodide	[*7783-96-2*]	2.6×10^{-6}	8.5×10^{-17} [d]
fluoride	[*7775-41-9*]	1.8×10^{3}	
chlorate	[*7783-92-8*]	9.0×10[b]	
bromate	[*7783-89-3*]	1.6[b]	
iodate	[*7783-97-3*]	4.4×10^{-1} [b]	3.1×10^{-8} [b]
nitrate	[*7761-88-8*]	2.16×10^{2} [c]	
nitrite	[*7783-99-5*]	4.2	
oxide	[*20667-12-3*]	2.2×10^{-2}	
selenate	[*7784-07-8*]	1.2[b]	4×10^{-9} [b]
sulfate	[*10294-26-5*]	8.3	1.2×10^{-5} [b]
sulfide	[*21548-73-2*]	1.4×10^{-4} [b]	1.0×10^{-50} [b,e]
			3.8×10^{-52} [f]
			5.9×10^{-52} [g]
sulfite	[*13465-98-0*]		2×10^{-14} [b]
thiocyanate	[*1701-93-5*]	1.3×10^{-4} [b]	1.0×10^{-12} [b]

[a] Unless otherwise stated.
[b] At 20°C.
[c] Per 100 g H_2O.
[d] At 25°C.
[e] Ref. 4.
[f] Ref. 5.
[g] Ref. 6.

Silver Chromate. Silver chromate, Ag_2CrO_4, is prepared by treating silver nitrate with a solution of a chromate salt or by heating a suspension of silver dichromate [*7784-02-3*].

Silver Cyanide. Silver cyanide, AgCN, forms as a precipitate when stoichiometric quantities of silver nitrate and a soluble cyanide are mixed. Silver(I) ion readily forms soluble complexes, ie, $Ag(CN)_2^-$ or $Ag(CN)_3^{2-}$, in the presence of excess cyanide ion.

Silver Halides and Other Halogen Salts. All silver halides are reduced to silver by treating an aqueous suspension with more active metals, such as magnesium, zinc, aluminum, copper, iron, or lead. Alternatively, the dry salts are reduced by heating with turnings or powders of these metals. Photolyzed silver halides are also reduced by organic reducing agents or developers, eg, hydroquinone, *p*-aminophenol, and *p*-phenylenediamine, during photographic processing (see Photography).

Silver Chloride. Silver chloride, AgCl, is a white precipitate that forms when chloride ion is added to a silver nitrate solution. The order of solubility of the three silver halides is $Cl^- > Br^- > I^-$. Because of the formation of complexes, silver chloride is soluble in solutions containing an excess of chloride ions and in solutions of cyanide, thiosulfate, and ammonia. It is insoluble in nitric and dilute sulfuric acid; treatment with concentrated sulfuric acid gives silver sulfate.

Silver chloride crystals have a face-centered cubic structure with a distance of 0.28 nm between each ion in the lattice. Silver chloride is the most ionic of the halides; it melts at 455°C and boils at 1550°C. Silver chloride is very ductile and can be rolled into large sheets. Individual crystals weighing up to 22 kg have been prepared (7).

The silver ion in silver chloride can be readily reduced by light; however, currently it is not used to any great extent in photography. With sufficient light intensity and time, silver chloride decomposes completely into silver and chlorine.

Silver Bromide. Silver bromide, AgBr, is formed by the addition of bromide ions to a solution of silver nitrate. The light yellow to green-yellow precipitate is less soluble in ammonia than silver chloride, but it easily dissolves in other complexing agents.

Silver bromide crystals, formed from stoichiometric amounts of silver nitrate and potassium bromide, have a cubic structure with interionic distances of 0.29 nm; however, if an excess of either ion is present, octahedral crystals tend to form. The yellow color of silver bromide has been attributed to ionic deformation, an indication of its partially covalent character. Silver bromide melts at 434°C and dissociates when heated above 500°C.

Silver bromide is significantly more photosensitive than silver chloride, which results in the extensive use of silver bromide in photographic products. The crystal structure of photographic silver bromide grains is often octahedral.

Silver Iodide. Silver iodide, AgI, precipitates as a yellow solid when iodide ion is added to a solution of silver nitrate. It dissolves in the presence of excess iodide ion, forming an AgI_2^- complex; however, silver iodide is only slightly soluble in ammonia and dissolves slowly in thiosulfate and cyanide solutions.

Silver iodide exists in one of three crystal structures depending on the temperature, a phenomenon frequently referred to as trimorphism. Below 137°C, silver iodide is in the cold cubic, or γ form; at 137–145.8°C, it exists in the green-yellow colored hexagonal, or β form; above 145.8°C, the yellow cubic or α form of silver iodide is the stable crystal structure. Silver iodide decomposes into its elements at 552°C.

Although silver iodide is the least photosensitive of the three halides, it has the broadest wavelength sensitivity in the visible spectrum. This feature makes silver iodide particularly useful in the photographic industry. It resists reduction by metals, but is reduced quantitatively by zinc and iron in the presence of sulfuric acid.

Silver Fluoride. Silver fluoride, AgF, is prepared by treating a basic silver salt, such as silver oxide or silver carbonate with hydrogen fluoride. It can exist as an anhydrous salt, a dihydrate [*72214-21-2*] (below 42°C), and a tetrahydrate [*22424-42-6*] (below 18°C). The anhydrous salt is colorless, but the dihydrate and tetrahydrate are yellow. Ultraviolet light or electrolysis decompose silver fluoride to silver subfluoride [*1302-01-8*], Ag_2F, and fluorine.

Halates. Silver chlorate, $AgClO_3$, silver bromate, $AgBrO_3$, and silver iodate, $AgIO_3$, have been prepared. The halates may decompose explosively if heated.

Perhalates. Silver perchlorate [*7783-93-9*], $AgClO_4$, and silver periodate [*15606-77-6*], $AgIO_4$, are well known, but silver perbromate [*54494-97-2*], $AgBrO_4$, has only recently been described (8). Silver perchlorate is prepared from silver oxide and perchloric acid, or by treating silver sulfate with barium perchlorate. Silver perchlorate is one of the few silver salts that is appreciably soluble in organic solvents such as glycerol, toluene, and chlorobenzene.

Silver Tetrafluoroborate. Silver tetrafluoroborate [*14104-20-2*], $AgBF_4$, is formed from silver borate and sodium borofluoride or bromine trifluoride. It is soluble in organic solvents.

Silver Nitrate. Silver nitrate, $AgNO_3$, is the most important commercial silver salt because it serves as the starting material for all other silver compounds. It is prepared by the oxidation of silver metal with hot nitric acid. Nitrogen oxides, NO and NO_2, are the by-products, and they are vented to the atmosphere or are scrubbed out of the fumes with an alkaline solution. Heavy metal impurities, such as copper, lead, and iron, are precipitated by increasing the pH of the solution to 5.5–6.5 with silver oxide and then boiling. The solution containing silver nitrate is made slightly acid, heated, evaporated, and then poured into pans to cool and crystallized. The crystals are washed, centrifuged, and dried. They can be further purified by recrystallization from hot water.

The Kestner-Johnson dissolver is widely used for the preparation of silver nitrate (9). In this process, silver bars are dissolved in 45% nitric acid in a pure oxygen atmosphere. Any nitric oxide, NO, which is produced is oxidized to nitrogen dioxide, NO_2, which in turn reacts with water to form more nitric acid and nitric oxide. The nitric acid is then passed over a bed of granulated silver in the presence of oxygen; most of the acid reacts. The resulting solution contains ca 840 g silver per liter (10). This solution can be further purified with charcoal (11), alumina (12), and ultraviolet radiation (13).

The manufacture of silver nitrate for the preparation of photographic emulsions requires silver of very high purity. At the Eastman Kodak Company, the principal U.S. producer of silver nitrate, silver bars (99.95% pure) are dissolved in 67% nitric acid in three tanks connected in parallel. Excess nitric acid is removed from the resulting solution, which contains 60–65% silver nitrate, and the solution is filtered. This solution is evaporated until its silver nitrate concentration is 84%; it is then cooled to prepare the first crop of crystals. The mother liquor is purified by the addition of silver oxide and returned to the initial stages of the process. The crude silver nitrate is centrifuged and recrystallized from hot, demineralized water. Equipment used in this process is made of ANSI 310 stainless steel (14).

Silver nitrate forms colorless, rhombic crystals. It is dimorphic and changes to the hexagonal rhombohedral form at 159.8°C; it melts at 212°C to a yellowish liquid which solidifies to a white, crystalline mass on cooling; an alchemist's name, lunar caustic, is still applied to this fused salt. In the presence of a trace of nitric acid, silver nitrate is stable to 350°C. It decomposes at 440°C to metallic silver, nitrogen, and nitrogen oxides. Solutions of silver nitrate are usually acidic, with a pH of 3.6–4.6. Silver nitrate is soluble in ethanol and acetone.

In the absence of organic matter, silver nitrate is not photosensitive. It is easily reduced to silver metal by glucose, tartaric acid, formaldehyde, hydrazine, and sodium borohydride.

Silver Nitrite. Silver nitrite, $AgNO_2$, is prepared from silver nitrate and a soluble nitrite, or silver sulfate and barium nitrite.

Organic Acid Salts. Slightly soluble or insoluble silver salts are precipitated when mono- and dicarboxylic aliphatic acids or their anions are treated with silver nitrate solutions. Silver behenate [*2489-05-6*], $C_{22}H_{43}O_2Ag$, silver laurate [*18268-45-6*], $C_{12}H_{23}O_2Ag$, and silver stearate [*3507-99-1*], $C_{18}H_{35}O_2Ag$, are used in commercial applications (see below). Silver oxalate [*533-51-7*], $C_2O_4Ag_2$, decomposes explosively when heated.

Silver Oxide. Silver oxide, Ag_2O, a dark brown-to-black material, is formed when an excess of hydroxide ion is added to a silver nitrate solution. Aqueous suspensions of this compound are basic. Silver oxide can also be prepared by heating finely-divided silver metal in the presence of oxygen. Anhydrous silver oxide is difficult to prepare; reagent-grade material may contain up to 1 wt % water. Silver oxide is soluble in most of the silver-complexing agents; however, if a suspension of silver oxide in excess hydroxide is treated with ammonium hydroxide, highly explosive fulminating silver, or Bertholet's silver (15), may be produced.

When heated to 100°C, silver oxide decomposes into its elements, and it is completely decomposed above 300°C. Silver oxide and sulfur form silver sulfide. Silver oxide absorbs carbon dioxide from the air, forming silver carbonate.

Silver Permanganate. Silver permanganate [*7783-98-4*], $AgMnO_4$, is a violet solid formed when a potassium permanganate solution is added to a silver nitrate solution. It decomposes upon heating, exposure to light, or by reaction with alcohol.

Silver Phosphates. Silver phosphate [*7784-09-0*], or silver orthophosphate, Ag_3PO_4, is a bright yellow material formed by treating silver nitrate with a soluble phosphate salt or phosphoric acid. Silver pyrophosphate [*13465-97-9*], $Ag_4P_2O_7$, is a white salt prepared by the addition of a soluble pyrophosphate to silver nitrate. Both the phosphate and the pyrophosphate are light-sensitive; silver pyrophosphate turns red upon exposure to light.

Silver Selenate. Silver selenate, Ag_2SeO_4, is prepared from silver carbonate and sodium selenate.

Silver Sulfate. Silver sulfate, Ag_2SO_4, is prepared by treating metallic silver with hot sulfuric acid. Alternatively, a solution of silver nitrate is acidified with sulfuric acid and the nitric acid is evaporated; a solution of silver sulfate is obtained. Silver sulfate is more soluble in sulfuric acid than in water because of the formation of silver hydrogen sulfate [*19287-89-9*], $AgHSO_4$.

Silver sulfate decomposes above 1085°C into silver, sulfur dioxide, and oxygen. This property is utilized in the separation of silver from sulfide ores by direct oxidation. Silver sulfate is reduced to silver metal by hydrogen, carbon, carbon monoxide, zinc, and copper.

Silver Sulfide. Silver sulfide, Ag_2S, forms as a finely divided black precipitate when solutions or suspensions of most silver salts are treated with an alkaline sulfide solution or hydrogen sulfide. Silver sulfide has a dimorphic crystal structure; transition from the rhombic (acanthite) to the cubic (argentite) form occurs at 175°C. Both crystal structures are found in nature.

Silver sulfide is one of the most insoluble salts known. It is not solubilized by nonoxidizing mineral acids, but it is soluble in concentrated nitric acid, concentrated sulfuric acid, and alkaline cyanide solutions.

Silver and sulfur combine even in the cold to form silver sulfide. Silver tarnish is an example of the ease with which silver and sulfur compounds react. It is removed by silver polishes that contain silver complexing agents, such as chloride ion or thiourea.

Silver sulfide is exceptionally stable in air and sunlight, but decomposes when heated to 810°C. Moss silver (filiform silver), consisting of long hairlike growths of pure silver, is formed when silver sulfide is heated for a prolonged period at elevated temperatures below 810°C.

Silver Sulfite. Silver sulfite, Ag_2SO_3, is obtained as a white precipitate when sulfur dioxide is bubbled through a solution of silver nitrate. Silver sulfite is unstable to light and heat, and solutions decompose when boiled.

Silver Thiocyanate. Silver thiocyanate, AgSCN, is formed by the reaction of stoichiometric amounts of silver ion and a soluble thiocyanate.

Silver Thiosulfate. Silver thiosulfate [*23149-52-2*], $Ag_2S_2O_3$, is an insoluble precipitate formed when a soluble thiosulfate reacts with an excess of silver nitrate; however, in order to minimize the formation of silver sulfide, the silver ion can be complexed before the addition of the thiosulfate solution.

Silver(I) Complexes

Silver ions form a number of complexes with both π-bonding and non-π-bonding ligands. Linear polynuclear complexes are known. The usual species are AgL and AgL_2, but silver complexes up to AgL_4 have been identified. Many of these complexes have commercial application (see below).

Ammonia and Amine Complexes. In the presence of excess ammonia, silver ion forms the complex ammine ions $Ag(NH_3)_2^+$ and $Ag(NH_3)_3^+$. To minimize the formation of fulminating silver, these complexes should not be prepared from strongly basic suspensions of silver oxide. Highly explosive fulminating silver, believed to consist of either silver nitride or silver imide, may detonate spontaneously when silver oxide is heated with ammonia or when alkaline solutions of a silver–amine complex are stored. Addition of appropriate amounts of HCl to a solution of fulminating silver renders it harmless. Stable silver complexes are also formed from many aliphatic and aromatic amines, eg, ethylamine, aniline, and pyridine.

Cyanide Complexes. Insoluble silver cyanide, AgCN, is readily dissolved in an excess of alkali cyanide. The predominant silver species present in such solutions is $Ag(CN)_2^-$, with some $Ag(CN)_3^{2-}$ and $Ag(CN)_4^{3-}$. Virtually all silver salts, including the insoluble silver sulfide, dissolve in the presence of excess cyanide because the dissociation constant for the $Ag(CN)_2^-$ complex is only 4×10^{-19}.

Halide Complexes. Silver halides form soluble complex ions, AgX_2^- and AgX_3^{2-}, with chloride, bromide, and iodide. The relative stability of these complexes is $I^- > Br^- > Cl^-$. As an example of the effect that complex formation has on solubility, the solubility of silver chloride in 1 *N* HCl is 100 times greater than it is in pure water.

Olefin Complexes. Silver ion forms complexes with olefins and many aromatic compounds. As a general rule, the stability of olefin complexes decreases as alkyl groups are substituted for the hydrogens bonded to the ethylene carbon atoms (16).

Sulfur Complexes. Silver compounds other than the sulfide dissolve in excess thiosulfate. Stable silver complexes are also formed with thiourea. Next to the cyanide complexes, these sulfur complexes of silver are the most stable. In photography, solutions of sodium or ammonium thiosulfate fixers are used to solubilize silver halides present in processed photographic emulsions. When insoluble silver thiosulfate is dissolved in excess thiosulfate, various silver complexes form. At low thiosulfate concentrations, the principal silver species is $Ag_2(S_2O_3)_2^{2-}$; at high thiosulfate concentrations, species such as $Ag_2(S_2O_3)_6^{10-}$ are present. Silver sulfide dissolves in alkali sulfide solutions to form complex ions such as $Ag(S_4)_2^{3-}$ and $Ag(HS)S_4^{2-}$. These ions are found in hot springs and waters highly charged with H_2S (17). Silver forms stable, slightly ionized salts with aliphatic and aromatic thiols.

Other Oxidation States

Silver(II) Compounds. Silver(II) is stabilized by coordination with nitrogen heterocyclic bases, such as pyridine and dipyridyl. These cationic complexes are prepared by the peroxysulfate oxidation of silver(I) solutions in the presence of an excess of the ligand. An extensive review of the higher oxidation states of silver has been published (18).

Silver(II) Fluoride. Silver(II) fluoride [*7783-95-1*], AgF_2, is a brown-to-black, hygroscopic material obtained by the treatment of silver chloride with fluorine gas at 200°C (19) or by the action of fluorine gas on silver metal. It is a strong oxidizing agent and strong fluorinating agent. When heated above 450°C, it decomposes into AgF and fluorine. It also decomposes in aqueous solutions unless stabilized with concentrated HNO_3.

Silver(II) Oxide. Silver(II) oxide [*1301-96-8, 35366-11-1*], AgO, is prepared by persulfate oxidation of Ag_2O in basic medium at 90°C or by the anodic oxidation of solutions of silver(I) salts. This black oxide is stable to 100°C; it dissolves in acids and evolves oxygen and gives some Ag^{2+} in solution. Although it was previously believed that this oxide is a peroxide, chemical evidence has ruled out this possibility. Because it is diamagnetic, its most likely formula is $Ag(I)Ag(III)O_2$ (16).

Silver(II) oxide is a strong oxidant; its reactions in alkaline medium have been studied extensively (16). It decomposes in aqueous solution unless stabilized with concentrated nitric acid.

Silver(III) Compounds. No simple silver(III) compounds exist. When mixtures of potassium or cesium halides are heated with silver halides in a stream of fluorine gas, yellow $KAgF_4$ [*23739-18-6*] or $CsAgF_4$ [*53585-89-0*], respectively, are obtained. These compounds are diamagnetic and extremely sensitive to moisture (18). When Ag_2SO_4 is treated with aqueous potassium persulfate in the presence of ethylenedibiguanidinium sulfate, the relatively stable Ag(III)–ethylenebiguanide complex is formed.

Economic Aspects

The cost of various silver compounds is a function of the silver market price. In 1980, the estimated usage of silver in the United States was 3730 metric tons (120×10^6 troy oz) (20). This silver is derived from silver mined within the United States; silver recycled or reclaimed from secondary sources, eg, coinage, flatware, jewelry, and photographic materials; and imported silver. In 1980, Canada, Mexico, and Peru, the principal exporters of silver to the United States, shipped 1670 t (53.8×10^6 troy oz) as silver bullion and silver compounds. The U.S. imports and exports are shown in Table 2.

New silver accounts for only a portion of the silver used in the United States; recycled silver makes up the difference. Its availability is dependent on the market price; as the market price increases, so does the flow of recycled silver. The New York price reached a then all-time high of $1543/kg ($48.00/troy oz) on Jan. 21, 1980; however, this was primarily the result of speculation rather than of an increase in silver demand by industry. The price fell to $347.30/kg ($10.80/troy oz) four months later as the pressure of speculative activity in the silver market lessened. The New York price range of silver at the end of June 1982 was $163.99/kg ($5.10/troy oz). Comprehensive reviews of the silver market are published yearly (20). New York prices between 1971 and 1981 are given in Table 3. The distribution of industrial uses is given in Table 4. The U.S. consumption by uses is given in Table 5.

Analytical Test Methods

Qualitative. The classic method for the qualitative determination of silver in solution is precipitation as silver chloride with dilute nitric acid and chloride ion. The silver chloride can be differentiated from lead or mercurous chlorides, which also may precipitate, by the fact that lead chloride is soluble in hot water but not in ammonium hydroxide, whereas mercurous chloride turns black in ammonium hydroxide; silver chloride dissolves in ammonium hydroxide because of the formation of soluble silver–ammonia complexes mentioned earlier. A number of selective spot tests (21) include the reactions with *p*-dimethylamino-benzylidenerhodanine, ceric ammonium nitrate, or bromopyrogallol red. Silver is detected by x-ray fluorescence and arc-emission spectrometry; two sensitive arc-emission lines for silver occur at 328.1 and 338.3 nm.

Quantitative. Classically, silver concentration in solution has been determined by titration with a standard solution of thiocyanate. Ferric ion is the indicator; the deep red ferric thiocyanate color appears only when the silver is completely precipitated. Gravimetrically, silver is determined by precipitation with chloride, sulfide, or 1,2,3-benzotriazole. Silver can be precipitated as the metal by electrodeposition or chemical reducing agents. A colored silver diethyldithiocarbamate complex, extractable by organic solvents, is used for the spectrophotometric determination of silver complexes.

Highly sensitive instrumental techniques, such as x-ray fluorescence, atomic absorption spectrometry, and inductively coupled plasma optical emission spectrometry, have wide application for the analysis of silver in a multitude of materials. In order to minimize the effects of various matrices in which silver may exist, samples are treated with perchloric or nitric acid. Direct-aspiration atomic absorption (22)

Table 2. U.S. Silver Imports and Exports, Jan.–Dec. 1980, Metric Tons[a]

Country	Ore and concentrates[b]	Waste and scrap	Doré and precipitates	Bullion, refined	Total[c]
Imports					
Argentina	1.5	[d]		5.32	6.78
Belgium–Luxembourg			0.9	24.8	25.8
Brazil	2.3			1.65	3.92
Canada	16	42.3	6.19	902	967
Chile	5.3	2	18.6	18.2	44.1
Dominican Republic		0.03	3.67	5.13	8.83
France		1.52	0.03	39.6	41.1
Germany, Federal Republic of		[d]	3.14	1.49	4.63
Honduras	20.5				20.5
Hong Kong	[d]	3.98	8.4	6.47	18.8
Japan	0.9		1.65	25.2	27.7
Korea, Republic of			7.56	11.7	18.8
Mexico	18	1.71	6.94	429	456
Netherlands				17.8	17.8
Panama		1.34	0.47	4.45	6.25
Peru	207	0.22	11.6	345	564
Poland				6.5	6.5
Switzerland		0.03	0.16	11.9	12.1
United Kingdom	24.5	1.9	0.62	114	141
Yugoslavia			0.31	35	35.4
other	6	5.75	0.65	8.64	21.1
Total[c]	*302*	*60.8*	*70.9*	*2014*	*2448*
Exports					
Belgium–Luxembourg	0.19	235	49.9	1.28	286
Canada	5.54	97.9	8.09	156	268
France	[d]	33.4		18.9	52.3
Germany, Federal Republic of	0.124	66.1	2.8	105	174
Japan	1.28	4.57	5.26	131	142
Netherlands		1.31		75.6	76.9
Spain		52.5			52.5
Switzerland	0.218	23.2	0.093	469	493
United Kingdom	2.05	134	3.86	809	949
other	0.124	7.15	0.404	13	20.7
Total[c]	*9.53*	*655*	*70.4*	*1779*	*2514*

[a] To convert metric tons to troy ounces, multiply by 32,154.
[b] Includes silver content of base metals and matte imported for refining.
[c] Data may not add to total shown because of independent rounding.
[d] Less than 16 g.

and inductively coupled plasma (23) have silver detection limits of 10 and 7 μg/L, respectively. The use of a graphite furnace in an atomic-absorption spectrograph lowers the silver detection limit to 0.2 μg/L.

The instrumental methods described above are useful for the determination of the total silver in a sample, but such methods do not differentiate the various species of silver that may be present. A silver-ion-selective electrode measures the activity of the silver ions present in a solution. These activity values can be related to the concentration of free silver ion in the solution. Commercially available silver-ion-selective electrodes measure Ag^+ down to 10 μg/L, and special silver ion electrodes measure free silver ion at 1 ng/L (24) (see Ion-selective electrodes).

Table 3. New York Silver Quotations[a]

Year	High		Low		Average	
	$/kg	$/troy oz	$/kg	$/troy oz	$/kg	$/troy oz
1981	528.94	16.45	255.63	7.95	338.26	10.52
1980	1543.41	48.00	347.27	10.80	663.39	20.63
1979	900.32	28.00	191.67	5.96	356.71	11.09
1978	202.44	6.29	155.27	4.82	173.66	5.40
1975	167.85	5.22	125.72	3.91	142.07	4.41
1971	56.33	1.75	41.41	1.29	49.70	1.54

[a] Courtesy of Handy & Harman, New York (20).

Table 4. U.S. Industrial Distribution of Silver Compounds, 1980[a]

Industry	Percent
photographic manufacturing	41.2
electrical contacts	21.6
brazing alloys and solders	7.4
sterling silverware	6.8
batteries	5.3
jewelry	4.1
electroplating	3.5
coins and medallions	3.5
catalysts	3.3
dental and medical supplies	1.6
mirrors	0.5
bearings	0.2
other	1.0

[a] Ref. 20.

Health and Safety Factors

Silver compounds that generate significant quantities of free silver ion in solution, eg, silver nitrate, may cause adverse acute health effects. Silver compounds whose counterions are inherently toxic, eg, silver arsenate and silver cyanide, also can cause such effects. The reported rat oral LD_{50} values for silver nitrate, silver arsenate [*13510-44-6*], and silver cyanide are 500–800 mg/kg (25), 200–400 mg/kg (25), and 123 mg/kg (26), respectively. Silver compounds or complexes in which the silver ion is not biologically available, eg, silver sulfide and silver thiosulfate complex, are considered to be without adverse health effects.

Toxic effects from chronic exposure to silver and silver compounds seem to be limited to the skin without evidence of health impairment (25,27–28). Argyria and argyrosis have resulted from therapeutic and occupational exposures to silver and its compounds. These disorders are characterized by either localized or general deposition of a silver–protein complex in parts of the body, and imparts a blue-gray discoloration to the areas affected. In generalized argyria, characteristic discoloration may appear on the face, ears, forearms, and under the fingernails (29). Although the exact quantities of silver required for the development of argyria are not known, estimates based on therapeutic exposure suggest that the gradual accumulation of 1–5 g leads to gen-

Table 5. U.S. Consumption of Silver by Use, Metric Tons [a]

		1980[d]				1981
Use[b]	1980[c]	First quarter	Second quarter	Third quarter	Fourth quarter	First quarter
electroplated ware	137.0	33.4	50.7	21.8	31.0	53.0
sterling ware	282.5	76.4	80.8	60.8	64.4	49.0
jewelry	183.3	47.6	56.6	34.7	44.4	60.0
photographic materials	1549.5	381.6	438.4	279.4	450.1	388.2
dental and medical supplies	68.3	17.8	15.8	18.6	16.1	17.8
mirrors	20.9	7.2	8.7	2.0	3.1	3.9
brazing alloys and solders	280.3	73.3	80.5	63.8	62.8	91.1
electrical electronic products:						
batteries	185.9	52.7	48.7	41.4	43.1	42.4
contacts and conductors	863.3	211.4	269.7	171.2	211.0	269.9
bearings	20.1	2.5	2.5	13.1	2.0	2.0
catalysts	94.1	31.0	5.0	30.3	27.8	23.7
coins, medallions, and commemorative objects	146.0	39.5	41.1	32.8	32.6	30.9
miscellaneous[e]	62.8	4.7	24.9	8.2	25.0	19.2
Total net industrial consumption[f]	3893.8	979.1	1123.3	778.0	1013.4	1051.1
coinage	2.2	0.1	[g]	0.8	1.2	2.3
Total consumption	3896.0	979.2	1123.3	778.8	1014.6	1053.4

[a] To convert metric tons to troy ounces, multiply by 32,154.
[b] Use as reported by converters of refined silver.
[c] Preliminary.
[d] Revised.
[e] Includes silver-bearing copper, silver-bearing lead anodes, ceramic paints, etc.
[f] Data may not add to total shown because of independent rounding.
[g] Less than 16 g.

eralized argyria (30). Localized argyria can occur from the prolonged handling of metallic silver, which causes silver particles to be embedded in the skin and subcutaneous tissues via sweat-gland pores, or from the application of silver compounds to abraded skin areas (31). In argyrosis, silver is deposited primarily in the cornea and conjunctiva; however, this does not appear to cause visual impairment (25,28).

Occupational argyria is now uncommon because of effective industrial-hygiene practices. In 1980, the ACGIH adopted a TLV for airborne silver metal particles of 100 $\mu g/m^3$, and proposed a TLV of 10 $\mu g/m^3$, as silver, for airborne soluble silver compounds (32). These values were selected to protect against argyria and argyrosis from industrial exposures to silver and silver compounds. Argyria caused by therapeutic use of silver compounds is now extremely rare because of the availability of silver-free antiseptics and antibiotics (33).

In 1975, the U.S. Public Health Service adopted an interim primary drinking-water standard for silver of 50 $\mu g/L$ (34) designed to protect against argyria in the general population. The Safe-Drinking-Water Committee of the NRC reviewed this interim standard and concluded that the natural concentrations of silver in waters are so low that consideration should be given to removing silver from the list of substances included in the primary drinking-water standards (35). In 1980, the EPA published ambient-water-quality criteria for silver. They concluded that an upper limit of 50 $\mu g/L$ in natural waters would provide adequate protection against adverse health effects (36).

Environmental Impact

The impact that a silver compound has in water is a function of silver-ion concentration generated by that compound, not the total silver concentration (24,37–39). In a standardized, acute aquatic bioassay, fathead minnows were exposed to various concentrations of silver compounds for a 96-h period, and the concentration of total silver lethal to half of the exposed population (96-h LC_{50}) was determined. For silver nitrate, the value obtained was 16 $\mu g/L$. For silver sulfide and silver thiosulfate complex, the values were >240 and >280 mg/L, respectively, the highest concentrations tested (24).

The chronic aquatic effects which relate silver speciation to adverse environmental effects were studied on rainbow trout eggs and fry. The maximum acceptable toxicant concentration (MATC) for silver nitrate, as total silver, was reported to be 90–170 ng/L (40). Using fathead minnow eggs and fry, the MATC, as total silver, for silver thiosulfate complex was reported as 21–44 mg/L, and for silver sulfide as >11 mg/L, the maximum concentration tested (24).

The EPA has noted that free ionic silver readily forms soluble complexes or insoluble materials with dissolved and suspended material present in natural waters, such as sediments and sulfide ions (41). The hardness of water is sometimes used as an indicator of its complex-forming capacity. Because of the direct relationship between the availability of free silver ions and adverse environmental effects, the 1980 ambient freshwater criterion for the protection of aquatic life is expressed as a function of the hardness of the water in question. The maximum recommended concentration of total recoverable silver is thus given by the expression (42):

$$e^{(1.72\ [\ln\ (\text{hardness})]\ -\ 6.52)}$$

in ppb. For example, at hardnesses of 50, 100, and 200 mg/L, as $CaCO_3$, the concentration of total recoverable silver should not exceed 1.2, 4.1, and 13 $\mu g/L$, respectively.

In the manufacture of photographic materials, silver is originally present as a halide. When light-exposed photographic films and papers are processed, the silver halide that has not been affected by light is normally removed by solubilization as a thiosulfate complex using a thiosulfate-containing fixing bath. Before disposing of exhausted fixing baths, most of the silver is recovered, frequently by metallic exchange or by electrolytic reduction. The resulting concentrations of silver thiosulfate complex in the final effluents are 0.1–20 mg/L, as total silver (24).

In secondary-waste-treatment plants receiving silver thiosulfate complex, microorganisms convert this complex predominantly to silver sulfide and some metallic silver (see Wastes, industrial). These silver species are substantially removed from the treatment-plant effluent at the settling step (43–44). It has also been reported that silver entering municipal secondary-waste-treatment plants tends to bind quickly to sulfide ions present in the system and precipitate into the treatment-plant sludge (45). Thus, silver discharged to secondary-waste-treatment-plants or into natural waters is not present as the free silver ion but rather as a complexed or insoluble species (see also Recycling, nonferrous metals).

Uses

Analysis. The ability of silver ion to form sparingly soluble precipitates with many anions has been applied to their quantitative determination (see Analytical methods). Bromide, chloride, iodide, thiocyanate, and borate are determined by the titration of solutions containing these anions with standardized silver nitrate solutions in the presence of a suitable indicator. These titrations use fluorescein, tartrazine, rhodamine 6-G, and phenosafranine as indicators (46).

Silver diethyldithiocarbamate [*1470-61-7*] is a commonly used reagent for the spectrophotometric measurement of arsenic in aqueous samples (47) and for the analysis of antimony (48). Silver iodate is used in the determination of chloride in biological samples such as blood (49).

Combination silver–silver salt electrodes have been used in electrochemistry. The potential of the common Ag/AgCl (saturated)–KCl (saturated) reference electrode is +0.199 V. Silver phosphate is suitable for the preparation of a reference electrode for the measurement of aqueous phosphate solutions (50). The silver–silver sulfate–sodium sulfate reference electrode has also been described (51).

Batteries. Primary (nonrechargeable) batteries containing silver compounds have gained in popularity in miniaturized electronic devices. The silver oxide–zinc cell has a cathode of Ag_2O or AgO. These cells are characterized by a high energy output per unit weight and a fairly constant voltage, ca 1.5 V, during discharge. Originally used almost exclusively for military applications, satellites, and space probes, silver oxide–zinc batteries are used today as power sources for wrist watches, pocket calculators, and hearing aids. Silver batteries have excellent shelf stability, with 90% of the original capacity retained after one year of storage at 21°C (see Batteries and electric cells). Silver chromate is one of several oxidizing agents that can be used in lithium primary batteries (52).

Silver sulfide, when pure, conducts electricity like a metal of high specific resistance, yet it has a zero temperature coefficient. This metallic conduction is believed to be due to the fact that a few silver ions exist in the divalent state, thus providing free electrons to transport current. A Japanese patent (53) describes the use of silver sulfide as a solid electrolyte in batteries.

Catalysts. Silver and silver compounds are widely used in research and industry as catalysts for oxidation, reduction, and polymerization reactions. Silver nitrite has been reported as a catalyst for the preparation of propylene oxide (qv) from propylene (54), and silver acetate has been reported as being a suitable catalyst for the production of ethylene oxide (qv) from ethylene (55). The solubility of silver perchlorate in organic solvents makes it a possible catalyst for polymerization reactions, such as the production of butyl acrylate polymers in dimethylformamide (56) or the polymerization of methacrylamide (57). Similarly, the solubility of silver tetrafluoroborate in organic solvents has enhanced its use in the synthesis of 3-pyrrolines by the cyclization of allenic amines (58).

Silver carbonate, alone or on Celite, has been used as a catalyst for the oxidation of methyl esters of D-fructose (59), ethylene (60), propylene (61), trioses (62), and α-diols (63). The mechanism of the catalysis of alcohol oxidation by silver carbonate on Celite has been studied (64).

Silver sulfate has been described as a catalyst for the reduction of aromatic hydrocarbons to cyclohexane derivatives (65). It is also a catalyst for oxidation reactions, and as such has long been recommended for the oxidation of organic materials during the determination of the COD of wastewater samples (66–67) (see Wastes).

Cloud Seeding. In 1947, it was demonstrated that silver iodide could initiate ice crystal formation because, in the β-crystalline form, it is isomorphic with ice crystals. As a result, cloud seeding with silver iodide has been used in weather modification attempts, such as increases and decreases in precipitation (rain or snow), the dissipation of fog. Optimum conditions for cloud seeding are present when precipitation is possible but the nuclei for the crystallization of water are lacking.

Silver iodide crystals, or smoke, for cloud seeding are produced predominantly by ground-based steady-state generators; short-term (5–10 min) flares are also used. In one study, ground-based generators produced an average of 255 g silver iodide crystals per hour; each generator is designed to cover a 259-km^2 (100-mi^2) target area. Cloud seeding techniques and results are reviewed in ref. 68.

Electroplating. Most silver-plating baths employ alkaline solutions of silver cyanide. The silver cyanide complexes that are obtained in a very low concentration of free silver ion in solution produce a much firmer deposit of silver during electroplating than solutions that contain higher concentrations. An excess of cyanide beyond that need to form the $Ag(CN)_2^-$ complex is employed to control the Ag^+ concentration. The silver is added to the solution either directly as silver cyanide or by oxidation of a silver-rod anode. Plating baths frequently contain 40–140 g of silver cyanide per liter (69) (see Electroplating).

Medicinal Preparations. Silver nitrate is used in medicine in the form of a stick, usually containing 1–3% silver chloride, or in solutions of varying concentrations. Uses of silver in medicine today are much reduced because of the availability of a broad spectrum of other remedies; however, silver preparations are not likely to cause sensitization. An example of a procedure still in use is the drop of 1% silver nitrate required in many states to be placed in the eyes of newborn infants as a prophylactic against ophthalmia neonatorum. When applied topically, silver sulfadiazene [*22199-08-2*] has proven effective in preventing infections in burn victims. Aqueous solutions containing 10–20% silver nitrate, or a solid stick, are highly corrosive and can be applied locally to remove warts or cauterize wounds.

Mirrors. The use of silver for the production of mirrors results in a highly reflective coating. The mirror is produced by the reaction of two separate solutions on the glass surface to be coated; one contains silver ammonia complex and the other an organic reducing agent, eg, formaldehyde, sodium potassium tartrate, sugar, or hydrazine. After rinsing, the silver coating can be protected by copper plating or silicone coating (70).

Photography. The largest single user of silver and its compounds is the photographic industry (71–72). Silver nitrate and a halide salt of an alkali metal or an ammonium halide give a light-sensitive silver halide. The silver halide can account for up to 30–40% of the total emulsion weight; gelatin is the other constituent (see Emulsions; Photography). Many different silver halide emulsions are manufactured; the ratios of the halides and preparation details are adjusted according to the specific properties and applications desired. For photographic papers and other emulsions of low sensitivity, silver chloride, chlorobromide, or bromide is employed. Emulsions of high sensitivity are primarily bromide plus up to 10 mol % of silver iodide. Pure silver iodide emulsions are not commercially important and silver fluoride has no comparable photographic use.

Heat-processed photographic systems have been described that utilize silver behenate, silver laurate, and silver stearate. These silver salts are coated on paper in the presence of organic reducing agents (73–74).

New methods of image-recording seek to avoid the high cost of silver; however, continued research has not led to systems that are able to offer the same combination of high sensitivity, high image density, exceptional resolution, permanence, and tricolor recording. In color photography, dyes comprise the finished image, and the emulsion silver is removed during processing. Silver ions present in the photographic fixing solution as a silver thiosulfate complex can be recovered by metallic replacement, electroplating, or sulfide precipitation.

Other Uses. Photochromic glass contains silver chloride (75) and silver molybdate [*13765-74-7*] (76) (see Chromogenic materials). An apparatus for the detection of rain or snow coated with silver nitrate has been described (77). Treatment with silver thiosulfate complex has been reported as dramatically increasing the post-harvest life of cut carnations (78). Silver sulfate has been used in the electrolytic coloring of aluminum (79). Silver sulfate also imparts a yellowish-red color to glass bulbs (80).

BIBLIOGRAPHY

"Silver Compounds" in *ECT* 1st ed., Vol. 12, pp. 438–443, by F. A. Meier, The American Platinum Works; "Silver Compounds" in *ECT* 2nd ed., Vol. 18, pp. 295–309, by Thomas N. Tischer, Eastman Kodak Company.

1. I. C. Smith and B. L. Carson, *Silver*, Vol. 2 of *Trace Metals in the Environment*, Ann Arbor Science, Ann Arbor, Mich., 1977, p. 12.
2. J. F. Kopp and R. C. Kroner, *Trace Metals in Waters of the United States*, U.S. Department of the Interior, FWPCA, Cincinnati, Ohio, 1970, p. 8.
3. *The United States Pharmacopeia XX* (*USP XX–NF XV*), The United States Pharmacopeial Convention, Inc., Rockville, Md., 1980, p. 723.
4. T. R. Hogness and W. C. Johnson, *Qualitative Analysis and Chemical Equilibrium*, 4th ed., Holt, Rinehart & Winston, Inc., New York, 1954, p. 565.
5. S. F. Ravitz, *J. Phys. Chem.* **40**, 61 (1936).
6. A. F. Kapistinski and I. A. Korschunov, *J. Phys. Chem.* (*U.S.S.R.*) **14**, 134 (1940).
7. N. R. Nail, F. Moser, P. E. Goddard, and F. Urback, *Rev. Sci. Instrum.* **28**, 275 (1957).
8. U.S. Pat. 4,022,811 (May 5, 1977), K. Baum, C. D. Beard, and V. Grakaukas (to U.S. Department of the Navy).
9. U.S. Pats. 2,581,518 and 2,581,519 (Feb. 17, 1948), T. Critchley (to Johnson and Sons' Smelting Works, Enfield, England).
10. Ref. 1, p. 203.
11. U.S. Pat. 2,543,792 (Nov. 2, 1949), M. Marasco and J. A. Moede (to E. I. du Pont de Nemours & Co., Inc.).
12. U.S. Pat. 2,614,029 (Feb. 21, 1951), J. A. Moede (to E. I. du Pont de Nemours & Co., Inc.).
13. U.S. Pat. 2,940,828 (Oct. 29, 1957), J. A. Moede (to E. I. du Pont de Nemours & Co., Inc.).
14. *Chem. Eng.* **59**, 217 (Oct. 1952).
15. F. Raschig, *Ann. Chem.* **233**, 93 (1886).
16. F. A. Cotton and G. Wilkinson, *Advanced Inorganic Chemistry*, Interscience Publishers, a division of John Wiley & Sons, Inc., New York, 1962, p. 642.
17. Ref. 1, p. 448.
18. J. McMillan, *Chem. Rev.* **62**, 65 (1962).
19. H. F. Priest, *Inorg. Synth.* **3**, 176 (1950).
20. S. Dunn, ed., *The Silver Market—1980 and 1981*, Handy & Harman, New York, 1981, 1982.
21. F. Feigl, *Spot Tests in Inorganic Analysis*, 5th ed., Elsevier Publishing Company, New York, 1958, pp. 58–64.

22. *Methods for Chemical Analysis of Water and Wastes*, EPA-600/4-79-020, U.S. Environmental Protection Agency, Cincinnati, Ohio, 1979, pp. 272.1-1 and 272.2-1.
23. U.S. Environmental Protection Agency, *Fed. Regist.* **44**(233), 69559 (1979).
24. H. B. Lockhart, Jr., *The Environmental Fate of Silver Discharged to the Environment by the Photographic Industry*, Eastman Kodak Company, Rochester, New York, 1980.
25. Unpublished data, Health, Safety and Human Factors Laboratory, Eastman Kodak Company, Rochester, New York, 1980.
26. *Registry of Toxic Effects of Chemical Substances*, National Institute of Occupational Safety and Health (NIOSH), Cincinnati, Ohio, 1978.
27. K. D. Rosenman, A. Noss, and S. Kon, *J. Occup. Med.* **21**, 430 (1979).
28. A. P. Moss, A. Sugar, and N. A. Hargett, *Arch. Ophthalmol.* **97**, 905 (1979).
29. E. Browning, *Toxicity of Industrial Metals*, Butterworths, London, 1961, pp. 262–267.
30. W. R. Hill and D. M. Pillsbury, *Argyria in the Pharmacology of Silver*, Williams & Wilkins Company, Baltimore, Md., 1939.
31. W. R. Buckley, *Arch. Dermatol.* **88**, 99 (1963).
32. *Threshold Limit Values for Chemical Substances and Physical Agents in the Workroom Environment with Intended Changes for 1980*, American Conference of Governmental Industrial Hygienists, Cincinnati, Ohio, 1980, p. 28.
33. Ref. 1, p. 264.
34. U.S. Environmental Protection Agency, *Fed. Regist.* **40**(248), 59566 (Dec. 24, 1975).
35. National Research Council, *Drinking Water and Health*, National Academy of Sciences, Washington, D.C., 1977, p. 292.
36. U.S. Environmental Protection Agency, *Ambient Water Quality Criteria for Silver*, PB-81-117822, Washington, D.C., 1980, p. C-128.
37. C. J. Terhaar, W. S. Ewell, S. P. Dziuba, and D. W. Fassett, *Photogr. Sci. Eng.* **16**, 370 (1972).
38. C. J. Terhaar, W. S. Ewell, S. P. Dziuba, W. W. White, and P. J. Murphy, *Water Res.* **11**, 101 (1977).
39. C. F. Cooper and W. C. Jolly, *Water Resour. Res.* **6**, 88 (1970).
40. P. H. Davies, J. P. Goettl, Jr., and J. R. Sinley, *Water Res.* **12**, 113 (1978).
41. M. A. Callahan, M. W. Slimack, N. W. Gabel, I. P. May, C. F. Fowler, J. R. Freed, P. Jennings, R. L. Durfee, F. C. Whitmore, B. Maestrik, W. R. Mabey, B. R. Holt, and C. Gould, *Water-Related Environmental Fate of 129 Priority Pollutants, Volume 1*, EPA—440/4-79/029a, U.S. Environmental Protection Agency, Washington, D.C., 1980, pp. 17-1–17-11.
42. Ref. 36, p. B-13.
43. C. C. Bard, J. J. Murphy, D. L. Stone, and C. J. Terhaar, *J. Water Poll. Contr. Fed.* **48**, 389 (1976).
44. J. B. F. Scientific Corporation, *Pathways of Photoprocessing Chemicals in Publicly-Owned Treatment Works*, National Association of Photographic Manufacturers, Inc., Harrison, N.Y., 1977.
45. H. D. Feiler, P. J. Storch, and A. Shattuck, *Treatment and Removal of Priority Industrial Pollutants at Publicly-Owned Treatment Works*, EPA-400/1-79-300, U.S. Environmental Protection Agency, Washington, D.C., 1979.
46. J. Bassett, R. C. Denney, G. H. Jeffery, and J. Mendham, eds., *Vogel's Textbook of Quantitative Inorganic Analysis*, 4th ed., Longman, Inc., New York, 1978, pp. 279–288.
47. E. P. Welsh, *Geol. Surv. Open File Rep.* (*U.S.*), 79 (1979).
48. Y. Yamamoto, M. Kanke, and Y. Mizukami, *Chem. Lett.* **7**, 535 (1972).
49. W. Sendroy, *J. Biol. Chem.* **120**, 335 (1937).
50. H. Tischner, E. Wendler-Kalsch, and H. Kaesche, *Corrosion* (*Houston*) **39**, 510 (1980).
51. D. A. Shores and R. C. John, *J. Appl. Electrochem.* **10**, 275 (1980).
52. P. Cignini, M. Scovi, S. Panero, and G. Pistoia, *J. Power Sources* **3**, 347 (1978).
53. Jpn. Pat. 80 119,366 (March 7, 1979), (to Citizen Watch Company, Ltd.).
54. Jpn. Pat. 80 85,574 (June 27, 1980), (to Showa Denko K.K.).
55. Fr. Pat. 5388 (April 28, 1978), J. M. Cognion and J. Kervennal (to Produits Chimiques Ugine Kuhlman).
56. M. Sahan and C. Senvar, *Chim. Acta Turc.* **8**, 55 (1980).
57. S. P. Manikam, N. R. Subbaratnam, and K. Venkatardo, *J. Polym. Sci. Polym. Chem. Ed.* **18**, 1679 (1980).
58. A. Claesson, C. Sahlberg, and K. Luthman, *Acta Chem. Scand. Ser. B* **B33**, 309 (1979).
59. H. Hammer and S. Morgenlie, *Acta Chem. Scand. Ser. B* **B32**, 343 (1978).
60. U.S. Pat. 4,102,820 (July 25, 1978), S. B. Cavitt (to Texaco Development Corporation).

61. Ger. Pat. 2,312,429 (Sept. 27, 1973), C. Piccinini, M. Morello, and P. Rebora (Snam Progetti S.P.A.).
62. S. Morgenlie, *Acta Chem. Scand.* **27,** 3009 (1973).
63. J. Bastard, M. Fetizon, and J. C. Gromain, *Tetrahedron* **29,** 2867 (1973).
64. F. J. Kakis, M. Fetizon, N. Douchkine, M. Golfier, P. Mourgues, and T. Prange, *J. Org. Chem.* **29,** 523 (1974).
65. U.S. Pat. 4,067,915 (July 17, 1980), S. Matsuhira, M. Nishino, and Y. Yasuhara (to Toray Industries, Inc.).
66. M. C. Rand, A. E. Greenberg, and M. J. Taras, eds., *Standard Methods for the Examination of Water and Wastewater*, 14th ed., American Public Health Association, Washington, D.C., 1976, pp. 550–54.
67. R. Wilson, *A Study of New Catalytic Agents to Determine Chemical Oxygen Demand*, PB-270965, National Technical Information Service (NTIS), Washington, D.C., 1977.
68. Ref. 1, pp. 220–225.
69. M. A. Orr in A. Butts and C. D. Coxe, eds., *Silver-Economics, Metallurgy and Use*, D. Van Nostrand Company, Inc., Princeton, N.J., 1967, p. 185.
70. R. D. Pohl in ref. 69, p. 193.
71. B. H. Carroll, G. C. Higgins, and T. H. James, *Introduction to Photographic Theory, The Silver Halide Process*, John Wiley & Sons, Inc., New York, 1980.
72. T. H. James, ed., *Theory of the Photographic Process*, MacMillan, Inc., New York, 1977.
73. Ger. Pat. 2,855,932 (July 5, 1979), K. Akashi, M. Akiyama, T. Shiga, T. Matsui, Y. Hayashi, and T. Kimura (Asahi Chemical Industry Company, Ltd.).
74. Jpn. Pat. 79 76926 (May 25, 1979), (to Fuji Photo Film Co., Ltd.).
75. U.S. Pat. 3,252,374 (Feb. 15, 1962), S. D. Stookey (to Corning Glass Works).
76. Belg. Pat. 644,989 (Sept. 10, 1964), L. C. Sawchuk and S. D. Stookey (to Corning Glass Works).
77. Jpn. Pat. 80 19,375 (May 26, 1980), (to Mitsui Toastsu Chemicals, Inc.).
78. G. G. Dimallo and J. Van Staden, *Z. Pflanzenphysiol.* **99,** 9 (1980).
79. Jpn. Pat. 78 04,504 (Feb. 17, 1978), T. Abe, T. Uchiyama, and T. Ohtsuka (to Showa Aluminium K.K.).
80. Jpn. Pat. 78 03,412 (Jan. 13, 1978), M. Sangen (to Matsushita Electronics Corporation).

HAINES B. LOCKHART, JR.
Eastman Kodak Company

SIMULATION AND PROCESS DESIGN

Simulation is the use of one system to imitate the actions of another, ie, to "simulate" it. Simulation may be important because the second or subject system does not yet exist (the design problem), because the system is too time consuming, dangerous or costly to operate itself to develop the necessary information (the cost problem), or because not enough is yet known about it (the information problem).

Although simulation in theory embraces the use of any kind of physical system to imitate the one in question, the term has come almost universally to mean the use of a digital computer in exercising a mathematical model of the system under study. It has been formally defined as "the process of designing a computerized model of a system (or process) and conducting experiments with this model for the purpose either of understanding the behavior of the system or of evaluating various strategies for the operation of the system" (1).

Therefore the principal steps in any simulation study are as follows:

(*1*) A precise formulation of the mathematical model of the system to be studied to the complexity needed for the study or to the limits of the available computer capability in speed of operation and available computing time. All simplifying assumptions used in developing the mathematical model detract to some degree from the faithfulness of the simulation results.

(*2*) Conversion of the mathematical model into a computer program for running on the available digital computer system. There is a wide choice of possibilities here including several special programming languages developed especially for simulation projects. The resulting computer program must then be verified, ie, it must be debugged and tested for logical accuracy versus the original model and refined and modified as necessary (2).

(*3*) Validation of the model and its representation in the computer program. That is, can the simulated system be proven to be a reasonable facsimile of the real system that it was intended to represent.

(*4*) Exercise of the model on the computer including the design of the experimental program to be carried out (ie, the choice of the several conditions to be tested).

(*5*) Interpretation and use of the simulation results to achieve the ultimate purpose of the intended study such as a new process design, a new production schedule, determination of the safe limits of operation of the process, etc.

(*6*) Documentation of the study including the recording of project activities and results as well as the documenting of the mathematical model, the computer program and their uses.

Figure 1 presents this sequence as a flow chart (1). In this figure strategic and tactical planning represent the establishment of the overall objectives and the detailed procedures respectively of the experimental program for exercising the system model on the computer (Item *4* above). These are discussed more fully in the last section of this article (see also Operations planning).

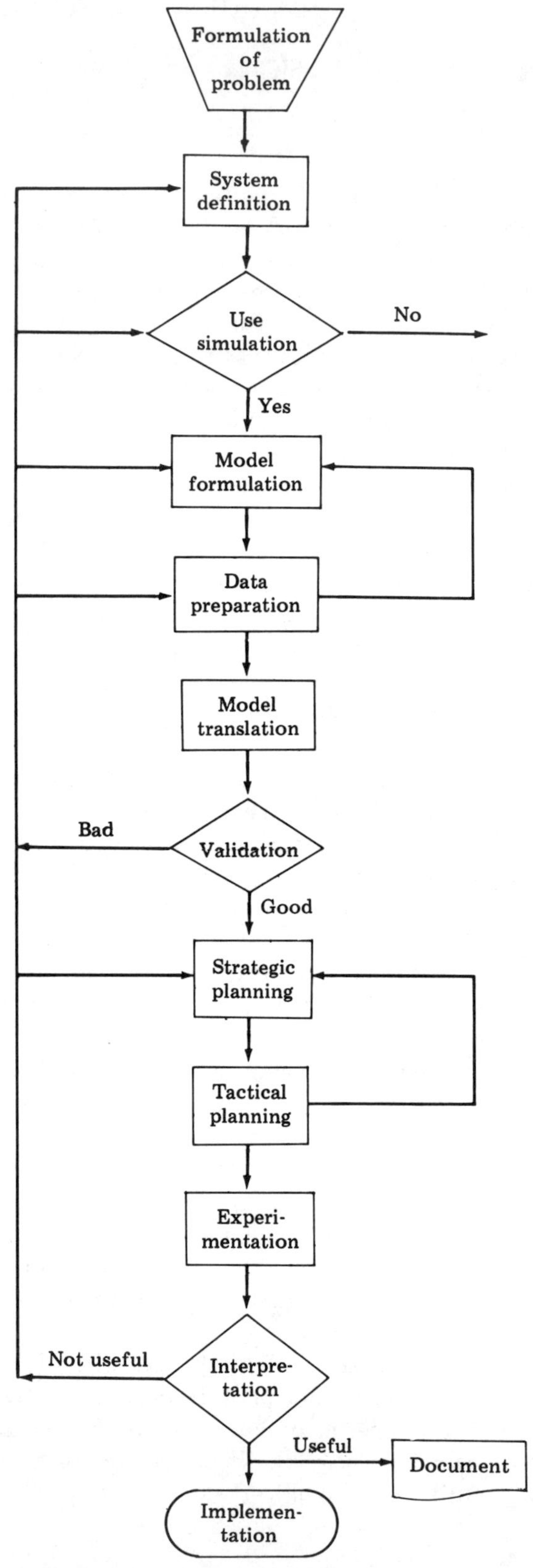

Figure 1. Steps in the simulation techniques (1).

The Advantages and Disadvantages of Simulation

Using simulation offers several advantages (3). Simulation makes it possible to investigate and experiment with the complex internal interactions of the system being studied whether a single process, an overall plant, or a complete company. It allows investigation of the sensitivity of a system to small changes in internal parameters or environmental conditions and thus helps determine the accuracy to which these factors must be known. Simulation also allows the testing of the model system in regimes of operation that would be too costly, too dangerous, too time consuming, or beyond the operating ranges of any one particular physical system, ie, systems sizes, construction materials, etc. Because it obviously causes no disturbances in the real system being studied, investigations at the limits of operation of that real system can be made with impunity without danger to the system or its human operators.

The process of modeling the desired system for a simulation may often be more valuable than the actual simulation itself in terms of what is learned about the system. Such knowledge often suggests changes in the system that can then themselves be tested by the same simulation. Simulation is an excellent teaching tool for plant operating personnel to allow them to become acquainted with a new process before they have to start up a new plant. Any problems foreseen with the new plant can then be anticipated and their remedies used as plant operating practice.

Time compression may be required for systems or processes with long time frames. Simulation affords complete control over time since a phenomenon may be speeded up or slowed down at will. Analysis of urban-decay problems, for example, is in this category.

For certain types of stochastic or random-variable problems, the sequence of events may be of particular importance. Statistical information about expected values or moments obtained from plant experimental data alone may not be sufficient to completely describe the process. In these cases, computer simulations with known statistical inputs may be the only satisfactory way of providing the necessary information. These problems are more likely to arise with discrete manufacturing systems or solids-handling systems rather than the continuous, fluid-flow systems usually encountered in chemical engineering studies. However, there are important examples of such stochastic events or data in the literature: ref. 4 provides data on the spectral content of the disturbances of several processes common to the chemical and paper industries; a stochastic polymerization model is developed in ref. 5; the statistics of plant process unit failure are discussed in ref. 6; and ref. 7 describes modeling of coal-pyrolysis reactors.

At the same time some important drawbacks or disadvantages of a simulation based study should be kept in mind. For example, exercise of a simulation model can be very expensive in terms of manpower and computer time if a proper design of experiments is not carried out. An extensive development period may be necessary in the production of the model itself, and the parameters of the model (kinetic coefficients, heat transfer coefficients, etc) may be difficult to determine thus necessitating very extensive plant tests on the real system with associated extensive data collection, reduction, and interpretation time and costs to obtain this required data. Hidden critical assumptions in the development of the model may cause it to deviate from the real system thus leading to erroneous results (see Design of experiments).

One of the primary reasons for using simulation is that many systems of interest cannot be studied or evaluated adequately by standard mathematical or operations-research techniques. These problems may arise for any number of reasons. There may, for example, be inherent stochastic, or random, processes in the system that can be analyzed statistically only by simulation. Complex interactions of the system variables may not be possible to model precisely using systems of equations solvable by closed-form mathematical techniques. The shear size and mathematical intractability of the resulting system of equations of variables and constraints may not be amenable to solution on the currently available machines any other way than by simulation.

Of course, direct experimentation on the physical system itself could perhaps supply the information needed but only with the appearance of other difficulties. Thus, direct experimentation upon the real-life system can eliminate many of the difficulties in obtaining a good match between the model and actual conditions; however, the disadvantages of direct experimentation are also sometimes great and the comparison will ultimately favor the simulation approach over direct experimentation (8). Experimentation could seriously disrupt company operations especially if many alternatives must be explored. It may be difficult to maintain the same operating conditions for each replication or run of the experiment and more time consuming or costly to obtain the same sample size (and therefore statistical significance).

The Development of Computer Simulation

Simulation was originally carried out with physical models of the system under study such as ship towing of tanks and ship-hull models, wind tunnels and aircraft models, chemical industry pilot plants, etc. With the development of the analogue computer in the late 1940s and 1950s simulation using such computers became the primary medium for such studies (9–10) particularly for aircraft and missile-system studies.

However, the difficulty of programming the analogue computer resulted in their almost total replacement by digital computers by the 1970s. The development of special programming techniques for digital computers and the tremendous rate of development of the digital computer field itself through the integrated circuit speeded the conversion until today almost all simulation is conducted on digital computers rather than on analogue computers. Hybrid computers, a combination of an analogue computer and a digital computer in one unit, attempted to combine the best features of both (11) but unfortunately did not adequately solve the programming problem of the analogue portion of the duo. Thus, these too have been effectively superseded by the digital computer (see Computers).

Simulation Models. The heart of every simulation study is the mathematical model of the system under investigation. Since the study usually involves the exercise of this model by a digital computer, the accuracy of the results obtained from the study will only be as good as the faithfulness of the model that is developed for it.

Because of the nature of the problems posed and the solution methods used, simulation studies must often be carried out on dynamic systems, ones for which the time history of the system's behavior is important. Such simulations require the exercise of the system mathematical models as the solutions of sets of differential equations or their equivalents in finite difference equations, matrix equations or statistical expressions.

Static or steady-state systems, on the other hand, are generally modeled as sets of algebraic equations. Although solution of these sets of equations is thus usually more straightforward, the greater ease of solution is usually compensated by the study of much more complex and extensive systems, each to the limits of the available computing power, particularly in carrying out any design-optimization studies. These static or steady-state systems are extremely important for engineering design studies, particularly of process systems, as discussed below.

Much laboratory instruction in the university today is carried out by simulation rather than through experiments on representative physical equipment. Likewise, a large percentage of university research on the behavior of physical systems is really carried out mainly by computer simulation instead of through laboratory experiments. The physical systems themselves are used primarily to obtain parameter values for the mathematical models or to validate the results of simulation. Industry is also adopting this trend. Thus, despite the many words of caution that have been written concerning the use of simulation instead of direct physical experimentation, there is a growing tendency to use simulation to replace direct experimentation in all fields of science and engineering.

All simulation models are in a sense input–output models (12): they yield the output of the system given the input to its interacting subsystems. Simulation models are therefore "run" rather than "solved" to obtain the desired information or results. They are incapable of generating a solution on their own in the sense of analytical models; they can only serve as a tool to analyze the behavior of a system under conditions specified by the experimenter. Thus simulation, including the necessary inherent model building, is not a theory but a methodology of problem solving. Furthermore, simulation is only one of several valuable problem-solving approaches available for the system analyst.

The mathematical model for any system to be studied can, of course, be programmed specifically for simulation on the digital computer and the resulting program operated to duplicate any reasonable set of operating conditions, but simulation is so important that many general-purpose problem-oriented languages or programs have been developed especially for this field. The best of these incorporate the most applicable numerical analysis techniques. Within any one program, they can handle a wide variety of different problems from a wide spectrum of industries. The relative ease of use of these languages and their importance in the computer-aided design process give them particular importance for simulation and process design. Any one language is usually specialized in its use to either dynamic or steady-state physical systems.

Dynamic Simulation Techniques

Classification of Dynamic Systems. The models developed for simulation studies, particularly those for dynamic system studies, may be divided in two different ways—according to whether the system being modeled is continuous or discrete, or according to whether the output is completely determined by the input and initial values of the system (ie, a deterministic system) or whether there is some inherent randomness in the system or in the external upsets to which it is subjected (a stochastic one).

Continuous systems are those whose dependent variables can assume any value over a specific range or interval, as for example, fluid flow in a pipe or temperature in a furnace. Discrete systems are those whose variables can assume only discrete or integer values over a finite, or possibly countably infinite, set. An example of a discrete system is a factory building automobiles or any other product that can be treated only as units.

Discrete systems are idealized as network flow systems and are characterized by the following three factors (13): (*1*) The system contains components (or elements or subsystems) each of which performs definite and prescribed functions. (*2*) Items flow through the system, from one component to another, requiring the performance of a function at a component before the item can move on to the next component. (*3*) Components have a finite capacity to process the items and therefore, items may have to wait in queues before reaching a particular component.

The main objective in studying discrete systems is to examine their behavior and to determine the capacity of the system, ie, how many items will pass through the system in a given period of time as a function of the structure of the system. A typical chemical engineering example of this type of problem is that involved in plant production scheduling, such as for polymer production (14).

The analytical techniques used to solve such problems are queueing theory and stochastic processes. The computations in this type of simulation consist, to a large extent, in keeping track of where individual items are at any particular time, moving them from queue to component, timing the necessary processing or functional transformations and removing and transporting the items to other components or waiting lines. The result of a discrete simulation "run" is a set of statistics describing the behavior of the simulated system during the run.

Figure 2 illustrates the difference between a deterministic and a stochastic system (3). It should be pointed out that any particular system may contain both deterministic and stochastic elements and can be effectively modeled with both elements included. Further, continuous systems may undergo discrete jumps or changes in their operating variables such as a step change in the set points of controllers. Furthermore, even systems that are physically completely deterministic may often be modeled as stochastic ones in order to give a smaller resulting model and to allow the use of statistical techniques in the evaluation of simulation results.

Continuous-system modeling from basic chemical and physical principles is well treated by references 15–17. Most texts in automatic control also contain considerable discussion of the principles of continuous-systems modeling (18–20). Indeed, except

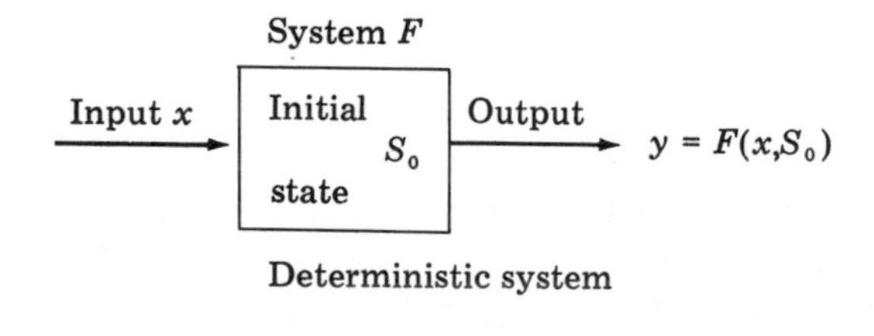

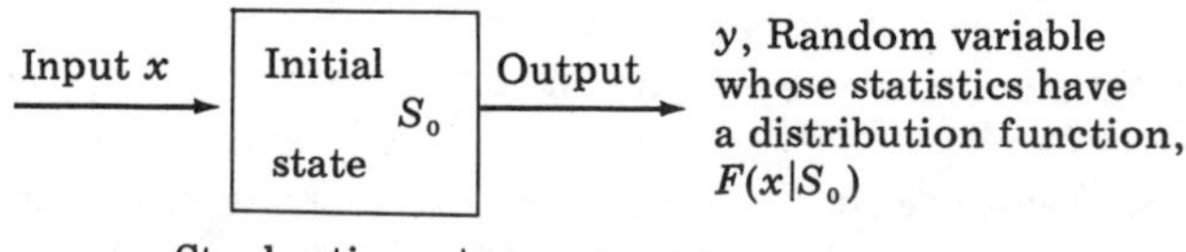

Figure 2. Comparison of deterministic and stochastic systems (3).

where analytical techniques are applicable, dynamic simulation is an indispensable tool in automatic-control studies, regardless of their field of application. The modeling of discrete systems is well covered by references 21 and 22.

Model Validation. The single most vital step of mathematical modeling and even of the simulation study itself is that of model validation in order to assure credibility of the final results. Where the existence of the real system being modeled permits, a comparison of simulation output results with historical data from the real system for the same input is an effective method of validation. However, if the use of the model is to be predictive of new operating regimes, this method of validation alone is not sufficient, and the model must be used with caution because of the problem of extrapolation and the possible existence of system nonlinearities that are not included explicitly in the model or detected by the insufficient validation. Another problem is that of determining when either the simulated or the real system is at a true equilibrium or steady state in its behavior. This is particularly true of systems subject to stochastic variations in input or behavior. Also, if the output of the model is in the form of a time series, the task of performing the necessary regression or spectral analysis can present additional difficulties. Methods of model verification are discussed in references 1 and 23–25.

Programming Languages and Systems for Dynamic Simulations. The importance of dynamic simulation as a technique is reflected in the fact that over 20 special simulation languages for dynamic systems are listed in a recent survey (26). A classification of some of the more important of the dynamic simulation languages is presented in Figure 3. Among the continuous-change languages, the block-oriented languages are especially important because their block structures allow process modeling to be done directly in the computer language, thus eliminating the mathematical equation-writing step. Further, for both continuous- and discrete-change languages, the block structure is so well known by users that in most cases it can be used directly for documentation of the system—a most important step (27,29).

In addition to the distinct language types, there are also numerous versions and extensions of them (12–13,26,61).

Besides the two classes of distinct computer programming languages, continuous- and discrete change, there is also a group of others that are themselves really self-contained methods of analysis and modeling techniques. These languages are composed of groups of subroutines programmed in a common base language such as FORTRAN. A particularly important example of this type of system is GERTS (62) which is useful in the modeling and analysis of any stochastic network problem such as an assembly line or conveyor system.

With such a wide variety of possibilities, the choice of the best language for a particular simulation task becomes difficult indeed. Aside from the consideration of a particular language's availability on the specific digital computer to be used, which is itself an important consideration, a flow-chart technique that gives an excellent step-by-step method of selection from among the available languages has been developed (Fig. 4).

Dynamic simulations are used in the chemical processing industry for several types of problems. The kinetic parameters for mathematical models of complex chemical reaction systems have been determined through application of parameter identification techniques against the composition-change history of sample real reactions (63). These kinetic-model parameters were then used in further simulation runs to optimize the design of a chemical reactor or its operation, ie, to improve the production rate or

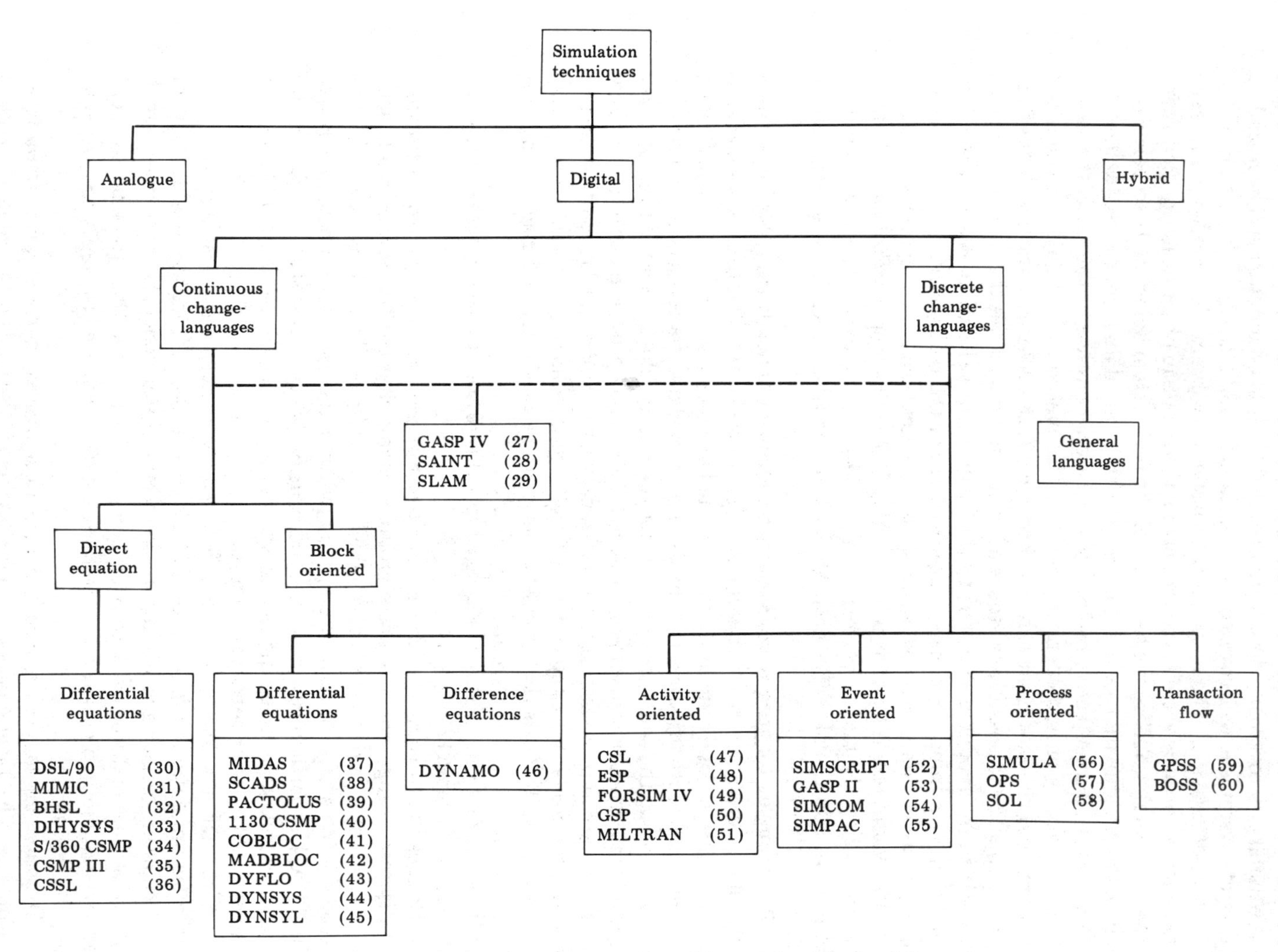

Figure 3. A language classification scheme, updated from ref. 1.

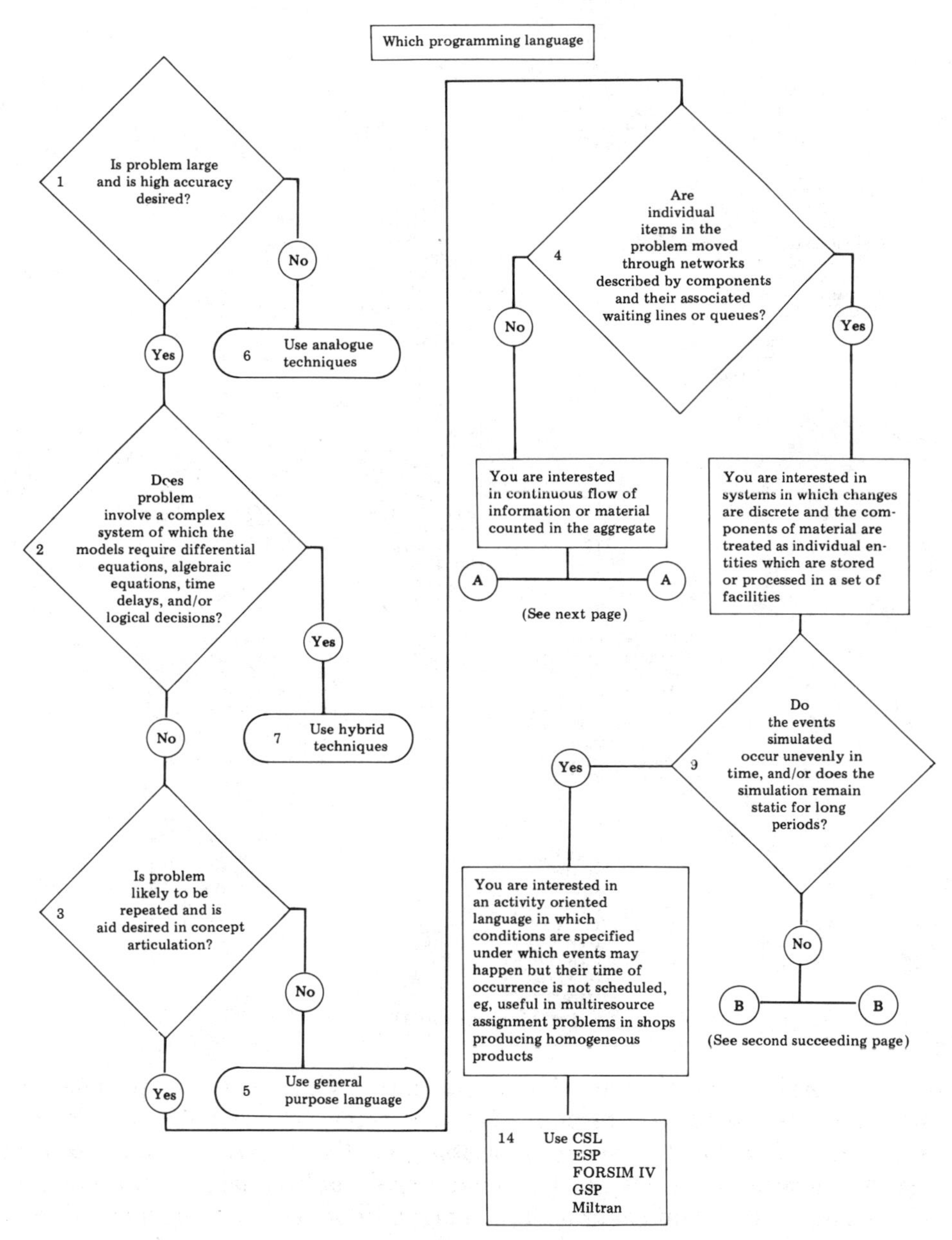

Figure 4. Steps in the choice of simulation techniques and languages (12).

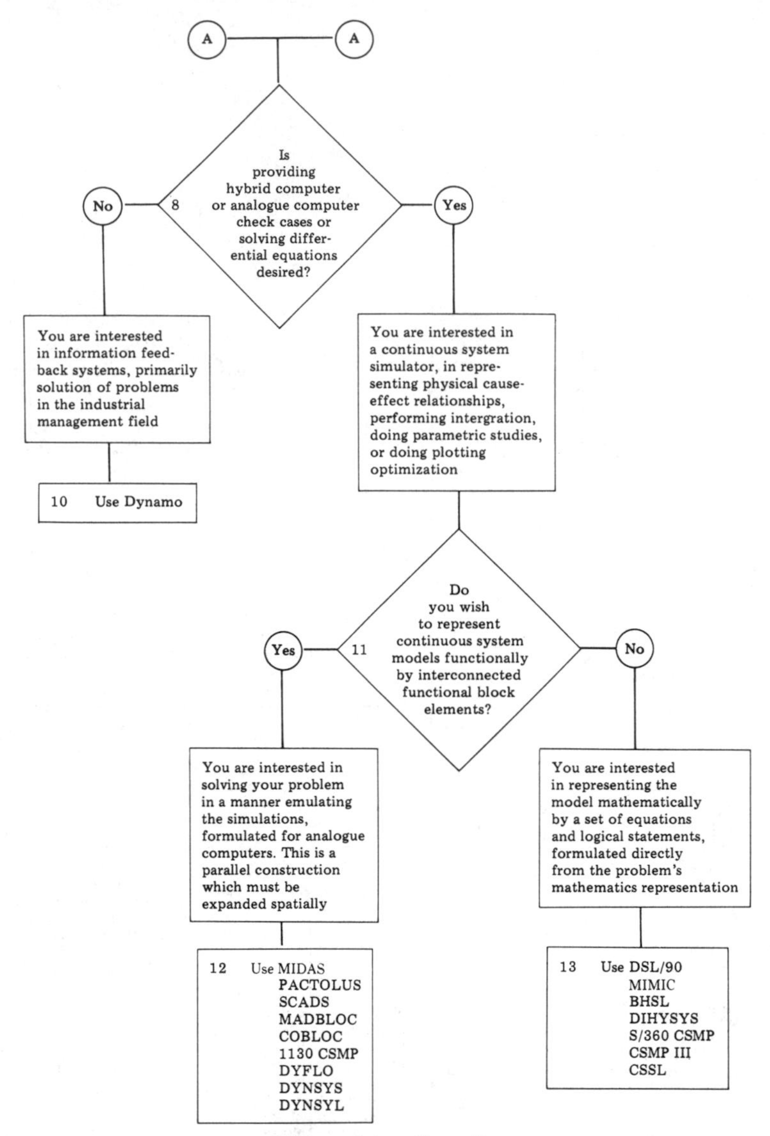

Figure 4. *(continued).*

throughput of a given chemical reaction, to maximize yield, or to design the most appropriate control system for the reactor (64) (see Reactor technology). Simulation has also been used to design and prove the applicability of other complex advanced control systems for chemical processes, such as feedforward, multivariable, noninteracting control for large distillation columns (65). The optimum cycles for regeneration of catalysts, for the cleaning of cracking-furnace tubes, for the relining of furnaces, for sootblowing of boiler tubes and for other chemical and petroleum plant processes that have complex and time-varying life cycles can be approached by simulation, as can the scheduling of multiple process units operating in parallel, such as multiple batch-reactor systems that produce a variety of products.

The literature presents many examples of the use of one or more dynamic simu-

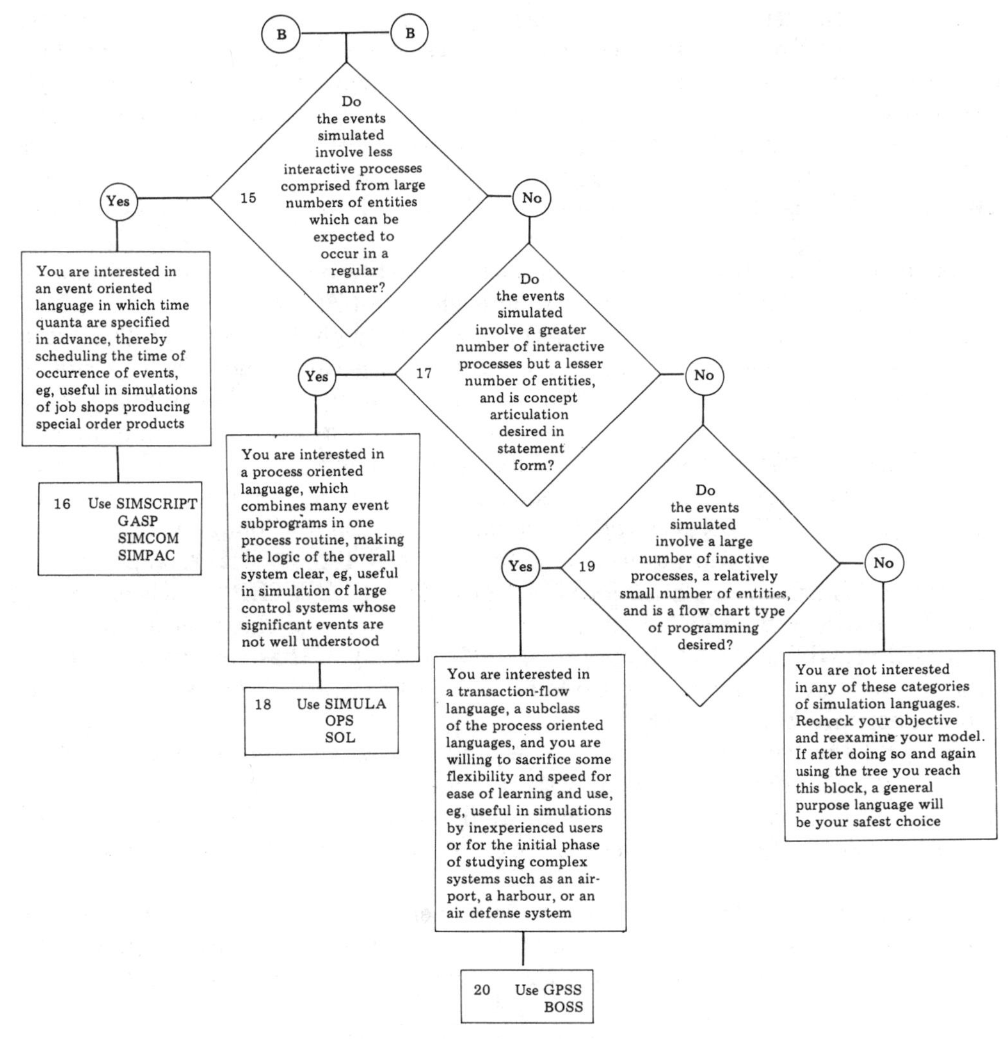

Figure 4. (*continued*).

lation package programs to study the following chemical engineering processes: distillation-column design, startup, and control (51,66); extraction-column design (45); batch-reaction control systems (67); polymerization-reaction systems (6,14,68); compressor and gas-flow systems (69); gas-turbine power systems (70); waste-heat boilers (71); nuclear power-plant system for control studies (72–73); and mixer–settler systems (74).

Steady-State Simulations

A principal use of simulation in chemical engineering has been that of process design and process operation studies. Although these have involved the use of both steady-state and dynamic simulations, the former has by far predominated. In the steady-state case, the principal application has been the optimization of the design of individual process units and of the sizing of large plant trains or even of complete plants and in the determination of the best operating conditions for existing process systems. The literature is replete with papers describing the use of these techniques for these purposes. An excellent paper describes the Monsanto Chemical Company's experience in this area (75). The contribution of simulation here is its capability of trying many cases of different values of the primary design variables of the process under consideration and in this way assuring that a nearly optimal design has been achieved even when an analytical solution of the optimization routine itself is impossible.

Simulation Programs. The great importance of these types of computations to chemical engineering has resulted in the development of a whole group of special languages or computer programs for this purpose. Although they are generally classed as steady-state simulation programs, they really comprise much more than the simple simulation of a chemical process or processes, as will be discussed below. Some of the most important of these programs, which are also known generally as material and energy-balance or flow-sheet programs, are listed in Table 1.

The most widely used of the steady-state simulation programs is probably FLOWTRAN (76–77) because the developer, the Monsanto Company, has made it widely available to the educational community through the CACHE (Computer Aids for Chemical Engineering Education) Committee (77). FLOWTRAN has been greatly expanded in the new ASPEN (Advanced System for Process Engineering) program developed at MIT under DOE sponsorship (78). Many of these programs were developed originally as university research projects (83–84,91,93). Others came originally from industrial sources (76,92,94).

Several excellent reviews treat these programs in considerable detail (95–98). These languages are distinguished by their inclusion of routines incorporating computations of the physical properties of the fluids and materials being processed as well as of other routines to calculate the relative economics of the several design cases being considered in the simulation.

Figure 5 outlines the steps in the development of a new chemical process (99). The shaded blocks indicate those steps that can be carried out on the computer. The blocks within the dashed border I are the steps normally encompassed in one of the

Table 1. A List of Some of the More Important Steady-State Process Simulator Programs

Program	References
FLOWTRAN	76, 77
ASPEN, ASPEN PLUS	78–79
GEMS	80–81
MASSBAL	82
PACER 245	83
CHESS	84–86
FLOWPACK	87
PROVES	88–89
CONCEPT–MARK III	90
SLED	91
CHEVRON	92
GEMCS	93
GMB	94

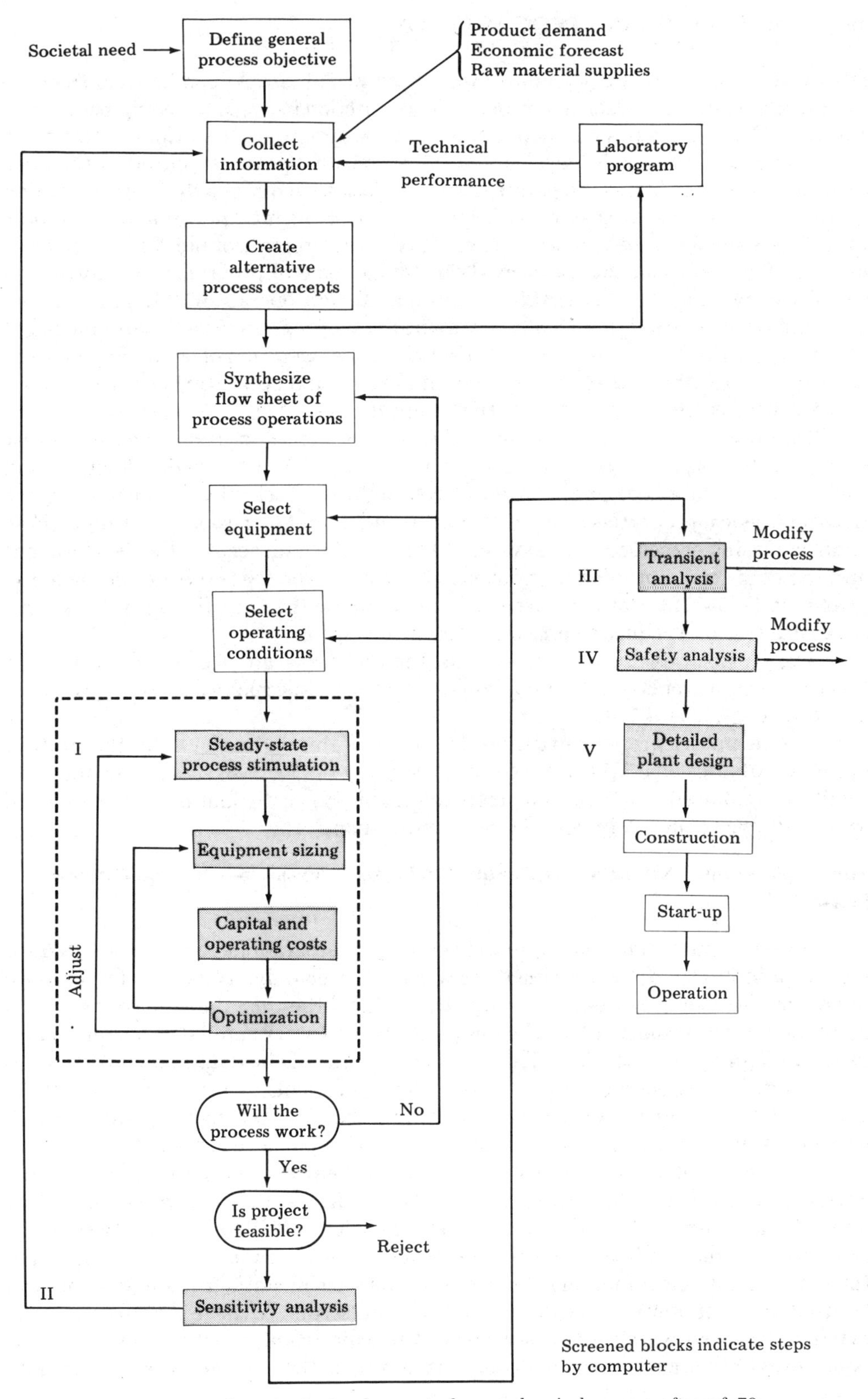

Figure 5. Steps in the development of a new chemical process, after ref. 79.

steady-state simulation design languages listed in Table 1. As can be seen, they use the process-system simulation to provide the connection between the equipment-sizing and capital-cost programs on the one hand and the optimization routines on the other. On the diagram of Figure 5, blocks II and IV can usually also be carried out by the same programs as block I but as a separate analysis. Block III requires a dynamic simulation system such as those discussed earlier in Figure 4 or the incorporation of a dynamic as well as a steady-state simulation capability in the program of block I. The computations of block V can also be carried out by the general programs but more often specific programs are developed for each separate unit operation (100–101).

It should be noted that steady-state simulation programs like those of block I of Figure 5 can also be used for the optimization of the operation of an already existing process by using the sizes of this equipment as input and optimizing on the operating cost features of the program rather than capital costs.

Two main problem areas arise in the development and use of the generalized steady-state design programs listed in Table 1. These relate to the development and use of first, a data base and/or a set of computational schemes for expressing the thermophysical properties of the materials (solid, liquid or gaseous, including phase changes) being processed in the system (90,102–106); and, second, the development and use of, a set of optimization routines that can handle the resulting nonlinear expressions (107–112). Both are necessary to automate the overall design procedures adequately and to reduce required human intervention.

The difficulties involved in establishing and using an effective data base for thermophysical property data and a proposal for improving such systems are presented in references 113 and 114.

The literature presents examples of the use of these techniques for the study of ammonia plants (77,107,112), a refinery glass plant (94), a flash separation unit (76), a sulfuric acid plant (96,115), a polymerization unit (86), a distillation system (95) and in overall project development (88–89), among many others.

The Application of Simulation Techniques to Problem Solution—The Experimental Phase

Once the model and corresponding computer program have been validated for the purpose intended, the investigator's work has often only just begun. The exercise of this model to obtain the final results desired and the final data analysis can be a main part of the project. Thus, the design of the necessary simulation experimental program is vital to the project. This is especially true of the dynamic simulation investigated. In the case of steady-state simulation (particularly in the use of one of the packaged programs), the exercise of the model is handled through a gradient search or other similar routine included in the optimization package (107–112).

The design of a computer simulation experiment is essentially a plan for purchasing a quantity of information that may be acquired at a varying price, depending upon the manner in which the data are obtained. The effective use of experimental resources is profoundly affected by the choice of design, first because the design of the experiment determines in great measure the form of statistical analysis that can be used appropriately to analyze the results; and second because the success of the experiment in answering the questions of the experimenter, without excessive expenditures of time and resources, depends largely upon the right choice of experimental design.

Computer simulation experiments are expensive in terms of time and labor of the experimenter as well as the cost of machine time. Since effort spent on one investigation is unavailable for another, it is important that the researcher plan to obtain as much information as possible from each experiment. The primary purpose of conducting simulation studies is to learn as much as possible about the behavior of the system being simulated at the lowest possible cost. To do so, we must plan and design carefully not only the model but also how it is to be run or used.

The running of a simulation experiment is the process of exercising or running the model to observe and analyze the resulting information. The experimental design selects a particular approach to gathering the information needed in order to draw valid inferences (see Design of experiments).

Differences between Experimentation and Simulation. Although the underlying objectives of computer simulation experiments are essentially the same as those for conducting physical experiments, some differences must be considered. Among the more important of these are the difficulties in defining a single datum point or sample; the ease with which experimental conditions can be repeated or reproduced; the ease of stopping and resuming experimentation; the presence or absence of correlation between subsequent data points; and the ability to control stochastic variability which in physical experiments is beyond the control of the experimenter.

In determining how to run a model and analyze the results, one of the first issues to be decided is what shall be considered a single datum point or sample. There are several possibilities, including: (*1*) a complete run of the model. This may entail considering the mean or average value of the response variable for the entire run as being the datum point. (*2*) A fixed time period during the run in terms of simulated time. For example, the model might be run for n time periods, when n is measured in hours, days, weeks, etc, and the mean or average value of the response variable for each time period considered a datum point. (*3*) Each transaction considered a separate sample. For example, turn-around time for each job or the total time in the system of each customer is considered a separate datum point. (*4*) Transactions aggregated into groups of fixed size. For example, we might take the turn-around time of each set of 25 jobs flowing through the system, and then use the mean time of the group as a single datum point.

Each of these has advantages and disadvantages. Depending upon the model, the first definition of a single datum point or response may result in excessive use of computer time owing to run lengths, or in having to discard the early part of each run owing to startup transient conditions. The second method, fixed-time increments, introduces the possibility of correlating data samples from one time increment to the next, as do the last two.

The ease with which experimental conditions can be repeated or reproduced in a computer model is often a distinct advantage of computer simulation over physical experimentation. If we are interested in comparing two alternatives in a relative manner only, we can run the model in such a way that each alternative is compared under identical conditions (same sequence of events). This is accomplished by repeating or using the same series of random numbers for each alternative, which reduces the residual variation in mean performance of the alternatives and requires considerably smaller sample sizes to establish statistically significant differences in the response. On the other hand, if we are interested in evaluation of absolute system performance, we can use a new stream of random numbers for each run.

Another difference between experimentation with computer models and that with physical systems is the ease with which the experimentation can be stopped and resumed. This facility allows the use of sequential or heuristic experimental methods that might not be feasible in a real-world system. With the computer model, it is always possible to stop the experiment while the results are analyzed and a decision made whether to change the parameters or continue as before. Being able to put the model back on the shelf, so to speak, while we think about what is occurring can be a distinct advantage not readily available in the real world. Again, the question of starting conditions may turn this advantage into a disadvantage.

Analysis of computer simulation experiments often presents some difficult problems because outputs are sometimes auto- or serially correlated. Autocorrelation arises when the observations in the output series are not independent of each other as is assumed by many experimental designs.

In many simulation models, the value of one output observation depends upon the value of the previous observation or upon some other past observation. Thus, not as much information is contained in that observation as would be if the two were completely independent. Since most experimental designs described in the literature assume independence of observations, many common statistical techniques are not directly applicable to autocorrelated simulation results.

In a well-conducted study, there are two areas of interface between the experimental planning design and the total computer simulation process. Figure 6 illustrates these two interface areas with a block diagram representing the overall simulation study process from the genesis of the problem to the final documentation and implementation of the study results. The two blocks representing the experimental planning–design functions are shown in heavy outline. Once the experimental objectives and a system definition have been established and the decision to employ computer simulation made, careful preliminary experimental planning at an early stage in the development of the computer model is necessary. One should have a fairly detailed idea of the experimental plan early so that the model itself can be better planned to provide efficient generation, and possibly partial analysis, of the desired experimental data. Since computer time is expensive, knowledge of the magnitude and special requirements of the desired data output may have a significant impact on the concept and details of the model. After the model is designed and coded is a poor time to find out that the operating philosophy and output formats need to be changed to meet the experimental data-generation needs.

As indicated in Figure 6, the second and principal function of experimental design is to provide the final strategic and tactical plans for executing the experiment. Here the project constraints on time and cost must be updated to the current conditions and imposed upon the design. Even though careful planning and budget control may have been exercised from the beginning of the project, now is the time to take a good hard, realistic look at the resources remaining and how best to utilize them. Whether the objectives of a given study are effectively and efficiently accomplished depends to a significant degree upon the care and skill exercised in experimental design. The larger and more complex the simulation, the more critical this phase becomes.

There are very few studies that do not have resource limitations imposed upon them in terms of schedule, dollars, and computer availability. In most situations these constraints place severe restrictions on the experimental design and often override academic statistical considerations.

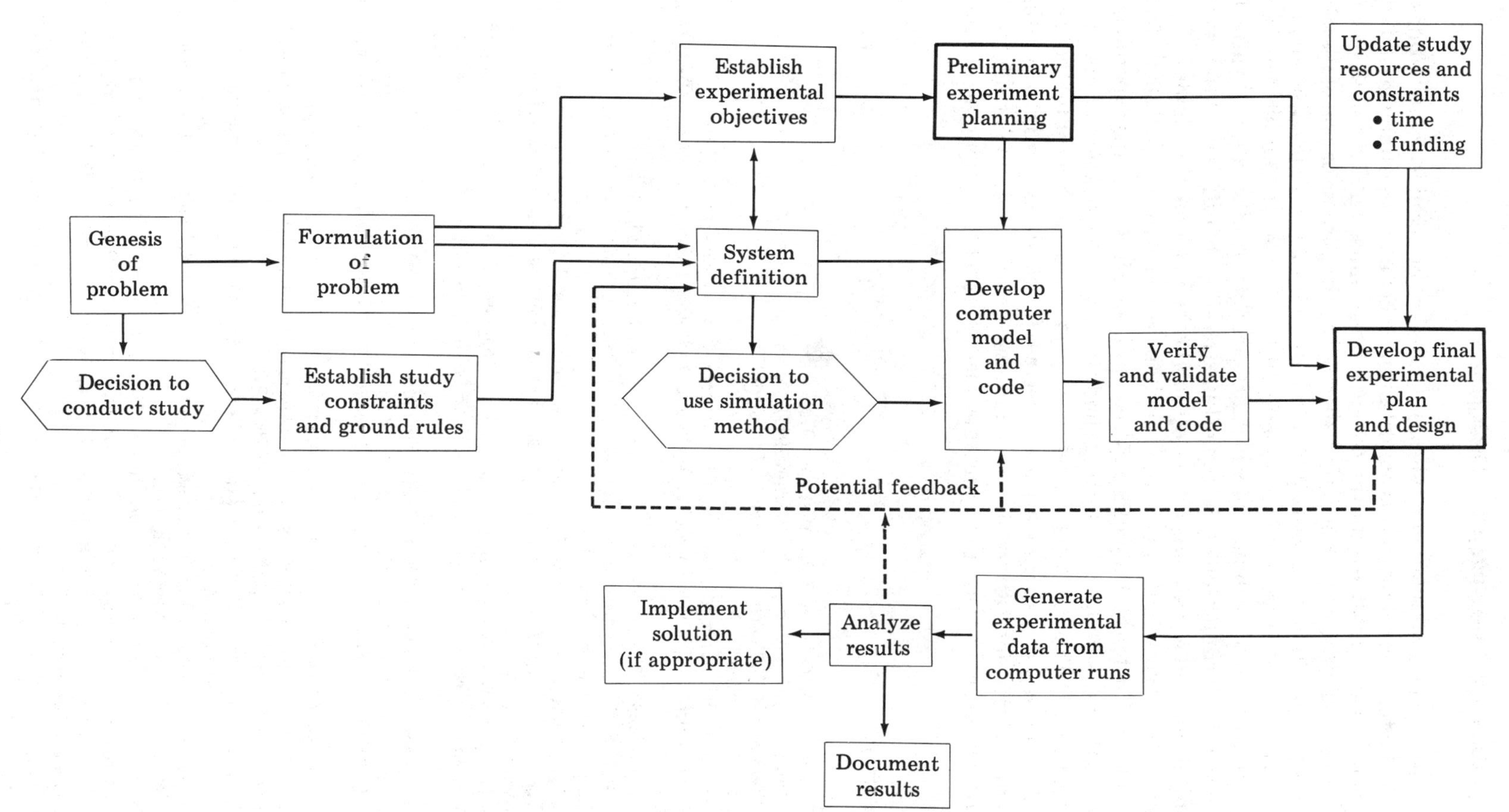

Figure 6. Overall simulation process (35).

In most complex simulation studies, the number of possible combinations of factors and factor levels of interest is almost infinite; hence, a large number of design trade-offs are made to stay within the resource constraints. The type of design the experimenter should choose is again largely dictated by the purpose or goal of the study and the type of statistical analysis required to fulfill those goals. Depending upon the specific purpose of the experimenter, several different types of analysis may be required. Among the more common of which are comparison of means and variances of alternatives; determination of the importance or effect of different variables and their limitations; and a searching for the optimal values of a set of variables.

Designs to accomplish the first type of analysis are generally so-called single-factor experiments and are fairly straightforward, eg, the main concerns of the experimenter are such matters as sample size, starting conditions, and the presence or absence of autocorrelation. The second type of analysis is one toward which most textbooks on design and analysis of experiments are directed. These designs primarily utilize analysis of variance and regression techniques for the interpretation of the results. The third type of analysis usually requires sequential or search techniques of experimentation.

BIBLIOGRAPHY

1. R. E. Shannon, *AIIE Trans.* **7**(3), 289 (Sept. 1975).
2. E. F. Miller, Jr., ed., *Computer* **11,** 10 (Apr. 1978).
3. G. Adkins and U. W. Pooch, *Computer* **10,** 12 (Apr. 1977).
4. D. A. Spitz, *Instrum. Technol.* **26,** 81 (Sept. 1979).
5. J. J. Davis and R. I. Kermode, *Ind. Eng. Chem. Process Des. Develop.* **14,** 459 (1975).
6. B. W. Overturf, G. V. Reklaitis, and J. M. Words, *Ind. Eng. Chem. Process Des. Dev.* **17,** 161 (1978).
7. F. Kaijchan and G. V. Reklaitis, *Ind. Eng. Chem. Process Des. Dev.* **19,** 15 (1980).
8. N. N. Barish, *Economic Analysis for Engineering and Managerial Decision Making*, McGraw-Hill Book Company, Inc., New York, 1962.
9. C. L. Johnson, *Analog Computer Techniques*, McGraw-Hill Book Company, Inc., New York, 1956.
10. W. W. Soroka, *Analog Methods in Computation and Simulation*, McGraw-Hill Book Company, Inc., New York, 1954.
11. G. A. Bekey and W. J. Karplus, *Hybrid Computation*, John Wiley & Sons, Inc., New York, 1968.
12. R. E. Shannon, *Systems Simulation, The Art and Science*, Prentice-Hall, Inc., Englewood Cliffs, N.J., 1975.
13. D. Tiechroew, J. F. Lubin, and T. D. Truitt, *Simulation* **9,** 180 (Oct. 1967).
14. J. R. Ford, *Chem. Eng. Prog.* **77**(9), 74 (Sept. 1981).
15. R. G. E. Franks, *Mathematical Modeling in Chemical Engineering*, John Wiley & Sons, Inc., New York, 1967.
16. W. L. Luyben, *Process Modeling*, McGraw-Hill Book Company, Inc., New York, 1973.
17. C. L. Smith, R. W. Pike, and P. W. Murrill, *Formulation and Optimization of Mathematical Models*, International Textbook Company, Scranton, Pa., 1970.
18. D. R. Coughanour and L. B. Koppel, *Process Systems Analysis and Control*, McGraw-Hill Book Company, Inc., New York, 1965.
19. T. H. Lee, G. E. Adams, and W. M. Gaines, *Computer Process Control: Modeling and Optimization*, John Wiley & Sons, Inc., New York, 1968.
20. W. H. Ray, *Advanced Process Control*, McGraw-Hill Book Company, Inc., New York, 1980.
21. F. F. Martin, *Computer Modeling and Simulation*, John Wiley & Sons, Inc., New York, 1968.
22. T. H. Naylor, J. L. Balintfy, D. S. Burdick, and K. Chu, *Computer Simulation Techniques*, John Wiley & Sons, Inc., New York, 1966.
23. T. H. Naylor, *Computer Simulation Experiments with Models of Economic Systems*, John Wiley & Sons, Inc., New York, 1971.

24. G. S. Fishman and P. J. Kiviat, *Digital Computer Simulation: Statistical Considerations, Memorandum RM-5387-PR*, The Rand Corporation, Santa Monica, Calif., Nov. 1967.
25. R. Chattergy and U. W. Pooch, *Computer* **10**(4), 40 (Apr. 1977).
26. J. E. Sammet, *Commun. A.C.M.* **17**(7), 601 (July 1972).
27. A. A. B. Pritsker, *The GASP IV Simulation Language*, John Wiley & Sons, Inc., New York, 1974.
28. D. B. Wortman, S. D. Duket, R. L. Hann, and G. P. Chubb, *The SAINT User's Manual, AMRL-RT-77-62*, The Aerospace Medical Research Laboratory, Wright-Patterson Air Force Base, Ohio, 1978.
29. D. D. Pegden and A. A. B. Pritsker, *Introduction to Simulation and SLAM*, John Wiley & Sons, Inc., New York, 1979.
30. Y. Chu, *Digital Simulation of Continuous Systems*, McGraw-Hill Book Company, Inc., New York, 1969, pp. 345–417.
31. H. E. Peterson, F. J. Sansom, and L. M. Warshawsky, *MIMIC–A Digital Simulation Program, SESCA Internal Memo 65-12*, Wright-Patterson Air Force Base, Ohio, May 1965.
32. J. C. Strauss, *AFIPS Nat. Comput. Conf. Expo. Conf. Proc.* **29**, 603 (1966).
33. J. Leon, C. O. Alford, and J. L. Hammond, Jr., *Proceedings of the 1970 Summer Simulation Conference*, June 1970, pp. 42–48.
34. R. D. Brennan, *Proceedings of the IFIP Working Conference on Simulation Programming Languages*, North Holland Publishing Company, Amsterdam, 1968, pp. 371–396.
35. *Continuous Simulation Modeling Program III (CSMP III)*, Program Reference Manual SH 19-7001-2, Program Number 5734-X59, 4th ed., Data Processing Division, IBM Corporation, White Plains, New York, Dec. 1975 .
36. *Simulation*, **9**(6), 281 (Dec. 1967).
37. R. J. Harnett, F. J. Sansom, and L. M. Warshawsky, *Simulation* **3**(4), 41 (Oct. 1964).
38. J. D. Strauss and W. L. Gilbert, *SCADS, A Programming System for the Simulation of Combined Analog Digital Systems*, 2nd ed., Carnegie Institute of Technology, Pittsburgh, Pa., Mar. 1964.
39. R. D. Brennan and H. Sano, *Proceedings of the Fall Joint Computer Conference*, 1964, pp. 299–312.
40. *1130 CSMP Program Reference Manual, IBM Report 1130-CX-13X*, 3rd ed., Data Processing Div., IBM Corp., White Plains, New York, Oct. 1969.
41. J. J. Skiles, R. M. Janoski, and R. L. Schaefer, *IEEE Trans. Electron. Comput.* **EC-15**(2), 78 (Feb. 1966).
42. V. C. Rideout and L. Tavernini, *Simulation* **4**(1), 20 (Jan. 1965).
43. R. G. E. Franks, *Modeling and Simulation in Chemical Engineering*, Wiley-Interscience, New York, 1972.
44. S. Babrow, J. W. Ponton and A. I. Johnson, *Can. J. Chem. Eng.* **49**, 391 (1971).
45. G. K. Patterson and R. B. Rozsa, *Comput. Chem. Eng.* **4**, 1 (1980).
46. A. L. Pugh, *DYNAMO II User's Manual*, M.I.T. Press, Cambridge, Mass., 1970.
47. J. N. Buxton and J. G. Laski, *Comput. J.* **5**, 194 (1964).
48. J. W. J. William, *Comput. J.* **6**, 328 (1964).
49. E. Famolari, *FORSIM IV Users Guide, SR-99*, The Mitre Corporation, Cambridge, Mass., Feb. 1964.
50. K. D. Tocher, *Handbook of the General Simulation Program Vols. I–II*, Report 77/ORC3/Tech and Report 88/ORC3/Tech, Dept. of Operations Research and Cybernetics, The United States Steel Companies, Ltd., Sheffield, UK, undated.
51. *MILTRAN Programming Manual, Report No. ESD-TDR-64-320*, Gulton Systems Research Group, Inc., Mineola, N.Y., June 1964.
52. H. M. Markowitz, B. Hausner, and H. W. Karr, *SIMSCRIPT: A Simulation Programming Language*, Prentice-Hall Inc., Englewood Cliffs, N.J., 1963.
53. P. J. Kiviot, *GASP–A General Activity Simulation Program P-2864*, The Rand Corporation, Santa Monica, Calif., 1964.
54. *SIMCON User's Guide, TR-65-2-149010*, Information Systems Operations, General Electric Company, Schenectady, N.Y., 1964.
55. R. P. Bennett, P. R. Cooley, S. W. Hovey, C. A. Kribs, and R. Lackner, *SIMPAC User's Manual, TM-602/000/00*, Systems Development Corporation, Los Angeles, Calif., Apr. 1962.
56. O. J. Dahl and L. Nygaard, *The SIMULA Language*, Norwegian Computer Center, Oslo, Norway, May 1965.

57. M. Greenburger, M. M. Jones, J. H. Morris, Jr., and D. N. Ness, *On-Line Computation and Simulation OPS-3*, The MIT Press, Cambridge, Mass., 1965.
58. D. C. Knuth and J. L. McNiley, *IEEE Trans. Electron. Comput.* **EC-13**(4), 401 (1964).
59. T. Schriber, *Simulation Using GPSS*, John Wiley & Sons, Inc., New York, 1974.
60. P. F. Roth, *Proceedings of the Fourth Conference on Applications of Simulation*, New York, Dec. 1970, pp. 244–250.
61. P. J. Kiviat, *Simulation* **8**(2), 65 (Feb. 1967).
62. A. A. B. Pritsker, *Modeling and Analysis Using Q-GERT Networks*, John Wiley & Sons, Inc., New York, 1977.
63. W. E. Stewart, ed., *Dynamics and Modeling of Reactive Systems*, Vol. 44, Mathematical Research Center Series, Academic Press, New York, 1980.
64. L. Lapidus and N. R. Amundson, eds., *Chemical Reactor Theory–A Review*, Prentice-Hall, Englewood Cliffs, N.J., 1977.
65. F. G. Shinskey, *Distillation Control for Productivity and Energy Conservation*, McGraw-Hill Book Company, New York, 1977.
66. M. J. Bush, J. Pulido, A. I. Johnson, and W. Y. Svrcek, *Comput. Chem. Eng.* **2,** 161 (1978).
67. R. J. Lackmeyer and D. G. Kempfer, *Proceedings of the 1977 Joint Automatic Control Conference, Vol. 2*, San Francisco, Calif., June 22–24, 1977, pp. 345–356.
68. J. R. Fowler and D. J. Harvey, *Chem. Eng. Prog.* **74**(1), 61 (Jan. 1978).
69. S. Ochiai, *Instrum. Technol.* **27**(8), 33 (Aug. 1980).
70. J. H. Tarey and T. M. Houlihan, *Proceedings of the 1979 Joint Automatic Control Conference*, Denver, Colo., June 17–20, 1979, pp. 344–347.
71. J. M. Davidson, *Proceedings of the 1978 Joint Automatic Control Conference*, Vol. 3, Philadelphia, Pa., Oct. 18–20, 1978, pp. 255–270.
72. R. P. Broadwater and R. A. Smoak, *Proceedings of the 1978 Joint Automatic Control Conference*, Vol. 4, Philadelphia, Pa., Oct. 18–20, 1978, pp. 127–140.
73. R. W. McNamara, M. R. Ringham, G. C. Bramblett, and L. C. Southworth, *Proceedings of the 1977 Joint Automatic Control Conference*, Vol. 1, San Francisco, Calif., June 22–24, 1977, pp. 345–350.
74. M. Lebhaber, P. Blumberg, and E. Kehat, *Ind. Eng. Chem. Process Des. Develop.* **13**(1), 39 (1974).
75. E. M. Rosen and D. J. Kaufman, *Comput. Ind.* **2**(1), 41 (1981).
76. E. M. Rosen and A. C. Pauls, *Comput. Chem. Eng.* **1,** 11 (1977).
77. J. D. Seader, W. D. Seider, and A. C. Pauls, *FLOWTRAN Simulation, An Introduction*, 2nd ed., CACHE Committee, Ulrich's Bookstore, Ann Arbor, Mich., 1977.
78. L. B. Evans and co-workers, *ASPEN, An Advanced System for Process Engineering*, Preprints, 12th Symposium on Computer Applications in Chemical Engineering, Montreaux, Switzerland, Apr. 8–11, 1979.
79. L. B. Evans, P. W. Gallier, H. I. Brett, and co-workers, *ASPEN PLUS*, Aspen Technology, Inc., Cambridge, Mass., 1981.
80. K. Thomas, *Tappi* **62**(2), 51 (Feb. 1979).
81. M. C. Uy and L. S. Adler, *paper presented at the 89th National Meeting, American Institute of Chemical Engineers*, Portland, Ore., Aug. 17–20, 1980.
82. C. Shewchuk and co-workers, *MASSBAL*, SACDA Engineering, University of Western Ontario, London, Ontario, Canada, undated.
83. *PACER 245 User Manual, Digital Systems Corp.*, Hanover, N.H., 1971.
84. R. L. Motard and H. M. Lee, *CHESS, Chemical Engineering User's Guide*, 3rd ed., University of Houston, Houston, Texas, 1971.
85. F. L. Worley, Jr. and R. L. Motard, *Chem. Eng. Computations*, Vol. 2, American Institute of Chemical Engineers Workshop Series, AIChE, New York, 1972.
86. L. D. Gaines and J. L. Gaddy, *Ind. Eng. Chem. Process Des. Dev.* **15**(1), 207 (1976).
87. *FLOWPACK–User's Manual*, Imperial Chemical Industries, Ltd., Runcorn, UK, 1970.
88. I. V. Klumpar, *Chem. Eng.* **77**(1), 107 (1970).
89. *Ibid.*, (14), 76 (1970).
90. *CONCEPT–Mark III*, Computer Aided Design Center, Cambridge University, Cambridge, UK, 1973.
91. H. D. Nott, *SLED: Simplified Language for Engineering Design–A Computerized System for the Design of Chemical Processes*, Ph.D. Dissertation, University of Michigan, Ann Arbor, Mich., 1971.
92. A. E. Ravicz and R. L. Norman, *Chem. Eng. Prog.* **60**(5), 71 (1964).

93. A. I. Johnson and T. Toong, *The Modular Approach to Systems Analysis and Design* (*GEMCS Manual*), McMaster University, Montreal, Canada, 1968.
94. R. A. Russell, *Comput. Chem. Eng.* **4,** 167 (1980).
95. R. L. Motard, M. Shacham, and E. M. Rosen, *AIChE J.* **21**(3), 417 (May 1975).
96. E. Kehat and M. Shacham, *Process Technol.* **18**(1/2), 35; (3) 115; (4/5), 181, (1973).
97. J. R. Flower and B. D. Whitehead, *Chem. Eng.* (*London*) **272,** 208; **273,** 271 (1973).
98. J. N. Peterson and co-workers, *Chem. Eng.* **85**(13), 145 (1978).
99. L. B. Evans and W. D. Seider, *Chem. Eng. Prog.* **72**(6), 80 (June 1976).
100. M. T. Fleischer and D. M. Prett, *Chem. Eng. Prog.* **77**(2), 72 (Feb. 1981).
101. R. V. Elshout and E. C. Hohmann, *Chem. Eng. Prog.* **75**(3), 72 (Mar. 1979).
102. R. C. Reid and L. B. Evans, *Design Data for Industry, Property Reduction with Computer System*, AIChE Today Series, New York, 1970.
103. R. V. Hughson and E. H. Steymann, *Chem. Eng.* **78**(14), 66 (1971); **80**(19), 121; (21), 127 (1973).
104. *PPDS-Physical Property Data System*, The Institution of Chemical Engineers, London, England, 1971.
105. *C. B. M.-Calcul des Belans de Matiers'-System*, Solvay and Cie, S. A., Brussels, Belgium, 1973.
106. *Uhde Thermophysical Properties Program Package*, Friedrich Uhde GMBH, Dortmund, FRG, 1973.
107. A. L. Parker and R. R. Hughes, *Comput. Chem. Eng.* **5**(3), 123 (1981).
108. P. Friedman and K. L. Pinder, *Ind. Eng. Chem. Process Des. Dev.* **11**(4), 512 (1972).
109. L. D. Gaines and J. L. Gaddy, *Ind. Eng. Chem. Process Des. Dev.* **15**(1), 206 (1976).
110. S. H. Ballman and J. L. Gaddy, *Ind. Eng. Chem. Process Des. Dev.* **16**(3), 377 (1977).
111. F. J. Doering and J. L. Gaddy, *Comput. Chem. Eng.* **4**(2), 113 (1980).
112. L. T. Biegler and R. R. Hughes, *Chem. Eng. Prog.* **77,** 76 (Apr. 1981).
113. L. B. Evans, *AIChE J.* **23,** 658 (Sept. 1977).
114. W. D. Seider, L. B. Evans, B. Joseph, E. Wong, and S. Jirapongphan, *Ind. Eng. Chem. Proc. Des. Dev.* **18,** 292 (1979).
115. T. K. Pho and L. Lapidus, *AIChE J.* **19,** 1170 (1973).

THEODORE J. WILLIAMS
Purdue University

SIMULTANEOUS HEAT AND MASS TRANSFER

Heat transfer and mass transfer occur simultaneously whenever a transfer operation involves a change in phase or a chemical reaction. Of these two situations, only the first is considered here because in reacting systems the complications of chemical reaction mechanisms and paths are usually primary (see Heat-exchange technology). Even in processes involving phase changes, design is frequently based upon the heat-transfer process alone; mass transfer is presumed to add no complications. However, as more is learned about such processes as the boiling and condensing of multicomponent mixtures, it is becoming clear that mass transfer effects do influence and even limit the process rate (see Mass transfer).

In processes where a condensing vapor or vapor from a liquid phase moves through an inert gas (eg, condensation in the presence of air, drying, humidification, crystallization, and boiling of a multicomponent liquid), mass transfer as well as heat transfer effects are important. In Drying, Evaporation, Crystallization, Air conditioning, Heat-exchange technology, and Mass transfer, such processes are discussed but with primary emphasis on either the heat transfer or the mass transfer taking place. Here the interactions between heat and mass transfer in such processes are discussed, and applications to humidification, dehumidification, and water cooling are developed. The same principles are applicable to the operations listed above.

Condensation and Vaporization as Effected by Simultaneous Heat and Mass Transfer

Consider the interphase transfer that occurs when one or more components change phase in the presence of inert or less active components. The transferring component must be transported through its original phase to the boundary and must then escape into the second phase. The phase change involves a heat effect. Energy diffuses to or from the boundary to balance the phase-change heat effect. The boundary temperature is influenced by the rate of heat transfer, and this determines the fugacity of the diffusing component at the boundary. Thus, to describe the process, rate equations for heat transfer and for mass transfer must be written along with material balances for the components present and an energy balance. Appropriate boundary conditions must be applied, and the resulting set of differential and algebraic equations solved.

This process has been used for various situations (1–14). For the condensation of a single component from a binary gas mixture, the gas-stream sensible heat and mass-transfer equations for a differential condenser section take the forms

$$G \cdot C_p \frac{dT_G}{dA} = -h_g \cdot (T_g - T_s) \frac{\epsilon}{e^{\epsilon} - 1} \tag{1}$$

$$\frac{dV}{dA} = -k_g \cdot (P_g - P_s) \tag{2}$$

Here, no condensation is taking place in the bulk gas phase. If fogging does occur, these equations become

$$G \cdot C_p \frac{dT_G}{dA} = -h_g \cdot (T_g - T_s) \frac{\epsilon}{e^{\epsilon} - 1} + \lambda \frac{dF}{dA} \tag{3}$$

$$\frac{dV}{dA} = -k_g \cdot (P_g - P_s) - \frac{dF}{dA} \tag{4}$$

The term $\epsilon/(e^{\epsilon} - 1)$ which appears in equations 1 and 3 was first developed to account for the sensible heat transferred by the diffusing vapor (1). The quantity ϵ represents the group $M_i{\cdot}C_{p_i}/h_g$ and may be thought of as the fractional influence of mass transfer on the heat-transfer process. The last term of equation 3 is the latent heat contributed to the gas phase by the fog formation. The vapor loss from the gas phase through both surface and gas phase condensation can be related to the partial pressure of the condensing vapor by using Dalton's law and a differential material balance.

The effect of latent and sensible heat transferred to the surface from the condensing vapor on the coolant temperature is

$$L \cdot M_L \cdot C_w \frac{dT_w}{dA} = \pm h_o \cdot (T_s - T_w) \tag{5}$$

The sign is negative for countercurrent flow.

Assuming a linear relation between surface temperature and corresponding vapor pressure of the condensable component allows a heat balance to be written from gas phase to the surface:

$$h_o \cdot (T_s - T_w) = h_g \cdot (T_g - T_s) \frac{\epsilon}{1 - e^{\epsilon}} + k_g \cdot \lambda(P_g - P_s) \tag{6}$$

Combining equation 6 with the heat- and mass-transfer rate expression gives

$$w \cdot C_w \frac{dT_w}{dA} = e^{\epsilon} \cdot G \cdot C_p \frac{dT_g}{dA} + \frac{V' \cdot \lambda P}{(P_t - P_{go})(P_t - P_{gf})} \cdot \frac{dP_g}{dA} \tag{7}$$

Equations 6 and 7 are not affected by fogging because the latent heat thus obtained is retained as sensible heat in the gas phase.

These basic relations have been solved for a wide range of cooler–condenser conditions and for different complexities of systems. A design procedure based on the assumption that the mixture is saturated throughout the condensation process has been developed (2). This assumption has later been shown to depend upon the rate of diffusion of the condensing component; some cases with rapidly diffusing components tend to superheat, and others with slowly diffusing vapors tend to subcool. The same approach extended to superheated mixtures has been used to develop the following equation for calculating T and partial pressures during condensation (3):

$$\frac{dP_g}{dT_g} = \frac{P_t - P_g}{(Le)^{2/3} \cdot P_{BM}} \cdot \frac{P_g - P_s}{T_g - T_s} \cdot \frac{e^{\epsilon} - 1}{\epsilon} \tag{8}$$

This relation was tested experimentally for water condensation in various gases and found to be acceptable (4). It has also been solved via analogue computers (5), and in another instance, for a set of conditions ranging from superheating to fogging (6). The use of standard j-factor correlations (see also Heat-exchange technology, heat transfer; Mass transfer) for coefficients of heat and mass transfer have been incorporated into the solution method (4,7–8). Experimental verification has been supplied by workers in the field of absorption (15–17), condensation (18–19), liquid–liquid extraction (8,20) and distillation (21), and in laboratory experiments where free convection played a significant role (21,22).

In considering the effect of mass transfer on the boiling of a multicomponent mixture, both the boiling mechanism and the driving force for transport must be examined (23–26). Moreover, the process is strongly influenced by the effects of convective flow on the boundary layer. In ref. 26 both effects have been taken into consideration to obtain a general correlation based on mechanistic reasoning that fits all available data within ca ± 15%.

The boiling mechanism can conveniently be divided into macroscopic and microscopic mechanisms. The macroscopic mechanism is associated with the heat transfer affected by the bulk movement of the vapor and liquid. The microscopic mechanism is that involved in the nucleation, growth, and departure of gas bubbles from the vaporization site. Both of these mechanistic steps are affected by mass transfer.

The final correlation for the overall boiling heat-transfer coefficient (26) is a direct addition of the macroscopic and microscopic contributions to the coefficient:

$$h_T = h_{\text{mac}} + h_{\text{mic}}$$

$$= \left[0.023(Re_{L\text{-only}})^{0.8}(Pr_L)^{0.4}\frac{k_L}{D_i}\right]\left[\frac{\left(\frac{dP}{dz}\right)_{2\phi}}{\left(\frac{dP}{dz}\right)_{L\text{-only}}}\right]^{0.444} f(Pr_L)\left[\frac{\Delta\tilde{T}}{\Delta T_s}\right]_{\text{mac}}$$

$$+\ 0.00122\,\frac{k_L^{0.79}\cdot C_{P_L}{}^{0.45}\cdot \rho_L^{0.49}\cdot g_c^{0.25}}{\sigma^{0.5}\mu_L^{0.29}\lambda^{0.24}\rho_V^{0.24}}\,\Delta T_s^{0.24}\cdot\Delta P_s^{0.75}\cdot S_{\text{binary}}\cdot Re_{2\phi} \quad (9)$$

where

$$f(Pr_L) = \left[\frac{Pr_L + 1}{2}\right]^{0.444} \quad (10)$$

$$\left(\frac{\Delta\tilde{T}}{\Delta T_s}\right)_{\text{mac}} = 1 - \frac{(1-y_j)q}{\rho_{\text{avg}}\lambda k_Y \Delta T_s}\cdot\frac{\Delta T_s}{\Delta x_{jPB}} \quad (11)$$

$$k_Y = 0.023(Re_{2\phi})^{0.8}\cdot(Sc)^{0.4}\frac{k_L}{D}$$

$$Re_{2\phi} = Re_{L\text{-only}}\left[f(Pr_L)\left[\frac{\left(\frac{dP}{dz}\right)_{2\phi}}{\left(\frac{dP}{dz}\right)_{L\text{-only}}}\right]^{0.444}\right]^{1.25} \quad (12)$$

$$S_{\text{binary}}\cdot Re_{2\phi} = \frac{1}{1-\frac{C_{PL}(y_j-x_j)}{\lambda}\cdot\frac{\partial T}{\partial x_j}\left[\frac{\alpha}{D}\right]^{1/2}}\,S\cdot Re_{2\phi} \quad (13)$$

In the macroscopic heat-transfer term of equation 9, the first group in brackets represents the usual Dittus-Boelter equation for heat-transfer coefficient. The second bracket is the ratio of frictional pressure drop per unit length for two-phase flow to that for liquid phase alone. The Prandtl-number function is an empirical correction term. The final bracket is the ratio of the binary macroscopic heat-transfer coefficient to the heat-transfer coefficient that would be calculated for a pure fluid with properties identical to those of the fluid mixture. This term is built on the postulate that mass transfer does not affect the boiling mechanism itself but does affect the driving force.

Likewise; the microscopic heat-transfer term takes accepted empirical correlations for pure-component pool boiling and adds corrections for mass-transfer and convection effects on the driving forces present in pool boiling. In addition to dependence on the usual physical properties, the extent of superheat, the saturation pressure change related to the superheat, and a suppression factor relating mixture behavior to equivalent pure-component heat-transfer coefficients are correlating functions.

The Description of Gas–Vapor Systems

In engineering application, the transport processes involving heat and mass transfer usually occur in process equipment involving vapor–gas mixtures where the vapor undergoes a phase transformation such as condensation to or evaporation from a liquid phase. In the simplest case, the liquid phase is pure, consisting of the vapor component alone.

The system of primary interest, then is that of a condensable vapor moving between a liquid phase, usually pure, and a vapor phase in which other components are present. Some of the gas phase components may be noncondensable. A simple example would be water vapor moving through air to condense on a cold surface. Here the condensed phase is characterized by T and P and exists at $x = 1$. The vapor-phase description requires y as well as T and P. The nomenclature used in the description of vapor–inert gas systems is given in Table 1.

The humidity term and such derivatives as relative humidity and molal humid volume were developed for the air–water system, and their use is generally restricted to that system. Recently, however, these terms have also been used for other vapor–noncondensable gas phases.

For the air–water system, the humidity is easily measured with a wet-bulb thermometer. Air passing the wet wick surrounding the thermometer bulb causes evaporation of moisture from the wick. The balance between heat transfer to the wick and energy required by the latent heat of the mass transfer from the wick gives, at steady state,

$$-k_Y \cdot A \cdot (Y_1 - Y_w) \cdot \lambda_w = (h_c + h_r) \cdot A \cdot (T_1 - T_w) \tag{14}$$

$$T_1 - T_w = \frac{k_Y \lambda_w}{(h_c + h_r)} (Y_w - Y_1) \tag{15}$$

If radiant energy transfer can be prevented

$$T_1 - T_w = \frac{k_Y \cdot \lambda_w}{h_c} (Y_w - Y_1) \tag{16}$$

Thus, a measurement of the wet-bulb temperature T_w and the temperature T_1 allows the molal humidity Y_1 to be calculated since Y_w is known.

Another relationship between temperature and humidity results from considering the path of T and Y during an adiabatic saturation process. Here consider a countercurrent packed column with no heat flow from or to the surroundings and the liquids (water) stream recycled. The gas stream (air) passes once through the unit. In this case the liquid reaches a steady-state temperature. If the column is very tall, the gas exit temperature would reach the temperature of the recycled liquid. Figure 1 shows the physical arrangement and the nomenclature. Material and energy balances written for an envelope encircling the exit and entrance streams from this column using

Table 1. Definitions of Humidity Terms

Term	Meaning	Units	Symbol
humidity	vapor content of a gas	mass vapor per mass noncondensable gas	$Y' = Y\frac{M_a}{M_b}$
molal humidity	vapor content of a gas	moles vapor per mole noncondensable gas	Y
relative saturation or relative humidity	ratio of partial pressure of vapor to partial pressure of vapor at saturation	kPa/kPa, or mole fraction per mole fraction, expressed as percent	$100\frac{y}{y_s}$
percent saturation or percent humidity	ratio of concentration of vapor to the concentration of vapor at saturation with concentrations expressed as mole ratios	mole ratio per mole ratio, expressed as percent	$100\frac{Y}{Y_s}$
molal humid volume	volume of 1 mol of dry gas plus its associated vapor	m³/mol of dry gas	$V_h = (1 + Y) \times 0.0224\frac{T}{273} \times \frac{1.013 \times 10^5}{P}$
molal humid heat	heat required to raise the temperature of 1 mol of dry gas plus its associated vapor 1°C	J/(mol of dry gas·°C)	$c_h = c_b + Yc_a$
adiabatic-saturation temperature	temperature that would be attained if the gas were saturated in an adiabatic process	°C or K	T_{sa}
wet-bulb temperature	steady-state temperature attained by a wet-bulb thermometer under standardized conditions	°C or K	T_w
dew-point temperature	temperature at which vapor begins to condense when the gas phase is cooled at constant pressure	°C or K	T_d

enthalpies in terms of molal humid heats, latent heats, and liquid heat capacities yields

$$C_{h_1} \cdot (T_{sa} - T_1) = \lambda_{sa} \cdot (Y_1 - Y_{sa}) \tag{17}$$

In this equation the subscripts sa refer to the adiabatic saturation condition; point 1 is any initial condition. This is the adiabatic saturation equation that traces the path of a moist gas stream as it is humidified under adiabatic conditions.

For the air–water system, Lewis recognized that $C_h = h_c/k_Y$ based on empirical evidence. Thus, the adiabatic saturation equation is identical to the wet-bulb temperature line. In general, again based on empirical evidence (27),

$$\frac{h_c}{k_Y} = C_h \left(\frac{Sc}{Pr}\right)^{0.56} \tag{18}$$

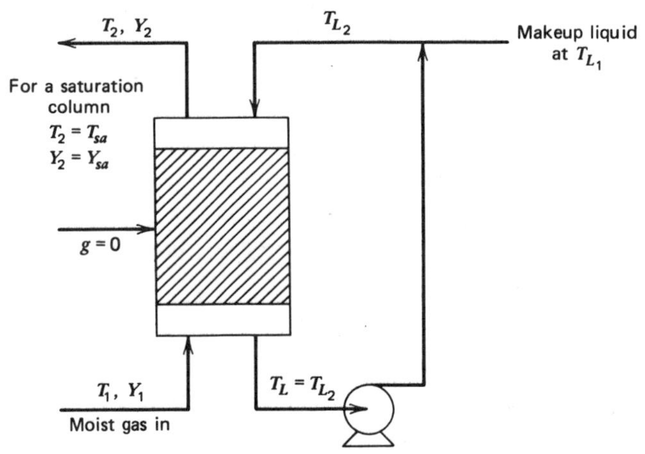

Figure 1. The adiabatic saturation process.

whereas, in general, $6.83 < h_c/k_Y < 7.82$, for air $Sc \cong Pr \cong 0.70$, and $h_c/k_Y = C_h = 6.94$.

A closer look at the Lewis relation requires an examination of the heat- and mass-transfer mechanisms active in the entire path from the liquid–vapor interface into the bulk of the vapor phase. Such an examination yields the conclusion that, in order for the Lewis relation to hold, eddy diffusivities for heat- and mass-transfer must be equal as must the thermal and mass diffusivities themselves. This equality may be expected for simple monatomic and diatomic gases and vapors. Air with small concentrations of water vapor fits these criteria closely.

The thermodynamic properties of a vapor–gas mixture, ie, two components one of which is condensable, are usually presented on a humidity diagram. Figure 2 is such a diagram for the air–water vapor phase at normal atmospheric pressure. Here humidity is plotted against temperature. Curves are given for saturated vapor and for constant values of relative humidity. Also plotted are lines of constant wet-bulb temperature, or adiabatic saturation lines. These are nearly straight with negative slopes slightly less than 30°. These lines are also lines of nearly constant enthalpy, as can be seen by rearranging the adiabatic saturation equation. The deviation is in the variations of humid heat and latent heat along the path. The chart shows values of the enthalpy at saturation as well as lines of constant enthalpy deviation. Thus, the enthalpy can be found by adding the enthalpy deviation to the enthalpy of the saturated gas phase at the wet-bulb temperature. The graph has a scale change at 0°C wet-bulb temperature. (A copy of this diagram covering a greater temperature span is given in Drying (Vol. 8, p. 79).)

Figure 3 gives the diagram in molar quantities. On that chart enthalpy deviations are not given. Enthalpy may also be calculated from the enthalpy of saturated air and of dry air using the percent saturation:

$$H = H_{\text{dry}} + (H_{\text{sat}} - H_{\text{dry}}) \cdot (\text{percent saturation}) \qquad (19)$$

Figure 3 gives percent humidity as the measure of vapor concentration whereas Figure 2 gives relative humidity (as percentage).

For systems other than air–water vapor or for total system pressures different than 101.3 kPa (1 atm), humidity diagrams can be constructed if basic phase-equilibria

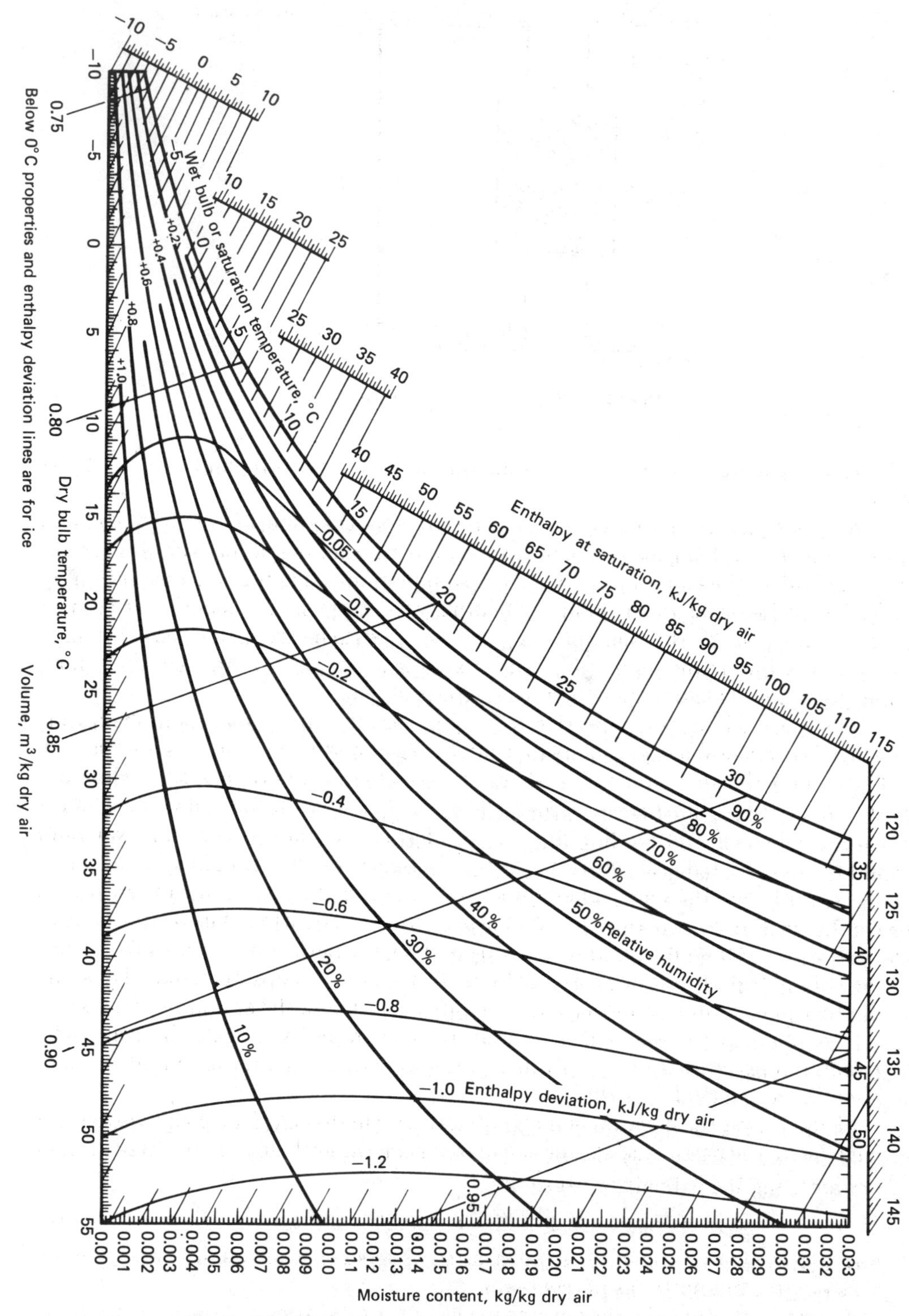

Figure 2. Psychrometric chart. Courtesy of Carrier Corporation.

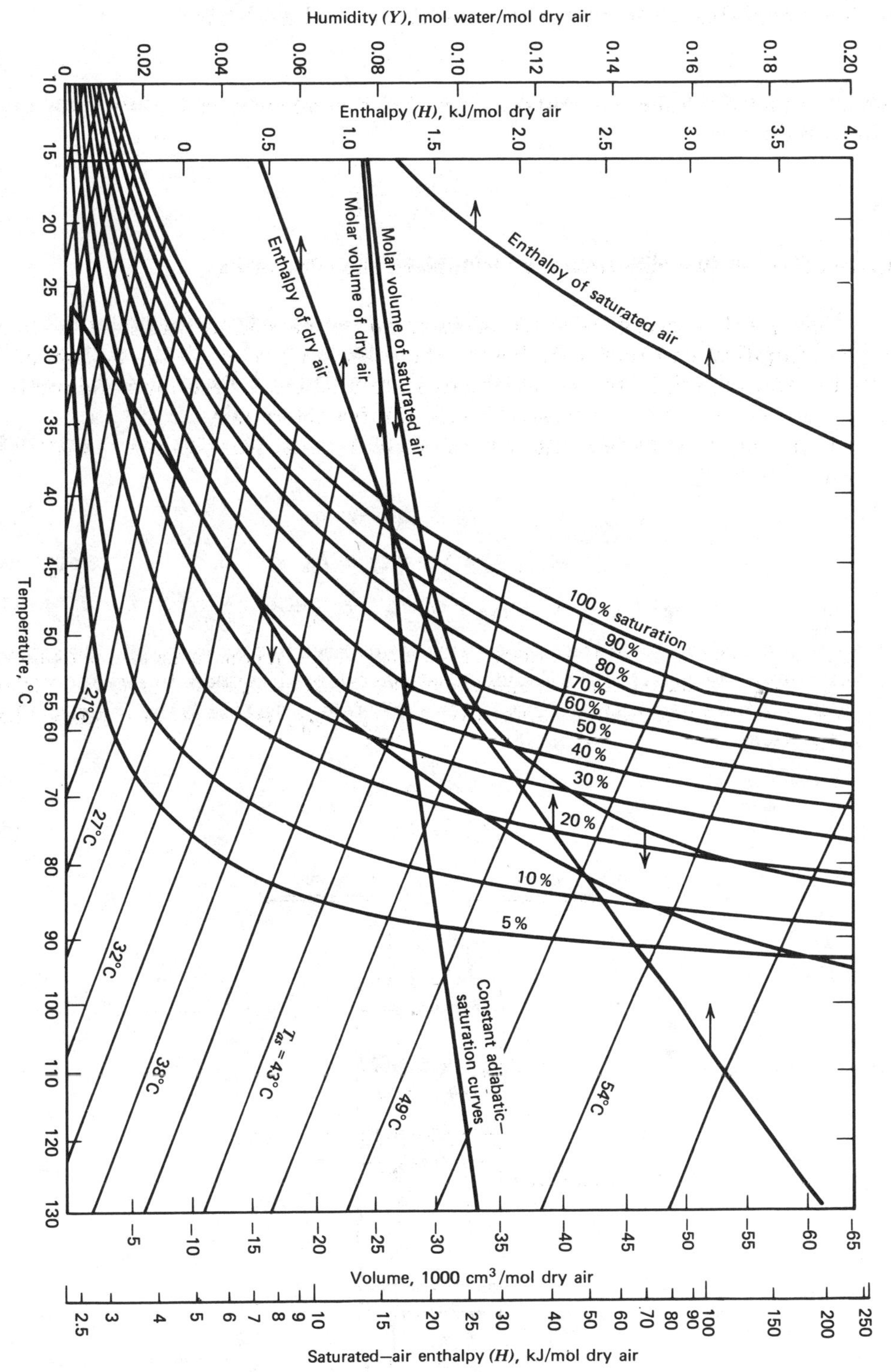

Figure 3. Humidity chart for the air–water system, molal quantities. To convert kJ to Btu, divide by 1.054; to convert cm^3 to ft^3, multiply by 35.31 × 10^6.

data are available. The simplest of these relations is Raoult's law

$$P \cdot y_s = p_i^\circ \cdot x_i \tag{20}$$

which, for a two-component system in which one component exists only in the vapor phase, reduces to

$$y_s = \frac{H_s}{1 + H_s} = \frac{p_i^\circ}{P} \tag{21}$$

Calculations for Humidification and Dehumidification Processes

Figure 4 shows the general arrangement and nomenclature for a humidification or dehumidification process. As before, the subscript 1 refers to the bottom of the column, and subscript 2 to the top. Steady state is assumed. Flow rates and compositions are given in molar terms, since this simplifies the results.

Total material, condensable component, and energy balances can be written for the entire column:

$$L_1 - L_2 = V_1 - V_2 \tag{22}$$

$$V' \cdot (Y_2 - Y_1) = L_2 - L_1 \tag{23}$$

$$L_2 \cdot H_{L_2} + V' \cdot H_{V_1} + q = L_1 \cdot H_{L_1} + V' \cdot H_{V_2} \tag{24}$$

Generally, q is small since the outside area is not large in comparison to the amount of heat being transferred, and the energy balance can be simplified. In these conditions it is also convenient to write balances over a differential section of the column. These balances yield

$$V' \cdot dY = dL \tag{25}$$

$$V' \cdot dH_V = d(L \cdot H_L) \tag{26}$$

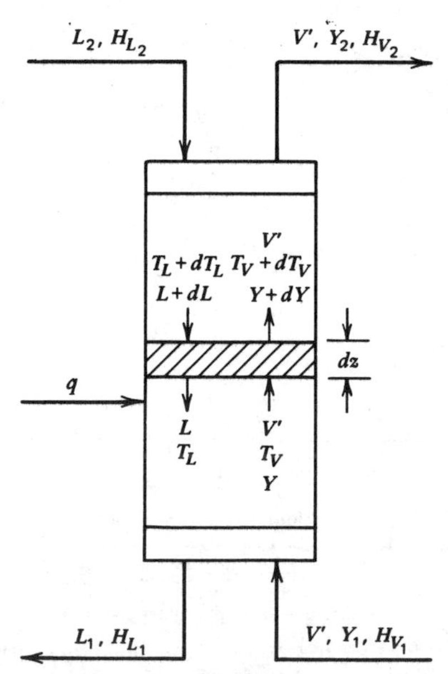

Figure 4. Arrangement and nomenclature for general humidification–dehumidification process.

If the amount of evaporation is small, the change in enthalpy in the liquid phase can be taken as a result of temperature change alone. Using an average liquid flow rate

$$V' \cdot dH = L_{avg} \cdot C_L \cdot dT_L \tag{27}$$

Similarly, the vapor enthalpy can be expressed in terms of humid heat and latent heat in relation to a base condition

$$V' \cdot dH = V' \cdot d[C_h \cdot (T_V - T_o) + Y \cdot \lambda_o] = V' \cdot C_h \cdot dT_V + V' \cdot \lambda_o \cdot dY \tag{28}$$

The energy transferred on both sides of the interface, as written above, can also be written in terms of the appropriate rate expressions. For the liquid phase

$$\frac{L_{avg}}{S} C_L \cdot dT_L = h_L \cdot a(T_L - T_i) \cdot dz \tag{29}$$

For the gas phase, energy transfers both as a result of a thermal driving force and as a by-product of vaporization. Thus,

$$\frac{V'}{S} C_h \cdot dT_V = h_c \cdot a(T_i - T_V) \cdot dz \tag{30}$$

and

$$\frac{V'}{S} \lambda_o \cdot dY = \lambda_o \cdot k_Y \cdot a(Y_i - Y) \cdot dz \tag{31}$$

Combining these two mechanisms for gas-phase transfer, as done in equation 28, yields

$$\frac{V'}{S} dH_V = h_c \cdot a(T_i - T_V) \cdot dz + \lambda_o \cdot k_Y \cdot a(Y_i - Y) \cdot dz \tag{32}$$

Rearranging equation 32 and defining the ratio $h_c \cdot a/(k_Y \cdot a \cdot C_h)$ as r, the psychrometric ratio, gives

$$\frac{V'}{S} dH_V = k_Y \cdot a[(C_h \cdot r \cdot T_i + \lambda_o \cdot \lambda_i) - (C_h \cdot r \cdot T_V + \lambda_o \cdot Y)] \cdot dz \tag{33}$$

For the air–water system, the Lewis relation shows that $r = 1$. Under these conditions the two parenthetical terms on the right hand side of equation 33 are enthalpies, and equation 33 becomes the design equation for humidification operations

$$\frac{V'}{S} dH_V = k_Y \cdot a(H_i - H_V) \cdot dz \tag{34}$$

or

$$\int_{H_{V1}}^{H_{V2}} \frac{V' \cdot dH_V}{k_Y \cdot aS(H_i - H_V)} = \int_o^z dz = z \tag{35}$$

The simplification of equation 33 to equation 34 is possible only where $r = 1$; that is, for simple monatomic and diatomic gases. For other systems the design equation can be obtained by a direct rearrangement of equation 33.

Although equation 35 is a very simple expression, it tends to be confusing. In this equation enthalpy difference appears as driving force in a mass-transfer expression. Enthalpy is not a potential, but rather an extensive thermodynamic function. Here it is used as enthalpy per mole and is a kind of shorthand for a combination of temperature and mass-concentration terms.

The integration of equation 35 requires a knowledge of the mass-transfer coefficient, $k_Y \cdot a$, and also of the interface conditions from which H_i could be obtained. Combining equations 27, 28, and 34 gives a relation balancing transfer rate on both sides of the interface

$$\frac{V'}{S} dH_V = h_L \cdot a(T_L - T_i) \cdot dz = k_Y \cdot a(H_i - H_V) \cdot dz \tag{36}$$

or

$$\frac{-h_L \cdot a}{k_Y \cdot a} = \frac{H_V - H_i}{T_L - T_i} \tag{37}$$

Thus, the enthalpy and temperature of the vapor–liquid interface are related to the liquid temperature and gas enthalpy at any point in the column through a ratio of heat and mass-transfer coefficients.

The integration can be carried out graphically or numerically using a computer. For illustrative purposes the graphical procedure is shown. In the plot of vapor enthalpy (H_V or H_i) vs liquid temperature (T_L or T_i) (Fig. 5), the curved line is the equilibrium curve for the system. For the air–water system, it is the 100% saturation line taken directly from the humidity diagram (Fig. 3).

The locus of corresponding T_L and H_V points, the operating line for the column, can be obtained by assuming V' and L_{avg} change little and integrating equation 27 across the length of the column.

$$\int_{H_{V1}}^{H_{V2}} V' \cdot dH = \int_{T_{L1}}^{T_{L2}} L_{avg} \cdot C_L \cdot dT \tag{38}$$

$$V' \cdot (H_{V_2} - H_{V_1}) = L_{avg} \cdot C_L \cdot (T_{L_2} - T_{L_1}) \tag{39}$$

or

$$\frac{H_{V_2} - H_{V_1}}{T_{L_2} - T_{L_1}} = \frac{L_{avg} \cdot C_L}{V'} \tag{40}$$

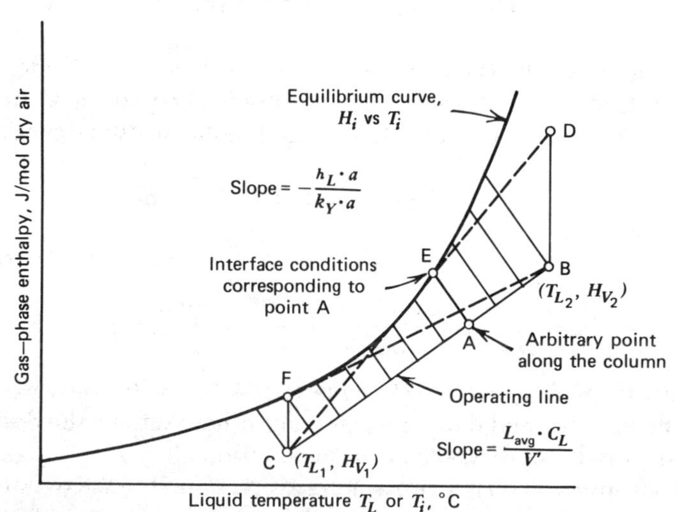

Figure 5. Adiabatic gas–liquid contacting, graphical representation. Conditions shown are those of a water cooling process. To convert J to Btu, divide by 1054.

Thus, the locus is a straight line, presuming the ratio on the right side of equation 40 is constant. The actual location of the line can be obtained if both end points are known or if one end point and the slope ($L_{avg} \cdot C_L/V'$) can be determined.

On Figure 5 the locus of interface points passes through points F and E. The operating line goes from B to C. In addition to specifying two points on the line itself (B, C) or the slope and one of these points, the column could be required to operate at some convenient gas-flow rate greater than minimum. Here the minimum gas-flow rate required to cool the liquid from 60°C to 30°C is given by line CD. Another possible limiting condition would be the flow of gas with the largest wet-bulb temperature possible to still allow water to cool from 60°C to 30°C no matter how large the column (line BF). A design and operating condition can be determined as an acceptable approach to either of these.

Once the operating line is set, interface conditions corresponding to any point on the operating line can be found if heat- and mass-transfer coefficients are available. Then a line of slope $-h_L \cdot a/k_Y \cdot a$ will connect a point on the operating line (point A, for example) with its corresponding interface condition (point E).

This information allows the integration of equation 35 to give the column height. The method is shown graphically in Figure 6, though again a numerical solution is possible.

Determination of the Gas-Phase Temperature. The development given above is in terms of interface conditions, bulk liquid temperature, and bulk gas enthalpy. Often the temperature of the vapor phase is important to the designer, either as one of the variables specified or as an important indicator of fogging conditions in the column. Such a condition would occur if the gas temperature equaled the saturation temperature, that is, the interface temperature. When fogging does occur, the column can no longer be expected to operate according to the relations presented here but is basically out of control.

Gas-phase temperatures have been obtained by an extension of the graphical method illustrated above (28). When equation 30 is divided by equation 34, one obtains

$$\frac{V' \cdot C_h \cdot dT_V}{V' \cdot dH_V} = \frac{h_c \cdot a \cdot (T_i - T_V) \cdot dz}{k_Y \cdot a \cdot (H_i - H_V) \cdot dz} \quad (41)$$

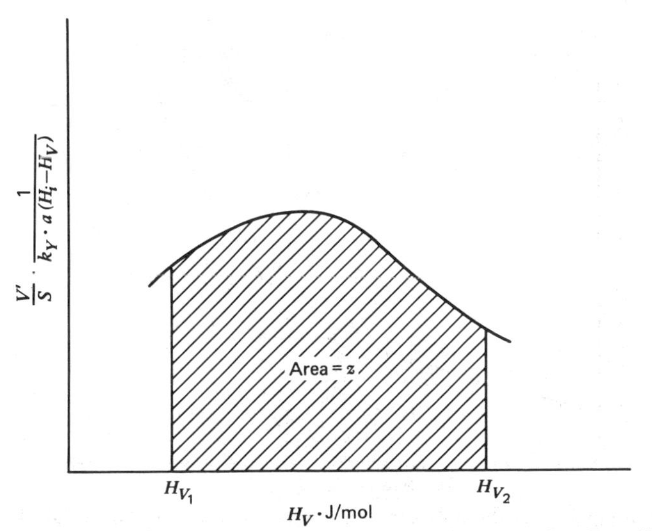

Figure 6. Integration of the design equation (eq. 35). To convert J to Btu, divide by 1054.

or

$$\frac{dT_V}{dH_V} = \frac{h_c \cdot a}{C_h \cdot k_Y \cdot a} \cdot \frac{(T_i - T_V)}{(H_i - H_V)} = \frac{T_i - T_V}{H_i - H_V} \tag{42}$$

The last expression of equation 42 is obtained by applying the Lewis relation ($C_h = h_c/k_Y$). If the differentials of equation 42 are replaced by finite differences

$$\frac{\Delta T_V}{\Delta H_V} \approx \frac{T_i - T_V}{H_i - H_V} \tag{43}$$

In effect, equation 43 states that the temperature and enthalpy values of the bulk gas phase continuously approach the interface condition at the same point in the column as that for which the gas-phase conditions apply. The graphical application is illustrated in Figure 7. The gas-phase temperature and enthalpy at the bottom of the column are usually known and are plotted at point G_0. The interface conditions at the bottom of the column are given at point H. Then by equation 43 the line GH is the path followed by the gas phase temperature. This path is followed until the interface condition shifts noticeably, perhaps to point J corresponding to a bulk liquid temperature at point I. Then the gas temperature line approaches point J. The interface conditions shift toward point Z, continuously changing the direction of the gas temperature line. The points at which the line changes slope depend on the intervals chosen along the operating line. Here G_1 corresponds to point I, G_2 to point K, G_3 to point M, G_4 to point O, etc.

In the example developed here fogging occurs at about the time the gas reaches the top of the column. That is far from inevitable and would not have occurred here if the operating line had terminated at point U.

Transfer Coefficients. The design method described depends on the availability of mass- and heat-transfer coefficients for its utility. Typically, $k_Y \cdot a$ and $h_L \cdot a$ are needed. These must be obtained from the standard correlations for mass and heat transfer, from data reported in the literature (29–36), or from data presented by

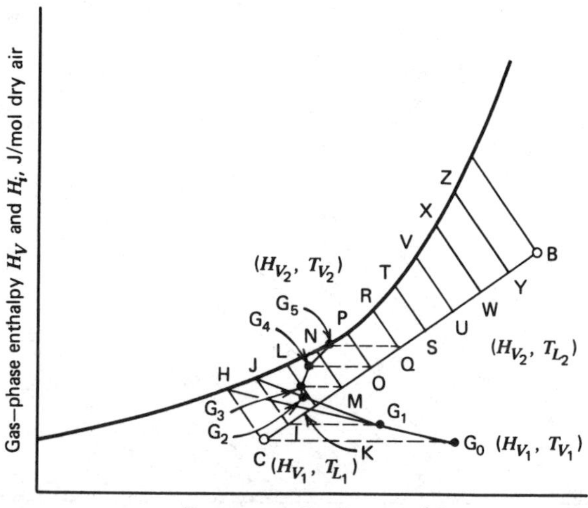

Figure 7. Determination of the bulk gas-phase temperature path. To convert J to Btu, divide by 1054.

equipment makers for particular packing (37–39). When this type of information is not available, it is possible to determine heat- and mass-transfer coefficients by a single test using the packing material of interest in a pilot-sized tower. If a steady state is obtained, measurement of air- and water-flow rates, water-inlet and -outlet temperatures, and air-inlet and -outlet wet- and dry-bulb temperatures comprises all the necessary information. Interface and operating lines on a T_L–H_V diagram, such as Figure 5, are directly obtained. Since column heights are known, the value of $-h_L \cdot a/k_Y \cdot a$ can be obtained by trial-and-error with the integration demonstrated in Figure 6 and adjustment of the slope of the operating interface condition line until z_{calc} equals the actual column height.

Overall Coefficients. Often overall coefficients of heat and mass transfer are available, rather than the film coefficients used earlier. In that case equation 35 can be rewritten

$$\int_{H_{V1}}^{H_{V2}} \frac{V' \cdot dH_V}{k_Y \cdot a \cdot S \cdot (H^* - H_V)} = \int_o^z dz = z \tag{44}$$

If $k_Y \cdot a$ is constant, this can be written

$$\frac{V'}{k_Y \cdot a \cdot S} \int_{H_{V1}}^{H_{V2}} \frac{dH_V}{(H^* - H_V)} = HTU{\cdot}NTU = z \tag{45}$$

Humidification and Dehumidification Equipment

The addition or removal of a condensable component to or from a noncondensable gas can be accomplished by direct contact between the vapor and the gas. This may be done in a countercurrent tower, usually packed as described in Mass transfer and in Absorptive separation. The direction of transfer depends upon the temperatures of the two streams. Such operations can also be done using spray ponds in which a grid of nozzles sprays liquid, usually water, into the gas phase, usually air. If the air is relatively dry, liquid evaporates into it both humidifying the air and cooling the liquid. If a large surface of water is available, the same process may be carried out through evaporation from the surface of lake or pond. Usually the purpose is the cooling of process water. As hot water is discharged into the pond, the surface temperature of the pond rises until evaporation balances the incoming thermal load. The planner must supply a large enough pond surface to allow evaporation to balance the thermal load at a manageable temperature rise. This area requirement may exceed the availability of land in the plant-site region.

Humidification processes also occur in spray contactors often used to scrub minor components from a gas stream. Here the gas passes through successive sprays of liquid. The liquid is often water but may be specially compounded to enhance absorption of the component to be removed.

Water-Cooling Towers. By far the most common and large-scale mode of humidification processing is in water-cooling towers. As supplies of cooling water become more strained, and as discharge-water temperatures are more closely controlled, water cooling and recirculation become a necessity more often. Two general types of direct-contact cooling towers are in use. The forced-draft tower depends on fans to move the air through the tower. Typically, the tower consists of a set of louvres and baffles over which the water falls, breaking into films and droplets. Air flow may be across this cascading liquid or countercurrent to it. Often both flow arrangements exist in the same tower. Figure 8 shows a cross-sectional view of a cooling tower. Here air flows across the cascading liquid, drawn by a fan located in the outlet duct. In other arrangements the fan could be placed to push the air through the tower.

There are several internal gridwork arrangements, all designed to enhance splashing and film formation in order to give a large water–air interface and allow a low pressure drop in the air stream passing through. The lattice members were traditionally wood treated to prevent biological and corrosion attack. More recently, different materials such as transite, various plastic laminates, and ceramics have been used. Packing design has also become more and more specialized, and proprietary designs are offered by most cooling-tower makers.

The thermal design of cooling towers follows the same general procedures presented above. With the availability of machine computation, integration of equation 35 is usually done numerically with mass-transfer coefficients, saturation enthalpies, etc, given in algebraic expressions. In mechanical-draft towers the air and water flow are both supplied by machines, and hence flow rates are fixed. Under these conditions the design procedure is straightforward.

Natural-Draft Cooling Towers. In a natural-draft cooling tower (see Fig. 9) the driving force for the air is provided by the buoyancy of the air column in a very tall stack. Stack heights of 100 m are common, and as power-plant sizes increases, the size of single towers is likely to increase also. In the absence of a fan, the air flow rate G is no longer an independent variable, but is dependent on the design and operating conditions of the tower. The governing equation for air flow now becomes

$$z_t \cdot \Delta\rho = N \frac{G^2}{2\,\rho g_c} \tag{46}$$

where $\Delta\rho$ is the average density difference between the outside air and the air in the stack, z_t is the height of the tower, and N is the resistance to air flow through the tower in velocity heads. N is specific for a given tower and can usually be expressed as a constant (40).

For a natural-draft tower, equations 46 and 44 must be solved simultaneously, introducing an expression for $\Delta\rho$ as a function of conditions inside and outside the tower. Up until 1955 this was a cumbersome procedure, and a number of approximate methods were devised to simplify the calculation. Now, however, the whole calculation can be programmed for computer handling. For a tower of given dimensions, the air flow can be guessed and the thermal performance and pressure loss calculated. From the thermal performance the density difference $\Delta\rho$ can be calculated and the left side of equation 46 is compared with the other side. When the correct air flow has been guessed, the equation is satisfied. Iterative procedures with rapid convergence can easily be devised. A recent research report on natural-draft cooling towers describes an iterative procedure of this type in considerable detail (41). Cooling-tower manufacturers express confidence in their present ability to design cooling towers that meet guaranteed performance and to predict off-design behavior.

Approximate Methods for Predicting Natural-Draft Cooling-Tower Performance. Though approximate methods are no longer needed for design work, they are still useful for rapid estimates of the effects of changing conditions on performance. In addition, a good grasp of the approximate theories leads to a better physical understanding of tower behavior.

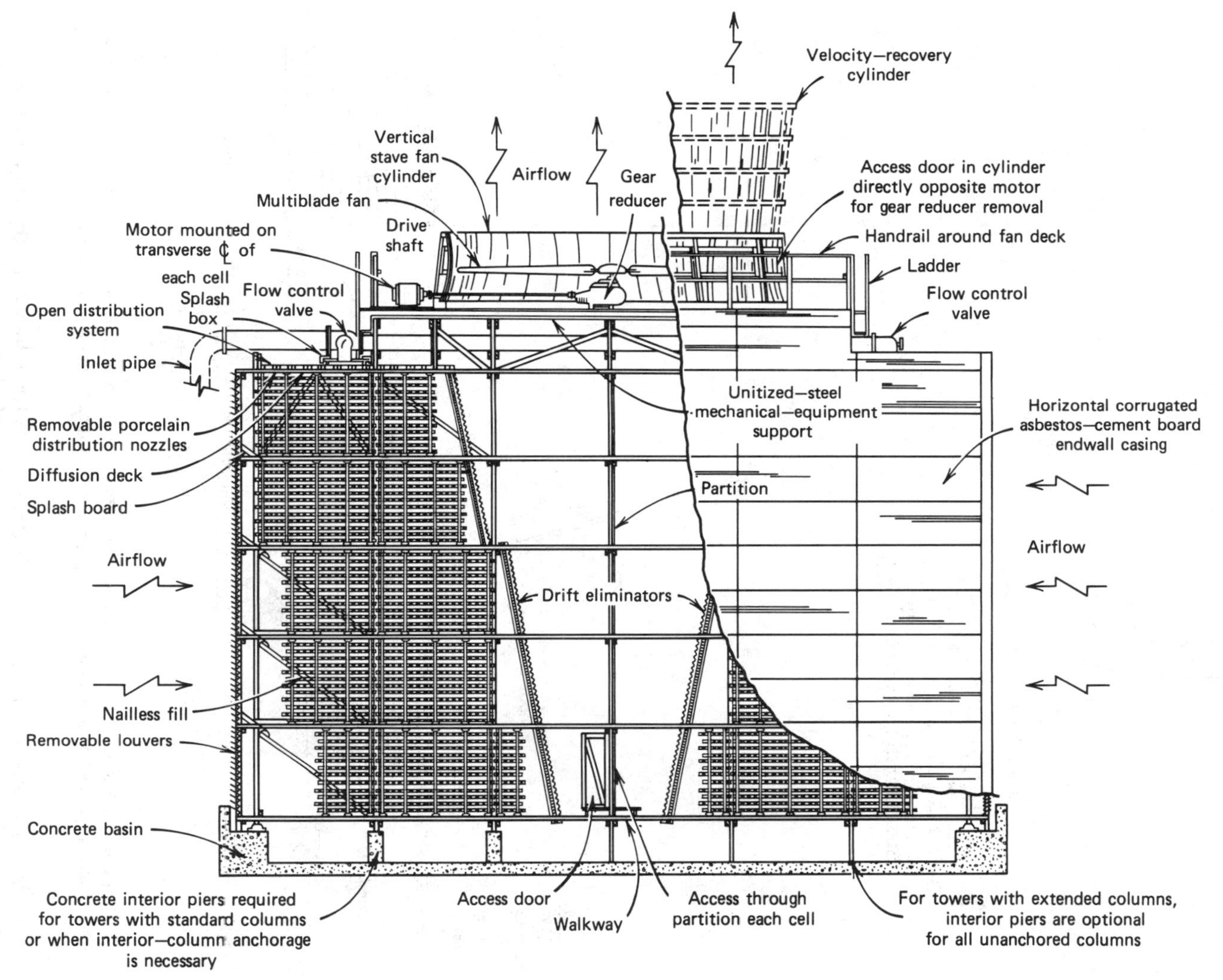

Figure 8. Transverse cross-sectional view of double-flow induced-draft cooling tower. Courtesy of The Marley Company.

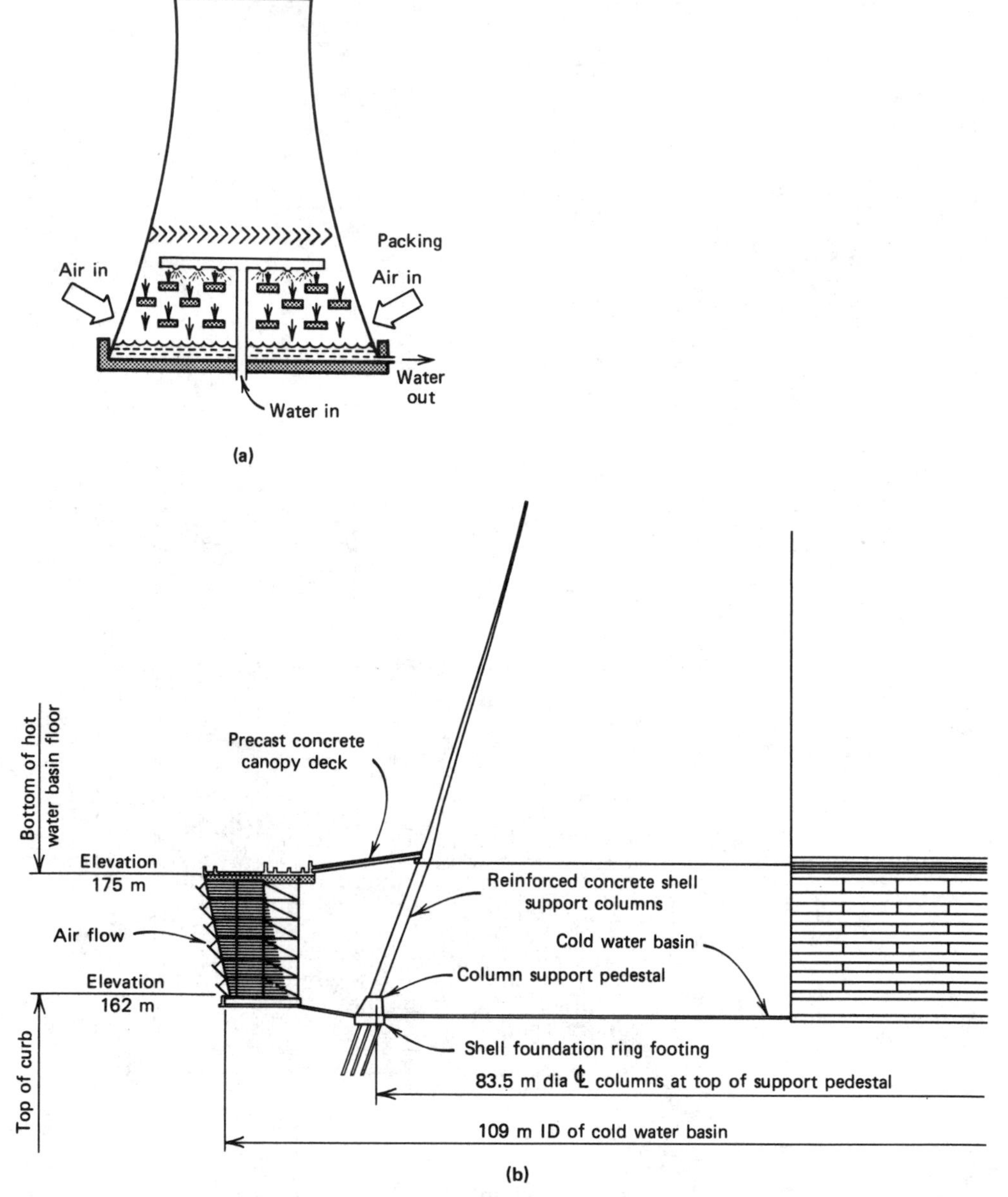

Figure 9. Natural draft cooling tower: (**a**) general tower drawing for countercurrent air–water flow arrangement; (**b**) sectional drawing showing arrangement for cross flow of air–water.

An approximate method for integrating equation 44 was originally proposed by Merkel and is discussed in later papers (40,42). Merkel's approximation leads to the equation

$$\frac{H_m^* - H_2}{T_{L_1} - T_{L_2}} = \frac{L}{2\,G} + \frac{L}{Kaz} \tag{47}$$

Using Merkel's approximation and knowing the desired thermal performance, the flow rates, and transfer coefficient, z can quickly be calculated. The difficulty with this method is that errors of ≥10% in z can arise if the cooling range $T_1 - T_2$ is larger than a few degrees.

Equations 46 and 47 have been combined to obtain rapid approximate methods for predicting cooling tower performance (40,43). The most interesting result is obtained from a rather simple analysis (44). If A is the cross-sectional area of the packing, then the liquid total flow rate $W_L = AL$, and the air flow rate $W_G = AG$. Substituting for G in equation 46 gives, with some rearrangement,

$$\frac{A\sqrt{z_t}}{(\sqrt{N/2})} = \frac{W_G}{\sqrt{\Delta\rho} \cdot \sqrt{\rho}} \tag{48}$$

which equals D, the duty coefficient of the tower. Let $\sqrt{N/2} = C^{3/2}$. Then

$$\frac{A\sqrt{z_t}}{C^{3/2}} = \frac{W_G}{\sqrt{\Delta\rho} \cdot \sqrt{\rho}} = D \tag{49}$$

Reference 40 shows that

$$\Delta\rho = 13.465 \times 10^{-5} \, (\Delta T_V + 0.3124 \, \Delta H) \tag{50}$$

Rearranging equation 50, applying the energy balance, and assuming air at standard conditions enters the tower yields (44):

$$\frac{W_L}{D} = 90.59 \, \frac{\Delta H}{\Delta T} \sqrt{(\Delta T + 0.3124 \, \Delta H)} \tag{51}$$

For most cooling towers in the UK, the exit air is saturated at a temperature close to the mean water temperature in the tower. Hence, if the water temperatures and the air inlet conditions are known, ΔH, ΔT_L, and ΔT_V can all be calculated, and W_L/D can be determined. It was found that the quantity C was approximately constant for these towers, ca 0.4–0.5 (40). If the value of C is known for a given tower, then the left side of equation 49 can be computed and, setting this equal to D, the allowable liquid flow rate can be found. Alternatively, with given W_L, T_{L_1}, and air-inlet conditions, the equations can be used to find T_{L_2}. A rapid estimate of the effects of off-design conditions can thus be made. Reference 44 presents a nomogram of these equations to facilitate the calculation.

Natural-draft cooling towers are extremely sensitive to air-inlet conditions, owing to the effects on draft. It can rapidly be established from these approximate equations that as the air-inlet temperature approaches the water-inlet temperature, the allowable heat load decreases rapidly. For this reason, natural-draft towers are unsuitable in many regions of the United States. Figure 10 shows the effect of air-inlet temperature on the allowable head load of a natural-draft tower. The inlet rh has been taken as 50%. Although the numerical values may be in error, the trend is correct.

Trends in Cooling-Tower Use and Development. Until recent years natural-draft cooling towers were rare in the United States, since ample water supplies were available for power-plant cooling, and natural-draft towers are best suited for large heat loads. Mechanical-draft towers were used in large numbers for industrial applications and occasionally for power plants. There has been a dramatic change in the situation. It appears that almost all new, large power plants require cooling towers. In the southern United States mechanical-draft towers will dominate the field. In the northern United

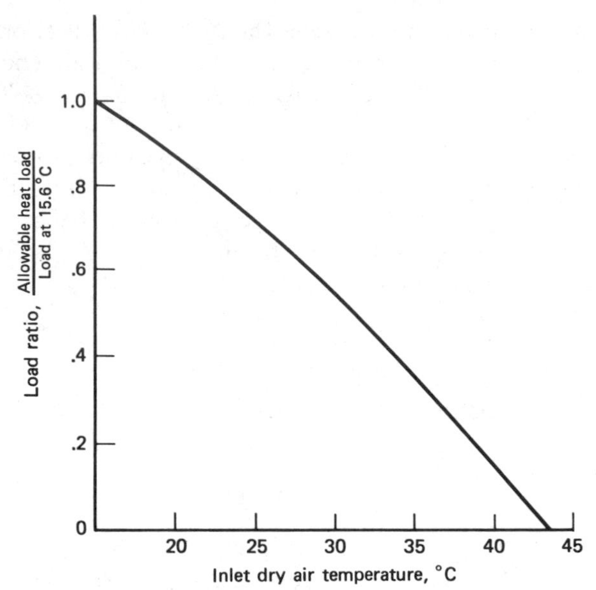

Figure 10. Effect of inlet dry air temperature on allowable load. Inlet relative humidity 50%. Water inlet temperature 43.3°C. Water exit temperature 32.2°C.

States the situation is not as clear. The issues involved in a decision are discussed in references 45–46.

The initial cost of a mechanical-draft cooling tower for a power plant is relatively low, ca $5.00/kW more than a direct-stream cooling system. However, the power required to run the fan is significant, and maintenance must also be considered. In a large power plant, many mechanical-draft towers are required, covering a large area of ground, and problems of water and power distribution become acute. In addition, the plume is discharged close to the ground and can be a source of fog.

The first cost of a natural-draft tower is substantially higher, ca $10/kW. However, the maintenance costs are much lower since there are no fans or electrical drives. The natural-draft tower occupies less ground space than the corresponding group of mechanical-draft towers. Because the plume is discharged ≥100 m from the ground, it is much less likely to cause local ground fog. As a result of these factors, the natural-draft tower is frequently more attractive, and may be chosen even if there is a slight overall cost disadvantage.

The economics of cooling towers has been discussed in several articles (44,47). A worthwhile evaluation of the optimum configuration must take into account the interaction between the tower performance and the plant performance. For example, in considering the additional expense of a cooling tower vs a direct-stream cooling system, it is desirable to optimize the whole plant for each system rather than add the cooling tower to a system optimized for run-of-the-river cooling. Consequently, it is not possible to produce general cost curves for cooling towers. Each installation must be evaluated separately.

A large natural-draft cooling tower may cost approximately $6,000,000, and the cost is divided almost equally among the foundation, the packing and water-distribution system, and the shell. Therefore, large cost reductions in any of these items will have a significant effect. New, light packings are being developed in a variety of materials and some cost improvements may be seen in the near future in this area. The tower shell is made of reinforced concrete; its hyperbolic shape is chosen mainly for structural rather than aerodynamic reasons. A hyperbolic shell requires less concrete than an equally strong cylindrical shell. It seems unlikely that important cost reductions in the shell or foundation can be made.

When all the expenses involved in using wet cooling towers on a power plant are considered, it appears that a mechanical-draft cooling tower system may raise the cost of generating electricity ca 3%, and a natural-draft tower ca 6% over the generating cost of a direct river-cooled power plant. These figures are approximate, but show the effects of a thermal-pollution regulatory program on the cost of generating electricity (see Thermal pollution).

Cooling-Tower Plumes. An important consideration in the acceptability of either a mechanical-draft or a natural-draft tower cooling system is the effect on the environment. The plume emitted by a cooling tower is seen by the surrounding community, and if it is a source of severe ground fog under some atmospheric conditions, trouble may follow. As has already been mentioned, the natural-draft tower is much less likely to produce fogging than is the mechanical draft tower. Nonetheless, it is desirable to devise techniques for predicting plume trajectory and attenuation.

Not only may the cooling-tower plume be a source of fog, which in some weather conditions can ice roadways, but it carries salts from the cooling water itself. These salts may come from salinity in the water, or they may be added by the cooling-tower operator to prevent corrosion and biological attack in the column. Efforts to combat bacteria and corrosion in cooling towers have gone through a long development and are now both complex and specific to different waters. Now cathodic and anodic inhibitors are used as well as biocides. Systems may include chromates, nitrites, orthophosphates, ferrocyanides, as cathodic inhibitors; zinc, nickel, lead, tin, copper, silicates as anodic inhibitors; and possibly added materials as biocides and pH controllers. These chemicals can be carried onto fields surrounding the cooling tower and seriously affect crop yield or ornamental plantings (see Water, industrial water treatment).

Thus, much work has been done on the modeling of wet cooling-tower plumes with the ultimate aim of determining their effect on the environment (48–50). The basic approach involves writing the equations of continuity, conservation of energy and of momentum, and the equations of motion for the conditions of the plume. These are then solved simultaneously using iterative and numerical methods on large computers. The accuracy of the results depends on whether the boundary conditions and simplifying assumptions are realistic. These are difficult to set since conditions change rapidly, ground configurations are seldom simple, and plume behavior is influenced by a host of casual, nonrepeated situations such as the passing of an airplane or the presence of cloud cover. Modeling owes much to meteorology and especially to the theory of cumulus clouds.

The recirculation of cooling water via a cooling tower ultimately removes process heat by evaporating water rather than by warming it, as would be the case with once-through systems. When water is especially scarce, it may be necessary to cool process water by transferring the heat to air through indirect heat transfer. This requires dry cooling towers, which have been built in a few dry regions of the United States. They usually take the form of natural-draft cooling towers with high surface heat-exchange areas replacing the usual gridwork. Since the air-to-circulating-water heat-transfer process passes heat through a solid surface, heat-transfer coefficients are low. Enormous heat-transfer areas are thus required. The cost of such towers may be tenfold that of wet towers, and the availability of tubing for heat transfer becomes a serious problem.

Nomenclature

A = interfacial area
a = interfacial area per unit column volume
C = heat capacity
D = molecular diffusivity for mass transfer
F = rate of fog formation, mol/time
G = mass flow rate of gas phase, kg/(h·m^2)
g_c = force–mass conversion constant
H = enthalpy
HTU = height of a transfer unit
h_c = convective heat transfer coefficient
h_r = coefficient for heat transfer by radiative mechanism
K_a = overall mass transfer coefficient per volume of contacting column
k_g = gas phase mass transfer coefficient in partial pressure driving force units
k_L = liquid phase thermal conductivity
k_Y = mass transfer coefficient in gas phase mole ratio units
L = liquid stream molar flow rate
Le = Lewis number
M = molecular weight
N = resistance to air flow in velocity heads
NTU = number of transfer units
P = total pressure
Pr = Prandtl number
$p°$ = vapor pressure
q = heat flux
Re = Reynolds number
r = psychrometric ratio
S = suppression factor $(\Delta Te/\Delta T)^{0.99}$
Sc = Schmidt number
T = temperature
$\Delta\tilde{T}$ = driving force for the binary macroscopic heat transfer
V = specific volume
V = gas phase molar flow rate
V' = molar flow rate of noncondensable component
w = total flow rate, mol/time
x = mole fraction in liquid phase
Y = mole ratio
Y' = mass ratio
y = mole fraction in gas phase
z = height of column
ϵ = Ackerman correction term, $\epsilon = m_i \cdot C_{p_i}/h_g$
λ = latent heat of vaporization
μ = viscosity
ρ = density
σ = surface tension

Subscripts

BM = mean value for noncondensing component
e = effective
g = gas phase
h = humid value, ie, including gas and vapor

i = interface condition
j = for the jth component (usually less volatile)
L = liquid phase
L_{only} = for the liquid-phase flow alone
mac = macroscopic contribution
mic = microscopic contribution
o = at reference condition
p = at constant pressure
s = at saturation
sa = at adiabatic saturation condition
V = in the vapor phase
w = for water, or at wet-bulb temperature
Y = the mole ratio driving force
1,2 = end points in the process
2ϕ = for two-phase flow

Superscripts

$*$ = bulk concentration in liquid phase but in gas-phase units, or vice versa
$'$ = mass rather than mole basis

BIBLIOGRAPHY

1. G. Ackerman, *Verh. Dtsch. Ing. Forschungsh.* **382,** 1 (1937).
2. A. P. Colburn and O. A. Hougen, *Ind. Eng. Chem.* **26,** 1178 (1934).
3. A. P. Colburn and T. B. Drew, *Trans. Am. Inst. Chem. Eng.* **33,** 197 (1937).
4. F. Stern and F. Votta, Jr., *AIChE J.* **14,** 928 (1968).
5. D. R. Coughanowr and E. O. Stensholt, *Ind. Eng. Chem. Proc. Des. Dev.* **3,** 369 (1964).
6. J. T. Schrodt, *Ind. Eng. Chem. Proc. Des. Dev.* **11,** 20 (1972).
7. J. T. Schrodt, *AIChE J.* **19,** 753 (1973).
8. R. L. Von Berg, *Recent Advances in Liquid–Liquid Extraction*, Pergamon Press, Oxford, England, 1971.
9. E. J. Hoffman, *AIChE J.* **17,** 741 (1971).
10. A. S. H. Jernqvist, *Br. Chem. Eng.* **11,** 1205 (1966).
11. F. Kayihan, O. C. Sandall, and D. A. Mellichamp, *Chem. Eng. Sci.* **30,** 1333 (1975).
12. W. R. Lindberg and R. D. Haberstroh, *AIChE J.* **18,** 243 (1972).
13. J. L. Manganaro and O. T. Hanna, *AIChE J.* **16,** 204 (1970).
14. G. L. Standart, R. Taylor, and R. Krishna, *Chem. Eng. Commun.* **3,** 277 (1979).
15. M. M. Dribika and O. C. Sandall, *Chem. Eng. Sci.* **34,** 733 (1979).
16. H. Hikita and K. Ishimi, *Chem. Eng. Com.* **3,** 547 (1979).
17. ABM Abdul Hye, *Simultaneous Heat and Mass Transfer from a Vertical, Isothermal Surface*, Ph.D. Thesis, University of Windsor, Canada, 1979.
18. T. Mizushina, M. Nakajima, and T. Oshima, *Chem. Eng. Sci.* **13,** 7 (1960).
19. J. T. Schrodt and E. R. Gerhard, *Ind. Eng. Chem. Fund.* **7,** 281 (1968).
20. L. W. Florschuetz and A. R. Khan, *Fourth International Heat Transfer Conference*, Paris, France, 1970.
21. O. C. Sandall and M. M. Dribika, *Inst. Chem. Eng. Sym. Ser.* **56,** 2.5/1 (1979).
22. H. J. Barton and O. Trass, *Can. J. Chem. Eng.* **47,** 20 (1969).
23. L. E. Scriven, *Chem. Eng. Sci.* **10,** 1 (1959).
24. F. Marshall and L. L. Moresco, *Int. J. Heat Mass Transfer* **20,** 1013 (1977).
25. R. A. W. Schock, *Int. J. Heat Mass Transfer* **20,** 701 (1977).
26. D. L. Bennett and J. C. Chen, *AIChE J.* **26,** 454 (1980).
27. C. H. Bedingfield and T. B. Drew, *Ind. Eng. Chem.* **42,** 1164 (1950).
28. H. S. Mickley, *Chem. Eng. Prog.* **45,** 739 (1949).
29. S. L. Hensel and R. E. Treybal, *Chem. Eng. Prog.* **48,** 362 (1952).
30. J. Lichtenstein, *Trans. ASME* **65,** 779 (1943).
31. R. L. Pigford and C. Pyle, *Ind. Eng. Chem.* **43,** 1649 (1951).

32. W. M. Simpson and T. K. Sherwood, *Refrig. Eng.* **52,** 535 (1946).
33. A. E. Surosky and B. F. Dodge, *Ind. Eng. Chem.* **42,** 1112 (1950).
34. F. P. West, W. D. Gilbert, and T. Shimizu, *Ind. Eng. Chem.* **44,** 2470 (1952).
35. J. Weisman and E. F. Bonilla, *Ind. Eng. Chem.* **42,** 1099 (1950).
36. F. Yoshida and T. Tanaka, *Ind. Eng. Chem.* **43,** 1467 (1951).
37. *Countercurrent Cooling Tower Performance*, J. F. Prichard Company, Kansas City, Mo., 1957.
38. *Technical Bulletins R-54-P-5, R-58-P-5*, Marley Company, Kansas City, 1957.
39. *Performance Curves*, Cooling Tower Institute, Houston, Texas, 1967.
40. H. Chilton, *Proc. IEE, Supply Sect.* **99,** 440 (1952).
41. J. R. Singham, *The Thermal Performance of Natural Draft Cooling Towers*, Imperial College of Science and Technology, Department of Mechanical Engineering, 1967.
42. H. B. Nottage, *ASHRAE Trans.* **47,** 429 (1941).
43. I. A. Furzer, *Ind. Eng. Chem. Proc. Des. Dev.* **7,** 555 (1968).
44. R. F. Rish and T. F. Steel, *Proc. ASCE J. Power Div.* **P05,** 89 (1959).
45. C. Waselkow, *National Conference on Thermal Pollution*, Federal Water Pollution Control Administration and Vanderbilt University, Aug. 1968.
46. W. R. Shade and A. F. Smith in ref. 45.
47. B. Berg and T. E. Larson, *Am. Power Conf.* **XXV,** 678 (1963).
48. K. G. Baker, *Chem. Proc. Eng.* 56 (Jan. 1967).
49. C. H. Hosler in *Cooling Towers*, CEP Technical Manual, 1972, p. 27.
50. D. B. Hoult, J. A. Fay, and L. J. Forney, *A Theory of Plume Rise Compared with Field Observations*, Fluid Mechanics Laboratory Publication No. 68-2, Massachusetts Institute of Technology Department of Mechanical Engineering, March 1968.

General References

Cooling Towers, CEP Technical Manual, American Institute of Chemical Engineering, New York, 1972.
R. E. Treybal, *Mass Transfer Operations*, 2nd ed., McGraw-Hill, Inc., New York, 1968, pp. 176–220.
R. H. Perry and C. H. Chilton, *Chemical Engineers Handbook*, 5th ed., McGraw-Hill, Inc., New York, 1973, pp. 12-17–12-21.
A. S. Foust and co-workers, *Principles of Unit Operations*, 2nd ed., John Wiley & Sons, Inc., New York, 1980, Chapt. 17, pp. 420–453.

LEONARD A. WENZEL
Lehigh University

SIZE ENLARGEMENT

Size enlargement concerns those processes that bring together fine powders into larger masses in order to improve the powder properties. Usually, they produce relatively permanent entities in which the original particles can still be identified, but this is not always necessary, as demonstrated in the formation of amorphous agglomerates by the cooling of melts, or by the production of weak and transient "instant" food agglomerates in which product strength need only be sufficient to withstand downstream handling, packaging, and transportation.

Size enlargement methods have been known for hundreds of years. The roots of the processes can be traced to such ancient techniques as the formation of clay bricks and other building materials, the hammering of implements from sponge iron, the manufacture of various items from precious metal powder, and the preparation of solid molded forms of medicinal agents (1). Agglomeration by heating and roasting techniques became established as a practical operation during the 19th century with the necessity to beneficiate and process fine coals and ores. In this century use of size enlargement has grown rapidly for a number of reasons. The application of high analysis nitrogen fertilizers in intensive agriculture led to development of noncaking and granulated, rather than powdered, products (2). The lower quality of available iron ore resulted in the need to upgrade the resource by grinding and rejection of the liberated impurities, followed by pelletization of the resulting fines into an acceptably coarse product (3). Environmental considerations have led to the recovery of many dusts and fine waste powders that can be recycled after size enlargement (4). In addition, modern high volume processing requires consistent feeds with good flow properties, requirements that for powders can often only be met through some form of agglomeration.

Many diverse industries benefit from the use of size enlargement, ranging from the relatively small scale, but important, requirements of pharmaceutical manufacturers to the tonnage requirements of the fertilizer and minerals processing industries. Benefits gained from size enlargement are as diverse as the industries in which the operation is used (1,5). Dusting losses and caking and lump formation are avoided with improved product appearance through the granulation of fertilizers. The wet granulation of pharmaceutical powders produces nonsegregating powder blends with consistent flow properties. The granulation then feeds high capacity tableting devices which yield tablets of a defined and consistent dosage. Objects with useful structural forms and shapes are produced in powder metallurgy and ceramic forming. The pelleting of carbon black increases bulk density and improves handling qualities. Control of powder properties such as solubility, porosity, and heat transfer capability can be attained through agglomeration procedures. Agglomeration in liquid suspension, such as the selective oil agglomeration of coal in water suspension, not only recovers the fine particles from the liquid but does so selectively in that unagglomerated impurity particles remain in suspension (see Carbon, carbon black; Ceramics, forming processes; Coal conversion processes, desulfurization; Coffee; Drying; Fertilizers; Iron; Pharmaceuticals; Plastics processing; Powder metallurgy).

Particle-Bonding Mechanisms

The mechanisms by which particles bond together and grow into agglomerates are affected by the specific size enlargement method. Nevertheless there are certain aspects of the bonding process that are essentially independent of the equipment and method. These aspects are described here in general terms before more detailed examination of the various size enlargement processes (see also Powder coating).

A classification of bonding mechanisms based on the fundamental nature of the interparticle bonds has been widely adopted in the literature and will be used here, together with the theoretical model used to estimate agglomerate strength (6). Rumpf's classification into five categories of particle–particle bridging is summarized in Table 1 together with references from the literature to examples in each category. More than one bonding mechanism is likely to act in a given process. For example, in bonding by tar deposited through solvent evaporation, it is likely that oxidative hardening will also occur. In sintering ores, bonding through chemical reaction also contributes to strength. With very fine powders it is difficult to determine whether bonding through long-range forces or vapor adsorption predominates. Although mechanical interlocking of particles influences agglomerate strength, its contribution is generally considered small in comparison to other mechanisms.

Theoretical Strength of Agglomerates. Based on statistical-geometrical considerations, Rumpf developed the following equation for the mean tensile strength of an agglomerate in which bonds are localized at the points of particle contact (6):

$$\sigma_T \approx \frac{9}{8}\left(\frac{1-\epsilon}{\epsilon}\right)\frac{H}{d^2} \tag{1}$$

where σ_T is the mean tensile strength per unit section area, Pa, ϵ is the void fraction of the agglomerate, d is the diameter of the (assumed) monosized spherical particles, m, and H is the tensile strength, N, of a single particle–particle bond. (To convert Pa to psi, multiply by 0.145×10^{-3}.)

In a second main class of agglomerate bonding particles are imbedded in or surrounded by an essentially continuous matrix of binding material rather than having bonding localized at points of particle contact. An important example of this second type of binding is the case in which particles are held together by mobile low viscosity liquids where adhesion occurs as a result of interfacial tension at the liquid surface and the pressure deficiency (capillary suction) created within the liquid phase by curvature at the liquid surface (20). The various regimes of low viscosity liquid which can exist in an agglomerate are shown in Figure 1.

In the pendular state, shown in Figure 1**a**, particles are held together by discrete lens-shaped rings at the points of contact or near-contact. For two uniformly sized spherical particles the adhesive force in the pendular state for a wetting liquid (contact angle zero degrees) can be calculated (17,21) and substituted for H in equation 1 to yield:

$$\sigma_T = \frac{9}{4}\left(\frac{1-\epsilon}{\epsilon}\right)\frac{\gamma}{d} \tag{2}$$

where γ is the liquid surface tension, N/m.

When the void space in an agglomerate is completely filled with a nonviscous liquid (Fig. 1**c**), the capillary state of wetting is reached, and the tensile strength of

Table 1. Classification of Binding Mechanisms According to Rumpf[a]

Class	Mechanism	Representative examples	References
solid bridges	sintering, heat hardening	induration of iron-ore pellets	3
		sintering of compacts in powder metallurgy	8
	chemical reaction, hardening binders, "curing"	cement binder for flue-dust pellets	9
		ammoniation–granulation of mixed fertilizers	10
		oxidation of tar binders	11
	incipient melting due to pressure, friction	briquetting of metals, plastics	12–13
	deposition through drying	crystallization of salts in fertilizer granulation	14
		deposition of colloidal bentonite in dry iron-ore balls	3
immobile liquids	viscous binders, adhesives	sugars, glues, gums in pharmaceutical tablets	15
	adsorption layers	instantizing food powders by steam condensation	16
		humidity effects in flow of fine powders	
mobile liquids	liquid bridges (pendular state)	flocculation of fine particles in liquid suspension by immiscible liquid wetting	7
		moistening–mixing of iron-ore sinter mix	3
	void space filled or partly filled with liquid (capillary and funicular states)	balling (wet pelletization of ores)	3
		soft plastic forming of ceramic powders	
intermolecular and long-range forces	Van der Waals forces electrostatic forces magnetic forces	adhesion of fine powders during storage, flow and handling	17
		spontaneous dry pelletization of fine powders (eg, carbon black, zinc oxide)	
mechanical interlocking	shape-related bonding	fracturing and deformation of particles under pressure	18
		fibrous particles, eg, peat moss	19

[a] Refs. 6–7.

the wet particle matrix arises from the pressure deficiency in the liquid network due to the concave liquid interfaces at the agglomerate surface. This pressure deficiency can be calculated from the Laplace equation for circular capillaries to yield, for liquids which completely wet the particles:

$$\sigma_T = C\left(\frac{1-\epsilon}{\epsilon}\right)\frac{\gamma}{d} \tag{3}$$

where the parameter C has a theoretical value of 6 for uniform spherical particles and ranges between 6.5–8 for nonuniform particles (20,22).

It is evident, by comparing equations 2 and 3, that tensile strength in the pendular

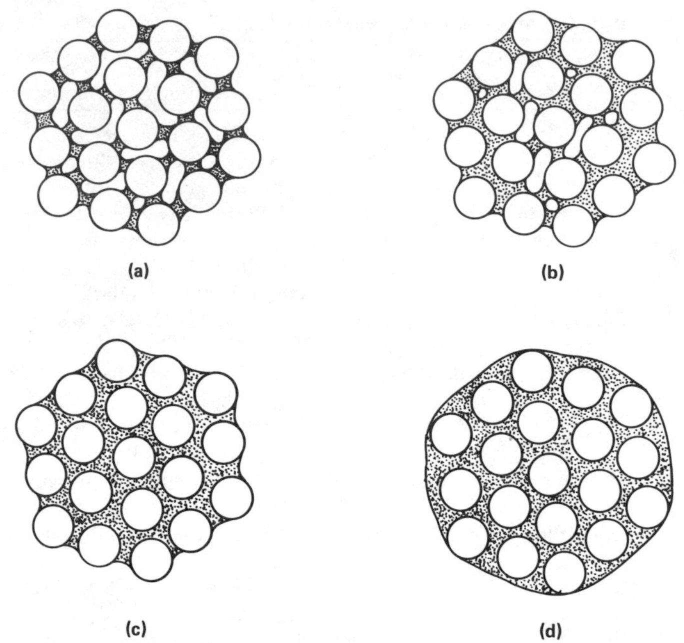

Figure 1. States of liquid content in bonding by low viscosity liquids; (**a**) pendular (**b**) funicular (**c**) capillary (**d**) particles surrounded by liquid droplet (7).

state is about one-third that in the capillary state. Intermediate liquid contents in the funicular state (Fig. 1**b**) yield intermediate values that can be approximated as:

$$\sigma_T = sC\left(\frac{1-\epsilon}{\epsilon}\right)\frac{\gamma}{d} \tag{4}$$

where s is the fractional filling of the agglomerate voids.

Measuring and Correlating Agglomerate Strength. In practice, simple and quick test methods are most often used to assess the quality of bonding and other desirable properties of product agglomerates. Compression, impact, abrasion, and other types of tests are widely accepted in industry to characterize agglomerate quality in relation to subsequent handling and processing. The mode of agglomerate failure is complex in these tests making it difficult to relate the results directly to theory.

Compression tests, in which agglomerates are crushed between parallel platens, are probably most universal. To obtain reproducible and accurate results, the rate of loading and method of load application must be strictly controlled. A variety of commercial testers are available to allow this needed control over the compression process. Several means of distributing the load uniformly at the point of contact are used including covering the platen surface with compressible board. With pliable agglomerates especially, the fracture load is highly dependent on the rate of loading (23–24).

During compression of spherical agglomerates, flattening takes place at the points of contact as particle–particle bonds fail locally and particles are driven into adjoining voids. Small, dense, wedgelike elements form in the regions adjacent to the platens and the agglomerate fails in tension along a circumferential crack joining the poles of loading (double-cone failure) (24–25). Internal frictional effects must be overcome in this compression process, in addition to the tensile strength of the particle matrix given by equations 1–4. For corresponding tensile and compression measurements on wet limestone pellets in the range for which the mean tensile strength is σ_T 20 kPa (3 psi)–83 kPa (12 psi), Rumpf found the ratio of tensile to compressive strength to be $\sigma_T/\sigma_C \approx 0.5$–$0.77$ (6).

For approximately spherical agglomerates, compression strength is calculated as follows:

$$\sigma_C = L/(\pi D^2/4) \tag{5}$$

where L is the compression force at failure in N, and D is the agglomerate diameter in m. Some typical compressive strengths of various agglomerates are indicated in Figure 2.

According to equation 5, compression test data can usually be correlated through a log–log plot of load at failure against pellet diameter to yield a straight line whose slope is often, but not always, equal to 2 (26–27). The intercept of such a plot is related to the compressive strength, σ_C. Because ultimate failure occurs in tension as explained above for the double-cone failure mechanism, this compressive strength factor can subsequently be correlated with bonding mechanism through equations 1–4 to account for the effect of such parameters as particle diameter, agglomerate porosity, and the strength of particle–particle bonding (20,28). Alternatively, empirical relationships can be developed to account for the effects of these parameters describing the particle matrix (29–30).

Several other tests of agglomerate quality are done routinely, the details de-

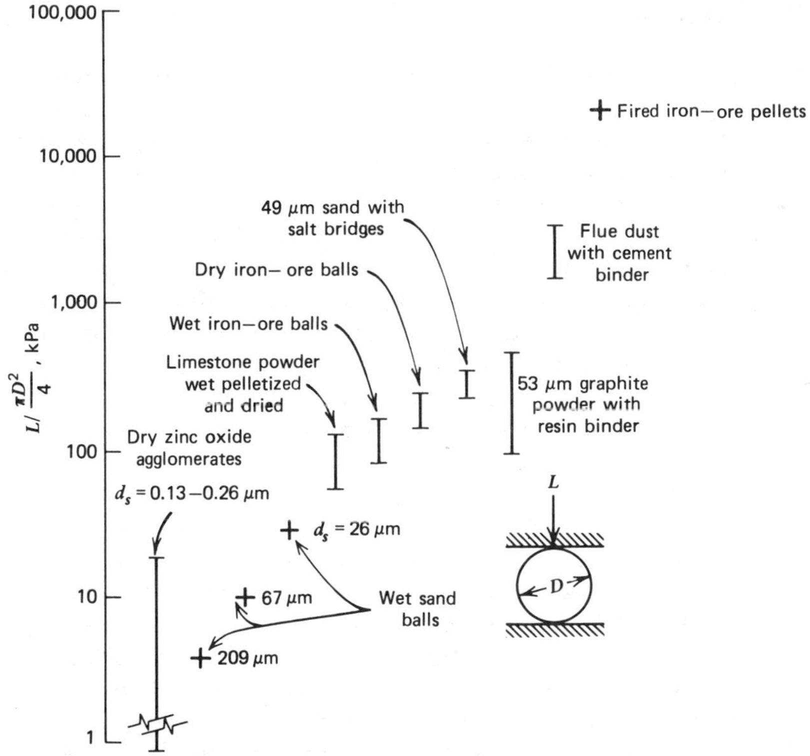

Figure 2. Compression strength of agglomerates formed into spherical shape by tumbling (7). To convert kPa to psi, multiply by 0.145.

pending on the practice accepted in a specific industry. For iron ore, a drop test is used on wet agglomerates as a measure of their ability to withstand handling up to the point in the process at which they are dried and fired (3). A test might consist of dropping a number of agglomerates from a height of 300 mm or 450 mm onto a steel plate. The average number of drops required to cause fracture is the drop number. In the ASTM tumbler test for iron-ore pellets (E 279), an 11.3 kg sample in the size range 6.4–38.1 mm is placed in a tumbler drum ca 910 mm diameter by 460 mm long and rotated at 24 rpm for a total of 200 revolutions. The drum is equipped with two equally spaced lifters 51 mm high. The abrasion index is given by the weight percentage of plus 6.4 mm material surviving the test and the dust index by the yield of minus 0.6 mm (30 mesh) material. Other strength tests such as the Linder rotating-furnace procedure, are used with iron-ore pellets in an attempt to determine their reducibility and breakdown under reducing conditions simulating those of a blast furnace (31).

Laboratory tests to assess the caking tendency of fertilizer granules consist of two parts (32). The first entails cake formation in a compression chamber under controlled conditions of air flow, humidity, temperature, etc. In the second part the cake is removed and its crushing strength determined as a measure of the degree of caking. In the tablet disintegration test outlined in USP XX, pharmaceutical tablets are contained in a basket-rack assembly which is immersed in a suitable dissolution fluid in a container and raised and lowered to agitate the tablets at a constant rate of 29–32 cycles per minute for a specified period of time (33). Acceptable tablets disintegrate completely by the end of the test period.

Types of Size Enlargement

Classification of size enlargement methods reveals two distinct categories (1,7). The first is forming-type processes in which the shape, dimensions, composition, and density of the individual larger pieces formed from finely divided materials are of importance. The second is those processes in which creation of a coarse granular material from fines is the objective and the characteristics of the individual agglomerates are important only in their effect on the properties of the bulk granular product.

Four principal mechanisms are used to bring fine particles together into larger agglomerates: agglomeration by tumbling and other agitation methods; pressure compaction and extrusion methods; heat reaction, fusion, and drying methods; and agglomeration from liquid suspensions.

Agglomeration by Tumbling and Other Agitation Methods

When fine particles, usually in a moist state, are brought into intimate contact through agitation binding forces come into action to hold the particles together as an agglomerate. Capillary binding forces caused by wetting with water or aqueous solutions is the most common binding mechanism, but others such as intermolecular forces developed in extremely fine dry powders may also be used to form the agglomerates. Several different forms of agitation may be used. The rolling cascading action of disk and drum devices produces rounded or roughly spherical agglomerates; other types of mixers, described below, generally yield more irregular shapes.

Agglomerate growth can occur through a number of mechanisms such as coalescence, crushing and layering, layering of fines, and abrasion transfer (34–35). More than one mechanism may occur simultaneously in a given process. Agglomerates of the order of 1–3 mm diameter formed by coalescence of fine feed particles, or recycled undersize product agglomerates act as seed nuclei for the process. The nuclei grow to larger sizes by the addition of layers of fines supplied continuously to the process.

To survive and grow in an agitated system, agglomerates must be able to withstand the destructive forces generated by the moving charge of powder. Equation 3 indicates that for given agitation conditions there is a maximum particle size that can produce sufficient tensile strength in the particle matrix to form agglomerates satisfactorily. The addition of fines to a given size distribution of feed material usually improves agglomeration not only because it reduces the mean particle size but also because it improves packing and reduces voids in the system. Similarly, more intensive agitation conditions generally reduce the size of agglomerates produced from a given material. For wet agglomeration on disk pelletizers top size is usually 300–600 μm (ca 30–50 mesh) and at least 25% of the feed powder should be finer than 75 μm (ca 200 mesh) (36). In iron-ore balling, a grind with 40–80% of the material below 44 μm (325 mesh) is normally used.

Agglomerates formed by wet pelletization in balling drums and disks are generally considered to have their internal pores saturated with binding liquid, that is, to be in the capillary state of wetting (see Fig. 1**c**) (34,37). The weight fraction for the theoretical liquid content of such agglomerates is given by equation 6

$$W = \frac{\epsilon \rho_L}{\epsilon \rho_L + (1 - \epsilon)\rho_S} \tag{6}$$

where W is the weight fraction of liquid (wet basis); ϵ is the void fraction in the agglomerates; and ρ_L, ρ_S are liquid and particle densities, respectively.

Equation 6 has been fitted to a wide variety of literature data in which solid density ranged from about 1–6 g/cm^3 and liquid densities were generally close to 1 g/cm^3 (37). The following relationships were found:

For average feed-particle diameters <30 μm:

$$W = \frac{1}{1 + 1.85 \dfrac{\rho_S}{\rho_L}} \tag{7}$$

For average feed-particle diameters >30 μm:

$$W = \frac{1}{1 + 2.17 \dfrac{\rho_S}{\rho_L}} \tag{8}$$

These relationships predict the binding liquid content for wet agglomeration with an accuracy of only ca 30%. Typical values of moisture content required for balling a variety of materials are listed in Table 2. Very accurate information on the optimum liquid content to agglomerate a particular feed material must be obtained from experimental tests.

Drum and Inclined-Disk Agglomerators. Although a wide variety of agitation equipment is used industrially to produce agglomerates, rotary drums or cylinders and inclined disks or pans are the most important equipment in terms of tonnages produced. Fertilizer granulation and iron-ore balling represent two of the largest applications (see Fertilizers; Iron).

Table 2. Moisture Requirements for Balling a Variety of Materials[a]

Raw material	Approximate size analysis of raw material, less than indicated μm (mesh)[b]	Moisture content of balled product, wt % H_2O
precipitated calcium carbonate	75 (200)	29.5–32.1
hydrated lime	45 (325)	25.7–26.6
pulverized coal	300 (48)	20.8–22.1
calcined ammonium metavandate	75 (200)	20.9–21.8
lead–zinc concentrate	850 (20)	6.9–7.2
iron pyrite calcine	150 (100)	12.2–12.8
specular hematite concentrate	106 (150)	9.4–9.9
taconite concentrate	106 (150)	9.2–10.1
magnetic concentrate	45 (325)	9.8–10.2
direct-shipping open-pit ores	1680 (10)	10.3–10.9
underground iron ore	6.4 mm	10.4–10.7
basic oxygen-converter fume	1 μm	9.2–9.6
raw cement meal	106 (150)	13.0–13.9
utilities–fly ash	106 (150)	24.9–25.8
fly ash–sewage-sludge composite	106 (150)	25.7–27.1
fly ash–clay-slurry composite	106 (150)	22.4–24.9
coal–limestone composite	150 (100)	21.3–22.8
coal–iron-ore composite	300 (48)	12.8–13.9
iron ore–limestone composite	150 (100)	9.7–10.9
coal–iron ore–limestone composite	1180 (14)	13.3–14.8

[a] Courtesy of Dravo Engineers and Constructors.
[b] Tyler equivalent scale.

Drum agglomerators (Fig. 3) consist of an inclined rotary cylinder powered by a fixed- or variable-speed drive. Feed material containing the correct amount of liquid-phase agglomerates under the rolling, tumbling action of the rotating drum. The pitch of the drum (up to 10° from the horizontal) assists material transport down the length of the cylinder. A retaining ring is often fitted at the feed end to prevent spill-back of feed. A dam ring may also be used at the exit to increase the depth of material

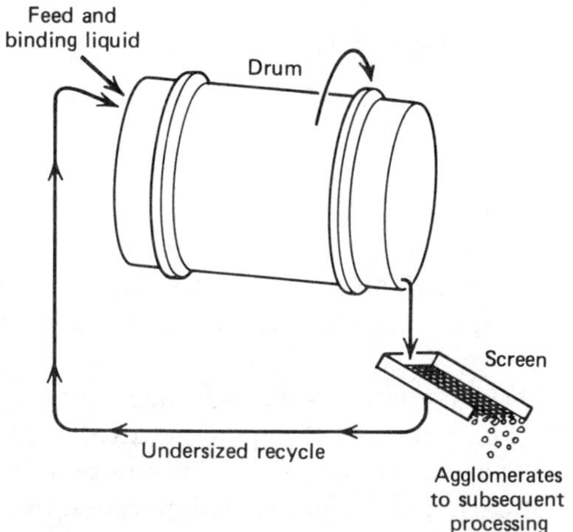

Figure 3. Schematic of a drum agglomerator.

and residence time in the drum. Liquid phase may be introduced either before or immediately after the solids enter the cylinder. With iron ores, the feed is usually premoistened wet filter cake, but water sprays may also be located inside the drum for moisture addition to aid control. In mixed fertilizer granulation, various solutions and slurries may be used, but water is the usual wetting agent. Various types of internal scrapers are in use to limit build-up of material on the inside surface and to provide a uniform layer to promote the correct rolling, tumbling action in the drum. Rubber flaps and liners as well as external knockers are used to limit build-up in fertilizer processing.

The drums used for fertilizer granulation range from 1.5 m in diameter by 3 m long, with installed power of 11 kW, and rotation rates of 10–17 rpm that process 8 metric tons per hour (t/h), up to 3 m diameter by 6 m long, with installed power of 112 kW, rotation rates of 7–12 rpm and a capacity of 51 t/h. These capacities exclude recycle so that the actual throughput of the drums may be much higher. Capacity depends primarily on the quantity of undersize and crushed oversize material recycled to the drum during continuous operation. The recycle ratio (amount recycled/output) varies from less than 1 (greater than 50% output) to 5 or 6 for hard-to-granulate grades of fertilizer. The grade being granulated, differing formulations for the same grade, differing plant operations, ambient temperatures, and skill all influence capacity in fertilizer granulation (38).

A typical drum used to ball iron ore is 3 m in diameter by 9.5 m long, is driven at 12–14 rpm by a 45 kW drive and produces 60 t/h product (39).

An inclined-disk agglomerator consists of a tilted rotating plate equipped with a rim to contain the agglomerating charge (Fig. 4). Solids are fed continuously from above onto the central part of the disk and product agglomerates discharge over the rim. Moisture or other binding agents can be sprayed on at various locations on the plate surface. Adjustable scrapers and plows maintain a uniform protective layer of product over the disk surface and also control the flow pattern of material on the disk. Plate angle can be adjusted from 40–70° to the horizontal to obtain the best results and both constant-speed and variable-speed motors are available as disk drives. Dust covers can be fitted when required. Characteristics of some of the range of inclined disks offered by one manufacturer are given in Table 3.

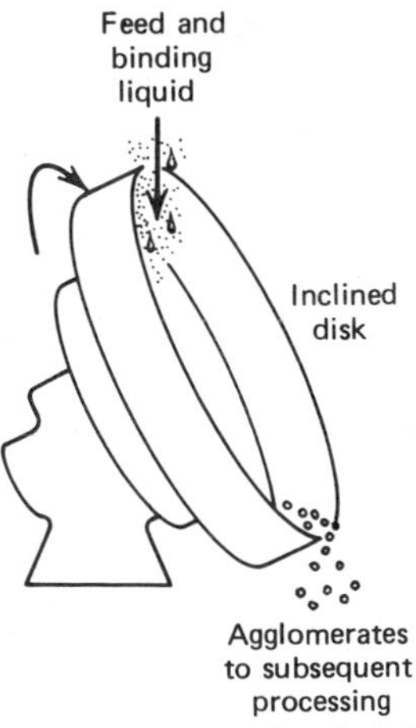

Figure 4. Schematic of an inclined-disk agglomerator.

Table 3. Characteristics of a Range of Inclined Disks[a]

Diameter, m	Depth, cm	Motor, kW	Rpm	Approximate capacity[b], t/h
0.41	8.9	0.19	12–36 variable	na
0.91	20.3	0.75	9–27	0.3
1.40	22.9	2.2	6.7–20.2	0.9
1.83	27.9	3.7	8.1–16.2	1.8
2.44	33.0	11.2	7.5–15	4.1
3.05	39.4	18.6	12.8 fixed	6.4
3.66	44.5	29.8	11.9	11
4.27	49.5	44.7	11.3	15
4.88	55.9	55.9	10.7	24
5.49	61.0	74.6	10.4	32
6.10	66.0	93.3	10.0	40
7.01	76.2	111.9	8.0	53
7.62	76.2	149.2	6.0	63

[a] Courtesy of Feeco International, Inc., 1980.

[b] Approximate capacity based on dry dust at 961 kg/m^3. Capacity depends on the type of material and desired product. Above rates are average for nominal 1.3-cm pellets.

Operation and control of tumbling agglomerators is affected primarily by the character of the feed powder (size distribution, solubility), by its optimum liquid content for agglomeration, and by retention time in the device (3,40–41). An important parameter for drum agglomeration is the amount of undersize product that must be recycled. A disk agglomerator requires experimental adjustment in angle of inclination, speed of rotation, and position of spray(s) and scraper(s). Temperature of operation can be a significant variable for soluble ingredients.

A most important feature of the inclined-disk agglomerator is its size separating ability. Feed particles and the smaller agglomerates sift down to the bottom of the tumbling load where, because of their high coefficient of friction, they are carried to the highest part of the disk before rolling downward in an even stream. Larger agglomerates remain closer to the top of the bed, where they travel shorter paths. In continuous operation the largest agglomerates are discharged from the top of the bed over the rim, while smaller ones and feed fines are retained for further growth. Because rotary drums do not possess the inherent classifying action of the inclined disk, agglomerates of a wide size distribution are discharged. Drums are therefore operated in closed circuit with screens to recycle the undersized (and crushed oversized, if present) material.

The inherent classifying action of inclined disks offers an advantage in applications that require accurate agglomerate sizing. Other advantages claimed for the inclined disk include less space requirements and lower cost than drums, as well as sensitivity to operating controls and easy observation of the agglomeration process. These latter features lend versatility in agglomerating a wide variety of materials of different degrees of ease of agglomeration.

Advantages claimed for the drum compared with the disk agglomerator are greater capacity, longer residence time for difficult materials and less sensitivity to upsets due to the damping effect of a larger recirculating load. Dusty materials and simultaneous processing steps (chemical reaction or drying during agglomeration) can be handled more easily in a drum.

Many variations on the design of the basic inclined-disk and drum agglomerators are in use. A well-known addition to the inclined disk is a separate reroll ring beyond the rim of the main disk. Normally disks are relatively shallow in depth with a rim height ca 20% of the disk diameter, but deeper configurations are also in use, including those with multi-stepped sidewalls, deep pan or deep drum designs and a cone pelletizer. Drum agglomerators may incorporate internal baffles (42), lifting blades, or independent paddle shafts in certain applications (43).

Mixer Agglomerators. Mixers are used by various industries in which, by contrast with drum or disk equipment, internal agitators of several designs provide a positive rubbing and shearing action to accomplish both mixing and size enlargement. Horizontal-pan mixers, pugmills and other types of intensive agitation devices are used. The positive cutting-out action of such equipment can handle plastic and sticky powder feeds and its kneading action is claimed to produce denser and stronger granules than the tumbling methods although agglomerates of more irregular shape usually result (see Mixing and blending).

Pug mixers (blungers, pugmills, paddle mixers) such as that shown in Figure 5 have been widely used in the granulation of fertilizer materials and for the mixing, moistening, and microagglomeration of sinter-strand feed in both the ferrous and nonferrous metallurgical industries. These mixers consist of a horizontal trough with a rotating shaft to which mixing blades or paddles of various designs may be attached. The vessel may be of a single-trough design although a double-trough arrangement is most popular. Twin shafts rotate in opposite directions throwing the materials forward and to the center as the pitched blades on the shaft pass through the charge. Construction is robust with the body of heavy plate [6.4 or 9.5 mm thick] and hardened agitators or tip inserts. Operational features include fume hoods, spray systems, and stainless steel construction. Provision can be made to feed materials at different points along the mixer as well as at the end to ensure that the entire mixing length is used

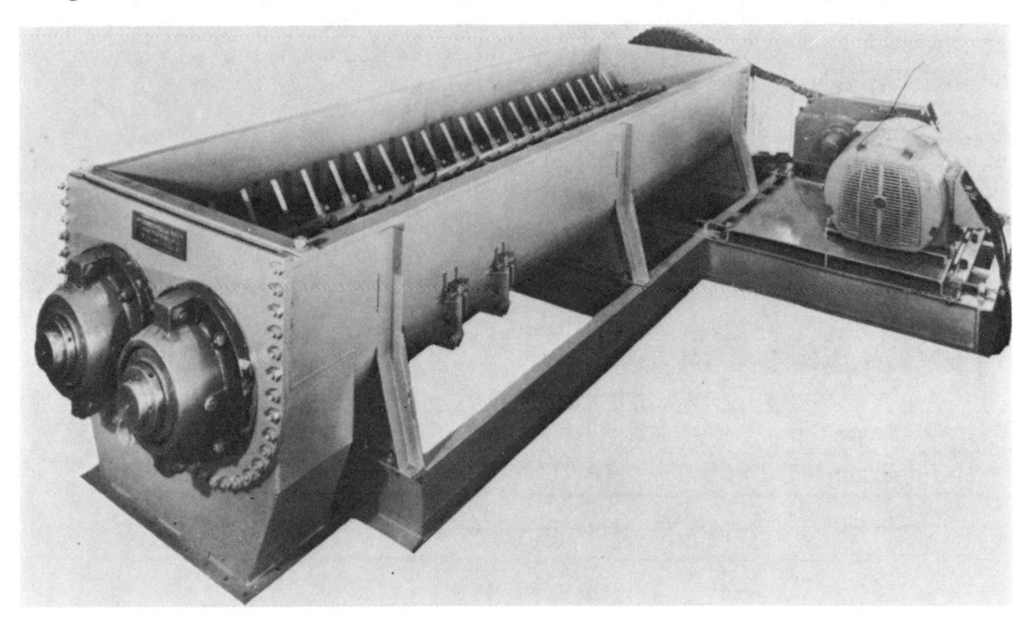

Figure 5. Double-shaft mixer used in fertilizer granulation. Courtesy of Edw. Renneburg and Sons Co.

and to add processing versatility. Capacities of these mixers cover a broad range from typical levels of 20–30 up to several hundred tons per hour.

An intensive countercurrent pan mixer can be used to homogenize feed powders while adding binding liquid to help pregranulate extreme fines before pelletizing. This mixer consists of a rotating mixing pan and two eccentrically mounted mixing stars which operate in the direction opposite to the pan rotation. This equipment operates in a batch mode to handle typically about 30 t/h of fertilizer materials.

Shaft mixers operating at very high rotational speeds are also used to granulate extreme fines, such as clays and carbon black, which may be highly aerated when dry, and plastic or sticky when wet. These machines are generally single-shaft devices in which the paddles are replaced by a series of pins, pegs or blades. The peg granulator (44) used to agglomerate ceramic clays in the china clay industry and the pinmixer (45) used in the wet pelleting of carbon black are representative examples. As shown in Figure 6, these mixers consist of a cylindrical shell within which rotates a shaft carrying a multitude of cylindrical rods (pegs or pins) arranged in a helix. The shaft rotates at speeds critical to machine performance. Wet feed or dry feed that is immediately moistened, enters the machine at one end and emerges as pellets at the opposite end.

Powder Clustering. Many applications of size enlargement require only relatively weak, small, cluster-type agglomerates to improve behavior of the original powder in flow, wetting, dispersion or dissolution. Tableting feeds in pharmaceutical manufacture, detergent powders, and "instant" food products are examples. In these cases, agglomeration is accomplished by superficially wetting the feed powder, often with less than 5% of bridging liquid in the form of a spray, steam, mist, etc. The wetting is carried out in a relatively dry state in standard or specialized powder mixers in which the mass becomes moist rather than wet or pasty. Equipment used (46) includes sigma-blade and heavy-duty planetary mixers (29), horizontal cylindrical vessels containing mixing and chopping tools (47), and rotary drums of special design (48) to form a constant-density falling curtain along the drum axis into which dispersed binding liquid is directed to form small agglomerates.

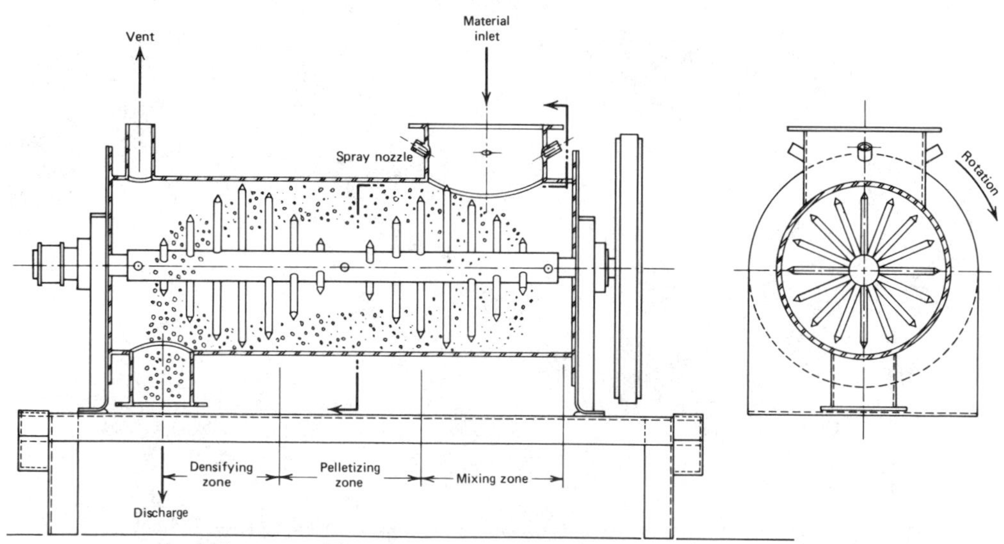

Figure 6. Pinmixer used in wet-pelleting carbon black (45).

Continuous-flow mixing systems are commonly used in the agglomeration of powdered food products. The Blaw-Knox Instantizer-Agglomerator (Fig. 7) is representative of these systems. In Figure 7 the feed powder is introduced to the moistening zone by a pneumatic conveyor and a rotary valve. The dry powder falls in a narrow stream between two jet tubes which inject the agglomerating fluid in a highly dispersed state. Steam, water, solvents, or a combination of these are used. Air at ambient temperature is also introduced through radial wall slots in the moistening chamber to induce a vortex motion. Control of this air flow controls the flow pattern and particle temperature. The reduced temperature serves to condense fluid onto the particles while the vortex motion induces particle collisions. Clustered material then drops through an air-heated chamber onto a conditioning conveyor where sufficient time is allowed to reach a uniform moisture distribution. The material then passes to an afterdryer, cooler, and sifter and is finally bagged (see Food processing).

Pressure Compaction and Extrusion Methods

The compression techniques of size enlargement produce agglomeration by application of suitable forces to particulates held in a confined space. The various methods in use differ in both the means of pressure application and the method used to confine the powder. Unidirectional compaction in punch and die assemblies and in molding presses makes use of a reciprocating punch or ram acting on the particulates in a closed die or mold. In roll-pressing equipment, the particulate material is compacted by squeezing as it is carried into the gap between two rotating rolls. In extrusion systems the particulates undergo shearing and mixing as they are consolidated while being forced through a die or orifice under the action of a screw or roller.

Tableting, pressing, molding, and extrusion operations are commonly used to produce agglomerates of well-defined shape, dimensions and uniformity in which the

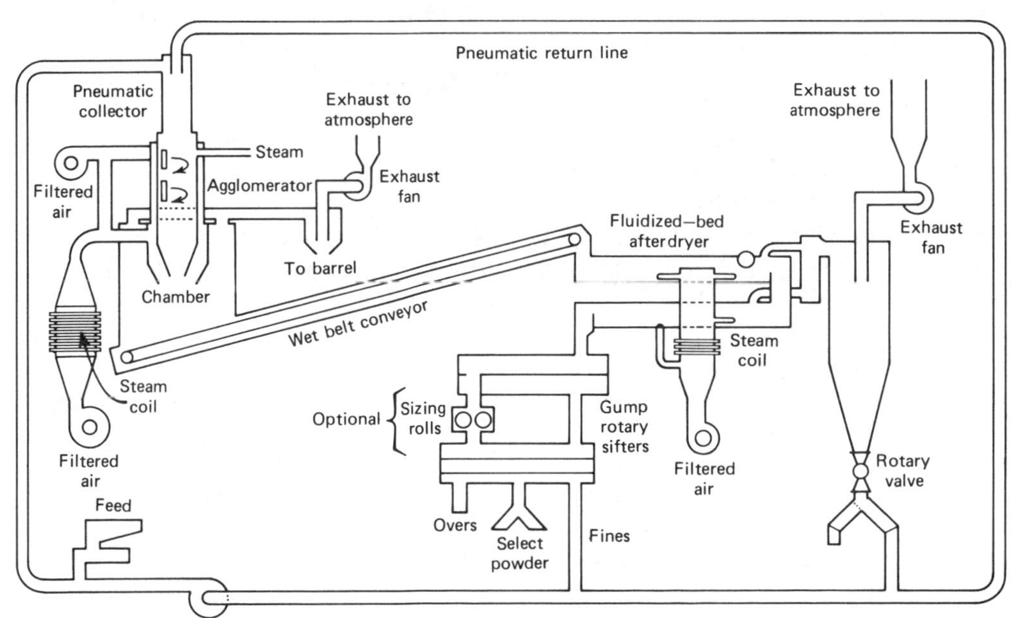

Figure 7. Blaw-Knox Instantizer-Agglomerator. Courtesy of Blaw-Knox Food and Chemical Equipment Co.

properties of each item are important and output is measured in pieces per hour (see Pharmaceuticals; Ceramics (forming processes); Plastics processing; Powder metallurgy).

The compaction process of void reduction may be considered to occur by two essentially independent mechanisms (49). The first is the filling of the holes of the same order of size as the original particles. This occurs as particles slide past one another; it is distinguished by the voids being filled with original particles that have undergone only slight size modification by fracture or by plastic deformation. The second process consists of the filling of voids that are substantially smaller than the original particles by plastic flow or by fragmentation. Many quantitative relationships have been suggested to represent this compaction process, usually in cylindrical die cavities (50–54). A successful theoretical design analysis for roll presses has been developed using small-scale laboratory measurements of the flow and compression properties of the feed powder (55–58).

The success of the compaction operation depends partly on the effective utilization and transmission of applied forces and partly on the physical properties and condition of the mixture being compressed. Friction at the die surface opposes the transmission of the applied pressure in this region, results in unequal distribution of forces within the compact and hence leads to density and strength maldistribution within the agglomerate (59). Lubricants, both external ones applied to the mold surfaces and internal ones mixed with the powder, are often used to reduce undesirable friction effects (60). For strong compacts external lubricants are preferable as they do not interfere with the optimum cohesion of clean particulate surfaces. Binder materials may be used to improve strength and also to act as lubricants.

Compacting Presses. In the automotive industry and other metal-working industries, coarse scrap-metal particulates are compressed and recycled to melting operations through piston-type briquetting presses (61–62). Feed materials are typically cast iron and steel borings or turnings, which tend to bond under pressure at least partially by mechanical interlocking. A reciprocating hydraulic press working on such materials might use a 56 kW hydraulics actuating pump with a rating of 318 t on a 12.7 cm diameter die. Briquettes of cylindrical shape 7.6 cm in length are produced at ca 3–4 t/h. Such compacting presses are not suited to larger tonnages when a small briquette is required. Their reciprocating nature is a disadvantage since this produces nonuniform loads on the drive motors.

Roll Briquetting and Compacting Machines. In roll presses, particulate material is compacted by squeezing as it is carried into the gap between two rolls rotating at equal speed. This is probably the most versatile method of size enlargement since most materials can now be agglomerated by this technique with the aid of binders, heat, and/or very high pressures if needed (63). Simplified briquetting and compaction flow diagrams are shown in Figure 8. In briquetting machines, pillow shapes are formed by matching indentations in the rolls. Precise design of these pockets based on practical experience is important to ensure optimum briquette density, minimum incidental "feather" (fines) production, and dependable pocket release of finished briquettes. In compaction machines the agglomerated product is in sheet form as produced by smooth or corrugated rolls. The compacted product can remain in sheet form or can be granulated into the desired particle size on conventional size reduction equipment.

Roll presses consist of the frame, the two rolls that do the pressing and the asso-

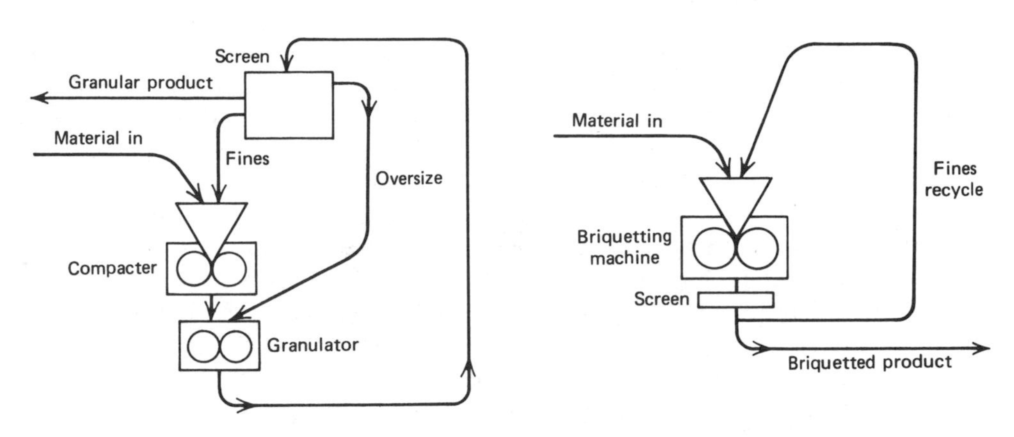

Figure 8. Simplified compacting and briquetting systems. Courtesy of Bepex Corporation.

ciated bearings, reduction gear, and fixed or variable-speed drive (see Fig. 9). Spacers between bearing housings prevent roll contact and allow adjustment of roll spacing.

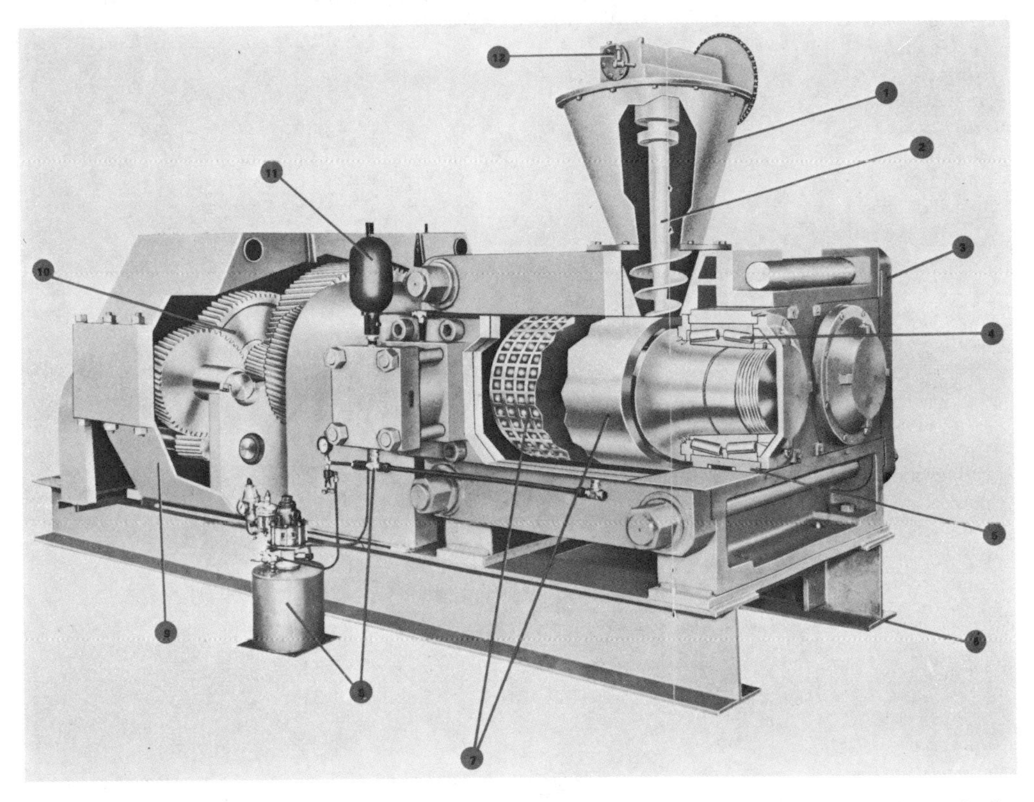

Figure 9. Cutaway view of a briquetting–compacting machine. 1. Predensifying feeder, 2. feeder screw, 3. machine housing, 4. antifriction bearing, 5. machined bearing block, 6. base frame, 7. pocketed or corrugated rolls, 8. hydraulic system, 9. speed reducer, 10. gears, 11. hydraulic accumulator, 12. feeder drive. Courtesy of Bepex Corporation.

The frame of the press is designed so that all forces are absorbed internally. The rolls are forced together by a hydraulic system which may incorporate a safety valve to prevent overpressure if foreign material intrudes between the roll faces. The rolls consist of a continuous roll shaft, the roll body, and attached molding equipment. The molding surface may be either solid or divided into segments. Segmented rolls are preferred for hot briquetting, as the thermal expansion of the equipment can be controlled more easily. Segmented rolls can be made from harder materials more resistant to wear than can one-piece rolls.

For fine powders that tend to bridge or stick and are of low bulk density, some form of forced feed, such as the tapered screw feeder shown in Figure 9 must be used to de-aerate, pre-compact, and pressurize the feed into the nip.

Capacity data for a variety of materials using a range of roll-press sizes is given in Table 4.

Table 4. Some Typical Capacities (t/h) for a Range of Roll Presses[a]

Roll diameter, cm	25.4	40.6	30.5	26.2	33.0	52.1	71.1	91.4
Maximum roll face width, cm	8.3	15.2	10.2	15.2	20.3	34.3	68.6	25.4
Roll separating force, t	23	45	36	45	68	136	272	327
Carbon								
coal, coke		1.8	0.91		2.7	5.4	23	
charcoal			7.3			12		
activated					2.7	15		
Metals and ores								
alumina					4.5	9.1	25	
aluminum				1.8	3.6	7.3	18	
brass, copper	0.5			1.4	2.7	5.4	15	
steel mill waste					4.5	9.1		
iron				2.7	5.4	14	33	
nickel powder					2.3	4.5		
nickel ore						18	33	
stainless steel				1.8	4.5	9.1		
steel								23
bauxite		1.4				9.1	18	
ferro-metals						9.1		
Chemicals								
copper sulfate	0.5	1.4		0.91	2.7	5.4	14	
potassium hydroxide				0.9	3.6	7.3		
soda ash	0.5				2.7	5.4	14	
urea	0.23					9.1		
dimethyl terephthalate	0.23				1.8	5.4		
Minerals								
potash						18	73	
salt				1.8	4.5	8.2		
lime					3.6	7.3	14	
calcium sulfate							12	33
fluorspar						4.5	9.1	25
magnesium oxide						1.4	4.5	
asbestos						1.4	2.7	
cement						4.5		
glass batch						4.5	11	

[a] Courtesy of Bepex Corporation, 1980.

Pellet Mills. Pellet mills differ from roll briquetting and compacting machines in that the particulates are compressed and formed into agglomerates by extrusion through a die rather than by squeezing as they are carried into the nip between two rolls. Several types of equipment that use the extrusion principle are available. The die may be a horizontal perforated plate with rollers acting on its upper surface to press material through the plate. Rolls may be either side-by-side with material extruded through one or both of the rolls; or one or more small rolls may be fitted inside a larger die roll. In yet another design, two inter-meshed gears are used and material is extruded through die holes located in the gear root.

The action of the roller and die assembly to produce a shearing and mixing action yields a plastic mix to be pushed through the die. Binders, plasticizing agents, and lubricants may be used to facilitate the process. Probably the most popular design of pellet mills in use is that shown in Figure 10 which utilizes a ring-type die and two or three rollers mounted in a vertical plane. Power is applied to the die to rotate it around the roller assembly which has a fixed axis. Capacities range from less than one ton per hour up to 25 or more tons per hour operating from 75–500 rpm with drive power of 7.5–185 kW. Dies with hole sizes from 2 or 3 mm–30 mm are used. Capacities vary greatly depending on speed, hole and die size, moisture content, and other characteristics of the material being pelleted. Scores of materials can be pelleted, from catalysts and carbon materials to rubber crumb, wood pulp and bark, and many chemicals.

Heat Reaction, Fusion and Drying Methods

These methods of size enlargement depend on heat transfer to accomplish particle bonding. Heat may be transferred to the particle agglomerates, as in the drying of a concentrated slurry or paste, the fusion of a mass of fines, or chemical reaction between particles at elevated temperatures. Alternatively, heat may be removed from the material to cause agglomeration by chilling as in the solidification–crystallization of a melt or concentrated suspension. Heat transfer may be direct from a heat-transfer fluid or indirect across a heat-transfer surface. As a consequence, heat transfer and

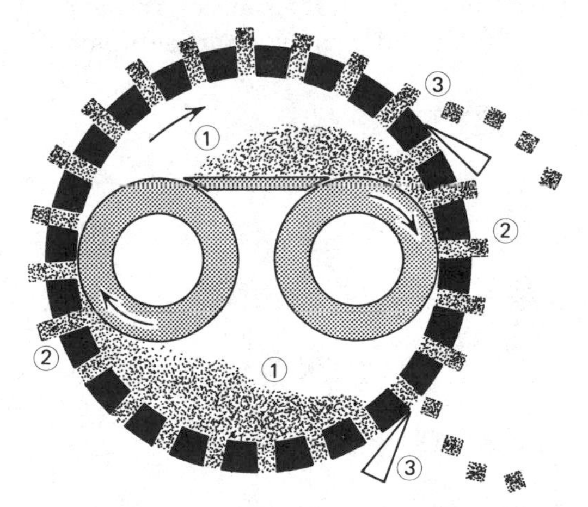

Figure 10. Operating principle of a common design of pellet mill. Courtesy of California Pellet Mill Co. 1. Loose material is fed into pelleting chamber. 2. Rotation of die and roller pressure forces material through die, compressing it into pellets. 3. Adjustable knives cut pellets to desired length.

drying equipment used is quite varied and includes packed-bed systems of particulates and aggregates, dispersed particle–fluid systems, and heating and cooling on moving surfaces. An equally wide variety of preagglomeration equipment is used to preform powders and pastes into agglomerates suitable for drying, firing, or chilling. Included are balling devices and pellet mills as described above and extruders using rollers, bars, or wiper blades to force plastic pastes through perforated plates or grids to preform the paste into rods and other shapes (64).

Sintering and Pelletizing. In extractive metallurgy (qv), sintering and pelletizing processes have been developed to allow processing of fine ores, concentrates, and recyclable dusts (3). In this connection, sintering refers to a process in which fuel (5–10% coke fines) is mixed with an ore and burned on a grate. A cake of hardened porous material is the resulting agglomerate. In the pelletizing process as applied on a large scale to iron ores, discrete "green" balls produced in balling drums, disks or cones are dried and hardened by passing combustion gases through a bed of the agglomerates, without fusing the agglomerates as in sintering.

Four separate sequential processes take place during these high temperature operations: drying, preheating, firing or high temperature reaction, and cooling. Ceramic bond formation and grain growth by diffusion are the two important mechanisms for bonding at the high temperatures (1093–1371°C) employed. Simultaneous useful processes may also occur, such as the elimination of sulfur and the decomposition of carbonates and sulfates. The highest tonnage applications are in the beneficiation of iron ores but nonferrous ores may also be treated (65).

A traveling-grate sintering machine is depicted in Figure 11. The machine consists of a strong frame of structural steel supporting two pallet driving gears and steel tracks or guides. Traveling on the track is an endless line of pallets with perforated bottoms or grates. The train of pallets passes first under a feeding mechanism which lays down a uniform layer of charge, then under an igniter, and finally over a series of wind boxes exhausted by fans. Each wind box is approximately equal in length and width. The speed of the train of pallets and the volume of air drawn through the charge are controlled so that the combustion layer reaches the grate just as the pallet passes off the machine. The sintering machine is a relatively small part of the equipment needed for a complex sintering plant. Auxiliary devices include conveying and storage equipment, mixing, nodulizing and proportioning equipment, fans, dust collectors, etc.

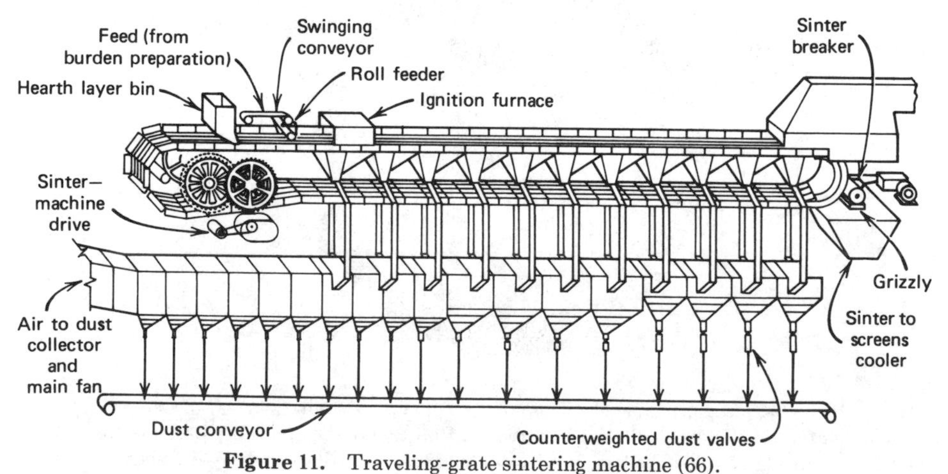

Figure 11. Traveling-grate sintering machine (66).

The capacity of a sintering strand is related directly to the rate at which the burning zone moves downward through the bed (67). This rate is controlled by the air rate through the bed, with the air acting in its function as the heat-transfer medium. The permeability of the sinter bed is thus very important in determining these rates and must be kept high for high throughput. Bed permeability is improved by pre-agglomeration of the feed material, proper feeding of material to the sinter strand, and the use of 20–30% of recycled product material to support the needed coarse structure of the bed. A modern sinter strand might be 4 m wide by 61 m long with a capacity of 7200 t/d of ferrous sintered product.

In the pelletizing process for iron ore, wet balls 9.5–13 mm in diameter are dried and heat hardened on traveling-grate machines adapted from the downdraft sintering grate. Installations now differ in air-flow arrangements used to accomplish the process steps of drying, firing, and cooling (3). Pelletizing grates differ from sintering grates in a number of other ways. Pelletizing grates possess a multiplicity of windboxes divided into large groupings to allow recovery of sensible heat (eg, air from the cooling section may be used for drying or combustion purposes). This improves fuel efficiency and adds flexibility to the processing steps. In pelletizing, pellets are held for a long period, relative to that used in sintering, at closely controlled temperatures to effect hardening. In addition, pellet cooling is usually done on the same machine used for heat hardening whereas sinter is cooled in separate equipment.

In addition to the common straight-grate machine, a circular-grate machine is also in use (68). The drying and firing of pellets may also be done in shaft furnaces or in a combined system using a traveling grate followed by final firing in a kiln (69).

Drying and Solidification on Surfaces. In this type of equipment, granular products are formed directly from fluid pastes and melts, without intermediate preforms, by drying or solidification on solid surfaces. Surfaces formed by single or double drums are common. Drum dryers consist of one or more heated metal rolls on which solutions, slurries, or pastes are dried in a thin film. Drying takes place in less than one revolution of the slowly revolving rolls, and a doctor blade scrapes the product off in flake, chip, or granular form. A wide variety of products, such as cereals, fruits, starches, vegetables, meat and fish products, inorganic and organic chemicals, and pharmaceuticals are beneficiated in this way. In drum flakers, a thin film of molten feed is applied to the polished external surface of a revolving, internally cooled drum. Virtually any molten material that will solidify rapidly with cooling can be treated in this method. Although ambient water is normally the cooling liquid, chilled water or other coolants may be used if lower temperatures are needed. The cooled solid is scraped from the drum as a flaked or granular product. Many organic and inorganic food and chemical products are flaked in this way, including caustics, resins, detergents, sugars, waxes, pharmaceuticals, and explosives.

Molten materials can also be cooled to solid products on endless-belt systems as shown in Figure 12. Some typical materials treated, product and feed characteristics, and capacities of belt cooling systems are given in Table 5.

As illustrated in Figure 12, a number of different feeding and discharge arrangements as well as surface sizes and speeds lend versatility to these methods based on drying and solidification on surfaces. For mobile slurries and fusible and/or soluble particles, these methods offer alternatives to more traditional size enlargement techniques.

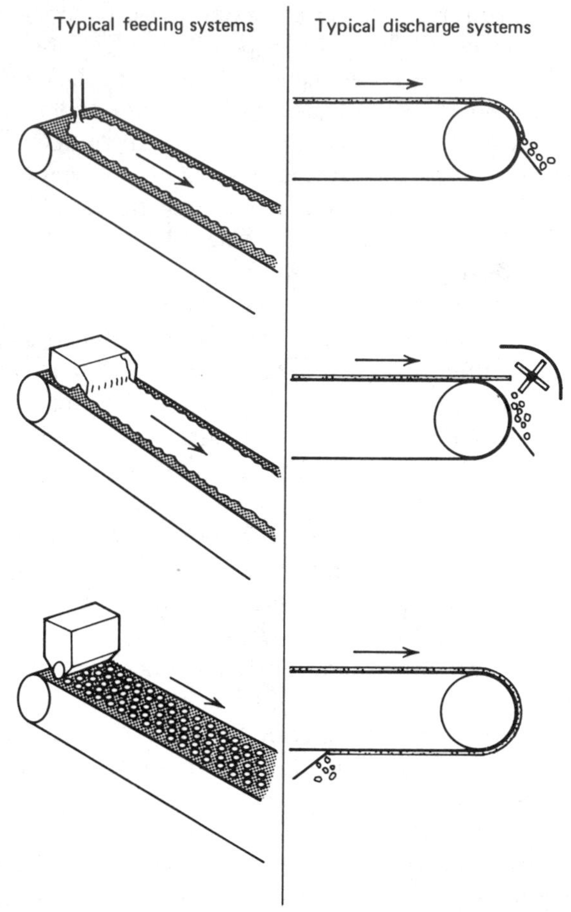

Figure 12. Typical feeding and discharge systems for endless-belt cooling of molten materials. Courtesy of Sandvik Conveyor, Inc.

Suspended-Particle Techniques. In these methods of size enlargement, granular solids are produced directly from a liquid or semiliquid phase by dispersion in a gas to allow solidification through heat and/or mass transfer. The feed liquid, which may be a solution, gel, paste, emulsion, slurry, or melt, must be pumpable and dispersible. Equipment used includes spray dryers, prilling towers, spouted and fluidized beds, and pneumatic conveying dryers, all of which are amenable to continuous, automated, large-scale operation. Because attrition and fines carryover are common problems with this technique, provisions must be made for recovery and recycling.

In the suspension methods, agglomerate formation occurs by hardening of feed droplets into solid particles by layering of solids deposited from the feed onto existing nuclei and by adhesion of small particles into aggregates as binding solids from the dispersed feed are deposited. The product size achievable in these methods is usually limited to ca 5 mm and is often much smaller (see Drying).

In spray drying, the largest particles produced are normally ca 1 mm diameter. Through design and operation of the dryer, however, larger agglomerated particles can be produced, and this technique is used to produce granular dried products in the pharmaceuticals and ceramics industries as well as coarse food powders with "instant"

Table 5. Product Characteristics and Capacity Data for Some Materials Treated in Belt Cooling Systems[a]

Product	Thickness, mm	Feed temp, °C	Discharge temp, °C	Capacity, kg/(h·m²)
resins				
phenolic	1.6	135	43	225
phenolic	1.2–1.3	138	33	277
sulfur	6.4	143	66	269
asphalt	3.2	218	52	90
urea	2.4	191	60	190
ammonium nitrate	1.6	204	71	439
chlorinated wax	1.6	149	38	303
sodium acetate	3.2	82	38	183
hot-melt adhesive	11.1	166	39	70
wax blend	0.6	132	29	129
epoxy resin	1.0	177	38	195

[a] Courtesy of Sandvik Conveyor, Inc. (1980).

properties. Although the variables of spray dryer design and operation all interact to influence product characteristics, a number of these have important effects on product size and size distribution (70). In general, the product size is increased by decreasing the intensity of atomization and of spray-air contact and by lowering the exit temperatures from the dryer. Higher liquid-feed viscosity and feed rate as well as the presence of natural or added binders also favor larger product size.

The flow sheet in Figure 13 shows one example of a system designed to yield agglomerated products. Coarse spray-dried instant food powders are produced directly from liquids in this system. Two stages of agglomeration take place. The initial stage

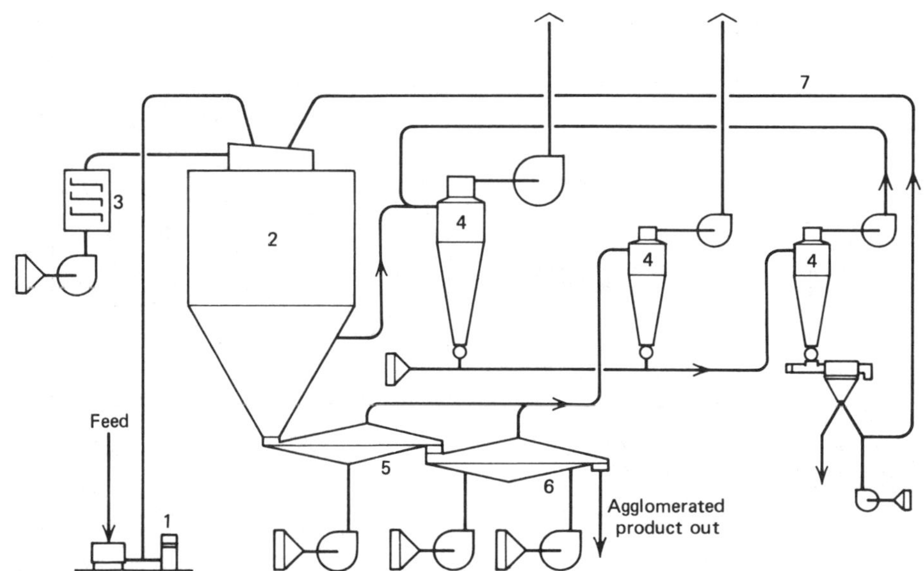

Figure 13. Spray-dryer system designed for production of agglomerated food powders with "instant" properties (71). 1. Liquid-feed system. 2. Spray-dryer chamber. 3. Drying air heater. 4. Cyclones for fines recovery. 5. Vibrofluidizer as after-dryer. 6. Vibrofluidizer as after-cooler. 7. Fines return to drying chamber.

occurs in the atomization zone of the spray dryer. There relatively cool air is used to retard the evaporation rate and enhance the agglomeration of fines. Further agglomeration is achieved by operating the spray dryer so that the powder is still moist when it leaves the dryer chamber. The agglomerated powder passes out of the bottom of the drying chamber to a vibrating fluid bed where drying is completed, and finally into a second fluid bed for cooling.

Spray cooling or solidification, more commonly known as prilling, is similar to spray drying in that liquid feed is dispersed into droplets at the top of a chamber. These congeal into a solid granular product as they travel down the chamber. The method differs from spray drying in that the liquid droplets are produced from a melt which solidifies primarily by cooling in the chamber with little, if any, drying. Product size is also generally larger (up to 3-mm dia) than in spray-dried materials. As a result of this relatively large prill size, the process is generally carried out in narrow but very tall towers to ensure that the prills are sufficiently solid when they reach the bottom. Because of the melt feed requirement, prilling is normally limited to materials of low melting point that do not decompose on fusion. Urea and ammonium nitrate fertilizers are traditionally treated by prilling (72–73) (see Fertilizers).

Fluid-bed (74–76) and spouted-bed granulation (77–78) both accomplish size enlargement and drying simultaneously by spraying feed liquids (solution, slurry, paste, melt) onto suspended layers of essentially dry particles. The seed-bed particles grow either by coalescence of two or more particles held together by a deposited binder material, or by layering of solids onto the surface of individual particles. Since multiple layers can be deposited, these granulation systems can produce larger granules than spray dryers. The two methods differ in the way the growing bed particles are agitated (Fig. 14). In a fluid bed, a suitable distributor such as a perforated plate passes hot gas uniformly into the base of the particle bed to suspend the particles. In the

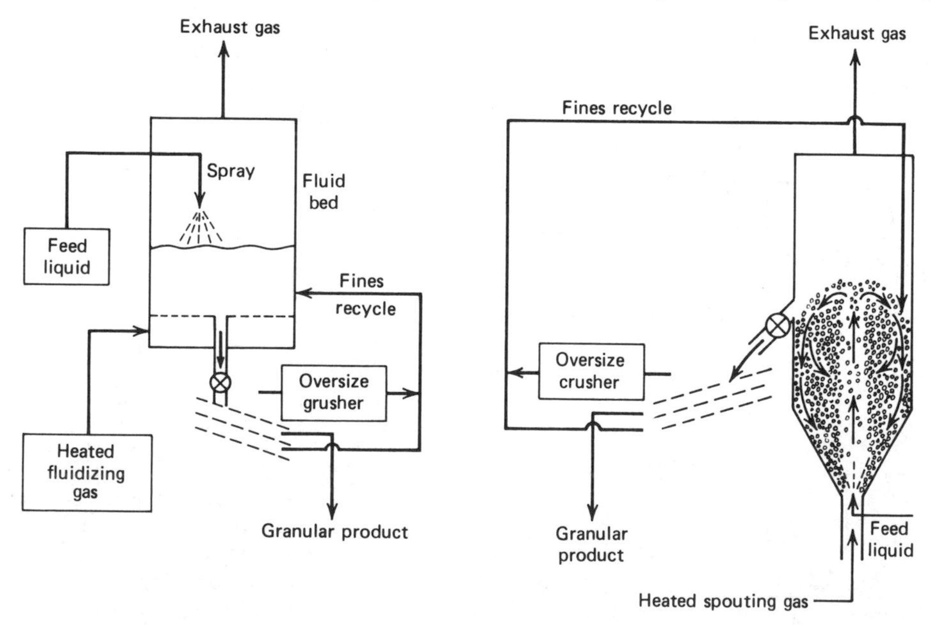

Figure 14. Schematics of fluid- and spouted-bed granulation systems.

spouted-bed hot gas is injected as a single jet into the conical base of the granulation unit causing the bed material to circulate in a fountain-like fashion. Spouted beds were originally developed as an alternative to fluidized beds as a means of contacting gas with solids. The operation of fluidized beds becomes less effective when particles greater than ca 1 mm diameter are to be treated. The gas–solids contacting efficiency is impaired at larger particle sizes as more and more gas bypasses in the form of large bubbles. In contrast, ca 1 mm is the minimum particle size for which spouting appears to be practical. Thus, spouted-bed granulators allow larger granules to be formed than do fluid-bed granulators (see Fluidization).

Agglomeration from Liquid Suspensions

Size enlargement of particles contained in liquids is a frequent aid to other operations such as filtration, dewatering, settling, etc. Flocculation procedures are the traditional means used to promote such size enlargement (see Flocculating agents); the product is usually in the form of loose aggregates of an open network structure. The present discussion is concerned with less conventional techniques in which stronger bonding and specialized equipment are used to form larger and more permanent agglomerates in liquid suspensions. Table 6 gives examples of such agglomeration processes, in which the objectives include not only the removal or capture of particles in suspension but also the production of granular (including spherical) material, displacement of as much suspending liquid as possible from the product, and the selective agglomeration of one or more components of a multiparticle mixture.

Agglomeration by Competitive Wetting. Fine particles in liquid suspension can readily be formed into large dense agglomerates of considerable integrity by adding a second or bridging liquid under suitable agitation conditions (88). This second liquid should be effectively immiscible with the suspending liquid and must wet preferentially the solid particles to be agglomerated. A simple example is the addition of oil to an aqueous suspension of fine coal. The oil readily adsorbs preferentially on the carbon

Table 6. Some Important Agglomeration Processes Carried Out in Liquid Systems

Process objective	Material treated and process used	References
sphere formation and production of coarse granular products	nuclear fuel and metal-powder production by sol-gel processes	79–80
	manufacturing of small spheres from refractory and high melting point solids (eg, tungsten carbide) by immiscible liquid wetting	81–82
removal and recovery of fine solids from liquids	removal of soot from refinery waters by wetting with oil	83
	recovery of fine coal from preparation-plant streams to allow recycling of water	84
displacement of suspending liquid	dewatering of various sludges by flocculation followed by mechanical drainage on filter belts, in revolving drum, etc	85–86
	displacement of moisture from fine coal by wetting with oil	84
selective separation of some components in a mixture of particles	removal of ash-forming impurities from coal and from tar sands by selective agglomeration	84, 87

particles and forms liquid bridges between these particles by coalescence during the collisions produced by agitation. Inorganic impurity (ash) particles are not wetted by the oil and remain in unagglomerated form in the aqueous slurry (see Coal conversion processes, desulfurization).

The agglomeration phenomena that occur as progressively larger amounts of bridging liquid are added to a solids suspension are depicted in Figure 15. The general relationships shown are not specific to a given system and relate equally well to siliceous particles suspended in oil and collected with water, or to coal particles suspended in water and agglomerated with oil. Of course, in the real world of separation of valuable particles from associated gangue particles, the colloid and surface chemistry involved is usually quite complex. As in the flotation process, selective agglomeration by immiscible liquids depends on the relative wettability of surfaces, and the same fundamentals of surface chemistry apply to the conditioning of particles to yield the required affinity for the wetting liquid (see Flotation). Table 7 summarizes a number of examples of ores, fossil fuels, and other particle mixtures in which particle conditioning and selective agglomeration are used to effect separation. A most useful feature of the agglomeration technique is its ability to work with extreme fines. Even particles of less than nanometer size (ca 10^{-10} m) can be treated, if appropriate, so that ultrafine grinding can be applied to materials with extreme impurity dissemination to allow recovery of agglomerates of higher purity (106).

The sol-gel process is a related technique which has been actively developed for the preparation of spherical oxide fuel particles, up to ca 1 mm diameter, for nuclear reactors (79–80). In agglomeration by immiscible liquid wetting, small amounts of a bridging phase adsorbed on the particles coalesce to draw the particles into larger entities. In the sol–gel process, fine particles are initially suspended in an excess of a bridging phase, the suspension is formed into spherical droplets and the excess bridging phase is removed to solidify the droplets into a particulate product. For example, an aqueous sol of colloidal particles such as thoria can be dispersed into droplets in a stream of immiscible water-extracting fluid, such as 2-ethyl-1-hexanol. As water is removed from the sol droplets, the gel formed solidifies and densifies the spherical

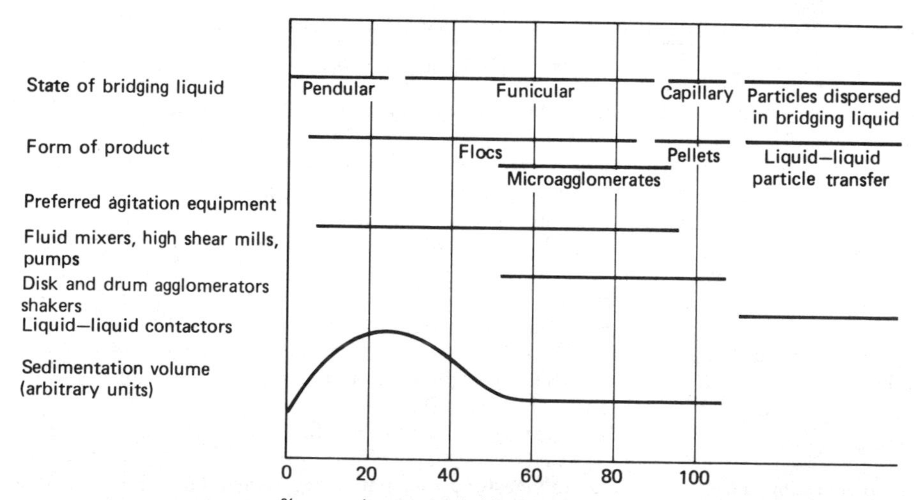

Figure 15. Phenomena that take place as increasing amounts of an immiscible wetting liquid are added to a suspension of fine particles (88).

Table 7. Examples of Surface Conditioning and Selective Agglomeration by Immiscible Liquid Wetting[a]

Material treated and objective	Suspending liquid	Conditioning agents and/or other additives	Collecting liquid	References
barite in mill tailings; upgrading	water	sodium dodecyl sulfate, sodium silicate, acid wash of agglomerates	still bottoms	89
brine (NaCl) solution, removal of colloidal iron oxides	saturated NaCl solution	tall oil	crude oil, sulfonated petroleum oil	90
coal, bituminous, separation of pyrite and other ash constituents	water	iron-oxidizing bacteria, alkali	petroleum distillates	91
coal, bituminous, balling of ash constituents	petroleum distillates	balling nuclei added	water	92
coal, subbituminous and lignitic	water	coke-oven tar, pitch, petroleum crudes, and fractions	hydrocarbons	84
germanium in carbonaceous sandstone	water	coal tar, Na_2CO_3 and sodium silicate	petroleum still bottoms	93
Glauber's salt ($Na_2SO_4.10H_2O$), removal of siliceous matter	saturated Na_2SO_4 solution	various amines	hydrocarbon	94
gold ore for upgrading	water	*sec*-butyl xanthate	still bottoms	95
graphite-sulfur mixture	light petroleum distillate			
microagglomeration and removal of graphite		tannic acid in collector	5% aqueous tannic acid	96
microagglomeration and removal of sulfur		Aerofloat 15	dilute NaOH solution	96
graphite-zinc sulfide-calcium carbonate mixture	water			
microagglomeration and removal of graphite		none	nitrobenzene	96
microagglomeration and removal of ZnS		*sec*-butyl xanthate, $CuSO_4$, NaOH	nitrobenzene	96
microagglomeration and removal of $CaCO_3$		oleic acid	nitrobenzene	96
ilmenite concentrate; removal of complex silicates	water	oleic acid, sodium silicate, pH adjuster	petroleum distillates	97
iron ores; removal of phosphatic and siliceous matter	water, $CaCl_2$ solution	various fatty acids, bases and acids for pH adjustment	crude or semirefined viscous petroleum oils	98–99
marl ($CaCO_3$) deposits, removal of siliceous matter	water	fatty acid after heat treatment	kerosene	100
methyl methacrylate suspension; balling	light petroleum distillate		aqueous chloral hydrate	101
shale; removal of $CaCO_3$	water	sodium oleate	oxidized crude oil	102
tar sands, agglomeration of bitumen	water	alkali	hydrocarbons present in sands	103
tar sands, extraction of bitumen and agglomeration of sands	petroleum oils		water with alkali	87
tin ore for upgrading	water	tall oil, acids and bases for pH adjustment	viscous petroleum oil	104–105

[a] Ref. 88.

agglomerates of the colloidal particles. Drying, calcining, and sintering of the recovered microspheres complete the process.

Pellet Flocculation. Flocculation procedures lead to aggregates with better settling, drainage, and filtration characteristics than the untreated particles. On the other hand, flocculated particles form a loose, bulky layer with relatively large pores so that a larger proportion of suspending medium is often retained than would be the case with untreated particles. To overcome this problem, pellet flocculation (85,107) and related techniques (86,108) have been developed. These combine the use of relatively large amounts of polymeric flocculants (ca 1 kg per ton of solids) with gentle rolling-mixing action to consolidate settled flocs into more compact agglomeratelike sludges of reduced liquid content. The agglomerates thus formed contain more interparticle bridges than those with lower polymer levels; they are able to grow to a larger size and therefore are more easily separated from the liquid phase; and they are strong and pliable enough to allow entrapped liquid to be squeezed out under mechanical working.

In a type of equipment developed in Japan to accomplish pellet flocculation, a horizontal drum revolves slowly at ca 1 m/min peripheral speed to dewater sludge. The drum interior is made up of three sections for successively pelletizing, decanting, and consolidation of the solids. Polymeric flocculant is added to the suspension upstream of the drum, together with auxiliary agglomerating agents such as calcium hydroxide or sodium silicate. Voluminous flocs formed ahead of the drum are rolled into denser sediment in the pelletizing section. These are then pushed into the decanting section by a guide baffle; there water is removed through intermittent slits in the drum wall. In the final consolidating section, the agglomerates are gently tumbled and rolled into a denser form, and water again escapes through wall slits. Other types of equipment in use include rotary-drum flocculation, rolling in cells formed by woven filter cloths and drainage in gravity and pressure-belt units (86,108).

Nomenclature

C = parameter in equation 3
D = agglomerate diameter, m
d = diameter of monosized spherical particles, m
H = tensile strength of a single particle-particle bond, N
L = compression force at failure, N
s = fractional filling of agglomerate voids
W = weight fraction of liquid (wet basis)
γ = liquid surface tension, N/m
ϵ = void fraction of agglomerate
ρ_L = liquid density, g/cm^3
ρ_S = particle density, g/cm^3
σ_C = compressive strength, Pa
σ_T = mean tensile strength per unit section area, Pa

BIBLIOGRAPHY

1. E. Swartzman, *Trans. Can. Inst. Min. Metall.* **57,** 198 (1954).
2. J. O. Hardesty, *Superphosphate: Its History, Chemistry and Manufacture*, U.S. Department of Agriculture, Washington, D.C., 1964, Chapt. 11.
3. D. F. Ball, J. Dartnell, J. Davison, A. Grieve, and R. Wild, *Agglomeration of Iron Ores*, Heinemann, London, UK, 1973.
4. W. H. Engelleitner, *Miner. Process.* **15**(7), 4 (1974).
5. J. E. Browning, *Chem. Eng.* **74**(25), 147 (1967).
6. H. Rumpf in W. A. Knepper, ed., *Agglomeration*, Interscience Publishers, a division of John Wiley

& Sons, Inc., New York, 1962, pp. 379–414.
7. C. E. Capes, *Particle Size Enlargement*, Elsevier, Amsterdam, 1980.
8. H. H. Hausner in ref. 6, pp. 55–91.
9. D. S. Cahn, *Trans. Soc. Mining Eng. AIME* **250,** 173 (1971).
10. T. P. Hignett in V. Sauchelli, ed., *Chemistry and Technology of Fertilizers*, Reinhold, New York, 1960, Chapt. 11, pp. 269–298.
11. B. K. Mazumdar, J. M. Sanyal, B. N. Bose, and A. Lahiri, *Indian J. Technol.* **7,** 212 (July 1969).
12. P. L. Waters, *Technical Communication No. 51*, CSIRO, Division of Mineral Chemistry, N.S.W., Australia, May 1969.
13. W. Pietsch, *Chem. Eng. Prog.* **66,** 31 (1970).
14. A. C. Herd, *N.Z.J. Sci.* **17,** 161 (1974).
15. R. E. King in *Remington's Pharmaceutical Sciences*, 14th ed., Mack Publishing Co., Easton, Pa., 1970, pp. 1649–1680.
16. J. D. Jensen, *Food Technol.*, 60 (June 1975).
17. W. B. Pietsch, *J. Eng. Ind.* **91B**(2), 435 (May 1969).
18. A. R. Cooper, Jr. and L. E. Eaton, *J. Am. Ceram. Soc.* **45**(3), 97 (1962).
19. U.S. Pat. 3,844,759 (Oct. 29, 1974), M. M. Ruel and A. F. Sirianni (to Canadian Patents and Development Limited).
20. D. M. Newitt and J. M. Conway-Jones, *Trans. Inst. Chem. Eng.* **36,** 422 (1958).
21. R. A. Fisher, *J. Agric. Sci.* **16,** 492 (1926).
22. P. C. Carman, *Soil Sci.* **52,** 1 (1941).
23. D. S. Kahn and J. M. Karpinski, *Trans. Amer. Inst. Min. Eng.* **241,** 475 (1968).
24. P. C. Kapur and D. W. Fuerstenau, *J. Am. Chem. Soc.* **50,** 14 (1967).
25. K. Shinohara and C. E. Capes, *Powder Technol.* **24,** 179 (1979).
26. C. E. Capes, *Powder Technol.* **5,** 119 (1971–1972).
27. C. E. Capes and R. D. Coleman, *Metall. Trans.* **5,** 2604 (1974).
28. C. E. Capes, *Powder Technol.* **4,** 77 (1970–1971).
29. L. Lachman, H. A. Lieberman, and J. L. Kanig, eds., *The Theory and Practice of Industrial Pharmacy*, Lea and Febiger, Philadelphia, Pa., 1970.
30. B. Hassler and P. G. Kihlstedt, Cold Bonding Agglomeration, private communication, 1977.
31. R. Linder, *J. Iron Steel Inst.* **189,** 233 (1958).
32. J. B. Bookey and B. Raistrick in ref. 10, Chapt. 18.
33. *United States Pharmacopeia*, XX (USP XX–NF XV), The United States Pharmacopeial Convention, Inc., Rockville, Md., 1980, pp. 958–959.
34. C. E. Capes and P. V. Danckwerts, *Trans. Inst. Chem. Eng.* **43,** 116 (1965).
35. K. V. S. Sastry and D. W. Fuerstenau in K. V. S. Sastry, ed., *Agglomeration 77*, AIME, New York, 1977, pp. 381–402.
36. W. G. Engelleitner, *Ceram. Age* **82**(12), 24, 25, 44, 45 (1966).
37. C. E. Capes, R. L. Germain, and R. D. Coleman, *Ind. Eng. Chem., Process Des. Dev.* **16,** 517 (1977).
38. G. M. Hebbard, The A. J. Sackett and Sons Co., private communication, July 14, 1977.
39. R. A. Koski in ref. 35, pp. 46–73.
40. G. C. Carter and F. Wright, *Inst. Mining Met. Soc. Proc. Adv. in Extractive Metall.*, 89 (1967).
41. W. B. Pietsch, *Aufbereit. Tech.* **7**(4), 144 (1966).
42. H. T. Stirling in ref. 6, pp. 177–207.
43. G. R. Hornke and R. E. Powers, *paper presented at the AIME Blast Furnace Coke Oven, and Raw Materials Conference*, St. Louis, Mo., April 6, 1959.
44. R. E. Brociner, *Chem. Eng. (London)* **220,** CE 227 (1968).
45. J. A. Frye, W. C. Newton, and W. C. Engelleitner, *Proc. Inst. Briquet. Agglom. Bien. Conf.* **14,** 207 (1975).
46. J. J. Fischer, *Chem. Eng.* 88 (Feb. 5, 1962).
47. *Ceram. Age* **86**(3), 15, 16, 18 (1970).
48. C. A. Sumner, *Soap Chem. Spec.* **51,** 29 (July, 1975).
49. A. R. Cooper, Jr. and L. E. Eaton, *J. Am. Ceram. Soc.* **45**(3), 97 (1962).
50. C. L. Huffine, Ph.D. Thesis, Columbia University, 1953.
51. R. S. Spencer, G. D. Gilmore, and R. M. Wiley, *J. Appl. Phys.* **21,** 527 (1950).
52. R. W. Heckel, *Trans. AIME* **221,** 671 (1961).
53. M. J. Donachie, Jr. and M. F. Burr, *J. Met.* **15,** 849 (1963).

54. W. D. Jones, *Fundamental Principles of Powder Metallurgy*, E. Arnold, London, UK, 1960.
55. J. R. Johanson, *Proc. Inst. Briquet. Agglom. Bien. Conf.* **9,** 17 (1965).
56. J. R. Johanson, *Trans. Am. Inst. Mech. Eng. Ser. E. J. Appl. Mech.* **32,** 842 (Dec. 1965).
57. J. R. Johanson, *Proc. Inst. Briquet. Agglom. Bien. Conf.* **11,** 135 (1969).
58. *Ibid.*, **13,** 89 (1973).
59. D. Train, *Trans. Inst. Chem. Eng.* **35,** 258 (1957).
60. K. R. Komarek, *Chem. Eng.* **74**(25), 154 (1967).
61. W. W. Eichenberger in ref. 55, p. 52.
62. W. F. Bohm in ref. 45, p. 219.
63. W. Pietsch, *Roll Pressing*, Heyden, London, UK, 1976.
64. I. H. Gibson in A. S. Goldberg, ed., *Powtech '75*, Heyden, London, UK, 1976, pp. 43–49.
65. T. E. Ban, C. A. Czako, C. D. Thompson, and D. C. Voletta in ref. 6, pp. 511–540.
66. R. H. Perry and C. H. Chilton, eds., *Chemical Engineers' Handbook*, 5th ed., McGraw-Hill, New York, 1973, Sect. 8.
67. R. L. Bennett and R. D. Lopez, *Chem. Eng. Prog. Symp. Ser.* **59**(43), 40 (1963).
68. N. R. Iammartino, *Chem. Eng.*, 76 (May 26, 1975).
69. A. A. Dor, A. English, R. D. Frans, and J. S. Wakeman, *paper presented at the 9th International Mineral Processing Congress*, Prague, 1970, pp. 173–238.
70. K. Masters, *Spray Drying Handbook*, 3rd ed., George Godwin, London, and Halsted Press, a division of John Wiley & Sons, Inc., New York, 1979.
71. K. Masters and A. Stoltze, *Food Eng.* **64** (Feb. 1973).
72. *Can. Chem. Process Ind.* **28,** 299 (May 1944).
73. *Horton Thermaprill Process*, *Bull. No. 8305P*, HPD Incorporated, Glen Ellyn, Ill., 1978.
74. J. W. D. Pictor, *Process Eng.*, 66 (June 1974).
75. W. L. Davies and W. L. Goor, *J. Pharm. Sci.* **60,** 1869 (1971); **61,** 618 (1972).
76. S. Mortensen and S. Hovmand in D. L. Keairns, ed., *Fluidization Technology*, Vol. II, Hemisphere Publishing Corp., Washington, D.C., 1976, pp. 519–544.
77. K. B. Mathur and N. Epstein, *Spouted Beds*, Academic Press, New York, 1974.
78. U.S. Pat. 3,231,413 (Jan. 25, 1966), Y. F. Berquin (to Potasse et Engrais Chimiques).
79. J. L. Kelly, A. T. Kleinsteuber, S. D. Clinton, and O. C. Dean, *Ind. Eng. Chem. Process Des. Dev.* **4,** 212 (1965).
80. M. E. A. Hermans, *Powder Metall. Int.* **5**(3), 137 (1973).
81. U.S. Pat. 3,368,004 (Feb. 6, 1968), A. F. Sirianni and I. E. Puddington (to Canadian Patents and Development Limited).
82. C. E. Capes, R. D. Coleman, and W. L. Thayer, *paper presented at the 1st International Conference on the Compaction and Consolidation of Particulate Matter*, London, UK, 1972.
83. F. J. Zuiderweg and N. van Lookeren Campagne, *Chem. Eng.* (*London*) (220), CE223 (1968).
84. C. E. Capes, A. E. McIlhinney, R. E. McKeever, and L. Messer, *paper presented at 7th International Coal Preparation Conference*, *Proc.*, Sydney, Australia, 1976.
85. M. Yusa, H. Suzuki, and S. Tanaka, *J. Am. Water Works Assoc.* **67,** 397 (1975).
86. *Flocpress*, Bull. DB845, Infilco Degremont Inc., Richmond, Va., Sept. 1976.
87. B. D. Sparks and F. W. Meadus, *Energy Processing* **72,** 55 (1979).
88. C. E. Capes, A. E. McIlhinney, and A. F. Sirianni in ref. 35, pp. 910–930.
89. F. W. Meadus and I. E. Puddington, *CIM Bull.* **66**(734), 123 (1973).
90. U.S. Pat. 3,399,765 (Sept. 3, 1968), I. E. Puddington and J. R. Farnard (to National Research Council, Ottawa).
91. C. E. Capes, A. E. McIlhinney, A. F. Sirianni, and I. E. Puddington, *CIM Bull.* **66**(739), 88 (1973).
92. U.S. Pat. 4,033,729 (July 5, 1977), C. E. Capes, R. J. Germain, A. E. McIlhinney, I. E. Puddington, and A. F. Sirianni (to Canadian Patents and Development Limited).
93. J. R. Farnand and I. E. Puddington, *CIM Bull.* **62**(683), 267 (1969).
94. U.S. Pat. 3,268,071 (Aug. 23, 1966), I. E. Puddington, H. M. Smith, and J. R. Farnard (to National Research Council, Ottawa).
95. J. R. Farnand, F. W. Meadus, E. C. Goodhue, and I. E. Puddington, *CIM Bull.* **62**(692), 1326 (1969).
96. J. R. Farnand, H. M. Smith, and I. E. Puddington, *Can. J. Chem. Eng.* **39,** 94 (1961).
97. B. D. Sparks and R. H. T. Wong, *CIM Bull.* **66**(729), 73 (1973).

98. A. F. Sirianni, R. D. Coleman, E. C. Goodhue, and I. E. Puddington, *Trans. Can. Inst. Min. Metall.* **71,** 149 (June 1968).
99. B. D. Sparks and A. F. Sirianni, *Int. J. Miner. Process.* **1,** 231 (1974).
100. B. D. Sparks and F. W. Meadus, *CIM Bull.* **67**(747), 111 (1974).
101. U.S. Pat. 3,471,267 (Oct. 7, 1969), C. E. Capes, J. P. Sutherland, and A. E. McIlhinney (to Canadian Patents and Development Limited).
102. R. D. Coleman, J. P. Sutherland, and C. E. Capes, *J. Appl. Chem.* **17,** 89 (April 1967).
103. B. D. Sparks, F. W. Meadus, and I. E. Puddington, *CIM Bull.* **64**(710), 67 (1971).
104. J. R. Farnand, F. W. Meadus, P. Tymchuk, and I. E. Puddington, *Can. Metall. Q.* **3,** 123 (1964).
105. F. W. Meadus, A. Mykytiuk, I. E. Puddington, and W. D. MacLeod, *Trans. Can. Inst. Mining Metall.* **69,** 303 (Aug. 1966).
106. A. F. Sirianni and I. E. Puddington, *Can. J. Chem. Eng.* **42,** 42 (1964).
107. M. Yusa and A. M. Gaudin, *Am. Ceram. Soc. Bull,* **43,** 402 (1964).
108. *DCG for Gravity Sludge Dewatering*, The Permutit Company, Paramus, N.J., 1972.

General Reference

Ref. 7 is also a general reference.

Note

This article has been issued as NRCC No. 20,388.

C. E. CAPES
National Research Council of Canada

SIZE MEASUREMENT OF PARTICLES

Particle size influences the combustion efficiency of powdered coal and sprayed liquid fuels, the setting time of cements, the flow characteristics of granular materials, the compacting and sintering behavior of metallurgical powders, and the hiding power of paint pigments. The size of particles in the atmosphere affects the color of a sunset; the probability of an explosion in a dusty workplace is related to dust-particle size; and the glossiness of paper is established by the size and shape of particles pressed upon it during manufacture. These examples only suggest the intimate involvement of particle size in energy generation, industrial processes, resource utilization, pollution, and natural phenomena. Many other illustrations could be drawn from such diverse topics as cosmology, soil science, ocean sediments, and stratospheric micrometeorites.

A particle is a small object having discrete physical boundaries. Typically, the term is limited in technical usage to objects having dimensions from ca 10^4–10^{-1} μm. The upper limit reaches the region where direct evaluation may be made without magnification, whereas the lower limit overlaps colloidal dimensions where surface or interfacial phenomena become equally as important as size. Particles are sometimes thought of exclusively as bits of solid matter, but the term applies with equal validity to liquid droplets and is sometimes employed when referring to tiny gas bubbles in liquids.

The size of a particle is best expressed in terms of a single, linear dimension, eg, a 50-μm particle, but a problem arises in selecting the appropriate dimension. Only in the case of spheres, cubes, and other fixed shapes can a single dimension adequately characterize particle size, and even then, the shape must be included. A unique description of size can be achieved through the mass, volume, or surface area of a particle. For example, an irregularly shaped particle might be said to be equivalent in size to a sphere having the same surface area as the particle. However, these parameters can rarely be measured individually, especially those of finer particles. Little, if any, advantage is gained by such measurements, since the behavior of a particle also depends on its shape.

The common solution to this problem is to characterize particle action in terms of an equivalent simple shape, generally that of a sphere, and thus effectively combine the parameters of size and shape. Thus, a particle can be described as "behaving as though it were a sphere of diameter d." This simplification is not without important consequences, however, since the definition of size depends on the method of measurement. An elongated particle sized by passing it through a sieve acts as though it were a sphere of some particular diameter. If the same particle were sized by determining how fast it fell through a liquid under the influence of gravity, it might behave as a sphere of significantly different diameter. Hence, it is necessary to specify both the size of a particle and the method by which that size was determined. Differences between equivalent sizes are ascribed to particle shape, and ratios of the equivalent sizes as obtained by different methods are sometimes referred to as shape factors.

Using the equivalent spherical diameter implies that all particles in a given system have the same shape. Obviously this is not strictly true for irregular particles but it is usually a reasonable assumption. Variations in particle shape manifest themselves in apparent size variations where this is the case.

Choice and application of sizing methods should take into account the uncertainty in the definition of particle size. Direct comparisons are valid only when sizes are determined by the same technique, eg, sieving versus sieving, microscopy versus microscopy, etc. Furthermore, if a system contains a broad range of sizes, the same technique should be employed for all sizes if at all possible. If impossible, and more than one method must be used, the ranges covered should overlap considerably to permit blending the data sets. Finally, choice of method should include consideration of the application for which size information is desired. Particles to be employed in a liquid suspension, for example, are most appropriately evaluated in a liquid, eg, by a sedimentation method.

Data Representation

Systems composed of identical particle sizes are extremely rare. Normally, a distribution of individual sizes is encountered. Frequently the range of values extends over several orders of magnitude. Assuming an appropriate diameter d can be assigned to the size of individual particles in a representative sample, the distribution may be characterized in terms of a single mean diameter, presented as a spread of values in a table or on a graph for visual examination, or described mathematically.

Mean Diameter. Mean diameters represent a continuum of particle sizes by a single value. They are frequently used for convenience sake where one characteristic may be of paramount concern, eg, surface area in relation to combustion properties; focusing on a particular mean, in this case a surface mean diameter, has merit.

Mean diameters are commonly employed in terms of the number of particles having a particular diameter (termed length for consistency) d. The descriptive designations given involve two and only two characteristic parameters of a particle system, ie, number and volume, surface and volume, number and surface, etc (see Table 1).

Mean diameters have special significance, as illustrated by a few examples using a system consisting of 10 spherical (or equivalent sphere) particles with diameters (lengths) of 1, 2, 3 . . . 10. The sum of these 10 diameters is 55. This number divided by 10 gives a number–length mean diameter of 5.5, which reveals that, as far as number and length are concerned, this distribution of unequally sized spheres is entirely equivalent to one of 10 equally sized particles having a diameter of 5.5.

The total surface area of the 10 unequal spheres is $(1)^2\pi + (2)^2\pi + (3)^2\pi + \ldots$

Table 1. Mean Diameters

Descriptive designation	Common designation	Mathematical expression[b]
number–length	arithmetic, number	$\Sigma nd/\Sigma n$
number–surface	surface diameter	$(\Sigma nd^2/\Sigma n)^{1/2}$
number–volume	volume diameter	$(\Sigma nd^3/\Sigma n)^{1/3}$
length–surface	length	$\Sigma nd^2/\Sigma nd$
length–volume	volume diameter	$(\Sigma nd^3/\Sigma nd)^{1/2}$
surface–volume	surface	$\Sigma nd^3/\Sigma nd^2$
volume (or weight)–moment mean[a]	volume	$\Sigma nd^4/\Sigma nd^3$

[a] Value is identical for volume and weight if all particles in system are of identical density.

[b] n = number; d = diameter.

$\pi(10)^2 = 385\pi$. Thus, ten equal-size spheres must have a diameter of $\sqrt{(385\pi)/(10\pi)}$ = 6.21 to have this total surface area and to be equivalent in both number and surface area. To achieve length–surface mean diameter equivalence, on the other hand, requires that number equivalence be abandoned. The length–surface mean diameter for particles having a total surface of 385π and a total length of 55 must be $385\pi/55\pi$ = 7. The number of equal-size spheres can now only be 55/7 = 7.857. In other words, 7.857 particles each of diameter 7 have the same total length and total surface area as the unequally sized set of 10 and are thus equivalent according to length and surface area.

Tabular. Presentation of particle-size information in useful tabular form requires data classification. With a narrow range of sizes, it may be sufficient to group the individual measurements in linear size intervals, eg, 0–1, 1–2, 2–3, etc, μm (or cm, m, etc) and then list the intervals as a function of the corresponding percent, or fraction, of the whole that each interval represents.

Grouping into linear size intervals results in a serious loss of information and is unsatisfactory when the range of sizes is large because the difference between 1 and 2 μm, for example, is much more significant than the difference between 100 and 101 μm. In this case, data are best classified on a geometric scale, eg, 1–2, 2–4, 4–8, 8–16, etc, which attaches the same importance to the difference between 1 and 2 μm as between 1 and 2 cm. Geometric intervals are equivalent to linear intervals on a logarithmic scale. Typical particle-size data are given in Table 2; the percent represented by each interval can be based on the total number of particles measured, the total sample weight, total sample volume, or any other basis upon which data might be accumulated. Counting techniques (microscopic examination, for example) yield numbers directly, whereas sieving and sedimentation give bulk-based data.

Graphical. Although a tabular presentation offers the ultimate in precision since all data are included, it is inconvenient to compare tables; a graph usually offers advantages. For example, a histogram provides a simple means for graphically presenting particle-size data. Normally, the percentage of particles in a given size interval, called the frequency, is plotted against size as shown in Figure 1. Histograms are particularly useful for comparisons among similar distributions. Information on sizes within the individual intervals is not utilized.

The cumulative curve affords a more useful graphical presentation. The data of Table 2 are readily converted to a plot representing either the percentage of coarser or the percentage of finer material. Since the 2.3% in the 1–2 μm interval is finer than 2 μm, the amount of coarser material is 97.7%. Other points are similarly derived; the data of Table 3 are plotted in Figure 2. However, cumulative plots tend to conceal detail

Table 2. Typical Particle-Size Distribution

Diameter range, μm	Percent, %
1–2	2.3
2–4	4.7
4–8	14.1
8–16	19.8
16–32	21.9
32–64	22.2
64–128	12.6
128–256	2.4

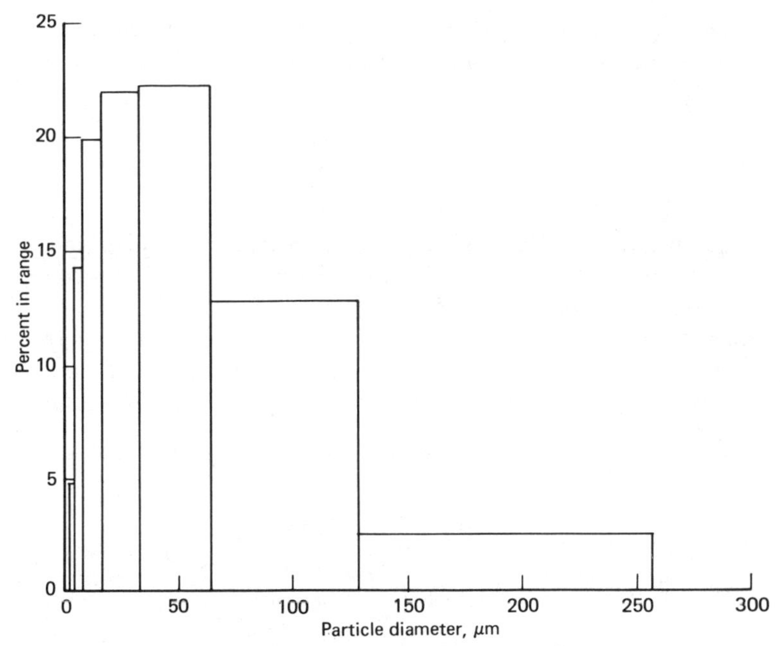

Figure 1. Histogram of the data from Table 2.

because of a smoothing effect. This effect can be minimized by employing greater numbers of smaller increments when justified by the data.

Mathematical. The maximum of useful information is revealed when particle-size data are summarized by a mathematical expression. Its application is simplified if it can be linearized. A linearized function also permits ready graphical representation and offers maximum opportunities for interpolation, extrapolation, and comparison among particle systems. Furthermore, still more useful information can be revealed if the parameters of the function can be related to properties of the particle-size systems or the process that produced it.

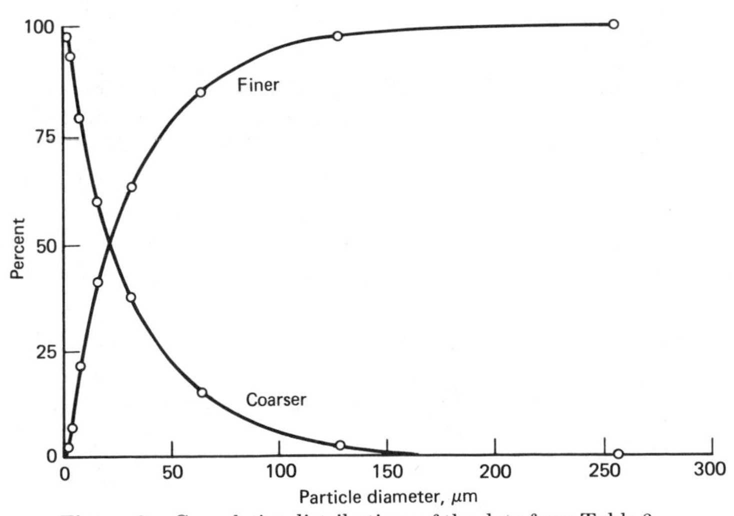

Figure 2. Cumulative distributions of the data from Table 3.

Defining $Y_n(x)$ as the total number of particles finer than size x, and $Y_m(x)$ as the total mass of particles finer than size x, the particle-size density functions

$$y_n(x) = \frac{dY_n}{dx} \quad (1)$$

$$y_m(x) = \frac{dY_m}{dx} \quad (2)$$

may be written where $y_n(x)dx$ and $y_m(x)dx$ represents the number and mass fractions, respectively, of particles having sizes between x and $x + dx$. If these density functions are employed as the ordinate on a plot, they would produce a histogram in which the size intervals are infinitesimally small. As noted above, the volume fraction $Y_v(x)$ is identical to the mass fraction when the density of the particles is constant and independent of size. Surface area and length fractions could be defined also but they are rarely utilized.

Inversion of equations 1 and 2 gives

$$Y_n(x) = \int_0^x y_n(x)dx \quad (3)$$

$$Y_m(x) = \int_0^x y_m(x)dx \quad (4)$$

from which it follows

$$\int_0^\infty y_n(x)dx = \int_0^\infty y_m(x)dx = 1. \quad (5)$$

An expression for the distribution function $Y(x)$ is required to utilize equations 3–5. Although there is no universally applicable distribution law, there are numerous empirical laws that give a reasonable fit with experimental size distribution data (1–19). These can be arranged in the form of algebraic expressions with one to three adjustable parameters. Only the more frequently useful expressions are included here.

The first, referred to as the Gaudin-Schuhmann distribution (1–2) often gives a remarkably good fit to data generated for crushed minerals by sieving. If $Y_m(x)$ is the weight of particles finer than size x, this distribution function is expressed by

$$Y_m(x) = \left(\frac{x}{k}\right)^{b} \quad (6)$$

where k is a size modulus and b a distribution modulus. When the spread of particle sizes conforms to equation 6, a straight line is obtained by plotting the cumulative weight percent of particles finer than each stated size versus that size on log–log paper. The size at which the straight line (extrapolated) crosses $Y_m = 1$ (100%) is the size modulus k and the slope of the straight line is the distribution modulus b. The size modulus k depends on the extent of grinding. In many instances the distribution modulus b appears to be a constant for a given grinding machine; typical values lie between 0.5 and 1.5.

The Rosin-Rammler empirical expression describes skewed distributions of particle sizes, ie, spreads of data where significant quantities of particles (either larger or smaller) exist removed from the region of predominant size (3). This mass (volume) distribution function may be expressed as

$$Y_m(x) = 1 - \exp\left[-\left(\frac{x}{m}\right)^{n}\right] \quad (7)$$

where n and m are adjustable parameters. The equation can be inverted to the form

$$\log \log \left(\frac{1}{1 - Y_m}\right) = n \log x - n \log m - \log 2.303 \tag{8}$$

where, when applicable, a straight line results if $\log \log [1/(1 - Y_m)]$ is plotted versus $\log x$. The slope of the line is n and m is the size at which the line crosses $Y_m = 63.21\%$.

A three-constant empirical expression known as the Nukiyama-Tanasawa distribution equation is particularly applicable to liquid atomization (4). It can be written in the form

$$\log \left(\frac{y_n}{x^2}\right) = \log p - \frac{qx^r}{2.303} \tag{9}$$

where p, q, and r are constants. Since the equation is linear, a proper selection of r should give a straight line. The slope of the resulting line is $q/2.303$. The value of p can then be determined from equation 9, once r and q are established. The value of r appears to be nearly constant for a particular nozzle over a wide range of operating conditions.

The normal or Gaussian distribution equation is very commonly utilized in a great variety of statistical applications. It is appropriate for particle-size data only in the case of very narrow distributions or in those instances where a sharp cut has been extracted from a wider range of sizes. However, when the normal distribution is applied to particle-size data on a logarithmic rather than a linear scale, an expression of considerable general utility is obtained (5). This logarithmic expression is easily used and permits ready computation of mean diameters. These features, along with the fact that it frequently well describes experimental data, make the logarithmic expression the most widely employed distribution functions in particle-size work. A word of caution is necessary. The attributes of the logarithmic distribution equation are so attractive that there is a tendency to force its use when prudence might suggest otherwise. Computer programs have been written to aid the fitting of data to both normal and log-normal distribution functions (20).

The logarithmic distribution function is expressed by

$$y(x) = \frac{2.303}{x\sqrt{2\pi} \log \sigma_g} \exp \left[-\frac{(\log x - \log \bar{x})^2}{2 \log \sigma_g}\right] \tag{10}$$

where σ_g is the geometric standard deviation and $\bar{x}$ is the geometric mean size. The geometric standard deviation relates to the spread of sizes of about the geometric mean size. Equation 10 is directly applied when cumulative data are plotted on log-probability paper (Fig. 3). This paper is scaled in such a way that all log normally distributed data give a straight line plot of slope σ_g with $\bar{x}$ as the 50% point. Particle data may be obtained according to number, surface, mass, etc, giving the corresponding geometric mean sizes $\bar{x}_N$, $\bar{x}_S$, $\bar{x}_M$, etc. The geometric standard deviation σ_g has a single value for any one material regardless of the basis of data acquisition (assuming the data were gathered correctly). Thus, number, area, and mass distribution yield parallel-plotted lines. The value of σ_g is simply

$$\sigma_g = \frac{x_{50}}{x_{15.87}} = \frac{x_{84.13}}{x_{50}} \tag{11}$$

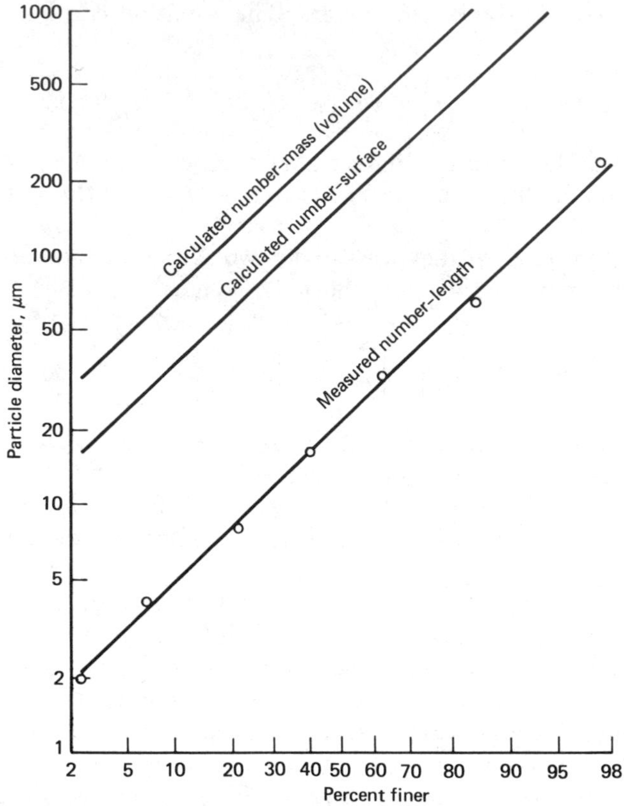

Figure 3. Log-probability distributions of the data from Table 3.

where x_{50} is the particle size for which Y is 50% (ie, $\bar{x}$), $x_{15.87}$ the size of 15.87%, and $x_{84.13}$ the size of 84.13%.

All other mean diameters are readily calculated once any one geometric mean diameter and the geometric standard deviation are established for a particle system. The logarithm of any mean diameter is the sum of the logarithm of the geometric mean diameter on a count basis and the product 1.1513 (order of the mean) $\log {}^2\sigma_g$. The order of the mean is defined as the sum of the powers of diameter in its definition (6), as given in Table 1. For example, the surface–volume mean diameter is obtained by

$$\log d_{\mathrm{SV}} = \log \bar{d}_{\mathrm{NL}} + 1.1513\,(2+3)\log {}^2\sigma_g \tag{12}$$

where L = length. If the geometric mean diameter has not initially been determined by the counting method, its value has to be computed from whatever version of equation 12 is applicable and then used with another form of the equation to calculate the desired mean.

The data of Table 3 are plotted on log-probability paper in Figure 3; a straight line satisfactorily fits the data points. The geometric mean diameter $\bar{d}$ (50% point) is 21.0 μm and the geometric standard deviation σ_g ($d_{84.13}/d_{50}$) is 3.19. If these data has been determined by microscopically measuring a number of individual particle diameters, the mean diameter is d_{NL} by definition, and equation 12 can be applied

Table 3. Cumulative Particle-Size Data

Diameter, μm	Cumulative coarser, %	Cumulative finer, %
2	97.7	2.3
4	93.0	7.0
8	78.9	21.1
16	59.1	40.9
32	37.2	62.8
64	15.0	85.0
128	2.4	97.6
256	0.0	100.0

to find other mean diameters as indicated below:

$$\log d_{NS} = \log 21.0 + 1.1513\ (2 + 1) \log^2 3.19$$

$$d_{NS} = 158\ \mu m$$

$$\log d_{NV} = \log 21.0 + 1.1513\ (3 + 1) \log^2 3.19$$

$$d_{NV} = 310\ \mu m$$

The corresponding distributions are plotted as in Figure 3 using these geometric mean sizes and the common geometric standard deviation, ie, slope of the line. The substantial shift in magnitude revealed by these numbers is evidence of the difficulty likely to be encountered when results from different measurement methods are compared. A mass mean diameter of 310 μm, for example, appears improbable from a cursory examination of Figure 3. The figure shows that in reality such a size can appear with ca 1% frequency, and it follows that one particle of this size contains a mass roughly equivalent to 99 smaller ones following the distribution given. A weighing technique thus strongly emphasizes a physical property different from one involving counting. The effect is especially pronounced when the distribution spans three orders of magnitude as is the case here.

If, on the other hand, the data of Table 3 were acquired by sieving, the geometric mean diameter (50% size) on Figure 3 would be d_{LV} by definition, and d_{NL} is found from

$$\log d_{NL} = \log \bar{d}_{LV} - 1.1513\ (3 + 1) \log^2 \sigma_g$$

All other diameters follow from application of forms of equation 12.

The log-probability function frequently fails to describe adequately a particle-size distribution because of deviations near the upper extreme of the size range. Probability theory merely requires that very small, as well as very large, particles occur with low frequencies. A smallest size of zero is inescapable, but the large-particle-size end may be open. This situation had led to a special treatment for upper-limit deviating data which, in effect, sets a maximum size and considers greater particles to be abberations (6). The function $d/(d_{max} - d)$ is plotted versus frequency of occurrence on log-probability paper. The term d_{max} is determined by trial and error to give the best straight line; it can usually be found in two or three trials. Expressions comparable to those derived for log-probability use and to the several forms of equation 12 are then employed for calculating mean diameters and for the interconversion of mean sizes.

Methods

Different methods have been developed in response to the variability of needs for particle-size measurement. Sometimes *in situ* analysis is required, whereas in other situations measurement a distance away from the point of origin is acceptable. Industry frequently needs size information in real time for process control. In other cases, analysis following the finished product is satisfactory. Some methods rapidly measure individual particles; others measure the material in bulk. Some methods are suitable only for analyzing dry particles, airborne particles, particles suspended in a liquid, or particles under partial vacuum.

Particle size may be inferred from the interaction of particles with many forms of radiant energy. Their images rather than the particles themselves are analyzed in other systems. The influence of gravity, the effect of imposed centrifugal force fields, and the response under various flow situations form the basis for other techniques. Passage along or through restrictions provide yet other means. Some techniques determine particle mass, directly, some volume, some cross-section, and some average diameter. Submicrometer sizes exclusively are resolved by certain techniques, whereas others are applicable to large particles only.

Sieving. The most widely employed sizing method determines particle size by the degree to which a powder is retained on a series of sieves with different opening dimensions. This technique is straightforward and requires simple equipment, but without attention to detail it can lead to erroneous results (21). The sieves, particularly those with the finer meshes, are damaged by careless handling. They tend to become clogged with irregularly shaped particles unless agitated, but distorted if agitated too much (22). Furthermore, it is always a concern to determine when all particles that might pass through a sieve have done so. Nevertheless, attempts to automate the procedure have not met with notable success (23–27).

A typical sieve is a shallow pan with a wire mesh bottom or an electroformed grid. Opening dimensions in any mesh or grid are generally uniform within a few percent. Sieves are available with openings from 5 μm in several series of size progression. Woven wire-mesh sieves have approximately square openings whereas electroformed sieves have round, square, or rectangular openings. Wire sieves tend to be sturdier and less expensive, and have a greater proportion of open area. They are much more frequently employed than are electroformed sieves except in the very fine particle range where only electroformed sieves are available.

Dry-sieving is typically performed with a stack of sieves with openings diminishing in size from the top downward. The lowest pan has a solid bottom to retain the final undersize. Powders are segregated according to size by placing the powder on the uppermost sieve and then shaking the stack manually, with a mechanical vibrator (28–29), or with air pulses of sonic frequency (30–31) until all particles fall onto sieves through which they are unable to pass or into the bottom pan. The unit powered by sonic energy shown in Figure 4 confines the sample with very flexible diaphragms, ensuring against loss of fine particles. In another device, sieves are employed one at a time within a container from which passing particles are captured by a filter. Agitation on the sieve is provided by a rotating air jet (32). The material retained by the sieve is recovered and recycled with the next coarser sieve until all of the powder is exposed to the desired series of sieves or all material passes.

Wet-sieving is performed with a stack of sieves in a similar manner except that

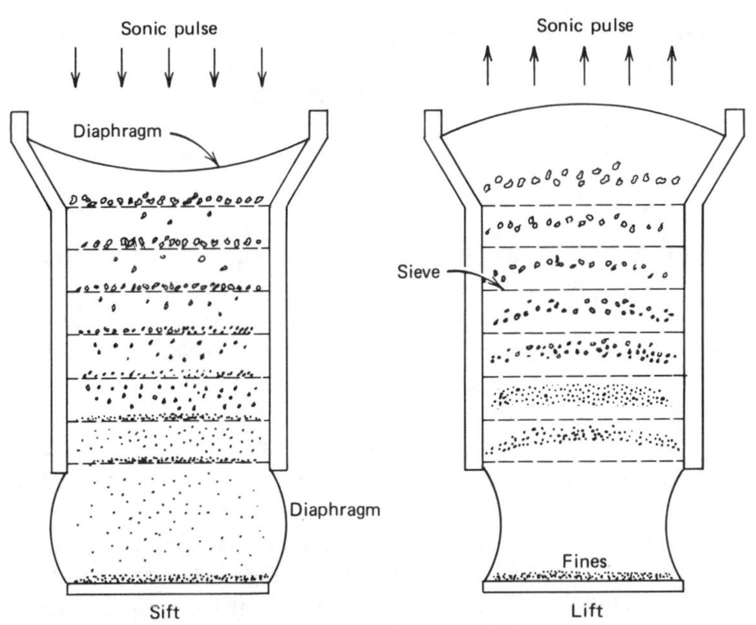

Figure 4. Dry siever employing sonic frequency.

water or another liquid that does not dissolve the material, is continually applied to facilitate particle passage. A detergent is frequently added to promote particle dispersion. This enhanced dispersion is almost essential for fine-particle analysis, because under dry conditions, electrostatic and surface forces greatly impede clean passage and isolation of sizes. A partial vacuum is sometimes applied to fine-particle sieving (33–34). Ultrasonic energy dislodges irregular particles trapped in sieve openings provided it is used at moderate density; a maximum of 0.45 W/cm^2 at a frequency of 40 kHz has been recommended (22) (see Ultrasonics).

The particle mass retained by each sieve is determined by weighing, after drying when necessary, and each fraction is designated by the sieve size it passed and the size on which it was retained. Mass fractions of the particles are then presented in tabular or graphical form as described above. Graph paper termed promesh combines a log-probability and standard-sieve-series mesh scale. It is particularly useful in presenting sieving results (35).

Sieve analysis is employed by the minerals industry to assess ore crushing for mineral release; in heavy construction work to evaluate soils, sand, and gravel for foundation stability; in powder-metallurgical operations for porosity control; by the ceramic and glass industries for feedstock evaluation; and in agriculture for grading seed quality and uniformity (see also Gravity concentration; Size reduction; Powder metallurgy).

Microscopic Examination. Examining particles one at a time with the aid of magnification is an obvious means for assessing size parameters. This technique is, however, tedious and time-consuming. To be analyzed individually, the particles must be thoroughly dispersed and hundreds of examinations may be required to establish statistical significance. Results are given in terms of numbers of particles, whereas data on a mass or volume basis are generally preferred. Several magnifications are usually required to span a range of particle sizes; these magnifications must be accurately known.

The procedure is facilitated when magnified particle images are photographed before examination. In addition, automated electronic image analyzers may be employed in conjunction with either microscopic images or photomicrographs. However, microscopic examination, both optical and electron, must be considered the standard method; it is frequently used to choose between one technique and another. It is neither a method suitable for routine applications nor one which easily gives precise results.

Optical microscopy is used for particles larger than ca 1.0 μm in diameter. The particles need to be spread evenly on a microscope slide with considerable free space between. The field of view should contain a maximum of 30 particles. Several techniques may be employed for slide preparation (15,36). Irregular particles tend to rest on a horizontal surface in their most stable position. This inevitably contributes a bias when the particles are in the shape of thin plates, for example. Only a detailed evaluation of shape can eliminate this bias; shape is too frequently neglected.

In particle-image measurement, four diameter definitions are currently applied: the length established by the movement of a hairline in traversing the extremities of the particle image in one direction, regardless of the particle position (37); the length across the midsection of the particle, again without regard to the particle position (38); the diameter of a circle that is judged to be equivalent in cross-sectional area to the particle (39–41); and the magnitude of the shift in particle image without regard to particle position that would bring two images of the same particle into contact (42–43). Obtaining the first two diameters requires an optical micrometer, the third an optical graticule, and the fourth an image-shearing eyepiece (see Fig. 5).

Particle analysis is speeded by evaluating photomicrographs of particle images printed on translucent paper (44–45). The image of an adjustable iris is projected onto the photomicrograph and adjusted until its cross-sectional area seems to be equivalent to that of a particle. The data are recorded. This method is useful to measure particle diameters according to the third definition above. Image-shearing eyepieces with an automatic recording feature are available.

Truly automatic image analyzers extract information directly from negatives, photomicrographs, or from microscopic (both optical and electron) images by a scanning technique. The scanner converts the optical image into a video signal wherein amplitude represents the original image. A decision-making process, called thresholding, establishes the particle boundary (46). This boundary is taken as the point where signal amplitude is midway between the optical density of the background in the vicinity of the particle and the maximum optical density of the particle image. A small error in boundary location is insignificant with large particles but can become critical with small particles (see Fig. 6). The outlining of the particle by the scanning signal permits a computer to calculate two-dimensional parameters such as size by area or diameter, projected length, area including holes, area excluding holes, and distributions according to each (47–52).

Electron microscopy is applicable to particles with diameters from ca 0.002–15 μm; the upper limit is set by the viewing field of the instrument. Particles are usually incorporated into very thin membranes or deposited on them (53). Fitting data from both optical and electron microscope measurements of a powder with a wide spread of particle sizes may present serious problems, which are alleviated by maximizing size-overlap.

Microscopic procedures must be conducted with randomly selected fields to avoid sampling bias. A series of slide preparations should be used in selecting the fields. The

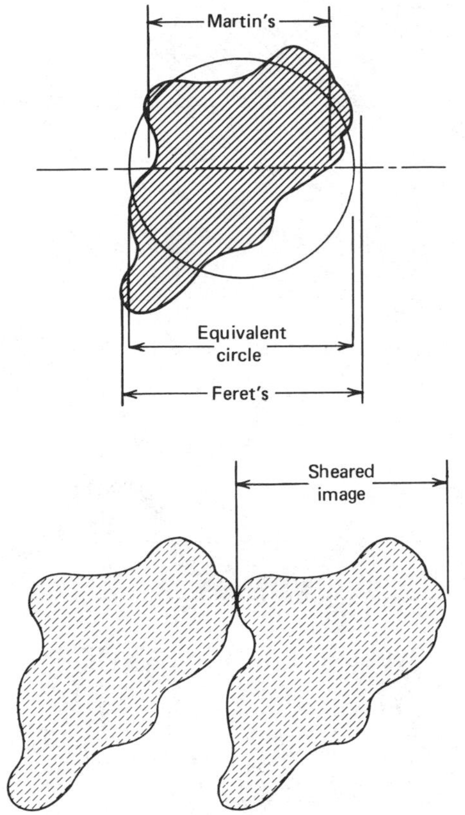

Figure 5. Particle-image definitions (37–38).

number of particles to measure depends on their size range. A number of measurements have to be made to achieve 99% certainty that the average size, defined as $(d_{100} - d_0)/d_{50}$, and the size distribution, defined as $(d_{50} - d_0)/d_{10}$ or $(d_{100} - d_{50})/d_{90}$, are correct within 10% (54) (see Table 4). Other guidelines have been offered (55–56) (see Sampling). The success of image analysis depends on the exactness with which the magnification is determined (57). The diameter data are reduced by arranging the individual measurements into size ranges in order to prepare a tabulation or plot. The number data are next converted into percentages. The reduced data are then treated to yield surface-, volume-, or mass-based size distributions as described previously.

Microscopy is applied when particle identification and, perhaps, shape evaluation are important in addition to size. Such uses include pollution or contamination assessment and forensic studies (see Air pollution control methods; Forensic chemistry).

Sedimentation. Measurement of the rate at which particles move under gravitational or centrifugal acceleration in a quiescent liquid provides a most important means for determining particle sizes. Gas-phase sedimentation has been investigated (58), but the difficulties of achieving true particle dispersion and the effect of electrostatic charging have restricted application. In liquid-phase sedimentation, the particles may initially be distributed uniformly throughout a liquid or they may be concentrated in a narrow band or layer at the liquid's surface. Their movement may be established from the accumulation rate at the base of the liquid container or from the change of concentration with time at other levels (see Sedimentation).

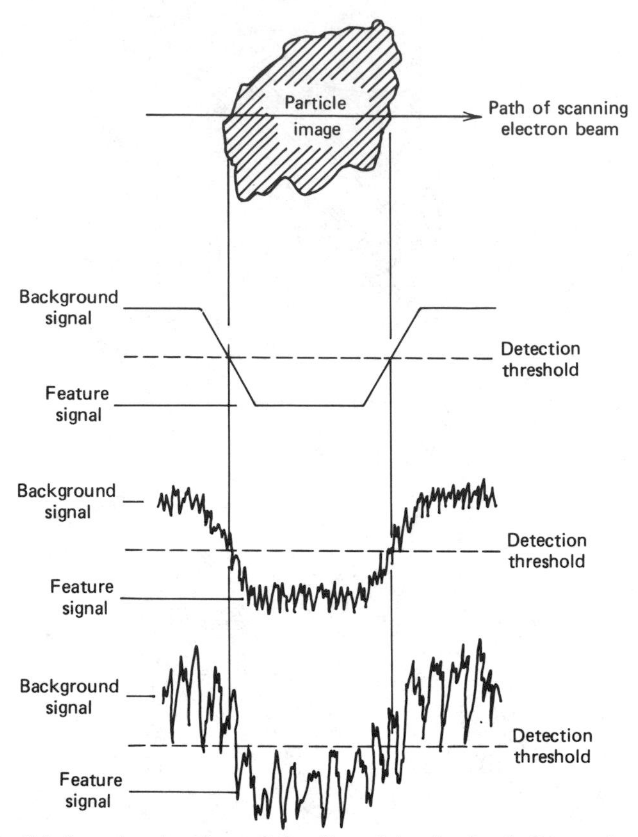

Figure 6. Particle-boundary location and the effect of signal noise. A, Correct size with ideal noise-free scanner signal. B, Well-defined size with low noise signal. C, Poorly defined size with high noise signal.

The upper-size limit for sedimentation methods is established by the value of the particle Reynolds number, $dv\rho_p/\eta$, where d is particle diameter, ρ_p particle density, and η liquid viscosity; particle velocity v is determined from Stokes law as given below. The particle Reynolds number should not exceed 0.3 (59). The lower-size limit is set by diffusion phenomena at a diameter of ca 0.2–0.3 μm for monodisperse particles (60–61), and lower for heterodisperse particle systems (62). The particle size determined by sedimentation techniques is an equivalent spherical diameter, ie, the diameter of a sphere of the same density as the irregularly shaped particle that exhibits an identical free-fall velocity. Thus it is an appropriate diameter upon which to base particle behavior in other fluid-flow situations. Unlike definitions employed in microscopy and sieving, it is rigorous and unequivocal.

In a simple sedimentation method, a pan is suspended from a sensitive balance. The particles accumulate in this pan as a function of time as they settle under gravity from a well-dispersed suspension (63–67). The technique is undesirable from a mathematical viewpoint because differentiation of the resulting accumulated weight–time curve is required to reduce the data to size-distribution information. A still simpler method employs a hydrometer to determine suspension density as a function of depth and time over intervals up to 48 h. Its application results in an integral or cumulative size distribution. Both methods are generally too slow for current needs but the latter technique especially is illustrative of modern sedimentation methods.

Table 4. Number of Particles that have to be Measured to Attain 99% Certainty that Results are Correct Within ±10%

Number of particles	Average size[a]	Size distribution[b]
20	0.60	
35	0.88	0.25
50	1.25	0.30
75	1.53	0.40
100	1.84	0.49
150	2.37	0.65
200	2.75	0.77
250		0.90
300		1.00
400		1.16
500		1.33
600		1.48
700		1.62
800		1.75
900		1.89
1000		2.00

[a] $(d_{100} - d_0)/d_{50}$.
[b] $(d_{50} - d_0)/d_{10}$ or $(d_{100} - d_{50})/d_{90}$.

Hydrometer. The concentration of particles remaining in suspension is readily calculated from density measurements made with a hydrometer. The percentage of particles remaining in suspension at the measured depth, termed the weight percent of finer material, P_f, is given by

$$P_f = 100 \frac{C_f}{C} \tag{13}$$

where C_f is the concentration after time t at the distance h from the upper surface of the suspension to the detection depth of the hydrometer and C the concentration when the suspension is uniformly mixed ($t = 0$). The diameter d_f of the particles that have fallen the distance h in time t is given by Stokes law.

$$d_f = \left[\frac{18\,\eta}{(\rho_p - \rho)g} \cdot \frac{h}{t}\right]^{1/2} \tag{14}$$

where ρ is the liquid density and g the acceleration of gravity. All other particles of size d_f are below the depth h at time t because they started at a lower depth. Particles of greater diameter are also at lower depths because of their greater settling velocity. A cumulative distribution of sizes is thus obtained by securing readings of C_f, h, and t and calculating corresponding values of P_f and d_f from equations 13 and 14. Any alteration of either liquid density or viscosity produced by surface-active agents employed as dispersing aids must be taken into account. Details of the hydrometer technique have been fully described (15,36,68–69).

X-Ray Methods. Prominent among the difficulties presented by the hydrometer technique are mechanical disturbance of the settling suspension caused by insertion and removal of the hydrometer, imprecision as to the actual depth of measurement, and long-term temperature alterations (70). Using a collimated beam of x rays permits particle concentration detection as a function of mass without physically disturbing the suspension; sharply defining the vertical dimensions of the beam permits shortening the distance over which particles must be allowed to fall and hence the time of fall; and moving the suspension container vertically downward relative to the detection system according to a prescribed schedule further reduces the distance of fall and time of analysis. These innovations constitute the main features of an x-ray particle-size analyzer (60,71–73). The relationship between the fraction of x-rays transmitted and the mass concentration of particles of atomic weight >12 is expressed as

$$\ln T = -\Delta\mu L C_f \tag{15}$$

where ln T is the natural logarithm of the transmittance relative to the suspending liquid, $\Delta\mu$ the difference in the x-ray mass-absorption coefficient of the solids and the liquid, and L the distance through the suspension in the direction of transmission. The weight percent of particles finer than size d_f, following from equation 13, is

$$P_f = 100 \frac{\ln T_f}{\ln T} \tag{16}$$

where T_f is the transmittance after time t at the distance h and T the transmittance at the starting time.

The instrument, shown in Figure 7, thus determines the mass concentration of particles remaining in suspension at various depths as a function of time. The logarithm of the transmitted x-ray intensity is electronically generated, scaled, and presented linearly as the cumulative percentage of finer mass on the Y axis of an X–Y recorder. The position of the cell in which sedimentation occurs is continuously changed with the result that the effective sedimentation depth is inversely proportional to the elapsed time. Cell movement is synchronized with the X axis of the X–Y recorder to indicate directly the equivalent spherical diameter corresponding to the elapsed time and the instantaneous sedimentation depth as related by equation 14. Particle diameters from 0.2–130 μm can be measured, depending on x-ray absorption and particle density. The time required for an analysis is inversely proportional to particle density and the maximum and minimum particle diameters.

Another version employs light instead of x-ray transmission to determine concentration (36,74–79). The technique is generally referred to as turbidimetry or photoextinction. Light can be used for low particle concentrations and detection of low molecular weight particles. However, the ability of a particle to scatter light is a complex function of its size and other variables and is predictable only for particles of simple shape. Furthermore, particles of different sizes may scatter light similarly. Consequently such measurements are most useful for comparing different samples of the same material.

X-ray sedimentation analysis is suitable for both quality control and research work and gives quick well-defined results. Thus, it is employed in the evaluation of a variety of materials including soils, sediments, pigments, fillers, phosphors, clays, minerals, and photographic halides. Sedimentation detected by light transmission is employed for size evaluation of organic particles and for comparison of powders identical in other respects.

Centrifugation. Centrifugal force in sedimentation analysis permits evaluation of smaller diameters but adds mechanical complexity. The first commercial application of centrifugal force introduced a two-layer technique (80–81). A thin layer of suspension is established above a clear liquid filling most of a centrifuge tube. The layer

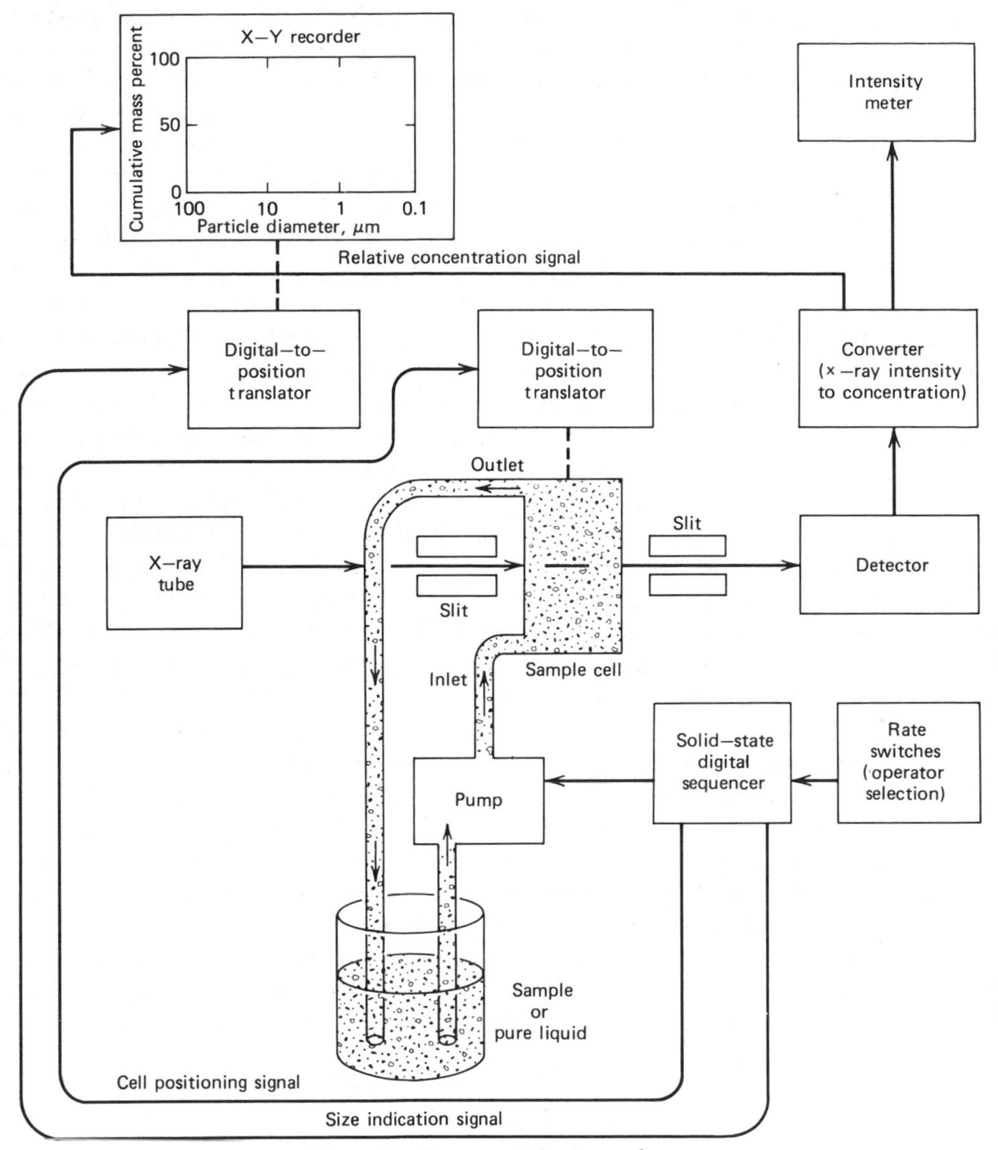

Figure 7. X-ray particle-size analyzer.

containing the particles must be formed of a liquid slightly less dense than the other liquid but completely miscible with it. The tube is transferred to a suitable centrifuge and spun at a series of speeds from 300–4000 rpm. The particles, largest first, are deposited at the bottom of the tube. The centrifuge is stopped after a certain time at each speed and the depth of the deposit is measured. Particle size distribution is calculated from the quantity of each deposit increment, the distance of sedimentation, the magnitude of the centrifugal force, and other pertinent parameters (see also Centrifugal separation).

Difficulties with this method include the uncertainty of the uniformity with which each deposit is packed, disturbances due to starting and stopping the centrifuge, and streaming. Streaming is a phenomenon unique to the two-layer method in which filaments of suspension break through the interface between the suspension and the clear liquid and carry down particles much faster than they normally would fall. It can be overcome by employing a very dilute suspension and placing a second layer of intermediate density and interfacial tension beneath the suspension layer (82).

Disk centrifuges of the general arrangement shown in Figure 8 utilize homogeneous suspensions and the two-layer technique (36,83–88). They operate with both a vertical and horizontal axis of rotation. The axis of rotation must be vertical with the entry space to accommodate homogeneous suspension. Any liquid within the disk cavity arranges itself to have an essentially flat air–liquid interface as shown when rotated at relatively high speed. Particles within the liquid migrate radially outward. Their rate of accumulation from a homogeneous suspension at the rim or the change in concentration at a particular radius from the center of rotation can be measured. The same is true when a thin layer of suspension is injected onto clear liquid already established in the disk. The interfacial region is somewhat perturbed by such an injection but only a small portion of clear liquid is affected. Figure 8 shows the inclusions of a cusp in the rotor face and a method of sample injection to minimize interface disturbance in a two-layer operation. In the case of the disk cavity being filled initially with homogeneous suspension, acceleration and Coriolis forces cause undesirable disturbances. Such disturbances are minor, however, with very small particles for which centrifuges are primarily used. Ordinary white and laser light and x-rays have been employed as concentration detectors.

The disturbances arising from acceleration have largely been eliminated in a centrifuge that employs a single cylindrical chamber rotated with its axis at right angle to and about three chamber heights away from the axis of rotation (89–90). The chamber is filled with a homogeneous suspension of the particles to be examined. Pressure measuring taps at either end of the chamber connect to a sensitive differential pressure detector. As sedimentation proceeds, the pressure difference caused by the presence of suspended particles diminishes in relation to their equivalent spherical diameter permitting the calculation of the particle-size distribution from the pressure-time data. A similar centrifuge has been built in which the mass of collecting sediment is determined as a function of time by a sensitive weighing device (91).

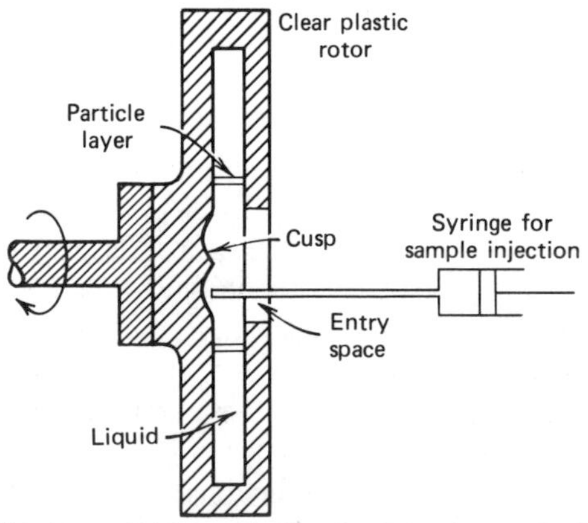

Figure 8. Disk profile showing injection method.

Sensing-Zone Methods. Sensing-zone methods produce size information as a result of particles being directed, one at a time, into a region where their presence affects a measurable condition, such as electrical conductivity; magnetic flux; light extinction, diffraction, and scattering; acoustic response; and thermal stability. Such systems require calibration, and their output decreases as the one-at-a-time condition is changed by the coincidental arrival of two or more particles (92).

The number and size of particles in an electrically conducting liquid may be determined by passing the particles through a small, short, nonconducting aperture on either side of which is an immersed electrode (50,93–96). The passage of each particle changes the resistance between the electrodes resulting in a voltage pulse of short duration proportional in amplitude to particle size. These pulses are amplified and fed into a threshold circuit with a variable level. A pulse is counted when it equals or exceeds this level. The distribution of sizes is established by varying the threshold level (see Fig. 9).

The resistance change caused by the particle presence in the aperture is linearly related to particle volume, provided the particle is substantially smaller than the aperture and shaped not greatly different from a sphere. Particle resistivity is rarely a problem. Because of interfacial effects, particles normally behave in a weak electrolyte as if they had infinite resistance whether conducting or not. Aperture diameters are available from 10–2000 μm which sets the practical range of particle size measurements from ca 0.5–250 μm. However, a single aperture is appropriate for a diametric range of only ca 20–1. Thus, samples that contain a wide range of sizes may have to be fractionated and several apertures of different sizes may be needed. For determination of large particles, a different aperture arrangement permits measurement of particles up to 1.5 mm in diameter (97), whereas particle diameters as small as 0.01 μm can be measured with a relatively long but tiny orifice (0.3 μm diameter) (98).

The sensing-zone method is employed most widely in medical applications involving blood cells, bacteria, bacteriophages, etc; it is also extensively used by the food, beverage, and pharmaceutical industries.

Ferromagnetic particles are sized similarly in a device that substitutes a small

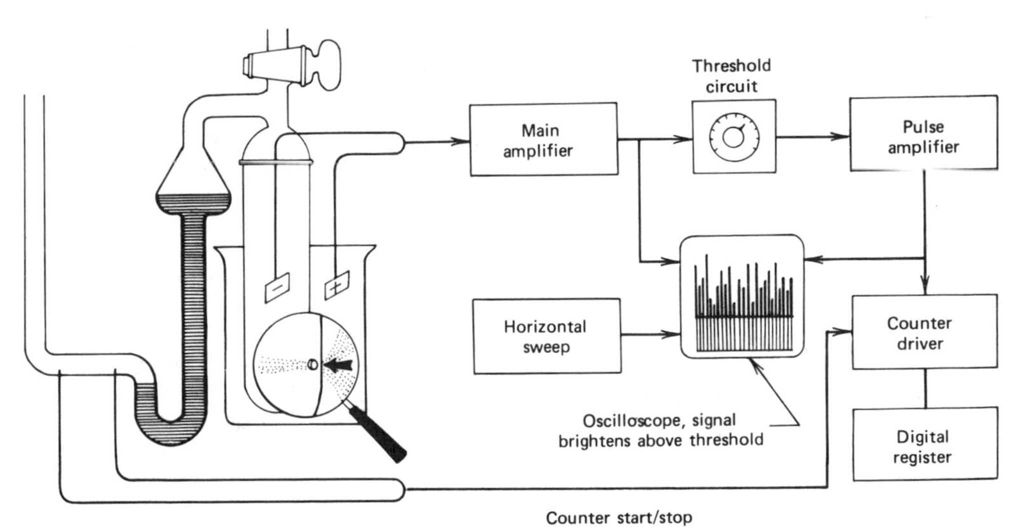

Figure 9. Schematic arrangement of conductometric apparatus.

search coil for the aperture (99). Particles passing through the coil give rise to magnetic flux pulses which, after calibration, are analyzed in a multichannel pulse-height analyzer.

Analysis of the light absorbed, diffracted, refracted, and scattered from a restricted beam of white light, or from a laser beam as either liquid- or gas-borne particles pass through, provides the basis for another type of particle-sizing instrumentation (100–110). The resulting light pulses are detected at various angles with the direction of illumination. A photodetector picks up the individual pulses and converts them into electrical signals with amplitudes that are functions of the optical characteristics and size of the particles, light wavelength, and the design of the optical system. The signals are amplified, classified into groups by a pulse-height amplifier, and counted, yielding a number-based size distribution. Diameters from submicrometer to over 100 μm may be measured, depending on the instrument (see Fig. 10).

Instruments designed to distinguish size on the basis of the light extinguished by particles passing one at a time through a small lighted orifice are particularly useful for the determination of particle contamination in lubricating oils and hydraulic fluids since no special condition, eg, electrical conductivity, is required (107–108). These instruments are also used to evaluate toner powders for dry-copying machines. Diffraction-based instruments, employed in the cement and ceramic industries, permit rapid analysis, and some are used in on-line applications. Light-based instruments have to be calibrated with particles of known size and optical properties similar to the test sample because of the complexity of light scattering phenomena.

Scattering by any isotropic sphere of diameter d and refractive index m relative to that of the medium when illuminated by unpolarized light of intensity I is given, according to Mie theory of light scattering by

$$I_0 = \frac{\lambda^2 I}{8\,\pi L^2} i_1(\alpha, m, \theta) + i_2(\alpha, m, \theta) \tag{17}$$

The scattered intensity I_0 is measured at the angle θ from the forward direction of incident-beam propagation. The wavelength of the incident light is λ; L is the distance

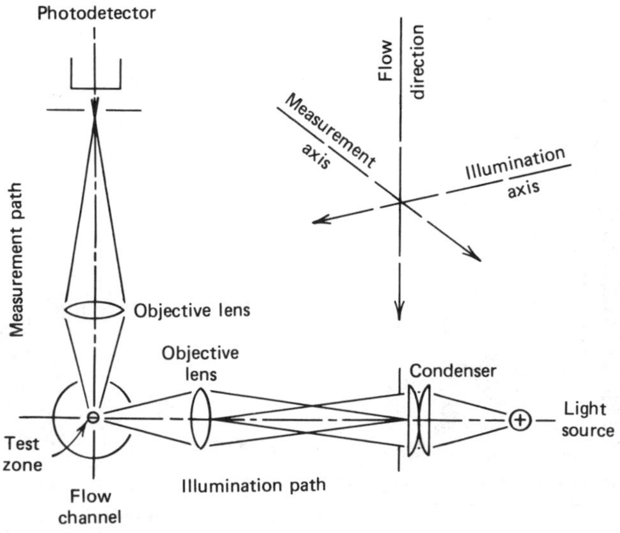

Figure 10. Typical light scattering system.

from the particle to the point of intensity measurement; and $\alpha = \pi/d\lambda$. The polarization intensity function i_1 is proportional to the scattered-light polarization component with the electric field perpendicular to the plane of measurement, whereas i_2 is proportional to the component with orthogonal polarization. Both functions have the form of an infinite series of half-order Bessel functions and Legendre polynomials.

Expressions for light scattering from particles that are not spherical have been worked out only for special cylinders, disks, and ellipsoids. Irregularly shaped particles such as quartz and coal fragments scatter with the same overall characteristics as spheres because of averaging over orientation. The fine structure exhibited by the scattering pattern of monodisperse spheres is, of course, not revealed in these cases. The effect of particle size on the scattering of light is shown in Figure 11.

True size data are only obtained when the complete scattering pattern of a particle is analyzed in detail. An instrument is used that employs electrostatic levitation to retain a particle in an incident laser beam for several seconds (111). During this time, a detector is rotated about the particle from nearly forward to back (0–180°) with respect to the incident radiation, while recording the angular variation of the scattered light intensity. This differential scattering pattern is then analyzed for particle size by computer in accordance with Mie theory. The technique can be speeded by employing an array of light guides to record the angular intensity variation (112).

A relatively large airborne particle passing rapidly through a gradually tapering tube that abruptly expands into a larger cavity generates an audible click that can be picked up by a microphone located in the tapered section (113–114). The acoustic response is only roughly proportional to particle size (115). However, an acoustic pressure wave that is characteristic of particle diameter between about 25 and 1500 μm is created when a thin orifice is employed through which particles are drawn by

Sphere smaller than 0.1 the light wavelength

Sphere about 0.25 the light wavelength

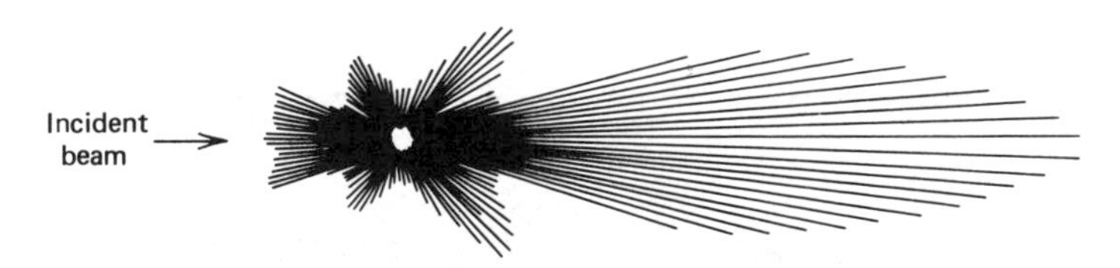

Sphere larger than the light wavelength

Figure 11. Angular intensity of scattered light.

a partial vacuum (116). Because of its greater density, such a particle lags behind the fluid adjacent to it as both are drawn toward the orifice. The velocity difference between particle and fluid attains a maximum as the particle reaches the plane of the orifice, which gives rise to the detectable acoustic pressure wave.

Other sensing-zone techniques include the momentary emission of visible light resulting when an airborne particle encounters an intense laser beam (see also Lasers). The time dependence of the light emission following exposure is related to particle size because of thermal diffusion within the particle and subsequent radiative cooling effects (117). When a particle, particularly a liquid droplet, encounters a heated wire such as in a hot-wire anemometer, a transient temperature decrease results in a size-related signal (114).

Fluid Dynamic. Until recently, techniques utilizing liquid- or gas-flow phenomena have been intended more for physical separation or classification than for size analysis. However, size-distribution information is obtained when fractions are collected and, perhaps, weighed. A noteworthy hydrodynamic unit incorporates a helical path in which radial stratification of particles occurs (118). There are also devices utilizing a series of small hydroclones (119–120). In another unit a cross-flow of air produces size separation (121), whereas in others air flow coupled with centrifugal force produces separation (122–123). Separation according to size is frequently not as sharp as might be desired except for relatively large particles.

Devices employing these techniques are found primarily in mineral dressing and deep-well operations, such as those used in oil exploration requiring drilling muds.

Discrete particle-size separation results from a transport phenomenon generally described as hydrodynamic chromatography (124–131), and a commercial instrument utilizing this principle is being developed (132). When, into a liquid flowing through a long capillary tube (length-to-diameter ratio >150,000), or through a column packed with uniform spheres, particles are injected much smaller in diameter than the capillary or packed spheres, they leave the tube or column in order of size (largest first and smallest last) and ahead of the liquid in which they were injected. This effect is illustrated in Figure 12 with a mixture consisting of particles of two sizes in a liquid of composition different from the primary liquid. A detector to resolve particle concentration at the exit as a function of egress time and computational capability provides size-distribution data.

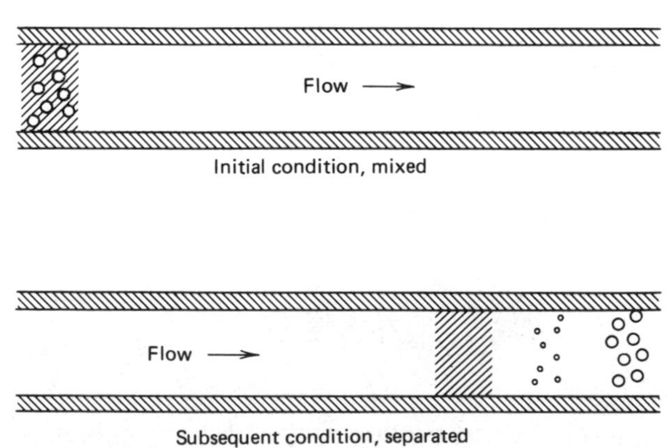

Figure 12. Hydrodynamic chromatography effect in a capillary tube.

Hydrodynamic chromatography is expected to have wide application for the determination of submicrometer particles such as in latices, carbon black, colloidal silica, paint and photographic pigments, lakes and dyes, food colors, natural and artificial blood, and metallic fumes.

Hydrodynamic separation theory is not entirely resolved although a probable predominant mechanism can be identified. Liquid, whether flowing laminarly through an empty tube or within the interstitial space among packed spheres, takes on a parabolic velocity profile in all flow channels with the maximum at the center and zero at the confining surface. A particle conveyed by liquid cannot experience a liquid velocity less than that prevailing one particle radius from the confining surface; hence particles must move with greater mean velocity than the liquid, and large particles must move on the average faster than small ones. Other influences include the ionic strength of the eluting liquid. This significantly alters the degree of size separation, possibly because of electrical double-layer effects associated with the particles and the capillary or packing-sphere surfaces.

Analysis of particles of 1–50 μm diameter appears feasible with capillaries, whereas packed columns extend the range from about 1 μm well into colloidal dimensions. Time from sample injection at the head of a capillary tube or a packed column to sample exit and report generation is likely to require 5–10 minutes, but, because overall sample spread is small, other injections can be made before the preceding analysis is completed. Thus particle-size distributions may possibly be obtained at a rate of one per minute.

Another broad class of size-separation techniques, termed field-flow fractionation (133–138), may also result in commercial instrumentation. Submicrometer particle separation is achieved here by applying a field of force at right angles to a liquid migration carrying the particles. Electrical, thermal, magnetic, gravitational, and centrifugal fields have been explored; centrifugal forces are the most promising. The flow channel in a field-flow fractionation centrifuge is circular and very narrow with respect to the axis of rotation. The field may be as great as 15,000 g. The liquid flowing through the channel establishes the characteristic parabolic velocity profile, whereas the external force field pushes the particles toward the slower-moving liquid region near the outer wall. A buildup of sample near the wall is resisted by normal diffusion in the opposite direction, resulting in the establishment of a particle zone wherein largest particles are nearest the wall and smallest farthest away. Field-flow fractionation appears to have the greatest promise of application with colloids and macromolecules.

Other Methods. Techniques to derive cut or average diameter rather than a distribution of sizes have not been discussed here because many of the preceding methods can yield this information. Among systems that can give only cut or average diameter is an elutriator-type device in which particles are introduced into an open vertical column through which an air stream passes upward (139–140). The smaller particles escape from the top, the larger discharge from the bottom. The mass-flow rates of both streams are measured at the exit. When the air flow is adjusted in such a way that the discharges are equal, the 50% cut size is defined; its diametric value is determined from drag laws and the elutriating air velocity. Equipment of this type is only suited for on-line control where the installation can be adjusted to process conditions.

In other devices, the average diameter of submicrometer particles is deduced from their random movement caused by Brownian motion (141–142). The frequency of this motion is inversely proportional to particle size, ie, the smaller the particle the faster its motion. Particle motion gives rise to a continual fluctuation of the light scattered at 90° from a low-power laser beam. This fluctuation is analyzed by mathematical autocorrelation techniques. The development of a result typically requires 2 minutes.

Ultrasonic energy attenuation provides particle-size information in a device that utilizes two pairs of receiving and detecting transducers (143). Whereas one pair operates at a frequency selected to accentuate viscous loss of input energy, the other operates at a frequency to emphasize scattering loss (see Ultrasonics). Viscous loss arises from the relative movement of particles and liquid. Large particles tend to lag the liquid movement and contribute more to the loss than small particles which tend to move more nearly in phase with the liquid. The scattering loss arises from the absorption of a certain amount of energy by every particle with subsequent reradiation. The defining mathematical interrelationships are complex, but a frequency can usually be found experimentally where ultrasonic energy attenuation is little influenced by particle size distribution and primarily determined by solids concentration. Changes in energy absorption are primarily related to changes in size distribution at another frequency. Analyzing the two signals of different frequency thus makes it possible to deduce both particle concentration and the percentage composition above or below a particular particle diameter. The system requires frequent calibration and adjustment since it is also sensitive to particle density. These techniques are applied mainly in mineral-dressing operations, particularly in gold-ore preparations.

BIBLIOGRAPHY

"Size Measurement of Particles" in *ECT* 1st ed., Vol. 12, pp. 472–497, by K. T. Whitby, University of Minnesota; "Size Measurement of Particles" in *ECT* 2nd ed., Vol. 18, pp. 310–324.

1. A. M. Gaudin, *Trans. Am. Inst. Min. Metall. Pet. Eng.* **73,** 253 (1926).
2. R. Schuhmann, Jr., *Min. Technol.* **4,** 1 (July 1940).
3. P. Rosin and E. Rammler, *J. Inst. Fuel* **7,** 29 (1933).
4. S. Nukiyama and Y. Tanasawa, *Trans. Soc. Mech. Eng. (Japan)* **5,** 1 (1939).
5. T. Hatch and S. P. Choate, *J. Franklin Inst.* **207,** 369 (1929).
6. R. A. Mugele and H. D. Evans, *Ind. Eng. Chem.* **43,** 1317 (1951).
7. H. C. Lewis, D. G. Edwards, M. J. Goglia, R. I. Rice, and L. W. Smith, *Ind. Eng. Chem.* **40,** 67 (1948).
8. W. Weibull, *J. Appl. Mech.* **18,** 293 (1951).
9. P. S. Roller, *J. Franklin Inst.* **223,** 609 (1937).
10. F. Kottler, *J. Franklin Inst.* **250,** 339 (1950).
11. *Ibid.*, **251,** 499 (1951).
12. F. Kottler, *J. Phys. Chem.* **56,** 442 (1952).
13. J. M. DallaValle, *Micromeritics, The Technology of Fine Particles*, Pitman, New York, 1948.
14. G. Herdan, *Small Particle Statistics*, Elsevier, Amsterdam, 1953.
15. C. Orr, Jr. and J. M. DallaValle, *Fine Particle Measurements*, Macmillan, New York, 1959.
16. C. Orr, Jr., *Particulate Technology*, Macmillan, New York, 1966.
17. A. D. Randolph and M. A. Larson, *Theory of Particulate Processes*, Academic Press, New York, 1971.
18. F. H. Steiger, *Chemtech* **1,** 225 (1971).
19. J. K. Beddow, *Particulate Science and Technology*, Chemical Publishing Co., Inc., New York, 1980.
20. B. B. Spencer and B. E. Lewis, *Powder Technol.* **27,** 219 (1980).
21. K. Leschonski, *Powder Technol.* **24,** 115 (1979).

22. J. Hidaka and S. Miwa, *Powder Technol.* **24,** 159 (1979).
23. K. Schönert, W. Schwenk, and K. Steier, *Aufbereit. Tech.* **15,** 368 (1979).
24. B. H. Kaye and N. I. Robb, *Powder Technol.* **24,** 125 (1979).
25. B. H. Kaye and M. R. Jackson, *Powder Technol.* **1,** 43 (1967).
26. R. W. Bartlett and T. H. Chin, *Trans. Am. Inst. Min. Metall. Pet. Eng.* **256,** 323 (1974).
27. C. Orr, D. K. Davis, and R. W. Camp, *Powder Technol.* **24,** 143 (1979).
28. K. T. Whitby, *Symposium on Particle Size Measurement*, ASTM Special Technical Publication No. 234, 1959, p. 3.
29. J. E. English, *Filtr. Sep.* **11,** 195 (1974).
30. H. O. Suhm, *Powder Technol.* **2,** 356 (1968–1969).
31. C. W. Ward, *Powder Technol.* **24,** 151 (1979).
32. B. J. Wahl and P. Larouche, *Am. Ceram. Soc. Bull.* **43,** 377 (1964).
33. J. D. Zwicker, *Am. Ceram. Soc. Bull.* **45,** 716 (1966).
34. H. B. Carroll and I. B. Akst, *Rev. Sci. Instrum.* **37,** 620 (1966).
35. P. J. Falivene, *Chem. Eng.* **88**(4), 87 (1981).
36. T. Allen, *Particle Size Measurement*, Chapman and Hall, London, England, 1975.
37. L. R. Feret, *Assoc. Int. l'Essai Mater. Group D, Zurich* **2,** (1931).
38. G. Martin, C. E. Blythe, and H. Tongue, *Trans. Ceram. Soc.* **23,** 61 (1924).
39. H. S. Patterson and W. Cawood, *Trans. Faraday Soc.* **32,** 1084 (1936).
40. G. L. Fairs, *Chem. Ind.* **62,** 374 (1943).
41. *Methods for the Determination of the Particle Size of Powders, Part IV, Optical Microscope Method*, British Standards 3406, 1961.
42. V. Timbrell, *Nature* **170,** 318 (1942).
43. J. Dyson, *J. Opt. Soc. Am.* **50,** 754 (1960).
44. F. Endtler and H. Gabauer, *Optik* **13,** 97 (1956).
45. P. Falcon-Uff and K. F. Leverington, *Proceedings of the Conference on Particle Size Analysis*, The Society for Analytical Chemistry, London, England, 1967, p. 45.
46. J. S. Glass, *Chem. Eng. Prog.* **68,** 58 (1972).
47. I. A. Cruttwell, *Microscope* **22,** 27 (1974).
48. V. N. Ryland and R. E. Stevens, *Ind. Res.* **19,** 91 (1977).
49. H. P. Hougardy, *Microscope* **24,** 7 (1976).
50. C. Orr in I. M. Kolthoff, P. J. Elving, and F. H. Stross, eds., *Treatise on Analytical Chemistry*, Part III, Vol. 4, Wiley-Interscience, New York, 1977, p. 273.
51. W. K. Witherow, *Opt. Eng.* **18,** 249 (1979).
52. H. Schwarz and H. E. Exner, *Powder Technol.* **27,** 207 (1980).
53. A. M. Glauert, *Practical Methods in Electron Microscopy*, North-Holland, Amsterdam, 1977.
54. D. W. Montgomery, *Rubber Age* **45,** 759 (1964).
55. R. R. Irani and C. F. Callis, *Particle Size*, John Wiley & Sons, Inc., New York, 1963.
56. W. D. Ross, *Filtr. Sep.* **10,** 587 (1973).
57. J. A. Davidson and H. S. Haller, *J. Colloid Interface Sci.* **47,** 459 (1974).
58. F. S. Eadie and R. E. Payne, *Iron Age* **174,** 99 (1954).
59. C. E. Lapple, *Fluid and Particle Mechanics*, University of Delaware, Newark, Del., 1956, p. 286.
60. W. P. Hendrix and C. Orr, Jr. in M. J. Grover and J. L. Wyatt-Sargent, eds., *Particle Size Analysis 1970*, The Society for Analytical Chemistry, London, England, 1972, p. 133.
61. D. W. Moore and C. Orr, Jr., *Powder Technol.* **8,** 13 (1973).
62. S. Berg, *The Analyst* **97,** 585 (1972).
63. W. Z. Bostock, *J. Sci. Instrum.* **29,** 209 (1952).
64. G. J. Rabaten and R. H. Gale, *Anal. Chem.* **28,** 1314 (1956).
65. E. S. Palik, *Ceram. Age* **78**(8), 43 (1962).
66. *The Determination of Particle Size. I. A Critical Review of Sedimentation Methods*, The Society for Analytical Chemistry, London, England, 1968.
67. S. Mori, T. Hara, and K. Aso, *paper presented at the Fourth Joint Meeting of the Mining and Metallurgical Institute of Japan (MMIJ) and the American Institute of Mining, Metallurgical, and Petroleum Engineers (AIME)*, Tokyo, Japan, Nov. 1980.
68. E. E. Bauer, *Eng. News-Rec.* **118,** 662 (1937).
69. R. F. Conley, *Powder Technol.* **3,** 102 (1969).
70. B. H. Kaye, *Paint Oil Colour J.* **147,** 735 (1965).
71. J. P. Olivier, G. K. Hickin, and C. Orr, Jr., *Powder Technol.* **4,** 257 (1970–1971).

72. P. Sennett, J. P. Olivier, and G. K. Hickin, *Tappi* **57,** 92 (1974).
73. R. K. Herrmann, *Keram. Z.* **31,** 275 (1979).
74. C. C. McMahon, *Cer. Bull.* **49,** 794 (1970).
75. W. A. Buerkel in A. R. Poster, ed., *Handbook of Metal Powders*, Reinhold, New York, 1966, Chapt. 3, p. 20.
76. K. C. Yang and R. Hogg, *Anal. Chem.* **51,** 758 (1979).
77. T. Allen, *Powder Technol.* **2,** 133 (1968–1969).
78. *Ibid.*, p. 141.
79. T. Allen in ref. 60, p. 167.
80. K. T. Whitby, *Heat. Piping Air Cond.* **61,** 33 (1955).
81. *Ibid.*, p. 449.
82. B. Scarlett, M. Rippon, and P. J. Lloyd, *Proceedings of the Conference on Particle Size Analysis*, The Society for Analytical Chemistry, London, England, 1967, p. 242.
83. E. Atherton, A. C. Cooper, and M. R. Fox, *J. Soc. Dyers Colour.* **80,** 521 (1964).
84. M. J. Groves, B. H. Kaye, and B. Scarlett, *Br. Chem. Eng.* **9,** 742 (1964).
85. B. H. Kaye and M. R. Jackson, *Powder Technol.* **1,** 81 (1967).
86. M. J. Groves, H. S. Yalabik, and J. A. Tempel, *Powder Technol.* **11,** 245 (1975).
87. W. J. Wnek, *Powder Technol.* **19,** 129 (1978).
88. M. J. Groves and H. S. Yalabik, *J. Pharm. Pharmacol.* **26,** 77P (1974).
89. P. Krischker, *TIZ-Fachber.* **103,** 327 (1979).
90. V. O. Lauer, *Chem. Tech.* **9,** 303 (1980).
91. A. Bürkholz, *Staub-Reinhalt. Luft.* **30,** 1 (1970).
92. J. F. Pisani and G. H. Thomson, *J. Phys. E* **4,** 359 (1971).
93. W. H. Coulter, *Proc. Natl. Electron. Conf.* **12,** 1034 (1956).
94. H. E. Kubitschek, *Nature* **182,** 234 (1958).
95. H. E. Kubitschek, *J. Res. Natl. Bur. Stand. Sect. A.* **13,** 128 (1960).
96. J. G. Harfield and W. M. Wood, in ref. 60, p. 293.
97. J. G. Harfield, B. Miller, R. W. Lines, and T. Godin in M. J. Groves, ed., *Particle Size Analysis*, Heydon, London, England, 1978, p. 378.
98. R. W. DeBlois and R. K. A. Wesley, *Report No. 76CRD192*, General Electric Co., New York, Nov. 1976.
99. G. Fratucelli and G. Giovanelli, *Rev. Sci. Instrum.* **42,** 663 (1971).
100. A. C. Thomas, A. N. Bird, R. H. Collins, and P. C. Rice, *Instrum. Soc. Am.* **8,** 52 (1961).
101. W. Kaye, *J. Colloid Interface Sci.* **44,** 384 (1973).
102. K. T. Whitby and R. A. Vomela, *Environ. Sci. Technol.* **1,** 801 (1967).
103. R. Davies and H. E. Turner in ref. 97, p. 238.
104. G. C. West in ref. 97, p. 347.
105. J. Cornillault, *Appl. Opt.* **11,** 265 (1972).
106. J. Cornillault and P. Evrard, *Cem. Technol.* **6,** 178 (1975).
107. A. L. Wertheimer and W. L. Wilcock, *Appl. Opt.* **15,** 1616 (1976).
108. H. N. Frock and R. L. Walton, *Cer. Bull.* **59,** 650 (1980).
109. H. Breuer, *Staub-Reinhalt. Luft.* **36,** 6 (1976).
110. R. Weinmann, *Powder Metall. Int.* **10,** 94 (1978).
111. P. J. Wyatt and D. T. Phillips, *J. Colloid Interface Sci.* **39,** 3 (1972).
112. I. K. Ludlow and P. H. Kaye, *J. Colloid Interface Sci.* **69,** 571 (1979).
113. G. Langer, *J. Colloid Sci.* **20,** 602 (1965).
114. R. Davies, *Am. Lab.* **6**(2), 47 (1974).
115. R. F. Karuhn, *paper presented at the First International Conference on Particle Technology*, Chicago, Ill., Aug. 1973, p. 203.
116. G. M. Bragg and B. C. Morrow, *J. Phys. E* **14,** 26 (1981).
117. R. W. Weeks and W. W. Duley, *J. Appl. Phys.* **45,** 4661 (1974).
118. A. B. Holland-Batt, *Powder Technol.* **11,** 11 (1975).
119. D. F. Kelsall and J. C. H. McAdam, *Trans. Inst. Chem. Eng.* **41,** 84 (1963).
120. C. B. Daellenbach and W. M. Mahan, *U.S. Bur. Mines Rep. Invest.*, No. 7879 (1974).
121. K. L. Metzger and K. Leschonski, *Chem. Ing. Tech.* **48,** 565 (1976).
122. W. Stöber and H. Flachbart, *Environ. Sci. Technol.* **3,** 1280 (1969).
123. Z. Tanaka, H. Takai, N. Okada, and K. Iinoya, *J. Chem. Eng. (Japan)* **4,** 167 (1971).
124. H. Small, *J. Colloid Interface Sci.* **48,** 147 (1974).

125. H. Small, F. L. Saunders, and J. Solc, *Adv. Colloid Interface Sci.* **6,** 237 (1976).
126. A. M. McHugh, C. Silebi, G. W. Poehlein, and J. W. Vanderhoff, *Colloid Interface Sci.* **IV,** 549 (1976).
127. H. Small, *Chemtech* **7,** 196 (1977).
128. M. E. Mullins and C. Orr, *Int. J. Multiphase Flow* **5,** 79 (1979).
129. C. Orr, Jr., in ref. 97, p. 77.
130. R. J. Noel, K. M. Gooding, F. E. Regnier, D. M. Ball, C. Orr, and M. E. Mullins, *J. Chromatogr.* **166,** 373 (1978).
131. C. A. Silebi and A. J. McHugh, *J. Appl. Polym. Sci.* **23,** 1699 (1979).
132. J. P. Olivier and C. E. Brown, *Pittsburgh Conference on Analytical Chemistry and Applied Spectroscopy*, Abstract No. 372, Atlantic City, N.J., 1980.
133. J. C. Giddings, *J. Chem. Educ.* **50,** 667 (1973).
134. J. C. Giddings, *Sep. Sci.* **8,** 567 (1973).
135. F. J. F. Yang, M. N. Myers, and J. C. Giddings, *Anal. Chem.* **46,** 1924 (1974).
136. J. C. Giddings, L. K. Smith, and M. N. Myers, *Anal. Chem.* **48,** 1587 (1976).
137. J. C. Giddings, K. D. Caldwell, J. F. Moellmer, T. H. Dickinson, M. N. Myers, and M. Martin, *Anal. Chem.* **51,** 30 (1979).
138. J. J. Kirkland, W. W. Yau, W. A. Doerner, and J. W. Grant, *Anal. Chem.* **52,** 1944 (1980).
139. T. Tanaka, *Miner. Process.* **10,** 20 (March 1969).
140. *Ibid.*, 20 (July 1969).
141. R. W. Lines, *Polym. Paint Colour J.* **171,** 214 (1981).
142. M. J. Groves, *J. Dispersion Sci. Technol.* **1,** 97 (1980).
143. A. R. Atkins and A. L. Hinde, *ISA Trans.* **14,** 318 (1975).

General References

G. Herden, *Small Particle Statistics*, 2nd ed., Academic Press, Inc., New York, 1960.
P. Somasundaran, ed., *Fine Particles Processing*, American Institute of Mining, Metallurgical, and Petroleum Engineers, Inc., New York, 2 vols., 1980.
B. Kaye, *Direct Characterization of Fine Particles*, John Wiley & Sons, Inc., New York, 1981.
T. Allen, *Particle Size Measurement*, 3rd ed., Chapman and Hall, London, UK, 1981.
R. D. Cadle, *The Measurement of Airborne Particles*, Wiley-Interscience, New York, 1975.
W. C. McCrone, *The Particle Atlas*, Ann Arbor Science Publishers, Ann Arbor, Mich., 6 vols., 1980.
W. C. McCrone, *The Asbestos Particle Atlas*, Ann Arbor Science Publishers, Ann Arbor, Mich., 1980.
W. C. McCrone, *Polarized Light Microscopy*, Ann Arbor Science Publishers, Ann Arbor, Mich., 1979.

CLYDE ORR
Micrometrics Instrument Corp.

SIZE REDUCTION

Size reduction, or comminution, is the process whereby particulate materials are mechanically reduced in size in order to increase their value. Size reduction to produce a material suitable for use is an important operation in many process industries, eg, cement, grain, stone, fertilizer, and coal. In others, eg, mineral ores, size reduction is an intermediate step in the separation of valuable constituents from waste materials (see Extractive metallurgy; Flotation; Gravity concentration).

Comminution processes are energy intensive. The NRC Committee on Comminution and Energy Consumption estimates that the United States in 1978 utilized ca 29×10^9 kW·h electrical energy (or 2% of total production) to comminute ca 10^9 metric tons of material (1). Minerals accounted for 53% of this energy consumption; cement, 26%; coal, 14.4%; and food grains, 6.6%.

Size-reduction devices are inherently inefficient and are costly to operate (frequently ≥25% of total raw-material processing costs). It has been estimated that <1% of the energy input to tumbling mills, for example, manifests itself as new surface. Equipment for a particular application must be chosen with regard to energy efficiency. The choice is dictated largely by the nature of the material to be comminuted and by the size desired. In the mineral industries, coarse size reduction (from large sizes to 1 cm) is normally achieved by explosive shattering followed by crushing, where individual pieces are broken by compression or impact. Fine size reduction (from 1 cm to 10 μm) is typically achieved by grinding in tumbling mills, where breakage occurs by probabilistic impact in a loosely tumbling mass of grinding media. Ultrafine size reduction (<10 μm) is achieved mainly by pulverizing, where size reduction results from abrasion and attrition.

A detailed review of size reduction is given in ref. 2, whereas ref. 3 expands on the practical aspects and ref. 4 on theoretical details.

Theory

Solids may be homogeneous or heterogeneous, crystalline or amorphous, hard or soft, brittle or plastic. However, most of the commercially important materials are heterogeneous, flawed but essentially brittle solids that fail catastrophically when stressed to a sufficient degree.

The resistance of these materials to breakage can be measured by stressing a sample under closely controlled conditions of compression, tension, or shear. Measurements of strain for a range of stresses are plotted in a diagram of the general form shown in Figure 1. Such diagrams are typically characterized by an initial region AB, where strain is directly proportional to stress; an intermediate region BC, where the stress–strain relationship is no longer constant but still increases; a second intermediate region CD, where strain continues even under apparently decreasing stress; and, finally, region DE, where the solid changes shape with no change in stress to break at E.

According to Hooke's law of elasticity, strain is directly proportional to the stress that caused it: ie, stress = constant × strain. Point B is the elastic limit for the material, and the slope of the straight line AB is Young's modulus E. Young's modulus is usually, but not always, the same for compression and tension, and is frequently used to characterize a material. Point C is the yield point, and the corresponding stress is

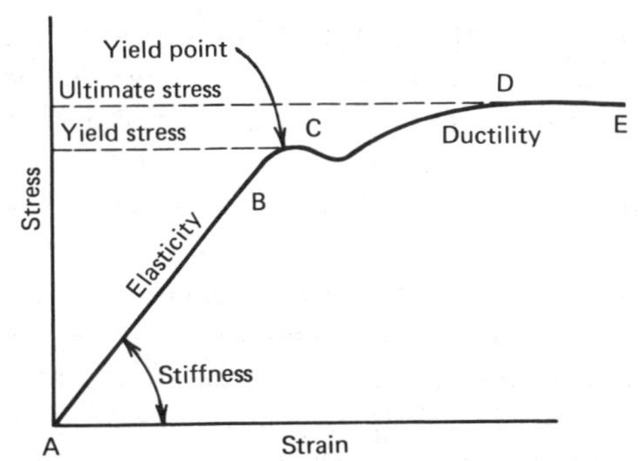

Figure 1. Stress–strain diagram for a solid (4).

known as the yield stress. The ultimate stress corresponds to point E. The region CE gives a measure of the ductility of the sample. The higher the stress at the elastic limit, the stronger the material. The less the difference between the strain at the elastic limit and that at the ultimate yield point, the more brittle is the material. Definitions of brittleness vary. Here a brittle material is defined as one that exhibits very little ductility.

Comminution machines work through compression or impact. However, when bodies are subjected to compression, tensile forces are set up at right angles to the imposed compressive stress. The relationship between the longitudinal strain and the concomitant lateral strain is a constant for a given body:

$$\frac{\text{lateral strain}}{\text{longitudinal strain}} = \text{constant } (G)$$

where G is Poisson's ratio, which varies from nearly zero to a max of 0.5. Stress is proportional to strain within the elastic region; therefore, a compressive stress of magnitude P_1 causes a tensile stress of magnitude GP_1 at right angles to it. Orders of precedence for the type of fracture likely to occur in a homogeneous material can be developed on the relative magnitudes of its ultimate tensile, compressive, and shear

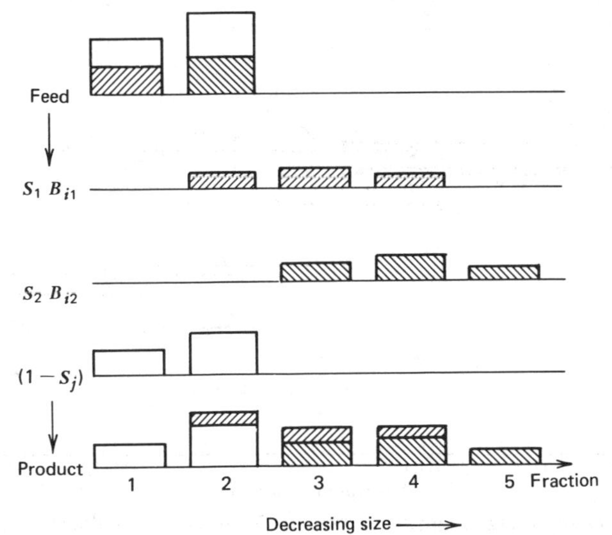

Figure 2. Schematic representation of the comminution process.

strengths. Although very few of the materials subjected to comminution are homogeneous, most naturally shaped particles break under tension or shear. Ultimate shear and tensile strengths are usually very similar, and it is not particularly important which predominates from an energy viewpoint.

Various investigators have attempted to account for the dissipation of this energy as the material fractures (5–6). It can take the form of kinetic energy, sound, potential strain, heat, light, and the energy of the new surfaces generated by fracture. Of these various forms, only the increase in surface energy bears a direct relationship to the ultimate objective of comminution, namely, the reduction of particle size. A number of investigators have concluded that the new surface energy constitutes a very small proportion of the actual energy input to the typical size-reduction process (<1% for grinding mills); therefore, it is difficult to relate specific surface measurements to energy requirements for comminution. Nevertheless, good correlations are frequently observed (7).

The only way to determine exactly the degree of size reduction that can be achieved in a particular comminution device is to carry out full-scale tests on a representative sample of the material. Nevertheless, considerable success has been achieved in empirically predicting the performance of full-scale equipment from laboratory and pilot-scale tests. More recent methods of analysis show promise of developing into true representations of comminution processes. However, these have not as yet supplanted the empirical laws, principally because of an insufficient data base.

Much of the earlier literature on comminution was devoted to fruitless attempts to relate new surface to the energy expended to produce that surface. The first of these

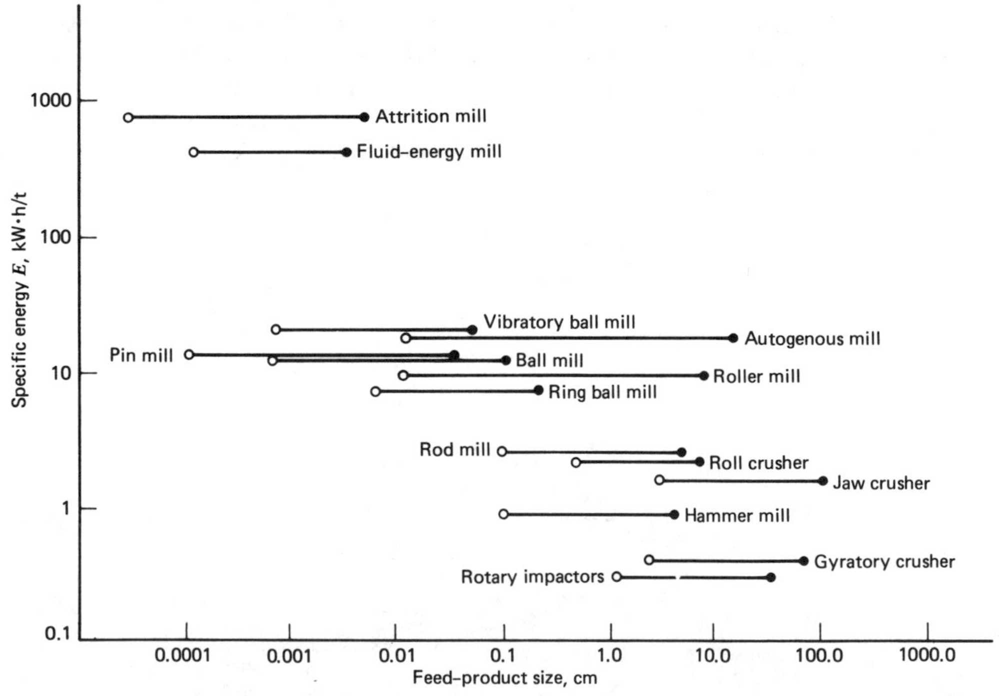

Figure 3. Reported average energy requirements for several size-reduction devices (1). ●, typical feed size; ○, typical product size.

energy–size relationships was developed by von Rittinger who, in 1867, postulated that the energy consumed in size reduction, E, is proportional to the new surface area produced (8). The relationship is typically expressed as:

$$E = K_1 \left(\frac{1}{d_p} - \frac{1}{d_f} \right) \tag{1}$$

where K_1 is a constant and d_p and d_f are the product and feed diameters, respectively. Countering this theory, Kick postulated, in 1885, that the energy required to produce "analogous changes in configuration in geometrically similar bodies of equal technological state varied as the volume or weight of these bodies" (9). This is expressed as:

$$E = K_2 \log \left(\frac{d_f}{d_p} \right) \tag{2}$$

where K_2 is a constant.

Although various authors have demonstrated the limited applicability of these two relationships, both theories present difficulties in any real situation and neither is utilized to any extent today. They give rise to widely different estimates of energy requirements but it was not until 1952 that Bond introduced his "third theory of comminution," based on a detailed analysis of a large number of operating and laboratory data (10). This relationship is expressed as:

$$W = W_i \left(\frac{10}{\sqrt{P}} - \frac{10}{\sqrt{F}} \right) \tag{3}$$

where W is the required work input, P and F are, respectively, the 80% passing sizes of the product and feed particles, and W_i is the "work index," representing the power required to comminute an infinitely large particle to 80% passing 100 μm (in kW·h/t). This relationship is widely used for the design of grinding circuits because a very large body of information on W_i values exists. In addition, experimental laboratory methods have been developed to determine work indexes for a particular material and comminution device. This parameter can then be used, along with equation 3, to size ball

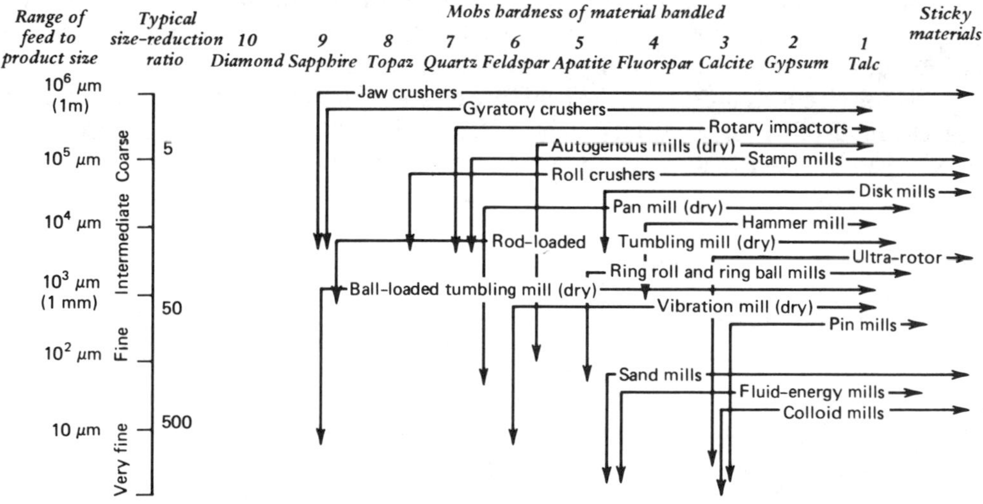

Figure 4. Typical reduction ratios (4). Courtesy of *Crushing and Grinding.*

and rod mills (11). It can also be used for crusher specification (12), although these machines are more typically selected from manufacturers' catalogues (13).

Comminution Mechanics

The principal limitation of the energy relationships given in the previous section is that they fail to account for the mechanics of the comminution device. Any such device must provide a means for moving material into a stress-application zone, applying stress to the material and then permitting the material to exit. Thus, a detailed study of a given comminution device must include the breakage process and machine dynamics.

The Breakage Process. If the spectrum of particle sizes existing in a particular comminution device is arbitrarily divided into a finite number of narrow-size intervals, the comminution process can be treated as a process of transforming particles from a given size interval into a mass distribution spread throughout a set of finer size intervals. The two fundamental parameters of the process can then be defined as: S_j, the specific rate, with respect to time or energy input, at which particles in size class j are broken; and b_{ij}, the mass fraction of the fragments produced by breaking size j that fall into size class i. It is often convenient to express b_{ij} in its cumulative form B_{ij}, which is the mass fraction produced by breaking size j into particles that are smaller than size i.

The set of S values thus describes the kinetics of the process and takes account of the rate at which particles are subjected to fruitful stress, ie, the frequency of stressing and the probability that this stress leads to fracture.

The b or B values characterize the products of the breakage event; ie, the fundamental product size distribution produced by the single-particle fracture processes described above. Figure 2 is a schematic representation of the comminution process illustrating how a feed material of size classes 1 and 2 is broken to give a product of size classes 1–5.

The population-balance models developed using these concepts generally utilize a first-order hypothesis (in a manner analogous to that of a first-order chemical reaction), which states that the rate of breakage of any size is proportional to the mass of that size present in the comminution zone of the particular device. There is no self-evident reason why this should be so, but the first-order assumption is valid to a first approximation for a wide variety of practical systems.

If the size distribution of the contents of the comminution device is given by M_i (ie, the mass of all particles in the device that fall into size class i), then the first-order model can be described by the following set of equations:

Size interval 1

$$\frac{dM_1}{dt} = -S_1M_1$$

Size intervals i = 2 to n − 1

$$\frac{dM_i}{dt} = \sum_{j=1}^{i-1} S_jb_{ij}M_j - S_iM_i$$

Size interval n

$$\frac{dM_n}{dt} = \sum_{j=1}^{n-1} S_jb_{ij}M_j$$

Table 1. United States and Canadian Comminution Equipment Manufacturers

Manufacturer	Equipment type[a]
Aerofall Mills, Ltd., Mississauga, Ontario, Canada	9, 10, 11, 12, 15
Aggregates Equipment, Inc., Leola, Pa.	1, 2, 4, 5, 6, 7
Air Products & Chemicals, Inc., Allentown, Pa.	9, 15
Allied Steel & Tractor Products, Solon, Ohio	5
Allis-Chalmers Corporation, Milwaukee, Wis.	1, 2, 4, 5, 6, 9, 10, 11, 12, 13, 14
Alpine American Corporation, Natick, Mass.	3, 4, 5, 9, 15
American Pulverizer Company, St. Louis, Mo.	4, 5, 7, 8, 15
Anixter Brothers, Inc., Skokie, Ill.	1, 2, 3, 4, 5, 6, 7, 8, 15
Auto-Weigh, Inc., Oak Brook, Ill.	6, 7
Babcock and Wilcox Company, New York, N.Y.	15
Barber-Greene Company, Aurora, Ill.	1, 2, 6, 7, 9, 10, 11, 12, 15
Bepex Corporation, Rosemont, Ill.	15
BICO, Inc., Burbank, Calif.	6, 15
Birdsboro Corporation, Birdsboro, Pa.	6
Clearfield Machine Company, Clearfield, Pa.	11
CMI Corporation, Oklahoma City, Okla.	15
Columbia Steel Casting Company, Inc., Portland, Ore.	1, 2, 5, 6, 7
Combustion Engineering, Inc., Chicago, Ill.	3, 4, 5, 7, 9, 11, 13, 15
Crossley-Economy, East Palestine, Ohio	9, 15
Dominion Engineering Works, Ltd., Montreal, Quebec, Canada	9, 11
Donaldson Company, Majac Div., Minneapolis, Minn.	4, 11, 15
Dowty Corporation, Zelienople, Pa.	1, 2, 3, 4, 5, 6, 7, 8
Dresser Industries, Inc., Houston, Texas	4, 5, 7, 15
Eagle Crusher Company, Galion, Ohio	2, 4, 5, 6, 7, 15
Entoleter Company, New Haven, Conn.	15
Exolon Company, Tonawanda, N.Y.	7
Ferro-Tech, Inc., Wyandotte, Mich.	5
Frog, Switch & Mfg. Co., Carlisle, Pa.	4
Fuller Company (GATX), Bethlehem, Pa.	2, 6, 7, 9, 10, 12, 13
Gruendler Crusher & Pulverizer Co., St. Louis, Mo.	4, 5, 6, 7, 8, 9, 15
Halbach & Braun Industries, Washington, Pa.	4, 5, 7
Hammermills, Inc., Cedar Rapids, Iowa	1, 4, 5, 6, 7, 15
Hazemag USA, Inc., Pittsburgh, Pa.	4, 5, 7, 15
Hazen Research International, Inc., Golden, Colo.	9, 10
Hewitt Robins Div., Litton Systems, Inc., Columbia, S.C.	1, 2, 4, 5, 6, 7, 15
Humboldt Wedag Div., Deutz Corp., Atlanta, Ga.	4, 5, 6, 7, 8, 9, 10, 15
Iowa Mfg. Co., Cedar Rapids, Iowa	1, 2, 4, 5, 6, 7, 15
Joy Mfg. Co., Pittsburgh, Pa.	4, 5, 6, 7, 9, 10
Koppers Company, Inc., York, Pa.	4, 5, 6, 7, 8, 9, 10, 11, 12, 15
Kue-Ken Div., Process Technology Corp., Oakland, Calif.	1, 6
Lippmann-Milwaukee, Inc., Milwaukee, Wis.	4, 5, 6, 7
Logan Corporation, Huntington, W.Va.	4, 5, 6
Long-Airdox Company, Oak Hill, W.Va.	4
Machinery Center, Inc., Salt Lake City, Utah	1, 5, 6, 7, 9, 10
Manufacturers Equipment Company, Middletown, Ohio	7
McLanahan Corporation, Hollidaysburg, Pa.	7
McNally Pittsburg Mfg. Co., Pittsburg, Kan.	1, 2, 4, 5, 6, 7, 9, 10, 13
Minerec Corporation, Baltimore, Md.	11
Morgardshammar, Inc., Olmsted, Ohio	1, 2, 6, 9, 11
Morrison Knudsen, Boise, Idaho	15
Morse Bros. Machinery Company, Denver, Colo.	6, 15
Nelmaco, North Vancouver, British Columbia, Canada	1, 2, 4, 5, 6, 7, 9, 10, 12
Norton Chemical Process Products, Cranston, R.I.	9

Table 1 (*continued*)

Manufacturer	Equipment type[a]
Orenstein & Koppel Canada, Ltd., Dundas, Ontario, Canada	4, 5
Owens Mfg., Inc., Bristol, Va.	4
Pennsylvania Crusher Corporation, West Chester, Pa.	4, 5, 6, 7, 8
C. Peters, Inc., Dallas, Texas	13
PHB Material Handling Corporation, Hackensack, N.J.	1, 5
Polysius Corporation, Atlanta, Ga.	9, 10, 11, 12, 13
Portec, Inc., Pioneer Div., Minneapolis, Minn.	1, 4, 5, 6, 7, 15
Preiser Scientific, Inc., St. Albans, W.Va.	5, 6, 15
Pulva Corporation, Perth Amboy, N.J.	5, 7, 15
Pulverizing Machinery Company, Summit, N.J.	4, 5, 6, 11, 15
Rexnord, Inc., Milwaukee, Wisc.	1, 2, 3, 4, 5, 6, 7, 9, 10, 11, 12, 15
S&S Corporation, Cedar Bluff, Va.	4
A. J. Sackett & Sons Co., Baltimore, Md.	4, 5, 15
Schroeder Brothers Corporation, McKees Rocks, Pa.	7
Sepor, Inc., Wilmington, Calif.	1, 2, 3, 4, 5, 6, 7, 9, 10, 15
Simplicity Engineering Company, Durand, Mich.	4
F. L. Smidth & Co., New York, N.Y.	4, 5, 9, 10, 11, 15
Stedman Foundry & Machine Co., Aurora, Ind.	3, 4, 5, 7, 11, 15
Sturtevant Mill Company, Boston, Mass.	3, 4, 5, 6, 7, 8, 15
SWECO, Inc., Los Angeles, Calif.	9
Tema, Inc., Cincinnati, Ohio	9, 15
Toronto Coppersmithing International, Ltd., Scarborough, Ontario, Canada	9
U.S. Steel Corporation, Pittsburgh, Pa.	11
Universal Engineering Corporation, Cedar Rapids, Iowa	4, 5, 6, 7, 15
Universal Road Machinery Company, Kingston, N.Y.	6
Vibra Screw, Inc., N.J.	4, 5, 15
Voest-Alpine International Corporation, New York, N.Y.	1, 2, 3, 4, 5, 6, 7, 8
Vortec Products Company, Compton, Calif.	4, 15
Vulcan Iron Works, Inc., Wilkes-Barre, Pa.	9
Weserhuette, Inc., St. Louis, Mo.	2, 4, 5, 6
Western States Machinery Company, Grand Junction, Colo.	9
Williams Patent Crusher & Pulverizer Co., St. Louis, Mo.	4, 5, 7, 8, 10, 11, 15
Young Industries, Inc., Muncy, Pa.	5

[a] 1, cone; 2, gyratory; 3, disk; 4, impact; 5, hammer; 6, jaw; 7, roll; 8, ring-roll; 9, ball; 10, rod; 11, autogenous; 12, pebble; 13, roller; 14, vibratory; 15, pulverizers.

This set of equations gives the general form of the batch comminution model. Analytical solutions were first given in 1965 (14). The numerous forms of the basic equation of the breakage process given in the literature are collected and compared in ref. 15.

A main advantage of the kinetic- or population-balance approach to modeling the breakage processes is the relative ease with which these equations can be incorporated into overall models of comminution devices, as has been done extensively for ball mills. The simple first-order model is well suited to this application since the energy input to these systems is relatively constant and interaction effects appear to be minimal.

This approach has also been used, but with less success in the analysis of rod mills,

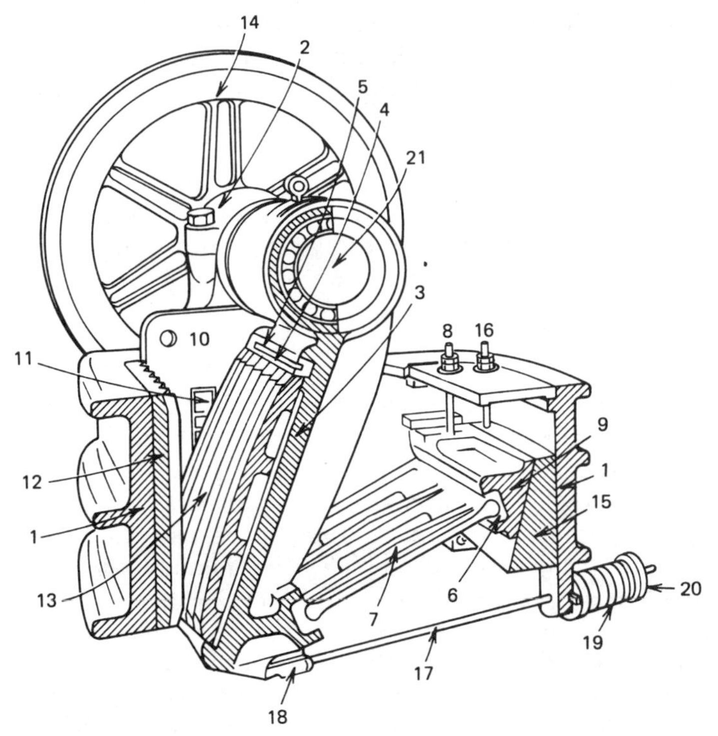

Figure 5. Single-toggle jaw crusher: 1, main frame; 2, main journal caps; 3, 4, and 5, swingstock, wedge, and bolts; 6, 7, 8, and 9, toggle cushion, toggle, bolts, and blocks; 10 and 11, cheek plates and bolts; 12, fixed jaw face; 13, swing jaw face; 14, flywheel; 15 and 16, adjusting wedge and bolts; 17, 18, 19, and 20, drawback rod, bolt, spring, and coverplate; 21, eccentric shaft. Courtesy of N.D. Engineering, Ltd.

autogenous and semiautogenous mills, cone crushers, hammer mills, and roller mills. Detailed information on S and B functions and how these functions are affected by changes in operating conditions is given in refs. 15–17.

So far these models have been of limited use in the design and operation of industrial comminution systems because of an inadequate data base and a lack of good scale-up laws. However, because of significant progress, it seems likely that this approach will supplant traditional empirical approaches within the decade, especially for devices other than ball mills where the empirical approach has not been very successful.

These models also offer considerable promise for circuit simulation and the optimal design of comminution circuits. Suitable simulation programs exist, but tend to be restricted in their utility because of inadequate parametric information on the performance of the individual units making up the circuit (18).

Machine Dynamics. It cannot be overemphasized that a practical commercial comminution device must perform two functions: material breakage and material transport. In most situations the device must be capable of handling high tonnage throughputs in order to be priced competitively, and must operate efficiently to minimize operating costs. The market success of a given device is largely dependent on its ability to perform these functions satisfactorily.

The capacity of a given machine may be controlled by the breakage process (as typically happens in crushers) or by the transport of material through the device.

Various transport mechanisms exist ranging from simple gravity flow, such as occurs in jaw, gyratory, and roll crushers, to the complex combination of gravity, mechanical, and fluid flow that exists in wet tumbling mills and air-swept tumbling mills. Generally, the relative energy consumption for material transport increases as the complexity of this function increases.

The relative efficiency of comminution equipment tends to decrease as the size of material decreases because of the difficulties of effectively applying stress to fine materials. Operating ranges and specific energy requirements for a variety of common comminution devices are given in Figure 3.

Equipment

The choice of comminution equipment type and the circuit in which it is to operate depend on the task to be performed. A clear and concise classification of comminuting equipment is not possible because all devices have developed over time with compromises among principle, cost, and attempts to maximize a particular effect. In general, however, equipment can be catagorized on the basis of primary comminuting action (3–4), namely, compression or nipping, blow or impact, tumbling or projection, cutting or shredding, and attrition.

Compression or nipping devices include jaw, gyratory, and roll crushers, disk pan, and roller mills. In each of these, particles are broken by compression (low speed impact) between two surfaces. These devices are further subdivided into those in which the gap between the crushing surfaces is controlled (eg, the jaw crusher) and the force is allowed to vary, and those in which a fixed force is used (eg, the roller mill) and the

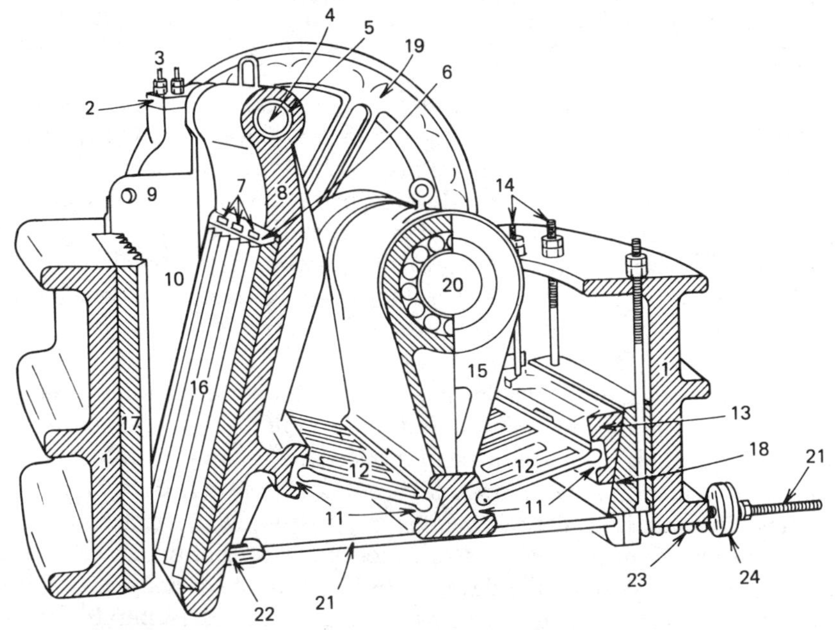

Figure 6. Double-toggle jaw crusher: 1, main frame; 2, 3, 4, 5, 6, and 7, cap, cap bolts, shaft, bushes, wedge, and wedge bolts for 8, swingstock; 9 and 10, cheek plates and bolts; 11, 12, 13, and 14, toggle cushion, toggles, block, and bolts; 15, pitman; 16, swing jaw face; 17, fixed jaw face; 18, adjusting wedge; 19, flywheel; 20, eccentric shaft; 21, 22, 23, and 24, drawback rod, bolts, spring, and coverplate. Courtesy of N.D. Engineering, Ltd.

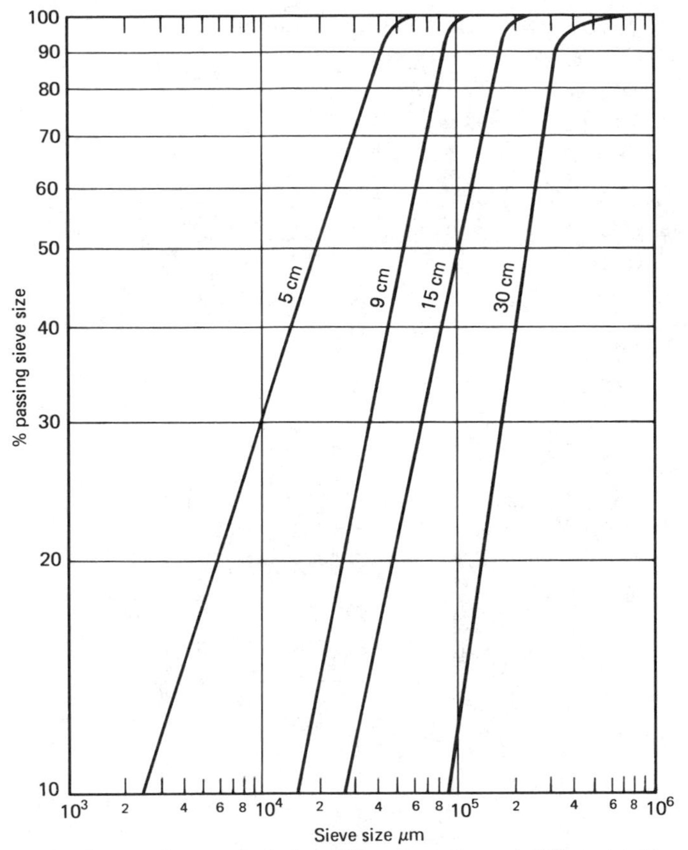

Figure 7. Jaw crusher product size; different crushers at different settings (4).

gap between the crushing surfaces is permitted to vary. The former find their greatest use in coarse particle (>2 cm) comminution; the latter more typically are used to produce fine materials, often with very high ratios of reduction.

Impact mills operate by subjecting particles to sudden high stress through impact. In hammer mills the particles are impacted with swinging hammers. The kinetic energy generated by each impact may also give rise to breakage of particles against the walls of the mill. Particles are broken in vibratory mills by high speed compression between two surfaces (balls).

Tumbling mills are the most important comminution devices, both in terms of installed capacity and overall energy consumption. These mills are a special subclass of impact devices where particles are broken by the action of a tumbling mass of loose grinding media. The media may be steel or ceramic balls or rods (as in ball and rod mills), or they may be large pieces of the material being comminuted (as in autogenous mills). The motion of the charge in tumbling mills is extremely complex, and the mechanics of breakage in these devices is still not fully understood. The energy input to the mill primarily goes into lifting the charge, ie, into potential energy (19). This energy is then converted into kinetic energy in a highly active zone where it is either utilized to break particles present in the zone or dissipated as sound and heat. Optimum performance is obtained by judicious choice of ball load and mill speed (11).

Cutting and shredding devices operate through the cutting action of a knife blade

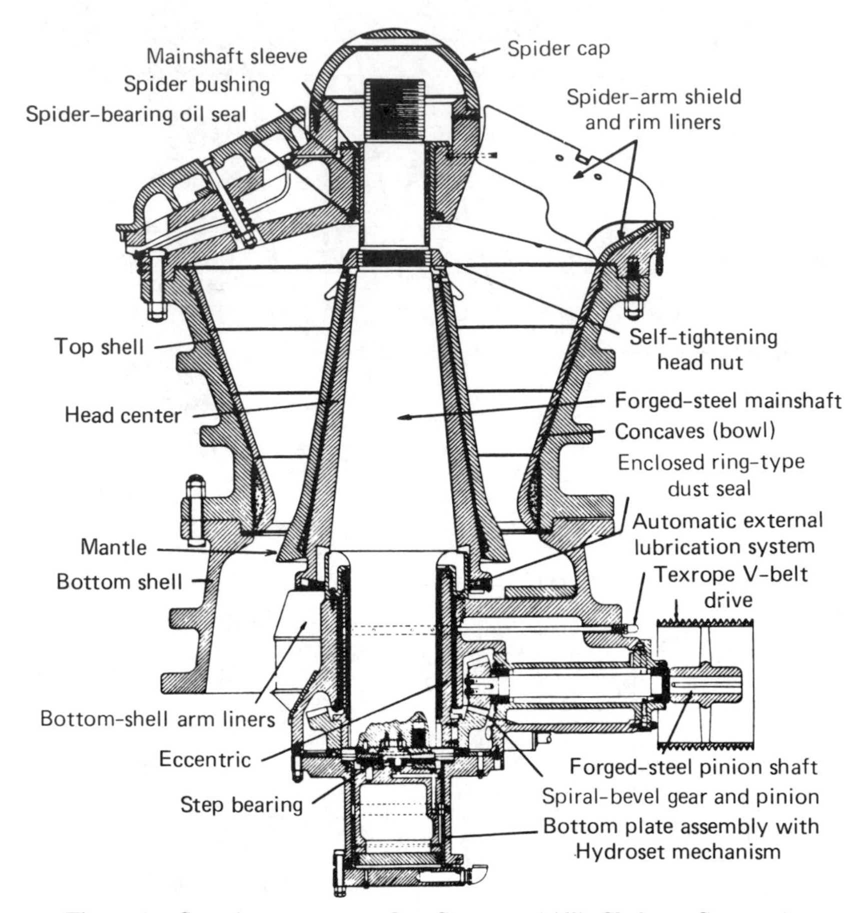

Figure 8. Superior gyratory crusher. Courtesy of Allis-Chalmers Corporation.

that can be either moving or stationary. Machines of this type are used principally for nonbrittle materials, eg, ductile metals and organic material.

Attrition mills are generally used for ultrafine grinding. This group, which includes stirred-ball mills and colloid mills, is comprised of comparatively low-capacity machines for specialized applications.

Several other specialized methods of comminution are used or have been proposed but cannot be classified conveniently in the above categories. Some of these methods are used on a large scale, whereas others are not fully developed and usually apply to particular materials or situations. These techniques include weathering, thermal shock, ultrasonics, electrical methods, pressure alteration, chemical methods, and explosion.

Manufacturers of comminution equipment in the United States and Canada are given in Table 1.

Selection Criteria. The selection of comminuting equipment relies on a careful evaluation of many elements in order to produce the desired product size at the lowest cost. Material characteristics, eg, hardness, compressive strength, coefficient of friction, abrasiveness, moisture content, flow characteristics, slabbing and blocking, and the tendency to agglomerate under pressure, must be assessed. Whether the operation is to be conducted wet or dry must also be determined.

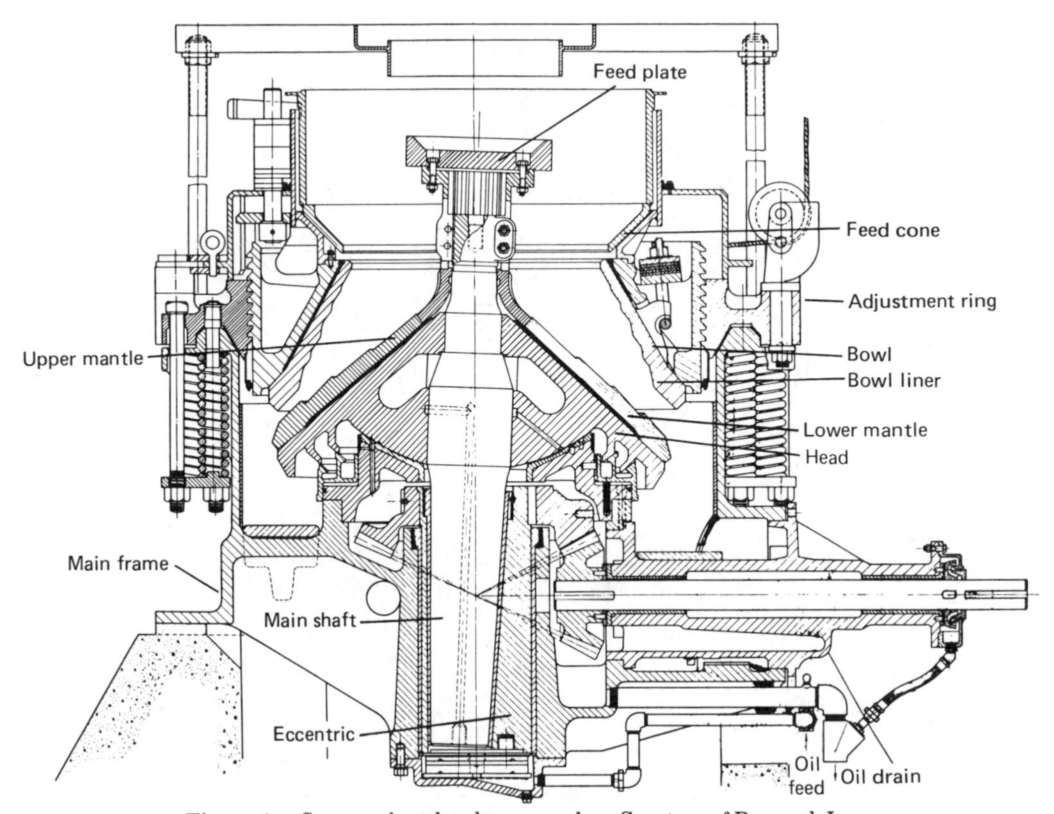

Figure 9. Symons short-head cone crusher. Courtesy of Rexnord, Inc.

Feed Size. The size of the feed material often determines the choice of the primary comminuting device. The top size of the feed material determines the feed opening required, whereas the size distribution determines the flow through the comminuting machine and affects the reduction ratio.

Reduction Ratio. The reduction ratio relates feed size to product size and is usually specified as $R_R = F_{80}/P_{80}$. All comminuting equipment operates within a range of possible reduction ratios (Fig. 4). Therefore, knowing the desired reduction ratio allows the selection and proper arrangement of equipment into a flow sheet.

Comminution Rate. The rate, eg, t/h, at which a material is to be comminuted obviously influences equipment selection. Several types of comminuting equipment have generally the same reduction characteristics, but factors influencing throughput and equipment size may vary.

Compression Devices. ***Jaw Crushers.*** Jaw crushers are used for hard, abrasive materials and will crush virtually any mineral. In contrast to other types, there is no rotary motion in the crushing cycle; crushing is performed by compression between two massive jaws and there is no rubbing action to consume energy and reduce capacity. Jaw crushers can handle material of widely varying sizes and also material with a high clay content. The jaw is either single toggle (Fig. 5) or double toggle (Fig.6). The latter is more expensive and is normally used only for tough, hard, abrasive material.

Jaw crushers are rated according to their receiving area, which is the width of the plates times the gape (the distance between the plates). The maximum feed size to

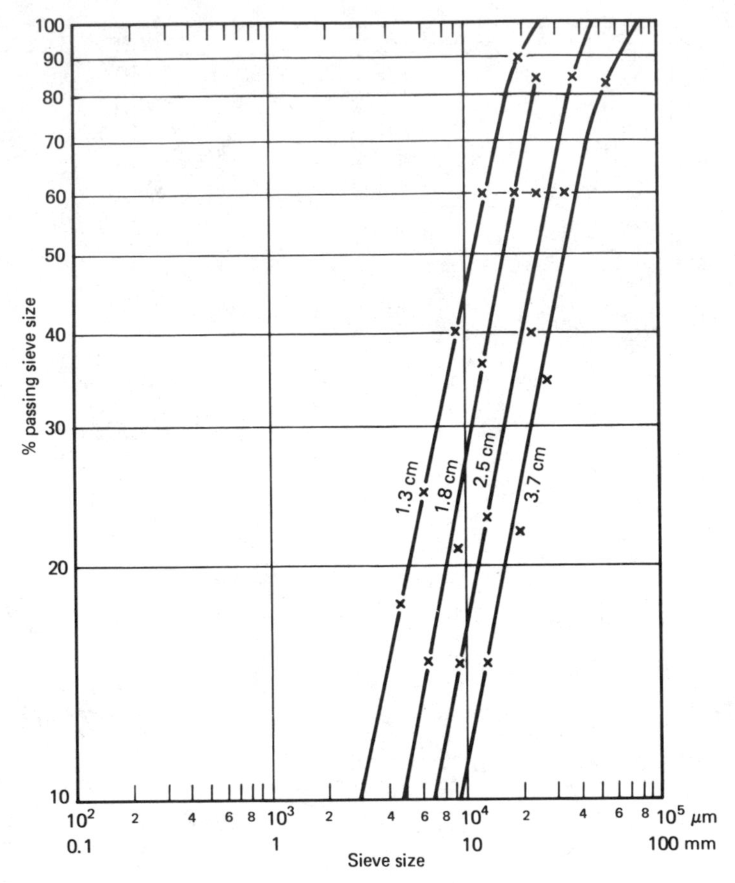

Figure 10. Cone crusher performance for indicated closed side settings in open circuit (4).

a particular jaw crusher is ca 0.9 × gape. The product size depends on the set, which is determined by the distance between the plates at the discharge point. A plot of product size for different settings is shown in Figure 7.

Gyratory Crushers. Gyratory crushers consist of a solid cone set on a revolving shaft located within a hollow shell which may have vertical or conical sloping sides. The solid cone revolves eccentrically with a motion similar to a series of infinite jaws opening and closing, but with a thrust motion. Gyratory crushers are high capacity devices and can be used as primary, secondary, or tertiary crushers. A typical cross-section of a primary crusher is shown in Figure 8, whereas Figure 9 illustrates a design more typical of secondary and tertiary crushers. Gyratory crushers are rated by the bowl diameter at the discharge point and the gape, the bowl diameter at the feed point and the gape, the circumference of the bowl at the feed point and the gape, the maximum diameter of the head and the gape. Crusher capacity is determined by the shape and angle of the cone and the concavity of the walls, as well as by speed, set, throw (eccentricity) and applied power. Product size is determined by the closed side setting, which is the shortest distance between the bowl and cone at any setting. Typical performance curves are given in Figure 10.

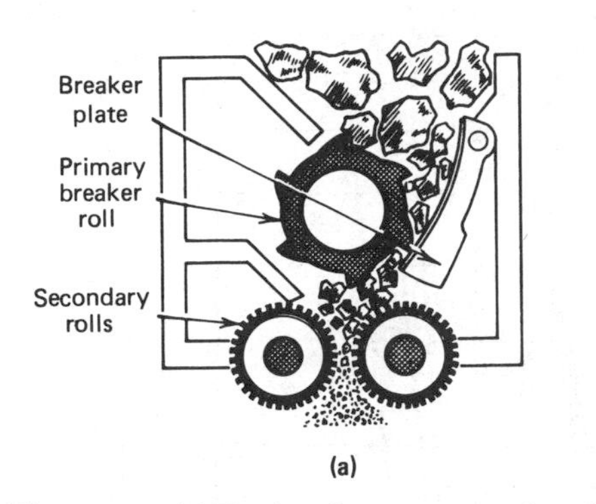

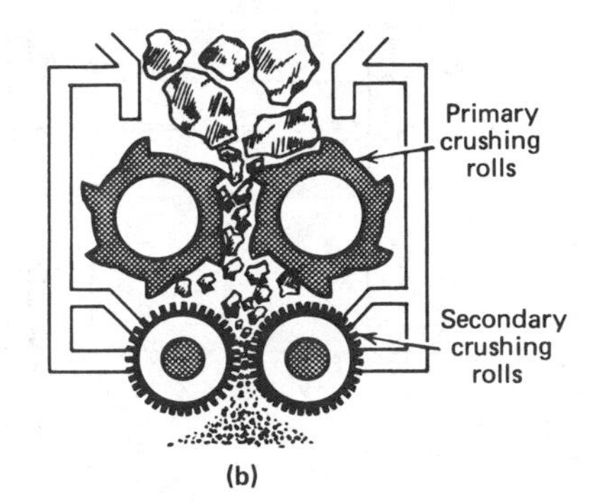

Figure 11. (**a**) Triple-roll two-stage crusher; (**b**) Quadroll two-stage crusher. Courtesy of McLanahan Corporation.

Roll Crushers. Roll crushers are of many types and can consist of a single roll or six or more rolls. The crushing action results from a cylinder rotating on its axis and nipping material between its surface and another surface, ie, either a fixed surface or another roller. The rolls can be smooth, corrugated, grooved, or beaded; they can have intermeshing teeth, fingers, or lugs; and they may be fluted, waffled, or serrated. Roll crushers produce a very narrow size distribution of material; they also exhibit small reduction ratios, typically around four. They can handle wet and sticky material and usually have relatively low power consumption.

Typical roll-crusher configurations are shown in Figures 11(**a**) and (**b**). Product size is determined by the distance between the crushing surfaces. However, this distance accounts only for ca two thirds of the product size. Typical product size-distribution curves are given in Figure 12.

Ring-Roll and Ring-Ball Mills. Ring-roll and ring-ball mills handle material of intermediate hardness which is <1 cm. These mills are capable of producing material <100 μm. They consist of either a number of rolls or balls pressed by mechanical or gravitational means onto a ring which is driven. An air sweep which entrains the finer particles is always incorporated, thereby removing them from the comminuting area. Closed-circuit grinding is often employed, ie, the mill discharge is classified and the coarse material is returned to the mill. Ring-roll and ring-ball mills do not handle wet or sticky material unless the circulating air is heated, thereby providing a drying action. Roller mills are usually restricted to materials with a Mohs hardness of 5 or below; harder materials result in excessive wear. Considerable stress is exerted on the material and breakage after leaving the mill is common.

The configurations shown in Figure 13 illustrate the most common arrangements of roller-bowl action. When the roller force is applied horizontally, wet material is handled more effectively than by other arrangements. Figure 14 shows the rollers positioned at an angle to the vertical axis and running in a grooved track, which is a wearing ring. The grinding force or load is applied by a hydraulic cylinder. In Figure 15 grinding is caused by balls held by a stationary upper ring while the bottom ring rotates. This configuration also handles wet material very well. The rollers are vertical and the table is rotated in another variation shown in Figure 16.

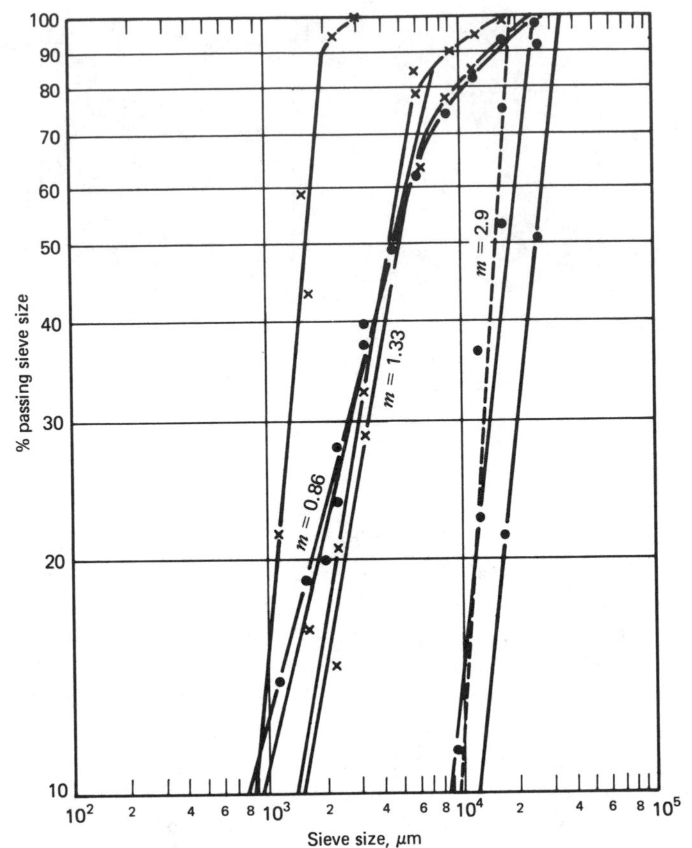

Figure 12. Product size distributions from roll crushers (4); the Gaudin m-values denote the slope.

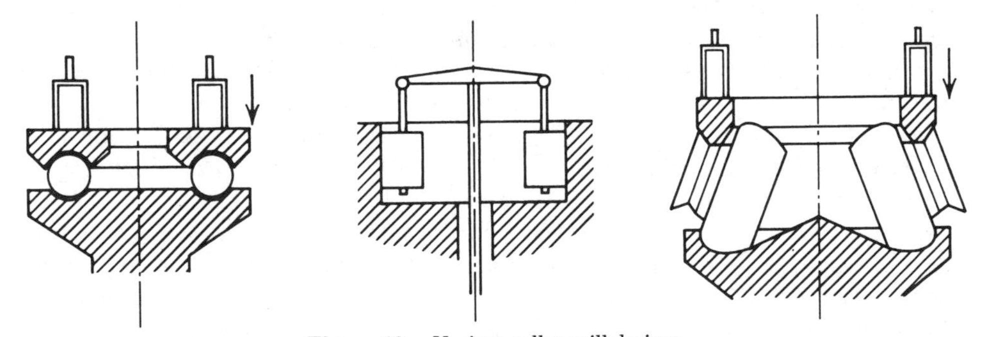

Figure 13. Various roller-mill designs.

The product from a roller mill varies as a function of the feed rate, airflow through the mill, material hardness, etc. A typical example of how product size distribution is effected in a Raymond mill is given in Figure 17.

Impact Mills. Impact mills feature a fast-moving part that transfers a portion of its kinetic energy to the material by direct contact. This transfer of energy sets up internal stresses that cause the material to shatter, resulting in the primary breakage action. Secondary breakage occurs when the material strikes a breaker plate or a stationary surface. Impact mills resemble roll crushers in their broad design, but differ

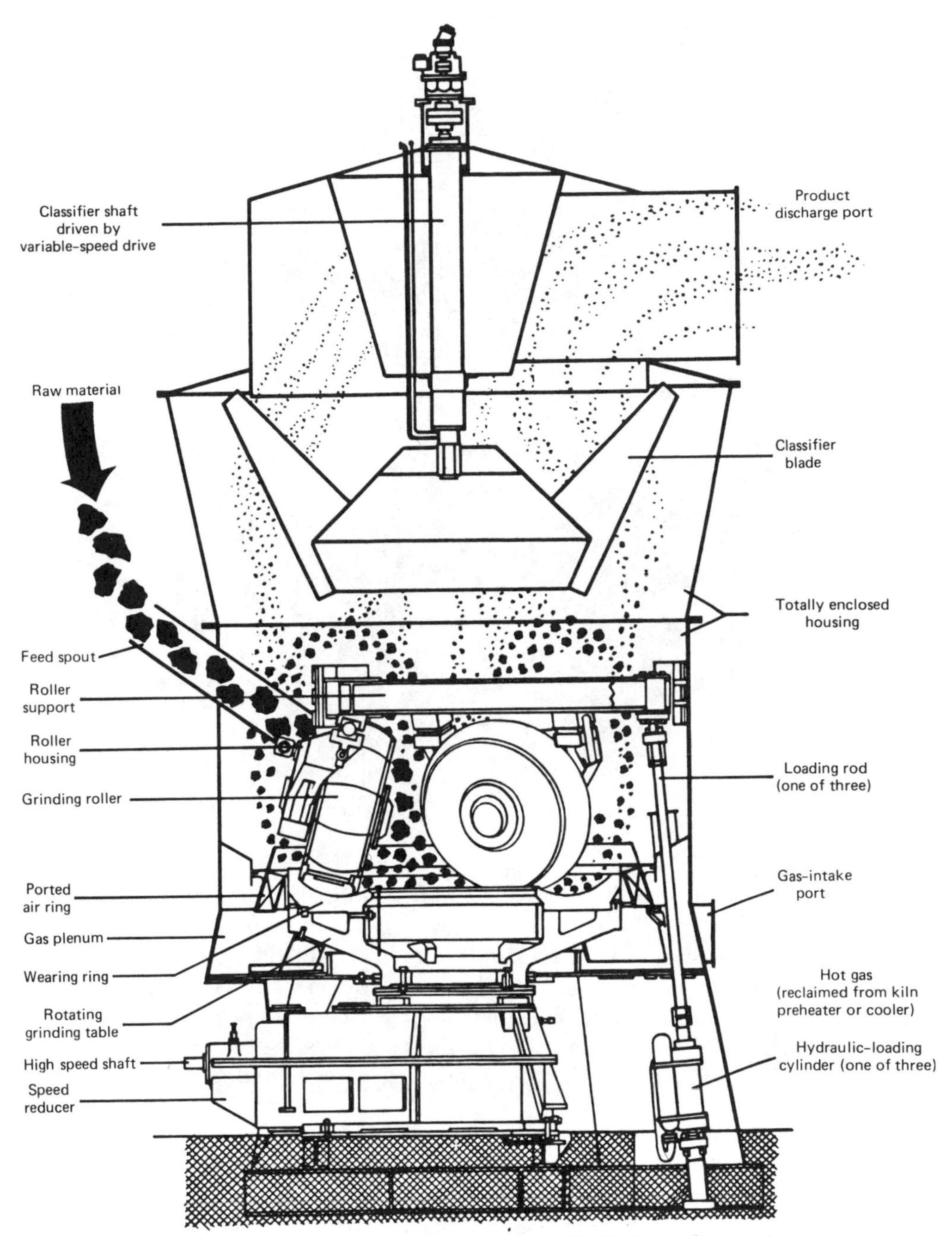

Figure 14. Pfeiffer roller mill. Courtesy of Allis-Chalmers Corporation.

in operation as well as the product. The impact device rotates several times faster, and the product from an impact mill does not retain stresses that result in flaking or breakage after leaving the mill.

Numerous types of mills can be classified as impact devices, eg, hammer, sledge, vibration, pin, centrifugal, cage, swing, and stamp mills. The type of mill that imparts

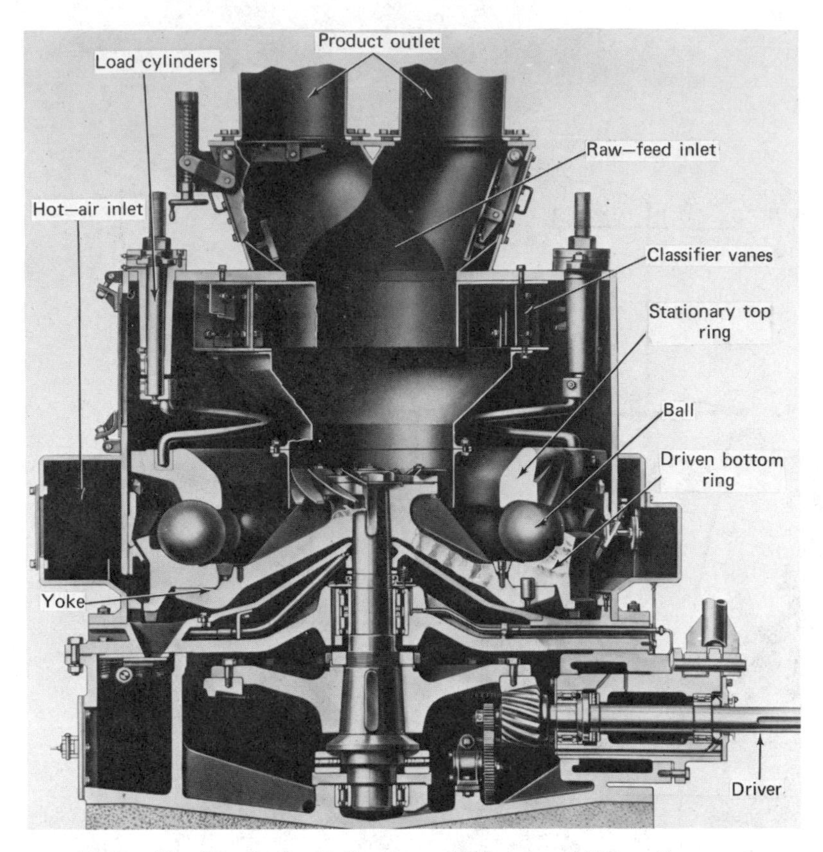

Figure 15. Ring-ball mill. Courtesy of Babcock and Wilcox Company.

energy to the particle by means of a fluid is not considered to be an impact mill because, as opposed to a hammer mill, the initial transfer of kinetic energy causes very little breakage.

Impact machines have the potential for very high reduction ratios, as high as 40:1; ratios of 10:1 and 20:1 are common. In addition, impact machines permit good control of the product size distribution. The machines can therefore be designed for a wide range of applications. Wear of impact machines is generally substantial, and most of the wear surfaces are designed for easy removal and hardening. Considerable maintenance is required.

Impact mills are generally divided into those that rotate and those where the energy is transferred by other means, such as vibration and stamping. Rotating mills are available in numerous configurations, eg, fixed hammers, pivoted hammers, rotary disks, cages, drums, and pins set in disks or drums. Such mills usually have additional breaking devices, such as plates, concentric cages, bars, or grates. The various hammer shapes and forms, each claiming certain benefits for specific applications, are too numerous to describe.

The nonreversible hammer mill depicted in Figure 18 is used for primary or secondary reduction of dry, friable, low abrasive materials, particularly when uniform product gradation is important and fines are not objectionable.

Size reduction starts upon impact, ie, when the hammer strikes the material as

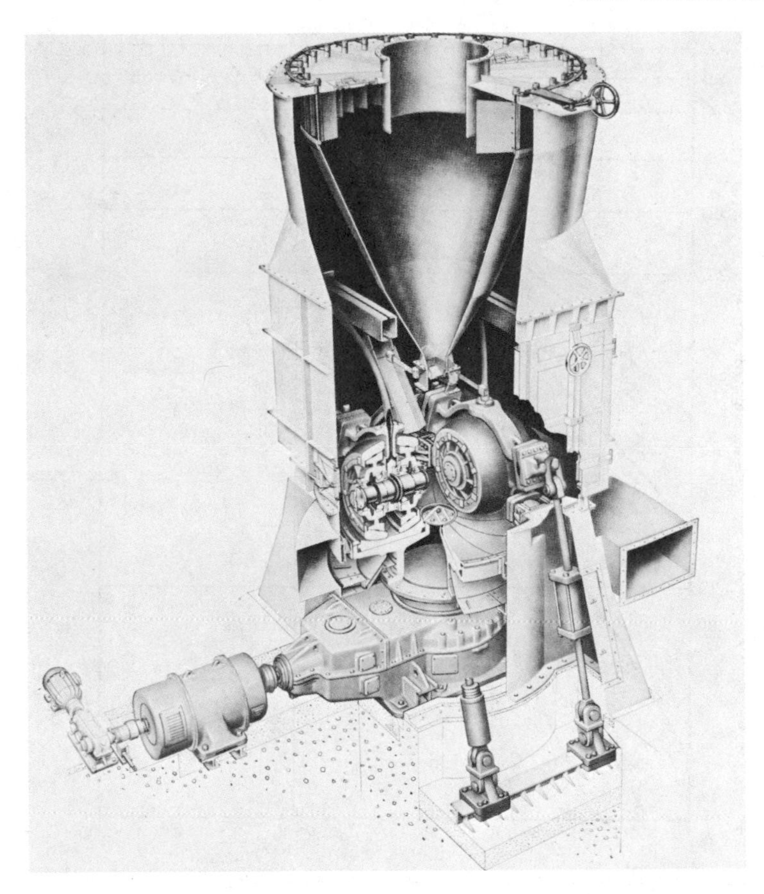

Figure 16. Polysius two-stage roller mill. Courtesy of Polysius Corporation.

it enters the crushing zone. Shattered fragments are swept down into the final crushing zone for further reduction at the pinch points between the hammers and screen bars. Oversize material remains in the machine until it is reduced sufficiently to pass through the screen-bar openings.

Some hammer mills have adjustable cages, a feature that permits changing the product size. The tighter the clearance between the screen bar and hammers, the smaller the particle size of the crushed product. However, for large size changes the screen-bar openings have to be changed.

A reversible hammer permits utilizing each side of both the hammer and grates without manually changing them. The feature allows longer operating time before maintenance is required. In the version shown in Figure 19, a larger number of hammers on the rotor comminutes coal, coke, and similar friable materials.

The impactor shown in Figure 20 has adjustable impact aprons for gap setting which allows good product-size control. Instead of swinging hammers, slide-in blow bars are stationary with respect to the rotor. The impact aprons are designed to retract under load if an uncrushable object enters the mill.

The variation of the single-rotor hammer mill shown in Figure 21 has twin rotors with hammers. It is used as a secondary crusher, and material breakage is accomplished solely by the impact of the material against the hammers. There are no impact plates,

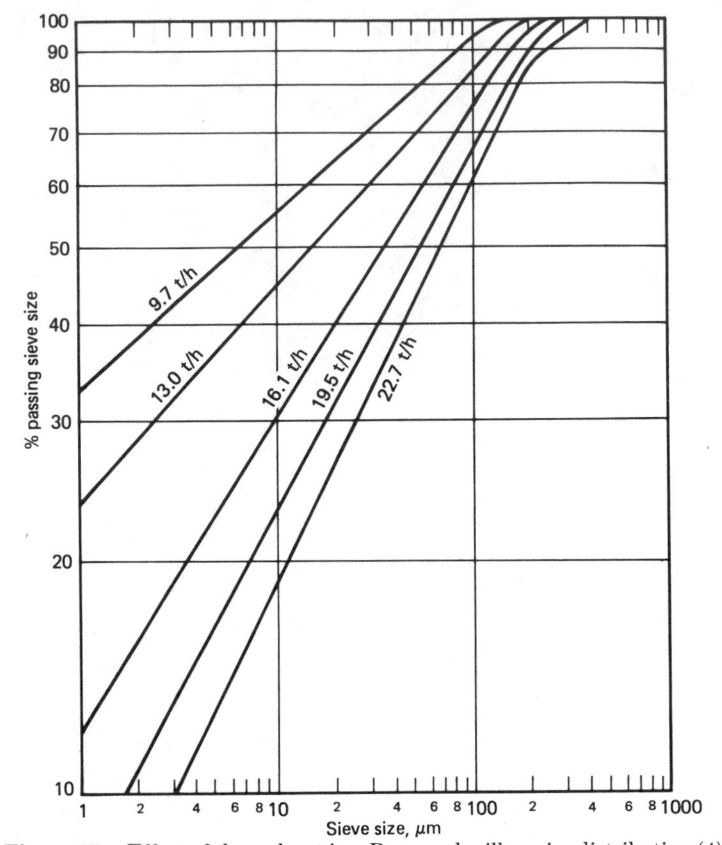

Figure 17. Effect of throughput in a Raymond mill on size distribution (4).

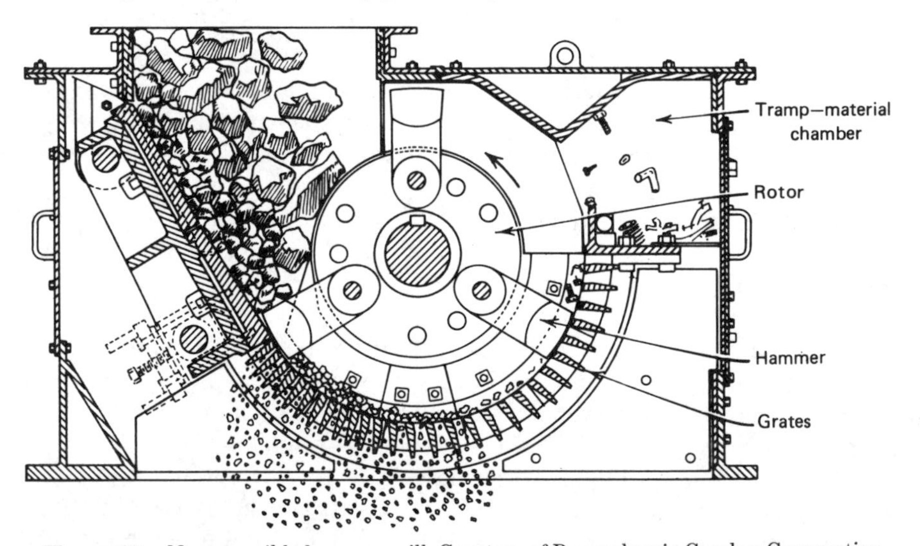

Figure 18. Nonreversible hammer mill. Courtesy of Pennsylvania Crusher Corporation.

grates, or other stationary impact areas. This mill handles material that is wet and sticky and can be equipped with burners to provide heat for drying.

In another version, rows of ring hammers crush by impact and compression (see

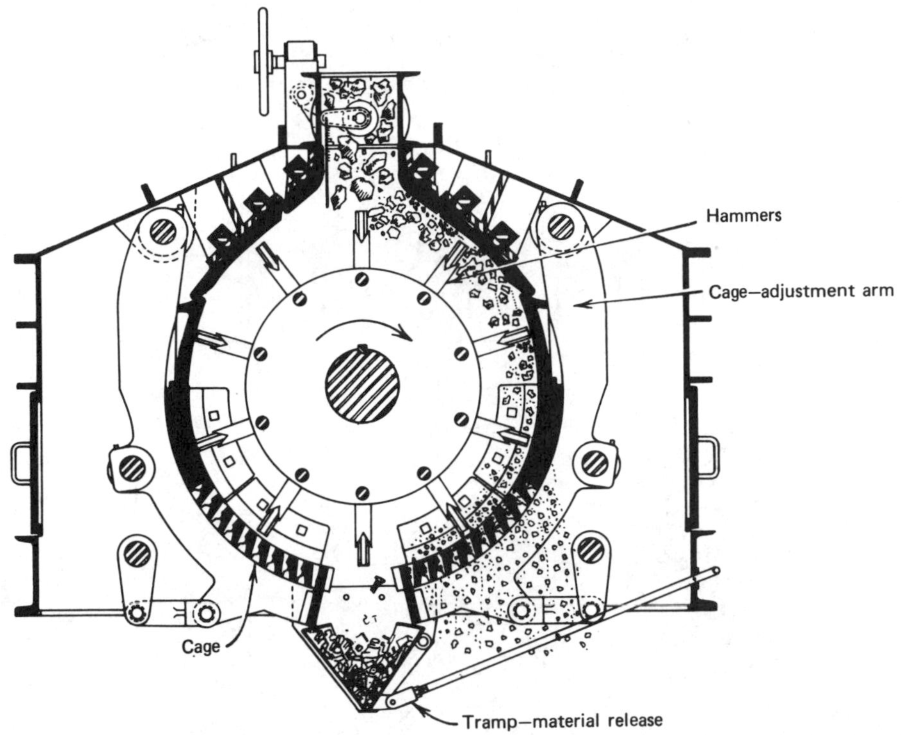

Figure 19. Reversible hammer mill for friable material. Courtesy of Pennsylvania Crusher Corporation.

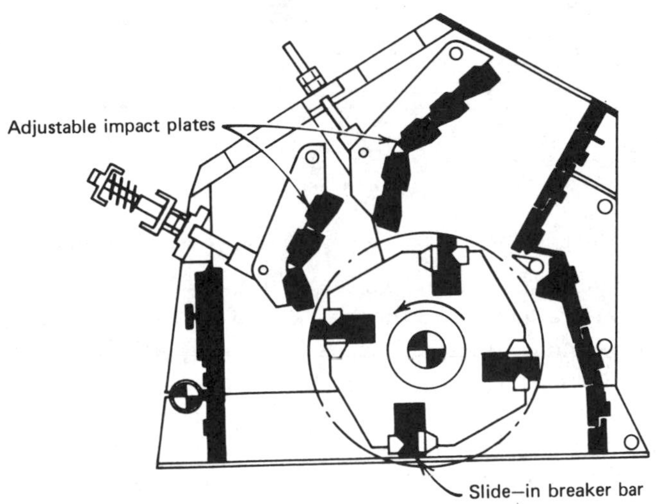

Figure 20. Impact crusher with blow bars. Courtesy of Hazemag USA, Inc.

Fig. 22). The product size is determined by the grate openings and adjusted by changing the clearance between the grate and the ring-hammer path.

An impact mill that imparts kinetic energy to the material by centrifugal force is shown in Figure 23. The mill utilizes either nonintermeshing impactor pins (Fig. 23(**a**)) or intermeshing rotor pins (Fig. 23(**b**)). Feed material flows down through the center, and the centrifugal force throws the material outward where the particles are

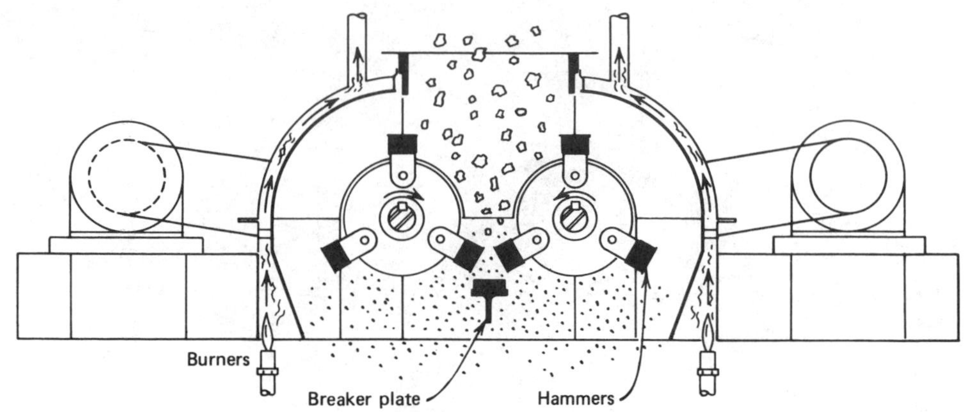

Figure 21. Twin-rotor impactor. Courtesy of Pennsylvania Crusher Corporation.

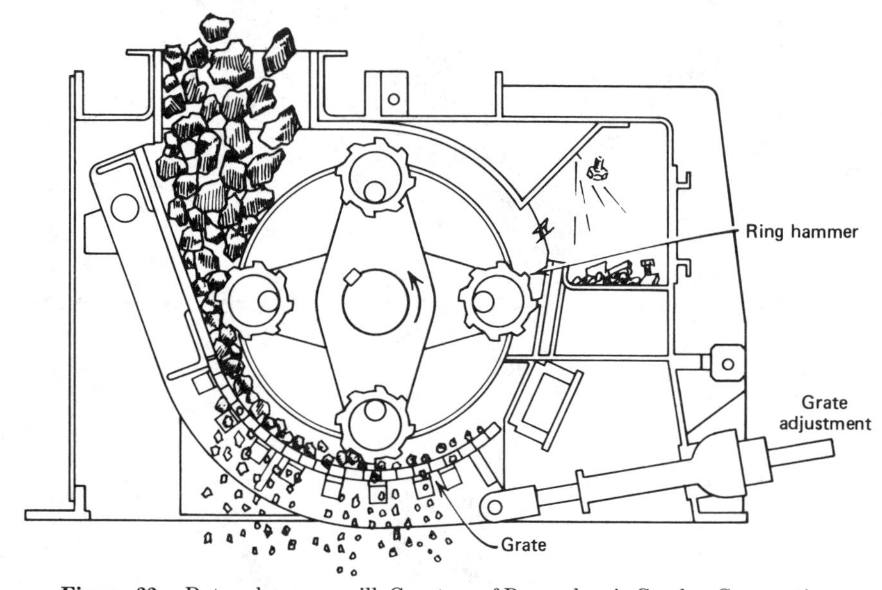

Figure 22. Rotary hammer mill. Courtesy of Pennsylvania Crusher Corporation.

struck by the rotor pins at the initial breakage point. After being struck by the impactor pins, the particles are accelerated to the tip speed of the rotor and hurled against the impact ring where secondary breakage occurs. To produce an even, fine-sized product, the multiple-row intermesh rotor design is used (Fig. 23(**b**)).

Vibration mills are also classified as impact devices. They have either a torus-shaped or cylindrical-shaped shell. Loose bodies are contained within the shell and caused to vibrate; the medium and the material impact against one another and the shell, causing breakage. Unlike in tumbling mills, the medium only moves a few millimeters, and the general path is a complex spiral determined by the amplitude and frequency of vibration. The mills can be run dry, wet, batch, or continuous, and produce a very uniform, fine product of 2–3 μm. These mills require only 10–15% of the power required by other devices to achieve the same reduction ratios. However, the capacity of vibratory mills is small, <5 t/h, and the feed size must be <0.25 mm.

The cylinder-type mill is essentially a tube that contains loose balls. The tube is vibrated by a horizontal and vertical component at right angles to the axis of the

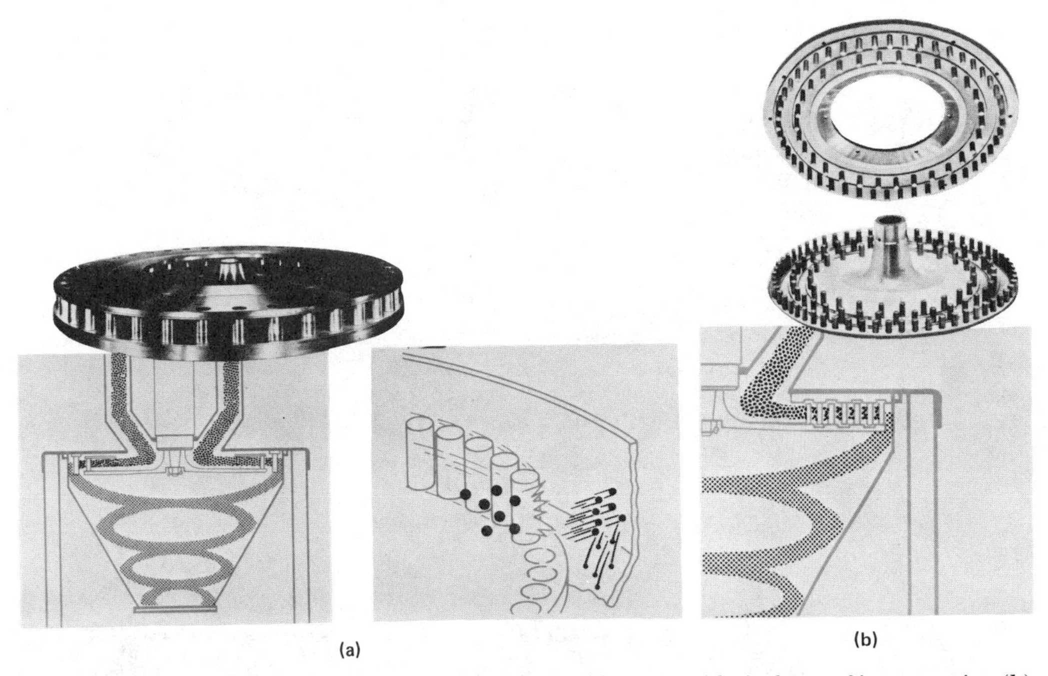

Figure 23. Centrifugal impact mill: (**a**) nonintermeshing rotor with single row of impactor pins; (**b**) intermeshing rotor design with three rows of pins. Courtesy of Entoleter Company.

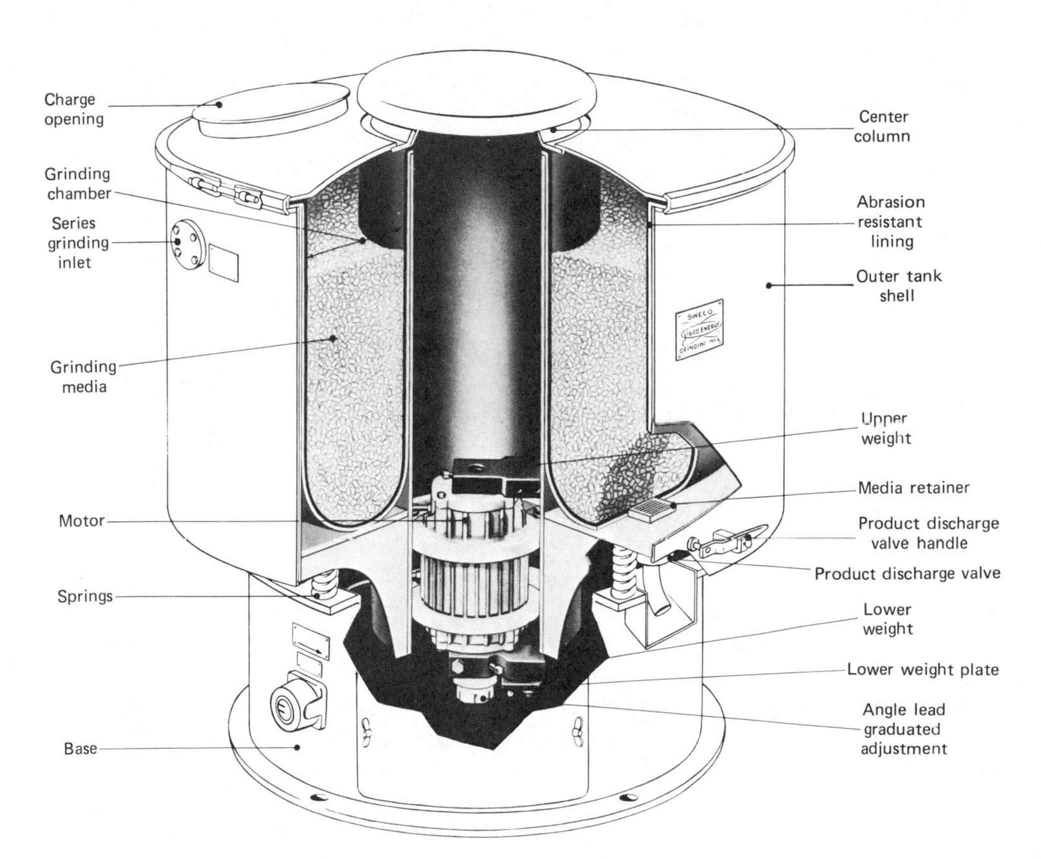

Figure 24. Vibro-Energy mill. Courtesy of Sweco, Inc.

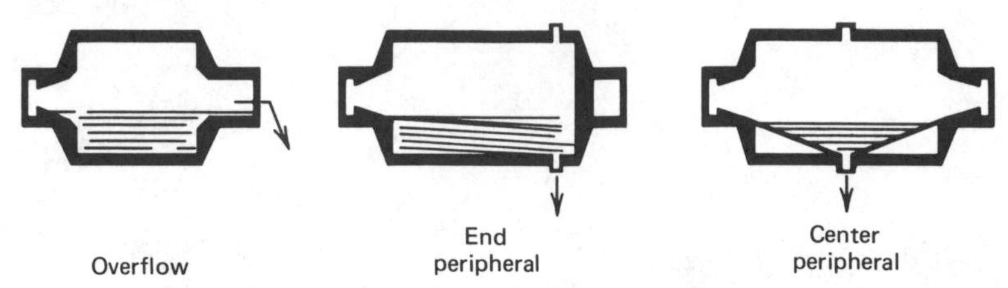

Figure 25. Rod-mill discharge arrangements.

	Overflow	*End peripheral*	*Center peripheral*
Process	wet only	wet or dry	wet or dry
Max reduction ratio	(15–20):1	(12–15):1	(4–8):1
Typical grinds, μm (mesh)	1680–420 (10–35)	4760–1410 (4–12)	6730–3360 (3–6)
Capacity	normal	normal	double
Typical speed, % critical	60–65	65–70	65–70

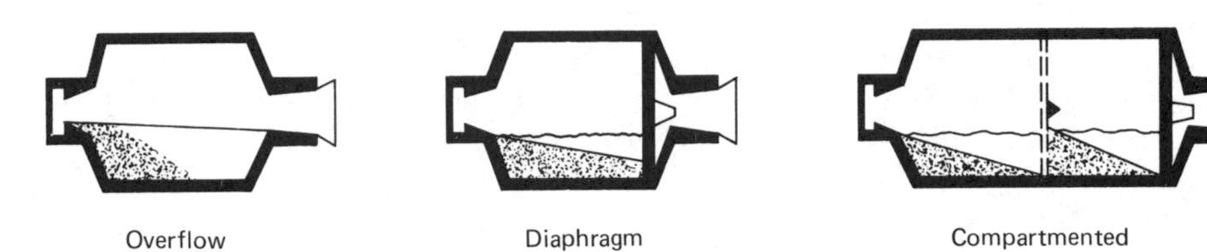

Figure 26. Ball-mill discharge arrangements.

	Overflow	*Diaphragm*	*Compartmented*
Process	wet only	wet or dry	wet or dry
Circuit	usually closed	closed	open or closed
Typical grinds μm (mesh)	fine 74 (200)	intermediate, wet 210–149 (65–100) fine, dry 44 (325)	fine 105–44 (150–325)
Max feed size	1.68–1.19 mm (10–14 mesh)	−1.25 cm	−1.25 cm
Length-to-dia ratio	(1–1.5):1	(1–1.5):1	open (3.5–5.0):1 closed (2.5–3.5):1
Typical speed, % critical	65–70	68–78	wet 65–75 dry 70–78
Charge, as % of mill volume	40–45	35–50	30–50

tube. Sometimes the tube is also given a swing about the center of the tube in a vertical plane.

A toroidal vibration mill is shown in Figure 24. The medium in this case consists of closely packed, small ceramic cylinders. The vibration is imparted in three dimensions and at an amplitude just sufficient so as not to disrupt the alignment of the cylinders assembled with their main axes vertical.

Tumbling Mills. Tumbling mills are characterized by a cylinder which contains grinding medium, rotating about a horizontal, or nearly horizontal, axis. The medium can be rods with one dimension several times the other, balls, or the material to be ground itself. The comminuting action takes place by the material being caught be-

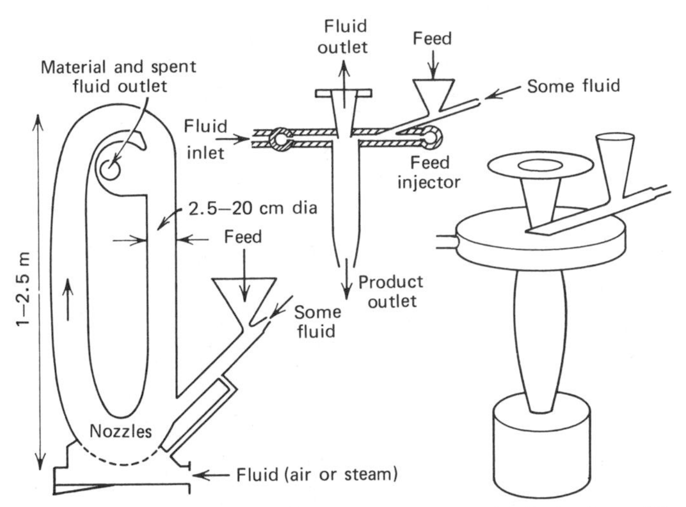

Figure 27. Fluid-energy mills showing torus and cylinder types (4).

tween medium elements while being elevated by the cylinder wall; by rolling-action encounters between the medium and the material; and by impacts from the media. Tumbling mills are commonly classified according to the medium used to effect breakage, eg, rod mills, ball mills, or autogenous mills. In addition, tumbling mills are also classified according to their means of discharge. A classification of tumbling mills is given below:

Rod mills	*Ball mills*	*Special mills*
overflow	overflow	pebble
end peripheral	diaphragm	autogenous
center peripheral	air-swept	combinations
	compartment	

Each type has unique design characteristics that bear on selection and operation. Figure 25 illustrates the types of processes (wet or dry), maximum reduction ratios, typical product size, and recommended operating speed for rod mills.

Rod mills operate most effectively in open circuit with a classifying action that minimizes the tramp (undesirable) oversize and the need for close-circuiting equipment. Recommended feed size is 100% passing 1.9 cm; reduction ratios are critical. Rod mills rotate at a relatively low speed to minimize rod and linear wear, particularly on hard, abrasive materials. These coarse grinding units have a minimum ratio of length to diameter of 1.25:1 to avoid rod tangling. The rods are 15.24 cm shorter than the actual grinding length of the mill. Overflow rod mills of up to 4.5 m in diameter have been utilized effectively in the mining industry.

Similar information for various ball mills is given in Figure 26. The overflow ball mill is preferred in the mining industry (after overflow rod mills) with a feed of ca 1.68–1.19 mm (10–14 mesh) for fine grinding. Overflow mills normally operate in closed circuits at speeds well below critical to minimize the wear. By utilization of scoop feeders and discharge ball cones, or reversed ball spirals, they operate at up to 45% volume ball charge.

The diaphragm, or grate-type, ball mill is operated either wet or dry for intermediate (ca 0.14–3.1 mm = 48–100 mesh) wet-grinding operations, and, in the cement

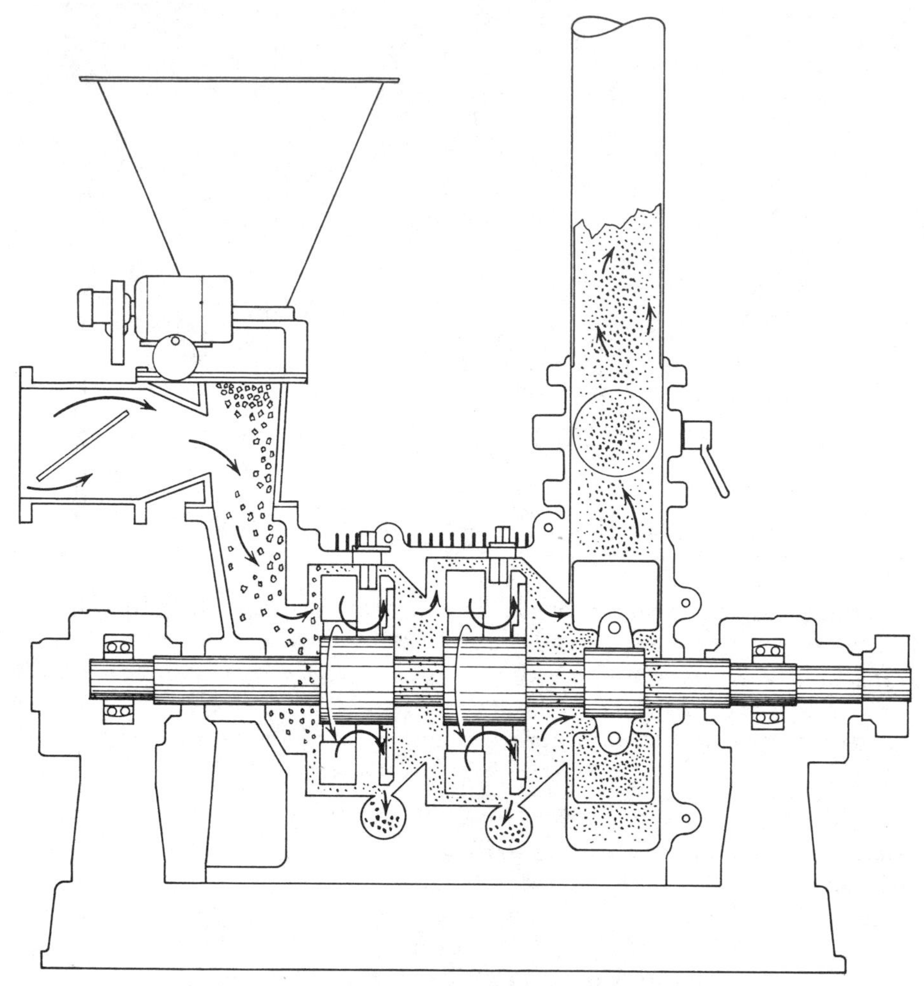

Figure 28. Supermicron mill. Courtesy of Vibra Screw, Inc.

industry, for the grinding of both raw material and finished cement. This mill is employed for semi-impact grinding, utilizing large balls (7.5–8.75 cm dia) and operating in the range of 70–78% critical speed. It carries a 40–50% ball charge and is not used for highly abrasive material.

Compeb mills have two or more compartments containing balls, pebbles, or rods and combine two or more grinding stages in one unit. These mills are particularly popular in the cement industry but can be used wherever stage milling is required without intermediate separation. Recently, however, peripheral discharge units have reappeared in the cement industry to permit the removal of fines with a minimum of overgrinding between the stages of the compartmented mill.

Autogenous mills rely on the observation that a large piece of rock will break a smaller piece of rock. Therefore, the material to be comminuted must contain large lumps. Semiautogenous milling is also practiced where some media, such as large-diameter balls, are added to the mill to assist the breaking action. Autogenous mills, like other tumbling mills, are operated either wet or dry.

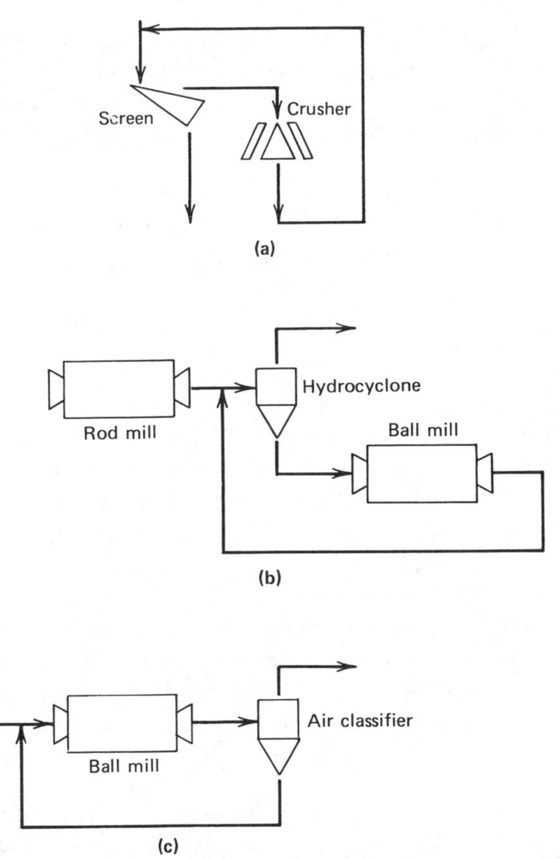

Figure 29. Typical closed-circuit comminution designs: (**a**) scalping screen followed by cone crusher in closed circuit; (**b**) rod mill followed by ball mill in reverse closed circuit with hydrocyclone; (**c**) ball mill in normal closed circuit with air classifier.

Attrition Mills. In attrition machines size reduction is effected by particles breaking each other after having acquired energy to do so from a solid or fluid impeller. The principal task of attrition mills is to produce very fine material ($<10\ \mu m$). They permit control of the product within a very narrow size range. Attrition mills typically handle small quantities of material and are therefore used only in special applications requiring closely controlled fine grinding. Figure 27 illustrates fluid-energy mill, where the fluid provides the energy for attrition, whereas the attrition mill shown in Figure 28 relies on the impact of the impeller moving the material through the mill (see Mixing and blending).

Other Comminuting Methods. Thermal size reduction is achieved by raising or lowering the temperature of a material causing differential thermal expansion and resulting in internal stresses that cause fractures. Thermal conditions can be achieved by steam, liquid nitrogen, plasma, electrical or r-f energy, fire, etc.

Electrical methods can initiate material fracture. Several have been tried without great success, including electrohydraulic comminuting, which requires the sudden release of an electrical discharge in a conducting fluid; high frequency electromagnetic waves impinging on materials; and electrical shock using wave amplitudes and frequencies to cause internal stresses and fracture.

Vacuum comminuting consists of subjecting particles to sudden pressure changes thereby causing fracture, primarily along grain boundaries.

Resonant vibration causes particles containing mixed compositions to be excited or to vibrate at different frequencies and thus fracture. This action can be generated by ultrasonics (qv).

Chemical methods rely on chemicals that are absorbed or that react with the material and thus weaken the material bond. The material can then be subjected to temperature changes or ultrasonic, electrical, or vacuum conditions to cause fracture.

Comminution Circuit Design

Although sizing classification is not a unit operation in which size reduction occurs, comminution and sizing-classification are often used in combination because the typical comminution device produces a mixture of unfinished (coarse) and finished (fine) material. If finished material can be removed as formed, overall energy efficiency is increased because energy is not wasted in producing material finer than required. (see also Size measurement of particles).

A sizing-classification device separates an input particle stream into a stream of finer particles and a stream of coarse particles (see also Size separation, Supplement Volume). The fine stream exits the comminution circuit, whereas the coarse stream is recycled for further size reduction. The sizing-classification unit can be an integral part of the comminution device, as are the screens in some hammer mills, or separate, as are the air classifiers in closed-circuit ball-milling. Typical arrangements are given in Figure 29.

Closed-circuit comminution systems can save a significant amount of energy. The extra power required for open-circuit systems is a function of the degree of product control. For ball-milling, open-circuit inefficiency factors range from 103.5% for a product-size control figure of 50% passing the specified size, to 170% for a control figure of 98% (11).

Comminution–classification circuits are often equipped with environmental-control systems. Circulating air can be heated to dry or heat feed materials or can be cooled to maintain circuit temperatures below the softening points of the material. Inert atmospheres minimize explosion risk and additives can cause a chemical reaction or modify the physical attributes of the system, eg, the reduction of pulp viscosity in wet ball milling (21).

Applications in the Chemical Process Industries

Size reduction in the chemical process industries is used mainly in the preparation of mineral feed stocks, eg, limestone plus lime, phosphate rock, coal (as a chemical feedstock), silica plus feldspars, etc. With impure minerals, the size-reduction step may be incorporated into a mineral-processing plant to upgrade the ore to meet user specifications. Lesser but still important applications exist in coal pulverization for fuel and in the preparation of suitably sized products for sale (see Coal). Size reduction is rarely an integral part of a chemical process. More typically it provides an appropriately sized intermediate, whether for increased reaction rates, for more rapid heat transfer, or to improve ease of handling, eg, the cutting of nylon polymer into small chips for blending before spinning.

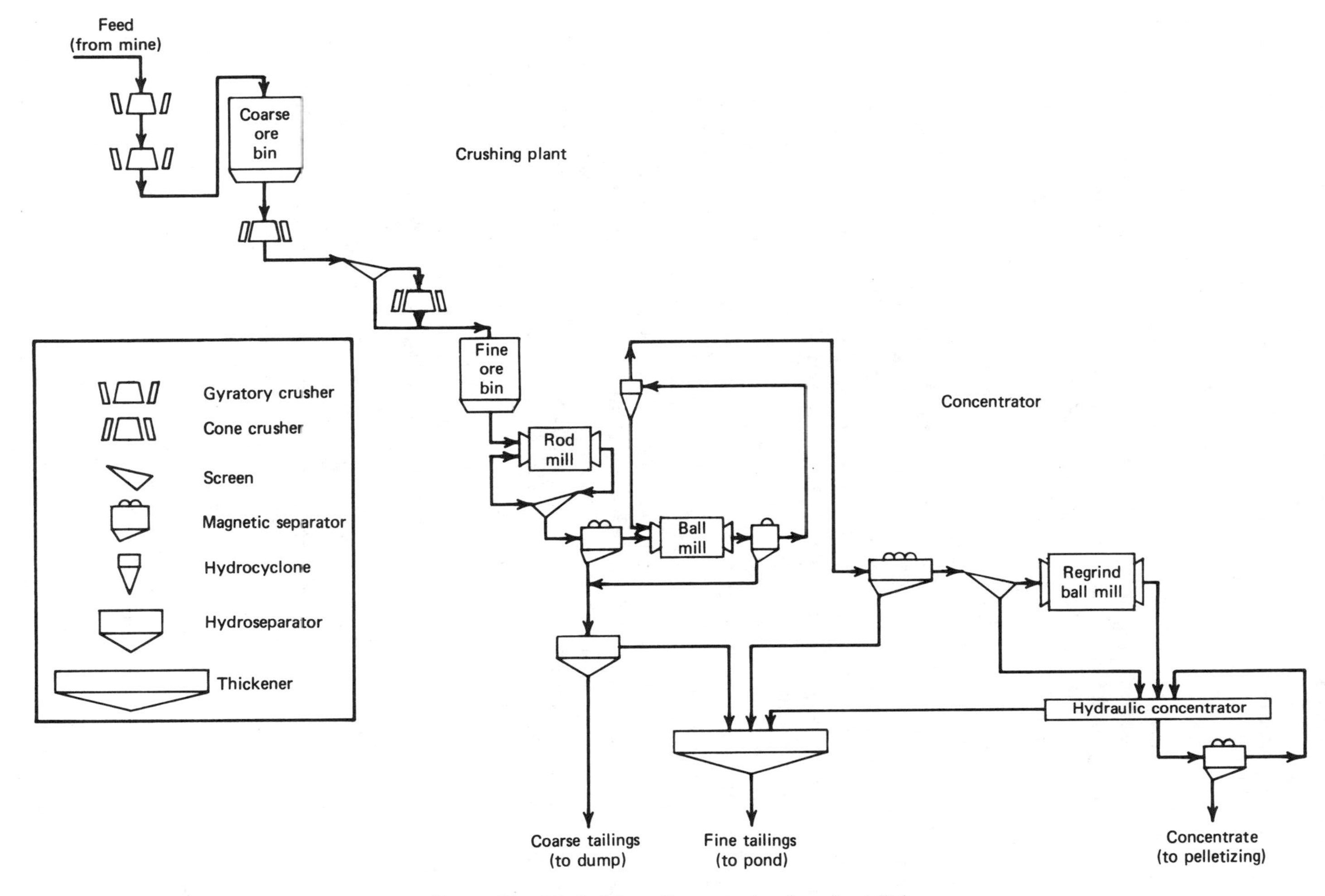

Figure 30. Typical taconite processing flow sheet (21).

Ores and Minerals. The taconite (low grade iron orc) industry of the Iron Range in Minnesota and the Upper Michigan Peninsula provides an excellent example of a large-scale minerals processing plant that utilizes staged comminution with intermediate size separation and beneficiation. These ores contain 20–40% Fe; the remainder is a hard cherty gangue. After staged size reduction, they are upgraded by magnetic separation and flotation to give a blast furnace feed of 64% Fe (see Magnetic separation). A flow sheet for a typical plant is given in Figure 30. In other plants, large autogeneous mills perform the crushing operation. Typical North American practice for a wide range of metallic and nonmetallic commodities is given in ref. 20.

Industrial Nonmetallic Minerals. The processing of limestone is largely one of reducing run-of-mine stone to desired sizes (22). Primary crushing (to a max size of 7.5–15 mm) is carried out in open circuit with jaw crushers, gyratories, or impact breakers. Secondary and tertiary crushing (to 2 mm) is carried out in open or closed circuit with cone crushers, hammer mills, or cage mills and occasionally roll crushers or rod mills. Fine grinding or pulverizing is effected in hammer mills, cage mills, roller mills, and rod, ball, tube, pebble, and vibrating mills. Grinding may be carried out wet or dry, in open or closed circuit, depending on the type of device and the use of the product. Although more efficient, wet grinding is falling into disfavor when a dry product is required, eg, in calcining, except where waste heat is available in adequate quantities (see Lime and limestone).

Other minerals of importance to the chemical industries include phosphate rock for fertilizers; silica and feldspars for glass, ceramics, and refractories; and sylvinite (γNaCl.KCl) for the preparation of KCl, principally for fertilizers. Processing of these minerals generally follows standard mining industry practice (4,20).

Cement. The materials required for portland cement manufacture are lime, alumina, and silica obtained from limestone, shale, clay, or cement rock. Run-of-mine stone is first crushed in jaw crushers, gyratories, impact crushers, or toothed rolls, then blended in appropriate proportions and ground to 75–90% passing 75 μm in wet or dry circuits. Modern plants prefer dry grinding in order to minimize fuel requirements for drying before calcination. Grinding is carried out, frequently in a single stage, in roller mills, ball mills, tube mills (long ball mills), and compartmented ball mills (2–3 compartments charged with different ball sizes). The material is dried in a separate dryer or in a mill-classifier closed circuit, employing waste heat from the cement kiln (see Cement).

Cement clinker from the kiln must be dry-ground to fine sizes (10–50 μm) to produce an acceptable product. Single-stage ball mills in closed circuit with air classifiers or roller mills with integral classifiers are used to yield a product with surface areas, as determined by the Blaine air permeability analysis, of 3200–6000 cm^2/g.

Pulverized Coal. The principal uses of coal (mainly bituminous) are combustion for energy and carbonization for coke, tar, coal, chemicals, and gas production. Pulverized coal is used increasingly in suspension firing in power-plant installations because of its high thermal efficiency. In smaller-scale chemical and metallurgical processes it replaces gas and oil. In the latter application, the coal is blown directly into the furnace or kiln as it is pulverized. The size-reduction equipment is generally small and self-contained. Ball mills and ring-roller and bowl mills, usually in closed circuit with air classifiers, are used in most applications. A recent innovation is the use of fluid-energy mills to provide ultrafine coal for the same purpose. In the Coja mill, steam is used as the conveying and separating medium to minimize the risk of spontaneous combustion and possible explosion.

Dispersion of Powders in Liquids. The term dispersion as used here refers to the complete process of incorporating a powder into a liquid medium in such a manner that the final product consists of fine particles distributed throughout the medium. Particle sizes in these processes are <1 μm. Dispersion processes are used in the production of adhesives, paint, paper, pharmaceuticals, pigments and dyestuffs (for intermediates), plastics, rubber, inks, etc. The basic types of equipment include low shear-rate devices, eg, pug mills and blade mixers; high shear-rate devices (as typified by high-speed impellers), both of which are primarily mixers; and ball mills and roll mills, where size reduction and agglomerate breakdown occurs in addition to mixing (see Mixing and blending).

In the paint industry size-reduction processes are used in the preparation of finely divided pigments and in the dispersion process itself. A good example is titanium dioxide, the most important white pigment. It is manufactured from ground ilmenite or rutile via sulfate digestion or chlorination, then reground in pebble or roller mills with air classification. In the paint process itself, the pigment is mixed with various tints, thinners, resins, and oils and then dispersed in tumbling, vibratory, or planetary ball mills, attrition mills, sand mills, or specialized roll mills (23). The action of all these mills is one of breakdown of agglomerates rather than particle size reduction, though this occurs also (see Paint).

BIBLIOGRAPHY

"Size Reduction" in *ECT* 1st ed., Vol. 12, pp. 498–520, by L. T. Work, Consulting Engineer; "Size Reduction" in *ECT* 2nd ed., Vol. 18, pp. 324–365, by Clyde Orr, Jr., Georgia Institute of Technology.

1. NRC Committee on Comminution and Energy Consumption, *Comminution and Energy Consumption*, Report NMAB-364, National Academy Press, Washington, D.C., 1981.
2. R. H. Perry and C. H. Chilton, eds., *Chemical Engineers' Handbook*, 5th ed., McGraw-Hill, Inc., New York, 1973.
3. A. F. Taggart, ed., *Handbook of Mineral Dressing: Ores and Industrial Minerals*, John Wiley & Sons, Inc., New York, 1945.
4. G. C. Lowrison, *Crushing and Grinding*, CRC Press, Cleveland, Ohio, 1974.
5. H. Rumpf, *Chem. Ing. Tech.* **37**(3), 187 (1965).
6. K. Schonert, *The Role of Fracture Physics in Understanding Comminution Phenomena*, AIME Preprint 71-B-115, AIME, New York, 1971.
7. B. H. Bergstrom, *Proceedings of the 5th Rock Mechanics Symposium*, Pergamon Press, Inc., New York, 1963, pp. 155–172.
8. P. R. von Rittinger, *Lehrbuch der Aufbereitungskunde*, Berlin, Germany, 1867.
9. F. Kick, *Das Gesetz der proportionalen Widerstande und Seine Anwendung*, Leipzig, 1885.
10. F. C. Bond, *Min. Eng.* **4,** 484 (1952).
11. C. A. Rowland and D. M. Kjos in *Mineral Processing Plant Design*, 2nd ed., SME/AIME, New York, 1980, Chapt. 12, pp. 239–278.
12. M. D. Flavel and H. W. Rimmer, *Particle Breakage Studies in an Impact Crushing Environment*, AIME Preprint, New York, 1981; *Pit Quarry* 75 (Sept. 1981).
13. J. C. Motz in ref. 11, Chapt. 11, pp. 203–238.
14. K. J. Reid, *Chem. Eng. Sci.* **20,** 953 (1965).
15. L. G. Austin, *Powder Technol.* **5,** 1 (1971/72).
16. J. A. Herbst, G. A. Grandy, and D. W. Fuerstenau, *Proceedings of the 10th International Mineral Processing Congress*, IMM, London, 1974, pp. 23–45.
17. A. J. Lynch, *Mineral Crushing and Grinding Circuits*, Elsevier, Amsterdam, the Netherlands, 1977.
18. A. L. Mular and J. A. Herbst in ref. 11, Chapt. 14, pp. 306–338.
19. R. Hogg and D. W. Fuerstenau, *Trans. AIME* **252,** 418 (1972).

20. N. Arbiter, ed., *Milling Methods in the Americas*, Gordon & Breach, Science Publishers, Inc., New York, 1964.
21. R. R. Klimpel, *Int. Symp. Fine Part. Process.* **12,** 1129 (1980).
22. R. S. Boynton, *Chemistry and Technology of Lime and Limestone*, 2nd ed., Wiley-Interscience, New York, 1980.
23. G. D. Parfitt, ed., *Dispersion of Powders in Liquids*, 2nd ed., Halsted Press, a division of John Wiley & Sons, Inc., New York, 1973, pp. 221–266.

B. P. FAULKNER
H. W. RIMMER
Allis Chalmers Corporation

SLAGCERAM. See Glass ceramics.

SLIMICIDES. See Industrial microbial agents.

SMOKES, FUMES, AND SMOG. See Air pollution.

SOAP

Although soap was first produced nearly 5000 years ago from wood ashes and animal fats, large-scale commercial production of soap did not start until the early eighteenth century. Prior to that soap manufacture was essentially a cottage industry, with literally thousands of small-scale manufacturers throughout the world. Until 1938 all soap was produced by the kettle process, which involves boiling fats and oils with caustic. Since then several commercially feasible continuous systems have been developed. These have largely replaced the kettle process among the major producers of soap in the U.S., but a significant amount of soap is still made in kettles.

Soaps are the alkali metal salts of long-chain monocarboxylic acids represented, for example, by sodium palmitate [*408-35-5*], $CH_3(CH_2)_{14}COONa$. Sodium is the common cation of hard soaps and is combined with the carboxylic (fatty) acids from animal fats (see Fats and fatty oils) and vegetable oils (qv). The surface activity of soaps results from the soap molecule having a hydrophobic group (the hydrocarbon chain) and a hydrophilic carboxylate group at opposite ends (see also Emulsions). In aqueous solution, this results in the formation of oriented monolayers at the surface with the hydrophobic tails pointing outward. In the body of the solution, the soap molecules aggregate into micelles. These phenomena are responsible for the dispersing and emulsifying powers that make soap solutions useful in soil removal. In hard water, soaps form a precipitate of calcium and magnesium salts, thus reducing their activity; bathtub ring is a manifestation of such precipitation. A host of synthetic surfactants has been developed whose solutions possess surface activity, but which are not prone to calcium precipitation. Hence, soaps have lost favor to the best of these synthetic detergents in washing applications, notably laundry and dishwashing, where the dilute solutions used may impose high calcium-to-soap ratios (see Drycleaning and laundering). On the other hand, soaps are still preferred for facial and body cleansing where the soap-to-water ratio is high. Only about 20% of the toilet-bar market is accounted for by detergent bars, and of these the most important contain soap as well as synthetic detergents. Dollar values for soap and detergents produced in selected years in the United States from 1947 to 1977 are given in Table 1 (see Surfactants and detersive systems).

Raw Materials

Fats and Oils. The naturally occurring fats and oils used in soapmaking are triglycerides with three fatty acid groups randomly esterified with glycerol. Each fat contains a number of long, saturated or unsaturated fatty acid molecules with an even number of carbon atoms ranging generally from C_{12} to C_{18} (see Carboxylic acids). The component fatty acids found in the most important soap stocks, ie, beef tallow and coconut oil, are listed in Table 2. A triglyceride mixture is converted to soap by saponification with aqueous caustic soda, simultaneously releasing glycerol in the following manner:

$$\begin{array}{l} RC(=O)OCH_2 \\ RC(=O)OCH \\ RC(=O)OCH_2 \end{array} + 3NaOH \longrightarrow 3RC(=O)ONa + \begin{array}{l} CH_2OH \\ CHOH \\ CH_2OH \end{array}$$

Thus, the properties of the resulting soap are determined by the amounts and com-

Table 1. Value of Products Shipped by All Producers, 10^6\$ [a]

Year	Nonhousehold detergents	Household detergents	Nonhousehold soaps	Household soaps
1947	na	93.2	101.8	706.0
1954	41.2	504.1	49.0	331.0
1958	66.9	813.1	73.7	288.5
1963	115.3	1029.4	58.7	296.3
1967	140.8	1329.7	51.9	332.0
1972	na	1633.9	na	412.0
1975	na	2368.5	na	584.0
1977	na	2817.2	na	707.0

[a] Ref. 1.

Table 2. The Principal Fatty Acids in Coconut Oil and Beef Tallow[a]

Acid	Formula	Melting point, °C	Double bonds	% in coconut	% in tallow
octanoic	$C_8H_{16}O_2$	16.5	0	8.0	
decanoic	$C_{10}H_{20}O_2$	31.3	0	7.0	
lauric	$C_{12}H_{24}O_2$	43.6	0	48.0	
myristic	$C_{14}H_{28}O_2$	53.8	0	17.5	2.0
palmitic	$C_{16}H_{32}O_2$	62.9	0	8.8	32.5
stearic	$C_{18}H_{36}O_2$	69.9	0	2.0	14.5
oleic	$C_{18}H_{34}O_2$	14.0	1	6.0	48.3
linoleic	$C_{18}H_{32}O_2$	−11.0	2	2.5	2.7

[a] Ref. 2.

positions of the component fatty acids in the starting fat mixture. In general, chain lengths of less than 12 carbon atoms are undesirable because their soaps are irritating to the skin; conversely, saturated chain lengths greater than 18 carbon atoms form soaps too insoluble for ready solution and sudsing. Similarly, too large a proportion of unsaturated fatty acids (eg, oleic, linoleic, linolenic) yield soaps susceptible to undesirable atmospheric oxidative changes. For these reasons and because of the economics of price and availability, the number of fats and oils suitable for soapmaking is limited.

Tallow. Tallow is the principal animal fat in soapmaking. It is a by-product of the meat-processing industry obtained by rendering the body fat from cattle and sheep. Tallows from different sources may vary considerably in color (both as received and after bleaching), titer (solidification point of the fatty acids), free fatty acid content, saponification value (alkali required for saponification), and iodine value (measure of unsaturation). The better grades of tallow, judged principally from titer and color after laboratory bleaching, are used for the preparation of fine toilet soaps; poorer grades are used in laundry soaps. Beef or mutton fat with a titer ≥4.4°C is generally classed as tallow (see Meat products).

Lard. Inedible lard (rendered hog fat) can be used as a limited replacement for tallow after partial hydrogenation to reduce its unsaturation.

Coconut Oil. Coconut oil, the most important vegetable oil used in soapmaking, is obtained by crushing and extracting the dried fruit (copra) of the coconut palm tree.

Palm Oil. Palm oil often serves as a partial substitute for tallow, particularly abroad. It is obtained from a pulp of the outer fleshy fruit of the tropical palm tree (*Elaeis guineesis*). Crude palm oil has an orange-red color and is normally air-bleached prior to saponification. This practice leads to dark-colored soap. Solvent extraction gives a useful by-product, ie, carotene, and an oil that yields light-colored soap.

Palm Kernel Oil. Palm kernel oil is extracted from the center nuts of the same fruit cluster that yields palm oil. Since its component fatty acids resemble in kind and amount those of coconut oil, palm kernel oil can be used as a partial substitute for coconut oil. The main difference between the two oils is that palm kernel oil contains ca 10% more oleic acid than coconut oil.

Miscellaneous Oils. Relatively small quantities of olive, castor, and babassu oil may be used.

Foots from Oil Refining. The crude soap or "foots" obtained from refining edible oils (cottonseed, soybean, etc), as well as the foots obtained by the soapmaker in refining fats and oils (see below under Refining), may be used in limited amounts for the preparation of low grade soaps (see Soybeans and other seed proteins; Vegetable oils).

Natural Fatty Acids. A modern trend, particularly in some continuous saponification operations, is to hydrolyze the oils listed above to fatty acids and glycerol prior to saponification. The fatty acids may then be vacuum-distilled to improve the quality of the resulting soap. Saponification or neutralization proceeds so readily that sodium carbonate can be partially or wholly substituted for caustic soda.

Nonfatty Soap Stocks. Rosin (see Terpenoids), and to a lesser extent tall oil (qv), and cycloparaffinic (naphthenic) acids are used in the preparation of laundry or textile-scouring soaps.

Alkalies. Aqueous sodium hydroxide solution (50 wt %) is commonly used to saponify fats, oils, and fatty acids for the preparation of hard soaps to be finished as bars, flakes, or beads. Aqueous potassium hydroxide is used for the preparation of soft soaps, ie, liquids or pastes, since the potassium soaps are more water-soluble than the sodium soaps. Blends of the two alkalies are occasionally used to achieve special properties.

Processing of Fats and Oils

Refining. Depending on the quality of the crude fats or oils, they may be refined and bleached before saponification. An amount of aqueous sodium hydroxide equivalent to the free fatty acid content of the oil is added with slow stirring to the hot oil. The precipitated crude soap or foots occludes some unwanted color bodies, and foots removal is accomplished by centrifugation or by filtration of the neutral oil after settling. The foots may be transferred to a kettle of low quality laundry soap or acidulated to release the fatty acids for purification by distillation.

Bleaching. Other coloring matter can be removed by agitating the hot neutral oil with 1–2% bleaching clays or activated carbon, followed by filtration in plate-and-frame presses.

Hydrogenation. The soapmaking qualities of the highly unsaturated triglycerides contained in some oils can be vastly improved by hydrogenation. Elimination of some unsaturation hardens the stock and improves its odor as well as stability. However, during hydrogenation some of the unsaturates are converted from the cis to trans form, which decreases the solubility of the resultant soap.

Fat and Oil Blending

The approximate percentages of the main fatty acids esterified with glycerol in coconut oil and beef tallow, the principal stocks used in most soapmaking, are listed in Table 2. Coconut oil contains large proportions of the relatively short-chain lauric and myristic acids, whereas tallow is rich in the longer-chain palmitic, stearic, and oleic acids. Soaps made from each of these stocks have desirable and complementary features. In general, the short-chain laurate and myristate soaps from coconut oil and the unsaturated oleate from tallow supply quick solution and foaming which is sustained by the palmitate and stearate soaps.

A sodium soap prepared from a blend of ca 20 wt % coconut oil and 80 wt % tallow exhibits good properties over a broad range of solution temperatures, as well as consumer acceptance when incorporated in a toilet bar. Limited substitutions for coconut oil and tallow are possible if they offer similar fatty acid distributions. Hence, most good-quality toilet soaps are saponified from a blend, whether by way of the mixed oils or the fatty acids therefrom.

Soap Phases

There are three basic operations in soapmaking, whether by the kettle process or the newer continuous processes: saponification, the chemical reaction in which fats and oils react with alkali to form soap and glycerol; washing to remove glycerol and salt; and fitting to separate dark colored impurities from the soap and leave it in a form suitable for further processing.

When manufacturing soap by the kettle process, the skilled soap-boiler relies heavily on appearance, flow properties, feel, and even taste in order to obtain the desired separations at each stage of the process. These separations can be best understood, however, in terms of the phase relationships.

Ternary Soap–Salt–Water Systems. Twenty years of experimentation by McBain and various associates established that a number of different phases can coexist in the three-component system of a soap kettle and that the various equilibria between the phases are rigorously governed by the phase rule. McBain's phase diagram of the ternary system sodium stearate–sodium chloride–water at 90°C (Fig. 1) depicts the equilibria involved during washing and fitting operations within the limits of experimental difficulties. This diagram, involving a single pure soap, sodium stearate [*822-16-2*], is only slightly different from that of a blended commercial soap. The following five areas of variable compositions outline the regions of stability of five phases in addition to the brine: curd soap, kettle-wax soap, neat soap, middle soap, and nigre. Not too clear on the diagram is the fact that the nigre phase, containing substantial amounts of soap, is continuous along the abscissa with a "lye" of very low soap and high salt content. Two-phase areas represent overall compositions where two immiscible phases separate. The most important two-phase separations in kettle boiling are the kettle wax–lye and neat–nigre separations. Within a number of triangular regions, three-phase separations occur where the compositions of the separating phases are fixed by the apexes of the triangles. This ternary diagram applies equally well to both the continuous and kettle processes since the desired equilibria and phase separations are only possible within the composition limits appropriate to the chosen soap formula.

Binary Soap–Water Systems. The product of the boiling and continuous procedures, starting with either fat or fatty acid, is neat soap of ca 69% soap and 30% water with a small residual salt content (ca ≤0.5%). Hence, the compositions that are finished to bar, flake, or bead products are essentially two-component systems of soap and water. The equilibria in these binary systems have also been determined. The phase diagram for sodium palmitate and water is shown in Figure 2. The diagrams for other soap and water systems, including a commercial soap, are generally similar (4); low salt content only slightly distorts the phase boundaries (5).

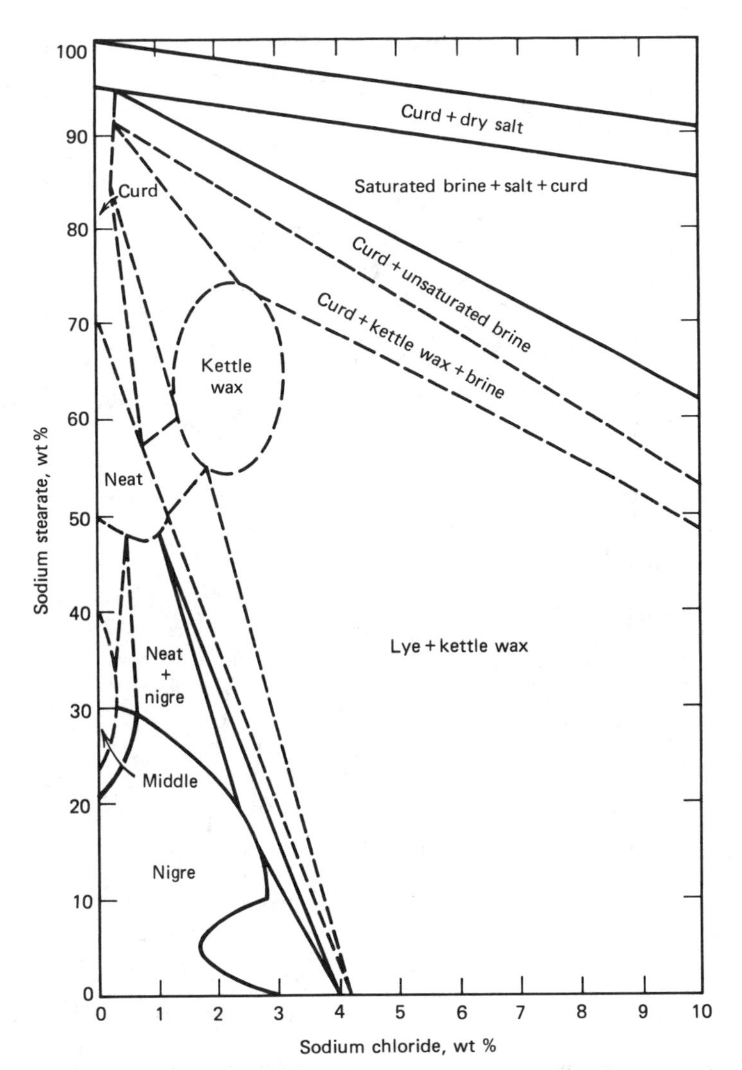

Figure 1. Phase diagram of the ternary system sodium stearate–sodium chloride–water at 90°C (3). The triangles indicate heterogeneous equilibria between the three condensed phases at the apexes of the triangle. The regions between the triangles should be filled with tie lines, each connecting two condensed phases, with the tie line nearest any triangle almost parallel to the nearest boundary of that triangle.

On cooling, molten neat soap of 70% soap content begins to separate at 83°C (see Fig. 2) as a solid-curd phase. The boundary containing this point is known as the T_c curve and represents the temperatures at which the solid-curd phase first separates various soap and water compositions on cooling. The solid-soap area below the T_c curve is the commercially important region in which bar soaps are processed. Yet this region in Figure 2 and similar phase diagrams shows only a heterogeneous mixture of curd soap of indefinite composition and a dilute isotropic solution. Evidence that soap-processing methods can produce marked differences in soap properties and crystal structures in this area is discussed below.

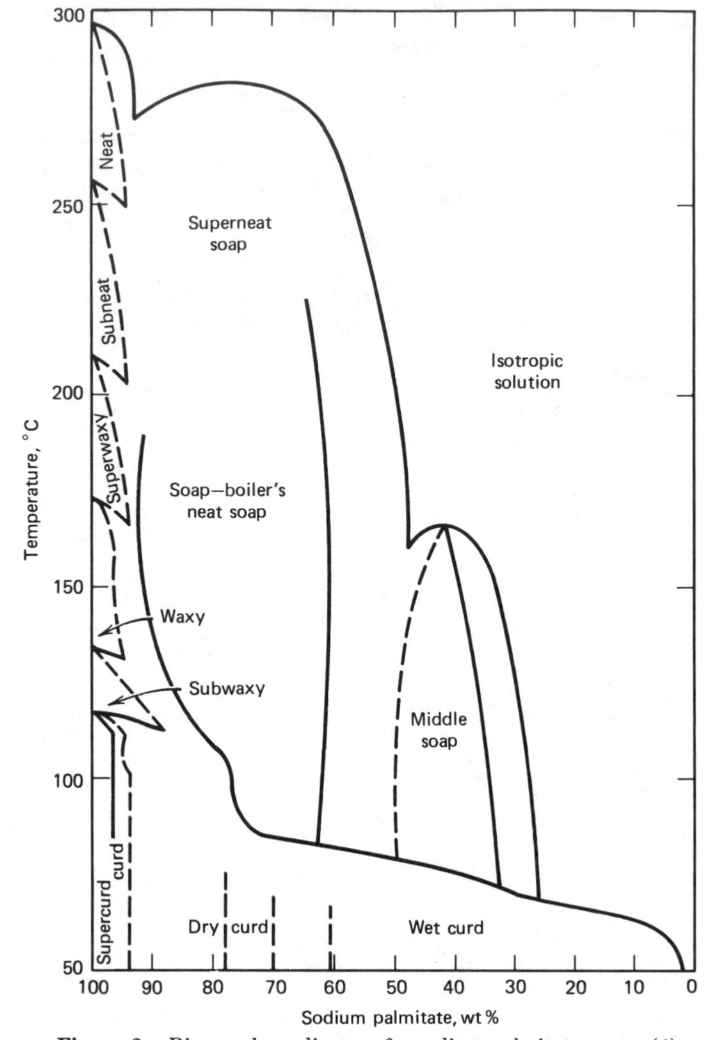

Figure 2. Binary phase diagram for sodium palmitate–water (4).

Solid-Soap Phases. In 1940 a patent was issued for a process for producing a floating soap by intense working of a semifluid soap-and-water mixture with air at elevated temperatures (71–107°C) followed by quick cooling without further agitation (6). Novel features, which distinguished this soap from prior art-milled or framed soaps of the same composition, were claimed. Hence, differences in the properties of solid soaps at room temperature, dependent on their processing histories, were noted.

Similarly, the room-temperature properties of soap-and-water compositions were discovered to be altered by working above or below a critical temperature characteristic of each composition, followed by rapid cooling without further agitation (7). In addition to having different use properties, soaps processed on either side of the critical-temperature boundary showed different x-ray-diffraction patterns at room temperature. Hence, demonstrable differences in the crystalline structures of identical compositions were shown.

Four different phases in solid hydrous soaps were characterized by different x-ray-diffraction patterns (8–9). Three of these phases were encountered in samples at room temperature of commercially important soap–water compositions. Phase transformations were induced by temperature or composition changes, particularly when assisted by the application of mechanical energy in the form of milling, plodding, or extrusion. Just as in the floating-soap patents, soaps of the same compositions but in different phases exhibited different solution and lathering properties. However, it was noted that the different processing conditions necessary to produce the different phases might equally well affect the soap's properties. Again, some detail indicative of structural difference was noted in the solid regions below the T_c boundaries of the phase diagrams.

By 1945 ten phases had been identified by x-ray short spacings in a number of individual pure soap-and-water compositions at room temperature (10–11). The soaps were vigorously worked at fixed temperatures and compositions and then cooled to room temperature without further agitation before x-ray irradiation (12). The phases determined at room temperature thus descended on cooling from the worked equilibrium mixtures; hence, working temperature–composition plots of the data were called descendant phase maps.

The phase maps obtained from data at room temperature are not to be confused with phase diagrams depicting the equilibria prevailing at the actual working temperatures. However, inspection of the phase map for the sodium myristate [*822-12-8*]–water system (12), shown in Figure 3, reveals striking similarities to certain features of McBain's phase diagrams in Figures 1 and 2. The nearly horizontal boundary between the μ and ζ fields (of different x-ray-diffraction patterns) of Figure 3 agrees fairly well with McBain's temperature of ready solubility of 60°C for sodium myristate (5).

The temperature of ready solubility (T_s) is defined as the temperature at which soap shows a sudden sharp increase in water solubility. In Figure 2, for example, ca

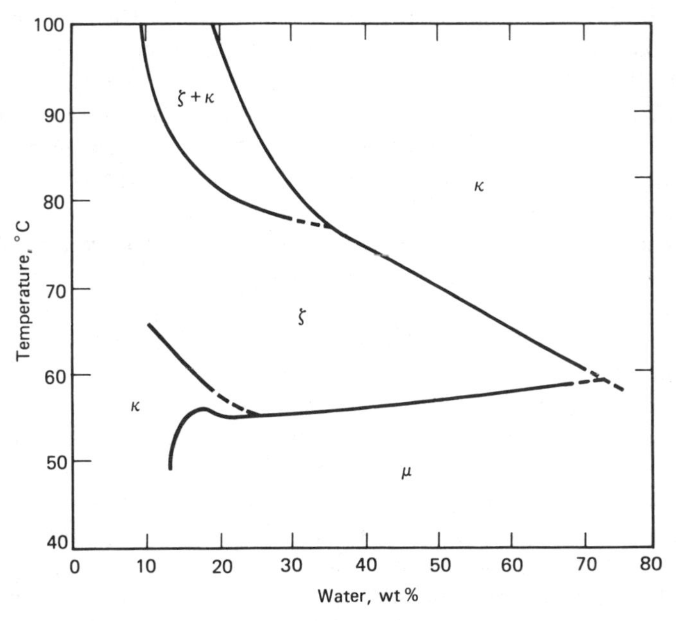

Figure 3. Phase map for the sodium myristate–water system.

5% of sodium palmitate dissolves to a clear isotropic solution at 63°C, but ca 25% is dissolved at 70°C. Similarly, the upper boundary of Figure 3 between the ζ and κ fields agrees fairly well with McBain's T_c boundary in the sodium myristate–water diagram (13), which is shown at 70°C at 50 wt % water and 84°C at 30 wt % water.

Thus, there is ample evidence that processing at fixed compositions on either side of McBain's boundaries produces phase discontinuities that lastingly affect the crystalline structures of subsequently solidified soap. Such structural changes can reasonably be expected to affect the properties of the soaps.

The phase changes described were generally induced in fixed soap–water compositions upon cooling from different working temperatures. However, a transparent soap was produced from an ordinary soap stock containing no additives by conventional milling and plodding at ca 38°C and at 20 wt % water content (14). At ca ±2% water, the same soap stock after the same milling and plodding yielded opaque soaps. Hence, the existence of another boundary in solid soaps was indicated, this time a critical composition instead of a critical temperature boundary. Similarly, a critical water boundary for translucency was identified at ca 15% for a modified soap stock (13).

Thus, solid-soap phases in the commercially important low water–low temperature area lack the mobility to establish equilibria spontaneously, in contrast to the hot-fluid phases of the kettle and the high temperature parts of the binary diagrams. Instead, vigorous working is necessary to establish equilibrium in relatively cold, concentrated systems. Recognition of this fact and the application of x-ray diffraction have indicated fundamental differences in solid-soap structures.

Batch Manufacturing Processes

Full-Boiled Kettle Method. Much of the world's soap production still begins in open steel kettles or pans capable of processing batches of 1–1000 kg soap. The pans are circular or square in cross section, but taper to cones at the bottom, and are generally lagged to conserve heat. Open steam coils in the cone section supply heat and agitation; closed steam coils may also be present to supply heat without adding condensed steam. The bottom layer in any stage of the boiling procedure is removed from the cone and transferred to storage tanks or other kettles via pumps, valves, and pipelines. Similarly, the upper layer is removed through an adjustable swing pipe located part way down the kettle.

This process supplies completely saponified neutral soap containing ca 30 wt % water, a composition which, when hot, can be pumped to various finishing operations. The soap is washed free of soluble colored and odorous impurities and in subsequent processes is separated as far as possible from glycerol and from the salt added to precipitate the soap.

Saponification Charge. Generally, part of the weighed coconut-oil fraction of the fat charge is first run into the kettle with a little water and salt. There it is heated and agitated with live steam. Saponification is started by the addition of (usually) 50 wt % caustic soda solution and the slow addition of the remaining coconut oil. Saponification comprises the following three stages: a slow incubation period, a rapid exothermic stage, and gradual completion. The incubation period is largely a result of poor contact between the starting oil and water phases and can be shortened by starting a boil on a preceding nigre (see Fitting Change), whose soap content helps emulsify the mix. If this is done, the entire charge of coconut oil and tallow is usually added when the kettle is charged. During the slow incubation period, the caustic must be added slowly since a large excess of unreacted alkali renders the oil and water phases even less miscible. When enough soap has formed to start emulsification, the exothermic stage begins and caustic is consumed rapidly. Steam control is necessary at this stage to prevent boiling over. A small amount of dry salt is usually added here to maintain a slight excess of electrolyte since the caustic soda content is rapidly diminishing. Some electrolyte excess is necessary to prevent "bunching," an undesired appearance of viscous middle soap (see Fig. 1) which forms at low electrolyte–high water concentrations. When the coconut-oil fraction is largely saponified, as judged by the lessening heat of reaction and the alkali consumption, the tallow fraction is pumped in and saponification is continued with more alkali. The saponification stage must finish with a bare excess ($\leq$0.1%) of unreacted alkali since the latter, which accumulates in the subsequent lye, complicates the glycerol-recovery process. Hence, the last, small amounts of caustic soda are added slowly with testing for alkali consumption as the pan contents are expanded by steaming to the top of the kettle on a "closed grain." The grain is determined by the amount of electrolyte present. Small amounts of salt and unreacted alkali produce a soft or closed grain, the condition required during saponification since it permits maximum interaction between fat and alkali. A closed grain is readily recognized by its smooth appearance. Excess alkalinity can be simply determined by titrating a soap sample dissolved in hot neutral alcohol to the phenolphthalein end point with standard acid solution.

After completion of the saponification, dry salt is added to the boiling soap until a hard or open grain is obtained. This condition corresponds to the two-phase separation of kettle wax and lye shown in Figure 1; it is recognized by the heterogeneous appearance of soap curds and thin aqueous liquid. However, the degree of graining is a critical factor in assessing the volume of lye and the amount of soap left in the lye, a determination that requires considerable experience. After a short boil to equilibrate the two phases, steaming is stopped, and the pan contents are allowed to separate for several hours. The lye is withdrawn and transferred to the glycerol-recovery system.

Wash Changes. Since only 30–45% of the available glycerol is removed in the saponification lye, two or more washes are made to extract as much of the remainder as economics dictate. The grained soap remaining in the pan is boiled up and water added. Since most of the salt previously added was removed with the saponification lye, the soap closes again and forms a homogeneous solution. After a short boil, the soap is grained with salt and another glycerol-containing lye is removed after each wash.

Fitting Change. After the last wash lye is removed, the final fit to neat soap is begun. The grained soap from the last wash is boiled and water gradually added to the desired fit. This composition corresponds to the two-phase neat–nigre area of Figure 1, so that on settling, a thin nigre, dark with unwanted impurities, separates below the neat-soap layer. This is another critical operation since the type of fit determines the proportions of neat and nigre. Soap boilers use a trowel to determine the fit. The trowel is heated in the soap mass, then quickly withdrawn so that a thin film of soap covers it. When the trowel is tilted, the soap film should slide off slowly as a sheet, leaving the trowel dry. If the film breaks up and leaves the trowel wet, the fit is too open; if the film does not slide off, the fit is insufficiently open. After several days of settling, the upper neat layer is removed through the swing pipe to storage tanks.

The bottom nigre layer may contain as much as one third of the fatty matter originally charged into the pan as dark-colored soap of ca 30–35% fatty acid content. Thus, it is essential to conserve the soap content while disposing of the separated impurities by various means. One procedure is to pump ca 25% of the nigre to an accumulation pan where, when enough has been collected, a fitting change is made and a secondary nigre is removed. The latter is usually disposed of by acidulation and the low grade fatty acids thus obtained are sold. The upper layer is used to start another boil of low quality soap. The remainder of the original nigre in the starting pan is washed with salt water to remove soluble impurities and settled to remove particulate impurities in the brine, which is discarded. A new boil of the same quality is started on the nigre whose soap content assists the new saponification. Alternatively, the whole of the original nigre may be transferred to the start of a new boil of lesser-quality soap.

Several variations of the soap-boiling sequence are practiced. In the direct-wash system outlined above, fresh water is used for all washes. The system requires a minimum number of kettles but produces a relatively large volume of lye for glycerol processing. In the counterwash system only the saponification lye is transferred directly to the glycerol-recovery units; the remaining lyes are worked backward through the next earlier changes in other boils. This system requires more kettles but reduces the volume of lye.

Another variation concerns the degree of completion of saponification sought in the first stage; complete saponification may be sought or, in the interest of obtaining a neutral lye, only 90–95% of the fat may be saponified. In the latter case, a strong change is added after several wash lyes are removed for glycerol recovery. An excess of caustic soda is added in the strong change to complete saponification and the alkali-rich lye is used in the saponification stage of a new boil. In the counterwash system, saponification is generally not complete in the first stage, but extra alkali added in later stages accumulates in lyes that are recycled as described.

Bleaching. The washing-stage lyes remove considerable color bodies, but some further improvement in the color of the soap can be obtained by adding sodium hydrosulfite or sodium hypochlorite to the final washing stage. After a short boil, the soap is grained out as usual (see Bleaching agents).

Rosin. When rosin is incorporated in the soap formula, it is added to the pan in a change just prior to the fit. The rosin acids are neutralized with caustic and, after graining, the lye may be discarded since rosin contributes no glycerol.

A boil requires approximately one week from start to finish. Soap producers maintain a steady supply of neat soap for the finishing operation in holding tanks by staggering the start-up of different boils.

Cold-Process Saponification. This is the simplest of the batch saponification procedures and requires a minimum of equipment. Since neither lyes nor nigre are separated, the glycerol and impurities from the fats remain in the soap. The fat charge is simply melted in a vessel equipped with a mechanical stirrer, and the calculated amount of caustic soda solution is added with vigorous stirring. After emulsification and thickening, the mass is poured into frames where saponification is completed during cooling and solidification. The sides of the frame are then removed, and the soap slab is cut into bars.

Semiboiled Saponification. This is similar to the cold process, but a higher temperature is used to speed saponification and permit adjustment of the alkali content before framing. The fat charge and alkali (which may be caustic potash for soft soaps) are thoroughly mixed at 70–80°C until the soap becomes smooth. Just prior to framing, the soap may be perfumed and small amounts of sodium silicate or other builders may be added to laundry soaps and fine sand, pumice, etc, to abrasive hand soaps (see Abrasives).

Soap from Fatty Acids. Fatty acids obtained from natural oils by hydrolysis (see Raw Materials) or by acidulation of foots are readily neutralized with caustic alkalies or carbonates to form soaps. Glycerol recovery is eliminated and, since the fatty acids are usually distilled before use, nigre separations may also be avoided so that neutralization can be carried out in crutchers, ie, steam-jacketed vessels containing a sweep agitator, rather than in kettles.

Continuous Processes

Efforts to shorten the processing time of the full-boiled kettle procedure have resulted in a number of continuous processes for soap manufacture. These include both continuous splitting, distillation and neutralization of fatty acids, and continuous saponification of neutral fats, followed by continuous washing and fitting with attendant lye and nigre separations. Some of the processes described (eg, Sharples and Mon Savon) are no longer commercially available but are still in use, at least in part, at some locations.

Procter & Gamble Process. In 1938, the Procter & Gamble Company began the manufacture of soap by a continuous process which converts raw fats to finished soaps in a matter of hours (15). The blended fats, containing zinc oxide as a catalyst, react countercurrently with water in a 20-m high stainless-steel hydrolyzing tower, maintained at 232–260°C and 4.1–4.8 MPa (600–700 psi), to yield a continuous output of fatty acid from the top and crude glycerol from the bottom. The fatty acids are distilled in vacuum and neutralized in a continuous process with proportioned caustic soda solution containing salt in a high-speed mixer. The product is neat soap ready for the various finishing operations.

Sharples Soap Process. The Sharples system converts fats directly to soaps and uses centrifuges to separate lyes for washing and glycerol recovery (16–18). Two stages of saponification (similar to the saponification change and the strong change of kettle boiling) are used, each making use of a mixer and a centrifuge. Washings with brine solutions are accomplished in similar mixer–centrifuge stages. The balance between soap, salt and water in the final mixer can be adjusted to yield a neat–nigre separation in the final centrifuge; such fitting may not be necessary, however, since the earlier centrifugal separations remove most of the impurities found in the settled nigres. Thus, the finish can be done in the neat soap–lye region of the phase diagram.

Mon Savon Process. This process continuously converts fats and oils directly into soap and provides for the recovery of glycerol (19). Hot, proportioned amounts of fat and caustic soda are first emulsified in a homogenizer to speed saponification. The emulsion is fed continuously onto the hot inner wall of a reactor where saponification is completed. Washings and lye are removed by countercurrent flows of crude soap and hot brine solutions through a multistage washing tower. Each unit of the tower is divided into mixing and settling zones. The soap discharged from the washing tower is fitted with caustic soda solution to the neat–nigre composition and settled in pans. The separated nigre is recycled through the washing tower.

De Laval Process. In the De Laval Centripure process, the fats and caustic soda solution are proportioned countercurrently into a vertical reactor, through which a large proportion of the saponified fat is continuously recirculated (20). The presence of the soap speeds saponification by emulsifying the incoming raw materials. A viscosity probe in the recirculating line automatically regulates the addition of caustic soda. An amount of fully saponified soap equal to the raw feeds emerges from the top of the reactor for continuous countercurrent washing and brine solution, followed by centrifugal separation of soap and lye. Additions of electrolyte to the fitting stage are controlled by the viscosity. Approximately 90% of the nigre separated in the last centrifuge is returned to the final washing stage.

Mazzoni Processes. Automated continuous Mazzoni systems are available for the saponification of fats, including the washing of the soap and the recovery of glycerol (SCN process). The SC process provides for the continuous neutralization of fatty acids by caustic soda, whereas the SCC process uses sodium carbonate for neutralization.

In the SCN process proportioned amounts of fat, caustic soda, and brine solution are metered into a four-stage reaction autoclave where heat, pressure, and recirculation bring saponification to 99.5% completion (21). An amount of reaction mixture equivalent to the combined feeds is continuously cooled and passed into a static separator for phase separation and removal of the spent lye. The soap is washed with countercurrent lyes in several stages of mixers plus static separators. In the last stage, the soap is washed with fresh brine and separated by centrifugation into soap and lye phases.

In the SC process, preheated fatty acids are proportioned by duplex pumps into a multistage centrifugal mixer, where they are mixed with a recirculating soap stream (22). An electrode in the soap stream monitors the caustic feed to maintain a constant preset pH. An amount of soap equal to the feed volumes is constantly withdrawn from this system to a holding tank. Sodium chloride is either added with the caustic soda or metered in separately. The SCC process is similar, but the sodium carbonate is used for most of the fatty acid neutralization and sodium hydroxide is used only for the final adjustment.

Crosfield Process. J. Crosfield and Sons, a Unilever subsidiary in the UK, began the continuous production of soap for spray-dried products in 1962 (23). The saponification unit consists of three interconnected vertical reactors with internal baffles. Metered quantities of fat and caustic soda solution are recycled through two of these reactors to maintain a high soap concentration, which emulsifies the incoming feed. Saponification control is maintained automatically by regulating the feed of one reactant by means of a viscosity indicator in the recirculating loop. A bleed from the last reactor feeds a liquid–liquid extracting unit for countercurrent washing with hot brine (see Extraction). Fitting is done in a shallow tank with a pump-operated recirculation loop. Another viscosity control adjusts the electrolyte concentration for a neat–nigre separation obtained in a bank of centrifuges.

Armour Process. In 1964 the Armour Company began producing soap by an automated continuous process combining units of the Mazzoni and the De Laval systems (24). The fat mixture is first hydrolyzed into fatty acids and water in splitting towers at high temperature and under high pressure. After distillation, the fatty acids are neutralized in a De Laval Centripure reactor and the neat soap is pumped directly into a holding tank. The finishing units consist of Mazzoni vacuum dryers, amalgamators where perfume, dye, and bacteriostat are mixed in, and Mazzoni Triplex plodders.

Jet Saponification. The Unilever process for the jet saponification of fats conserves steam by withholding it during the exothermic stage of the reaction (25). A three-way jet uses steam to atomize proportioned feeds of fat and of caustic soda solution containing salt, thus producing the emulsion necessary to start saponification. The emulsion is sprayed into the first compartment of a receiving vessel where saponification takes place nearly to completion without benefit of added heat or agitation. After holding for a time, determined by the height of a weir, the contents spill over into a second compartment where live steam may be added to complete saponification. The crude neat soap is washed countercurrently. Separation of neat soap from lye and fitting operations are carried out by means of centrifuges.

High Caustic–High Solids Saponification. The high caustic–high solids saponification process can result in production of 86% soap, thereby eliminating the need for drying (26–28). Fatty acids and 50% caustic soda react in an Alfa-Laval neutralization column. Viscosity measurement is used to control the neutralization (see Rheological measurements).

Finishing Operations

The result of both the kettle and continuous saponification procedures is neat soap containing ca 30% water. Hence, the following finishing steps are customary, in one form or another, to convert neat soap to useful products.

Crutching. Incorporation of various additives, eg, builders, into the soap may be accomplished using vessels called crutchers. These are cylindrical vessels usually containing dual agitators. A screw agitator lifts the mixture from the bottom of the crutcher, and a sweep agitator provides lateral mixing.

Framing. Bar soaps can be produced by allowing neat soap to cool in rectangular frames. The solidified soap is then cut into cakes. These cakes can be stamped following partial drying of the surfaces. Addition of perfumes to framed soaps is restricted by the fact that the soap is hot when poured into the frames.

Drying for Bars or Flakes. The 30% water content of neat soap must be reduced to 10–15% before the soap can be shaped into bars and to 5–10% before it is cut into flakes. The drying is generally accomplished in cabinet, flash, or vacuum drying units (see Drying). Prior to cabinet drying, the molten neat soap from storage is dropped onto a chill roll, and ribbons of the solidified soap are scraped off the roll and dropped on a wire-mesh conveyor. The latter moves through a hot-air cabinet where the residence time and the temperature of the air are adjusted to give the desired degree of drying. In flash drying, the neat soap is superheated under pressure in a heat exchanger, then released through an orifice onto a chill roll.

Vacuum spray drying has largely replaced both of the above methods. In a vacuum spray dryer, the neat soap is pumped through a heat exchanger. The hot neat soap is then sprayed through a rotating nozzle onto the cold inner wall of a vacuum chamber from which it is scraped mechanically. The dried soap falls onto a screw and is extruded through a multi-hole plate and cut into pellets. A major advantage of this type of drying is greater uniformity. This minimizes formation of hard particles that give finished bar soaps a rough feel.

Mixing and Milling. Pigments, dies, perfumes, germicides, antioxidants, etc, may be added to bar and flake soap products. These ingredients are generally coarsely worked into the dried soap pellets or flakes in a batch mixer equipped with a helical agitator. The mixture is then dumped onto a three- or five-roll mill where it is squeezed through the first two rolls, picked up by the faster roll, and passed on. The soap is scraped off the last roll as ribbons by a series of staggered knives. Each mixer batch of soap is generally milled twice. The milling operation completes mixing of the additives with the soap, and the compression and shearing action grind out hard particles that would give the soap a rough feel.

Alternatively, the mixer contents may be homogenized in cylindrical refiners where an internal screw forces the soap through small holes in an end plate. Pellets cut from this plate may be fed to a second refiner or to a mill.

Plodding of Bar Soaps. Mill ribbons or refiner pellets are fed through a hopper into a plodder where an internal screw forces the soap into a compression area that ends in a tapered outlet fitted with a die through which a log of soap is continuously extruded. Cooling water can be circulated through a jacket surrounding the plodder barrel. The plodder die can be heated to obtain a smooth surface on the extruded soap. The log is then cut, cooled, stamped, wrapped, and cartoned. All of these operations are conducted continuously by machines capable of producing >300 bars per minute.

Flakes. In the manufacture of flakes, soap from the mixer is passed through finishing rolls adjusted to a close tolerance to give a thin, glossy film of soap. The flake shapes are marked by rotating cutters and stripped from the final roll by a knife.

Spray-Dried Powders. In the manufacture of spray-dried powders, additives are suspended or dissolved in hot neat soap in a crutcher before drying. Additives may include inorganic builders, optical brighteners (see Brighteners, fluorescent), and dyes. When the additives are thoroughly dispersed or dissolved, the mix is pumped to a holding tank and then through a high pressure pump to atomizing nozzles near the top of a spray tower. Hot air circulated through the tower evaporates water from the spray and forms beads. Powder from the base of the tower is screened to remove off-size material, which is returned to the crutcher. Then the powder can be conveyed through a zone for mist perfuming before packaging.

Heavy-duty soap powders containing inorganic builders, eg, sodium tripolyphosphate, sodium pyrophosphate, sodium perborate, and sodium silicate, have been almost completely replaced in the United States by heavy-duty synthetic detergents containing the same additives.

Specialty Soaps

Deodorant Soaps. Modern antibacterial agents have supplanted the phenol and cresylic acid bar-soap formulations and are in great demand by consumers. The deodorant segment accounts for 50% of all bar soap sales. Antibacterial agents (qv) used in soap bars are 3,4,4′-trichlorocarbanilide (**1**) (TCC) and 2-hydroxy-2′,4,4′-trichlorodiphenyl ether (**2**) (Irgasan DP-300) (see Disinfectants and antiseptics). The functional properties of these agents are the effective suppression of the growth of skin bacteria responsible for body odor at low concentrations (29) and a substantivity to the skin such that rinsing away is resisted.

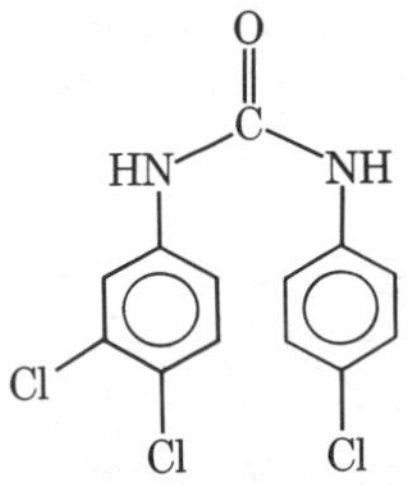

(1)

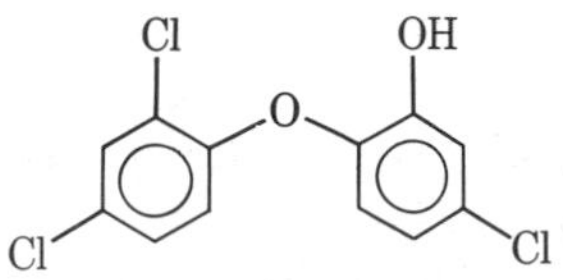

(2)

Superfatted Soaps. Many toilet soaps are superfatted with 2–10% excess of unsaponified oil, fatty acid, or lanolin to produce a soft cold-cream effect by leaving a residual film on the skin after washing and rinsing. These soaps also exhibit a richer, more dense lather than nonsuperfatted soaps. Other superfatting agents that have been used include mineral oil, fatty esters, and fatty alcohols (30–31).

Liquid Soaps. Liquid soaps of the type generally found in public washrooms and shampoos are formulated with the more soluble potassium, ammonium, or triethanolamine soaps of coconut, olive, and other low titer oils (see Alkanolamines). Synthetic detergents have replaced soap in most shampoo formulations and in the liquid "soaps" marketed for household use.

Floating Soaps. The original floating soaps were made by beating air into molten neat soap in an open crutcher, followed by solidification in frames. These coarse air dispersions have been vastly improved by the newer methods of incorporating air under pressure in closed kneading units (6–7). Moreover, the lower water content of the new soaps (ca 20% instead of 30%) affords sufficient firmness to permit extrusion into logs for cutting and stamping after in-process cooling.

Marbleized Soaps. Marbleized soap bars are produced either by mixing different-colored soap pellets in a plodder (32–33) or by injecting a liquid into the soap as it passes through the plodder (34–41).

Transparent Soaps. Additions of alcohol, sugar solution, and glycerol to the hot soap in the semiboiled method inhibit the growth of soap crystallites during frame cooling and promote a glossy, transparent condition. Milled translucent soaps are also made without benefit of these additives by mechanically working the soap at controlled temperatures (see Solid-Soap Phases, above).

Scouring Soaps. Slightly abrasive soaps manufactured in bar, paste, or powder forms contain a finely powdered insoluble material such as pumice to assist in particular cleaning jobs. Alkaline builders, eg, sodium silicate, sodium carbonate, or trisodium phosphate, are frequently incorporated. Soap contents in such products are generally low (5–10%) since no great amount of sudsing is desired. Mechanics' hand soaps are examples of such formulations.

Shaving Creams. Soaps for shaving are generally formulated by saponifying a mixture of coconut oil and stearic acid with mixed caustic potash and soda. Stearic acid (4–8%) may be left unneutralized to supply a pearly luster and 5–10% glycerol may be added for body (see Cosmetics).

Analysis

In the manufacture of soaps, certain analytical tests are performed during and after the various processing operations. The American Oil Chemists' Society publishes a manual of sampling and analytical testing methods for soaps and soap products (42). Some of the more important analytical tests made on soaps are for total fatty acids, color of fatty acids, free alkalinity, salt, and glycerol. The lyes are tested for alkalinity, salt, and glycerol.

Total fatty acids. The sample is hydrolyzed with acid, the fatty acids are extracted with ether, which is evaporated, and the residue is weighed.

Color of soap is usually closely related to the color of the washed fatty acids split from the sample. The fatty acid color is compared with standard colors. For the light-colored fatty acids, a 13.3-cm column is compared with color standards in a Lovibond tintometer. Darker fatty acids are compared with standard color disks, which conform to specifications of the Fat Analysis Committee (FAC) of the AOCS.

Free alkalinity. A sample is dissolved in alcohol and titrated to a phenolphthalein end point with standard acid. The result is usually expressed in terms of Na_2O.

Salt is determined by titration with silver nitrate, using potassium chromate as an indicator.

Glycerol. Analysis is accomplished by hydrolyzing the soap with mineral acid and determining the glycerol content of the aqueous phase by oxidation with potassium dichromate or sodium periodate.

Health and Safety Factors

In the soap industry, the handling of strong caustic soda solutions is probably the greatest safety hazard. Caustic soda is usually stored and used in processing operations as a 1.53 g/cm^3 (50° Bé) solution (50 wt % NaOH). At this strength it is corrosive and may cause serious eye injuries and body burns if not washed off quickly with water. Goggles and protective clothing are worn where danger from caustic soda exists.

In some flake- and powder-processing operations, high dust concentrations may prove irritating to the mucous surfaces of the nasal passages and throat. Such irritation causes mild to severe discomfort. The discomfort arises from irritation, dryness, and soreness of mucous surfaces, sometimes accompanied by excessive mucous discharge. Workers in dust areas wear masks to prevent irritation.

Uses

Table 3 lists the quantity and value of soap shipped by all U.S. producers during 1977 by product class. This table shows that the principal use of the soluble alkali soaps is in toilet bars. However, their detersive and emulsifying properties are also utilized in a number of other areas. The vast array of anionic, cationic and nonionic synthetic detergents available today has undoubtedly restricted, but not eliminated, the role of soaps in some of these special applications.

Textile mills consume soap in kier-boiling cotton (qv), scouring wool (qv), and degumming silk (qv) to remove impurities prior to finishing operations and to assist

Table 3. Quantity and Value of Shipments by All Producers, 1977 [a]

Soap	Quantity	Value, 10^6\$
Nonhousehold		
chips, flakes, granulated, powdered, sprayed, including washing powders	68.2×10^6 t	44.5
liquid (potash and others, excluding shampoos)	49.6×10^6 L [b]	52.1
other nonhousehold soaps, including mechanics hand soap	43.9×10^6 t	43.0
Household		
toilet soaps	381.0×10^6 t	614.5
chips, flakes, etc	32.4×10^6 t	41.1
laundry and other household bars	na	na
other household soaps	na	23.2
soap, not specified by kind	na	8.2
shampoos	na	87.1
shaving creams	na	96.2

[a] Ref. 43.
[b] To convert L to gal, divide by 3.785.

the level application of softening agents used to improve fabric feel (see Quaternary ammonium compounds).

Soaps play a key role in emulsion-polymerization processes used in the rubber and plastic industries. In producing SBR (styrene-butadiene rubber), for example, soaps maintain an intimate emulsion between the dispersed monomers and the aqueous phase during the polymerization. Partially hydrogenated tallow soaps are generally formed *in situ* by dissolving the fatty acids in the monomers and adding the sodium or potassium alkali to the water phase. Soaps function similarly in the emulsion polymerizations of acrylic and vinyl monomers.

Soaps are widely used in the cosmetic industry to emulsify a variety of skin cleaners and conditioners. A partial listing of the latter includes vegetable oils, fatty acids, waxes, and mineral oil. The soaps are generally prepared *in situ*, and the kind and amount of soap largely determines the liquid, paste, or gel forms of the end products.

Sodium and lithium soaps are extensively used to thicken mineral oil in the manufacture of lubricating greases (see also Driers and metallic soaps).

Leather (qv) is degreased by hot soap solutions during processing. Finished saddles and leather goods are maintained by application of a saddle soap which contains a few percent of emulsified beeswax or carnauba wax .

Soaps are still used as wetting and spreading agents to improve the dispersal of insecticidal and fungicidal components of agricultural sprays (see Fungicides; Insect control technology).

Ammonia (qv) and alkanolamines, eg, mono- and triethanolamines, monoisopropanolamine, and 2-amino-2-methylpropanol (AMP), are used to neutralize fatty acids to form specialty soaps. These soaps are good emulsifiers and are diversely employed in cosmetic preparations, soluble cutting oils used in textile processing, furniture, floor, and automobile polishes, and emulsion paints.

BIBLIOGRAPHY

"Soap" in *ECT* 1st ed., Vol. 12, pp. 553–598, by G. W. Busby, Lever Brothers Company; "Soap" in *ECT* 2nd ed., Vol. 18, pp. 415–432, by F. V. Ryer, Lever Brothers Co., Inc.

1. *1954, 1967, 1977 Census of Manufacturers*, Vol. II, U.S. Government Printing Office, Washington, D.C., Product Statistics, Tables 6A and 6C.
2. W. H. Mattil, *Oil Soap* **21,** 198 (1944).
3. J. W. McBain and W. W. Lee, *Ind. Eng. Chem.* **35,** 917 (1943).
4. J. W. McBain and W. W. Lee, *Oil Soap* **20,** 17 (1943).
5. J. W. McBain, R. D. Vold, and K. Gardiner, *Oil Soap* **20,** 221 (1943).
6. U.S. Pat. 2,215,539 (Sept. 24, 1940), J. W. Bodman (to Lever Brothers Co., Inc.).
7. U.S. Pats. 2,295,594; 2,295,595; 2,295,596 (Sept. 15, 1942), V. Mills (to Procter & Gamble).
8. R. H. Ferguson, F. B. Rosevear, and R. C. Stillman, *Ind. Eng. Chem.* **35,** 1005 (1943).
9. R. H. Ferguson, *Oil Soap* **21,** 6 (1944).
10. M. J. Buerger, L. B. Smith, F. V. Ryer, and J. E. Spike, Jr., *Proc. Natl. Acad. Sci. U.S.* **31,** 226 (1945).
11. R. D. Vold, J. D. Grandine, and H. Schott, *J. Phys. Chem.* **56,** 128 (1952).
12. M. J. Buerger, L. B. Smith, and F. V. Ryer, *J. Am. Oil Chem. Soc.* **24,** 193 (1947).
13. U.S. Pat. 3,274,119 (Sept. 20, 1966), S. Goldwasser and F. V. Ryer (to Lever Bros. Co., Inc.).
14. U.S. Pat. 2,970,116 (Jan. 31, 1961), W. A. Kelly and H. D. Hamilton (to Lever Brothers Co., Inc.).
15. G. W. McBride, *Chem. Eng.* **54,** 94 (1947).
16. U.S. Pats. 2,300,749; 2,300,750 (Nov. 3, 1942), A. T. Scott (to Sharples Corporation).
17. U.S. Pat. 2,300,751 (Nov. 3, 1942), A. T. Scott and L. Sender (to Sharples Corporation).
18. U.S. Pat. 2,336,893 (Dec. 14, 1943), A. T. Scott (to Sharples Corporation).
19. F. Lachampt and R. Perron in R. T. Halman, W. O. Lundberg, and T. Malkin, eds., *The Chemistry of Fats and Other Lipids*, Vol. 5, Pergamon Press, Inc., New York, 1958, p. 34.
20. F. T. E. Palmqvist and F. E. Sullivan, *J. Am. Oil Chem. Soc.* **36,** 173 (1959).
21. A. L. Schulerud, *J. Am. Oil Chem. Soc.* **40,** 1609 (1963).
22. A. Lanteri, *Seifen Oele Fette Wachse* **84,** 589 (1958).
23. F. V. Wells, *Soap Chem. Spec.* **38,** 49 (1962).
24. *Soap Chem. Spec.* **40,** 67 (1964).
25. U.S. Pat. 2,566,359 (Sept. 4, 1954), R. V. Owen (to Lever Brothers Co., Inc.).
26. D. Osteroth, *Fette Seifen Anstrichm.* **74,** 411 (1972).
27. E. Jungermann, *J. Am. Oil Chem. Soc.* **50,** 475 (1973).
28. E. Jungermann, *Tenside Deterg.* **13,** 9 (1976).
29. E. Jungermann, J. Brown, Jr., F. Yackovich, and D. Taber, *J. Am. Oil Chem. Soc.* **44,** 232 (1967).
30. U.S. Pat. 3,598,746 (Aug. 10, 1971), T. J. Kanieki, T. J. Hassapis, and T. Liebman (to Armour-Dial, Inc.).
31. D. Osteroth and W. Heers, *Seifen Oele Fette Wachse* **97,** 495 (1971).
32. Ital. Pat. 790,992 (Nov. 2, 1967), A. Busto (to Costruzioni Miche G. Mazzoni).
33. U.S. Pat. 3,673,294 (June 27, 1972), R. G. Matthaei (to Unilever, Ltd.).
34. U.S. Pat. 3,663,671 (May 16, 1972), R. W. Meye and G. Thor (to Henkel and Cie.).
35. Brit. Pat. 1,148,273 (Apr. 10, 1969), R. E. Compa and M. L. Liebowitz (to Colgate-Palmolive Company).
36. Ger. Pat. 2,210,385 (Sept. 14, 1972), E. H. Evans, P. F. Humphreys, E. Schoenig, and H. Brueckel (to Unilever, N.V.).
37. U.S. Pat. 3,485,905 (Dec. 23, 1969), R. E. Compa and M. Liebowitz (to Colgate-Palmolive Company).
38. U.S. Pat. 3,676,538 (July 11, 1972), C. B. Patterson (to Purex Company).
39. Ger. Pat. 2,206,221 (May 23, 1972), R. G. Matthaei (to Unilever, N.V.).
40. U.S. Pat. 3,769,225 (Oct. 30, 1973), R. G. Matthaei (to Lever Brothers Co., Inc.).
41. Ger. Offen. 2,054,865 (June 24, 1971), R. G. Matthaei (to Unilever, Ltd.).
42. *Official and Tentative Methods of the American Oil Chemist's Society*, American Oil Chemist's Society, Chicago, Ill. (Annual Publication).

43. *1977 Census of Manufacturers*, Vol. II, U.S. Government Printing Office, Washington, D.C., Product Statistics, Table 6A.

General Reference

D. Swern, ed., *Bailey's Industrial Oil and Fat Products*, 4th ed., Vol. 1, John Wiley & Sons, Inc., 1979.

FREDERICK S. OSMER
Lever Brothers Co., Inc.

SODA. See Alkali and chlorine products.

SODIUM AND SODIUM ALLOYS

SODIUM

Sodium [*7440-23-5*], an alkali metal, is the second element of group IA of the periodic table, at wt 22.9898. The chemical symbol Na is derived from the German (or Latin) Natrium. Commercial interest in the metal derives from its high chemical reactivity, low melting point, high boiling point, good thermal and electrical conductivity, and low cost.

Sir Humphry Davy first isolated metallic sodium in 1807 by the electrolytic decomposition of sodium hydroxide. Later, the metal was produced experimentally by thermal reduction of the hydroxide with iron. In 1855, commercial production was started by the Deville process in which sodium carbonate is reduced with carbon at ≥1100°C. In 1886, Castner developed a process for the thermal reduction of sodium hydroxide with carbon and later made sodium on a commercial scale by the electrolysis of sodium hydroxide (1–2). The Downs process for the electrolytic decomposition of fused sodium chloride was patented in 1924 (2–3). It has been the preferred process since installation of the first electrolysis cells at Niagara Falls in 1925 and is widely used throughout the world.

Sodium was first used commercially to make aluminum by the reduction of sodium aluminum chloride. Other uses developed over the years, but at present the principal applications are for the manufacture of lead antiknock agents and titanium metal and as a heat-transfer medium. Small amounts are used to produce sodium hydride, sodamide, and sodium peroxide, and in the preparation of organic compounds and pharmaceuticals.

Sodium is not found in the free state in nature because of its high chemical reactivity. It occurs naturally as a component of many complex minerals and of such simple ones as sodium chloride, sodium carbonate, sodium sulfate, sodium borate, and sodium nitrate. Soluble sodium salts are found in seawater, mineral springs, and salt lakes. Principal U.S. commercial deposits of sodium salts are the Great Salt Lake, Searles Lake, and the rock salt beds of the Gulf Coast, Virginia, New York, and Michigan (see Chemicals from brine). Sodium-23 is the only naturally occurring isotope. The six artificial radioisotopes (qv) are listed in Table 1.

Physical Properties

Sodium is a soft, malleable solid which is readily cut with a knife or extruded as wire. It is commonly coated with a layer of white sodium monoxide, carbonate, or hydroxide depending on the degree and kind of atmospheric exposure. In a strictly anhydrous inert atmosphere, the freshly cut surface has a faintly pink, bright metallic luster. Liquid sodium in such an atmosphere looks much like mercury. Both liquid and solid oxidize in air but traces of moisture appear to be required for the reaction to proceed. Oxidation of the liquid is accelerated by an increase in temperature.

Only body-centered cubic crystals, lattice constant 428.2 pm at 20°C, are reported for sodium (4). The atomic radius is 185 pm, the ionic radius 97 pm, and the electronic configuration $1s^22s^22p^63s^1$ (5). Other physical properties of sodium are given in Table 2. For greater detail and properties not included here see ref. 5 which includes 152 references plus a reading list.

Sodium is paramagnetic. The vapor is chiefly monatomic, although the dimer and tetramer are reported (6). Thin films are opaque in the visible range but transmit in the ultraviolet at ca 210 nm. The vapor is blue, but brilliant green is frequently observed when working with sodium at high temperature, presumably because of mixing of the blue with yellow from partial burning of the vapor.

At 100–300°C sodium readily wets and spreads over many dry solids, eg, sodium chloride or aluminum oxide. In this form it is highly reactive (7). It does not easily wet stainless or carbon steels. Wetting of such structural metals is influenced by the cleanliness of the surface, the purity of the sodium, temperature, and the time of exposure. Wetting occurs more readily at ≥300°C and, once attained, persists at lower temperatures (5).

Sodium Dispersions. Sodium is easily dispersed in inert hydrocarbons, eg, white oil or kerosene, by agitation or with a homogenizing device. Addition of oleic acid and other long-chain fatty acids, higher alcohols and esters, and some finely divided solids, eg, carbon or bentonite, accelerate dispersion and produce finer particles. Above 98°C, the sodium is present as liquid spheres. On cooling to lower temperatures, solid spheres of sodium remain dispersed in the hydrocarbon and present an extended surface for reaction. Dispersions may contain as much as 50 wt % sodium. Sodium in this form

Table 1. Radioisotopes of Sodium

Isotope	CAS Registry No.	Half-life
sodium-20	[*14809-59-7*]	0.4 s
sodium-21	[*15594-24-8*]	23.0 s
sodium-22	[*13966-32-0*]	2.58 yr
sodium-24	[*13982-04-2*]	15.0 h
sodium-25	[*15760-13-1*]	60.0 s
sodium-26	[*26103-12-8*]	1.0 s

Table 2. Physical Properties of Sodium[a]

Property	Value	Property	Value
ionization potential, V	5.12	specific heat, kJ/(kg·K)[b]	
melting point, °C	97.82	solid	
heat of fusion, kJ/kg[b]	113	at 20°C	2.01
volume change on melting, %	+2.63	at mp	2.16
boiling point, °C	881.4	liquid	
heat of vaporization at bp, MJ/kg[b]	3.874	at mp	1.38
density of the solid, g/cm^3		at 400°C	1.28
at 20°C	0.968	at 550°C	1.26
at 50°C	0.962	electrical resistivity, μΩ·cm	
at mp	0.951[c]	solid	
density of the liquid, g/cm^3		at 20°C	4.69
at mp	0.927	at mp	6.60[c]
at 400°C	0.856	liquid	
at 550°C	0.820	at mp	9.64
viscosity, mPa·s (= cP)		at 400°C	22.14
at 100°C	0.680	at 550°C	29.91
at 400°C	0.284	thermal conductivity, W/(m·K)	
at 550°C	0.225	solid	
surface tension, mN/m (= dyn/cm)		at 20°C	1323
at mp	192	at mp	1193[c]
at 400°C	161	liquid	
at 550°C	146	at mp	870
		at 400°C	722
		at 550°C	648

[a] Ref. 5.
[b] To convert J to cal, divide by 4.184.
[c] Estimated.

is easily handled and reacts rapidly. For some purposes the presence of the inert hydrocarbon is a disadvantage.

High Surface Sodium. Liquid sodium readily wets many solid surfaces. This property may be used to provide a highly reactive form of sodium without contamination by hydrocarbons. Powdered solids with a high surface area per unit volume, eg, completely dehydrated activated alumina powder, provide a suitable base for high surface sodium. Other powders, eg, sodium chloride, hydride, monoxide, or carbonate, can also be used.

The solid to be coated with sodium is placed in a vessel equipped with a stirrer, filled with pure, dry nitrogen or another inert gas, and heated to 110–250°C. Clean sodium is added with stirring. If enough sodium is added to provide at least a monolayer, it is rapidly distributed over the entire available surface. Depending on the available surface, up to 10 wt % or more can be added without changing the free-flowing character of the system (7–8).

Chemical Properties

Sodium forms unstable solutions in liquid ammonia where a slow reaction takes place to form sodamide and hydrogen as follows:

$$Na + NH_3 \rightarrow NaNH_2 + \tfrac{1}{2} H_2$$

Iron, cobalt, and nickel catalyze this reaction; the rate depends upon temperature and sodium concentration. At −33.5°C, 0.251 kg sodium is soluble in 1 kg ammonia. Concentrated solutions of sodium in ammonia separate into two liquid phases when cooled below the consolute temperature of −41.6°C. The compositions of the phases depend on the temperature; at the peak of the conjugate solutions curve, the composition is 4.15 atom % sodium. The density decreases with increasing concentration of sodium. Thus, in the two-phase region the dilute bottom phase, low in sodium concentration, has a deep-blue color, whereas the light top phase, high in sodium concentration, has a metallic bronze appearance (9–13).

At high temperature sodium and its fused halides are mutually soluble (14). The consolute temperatures and corresponding Na mol fractions are given in Table 3. Nitrogen is soluble in liquid sodium to a very limited extent but sodium has been reported as a nitrogen-transfer medium in fast-breeder reactors (5) (see Nuclear reactors).

The solubility–temperature relationships of sodium, sodium compounds, iron, chromium, nickel, helium, hydrogen, and some of the rare gases are important in the design of sodium heat exchangers, especially those used in liquid-metal fast-breeder reactors (LMFBR). The solubility of oxygen in sodium is particularly important because of its marked effect on the corrosion of containment metals and because of problems of plugging narrow passages. This solubility is given as

$$\log S \text{ (ppm wt O)} = 6.239 - \frac{2447}{T(\text{K})}$$

from ca 400 to 825 K in ref. 5, which also presents solubility data on many other materials. Because metallic calcium is always present in commercial sodium, and to a lesser extent in reactor-grade sodium, the solubilities of calcium oxide and nitride in sodium are critical to the design of heat-transfer systems. These compounds are substantially insoluble at 100–120°C (15) (see Heat-exchange technology).

Sodium is soluble in ethylenediamine (16–17), but solubility in other amines such as methyl- or ethylamine may require the presence of ammonia. Sodium is insoluble in most hydrocarbons and is readily dispersed in kerosene or similar liquids toward which it is chemically inert. Such dispersions provide a reactive form of the metal.

In 1932 a new class of complexes of ethers, sodium, and polycyclic hydrocarbons was discovered (18). Sodium reacts with naphthalene in dimethyl ether as solvent to form a soluble, dark green, reactive complex. The solution is electrically conductive. The reaction has been described as follows (8):

Na· + (naphthalene) $\xrightarrow[\text{ether}]{\text{active}}$ (H, Na)·(ether)$_x$ ⟷ (H, Na)·(ether)$_x$

Table 3. Mutual Solubility of Sodium and Its Fused Halides

Compound	Consolute temperature, °C	Na mol fraction
Na/NaF	1182	0.28
Na/NaCl	1080	0.50
Na/NaBr	1025	0.52
Na/NaI	1033	0.50

The addition product, $C_{10}H_8Na$, called naphthalenesodium or sodium naphthalene complex, may be regarded as a resonance hybrid. The ether is more than just a solvent that promotes the reaction: stability of the complex depends on the presence of the ether, and sodium can be liberated by evaporating the ether or by dilution with an indifferent solvent, such as ethyl ether. A number of ether-type solvents are effective in complex preparation, such as methyl ethyl ether, ethylene glycol dimethyl ether, dioxane, THF, and others. Trimethylamine also promotes complex formation. This reaction proceeds with all alkali metals. Other aromatic compounds, eg, diphenyl, anthracene, and phenanthrene, also form sodium complexes (16,19).

Sodium reacts with many elements and substances (5,16,19) and forms well-defined compounds with a number of metals; some of these alloys are liquid below 300°C. When heated in air, sodium ignites at ca 120°C and burns with a yellow flame, evolving a dense white acrid smoke. With air or oxygen a monoxide or peroxide is formed. With limited oxygen supply and below 160°C, the principal reaction product is sodium monoxide, Na_2O. At 250–300°C with adequate oxygen, sodium peroxide, Na_2O_2, is formed along with very small amounts of superoxide, NaO_2 (see Peroxides). Sodium superoxide is made from sodium peroxide and oxygen at high temperature and pressure. Sodium does not react with extremely dry oxygen or air beyond the possible formation of a surface film of transparent oxide (20).

The reaction of sodium and water according to the equation

$$Na + H_2O \rightarrow NaOH + \tfrac{1}{2} H_2 + 141 \text{ kJ/mol (98.95 kcal/mol)}$$

has been extensively studied as it relates to the generation of steam in sodium-cooled breeder reactors (5). Under ordinary circumstances, this reaction is very rapid. The liberated heat melts the sodium and frequently ignites the evolved hydrogen if air is present. In the absence of air and with a large excess of either reactant, the reaction may be relatively nonviolent. Thus, dry steam may be used to clean equipment contaminated with sodium residues, but precaution must be taken to exclude air, avoid condensation, and design equipment to drain without leaving pockets of sodium. Such sodium may become isolated by a layer of solid sodium hydroxide and can be very hazardous.

Hydrogen and sodium do not react at room temperature, but at 200–350°C sodium hydride is formed (21–22). The reaction with bulk sodium is slow because of the limited surface available for reaction, but dispersions in hydrocarbons and high surface sodium react more rapidly (7). With the latter, reaction is further accelerated by surface-active agents such as sodium anthracene-9-carboxylate and sodium phenanthrene-9-carboxylate (23–25).

There is very little evidence for the direct formation of sodium carbide from the elements (26–27), but sodium and graphite form lamellar intercalation compounds (16,28–30). At 500–700°C, sodium and sodium carbonate produce the carbide, Na_2C_2, but above 700°C free carbon is also formed (31). Sodium reacts with carbon monoxide to given sodium carbide (31), and with acetylene to give sodium acetylide ($NaHC_2$) and sodium carbide (disodium acetylide, Na_2C_2) (8).

Nitrogen and sodium do not react at any temperature under ordinary circumstances but are reported to form the nitride or azide under the influence of an electric discharge (14,32). Sodium silicide, NaSi, has been synthesized from the elements (33–34). When heated together, sodium and phosphorus form sodium phosphide, but in the presence of air with ignition sodium phosphate is formed. Sulfur, selenium, and tellurium form the sulfide, selenide, and telluride, respectively. In vapor phase, sodium forms halides with all halogens (14). At room temperature, chlorine and bromine react rapidly with thin films of sodium (35), whereas fluorine and sodium ignite. Molten sodium ignites in chlorine and burns to sodium chloride.

At room temperature, little reaction occurs between carbon dioxide and sodium, but burning sodium reacts vigorously. Under controlled conditions, sodium formate or oxalate may be obtained (8,16). On impact, sodium is reported to react explosively with solid carbon dioxide. In addition to the carbide-forming reaction noted above, carbon monoxide reacts with sodium at 250–340°C to yield sodium carbonyl, $(NaCO)_6$ (36–37). Above 1100°C (temperature of the Deville process), carbon monoxide and sodium do not react. Sodium reacts with nitrous oxide to form sodium oxide and burns in nitric oxide to form a mixture of nitrite and hyponitrite. At low temperature, liquid nitrogen pentoxide reacts with sodium to produce nitrogen dioxide and sodium nitrate.

Phosphorus trichloride and pentachloride form sodium chloride and sodium phosphide, respectively, with sodium. Phosphorus oxychloride, $POCl_3$, when heated with sodium explodes. Carbon disulfide reacts violently, forming sodium sulfide. Sodium amide (sodamide), $NaNH_2$, is formed by the reaction of ammonia gas with liquid sodium. Solid sodium reacts only superficially with liquid sulfur dioxide but molten sodium and gaseous sulfur dioxide react violently. Under carefully controlled conditions, sodium and sulfur dioxide yield sodium hydrosulfite, $Na_2S_2O_4$ (38). Dry hydrogen sulfide gas reacts slowly with solid sodium, but in the presence of moisture the reaction is very rapid. The product is sodium sulfide.

Sodium reacts with dilute acids about as vigorously as it reacts with water. The reaction with concentrated sulfuric acid may be somewhat less vigorous.

At 300–385°C sodium and sodium hydroxide react according to the following equilibrium:

$$2\,Na + NaOH \rightleftharpoons Na_2O + NaH$$

The reaction is displaced to the right by dissociation of sodium hydride with liberation of hydrogen. This dissociation is favored by conducting the reaction under vacuum or by sweeping the reaction zone with an inert gas to remove the hydrogen (21–22). In this manner sodium monoxide substantially free of sodium and sodium hydroxide is produced. In the more complicated reaction between sodium metal and anhydrous potassium hydroxide, potassium metal and sodium hydroxide are produced in a reversible reaction (39–40):

$$Na + KOH \rightleftharpoons K + NaOH$$

Superimposed on this simple equilibrium are complex reactions involving the oxides and hydrides of the respective metals. At ca 400°C the metal phase resulting from the reaction of sodium and potassium hydroxide contains an unidentified reaction product that precipitates at ca 300°C (15).

Data on the free energy of formation (41–42) indicate that sodium reduces the oxides of group IA elements except lithium oxide but does not reduce oxides of group IIA elements except mercury, cadmium, and zinc oxides. Many other oxides are reduced by metallic sodium. In some cases reduction depends on the formation of exothermic complex oxides. Iron oxide is reduced by sodium below ca 1200°C; above this temperature the reaction is reversed. Sodium reduces most fluorides except the fluorides of lithium, the alkaline earths, and some lanthanides. It reduces most metallic chlorides, although some of the group IA and group IIA chlorides give two-phase equilibrium systems consisting of fused salt and alloy layers (40). Some heavy metal sulfides and cyanides are also reduced by sodium.

Sodium reacts with many organic compounds, particularly those containing oxygen, nitrogen, sulfur, halogens, carboxyl, or hydroxyl groups. The reactions are violent in many cases; for example, with organic halides. Carbon may be deposited or hydrogen liberated, and compounds containing sulfur or halogens usually form sodium sulfide or sodium halides. Alcohols give alkoxides (see Alkoxides, metal); primary alcohols react more rapidly than secondary or tertiary. The reactivity decreases with increasing number of alcohol carbon atoms.

Organosodium compounds are prepared from sodium and other organometallic compounds or active methylene compounds by reaction with organic halides, cleavage of ethers, or addition to unsaturated compounds. Some aromatic vinyl compounds and allylic compounds also give sodium derivatives.

Sodium does not react with anhydrous ethyl ether but may react with higher ethers or mixed ethers. Organic acids give the corresponding salts with evolution of hydrogen or decompose. Pure, dry, saturated hydrocarbons, eg, xylene, toluene, and mineral oil, do not react with sodium below the hydrocarbon-cracking temperature. With unsaturated hydrocarbons, sodium may add at a double bond or cause polymerization.

Sodium amalgam or sodium and alcohol are employed for organic reductions. Sodium is also used as a condensing agent in acetoacetic ester and malonic ester syntheses and the Wurtz-Fittig reaction. For a more detailed discussion of the organic reactions of sodium, see reference 16.

Manufacture

Thermal Reduction. Metallic sodium is produced by thermal reduction of several of its compounds. The earliest commercial processes were based on the carbon reduction of sodium carbonate (43–46) or sodium hydroxide (1,8,47):

$$2\,C + Na_2CO_3 \rightarrow 2\,Na + 3\,CO$$

$$2\,C + 6\,NaOH \rightarrow 2\,Na_2CO_3 + 3\,H_2 + 2\,Na$$

Sodium chloride is reduced by ferrosilicon in the presence of lime:

$$4\,NaCl + 3\,CaO + (Fe)Si \rightarrow 2\,CaCl_2 + CaO.SiO_2 + 4\,Na + (Fe)$$

This process was operated briefly in vacuum retorts by Union Carbide in 1945 (48). The chloride is also reduced by calcium carbide at 800–1200°C under vacuum (49):

$$2\,NaCl + CaC_2 \rightarrow 2\,C + CaCl_2 + 2\,Na$$

A number of other thermal reductions are described in the literature (8), but it is doubtful that any have been carried out on commercial scale.

Electrolysis of Fused Sodium Hydroxide. The first successful electrolytic production of sodium was achieved with the Castner cell (2):

Cathode $$4\,Na^+ + 4\,e \rightarrow 4\,Na$$

Anode $$4\,OH^- - 4\,e \rightarrow 2\,H_2O + O_2$$

The water formed at the anode diffuses to the cathode compartment where it reacts with its equivalent of sodium:

$$2\,H_2O + 2\,Na \rightarrow 2\,NaOH + H_2$$

The net change is represented as follows:

$$2\,NaOH \rightarrow 2\,Na + H_2 + O_2$$

Since the water reacts with half of the sodium produced by the electrolysis, the current yield can never be more than 50% of theoretical. Other reactions in the cell lower this yield still more.

The Castner cell was so simple in design and operation that over the years only minor changes were made. A section of a cell used in England in the early 1950s is shown in Figure 1. The fused caustic bath is contained in the cast-iron outer pot which rests in a brick chamber. The cylindrical copper cathode is supported on the cathode stem which extends upward through the bottom of the cell. The cathode stem is sealed

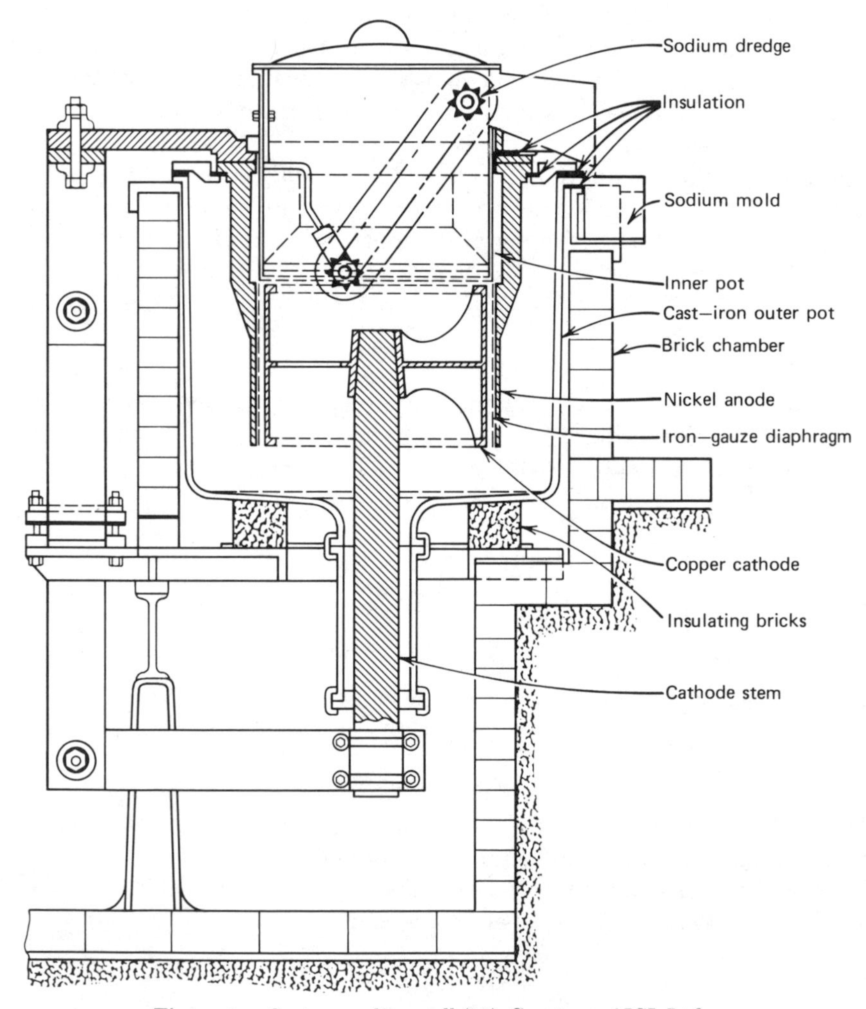

Figure 1. Castner sodium cell (50). Courtesy of ICI, Ltd.

and insulated from the outer pot by a frozen portion of the bath. The cylindrical nickel anode concentric with the cathode, is supported from the rim of the outer pot. The cylindrical iron-gauze diaphragm, located in the 2.5-cm annular space between the electrodes, is suspended from the inner pot. Because of the difference in density, sodium rises in the hydroxide bath and collects on its surface in the inner pot; the latter is electrically insulated from the top anode ring by which it is supported. The inner pot is closed by a cover which maintains an atmosphere of hydrogen over the sodium to prevent burning. No practical way has been found to collect the hydrogen, which is vented, as is the oxygen liberated at the anode. A perforated hand ladle was used to remove sodium from early Castner cells (51). Later, the sodium was removed with mechanically driven iron-gauze buckets. The cell shown in Figure 1 can hold approximately one ton of molten bath, consisting primarily of mercury-cell caustic soda with up to 10% each of sodium chloride and sodium carbonate. Some salt is added initially to improve the bath conductivity, but the carbonate is an unwanted impurity. After several months of operation, the chloride, carbonate, and other impurities attain concentrations that seriously impair the efficiency of the cell; the bath is then renewed. Operating characteristics are given in Table 4.

Small cells are heated externally to maintain operating temperature, but large cells are heated by the electrolysis current. Of the many ingenious systems proposed to prevent the reaction of sodium with the water produced at the anode, none are known to have been applied commercially. A study of the effect of sulfate on the operation of the Castner process under industrial conditions published in Moscow in 1972 suggests that the process may still be in operation in the USSR (52).

Electrolysis of Fused Sodium Chloride. Although many cells have been developed for the electrolysis of fused sodium chloride (8,53–58), the Downs cell (3) has been most successful. In cells in general use by 1945, a single cylindrical anode constructed of several graphite blocks is inserted through the center of the cell bottom and surrounded by an iron-gauze diaphragm and a cylindrical iron cathode. In the 1940s, the single anode and cathode were replaced with a multiple electrode arrangement consisting of four anodes of smaller diameter in a square pattern, each surrounded by a cylindrical diaphragm and cathode as shown in Figure 2. Without substantially increasing the overall cell dimensions, this design increases the electrode area per cell and allows increased amperage.

The cell consists of three chambers. The upper chamber is outside the chlorine dome and above the sodium-collecting ring. The other two chambers are the chlo-

Table 4. Operating Characteristics of the Castner Cell

Property	Value
bath temperature, °C	320 ± 10
cell current, kA	9 ± 0.5
cell voltage, V	4.3–5.0
cathode current density, kA/m^2	10.9
current efficiency, %	40[a]
sodium produced	
g/(A·h)	0.4
g/(kW·h)	90

[a] Approximate value.

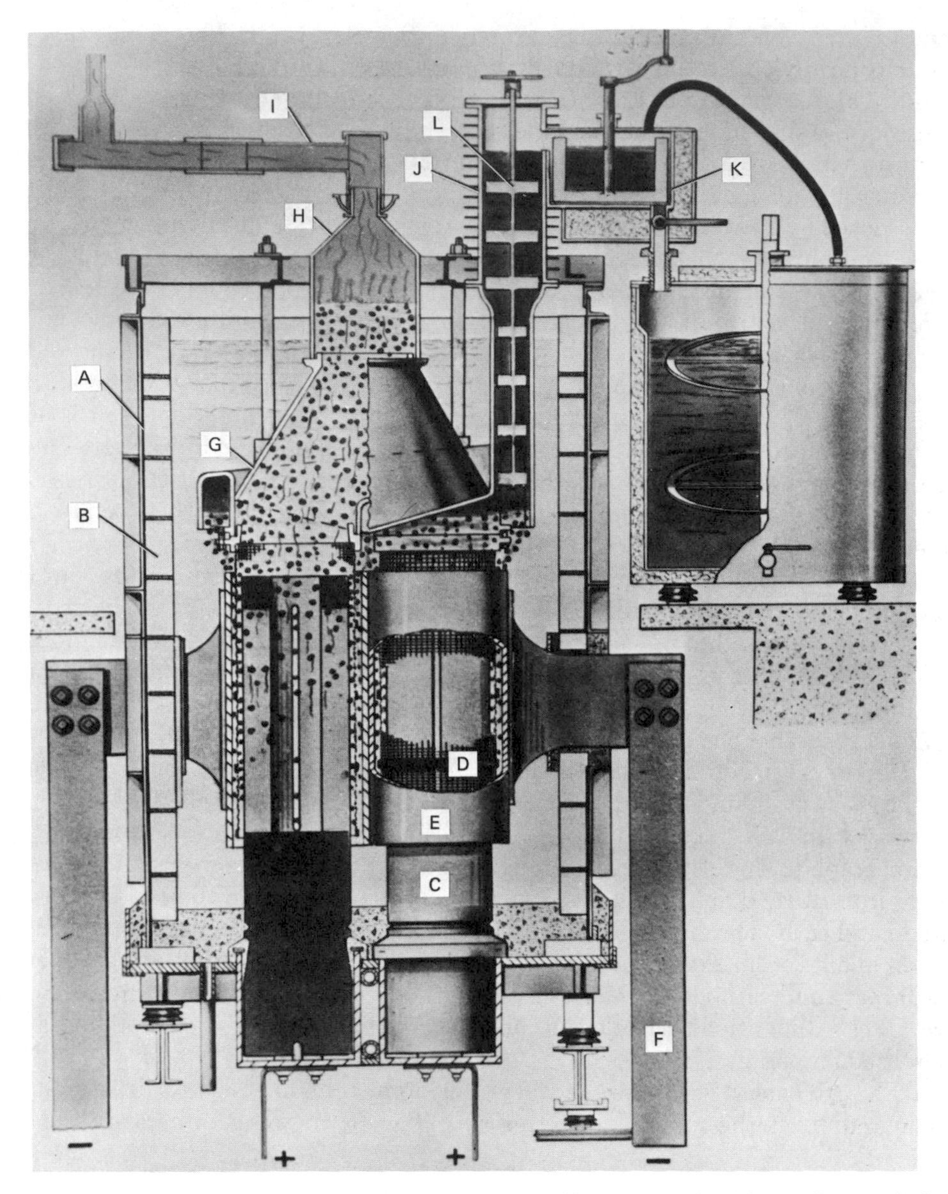

Figure 2. Downs cell. A, the steel shell, contains the fused bath; B is the brick lining. C, four cylindrical graphite anodes, project upward from the base of the cell, each surrounded by D, a diaphragm of iron gauze, and E, a steel cathode. The four cathode cylinders are joined to form a single unit supported on cathode arms projecting through the cell walls and connected to F, the cathode bus bar. The diaphragms are suspended from G, the collector assembly, which is supported from steel beams spanning the cell top. For description of H–K, see text.

rine-collecting zone inside the dome and diaphragm, and the sodium-collecting zone outside the diaphragm and under the sodium-collecting ring. This arrangement prevents recombination of the sodium and chlorine. The collector is a complex assembly of inverted troughs and chambers arranged to collect the products in separate compartments as they rise through the bath. The chlorine emerges through the nickel dome

H and is removed through the chlorine line I to a header. Sodium is channeled to a riser pipe J which leads to a discharge point above the cell wall. The difference in level between the overflowing sodium and the cell bath is due to the roughly 2:1 density ratio of the fused bath and liquid sodium. The upper end of the riser pipe is fitted with fins that cool the sodium and thereby precipitate dissolved calcium. The sodium, still containing some calcium, electrolyte, and oxide, overflows into a receiver, K. The calcium precipitated in the riser pipe tends to adhere to the wall from where it is dislodged by the scraper, L, and returned to the base of the riser. The cell is fitted with an insulated cover to conserve heat and protect the operators, but a small area is left uncovered for visual observation, bath-level regulation, and salt feed. Fine, dry crystalline salt is fed to the open bath through a feed chute (not shown).

The cell bath in early Downs cells (8,14) consisted of about 58 wt % calcium chloride and 42 wt % sodium chloride. This composition is a compromise between melting point and sodium content. Additional calcium chloride would further lower the melting point at the expense of depletion of sodium in the electrolysis zone with resulting complications. With the above composition the cells operate at 580 ± 10°C, well below the temperature of highest sodium solubility in the bath. Calcium chloride causes problems because of the following equilibrium (40):

$$2\,Na + CaCl_2 \leftrightarrows 2\,NaCl + Ca$$

The alloy phase contains about 5 wt % calcium at cell conditions, an amount intolerable for most industrial uses. The bulk is removed by precipitation in the cooled riser pipe. Any precipitated calcium that adheres to the walls of the riser must be scraped off to prevent plugging. The precipitate drops to the bath–metal interface where it reacts to reform calcium chloride and sodium according to the above equilibrium. Calcium remaining in the sodium is largely removed by filtration at ca 110°C. The filtered sodium contains <0.04 wt % calcium. The filtration operation produces a filter cake of calcium, sodium, chlorides, and oxides. Several methods of recovering metallic calcium or calcium alloys from the filter cake have been proposed, but none has been commercialized.

Characteristics of Downs cells as operated in the United States have not been published; data from the UK (2) and the FRG (58) are given in Table 5.

Salt substantially free of sulfate and other impurities is the cell feed. This grade may be purchased from commercial salt suppliers or made on site by purification of crude rock salt. Dried calcium chloride or cell bath from dismantled cells is added to the bath periodically as needed to balance calcium removed with the sodium. The heat

Table 5. Characteristics of Downs Cells from the UK and the FRG

Property	UK	FRG
bath temperature, °C	580 ± 15	590 ± 5
cell current, kA	25–35	24–32
cell voltage, V	7[a]	5.7–6.0
cathode current density, kA/m^2	9.8	9.8
current efficiency, %	75–80	78
cell life, d	500–700	300–350
diaphragm life, d	20–100	20–30

[a] Approximate value.

required to maintain the bath in the molten condition is supplied by the electrolysis current. Other electrolyte compositions have been proposed in which part or all of the calcium chloride is replaced by other salts (59–62). Such baths offer improved current efficiencies and production of crude sodium containing relatively little calcium.

Cell life is determined by the loss of graphite from the anodes. Oxygen released at the anode by electrolysis of oxides or water in the bath reacts with the graphite to form CO and CO_2. In time, erosion of the anode increases the interelectrode spacing with corresponding increases in cell voltage and temperature. At this stage the cell is replaced.

A dimensionally stable anode consisting of an electrically conducting ceramic substrate coated with a noble metal oxide has been developed (53). Iridium oxide, for example, resists anode wear experienced in the Downs and similar electrolytic cells. (see Metal anodes).

Other commercial cells designed for the electrolysis of fused sodium chloride include the Danneel-Lonza cell used before World War I. It had no diaphragm and the sodium was confined to the cathode zone by salt curtains (ceramic walls). The Seward cell was also operated in the United States for a short time before World War I. It utilized the contact-electrode principle, with the cathode immersed only a few millimeters in the electrolyte. The Ciba cell was used over a longer period of time. It was an adaptation to sodium chloride of the Castner cell for the electrolysis of fused caustic. A mixture of sodium chloride and other chlorides, molten at 620°C, was electrolyzed in rectangular or oval cells heated only by the current. Several cells have been patented for the electrolysis of fused salt in cells with molten lead cathodes (63). However, it is difficult to separate the lead from the sodium (see Electrochemical processing).

Electrolysis of Amalgam. Sodium in the form of amalgam as made by the electrolysis of sodium chloride brine in mercury cathode cells is much cheaper than any other form of the metal but commercial use of amalgam is restricted largely to production of caustic soda (see Alkali and chlorine products). Many efforts have been made to develop processes for recovering sodium from amalgam (64–67). Recovery by electrolysis with the amalgam serving as anode has been the favored approach. The electrolytes were generally low melting sodium-salt combinations, although liquid ammonia (68) and organic solvents (69) have been reported. The addition of lead before electrolysis with a low melting salt electrolyte has been patented (70–71).

Sodium was made from amalgam in Germany during World War II (66). The only other commercial application appears to be the Tekkosha process (72–74). In this method, preheated amalgam from a chlor-alkali cell is supplied as anode to a second cell operating at 220–240°C. This cell has an electrolyte of fused sodium hydroxide, sodium iodide, and sodium cyanide and an iron cathode. Operating conditions are given in Table 6.

The sodium produced contains 0.1–0.5 wt % mercury. This mercury is converted to calcium amalgam by treating the crude sodium with powdered anhydrous calcium chloride:

$$Hg + 2\,Na + CaCl_2 \rightarrow CaHg + 2\,NaCl$$

The residual salts ($CaCl_2$, NaCl) and the calcium amalgam are removed by cooling and filtration. The literature does not disclose the mercury content of the refined sodium nor the method for recovering mercury from the filtered solids.

Table 6. Tekkosha Fused-Salt Electrolysis Cell[a]

Property	Value
voltage, V	3.0–3.1
current, kA	60
current density, kA/m^2	4
current efficiency, %	96–98

[a] Refs. 72–74.

The Tekkosha process offers the advantages of moderate temperature, minimum corrosion, simple operation, high efficiency, low labor cost, good working conditions, and process adaptability. In the United States, these advantages would be largely offset by the environmental problems inherent in handling of mercury and the need to produce some caustic soda to balance the in-process sodium inventory.

Electrolysis Based on Cationically Conducting Ceramics. Searching for a method for using sodium and sulfur as reactants in a secondary battery, the Ford Motor Company developed a polycrystalline β-alumina ceramic material that selectively transports sodium cations when subjected to an electric field (45,75) (see Batteries and electric cells; Ceramics, electrically conducting). This ceramic, or any of its many variants, is useful as a diaphragm or divider in a two-compartment cell. In one compartment the sodium is in contact with the ceramic, whereas in the other a suitable liquid electrolyte is in contact with the opposite side of the ceramic. Thus, the sodium is in electrochemical but not physical contact with the liquid electrolyte. Many low melting electrolytes can be used that are otherwise incompatible with sodium; eg, sodium polysulfides, sodium tetrachloroaluminate ($NaAlCl_4$), sodium hydroxide, and mixtures of sodium chloride and zinc chloride or sodium nitrite and nitrate. Because sodium is not in contact with the liquid electrolyte, the various reactions that usually lower the current efficiency of commercial cells do not occur. Cells based on this principle generally operate close to 100% current efficiency. Sodium of exceptional purity is produced at satisfactory operating conditions. However, because a ceramic of predictable properties and long service life has not yet been developed, these cells have not been commercialized. Research on sodium–sulfur batteries is continuing and the literature is extensive and growing (76–83). Solid electrolytes other than sodium beta-alumina are also reported (84–85).

Energy Requirements. The energy requirements of several sodium processes are compared in ref. 74 (see Table 7). The data contain some ambiguities because of the allocation of energy to the coproduction of chlorine. An independent calculation shows a somewhat lower energy consumption for the Downs process (86).

Table 7. Sodium Process Energy Requirements[a]

Process	Total energy, MJ/kg Na[b]
Downs, fused NaCl	107[c]
Castner, fused NaOH	328
Tekkosha, double electrolysis	80
sodium lead, evaporation	55

[a] Ref. 74.
[b] To convert MJ/kg to Btu/lb, multiply by 430.2.
[c] Ref. 86 gives a value of 97 MJ/kg.

Specifications, Shipping

Sodium, generally about 99.95% Na, is available in two grades:

	Calcium, wt %	*Chlorides, wt %*
regular	0.040	0.005
reactor	0.001	0.005

Reactor grade is packed in specially cleaned containers and in some cases with special cover atmospheres.

Sodium is usually shipped in 36- or 54-t tank cars. Smaller amounts are shipped in tank trucks. Sodium is also available in 104- and 190-kg drums and in bricks (ca 0.5–5 kg). A thin layer of oxide, hydroxide, or carbonate is usually present. Sodium is also marketed in small lots as a dispersion in an inert hydrocarbon.

Economic Aspects

Because of the relatively high demand for gasoline antiknock agents, U.S. production of sodium has for many years been 70–85% of world production (see Table 8). As lead compounds are phased out of gasoline in the United States, this situation is likely to change. Data on U.S. production and use distribution are shown in Table 9. Price development is given in Table 10.

Table 8. World Sodium Production Capacities[a]

Country	Company	Capacity, 1000 t/yr
U.S.	DuPont	63
	Ethyl Corporation	63
	RMI Incorporated	30
FRG	Degussa	9
UK	ICI	15
	Associated Octel	18
France	Ugine Kuhlmann	9
Japan	Nippon Soda	8
	Toyo Soda	13
USSR		9[b]

[a] Ref. 87.
[b] Estimate.

Table 9. U.S. Sodium Production and Uses, Thousand Metric Tons[a]

Use	1959	1963	1967	1972	1978	1980[b]
gasoline additives	72	93	124	121	101	78
metals reduction	5	5	12	6	14	13
all others	24	16	13	19	11	10
Total	101	114	149	146	126	101

[a] Refs. 88–89.
[b] Estimates.

Table 10. U.S. Sodium Prices

Year	$/kg	Year	$/kg
1890	4.40	1974	0.41
1906	0.55	1977	0.73
1946	0.33	1980	1.10
1953	0.35	1981	1.32
1968	0.38	1982	1.48[a]

[a] Estimate.

Analytical Methods

Sodium is identified by the intense yellow color that sodium compounds impart to a flame or spectroscopically by the characteristic sodium lines. The latter test is extremely sensitive and, since many materials contain traces of sodium salts as impurities, it is not conclusive evidence of the presence of sodium in any considerable quantity.

The alkali metals are commonly separated from all other elements except chlorine before gravimetric determination. In the absence of other alkalies, sodium may be weighed as the chloride or converted to the sulfate and weighed. Well-known gravimetric procedures employ precipitation as the uranyl acetate of sodium–zinc or sodium–magnesium. Quantitative determination of sodium without separation is frequently possible by emission or atomic-absorption spectrometric techniques.

Metallic sodium is determined with fair accuracy by measuring the hydrogen liberated on the addition of ethyl alcohol. Sodium amalgam is analyzed by treating a sample with a measured volume of dilute standard acid. After the evolution of hydrogen stops, the excess acid is titrated with a standard base. Total alkalinity is calculated as sodium.

Calcium in commercial sodium is usually determined by permanganate titration of its oxalate. The trace amounts of calcium present in reactor-grade sodium, as little as 0.5 ppm, are determined by atomic absorption spectrometry. Chloride is determined as silver chloride by a turbidimetric method in which glycerol stabilizes the suspended precipitate. Sodium oxide is separated from sodium by treatment with mercury. The oxide, which is insoluble in the amalgam formed, can be separated and determined by acid titration.

Methods for the determination of impurities in sodium are given in references 5, 8, 20, and 90.

Health and Safety Factors

The safe handling of sodium requires special consideration because of its high reactivity. With properly designed equipment and procedures, sodium is used in large and small applications without incident. The hazards of handling sodium are no greater than those encountered with many other industrial chemicals (5).

Direct contact of sodium with the skin can cause deep, serious burns caused by the action of sodium with the moisture present and the subsequent corrosive action of the caustic formed. Sodium can cause blindness on contact with the eyes. For these reasons, goggles, face shield, gloves, and flame-retardant protective clothing are recommended when working with sodium (91).

Perhaps the greatest hazard presented by metallic sodium stems from its extremely vigorous reaction with water to form sodium hydroxide and hydrogen with the evolution of heat (5,14,92–93). In the presence of air this combination usually results in explosion; in a closed system with an inert atmosphere present, the hydrogen evolved can cause a rapid increase in pressure. In the absence of air the rate of reaction is substantially equal to the rate of mixing the reactants, and the reaction does not generally cause mechanical damage to heat-transfer equipment. In the presence of air, the results cannot be predicted.

Another hazard arises from the reaction of air and sodium. Liquid sodium ignites at ca 120°C, although under some conditions dispersed or high surface sodium may ignite at much lower temperatures (7). A small local sodium fire can be put out by submerging the burning mass in the remaining pool of liquid sodium with an iron blade if the bulk of the sodium has not reached the ignition point. Larger fires are more difficult to handle. The common fire extinguishers, ie, water, CO_2, CCl_4, etc, only aggravate the situation by introducing additional explosion or reaction hazards. If the vessel containing the burning sodium can be flooded with nitrogen or closed to exclude air, the fire subsides and the system can be cooled. Fires that cannot be extinguished by excluding air may be quenched by large quantities of dry salt or other dry, cold, inert powder (94). Dry soda ash is excellent for this purpose but tends to become damp in storage and must be carefully protected from contact with the air. Equipment should be designed to confine any possible sodium fire.

In recent years the techniques for handling sodium in commercial-scale applications have been greatly improved (5,20,92,95–96). Contamination by sodium oxide is kept at a minimum by completely welded construction. Residual oxide is removed by cold traps or micrometallic filters. Leakfree electromagnetic or mechanical pumps and meters work well with clean liquid sodium. Corrosion of stainless- or carbon-steel equipment is minimized by keeping the oxide content low.

In the laboratory, sodium is best handled in a glove box filled with nitrogen or another inert gas. When sodium is handled on the bench top, water and aqueous solutions must be excluded from the area. Tools for cutting or handling sodium must be clean and dry. Contact of sodium with air should be kept to a minimum, since moisture in the air reacts rapidly with sodium. A metal catch pan under the equipment is essential to contain any spills or fires. Provision should be made for safe removal of sodium residues from equipment and for cleaning the apparatus. Residue and sodium scrap can be destroyed by burning in a steel pan in a well-ventilated hood. Equipment may be cleaned by being opened to the air and heated until any sodium present is oxidized, or by purging thoroughly with nitrogen, then slowly admitting dry steam to the system while maintaining the nitrogen purge. The burning of sodium as part of any cleaning procedure produces an irritating and hazardous smoke of sodium oxide. This should be collected by an appropriate hood or duct and scrubbed. Dilute aqueous sodium hydroxide is a satisfactory scrubbing liquid.

Other methods for safely cleaning apparatus containing sodium residues or disposing of waste sodium are based on treatment with bismuth or lead (97), inert organic liquids (98–100), or by reaction with water vapor carried in an inert gas stream (101).

Most reactions of sodium are heterogeneous, occurring on the surface of solid or liquid sodium. Such reactions are accelerated by extending the sodium surface exposed. The sodium is dispersed in a suitable medium (102) or spread over a solid powder of high surface area (7–8). Dispersions in inert hydrocarbons may be briefly exposed to air and present no special hazards as long as the hydrocarbon covers the dispersed sodium, but high surface sodium reacts very rapidly with air and cannot be exposed without risk of fire. Dispersions of sodium spilled on cloth or other absorbent material may ignite quickly.

Uses

The production of tetraethyllead and tetramethyllead antiknock agents for gasoline is the largest outlet for sodium in the United States (see Lead compounds; Organometallics). Sodium is also used for the production of other organometallic compounds.

The manufacture of refractory metals such as titanium, zirconium, and hafnium by sodium reduction of their halides is a growing application (103–108). A typical overall reaction is the following:

$$TiCl_4 + 4\ Na \rightarrow Ti + 4\ NaCl$$

Sodium reduction processes are also described for tantalum (109), silicon (110–112), magnesium (113), and other metals.

Metallic potassium and potassium–sodium alloys are made by the reaction of sodium with fused KCl (8,92) or KOH (8,15). Calcium metal and calcium hydride are prepared by the reduction of granular calcium chloride with sodium or sodium and hydrogen, respectively, at temperatures below the fusion point of the resulting salt mixtures (114–115).

Manufacture of sodium peroxide, once an important chemical, has declined in recent years (116–120).

Sodium hydride, made from sodium and hydrogen, is employed as catalyst or reactant in numerous organic reactions and for the production of other hydrides, eg, sodium borohydride. Sodium is used indirectly for the descaling of metals such as stainless steel and titanium (38). Sodium and hydrogen are fed to a molten bath of anhydrous caustic to generate sodium hydride, which dissolves in the melt and is the effective descaling agent (see Metal treatments).

Many sodium compounds are made from sodium. Until 1961 sodium cyanide was made in the United States by the Castner process:

$$Na + C + NH_3 \rightarrow NaCN + 1.5\ H_2$$

Today, however, a process based on the neutralization of aqueous NaOH with HCN gas is preferred.

Sodium is employed as a reducing agent in numerous preparations, including the manufacture of dyes, herbicides (121), pharmaceuticals, high molecular weight alcohols (122), perfume materials (123), and isosebacic acid (124–125).

Sodium is a catalyst for many polymerizations; the two most familiar are the polymerization of 1,2-butadiene (the Buna process) and the copolymerization of styrene–butadiene mixtures (the modified GRS process). The alfin catalysts, made from sodium, give extremely rapid or unusual polymerizations of some dienes and of styrene (126–130) (see Elastomers, synthetic).

Naphthalene sodium prepared in dimethyl ether or other appropriate solvent, or metallic sodium dissolved in liquid ammonia or dimethylsulfoxide are used to treat polyfluorocarbon and other resins to promote adhesion (131–133).

Sodium, usually in dispersed form, is used to desulfurize a variety of hydrocarbon stocks (134). The process is most useful for removal of small amounts of sulfur remaining after hydrodesulfurization.

Sodium as an active electrode component of primary and secondary batteries offers the advantages of low atomic weight and high potential (135–136). In addition to the secondary battery developed by the Ford Motor Company discussed above, a remarkable primary cell has been developed by Lockheed Aircraft Corporation in which sodium metal and water (in the form of aqueous sodium hydroxide) are the reactants (137–140). No separators or diaphragms are used, the counterelectrode is mild steel and the interelectrode distance is very short. The unexpected discovery which makes this cell possible is that, given an external circuit of reasonable resistance, hydrogen is released on the iron counterelectrode rather than on the sodium surface. Thus, sodium dissolves as NaOH in a vigorous but nonviolent manner and the released electrons traverse the external circuit to discharge hydrogen ions at the iron electrode. Concentration cells based on amalgam of differing sodium content that are regenerated thermally have been described (141–142).

Because of its electrical conductivity, low density, low cost, and extrudability, cables are made of sodium sheathed in polyethylene (143). An earlier application used sodium-filled iron pipe as a conductor (144), whereas a more recent patent describes a conductor of sodium contained in aluminum, copper, or steel tubing (145). A corrugated flexible thin-walled copper tube filled with sodium and particularly well adapted for use in gas-insulated high voltage transmission lines has been developed by I.T.E. Imperial Corporation (146). Sodium-conductor distribution cables offer both economic and energy-saving advantages (147).

Sodium is used as a heat-transfer medium in primary and secondary loops of liquid-metal fast-breeder power reactors (5,148–150). Low neutron cross-section, short half-life of radioisotopes produced, low corrosiveness, low density, low viscosity, low melting point, high boiling point, and high thermal conductivity make sodium attractive for this application (37). Sodium has also been suggested for heat transfer in advanced solar-energy collectors (151–152). A comparison of sodium with other working fluids in heat pipes is given in refs. 153 and 154 (see Heat-exchange technology, heat pipe). Small amounts of sodium have been used for many years to cool exhaust valves of heavy-duty internal combustion engines (8).

In metallurgical practice, sodium uses include preparation of powdered metals; removal of antimony, tin, and sulfur from lead; modification of the structure of silicon–aluminum alloys; application of diffusion alloy coatings to substrate metals (155–156); cleaning and desulfurizing molten steel (157); nodularization of graphite in cast iron; deoxidation of molten metals; heat treatment; and the coating of steel with aluminum or zinc.

Sodium vapor lamps, in use for many years, continue to be improved, both with respect to efficiency and color of emitted light. These lamps, however, contain only a few milligrams of sodium each.

SODIUM ALLOYS

Sodium is miscible with many metals in liquid phase and forms alloys or compounds. Important examples are listed in Table 11; phase diagrams are given in refs. 4–5, 14, 32.

Table 11. Metal–Sodium Systems[a]

Metal	Alloy formation	Compound formation	Consolute temperature, °C
barium	x	x	miscible
calcium	x		ca 1200
lead	x	x	
lithium[b]	x		306
magnesium	x		>800
mercury	x	x	
potassium	x	x	miscible
rubidium	x		miscible
tin	x	x	
zinc	x	x	>800

[a] Refs. 4–5, 14, 34.
[b] Refs. 158–159.

The brittleness of metals is frequently increased by the addition of sodium to form alloys. The metals vary in their ability to dilute the natural reactivity of sodium. Most binary alloys are unstable in air and react with water. Ternary and quaternary alloys are more stable.

Sodium–potassium alloy is easily prepared by melting the clean metals in an inert atmosphere or under an inert hydrocarbon, or by the reaction of sodium with molten KCl, KOH, or solid K_2CO_3 powder.

Alloys of lead and sodium containing up to 30 wt % sodium are obtained by heating the metals together in the desired ratio, allowing a slight excess of sodium to compensate for loss by oxidation. At ca 225°C the elements react and generate enough heat to cause a rapid temperature rise. External heating is discontinued and the mixture is cooled and poured into molds. The brittle alloys can be ground to a powder and should be stored under a hydrocarbon or in airtight containers to prevent surface oxidation. The 30 wt % sodium alloy reacts vigorously with water to liberate hydrogen which provides a convenient laboratory source of this gas. An alloy containing 10 wt % sodium may be used in controlled reactions with organic halogen compounds that react violently with pure sodium. Sodium–lead alloys that contain large amounts of sodium are used to dry organic liquids.

Sodium–lead alloys that contain other metals, eg, the alkaline-earth metals, are hard even at high temperatures, and are thus suitable as bearing metals. Tempered lead, for example, is a bearing alloy that contains 1.3 wt % sodium, 0.12 wt % antimony, 0.08 wt % tin, and the remainder lead. The German Bahnmetall, which was used in axle bearings on railroad engines and cars, contains 0.6 wt % sodium, 0.04 wt % lithium, 0.6 wt % calcium, and the remainder lead, and has a Brinell hardness of 34 (see Bearing materials).

Up to ca 0.6 wt % sodium dissolves readily in mercury to form amalgams which are liquid at room temperature (160). The solubility of sodium in mercury is ca 1 wt % at 70°C (160) and 2 wt % at 140°C (34). Alloys containing >2 wt % sodium are brittle at room temperature. Sodium-rich amalgam may be made by adding mercury dropwise to a pool of molten sodium; mercury-rich amalgam is prepared by adding small, clean pieces to sodium to clean mercury with agitation. In either case an inert atmosphere must be maintained, and the heat evolved must be removed. Solid amalgams are easily broken and powdered, but must be carefully protected against air oxidation. Amalgams are useful in many reactions in place of sodium because the reactions are easier to control (160).

Sodium amalgam is employed in the manufacture of sodium hydroxide; sodium–potassium alloy, NaK, is used in heat-transfer applications; and sodium–lead alloy is used in the manufacture of tetraethyllead and tetramethyllead.

Sodium does not form alloys with aluminum but is used to modify the grain structure of aluminum–silicon alloys and aluminum–copper alloys. Sodium–gold alloy is photoelectrically sensitive and may be used in photoelectric cells. A sodium–zinc alloy, containing 2 wt % sodium and 98 wt % zinc, is used as a deoxidizer for other metals.

BIBLIOGRAPHY

"Alkali Metals, Sodium" in *ECT* 1st ed., Vol. 1, pp. 435–447, by E. H. Burkey, J. A. Morrow, and M. S. Andrew, E. I. du Pont de Nemours & Co., Inc.; "Sodium" in *ECT* 2nd ed., Vol. 18, pp. 432–457 by Charles H. Lemke, E. I. du Pont de Nemours & Co., Inc.

1. A. Fleck, *Chem. Ind. (London)* **66,** 515 (1947).
2. D. W. F. Hardie, *Ind. Chem.* **30,** 161 (1954).
3. U.S. Pat. 1,501,756 (July 15, 1924), J. C. Downs (to Roessler and Hasslacher Chemical Company).
4. C. J. Smithells, *Metals Reference Book*, 2nd ed., Vols. I and II, Interscience Publishers, Inc., New York, 1955.
5. O. J. Foust, ed., *Sodium–NaK Engineering Handbook*, Gordon and Breach, Science Publishers, Inc., New York, 1972.
6. J. P. Stone and co-workers, *High Temperature Properties of Sodium, NRL Report 6241*, U.S. Naval Research Laboratory, Defense Documentation Center, AD 622191, Washington, D.C., Sept. 24, 1965.
7. *High Surface Sodium*, U.S. Industrial Chemicals Company, New York, 1953 (now RMI Company, Niles, Ohio).
8. M. Sittig, *Sodium: Its Manufacture, Properties and Uses*, Reinhold Publishing Corporation, New York, 1956.
9. P. B. Dransfield, *Chem. Soc. (London), Spec. Pub. No. 22*, 222 (1967).
10. C. A. Kraus and W. W. Lucasse, *J. Am. Chem. Soc.* **44,** 1949 (1922).
11. C. A. Kraus, *J. Chem. Educ.* **30,** 83 (1953).
12. J. F. Dewald and G. Lepoutre, *J. Am. Chem. Soc.* **76,** 3369 (1954).
13. P. D. Schettler, Jr., P. W. Doumaux, and A. Patterson, Jr., *J. Phys. Chem.* **71**(12), 3797 (1967).
14. J. W. Mellor, *The Alkali Metals*, Vol. II, Suppl. II of *Comprehensive Treatise on Inorganic and Theoretical Chemistry*, John Wiley & Sons, Inc., New York, 1961, Part. I.
15. Unpublished data, E. I. du Pont de Nemours & Co., Inc., Wilmington, Del.
16. R. E. Robinson and I. L. Mador "Alkali Metals" in N. M. Bikales, ed., *Encyclopedia of Polymer Science and Technology*, Vol. 1, Interscience Publishers, a division of John Wiley & Sons, Inc., New York, 1970, pp. 639–658.
17. S. B. Windwer, *Solutions of Alkali Metals in Ethylenediamine*, L. C. Card No. Mic 61-707, University Microfilms, Ann Arbor, Mich., 1960, 79 pp.
18. N. D. Scott, J. F. Walker, and V. L. Hansley, *J. Am. Chem. Soc.* **58,** 2442 (1936).
19. T. P. Whaley, *Sodium, Potassium, Rubidium, Cesium and Francium*, Chapt. 8 in A. F. Trotman-Dickenson, ed., *Comprehensive Inorganic Chemistry*, Pergamon Press, Oxford, England, 1973.
20. J. W. Mausteller, F. Tepper, and S. J. Rodgers, *Alkali Metal Handling and Systems Operating Techniques*, Gordon and Breach, Science Publishers, New York, 1967.
21. A. C. Wittingham, *Liquid Sodium–Hydrogen System: Equilibrium and Kinetic Measurements in the 610–667 K Temperature Range*, NTIS Accession No. RD/B/N-2550, National Technical Information Service, Washington, D.C.; abstract, NSA 31 02, No. 05224, Aug. 1974.

22. D. D. Williams, *A Study of the Sodium–Hydrogen–Oxygen System*, U.S. Naval Research Laboratory Memorandum Report No. 33, Washington, D.C., June 1952.
23. V. L. Hansley and P. J. Carlisle, *Chem. Eng. News* **23**(2), 1332 (1945).
24. P. V. H. Pascal, *Nouveau Traité de Chimie Minérale*, Vol. II, Masson et Cie., Paris, France, 1966.
25. T. P. Whaley and C. C. Chappelow, Jr. in T. Moeller, ed., *Inorganic Syntheses*, Vol. 5, McGraw-Hill Book Company, Inc., New York, 1957, pp. 10–13.
26. E. W. Guernsey and M. S. Sherman, *J. Am. Chem. Soc.* **47,** 1932 (1925).
27. U.S. Pat. 2,802,723 (Aug. 13, 1957), C. H. Lemke (to E. I. du Pont de Nemours & Co., Inc.).
28. W. C. Sleppy, *Inorg. Chem.* **5**(11), 2021 (1966).
29. R. C. Asher and S. A. Wilson, *Nature* **181,** 409 (1958).
30. A. Hérold, *Bull. Soc. Chim. France*, 999 (1955).
31. U.S. Pat. 2,642,347 (June 16, 1953), H. N. Gilbert (to E. I. du Pont de Nemours & Co., Inc.).
32. G. J. Moody and J. D. R. Thomas, *J. Chem. Educ.* **43**(4), 205 (1966).
33. E. Hohmann, *Z. Anorg. Allgem. Chem.* **251,** 113 (1948).
34. M. Hansen, *Constitution of Binary Alloys*, McGraw-Hill Book Company, Inc., New York, 1958.
35. M. J. Dignam and D. A. Huggins, *J. Electrochem. Soc.* **114**(2), 117 (1967).
36. U.S. Pat. 2,858,194 (Oct. 28, 1958), H. C. Miller (to E. I. du Pont de Nemours & Co., Inc.).
37. "The Alkali Metals," *An International Symposium, Nottingham, England, July 1966, Special Publication No. 22*, The Chemical Society, London, 1967.
38. H. N. Gilbert, *Chem. Eng. News* **26,** 2604 (1948).
39. M. I. Klyashtornyi, *Zh. Prikl. Khim.* **31,** 684 (1958).
40. E. Rinck, *Ann. Chim. (Paris)* **18,** 395 (1932).
41. A. Glassner, *The Thermochemical Properties of the Oxides, Fluorides, and Chlorides to 2500 K*, Argonne National Laboratory Report ANL-5750, U.S. Government Printing Office, Washington, D.C., 1957.
42. D. R. Stull and H. Prophet, project directors, *JANAF Thermochemical Tables*, 2nd ed., National Standards Reference Data Series, U.S. National Bureau of Standards, no. 37, June 1971, available from U.S. Government Printing Office, Superintendent of Documents, Washington, D.C.
43. G. L. Clark, ed., *Encyclopedia of Chemistry*, 2nd ed., Reinhold Publishing Corporation, New York, 1966, p. 997.
44. U.S. Pat. 2,391,728 (Dec. 25, 1945), T. H. McConica and co-workers (to Dow Chemical Company).
45. Ger. Offen. 2,243,004 (March 15, 1973), V. M. Chong (to Sun Research and Development Company).
46. Ger. Offen. 2,252,611 (May 10, 1973), E. L. Mongan, Jr. (to E. I. du Pont de Nemours & Co., Inc.).
47. U.S. Pat. 2,789,047 (Apr. 16, 1957), C. H. Lemke (to E. I. du Pont de Nemours & Co., Inc.).
48. D. J. Hansen, private communication, Union Carbide Corporation, Niagara Falls. N.Y., April 19, 1968.
49. Ger. Offen. 2,044,402 (March 18, 1971), C. Gentaz and G. Bienvenue (to Battelle Memorial Institute).
50. T. Wallace, *Chem. Ind. (London)*, 876 (1953).
51. H. N. Gilbert, *J. Electrochem. Soc.* **99,** 3050 (1952).
52. L. A. Kal'man, P. T. Merenkov, and I. G. Grechko, *Khim. Promst. (Moscow)* **48**(10), 789 (1972).
53. U.S. Pat. 4,192,724 (March 11, 1980), T. Minami and S. Toda (to Chlorine Engineers Corporation, Ltd., Tokyo).
54. U.S. Pat. 3,507,768 (April 21, 1970), E. I. Adaev, A. V. Blinov, G. M. Kamaryan, V. A. Novoselov, V. N. Suchkov, and L. M. Yakimenko.
55. C. L. Mantell, *Electrochemical Engineering*, 4th ed., McGraw-Hill Book Company, Inc., New York, 1960.
56. H. E. Batsford, *Chem. Metall. Eng.* **26.** 888 (1922).
57. *Ibid.*, 932 (1922).
58. W. C. Gardiner, *Office of Technical Services Report PB-44761*, U.S. Department of Commerce, Washington, D.C., 1946; Field Information Agency, Technical (FIAT) Final Report 820.
59. U.S. Pat. 2,850,442 (Sept. 2, 1958), W. S. Cathcart and co-workers (to E. I. du Pont de Nemours & Co., Inc.).
60. U.S. Pat. 3,020,221 (Feb. 6, 1962), W. H. Loftus (to E. I. du Pont de Nemours & Co., Inc.).
61. USSR Pat. 320,552 (Nov. 4, 1971), E. I. Adaev and co-workers.
62. U.S. Pat. 3,712,858 (Jan. 23, 1973), F. J. Ross (to E. I. du Pont de Nemours & Co., Inc.).
63. *Chem. Eng.* **69**(6), 90 (March 1962).

64. U.S. Pat. 2,148,404 (Feb. 21, 1939), H. N. Gilbert (to E. I. du Pont de Nemours & Co., Inc.).
65. U.S. Pat. 2,234,967 (March 18, 1941), H. N. Gilbert (to E. I. du Pont de Nemours & Co., Inc.).
66. W. C. Gardiner, *Office of Technical Services Report PB-44760*, U.S. Department of Commerce, Washington, D.C., 1946; Field Information Agency, Technical (FIAT) Final Report 819.
67. U.S. Pat. 3,265,490 (Aug. 9, 1966), S. Yoshizawa and co-workers (to Tekkosha Company Ltd., Tokyo).
68. Fr. Pat. 1,457,562 (Nov. 4, 1966), (to Showa Denko K. K.).
69. Jpn. Pat. 71/03846 (Jan. 30, 1971), T. Ohshiba (to Showa Denko K. K.).
70. Jpn. Pat. 74/28322 (July 25, 1974), N. Watanabe and M. Tomatsuri (to Tekkosha Company, Ltd.).
71. N. Watanabe, M. Tomatsuri, and K. Nakanishi, *Denki Kagaku* **38**(8), 584 (1970).
72. T. Yamaguchi, *Chem. Econ. Eng. Rev.* **4**(1), 24 (Jan. 1972).
73. T. Nakamura and Y. Fukuchi, *J. Metall.* **24**(8), 25 (Aug. 1972).
74. L. E. Vaaler and co-workers, *Final Report on a Survey of Electrochemical Metal Winning Processes*, ANL/OEPM-79-3, prepared by Battelle Columbus Laboratories for Agronne National Laboratory, March 1979, Available from National Technical Information Service, Washington, D.C.
75. U.S. Pat. 3,488,271 (Jan. 6, 1970), J. T. Kummer and N. Weber (to Ford Motor Company).
76. U.S. Pat. 4,108,743 (Aug. 22, 1978), R. W. Minck (to Ford Motor Company).
77. Brit. Pat. 1,155,927 (June 25, 1969), A. T. Kuhn and S. F. Mellish (to Imperial Chemical Industries, Ltd.).
78. Brit. Pat. 1,200,103 (March 31, 1967), A. T. Kuhn (to Imperial Chemical Industries, Ltd.).
79. U.S. Pat. 4,089,770 (May 16, 1978), C. H. Lemke (to E. I. du Pont de Nemours & Co., Inc.).
80. U.S. Pat. 4,133,728 (Jan. 9, 1979), S. A. Cope (to E. I. du Pont de Nemours & Co., Inc.).
81. U.S. Pat. 4,203,819 (May 20, 1980), S. A. Cope (to E. I. du Pont de Nemours & Co., Inc.).
82. S. Yoshizawa and Y. Ito, *Extended Abstracts of the 30th Meeting of the International Society of Electrochemistry*, Trondheim, Norway, Aug. 1979, pp. 38–40.
83. Jpn. Pat. 77/135,811 (Nov. 14, 1977), J. Koshiba and co-workers (to Toyo Soda Manufacturing Company, Ltd.).
84. U.S. Pat. 4,097,345 (June 27, 1978), R. D. Shannon (to E. I. du Pont de Nemours & Co., Inc.).
85. C. A. Levine, R. G. Heitz, and W. E. Brown, *Proceedings of the 7th Intersociety Energy Conversion and Engineering Conference*, American Chemical Society, Washington, D.C., 1972, pp. 50–53.
86. Private communication, E. I. du Pont de Nemours & Co., Inc. to Battelle Columbus Laboratories, June 1975.
87. K. Yajima, *Yoyuen* **21**(1), 147 (1978).
88. *Sodium* in *Chemical Products Synopsis*, Mannsville Chemical Products, Mannsville, N.Y., Dec. 1978.
89. N. M. Levinson in *Chemical Economics Handbook, Sodium Metal*, 770.1000A-D, Stanford Research Institute, Menlo Park, Calif., Oct. 1979.
90. C. H. Lemke, N. D. Clare, and R. E. DeSantis, *Nucleonics* **19**(2), 78 (Feb. 1961).
91. *Sodium Material Safety Data Sheet*, E. I. du Pont de Nemours & Co., Inc., Wilmington, Del., 1980.
92. D. D. Adams, G. J. Barenborg, and W. W. Kendall, "Handling and Uses of the Alkali Metals," *Adv. Chem. Ser.* **19,** 92 (1957).
93. L. F. Epstein in C. M. Nicholls, ed., *Progress in Nuclear Energy*, Ser. IV, Vol. 4, Pergamon Press, Inc., New York, 1961, pp. 461–483.
94. P. Menzenhauer, G. Ochs, and W. Peppler, *Kernforschungszent Karlsruhe* (*Ber.*), KFK 2525 (1977).
95. *Chem. Eng. News* **34**(17), 1991 (April 23, 1956).
96. *Chem. Eng.* **65**(12), 63 (June 16, 1958).
97. U.S. Pat. 4,032,615 (June 28, 1977), T. R. Johnson (to U.S. Energy Research and Development Administration).
98. Jpn. Pat. 79/114,473 (Sept. 6, 1979), Y. Nishizawa (to Mitsubishi Atomic Power Industries, Inc.).
99. U.S. Pat. 3,729,548 (April 24, 1973), C. H. Lemke (to E. I. du Pont de Nemours & Co., Inc.).
100. U.S. Pat. 3,459,493 (Aug. 15, 1969), F. J. Ross (to E. I. du Pont de Nemours & Co., Inc.).
101. H. P. Maffei, C. W. Funk, and J. L. Ballif, *Sodium Removal Disassembly and Examination of the Fermi Secondary Sodium Pump, Report 1974, HEDL-TC-133*, available from National Technical Information Service, Washington, D.C.
102. I. Fatt and M. Tashima, *Alkali Metal Dispersions*, D. Van Nostrand Company, Inc., Princeton, N.J., 1961.

103. U.S. Pat. 2,890,111 (June 9, 1959), S. M. Shelton (to United States of America).
104. U.S. Pat. 2,828,119 (March 25, 1958), G. R. Findley (to National Research Corporation).
105. U.S. Pat. 2,890,112 (June 9, 1959), C. H. Winter, Jr. (to E. I. du Pont de Nemours & Co., Inc.).
106. Brit. Pat. 816,017 (July 8, 1959), (to National Distillers and Chemical Corporation).
107. U.S. Pat. 3,736,132 (May 29, 1973), H. H. Morse and co-workers (to United States Steel Corporation).
108. Brit. Pat. 1,355,433 (June 5, 1974), P. D. Johnston and co-workers (to Electricity Council).
109. Ger. Offen. 2,517,180 (Oct. 21, 1976), R. Haehn and D. Behrens (to Firma Hermann C. Starck, Berlin).
110. J. V. R. Heberlein, J. F. Lowry, T. N. Meyer, and D. F. Ciliberti, *Conference Proceedings of the 4th International Symposium on Plasma Chemistry*, Pt. 2, Zurich, Switzerland, 1979, pp. 716–722.
111. A. Sanjurjo, L. Nanis, K. Sancier, R. Bartlett, and V. Kapur, *J. Electrochem. Soc.* **128**(1), 179 (Jan. 1981).
112. D. B. Olson and W. J. Miller, *Report 1978, DOE/JPL/-954777-5, AeroChem-TN-99*, available from National Technical Information Service, Washington, D.C.
113. U.S. Pat. 4,014,687 (Mar. 29, 1977), N. D. Clare and C. H. Lemke (to E. I. du Pont de Nemours & Co., Inc.).
114. U.S. Pat. 2,794,732 (June 4, 1957), P. P. Alexander (to Metal Hydrides, Inc.).
115. U.S. Pat. 2,794,733 (June 4, 1957), P. P. Alexander and R. C. Wade (to Metal Hydrides, Inc.).
116. U.S. Pat. 1,796,241 (March 10, 1931), H. R. Carveth (to Roessler and Hasslacher Chemical Company).
117. U.S. Pat. 2,633,406 (March 31, 1953), D. S. Nantz (to National Distillers Products Corporation).
118. U.S. Pat. 2,671,010 (March 2, 1954), L. J. Governale (to Ethyl Corporation).
119. U.S. Pat. 2,685,500 (Aug. 3, 1954), R. E. Hulse and D. S. Nantz (to National Distillers Products Company).
120. I. I. Vol'nov, *Peroxides, Superoxides, and Ozonides of Alkali and Alkaline Earth Metals*, Plenum Publishers Corporation, New York, 1966.
121. *Eur. Chem. News (London)* **14**(336), 34 (1968).
122. U.S. Pat. 2,915,564 (Dec. 1, 1959), V. L. Hansley (to National Distillers and Chemical Corporation).
123. H. N. Gilbert, N. D. Scott, W. F. Zimmerli, and V. L. Hansley, *Ind. Eng. Chem.* **25,** 735 (1933).
124. M. Sittig, *Mod. Plast.* **32**(12), 150, 217 (1955).
125. U.S. Pat. 2,352,461 (June 27, 1944), J. F. Walker (to E. I. du Pont de Nemours & Co., Inc.).
126. A. A. Morton, *Ind. Eng. Chem.* **42,** 1488 (1950).
127. A. A. Morton in N. M. Bikales, ed., *Encyclopedia of Polymer Science and Technology*, Interscience Publishers, a division of John Wiley & Sons, Inc., New York, 1964, pp. 629–638.
128. U.S. Pat. 3,966,691 (June 29, 1976), A. F. Aalasa (to Firestone Tire and Rubber Company).
129. Jpn. Pat. 76/68491 (June 14, 1976), T. Kitsunai, Y. Mitsuda, and S. Sato (to Denki Kagaku Kogyo K. K.).
130. Ger. Offen. 2,802,044 (July 20, 1978), A. Proni and A. Roggero.
131. U.S. Pat. 2,809,130 (Oct. 8, 1957), G. Rappaport (to General Motors Corporation).
132. U.S. Pat. 2,789,063 (April 16, 1957), R. J. Purvis and W. R. Beck (to Minnesota Mining & Manufacturing Co.).
133. Brit. Pat. 1,078,048 (Aug. 2, 1967), R. S. Haines (to International Business Machines Corporation).
134. U.S. Pat. 3,565,792 (Feb. 23, 1971), F. B. Haskett.
135. R. D. Weaver, S. W. Smith, and N. L. Willmann, *J. Electrochem. Soc.* **109**(8), 653 (1962).
136. N. Weber and J. T. Kummer, *Proceedings of the 21st Annual Power Sources Conference*, U.S. Army, Electronics Command, Atlantic City, N.J., 1967, pp. 37–39.
137. U.S. Pat. 3,791,871 (Feb. 12, 1974), L. S. Rowley (to Lockheed Aircraft Corporation).
138. U.S. Pat. 4,053,685 (Oct. 11, 1977), L. S. Rowley and H. J. Halberstadt (to Lockheed Missiles and Space Company, Inc.).
139. H. J. Halberstadt, *Proceedings of the 8th Intersociety Energy Conversion and Engineering Conference*, American Institution of Aeronautics and Astronautics, New York, 1973, pp. 63–66.
140. R. R. Roll, *Proc. Symp. Batteries Traction Propul.*, 209 (1972).
141. I. J. Groce and R. D. Oldenkamp, *Adv. Chem. Ser.* **64,** 43 (1967).
142. L. A. Heredy, M. L. Iverson, G. D. Ulrich, and H. L. Recht in ref. 141, p. 30.
143. *Chem. Week* **101**(8), 79 (Aug. 19, 1967).
144. R. H. Boundy, *Trans. Electrochem. Soc.* **62,** 151 (1932).

145. Brit. Pat. 1,188,544 (Apr. 15, 1970), (to General Cable Corporation).
146. U.S. Pat. 4,056,679 (Nov. 1, 1977), T. F. Brandt (to I. T. E. Imperial Corporation).
147. *Assessment of Sodium Conductor Distribution Cable, DOE/ET-5041-1*, Westinghouse Research and Development Center, Pittsburgh, Pa., June 1979.
148. W. Peppler, *Chem. Ztg.* **103**(6), 195 (1979).
149. *Proceedings of the International Conference on Liquid Alkali Metals*, British Nuclear Energy Society, London, April 4–6, 1973.
150. M. H. Cooper, ed., *Proceedings of the International Conference on Liquid Metal Technology in Energy Production*, CONF-760503-P1 and P2, Champion, Pa., May 3–6, 1976.
151. J. J. Bartel, H. J. Rack, R. W. Mar, S. L. Robinson, F. P. Gerstle, Jr., and K. B. Wischmann, *1. Molten Salt and Liquid Metal, Sandia Laboratories Materials Task Group Review of Advanced Central Receiver Preliminary Designs*, Sept. 1979, SAND 79-8633, available from National Technical Information, Service, Washington, D.C.
152. A. B. Meinel and M. P. Meinel, *Applied Solar Energy*, Addison-Wesley Publishing Company, Reading, Mass., 1976.
153. J. W. H. Chi, *Proc. Top. Meet. Technol. Controlled Nucl. Fusion* **2**(2), 443 (1976).
154. J. E. Kemme, J. E. Deverall, E. S. Keddy, J. R. Phillips, and W. A. Rankin, *Temperature Control with High Temperature Gravity-Assist Heat Pipes*, Los Alamos Scientific Laboratory, 1975, available from National Technical Information Service, Accession No. CONF-750812-10.
155. U.S. Pat. 3,220,876 (Nov. 30, 1965), R. D. Moeller (to North American Aviation, Inc.).
156. U.S. Pat. 3,251,719 (May 17, 1966), F. Tepper and co-workers (to M.S.A. Research Corporation).
157. U.S. Pat. 3,598,572 (Aug. 10, 1971), J. C. Robertson (to Dow Chemical Company).
158. M. G. Down, P. Hubberstey, and R. J. Pulham, *J. Chem. Soc. Dalton Trans.* (14), 1490 (1975).
159. R. P. Elliot, *Constitution of Binary Alloys*, 1st Suppl., McGraw-Hill Book Company, New York, 1958.
160. R. B. MacMullin, *Chem. Eng. Prog.* **46,** 440 (1950).

General References

Reference 5 is a critical source of data on sodium properties, components, systems, handling and safety.
References 14 and 19 cover the inorganic chemistry of sodium.
References 8 and 16 survey sodium organic chemistry.
References 34 and 158 present phase diagrams of sodium with other metals.
Natrium, Vol. 2 of *Gmelins Handbuch der Anorganischen Chemie*, 8th ed., Verlag Chemie GmbH, Weinheim/Bergstr., FRG, 1965, pp. 401–627; covers inorganic chemistry of sodium.

CHARLES H. LEMKE
E. I. du Pont de Nemours & Co., Inc.
University of Delaware

SODIUM CARBONATE. See Alkali and chlorine products.

SODIUM COMPOUNDS

SODIUM HALIDES, SODIUM CHLORIDE

Sodium chloride [*7647-14-5*], NaCl, in this article referred to as salt, supplies sodium and chloride ions, both essential to animal life. It is one of the principal raw materials used in the chemical industry and the source of almost all industrial compounds containing sodium or chlorine. The largest use of salt, mainly in the form of aqueous solutions (brine), is in the electrolytic production of chlorine and sodium hydroxide (see Alkali and chlorine products; Chemicals from brine).

Salt is widely distributed throughout the world. It occurs in seawater and other saline waters and in dry deposits as rock salt or playas.

Most of the world's salt is contained in the oceans, and seawater is an important source of salt (see Table 1). Except for such highly saline waters as the Great Salt Lake and the Dead Sea, naturally occurring inland saline waters are minor salt sources.

A playa is a shallow basin in a desert plain. Leaching of the surrounding sediments, followed by evaporation, produces a deposit in the basin. Playas are not important salt sources but may contain significant amounts of other salines, eg, sodium sulfate.

The bedded salt deposits are true sedimentary rocks; they are thus referred to as rock salt or halite and present a very important source of salt. Bedded deposits were formed from evaporation of large inland seas that become separated from the oceans. Geologically, rock-salt deposits are very old, dating back to Cambrian and perhaps pre-Cambrian periods.

In North America huge quantities of rock salt are found in the Silurian basin covering parts of Michigan, Ontario, Ohio, Pennsylvania, and New York. Although these deposits date mostly back to the Silurian period, Devonian salt is found in the Michigan basin. The Permian basin covers parts of Kansas, Colorado, Oklahoma, New Mexico, and Texas, and may extend into northern Mexico. The Gulf Coast basin, dating to the pre-Jurassic period, covers parts of Texas, Arkansas, Louisiana, Mississippi, Alabama, and northeastern Mexico; it also extends out beneath the Gulf of Mexico. The salt deposits in the Isthmus of Tehuantepec in southern Mexico may be an extension of the same deposit. The Williston and Elk Point basins, covering parts of North and South Dakota, Montana, Saskatchewan, and Alberta, contain salt of the Silurian-Devonian period.

Table 1. Composition of Seawater

Component	Wt %	Component	Wt %
sodium chloride	2.68	potassium chloride	0.07
magnesium chloride	0.32	sodium bromide	0.008
magnesium sulfate	0.22	water	96.582
calcium sulfate	0.12		

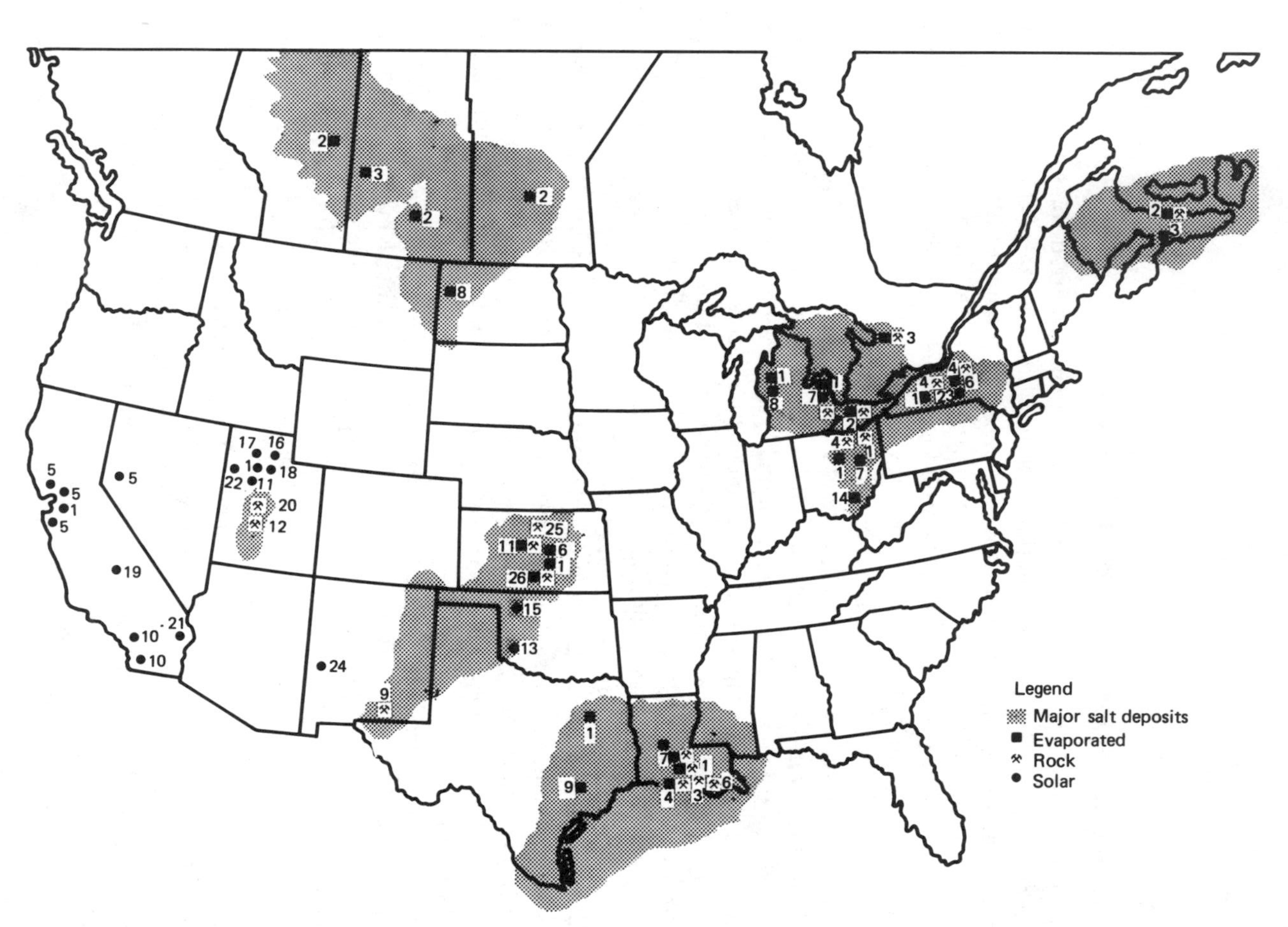

Figure 1. Principal salt deposits and dry-salt production sites, U.S. and Canada. Numerical key appears on page 207.

The North American rock-salt deposits are among the most extensive in the world (see Fig. 1), although significant deposits exist in South America, the UK, Europe, and the USSR.

Rock salt commonly occurs in layered strata of relatively pure salt separated by thin layers of calcium sulfate. The thickness of the salt strata is generally 5–40 m or even more. The deposits usually occur at depths of 150–1200 m.

The salt dome, a large vertical column of salt, is one of the most unusual and interesting types of rock-salt deposits. The top of a dome used for salt production is usually within 100 m of the earth's surface and may come to within 20 m. The depth may be well over 10 km. Some domes are estimated to have a volume of ca 80 km^3. Salt domes are believed to have been formed from bedded deposits by salt flowage. Halite is somewhat plastic, and the presence of surrounding dense rock caused the bedded salt to flow and pierce the overlying rock through zones of weakness. Although salt domes occur in different parts of the world, they are most numerous around the Gulf of Mexico bordering Texas, Louisiana, Mississippi, and Alabama, where 329 domes have been located and characterized (1). Deposits of sulfur and oil are commonly associated with salt domes.

The principal impurity in naturally occurring rock salt is calcium sulfate, generally 1–4%, with small amounts of calcium chloride and magnesium chloride. Salt domes contain 1–10% calcium sulfate, but mined dome salt is usually quite pure, with impurities of 1–3%. It is therefore preferred over other types of rock salt for regeneration of water-softening resins (see Ion exchange).

Salt also occurs as sylvinite, which consists of mixed crystals of potassium chloride and sodium chloride. It is mainly a source of potassium chloride. In dry mining, the potassium chloride is separated by flotation (qv), and the associated salt is waste product. However, when sylvinite is solution-mined, the salt may be recovered in high purity granulated form before production of granulated potassium chloride in a crystallizing evaporator system.

Some rock-salt deposits contain pure, crystallized salt, ranging from very small isolated patches to formations weighing several metric tons. This salt is extremely clear and free from defects; it is ca 99.98% pure sodium chloride.

Numerical key to Figure 1: 1, Morton: Newark, Calif.; Saltair, Utah; Hutchinson, Kans.; Grand Saline, Texas; Manistee, Mich.; Marysville, Mich.; Weeks Island, La.; Rittman, Ohio; Fairport, Ohio; and Silver Springs, N.Y.; 2, Canadian: Lindbergh, Alb.; Belle Plaine, Sask.; Nepawa, Man.; Windsor, Ont.; and Pugwash, N.B.; 3, Domtar: Unity, Sask.; Goderich, Ont.; Nappan, N.B.; and Cote Blanche, La.; 4, International: Avery Island, La.; Detroit, Mich.; Cleveland, Ohio; and Watkins Glen, N.Y.; 5, Leslie: Napa, Calif.; Newark, Calif.; Redwood City, Calif.; and Fallon, Nev.; 6, Cargill: Hutchinson, Kan.; Belle Isle, La.; and Lansing, N.Y.; 7, Diamond Crystal: Jefferson Island, La.; St. Clair, Mich.; and Akron, Ohio; 8, Hardy: Williston, N.D.; and Manistee, Mich.; 9, United: Carlsbad, N.M.; and Blue Ridge, Hockley, Texas; 10, Western: Santa Ana, Calif.; and Chula Vista, Calif.; 11, American: Grantsville, Utah; and Lyons, Kan.; 12, Albert Poulson: Redmond, Utah; 13, Acme: Erick, Okla.; 14, Excelsior: Pomeroy, Ohio; 15, Ezra S. Blackmon: Freedom, Okla.; 16, Great Salt Lake Minerals and Chemical: Ogden, Utah; 17, Lake Crystal: Ogden, Utah; 18, Lake Point: Lake Point, Utah; 19, Pacific Salt & Chemical Company: Trona, Calif.; 20, Redmond Clay: Redmond, Utah; 21, Standard Salt and Chemical Co.: Rice, Calif.; 22, Utah: Wendover, Utah; 23, Watkins: Watkins Glen, N.Y.; 24, Zuni: Quemado, N.M.; 25, Independent: Kanopolis, Kan.; 26, Carey: Hutchinson, Kan. Courtesy of The Salt Institute.

Properties

Sodium chloride is composed of 39.339 wt % sodium and 60.661 wt % chlorine. The ionic crystal has a cubic close-packed lattice, with alternate positions occupied by sodium ions and chloride ions; the unit cell is 0.5627 nm on edge. The lattice energy is 761.5 J/mol (182 cal/mol).

Pure sodium chloride is colorless. Unlike glass, it is transparent in the near- and mid-infrared regions, and can be used as cell windows in infrared spectrophotometry. The physical properties of sodium chloride are given in Table 2.

Salt is soluble in polar solvents, insoluble in nonpolar. The solubility in water and various solvents is given in Table 3. The aqueous solution is essentially neutral. Dissolution of salt in water is endothermic; the integral heat of solution is 3.75 kJ/mol (898 cal/mol) in 1000 g water at 25°C. The properties of saturated brine solution are given in Table 4. An extensive tabulation of the physical properties of salt is found in ref. 2.

The phase diagram for sodium chloride–water is shown in Figure 2. The eutectic temperature is −21.12°C; the composition of the eutectic mixture is 23.31 wt % salt and 76.69 wt % water, or 37.68 wt % $NaCl.2H_2O$ and 62.32 wt % water. At low temperatures, brines more concentrated than 23.31 wt % salt deposit large, transparent crystals of monoclinic sodium chloride dihydrate [*23724-87-0*]. These crystals, although similar to ice in appearance, are birefringent. They account for freezing of moist piles of highway deicing salt during storage in cold weather.

Table 2. Properties of Sodium Chloride

Property	Value
mp, °C	800.8
bp, °C	1465
density, g/cm^3	2.165
hardness	2.5
refractive index, n_D^{20}	1.544
specific heat, J/(g·°C)[a]	0.853
heat of fusion, J/g[a]	517.1
critical humidity at 20°C, %	75.3
heat of solution in 1 kg H_2O at 25°C, kJ/mol[a]	3.757

[a] To convert J to cal, divide by 4.184.

Table 3. Solubility of Salt

Solvent	Temperature, °C	g NaCl/100 g solvent
methanol	25	1.40
ethanol	25	0.065
formic acid	25	5.21
ethylene glycol	25	7.15
monoethanolamine	25	1.86
liquid ammonia	−40	2.15
water	0	35.7
water	120	39.8

Table 4. Properties of Saturated Brine

Property	Value
bp, °C	108.7
NaCl, %	
at bp	28.4
at 25°C	26.48
sp gr[a]	1.1978
specific heat, J/(g·K)[b]	3.28
vapor pressure, kPa[c]	
20°C	1.76
25°C	2.39
50°C	9.26
70°C	23.27
90°C	52.20

[a] Relative to water at 4°C.
[b] To convert J to cal, divide by 4.184.
[c] To convert kPa to mm Hg, multiply by 7.5.

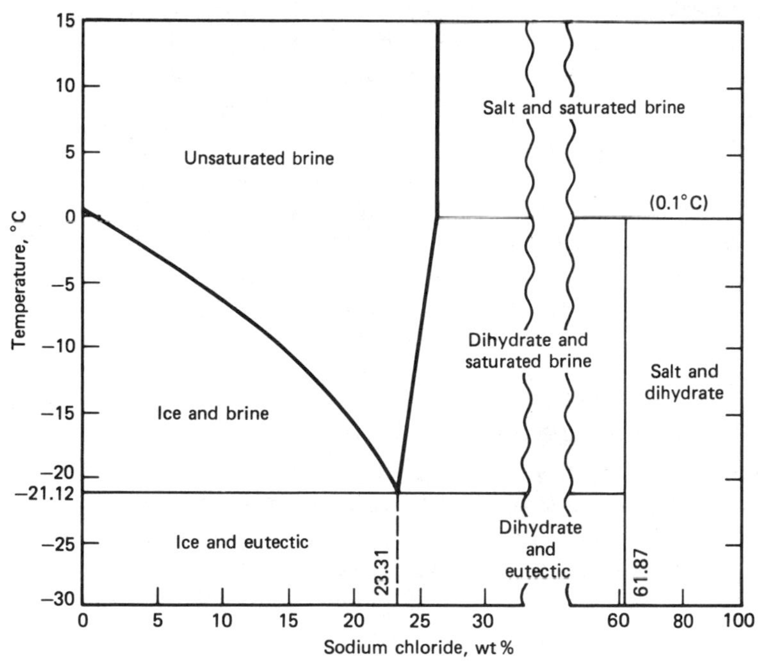

Figure 2. The system sodium chloride–water.

Processing

Shaft Mining. Salt is mined similarly to coal. A shaft is sunk into the rock-salt vein. With well-established techniques of undercutting, side shearing, drilling, blasting, loading, and transporting, the salt is removed from the deposit for further processing, which is primarily a materials-handling operation involving crushing, screening, bagging, and loading. Some of these operations may be carried out in the mine.

Some rock-salt fines (<2 mm or −10 mesh) are used as loose salt for animal feeding. Alternatively, the fines may be compacted to a coarser aggregate or pressed into blocks, ie, cattle salt licks. Rock-salt grades are designated coarse (6.7–2.4 mm or 3–8 mesh), medium (3.3–1.2 mm or 6–16 mesh), and fine or kiln dried (1.68–0.21 mm or 12–70 mesh), although neither designation nor screening is standardized. Louisiana (dome) salt includes an extra coarse grade (13–4.7 mm or 4 mesh) for regenerating water softeners. The most popular type of salt for snow and ice control is called coarse crushed or CC (ca 95–2.0 mm or 10 mesh).

Rock salt is mined by the room-and-pillar method. As the salt is removed, big empty spaces (rooms) are formed; pillars are left for support. In layered-strata deposits, the height of the room is generally limited to 8 m and the length to 15–20 m.

In salt domes, the height is less limited. A special type of room-and-pillar mining, referred to as quarrying, is used that permits room height and length each up to 30 m.

Working conditions in a salt mine are excellent. The air is dry and the temperature moderate at ca 15°C, although it may be as high as 27°C in warm regions. The working areas are well lit, and the mine is ventilated with forced draft. Safety records are excellent.

Solution Mining. Salt brine is obtained by pumping water into a rock-salt deposit. The salt is dissolved and the brine brought to the surface. In one technique, water is pumped into and brine out of a single well with a double-pipe arrangement. In general, water is pumped into the well in the outer annular space between the pipes and brine is removed via the inside pipe.

With another method, frequently used in layered-strata deposits, two holes, 100–300 m apart, are drilled into the deposit. The mass between the two wells is fractured with high pressure pumps in a process called hydrofracing. Water is then pumped into one well and brine flows out of the other. With the proper arrangement, one injection well can serve more than one brine-producing well.

Solution mining produces essentially saturated brine. Its principal impurity is calcium sulfate, which is usually associated with salt deposits. Raw brine contains up to 0.5 wt % calcium sulfate and smaller amounts of calcium and magnesium chlorides. Minor amounts of hydrogen sulfide may also be present.

Many electrolytic plants producing chlorine and caustic soda utilize brine obtained by solution mining. Furthermore, such brines serve as the raw material for salt produced by several different types of crystallizers or salt evaporators.

For ordinary grades of salt, no brine treatment is required other than hydrogen sulfide removal, if necessary, and settling to remove solids. Hydrogen sulfide is removed by aeration and chlorination. For special grades for food or chemical use, pretreatment of brine may be required, generally with sodium hydroxide or lime to remove magnesium and with sodium carbonate to remove calcium. The brine is usually treated on a batch basis and allowed to settle to produce a sparkling clear brine. The principal impurity left is soluble sodium sulfate, which in most cases presents no problem in the final purity of the crystalline salt product. However, if necessary, the sulfate can be removed by barium chloride, but only if the salt is not intended for food use.

In the United States, regular table salt is produced from brine without removal of calcium and magnesium and hence contains small quantities of calcium sulfate crystals. In some countries, virtually all brine is pretreated before crystallization in evaporators.

Crystallizing Evaporators. The multiple-effect crystallizing evaporator system, or vacuum-pan system, is the most common technique for the evaporation of water from brine to produce salt (see Evaporation). Salt produced in this manner is called evaporated, or granulated, salt. It is generally in the shape of well-formed cubes, and sizes are 0.84–0.21 mm (20–70 mesh).

Evaporators are either the calandria (internal heating-surface) type, shown in Figure 3, or the forced-circulation (external heat-exchanger) unit, shown in Figure 4. The principal advantage of the multiple-effect system is heat economy. The vapors produced in one effect are reused as a source of heat to boil brine at a lower pressure in a following effect. Thus, in a triple-effect set of evaporators, 1 kg steam can evaporate almost 3 kg water. In salt production, triple-, quadruple-, and quintuple-effect evaporators are commonly used. At the end of the system, vapor from the last effect may be condensed in a water-cooled barometric condenser to provide the vacuum for that evaporator.

The need for condenser water may be eliminated with mechanical vapor-recompression evaporators or thermocompression evaporators. In thermocompression, the vapor given off from the boiling liquid is compressed and used as steam for heating the evaporator from which it was withdrawn. Thermocompression evaporators are particularly attractive in countries where fuel costs are high and cheap hydroelectric power is available.

The older calandria evaporators are generally limited to low steam pressure, although calandrias can be designed to operate at high pressure. The capacity can be increased with a forced-circulation evaporator operating at higher pressure, and providing vapor at lower pressure for the existing calandria evaporators.

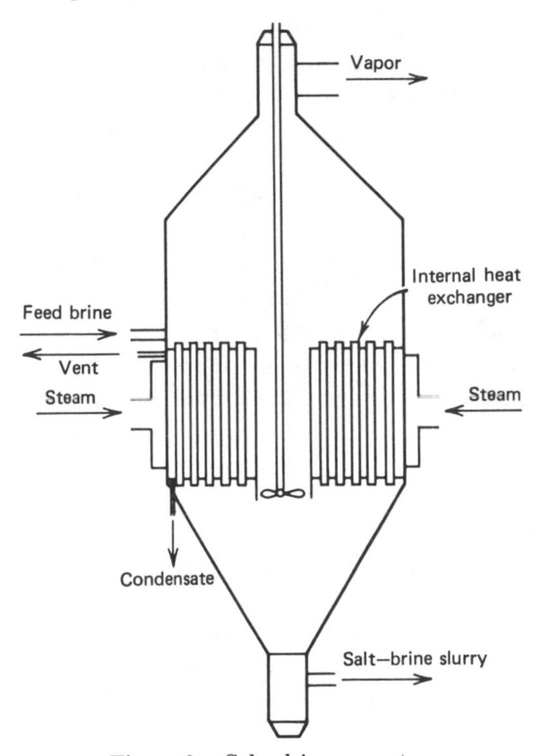

Figure 3. Calandria evaporator.

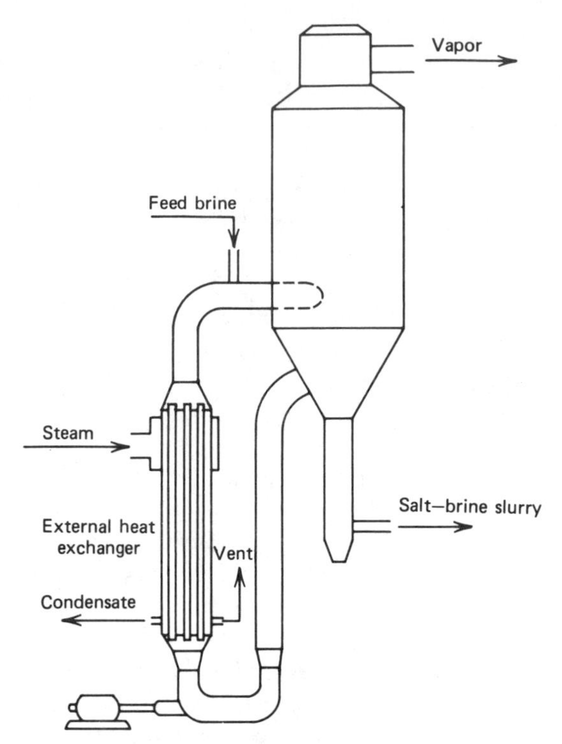

Figure 4. Forced-circulation evaporator.

The older evaporators were cast-iron vessels, as much as 8 m dia, with copper heating tubes. Modern evaporators up to 12 m diameter and 25–30 m high are constructed of Monel with copper–nickel or titanium heating tubes. A single modern evaporator may produce up to 50 metric tons of salt per hour.

Untreated brine may cause scaling of heating surfaces with calcium sulfate because of the inverse solubility of calcium sulfate with temperature. This problem is controlled by maintaining a suspension of very fine calcium sulfate seed crystals in the pans. Scaling is complicated by the fact that anhydrite ($CaSO_4$) forms in the hottest pans, whereas $5CaSO_4.Na_2SO_4.3H_2O$ may exist at intermediate temperatures, and gypsum ($CaSO_4.2H_2O$) in cooler pans. Anhydrite is the best agent for scale control; the other two forms may grow in size and substantially reduce the salt purity.

The salt is removed from the evaporators in a brine slurry and washed countercurrently with fresh brine to float off the calcium sulfate. It is then dewatered by centrifugation or by vacuum filtration and dried. The highest purity salt is washed with water on the filter.

Oslo or Krystal type evaporators produce coarse spherical crystals.

A special type of salt is obtained when a small amount of sodium ferrocyanide is added to the brine. This additive modifies the growth of the crystals, ie, it inhibits growth on the faces and promotes growth on the corners (3). The product is called dendritic salt; it has a lower bulk density and higher surface area than granulated salt and blends well with fine powders.

Grainers. With the grainer process, a special type of salt called flake salt is obtained. This process is very wasteful of heat and therefore has recently been phased out.

A grainer is a long, narrow, shallow pan. The brine is heated by coils. Crystals forming at the surface are supported by the surface tension of the brine. As their weight increases, the crystals sink and hollow pyramids, or hopper crystals, are formed. These hoppers drop to the bottom and are removed by a raking mechanism. In the process of removal, filtering, and drying, the hoppers break up into the flakes, which give this type of salt its name.

Salt with characteristics similar to flake salt can be produced synthetically. In one process, salt is fused and cast into molds, and the cooled salt is crushed and screened. In another process, granulated salt is compressed between smooth-faced rolls to form individual flat flakes. Low energy requirements make this process very attractive compared to grainer operations.

Alberger Process. This process produces a special type of salt which is a mixture of flake salt and fine cubic crystals (4). Brine is heated under pressure to 145°C, passed through a vessel where calcium sulfate is removed, and then flashed down to atmospheric pressure, producing fine cubic crystals. The slurry is fed to open circular evaporating pans, where flake salt is produced by surface evaporation. It is mechanically raked out of the pans, centrifuged, and dried in rotary or fluidized-bed dryers.

Recrystallizer Process. This process was developed primarily to convert rock salt to evaporated salt (5) (see Fig. 5). Rock salt is introduced into a circulating-brine system. The slurry is heated by direct injection of live steam to dissolve the salt. The saturated brine is settled, filtered, and sent to a forced-circulation evaporator where water is flashed off under reduced pressure and salt is crystallized. Under these conditions calcium sulfate does not crystallize, since its solubility increases during the flash cooling. The salt is dewatered and dried.

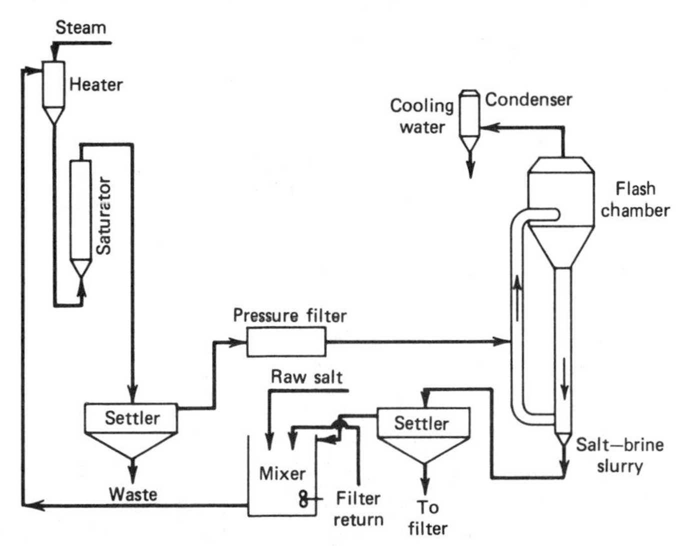

Figure 5. Recrystallizer process.

Solar Evaporation. Solar evaporation is the oldest method of salt recovery. In the United States and Canada only ca 5% of the total salt production derives from solar salt, although energy considerations are making this process more attractive. The San Francisco Bay area, southern California, and the Great Salt Lake are the only areas where solar salt is produced, although substantial quantities are imported from the Bahamas, Netherlands Antilles, and Mexico. In other parts of the world, solar salt evaporation is of great importance. Worldwide, solar salt accounts for close to half of the world's salt production; in many areas, it is the only type available.

Solar-salt production is limited to areas that combine favorable meteorological conditions, availability of acceptable land, and accessibility to markets. Very large and relatively level land areas with reasonably impervious soil are required.

Solar evaporation is basically a fractional crystallization process using the sun as the energy source. Seawater or other natural brine is brought to saturation in a series of large open concentrating ponds. In the lime ponds preceding the salt-crystallizing areas, substantial deposition of calcium sulfate occurs. Thereafter, further evaporation in crystallizing areas, or pans, results in formation of salt crystals that accumulate in a bed on the floor of the crystallizing pan. Except for calcium sulfate, the brine constituents other than salt remain in solution and are generally discarded as bitterns. The salt crop is recovered with special equipment in an operation called harvesting. The harvested salt is washed and stockpiled, and further processed in a plant by drying, crushing, and screening.

Solar-salt operations, with their large open ponds, depend highly upon favorable weather conditions; a particularly rainy year, or even one hurricane, may wipe out most of an entire season's production. For this reason, a year's salt production may be stockpiled as a rainy-day reserve. The stockpiled salt has time to drain, and this drainage plus additional washing by rainfall improve the salt's purity.

In some areas solar salt is redissolved, and the brine is processed in vacuum pans to food-grade evaporated salt. In the United States, some salt products are marketed as sea salt with implied nutritional or health benefits; some of these products are ordinary vacuum-evaporator salt. In practice, since solar evaporation is a fractional crystallization process, most minor components of seawater are not included. Total evaporation of seawater would produce a true sea salt, but it would not be acceptable for food uses.

In addition to sodium chloride, the bitterns from a solar operation contain a variety of magnesium and potassium compounds, plus other recoverable values. Magnesium chloride, potassium chloride, bromine, potassium sulfate, and sodium sulfate are successfully recovered from bitterns.

As an alternative to solar evaporation, an ion-exchange membrane process produces a salt-rich brine from seawater by means of electrodialysis; the salt is then crystallized by a conventional evaporator process. In Japan, this process accounted for production of 685,000 metric tons in 1977, or ca 50% of Japan's total salt production (6). Capacity has been increased substantially since that time.

Incoming seawater is filtered before being fed to the electrodialyzer. An electrodialyzer may contain as many as 3600 sheets of ion-exchange membranes between a set of electrodes and is capable of producing brine equivalent to 11,000 t salt per year. The membranes are selectively permeable to monovalent ions (see Membrane technology).

Prevention of Caking. At >75% rh, untreated salt absorbs moisture and generally cakes in storage. In addition, a small quantity of brine entrapped within the salt during the crystallization process may be slowly released later to contribute to caking. Thus a water-insoluble free-flow agent is generally added to table salt and some industrial grades. The most common free-flow agents are finely divided adsorbents, eg, sodium silicoaluminate and tricalcium phosphate; magnesium carbonate, calcium polysilicate, and silicon dioxide are also used. Concentrations are ca 0.5%, and up to 2% for very fine grades of salt.

Alternatively, water-soluble anticaking agents may be added, eg, sodium ferrocyanide or ferric ammonium citrate. They are used at very low concentrations of ≤13 ppm for salt for food use.

A humectant, eg, propylene glycol, glycerol, sorbitol, or calcium chloride, prevents drying and caking.

All these additives are approved food additives (see Food additives).

Economic Aspects

World salt production was ca 172 × 10^6 t in 1979, and 165 × 10^6 t in 1980; thirteen nations (the U.S., People's Republic of China, the USSR, the FRG, the UK, France, Canada, Mexico, Italy, Romania, Australia, Poland, and India) account for ca 83% of the world's salt output (see Table 5) (7). United States production accounted for ca 25% of 1979 world total.

Annual U.S. salt production and export–import data are given in Table 6 (8). In 1980, 47 U.S. companies operated 89 salt-producing plants in 17 states. Over 10^6 t salt was sold or used by each of 11 companies, representing 82% of the U.S. total. In 1980, the apparent U.S. salt consumption fell to 40.6 × 10^6 t, the lowest level since 1975.

Table 5. Worldwide Salt Production, 10^6 metric tons

Continent	1978	1979
Europe	65.7	72.6
North America	53.3	55.2
Asia	31.8	30.2
South America	5.4	5.2
Oceania	4.7	5.9
Africa	2.8	1.9

Table 6. U.S. Production, Exports, Imports, and World Production

	1976		1978		1980	
Use	10^5 t	10^6 $	10^5 t	10^6 $	10^5 t	10^6 $
U.S. production[a]	397.36		388.99		376.33	
sold or used by producers[a]	400.90	430.96	388.90	499.35	366.07	656.16
exports	9.13	10.33	7.04	9.80	7.54	12.83
imports for consumption	39.48	23.48	48.81	34.25	47.75	44.07
apparent consumption	431.24		430.67		406.3	
world production	1606.17		1669.19		1647.53	

[a] Excluding Puerto Rico, which produced an estimated 0.24 × 10^5 t/yr (1975–1980).

There are several reasons for this decrease: The 1980 demand for deicing salt decreased by 21% from the previous year as a result of a mild winter and imports of salt for highway use decreased 55%. Furthermore, production of chlorine and sodium hydroxide declined in 1980; the output of metallic sodium dropped because of reduced consumption of leaded gasoline additives.

A breakdown of U.S. production by recovery is given in Table 7 (8). Of total 1980 tonnage, ca 10% was evaporated (granulated) salt chiefly from crystallizing evaporators, 6% solar-evaporated, and 29% mined rock salt; 55% was used as brine, which is the principal raw material for heavy chemicals.

Louisiana, Texas, New York, Ohio, and Michigan are the leading states in amount of salt sold or used, and together they represent ca 85% of total U.S. tonnage (see Table 8) (8); salt prices, as of July 27, 1981, are given in Table 9.

Table 7. U.S. Salt Production, Method of Recovery [a]

Recovery method	1979		1980	
	10^5 t	10^6 \$	10^5 t	10^6 \$
evaporated				
bulk				
evaporated (granulated) [b]	33.80	229.662	32.54	274.188
solar-evaporated	19.09	25.575	21.17	36.516
pressed blocks	3.55	19.727	3.57	24.412
Total	*56.44*	*274.965*	*57.28*	*335.117*
rock				
bulk	135.09	148.205	106.52	172.039
pressed blocks	0.58	3.987	0.59	4.502
Total	*135.67*	*152.192*	*107.11*	*176.541*
salt in brine (sold or used as such)	223.90	111.195	201.68	144.507
Grand total	*416.01*	*538.352*	*366.07*	*656.164*

[a] Excludes Puerto Rico.
[b] Includes grainer salt.

Table 8. Salt Sold or Used by Producers in the United States, by State

State	1979		1980	
	10^5 t	10^6 \$	10^5 t	10^6 \$
Kansas [a]	17.24	61.184	14.26	64.276
Louisiana	128.89	113.167	114.87	132.182
Michigan	27.94	82.540	21.83	104.842
New York	57.94	77.751	49.98	99.395
Ohio	37.51	79.598	29.28	87.371
Texas	102.36	67.602	90.52	93.414
Utah	10.92	14.723	10.50	19.373
West Virginia	9.78	[b]	8.65	[b]
other states	22.86	41.787	26.19	55.311
Total	*415.44*	*538.352*	*366.08*	*656.164*
Puerto Rico	0.25	0.639	0.25	0.642

[a] Quantity and value of brine included under other states.
[b] Proprietary data; included under other states.

Table 9. Salt Prices, July 1981

Salt product	Price, $
Evaporated[a]	
common	
in 36-kg bags, per kg	0.08
bulk, per metric ton	55.13
chemical grade in 36-kg bags, per kg	0.09
Rock, medium or coarse[a]	
in 36-kg bags, per kg	0.05
bulk, per metric ton	55.13

[a] Carlots, truckloads, North, works.

In 1980, U.S. salt exports amounted to 753,700 t, mostly to Canada, with minor quantities to Saudi Arabia, Iraq, and Mexico. Salt imports, mainly from Canada and Mexico, decreased slightly to 4.8×10^6 t in 1980 as a result of reduced consumption.

The distribution of the 1979 U.S. consumption is given in Table 10. Production of chlorine, caustic soda, and soda ash accounts for ca 57% of total usage, mainly in the form of brine. Highway use of rock salt represents ca 19% of total tonnage.

Table 10. Distribution of U.S. Salt Consumption, 1979, Thousand Metric Tons

Consumer or use	Evaporated	Rock	Brine	Total
chlorine, caustic soda, and soda ash	505	1,650	21,613	23,768
all other chemicals	405	567	136	1,108
textile and dyeing	122	48		171
meatpackers, tanners, and casing manufacturers	235	260		495
dairy	71	6		77
canning	164	90		254
baking	99	9		108
flour processors, including cereal	64	23		86
other food processing	185	51		237
feed dealers	624	459		1,083
feed mixers	330	326		656
metals	64	259		323
rubber	[a]	8	[a]	90
oil	207	93		499
paper and pulp	[a]	122	[a]	176
water softener manufacturers and service companies	421	313		734
grocery stores	805	230		1,034
highway use	279	7,650		7,931
U.S. government	18	53		71
distributors, brokers, wholesalers	533	[a]	[a]	1,133
other uses	547	1,297	445	1,555
Total	*5,678*	*13,514*	*22,393*	*41,589*

[a] Proprietary data; included under other uses.

Analytical Methods

The most common impurities, depending on type of salt, are calcium sulfate, calcium chloride, magnesium chloride or magnesium sulfate, sodium sulfate, and water-insoluble material. Surface moisture is determined by drying, material insoluble in water by weighing, calcium and magnesium by EDTA titration, and sulfate gravimetrically; salt content is then expressed as difference (9). Other procedures use an assay based upon chloride titration, but accurate determination of salt purity by assay is difficult, especially for purified grades of salt.

Health and Safety Factors

In every mammal, sodium regulates the volume of blood and maintains the balance of fluids and pressure inside and outside the body's cells. It also plays an important role in nerve-impulse transmission, heart action, and the metabolism of carbohydrates and protein. Chloride ion is essential in maintaining the acid–base balance in blood and osmotic balance in tissues. Chloride ion is needed for activating certain essential enzymes and for the formation in the stomach of the hydrochloric acid necessary for the digestive process.

The daily human requirement for sodium is difficult to establish because the need fluctuates, depending upon sweating and elimination. Rather than recommend a specific daily amount, the NRC-NAS has issued an estimate of an adequate and safe sodium intake at 1.1–3.3 g/d for adults (10). At present, U.S. citizens consume 4–5 g sodium per day.

Although sodium is essential to the normal functioning of the human body, there has been recent concern about excessive sodium in the U.S. diet with regard to hypertension or high blood pressure. Hypertension afflicts >20% of the world population, according to estimates by a number of health professionals. Only in ca 5–10% of the cases can the actual cause of the hypertension be determined. The remaining 90–95% of the hypertensives are referred to as suffering from essential or primary hypertension (11).

Medical researchers have not established exactly what causes high blood pressure.

Since the sodium ion plays such a decisive role in the regulation of body fluids, it seems reasonable to postulate that it may be capable of influencing blood pressure. Although the body normally maintains a balance of sodium and other minerals at the proper concentration, evidence suggests that individuals who are genetically predisposed to hypertension may increase their risk with a high sodium diet.

Research on the possible role of sodium in essential hypertension started at least 60 years ago. It has not been established that excessive sodium actually causes hypertension, although sodium restriction often lowers the blood pressure of hypertensive persons. Moderate amounts of sodium generally have no effect on healthy persons.

High blood pressure can be reduced by a diet severely restricted in sodium. Limitation of sodium intake to ca 1.38 g/d (60 meq/d) may prevent the onset of hypertension (12), but limitation to 0.46 g/d may be required to reduce severe hypertensive pressure. Mild hypertension may be reduced by a sodium intake of 0.92–1.15 g/d (12). However, the value of low sodium diets has been questioned (13).

High blood pressure is familial, and, for unknown reasons, hypertension seems more prevalent in black people than in any other ethnic group.

Voluntary nutrition labeling by food manufacturers is expected to include sodium content in the future.

Salt is not the only source of sodium in the diet. Substantial sodium intake derives from bicarbonate of soda, monosodium glutamate, and a variety of seasonings. The natural sodium content in food items varies (see Mineral nutrients).

Toxicity. Except in the case of infants, acute oral toxicity of salt is hardly meaningful. The oral LD_{50} is 3,000 mg/kg in rats and 4,000 mg/kg in mice, indicating a toxic dosage of 200–280 g for a 70-kg adult.

Environmental Concerns

Mainly because of its large-scale usage, salt creates significant environmental problems. In the production of granulated salt, impurities concentrate in the brine. The waste brine must eventually be disposed of, usually by injection into deep wells. The calcium carbonate–magnesium hydroxide sludge from the treatment of raw brine and the high sodium sulfate brine from the crystallizers both present disposal problems.

In some areas, disposal of water-softener regenerant brines is a problem, especially for large-volume water-softener dealers. Similarly, contaminated brines, as obtained from pickle processing, must be disposed of. In some instances, waste brines are being treated and recycled.

The general public is mainly concerned with deicing salt for roads. Arguments relate to vegetation damage, contamination of waterways or shallow wells, auto corrosion, scaling of concrete surfaces, and corrosion of steel reinforcing bars on bridge decks (14).

Shade trees that are more salt resistant are now planted along roadways. Protective coverings over deicing salt piles and improved salt-spreading equipment have minimized wastage. Scaling of concrete occurs whatever deicer is used. Air-entrained concrete minimizes surface scaling and spalling of concrete.

Many substitutes for deicing salt have been suggested, but most are too expensive and unavailable in the large quantities needed. Urea is used on airport runways. Liquids such as ethylene glycol have been used for deicing of airplanes with minimum corrosion (see Antifreezes and deicing fluids). Protective coatings reduce automobile corrosion. For the time being, large amounts of salt will be used for deicing for lack of attractive, inexpensive alternatives (14–15).

Uses

In Food. Historically, salt has been used to flavor and preserve meat and fish. In food flavoring, it is second only to sugar in the amounts used. U.S. citizens consume an average of 10–12 g of salt per day, of which ca 3 g occur naturally in food, 3 g are added in cooking and at the table, and 4–6 g are added during commercial processing of food (16) (see Food additives).

Flavor Enhancer. Salt improves the flavor of many foods and contributes to palatability and acceptance. Without salt, bread tends to be bland, cheeses are bitter and tart, and processed meats lack texture and flavor (see Flavors and spices).

Preservative. Salt is an important preservative of meat, dairy products, pickles, olives, margarine, salad dressings, and many other foods. It retards the growth of microorganisms.

Color Developer. In ham, bacon, hot dogs, sauerkraut, and sausage, salt promotes the natural development of the color to which consumers are accustomed.

Binder. In sausage and other processed meat products, salt promotes formation of a binding gel consisting of meat, fat, and moisture.

Texture Aid. Salt improves the tenderness of cured meats such as ham by promoting the absorption of water by protein. It also imparts a smooth, firm texture to processed meats. Salt develops an even consistency in bread and other yeast-raised baked goods, as well as in sauerkraut and cheeses. It develops the characteristic rind on hard cheeses.

Fermentation Control Agent. In production of pickles, sauerkraut, summer sausage, cheese, and bread, salt controls the rate of fermentation. This assures consistency in color, flavor, and texture of the finished products.

Iodized Salt. Since 1924, table salt in the United States has been used as a carrier for iodine, a trace nutrient. It is sold at the same price as plain table salt. More than 50% of the table salt sold in the United States is iodized.

Iodine is used in the body for formation of thyroxine, an essential hormone (see Thyroid and antithyroid preparations). Iodine deficiency occurs in areas where excessive run-off has leached naturally occurring iodine compounds from the soil. The recommended dietary allowance for iodine is 150 μg for adults. Currently, there is some concern that this level is being substantially exceeded in the U.S. diet. Much of this excess appears to come from food-processing aids and iodine additives in animal feeds.

The only iodizing agent approved for table salt in the United States is potassium iodide; it is present at a concentration of $<0.01\%$. In the presence of moisture, potassium iodide is unstable, especially in an acid environment. Therefore, the iodizing solution added to salt contains sodium carbonate or bicarbonate for alkalinity, plus a stabilizer such as sodium thiosulfate or dextrose. Without stabilization, potassium iodide is oxidized by air to iodine and lost from the product.

Chlorinated water may liberate iodine from iodized salt, and some persons find iodine taste and odor objectionable. Potassium iodate or calcium iodate would be more stable than potassium iodide, but they are not approved for use in table salt.

In some countries, table salt is used as a carrier of fluoride ion for prevention of dental caries, whereas in India, salt has been evaluated as a carrier of iron and calcium.

Industrial Uses. Many important industrial chemical processes are based on sodium chloride (see Tables 11 and 12).

A growing use of salt is for the regeneration of cation-exchange resins in water softeners. Compressed evaporated salt (in the form of briquets, chunks, or blocks), Southern (dome) rock salt, and some solar salt are utilized. Northern rock salt is not extensively used because it contains substantial quantities of insoluble matter.

In agriculture, salt is a nutritional requirement for animals and is also used as a carrier for trace minerals, vitamins, and medication.

Salt-gradient solar ponds have a top layer of water and a bottom layer of concentrated brine; density differences prevent mixing. A pond may consist of a 0.2-m convective layer resulting from wind action, a 1-m-thick nonconvecting layer, and a

Table 11. Chemicals Produced by Reactions Using Salt as Raw Material

Process	Reaction
brine electrolytic cell[a]	$2\ NaCl + 2\ H_2O \rightarrow 2\ NaOH + H_2\uparrow + Cl_2$
brine electrolytic cell[b]	$NaCl + H_2O \rightarrow NaClO + H_2\uparrow$
	$NaCl + 3\ H_2O \rightarrow NaClO_3 + 3\ H_2\uparrow$
	$NaCl + 4\ H_2O \rightarrow NaClO_4 + 4\ H_2\uparrow$
ammonia–soda Solvay process	$2\ NaCl + CaCO_3 \rightarrow Na_2CO_3 + CaCl_2$
Downs sodium electrolytic cell	$2\ NaCl \rightarrow 2\ Na + Cl_2$
nitrosyl chloride process	$3\ NaCl + 4\ HNO_3 \rightarrow Cl_2 + 3\ NaNO_3 + NOCl + 2\ H_2O$
Mannheim furnace	$2\ NaCl + H_2SO_4 \rightarrow Na_2SO_4 + 2\ HCl$
Hargreaves process	$4\ NaCl + 2\ SO_2 + O_2 + 2\ H_2O \rightarrow 2\ Na_2SO_4 + 4\ HCl$
cyanamide electric furnace	$2\ NaCl + CaCN_2 + C \rightarrow CaCl_2 + 2\ NaCN$
niter-cake retort	$NaCl + H_2SO_4 \rightarrow NaHSO_4 + HCl$

[a] Chlor-alkali cell.
[b] Without separation of cell compartments.

Table 12. Industrial Applications of Salt

Process	Action
leather tanning	prevention of bacterial decomposition in hides
textile dyeing	standardization of dye batches; dye fixation
soapmaking	separation of soap from water and glycerol
pulp and paper manufacture	precipitation of waterproofing compositions; electrolytic generation of chlorine bleach
ceramics manufacture	surface vitrification of heated clays
rubber manufacture	salting out of rubber from latex
refrigeration	salt–ice mixtures for direct cooling; brine refrigerant
oil-well operation	inhibition of fermentation in drilling muds; increase density of drilling fluid; stabilization of rock-salt strata
pigment and dry-detergent formulation	filler; grinding agent in pigment

2–7-m bottom convective layer for heat storage and energy extraction. Solar energy warms the base of the pond, and heat is retained by brine in the deeper parts. This warm brine may be withdrawn and its heat used for space heating or other application. In Israel, the heat has been used to produce electric power.

Salt has been used experimentally to enhance sulfur dioxide removal by limestone during fluidized-bed combustion of coal. Salt caverns, especially in salt domes, are used to some extent for storage of petroleum, natural gas, and various organic chemicals. Storage of nuclear waste in salt caverns is controversial.

Salt Substitutes

Most commercial salt substitutes are based upon potassium chloride; other ingredients modify the principal deficiencies of potassium chloride, namely, the cooling sensation when it dissolves on the tongue and its characteristic bitter aftertaste.

Typical ingredients include potassium citrate, fumaric acid, ammonium chloride, monocalcium phosphate, potassium phosphates, citric acid, tartaric acid, cream of tartar (potassium bitartrate), choline bitartrate, glutamic acid, and potassium glutamate. A plain salt substitute contains <100 ppm sodium, whereas seasoned salt substitutes contain <200 ppm sodium.

In an approach different from the low sodium salt-substitute concept, a combination of salt and potassium chloride has been evaluated that provides only a partial reduction of sodium intake (17). It is claimed that the composition has a bitter taste to not more than 20% of the population.

Overall, no other compounds seem to impart the salty flavor characteristic of salt. The choice of substitute cations is limited to potassium, calcium, and magnesium. Calcium and magnesium compounds, however, tend to be hygroscopic which prevents the free flow required in a seasoning product. Lithium chloride has a salty flavor, but toxicity considerations preclude its use.

Generally, a wide variety of spices, herbs, and seasonings are used in low sodium diets to counteract the blandness of unsalted foods.

BIBLIOGRAPHY

"Salt" in *ECT* 1st ed., Vol. 12, pp. 67–82, by C. D. Looker, International Salt Company, Inc.; "Salt" in *ECT* 2nd ed., Vol. 18, pp. 468–484, by E. J. Kuhajek and H. W. Fiedelman, Morton International, Inc.

1. M. E. Hawkins and C. J. Jirik, *Salt Domes in Texas, Louisiana, Mississippi, Alabama and Offshore Tidelands, U.S. Bureau of Mines Information Circular 8813*, Washington, D.C., 1966.
2. D. W. Kaufmann, *Sodium Chloride*, Reinhold Publishing Corporation, New York, 1960, pp. 587–626, 668–669.
3. U.S. Pat. 2,642,335 (June 16, 1953), W. May and T. Scott (to Imperial Chemical Industries, Ltd.).
4. U.S. Pats. 351,082 (Oct. 19, 1886), 400,983 (April 9, 1889), and 443,186 (Dec. 23, 1890), H. Williams, J. L. Alberger, and L. R. Alberger.
5. U.S. Pats. 2,555,340 (June 5, 1951) and 2,876,182 (Mar. 3, 1959), C. M. Hopper and R. B. Richards (to International Salt Company).
6. H. Kawate, T. Seto, R. Komori, and Y. Nagasato in A. H. Coogan and L. Hauber, eds., *Fifth Symposium on Salt*, Vol. 2, Northern Ohio Geological Society, Cleveland, Ohio, 1980.
7. R. J. Foster in A. E. Schreck, ed., *Bureau of Mines Mineral Yearbook 1978–79*, U.S. Department of Interior, Washington, D.C.
8. D. S. Kostick in *Bureau of Mines Minerals Yearbook 1980–81*, U.S. Department of Interior, Washington, D.C.
9. *ASTM E 534-75* in *1979 Annual Book of ASTM Standards*, American Society of Testing and Materials, Philadelphia, Pa., pp. 1002–1009.
10. *Sodium Restricted Diets and the Use of Diuretics. Rational Complications and Practical Aspects of the Use*, Committee on Sodium-Restricted Diets, Food and Nutrition Board, National Academy of Sciences/National Research Council, Washington, D.C., 1979.
11. J. L. Marx, *Science* **194,** 821 (1976).
12. L. Tobian, *Am. J. Clin. Nutr.* **32,** 2739 (1979).
13. G. Kolata, *Science* **216,** 38 (1982).
14. D. M. Murray and U. F. W. Ernst, *An Economic Analysis of the Environmental Impact of Highway Deicing*, EPA 600/2-76-105, U.S. Environmental Protection Agency, distributed by National Technical Information Service, Springfield, Va., 1976.
15. R. Brenner and J. Moshman, *Benefits and Costs in the Use of Salt to Deice Highways*, The Institute for Safety Analysis, Washington, D.C., 1976.
16. *Dietary Salt—A Scientific Status Summary by the Institute of Food Technologists Expert Panel on Food Safety and Nutrition and the Committee on Public Information*, Institute of Food Technologists, Chicago, Ill., 1980.
17. U.S. Pat. 3,514,296 (May 26, 1970), R. L. Frank and O. Mickelsen (to Morton International, Inc.).

General References

Reference 2 is also a general reference.

W. E. Ver Planck, *Salt in California*, Bulletin 175, State of California, Division of Mines, San Francisco, Calif., 1958.

H. Borchert and R. O. Muir, *Salt Deposits*, D. Van Nostrand Company, Inc., Princeton, N.J., 1964.

A. C. Bersticker, ed., *First Symposium on Salt*, Northern Ohio Geological Society, Cleveland, Ohio, 1963.

J. L. Rau, ed., *Second Symposium on Salt*, Vols. 1 and 2, Northern Ohio Geological Society, Cleveland, Ohio, 1965.

J. L. Rau and L. F. Dellwig, ed., *Third Symposium on Salt*, Vols. 1 and 2, Northern Ohio Geological Society, Cleveland, Ohio, 1970.

A. H. Coogan, ed., *Fourth Symposium on Salt*, Vols. 1 and 2, Northern Ohio Geological Society, Cleveland, Ohio, 1974.

A. H. Coogan and L. Hauber, ed., *Fifth Symposium on Salt*, Vols. 1 and 2, Northern Ohio Geological Society, Cleveland, Ohio, 1980.

J. W. Mellor, *Comprehensive Treatise on Inorganic and Theoretical Chemistry*, Vol. II, Suppl. II, *The Alkali Metals*, John Wiley & Sons, Inc., New York, 1962, pp. 751–933.

G. L. Eskew, *Salt, The Fifth Element*, J. G. Ferguson and Associates, Chicago, Ill., 1948.

R. T. MacMillan, *Industrial Minerals and Rocks*, American Institute of Mining, Metallurgical and Petroleum Engineers, New York, 1960, pp. 715–721.

M. T. Halbouty, *Salt Domes—Gulf Region and Mexico*, Gulf Publishing Company, Houston, Texas, 1967.

S. J. Lefond, *Handbook of World Salt Resources* in R. W. Fairbridge, ed., *Monographs in Geoscience*, Plenum Publishing Corporation, New York, 1969.

R. B. Mattox, ed., *Saline Deposits*, Geological Society of America, Boulder, Col., 1968; symposium based on *International Conference on Saline Deposits*, Houston, Texas, 1962.

C. Moses, ed., *Sodium in Medicine and Health—A Monograph*, Reese Press, Inc., Baltimore, Md., 1980.

L. H. Gevantman, ed., *Physical Properties Data for Rock Salt*, *National Bureau of Standards*, *Monograph 167*, U.S. Government Printing Office, Washington, D.C., 1981.

K. D. Fisher and C. J. Carr, *Iodine in Foods: Chemical Methodology and Sources of Iodine in the Human Diet*, distributed by National Technical Information Service, Washington, D.C., 1974.

JOHN F. HEISS
Diamond Crystal Salt Co.

EUGENE J. KUHAJEK
Morton Salt

SODIUM HALIDES, SODIUM BROMIDE

The most common salt of hydrobromic acid is sodium bromide [*7647-15-6*], NaBr, a colorless compound that crystallizes in the cubic system.

Properties. Physical properties are given in Table 1. Sodium bromide has a somewhat bitter salty taste. It is moderately hygroscopic and has fairly high water solubility (see Table 2). Below 51°C, the solid phase in equilibrium with saturated aqueous solution is the dihydrate [*13466-08-5*]; above 51°C, it is the anhydrous form.

Preparation. Pure sodium bromide is obtained most simply by neutralizing sodium carbonate or hydroxide with hydrobromic acid, followed by evaporation and crystallization. Processes for the direct formation of sodium bromide from bromine, sodium hydroxide or carbonate, and a reducing agent in water have been described (1–5). Reducing agents include ferrous iron, ammonia, activated carbon, formic acid, and hydrazine. Some sodium bromide is recovered as a by-product from solutions obtained during bromination, bromo-oxidation, or hydrolysis of organic compounds.

Economic Aspects. Total production in the noncommunist world during 1981 was estimated to be slightly under 6000 metric tons; U.S. prices in 1982 were $2.10–2.50/kg in 181-kg (400 lb) drums, fob works, for granular 99% material (commercial or technical grade). Also available were a reagent grade meeting ACS specifications and 35–50% aqueous solutions.

Table 1. Properties of Sodium Bromide

Property	Value
mol wt	102.89
mp, °C	747
lattice constant, nm	0.5977
refractive index, n_D^{25}	1.6141
sp gr $_4^{25}$	3.200
heat of fusion, J/g[a]	253.6
heat capacity, J/(kg·K)[a]	
at 25°C	498
at 200°C	527

[a] To convert J to cal, divide by 4.184.

Table 2. Solubility of Sodium Bromide

Solvent	Temperature, °C	g NaBr/100 g solution
water	0	44.5
	20	47.6
	30	49.6
	50	53.8
	100	54
	140	56
methanol	25	14.8
ethanol, 95%	25	3.8

Health and Safety Factors. Sodium bromide dust and solutions are moderately irritating to eyes and skin; prolonged exposures should be avoided. Internally in large doses, it causes depression of the central nervous system. Continued intake may lead to mental deterioration and acne-type skin eruptions (6). The LD_{50} orally in rats has been reported as ca 3.5 g/kg body weight (7). In more recent rat studies involving performance testing, response rates during the tests increased (ie, improved) with daily dosage of 100 mg/kg, but decreased with 300 mg/kg. When the dosages were stopped after six weeks of administration, response rates quickly returned to normal (8). In semichronic toxicity testing on rats, 19,200 ppm in the diet for 90 d caused motor incoordination, growth retardation in males, and increased adrenal and thyroid gland weights in both sexes (9). Incorporation of either sodium, potassium, or ammonium bromide at 10 mg Br^-/kg dry feed improves the nutrient utilization and body weight of broiler chickens (10) (see Pet and other livestock feeds; Mineral nutrients).

Uses. Sodium bromide is used in photographic processes both for the preparation of light-sensitive silver bromide emulsions and as a restrainer in developers (see Photography). Potassium and ammonium bromides have been making inroads into photographic applications. Other uses include swimming pools and health spas, and formulations of bleaching and disinfecting agents; in these cases, it becomes a source of free bromine, hypobromite, or bromamine when oxidized by chlorine, hypochlorite, or other oxidant (see Bleaching agents; Water, treatment of swimming pools, spas, and hot tubs). Similarly, it can be employed for brominations in organic syntheses.

Numerous applications of sodium bromide as a catalyst for the partial oxidation of hydrocarbons and other organic compounds have been described; many are patented. Cobalt and copper compounds may serve as co-catalysts (11). Another growing use, along with that of calcium bromide, is for increasing the density of aqueous oil-well drilling fluids (see Petroleum). Use as an electrolyte component in a low temperature sodium–halogen battery has been reported (12). Sodium bromide is widely employed as a sedative, hypnotic, and anticonvulsant (see Hypnotics, sedatives, anticonvulsants).

BIBLIOGRAPHY

"Sodium Halides" under "Sodium Compounds" in *ECT* 1st ed., Vol. 12, pp. 603–605, by J. A. Brink, Jr., Purdue University; "Sodium Bromide" under "Sodium Compounds" in *ECT* 2nd ed., Vol. 18, pp. 484–485, by V. A. Stenger, The Dow Chemical Company.

1. U.S. Pat. 1,775,598 (Sept. 8, 1930), J. H. van der Meulen.
2. U.S. Pats. 1,843,355 (Feb. 2, 1932); 1,916,457 (July 4, 1933), A. S. Behrman.
3. U.S. Pat. 1,863,375 (June 14, 1932), C. W. Jones (to The Dow Chemical Company).
4. U.S. Pat. 2,269,733 (Jan. 13, 1942), E. P. Pearson (to American Potash and Chemical Company).
5. Jpn. Kokai 72 33,992 (Nov. 10, 1972), T. Higuchi (to Kyowa Chemical Industries Company, Ltd.).
6. T. O. Soine and C. O. Wilson, *Roger's Inorganic Pharmaceutical Chemistry*, 8th ed., Lea & Febiger, Philadelphia, Pa., 1967, pp. 213–216.
7. P. K. Smith and W. E. Hambourger, *J. Pharmacol. Exp. Ther.* **55**, 200 (1935).
8. M. W. Oglesby, J. Rosenberg, and J. C. Winter, *Psychopharmacologia* **32**(1), 85 (1973).
9. M. J. Van Logten and co-workers, *Toxicology* **1**(4), 321 (1973); **2**(3), 257 (1974).
10. V. F. Lemesh and R. E. Gushcha, *Vestn. Akad. Nauk Beloruss. SSSR* (1), 97 (1974); *Chem. Abstr.* **81:** 2878 y (1974).
11. Jpn. Kokai 80 33,430 (Mar. 8, 1980), Y. Kamiya and T. Okada (to Nippon Petrochemicals Company, Ltd.).
12. U.S. Pat. 3,918,991 (Nov. 11, 1975), H. G. Hess (to General Electric Company).

V. A. STENGER
Dow Chemical U.S.A.

SODIUM HALIDES, SODIUM IODIDE

Sodium iodide [*7681-55-2*], NaI, occurs as colorless crystals or as a white crystalline solid. It has a salty and slightly bitter taste. In moist air, it gradually absorbs as much as 5% water, which causes caking. It slowly becomes brown when exposed to air because iodine is liberated. Water solutions are neutral or slightly alkaline and gradually become brown for the same reason. Aqueous solutions are stabilized by raising the pH to 8–9.5.

Properties. Sodium iodide crystallizes in the cubic system; physical properties are given in Table 1 (1). Sodium iodide is soluble in methanol, ethanol, acetone, glycerol, and several other organic solvents; solubility in water is given in Table 2.

Below 65°C, sodium iodide is present in aqueous solutions as hydrates containing varying amounts of water. When anhydrous NaI is dissolved in water, heat is liberated because of hydrate formation; eg, ΔH = 174.4 kJ/mol (41.7 kcal/mol) when the dihydrate is formed. At room temperature, sodium iodide crystallizes from water as the dihydrate, $NaI.2H_2O$ [*13517-06-1*], in the form of colorless prismatic crystals.

Manufacture. Bulk production of USP and reagent grades is based on the reaction of sodium carbonate or hydroxide with an acidic iodide solution. After removal of chemical impurities, the solution is filtered and concentrated. Evaporation gives the anhydrous form; controlled cool-down provides either a di- or pentahydrate [*81626-33-7*].

Table 1. Physical Properties of Sodium Iodide[a]

Property	Value
mol wt	149.895
mp, °C	651
bp, °C	1304
d_4^{25}, g/cm^3	3.667
specific heat, J/(kg·K)[b]	
at 0°C	350
at 50°C	360

[a] Ref. 1.
[b] To convert J to cal, divide by 4.184.

Table 2. Solubility of Sodium Iodide

Temperature, °C	g NaI/100 g H_2O
0	158.7
20	178.7
40	205.0
60	256.8
70	294
80	296
100	302
140	321

Economic Aspects. The price development and U.S. volume of sodium iodide is given in Table 3. The increase over the past decade reflects the price increase of crude iodine.

Table 3. Price Development and U.S. Production of Sodium Iodide

Year	NaI, USP, $/kg	U.S. production, t/yr	Iodine, $/kg
1970	2.60	22.5	1.30
1975	5.16	29	2.59
1978	5.76		3.10
1979	6.62	27	3.63
1980	11.95		6.35
1981	12.98	32.6	7.26

Uses. The principal use of sodium iodide is as an expectorant in cough medicines (see Expectorants, antitussives, and related agents). A small amount is used in the wet extraction of silver, iodized salt, animal feeds to prevent hoofrot, photographic chemicals, scintillation crystals, and the manufacture of organic chemicals. It has also been used in cloud seeding.

Sodium iodide USP XX contains 99–101.5% NaI, calculated on the anhydrous basis (2). It is used interchangeably with potassium iodide as a therapeutic agent, except where sodium ion is contraindicated. Intravenous sodium iodide formulations have been used for a variety of diseases from thyroid deficiency to neuralgia. However, these solutions are no longer listed in the NF XV (2), which indicates that their therapeutic value has not been satisfactorily demonstrated.

Veterinary uses include the treatment of horses, cattle, sheep, swine, and dogs for various afflictions (see Veterinary drugs).

BIBLIOGRAPHY

"Sodium Halides" under "Sodium Compounds" in *ECT* 1st ed., Vol. 12, pp. 603–604, by J. A. Brink, Jr., Purdue University; "Sodium Iodide" under "Sodium Compounds" in *ECT* 2nd ed., Vol. 18, pp. 485–486, by F. N. Anderson, Mallinckrodt Chemical Works.

1. J. C. Bailar, Jr., H. J. Emelius, R. Nyholm, and A. F. Trotman-Dickenson, eds., *Comprehensive Inorganic Chemistry*, Pergamon Press, Inc., Elmsford, N.Y., 1973, Vol. 1, p. 402; Vol. 2, p. 1107.
2. *The United States Pharmacopeia XX* (*USP XX–NF XV*), The United States Pharmacopeial Convention, Inc., Rockville, Md., 1980, p. 732.

P. H. MERRELL
E. M. PETERS
Mallinckrodt, Inc.

SODIUM NITRATE

Sodium nitrate [*7631-99-4*], $NaNO_3$, occurs in nature in deposits associated with sodium and potassium chlorides, potassium nitrate, sodium sulfate, magnesium chloride, and other salts. There are deposits of sodium nitrate in different parts of the world, eg, Chile, Egypt, South Africa, Mexico, and the United States. Only the Chilean reserves are exploited commercially, and they are referred to as Chilean saltpeter or Chilean nitrate.

Chilean nitrate is the only inorganic nitrogen fertilizer produced today. In 1880–1910, it accounted for 60% of world nitrogen production. Synthetic sodium nitrate production started early in the twentieth century and peaked in the 1930s. Since then, it has declined steadily to a very low proportion of the world nitrogen industry. This decrease resulted from the introduction in the market of synthetic, less expensive fertilizers, ie, ammonia and its derivatives, mainly urea and ammonium nitrate (see Fertilizers; Ammonium compounds). The main industrial market for sodium nitrate during its beginnings was the manufacture of nitric acid and explosives, products which have since been produced from less expensive raw materials, although some sodium nitrate is still used in explosives (see Explosives and propellants).

The highest production levels attained by the Chilean nitrate industry were from 1910 to 1920, ca 2.5×10^6 metric tons per year. Synthetic sodium nitrate production peaked in the middle of the 1930s with a total world production of ca 8×10^5 t/yr. During this period, the Chilean production decreased to 1.5×10^6 t. Recently, Chile's production has averaged 6.3×10^5 t and synthetic production has not exceeded 1.5×10^5 t.

Deposits

Commercially valuable Chilean deposits of sodium nitrate are in the northern part of the country between the southern latitudes 19° and 26°. They occur on the eastern slopes of the Chilean coastal range at 1070–2290 m above sea level. The deposits are scattered in an area 640 km in the north–south direction and are 8–64 km wide. The ore bodies are very heterogeneous, varying in size, thickness of the main body and overburden, composition, and hardness. They are 0.3–1.2 m below an overburden of sandy soil and vary in thickness from 0.15 to 4.3 m.

Caliche, the ore from which the Chilean nitrate is extracted, is a conglomerate of inert detritus cemented by salts. Sodium nitrate and sodium chloride are the main components of the saline fraction. The inert material is composed of volcanic rocks interbedded with limestone, clays, lime, and sand. The ore body is most commonly a stratified-layer type, with the varying salt content defining the strata; however, the physical characteristics of the ore remain fairly constant. Thus, the concept of exploitable caliche is very broad, and the economic feasibility of exploitation is related to the layer thickness. In Figure 1, the different layers in a nitrate ore are shown.

Typical analyses of samples being mined during the history of the nitrate industry are given in Table 1 (1). The ore composition has degraded considerably since the early days of the industry, when it was reported that ores of up to 50 wt % sodium nitrate were mined.

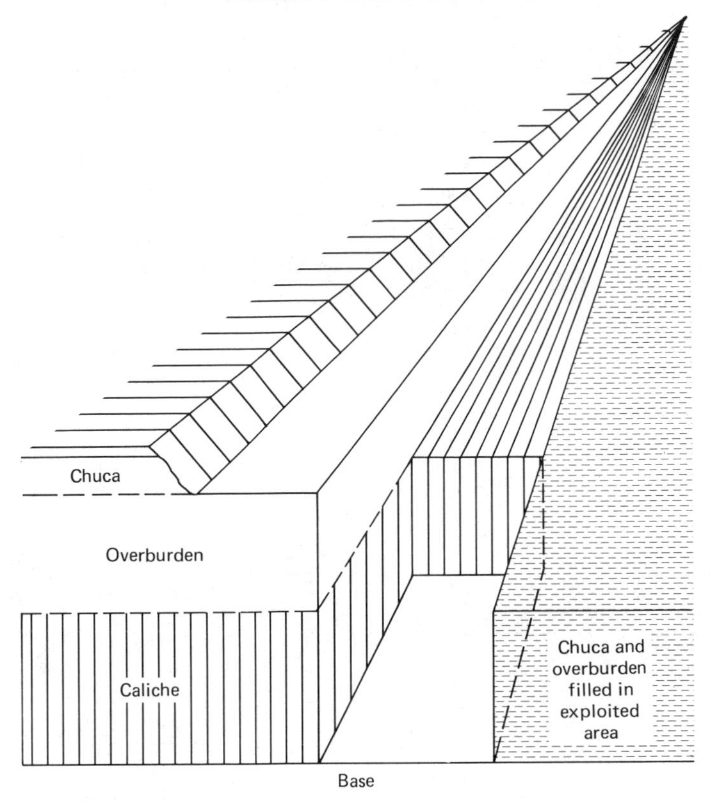

Figure 1. Schematic of layers in a nitrate ore bed.

Table 1. Analyses of Nitrate Ore Samples, wt % [a]

Component	Mining period		
	Up to 1920	1960–1970	1970–1980
$NaNO_3$	20–50	7–9	6.5–8
NaCl	20	10–15	12–16
Na_2SO_4	12–15	12–16	10–15
I_2	0.03	0.03	0.03
$Na_2B_4O_7$	0.4–0.6	0.4–0.6	0.4–0.6
K	1.0–2.0	0.4–1.5	0.3–1.2
$KClO_4$	0.03	0.03	0.03
Mg	0.2–0.8	0.2–0.8	0.2–1.2
Ca	0.4–3.3	0.4–2.3	0.5–2.4
Li	0.03	0.03	0.03
H_2O	1.0–2.0	1.1–1.8	1.1–1.8
clay, sand, and insolubles	6–45	53–68	54–69

[a] Ref. 1.

Properties

Selected physical and chemical properties of sodium nitrate are listed in Table 2. At room temperature, sodium nitrate is an odorless and colorless solid, moderately hygroscopic, saline in taste, and very soluble in water, ammonia, and glycerol. Detailed

Table 2. Selected Properties of Sodium Nitrate

Property	Value
mol wt	84.99
crystal system	trigonal, rhombohedral
mp, °C	308
refractive index, n_D^{20}	
trigonal	1.587
rhombohedral	1.336
density (solid), g/cm³	2.257
solubility in H_2O, molality (±2%)	
at −17.5°C	7.4
at 0°C	8.62
at 40°C	12.39
at 80°C	17.42
at 120°C	24.80
specific conductivity (at 300°C), S/cm	0.95
viscosity[a] (η), mPa·s (= cP)	
$\eta = 25.0987 - 6.0544 \times 10^{-2}\, T + 3.8709 \times 10^{-5}\, T^2$ [b]	
at 590 K	2.85
at 730 K	1.53
heat of fusion, J/g[c]	189.5
heat capacity, J/g[c]	
solid state	
at 0°C	1.035 ± 0.005
at 100°C	1.23 ± 0.006
liquid state	
at 350°C	1.80 ± 0.02
aqueous solution, p = wt % solute	
$C_p = 4.175 - 37.42 \times 10^{-3}\, p + 216.6 \times 10^{-6}\, p^2$ [d]	4.138–3.045

[a] Measurement method: capillary.
[b] Precision = ca ±0.6%; uncertainty = ca ±3%.
[c] To convert J to cal, divide by 4.184.
[d] Range 1–39%.

physical and chemical properties are given in refs. 2–3 and in the *Handbook of Chemistry and Physics.*

Manufacture and Processing

The manufacture of natural sodium nitrate is carried out by its extraction from the ore with a brine followed by fractional crystallization.

Shanks Process. In the Shanks process, the ore was crushed to 3.8–5.1-cm pieces and loaded into large steel vats, each with a capacity of up to 75 t. The leaching solution consisted of water and a mother-liquor brine with ca 450 g/L sodium nitrate. During the leaching process, which is carried out at 70°C, this brine was concentrated to 700–750 g/L sodium nitrate and was pumped to a preliminary cooling pan to allow the temperature to drop no more than 5°C. Sodium chloride then crystallized and the slimes were allowed to settle after the addition of a coagulant, eg, wheat flour. The clear liquor was added to $NaNO_3$-crystallizing pans, where the temperature was allowed to drop overnight to ambient temperature. The brine or mother liquor was then pumped and sent to the last stages of the leaching cycle. The crystals were piled in the crystallizing pan to allow drainage of the entrained mother liquor. Finally, the crystals were transferred to metallic drying floors, where the remaining mother liquor was evaporated. The final product was a crystalline material containing the following: $NaNO_3$ (including up to 5 wt % KNO_3), 95.5 wt %; NaCl, 0.75 wt %; Na_2SO_4, 0.45 wt %; other salts, 1.0 wt %; and H_2O, 2.3 wt %.

The Shanks process made possible the recovery of ca 60% of the sodium nitrate in the ore. Fuel consumption was ca 0.154 tons of fuel oil per ton of $NaNO_3$. This process was considered obsolete in the early 1960s and there are no plants based on this system.

Guggenheim Process. Because of the inefficiency in ore extraction and fuel consumption of the Shanks process, in 1918 the U.S. mining company Guggenheim Brothers developed a low temperature leaching process. The process was developed based on the following facts: Leaching of caliche at 40°C yields a fairly good extraction. Also, the sodium nitrate concentration can be as high as 450 g/L, which is more than 50% of the total dissolved solids, and it can be easily removed through crystallization. If the leaching solution contains a certain level of protective salts, eg, $MgSO_4$ and $CaSO_4$, the sparingly-soluble double salt $NaNO_3.Na_2SO_4.H_2O$ (darapskite) present in the caliche is broken up by magnesium action, thereby increasing $NaNO_3$ extraction.

The Guggenheim nitrate process soon demonstrated that caliche ores as low as 7 wt % in nitrate content could be economically exploited. In Chile, the María Elena and Pedro de Valdivia plants have a joint capacity of 750,000 t and are processing ores with 7.5–7.7 wt % $NaNO_3$ concentrations. A flow sheet of the processing operations there is given in Figure 2.

A mining section is ca 2000 m long and 4–5 m wide and follows the pattern of Figure 1. The mining process consists mainly of stripping and rolling the chuca with bulldozers and filling previously mined areas. Then the overburden is loosened by drilling and blasting and is removed with draglines to expose the ore body. The overburden is also sent to fill previously mined areas (see Fig. 1). The ore body is then blasted and loaded by shovels or draglines into railway cars or trucks with 25–35-t capacities. The cars are hauled to the plant by battery-powered locomotives to central stations. The trucks dump the ore in piles at these central stations, where it is transferred to railway cars. Subsequently, the cars are hauled to the plants by electric- or diesel-powered locomotives.

At the plant the cars dump the ore into rotary dumpers. The ore is transferred to primary jaw- or gyratory-type crushing units. The feed is heterogeneous, ranging from fine particles to chunks of 3–5 t. Crushing is carried out in three stages. Of the resulting product, 20–25% is <9.5-mm dia. This portion of the feed is known as fines. The remaining 75–80% is 9.5–19 mm in diameter and is called the coarse fraction. Both fractions are sent to different leaching units. The removal of the fines from the coarse fraction is essential, as leaching takes place by downward percolation in big vats. This percolation becomes very difficult because of the presence of clay in the fines which swell when combined with rich brines. The fines fraction is sent to the filter plant (see Fig. 2).

The vats are built of reinforced concrete, since the solutions are highly corrosive.

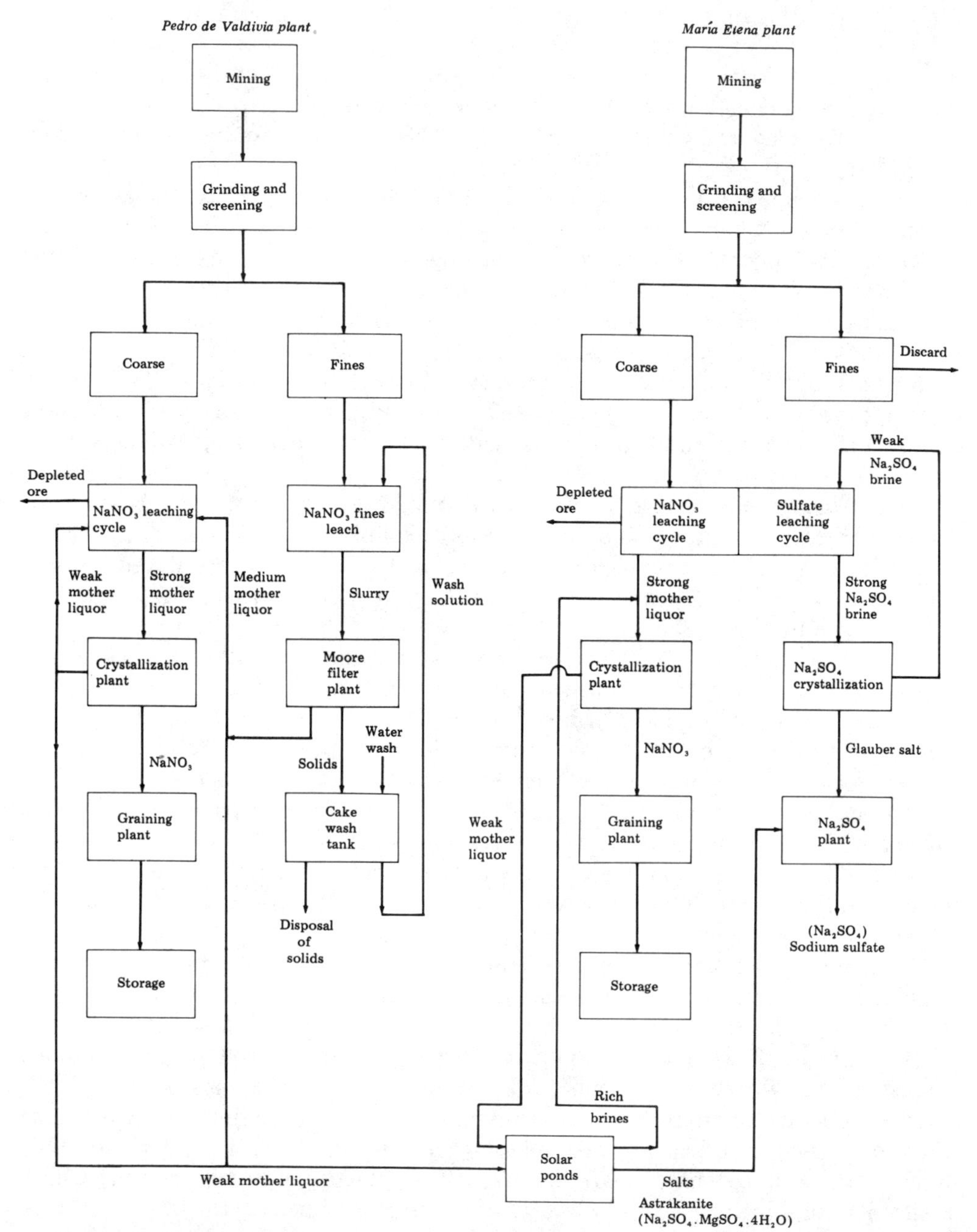

Figure 2. Processing operations flow sheet of the María Elena and Pedro de Valdivia plants.

The vats are ca 49 m long, 34 m wide, and 6 m tall. Each has an ore capacity of 10,000 t of crushed ore. The bottoms of the vats contain filter beds to ensure clear solution discharge.

Usually the leaching cycle involves 8–10 vats, of which 4 are used at any given time in the leaching process. While one vat is being unloaded, another is being loaded. The remaining vats are washed with water and are drained. The resulting brines (weak brines) from the vats are used as water makeup because of losses in the system, entrained water in the crystallized salts, leakage, and evaporation. The remaining weak brines are sent to solar ponds where they are concentrated until they reach the same composition as the brine coming from the leaching stage.

The average leaching temperature is 40°C, and the mother-liquor concentrates contain ca 320 g/L to a final concentration of 440–450 g/L soln. The leaching cycle lasts 40 h, and the total vat cycle is ca 168 h. The final tailings contain 0.75–2.00 wt % sodium nitrate (mainly entrained) and are removed from the vats with 5-t clam-shell grab buckets suspended from a movable crane. The buckets dump the tailings into trucks that haul them to tailings disposal areas or botaderos.

Filter Plant. The fines produced during crushing are treated in the filter plant. They are mixed with liquor containing 150–200 g/L of nitrate, with the mixing ratio being two tons of solids per cubic meter of liquor, and the leaching is carried out at 40–50°C. The mixture is air-agitated and then filtered in Moore vacuum filters. Each filter has ca 560 m^2 of filtration area. The filtrate contains 350 g/L of sodium nitrate and is sent to the last stages of the leaching plant, where it is again filtered in the vats and combined with rich leaching solution from that vat.

As the cake, which is mainly silt and clays, builds up to 3.8 cm in thickness on the filter leaves, the filter basket is transferred from the cake-forming tank to a washing tank. There the cake is washed by displacement with water or weak brines taken from the leaching plant. The resulting brine is used as the leaching solution in the sodium nitrate extraction from the fines. This new countercurrent leaching system was first used in 1977 and is more efficient than the early system involving the same units (see Modified Guggenheim Process). After washing, the filter basket is transferred to a drying tank where the cake is pushed off the filter leaves by compressed air. As the cake drops to the bottom of the tank, it is mixed with water and the resulting slurry is pumped to the disposal lagoon or botaderos.

Crystallizing Plant. The rich liquor from the leaching vats containing 450 g/L of sodium nitrate is sent to crystallizers where it is chilled from 40°C to 0 or 5°C. The crystallizing system comprises 20 vertical shell-and-tube units in series with the rich brine inside the tubes. Each crystallizer has more than 500 tubes 7.6 cm in diameter and 4.9 m long. Mother liquor in the first 14 crystallizers and then liquid ammonia are used as coolants. At the exit of the last crystallizers, the mother liquor and the sodium nitrate crystals are pumped out as a slurry (crystal size, 325 μm (48 mesh)) to continuous centrifuges. In these the slurry is spun out and the resulting cake is rinsed with water to displace entrained brines. The discharged crystals are white and contain 5–7 wt % moisture.

Graining Plant. The sodium nitrate crystals are transferred from the draining piles by conveyor belts to the graining plant, which consists of large, oil-fired, reverberatory-type furnaces. The crystals are heated to 315–325°C and change to liquid form. The molten sodium nitrate is pumped to large spray chambers, which are 30.5 m tall in Pedro de Valdivia and 45.7 m tall in María Elena, where it passes through the spray nozzles 1.6 cm in diameter. The spray of molten sodium nitrate solidifies as prills, which are subsequently transferred while warm by conveyors to a screening plant. The oversized and undersized products are sent back to the furnaces for remelting. The screened product is cooled in heat exchangers, where leaching brines are used as the coolant to recuperate additional heat for that process. As the prills come from cooling at 35°C, they are passed by belt conveyors to rail cars and finally to storage silos at the shipping port.

The Guggenheim process recovers 80–85 wt % of the $NaNO_3$ contained in the coarse fractions with a total fuel consumption of 0.100 tons of fuel oil per ton of sodium nitrate produced. Sodium sulfate and iodine are also recovered in association with the Guggenheim process. Sodium sulfate is recovered from the weak nitrate brines through a fractional crystallization where sodium sulfate decahydrate is obtained. Iodine is obtained by converting the sodium iodate in the brines into sodium iodide and finally mixing the iodide and iodate brines yielding iodine.

Modified Guggenheim Process. During the late 1950s and early 1960s, it was recognized that sodium nitrate and other compounds could be recovered from weak brines through concentration by solar evaporation. It was also discovered that potassium concentration was constant in the rich brines coming from the leaching stages despite changes in the potassium concentration in the ore and that the leaching brine was constantly recycled into the vats. Since then numerous studies show a potassium solid–liquid equilibrium between different clays present in the ore and the brines. Common practice was to send the brines from the leaching stage, as in the normal Guggenheim system, to the crystallizers and then to transfer the mother liquor to solar ponds. As water is eliminated in the ponds, a higher amount of makeup water is needed, so a totally countercurrent leaching system was developed. This enabled an increase in leaching yields of sodium nitrate, potassium nitrate, and iodine. These ideas were tested in the laboratory in the late 1960s and put into full scale operation in 1976 at the Chilean María Elena–Coya complex.

In the new process the ore is leached countercurrently with water in vats. From the weak sodium nitrate brines, sodium sulfate is extracted by a chilling process that yields Glauber's salt. As leaching proceeds, iodine concentration increases until it reaches 1 g/L, at which time it is removed from the vats and iodine is extracted. The depleted iodine brine is returned to the leaching system until either the sodium nitrate concentration reaches 420–440 g/L or the iodine concentration again reaches 1 g/L. The rich sodium nitrate solution is passed to the crystallizers and the mother liquor to the solar-pond system. Whenever the brines concentration in the ponds is greater than 420 g/L, the brines are sent to the crystallizers again (see Fig. 2).

The continuous evaporation process of the brines in the ponds results in concentration increases of potassium nitrate and iodine. Potassium nitrate is removed from the brine through crystallization, and a natural product containing 30 wt % KNO_3 and 70 wt % $NaNO_3$ is obtained. Iodine is removed in the iodine plant by the same process as for the rich vat solutions in the Pedro de Valdivia plant.

The salts that deposit in the solar ponds are constituted mainly of astrakanite ($Na_2SO_4.MgSO_4.4H_2O$) and sodium chloride. A new system was developed in 1976 that has permitted the recovery of sodium sulfate from these waste salts (see Chemicals from brine).

Synthetic Process. As of today, synthetic sodium nitrate is produced by neutralization of nitric acid with soda ash or caustic soda (4). The sodium alkali used is chosen based upon the relative prices of the two chemicals. The chemical reactions are as follows:

$$2\,HNO_3 + Na_2CO_3 \rightarrow 2\,NaNO_3 + CO_2 + H_2O$$
$$2\,HNO_3 + 2\,NaOH \rightarrow 2\,NaNO_3 + H_2O$$

After the reaction, the water evaporates leaving behind the sodium nitrate. From this point the sodium nitrate is melted and prilled in processes very similar to those of the Guggenheim process.

Production and Consumption

Sodium nitrate is predominantly supplied from natural sources exploited in Chile. From a worldwide total of approximately 8.25×10^5 t/yr, 6.5×10^5 t/yr is provided by the Chilean industry. The product is offered mainly in two grades: technical and agricultural. The latter is provided entirely by Chile and the former mainly by SQM (Chile) and a plant at Lake Charles, Louisiana, owned by the Olin Corporation. Total world consumption of the technical grade is ca 3.25×10^5 t/yr and that of the agricultural grade is ca 5×10^5 t/yr. In Table 3, typical analyses for each grade are given.

Technical Grade. Production of technical-grade sodium nitrate is mainly from Olin and a refined grade is provided by SQM. The former supplies ca 8×10^4 t mainly to the U.S. market, and SQM (Chile) supplies ca 1.7×10^5 t of the refined grade. The estimated remaining 7.5×10^4 t is produced as a by-product in Europe by many different companies.

Consumption by the United States and Canada is ca 50% of world consumption, European consumption is ca 25%, and the rest is distributed to Latin America and Asia. The main markets are explosives, fiberglass, glass, and charcoal-briquette manufacture.

Recent SQM prices (fob warehouse) for the technical grade are 1975, $130/t; 1976, $130/t; 1977, $130/t; 1978, $136/t; 1979, $143/t; 1980, $161/t; and 1981, $176/t. Olin's prices have been ca 10% higher than those of SQM because of the lower impurities content of the former's product.

Agricultural Grades. Annual production of sodium nitrate from natural sources has been declining steadily since the peak year of 1929, when production reached 3.1×10^6 t. Today production is ca 6.5×10^5 t and represents both technical and agricultural grades. During the 1940s, Egypt produced ca 1.13×10^5 t/yr of natural sodium nitrate from 7 wt % $NaNO_3$ ore. Since that time, Egypt has stopped production.

The decline in the use of sodium nitrate in agriculture stems directly to unfavorable competition with the synthetic nitrogen-containing products, eg, ammonia, ammonium nitrate, and urea, which have become available at much lower costs per unit of contained nitrogen. The main reasons for the lower cost are as follows: there is a higher content of nitrogen in the synthetic fertilizers compared to sodium nitrate; synthetic nitrogen-containing product plants are located closer to the markets; and inexpensive by-product gas can be used as fuel for processing the synthetic products.

Sodium nitrate consumption has continued in some markets where its marginal productivity is higher than that of the synthetic products and where sodium is needed in the soil or by plants, eg, beet roots, tobacco, and cotton. However, the amount of sodium nitrate used in the nitrogen fertilizer market has gradually diminished: 66% in 1900, 46% in 1920, 8% in 1940, 3% in 1960, and 0.15% in 1980.

Prices of agricultural-grade sodium nitrate for the U.S. market have been: 1960, $49/t; 1965, $49/t; 1970, $49/t; 1975, $139/t; 1980, $132/t; and 1981, $149/t.

Table 3. Typical Sodium Nitrate Analyses[a]

Component	Agricultural Chilean nitrate, wt %	Industrial Chilean nitrate, wt %	Synthetic sodium nitrate (Olin Corporation), wt %
sodium nitrate	97.10[b]	98.15[b]	99.5 min
nitrogen	16.18	16.23	16.40 min
sodium nitrite	0.0031	0.0024	0.025 max
potassium nitrate	0.816	0.633	na
sodium chloride	1.02	0.761	0.25 max
sodium sulfate	0.38	0.313	0.30 max
sodium borate	0.097	0.051	na
silicon	0.0037	0.0044	0.0085[c] max
calcium oxide	0.034	0.061	0.03 max
magnesium oxide	0.193	0.093	0.002 max
aluminum oxide	0.036	0.0075	0.001 max
manganese	na	na	0.0001 max
copper	0.0001	na	0.0001 max
iodine	0.034[d]	0.079[d]	0.0001 max
iron oxide (Fe_2O_3)	0.0003	0.0012	0.0025 max
water	0.159	0.067	0.02 max
hot-water insolubles	0.057	0.097	0.03 max
alkalinity	0.111[e]	0.035[e]	0.04[f] max
potassium perchlorates including chlorate	0.018	0.022	na
chromium	na	0.0007	na
phosphate (P_2O_5)	<0.001	<0.001	na
arsenic	<0.0001	<0.0001	na
sulfide	none	none	na
lead	<0.0001	na	na
Screen analysis			
through 470 μm (35 mesh)	0.03	0.15	na
on 3360 μm (6 mesh) max	na	na	1
through 3360 μm on 840 μm (20 mesh) min	na	na	80
2380 μm (8 mesh)	7.35	28.10	na
through 2380 μm on 1400 μm (14 mesh)	89.66	71.54	na
through 1400 μm on 840 μm (20 mesh)	2.63	0.13	na
through 840 μm on 625 μm (28 mesh)	0.22	0.07	na
through 625 μm on 470 μm (35 mesh)	0.11	0.01	na
through 250 μm (60 mesh), on 149 μm (100 mesh) max	na	na	3
through 149 μm max	na	na	1.5

[a] Courtesy of Chilean Nitrate Corporation and Olin Corporation.
[b] By difference.
[c] As SiO_2.
[d] As potassium iodate.
[e] As Na_2O.
[f] Total methyl orange alkalinity as Na_2CO_3.

Standards and Specifications

The specifications for reagent-grade sodium nitrate are given in Table 4. Food Chemicals Codex specifications for sodium nitrate are given in Table 5. U.S. Federal specifications for technical-grade sodium nitrate are listed in Table 6.

Health and Safety Factors

The acceptable daily intake by human adults for nitrates (suggested by the World Health Organization) is 5 mg/kg in addition to naturally occurring nitrates. However, large doses of nitrates, including sodium nitrate, are lethal. Accidental ingestion of ca 8–15 g or more of sodium nitrate or potassium nitrate causes severe abdominal pain, bloody stools and urine, weakness, and collapse. Victims of sodium nitrate or potassium nitrate poisoning contract severe gastroenteritis.

Table 4. ACS Specifications for Reagent-Grade Sodium Nitrate

Specification	Value
insoluble matter, wt %	≤0.005
pH of a 5 wt % solution (at 25°C)	5.5–8.3
total chlorine, wt %	≤0.001
iodate (IO_3), wt %	ca 0.0005
nitrite (NO_2), wt %	ca 0.001
phosphate (PO_4), wt %	≤0.0005
sulfate (SO_4), wt %	≤0.003
calcium, magnesium, and R_2O_3 precipitate, wt %	≤0.005
heavy metals (as Pb), wt %	≤0.0005
iron (Fe), wt %	≤0.0003

Table 5. *Food Chemicals Codex* Specifications for Sodium Nitrate[a]

Specification	Value
assay, wt % $NaNO_3$ after drying	99.0
limits of impurities	
arsenic (as As), wt %	0.0003
heavy metals (as Pb), wt %	0.001
total chlorine, wt %	ca 0.2 (passes test)

[a] Ref. 5.

Table 6. Federal Government Specifications for Technical-Grade Sodium Nitrate[a]

Specification	Value
purity of $NaNO_3$	97 min
granule size, % passing through USS sieve sizes	
no. 4 min (<4.76 nm)	100
no. 30 max (>0.59 nm)	10
moisture content and other volatile matter, wt %	0.5

[a] Ref. 6.

Outbreaks of poisoning from the ingestion of meats containing sodium nitrate and sodium nitrite have occurred from the accidental incorporation of excessive amounts of nitrate–nitrite mixtures, ie, 0.5 wt % nitrite as compared to maximum ingredient specifications of 0.05 and 0.02 wt % of nitrates and nitrites, respectively. Of the few studies conducted on nitrate consumption by humans, one study on infants showed that a daily intake for a week of ca 16–21 mg/kg body wt of NO_3^- from spinach produced no signs of methemoglobin formation or other toxic effects.

Nitrate toxicity in ruminants is considered to depend on the ability of nitrate to reduce to nitrite since nitrites are considerably more toxic. The lethal dose of nitrites is estimated to be 1 g in adults. Toxicity studies on animals show that rats tolerate 1 wt % sodium nitrate in the diet for 2 yr with no effects and 5 wt % with only slight growth depression. Dogs show no adverse effects at 2 wt % sodium nitrate in the diet for 105–125 d. The no-effect daily dose appears to be ca 500 mg/kg (see also *N*-Nitrosamines). For further information see reference 7.

Uses

Sodium nitrate is used primarily in agriculture as a fertilizer. The process of growing and harvesting crops produces a steady drain on all of the nutrients in the soil. A lack of nitrogen is generally said to be the first limiting factor in the growth of crops. Nitrogen primarily functions to promote root, stem, leaf, and fruit growth. The easily assimilated nitrogen in the nitrate is used by the plant in the formation of proteins by chemical condensation of carbohydrates and inorganic nitrogen. The sodium in sodium nitrate has value as a corrector for soil acidity, an agent for liberating soil phosphate, and a partial potassium substitute for plants. The main advantages of nitrogen from sodium nitrate over nitrogen from ammonia (qv) are quicker availablity of nitrogen, no increase in the acidity of the soil, liberation of soil phosphate, and substitute of sodium for potassium via potassium liberation in the soil. Sodium nitrate is particularly effective as a nitrogen source for cotton, tobacco, and vegetable crops (see also Plant-growth substances).

The main industrial use of sodium nitrate is in the manufacture of explosives. Its primary use is as an oxidizer in water-based slurry blasting agents and explosives. Typical slurry explosives contain 10–15 wt % sodium nitrate. Dynamite may also contain sodium nitrate as an energy modifier. Typically, 40% dynamite contains 44 wt % sodium nitrate, and 40 wt % ammonia dynamite contains 33 wt % sodium nitrate. Also, 40% gelatin and 40% ammonia gelatin dynamites typically contain 52 and 49 wt % sodium nitrate, respectively (see Explosives and propellants).

Sodium nitrate is also used in the manufacture of glass (qv), fiber glass, enamels, and porcelain as an oxidizing and fluxing agent (see Enamels, porcelain or vitreous). In procelain-enamel frits used for metal coating, the amount of sodium nitrate in a batch generally varies with the various metal bases to be coated. Typically the following contain the corresponding quantities of sodium nitrate in the batch: sheet steel ground coat, 3.8 wt %; titania cover enamel, 7.8 wt %; high lead cast iron, 4.8 wt %; low lead cast iron, 6.7 wt %; cast iron ground coat, 4.0 wt %; cast iron cover coat, 4.5 wt %.

Another large industrial application is as an ingredient in the production of charcoal briquettes. The amount of sodium nitrate used in charcoal-briquette manufacture depends on the type of wood and coal used in a particular production run. Typically charcoal briquettes contain ca 2–3 wt % sodium nitrate.

In the field of health and nutrition, sodium nitrate has applications in certain antibiotic and pharmaceutical production. In curing beef, bacons, and other meats, sodium nitrate is used as a preservative (see Food additives; *N*-Nitrosamines).

In heat-treatment baths for alloys and metals, sodium nitrate is used either alone or in mixtures with other salts. Other uses of sodium nitrate are as a raw material in the production of potassium nitrate, sodium nitrite, sulfuric acid, dyestuff intermediates, and paint pigments. The use of sodium nitrate as a heat-exchange medium is potentially a very large one. Sodium nitrate and other molten salts are being used increasingly in solar-energy technology as heat-transfer and sensible-heat-storage fluids (see Heat-exchange technology). As solar-energy production increases, the demand for sodium nitrate is expected to increase significantly.

BIBLIOGRAPHY

"Sodium Nitrate" in *ECT* 1st ed., under "Sodium Compounds," Vol. 12, pp. 605–606, by J. A. Brink, Jr., Purdue University; "Sodium Nitrate" in *ECT* 2nd ed. under "Sodium Compounds," Vol. 18, pp. 486–498, by L. C. Pan, Chemical Construction Corporation.

1. Grossling and Ericksen, *Computer Studies of the Composition of Chilean Nitrate Ores*, U.S. Geological Survey, Dec. 1970.
2. *International Critical Tables*, McGraw-Hill Book Co., Inc., New York, 1928.
3. G. J. Janz, C. B. Allen, N. P. Bansal, R. M. Murphy, and R. P. T. Tomkins, *Physical Properties Data Compilations Relevant to Energy Storage. II. Molten Salts: Data on Single and Multi-Component Salt Systems*, U.S. Department of Commerce, National Bureau of Standards, Washington, D.C., Apr. 1979, pp. 142–154.
4. V. Sauchelli, ed., *Fertilizer Nitrogen, Its Chemistry and Technology*, Reinhold Publishing Corporation, New York, 1964.
5. *Food Chemical Codex*, 3rd ed., National Academy Press, Washington, D.C., 1981, p. 292.
6. *Sodium Nitrate, U.S. Military Technical Specification*, MIL-S-322C, U.S. Government Printing Office, Washington, D.C., Feb. 5, 1968.
7. D. W. Fasset, "Nitrates and Nitrites," in *Toxicants Occurring Naturally in Foods*, National Academy of Sciences, Washington, D.C., 1973.

SAMUEL MAYA
Beecham Products

MARTIN LABORDE
CODELCO

SODIUM NITRITE

Properties

Sodium nitrite [*7632-00-0*], $NaNO_2$, is a white-to-pale straw-colored compound with a mol wt of 69.00. It is supplied commercially in two forms, dry and liquid, and three grades, technical, USP, and food. The anhydrous form is odorless and hygroscopic, and its tendency to lump and cake is overcome by the addition of an anticaking agent. Dry sodium nitrite is packed in polyethylene-lined paper bags and fiber drums. Sodium nitrite seldom occurs in nature and is manufactured mainly by dissolving oxides of nitrogen in aqueous alkaline solutions.

Anhydrous sodium nitrite has a specific gravity of 2.168 0°C/0°C and a heat of formation of −355.85 to −369.24 kJ/mol (−85.05 to −88.25 kcal/mol) (1). It melts at 285 ± 5°C (1) and starts to decompose at ca 320°C into sodium oxide and nitrogen or nitrogen oxide. Its crystal structure is body-centered orthorhombic with the unit cell dimensions a = 0.355 nm, b = 0.556 nm, and c = 0.557 nm (2). Crystalline sodium nitrite exhibits a transition point at 158–165°C with a heat of transition of 1192 kJ/mol (285 kcal/mol) (1). This transition point is evident in the physical properties of the crystal, eg, specific heat. The specific heat of sodium nitrite increases with temperature from ca 0.98 J/g (0.23 cal/g) at 60°C to ca 1.16 J/g (0.28 cal/g) at 200°C. A peak value at the transition point is 2.29 J/g (0.55 cal/g) (1).

Sodium nitrite is an oxidizing agent that does not deteriorate in air at ambient temperature. It functions as a reducing agent toward such powerful oxidizing agents as dichromates, permanganates, and chlorates. In the presence of acids, it forms nitrous acid which is used in organic synthesis, eg, in the diazotization and nitrosations of aromatic amines (see Azo dyes).

Sodium nitrite is hygroscopic and very soluble in water, but it is relatively insoluble in most organic solvents. The heat of solution has been reported as 14.9 kJ/mol (3.56 kcal/mol) at 20°C and 15.3 kJ/mol (3.66 kcal/mol) at 14°C (2). The solubility diagram for the sodium nitrite–water systems is presented in Figure 1 (3). At −19.5°C the solution forms a eutectic containing 28.1 wt % sodium nitrite. At −19.5°C to −5.1°C a hemihydrate [*82010-95-5*], $NaNO_2.\frac{1}{2}H_2O$ has been reported. Aqueous solutions of sodium nitrite are mildly basic; a 1 wt % solution has a pH of ca 7.1.

Manufacture

The numerous methods for the preparation of sodium nitrite can be divided into two groups: reduction of sodium nitrate and absorption of oxides of nitrogen in an alkali.

The reduction of nitrates to nitrite can be carried out by heat, light, and ionizing radiation (2); addition of lead to fused $NaNO_3$ at 400–450°C (2); reaction of the nitrate in the presence of sodium ferrate and nitric oxide at 400 ± 10°C (2); contacting molten sodium nitrate with hydrogen (4); and electrolytic reduction of sodium nitrate in a cell having a cation-exchange membrane, rhodium-plated titanium anode, and lead cathode (5).

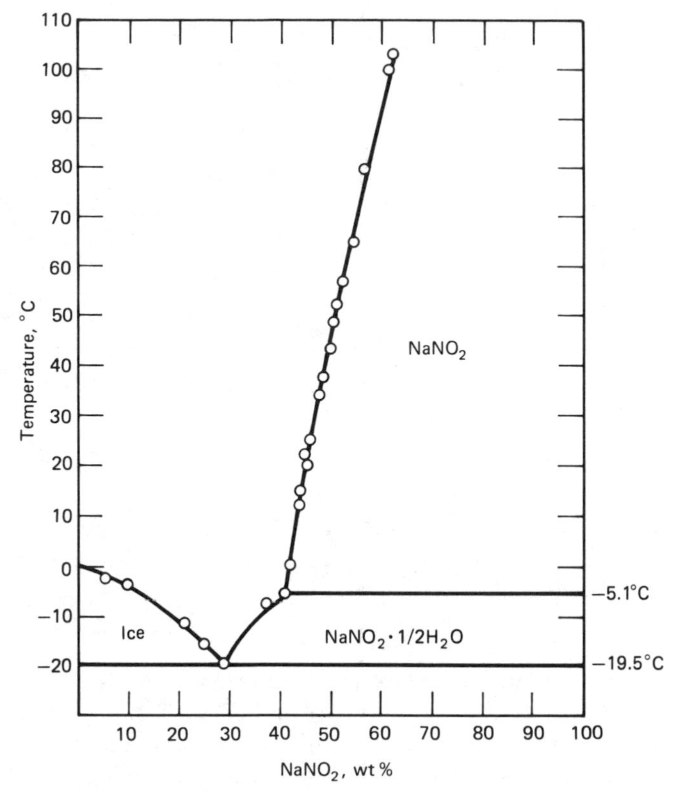

Figure 1. Solubility of sodium nitrite in water (3).

Industrially sodium nitrite is prepared by contacting oxides of nitrogen with aqueous alkaline solutions. Nitrous gases can be produced by several processes, eg, the arc process of nitrogen fixation, the air oxidation of ammonia, or as a waste gas in the production of nitric acid and oxalic acid (6–7). In order to maximize the formation of nitrite and to minimize the concentration of nitrate, it is common to use a mixture with an excess of nitric oxide (NO) over nitrogen dioxide as well as to adjust the process conditions (8). Thus, a solution of sodium nitrite with a low concentration of nitrate is prepared by rapidly cooling a gas mixture containing, in parts per volume, 0–10 parts O_2, 2–12 parts NO_x, 15–20 parts H_2O, and 58–83 parts inert gas from ca 900°C to 300–350°C and then dispersing it under the surface of an aqueous sodium hydroxide solution at 30–120°C (9). If the nitrogen oxide mixture contacts a cold surface, nitrates will form (9). The reaction between the gaseous oxides of nitrogen and sodium hydroxide or sodium carbonate produces a dilute solution which, after evaporation and drying, typically gives a product containing 99 wt % $NaNO_2$. The specifications for technical, USP, and food-grade sodium nitrite are given in Table 1.

Economic Aspects

U.S. consumption of sodium nitrite in 1980 was ca 5.4×10^4 metric tons. The market demand was supplied by a combination of 93% U.S. manufactured and 7% imported products. Future growth is estimated at 3%/yr through 1985. Prices of sodium

Table 1. Specifications for Technical, USP, and Food-Grade Sodium Nitrite

Property	Technical[a]	USP[b]	Food[c]
sodium nitrite, wt %, min	97.0	97.0	97.0
loss on drying, wt %, max		0.25	0.25
heavy metals (as Pb), wt %, max		0.002	0.002
arsenic (as As), ppm, max			3
lead (as Pb), ppm, max			10
sodium sulfate (as Na_2SO_4), wt %, max	0.2		
sodium chloride (as NaCl), wt %, max	0.2		
insolubles, wt %, max	0.5		
pH	8 ± 1		

[a] Ref. 10.
[b] Ref. 11.
[c] Ref. 12.

nitrite in 45-kg paper bag, fob works in 1968, 1975, and 1980 were $0.21/kg, $0.42/kg, and $0.66/kg, respectively.

Storage and Handling

Dry sodium nitrite is usually shipped in multiwall bags or fiber drums. Bulk shipments of the solid product in sparger cars and of solutions in tank cars and tank trucks are available when use, volume, and shipping costs make it desirable. Sodium nitrite is classified by the DOT as a hazardous substance and an oxidizer. Its containers must bear an appropriate oxidizer label.

Sodium nitrite should be stored in a tightly closed container in a cool, dry place away from combustible materials. Sodium nitrite is not flammable, but it decomposes relatively fast at elevated temperatures and increases the rate of burning of combustibles. In case of fire, the area should be flooded with water and the runoff should be kept from entering the sewers and natural waters. Spills of sodium nitrite should be kept away from acids and treated according to local regulations.

Under some conditions, sodium nitrite reacts violently with certain chemicals, particularly ammonia salts, thiocyanates, and strong reducing agents. When reacting with acids, it gives off toxic oxides of nitrogen, and with secondary amines it forms nitrosamines (see *N*-Nitrosamines).

Health and Safety Factors

Although small quantities of food-grade sodium nitrite are used in food processing, sodium nitrite is a moderately toxic chemical. Inhalation of large volumes of dust or liquid mist or direct ingestion can result in acute toxic effects including death. Sodium nitrite poisoning is characterized by loss of blood pressure and cyanosis caused by sodium nitrite's ability to dilate blood vessels and to produce methemoglobinemia, respectively. The oral LD_{50} for sodium nitrite is 85 mg/kg (rat) (13).

Subsequent to minor ingestion or inhalation exposure, irritation of the mouth, esophagus, nasal passages, and respiratory tract may result. Prolonged contact with the skin or contact with the eyes can result in irritation. Recommended first aid for exposure to sodium nitrite is

Exposure	*Treatment*
skin contact	wash with plenty of water; call a physician if irritation persists
eye contact	wash eyes with water for at least 15 min; get medical attention
inhalation	remove victim to fresh air source; administer artificial respiration if not breathing; get immediate medical attention
ingestion	induce vomiting if victim is conscious; get immediate medical attention

Employees who handle sodium nitrite should wear impervious gloves, safety glasses, and long-sleeved shirts. An appropriate respirator should be worn where product mists or dusts are present. Employees should thoroughly wash their hands and faces before eating or smoking.

Uses

Many industrial uses for sodium nitrite are mainly based on its oxidizing properties and its decomposition in acid solution to nitrous acid.

Dyes. Sodium nitrite is a convenient source of nitrous acid in the nitrosation and diazotization of aromatic amines. When primary aromatic amines react with nitrous acid, the intermediate diamine salts are produced which, on coupling to amines, phenols, naphthols, and other compounds, form the azo dyes. The color center of a dye or pigment is the —N=N— group and the attached groups modify the color. Many dyes and pigments have been manufactured with shades of the whole color spectrum (14).

Rubber. Sodium nitrite is an important material in the manufacture of rubber-processing chemicals (see Rubber chemicals). Rubber chemicals that utilize sodium nitrite are accelerators, retarders, and antioxidants. Accelerators greatly increase the rate of vulcanization and lead to marked improvement in the quality of rubber. Retarders, on the other hand, delay the onset of vulcanization and prevent scorching of rubber during processing. Antioxidants slow down the rate of oxidation by acting as chain terminators, transfer agents, and peroxide decomposers. Accelerators and retarders whose manufacture requires sodium nitrite as a source of nitrous acid are thiuram and *N*-nitrosodiphenylamine, respectively.

Heat Treatment and Heat-Transfer Salts. Combinations of sodium nitrite, sodium nitrate, and potassium nitrate are used in the preparation of molten salt baths and heat-transfer media. The most frequently used heat-transfer mixture is the triple eutectic 40 wt % $NaNO_2$, 7 wt % $NaNO_3$, and 53 wt % KNO_3, which has a melting point of 143°C and is used only in the liquid phase at temperatures $<550°C$. Among the principal advantages of this mixture are low melting point, low cost, high heat transfer, noncorrosive properties, and thermal stability. Recently the range of applications of this heat-transfer salt has been extended by the salt-dilution technique, which lowers its freezing point to 10°C (15) (see Heat-exchange technology, Heat-transfer media other than water).

Metallic Coatings. Sodium nitrite plays an important role in metal treatment and finishing operations, eg, phosphating and gold plating (see Metallic coatings). In phosphating solutions, sodium nitrite performs two functions, namely, that of an accelerator and oxidizer. As an accelerator sodium nitrite reduces processing time, and as an oxidizer it controls the buildup of ferrous ion in phosphating solutions. In gold plating, sodium nitrite is used in the preparation of gold–sulfite–nitrite complex, which is the source of gold in a gold-electroplating bath (16–17). This bath is nonpoisonous as opposed to cyanide baths.

Corrosion Inhibition. Sodium nitrite is an anodic inhibitor that forms a tightly adhering film over steel and prevents the dissolution of metal at anodic areas. Sodium nitrite is most effective in combating the corrosion of iron and steel. When used in the presence of nonferrous metals, other additives, eg, benzoate, borate, silicate, etc, may be required. Its applications as a corrosion inhibitor in open and closed recirculating cooling-water systems, boiler-water treatment, and in the protection of metals in process and storage are well known. It is also being used to reduce corrosion of steel reinforcing bars in concrete (18) (see Corrosion and corrosion inhibitors).

Meat Curing. Sodium nitrite is used as a curing and preserving agent in bacon, ham, frankfurters, etc. Brines containing sodium nitrite inhibit the growth of bacteria including *clostridium* botulism. The maximum allowable quantity of sodium nitrite in the brines and the residual remaining in the meat is under the jurisdiction of FDA and USDA (see Food additives).

Miscellaneous. Other applications of sodium nitrite include: inhibition of polymerization (19–21); textile dyeing (22–25); synthesis of aromatic fluorine compounds; synthesis of cyclohexanone oxime and adipic acid (26–27); production of blowing agents (28–30); synthesis of insecticides, herbicides, synthetic caffeine, drugs, and flavors; and detinning; etc.

BIBLIOGRAPHY

"Sodium Nitrite" in *ECT* 1st ed., under "Sodium Compounds," Vol. 12, p. 606, by J. A. Brink, Jr., Purdue University; "Sodium Nitrite" in *ECT* 2nd ed. under "Sodium Compounds," Vol. 18, pp. 498–502, by L. C. Pan, Chemical Construction Corporation.

1. "Natrium" in *Gmelins Handbuch der Anorganischen Chemie*, System 21, Vol. 3, Verlag Chemie, Weinheim, FRG, 1966.
2. J. W. Mellor, *Supplement to Mellor's Treatise on Inorganic and Theoretical Chemistry*, Vol. VIII, Supp. II, Part II, John Wiley & Sons, Inc., New York, 1967.
3. Seidel, *Solubilities of Inorganic and Metal Organic Compounds*, 4th ed., Vol. 2, American Chemical Society, Washington, D.C., 1965.
4. U.S. Pat. 2,294,374 (Sept. 1, 1945), J. R. Bates (to Houdry Process Corporation).
5. Ger. Offen. 2,940,186 (Apr. 24, 1980), M. Yoshida (to Asahi Chemical Industry Co., Ltd.).
6. U.S. Pat. 3,965,247 (June 22, 1976), C. F. Hecklinger and D. E. Crook (to Allied Chemical Company).
7. S. D. Deshpande and S. V. Vyas, *Ind. Eng. Chem. Prod. Res. Dev.* **18**(1), 69 (1979).
8. U.S. Pat. 2,032,699 (Mar. 3, 1936), J. W. Hayes and H. C. Britton (to Allied Chemical Company).
9. U.S. Pat. 4,009,246 (Feb. 22, 1977), M. M. Wendel (to E. I. du Pont de Nemours & Co., Inc.).
10. *Sodium Nitrite, U.S. Military Technical Specification*, MIL-S-24521, U.S. Government Printing Office, Washington, D.C., Sept. 2, 1975.
11. *The United States Pharmacopeia XX* (*USP XX–NF XV*), The United States Pharmacopeial Convention, Inc., Rockville, Md., July 1, 1980.
12. *Food Chemicals Codex*, 3rd ed., National Academy of Sciences–National Research Council, Washington, D.C., 1980.
13. *NIOSH Registry of Toxic Effects*, National Institute for Occupational Safety and Health, Washington, D.C., 1979.
14. T. C. Patten, *Pigment Handbook*, Vol. 1, John Wiley & Sons, Inc., New York, 1973.
15. *Oil Gas J.* **69**, 76 (1970).

16. S. Afr. Pat. 73 07,671 (Sept. 28, 1973), C. W. Bradford and H. Middleton (to Johnson Matthey and Co., Ltd.).
17. Ger. Offen. 2,615,631 (Oct. 28, 1976), C. W. Bradford and P. C. Hydes (to Johnson Matthey and Co., Ltd.).
18. J. T. Lundquist, Jr. and co-workers, *Mater. Perform.* **18**(3), 36 (1979).
19. G. S. Whitby, *Synthetic Rubber*, John Wiley & Sons, Inc., New York, 1954.
20. Br. Pat. 900,970 (July 11, 1962), (to E. I. du Pont de Nemours & Co., Inc.).
21. U.S. Pat. 3,714,008 (Jan. 30, 1973), T. Masaaki and co-workers (to Japan Atomic Energy Research Institute).
22. U.S. Pat. 3,597,144 (Aug. 3, 1971), J. A. Leddy (to Geigy Chemical Corporation).
23. Jpn. Pat. 71 06,117 (Feb. 16, 1971), Z. Munakata (to Daito Chemical Industry Co., Ltd.).
24. Jpn. Pat. 73 06,668 (Feb. 28, 1973), J. Yasuda and co-workers (to Mitsubishi Chemical Industrial Co., Ltd.).
25. U.S. Pat. 3,795,482 (Mar. 5, 1974), R. E. Yelin and co-workers (to FMC Corporation).
26. Sp. Pat. 384,210 (June 1, 1973), S. L. Kemichrom.
27. Fr. Addn. 2,054,701 (June 11, 1971), (to Société des Usines Chimiques Rhone-Poulenc).
28. Jpn. Kokai 73 60,769 (Aug. 25, 1973), S. Murakami and co-workers (to Eiwa Chemical Industrial Co., Ltd.).
29. Jpn. Kokai 74 00,369 (Jan. 4, 1974), S. Murakami and co-workers (to Eiwa Chemical Industrial Co., Ltd.).
30. Jpn. Kokai 74 02,868 (Jan. 11, 1974), S. Murakami and co-workers (to Eiwa Chemical Industrial Co., Ltd.).

JOHN KRALJIC
Allied Corporation

SODIUM SULFATES

The sulfates of sodium are commonly sold in four commercial forms: anhydrous sodium sulfate [*7757-82-6*], Na_2SO_4; technical-grade sodium sulfate, known as salt cake; sodium sulfate decahydrate [*7727-73-3*], $Na_2SO_4.10H_2O$, known as Glauber's salt; and sodium hydrogen sulfate [*7681-38-1*], $NaHSO_4$, known as niter cake. Both grades of sodium sulfate are significant items of commerce with a combined U.S. production of 1.18×10^6 metric tons in 1979. Glauber's salt sales have declined since 1960 to the point where it is a specialty item. Sodium hydrogen sulfate has numerous low volume uses making it, too, a specialty chemical with an annual U.S. production of $(7.0–7.5) \times 10^4$ t. Because of the multiplicity of sources and processes for sodium sulfates, a variety of names have developed:

Na_2SO_4

salt cake	generic name originating from Mannheim and Leblanc processes, implies lower purity
anhydrous	implies manufacture from Glauber's salt
thenardite	naturally occurring mineral

rayon cake	rayon by-product
chrome cake	dichromate by-product
phenol cake	phenol by-product
kaiseroda	German potash by-product
synthetic salt cake	sulfate substitute, mixture of Na_2CO_3 and molten sulfur

$Na_2SO_4.10H_2O$

Glauber's salt	most common name
mirabilite	naturally occurring mineral
sodium sulfate crystals	incorrect designation, occasionally used

$NaHSO_4$

sodium bisulfate	
sodium hydrogen sulfate	used interchangeably
sodium acid sulfate	
niter cake	nitric acid by-product, implies lower purity

Sodium sulfate has been important throughout the history of the chemical industry but more as a by-product or intermediate than as a leading product. Its abundance, both naturally and as a by-product of many chemical processes, has kept its price low. As a source of both alkali (Na_2O) and sulfur, it represents an inexpensive substitute for salt, soda ash, and sulfur under certain conditions; but for such uses, high energy input is required.

In the seventeenth century, Johann Glauber produced salt cake as a by-product of the reaction of salt and sulfuric acid to form hydrochloric acid.

$$2\ NaCl + H_2SO_4 \rightarrow Na_2SO_4 + 2\ HCl\uparrow$$

The reaction used by Glauber was important in the eighteenth and nineteenth centuries as the principal source of both soda ash and hydrochloric acid. In the Leblanc process, the salt cake produced in the Glauber reaction, frequently called the Mannheim process, reacts with carbon and limestone to produce soda ash. The LeBlanc process initiated some of the first air- and water-pollution legislation for regulation of HCl emissions and salt cake disposal (see Air pollution; Water, water pollution). During periods when salt was in low supply in the UK, salt cake was fed directly into the carbon-reduction step of the Leblanc process. The Leblanc process was replaced by the Solvay process for soda ash manufacture in the nineteenth century, eliminating Na_2SO_4 as an intermediate in soda ash production.

Salt cake was produced thereafter principally by the Mannheim process. The Hargreaves process was developed as a refinement to the Mannheim process for making hydrochloric acid and salt cake via sulfur dioxide and oxygen to replace sulfuric acid.

$$4\ NaCl + 2\ SO_2 + O_2 + 2\ H_2O \rightarrow 2\ Na_2SO_4 + 4\ HCl\uparrow$$

Both the Mannheim and Hargreaves processes are still used throughout the world.

Nitric acid was produced during the nineteenth century by the reaction of sulfuric acid and sodium nitrate.

$$H_2SO_4 + NaNO_3 \rightarrow NaHSO_4 + HNO_3\ (g)$$

The niter cake ($NaHSO_4$) was further used in a variation of the Mannheim process.

$$NaHSO_4 + NaCl \rightarrow Na_2SO_4 + HCl\ (g)$$

Again, salt cake was a by-product to be disposed of when nitric and hydrochloric acid needs exceeded those for salt cake. The $NaNO_3$ process was replaced in the 1920s by the Ostwald and similar processes in which ammonia is oxidized to form nitric acid.

In 1884, the kraft paper-pulp process was invented. Development and commercialization of the kraft process in the early twentieth century created a new important use for sodium sulfate. Since 1940 only kraft process plants have been built and have completely replaced those based on the older sulfite process. Increasing use of the kraft process has raised demand for salt cake. In the United States, about two thirds of the total sodium sulfate produced is used in paper-pulp manufacturing (see Pulp).

The chemical inertness and low price of sodium sulfate have led to its use as a filler in the production of household detergents since World War II (see Drycleaning and laundering, laundering). During the early and middle twentieth century, U.S. sulfate supply was dominated by by-product salt cake from the rayon industry. Rayon producers had to dispose of the sulfate by-product, whose production rate was set by the demand for rayon. Thus salt cake pricing depended on the demand for rayon. Since the 1960s rayon production declined, causing a decrease in by-product salt cake from that industry. Natural sulfate production increased to fill the market need.

Sulfate consumption in the United States is declining slowly as economic and environmental factors force kraft paper producers to recycle more of their effluents, thereby reducing salt cake consumption per unit of paper produced. Few new uses have developed to offset the decline (see Sulfur recovery).

Properties

Physical and chemical properties of the three most common forms of sodium sulfate are summarized in Table 1. The solubility of Na_2SO_4 in water from 0 to 360°C is shown in Figure 1 (6). The existence of a metastable heptahydrate [*13472-39-4*] form is well documented and shown as a dotted line in the figure. No naturally occurring heptahydrate has been reported. Published evidence of the existence of $Na_2SO_4.H_2O$ [*10034-88-5*] is not well documented. Industrial processes frequently involve NaCl to depress Na_2SO_4 solubility and to improve recovery efficiency. The reduction in Na_2SO_4 solubility in aqueous solutions saturated with NaCl is presented in Figure 1 as well.

The reactivity of Na_2SO_4 varies widely with temperature. In the solid state and close to ambient temperatures, Na_2SO_4 is relatively inert but does form acid compounds in the presence of sulfuric acid. Solubilities in the Na_2SO_4–H_2SO_4–H_2O system are outlined in Figure 2 (6). Of the various acid forms, only $NaHSO_4$ is of significant industrial importance. Sodium sulfate is the product of the reaction of a strong acid, ie, sulfuric acid, and a strong base, ie, sodium hydroxide. It is a neutral salt that undergoes no further reaction. In nature certain bacteria are capable of reducing sulfate to elemental sulfur.

At higher temperatures and particularly in the molten state, Na_2SO_4 is very reactive (2). The compound has a vapor pressure of 20 Pa (0.15 mm Hg) at its mp (882°C), and the vapor pressure rises to 133.3 Pa (1 mm Hg) at 1371°C. In pure form it can be completely vaporized. Common impurities, ie, silica, alumina, and iron oxide, promote decomposition above 1100°C to SO_2, SO_3, O_2, and Na_2O. The presence of carbon at 1000°C leads to Na_2S or SO_2, depending upon the amount of air. Sodium

Table 1. Properties of Sodium Sulfates

Properties	Sodium sulfate		Sodium hydrogen sulfate
	Anhydrous	Decahydrate	
molecular weight	142.04	322.19	120.06
melting point, °C	882[a]	32.4[b]	315[b]
specific gravity[c]	2.664	1.464	2.435
refractive index[c]	1.464, 1,474, 1.485	1.394, 1.396, 1.398	1.459, 1.479
crystalline form	rhombic, monoclinic, or hexagonal	monoclinic	triclinic
specific heat[c], J/(g·K)[d]	1.05_{400}, 1.42_{1000}		
heat of formation[c], MJ/mol[d]	−1.384	−4.324	−1.126
enthalpy of melting, kJ/mol[d]	23.3[a]		
heat of solution (at ∞ dilution)[e], MJ evolved/mol[d]	1.17	−78.41	7.28
heat of crystallization from saturated soln[f], J/mol[d]	$-8.8_{50°C}$	$74.98_{25°C}$, $75.52_{11°C}$	
transition temperature, °C	242[a]		
heat of transition, kJ/mol[d]	6.9		

[a] Ref. 1.
[b] Ref. 2.
[c] Ref. 3.
[d] To convert J to cal, divide by 4.184.
[e] Ref. 4.
[f] Ref. 5.

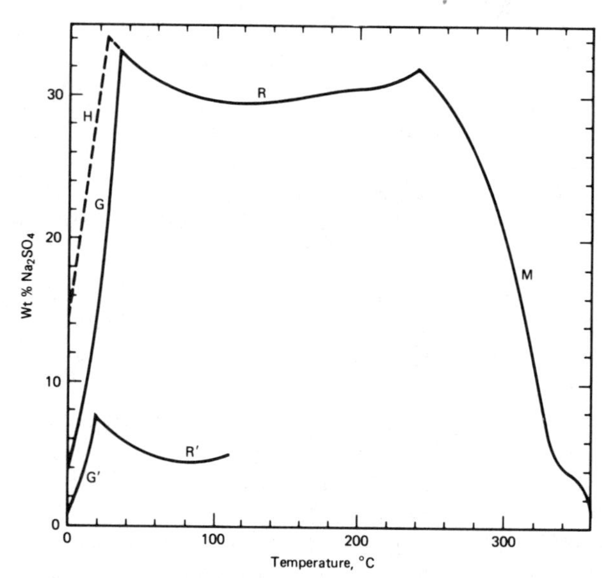

Figure 1. The solubility system Na_2SO_4–NaCl–H_2O from 0 to 360°C (1). Above boiling point the system is under sufficient pressure to maintain H_2O as a liquid. H, $Na_2SO_4.7H_2O$; G′, $Na_2SO_4.10H_2O$; R, Na_2SO_4 (rhombic); M, Na_2SO_4 (monoclinic); G, $Na_2SO_4.10H_2O$ (also saturated with NaCl); R′, Na_2SO_4 (rhombic, also saturated with NaCl); - - -, metastable form.

carbonate can be produced by this reaction. Reduction can also occur with hydrogen, starting at 800°C.

The decahydrate melts incongruently, ie, to a saturated sulfate liquid phase and

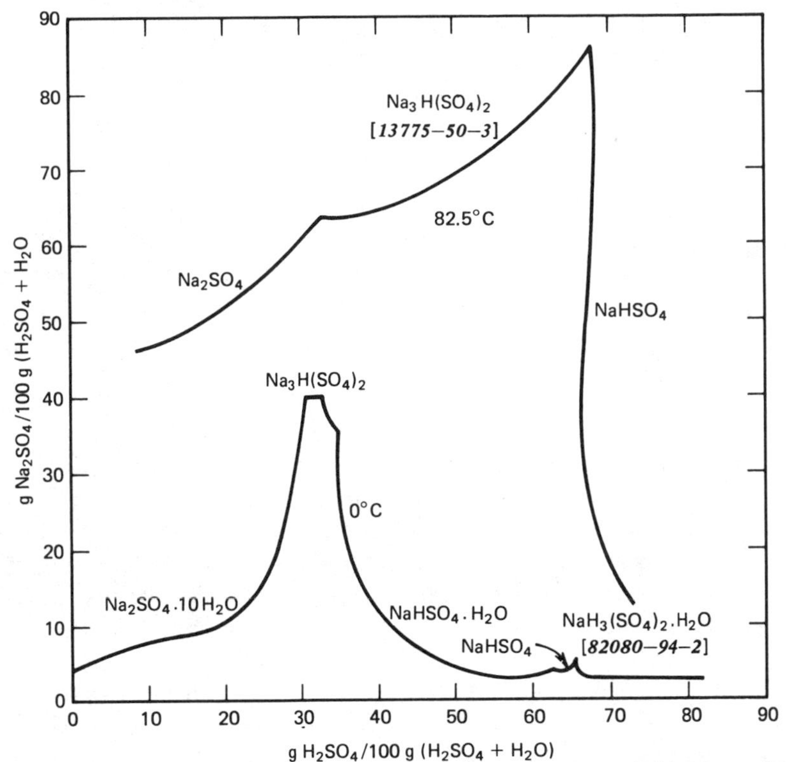

Figure 2. The solubility system Na_2SO_4—H_2SO_4—H_2O at 0 and 82.5°C (1). Metastable forms not shown.

an anhydrous sulfate solid, at 32.4°C when pure. The presence of other salts depresses the melting point to as low as 17.9°C in the case of NaCl (6).

Sodium sulfate crystallized from solution has an affinity for impurities, eg, iron and its compounds and various organics. Glauber's salt has no such affinity and rejects most impurities. It does, however, form solid solutions with $Na_2CO_3.10H_2O$. Because of this property, higher quality commercial grades of anhydrous Na_2SO_4 are usually made from Glauber's salt.

Sodium sulfate deposits occur naturally throughout the world and result from the evaporation of inland seas and lakes (2). In cold climates, eg, Canada and the USSR, deposits of mirabilite ($Na_2SO_4.10H_2O$) occur. In warmer areas, eg, South America, India, Mexico, and the western United States, the anhydrous form, ie, thenardite, occurs. In other evaporite deposits where other anions and cations are present, Na_2SO_4 occurs in compound minerals with other salts, the most common of which are listed in Table 2.

Manufacture and Processing

Both natural and by-product processes involve the same basic principles to obtain marketable sodium sulfate. Nearly all producers purify the sulfate starting material by crystallizing it as Glauber's salt. The Glauber's salt is then converted to the anhydrous form by melting or complete dehydration. Considerable variety in starting material and equipment result in process diversity.

Table 2. Common Minerals Containing Na_2SO_4

Mineral	CAS Registry No.	Chemical formula
aphthitalite	[*12274-47-4, 13932-19-9*]	$Na_2SO_4.3K_2SO_4$
glaserite	[*20741-74-6*]	
bloedite	[*15083-77-9*]	$Na_2SO_4.MgSO_4.4H_2O$
burkeite	[*12179-88-3*]	$2Na_2SO_4.Na_2CO_3$
glauberite	[*13767-89-0*]	$Na_2SO_4.CaSO_4$
hanksite	[*12180-10-8*]	$9Na_2SO_4.2Na_2CO_3.KCl$
loewite	[*16633-52-6*]	$Na_2SO_4.MgSO_4.2.5H_2O$
tychite	[*12188-92-0*]	$Na_2SO_4.2Na_2CO_3.2MgCO_3$
vanthoffite	[*15557-33-2*]	$3Na_2SO_4.MgSO_4$

Salt cake produced by the Hargreaves and Mannheim processes is an exception to the above generalization. In these processes, Na_2SO_4 is the residue after HCl is driven from the reaction mixture that is created when salt and sulfuric acid or niter cake are heated to 650°C; SO_2, C, and H_2O in the Hargreaves process replace H_2SO_4. The crude product is cooled and sold directly. Only three Mannheim and one Hargreaves plants are operated in the United States and account for only a minor portion of U.S. production. Such processes are widely used throughout the rest of the world, however.

A solution of Na_2SO_4 forms in the processes for spin-bath rayon, sodium dichromate, boric acid, phenol, iron from pyrites, and leather tanning. Usually the solution contains chemicals too valuable or too toxic to discard. The sulfate is removed by cooling to crystallize Glauber's salt, other impurities are removed as necessary and the solution is recycled for reuse in the process. Glauber's salt can be converted to Na_2SO_4 by any of the processes described below.

Sodium sulfate processing methods are summarized in Figure 3. The Glauber's salt starting material can be obtained by cooling a natural brine, a solution obtained by solution-mining a sulfate deposit or cooling a process stream as above. Glauber's salt can also be natural mirabilite, but insolubles in mirabilite usually make the resultant salt cake unmarketable. Impurities, eg, NaCl, are beneficial in reducing sulfate solubility. Texas sulfate brines are deliberately saturated with NaCl before crystallizing Glauber's salt (7–8). Removal of Glauber's salt dehydrates the liquor, thereby concentrating the noncrystallizing components. This technique is used at Searles Lake to preconcentrate the brine fed to triple-effect evaporators to improve borax and KCl recoveries (9).

Crystallization of Glauber's salt is accomplished by winter cooling in ponds (10–11), surface cooling in crystallizers using polished tubes (12), and vacuum cooling in modern crystallizers (9). Dewatering is accomplished by draining in piles (10) or filtration by conventional means (9,12). The mother liquor either is discarded or recycled for further pickup of sulfate or is reused in the process as in the rayon (qv) industry.

Glauber's salt is processed in a variety of ways depending on its purity and the quality of salt cake desired. Conversion is accomplished by complete dehydration in a rotary dryer (2), by a two-step process in which the anhydrous salt is produced by simple melting followed by salting out a salt-cake-quality product with NaCl (9), by evaporation involving submerged combustion (7), and by evaporation in a double-effect

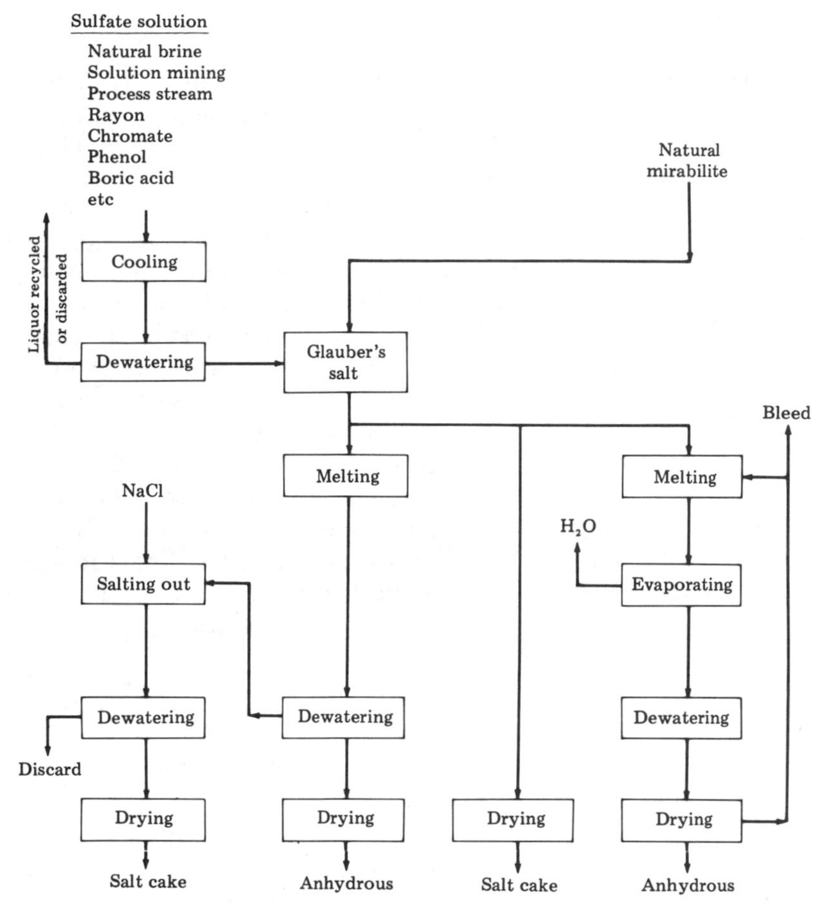

Figure 3. Natural Na_2SO_4 processing methods.

evaporator (12). Variations of these methods are possible depending on the specific liquor compositions and local conditions. For example, boiler-flue-gas scrubbing with a sodium base, eg, soda ash or caustic soda, could provide a Na_2SO_4 solution after air oxidation of the sulfite to sulfate.

Energy requirements for the different processes discussed above vary widely depending on local conditions, and compromises must be made. Winter crystallization of Glauber's salt in natural lakes or man-made ponds may require no energy input other than nature's. This advantage must be traded off against the harvesting and transportation costs. Capital costs for energy efficiency, eg, double- or triple-effect evaporation for dehydrating Glauber's salt, must be balanced against the low capital cost required for simple melting with its higher energy cost and lower recovery efficiency. No generalities can be made concerning energy costs for Na_2SO_4 manufacture, except that costs must be kept low to remain competitive.

Sodium bisulfate as niter cake is no longer produced as a by-product in nitric acid production in the United States by the reaction of sodium nitrate and sulfuric acid. That process has been supplanted by the Ostwald and similar processes in which no $NaHSO_4$ is formed. The principal U.S. process for $NaHSO_4$ involves the reaction of sulfuric acid and salt cake under anhydrous conditions.

Economic Aspects

Trends in sodium sulfate production and consumption in the United States appear to be leading the rest of the world and foretell a declining market (13). The greatest U.S. production occurred from 1965 through 1974 when annual production at ca 1.3×10^6 metric tons and annual consumption at ca 1.5×10^6 t peaked (Fig. 4). Worldwide production peaked in 1975 and 1976 at 4.4×10^6 t/yr. Production outside of the United States decreased 6.4% from 1977 to 1979 below the peak years. Whether this drop represents a continuing downward trend can only be determined by future performance. U.S. consumption declined rapidly by 4.8%/yr since 1974. Production, however, has decreased only 1%/yr. The difference is explained by the change in the export–import balance. Imports decreased markedly whereas exports increased. With imports approaching a balance with exports, the declining consumption should have an adverse impact on production if the decline continues.

The U.S. decline can be attributed to the rapid escalation of fluid fuel prices since 1973 and increasingly strict regulations governing air and water pollution. The kraft paper industry, representing approximately two thirds of the total U.S. sulfate consumption, is the main contributor to the U.S. consumption decline. Pressures on the kraft producers to clean their effluents to meet new regulations and to reduce unit

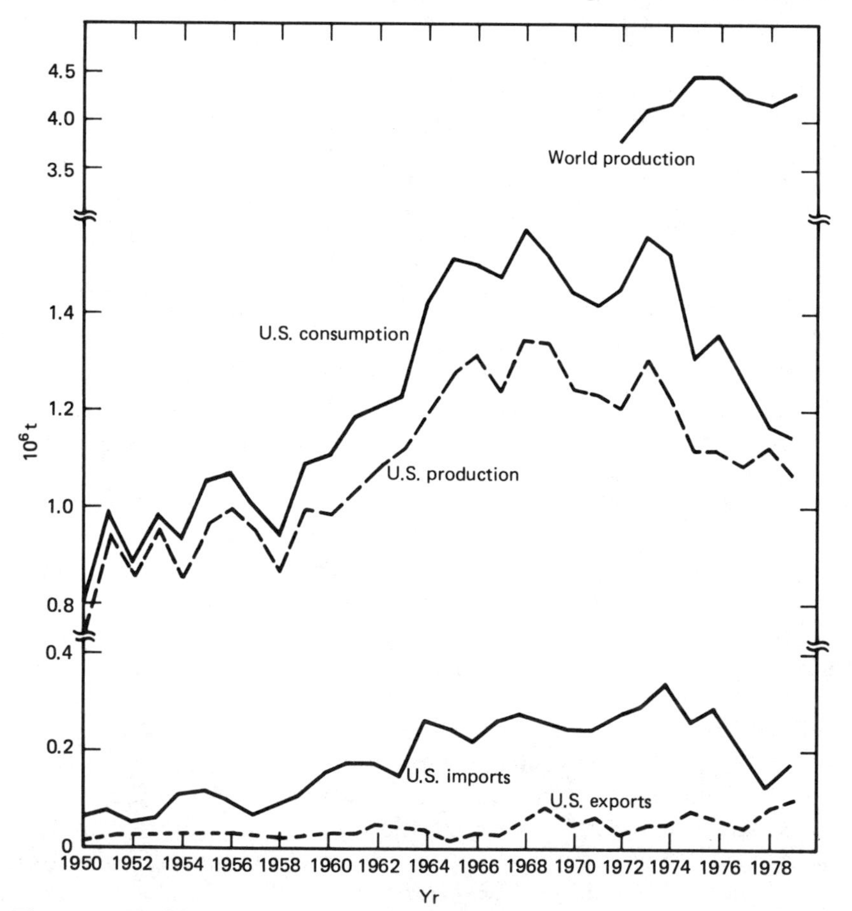

Figure 4. Na_2SO_4 production and consumption for the period 1950 through 1979 (13).

energy costs to offset fuel price increases have provided the impetus. They have responded by improving their internal process efficiencies and by recovering and recycling sulfate which previously was discarded (14–15). These improvements have resulted in decreases in the amount of salt cake required per ton of paper produced.

It is unlikely that the present rate of decline will continue for long. The kraft producers will soon attain a new minimum salt cake consumption rate per ton of paper. With the slow but continuing expansion of detergent and paper production, total U.S. salt cake consumption will start to rise again to match the expansion rate of these industries.

Salt cake pricing, as shown for natural sulfate in Figure 5, has followed an unusual trend with prices skyrocketing at just the time when U.S. consumption was declining. The price increases accelerated the drop in consumption. Downward price adjustments can be expected in the near future if production rates decline further.

Annual U.S. sodium hydrogen sulfate production has remained relatively constant at $(7.0–7.5) \times 10^4$ t (16). Nearly all of the production of sodium bisulfate from the three main U.S. producers (DuPont, Allied, Monsanto) goes into the manufacture of consumer products.

Health and Safety Factors

Environmental concerns have influenced Na_2SO_4 manufacturing and its uses. Most by-product sulfate producers have been forced to recover Na_2SO_4 because of

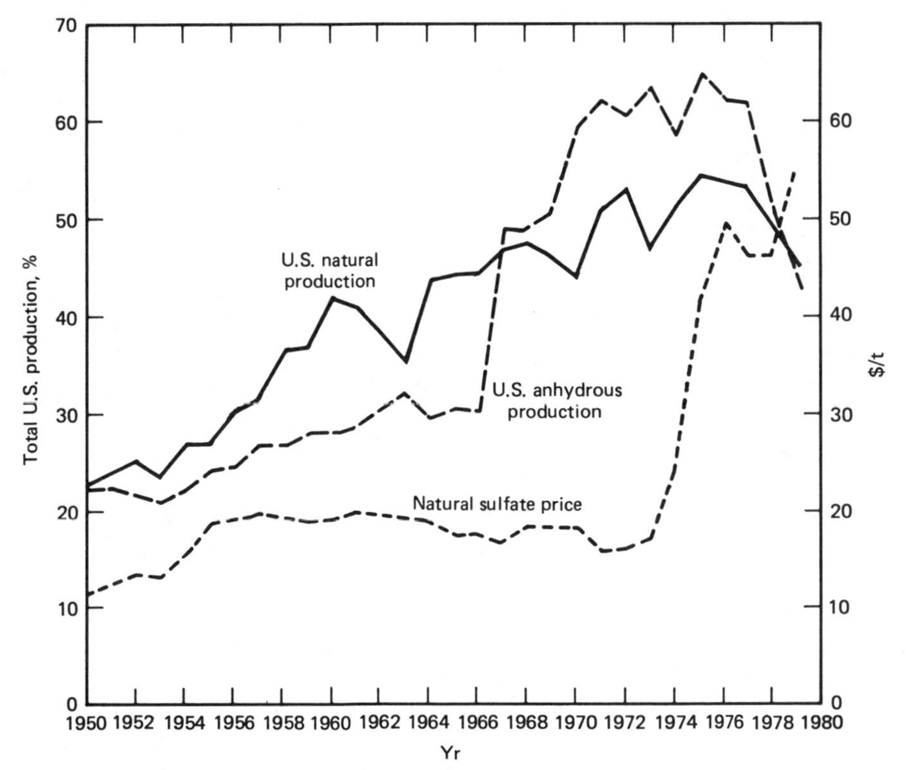

Figure 5. Na_2SO_4 prices and production in the United States from 1950 through 1979 (13).

regulations concerning discharge of sulfate-containing liquors into waterways. Paper producers have been similarly restricted. Sodium sulfate occurs as a particulate contributing to the smog over Los Angeles (17). Most SO_2 emitted in flue gas combines with basic gases, eg, NH_3, to form sulfates other than Na_2SO_4 (18). In areas where Mannheim or Hargreaves processes are in operation, particulate Na_2SO_4 might occur because of its high vapor pressure at furnace temperatures. Acid rain containing sulfuric acid can react with a sodium-containing base to produce Na_2SO_4. In general, however, Na_2SO_4 does not appear to be an environmentally dangerous material.

Sodium sulfate in moderation is used medically as a diuretic (qv) and cathartic for humans and for animals (19) (see Gastrointestinal agents). It also is a component of consumer products, eg, laxatives and antacids, and is a neutral salt that is used extensively in domestic laundry detergents (see Surfactants and detersive systems). Its dust is classified as a nuisance but is not classified as toxic. In general, Na_2SO_4 is a relatively safe chemical in pure form. Individual producers should be consulted concerning impurities that may have been introduced in their particular process.

Sodium bisulfate is mildly acidic and appropriate precautions should be taken when using it.

Uses

Sodium sulfate is used industrially because of its inertness at low temperatures and its reactivity at high temperatures. The principal uses for Na_2SO_4 are paper, 67%; detergents, 26%; glass and all others, 7% (13). Consumption in the paper industry is declining, and uses for detergents are increasing slowly. New uses have not developed sufficiently during the 1970s to expand the market.

In the kraft or sulfate paper process, a mixture of sodium sulfide and sodium hydroxide is used to digest wood chips. This process is more effective and can be used for a wider range of wood types than can the competing pulping processes (see Pulp). The sulfide and hydroxide are generated starting with salt cake as the raw material. The low price of salt cake makes it a cheaper raw material, although a few instances where pulp mills have used Na_2S and NaOH have been reported (2). The breakdown of Na_2SO_4 at the high temperatures in the black-liquor furnaces makes it uniquely suitable for paper processing. Some attempts have been made to sell mixtures of Na_2SO_4 and Na_2CO_3 in competition with salt cake, but the higher pricing of soda ash makes that approach uneconomical except under unusual conditions.

The high temperature properties of Na_2SO_4 are also advantageous in glassmaking, where it helps speed up the melting process, reduces the tendency for alkaline gas bubbles and seeds to form, and provides a less expensive source of Na_2O than soda ash (2). Sodium sulfate improves the boiling and working properties of high silica glasses. Use in the glass industry is declining because of some disadvantages: release of SO_2 from the melt creates an air-pollution problem, the presence of sulfate increases the corrosive attack of refractories, and the reactions involved increase iron contamination which requires addition of bleaching agents.

The low temperature, inert properties of Na_2SO_4 make it desirable as a principal component of household laundry detergents. Such detergents contain as much as 75 wt % Na_2SO_4 as a diluent and builder for the concentrated detergents. Consumers apparently prefer to add a cupful of solid rather than a smaller amount of powder or liquid concentrate. Sodium sulfate is the least expensive inert material available.

Sodium chloride is not suitable because of its astringent property. The sulfate also improves the detergency of surfactants. Much emphasis is placed by detergent makers on the whiteness and particle size of the Na_2SO_4 used to improve the appearance and handling properties of their detergent products. Anhydrous Na_2SO_4 is the preferred form because of its higher purity and whiteness.

In the dye industry both Na_2SO_4 and $NaHSO_4$ are used to dilute or standardize the dyes and to adjust pH. Sodium sulfate is used in cattle feed, in cellulose-sponge manufacture as Glauber's salt, as a cement and plaster hardener, and as an aid in metallurgical refining (see Pet and livestock feeds). None of these uses are of significance in terms of tonnage (2).

Recent interest in solar energy (qv) has created renewed interest in the high heat of crystallization of Glauber's salt as a means of storing energy (20–21). The interest thus far has not led to any commercial developments requiring significant sulfate tonnages. A proposal for using molten Na_2SO_4 for off-peak storage of electrical energy was made but not commercialized (22).

As a solid, $NaHSO_4$ provides a convenient acid for such household uses as toilet-bowl cleaning, automobile-radiator cleaning, and swimming-pool-pH adjusting (see Water, treatment of swimming pools, hot tubs, and spas). Industrially it is used for metal pickling, as a dye-reducing agent, for soil disinfecting, and as a promoter in hardening certain types of cement (see Metal surface treatments).

BIBLIOGRAPHY

"Sodium Sulfates" under "Sodium Compounds" in *ECT* 1st ed., Vol. 12, pp. 607–609, by J. A. Brink, Jr., Purdue University; "Sodium Sulfates" under "Sodium Compounds" in *ECT* 2nd ed., Vol. 18, pp. 502–510, by Joseph J. Jacobs, Jacobs Engineering Co.

1. L. Denielou, J.-P. Petitet, and C. Tequi, *Thermochim. Acta* **9,** 135 (1974).
2. *Mellor's Comprehensive Treatise on Inorganic and Theoretical Chemistry*, Vol. II, Suppl. II, John Wiley & Sons, Inc., New York, 1961.
3. J. A. Dean, ed., *Lange's Handbook of Chemistry*, 12th ed., McGraw-Hill, Inc., New York, 1979.
4. R. H. Perry and C. H. Chilton, eds., *Chemical Engineers' Handbook*, 5th ed., McGraw-Hill, Inc., New York, 1973, Sect. 3, p. 149.
5. B. V. Gritsus, E. I. Akhymov, and L. P. Zhiliua, *Zh. Prikl. Khim.* **45**(11), 2556 (1972).
6. W. F. Linke, *Solubilities of Inorganic and Metal-Organic Compounds*, 4th ed., Vol. II, American Chemical Society, Washington, D.C., 1965, p. 982.
7. W. I. Weisman and R. C. Anderson, *Min. Eng.* **5,** 711 (1953).
8. W. I. Weisman, *Chem. Eng. Prog.* **60**(11), 47 (1964).
9. G. Parkinson, *Eng. Min. J.* **178**(10), 71 (1977).
10. W. A. MacWilliams and R. G. Reynolds, *Can. Min. Metall. Bull.* 115 (June, 1973).
11. U.S. Pat. 3,556,596 (Jan. 19, 1971), W. A. MacWilliams and L. Bithel (to Chemcell Ltd.).
12. C. H. Chilton, *Chem. Eng.* **65,** 116 (Aug. 11, 1958).
13. "Sodium and Sodium Compounds" in *Bureau of Mines Minerals Yearbook*, U.S. Department of the Interior, Washington, D.C., 1977.
14. *Industrial Minerals Glass Survey '77* (125), 81 (Feb. 1978).
15. *Chem. Week* **120,** 37 (Mar. 2, 1977).
16. R. N. Shreve and J. A. Brink, Jr., *Chemical Process Industries*, 4th ed., McGraw-Hill, Inc., New York, 1977, pp. 200–201.
17. R. D. Cadle and co-workers, *Arch. Ind. Hyg. Occup. Med.* **2,** 698 (1950).
18. B. R. Appel and co-workers, *Env. Sci. Technol.* **12**(4), 418 (1978).
19. M. Windholz, ed., *The Merck Index*, 9th ed., Merck & Co., Inc., Rahway, N.J., 1976.
20. M. Telkes, *ASHRAE J.* **14,** 38 (1974).
21. *Wall Street J.* **190**(121), 11 (Dec. 21, 1977).

22. A. Verma and Y. J. Hu, *Can. J. Chem. Eng.* **55**, 105 (1977).

General References

"Sodium Sulfate," *Chem. Mark. Rep.* **213**(5), 9 (Jan. 30, 1978); **207**(8), 9 (Feb. 24, 1975).
J. A. Kent, ed., *Riegel's Handbook of Industrial Chemistry*, 7th ed., Van Nostrand Reinhold, New York, 1974, pp. 132–134. Description of the Mannheim furnace.
D. M. Considine, ed., *Chemical and Process Technology Encyclopedia*, McGraw-Hill Book Co., New York, 1974.
F. A. Lowenheim and M. K. Moran, eds., *Faith, Keyes and Clark's Industrial Chemicals*, 4th ed., John Wiley & Sons, Inc., New York, 1975. Good summary of process energy requirements and list of manufacturers.

THOMAS F. CANNING
Kerr-McGee Chemical Corporation

SODIUM SULFIDES

Sodium sulfide [*1313-82-2*] (Na_2S, mol wt 78.05), sodium hydrosulfide [*16721-80-5*] (NaHS, mol wt 56.06), and sodium tetrasulfide [*12034-39-8*] (Na_2S_4, mol wt 174.24) have diverse, valuable industrial uses. These sulfides to some extent are interchangeable in the paper-pulping and leather industries (see Pulp; Leather). Recent environmental emphasis has created a use in fixation of heavy metals, but the required recovery of the sulfides from effluents has reduced the market overall.

Sodium Sulfide

Properties. Pure sodium sulfide (sodium sulfuret) is a white, crystalline solid (mp 1180°C, sp gr 1.856) (see also Sulfur compounds). Commercial material is white to light yellow or pink. Figure 1 shows the boiling points, densities, and freezing points for different strength solutions (1). Heat of formation for the crystalline state is −373 kJ/mol (−89 kcal/mol), and the heat of solution is −63.5 kJ/mol (−15.2 kcal/mol) (2). It crystallizes from aqueous solutions as the nonahydrate [*1313-84-4*], $Na_2S.9H_2O$. In air, sodium sulfide slowly converts to sodium carbonate and sodium thiosulfate and is deliquescent. Reactions with strong oxidizing agents give elemental sulfur.

Manufacture. The oldest manufacturing process for Na_2S is not used in the United States but was part of the LeBlanc soda ash process. Sodium sulfate was reduced by powdered coal in a furnace under extreme conditions of 900–1000°C.

$$Na_2SO_4 + 4\,C \rightarrow Na_2S + 4\,CO \quad (1)$$

$$Na_2SO_4 + 4\,CO \rightarrow Na_2S + 4\,CO_2 \quad (2)$$

Two processes are practiced currently: in one, the reduction of barytes ore ($BaSO_4$)

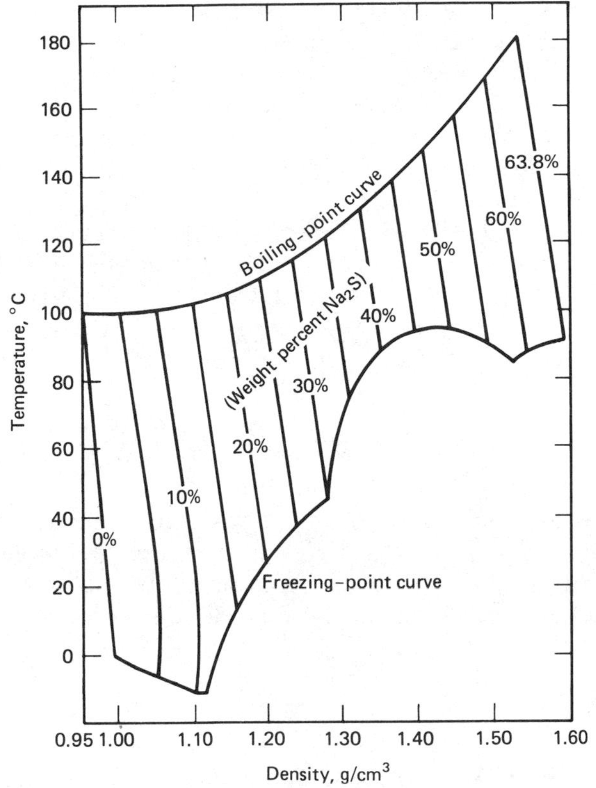

Figure 1. Density of sodium sulfide solutions in water. To convert g/cm³ to lb/gal, multiply by 8.344.

with coal in a rotary kiln at ca 800°C gives a crude black ash. Leaching dissolves barium sulfide, which then is converted to barium carbonate by addition of sodium carbonate. A weak sodium sulfide solution is produced, which is concentrated and crystallized or flaked. Iron contamination is reduced by use of nickel or nickel-clad equipment.

The second process involves two steps. Sodium hydrosulfide obtained from the reaction of hydrogen sulfide with caustic soda reacts further with caustic soda to yield Na_2S. Concentration by evaporation to 60 wt % is practiced unless concentrated caustic soda is used in the second step.

$$H_2S + NaOH \rightarrow NaHS + H_2O \quad (3)$$

$$NaHS + NaOH \rightarrow Na_2S + H_2O \quad (4)$$

Hydrogen sulfide can be obtained by the reaction of hydrogen and sulfur, as a by-product of the carbon disulfide-from-methane process, or from desulfurizing petroleum fractions. With high purity raw materials, a chemical-grade Na_2S is produced and is suitable for dyestuffs, polymers, and leather dehairing. If the caustic soda has heavy-metal impurities, filtration of the sodium hydrosulfide may be necessary.

Sodium sulfide is marketed as 30–34 wt % fused crystals and 60–62 wt % flakes (3). Each container has a corrosive label and a product label stating that the product causes severe burns to eyes or skin, and that contact with acid liberates poisonous hydrogen sulfide gas. The material is nonflammable, noncombustible, and nonexplosive.

Economic Aspects. U.S. production of sodium sulfide increased rapidly from 1965 through 1972 and then decreased through 1974 (4). Because only three producers remained, 1974 was the last year that the U.S. Bureau of Census released production figures. Current estimates indicate that the market has declined to levels of the 1950s. Production data on a 100 wt % Na_2S basis and the value of the product on the same basis are listed in Table 1. Increased sulfur and caustic soda prices have inflated the sales value of sodium sulfide since 1974.

Uses. Uses of sodium sulfide are listed in Table 2. Many of the applications involve either sodium sulfide or sodium hydrosulfide to meet a sulfidity need. Sulfidity is the sodium sulfide fraction of active alkali from Na_2S and NaOH. Sodium sulfide provides stronger alkalinity than the hydrosulfide but less sulfidity. In some leather-dehairing operations, the stronger alkalinity as well as the sulfidity is preferred before tanning.

In froth flotation, the mining industry uses Na_2S to form insoluble metal sulfides of copper, lead, and molybdenum. The same reaction is used to remove heavy metals from wastewaters of other industries (see Flotation).

In the kraft wood-pulping process, the sulfide is used in synthetic cooking liquor. In the dyestuffs industry it is used as a solvent for water- insoluble dyes and as a reducing agent. Other uses include the preparation of lubrication oils (see Lubrication and lubricants) and the production of polysulfide elastomers and plastics (see Polymers containing sulfur).

Sodium Hydrosulfide

Properties. Pure sodium hydrosulfide (sodium sulfhydrate, sodium hydrogen sulfide, sodium bisulfide) is a white, crystalline solid (mp 350°C, sp gr 1.79). It is highly soluble in water, alcohol, or ether. The heat of formation is −237.6 kJ/mol (− 56.79 kcal/mol), and the heat of solution is 15.9 kJ/mol (3.8 kcal/mol) (2). Figure 2 shows

Table 1. U.S. Production and Sales Value of Sodium Sulfide [a]

Year	Production, 1000 metric tons	Sales value, $/t
1965	39	225
1970	68	246
1972	80	252
1974	65	262
1980	19 estd	506 estd

[a] Ref. 4.

Table 2. Estimates of Sodium Sulfide Uses in the United States in 1980

Use	%
mining (flotation)	33
leather depilatory	20
chemicals and dyestuffs	19
export	18
other	10

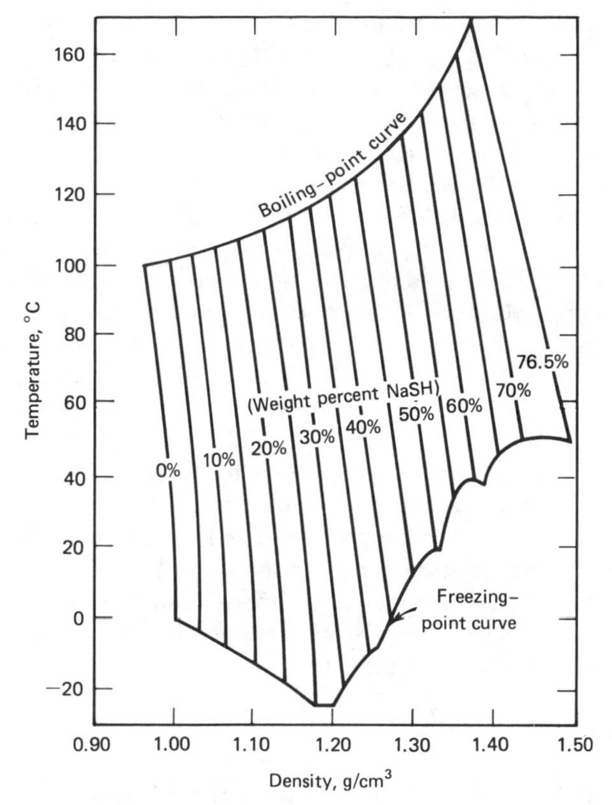

Figure 2. Density of sodium hydrosulfide solutions in water. To convert g/cm^3 to lb/gal, multiply by 8.344.

the graph of boiling points, densities, and freezing points for solutions of different strengths (1). The commercial product occurs in different shades of yellow and is highly deliquescent. Exposure to air converts it to sodium thiosulfate and sodium carbonate. In the presence of organic matter, combustion can occur. Heating releases hydrogen sulfide, which is a toxic gas. The bulk density of 70–72 wt % flake is 0.64 g/cm^3 (1).

Manufacture. Sodium hydrosulfide was first produced in the United States in 1938 after being introduced in Germany a few years earlier. This may account for the use of sodium sulfide in some traditional processes where NaHS would be more cost-effective, eg, in tanneries.

Processes to make NaHS are closely related to the source of the hydrogen sulfide which reacts with caustic soda in equation 3. With the carbon disulfide process as the source, the carbon disulfide can be separated either from the hydrogen sulfide by absorption or from the NaHS after the H_2S reacts. Other sources of H_2S are mentioned under the manufacture of sodium sulfide. The reaction of H_2S and caustic soda gives high yields based on both reactants.

Sodium hydrosulfide is marketed as 70–72 wt % flakes and 44–60 wt % liquor in the high purity grades and as 10–40 wt % liquor from recovered caustic wash in the oil-refining desulfurization processes (See Petroleum, refinery processes).

Shipment labeling is the same as for sodium sulfide. The product is shipped either as flake in drums or as solutions in tank cars or tank trucks.

Economic Aspects. Production and sales values for sodium hydrosulfide on a 100 wt % basis since 1965 are listed in Table 3 (3). These figures, as well as those in Table 4, exclude the low purity material recovered by oil refiners and believed to be sold to pulp mills. Reliable data are not available for oil-refiners' production, but it is believed to be 2–4 times that of the chemical industry.

Estimates of uses in the United States are given in Table 4 (3,5). Production of NaHS peaked in 1966 at 4.1×10^4 metric tons (4). Declining markets for dyes used on cotton and rayon and for leather hide dehairing have reduced demand, and efforts to curb air and water pollution have encouraged recovery and reuse. Froth flotation to beneficiate low grade minerals has provided a growing market and has become a large use. A new engineering plastic, poly(*p*-phenylene sulfide), is made by reaction of *p*-dichlorobenzene and sodium sulfide (6) (see Polymers containing sulfur).

The amount of low purity material recovered by oil refiners has been ca (4.5–9.0) $\times 10^4$ t/yr on a 100-wt % basis (7). It can be a mixture of both sulfide and sulfhydrate, ie, 15 wt % and 85 wt % respectively. Capacity for high purity NaHS in the United States is ca 7.0×10^4 t/yr, although this volume varies with Na_2S production (5).

The lower freezing points of solutions of sodium hydrosulfide provide an advantage over those of sodium sulfide in shipping by tank truck and tank car. Recently systems have been designed to enable customers to make their own sodium sulfide solutions by reaction of NaHS and NaOH. This permits advantages to the customer of bulk shipments and sulfidity control (8).

Sodium Polysulfides

Sodium tetrasulfide, Na_2S_4, is prepared by the reaction of sodium sulfide with sulfur.

$$Na_2S + 3\,S \rightarrow Na_2S_4 \qquad (5)$$

The 40 wt % solution is dark red to near black, solidifies at −33 to 10°C, boils at 115°C, and has a specific gravity at 15.5°C of 1.335 (1). Sodium tetrasulfide is used in leather

Table 3. U.S. Production and Sales for Sodium Hydrosulfide [a]

Year	Production, 1000 t	Sales, $/t
1965	40.5	97
1970	26.7	150
1975	21.5	270
1980	28.6	500

[a] Ref. 3.

Table 4. Estimate of Sodium Hydrosulfide Uses in the United States in 1980

Use	%
dyestuffs	20
leather depilatory	15
paper pulping	20
mining (flotation)	30
plastics and chemicals	15

processing to solubilize globular proteins and for many of the applications listed previously for sodium sulfide and sodium sulfhydrate. Commercial polysulfide elastomers, eg, Thiokols, are made by condensation of alkaline polysulfides and organic chloride compounds (9).

Other polysulfides have been studied from di- through pentasulfides, but there are questions as to the actual structures. No commercial polysulfide of significance is produced other than the tetrasulfide. Two producers of Na_2S_4 and one of sodium polysulfide are listed for the United States: PPG Industries, Inc., New Martinsville, West Virginia, Van de Mark, Lockport, N.Y., and FMC Corp., Modesto, Calif., respectively.

Analysis

A double-end-point, acid–base titration can be used to determine the amount of Na_2S and either NaHS or NaOH present; neither NaHS nor NaOH are ever present (see Hydrogen-ion concentration). Standardized hydrochloric acid is the titrant, and the indicators are thymolphthalein and bromophenol blue (1). The sample is mixed with 10 wt % $BaCl_2$ solution to precipitate carbonates and sulfates. The first titration against thymolphthalein to the greenish-yellow end point determines the amount of NaOH and one half of the Na_2S. The yellow second end point obtained with bromophenol blue indicates completion of titration of the NaHS and the other half of the Na_2S.

Sodium thiosulfate and sodium thiocarbonate interfere quantitatively and are determined separately. This method requires a well-ventilated hood because of H_2S liberation. Impurities typically include thiosulfate, thiocarbonate, carbonate, nickel, and iron.

Analysis of sodium tetrasulfide requires sulfide determination, liberation of hydrogen sulfide, and determination of free sulfur by a gravimetric procedure.

Health and Safety Factors

The sodium sulfides are similar to caustic alkalies as corrosive substances on animal tissues. Contact with these as solids or liquids irritates and burns the eyes, skin, and the respiratory tract. Ingestion and reaction with gastric acids can result in decomposition to hydrogen sulfide with subsequent systemic poisoning. No threshold-limit values (TLVs) have been adopted by the ACGIH (1976). A more serious hazard involving the sulfides is the generation of toxic hydrogen sulfide gas by their reaction with any acid. The properties of hydrogen sulfide, including explosion limits in air, toxicity, and safety precautions, are described elsewhere (see Sulfur compounds).

Heating solutions of sodium sulfides increases the evolution of hydrogen sulfide, creating a potentially dangerous condition in storage, process, and shipping equipment, especially when these products are involved in a fire. Respiratory protection should be provided in such cases. Spills should be absorbed with inert materials and not allowed to enter sewers where acids may be present. These sulfides should not be stored in zinc, aluminum, or copper containers.

Safety recommendations include use of eye and face protection, gloves, rubber boots, and impervious clothing. Self-contained breathing apparatus and safety eyewash and shower facilities should be available. First aid for eye or skin contact is flushing with copious amounts of water for 15 min. In case of ingestion, water or milk should be taken, but vomiting should not be induced; a physician should be consulted. Use of a continuous monitor and alarm for hydrogen sulfide is recommended by NIOSH (10).

BIBLIOGRAPHY

"Sodium Sulfides" in *ECT* 1st ed., Vol. 12, pp. 609–611, by J. A. Brink, Jr., Purdue University; "Sodium Sulfides" in *ECT* 2nd ed. under "Sodium Compounds," Vol. 18, pp. 510–515, by J. S. Sconce, Hooker Chemical Corporation.

1. *Sulfur Chemicals*, PPG Industries, Inc., Pittsburgh, Pa., 1979.
2. D. D. Wagman, *Selected Values of Chemical Thermodynamic Properties*, Circular 500, National Bureau of Standards, Washington, D.C., pp. 456–460.
3. *Sulfur Products, Chemical Economics Handbook*, 780.4000A-H, Stanford Research Institute International, Menlo Park, California, Dec., 1978.
4. *Current Industrial Reports*, U.S. Bureau of the Census, Washington, D.C., M28A.
5. "Sodium Hydrosulfide: A Chemical Profile," *Chem. Mark. Rep.*, 9 (Feb. 2, 1981).
6. U.S. Pat. 3,354,129 (Nov. 21, 1967), J. T. Edmonds and co-workers (to Phillips Petroleum).
7. *Directory of Chemical Producers—United States*, Stanford Research Institute International, Menlo Park, Calif., 1981.
8. *Paper J.* **160,** 35 (May 1, 1976). (Excerpts from Business Data Base, Predicasts, Inc., Cleveland, Ohio).
9. M. B. Berenbaum in N. Gaylord, ed., *High Polymers Series*, Vol. XIII, Wiley-Interscience, New York, 1962, pp. 43–224.
10. *Occupational Exposure to Hydrogen Sulfide*, Public No. DHEW (NIOSH) 77-158, U.S. Department of Health, Education, and Welfare, Washington, D.C., May 1977.

CHARLES DRUM
PPG Industries, Inc.

SODIUM HYDROXIDE. See Alkali and chlorine products.

SODIUM SILICATE. See Silicon compounds.

SODIUM SULFITES. See Sulfur compounds.

SODIUM TRIPOLYPHOSPHATE. See Phosphoric acids and phosphates.

SOIL CHEMISTRY OF PESTICIDES

One of the main trends in the United States during the 1970s was an appraisal of the effect of new chemical technology on humans and their environment, and as a result, pesticides came under scrutiny. Soils were of special interest since they represent the ultimate reservoir for most pesticides. In reality, soils represent a dynamic environment in which a number of processes operate on any pesticide. The susceptibility of a pesticide molecule to these soil processes has in some circumstances dictated the future of that pesticide. This awareness of the significance of soil processes on the fate and behavior of pesticides and of hazardous chemicals has fostered a better understanding of the chemistry of pesticides in soils.

The most significant development in the chemistry of pesticides in the last decade has been the introduction of the synthetic pyrethroid insecticides. Although the use of many of the chlorinated hydrocarbon insecticides has been restricted in the United States, these insecticides continue to be an important series of compounds on a global basis. The largest class of pesticidal chemicals are the organic herbicides used for weed control.

Common and chemical names, structure numbers, Chemical Abstracts Service Registry Numbers, and classifications of pesticides and metabolites mentioned in the text are listed in Table 1 (see Fungicides; Herbicides; Insect control technology).

Insecticides

Principal classes of insecticides are the chlorinated hydrocarbons, organophosphates, carbamates, and synthetic pyrethroids. Studies of mode of action have led to the introduction of new compounds, eg, the juvenile hormone analogues and diflubenzuron, which is an inhibitor of the enzyme responsible for the synthesis of the skeletal material chitin. Of the insecticides, the chlorinated hydrocarbons are some of the most persistent in soils; in an extreme case that was related to high initial rates of application, 40–45% of the original material remained after 14 yr (1). The low solubility of most of the chlorinated hydrocarbon insecticides probably contributes to their persistence in soils. Other factors influencing their persistence are moisture, pH, and microbial activity.

DDT. Total worldwide production of DDT from its introduction until 1968 was greater than 1,600,000 metric tons (2). In 1971, registration of DDT for most uses in the United States was canceled, and environmental residues, which were common, are steadily declining (3). DDT is nearly immobile in soils. In water, the molecules of this extremely hydrophobic material tend to migrate to the air–water interface, where DDT evaporates with water molecules faster than would be predicted from its extremely low vapor pressure. The occurrence of DDT in untreated areas throughout the world may be a consequence of its ability to evaporate with water vapor.

Two processes that structurally alter DDT are microbial metabolism and photodecomposition. The course of degradation of DDT (**1**) in soils is not fully established. Certain organisms can dehydrohalogenate DDT to DDE (**2**). Resistant strains of insects can effect this conversion, which is also brought about easily by base or catalytically by iron. Many soil microorganisms can convert DDT to DDD (**3**) under anaerobic conditions. DDD is an important product of DDT degradation, but it has been difficult

Table 1. Chemical Designations of Pesticides or Metabolites Mentioned in Text

Common name	IUPAC name	Structure no.	CAS Registry No.	Type of pesticide[a]
Agent Orange	50:50 mixture of the *n*-butyl esters of 2,4-D and 2,4,5-T		[*94-80-4*] [*93-79-8*] or [*39277-47-9*]	H
alachlor	2-chloro-2′,6′-diethyl-*N*-(methoxymethyl)acetanilide		[*15972-60-8*]	H
aldrin	1,2,3,4,10,10,hexachloro-1,4,4a,5,8,8a-hexahydro-1,4-*endo-exo*-5,8-dimethanonaphthalene	**(5)**	[*309-00-2*]	I
allethrin	2-methyl-4-oxo-3-(2-propenyl)-2-cyclopenten-1-yl 2,2-dimethyl-3-(2-methyl-1-propenyl)cyclopropanecarboxylate	**(26)**	[*584-79-2*]	I
aminoparathion	*O*,*O*-diethyl *O*-*p*-aminophenyl phosphorothioate		[*3735-01-1*]	M
amitrole	3-amino-1,2,4-triazole	**(62)**	[*61-82-5*]	H
arsenic pentoxide	arsenic oxide (2:5)		[*1303-28-2*]	I, H, SS, V
arsenic trioxide	arsenic oxide (2:3)		[*1327-53-3*]	I, H, SS, V
atrazine	2-chloro-4-ethylamino-6-isopropylamino-*s*-triazine		[*1912-24-9*]	H
bensulide	*N*-(2-mercaptoethyl)benzenesulfonamide		[*741-58-2*]	H
Bordeaux mixture	mixture of cupric sulfate and cupric hydroxide		[*18939-61-2*] [*20427-59-2*]	F
γ-BHC	(see Lindane)			I
cacodylic acid	dimethylarsinic acid		[*75-60-5*]	H
calcium arsenate			[*7778-44-1*]	I, H, SS, V
captan	*N*-trichloromethylmercapto-4-cyclohexene-1,2-dicarboximide	**(70)**	[*133-06-2*]	F
carbaryl	1-naphthyl *N*-methylcarbamate	**(16)**	[*63-25-2*]	I
chloramben	3-amino-2,5-dichlorobenzoic acid		[*133-90-4*]	H
chlordane	1,2,4,5,6,7,8,8-octachloro-2,3,3a,4,7,7a-hexahydro-4,7-methano-1*H*-indene (principal constituent)		[*57-74-9*]	I
chloroxuron	3-[*p*-(*p*-chlorophenoxy)phenyl]-1,1-dimethylurea		[*1982-47-4*]	H
chlorpropham	isopropyl *m*-chlorocarbanilate	**(52)**	[*101-21-3*]	H
copper			[*7440-50-8*]	F
cypermethrin	cyano(3-phenoxyphenyl)methyl 3-(2,2-dichlorovinyl)-2,2-dimethylcyclopropanecarboxylate		[*52315-07-8*]	I

2,4-D	2,4-dichlorophenoxyacetic acid	**(33)**	[*94-75-7*]	H
dacthal	dimethyl tetrachloroterephthalate		[*1861-32-1*]	H
dalapon	2,2-dichloropropionic acid	**(60)**	[*75-99-0*]	H
2,4-DB	4-(2,4-dichlorophenoxy)butyric acid	**(32)**	[*94-82-6*]	H
DBCP	1,2-dibromo-3-chloropropane	**(76)**	[*96-12-8*]	N
DDD	1,1-dichloro-2,2-bis(*p*-chlorophenyl)ethane		[*72-54-8*]	I, M
DDE	1,1-dichloro-2,2-bis(*p*-chlorophenyl)ethene		[*72-55-9*]	I, M
DDT	dichlorodiphenyltrichloroethane: mixture of *p,p′*-DDT:1,1,1-trichloro-2,2-bis(*p*-chlorophenyl)ethane and *o,p′*-DDT: 1,1,1-trichloro-2-(*o*-chlorophenyl)-2-(*p*-chlorophenyl)ethane	**(1)**	[*50-29-3*] [*789-02-6*]	I
decamethrin	(1*R*-[1α(*S**),3α])-cyano(3-phenoxyphenyl)methyl 3-(2,2-dibromovinyl)-2,2-dimethylcyclopropanecarboxylate	**(29)**	[*52918-63-5*]	I
diazinon	*O,O*-diethyl *O*-2-isopropyl-6-methyl-4-pyrimidinyl phosphorothioate	**(19)**	[*333-41-5*]	I
dieldrin	1,2,3,4,10,10-hexachloro-6,7-epoxy-1,4,4a,5,6,7,8,8a-octahydro-1,4-*endo*-*exo*-5,8-dimethanonaphthalene	**(6)**	[*60-57-1*]	I
diflubenzuron	*N*-[(4-chlorophenyl)aminocarbonyl]-2,6-difluorobenzamide		[*35367-38-5*]	I
dimetan	5,5-dimethyl-3-oxo-1-cyclohexen-1-yl dimethylcarbamate		[*122-15-6*]	I
diquat	6,7-dihydrodipyrido[1,2-*a*:2′,1′-*c*]pyrazidinium dibromide	**(65)**	[*85-00-7*]	H
diuron	3-(3,4-dichlorophenyl)-1,1-dimethylurea		[*330-54-1*]	H
DSMA	disodium methanearsonate		[*144-21-8*]	H
endrin	1,2,3,4,10,10-hexachloro-6,7-epoxy-1,4,4a,5,6,7,8,8a-octahydro-1,4-*endo*-*endo*-5,8-dimethanonaphthalene		[*72-20-8*]	I
ethyl mercuric acetate			[*109-62-6*]	F
ETU	ethylenethiourea	**(75)**	[*96-45-7*]	M, P
fenac	2,3,6-trichlorophenylacetic acid	**(61)**	[*85-34-7*]	H
fenuron	1,1-dimethyl-3-phenylurea		[*101-42-8*]	H
fenvalerate	cyano(3-phenoxyphenyl)methyl 4-chloro-α-(1-methylethyl)phenylacetate	**(30)**	[*51630-58-1*]	I
ferbam	ferric dimethyldithiocarbamate	**(68)**	[*14484-64-1*]	F
glyphosate	*N*-(phosphonomethyl)glycine		[*1071-83-6*]	H
heptachlor	1,4,5,6,7,8,8-heptachloro-3a,4,7,7a-tetrahydro-4,7-methano-1*H*-indene	**(7)**	[*76-44-8*]	I
heptachlor epoxide	1,4,5,6,7,8,8-heptachloro-2,3-epoxy-2,3,3a,7a-tetrahydro-4,7-methano-1*H*-indene	**(8)**	[*1024-57-3*]	I, M

Table 1 (*continued*)

Common name	IUPAC name	Structure no.	CAS Registry No.	Type of pesticide[a]
kepone	decachlorooctahydro-1,3,4-metheno-2*H*-cyclobuta[*cd*]pentalene-2-one		[*143-50-0*]	I, AC
lead arsenate			[*7784-40-9*]	I, H, SS, V
lindane (γ-BHC)	γ-1,2,3,4,5,6-hexachlorocyclohexane	**(14)**	[*58-89-9*]	I
linuron	3-(3,4-dichlorophenyl)-1-methoxy-1-methylurea		[*330-55-2*]	H
malathion	*S*-[1,2-bis(ethoxycarbonyl)ethyl] *O*,*O*-dimethyl phosphorodithioate		[*121-75-5*]	I
mancozeb	coordination product of Zn^{2+} and manganese ethylenebis(dithiocarbamate)		[*8018-01-7*]	F
maneb	manganous ethylenebis(dithiocarbamate)		[*12427-38-2*]	F
MCPA	4-chloro-2-methylphenoxyacetic acid		[*94-74-6*]	H
mercurous chloride			[*10112-91-1*]	F
mercuric chloride			[*7487-94-7*]	F
metiram	mixture of [ethylenebis(dithiocarbamato)]zinc ammoniates with ethylenebis(dithiocarbamic acid)anhydrosulfides		[*12122-67-7*]	F
metobromuron	3-(*p*-bromophenyl)-1-methoxy-1-methylurea	**(47)**	[*3060-89-7*]	H
metolachlor	2-chloro-*N*-(2-ethyl-6-methylphenyl)-*N*-(2-methoxy-1-methylethyl)-acetamide		[*51218-45-2*]	H
mevinphos	methyl(*E*)-3-hydroxycrotonate dimethyl phosphate		[*298-01-1*]	I, AC
mirex	dodecachlorooctahydro-1,3,4-metheno-1*H*-cyclobuta[*cd*]pentalene		[*2385-85-5*]	I
monuron	3-(*p*-chlorophenyl)-1,1-dimethylurea		[*150-68-5*]	H
nabam	sodium ethylenebis(dithiocarbamate)	**(67)**	[*142-59-6*]	F
neburon	1-butyl-3-(3,4-dichlorophenyl)-1-methylurea		[*555-37-3*]	H
nitralin	4-(methylsulfonyl)-2,6-dinitro-*N*,*N*-dipropylaniline		[*4726-14-1*]	H
paraoxon	*O*,*O*-diethyl *O*-*p*-nitrophenyl phosphate	**(23)**	[*311-45-5*]	I, M
paraquat	1,1′-dimethyl-4,4′-bipyridinium salt	**(64)**	[*4685-14-7*]	H
parathion	*O*,*O*-diethyl *O*-*p*-nitrophenyl phosphorothioate	**(18)**	[*56-38-2*]	I
Paris green	cupric acetoarsenite		[*12002-03-8*]	I, H, SS, V
permethrin	(3-phenoxyphenyl)methyl 3-(2,2-dichlorovinyl)-2,2-dimethylcyclopropanecarboxylate	**(28)**	[*52645-53-1*]	I, AC, N
phenothrin	(3-phenoxyphenyl)methyl 2,2-dimethyl-3-(2-methyl-1-propenyl)-cyclopropanecarboxylate		[*26002-80-2*]	I

phorate	*O,O*-diethyl *S*-[(ethylthio)methyl] phosphorodithioate		[*298-02-2*]	I
photodieldrin	1,9,10,10,11-*exo*-12-hexachloro-4,5-*exo*-epoxy-8,3-7,6-*endo*-8,9-7,11-*exo*-pentacyclo[7.3.0.0^{2,6}.0^{3,8}.0^{7,11}]dodecane	**(72)**	[*13366-73-9*]	M
picloram	4-amino-3,5,6-trichloropicolinic acid		[*1918-02-1*]	H
PMA	phenylmercury acetate		[*62-38-4*]	F
propanil	3′,4′-dichloropropionanilide	**(53)**	[*709-98-8*]	H
propham	isopropyl carbanilate	**(55)**	[*122-42-9*]	H
pyrethrin I	2-methyl-4-oxo-3-(2,4-pentadienyl)-2-cyclopenten-1-yl 2,2-dimethyl-3-(2-methyl-1-propenyl)cyclopropanecarboxylate	**(24)**	[*121-21-1*]	I
pyrethrin II	2-methyl-4-oxo-3-(2,4-pentadienyl)-2-cyclopenten-1-yl 3-(3-methoxy-2-methyl-3-oxo-1-propenyl)-2,2-dimethylcyclopropanecarboxylate	**(25)**	[*121-29-9*]	I
semesan	(3-chloro-4-hydroxyphenyl)hydroxymercury	**(69)**	[*538-04-5*]	F
silvex	2-(2,4,5-trichlorophenoxy)propionic acid		[*93-72-1*]	H
simazine	2-chloro-4,6-bis(ethylamino)-*s*-triazine		[*122-34-9*]	H
sodium arsenate			[*13464-38-5*]	I, H, SS, V
sodium arsenite			[*13464-37-4*]	I, H, SS, V
strobane	terpene polychlorinates (65 wt % Cl)		[*8001-50-1*]	I
sulfur			[*7704-34-9*]	F
2,4,5-T	2,4,5-trichlorophenoxyacetic acid		[*93-76-5*]	H
TCA	trichloroacetic acid	**(59)**	[*76-03-9*]	H
TCDD	2,3,7,8-tetrachlorodibenzo-*p*-dioxin	**(74)**	[*1746-01-6*]	P
toxaphene	camphene polychlorinates (67–69% Cl)		[*8001-35-2*]	I
trifluralin	α,α,α-trifluoro-2,6-dinitro-*N,N*-dipropyl-*p*-toluidine	**(63)**	[*1582-09-8*]	H
vapam	sodium methyldithiocarbamate	**(71)**	[*137-42-8*]	F, N, I, H
zineb	zinc ethylenebis(dithiocarbamate)		[*12122-67-7*]	F

[a] H, herbicide; I, insecticide; F, fungicide; N, nematicide; AC, acaricide; M, metabolite; P, impurity; SS, soil sterilant; V, silvicide.

to ascertain the route of further breakdown after initial conversion to DDD (4). The production of nonextractable polar material from soils has been noted. Anaerobic breakdown of DDT by *Enterobacter aerogenes* microorganisms yields reduced dechlorinated compounds as well as oxidized derivatives and ultimately *p,p′*-dichlorobenzophenone (4) (5).

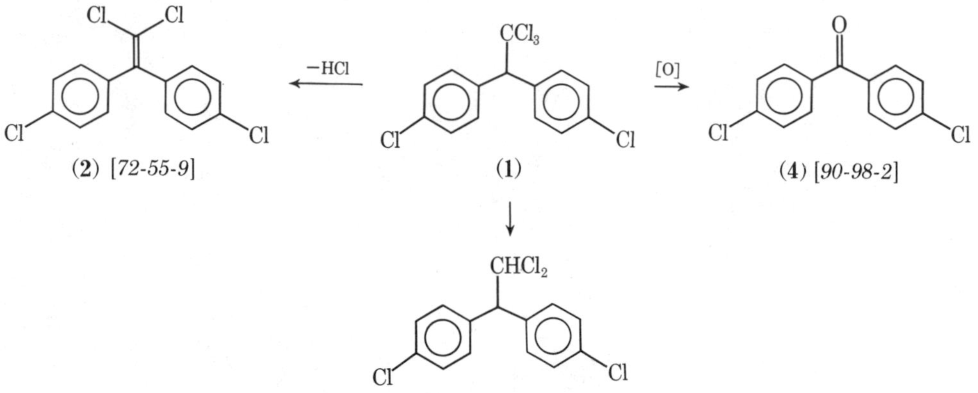

(2) [*72-55-9*] (1) (4) [*90-98-2*]

(3) [*72-54-8*]

Cyclodienes. Aldrin (**5**) is oxidized to dieldrin (**6**) and heptachlor (**7**) to heptachlor epoxide (**8**) in soils. Both epoxides are insecticidal and prolong biological activity. Epoxidation is probably mediated by soil microorganisms.

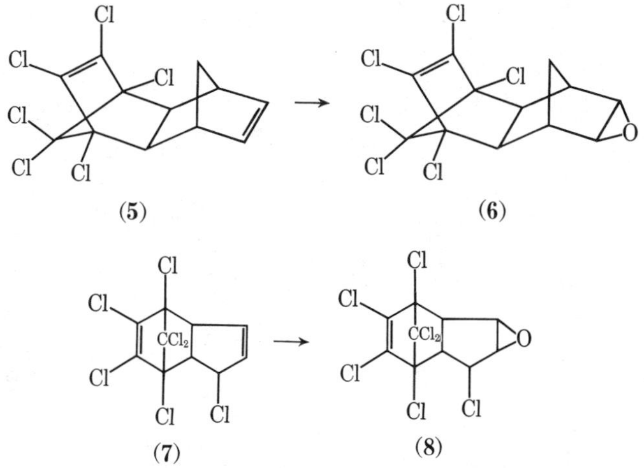

(**5**) (**6**)

(**7**) (**8**)

Dieldrin is degraded by a number of soil microorganisms. The main degradation products resulting from metabolism by *Pseudomonas* spp include aldrin, a ketone derived by rearrangement of the epoxide ring to a carbonyl function (**9**), a compound (**10**) formed by ring closure in structure (**9**), probably dihydroxydihydroaldrin (**11**), and tentatively an aldehyde (**12**) and an acid (**13**) formed by loss of carbon from the ring to which the epoxide function was formerly attached. *Enterobacter aerogenes* in an aerobic culture medium converts dieldrin to 6,7-*trans*-dihydroxydihydroaldrin, which is a specific configuration of structure (**11**) (5).

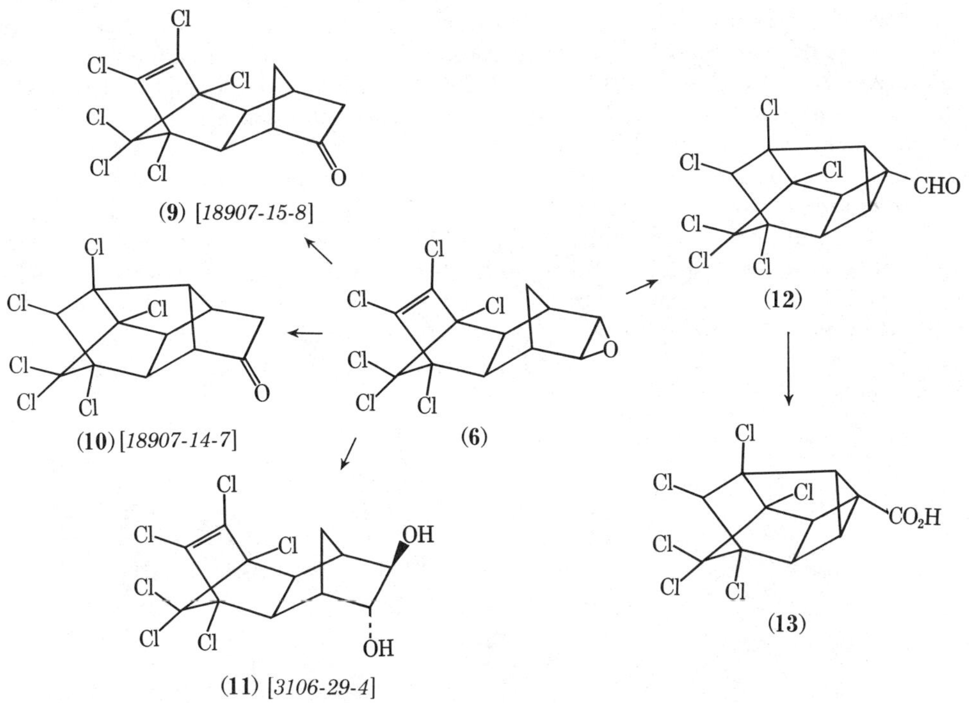

(9) [18907-15-8]

(10) [18907-14-7]

(6)

(12)

(13)

(11) [3106-29-4]

Lindane. Lindane (**14**) is less persistent in soil than DDT or the chlorinated cyclodiene insecticides. It degrades in soils to γ-pentachlorocyclohexene (**15**) probably by way of a microbial conversion (6).

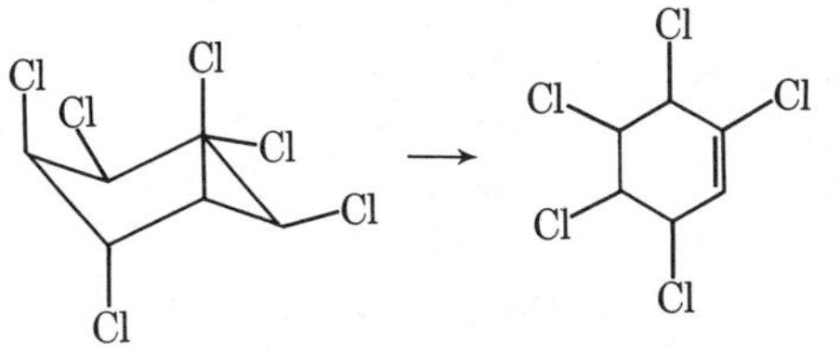

(**14**) (**15**) [54083-25-9]

Carbamates. Several methyl carbamates are sold as broad-spectrum insecticides. Of these, carbaryl (**16**) is the most extensively used. Little is published on the fate of these insecticides in soils; however, this scarcity of information may relate to their brief persistence. Methyl carbamate insecticides are seldom, if ever, detected in national soils-monitoring programs. Carbaryl ($t_{1/2}$ = 8 d in soils) probably hydrolyzes by initial hydrolysis of the ester linkage, forming 1-naphthol (**17**), methylamine, and carbon dioxide. Methyl isocyanate is a possible transient intermediate. Most methyl carbamates are unstable in alkaline media; consequently, localized areas of high pH in soils may facilitate hydrolysis.

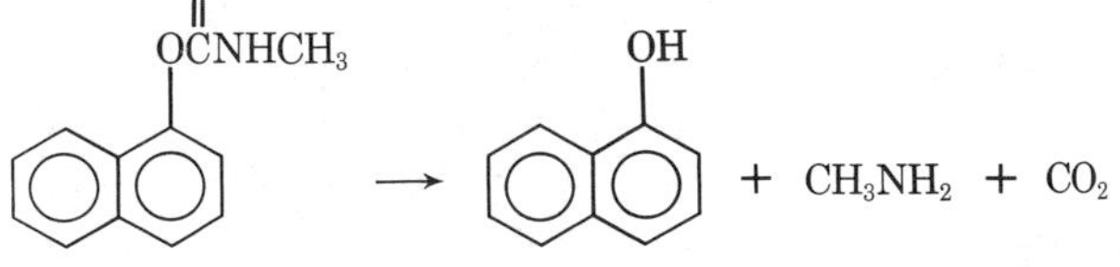

$$\text{(16)} \longrightarrow \text{(17)} + CH_3NH_2 + CO_2$$

(**16**) (**17**) [90-15-3]

Since there is no direct evidence for methyl carbamate metabolism by soil microorganisms, one can only speculate as to probable sites of attack from a study of related systems. Possible points of metabolic attack include hydrolysis of the ester linkage, oxidation of the *N*-methyl group, ring substitution, and ring cleavage. Based largely on the rapidity with which carbaryl breaks down in soil, esterase activity is the favored mode of degradation. The other reactions are generally slower in soils.

Phosphates. Like the carbamates, the organic phosphates are cholinesterase inhibitors. Parathion (**18**), diazinon (**19**), and phorate are used to control soil-borne insects. The organophosphorus compounds degrade fairly rapidly, with the rate increasing with soil moisture; greater soil pH favors hydrolysis. The products of hydrolysis of parathion are *O*,*O*-diethyl phosphorothioate (**20**) and *p*-nitrophenol (**21**). Hydrolysis of diazinon yields compound (**20**) and 6-hydroxy-2-isopropyl-1-methylpyrimidine (**22**):

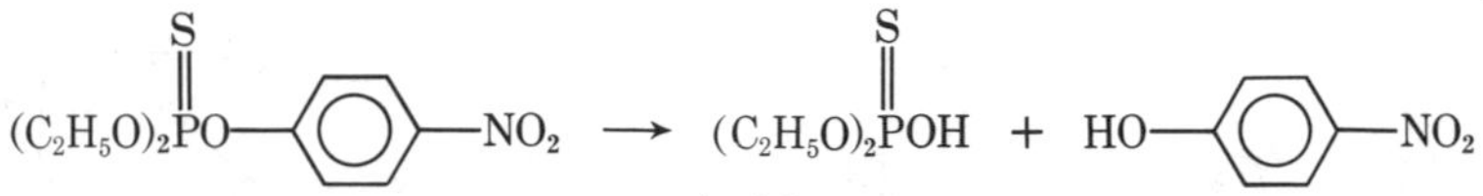

$$(C_2H_5O)_2\overset{S}{\overset{\|}{P}}O-C_6H_4-NO_2 \longrightarrow (C_2H_5O)_2\overset{S}{\overset{\|}{P}}OH + HO-C_6H_4-NO_2$$

(**18**) (**20**) *[2465-65-8]* (**21**) *[100-02-7]*

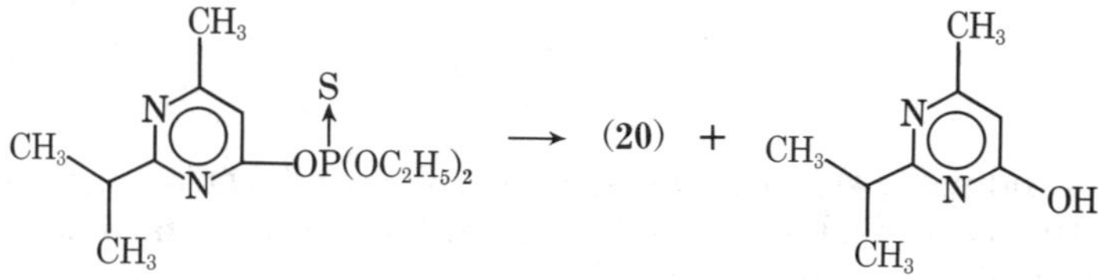

$$\text{(19)} \longrightarrow \text{(20)} + \text{(22)}$$

(**19**) (**22**)

The process is catalyzed by adsorption and occurs at degradation rates up to 11%/d. Thus, chemical hydrolysis may occur so rapidly as to preclude any extensive biological breakdown, especially at low pH. Diazinon and related pesticides are also readily hydrolyzed in the presence of Cu^{2+} at pH 5–6.

Soil microorganisms have also been implicated in phosphate insecticide degradation. Typically, initial chemical hydrolysis of the phosphate linkage is followed by metabolism of the resultant products. Malathion is rapidly metabolized by a soil fungus, *Trichoderma viride*, and a bacterium, *Pseudomonas* sp, which is isolated from soils receiving heavy applications of the insecticide (7). *Pseudomonas fluorescens* and *Thiobacillus thiooxidans* metabolize phorate and other phosphorothioates. Oxidation of parathion produces the more toxic paraoxon (**23**) (8).

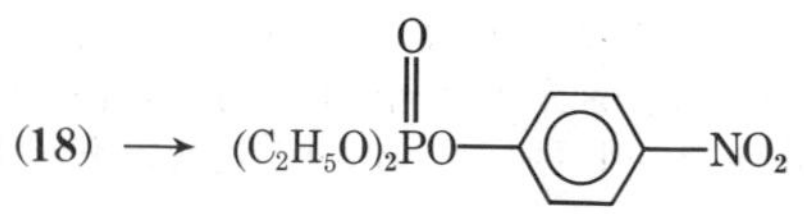

$$\text{(18)} \longrightarrow (C_2H_5O)_2\overset{O}{\overset{\|}{P}}O-C_6H_4-NO_2$$

(**23**)

The respective roles of soil microorganisms and that of adsorption in the overall disappearance of organophosphate insecticides in soils are unclear. Autoclaving, which is the common procedure for sterilizing soils, not only reduces the microbial population, but also alters the structure of many organic components in soils. Consequently, steam sterilization is not always a reliable indicator of microbial involvement, particularly if the initial reaction involves the participation of some labile organic compound in soils. For example, malathion and several other phosphate insecticides decompose much faster in gamma-radiation-sterilized soils than in autoclaved soils; a heat-labile, water-soluble substance that accelerates malathion degradation has been extracted from several nonautoclaved and radiation-sterilized soils (9). Thus, a third component, ie, soil organic matter, appears to be important in phosphate degradation. The finding of a specific organic catalyst for this class of pesticides illustrates an important concept in the soil chemistry of pesticides, ie, the participation of purely organic reactions in soils not mediated by soil microorganisms. The inhomogeneity of soil organic matter, however, has made the understanding of such reactions difficult.

Pyrethroids. The insecticidal activity of the extracts of pyrethrum (*Chrysanthemum cinesariaefolium*) plants has been exploited commercially for over a century. The active principles are the pyrethrins I (**24**) and II (**25**) and a number of related compounds.

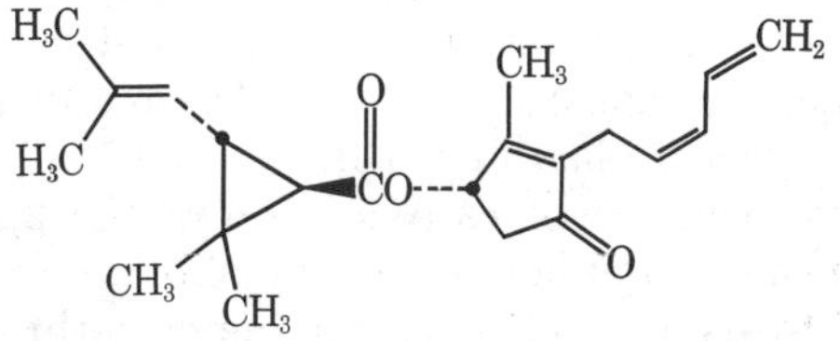

(**24**) 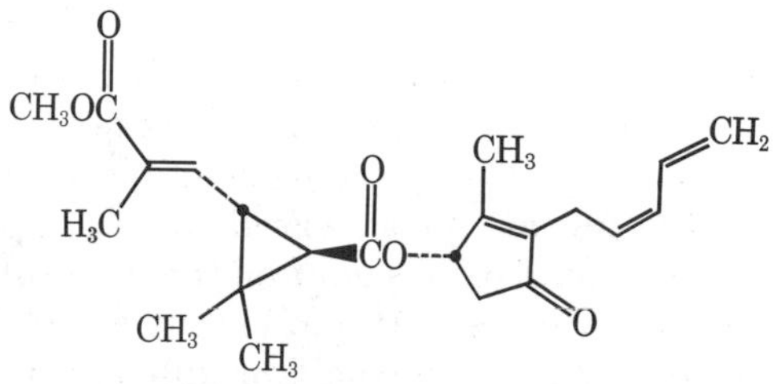 (**25**)

Although these naturally occurring compounds are effective insecticides, their biological and photochemical degradation is so rapid that their use as agricultural insecticides is impractical. Wartime limitations on the import of naturally occurring insecticides provided a stimulus for the search for new synthetic substitutes, eg, allethrin (**26**) (10). Modifications of the pyrethroid structure had led to compounds that are more economical and convenient to synthesize, more effective insecticides, and more stable towards light (11). There are several pyrethroids that show promise for control of insect pests in cotton (qv).

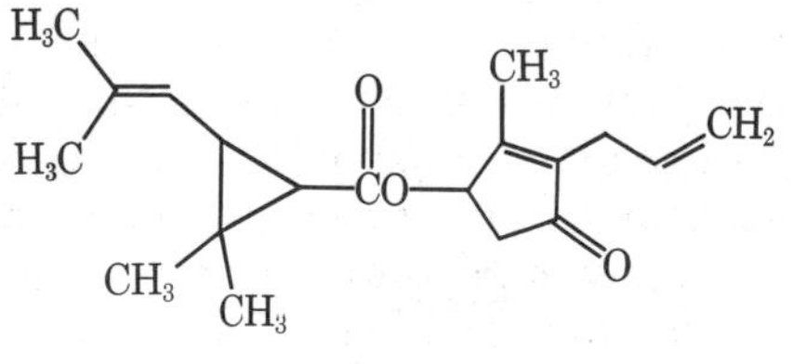

(**26**)

Photostabilization has been achieved by substitution in the chrysanthemic acid moiety (**27**) of a 2,2-dichlorovinyl group in place of the 2,2-dimethylvinyl group. The pyrethrolone portion of the molecule is replaced by 3-phenoxybenzyl alcohol to give the pyrethroid, permethrin (**28**).

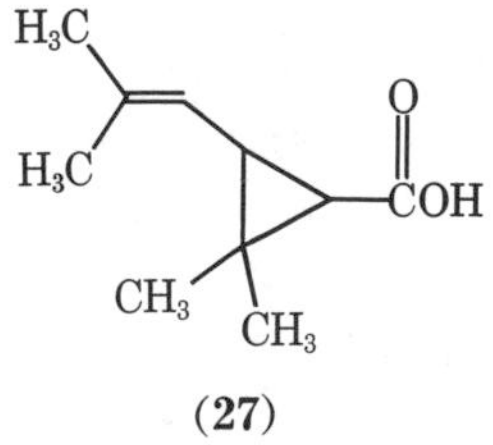

(**27**) 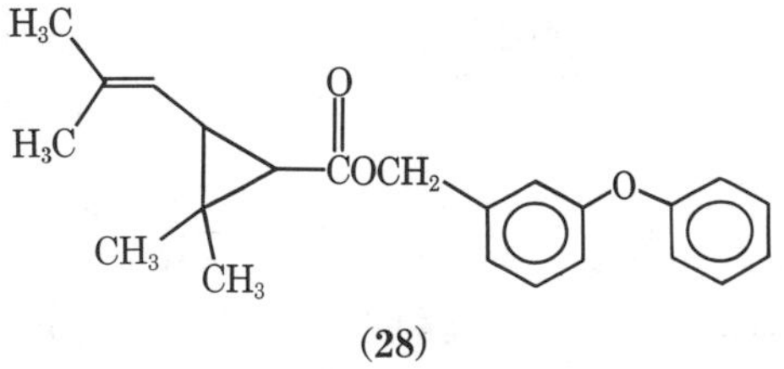 (**28**)

The half-life of this compound in sunlight is expressed in terms of hours rather than days and a number of halogenated analogues have since been developed. The selection of individual stereoisomers and the introduction of an α-cyano group in the phenoxybenzyl moiety enhance biological activity so that compounds, eg, decamethrin (**29**), are far more potent insecticides than the natural pyrethroids. The newer synthetic pyrethroids can be applied at very low rates (0.5–20 mg/m^2). Another commercially

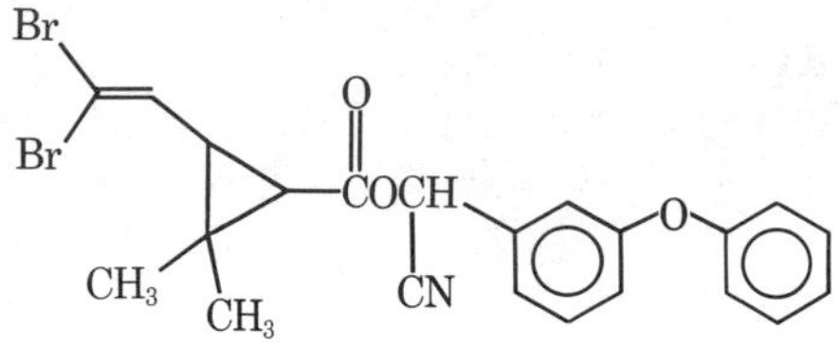

(**29**) 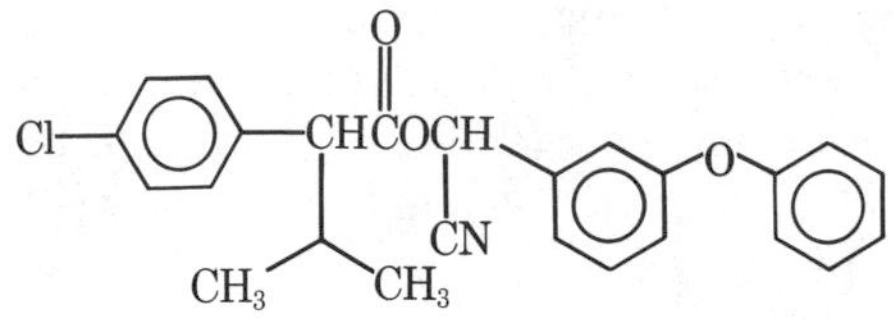 (**30**)

important photostabilized pyrethroid is a fenvalerate (**30**), which does not contain a dimethyl cyclopropyl ring. The structure and conformation of the synthetic pyrethroids have been discussed in relation to their biological activity (12). Decamethrin and fenvalerate photodegrade by a complex series of reactions (13).

Photodecomposition of the natural pyrethrins involves oxidation of the 2,2-dimethylvinyl residue and modification of the alcohol moiety (14). The photolability of the chrysanthemic acid residue is reduced by halogen substitution in permethrin, and photolysis in solution or soil brings about isomerization around the 1,3-bond of the cyclopropane ring and the ester is cleaved (15). Other minor reactions include reductive dechlorination, cyclopropane ring fusion, and oxidation of phenoxybenzyl alcohol (15).

The degradation of pyrethroids in soil is rapid and, in the case of permethrin, the initial step in biological degradation is primarily ester cleavage. This is followed by complete oxidation of the fragments to CO_2 (16). Other pyrethroids degrade by similar pathways (17).

Herbicides

More than 180 herbicides are manufactured in the United States and >6000 formulations are registered. Principal chemical classes include the phenoxyalkanoic acids, *s*-triazines, phenylcarbamates, phenylureas, dinitrobenzenes, benzoic acids, thiocarbamates, anilides, dipyridyls, chlorinated aliphatic acids, and uracils. Other classes, eg, the phenols, organoarsenical compounds, and diphenyl ethers, are also used for weed control. There are many important herbicides that are sole representatives of their chemical class, eg, picloram, glyphosate, and amitrole.

The presence of residual herbicides or their alteration products in soil may adversely affect subsequent crops. Many herbicides may be active only when applied to soils. There are a number of reviews of the chemistry of herbicides in soils (18–19).

Phenoxyalkanoic Acids. The phenoxyalkanoic acids formerly constituted the main group of herbicides used in the United States. The phenoxy herbicides are related by the following common structure:

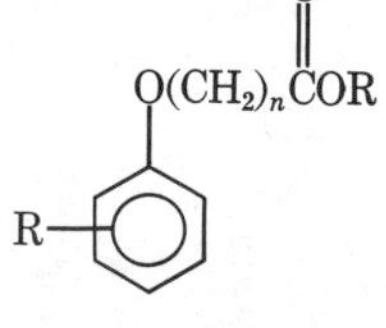

(**31**)

Microbial metabolism is the principal mechanism of phenoxyalkanoic acid breakdown within soils. Loss of 2,4-D occurs more quickly than the loss of either MCPA or 2,4,5-T, the latter of which is perhaps the most persistent phenoxy herbicide. Warm, moist, and well-aerated soil with ample organic matter favors the growth of microorganisms that degrade the phenoxys; herbicide persistence usually is fairly short in this environment. The rate of breakdown may also depend on the development of an adaptive population. Repeated applications of 2,4-D result in a microbial population that utilizes the herbicide as a carbon source; persistence after each addition becomes progressively shorter. An interesting corollary of this is the ability of MCPA-enriched soils to degrade other phenoxyacetic acids, sometimes faster than soils enriched in the test herbicides.

$OCH_2CH_2CH_2COOH$, Cl, Cl

(**32**)

(**34**) [*120-83-2*] (A) ← (**33**) (B) → (**41**) [*2639-79-4*] or [*2639-78-3*]

(**36**) [*2138-22-9*] (**35**) [*13673-92-2*] → (**39**) [*18529-78-7*]

[*20665-95-6*] → (**40**) [*6666-46-2*] → [*24740-88-3*] → acetate + dicarboxylic acid ← [*22752-94-9*]

The two main routes of phenoxyalkanoic acid metabolism in pure cultures of soil microorganisms appear to be: pathway A, initial attack and removal of the aliphatic side chain, followed by alteration of the aromatic ring; and pathway B, direct hydroxylation of the benzene moiety, often with product accumulation. Pathway A may involve β-oxidation, ie, progressive cleavage of two-carbon fragments along the aliphatic chain. Thus, 2,4-DB (**32**) is converted in soil to its two-carbon analogue 2,4-D (**33**), perhaps by a crotonic acid intermediate. By another mechanism, a *Flavobac-*

terium sp cleaves the ether bond in a 2,4-dichlorophenoxyalkanoic acid directly with formation of 2,4-dichlorophenol (**34**) and the fatty acid. The common metabolite in pathway A is the substituted phenol. Ring hydroxylation, a well-known mode of biological detoxication, apparently occurs next. In cultures and enzyme preparations of the soil-isolated bacterium *Arthrobacter* sp, 2,4-D is first converted to 2,4-dichlorophenol and then to 3,5-dichlorocatechol (**35**). By an analogous process, MCPA is cleaved to 4-chloro-2-methylphenol, then to an unidentified catechol. All steps require molecular oxygen. Other organisms may release chlorine during hydroxylation, affording 4-chlorocatechol (**36**) from 2,4-D. Cleavage of catechols may occur either adjacent to a hydroxyl yielding a muconic semialdehyde (**37**) or between the hydroxyl with formation of a muconic acid (**38**).

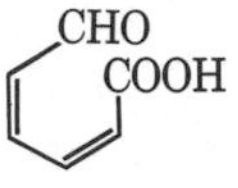

CHO COOH

COOH COOH

(**37**) [*69815-28-7*] 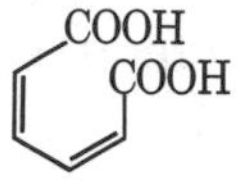 (**38**) [*505-70-4*]

Only muconic acids have been isolated from cultures containing the phenoxy herbicides. Further metabolism may proceed by lactone formation, eg, (**39**) and (**40**) observed with 2,4-D and MCPA, including loss of any carbon-4 halogen, and ultimate oxidation to carbon dioxide and, in chlorinated compounds, chloride.

Pathway B proceeds by hydroxylation of the aromatic ring. However, β-oxidation to a phenoxyacetic acid may occur initially. The fungus *Aspergillus niger* produces *ortho-* and *para-*hydroxylated phenoxyalkanoates from the unsubstituted acids; the same organism forms 5-OH-2,4-D [*2639-79-4*] (**41**) and 5-OH-MCPA [*2639-79-4*] as principal products of 2,4-D and MCPA, respectively. The formation of 6-hydroxy intermediates by certain bacteria has been postulated; subsequent ether cleavage would yield catechols directly. The relative importance in soils of the alternate metabolic pathways whereby phenoxy compounds are degraded is unclear. Soil conditions governing distribution of microbial populations may be the most influential factors.

The phenoxyalkanoic acids decompose in aqueous solution in sunlight. The main reactions occurring are the fission of the ether bond and sequential replacement of the halogen atoms by hydroxyl groups:

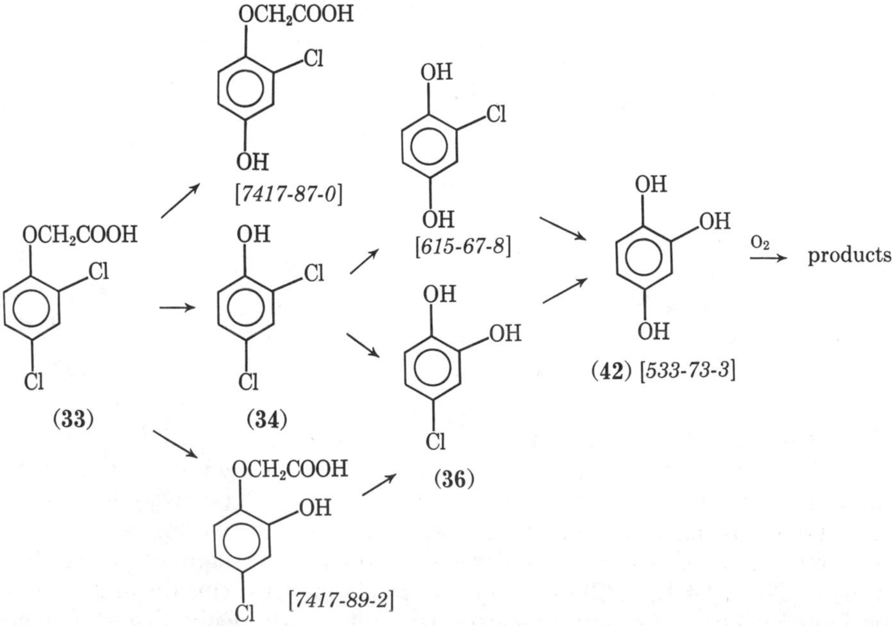

OCH_2COOH, Cl, OH [*7417-87-0*]

OH, Cl, OH [*615-67-8*]

OCH_2COOH, Cl, Cl (**33**)

OH, Cl, Cl (**34**)

OH, OH, Cl (**36**)

OH, OH, OH (**42**) [*533-73-3*] $\xrightarrow{O_2}$ products

OCH_2COOH, OH, Cl [*7417-89-2*]

A trihydroxybenzene (**42**) is the ultimate photolysis product, but it is susceptible to oxidation in air and can only be isolated as the acetate when an oxidation inhibitor, eg, sodium bisulfite, is added to the reaction mixture. Oxidation of compound (**42**) by air gives dark-colored polymeric material resembling humic acid.

***s*-Triazines.** Structurally herbicidal *s*-triazines are generally 2-substituted derivatives of the 4,6-bis(alkylamino)-*s*-triazines (**43**), where X may be Cl, SCH_3, or OCH_3 and R and R′ are alkyl groups.

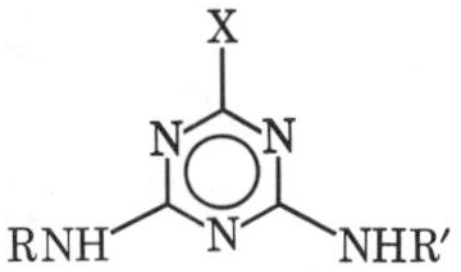

X
N N
RNH N NHR′

(**43**)

With a few exceptions, the *s*-triazine herbicides are considered to be slightly mobile, nonvolatile, and unaltered in solution by sunlight. Simazine adsorbed on filter paper is inactivated by sunlight. Chemical and, to a lesser extent, biological reactions play an important role in altering these herbicides in soils. The principal degradation product is the corresponding 2-hydroxy-4,6-bis(alkylamino)-*s*-triazine (**44**). The dehalogenation reaction in this case is a chemical reaction facilitated by protonation at the surface of a clay colloid.

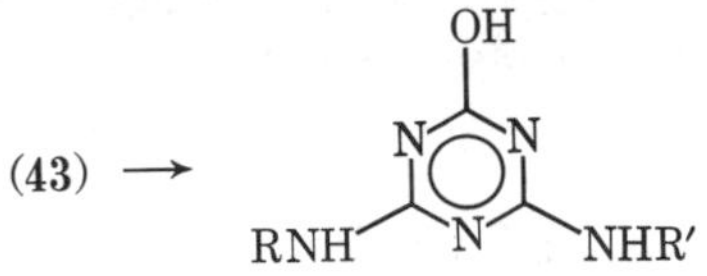

OH
(**43**) ⟶ N N
RNH N NHR′

(**44**)

Certain soil fungi, eg, *Aspergillus fumigatus*, *Rhizopus stolonifer*, and *Penicillium decumbens*, can dealkylate the side chain groups to produce the corresponding 2-chloro-4-amino-6-alkylamino-*s*-triazines (**45**).

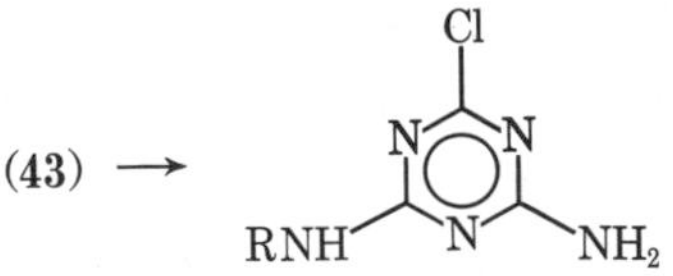

Cl
(**43**) ⟶ N N
RNH N NH_2

(**45**)

Detailed radioactivity studies with alkyl-^{14}C- and ring-^{14}C-labeled *s*-triazine herbicides indicate that chemical reactions are primarily responsible for degrading the herbicidal 2-chloro-*s*-triazines in most soils.

Phenylureas. Most phenylureas have the 1,1-dimethyl-3-phenylurea moiety (**46**) as part of their structure, where X and Y may be various substituents, eg, H, Cl, CF_3, and OC_6H_5. A few of the phenylureas are 1-methoxy-1-methyl-substituted compounds.

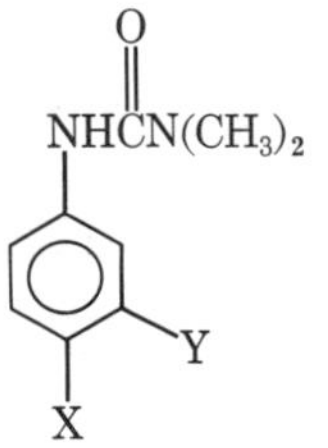

O
‖
$NHCN(CH_3)_2$
Y
X

(**46**)

In soils, the phenylureas behave somewhat like the *s*-triazines; ie, they are relatively immobile and nonvolatile. Degradation occurs under conditions most conducive for microbial activity, ie, warm, moist soils with high organic matter content. Although the causative organisms are difficult to isolate, the following pathway has been proposed for the degradation of the phenylurea herbicides in soil:

(46) → $C_6H_5NHC(O)NHCH_3$ [*1007-36-9*] → $C_6H_5NHC(O)NH_2$ [*64-10-8*] → $C_6H_5NH_2$ [*62-53-3*]

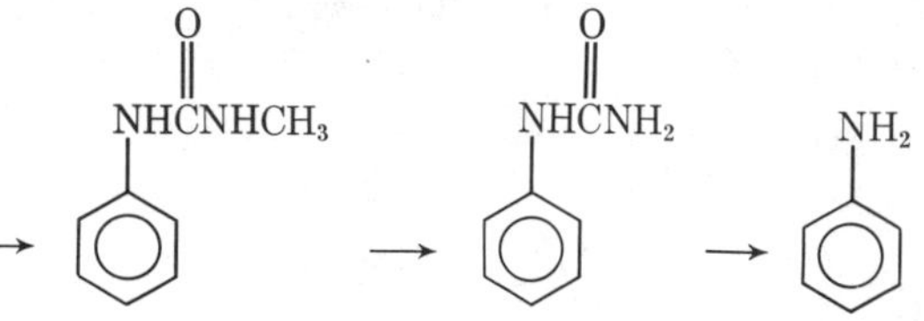

All of the corresponding products in the above pathway have been detected from the herbicides monuron, diuron, and chloroxuron. *N*-Dealkylation is a reaction common to many pesticides and is apparently mediated by mixed-function oxidase systems requiring molecular oxygen.

Urea herbicides readily undergo photolysis. Metobromuron (**47**), a representative urea herbicide, decomposes in aqueous solution on exposure to sunlight (20). The main product is 3-(*p*-hydroxyphenyl)-1-methoxy-1-methylurea (**48**). Present in lesser amounts are 3-(*p*-bromophenyl)-1-methylurea (**49**) and *p*-bromophenylurea (**50**). Substituted diphenyl derivatives may also form.

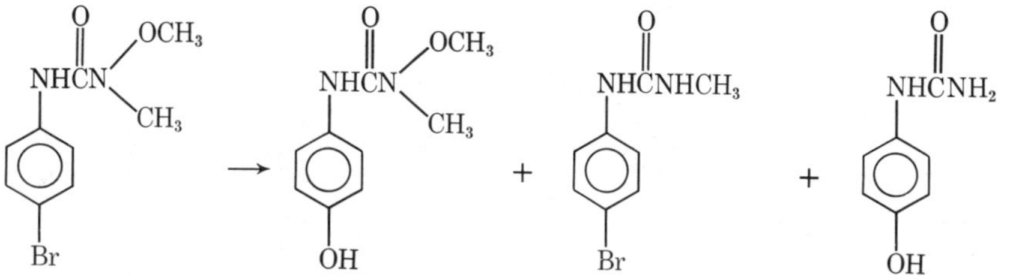

$p\text{-}BrC_6H_4NHC(O)N(OCH_3)CH_3$ (**47**) → $p\text{-}HOC_6H_4NHC(O)N(OCH_3)CH_3$ (**48**) [*20680-06-2*] + $p\text{-}BrC_6H_4NHC(O)NHCH_3$ (**49**) [*20680-07-3*] + $p\text{-}HOC_6H_4NHC(O)NH_2$ (**50**) [*1967-25-5*]

Chloroxuron decomposes on irradiation and degrades by sequential loss of *N*-methyl groups. Monuron, neburon, diuron, linuron, and fenuron are also degraded by sunlight.

Phenylcarbamates. The phenylcarbamate herbicides differ considerably from the methylcarbamate insecticides in soils. The herbicidal carbamates are lost from soils by volatilization and metabolism by soil microorganisms. This group has as its primary structure a phenylcarbamic acid esterified with various alcohols (**51**).

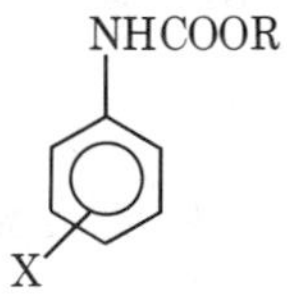

$X\text{-}C_6H_4NHCOOR$

(**51**)

Microorganisms, eg, *Pseudomonas striata*, *Flavobacterium*, *Azobacterium*, and

Achromobacter, metabolize these herbicides to anilines, alcohols, and carbon dioxide, as shown with chlorpropham (**52**).

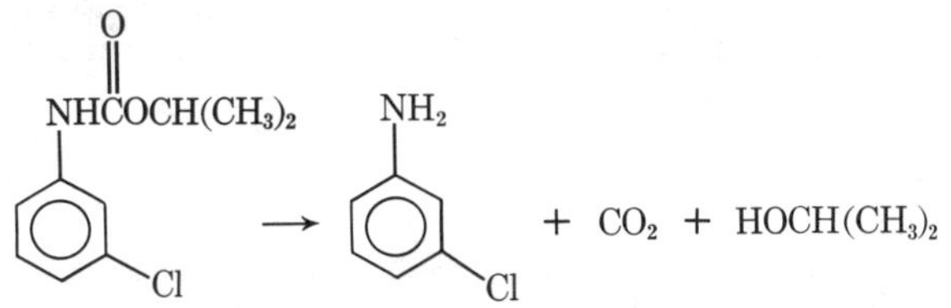

$$NHCOCH(CH_3)_2 \text{ (3-chlorophenyl)} \rightarrow NH_2 \text{ (3-chlorophenyl)} + CO_2 + HOCH(CH_3)_2$$

(**52**) [*108-42-9*]

The acylanilides are also aniline-based herbicides closely resembling the phenylcarbamates. An important member of this group is propanil (**53**), which is the principal herbicide used for weed control in rice. Soil microorganisms cleave propanil to yield 3,4-dichloroaniline and propionic acid. The dichloroaniline undergoes condensation to form 3,3′,4,4′-tetrachloroazobenzene (**54**) (TCAB). 3,3′,4,4′-Tetrachloroazobenzene and azobenzenes arising from several other soil-incorporated chlorinated anilines are likely to form when precursor concentration is high.

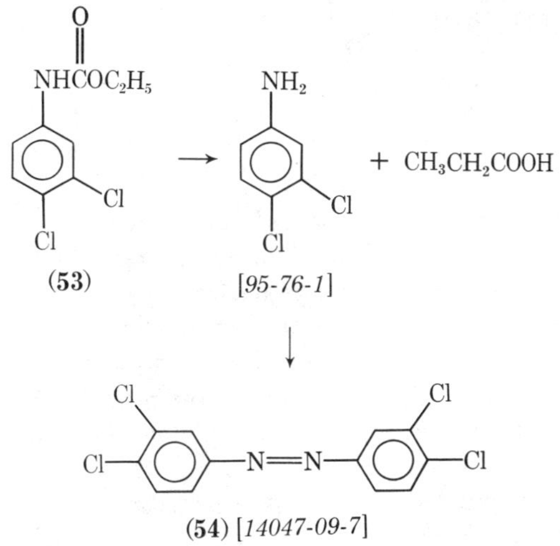

$$NHCOC_2H_5 \text{ (3,4-dichlorophenyl)} \rightarrow NH_2 \text{ (3,4-dichlorophenyl)} + CH_3CH_2COOH$$

(**53**) [*95-76-1*]

$$\downarrow$$

$$Cl_2C_6H_3{-}N{=}N{-}C_6H_3Cl_2$$

(**54**) [*14047-09-7*]

The microbial decomposition of aniline derivatives in soil has been the subject of intensive investigation. The initial formation of 4-chlorophenylhydroxylamine [*823-86-9*] when 4-chloroaniline [*106-47-8*] is incubated with cultures of *Fusarium oxysporum* suggests that such compounds are intermediates in the formation of azoxybenzenes and azobenzenes in soil. Hydroxylation of the aniline at the ortho position is also accompanied by acylation of hydroxyl or amine functions. This reaction is interesting as a route of microbial transformation, and anilines are transformed to acetyl and formyl derivatives in soil (21–22). The isolation of a triazene from soils after incubation with propanil suggests that the aniline-derived diazonium ions may be implicated in the formation of such condensed nitrogen-containing compounds (23); this pathway has been investigated (24).

The carbamate herbicide propham (**55**) photodecomposes by at least two pathways. Reaction of aniline (**56**) and phenyl isocyanate (**57**) yields *sym*-diphenylurea (**58**).

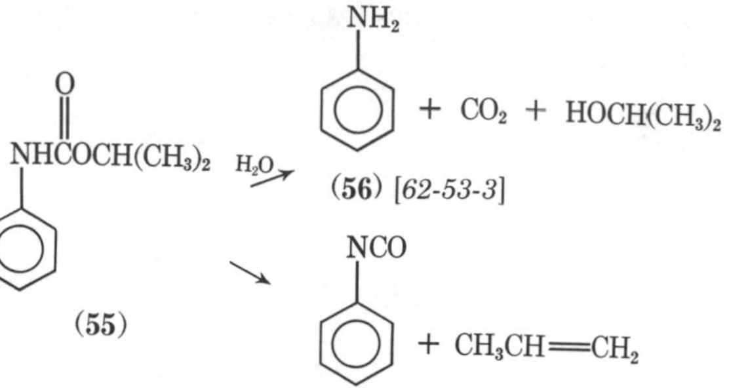

$$C_6H_5NHC(=O)OCH(CH_3)_2 \xrightarrow{H_2O} C_6H_5NH_2 + CO_2 + HOCH(CH_3)_2$$

(55) → (56) [62-53-3]

$$C_6H_5NHC(=O)OCH(CH_3)_2 \rightarrow C_6H_5NCO + CH_3CH{=}CH_2$$

(55) → (57) [103-71-9]

$$(56) + (57) \rightarrow C_6H_5NHC(=O)NHC_6H_5$$

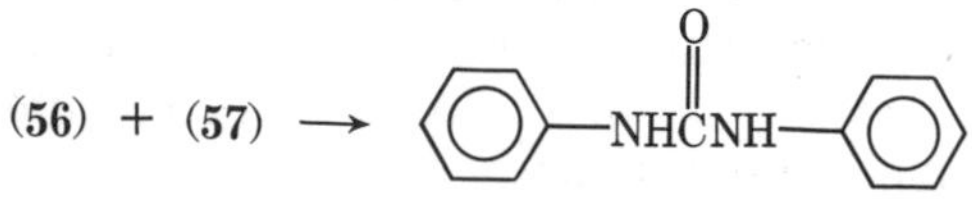

(58) [102-07-8]

Chlorinated Alkyl and Aryl Acids. Two chlorinated aliphatic acids, ie, TCA (**59**) and dalapon (**60**), are used extensively as herbicides. Microorganisms rapidly me-

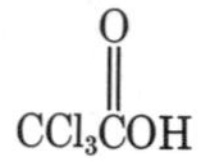

$$CCl_3C(=O)OH \quad (59) \qquad CH_3CCl_2C(=O)OH \quad (60)$$

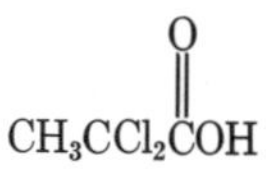

tabolize most members of this family of herbicides by replacing the halogen to form the corresponding hydroxy or keto acids. For example, dalapon yields pyruvic acid by enzymic hydrolysis in a system isolated from an *Arthrobacter* sp. Little is known about the metabolism of the chlorinated aromatic acids. Replacement of chlorine by the hydroxyl group commonly occurs when an aqueous solution of a chlorinated aromatic acid is exposed to sunlight. Reductive dehalogenation may also take place when chlorine is replaced by hydrogen; however, this reaction is more favored in hydrogen-donor solvents, eg, cyclohexane.

Loss of chlorine from the ring is preferred in certain orientations: Chloramben loses Cl preferentially from the 2 position. A chlorine atom that is substituted meta to a carboxylic acid group is more resistant to photolysis than an ortho or para chlorine substituent. In addition, chlorinated phenylacetic acids also undergo reaction in the side chain and the probable sequence of reactions is indicated with fenac (**61**).

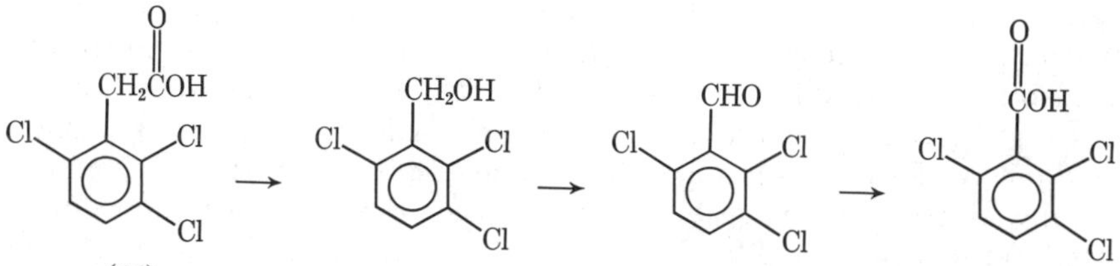

$$Cl_3C_6H_2CH_2C(=O)OH \rightarrow Cl_3C_6H_2CH_2OH \rightarrow Cl_3C_6H_2CHO \rightarrow Cl_3C_6H_2C(=O)OH$$

(61) → → [4659-47-6] → [50-31-7]

Amitrole. Amitrole has been used for a variety of industrial and agricultural weed-control applications. Cranberries were withdrawn from the market in 1959 when it was determined that they contained amitrole residues. In 1971, all registered uses of the herbicide on food were canceled.

Several adducts have been identified as plant metabolites. Amitrole rapidly disappears from soils. Transformations of amitrole (**62**) in soil are governed by the activity of microorganisms, adsorption on soil colloids, and chemical transformations. Free radicals present in soil organic matter probably play an important role in amitrole degradation (25). A number of model free-radical systems yield the following products:

$$(62) \longrightarrow NH_2CNH_2 \,(C{=}O) + NH_2CN$$

(**62**) → $NH_2C(O)NH_2$ [57-13-6] + NH_2CN [420-04-2]

Similar systems are probably responsible for amitrole degradation in soils.

Amitrole degradation by uv irradiation affords the same products that are obtained by riboflavin-sensitized photolysis at the wavelengths of visible light. Urea and cyanamide can be obtained from the reaction mixture; the photolysis parallels the reaction of amitrole with hydroxyl radicals generated by Fenton's reagent (26).

Dinitroanilines. Trifluralin (**63**), nitralin, and bensulide are important herbicides used in cotton and soybean cultivation. Photodecomposition, volatilization, and soil microorganisms contribute to the disappearance of these herbicides from their areas of application.

(**63**) [$N(C_3H_7)_2$, O_2N, NO_2, CF_3] → [2077-99-8] [HNC_3H_7, O_2N, NO_2, CF_3] → [445-66-9] [NH_2, O_2N, NO_2, CF_3]

Degradation proceeds by sequential removal of the *N*-alkyl groups, as shown with trifluralin. Under anaerobic field conditions, eg, in flooded soils, the nitro group in trifluralin is reduced to an amino group. Combinations of these two pathways occur in certain soils.

(**63**) → [2078-04-8] [$N(C_3H_7)_2$, O_2N, NH_2, CF_3]

Dipyridyls. The organic cations, paraquat (**64**) and the closely related diquat (**65**), are desiccants and contact herbicides. In soils they are tightly adsorbed to soil particles and are inactivated almost immediately. In soil enrichment cultures, however, bacteria have been isolated that carry out the following series of reactions:

$[CH_3{-}\overset{+}{N}C_5H_4{-}C_5H_4\overset{+}{N}{-}CH_3]\ 2\,Cl^- \longrightarrow$ (**64**) [1910-42-5]

$CH_3{-}NC_5H_4{-}C_5H_4N^+Cl^- \longrightarrow CH_3{-}\overset{+}{N}C_5H_4{-}COO^-$

[4397-88-0] [824-77-1]

Similar degradation products of paraquat have been identified in photochemical studies. Diquat decomposes in sunlight and affords a tetrahydropyridopyrazidinium salt (**66**) as a principal product.

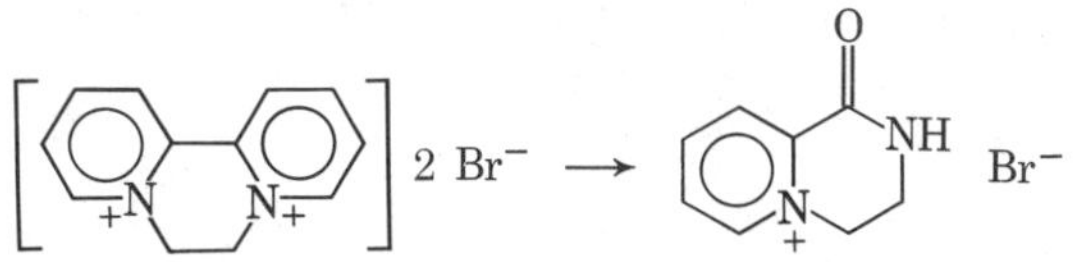

$$[\text{diquat}]\ 2\ Br^- \longrightarrow \text{(66)}\ Br^-$$

(**65**) (**66**)

Arsenical Pesticides

Inorganic arsenicals, eg, arsenic trioxide, As_2O_3; arsenic pentoxide, As_2O_5; sodium arsenate, Na_2HAsO_4; sodium arsenite, nominally Na_3AsO_3; lead arsenate, $Pb_3(AsO_4)_2$; calcium arsenate, $Ca_3(AsO_4)_2$; and Paris green, $3Cu(AsO_2)_2.Cu(C_2H_3O_2)_2$, have been used in agriculture for many years as insecticides, herbicides, soil sterilants, and silvicides. Arsenic is ubiquitous in nature and occurs naturally in soils; the average content in soils collected from various countries is ca 5 ppm. Quantities ranging from a trace to 14 ppm have been reported for most soils.

The extensive use of arsenic pesticides in early pest-control programs has increased the residual content in certain soils. For example, arsenic residues of 440 ppm have been reported in orchard soils with a history of lead arsenate applications. Beans exhibit toxic symptoms when grown on these soils (27). A more recent survey reports residues of up to 2500 ppm in soils treated with lead arsenate in the 1930s.

The inorganic arsenicals have largely been replaced by the less toxic organic arsenicals. Two important organic arsenicals in use are DSMA, $CH_3AsO(ONa)_2$, and cacodylic acid, $(CH_3)_2AsO(OH)$. They are considerably less toxic than the inorganic arsenicals; the respective LD_{50}s are 1800 and 830 mg/kg. DSMA is particularly effective against the perennial weed Johnsongrass and affords one of the most economical methods of control in areas of heavy infestation. Cacodylic acid is a contact herbicide which defoliates or desiccates a large variety of plants.

Rice is particularly sensitive to arsenic residues in soil (28). One theory to explain the greater sensitivity of rice to arsenic residues maintains that, under waterlogged conditions, soils are in a highly reduced condition and arsenate is reduced to the more phytotoxic arsenite. Another theory is based on the reduction of arsenic to arsine by certain soil fungi. The alkylarsines are foul-smelling gases and are highly toxic to mammals. Trimethylarsine [*593-88-4*], $(CH_3)_3As$, is produced by certain strains of molds in the presence of arsenic trioxide or sodium cacodylate (29). It has been suggested that arsenic is lost from soils as an alkyl arsine.

Arsenate residues in soils exist in the following five forms: water-soluble form, and iron, aluminum, calcium, and nonextractable arsenate salts. The amount of any one form varies, depending on the amount of cation present. The water-soluble arsenate is the most toxic to plants. The iron and aluminum salts are extremely insoluble and consequently are relatively less phytotoxic.

Fungicides

Fungicides are used at low rates as seed protectants and disinfectants on most crop areas. On smaller areas, however, fairly high rates are used to control diseases on citrus and deciduous fruits and to a lesser extent on other fruits and vegetables. Most fungicides can be conveniently placed in the following classes of compounds: sulfur, copper, mercury, carbamate, imidazoline, quinone, and guanidine. Unfortunately, there is little published literature on the fate of many organic fungicides in soils. However, there are several excellent references on fungicides and their properties in soils (30–32).

Sulfur and Copper. The inorganic fungicides, including sulfur and Bordeaux mixture, $CuSO_4 + Cu(OH)_2$, are among the oldest known pesticides. Elemental sulfur was first recommended for the control of powdery mildews in 1835 (30). In soils, elemental sulfur is oxidized slowly by chemical means. However, oxidation is rapid when initiated by microbes, eg, chemoautotrophic bacteria of the genus *Thiobacillus*, heterotrophic bacteria, fungi, and actinomycetes. The product, sulfate ion, is the dominant form absorbed by higher plants.

Copper, like arsenic, is a natural component in soils and ranges in concentration from 1 to 50 ppm (33). Organic soils tend to fix copper, and hence it does not move from its site of application. In France from 1920 to 1980, Bordeaux mixture has been applied annually to vineyard soils at a rate of ca 1120 kg/ha (2770 lb/acre) of copper sulfate without phytotoxic effects. In the United States, 36 kg/ha (80 lb/acre) of copper sulfate added over 32 yr to some Long Island potato soils caused no plant injuries.

Mercury. Mercurous chloride, Hg_2Cl_2, and the organic mercurials, including methyl- [*115-09-3*], ethyl- [*107-27-7*], and phenylmercury [*100-56-1*] derivatives, are used as seed disinfectants. The action of mercurous and mercuric chloride applied to soils results from the release of mercury vapor. Soil factors that increase the reduction of mercury salts to metallic mercury include greater soil moisture and organic matter contents, pH, and temperature.

Metallic mercury and trace amounts of phenylmercuric acetate (PMA) were detected in the air surrounding soil following PMA treatment (34). Ethylmercury acetate treatment affords approximately equal amounts of mercury vapor and the vapor of the ethylmercury compound. Methylmercury compounds give methylmercury vapor with a trace quantity of mercury vapor. After 30–50 d, a large portion of soil-applied organic mercurial compound is unaltered.

Thiocarbamates. Most commercially available thiocarbamate fungicides are derivatives of dithiocarbamic acid NH_2CS_2H. These derivatives may be classified into the following three groups: thiuram disulfides, $R_2NCS_2CS_2NR_2$; metallic dithiocarbamates, R_2NCS_2M (M = metal); and ethylenebis(dithiocarbamates), $(CH_2NHCS_2)_2M$. Nonvolatile fungicides nabam (**67**) and ferbam (**68**) are inactivated nonbiologically in soils, whereas semesan (**69**) is metabolized by soil microorganisms. Under similar conditions, captan (**70**) is recovered unchanged after 150 d (35). Nabam in soils produces volatile fungitoxic materials, eg, carbon disulfide [*75-15-0*], ethylene diisothiocyanate [*3688-08-2*], $(CH_2N{=}C{=}S)_2$, and carbonyl sulfide [*463-58-1*], COS. In addition, ethylenediamine [*107-15-3*], ethylene(thiocarbamoyl) disulfide [*3082-38-0*], and ethylenebis(thiocarbamoyl) monosulfide [*5782-83-2*] have been reported as degradation products of ethylenebis(dithiocarbamate) fungicides (36).

NaSC(=S)NH–CH₂CH₂–NHC(=S)SNa
(**67**)

$[(CH_3)_2NC(=S)S]_3$ Fe
(**68**)

HO–C₆H₃(Cl)–OHgOH
(**69**)

NSCCl₃
(**70**)

$CH_3NHC(=S)SNa$
(**71**)

Vapam (71) decomposes in soil to methyl isothiocyanate and N,N'-dimethylthiourea. Decomposition is apparently a chemical process favored by increased soil temperature and decreased moisture content. At 40°C and 6 wt % moisture, vapam degradation occurs after 1.5–2.5 h (37).

Processes Affecting Pesticides

The form, quantity, and location of pesticide residues within various soil compartments are influenced by a variety of physical, chemical, and biological processes. The relative importance of each process is affected by the unique nature of each chemical, by various soil characteristics, by climate, and by agronomic considerations, eg, the method of pesticide application and past history of pesticide usage on the land. Thus, persistence of a chemical represents a complex interaction. Since persistence affects not only a pesticide's efficacy but also its potential as an environmental problem, knowledge of the processes affecting pesticides in soils is important.

Adsorption. Adsorption refers to the soil surface retention or in some cases exclusion of pesticide molecules or ions. Adsorption differs from absorption, ie, the penetration of pesticides into the soil matrix; adsorption implies a binding mechanism. In practice, the processes are not distinguished during measurement, and some authors prefer to describe loss of pesticide from solution phases as sorption (38).

Adsorption is listed first among environmental processes affecting pesticides because it influences pesticide movement, photodecomposition, microbial or chemical decomposition, and plant uptake (see Adsorptive separation). By restricting leaching, strong adsorption diminishes the possibility of groundwater contamination but may increase the opportunity for pesticide loss by photodecomposition or volatilization if the chemical was originally surface-applied. Adsorption reduces pesticide concentration in the soil solution and probably limits the rate of microbial degradation and plant uptake. A number of reviews describing pesticide adsorption are available (38–44).

Pesticide adsorption is usually presented in the form of adsorption isotherms, ie, graphs relating equilibrium solution concentration to the amount of solute sorbed per unit mass of adsorbent. The Langmuir equation was an early theoretical treatment of gas–solid adsorption but only occasionally seems to be used for pesticides in aqueous soil systems. One of its assumptions, ie, that the energy of adsorption is constant and independent of the extent of surface coverage, appears particularly tenuous in light of the surface heterogeneity of soil. The Langmuir relationship successfully describes sorption of many inorganic ions in soil (38); modification to accommodate shifts in adsorption energy as adsorbate sites become occupied has led to several variations in the basic isotherm equation. When the heat of adsorption is assumed to be a logarithmic function of the surface coverage, the following isotherm occurs:

$$\ln S = k + N \ln C \tag{1}$$

where S is the amount of adsorbed solute, C is the solute concentration in the soil solution, and k and N are constants.

Equation 1 is conceptually identical to the empirically derived Freundlich equation (eq. 2), which is also described as equation 3.

$$S = KC^N \tag{2}$$

$$\log x/m = \log K + 1/n \log C \tag{3}$$

Equation 3, in which x/m represents the amount of pesticide adsorbed per unit mass of soil and K is the Freundlich equilibrium constant, has been very useful in evaluating relative adsorptivity among pesticides and soils. The exponential constant N or $1/n$ is often somewhat less than unity, implying a nonlinear sorption relationship. When this constant is 1, equation 2 reduces to a simple distribution equation $S = K_dC$. It has been indicated that such an equation may be valid for neutral organic molecules when their aqueous phase concentration is $<10^{-5}\,M$ or <50% of their water solubility (45).

Adsorption isotherms have not been too successful in the characterization of adsorption mechanisms. Solute adsorption has been classified into four types; the types and their variations are distinguished by isotherm shape (46). These may reflect some of the solute–solvent–adsorbent interactions. Thermodynamic calculations based on isotherms have had limited value in assessing adsorption mechanisms (see Adsorptive separation). Infrared spectroscopy is useful in some studies of pesticide–clay bonding, as is x-ray diffraction to measure layer silicate expansion following sorption of organics. Among mechanisms identified or postulated to account for pesticide–soil adsorption are weak van der Waals forces, hydrogen bonding, ligand exchange, charge transfer, entropy generation or hydrophobic bonding, strong ion exchange (qv), and strong chemisorption. The energy of adsorption is the sum of several such mechanisms.

The amount of soil organic matter and the amount and type of clay (in that order) are usually well-correlated with the amount of pesticide sorption to soil; secondary properties, eg, surface area and cation-exchange capacity, are also related to sorption. Adsorption among various soils is often compared after dividing K or K_d by the organic-carbon content in the soil. Although this normally reduces soil-to-soil variability, interpretations should be made cautiously, since changes in other factors, eg, soil pH, may affect individual chemicals in different ways. For example, organic acids become primarily neutral species and neutral amines become cations when pH is less than the pK; both groups are adsorbed more strongly at lower pH values.

The crystalline-layer silicate clays, eg, montmorillonite and vermiculite have high cation-exchange capacities (80–150 meq/100 g) and surface areas (600–800 m^2/g) and are especially associated with strong bonding of organic cationic pesticides. The nature of the exchangeable metallic cations bonded to clays may significantly modify pesticide adsorption (40). Other soil clays further contribute to adsorption as do amorphous oxides of iron and aluminum. This latter group is characterized by high surface area and development of a pH-dependent positive charge; both factors contribute to adsorption of organic acids.

Characteristic functional groups of pesticides that seem to be associated with enhanced adsorption include: R_3N^+, $CONH_2$, OH, NHCOR, NH_2, OCOR, and NHR (42). Protonation of any functional group increases adsorption, as does increasing molecular size.

Leaching. Leaching refers to the transport of solutes within soil, either through diffusion or by mass transport with percolating water. The subject has been reviewed and is distinguished from surface transport by runoff, which is a main contributor to pesticide residues in surface waters (42,47–52).

Although movement of most pesticides is probably restricted to <1 m within soil, residues in wellwater appear occasionally. In cases where a chemical is both persistent and inherently mobile, the chances of some movement to groundwater are increased. That situation is exemplified by the nematicide DBCP. Some leaching may be desirable in order for the pesticide to reach target organisms or to protect the chemical against surface-loss mechanisms. Knowledge of chemical and soil effects on pesticide transport is important not only for agronomic reasons, but also because it offers some basis on which to predict movement of hazardous wastes from land disposal sites. Three problems with such predictions, however, are the higher concentration, deeper initial distribution in soil, and exposure to substantially different soil solution chemistry as compared to typical pesticide regimes.

The diffusion process has been reviewed at length (50,53). Diffusion of nonvolatile pesticides in water, though only short-range, contributes to pesticide effectiveness. Diffusion coefficients usually increase as soil moisture content increases, the exceptions being for volatile chemicals which diffuse primarily in the vapor phase. Other factors associated with greater diffusion rates are higher temperature and lower soil bulk density. Diffusion is reduced by adsorption.

In a general transport equation, eg, equation 4

$$D\frac{d^2C}{dz^2} - \nu_o\frac{dC}{dz} - \left[1 + \frac{\rho K_d}{\theta}\right]\frac{dC}{dt} = 0 \qquad (4)$$

which describes movement of a nonreactive solute through a uniform porous medium under steady-state conditions, the mass transfer and diffusion components are combined as D the apparent dispersion coefficient. Other terms in equation 4 are solute concentration C; soil depth z; average pore-water velocity ν_o; soil bulk density ρ; distribution coefficient K_d; volumetric soil water content θ; and time t. As with movement of other solutes in porous media, pesticides moving into soil become redistributed, usually as a broad band. The concentration profile is strongly dependent on the pattern of water movement with time. One practical aspect of this is that, when surface drying occurs, soil water may move upward and bring some leached pesticide near to the surface. Soil characteristics causing reduced hydraulic conductivity, eg, high clay content or subsurface hardpan layers, also restrict pesticide movement. When the soil profile is discontinuous, water and pesticide percolation are likely to be "fingers" at points where the soil restriction to movement is less. This has obvious implications for pesticide-residue sampling.

Adsorption of pesticides has been correlated with restricted mobility. Among soil properties, organic-matter content, clay content, field moisture capacity, surface area, and cation-exchange capacity have all been negatively correlated with pesticide leachability. The last three properties are closely correlated with organic matter and clay contents, which may partially explain their predictive value for pesticide leaching. Organic matter is the most universally significant soil component with respect to pesticide retardation, but clay is more important for organic cations and perhaps for chemicals that may coordinate with mineral cations. Soil pH becomes important when pesticide dissociation occurs within the environmentally relevant pH range of ca 4–8.

The chemical characteristics that affect adsorptive behavior are also those related to inherent pesticide mobility. Organic cations are strongly sorbed despite very high water solubility; at the other extreme, very poorly soluble pesticides are also immobile. Among the more mobile chemicals are organic anions, eg, acids in soils for which $pH > pK_a$. Degradation is an important attenuating factor. Thus, for example, dalapon is highly mobile in laboratory leaching tests but is not a problem in actual use because it degrades rapidly. Glyphosate would likely be mobile except that it apparently complexes through the phosphonate group with Fe or Al in the soil (54). The relative mobilities of 82 pesticides have been compared (45).

Leaching is usually evaluated in the laboratory by soil column or soil thin-layer chromatographic methods. These studies may be scaled-up to the lysimeter stage and ultimately to field studies. Methods for studying leaching have been reviewed (47–49).

Photodecomposition. Decomposition of pesticides by light occurs not only through the direct absorption of light by the molecule that is degraded, but also by reaction with products of a photochemical process. Thus, reaction may occur as a consequence of light absorption by a molecule which subsequently transfers energy to a second molecular species that undergoes reaction, ie, photosensitization. Alternatively, reactive species, eg, hydroxyl radicals or singlet molecular oxygen, generated in a photochemical process may be responsible for degradation of pesticide molecules. Many pesticides absorb at 200–350 nm. Photochemical processes require the absorption of light; direct absorption of light by pesticides occurs if the absorption spectrum overlaps that of solar radiation. Although solar radiation is a poor source of uv light (the so called uv-B region of sunlight covers wavelengths of 280–320 nm) and only a very small amount of radiation of wavelengths <295 nm falls on the earth's surface, this region is important because such radiation is sufficiently energetic to dissociate chemical bonds or affect biological processes.

Photosensitized reactions may accelerate decomposition of some pesticides. Photooxidation may involve intermediates, eg, hydroxyl radicals, ozone, single molecular oxygen, or other active species. The influence of physical state is also important in determining the course of a photochemical reaction. Absorption of light may occur on soil or on leaf surfaces but not in the bulk of the soil medium. Molecules at these surfaces may be in the adsorbed, crystalline, or solution state. The physical state may influence the absorption spectrum of the compound and thus the rate and products of reaction. For example, aldrin (**5**) forms photodieldrin (**72**) and dieldrin (**6**) on photolysis in the solid state. Photolysis of dieldrin in an organic solvent brings about reductive dechlorination to compound (**73**).

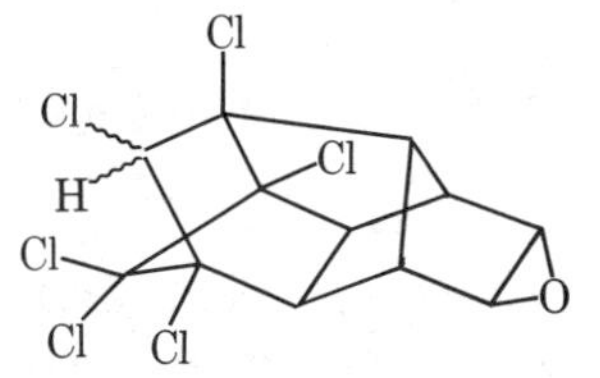

(**72**)

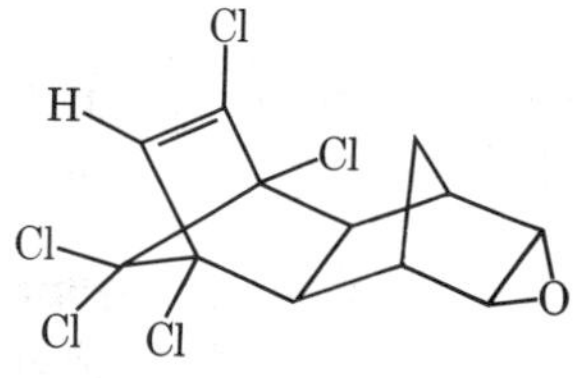

(**73**) [*18417-21-5*]

Studies of the photochemistry of pesticides on silica reveal that the acidic nature of silica changes the adsorption spectrum and it is therefore difficult to extrapolate results of experiments that are not conducted in the soil environment. The effect of soil and silica environments on photolysis has been compared, and the latter may favor photodecomposition of adsorbed materials because it allows passage of uv energy to a much greater extent than soil. Thus soil exerts a protective effect; photolysis of trifluralin and 2,3,7,8-tetrachlorodibenzo-*p*-dioxin [*1746-01-6*] (**74**) occurs to a lesser extent on soil than on silica (55).

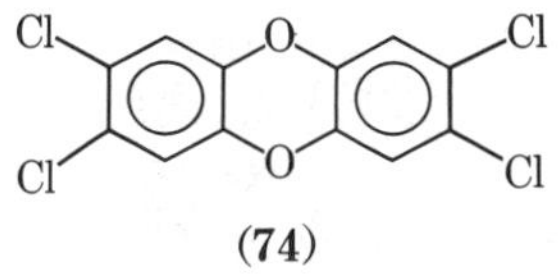

(**74**)

The complexity of the natural environment has restricted photochemical studies primarily to the investigation of the products of photolysis in solution or as solid films. Often the study of photodecomposition has been undertaken to determine the loss of biological activity following irradiation. The rate of bioactivity loss reflects the rate of photodecomposition but affords little other chemical information.

Many organic pesticides are complex molecules and contain a number of functional groups. Photochemical reactions of individual groups are often predictable by reference to previously investigated models, since even in a more complex molecular environment similar reactions may occur. However, the relative reactivity of functional groups becomes extremely difficult to predict. The replacement of a halogen atom by hydroxyl or occasionally by hydrogen in aqueous solution or by hydrogen in a hydrogen-donor solvent, eg, methanol, is frequently observed when halogen atoms are bonded to an aromatic ring system. Photoreactivity of halogen atoms attached to an isolated double bond is observed in the cyclodiene insecticides. New carbon–carbon bonds are formed at the site of departure of the chlorine atom, or reduction may occur in a hydrogen-donor solvent at greater dilution.

Photooxidation usually transforms amines and phenols into dark materials of high molecular weight. In aqueous solution, halogenated phenoxyaliphatic acids generate polyhydroxyphenols photolytically by ether fission and replacement of halogen by hydroxyl. Since polyhydroxyphenols are extremely susceptible to oxidation at low pH, the resultant phenols readily polymerize and can only be isolated if their further oxidation is inhibited. Polyphenols and many other classes of compounds generated by photolysis are very reactive and can therefore be isolated from the reaction mixture only with difficulty. Another general reaction is the removal of alkyl or alkoxy groups attached to nitrogen atoms.

Several texts on organic photochemistry are available and provide an excellent foundation for understanding the photochemistry of pesticides (56–57). However, predictions based on extrapolation from the photochemical behavior of a model system are of limited value until broader knowledge of the effect of many complicating factors can be accumulated.

Metabolism and Chemical Degradation. Microbiological and chemical reactions are important processes affecting pesticide persistence in soils. Enzyme systems in bacteria, actinomycetes, and fungi can cleave chemical bonds common to many pesticide molecules. The principal reactions associated with pesticide decomposition by soil microorganisms include dehalogenation, dealkylation, amide or ester hydrolysis, oxidation, reduction, ether fission, aromatic ring hydroxylation, and ring cleavage (58–60). The first five metabolic reactions are of particular importance.

Dehalogenation reactions lead to replacement of the halogen, usually Cl, by —OH under aerobic conditions and by —H under anaerobic conditions. Studies in soils and pure culture solutions show that the number, position, and type of halogen substitution affect the rate of decomposition of several pesticides. Important examples of this class of reaction include the dehydrohalogenation of lindane to γ-pentachlorocyclohexene and the reductive dehalogenation of DDT to DDD by *Aerobacter aerogenes*.

Dealkylation reactions lead to cleavage of the *N*-alkyl and *O*-alkyl bond. *N*-alkyl groups are common in herbicides, including secondary amines such as dialkyl-*s*-triazines, tertiary amines such as *N*,*N*-dialkylphenylureas, and dinitroanilines, and in quaternary ammonium salts such as paraquat. Sequential removal in soils of alkyl groups in phenylureas and dinitrotoluidines proceeds by means of the *N*-oxide and requires a NADPH(reduced nicotinamide adenine dinucleotide phosphate)-generating system. Methyl and ethyl ethers of several pesticides are split into phenols and aldehydes.

Amide and ester hydrolyses of pesticides have been most widely studied with phenylamide herbicides of the general structure C_6H_5NHCOR, including the phenylcarbamates and acylamides. Hydrolysis proceeds by cleavage of the amide or ester linkage to yield aniline, carbon dioxide, and alcohol from the phenylcarbamates and aniline and aliphatic acid from the acylanilides. *Pseudomonas striata* can catalyze the ester hydrolysis of chlorpropham; the responsible enzyme has been partially characterized. Rapid hydrolysis of the ester linkage in the synthetic pyrethroids, phenothrin, permethrin, cypermethrin, decamethrin, and fenvalerate is a main microbial route in soils.

Oxidation is an important reaction for many of the chlorinated insecticides containing an isolated double bond and leads to formation of the corresponding epoxides. Important examples include the microbial conversion of aldrin to dieldrin and heptachlor to heptachlor epoxide. Reductive reactions usually involve pesticides with a nitro substituent undergoing reduction to an amine. One of the most important reactions in this series involves the microbial conversion of parathion to aminoparathion.

The genetic basis for pesticide metabolism by soil microorganisms has been identified in small chromosomes, termed plasmids, some of which can be transferred from one bacterial cell to another. Plasmid involvement in the degradation of camphor was first reported in 1971 (61). Subsequent investigations revealed plasmids that confer the ability to degrade naturally occurring aromatic and aliphatic compounds, eg, octane, naphthalene, salicylates, and toluene (62). The first indication that plasmids were involved in the environmental degradation of a pesticide came from a report that strains of *Alcaligenes paradoxus*, a ubiquitous saprophytic soil bacterium, has a transmissible plasmid that confers the ability to degrade 2,4-D (63). The property can be transferred from one bacterial population to another. Work is in progress on the isolation and identification of other pesticide-degrading plasmids. It may be possible to construct bacterial strains through genetic engineering techniques with an even wider range of metabolic activities for recalcitrant pesticides (see Genetic engineering).

Purely chemical reactions are probably less well understood than the biochemical reactions mediated by soil microorganisms. One difficulty in distinguishing chemical from biochemical reactions has been obtaining a sterile soil system that has not undergone extensive physical and chemical alteration by autoclaving or other methods of sterilization. Despite these difficulties, several important pesticides are known to be converted by chemical hydrolysis. Examples include the solvolysis of metabolites of pyrethroid insecticides, hydrolyses of such phosphate insecticides as diazinon, malathion, dimetan, and mevinphos, and hydrolyses of the chloro-*s*-triazine herbicides, eg, atrazine, to their corresponding 2-hydroxy derivatives. Free-radical reactions leading to ring cleavage of amitrole occur in nondestructively sterilized soils.

Plant Uptake. Plant absorption and translocation of residual pesticides used in a current or previous cropping system has been a subject of considerable study. It represents one possible avenue of direct human exposure to pesticides in the diet. An early problem in determining the degree of risk from such residues, ie, lack of selective and sensitive methods, has largely been overcome by the use of gas–liquid chromatography.

A number of plant, soil, and pesticide properties affect the uptake process (64). Generally the plant variables that affect their residues are concentration of soil residues; plant species; growth pattern, eg, root crops vs aerial plants; oil content; and growth stage. Plant residues are directly related to soil residues and oil content.

The most important chemical property of the pesticide influencing plant contamination is water solubility. The more water-soluble the molecule, the more readily it reaches the root, passes through the epidermis, and is translocated throughout the plant. In contrast, nonpolar molecules tend to adsorb onto root surfaces rather than pass through the epidermis.

The main soil factor influencing plant contamination from root sorption, ie, organic matter content, is inversely correlated with plant sorption, particularly for nonpolar pesticides. This suggests that pesticide adsorption to the soil organic fraction has reduced uptake.

Under actual field conditions, it is impossible to separate contamination by root uptake and translocation from contamination by dust, drift, splashing, or volatilization. Data from ref. 64 clearly show that volatilization of DDT from soil surfaces onto the lower leaves and stems of cotton is a far more important contamination pathway than root uptake and translocation.

Problems

A number of unanticipated problems arose in the 1970s that merit special consideration in the soil chemistry of pesticides. Some of these were caused by an undetected toxicological symptom appearing after the compound had previously been registered, eg, with DBCP. Another area of concern was the detection of toxic impurities in certain pesticidal formulations that eventually appeared in soil. Some of the most important impurities that have received attention include hexachlorobenzene in dacthal, nitrosamines in dinitroanilines and some acid herbicides, ethylenethiourea [*96-45*-7] (ETU) in the ethylenebis(dithiocarbamate) (EBCD) fungicides, and the chlorinated dioxins in certain phenoxyalkanoic acid herbicides.

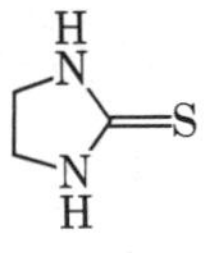

H N N H =S

(75)

$CH_2BrCHBrCH_2Cl$

(76)

Dioxin. Chlorinated dibenzo-*p*-dioxins comprise a large family of chlorinated aromatic hydrocarbons, including about 75 isomers. They have been identified as impurities in and as pyrolysis products of chlorophenols. The main dioxin of concern, ie, 2,3,7,8-tetrachlorodibenzo-*p*-dioxin (TCDD) (**74**), can form during the conversion of 2,4,5-trichlorophenol from 1,2,4,5-tetrachlorobenzene during synthesis of the herbicide 2,4,5-T. TCDD is extremely toxic (LD_{50} = 6 μg/kg in male guinea pig); it is fetotoxic and teratogenic to laboratory animals. Species differences occur, and it is uncertain whether toxic effects elicited in laboratory animals can be directly translated to man. Chloracne (a severe form of acne) is the most common symptom observed in industrial workers exposed to TCDD.

Interest continues in the potential exposure of military personnel to TCDD in Agent Orange, a defoliant used in Vietnam. This herbicide was formulated to contain a 50:50 mixture of the *n*-butyl esters of 2,4-D and 2,4,5-T. Of ca 48,500 t of all herbicides sprayed on South Vietnam from 1962 through February 1971 (65), ca 94 wt % was 2,4-D (53 wt %) and 2,4,5-T (41 wt %).

TCDD is very insoluble in water (0.2 ppb), which governs many of its properties in soils. The dioxin is relatively persistent ($t_{1/2} > 0.5$ yr, depending on initial concentration and climate), does not leach, is not taken up by plants, is rapidly photodecomposed in solution but is not appreciably altered on dry soil surfaces, and is not biosynthesized in soil from 2,4,5-trichlorophenol [*95-95-4*]. In controlled laboratory studies with aquatic ecosystems, TCDD accumulates in amounts of $(2–2.6) \times 10^4$ times the initial water concentration for snails, mosquito, fish, and daphnids and averages 4900 for duckweed, algae, and catfish (66). Because of the low concentrations of TCDD likely to be encountered in the environment, classical monitoring and detection techniques contributed little to our understanding of its accumulation in real world ecosystems. Most 2,4,5-T samples produced in the 1970s contained ≤0.1 ppm TCDD. A 110-mg/m^2 application of 2,4,5-T (typical rates are 30–110 mg/m^2) containing 0.1 ppm TCDD applied to soil would result in a maximum of only 0.1 ppt (parts per trillion) in the top 15 cm of the soil. Detection of concentrations of 10–50 ppt are feasible by rather elaborate cleanup methods followed by the use of specific ion monitoring with high resolution mass spectrometers interfaced with gas chromatography (67).

Recent studies involving highly complex instrumentation have made possible detection of TCDD in certain fish and bird samples collected in the Great Lakes region. The source of these residues is the subject of much current debate, and there is evidence suggesting that certain combustion processes may be the origin of the TCDD contamination (see Trace and residue analysis).

Ethylenethiourea. ETU (**75**) is a manufacturing, processing, and metabolic product of the EBDC fungicides, including nabam, mancozeb, metiram, maneb, and zineb. ETU became an environmental issue when toxicological studies indicated that it may be goiterogenic, tumorogenic, and teratogenic to laboratory animals. ETU residues appeared on certain agricultural commodities sprayed with EBDC fungicides. These residues may result from the presence of ETU in the pesticide formulation or from the subsequent transformation either as a plant metabolite or as a by-product formed during processing of the crop.

In soils ETU is rapidly oxidized to a number of products, including 2,4-imidazolidinedione [*461-72-3*] (hydantoin), 1-(4,5-dihydro-1*H*-imidazol-2-yl)-2-imidazolidinethione [*484-92-4*] (Jaffe's base), and 2-imidazolidinone [*120-93-4*] (ethyleneurea, EU). Soil microorganisms probably play a role in metabolism of ETU, and CO_2 is a principal product from nonsterile soils. Plants produce a number of unknown products in addition to EU and 2-imidazoline [*504-75-6*] from either soil or foliar treatment with the EBDC fungicides. Based primarily on the findings that ETU does not persist or bioaccumulate in plants, soils, or water, much of the initial concern about this product has lessened. Current management practices in agriculture have generally tended to reduce the usage of the EBDC fungicides for prescribed periods before harvest, thus reducing ETU residues in the crop. For an excellent review on the occurrence of ETU as a terminal residue resulting from agricultural use of the EBDC fungicides, see ref. 68.

DBCP. The highly effective nematicide DBCP (**76**) was synthesized following the discovery in the 1940s that a halogenated hydrocarbon mixture substantially improved pineapple growth. For ca 20 yr, DBCP aided growers of citrus, peaches,

Table 2. Estimated Worldwide and United States Retail Sales of Pesticides, 10^6 (U.S. 1980 $)[a]

Pesticide	1980	1982	1985
Noncommunist world			
herbicides	4,891	5,307	6,022
insecticides	3,916	4,228	4,764
fungicides	2,199	2,417	2,772
other[b]	599	654	758
Total	*11,605*	*12,606*	*14,316*
United States			
herbicides	2,171	2,418	2,760
insecticides	908	1,013	1,128
fungicides	226	240	276
other[b]	199	248	319
Total	*3,504*	*3,919*	*4,483*

[a] Ref. 73.
[b] Includes soil fumigants, defoliants and desiccants, plant-growth regulators, pheromones, attractants and viruses, and rodenticides.

grapes, cotton, and numerous other fruit and vegetable crops with no apparent environmental or toxicological problems (see Plant growth substances). In 1977, DBCP was discovered to cause temporary sterility among male production-plant workers and, at about the same time, the chemical was identified as a potential carcinogen (see Bromine compounds). Use of DBCP in California was suspended in 1977, but subsequent field testing as well as years of previous use seemed to demonstrate that DBCP could be used safely.

A survey in May 1979 revealed that 59 of 119 wells tested in California's San Joaquin Valley contained DBCP residues at levels of 0.1–39 ppb and averaging 5 ppb (69). DBCP had been used on these sandy soils from ca 1960 to 1977. Although residues were highest in shallower wells (0.3 ppb), DBCP was reported in two wells that were 180 m deep. As a result, the use of DBCP was suspended throughout the United States, except for Hawaii, in October 1979, and was canceled later (70). Apparently the DBCP that was not initially lost by volatilization was gradually leaching downward. This implies limited adsorption and degradation as well as significant hydraulic recharge.

If leaching to groundwater can be prevented, then it may be possible to restore the use of DBCP with specific geographical and other restrictions. A recent survey of groundwater samples in Florida, Georgia, and South Carolina failed to show the presence of DBCP (71).

Economic Aspects

Although the regulatory actions of EPA changed the pattern of compounds used for pest control, the industry grew substantially during the period 1970–1980 (see Fig. 1). The price of pesticides almost tripled during the 1970s, causing the value of U.S. production to surge 350%. U.S. consumption peaked in 1975, whereas exports nearly doubled and imports increased almost eightfold. The total production of pesticides increased 43%. By comparison, the total production of synthetic organic pesticides in 1945 was 15,400 metric tons; in 1955, 299,000 t; and in 1965, 398,000 t.

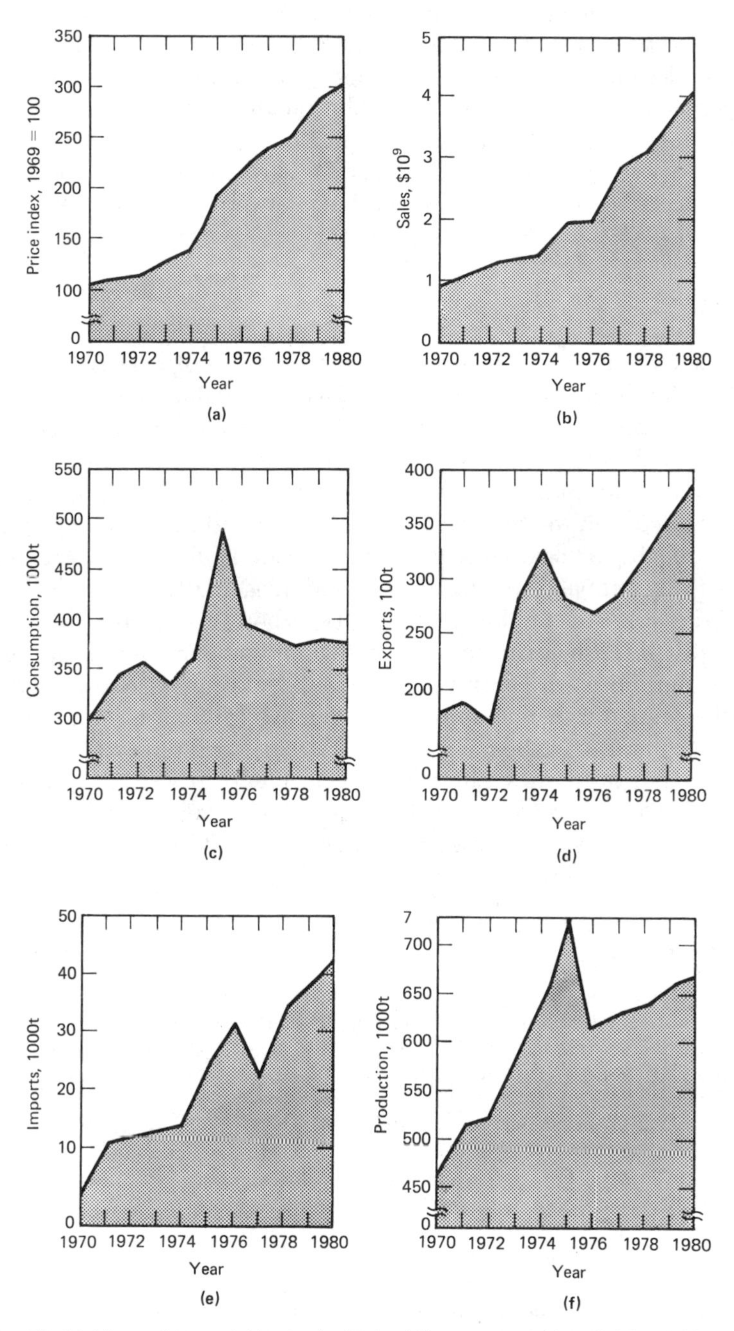

Figure 1. Herbicides and insecticides in the United States, 1970–1980. (**a**) Pesticide prices; (**b**) sales; (**c**) consumption; (**d**) pesticide exports; (**e**) pesticide imports; (**f**) production. Courtesy of the American Chemical Society (72).

A recent survey suggests that the 1980 sales of pesticides worldwide were \$11.5 $\times 10^9$, with projected sales of \$12.6 $\times 10^9$ in 1982 and \$14.3 $\times 10^9$ in 1985 (73). The distribution of sales between various pesticides on a worldwide and United States basis is shown in Table 2. The main markets for herbicides in the United States were (in U.S. 1980 dollars): corn, \$754 $\times 10^6$; soybeans, \$749 $\times 10^6$; cotton, \$142 $\times 10^6$; fruits, vegetables, and horticultural crops, \$116 $\times 10^6$; and wheat, \$97 $\times 10^6$. For insecticides, the main markets are fruits, vegetables, and horticultural crops, \$262 $\times 10^6$; corn, \$242 $\times 10^6$; and cotton, \$212 $\times 10^6$.

Regulatory Actions

The single most important event affecting the use of pesticides in the United States was the establishment of the EPA in 1970 (see Regulatory agencies). The Agency was granted broad administrative power under the Federal Environmental Pesticide Control Act of 1972 to regulate and control pesticides. A goal of the Agency was to remove from the marketplace those pesticides deemed most detrimental to the environment. Every decision to ban a pesticide led to considerable scientific debate, the most famous probably associated with the insecticide DDT. These regulatory actions by EPA drastically changed the pesticides used in agriculture.

Most of the important chlorinated-hydrocarbon insecticides, including aldrin, chlordane, DDD, DDT, dieldrin, endrin, heptachlor, kepone, lindane, mirex, strobane, and toxaphene, have been discontinued, suspended, or otherwise restricted. Generally, the insecticides tended to persist in soils; some moved from their application site and tended to bioaccumulate in the fatty tissue of aquatic and terrestrial organisms. The ubiquitous presence of many such residues was a factor leading to their ban. Other pesticides have been removed because of impurities or because they were uneconomical in light of projected sales. Many metal- or metaloid-containing pesticides, including arsenic, mercury, and thallium, have been restricted.

BIBLIOGRAPHY

"Soil Chemistry" in *ECT* 1st ed., Vol. 12, pp. 614–633, by E. R. Graham, University of Missouri; "Soil Chemistry of Pesticides" in *ECT* 2nd ed., Vol. 18, pp. 515–540, by P. C. Kearney, J. R. Plimmer, and C. S. Helling, U.S. Department of Agriculture.

1. R. G. Nash and E. A. Woolson, *Science* **157,** 924 (1967).
2. U.S. Tariff Commission, *United States Production and Sales of Pesticides and Related Products*, U.S. Government Printing Office, Washington, D.C., 1968.
3. G. W. Ware, B. J. Estesen, N. A. Buck, and W. P. Cahill, *Pestic. Monit. J.* **12,** 1 (1978).
4. F. A. Gunther and D. E. Ott, *Residue Rev.* **10,** 70 (1965).
5. G. Wedemeyer, *Appl. Microbiol.* **16,** 661 (1968).
6. W. N. Yule, M. Chiba, and H. W. Morley, *J. Agric. Food Chem.* **15,** 1000 (1967).
7. F. Matsumura and G. M. Boush, *Science* **153,** 1278 (1966).
8. D. G. Crosby, E. Leitis, and W. L. Winterlin, *J. Agric. Food Chem.* **13,** 204 (1965).
9. L. W. Getzin and I. Rosefield, *J. Agric. Food Chem.* **16,** 598 (1968).
10. M. L. Schechter, N. Green, and F. B. La Forge, *J. Am. Chem. Soc.* **71,** 3165 (1949).
11. M. Elliott, A. W. Farnham, N. F. Janes, P. H. Needham, and D. A. Pulman, *Nature* **244,** 456 (1973).
12. M. Elliott and N. F. Janes, *Chem. Soc. Rev.* **7,** 473 (1978).
13. R. L. Holmstead and J. E. Casida, *Am. Chem. Soc. Symp. Ser.* **42,** 137 (1978).
14. M. V. Bullivant and G. Pattenden, *Pestic. Sci.* **7,** 231 (1976).
15. R. L. Holmstead, D. G. Fullmer, and L. O. Ruzo, *J. Agric. Food. Chem.* **26,** 954 (1978).

16. D. D. Kaufman, S. C. Haynes, E. G. Jordan, and A. J. Kayser, *Am. Chem. Soc. Symp. Ser.* **42,** 147 (1977).
17. T. R. Roberts and M. E. Standen, *Pestic. Sci.* **8,** 305 (1977).
18. P. C. Kearney and C. S. Helling, *Residue Rev.* **25,** 25–44 (1969).
19. R. J. Hance, ed., *Interactions between Herbicides and Soil*, Academic Press, Inc., New York, 1980.
20. J. D. Rosen and R. F. Strusz, *J. Agric. Food Chem.* **16,** 568 (1968).
21. D. D. Kaufman, J. R. Plimmer, and U. I. Klingebiel, *J. Agric. Food Chem.* **21,** 127 (1973).
22. P. C. Kearney and J. R. Plimmer, *J. Agric. Food Chem.* **20,** 584 (1972).
23. J. R. Plimmer, P. C. Kearney, H. Chisaka, J. B. Yount, and U. I. Klingebiel, *J. Agric. Food Chem.* **18,** 859 (1970).
24. C. T. Corke, N. J. Bunce, A.-L. Beaumont, and R. L. Merrick, *J. Agric. Food Chem.* **27,** 644 (1979).
25. D. D. Kaufman, J. R. Plimmer, P. C. Kearney, J. Blake, and F. S. Guardia, *Weeds* **16,** 266 (1968).
26. J. R. Plimmer, P. C. Kearney, D. D. Kaufman, and F. S. Guardia, *J. Agric. Food Chem.* **15,** 996 (1967).
27. R. Dickens and A. E. Hiltbold, *Weeds* **15,** 299 (1967).
28. E. A. Epps and M. B. Sturgis, *Soil Sci. Soc. Am. Proc.* **4,** 215 (1939).
29. F. Challenger, *Chem. Ind.* **54,** 657 (1935).
30. E. G. Sharvelle, *The Nature and Uses of Modern Fungicides*, Burgess Publishing Company, Minneapolis, Minn., 1961.
31. A. L. Morehart and D. F. Crossman, *Del. Exp. Sta. Bull.* **357,** 1965, 26 pp.
32. C. A. I. Goring, *Ann. Rev. Phytopath.* **5,** 285 (1967).
33. F. A. Gilbert, *Adv. Agron.* **40,** 147 (1952).
34. Y. Kimura and V. L. Miller, *J. Agric. Food Chem.* **12,** 253 (1964).
35. D. E. Munnecke, *Phytopathology* **48,** 581 (1958).
36. W. Moje, D. E. Munnecke, and L. T. Richardson, *Nature* **202,** 831 (1964).
37. N. J. Turner and M. E. Corden, *Phytopathology* **53,** 1388 (1963).
38. C. C. Travis and E. L. Etnier, *J. Environ. Qual.* **10,** 8 (1981).
39. G. W. Bailey and J. L. White, *Residue Rev.* **32,** 29 (1970).
40. R. Calvet in R. J. Hance, ed., *Interactions between Herbicides and the Soil*, Academic Press, Inc., New York, 1980, pp. 1–30.
41. J. W. Hamaker and J. M. Thompson in C. A. I. Goring and J. W. Hamaker, eds., *Organic Chemicals in the Soil Environment*, Marcel Dekker, Inc., New York, 1972, pp. 49–143.
42. C. S. Helling, P. C. Kearney, and M. Alexander, *Adv. Agron.* **23,** 147 (1970).
43. S. U. Khan, *Pesticides in the Soil Environment*, Elsevier Scientific Publishing Company, New York, 1980, pp. 29–70.
44. R. E. Green in W. D. Guenzi, ed., *Pesticides in Soil and Water*, Soil Science Society of America, Madison, Wisc., 1974, pp. 3–37.
45. S. W. Karickhoff, D. S. Brown, and T. A. Scott, *Water Res.* **13,** 241 (1979).
46. C. H. Giles in *Sorption and Transport Processes in Soil*, S.C.I. Monograph No. 37, Society of Chemical Industry, London, 1970, pp. 14–32.
47. C. S. Helling, *Residue Rev.* **32,** 175 (1970).
48. C. S. Helling and J. Dragun in *Test Protocols for Environmental Fate and Movement of Toxicants*, Association of Official Analytical Chemists, Arlington, Va., 1981, pp. 43–88.
49. M. Leistra in ref. 19, pp. 31–58.
50. J. Letey and W. J. Farmer in ref. 44, pp. 67–97.
51. J. Letey and J. K. Oddson in ref. 41, pp. 399–440.
52. R. D. Waucope, *J. Environ. Qual.* **7,** 459 (1978).
53. J. W. Hamaker in ref. 41, pp. 341–397.
54. P. Sprankle, W. F. Meggitt, and D. Penner, *Weed Sci.* **23,** 229 (1975).
55. J. R. Plimmer, *Bull. Environ. Contam. Toxicol.* **20,** 87 (1978).
56. N. J. Turro, *Molecular Photochemistry*, W. A. Benjamin, Inc., New York, 1965.
57. J. G. Calvert and J. N. Pitts, Jr., *Photochemistry*, John Wiley and Sons, Inc., New York, 1966.
58. P. C. Kearney and D. D. Kaufman in *Degradation of Synthetic Organic Molecules in the Biosphere*, National Academy of Sciences, Washington, D.C., 1972, pp. 166–189.
59. C. M. Menzie, *Metabolism of Pesticides, Special Scientific Report—Wildlife No. 127*, U.S. Department of Interior, Bureau of Sport Fisheries and Wildlife, Washington, D.C., 1969, 487 pp.
60. C. M. Menzie, *Metabolism of Pesticides, Update II, Special Scientific Report—Wildlife No. 212*,

U.S. Department of Interior, Bureau of Sport Fisheries and Wildlife, Washington, D.C., 1978, 381 pp.
61. A. M. Chakrabarty and I. C. Cunsalus, *Bacteriol. Proc.* **46** (1971).
62. A. M. Chakrabarty, *Annu. Rev. Genet.* **10,** 7 (1976).
63. J. M. Pemberton and P. R. Fisher, *Nature* (*London*) **268,** 732 (1977).
64. R. G. Nash in ref. 44, pp. 257–313.
65. A. L. Young, J. A. Calcagni, C. E. Thalken, and J. W. Tremblay, *The Toxicology, Environmental Fate and Human Risk of Herbicide Orange and Its Associated Dioxin*, USAF OEHL Technical Report, Brooks Air Force Base, Texas, 1978, 400 pp.
66. A. R. Isensee and G. E. Jones, *Environ. Sci. Technol.* **9,** 668 (1975).
67. R. W. Baughman and M. S. Meselson, *Adv. Chem. Ser.* **120,** 92 (1973).
68. IUPAC, *Pure Appl. Chem.* **49,** 675 (1977).
69. S. A. Peoples, K. T. Maddy, W. Cusick, T. Jackson, C. Cooper, and A. S. Frederickson, *Bull. Environ. Contam. Toxicol.* **24,** 611 (1980).
70. U.S. Environmental Protection Agency, *Fed. Regist.* **44**(219), 65135 (1979).
71. J. M. Hosenfeld, J. Neils, K. Thomas, and J. E. Going, *Sampling and Analysis of Drinking Water for Residues of DBCP, Final Report, EPA Contract No. 68-01-5915*, Midwest Research Institute, Kansas City, Missouri, 1981, 66 pp.
72. W. J. Storck, *Chem. Eng. News* 10 (April 28, 1980).
73. *Farm. Chem.* **144,** 55 (1981).

P. C. KEARNEY
C. S. HELLING
J. R. PLIMMER
U.S. Department of Agriculture

SOLAR ENERGY

The sun is a fusion reactor at a distance of ca 150×10^6 km (see Fusion energy). Its probable age is $>4 \times 10^9$ yr and it is expected to radiate energy for another 4×10^9 yr. That radiation comprises a spectrum from 10 pm (gamma rays) to 1 km (radio waves). However, more than 99% of the sun's radiation lies within the optical range from 0.276–4.96 μm. Electromagnetic radiation from the sun amounts to ca 3.8×10^{26} W. Because of the earth's distance and small size, it receives only a tiny fraction of the sun's total radiation; this fraction nevertheless amounts to ca 1.7×10^{17} W. The energy flux reaching the earth's outer atmosphere exceeds human energy consumption 27,000-fold. About half of the energy is reflected back into space, but about 13,000 times as much energy strikes the surface of the earth as the entire world consumes in the form of fossil and nuclear fuels; ca 46,400 EJ (44×10^{18} Btu) of solar energy reaches the continental United States each year, compared to the 84 EJ (80×10^{15} Btu) of energy consumed (1 EJ = 0.95×10^{15} Btu or ca 1 quad) (see Fuels, survey).

Solar-energy technology comprises two distinct categories: thermal conversion and photoconversion. Thermal conversion takes place through direct heating, ocean waves and currents, and wind. Photoconversion includes photosynthesis, photochemistry, photoelectrochemistry, photogalvanism, and photovoltaics. Solar radiation is collected and converted by natural collectors, such as the atmosphere, the ocean, and plant life, as well as by man-made collectors of many kinds (Figs. 1 and 2) (see Fuels from biomass; Photovoltaic cells).

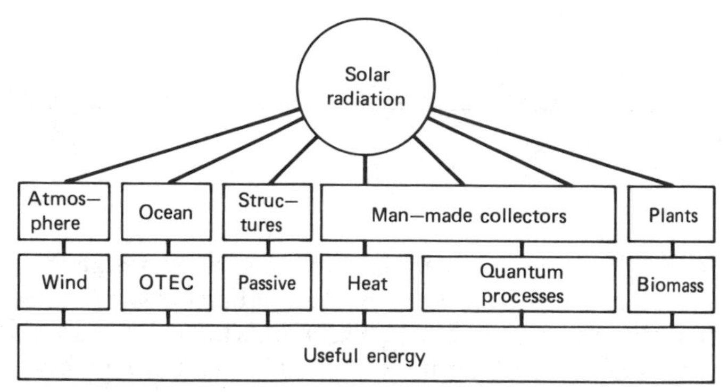

Figure 1. Natural and man-made solar collectors. OTEC = Ocean Thermal Energy Conversion.

The United States is making appreciable use of solar energy in the form of hydroelectric power, amounting to ca 3 EJ (2.85×10^{15} Btu) annually, approximately equal to the contribution of nuclear power plants (see Nuclear reactors). In addition, ca 2 EJ (1.9×10^{15} Btu) is derived from burning biomass, mostly wood and agricultural wastes. Thus, solar energy in the aggregate contributes ca 6% of total energy used in the United States. The aim of solar-energy development is to increase this minimal use of the vast solar-energy resource. The contributions of various energy sources in 1978 are shown in Table 1.

Solar Thermal Energy

The surface temperature of the earth is increased by ca 250°C over surrounding space by exposure to the sun. However, the sun radiates energy at an effective surface temperature of ca 6000 K and terrestrial solar concentrators capture significant amounts of this thermal energy. Solar collectors provide a heat range that permits applications from passive residential heating through medium temperatures for industrial process heat to the high temperatures needed for power-generating engines, and even for operating refractory furnaces and testing materials at very high temperatures.

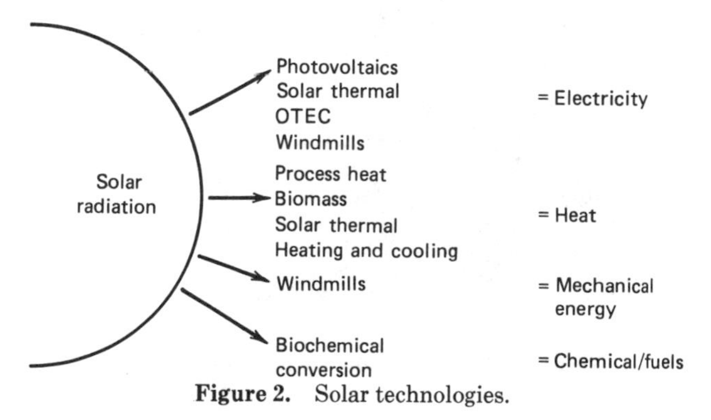

Figure 2. Solar technologies.

Table 1. U.S. Energy Consumption, 1978[a]

Source	Percent[b]
petroleum	47.5
natural gas	25.3
coal	18.2
nuclear	3.5
hydropower	3.1
biomass	2.3

[a] Ref. 1.
[b] Rounded.

Passive Heating. In the 5th century BC, buildings were designed and oriented to take advantage of solar heat during cold weather. Indians in the U.S. Southwest built passively heated residences into south-facing cliffs, which also shaded the structures from the summer sun (Fig. 3). Modern society is relearning this technique with better

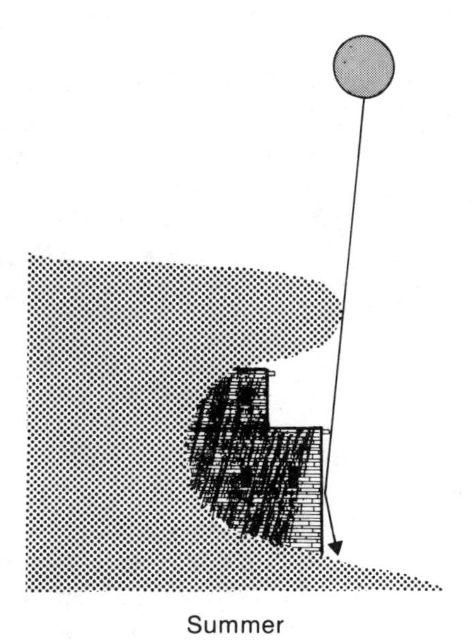

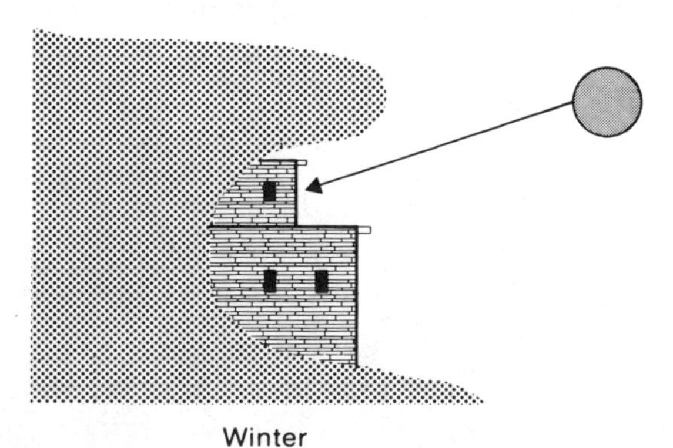

Figure 3. Cliff-dwelling passive solar heating.

engineering and materials. Passive design is a matter of orientation, glazing, shading, insulation, and heat storage. Interest in passive solar design generally has followed earlier exploitation of active solar heating. The concept of the building itself as a large-volume solar collector has been rediscovered and is being reestablished rapidly (see Fig. 4).

The energy in solar radiation is not a function of ambient air temperature but depends on the clarity of the sky and the angle at which the sun's rays are received. Even on an overcast day, 50% of maximum solar radiation may reach the surface. A square meter of insolated area provides ca 22.7 MJ (ca 21,500 Btu) on a clear day. Thus, a greenhouse, or other glazing, of 18.6 m^2 (200 ft^2) may intercept up to 422 MJ/d (4×10^5 Btu/d). A well-designed passively heated home may receive all the heat it needs from the sun. Surplus heat is stored in floors, walls, water tanks, or rock bins. Techniques such as Trombe walls add effectiveness to passive designs. A Trombe wall is a glazed section over a brick or concrete wall. This construction provides the building with warm air through vents, plus heat radiated from the wall itself (Fig. 5).

Passive cooling techniques include cross ventilation, insulation, shading, regenerative rockbed cooling, evaporation, and radiation to the night sky. However, these are not solar-driven processes, except for natural breezes. Passive desiccant cooling,

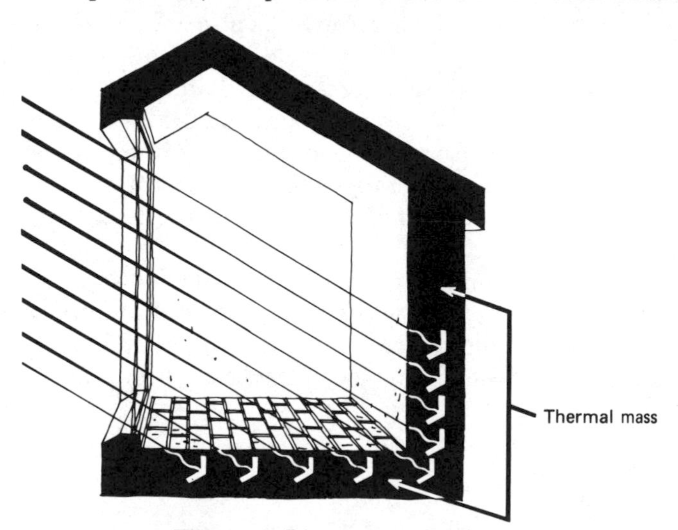

Figure 4. Direct passive heating.

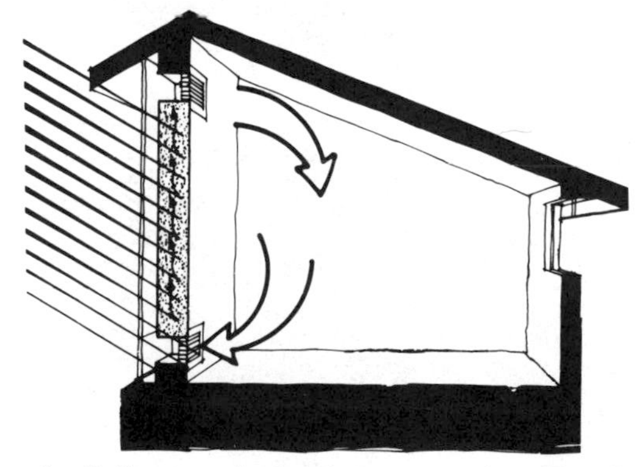

Figure 5. Indirect passive heating system using a Trombe wall.

which dates to the U.S. South of a century ago, has been adapted for solar regeneration of the desiccant. Typical desiccants include salt, silica gel, ethylene glycol, and molecular-sieve zeolites (see Drying agents; Molecular sieves).

Active Heating and Cooling. In active solar heating, the heat-exchange medium is moved from a solar collector to the space that is to be heated. The most common solar collectors are those long used for domestic water heating. Typically, a shallow rectangular box is insulated on the bottom and at the sides, fitted with a metal absorber plate inside the box, and covered with a glazing of glass or translucent plastic (see Fig. 6). Heat from the absorber plate is removed by a heat-exchange medium, eg, water, oil, antifreeze solution, or air. Air systems protect against freezing and no harm results from leaks (except system inefficiency). However, an additional heat exchanger is required if the air collector is to heat water (see Heat-exchange technology).

Water systems are much more common than air systems. They offer better heat-exchange performance but have the disadvantage of potentially damaging leaks and susceptibility to freezing. (Because the absorber plate is generally a good radiator, the water in the collector may freeze even at ambient temperatures well above 0°C.) Protection against freezing includes antifreeze in a separate loop (closed system), a drain-down mechanism, or heating the collector in cold weather either electrically or by pumping in warm water. The antifreeze loop seems to offer the best protection.

The absorber plate of a solar collector is painted dull black or otherwise treated to permit radiation absorptance. However, a good absorber is generally a good emitter as well. For this reason, much effort has been spent on developing selective surfaces by coating or other treatments in order to obtain high absorptance but low emittance. Proportional control varies the speed of fluid through a collector, whereas a high insolation rate speeds up the pump.

Current solar air-conditioning projects include solar-powered refrigeration systems of both Rankine-cycle vapor-compression types and absorption systems (Fig. 7). Other cooling systems include solar-regenerated desiccant cooling. In the system developed by the Institute of Gas Technology, solar heat is substituted for conventional

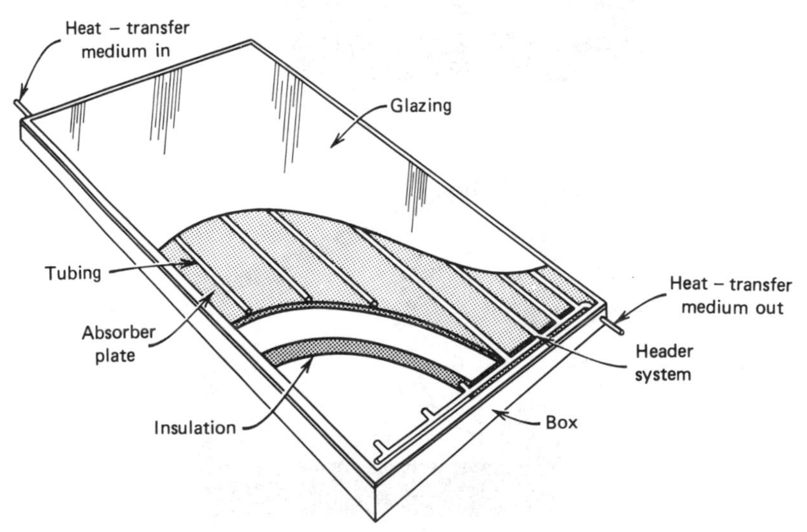

Figure 6. Diagram of solar collector.

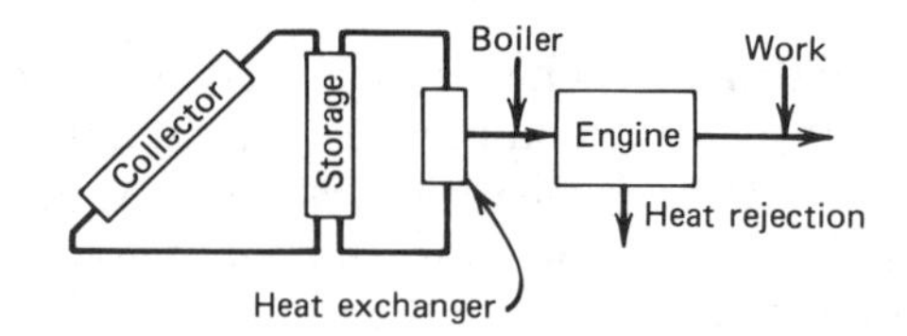

Rankine–cycle system

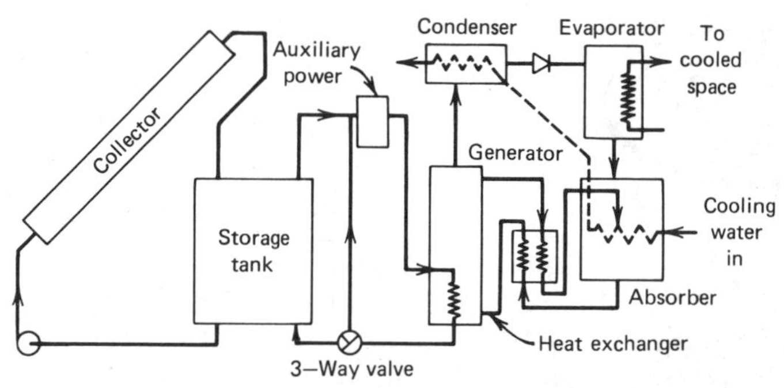

Absorption refrigeration system

Figure 7. Rankine-cycle and absorption solar-refrigeration systems. Courtesy of Marcel Dekker, Inc.

fuel in the desiccant-regeneration cycle. The DOE operates the National Desiccant Cooling Program; approaches include absorption and adsorption, and both open and closed cycles. Open-cycle adsorption systems seem to offer the best prospects (Fig. 8). In a typical desiccant cooling cycle, ambient air is adiabatically cooled, dehumidified, both sensibly and evaporatively cooled, and ducted to the living area. In the regenerative stage, air is evaporatively cooled, heated as it cools the supply airstream, heated again by solar collectors, and humidified. Zeolites have been employed in some experimental work, and small solar-refrigeration systems have been built. Simulation and analysis of desiccant cooling systems suggest that solar-regenerated systems can be cost-competitive with conventional vapor-compression or absorption systems, and provide a 20-yr payback on costs. Desiccant cooling seems best suited for regions with about equal heating and cooling loads and high humidity (see Air conditioning; Refrigeration).

Industrial Process Heat. Industrial and agricultural process-heat applications are often amenable to active solar heating. Food dryers and car washes represent the low temperature end of this spectrum, which extends upward to very hot water and other liquids in the food, textile, brewing, and other industries, and even to steam. The best flat-plate collectors produce temperatures up to ca 100°C, but at considerable loss

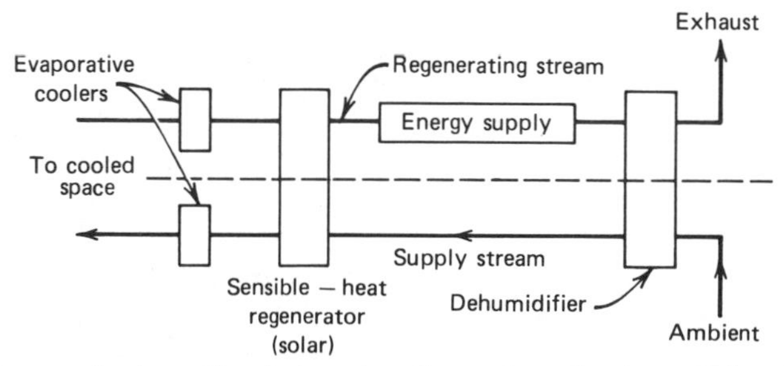

Figure 8. Open-cycle absorption desiccant cooling system. Courtesy of Marcel Dekker, Inc.

of efficiency at the high end. Thus, concentrating collectors are preferred at the medium temperatures. The types shown in Figure 9 are capable of producing steam for industrial process heat (see Power generation; Steam). Applications include textile processing, irrigation pumps, and domestic and commercial air conditioning. The manufacturing industry uses about 35% of the nation's energy and offers a large potential market for solar-energy systems. Approximate percentages for applications at various temperatures are given in Table 2.

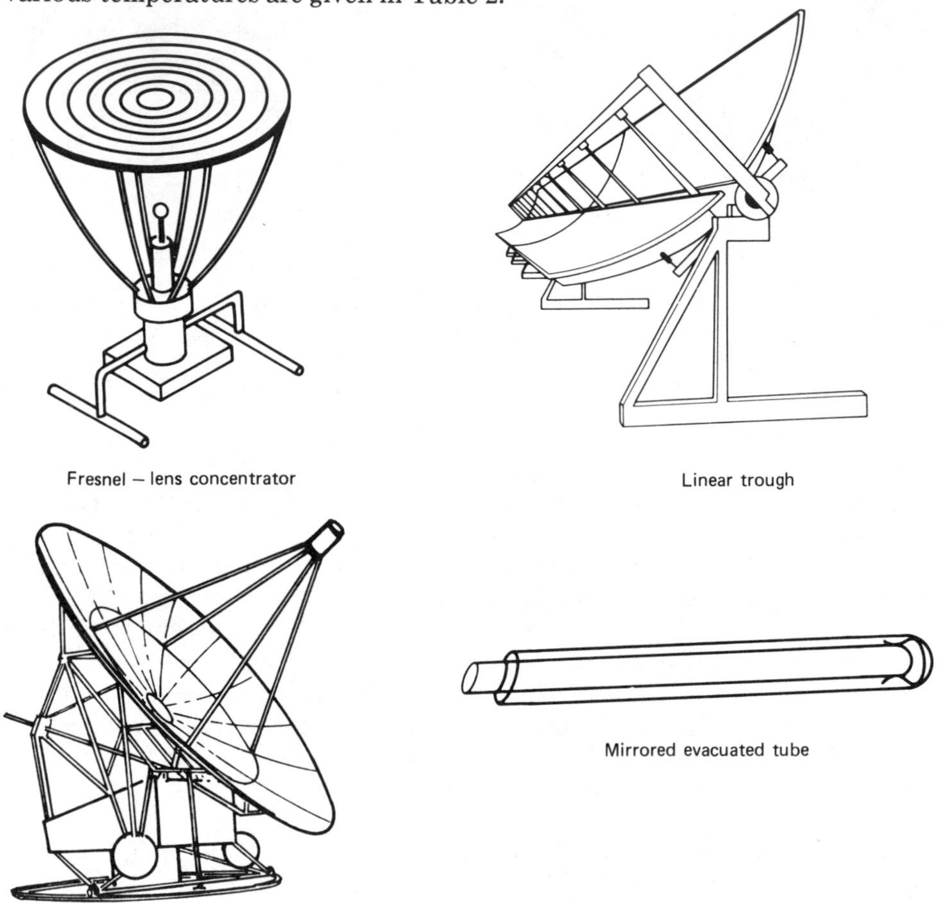

Figure 9. Concentrating heat collectors.

Table 2. Industrial Applications at Various Temperatures[a]

Temperature, °C	Percent of total
100	10
175	10
175–300	5
300–600	18
600–1100	12
1100	45

[a] Ref. 2.

These are industry-wide averages; requirements within a specific industry vary. For example, estimates for the oil-refining industry are: <300°C, 22%; 300–600°C, 63%; >600°C, 15%. Some oil companies have investigated the use of solar heat in secondary-recovery operations but found it is not cost-effective at prevailing oil or gas fuel prices. At the other end of the temperature range is a large-area shallow solar-pond collector used in a uranium ore processing demonstration which was almost cost-competitive with oil several years ago. Temperatures produced by various solar collector systems are given in Table 3.

Table 3. Temperatures Produced by Various Collectors

Collector type	Temperature, °C
solar ponds/flat plate	100
evacuated glass-tube collectors	50–175
line-focus concentrators	65–300
point-focus concentrators	200–800
central receiver	800–1400

Six DOE-funded industrial process-heat demonstration projects for low temperature (<38°C) applications included the drying of textiles, lumber, soybeans, and fruit, the curing of concrete block, and can washing (Fig. 10). Both liquid and air were used as working fluids. Installed costs (less prototype design costs) were $(250–870)/m^2 ($(23–81)/ft^2) of the collector area. These high first costs resulted from the fact that

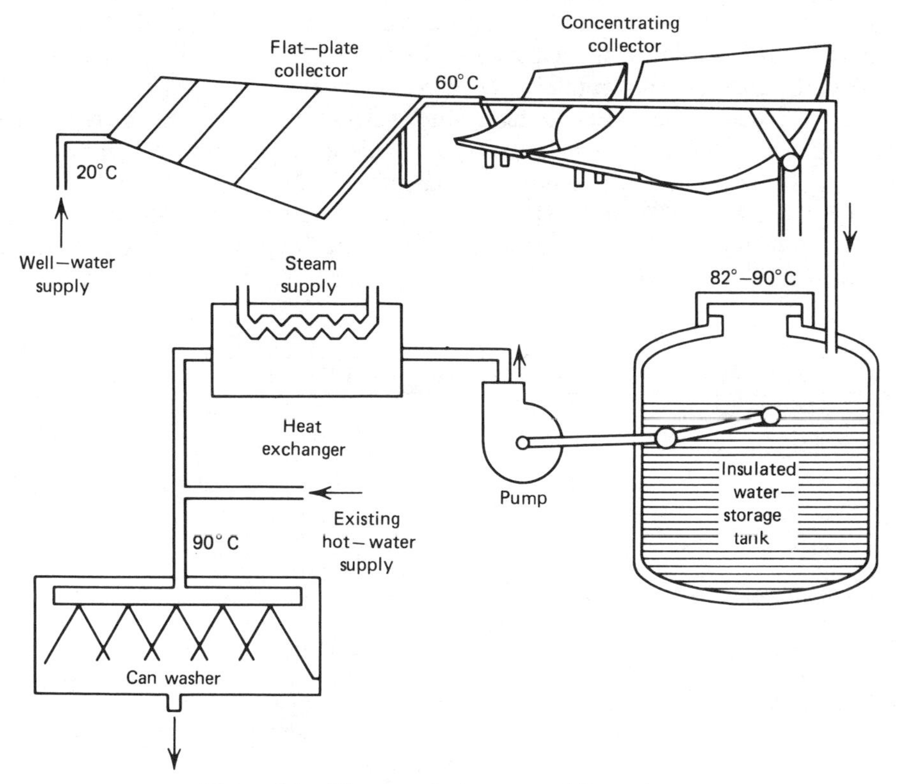

Figure 10. Diagram of solar can-washing system.

each system and application were unique and required special treatment throughout the projects. A cost of $(50–100)/m^2 (ca $(5–10)/ft^2) is a desirable goal. At present, several dozen industries in the United States have solar-energy process-heat installations on the drawing board or in operation. A recent journal article reported on four typical systems, including a brewery, food-dehydration plant, natural-gas-processing facility, and food-processing plant (3). Operating temperatures range up to 300°C. Although these are pilot plants, the economic projections are encouraging. One system was ranked as not cost effective, whereas the others have estimated payback times of 16–21.5 yr with projected fuel increases. Typical industrial applications are given in Table 4.

Power Generation. It is widely believed that solar energy is too diffuse for use as a power source; however, an area of a square meter normal to the sun intercepts ca 1 kW of solar radiation in clear weather. The south-facing portion of a house roof of ca 80 m^2 (860 ft^2) receives far more energy than the house consumes for heating, cooling, lights, and appliances. Thus, the 80×10^6 residences in the United States receive a peak total of 6.4×10^6 MW_p (photogenerated) of solar energy. A case can also be made for central solar power plants, and a DOE 10-MW pilot plant for solar thermal power is now being operated in southern California. A tower, ca 100 m tall, houses a boiler heated by solar radiation reflected from 1800 mirrors surrounding the tower. Radiant energy is delivered directly to the boiler, which converts water into high pressure steam at 595°C, typical of conventional fossil- or nuclear-fired power plants. Linear-trough and dish-type concentrators are also being developed to achieve high temperatures and thus high thermodynamic efficiencies. Costs on the order of $2.5/W are projected for commercial solar power plants, competitive with most fossil fuels. In small sizes, solar thermal power plants are relatively economical. An intermediate size solar ammonia-plant project is shown in Figure 11.

The DOE has funded 14 projects in which solar energy replaces conventional fuels at existing manufacturing or electric-power generating plants; five projects are in Texas, three in California, two each in Arizona and New Mexico, and one each in Nevada and Oklahoma. As these sites suggest, solar power plants function best in the Southwest where direct solar radiation is abundant. Typical average direct normal radiation is ca 7 (kW·h)/(m^2·d). Eight of the projects involve conversion of existing

Table 4. Typical Commercial Applications of Solar Energy in Operation [a]

Company	Location	Application	System type	Energy saved, GJ/yr [b]
Anheuser-Busch	Jacksonville, Fla.	pasteurization	trough	527
Campbell Soup	Sacramento, Calif.	can washing	trough	1900
Gold Kist	Decatur, Ala.	soybean drying	flat plate	843
J.A. LaCour Kiln Services	Canton, Miss.	lumber drying	flat plate	422
Johnson & Johnson	Sherman, Tex.	bleaching, sterilization	trough	1580
Lamanuzzi & Pantaleo Foods	Fresno, Calif.	food drying	flat plate	1050
Riegel Textile	La France, S.C.	dyeing	flat plate	422
Reedy Creek Utility	Lake Buena Vista, Fla.	air conditioning	trough	843
York Building Products	Harrisburg, Pa.	concrete-block curing	flat plate	422

[a] Ref. 4.

[b] To convert J to Btu, divide by 1054.

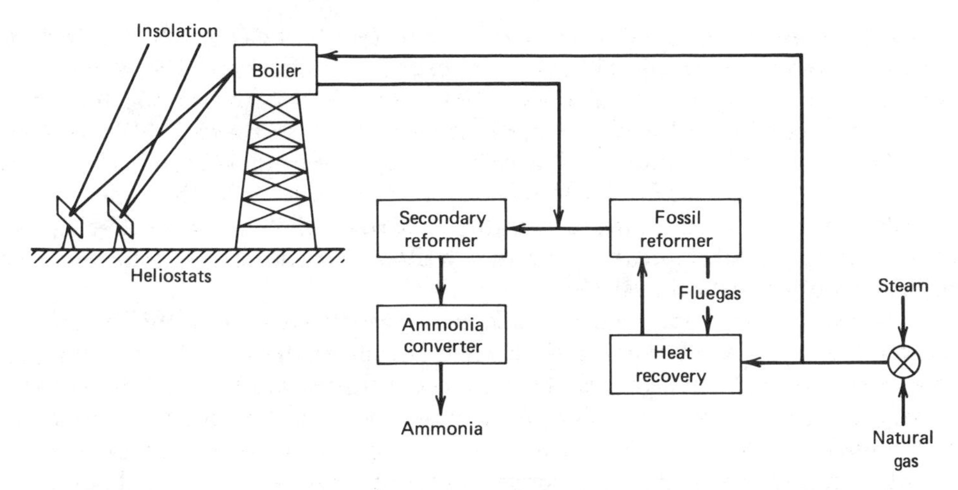

Figure 11. Ammonia-plant solar project.

oil or natural-gas electric-utility power plants to solar power plants. The other six include various industrial process-heat applications.

Other proposed projects include the ambitious orbiting solar-power satellite (SPS), comprising many square kilometers of arrays of photovoltaic cells for converting solar radiation to electricity. The electric-power output would be converted continuously to microwave frequencies and beamed to rectifying antenna arrays on earth. Here, the microwave radiation would be converted to alternating current at frequencies compatible with utility grids. Estimated to cost ca $\$10^{12}$, the SPS project would require assembly in space. Parts would be placed into low orbit by space shuttles, assembled, and boosted into the necessary 35,000-km synchronous orbit (perhaps using power generated by the satellite). In addition to serious logistics problems, there are environmental concerns that the microwave energy flux may be harmful to life. Political complications and vulnerability to possible military attack also have to be considered (see Microwave technology).

Solar Ponds. Solar collectors tend to be expensive, eg, a typical flat-plate collector costs $\$150/m^2$ (ca $\$14/ft^2$) and a tracking concentrator might be several times as much. Such costs make the solar pond very interesting. Solar ponds were first developed in Israel, perhaps because of the combination of ample solar energy, the Dead Sea, and the new nation's need for energy resources.

At the end of the 19th century, the phenomenon of the reverse-gradient lake was discovered. Water at the bottom of Lake Medve in Transylvania, Romania, had a temperature of ca 65°C, the reverse of the normal situation where warm water rises to the surface. When the temperature was measured at different levels, stratified, nonmixing layers of varying density were found. The denser lower levels were saltier, thus suppressing normal convection (5). Other natural reverse-gradient ponds have been found in Venezuela, in the state of Washington, and in Antarctica.

In Israel, this discovery was applied to a heat-producing solar pond. By 1959, the first man-made, reverse-gradient solar pond was completed and a specially designed turbine for producing electric power from water at 90°C was installed (6). However, solar electricity could not compete with the very low price of fossil fuels at that time, and the turbine was converted to be powered by fossil fuels. After 1973 and the growing uncertainty of oil supplies, Israel started again to develop solar ponds. The turbines were larger, and in 1979 a 150 kW generator began operation at a 7500 m^2 (ca 81,000 ft^2) solar pond. This was a test unit only; the Israeli program calls for a 5 MW turbine per 0.25 km^2 pond in 1982; a 20 MW per 1 km^2 system in 1983; and 20 MW per 4 km^2 system by 1985. The long-range goal is for 2 GW of power from solar ponds by the end of this century. The Dead Sea itself may be a giant solar pond by that time.

Solar ponds are constrained to operate in a horizontal plane, thus suffering some loss of intercepted radiation as compared to a tilted or tracking solar collector. However, construction costs are very low.

A nonconvecting solar pond typically comprises three layers: a thin, convecting top layer that serves as a buffer for wind; a middle, salt-gradient, nonconvecting layer; and a convecting bottom layer that stores most of the heat (Fig. 12). The bottom in nonconvecting as well as convecting ponds is covered with dark, heat-absorbing material. Thin, transparent membranes interposed between layers may be useful to prevent mixing. Experimental nonconvecting solar ponds include gelled ponds and viscosity-stabilized ponds; more research is needed in this area.

The United States has not yet matched the Israeli effort, but some impressive projects are operating and much larger ones are under study. In 1978, the city of Miamisburg, Ohio, built a solar pond, 50 × 55 m and 3 m deep, to heat an outdoor swimming pool in summer and a recreation hall in winter (7). The excavation was lined with chemically resistant polymer-coated polyester fabric and the pond was filled with 5.3×10^5 m^3 (1.4×10^8 gal) water. About 1100 metric tons of sodium chloride was added and a reverse-density gradient was formed by progressively increasing the concentration from zero at the surface to ca 18% at the heat-storage layer at the bottom. Heat is extracted from the pond through a copper heat-exchange system. In July 1979, the pond reached a temperature of ca 68°C. In winter, some heat is collected even with a layer of ice on the surface. By February, storage temperature decreases to ca 27°C. A total heat production of ca 991 GJ (940×10^6 Btu) is estimated, of which ca 823 GJ (780×10^6 Btu) is used for heating the pool in summer and the remaining 169 GJ (160×10^6 Btu) for heating buildings in winter; this is equivalent to the energy of 39.75 m^3 (250 bbl) oil. Total cost for the project was \$37/m^2 (ca \$3.40/ft^2) of pond area, which is far less than for conventional solar collectors. Average cost per GJ is ca \$6.00 (ca \$5.70 per 10^6 Btu), well below the present cost of oil. Performance and costs of several U.S. solar ponds are given in Table 5.

Exceeding the Israeli achievements in scope is the proposed Salton Sea solar-pond

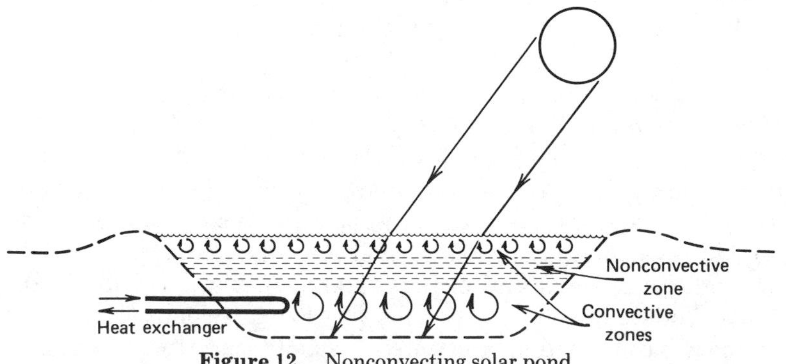

Figure 12. Nonconvecting solar pond.

Table 5. Performance of Solar Ponds[a]

Application			Pond			Av output, kW (W/m^2)	Energy cost, ¢/(kW·h) ($/$10^6$ Btu)
Process	Location	Latitude	Shape	Dimensions, m	Cost, $/$m^2$ ($/$ft^2$)		
winter space heating	Ohio	40° N	circular	50 dia	47.50 (4.41)	54 (27.5)	2 (5.86)
laundry-water preheating	Ohio	40° N	circular	100 dia	18.33 (1.70)	291 (37)	0.57 (1.67)
water heating for hotel	Hawaii	20° N	circular	100 dia	12.00 (1.11)	526 (67)	0.20 (0.59)
process heat for salt works	Texas	26° N	rectangular	500 × 2000	5.30 (0.49)	50[b] (50)	0.12 (0.35)

[a] Ref. 8.
[b] MW.

project in southern California (9) (see Fig. 13). Funded by DOE, DOD, the State of California, Southern California Edison, and the Israeli Ormat turbine firm, the total cost of the venture is $650,000 and Phase I, a feasibility study, began early in 1981. If findings are favorable, Phase 2, a 5 MW pilot plant, will be built. Phase 3 entails construction of a 600 MW power plant. The Salton Sea should be a good site for a large pond. With a surface area of 969 km^2, it has a salinity of 4%. Furthermore, it is in one of the sunniest regions in the world. Should the entire sea be converted to a solar pond, its output would be in the range of 5 GW. Other bodies of water in the United States that are being studied as potential solar-pond sites include Utah's Great Salt Lake and San Francisco Bay. The Great Salt Lake's 4365 km^2 suggests a potential output of 22 GW.

Shallow convecting ponds are also used for power generation, eg, those of Lawrence Livermore Laboratory (10). Typically 5 × 60 m, and 10 cm deep, these ponds have black, heat-absorbing bottoms, insulated sides, and plastic-glazed tops, as shown in Figure 14. Water is heated during the day in the ponds and stored at night in underground tanks. Such installations are nearly competitive with petroleum.

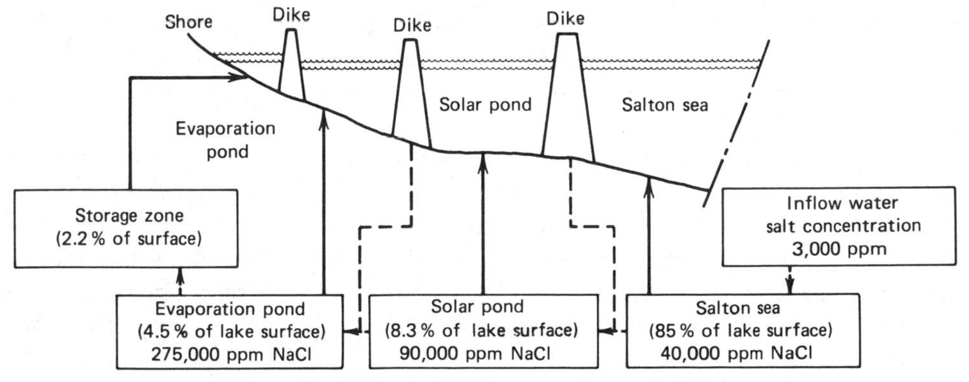

Figure 13. Diagram of Salton sea solar-pond project.

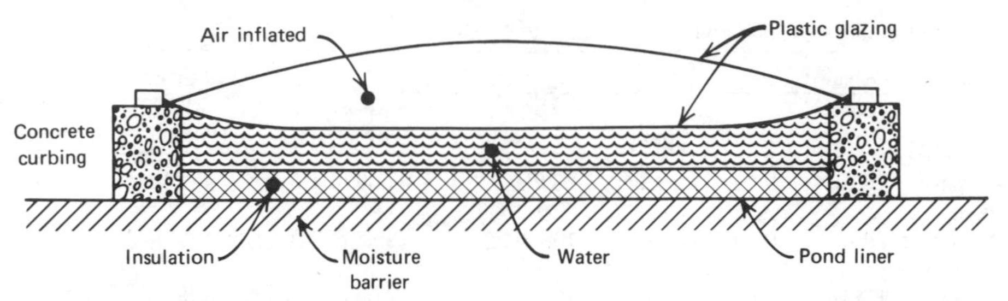

Figure 14. Shallow solar pond. Courtesy of Lawrence Livermore Laboratories.

Materials and Equipment. Because of the wide range of solar thermal devices and applications, a variety of materials is used. These may be categorized generally as glazing, heat absorbers, plumbing, insulation, supporting structures, reflective materials, heat-transfer fluids, and seals and sealants. Criteria for selection include cost, weight, lifetime, weathering characteristics, toxicity, flammability, outgassing properties, etc.

Glazing. Glazing is the first component of a solar collector to receive incoming solar radiation. Glass is the traditional material, and plastic substitutes have yet to match its stability, weatherability, and resistance to ultraviolet radiation. The Solar Energy Research Institute recently encouraged the manufacture of fusion-drawn glass by the Corning Glass Co. They produced a highly transmissive, very thin glazing material in an oxidizing atmosphere; it contains optically inactive iron(III) and has a transmissivity near the theoretical maximum. The 20-mm thickness permits bending for use in linear parabolic-mirror applications (see Glass). The advantages and disadvantages of glass and various plastic glazing materials are given in Table 6.

Reflective Materials. Reflective materials for heliostats or concentrators include silvered glass, polished metals (often anodized or otherwise treated), eg, aluminum such as Alzak or Coilzak, and reflective foils and films. Glass mirrors are subject to breakage, wind loading, scouring by sand and dirt, radiation damage, and degradation by moisture. Metals incur similar damage, and some metallic concentrating collectors have been covered with plastic reflective films to increase or restore reflectivity.

Absorber Plates. Absorber plates in solar collectors vary from blackened plastics for low temperature use, such as pool heating, to metals with selective surface coatings. Nonselective materials may be industrial flat-black paints or special solar paints with better ratios of absorptance to emittance (see Table 7).

Research on selective coatings was pioneered in Israel, where solar water heaters are used extensively. Today, numerous commercial coatings are available (see Table 8).

Absorber plates for low temperature applications are made from plastics but must resist ultraviolet degradation. Metals are used for higher temperatures and include aluminum, copper, steel, and stainless steel. Copper and stainless steel have good corrosion resistance, an important factor for water-carrying collectors. The absorber is often a composite of aluminum fins and copper-tube water passages. Some collectors are made entirely of aluminum or iron. In such cases, a noncorrosive or inhibited heat-exchange fluid is used, and galvanic action between dissimilar metals must be prevented (see also Table 9).

Table 6. Advantages and Disadvantages of Glazing Materials[a]

Glazing	Trade names	Advantage	Disadvantage
glass		excellent weathering resistance	low impact strength high density poor thermal-stress resistance
acrylic	Plexiglass Acrylite Flexigard	good insulator high impact strength good weathering resistance	low softening point distortion when heated surface abrasion
fluorinated ethylene–propylene	Teflon	high chemical stability excellent weathering resistance	high long-wave transmittance
glass fiber-reinforced polyester	Fiberglass Filon Lascolite	lightweight high strength high impact resistance high heat resistance near opacity to long-wave radiation high durability	poor weathering[b]
polycarbonate	Lexan Poly-Glaz	good insulator high initial impact strength of all plastics	low mar-resistance types do not weather well surface deterioration by radiation/moisture high cost
polyester	Mylar	surface hardness low cost	high long-wave transmittance uv degradation unless coated
polyethylene		light weight flexible low cost	short lifetime high long-wave transmittance wind and temperature sagging
poly(vinyl fluoride)[b,c]	Tedlar	high chemical stability high resistance to abrasion excellent weathering resistance low cost	high long-wave transmittance

[a] Ref. 8.
[b] PVF coating improves surface-erosion resistance.
[c] PVF.

Heat-Transfer Media. Heat-transfer media include liquids and gases; the former are more commonly used. Some solar collectors utilize air; some chlorofluorocarbons, including Freons. The latter may cycle between the liquid and gas phases during operation. Liquid heat-exchange fluids include water, hydrocarbon oils, glycols, and silicones. Choices are made on the basis of cost and properties such as heat capacity, specific gravity, thermal conductivity, viscosity, boiling and freezing points, and corrosion resistance (see Table 9). Water is favored because of its high specific heat, good thermal conductivity, and low viscosity. Glycols (qv) are slightly heavier than

Table 7. Absorptance and Emittance of Nonselective Absorber Coatings

Coating	Absorptance	Emittance	References
anodized aluminum	0.95	0.80	11
black paint	0.95–0.98	0.88–0.89	12–13
black paint	0.90	0.60	14
urethane resin over epoxy primer	0.97	0.92	15

Table 8. Absorptance and Emittance of Solar Absorber Coatings

Coating	Substrate	Absorptance	Emittance	References
black chrome[a] (Cr and CrO)	nickel-plated metals, copper, steel	0.91–0.96	0.07–0.16	11, 14, and 16
black copper[b] (CuO and Cu_2O)	copper, nickel, aluminum	0.81–0.93	0.11–0.17	16–17
black nickel[c] (NiO_2)	nickel, iron steel	0.89–0.96	0.07–0.17	16–17
iron oxide (Fe_2O_3)	iron, steel	0.85	0.08	16
stainless steel oxide (FeO)	stainless steel	0.89	0.07	11
selective paint[a]	any	0.90	0.30	16

[a] Minneapolis Honeywell.

Table 9. Properties of Heat-Transfer Media[a]

Characteristic	Water	Glycols	Silicones	Hydrocarbons
heat capacity at 82°C, kJ/(kg·K)[b]	4.18	3.55	1.59	2.00
specific gravity at 82°C	1.00	1.02	0.93	0.88
thermal conductivity at 82°C, W/(m·K)	0.64	0.41	0.14	0.12
viscosity at 82°C, mm^2/s (= cSt)	0.35	1–2	5–10	1–10
coefficient of thermal expansion, %/°C	0.005	0.017	0.033	0.017–0.044

[a] Ref. 8.
[b] To convert J to Btu, divide by 1054.

water; hydrocarbons and silicones, slightly lighter. Water suffers the disadvantages that it is liquid over a relatively small temperature range and is corrosive. However, distilled or deionized water (with appropriate inhibitors) is acceptable in metal passageways. The general pH range is given below.

Metal	*pH range*
aluminum	<5 or >8
galvanized steel	<7 or >12
stainless steel	<5 or >12

Copper must be protected from corrosive fluxes and certain chemicals; stainless steel from corrosive fluxes and chloride ions (8).

Collector Boxes. Collector boxes are made from a variety of materials, including wood, plastic, and metal. Wood, however, is combustible and susceptible to rotting, whereas plastics may not withstand high temperatures. Metals are safest and most durable; aluminum has the advantage of light weight.

Insulation. Insulation prevents loss of collected heat energy. Operating temperatures and environmental conditions dictate the type and amount of insulation required. Rigid polystyrene foam may be used in a solar swimming-pool heater, which has an upper temperature limit of ca 75°C. However, a glazed, medium temperature solar collector can attain very high temperatures and fiber-glass insulation or an equivalent is necessary. Characteristics of various materials are given in Table 10. Because of the outgassing of binder material, some manufacturers use binderless insulation, eg, wool, especially where toxicity cannot be tolerated (see Insulation). Urethane foams, eg, are highly toxic at elevated temperatures. Outgassing products of several insulation materials are given in Table 11.

Seals and Sealants. Seals and sealants are used to join materials, accommodate to thermal expansion, prevent leaks, and isolate the collector and system from the environment as shown in Figure 15. Glazing must be tightly yet flexibly in contact with collector boxes; insulation must adhere to box or absorber; joints must be secure in all weather conditions and against temperature and internal pressure; and pumps must be tightly sealed against leaks. Seal and sealant materials must withstand thermal cycling and undergo minimum outgassing. Elastomers are choice materials because of their rubberlike, elastic qualities. Their temperature limits and compatibility with heat-transfer fluids are given in Tables 12 and 13, respectively (see Sealants; Elastomers, synthetic).

Table 10. Thermal Conductivities and Upper Temperature Limits of Insulation Material[a]

Insulation	Thermal conductivity[b,c], mW/(m·K)	Upper temperature limit, °C
calcium silicate	55[c]	650
mineral fiber block	36	1040
mineral fiberboard	55	650
perlite	69[c]	815
refractory fiberboard	40[c]	650
glass fiberboard	36	345
cellular elastomer[d]	36	105
polystyrene	28	75
polyurethane	23	105
isocyanurate resin	24	120
phenolic resin	32[e]	135
urea–formaldehyde resin	35	132

[a] Ref. 8.
[b] At 24°C unless otherwise stated.
[c] At 93°C.
[d] Cellular elastomer on flexible foamed plastic is made from foamed resin combined with elastomer to produce a cellular flexible material (see Foamed plastics).
[e] At 21°C.

Table 11. Outgassing Products from Insulation Materials[a]

Insulation	Outgassing products
polystyrene[b]	styrene α-methylstyrene
isocyanurates[b]	carbon dioxide hydrogen chloride trichlorofluoromethane
polyurethane[b]	carbon dioxide hydrogen chloride trichlorofluoromethane tris(2-chloroethyl) phosphate
urea–formaldehyde[c]	formaldehyde
glass fiber[b]	carbon dioxide saturated alkanes

[a] Ref. 8.
[b] Ref. 18.
[c] Ref. 19.

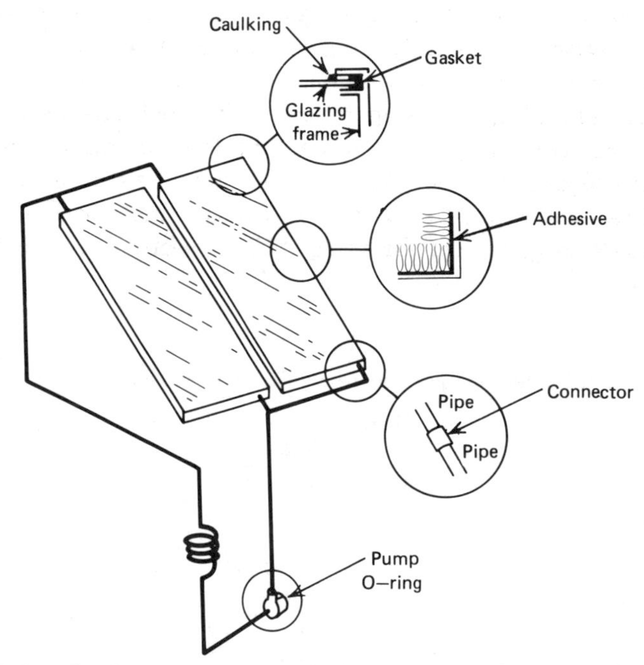

Figure 15. Seals and sealant materials in a solar heating system. Courtesy of *Solar Engineering*.

Heat Storage

If solar energy were received continuously and evenly, without difference between winter and summer, the design of solar-energy installations would be simple. However, insolation varies diurnally as well as seasonally, and it is generally necessary to provide some means of storing solar heat for nights and cloudy periods. It is seldom cost-effective to design a solar collector for the least amount of energy received. Although such an approach would guarantee heat even during rare, week-long cloudy periods,

Table 12. Temperature Limits of Elastomers[a]

Elastomer	ASTM designation[b]	Service temperature, °C Low	High	Heat resistance[c]
acrylate–butadiene	ABR	−30	175	
butadiene rubber	BR	−75	80	
styrene–butadiene (buna-S)	SBR	−55	95	F
isobutene–isoprene (butyl)	IIR	−55	120	G–E
ethylene–propylene–diene rubber	EPDM	−60	150	G–E
fluoroelastomer	FKM	−40	230	E
poly(chloromethyloxirane) (epichlorohydrin polymer)	CO, ECO 494	−40	165	
chlorosulfonated polyethylene (Hypalon)	CSM 3	−55	135	
polyisoprene, synthetic	IR 249+	−60	80	
natural rubber	NR	−60	80	
polychloroprene	CR	−55	120	G–E
acrylonitrile–butadiene (nitrile) rubber	NBR	−55	135	G–E
polyblend (PVC–NBR[d])	PVC–NBR	−20	120	
polysulfide (T-13[d])		−40	110	P–G
silicone (SI, FSI, PSI 35+, VSI, PVSI[d])		−95	290	E
polyurethane (U-138[d])		−40	120	

[a] From ref. 20.
[b] ASTM D 1418-79a.
[c] P = poor; F = fair; G = good; E = excellent.
[d] Commercial designation.

Table 13. Compatibility of Elastomers with Heat-Transfer Fluids[a]

Elastomer	ASTM designation[b]	Mineral oils[c]	Aromatic hydrocarbons[c] Oils	Alkylated	Ethylene glycol[c]	Silicone oils[c]
acrylate–butadiene	ABR	G	G	G	P	G
butadiene	BR	P	P	P	G–E	F–G
poly(ethylene-*co*-propylene)	EPM	P	P–F	P–F	E	E
ethylene–propylene–diene rubber	EPDM	P	P–F	P–F	E	E
chloroprene	CR	G	P–F	P–F	G–E	G–E
acrylonitrile–butadiene (nitrile) rubber	NBR	E	F–G	F–G	G–E	G–E
polysulfide (T-13[d])		E	E	E	F–E	P–F
silicone (SI 35+[d])		P–F	P–F	P–F	F–E	P–F

[a] From ref. 20.
[b] ASTM D 1418-79a.
[c] P = poor; F = fair; G = good; E = excellent.
[d] Commercial designation.

capital investment would be too high. It is more economical to employ a conventional standby system, with the sun providing some optimum fraction of the needed amount of energy.

Passive solar space heating presents a special and very interesting example of the need for heat storage. If a sizable greenhouse is attached to a conventional residence, or sufficient south-facing window area is provided, the residence may quickly become very hot by day but very cold at night. The incorporation of heat-storage areas, eg, masonry floors and walls, rock bins, or water tanks (commonly called thermal mass), solves the problem of these extremes. While the sun is shining, the house's thermal mass heats slowly, keeping air temperature moderate. At night, the thermal mass slowly radiates stored heat to keep the house warm (Fig. 16).

Sensible Heat Storage. Heat can be stored in water at ca 4 kJ/(kg·K) [1 Btu/(lb·°F)] increase in temperature. Rock stores heat, too, although not as much for a given volume. Both these media store sensible heat:

$$Q = MC_pT$$

where Q = heat stored, M = mass of storage material, C_p = specific heat of storage material, and T = temperature change of storage material.

Sensible heat storage in residences is common in Europe, especially in the Federal Republic of Germany and in the United Kingdom. Typical systems use refractory brick for storing heat at temperatures as high as 650°C (see Refractories). Effective and durable insulation is required to keep baseboard units at a safe outside temperature of ca 60°C. Electric heating rods are used, and thermostatically controlled fans circulate hot air into the room. Larger, central-heating storage furnaces have refractory-brick or cast-iron cores (21). Solar energy in residential applications provides only moderate temperatures, and a much larger volume of thermal mass is required.

A heat-storage system, called Annual Cycle Energy Storage (ACES), was developed several years ago at Oak Ridge National Laboratories (Fig. 17). In the basement of a house, 70 m^3 (18,500 gal) of water is stored. In winter, a heat-pump system heats the house, drawing on the stored water. By spring, the water is frozen and then used for cooling the house in summer. This system uses both sensible and phase-change heat storage. Analogous large-scale heat-storage systems for communities entail storing heat in aquifers, ponds, earth, or rock. Economies of scale are gained because the ratio of volume to surface area increases with size of the storage area (22).

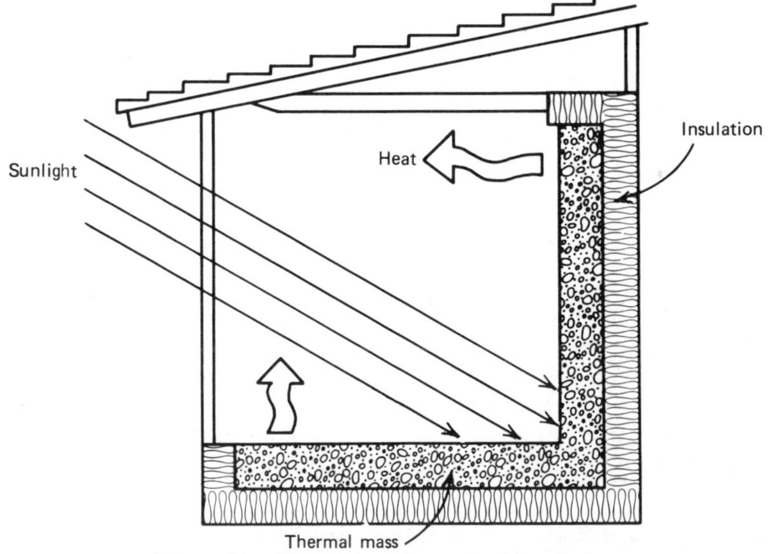

Figure 16. Heat storage in passive solar design.

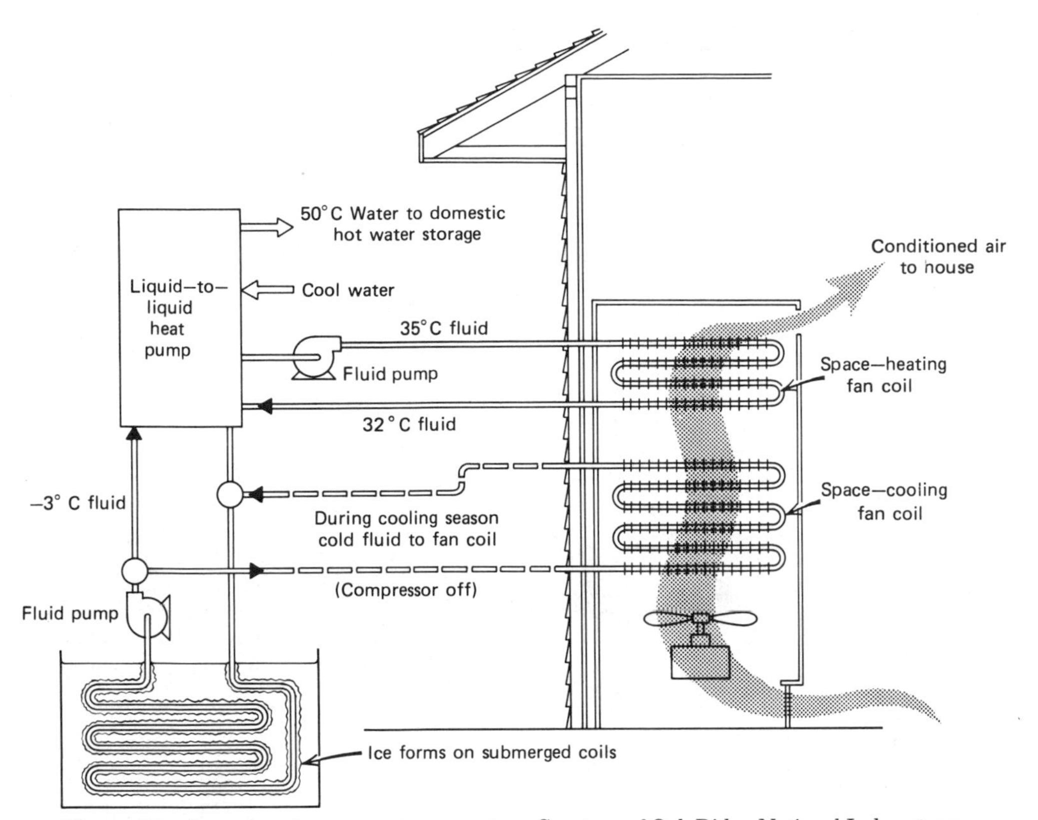

Figure 17. Annual-cycle energy storage system. Courtesy of Oak Ridge National Laboratory.

High specific heat is desirable in a storage medium. Furthermore, the thermal mass must provide continuous heat conduction. For example, a solid concrete wall provides better thermal storage than a hollow block wall filled with sand. If a block wall is filled with grouting, voids should be prevented. The properties of heat-storage material are given in Table 14.

Latent Heat Storage. Latent heat storage requires a phase change in the storage medium, ruling out materials such as brick, concrete, and rocks. Water is a common example of enhanced heat storage with a phase change. Melting ice absorbs great quantities of heat with no change in temperature. For cooling, this property can be put to good use, but the phase change temperature of water is not suitable for storing heat for later use. A material is needed with a high heat of fusion that changes phase at a higher temperature than water. Desirable characteristics include small change in volume and low vapor pressure. Sodium sulfate decahydrate, commonly called Glaubers salt, was one of the first materials used. The properties of typical phase-change materials (PCMs) are given in Table 15.

Storage of a GJ (ca 10^6 Btu) at a reasonable space-heating temperature requires ca 33 m^3 (1150 ft^3) of rock or 13 m^3 (460 ft^3 or ca 3450 gal) water. A phase-changing salt solution, however, can store a GJ (ca 10^6 Btu) in only 2.8 m^3 (24). The problems encountered in phase-change heat storage stem from crystallization and gradual loss of ability to change from liquid to solid after a number of phase-change cycles. Storage in shallow trays, mechanical mixing, rotating drums, or binding agents prevent crystallization or settling out. Compactness is desirable for retrofit heat-storage applica-

Table 14. Properties of Various Heat-Storage Media [a]

Storage medium	Heat capacity, kJ/(kg·K) [b]	Density, g/cm^3	Unit heat capacity, $J/(cm^3·K)$ [b] No voids	30% voids
water	4.18	1.0	3.6	
water 30%/glycol 70%	3.34	1.02	3.1	
scrap iron	0.45	7.82	3.4	2.3
magnetite	0.67	5.12	3.4	2.3
scrap aluminum	0.88	2.69	2.3	1.4
concrete	1.13	2.24	2.3	1.6
rock (basalt)	0.84	2.88	2.3	1.4
brick	0.84	2.24	1.8	1.3
rock salt	0.91	2.17	2.0	1.3
adobe	0.84	1.92	1.4	
wood (pine)	2.80	0.49	1.3	
paraffin oil	2.09	0.78	1.4	

[a] Ref. 23.
[b] To convert J to Btu, divide by 1054.

Table 15. Properties of Phase-Change Materials [a]

Material	Mp, °C	Heat of fusion, kJ/kg [b]	Heat-storage density, MJ/m^3 [c]	Approx. cost, $/kg
sodium sulfate decahydrate	31	251	372	0.02
sodium thiosulfate pentahydrate	49	209	335	0.16
barium hydroxide octahydrate	82	265	657	0.66
C_{14}–C_{16} paraffins	2–7	152	119	0.22
sodium fluoride·magnesium fluoride (1:1)	832	625	1370	1.10–0.55
sodium hydroxide	320	159	284	0.65

[a] Ref. 24.
[b] To convert kJ/kg to Btu/lb, divide by 2.32.
[c] To convert J to Btu, divide by 1054.

tions. However, care must be taken to prevent undesired interactions with other materials; for example, some PCM materials placed within hollow concrete blocks expand and cause fractures.

Because of the problems presented by phase-changing salts, some researchers have recommended paraffin waxes for solar heating and tetrahydrofuran for air conditioning. However, the Cabot Corporation test house built some years ago at MIT successfully used a eutectic salt mixture through 3500 cycles without appreciable degradation of performance. Trade journals advertise an energy rod that uses a phase-change compound capable of storing 385 MJ/m^3 (10,300 Btu/ft^3) at 27°C, and 485 MJ/m^3 (13,000 Btu/ft^3) at 49°C.

High Temperature Heat Storage. Solar power and industrial process-heat systems call for higher temperature heat storage, ie, >250°C, than residential or commercial applications do. Obviously, more insulation is required to maintain elevated temperatures for long periods. At very high temperatures, for example, for a turbine driven at 800°C, the containment vessel itself must be constructed of higher quality material, perhaps stainless steel. The insulating material must not only insulate the heat-transfer medium but also survive the elevated temperatures. High temperature storage media include oil/rock composites and phase-change salts. A rock-bed high temperature storage system is illustrated in Figure 18.

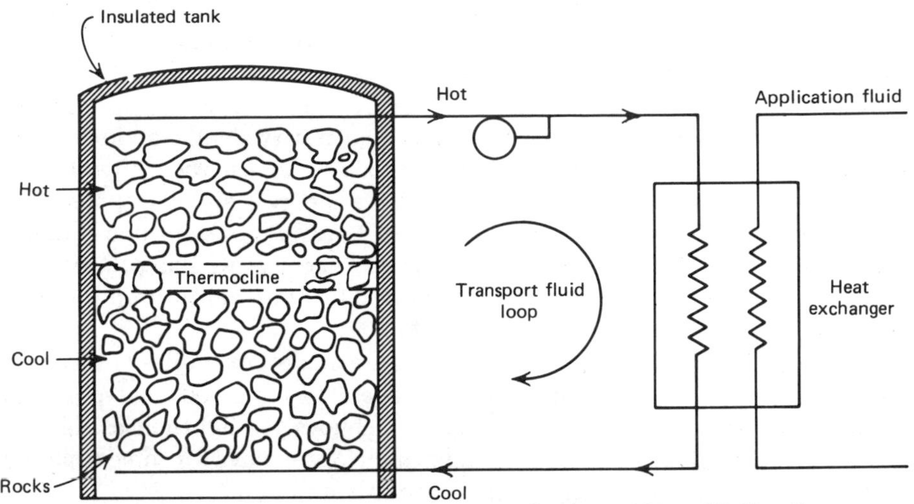

Figure 18. High temperature storage system. Courtesy of Marcel Dekker, Inc.

Reversible Thermochemical Reactions. Heat-storage materials used for reversible thermochemical reactions include hydrated, ammoniated, or methanolated salts, sulfuric acid, hydrogenated metals, and hydrated zeolites. Functioning as chemical heat pumps, these materials have estimated coefficients of performance ca 1.6 in the heating mode, and <1.0 in the cooling mode. Essentially, thermochemical energy storage is based on the concept: AB = A + B (Fig. 19). It offers high energy density and long-term storage at ambient temperatures. Products and reactants are easily transported, and therefore, endothermic and exothermic reactors can be separated by appreciable distances. However, thermochemical systems are extremely complex compared to both sensible and phase-change storage systems. At present, thermochemical storage is a long way from reality (25).

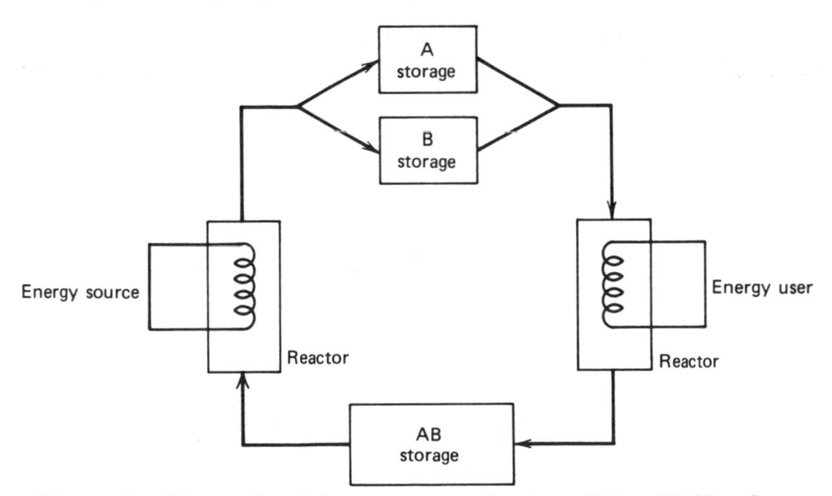

Figure 19. Thermochemical energy storage. Courtesy of Marcel Dekker, Inc.

Ocean Energy

More than 70% of the solar radiation reaching the earth falls on the ocean. For decades, ways of extracting power from the oceans have been sought by scientists and engineers in many countries. Mechanical, thermal, and chemical energy contained in the ocean represents an estimated 2 EJ (1.9×10^{15} Btu) annually (26).

Wave Power. The flux in the upper layer of ocean water influenced by winds is estimated at ca 100 EJ/yr (95×10^{15} Btu/yr). This quantity exceeds present U.S. use of energy but the prospect for converting more than a very small fraction of the potential is slim. Most experimentation in this field today is done in the United Kingdom, although several other nations are also active. Wave energy is highly site-specific, but the west coasts of North America, Scotland, and Norway represent power densities of roughly 10–100 kW/m wave front. Wave power increases with depth. An annual average of 91 kW/m crest length has been reported for Weather Station India in the northeast Atlantic (27).

Bell buoys, whistle buoys, and bilge pumps are modest examples of wave-power utilization. Japan has for more than a decade operated some 400 wave-power units generating 70–170 W. These serve as navigational buoys and lighthouses. Japan, the United Kingdom, the United States, Canada, and Norway have for some time sought methods to produce large amounts of power from waves.

There are several ways of extracting power from waves. The small Japanese point absorbers accept wave energy from all directions and are therefore three-dimensional. Vertical motion is converted to electric power inside the buoy, whereas two-dimensional devices accept energy in one direction only, the line along which wave trains move. Some, like the Salter Cam of the UK, use the rotary motion of waves. This design is claimed to be ca 80% efficient. Elongated wave-power devices may be linked with hinges, and power extracted by mechanisms at the hinge. In pneumatic devices, a displaced air column drives a turbine (see Fig. 20). Some wave machines are permanently moored to a pier or anchored to the ocean bottom to provide a fixed point against which wave motion works. Advanced concepts include wave-focusing devices and ships designed to extract propulsive or secondary power from the waves. Small boats have been driven by wave-power mechanisms. Several wave-power designs are shown in Figure 21.

In the United Kingdom, much time and money have been spent on wave-power experiments. The results have been technically successful but economic evaluations

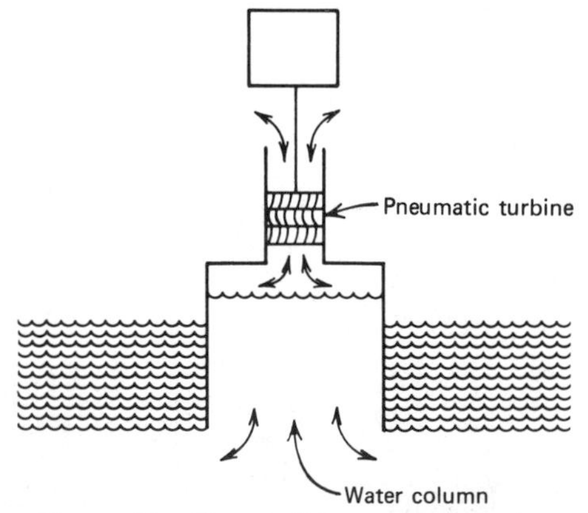

Figure 20. Pneumatic wave-power device.

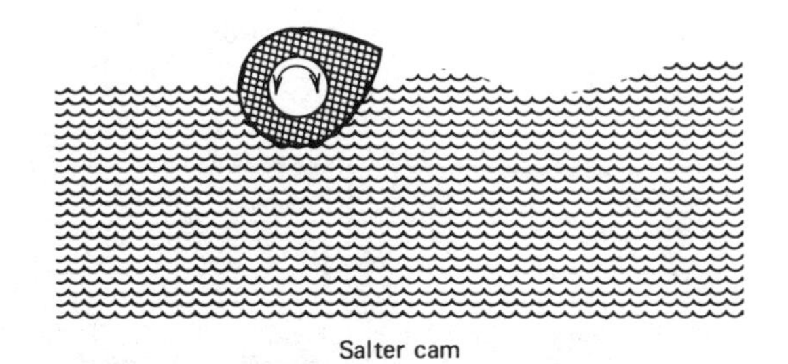

Salter cam

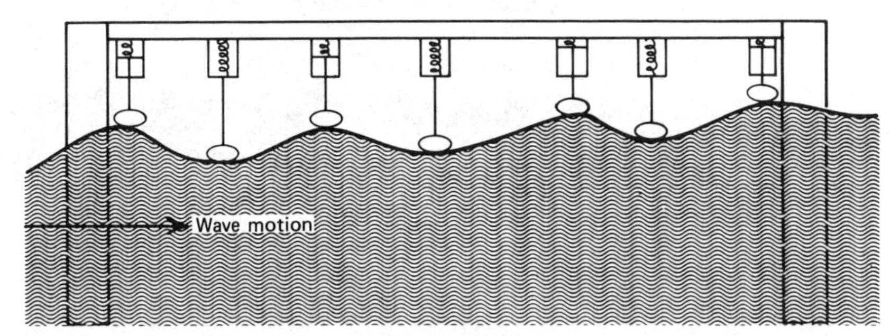

Pier—mounted device

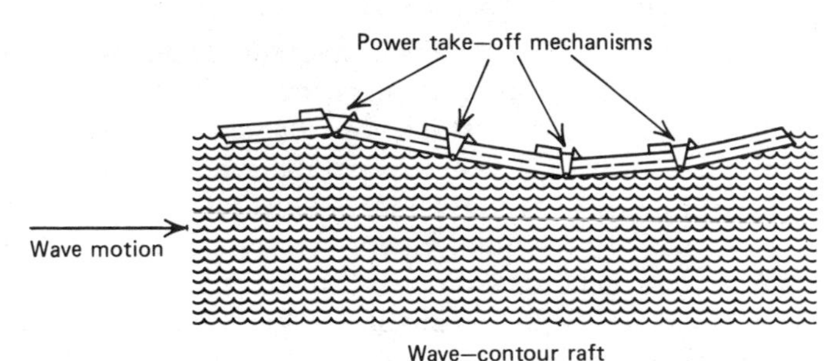

Wave—contour raft

Figure 21. Wave-power designs.

have been discouraging. Funding for continued effort has been directed to designs with better potential for cost-effectiveness. The basic problem is to design a device that produces power efficiently and yet withstands strong storms that have wrecked many installations. The recent Dam-Atoll design by Lockheed Corporation is shown in Figure 22. Waves sweep around the top of a 75 m dia inverted bowl and are driven into a swirl that flows into an 18 m cylindrical opening in the center. This whirlpool action turns a turbogenerator to produce electricity. The inventors claim that 500 to 1000 Dam-Atoll machines in a coastal area with 40 kW of wave power per meter would match the output of Hoover Dam (28).

Ocean Currents. Ocean gyres are sinuous reservoirs of kinetic energy that could be tapped for power. Driven by the prevailing global wind systems and modified directionally by the Coriolis effect, the currents rotate clockwise in the northern hemisphere and the reverse south of the equator. Of most interest to the United States is the Gulf Stream moving roughly parallel to the east coast. This current was studied by Benjamin Franklin in 1769 and later called by oceanographer Maury "a mighty river in the ocean." The Florida current, part of the Gulf Stream system, flows within 33 km of Miami and is estimated to contain more energy than all the rivers of the world; $(70–90) \times 10^6$ m^3 water per second [ca $(25–32) \times 10^8$ ft^3/s] is moved by this ocean

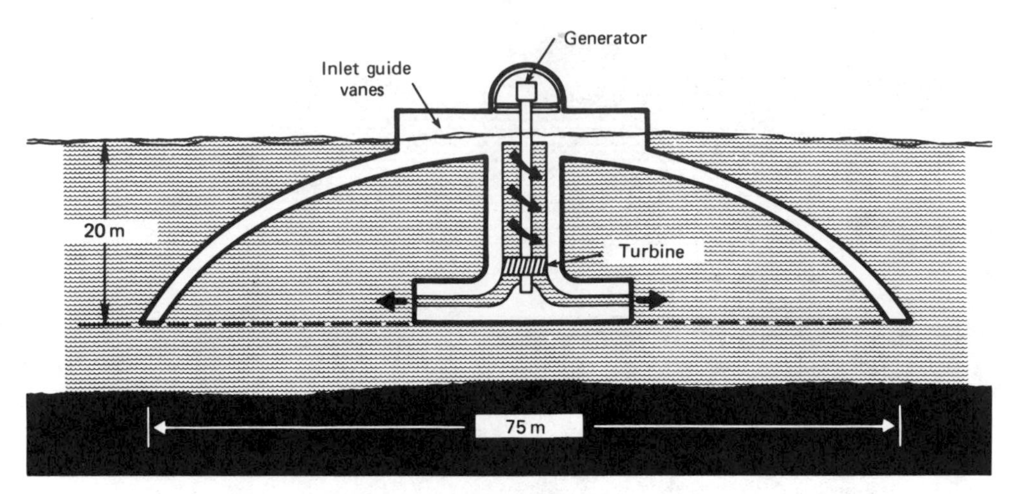

Figure 22. Dam–Atoll wave-power system.

current. Other currents include the Caribbean, North Equatorial, South Equatorial, Guiana, Humboldt, Japan, and Labrador. The annual principal ocean-current flux is estimated at ca 1 EJ (0.95×10^{15} Btu) (12).

Although appreciable use has not been made of strong river currents, many schemes have been proposed for tapping ocean currents. These include huge undersea turbines, as well as large parachutes attached to long cables on a system of pulleys. Aerovironment, Inc., of California has estimated that an average of 10 GW of power could be extracted continuously without appreciably affecting the flow of the Florida current. As a subcontractor to the Solar Energy Research Institute, Aerovironment has proposed an array of 242 moored submarine turbines, each 170 m in diameter (see Fig. 23). Extending 30 km across the stream and 60 km downstream, this array would produce a total of 10 GW of power and ca 50×10^{12} W·h/yr, a use factor of ca 57%. This quantity represents about half the annual electric-power needs of the state of Florida. Installed cost is estimated at ca $1200/kW with electric power delivered at ca 4¢/kW·h (29). Cost comparisons are given in Table 16.

Table 16. Cost Comparison for New Power Plants in the United States, Estimate 1978 [a]

Energy source	Plant cost with transmission, $/kW	Plant factor, %	Fuel cost, ¢/(kW·h)	Bus bar power cost, ¢/(kW·h)
combined cycle with oil	535	65	6.0	8.3
coal plus cleanup	1035	65	2.1	7.3
nuclear	1330	70	0.7	5.6
Coriolis force	1050	57	0	4.0

[a] Ref. 29.

Ocean Thermal Energy Conversion (OTEC). For ages, the oceans have absorbed solar radiation and have reached an equilibrium in which heat losses from the water are balanced by incoming radiation. The amount of heat energy contained in the ocean is estimated at ca 1000 EJ (ca 10^{18} Btu) annually (12).

The idea of exploiting thermal gradients was first mentioned by a French scientist in 1881, in a paper suggesting a heat engine operating on the modest temperature

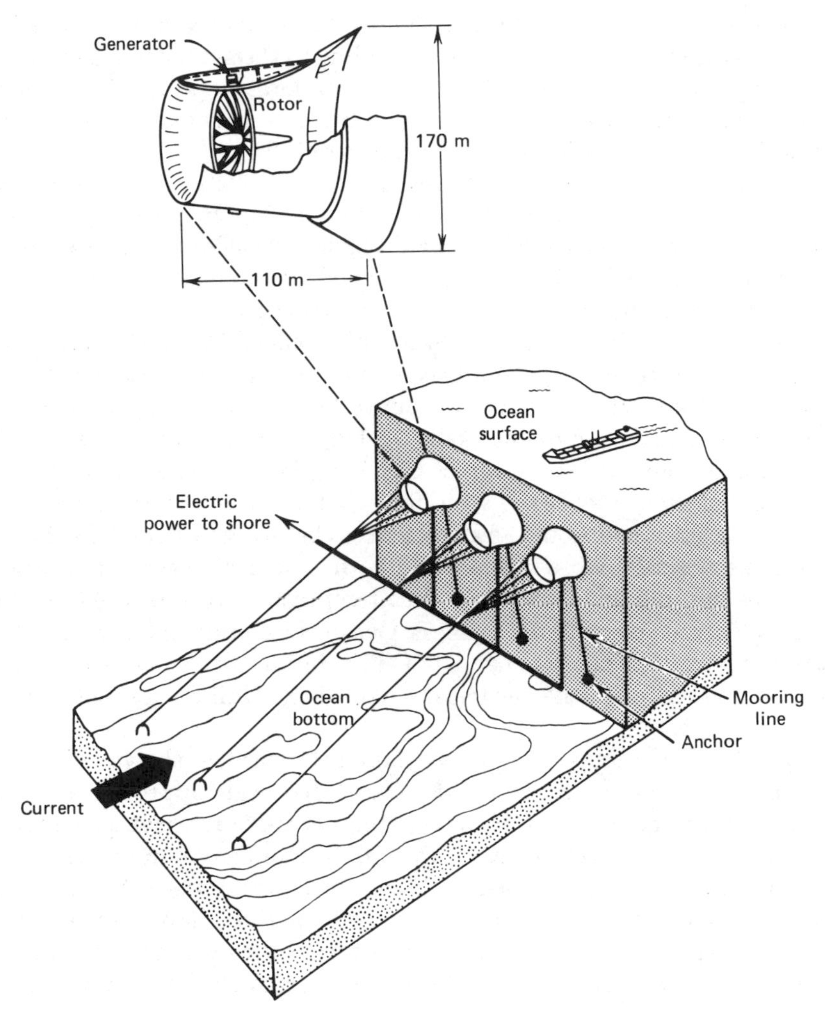

Figure 23. Coriolis ocean-current turbine. Courtesy of *Oceanus*.

difference between snow and river water (30). This idea was expanded to a heat engine that would operate on the temperature difference between surface water of the ocean and water at considerable depth.

The OTEC plant as a heat engine is limited in efficiency by the Carnot cycle:

$$\text{Eff} = \frac{T_{\text{in}} - T_{\text{out}}}{T_{\text{in}}}$$

where T = temperature in K. The OTEC plant operates on a temperature difference of ca 30°C at best. Assuming surface water at 27°C (300 K) and at 1000 m depth at 4°C (277 K), the maximum theoretical efficiency of an OTEC plant is

$$\frac{300 - 277}{300} \times 100\% = 7.67\%$$

This is a theoretical value which is reduced to ca 2–3% in practice. Nevertheless, the OTEC scheme is attractive because the ocean heat source is virtually limitless.

In 1929, a land-based sea-thermal energy plant was erected in Cuba. The cold water was pumped through a long pipe extending to the sea bottom and used to condense warm surface water flashed into low temperature steam in the boiler. An open-cycle engine produced ca 20 kW, but the pumping equipment used ca 60 kW. A floating OTEC plant was hardly more successful (31).

The French government started a sea-power plant at Abijan in North Africa in the 1940s, but abandoned the project before completion. By the early 1960s, private U.S. researchers had begun to design OTEC plants. Following the energy crisis of 1973 and the formation of the U.S. Energy Research and Development Administration (later DOE), the U.S. government became interested in OTEC.

In 1979, a small, privately developed test plant called MiniOTEC was successfully operated off Hawaii. Its input-output power balance bettered the earlier venture: gross output was 50 kW and the net electric output ca 12 kW. A closed-cycle design was used with heat exchangers and ammonia as the working fluid. A year later, DOE's OTEC-1, a 1 MW test bed mounted in a converted U.S. Navy tanker, began operating in 1200-m deep water off Hawaii (Fig. 24). A 4-module pilot plant of 40 MW output is planned next. Such an OTEC plant would be attractive to Hawaii and Puerto Rico. Both islands offer a cold-water source close to shore and both depend almost entirely on imported oil. Best sites are in equatorial waters but temperature differences of ca 20°C are considered sufficient for cost-effectiveness. Ocean areas offering this gradient abound, as shown in Figure 25. The proponents of OTEC foresee fleets of sizable plants in the Gulf of Mexico serving all the electrical needs of the Gulf States by the end of this century.

Design of OTEC plants is relatively straightforward, as shown in Figure 26. Although the open-cycle design is simple, the closed-cycle approach requires much smaller boilers and condensers. However, the necessity for heat exchangers of expensive, noncorrosive materials (eg, stainless steel and titanium) increases the cost and operational complexity of closed-cycle systems. Furthermore, the heat exchangers would be fouled by marine organisms. A number of methods have been devised to solve this problem, such as brushes or rubber balls forced through the heat exchanger periodically or chlorine introduced to inhibit marine growth. Environmental concerns

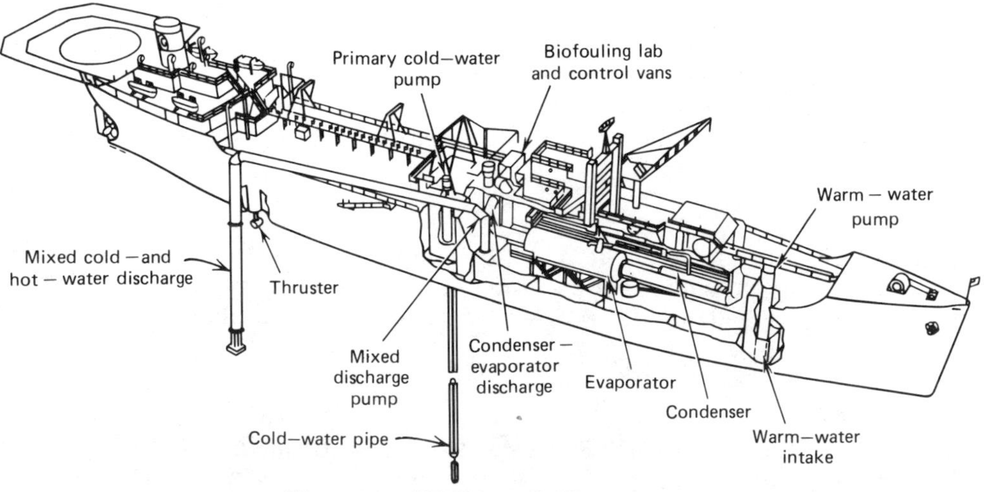

Figure 24. OTEC-1, a 1-MW test facility.

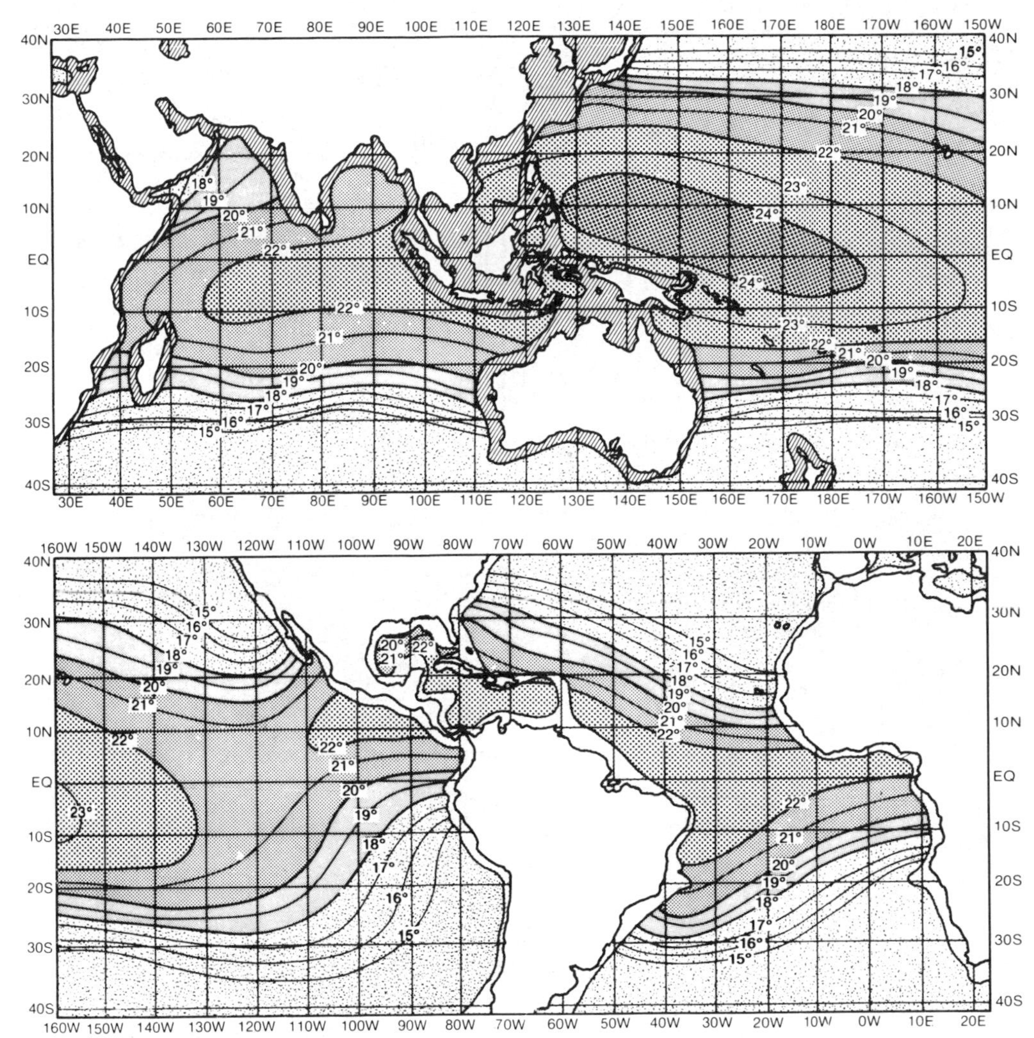

Figure 25. OTEC thermal resource ΔT (°C) between surface and 1000 m.

are raised by the possibility of leaks of ammonia or other working fluids. For these reasons, U.S. government research has turned to open-cycle designs. This effort is led by the Solar Energy Research Institute in Colorado.

A commercial OTEC plant would be ca 100 m across and would dangle a pipe 20–60 m dia hundreds of meters below it (Fig. 27). In addition to metals and plastics, concrete could be used for the OTEC plant and cold-water pipe. Much experience has been gained in building barges, floating docks, and other facilities, and it is felt that the state of the art is capable of fulfilling OTEC needs.

Both open- and closed-cycle OTEC systems raise nutrient-laden cold water which increases fish populations in the vicinity of the plant. It has been suggested that valuable minerals and chemicals might be recovered profitably as an adjunct to the production of power. During World War I, Haber suggested recovering the minute amounts of gold in seawater (32). His idea was impractical at the time because of high operational costs, but the greatly increased value of gold (and other metals), combined with reduced recovery costs in conjunction with the basic power-generation operation of the OTEC plant, may make this suggestion realizable.

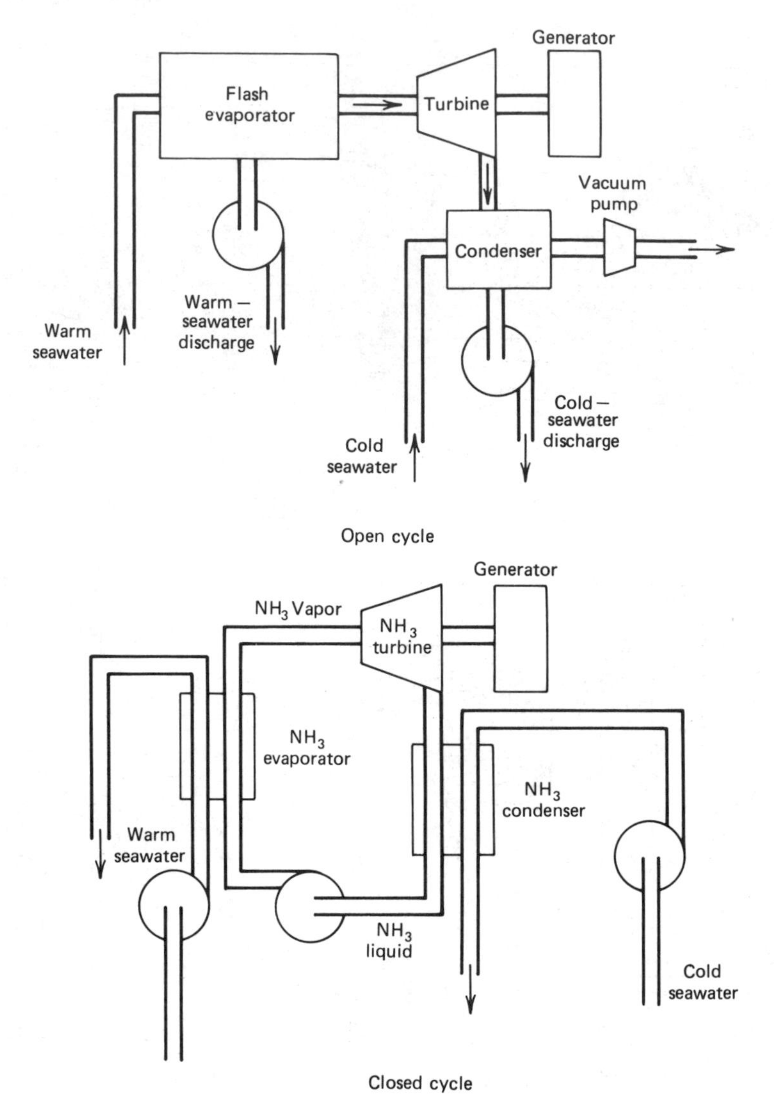

Figure 26. Open- and closed-cycle OTEC systems schematic.

The hybrid-cycle OTEC design combines features from both closed and open cycles. Warm seawater is flash-vaporized in partially evacuated evaporators. The vapor produced is then used to evaporate a second working fluid as in a closed-cycle system.

Variations of the basic open cycle use hydraulic effects rather than expansion to produce mechanical power. The mist-flow design uses convection to raise moisture in a low-pressure chamber. The condensate then falls and drives a hydraulic turbine. In the foam design, warm seawater is mixed with a foam-promoting biodegradable surfactant and introduced into an evacuated chamber. The warm foam rises, is condensed at the top of the chamber by cold seawater, and falls to a hydraulic turbine. A novel thermoelectric OTEC design was proposed in 1979 (33). Thermoelectric generation uses the Seebeck effect, in which the temperature gradient across a material drives charge carriers from the hot side of the thermocouple to the cold side. The

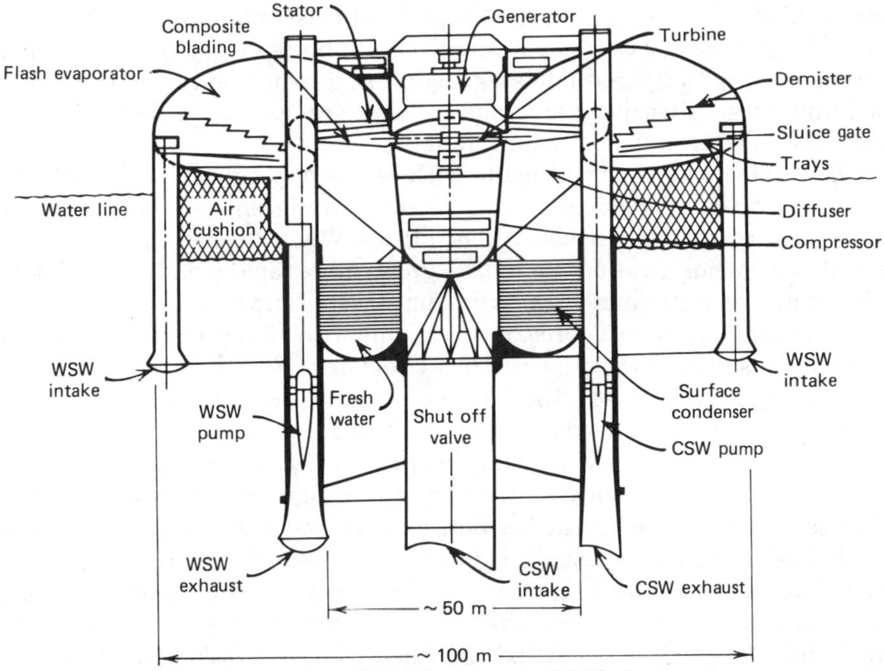

Figure 27. 100-MW open-cycle OTEC plant.

voltage produced is proportional to the temperature gradient. In operation, the thermoelectric OTEC plant would pump warm surface water over one side of the thermoelectric generator and cold water from the depths over the other (see Fig. 28) (see also Thermoelectric energy conversion).

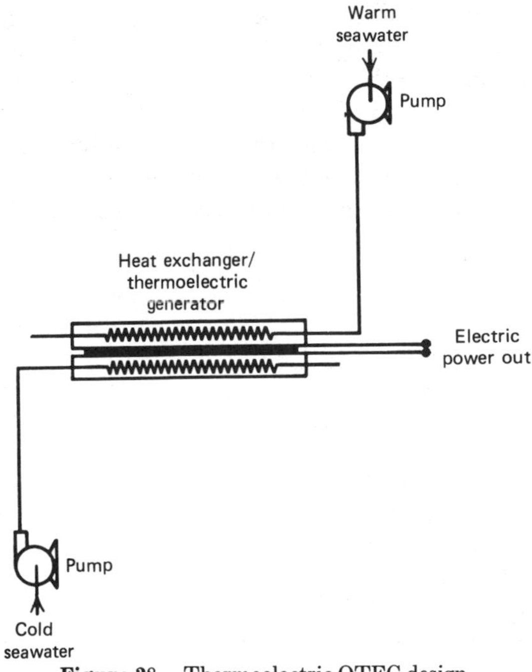

Figure 28. Thermoelectric OTEC design.

Marine Biomass. The ocean is the largest but least cultivated pasture on earth with ca 10^{10} metric tons of carbon per year fixed by photosynthesis (34). Some marine crops are harvested for hydrocolloids, eg, alginic acid, agar, and carrageenan, used by the food and drug industries (see Gums). Others are used as cattle feed, compost, fertilizer, and even for human consumption.

In the United States, experiments with seaweed in the late 1960s led to new techniques for growing *Chondrus crispus*, a red seaweed commonly called Irish moss, the source of carrageenan (see also Aquaculture). When grown suspended in water by vertical currents or aeration, *Chondrus* grows more rapidly and maintains a permanent vegetative, reproductive, nonfruiting stage offering the possibility of cultivating biomass as an energy source. Experiments with 50 species of seaweed native to Florida waters proved that the red alga *Gracilaria Tikvahiae* could produce an annual yield equivalent to >12.4 kg/m^2 under ideal conditions. Commercial yields of *Gracilaria* and other seaweeds are given in Table 17.

On the west coast of the U.S., experiments have been done with the giant kelp *Macrocystis pyrifera*. This alga grows to lengths of 50 m or more and has been harvested since 1910 for products such as potash, agar, and algin. Because marine algal tissue contains only small amounts of lignin and cellulose (which can present a problem in the production of fuels from biomass), and because plants like *Macrocystis* absorb ca 99% of sunlight (36), marine biomass was examined with increasing interest.

In the mid-1970s, the Ocean Food and Energy Farm (OFEF) project was funded by the Energy Research and Development Administration (ERDA, the forerunner of DOE), NSF, American Gas Association, U.S. Navy, and several other public and private organizations (35). The layout is shown in Figure 29. A huge, open-ocean kelp farm was planned, covering 4×10^4 hm^2 (ca 10^5 acre) ca 160 km off the southern California coast. Later, the project included the General Electric and United Aircraft Companies under the title Marine Farm Project.

A test module was installed in the ocean off Laguna Beach, which consisted of a 3 m upright buoy holding 103 kelp plants (see Fig. 30). These plants were fed nutrients that upwelled from a depth of ca 460 m. Although the adult kelp plants did not survive, they seeded the farm yielding an estimated 30,000 juvenile plants. Although many technical and economical problems remain to be solved, research on a full-scale kelp-farm project continues, with special emphasis on near-shore projects. An annual yield of ca 2 kg dry biomass per square meter has been estimated (see also Fuels from biomass).

Table 17. Yields from Commercial Seaweed Cultures [a]

Genus	Location	Yield, t/(hm^2·yr)
Laminaria (kelp)	People's Republic of China	4–7
Gracilaria	Taiwan	1.5–3
Eucheuma	Philippines	1.8
Gelidium	Japan	0.8–1.8
Porphyra (nori)	Japan	0.04–0.5
Porphyra (nori)	People's Republic of China	0.08–1.4
Undaria (wakami)	Japan	0.68

[a] Ref. 35.

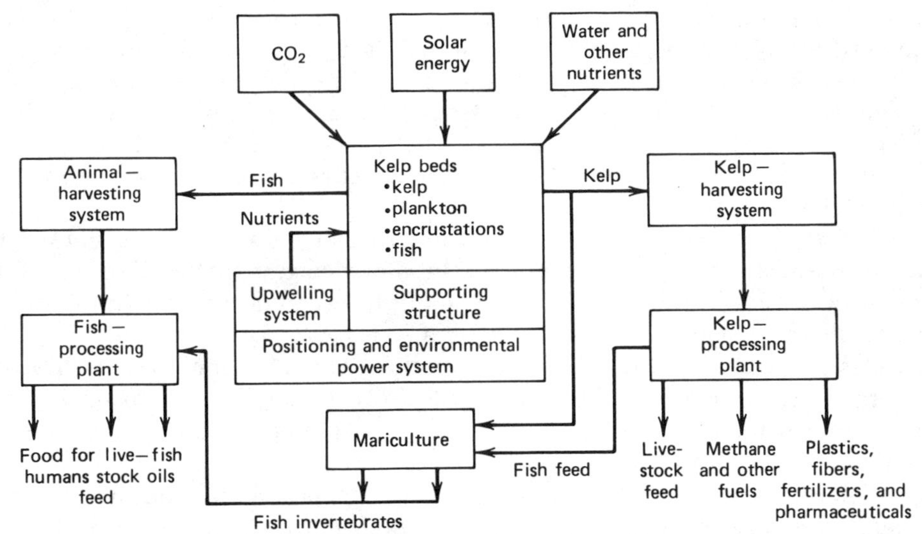

Figure 29. Ocean food and energy farm schematic. Courtesy of U.S. Naval Underseas Center.

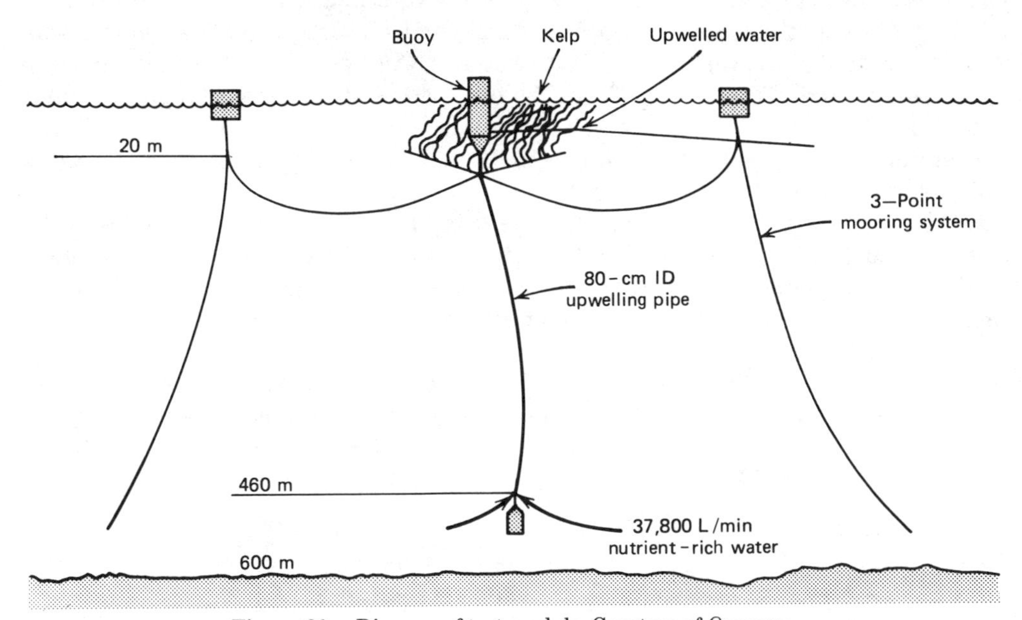

Figure 30. Diagram of test module. Courtesy of *Oceanus*.

Wind Energy

Wind power has been exploited to operate sailing vessels as early as 3000 BC. Crude wind machines were operating several centuries BC and European windmill designers were using complex aerodynamic techniques several centuries ago. During the 1940s, a large windmill generated 1.3 MW of electric power for a Vermont utility company. Machines of 2.5 MW output are operating today and designs as large as 5 MW are under development. Efforts of this magnitude obviously involve utility companies' participation.

Wind is not generally understood as the solar-thermal process it is; however, a small portion of solar energy in the atmosphere is transformed into wind. Factors involved in this transfer of heat energy to the atmospheric cycle include unequal heating of land and air masses, resulting atmospheric pressure changes, the rotation of the earth, and the diurnal and seasonal changes in solar-energy input. The atmosphere is indeed a huge heat engine fueled by solar radiation. Some of the kinetic energy thus generated is available as relative motion at the air–surface interface and can be exploited by wind machines (37).

Resource. The average solar flux reaching the earth's surface is ca 350 W/m^2. Estimates made at Lawrence Livermore Laboratory suggest that ca 2% of this flux is converted to wind and that 10% of the available low altitude wind might be tapped by windmills. This operation would yield ca 4000 EJ (3.8×10^{18} Btu) annually over the entire world, and 60 EJ (57×10^{15} Btu) over the continental United States. In 1979, U.S. energy consumption was ca 80 EJ (76×10^{15} Btu). A study at Princeton University suggests that a wind resource of 8% of the area of the contiguous states would provide 220 GW of electric power (27).

Mechanics. Because of the complexity of atmospheric circulation and the great irregularities in topography, wind velocity varies from calm to in excess of 67 m/s. Neither of these extremes is useful for present wind machines. Wind velocities of 5.4–17.9 m/s are attractive. A classification of wind-power density is given in Table 18. In a favorable wind regime, such as prevails on the Oregon coast, a wind machine could be expected to operate ca 88% of the time, with output at rated capacity about 33% of the time (27). Annual average wind speeds in the United States are 4.6 m/s (power density 143.8 W/m^2), ranging from 3.5 m/s (78.8 W/m^2) in Georgia to 7.3 m/s (399.8 W/m^2) in Wyoming.

Wind power varies with the cube of velocity; thus, higher speeds are desirable. The rated output of a given machine should be at ca 1.5 to 2 times the measured average wind. This is a compromise between a machine too small to take advantage of frequent strong winds and one so large it seldom operates at its maximum capacity (see Fig. 31).

Wind machines extract power from wind by converting its kinetic energy to rotary or other mechanical motion. Momentum theory dictates inescapable losses in the

Table 18. Classification of Average Wind-Power Density at 50 m Height[a]

	Average wind-power density and speed at 50 m height	
Class	Wind-power density, W/m^2	Speed, m/s[b]
1[c]	0–200	0–5.6
2	200–300	5.6–6.4
3	300–400	6.4–7.0
4	400–500	7.0–7.5
5	500–600	7.5–8.0
6	600–800	8.0–8.8
7	800–2000	8.8–11.9

[a] Ref. 38.
[b] Assuming Rayleigh speed distribution and sea-level conditions.
[c] Economic operation not feasible.

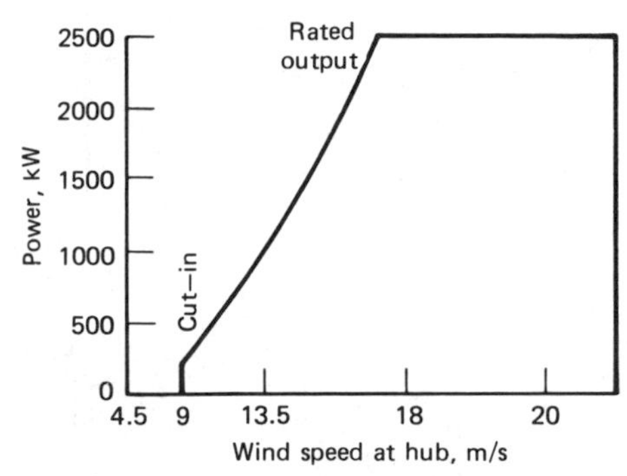

Figure 31. Typical cut-in/cut-out speed regime for wind machine. Courtesy of *Solar Age*.

process of extracting power from the wind. Assuming a wind machine of 100% mechanical efficiency, power extracted is expressed as follows:

$$\text{power} = 0.593\,\frac{\rho V_0{}^3}{2}$$

where: ρ = air density, V_0 = velocity, and $\rho V_0{}^3/2$ = ambient power density. The factor 0.593 is termed the Betz coefficient. Even though wind machines are unable to convert all the energy contained in the wind, they are still more efficient than many other energy converters; for example, two-bladed rotor designs have attained efficiencies of more than 40%.

Wind machines rotate about a vertical or a horizontal axis. Early Persian grain mills were vertical-axis machines; so are the modern Savonius, Darrieus, and giromill designs. However, most wind machines throughout history have been horizontal-axis designs. In the decades preceding rural electrification, 6×10^6 small horizontal-axis windmills operated in the United States, and all existing multi-megawatt machines are horizontal-axis designs. Most are two-bladed types, although there are sizable three-bladed wind machines and a company in the FRG is building a large, single-bladed design. Vertical-axis types include vortex generators, tracked airfoils, and Magnus rotors; horizontal-axis innovations include passive cyclic pitch turbines and venturi effects (see Figs. 32 and 33). Unusual concepts include solar-heated chimney machines and charged-particle extractors.

Technology. Small wind machines are derivative of earlier designs, improved by materials and electrical/electronic packaging. Wind machines are not especially sensitive to economies of scale; machines as small as a few kilowatts are often cost-effective, assuming a relatively long life with low maintenance costs. For a diagram of a typical wind-machine system, see Figure 34.

A recent directory lists 38 manufacturers of large and small wind machines (39). Not surprisingly, most large wind machines are designed and built by large industrial firms, eg, General Electric and Bendix, and aircraft companies like Hamilton Standard and Boeing. Such organizations bring aircraft and space technology to the field. As a result, new designs and unconventional materials are more evident in large than in small wind machines. In addition to wood, steel, and aluminum blades, materials include cement, plastics, fiber glass, and even carbon and boron fibers. Cost projections for large wind machines show them competitive with conventional electric power.

Because of the Public Utilities Regulatory Policies Act, utilities are required to

Figure 32. Vertical- and horizontal-axis wind-machine designs. Courtesy of Van Nostrand Reinhold Co.

Figure 33. Innovative wind-machine designs. Courtesy of Van Nostrand Reinhold Co.

purchase electric power from small power producers such as operators of modest-sized wind machines. For this reason, the operator of a small domestic wind-electric system may not need to provide battery or other storage. Instead, surplus electricity can be sold to the utility (properly conditioned to be compatible with utility voltage and frequency) and bought during periods of little or no wind. Properly managed, such arrangements can be of mutual benefit, saving the utility the expense of new plant capacity.

Large wind machines are obviously of greater interest to utilities geared to large power-plant operation. Pioneering the use of wind power are utilities like Southern

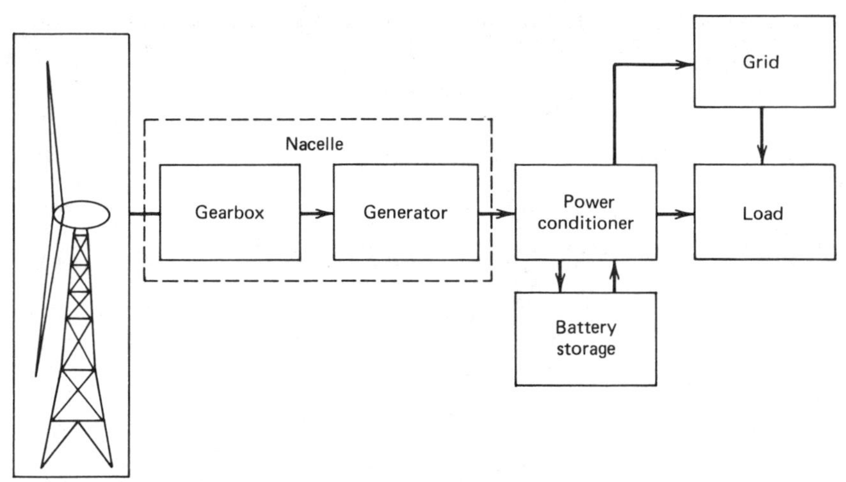

Figure 34. Diagram of wind-machine system.

California Edison and Pacific Gas and Electric, both of which are building arrays of wind machines in remote, windy areas. Prototype machines in the 2- and 3-MW range are in place and undergoing tests to determine capacity factor, grid compatibility, and maintenance costs. The Department of Interior, a large Federal producer of hydropower, has a project underway with the specific goal of supplementing the hydroelectric power supply from the Colorado River Project with large wind-machine arrays installed at Medicine Bow, Wyo.

At one time, all ships were sail-powered, and an interesting recent development in wind-power applications is that of sail-powered commercial shipping; eg, Japan's Shin-Aitoku Maru is a sail-assisted oil tanker. This innovation is a result of the huge increase in oil cost in the last decade.

Siting. Wind regimes vary regionally, and also on a micrometeorological scale within regions. The continental United States contains several regions with average wind-power densities at 50-m height >300 W/m^2, comparable to average solar radiation density. Regions of high wind-power density include the midwest, pacific northwest, Texas coast, northeast, and the northeast seacoast. Since most of the earth's surface is covered by ocean, much of the wind blows over open water. It has been proposed that large wind-machine arrays be sited offshore from the New England and the central seaboard states, taking advantage of high winds and relatively short underwater cable power-transmission lines. It has been established that the sites within 24 km of shore offer ca 20 EJ (19×10^{15} Btu) in addition to the 60 EJ (57×10^{15} Btu) estimated for the contiguous United States.

The economics of wind machines depend on a complex interrelationship of factors including not only the vagaries of wind but also the cost of conventional power, lifestyle, etc. Payback projections for wind machines/heat-pump systems are given in Table 19.

Hill crests and narrow canyons usually create stronger than normal winds. Over flat land, wind speed increases as the one-seventh power of height, giving high-mounted wind machines an advantage. Site selection for most efficient use of a wind machine is not an easy task. Although wind regimes are fairly well understood, careful monitoring is desirable for a specific site. The effect of altitude (air density) on windmill power output is shown in Table 20.

Table 19. Payback Periods For Wind-Power Installation For Residential Heat Pump and Electricity[a,b]

Location	Payback period, yr	Estimated 1979 rates, $/(kW·h)
Boston, Mass.	3.7	0.060
New York, N.Y.	4.2	0.086
Dodge City, Kan.	4.5	0.044
Brownsville, Texas	4.5	0.053
Cape Hatteras, N.C.	5.4	0.052
Bismarck, N.D.	5.8	0.048
Fort Worth, Texas	7.8	0.039
Columbia, Mo.	8.1	0.047
Caribou, Maine	8.1	0.042
Omaha, Neb.	9.1	0.039
Great Falls, Minn.	9.3	0.028
Charleston, S.C.	9.4	0.048
Miami, Fla.	9.5	0.042
El Paso, Texas	9.6	0.060
Albuquerque, N.M.	9.6	0.056
Madison, Wis.	11.3	0.042
Santa Maria, Calif.	12.4	0.049
Ely, Nev.	15.0	0.028
Lake Charles, La.	15.2	0.034
Washington, D.C.	15.6	0.036
Apalachicola, Fla.	16.9	0.045
Nashville, Tenn.	17.2	0.029
Fresno, Calif.	23.8	0.049
Phoenix, Ariz.	25.4	0.057
Seattle, Wash.	29.7	0.015
Medford, Ore.	56.7	0.030

[a] Ref. 40.
[b] Fuel escalation at 2%/yr, no waste electricity.

Table 20. Wind Energy vs Elevation[a]

Elevation, m	Relative wind energy, %
sea level	100
150	99
300	97
450	96
600	94
1500	86

[a] Ref. 41.

Limitations. Wind power presents a number of problems, including icing of blades, blade or tower failure, and alteration of surface wind regime (42). Noise and interference with TV reception are particularly objectionable in urban areas and have created legal obstacles to the installation of windmills. Residents of Boone, N.C., site of a large DOE wind-machine demonstration, complained about the ultrasound generated by the slowly whirling 57-m dia blades. This problem can be mitigated by siting large arrays of wind-electric generators in remote areas. High winds, hurricanes, and tornadoes also pose problems. Thunderstorms can create gusts of 67 m/s and downslope or foehn winds can be of the same magnitude. Although transmission-line towers in Guam have withstood typhoon winds of 85 m/s, such structures are not equipped with propeller blades up to 91 m in diameter. Several sizable wind machines of both horizontal- and vertical-axis design, have been damaged or destroyed by severe gusts.

Applications. Development and application of wind technology is taking place on two fronts: new machines provide power of up to 100 kW in rural areas and in small communities. In addition, machines of much higher capacity are producing power for electric-utility grids. Many of these installations are Federal demonstration projects, but in an increasing number of cases, utility companies cooperate with the wind-machine industry; eg, the Hawaiian Electric Co. is involved in a program to install an array of large wind machines producing 80 MW of electric power. Applications of wind machines of various power output are given in Table 21.

In 1980, the U.S. Congress passed the Wind Energy Systems Act, an 8-yr commercialization program for 700 MW produced by large machines and 100 MW produced by small machines. A long-range goal is the production of ca 2% of total energy needs or ca 2 EJ (1.9×10^{15} Btu) of wind energy by the year 2000; ca 220 GW of electric power might be produced by wind arrays covering 8% of the total area of the contiguous states, or 50% of those areas of highest wind potential. More than 300,000 2-MW machines would be required, spaced at intervals of about 1.6 km.

Table 21. Typical Applications for Wind-Energy Systems[a]

Rated output, kW	Application
2500	electric-power generation for utilities and industry
	pumping water for aqueducts and irrigation systems
	space and water heating and cooling for large buildings
1500	heating and cooling for water for aquaculture
	fertilizer and chemical production
	food processing
350	desalination of water
	electric-power generation for farms and industry
100	pumping water and other fluids and compressing air
	space heating and cooling of farm buildings
	heating and cooling of water for farms
40	drying and refrigeration of crops
	heating and cooling of commercial buildings and schools
	heating and cooling of household air space
10	heating and cooling of water for residences
	battery charging for electric vehicles
	generation of electricity for remote sites
1	generation of electricity for remote sites
	cathode protection for pipelines

[a] Ref. 40.

Photoconversion

Biological photosynthesis, the most natural form of photoconversion, produces biomass. However, photoconversion in its broadest sense includes all physicochemical processes that are driven by light, such as artificial photodecomposition of molecules, photovoltaic effects, photocatalysis, photosynthesis, photo-induced charge transfer, and photoisomerization. In the context of solar-energy conversion, photoconversion refers to the direct production of fuels and chemicals from sunlight and water, carbon dioxide, nitrogen, and simple organic compounds. Typical products are hydrogen, methane, alcohols, ammonia, and organic nitrogen compounds.

In general, photoconversion proceeds through the alteration of chemical bonds by the absorption of electromagnetic energy in the ultraviolet to the near-infrared regions. Of interest in solar-energy conversion and storage applications are wavelengths from ca 400–1200 nm. This range corresponds to ca 3.1 to 1.0 eV per photon. Chemical bonds have the same energy range and photoconversion processes should permit high conversion efficiencies. Photoconversion is a one-step process where light absorption and the ensuing chemical reactions occur simultaneously. Photoconversion systems are inorganic, organic, organometallic, or bionomic. Biomass conversion differs from photoconversion since it is a two-step process. Biomass is produced by sunlight (a photoconversion process), followed by the conversion of the biomass in the dark to fuels and chemicals.

Photobiological Processes. The natural photosynthetic process produces large amounts of biomass useful for fuel, but only at modest conversion efficiencies. For example, although the calculated theoretical conversion efficiency of electromagnetic energy to stored chemical energy in green plants is ca 10%, actual efficiencies of no more than ca 4–5% have been measured in the field, and these are only experimental, short-term conversion rates. Average annual efficiencies for a variety of crops range from <0.2 to 2%. Worldwide, the long-term conversion efficiency of photosynthesis is estimated at only ca 0.15% (see Table 22). The calculated theoretical maximum for

Table 22. Average Annual Photosynthetic Efficiencies[a]

Category	Utilization of sunlight, %
Crop	
cassava	0.8
kelp	2.0
napier grass	1.6
soybean	0.16
sugarcane	1.2
Area	
desert	0.02
forest, deciduous	0.4–0.6
forest, evergreen	0.8
grassland	0.8
reed swamps	1.1
Average total	
oceans	0.08
land	0.31
world	0.15

[a] Ref. 43.

photoconversion of incident solar radiation to usable chemical or electrical free energy is ca 32% (44).

Photobiological systems require the physical separation of dark reactions from light-driven reactions, as has been demonstrated for both bacteria and green plants, eg, by immobilization of the photosynthetic mechanism which produces charge separation. The immobilized system can be coupled to a photoelectrochemical cell, as in Figure 35, or be used to produce reduced substrates, which in turn produce hydrogen in the presence of an appropriate enzyme. Reaction-center complexes have been isolated from the photosynthetic bacterium *Rhodopseudomonas sphaeroides* R-26 and dried as an unoriented film on SnO_2 semiconductor electrodes. Both bacterial chromatophores and sphericles were used as coatings. (Chromatophores are the subcellular membrane vesicles from which reaction centers are isolated; *in situ* these reaction centers are perfectly oriented in the membrane. Sphericles are inside-out chromatophores with the reaction center oriented more than 80% in the opposite direction.)

Electron transfer between electrodes and chromatophores was detected in the external cell circuit (45). Photoeffects generated by chromatophore- and sphericle-coated tin oxide electrodes are given below. The electrolyte (pH = 7, E_h = 236–344 mV) contained 10 mM hydroquinone; liquid irradiance = 3 mW/cm^2.

	Photovoltage, mV		*Photocurrent, μA/cm²*	
Electrode	*peak*	*steady-state*	*peak*	*steady-state*
chromatophore-coated	85	70	0.52	0.26
sphericle-coated	27	23	0.16	0.10

Results with other biological components are given in Table 23.

The production of hydrogen by various phototropic organisms occurs in the presence of light with photosynthetic bacteria, cyanobacteria, or algae. The bacteria require reduced carbon substrates (eg, organic acids such as lactic acid), whereas water serves as the electron donor for algae. The primary hydrogen-producing enzymes are hydrogenase and nitrogenase (Fig. 36). Both photosynthetic and nonphotosynthetic organisms exhibit these activities (46).

In vivo cultures of photosynthetic bacteria offer the best prospect for efficient photobiological hydrogen production (46) (see Table 24). *In vivo* hydrogen production by green algae or cyanobacteria proceeds at much slower rates than production by photosynthetic bacteria.

In vitro H_2 generation, employing isolated spinach chloroplasts, ferredoxin,

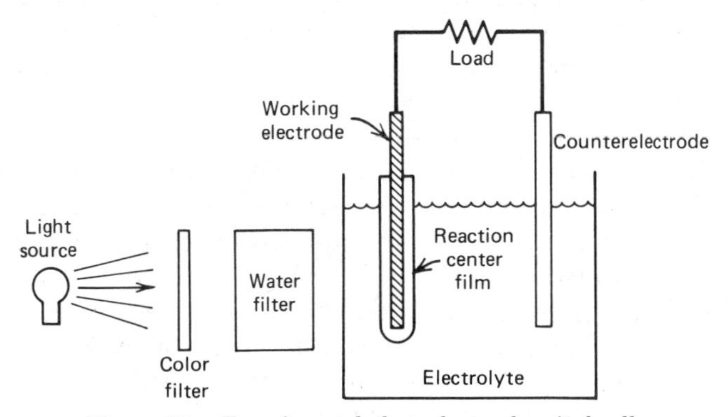

Figure 35. Experimental photoelectrochemical cell.

Table 23. Electrochemical Cells Employing Biological Components[a]

Biological component	Physical configuration	Light: Energy, W/m^2 [b]	Light: Wavelength	Conversion efficiency, %	Dark stability, days
immobilized algae	cells on SnO_2	500	red	very low	20
broken chloroplast membranes	membranes on platinized platinum	33.33	630 mm	2.45[c]	1–3
broken chloroplasts and PS I[d] particles	two half-cells separated by an ultrafiltration membrane, chloroplasts and PS I particles in solution	800	red	na	1–3
PS I particles	two half-cells separated by PS I-impregnated filter	8000	incandescent	0.13	10
purple membrane	two half-cells separated by bacteriorhodopsin incorporated into a stabilized BLM	200	incandescent red	6×10^{-5}	0.2
bacterial reaction centers	two half-cells separated by reaction centers incorporated into a planar lipid membrane		red artificial	very low very low	1 1

[a] Ref. 46.
[b] $1000\ W/m^2$ = 1 sun.
[c] Poised system.
[d] PS I = Photo System I.

Chromatium hydrogenase, and cysteine, was first reported in 1961 (47). Since then considerable improvements have been made. With water as a substrate, production rates have been increased by a factor of 6, and duration of the operation has been lengthened. Much remains to be accomplished in the photobiological production of electricity or fuels but encouraging progress is being made. The potential efficiency of such systems, and the increasing costs of conventional energy, justify a continued effort in this direction (see Photovoltaic cells; Hydrogen energy).

Photochemistry. Photochemical conversion mimics biological photosynthetic processes. Photochemical reactions are carried out in either homogeneous or heterogeneous media. Liquid-phase processes are preferred because of suitable reaction rates, ease of handling, and material transport. Aqueous systems are of particular interest because of the ready availability of water. In a photosensitized photochemical reaction, a sensitizer absorbs the light and transfers the absorbed energy to another reactant; the sensitizer is reversibly generated during the reaction cycle. In a direct photochemical process, at least one reactant absorbs light directly and is itself consumed in the overall reaction (see also Photochemical technology).

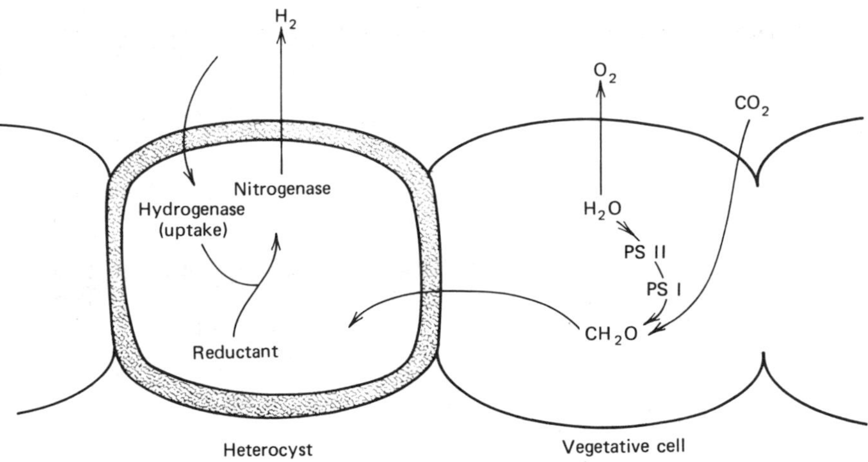

Figure 36. Hydrogen metabolism in cyanobacteria.

Table 24. Representative Rates of Hydrogen Photoproduction from Photosynthetic Bacteria[a]

Organism	Designation	Activity, mL H_2/(mg·h)[b]
Rhodopseudomonas capsulata	SCJ	168
Rhodopseudomonas capsulata	B10	124
Rhodopseudomonas sulfidophila	BSW8	106
Rhodopseudomonas viridis	NTHC 133	3
Rhodospirillum rubrum	S-1	146
Rhodopseudomonas palustris	EC	62
Rhodopseudomonas sulfidophila plus an unidentified marine species	BSW8	6
Rhodopseudomonas capsulata	W12 (B10 Nif)	3
Rhodopseudomonas capsulata	W52 (B10 Hup)	144
Rhodospirillum rubrum	S-1	4

[a] Ref. 46.
[b] Dry weight.

In light-driven homogeneous photoreactions with positive free-energy charges, the energy is driven uphill. The stored energy can be recycled. Desired are a free-energy difference of 1 to 2 eV, high energy density, high photon efficiency, and good visible-wavelength response. Researchers have set upper limits on photochemical conversion efficiencies, analogous to Carnot efficiencies of heat engines (33). Possible effects of temperature, concentration, and use of multiphoton absorbers are shown in Tables 25–27. No man-made photochemical system has yet approached such efficiencies, but they are a tantalizing goal.

Synthetic chloroplasts have been produced analogous to the natural chloroplasts in green plant cells (49). These structures split hydrogen from water and form carbohydrates by combining hydrogen with carbon from carbon dioxide in the atmosphere. The chloroplast substitutes are tiny spheres of oil floating in water which separate hydrogen from the water and release it as a gas. In addition to releasing hydrogen as a gas rather than uniting it with carbon, artificial chloroplasts offer a possible conversion efficiency of ca 30%, an order of magnitude better than that of natural photosynthesis.

Table 25. Optimum Wavelength Combinations and Corresponding Thermodynamic Efficiencies[a,b]

Number of absorbers	Optimum wavelengths, nm								Efficiency n_p, %
1								843	32.3
2							722	1318	43.6
3						581	836	1333	49.7
4					510	669	853	1340	52.6
5				505	649	831	1081	1355	54.4
6			490	620	778	997	1189	1368	54.9
7		479	583	715	840	1024	1187	1378	56.2
8	436	518	611	719	852	1090	1268	1397	57.0

[a] Ref. 48.
[b] Air mass = 1.2, T = 300 K.

Table 26. Maximum Thermodynamic Single-Photon Efficiencies for Solar Radiation at Various Intensities[a,b]

Intensity, W/m^2 [c]	Efficiency, % at absorber temperature, K				
	300	350	400	450	500
10^3	32.3	30.2	28.1	26.1	24.0
10^4	34.1	32.3	30.5	28.7	26.9
10^5	35.9	34.4	32.8	31.3	29.8
10^6	37.7	36.5	35.2	34.0	32.7
10^7	39.5	38.5	37.6	36.7	35.7

[a] Ref. 48.
[b] Air mass = 1.2, λ = 840 nm.
[c] 1000 W/cm^2 = 1 sun.

Table 27. Thermodynamic Efficiencies for Optimal Combinations of Two Absorber Wavelengths at Air Mass 1.2[a]

Intensity, W/m^2 [b]	Temperature, K	Absorber wavelength, nm		Efficiency n_p, %
		1	2	
10^3	300	722	1318	43.6
10^4	350	722	1325	43.6
10^5	400	722	1325	44.6
10^6	450	722	1327	46.4
10^7	500	722	1339	49.3

[a] Ref. 48.
[b] 1000 W/m^2 = 1 sun.

Other work in the area of chlorophyll analogues includes porphyrin-*p*-quinone structures (50). Porphyrin, a pigment module, mimics the action of chlorophyll in living plants and is the electron donor. *p*-Quinone is the electron acceptor. Although initial results show efficiency of less than 0.003%, investigators estimate that 10% or better is possible and envision large-area, low-cost plastic films in applications like rooftop electricity production (see Quinones).

Photoelectrochemistry. Photoelectrochemistry is based on the properties of photoactive semiconductor electrodes in contact with liquid electrolytes. Illumination of the semiconductor produces electrons and positive holes that separate in the electric field produced by the semiconductor–electrolyte interface. The holes are injected into the electrolyte from semiconductor anodes (*n*-type semiconductors) to produce oxidation reactions, whereas electrons are injected from semiconductor cathodes (*p*-type semiconductors) to produce reduction reactions. These oxidation and reduction reactions are called redox reactions (see Semiconductors).

Photoelectrochemical cells can be designed to produce either chemicals and fuels, or electricity. In the former case, so-called photoelectrosynthetic cells (Fig. 37), the oxidation and reduction reactions are different, resulting in a net change in the electrolyte that creates fuels or chemicals. For example, water can be oxidized to oxygen at the anode and reduced to hydrogen at the cathode, resulting in the overall photolytic splitting of water into hydrogen and oxygen. The hydrogen serves as both chemical feedstock and fuel. In the latter case, called electrochemical photovoltaic cells or liquid-junction solar cells, only one redox reaction occurs, and whatever is oxidized at the anode is reduced at the cathode; no net chemical change occurs in the electrolyte which serves only to move charges across the cell. Liquid-junction photovoltaic cells were discovered by Edmond Becquerel in 1839 (see Photovoltaic cells).

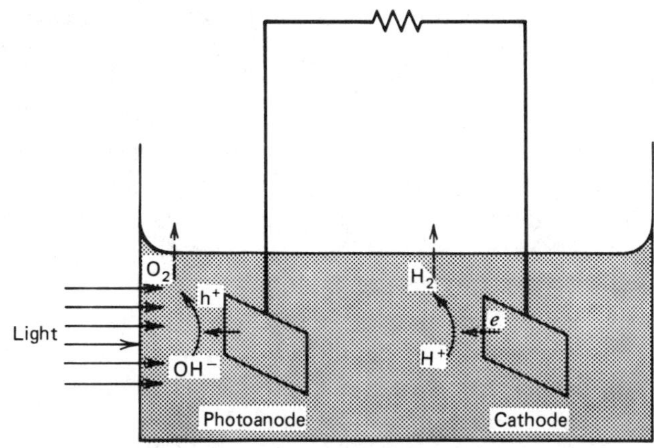

Figure 37. Photoelectrosynthetic cell.

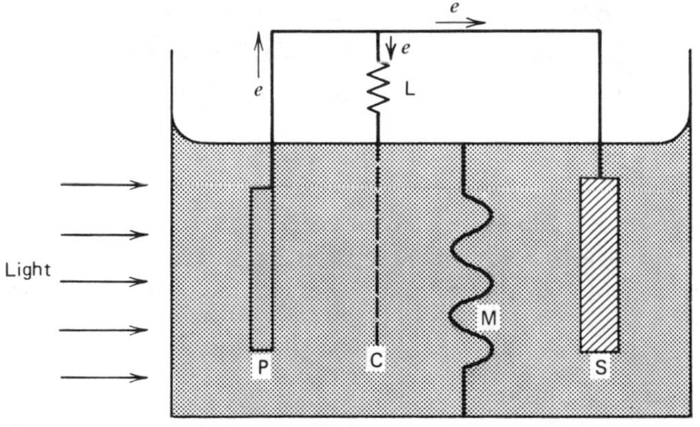

Figure 38. Photoelectrochemical cell with *in situ* storage. P, *n*-type photoelectrode. C, counterelectrode. S, storage electrode. M, semipermeable membrane. L, load.

Storage of electricity is also possible in an electrochemical photovoltaic cell with the help of a third storage electrode or the use of redox electrolytes in a redox battery (Fig. 38) (see Batteries and electric cells). Experimental, solar-powered, electrochemical, photovoltaic storage cells have been operated for months, providing an electric current day and night.

Liquid-junction solar cells offer several advantages over conventional solid-state photovoltaic cells. A redox electrolyte has a Fermi level or redox potential and thus can take the place of a *p*-type semiconductor or a metal to form the necessary junction. The junction is simple to make, and because the liquid electrolyte makes intimate contact with the semiconductor surface, there are fewer physical mismatches at the junction. This is especially beneficial when a polycrystalline photoelectrode is used. Furthermore, because the electrolyte can be transparent, there need be no absorption losses before light reaches the junction. Bell Laboratories researchers have produced liquid-junction gallium arsenide cells with efficiency in the range of 12%.

Liquid-junction cells are much larger and heavier than solid-state photovoltaic cells. Furthermore, they may require sealing if the electrolyte solution is affected by air. Corrosion is also a serious problem. Whereas solid semiconductors are stable with reported lifetimes >20 years, photoinduced chemical corrosion can occur in electrochemical photovoltaic cells. Various solutions have been proposed for these problems. A redox species is included in the electrolyte that preferentially traps the photoinduced charges before they corrode the electrode; eg, cadmium selenide as the semiconductor material in an alkaline solution of sulfur, and sulfide as the electrolyte (51). Oxidation of the semiconductor produces selenium which dissolves in the polysulfide electrolyte and is replaced by sulfide ions. Instead of corroding, the photoelectrode is constantly coated at the semiconductor/electrolyte junction with a cadmium sulfide layer. The photoelectrode described includes a titanium substrate. Titanium is kinetically sluggish for most electrochemical reactions and suppresses dark reduction currents and increases net photoanodic current.

Photoelectrosynthetic cells that drive reactions thermodynamically uphill are of particular interest since solar energy is thus stored as chemical energy. In 1976, workers at MIT and others discovered a $SrTiO_3$-based cell that requires no external bias (51). With a threshold energy just above 3.0 eV, this cell responds only to the ultraviolet. However, it converts this portion to storable energy at ca 25% efficiency (52).

The photolytic splitting of water in such cells, called photoelectrolysis, is of exceptional interest and intense research is in progress worldwide. Photochemical diodes of aqueous suspensions of *n*-type TiO_2/*p*-type GaP particles have been developed (53). These diodes yield hydrogen and oxygen when optically excited. Reduction of carbon dioxide and water to formic acid with photochemical diodes has also been reported. Photochemical diode powders have been reported as photosynthetic and photocatalytic materials (54). Gas/solid photochemistry may also have potential for converting light to fuel. Thus, much serious research is aimed at improving the present ca 1% efficiency for photoelectrolysis of water. Efficiencies for photoelectrolysis of ca 5% with stable InP photocathodes containing catalytic ions on the surface have been obtained (55).

Photogalvanism. Photoelectrochemical devices typically depend on the photoactivation of solid semiconductor electrodes to produce charge carriers. Photogalvanic processes, on the other hand, involve absorption of incident light by dye molecules in the cell electrolyte. The photoexcited dye molecules drive redox reactions in the electrolyte, and these redox products then transfer charge to the cell electrodes to produce an electric current. Photogalvanic devices convert light energy to electricity, or serve as storage cells recharged by light energy. Having no junction to prevent charge recombination as in a semiconductor, a photogalvanic device requires charge carriers in the form of redox species that back-react slowly enough to diffuse to electrodes. Chloroplasts in green plant cells evolve oxygen from water when illuminated in the presence of certain ferric compounds. In this reaction, hydrogen and oxygen were separated when water was illuminated by ultraviolet light. Photogalvanic reactions in ferrous-solution electrolytes were also reported (56).

The photoredox reaction of thionine TH^+ with the Fe^{3+}/Fe^{2+} couple in an acid solution has been extensively investigated. Leucothionine, TH_4^{2+}, is transparent in the visible and reacts relatively slowly with Fe^{3+}. Complete oxidation of leucothionine through the external circuit produces the cell's power (57). More recently, a totally illuminated device was described (rather than the first light and dark cell) with potentially greater efficiency (58). However, highest reported efficiency for such devices is 0.063%. This is an indication of the improvement needed before the photogalvanic approach is practical.

BIBLIOGRAPHY

1. A. R. Hoffman, *Domestic Policy Review of Solar Energy*, Solar Energy Policy Committee, U.S. Department of Energy, Washington, D.C., Aug. 25, 1980.
2. K. C. Brown, *The Use of Solar Energy to Produce Heat for Industry, TP-731-626*, Solar Energy Research Institute, Golden, Colo., April 1980.
3. K. C. Brown, *Power* **125,** 72 (March 1981).
4. *Chem. Week*, 33 (Dec. 3, 1980).
5. *Ann. Phys. (Leipzig)* **IV 7,** 408 (1902).
6. H. Tabor, *Electron. Power* **10,** 296 (Sept. 1964).
7. L. J. Wittenberg and M. J. Harris, *Sol. Eng.* **5,** 26 (April 1980).
8. *Solar Heating Materials Handbook*, U.S. Department of Energy NTIS, Springfield, Va., 1981.
9. *Sol. Eng.* **5,** 20 (April 1980).
10. A. F. Clark and W. C. Dickinson, *Solar Technology Handbook, Part A*, Marcel Dekker, Inc., New York, 1980, pp. 377–402.
11. A. C. Krupnick, M. L. Roberts, and M. H. Sharpe, *Natural-Oxide Solar-Collector Coatings*, Technical support package, brief No. MFS-23518, National Aeronautics and Space Administration, Marshall Space Flight Center, Ala., 1978.
12. *Nextel Brand Velvet Coating Series 101, Product Bulletin 201*, 3-M Company, St. Paul, Minn., 1973.
13. *PPG Standard Solar Collector General Information*, PPG Industries, Inc., Pittsburgh, Pa., 1976.
14. *Technical Data on the Chamberlain Manufacturing Corporation Flat Plate Solar Collector*, Chamberlain Manufacturing Corp., Elmhurst, Ill., 1976.
15. *Enersorb A, Super Koropon Primer*, technical data report, performance and durability testing, DeSoto, Inc., Des Plaines, Ill., 1977.
16. H. Y. B. Mar, *Sol. Eng.*, 31 (Aug. 1978).
17. E. Gore and A. S. Quershi, *Heat./Piping/Air Cond.* **49**(7), 43 (1977).
18. *Outgassing Products at 205°C in High Vacuum Determined by Mass Spectroscopy*, Sandia Laboratories, Albuquerque, N.M.
19. G. G. Allan, J. Dutloewoca, and E. J. Gilmartin, *Environ. Sci. Technol.* **14,** 1235 (1980).
20. G. A. Harper, ed., *Handbook of Plastics and Elastomers*, McGraw-Hill, New York, 1975.
21. J. G. Asbury, R. F. Giese, R. O. Mueller, and S. H. Nelson, *Proc. Am. Power Conf.* **39,** (1977).
22. T. B. Taylor, *Long-Term Storage of Heat and Cold*, Center for Energy and Environmental Studies, Princeton University, Princeton, N.J., June 1981.

23. H. G. Lorsch and K. W. Kaufman, *Central Storage for Solar Active Space Heating*, in *Proceedings of Solar Energy Storage Options*, CONF-790328-P1, U.S. Department of Energy, 1979; *Copper Brass Bronze Design Handbooks, Solar Energy Systems*, Copper Development Association, New York, 1978; A. Wilson, *N.M. Sol Energy Assoc. Bull.* 3(12), 10 (1978).
24. P. G. Grodzka in ref. 10.
25. R. W. Mar and T. T. Bramlette in ref. 10.
26. J. D. Isaacs and W. R. Schmitt, *Science* **207,** 265 (1980).
27. G. Thompson, *The Prospects of Wind and Wave Power in North America*, Center for Energy and Environmental Studies, Princeton University, Princeton, N.J., June 1981.
28. *Oceanus* **22**(4), 46 (Winter 1979/80).
29. P. B. S. Lissaman, *Oceanus* **22,** 23 (Winter 1979/80).
30. J. d'Arsonval, *La Revue Scientifique*, 370 (Sept. 17, 1881).
31. G. Claude, *Mech. Eng.* **52,** 1039 (1930).
32. F. Haber, *Z. Angew. Chem.* **40,** 303 (1927).
33. T. S. Jayadev, D. K. Benson, and M. S. Bohn, *Thermoelectric Ocean Thermal Energy Conversion*, TP-35-254, Solar Energy Research Institute, Golden, Colo., June 1979.
34. J. D. Issacs and W. R. Schmitt, *Science* **207,** 270 (1980).
35. J. H. Ryther, *Oceanus* **22**(4), 48 (Winter 1979/80).
36. W. J. North, *Chem. Technol.*, 294 (May 1981).
37. P. C. Putnam, *Power From The Wind*, Van Nostrand Reinhold Co., New York, 1974.
38. *Wind Energy Resource Atlas, Vol. 9, Southwest Region*, PNL-31-95, WIRA-9-VC-60, Pacific Northwest Laboratories, Richland, Wash., Nov. 1980.
39. *Wind Energy Equipment Manufacturers*, SP-451-569, Solar Energy Research Institute, Golden, Colo., Nov. 1980.
40. F. R. Eldridge, *Wind System Applications and Commercialization*, *Proceedings of the American Section/International Solar Energy Society*, Vol. 3.2, Phoenix, Ariz., June 2–6, 1980.
41. P. Gipe, *Sol. Age*, 34 (April 1981).
42. L. Coit, *Wind Energy: Legal Issues And Institutional Barriers*, TR-62-641, Solar Energy Research Institute, Golden, Colo., June 1979.
43. M. Seibert, J. S. Connolly, T. A. Milne, and T. B. Reed, *AIChE Symp. Ser.* **74,** 42 (1978).
44. M. Seibert, S. Lein, and A. F. Janzen, *Photobiological Production of Hydrogen and Electricity*, TP-623-815, Solar Energy Research Institute, Golden, Colo., Aug. 1980.
45. J. R. Bolton, *Science* **202,** 705 (1978).
46. M. Seibert and A. F. Janzen, *A Biological Solar Cell*, TP-332-476, Solar Energy Research Institute, Golden, Colo., 1980.
47. D. I. Arnon, A. Matsui, and A. Paneque, *Science* **134,** 1425 (1961).
48. R. V. Bilchak, J. S. Connolly, and J. R. Bolton, *Proceedings of American Section/International Solar Energy Society Annual Meeting*, Phoenix, Ariz., June 2–6, 1980.
49. M. Calvin, *Science* **184,** 375 (1974).
50. J. S. Connolly and J. R. Bolton, *paper presented at the 1982 meeting of the American Section of the International Solar Energy Society*, Houston, Texas, June 1982.
51. M. S. Wrighton, A. B. Ellis, P. T. Wolczanski, D. L. Morse, H. B. Abrahamson, and D. S. Ginley, *J. Am. Chem. Soc.* **98,** 2774 (1976).
52. M. S. Wrighton, *Chem. Eng. News* **58,** 29 (Sept. 3, 1979).
53. A. J. Nozik, *Ann. Rev. Phys. Chem.* **29,** 189 (1978).
54. A. J. Bard, *Science* **207,** 139 (1980).
55. A. Heller and R. G. Vademosky, *Phys. Rev. Lett.* **46,** 1153 (1981).
56. E. Rabinowich, *J. Chem. Phys.* **8,** 551 (1940).
57. N. H. Lichtin, *Chem. Technol.* 252 (April 1980).
58. W. D. K. Clark and J. A. Eckert, *Sol. Energy* **17,** 147 (1975).

General References

Proceedings of the 1980 Annual Meeting, American Section of the International Solar Energy Society, Vol. 3.1, Phoenix, Ariz., June 2–6, 1980.

W. C. Dickinson and P. M. Cheremisinoff, eds., *Solar Energy Technology Handbook, Parts A & B*, Marcel Dekker, Inc., New York and Basel, 1980.

Proceedings of the 1981 Annual Meeting American Section of the International Solar Energy Society, Vol. 4.1, Philadelphia, Pa., May 26–30, 1981.

Solar Thermal Energy

Proceedings of AAAS Symposium Efficient Comfort Conditioning: Heating and Cooling of Buildings, Feb. 1978.
Solar Energy for Agricultural and Industrial Process Heat, Program Summary, DOE/CS-0053, U.S. Department of Energy, Washington, D.C., Sept. 1978.
Proceedings: Solar Industrial Process Heat Conference, Vol. 1, SERI/TP-49-065, Solar Energy Research Institute, Golden, Colo., Oct. 18–20, 1978.
Proceedings of the American Section/International Solar Energy Society 5th Passive Conference, Amherst, Mass., 1980.

Heat Storage

Thermal Energy Storage, Solar Applications: An Overview, SERI/TR-34-089, Solar Energy Research Institute, Golden, Colo., March 1979.
Low-Temperature Thermal Energy Storage: A State of the Art Survey, SERI/RR-54-164, Solar Energy Research Institute, Golden, Colo., July 1979.
Survey of Sensible and Latent Heat: Thermal Energy Storage, SERI/RR-355-456, Solar Energy Research Institute, Golden, Colo., Oct. 1979.

Ocean Energy

Solar Sea Thermal Energy, Hearing before the subcommittee for Energy of the Committee on Science and Astronautics; U.S. House of Representatives, Ninety-Third Congress, Second Session, No. 41, May 28, 1974.
Proceedings of the 6th OTEC Conference, Washington, D.C., sponsored by U.S. Department of Energy, June 1979.
Proceedings of the 7th Ocean Energy Conference, Washington, D.C., sponsored by U.S. Department of Energy, June 1980.
Ocean Energy Conversion Systems: Annual Research Report, SERI/TR-634-1011, Solar Energy Research Institute, Golden, Colo., March 1981.
Proceedings of the 8th Ocean Energy Conference, Washington, D.C., sponsored by U.S. Department of Energy, June 1981.

Wind Energy

E. W. Golding, *The Generation of Electricity from Wind Power*, Halsted Press, a division of John Wiley & Sons, Inc., New York, 1976.
N. P. Cheremisinoff, *Fundamentals of Wind Energy*, Ann Arbor Science Publications, Ann Arbor, Mich., 1978.
F. R. Eldridge, *Wind Machines*, 2nd ed., Van Nostrand Reinhold Co., New York, 1980.

Photoconversion

J. R. Bolton, ed., *Solar Power and Fuels*, Academic Press, New York, 1977.
A. Mitsui, S. Miyachi, A. San Pietro, and S. Tammura, eds., *Biological Solar Energy Conversion*, Academic Press, New York, 1977.
L. A. Harris and R. H. Wilson, *Annu. Rev. Mater. Sci.* **8,** 99 (1978).
K. Rajeshwar, P. Singh, and J. DuBow, *Electrochim. Acta* **23,** 1117 (1978).
S. Wrighton, *Acc. Chem. Res.* **12,** 303 (1979).
H. Gerischer in *Solar Energy Conversion, Topics in Applied Physics*, Vol. 31, Springer-Verlag, Berlin, 1979.
M. Tomkiewicz and H. Fay, *Appl. Phys.* **18,** 1 (1979).
M. Archer, ed., *J. Photochem.* **10** (1979).
M. A. Butler and D. S. Ginley, *J. Mater. Sci.* **15** (1980).

A. Heller, *Acc. Chem. Res.* **14,** 154 (1981).
J. S. Connelly, ed., *Photochemical Conversion and Storage of Solar Energy*, Academic Press, New York, 1981.
A. J. Nozik, *Photoeffects at Semiconductor-Electrolyte Interfaces, ACS Symposium Series*, Vol. 146, American Chemical Society, Washington, D.C., 1981.

DAN HALACY
Solar Energy Research Institute

SOLDERS AND BRAZING ALLOYS

Solders

Soldering generally is used for making a mechanical, electromechanical, or electronic connection. This distinction is important because each application has its own specific requirements for solder alloys, fluxes, heating methods, and flux-residue removal. A mechanical joint made by a plumber in connecting two pieces of copper tubing is based on a tin–lead alloy as a filler metal to ensure that there are no leaks. A strong aggressive flux, a torch to heat the joint, and a damp cloth to remove excess flux residue are also used. Soldering the heating element contact to the external cord of a space heater may be considered as an electromechanical connection. In this case, a tin–lead-cored solder containing rosin flux and a soldering iron to melt the solder are used. The flux residue usually is not removed. Electronic connections are best exemplified by those used to attach components to the conductor paths of a printed wiring assembly by wave soldering. The solder is a tin–lead alloy. The flux is either rosin or a water-soluble organic flux. The heat is supplied by the molten solder. After soldering, the assembly is in most cases carefully cleaned to remove any residues which would cause malfunctioning of the electronic assembly (see Electrical connectors; also Welding).

Some of the more common solder alloys, their melting points, and their uses are listed in Table 1.

Joints. ***Design.*** A good solder joint should be one that provides visual inspectability, electrical conductivity, mechanical strength, ease of manufacture, and simplicity of repair. Each of these characteristics is determined by the design and selection of materials. Visual inspection of the solder joint is the most widely used nondestructive method of inspection, and proper design allows for visual inspection (See Nondestructive testing). Electrical conductivity is important in electronic assemblies, and joint design should maximize wire-to-terminal contact with the solder serving to fill all spaces and as a protection against atmospheric corrosion. Solders should never be relied on to provide mechanical strength. The joint should be designed to assure that any mechanical or thermal stresses are absorbed by the terminal-wire and not by the solder fillet. Design should take into account the solderability of materials used, space

Table 1. Common Solder Alloys

Composition, at %						Melting	
Sn	Pb	Cd	Bi	Ag	Sb	range, °C	Use
63	37					183	eutectic solder for electronic application
60	40					183–190	high quality solder
50	50					183–216	general-purpose solder, plumbing
40	60					183–238	wiping solder, radiator solder
30	70					183–255	machine and torch soldering
20	80					183–277	automotive-body solder
95					5	235–240	refrigeration soldering
62	36			2		179	soldering silver surfaces
1	97.5			1.5		309	high temperature soldering
15.5	32		52.5			90	fusible links
13	27	10	50			70	low melting solder

available for cleaning, type of heating used, as well as many other factors. Finally, the design should provide for ease of repair and replacement of parts.

Surface Preparation. It is essential to good soldering that the metals to be joined are compatible with the solder and that the surface be solderable under the conditions being used. Not all metals can be wet by solder. Copper, copper alloys, mild steels, nickel, etc, can be soldered. Certain stainless steels, titanium, molybdenum, and other metals cannot be soldered. For those metals that cannot be soldered, it is usually feasible to apply solderable coatings by electrodeposition or cladding.

Once the selection of metals that are compatible with solder has been made, it is necessary to prepare these surfaces for soldering. Parts being joined must be clean; oil, grease, dirt, and organic soils inhibit soldering and must be removed. This is usually accomplished by removing the contaminant with an appropriate solvent. Oxides, sulfides, carbonates, and other reaction products of the base metals are nonsolderable and must be removed by mechanical means, ie, abrasion, or chemical means, ie, bright dips. Surface preparation should be accomplished just prior to the soldering operation. The solderability of parts that have been stored for periods of time can be assessed by certain test methods (1–2).

Fluxes. Despite careful and thorough preparation of the surfaces being joined, it is always necessary to use a flux during soldering. A flux performs the following three functions: reacts with and removes surface compounds, eg, oxides and sulfides; reduces the surface tension of the molten solder alloy; and prevents oxidation during the heating cycle by providing a surface blanket to the base metal and solder alloy. Fluxes range in activity from inorganic acids and salts, which are the strongest, to rosin, which is the weakest. Classification by chemical composition is most commonly used.

Inorganic. Most inorganic fluxes are a combination of salts and acids dissolved in water with a wetting agent. Zinc chloride, ammonium chloride, hydrochloric acid, stannous chloride, and others are used. These fluxes are used for difficult-to-solder metals and usually in mechanical soldering where corrosion is not a problem. They should not be used in electrical or electronic soldering because the residues are too corrosive and generally are difficult to remove.

Organic. Organic fluxes are of the water-insoluble (rosin) and water-soluble (organic acid) kinds. The large majority of fluxes in use are organic. Rosin fluxes have been the most commonly used fluxes for electronic and electromechanical joining. They are available in core solder as well as in liquid form. Rosin is a complex mixture of isomeric acids; the principal component is abietic acid. Rosin flux is inert, noncorrosive, and nonconductive in the cold solid state but is active in removing tarnish films when hot. However, for many applications rosin is not suitable as a flux because it is too slow for modern, high-speed soldering and not aggressive enough for poorly solderable components. This deficiency has been overcome to some extent by adding activators to the rosin. These materials are generally amine hydrohalides, organic acids, or combinations of them. There are three types of rosins: nonactivated, mildly activated, and activated. Federal Specification QQ-S-571 characterizes rosin (R), rosin, mildly activated (RMA), and rosin, activated (RA) (3). For very high reliability equipment, only types R and RMA are permitted, because their flux residues are considered safe even if present after a cleaning operation. Type RA may be used if careful cleaning of the residues is provided and the assembly meets the cleanliness criteria of MIL-P-28809 Par 3.7 (4) (see Terpenoids).

Water-soluble organic fluxes, which have been used in mechanical soldering for many years, have recently achieved popularity in electromechanical and electronic soldering. The impetus for the interest in water-soluble fluxes lies in the fact that the residues can be removed with water rather than with the solvents required by rosin fluxes. Because they are more aggressive than rosin fluxes, they are useful where marginally solderable surfaces are being soldered. Water-soluble organic fluxes are mixtures of organic acids, amine hydrohalides, and surfactants dissolved in water or alcohol. They are available as neutral or acidic solutions and require very efficient cleaning systems if they are to be used effectively and safely on electronic assemblies.

Solder Selection. Most solder alloys are composed of combinations of tin and lead (see Tin and tin alloys; Lead alloys). The binary tin–lead eutectic composition is 63 at % tin and 37 at % lead, and its melting point is 183°C. For electronic assembly 60 Sn–40 Pb or 63 Sn–37 Pb solders are almost universally used. 50 Sn–50 Pb alloys are widely used in plumbing. 40 Sn–60 Pb is an inexpensive utility solder. 20 Sn–80 Pb is used as a body solder because of its large melting range. Military applications require that solders contain 0.2–0.5 at % antimony to reduce the possibility of transformation to a brittle phase at low temperatures. A number of alloys are used for special purposes. 95 Sn–5 Sb is used in refrigeration joints where tensile strength is important. Sequential soldering, in which a second joint is soldered in the vicinity of a previously soldered joint, must be accomplished with a lower melting alloy, eg, Sn–Pb–Cd or Sn–Pb–Bi, so as not to melt the first joint. Silver-fired or -coated parts are often soldered with Sn–Ag or Sn–Pb–Ag alloys to prevent the scavenging of silver by the molten alloy. These are but a few of the alloys in use. Additional alloys are described in refs. 3 and 5.

Solder alloys are made in many shapes and forms. Wire solder, in which the solder is supplied as solid or with a flux core, is used in making discrete joints. Bar solder is used in tinning pots, wiping solder pots, or wave or dip solder pots. Solder foil is used to produce stampings of special shapes, eg, washers and disks called preforms. These can be supplied with the flux on the inside (flux-filled) or on the outside (flux-coated) of the preform. Solder can be made as spheres and rings. One form of solder, which has been used for years by plumbers and has many industrial uses, is solder paste. In this form, powdered solder metal is suspended in a flux vehicle and placed on the part prior to soldering.

Heating. The choice of a heating method depends on a number of factors. For small runs, manual heating is generally used. Whatever method is used, it is important to bring all parts to be soldered to ca 50°C above the melting point of the solder. The most popular manual method is the electric soldering iron. The soldering bit is copper, usually coated with iron or an iron alloy to prolong the bit life. A variety of sizes and shapes of bits are available as well as soldering irons with different wattages. In addition to electrically heated soldering irons, gas-heated irons, resistance soldering tools, soldering guns, gas flames, and hot gas can be used.

Automatic or mass soldering is accomplished by a number of different methods. The most common method is wave soldering. The heat is applied usually to the mass soldering of printed wiring assemblies. An insulating, plastic-laminated board containing conductor paths is used as the support for the components. Components are mounted on the top side of the board with leads passing through holes where they are to be soldered to the circuitry (conductive paths) on the bottom of the board. The board is placed on a moving conveyor, passes over a flux applicator, a preheat unit, and a standing wave of molten solder. After cooling, the unit is ready for flux-residue removal. By this method it is possible to solder hundreds of connections in a few seconds with very little labor. There are many variations to the basic procedure. Printed circuit boards can be rigid or flexible. Components can be placed on one or both sides of the board. The boards can be soldered in a still pot of solder (drag soldering).

Oven soldering is another form of mass soldering. Solder as a preform or paste is placed on the assembly, and the units are placed in an oven for soldering. This can be a batch operation, or conveyor belts passing through the oven can be used. The source of heat in the oven may be gas, electrical, infrared heat lamps, etc.

Vapor-phase reflow soldering is a rapidly growing technique for special mass soldering applications (6). The source of heat is the boiling vapors of a stable organic liquid. The assembly, with components and solder preplaced, is lowered into a chamber of vapor whose temperature is above the liquidus of the solder being used. Soldering is rapid and controlled since the temperature can never exceed the boiling point of the liquid.

Cleaning. After soldering it is usually necessary to perform a final cleaning. This consists of removal of flux residues and oxides which have formed on areas unprotected by flux during soldering. Flux residues that are corrosive or whose presence would cause malfunctions, eg, short circuits or noise, must be removed. If the soldered joint is a mechanical connection, it is usually made with a water-soluble flux and the flux residue can be removed by total immersion in water or slightly acidified water. If the unit is an electromechanical or electronic assembly, total immersion in water may not be possible and flux-residue removal becomes an important aspect of the soldering process.

If flux-residue removal is required, the type of flux used determines the cleaning solvent to be used, and the nature of the assembly determines the method to be used. There are many types of cleaners for rosin fluxes. Flammable solvents, eg, alcohols and ketones, are excellent solvents for rosin and activators but are not used because of their associated fire hazard. Chlorinated solvents are excellent for rosin but are ineffective in removing the polar activators, and they present toxicity problems.

Combinations of alcohols and chlorinated hydrocarbons, which are nonflammable and have sufficient alcohol to remove activators, are commonly used. These bipolar solvents can be used as cold cleaners sometimes with ultrasonic energy or as vapor degreasers provided they are azeotropes or near azeotropes. The toxic vapors must be controlled by proper hooding and exhausts. Another method of removal of rosin residues is an aqueous process. Rosin can be converted to a soluble ammoniated soap by treatment with hot ammoniacal solutions. Special equipment is necessary to provide the combination of mechanical and chemical action necessary to solubilize the residues.

Water-soluble organic flux residues are usually removed with water. A well-designed and -monitored cleaning system is necessary for removal of water-soluble flux residues from electronic equipment.

Additional useful information on solders and soldering is given in refs. 7–9.

Health and Safety Factors. Solder is safe to use and presents no hazard when proper working conditions prevail and workers observe safety rules. Aside from the danger of handling hot molten metal, burning oneself with a hot soldering iron, etc, the greatest danger to health lies in the presence of lead or cadmium in the solder alloy. Lead is present in large amounts in most solder alloys and some special alloys contain cadmium. The precautions used in handling lead alloys apply also to cadmium-bearing solders.

Lead is absorbed through mucous membranes of the lung, stomach, or intestines and then enters the bloodstream. Excessive amounts of lead absorption can cause anemia, fatigue, headache, weight loss, and constipation. The chief sources of lead toxicity are lead fumes, lead dust, and lead compounds ingested by eating or drinking. Lead fumes and dust can be controlled by ventilation. Air monitoring for lead content is an effective method for determining the safety of the working environment. Establishment and enforcement of rules of hygiene, eg, prohibition of eating and smoking in the work area, thorough washing of hands before eating, etc, are recommended. Periodic blood tests for lead are a recognized means of monitoring the effectiveness of the safety programs. Additional health and safety information is available from the Lead Industries Association, Inc., New York (see also Lead compounds, industrial toxicology).

Brazing Alloys

Welding, brazing, and soldering are metal-to-metal metallic bonding processes. In welding, like or similar components are fusion-bonded at or just below their melting points. In most brazing, except for the brazing of aluminum and magnesium alloys, and almost all soldering, the components, alike or dissimilar, are molecularly bonded well below their melting points. In welding, the filler metal is either puddled into relatively wide gaps or the metal surfaces being joined are partially melted and bonded by fusion or by a combination of puddling and fusion. In brazing or soldering, the filler metals are drawn into closely fitted joints by capillary attraction, and they bond and solidify without melting the components. Appreciable alloying may or may not take place during brazing, and extensive alloying is to be avoided as it may result in an unfilled joint.

Joints. The brazing process is relatively simple (10). An unskilled workman can obtain a high proficiency after a short training period, or the process can be reliably and inexpensively automated. Certain steps, however, must be taken to produce good joints, ie, design, preparation of joint surfaces, fluxing, assembling and jigging, heating, and final cleaning.

Design. There are two basic types of joints, ie, butt and lap. Because the area of a butt joint is limited to the cross-section of the pieces being joined, the butt joint is not generally used except where lap joints cannot be accommodated by the shape of the parts. The lap joint can be adjusted in length for added strength if need be, and lap joints may help to jig the assembly.

To take full advantage of the capillary forces, close clearances of 0.025–0.13 mm should be employed. In shop terms this amounts to an easy slip fit on tubing or one flat part on top of another providing that the parts are mill-finished and not highly polished. If dissimilar metals that have different coefficients of expansion are involved, these must be accounted for, particularly with tubular assemblies, to obtain the proper clearance at the brazing temperature.

Surface Preparation. Oil, grease, dirt, oxides, and scale interfere with the formation of a molecular bond and must be removed prior to brazing. First the oil, grease, and dirt is removed by dipping the parts into a degreasing solvent, eg, trichlorethane, or by vapor degreasing or alkaline cleaning. Then the oxide or scale can be removed mechanically by abrasive cleaning or chemically with an acid pickle; however, no traces of acid may remain in crevices or blind holes (see Metal surface treatments, cleaning, pickling, and related processes). The next step is to flux and braze the parts as soon as possible while they are clean.

Fluxes. Fluxes are formulated to prevent the oxidation of the brazing alloy and the parts being joined below the brazing temperature and throughout the brazing cycle. This calls for fluidity and thorough wetting, chemical activity, and as little fuming as possible.

The fluxes used with Al–Si and Mg–Al brazing alloys function at 370–675°C and 480–650°C, respectively, and they contain fluorides and chlorides. These are very corrosive and are supplied as dry powders in air-tight containers, and they are converted into pastes with water, alcohol, etc, just before use. Use of an excessive amount should be avoided.

A number of fluxes are used with other brazing alloys. Two consist of borates, boric acid, fluoroborates, and fluorides and function at 565–870°C and 730–1150°C. The third type consists of borates, fluorides, and chlorides and functions at 565–870°C on aluminum–bronze and nickel alloys containing aluminum and titanium. The fourth consists of borates, boric acid, and borax and functions at 760–1205°C. These are supplied as powders, pastes, and, except for the third, liquids. Liberal application is advisable to ensure complete protection and to simplify final cleaning.

Protective atmospheres reduce the requirements for flux, but they are still desirable with brazing alloys and parts containing volatile elements, eg, cadmium or zinc, and with parts containing appreciable quantities of aluminum, beryllium, silicon, or titanium. The protective atmospheres range from the conventional, eg, wet or dry combusted gas and tank hydrogen, to the specific, eg, dissociated ammonia, deoxidized and dried hydrogen, purified inert gas, and vacuum.

Table 2. Compositions and Solidus, Liquidus, and Brazing Temperatures of Brazing Alloys[a]

Brazing alloy	Composition, at %																				
	Cu	P	Pb	Al	Zn	Ni	Fe	Mn	Si	Ag	Cd	Sn	Li	Au	Pd	Cr	B	C	Mg	Ti	Co	W
copper																						
1	99.90 min	0.075 max																				
2	99.00 min																					
3	86.50 min																					
copper–zinc																						
1	59.0				40[d]							0.63										
2	58.0				40[d]		0.73	0.25	0.10			0.95										
3	48.0	0.25 max			41[d]	10.0			0.15													
copper–phosphorus																						
1	95[d]	5.0																				
2	80[d]	5.0								15.0												
3	89[d]	6.0								5.0												
4	88[d]	6.75								5.0												
5	91[d]	7.00								2.0												
6	93[d]	7.25																				
7	87[d]	7.25								6.0												
silver–copper																						
1	15.0				16.0					45.0	24.0											
2	15.5				16.5					50.0	18.0											
3	22.0				17.0					56.0		5.0										
4	15.5				15.5	3.0				50.0	16.0											
5	26.0				21.0					35.0	18.0											
6	27.0				23.0					30.0	20.0											
7	30.0	0.025								60.0		10.0										
8	30.0				25.0					45.0												
9	28.0									71.5			0.5									
10	38.0				32.0					30.0												
11	34.0				16.0					50.0												
12	28.0									72.0												
13	30.0				28.0	2.0				40.0												
14	28.5					2.5				63.0		6.0										
15	40.0				5.0	1.0				54.0												
16	7.3									92.5			0.2									
17	42.0					2.0				56.0												
gold																						
1	20.0													80.0								
2						18.0								82.0								
3	62.5													37.5								
4	62.0					3.0								35.0								
5						36.0								30.0	34.0							
nickel[g]																						
1						70[d]			10.1							19.0						
2						73[d]	4.5		4.5							14.0	3.1	0.75				
3						73[d]	4.5		4.5							14.0	3.1					
4						82[d]	3.0		4.5							7.0	3.1					
5						91[d]	0.5 max		4.5								3.1					
6						92[d]	1.5 max		3.5								1.85					
7		11.0				88[d]																
8		10.1				75[d]	0.2 max		0.10 max							14.0						
9	4.5					65[d]		23.0	7.0													
cobalt																						
1						17.0	1.0		8.0							19.0	0.8	0.4			49[d]	4.0

							Other elements											Temperatures, °C		
Be	Pb	Al	Ti	Zr	S	P	Cu	Si	Fe	Ni	Mn	B	C	Cd	Zn	Each	Total	Sol.	Liq.	Brazing
	0.02 max	0.01 max															0.10 max	1082	1082	1093–1149
																	0.30 max[b]	1082	1082	1093–1149
																	0.50 max[c]	1082	1082	1093–1149
0.05 max		0.01 max															0.50[e]	888	899	910–954
					0.05 max	0.01 max											0.50[e]	866	888	910–954
0.05 max		0.01 max															0.50[e]	921	935	938–982
																	0.15	710	924	788–927
																	0.15	643	802	704–816
																	0.15	643	813	718–816
																	0.15	643	771	704–816
																	0.15	643	788	732–816
																	0.15	710	793	732–843
																	0.15	643	718	691–788
																	0.15	607	618	618–760
																	0.15	627	635	635–760
																	0.15	618	652	652–760
																	0.15	632	688	688–816
																	0.15	607	702	702–843
																	0.15	607	710	710–843
																	0.15	602	718	718–843
																	0.15	677	743	743–843
																	0.15	766	766	766–871
																	0.15	677	766	766–871
																	0.15	688	774	774–871
																	0.15	779	779	779–899
																	0.15	671	779	779–899
																	0.15	691	802	802–899
																	0.15	718	857	857–968
																	0.15	779	891	891–982
																	0.15	771	893	893–982
																	0.15	891	891	891–1010
																	0.15	949	949	949–1004
																	0.15	991	1016	1016–1093
																	0.15	974	1029	1029–1091
																	0.15	1135	1166	1166–1232
		0.05 max	0.05 max	0.05 max	0.02 max	0.02 max						0.3 max	0.10 max				0.50	1079	1135	1149–1204
		0.05 max	0.05 max	0.05 max	0.02 max	0.02 max							0.06 max				0.50	977	1038	1066–1204
		0.05 max	0.05 max	0.05 max	0.02 max	0.02 max							0.06 max				0.50	977	1077	1077–1204
		0.05 max	0.05 max	0.05 max	0.02 max	0.02 max							0.06 max				0.50	971	999	1010–1177
		0.05 max	0.05 max	0.05 max	0.02 max	0.02 max							0.06 max				0.50	982	1038	1010–1177
		0.05 max	0.05 max	0.05 max	0.02 max	0.02 max							0.10 max				0.50	982	1066	1010–1177
		0.05 max	0.05 max	0.05 max	0.02 max	0.02 max							0.08 max				0.50	877	877	927–1093
		0.05 max	0.05 max	0.05 max	0.02 max			0.10 max	0.2 max		0.04 max	0.01 max	0.10 max				0.50	888	888	927–1093
		0.05 max	0.05 max	0.05 max	0.02 max	0.02 max											0.50	982	1010	1010–1093
		0.05 max	0.05 max	0.05 max	0.02 max	0.02 max											0.50	1121	1149	1149–1232

Table 2 (*continued*)

Brazing alloy	Composition, at %																					
	Cu	P	Pb	Al	Zn	Ni	Fe	Mn	Si	Ag	Cd	Sn	Li	Au	Pd	Cr	B	C	Mg	Ti	Co	W
aluminum–silicon																						
1	0.25 max			91[c]	0.20 max		0.8 max	0.10 max	7.5													
2	0.30 max			88[c]	0.10 max		0.8 max	0.05 max	10.0										0.05 max			
3	0.30 max			86[c]	0.20 max		0.8 max	0.15 max	12.0										0.10 max			
4	4.00			84[c]	0.20 max		0.8 max	0.15 max	10.0							0.15 max			0.15 max	0.20 max		
5	0.25 max			89[c]	0.20 max		0.8 max	0.10 max	7.5										2.5			
6	0.25 max			87[c]	0.20 max		0.8 max	0.10 max	10.0										1.5[d]			
7	0.25 max			85[c]	0.20 max		0.8 max	0.10 max	12.0										1.5[d]			
magnesium–aluminum																						
1				9.0	2.0			0.15 min											88			
2				12.0	5.0														82			
silver–copper (vacuum grade)[f]																						
1	0.05 max									99.95								0.005 max				
2	28.0									72.0								0.005 max				
3	28.0					1.0				71.0								0.005 max				
4	27.0									68.0					5.0			0.005 max				
5	50.0									50.0								0.005 max				
6	99.95									0.05 max								0.005 max				

[a] Adapted from American Society of Metals (11).
[b] Cu powder: 0.30 at % pertains only to metallic elements; balance is essentially oxygen.
[c] Copper oxide paste: 0.50 at % pertains only to metallic elements. Maximums of nonmetallic compounds are: 0.4 at % chlorides; 0.1 at % sulfates; 0.3 at % insoluble nitric acid; 0.5 at % insoluble acetone; balance, oxygen.

Assembling and Jigging. In assembling the parts to be joined, it is advantageous to preplace the filler metal. Thin sheets or washers may be used between flat surfaces and rings made of rectangular strip or round wire on tubular members. It is preferable to place the alloy inside the joint in contact with the parts but away from the heat source; otherwise, as the filler metal is usually of lighter gauge than the parts being joined, it may melt away before the assembly is heated to the brazing temperature.

Joints should be designed to be as self-supporting as possible, because jigging sometimes requires considerable ingenuity. The jig should be simple and light with a minimum of contact area between the jig and the assembly. Poor heat conductors that expand very little on heating, eg, stainless steel, inconel, and ceramics, are excellent jig components.

Heating. The assembly must be heated broadly and uniformly in the vicinity of the joint to a temperature just above the flow point of the brazing alloy. If a torch is being used to provide the heat, the filler metal flows toward the source of the heat; the good conductor must be heated in preference to the poor conductor and the thick part in preference to the thin part.

Torch, furnace, induction, resistance, and salt-bath brazing are commonly used, and all methods can be used with many brazing alloys; however, some methods, ie, induction and resistance brazing, may call for particular brazing alloys.

Cleaning. The chloride fluxes for the Al–Si and Mg–Al brazing alloys and the borate fluxes must be completely removed immediately after brazing. Quenching in hot water where possible followed by scrubbing with fiber brushes in running or agi-

Other elements																		Temperatures, °C		
Be	Pb	Al	Ti	Zr	S	P	Cu	Si	Fe	Ni	Mn	B	C	Cd	Zn	Each	Total	Sol.	Liq.	Brazing
																0.05 max	0.15 max	577	613	599–621
																0.05 max	0.15 max	577	591	588–604
																0.05 max	0.15 max	577	582	582–604
																0.05 max	0.15 max	521	585	571–604
																0.05 max	0.15 max	559	607	599–621
																0.05 max	0.15 max	559	591	588–604
																0.05 max	0.15 max	559	579	582–604
0.0005							0.05 max	0.05 max	0.005 max	0.005 max							0.30	443	599	604–627
0.0005																	0.30	410	566	582–610
	0.002 max					0.002 max								0.001 max	0.001 max			961	961	961–1038
	0.002 max					0.002 max								0.001 max	0.001 max			779	779	779–899
	0.002 max					0.002 max								0.001 max	0.001 max			779	795	799–899
	0.002 max					0.002 max								0.001 max	0.001 max			807	810	810–927
	0.002 max					0.002 max								0.001 max	0.001 max			779	871	871–982
	0.002 max					0.002 max								0.002 max	0.002 max			1083	1083	1093–1149

[d] Balance.
[e] Includes Fe, Mn, Si, Pb, and Al.
[f] If determined, cobalt content is 0.1 at % max unless otherwise specified.
Sol. = solidus; Liq. = liquidus.

tated hot water, ultrasonic cleaning, or both, is conventional (see Ultrasonics). Where these are inadequate, chemical cleaners or pickles may have to be used, particularly with aluminum or magnesium assemblies, to remove the last traces of flux.

Health and Safety Factors. Zinc and cadmium volatilize from zinc- and cadmium-containing brazing alloys during brazing, and inhalation of their vapors or their oxides produces temporary and mild effects with respect to zinc or zinc oxide but severe, painful, and sometimes fatal effects with respect to cadmium or cadmium oxide. Precautions include proper ventilation and segregation and collection of the oxides of cadmium.

Cadmium is soluble in the inorganic acids present in food, and the salts are converted to cadmium chloride by gastric juices. Small quantities of cadmium chloride, if ingested, produce severe, painful, and perhaps fatal results. Therefore, cadmium-free brazing alloys or solders are recommended for use on food-handling equipment.

Alloys. The ten types of brazing alloys listed in Table 2 include sixty brazing alloys that have been used in considerable volume and that are readily available commercially (10–11).

Copper. Copper is a furnace brazing alloy that is extremely fluid on iron and iron alloys and generally requires 0.03–0.08 mm clearances and near-press to press fits (see Copper alloys). Its brazing temperature range (1093–1194°C) is fairly high, its service temperature is low (ca 200°C), and it is attacked by sulfur. It can be used on the few copper alloys, eg, 30 at % cupro–nickel, that are higher melting than it is.

One of its chief uses is in furnace brazing of low carbon, low alloy steels in reducing protective atmospheres without flux. It is used with flux on cast iron and carbon tool steels, but the brazing temperature of 1093–1194°C is questionable for cast iron and may be so for carbon tool steels. It is used in pure dry atmospheres on stainless steels during regular or solution annealings.

When copper is used on nickel alloys, it alloys with them and solidifies rather than flows. Therefore, it must be preplaced at, near, or in the joint.

Copper–Zinc. Most brazing processes can be used with copper–zinc alloys. These alloys are less fluid than copper and tolerate clearances of 0.05–0.1 mm. They require flux and if overheated produce porous joints caused by the volatilization of zinc. Their service temperature is ca 200°C, and their corrosion resistance is questionable on copper, iron, and nickel alloys.

Alloys 1, 2, and perhaps 3 are sometimes used on the copper alloys that are higher melting than they are. These include copper–silicon, copper–tin, and copper–nickel. Low carbon, low alloy steels and cast iron can be brazed with these alloys, but the temperatures involved may be too high with respect to cast iron.

Carbon tool steels are brazed with alloys 1, 2, and 3. When these are dip-brazed during the carburizing or hardening of the steels, a process involving rapid heating, minimum time at temperature, and minimum volatilization of zinc, these alloys can and do compete with silver–copper brazing alloys 15 and 17. Stainless steels and nickel alloys are also brazed with 1, 2, and 3 where their poor corrosion resistance is insignificant (see also Zinc and zinc alloys).

Copper–Phosphorus. Copper–phosphorus alloys 1–7 flow below their liquidus temperatures and braze by all processes. They are used on copper, on which they are self-fluxing; copper alloys, excluding aluminum–bronze, beryllium–copper, silicon–bronze, and cupro–nickel containing <10 wt % nickel; and to a limited extent for brazing silver–molybdenum or silver–tungsten contacts to copper. They are not used on iron or nickel alloys.

The fluidity of these alloys depends on their phosphorus content and to a lesser extent on their silver content. Thus, the 5 wt % P alloy 2 is somewhat less fluid and fillets better than the 6 wt % P alloy 3, which is less fluid than the 6.75–7.25 wt % P alloys 4–7. Conversely, the ductility of these alloys is inversely proportional to their phosphorus content. The 5 wt % P alloys have limited ductility; the 6.75–7.25 wt % P alloys have little or no ductility. Alloys 1–7 do not produce inherently brittle joints on joining plastic to tough copper and copper alloys, but lap joints are preferred to butt joints. Alloys 4–7 may be questionable if they are to be subject to critical impact or vibration. Alloys 1–7 have service temperatures of ca 150°C and are attacked by sulfur.

Silver–Copper. The 17 Ag–Cu brazing alloys have service temperatures of ca 200–370°C (see Silver and silver alloys). They stem from the Ag–Cu eutectic system and include temperature depressants, ie, cadmium, tin, and zinc, and wetting agents, ie, lithium, nickel, zinc, and occasionally manganese. They are adapted to all brazing processes and are used on the ferrous as well as the nonferrous metals, excluding aluminum and magnesium.

Of the four Ag–Cd–Cu–Zn alloys, 1 and 2 are eutectiferous and are the lowest melting and generally the most fluid of all the Ag–Cu alloys. Alloys 4 and 5 flow freely and are useful where joint clearances are large or fillets are desired. The copper–phosphorus alloys are preferred to alloys 1, 2, 4, and 5 on copper, but alloys 1–5 are preferred to the copper–phosphorus alloys and the other silver–copper alloys on all of the copper alloys except aluminum–bronze. Alloys 1–5 are used for torch and induction brazing of low carbon, low alloy steel and to braze stainless steel. Alloys 1 and

2, however, are preferred to alloys 4 and 5 on high speed tool steel and nickel alloys.

The Ag–Cu–Sn–Zn alloy 3 is the lowest melting cadmium-free alloy available to braze food-handling equipment. It is less fluid than the Ag–Cd–Cu–Zn alloys, but it is often used preferentially on leaded brass.

The Ag–Cu–Zn alloys 8, 10, and 11 differ principally in their zinc contents in the fact that alloys 8 and 11 liquate but 10 does not. They are used on copper and copper alloys but not on leaded brass, phosphor–bronze alloy 11, silicon–bronze, cupro–nickel alloys 10 and 11, aluminum–bronze or beryllium–copper. They are also used in torch brazing low carbon, low alloy steel and in brazing stainless steel.

Ag–Cd–Cu–Ni–Zn alloy 4 is used on silicon–bronze and is preferred on aluminum–bronze and in brazing beryllium–copper to steel. Alloy 4 and Ag–Cu–Ni–Zn alloy 13 are used in torch brazing low carbon, low alloy steel, but 4 is used in its induction brazing. Both alloys 4 and 13 are used in brazing cast iron, stainless steel, and carbide tool tips.

The remaining alloys, only one of which, ie, alloy 15, contains any zinc, are primarily furnace brazing alloys. Alloy 12, which is the Ag–Cu eutectic, is used on beryllium–copper and in general on electrical components requiring the highest electrical and thermal conductivity. Alloy 7 (Ag–Cu–Sn) is the lowest melting cadmium-and-zinc-free alloy of the seventeen listed and is occasionally used on copper, sometimes on leaded brass, and more frequently on cupro–nickel. In general, it is used on vacuum seals and on marine heat exchangers exposed to hot salt water, under which conditions zinc is objectionable. Alloy 14 (Ag–Cu–Ni–Sn) produces 400 series stainless-steel joints that do not crevice corrode. It is used on stainless steels 302–316, 430, 403, 410, and 440 A. Alloys 15 and 17 are basically Ag–Cu–Ni alloys and are used on stainless steels 302–316, 321, 347, 446, 403, 410, and 17-4 PH (precipitation hardenable), and they have service temperatures of ca 370°C. Lithium-containing Ag–Cu alloys 9 and 16 are used on thin stainless steels 302–316, 17-7 PH, PH 15-7 Mo, 17-4 PH, and AM 350. They do not interact with these steels at all.

Gold. Gold alloys 1, 3, and 4 are vacuum-tube step-brazing alloys (see Gold and gold compounds). Alloys 2 and 5 are high temperature brazing alloys and are fluid, but they interact very little with cobalt, iron, and nickel high temperature alloys. They are used in furnace brazing in dry protective atmospheres or in vacuum as well as in induction and resistance brazing. The joints produced are tough and corrosion-, heat-, and oxidation-resistant with service temperatures up to 815°C.

Nickel. Nickel alloys 1–9 include three types of alloys: the Ni–Cr–Si, Ni–Cr–Si–B, and Ni–Si–B alloys 1–6; the Ni–P and Ni–Cr–P alloys 7 and 8; and the Ni–Mn–Si–Cu alloy 9 (see Nickel and nickel alloys). All of these are brittle and with one or two exceptions are only available as powder, powder pastes, and transfer tapes. Concerning the exceptions, 7 is available as foil, and 4–6 are produced with boron-diffused boride coatings on ductile Ni–Cr–Si and Ni–Si strip or wire cores.

Nickel alloys 1–6 have extremely low vapor pressures and produce heat-, corrosion-, and oxidation-resistant joints. They are amenable to all brazing processes and are mainly used on stainless steels, although they can be used on other steels and nickel and cobalt alloys. Their service temperature is ca 650°C. Of these alloys, 1, 2, and 3 are the strongest and most heat-resistant. Alloys 4–6 are lower melting and alloys 5 and 6 fill gaps better. All of the boron-containing alloys, ie, 2–6, however, erode stainless steel, especially if long brazing cycles are involved.

Phosphorus alloys 7 and 8 are very fluid on stainless steel, but erode it far less than silicon–boron alloys 2–6. Prolonged furnace-brazing cycles may be used with alloy 8 on thin-walled assemblies. On nickel alloys, alloys 7 and 8 are preferred to the silicon–boron alloys although they erode nickel appreciably. The reaction of erosion, however, increases the remelt temperatures of alloys 7 and 8 as well as their oxidation and corrosion resistance.

Cobalt. Cobalt brazing alloy is compatible with cobalt alloys and is used in diffusion brazing of jet engine parts.

Aluminum–Silicon. Aluminum–silicon brazing alloys 1–4 are conventional brazing alloys, and alloys 5–7 are vacuum brazing alloys. The aluminum alloys that are brazed with 1–7 have liquidus temperatures of 613–657°C and solidus temperatures from 646°C to 584–571°C (brazeable) to 557–543°C (slightly brazeable) to 510–427°C (unbrazeable). It follows that in the use of alloys 1–7, certain things are desirable, eg, single- and double-brazing, alloy-clad, aluminum alloy sheet, precise temperature control (±2.8°C), continuous circulation of furnace atmospheres, preheating sizeable dip-brazing assemblies prior to dipping, and minimum time at temperature, ie, 2–6 min in furnace brazing. Under these conditions, alloys 1–7 produce strong joints and alloys 5–7 permit the brazing of heat exchangers and radiators without requiring the complete removal subsequent to brazing of the corrosive flux used with alloys 1–4.

The brazeable alloys include those containing minimal amounts of magnesium, eg, EC, 1100, 3003–3005, 5005, 5050, 6053, 6061–6063, and 6951, and casting alloys A612 and C612. These are furnace-, dip-, or torch-brazed and the heat-treatable alloys, eg, 6951, must be quenched following brazing if they are to be heat-treated subsequently.

Magnesium–Aluminum. Magnesium–aluminum brazing alloys 1 and 2 are used on AZ10, A1A, and M1A, which have respective solidus and liquidus temperatures of 632°C and 638°C (AZ10) and 648°C and 649°C (K1A and M1A). Alloy 2 is used on A231A and ZE10A, which have respective solidus and liquidus temperatures of 566°C and 627°C (A231A) and 593°C and 646°C (ZE10A).

Furnace, dip, or torch brazing of these alloys must be precisely controlled in the same manner as brazing of the brazeable aluminum alloys.

Silver–Copper (Vacuum-Grade). Silver–copper alloys 1–6 include two pure metals, ie, alloys 1 and 6, and two pure eutectics, ie, alloys 2 and 4. They are used in the step brazing of high-grade electron tubes and electronic devices. They must be used in pure atmospheres to maintain the purity of the brazing alloy and to ensure proper brazing and to guarantee the quality of the joint. Alloy 1 is used for metallizing ceramics to be used as semiconductors.

BIBLIOGRAPHY

"Solders and Brazing Alloys" in *ECT* 1st ed., Vol. 12, pp. 634–640, by C. H. Chatfield, Handy & Harman; "Solders and Brazing Alloys" in *ECT* 2nd ed., Vol. 18, pp. 541–549, by C. H. Chatfield, Handy & Harman.

1. *Solderability Test Standard RS-178-B*, Electronic Industries Association, Washington, D.C., Jan. 1973.
2. *Solderability Test Methods for Printed Wiring Boards*, Institute for Interconnecting and Packaging Electronic Circuits, Evanston, Ill., 1970.
3. *U.S. Federal Specification QQ-S-571*, Solder, May 5, 1972.
4. *U.S. Military Specification MIL-P-28809, Printed Wiring Assemblies*, U.S. Government Printing Office, Washington, D.C., March 21, 1975.

5. *ASTM B32, Standard Specification for Solder Metal*, American Society for Testing and Materials, Philadelphia, Pa., 1976.
6. *Vapor Phase Reflow Soldering*, VPS, 3M Company, St. Paul, Minn.
7. H. H. Manko, *Solders and Soldering*, 2nd ed., McGraw-Hill Book Company, New York, 1979.
8. *Solders and Soldering, A Primer*, Lead Industries Association, New York, N.Y.
9. *Soldering Manual*, 2nd ed., rev., American Welding Society, Inc., Miami, Fla., 1978.
10. *Welding and Brazing*, Vol. 6 of *Metals Handbook*, 8th ed., American Society for Metals, Metals Park, Ohio, 1971, pp. 593–702.
11. *Materials and Processing Databook "81," Metals Progress*, American Society for Metals, Metals Park, Ohio, mid-June 1981, pp. 150 and 151.

GEORGE SISTARE
Consultant

FREDRICK DISQUE
Alpha Metals, Inc.

SOLVENT RECOVERY

The recovery of valuable solvents has long been regarded as an economically rewarding practice. Ever since the surge in solvent manufacturing and solvent-utilizing facilities began around 1930, processes employing solvents have generally included solvent-recovery systems as part of the initial installation. However, in some highly innovative and extremely profitable processes the inclusion of solvent recovery would have reduced the overall profitability, and solvent recovery facilities were generally added only after competition reduced profits to a normal level. There were also operations that used inexpensive solvents in quantities sufficiently small that solvent recovery could not be economically justified.

The environmental and safety hazards associated with the discharge of organic vapors and liquids to air or sewers were also recognized early with the result that even marginally profitable solvent-recovery systems were often installed to comply with industry standards or with state or local regulations. However, the establishment of the EPA in 1970 and the ensuing air- and water-pollution legislation have brought these considerations more sharply into focus (see Air pollution; Water, water pollution). The EPA now has responsibility to survey current and achievable industrial practices, and to establish limitations on the discharge of polluting emissions to air and effluents to waters. These limitations are to become more stringent in a series of steps aimed at an expressed national goal of zero pollution. For new installations the EPA is authorized to establish guidelines for pollution-reducing equipment and control systems to be used either at the end of a process or within the process (see also Regulatory agencies).

It is difficult to speculate on the ultimate impact that such far-reaching powers are likely to have on the design of solvent-recovery systems. Certainly formulation, revision, and enforcement of the regulations are subjected to a variety of disparate influences, including those of large and small operating companies, industry associations, environmental activist groups, equipment manufacturers, agency personnel, cooperating state and local authorities, and the courts. Assuming a reasonable resolution of all these influences, it is expected that solvent-recovery designs will continue to be chosen by process design engineers on the basis of economic evaluation of a variety of possibilities. However, it will no longer be possible to discharge large, or even relatively small, amounts of solvents to the atmosphere, municipal sewers, streams, lakes, or oceans. Inclusion of solvent recovery in a process is indicated as long as the costs and credits associated with it are more favorable than the often considerable cost of eliminating solvent pollution at the end of the process.

The solvent that has seen the greatest recent increase in recovery and reuse is, of course, water. However, because of its special importance and unique purification technology, this subject is covered elsewhere (see Water, water reuse). This article deals with the recovery of nonaqueous solvents employed in the following industrial processes: the formation and drying of synthetic fibers and films, plastics and rubber products, smokeless powder, impregnated fabrics, adhesives, printing inks, paints, lacquers, enamels, and other deposited coatings; the solvent extraction of oils and fats from vegetable and animal products, of fossil fuel materials from coal, shale, tar, and oil sands, of metallic compounds from treated ores, and other solid–liquid extractions; the solvent-refining of mineral and vegetable oils, other liquid–liquid extractions, extractive and azeotropic distillations, and gas-absorption processes; the degreasing of fabricated parts, dry cleaning, and other washing operations; and polymerizations and other chemical reactions carried out in solution, and precipitation and recrystallization from solution.

In most of these applications the recovered solvent is recycled. There are, however, a number of applications where solvent separation takes place at a different place and time than solvent addition. In such cases the recovered solvent is reused in a different and usually a lower value application.

In a number of applications, organic liquids that are potentially solvents are purified and separated from processes in which they were not used primarily as solvents. Except for the use to which the organic liquid is put, such operations may resemble solvent-recovery systems in every respect. Among such closely related applications are the recovery of natural gasoline and light hydrocarbons from natural and casing-head gas; vapor-recovery operations in petroleum refining; recovery of ethanol from fermentation gases; recovery of organic liquids from wood distillation and coal-tar distillation operations; and recovery of by-products or unused reactants from chemical synthesis (see Gasoline; Petroleum; Ethanol; Coal; Tar and pitch).

The solvent flow in a typical recovery system is given in Figure 1. Fresh and recovered solvents join together and enter an industrial process. In the separations section the products are separated from recycle streams and a stream containing the bulk of the solvent for recovery. The last stream is recycled after one or more purification steps in which solvent is separated from the other gas, liquid, or solid components. Solvent that is not recycled leaves the system as a component of the product, by-product, or waste streams. If emissions, spills, and solvent leaving in degraded form are included, it is obvious from Figure 1 that solvent make-up must be equal to solvent loss. Thus, good solvent recovery means low solvent use, and therefore low activity in the manufacture of chemicals specifically for solvent applications.

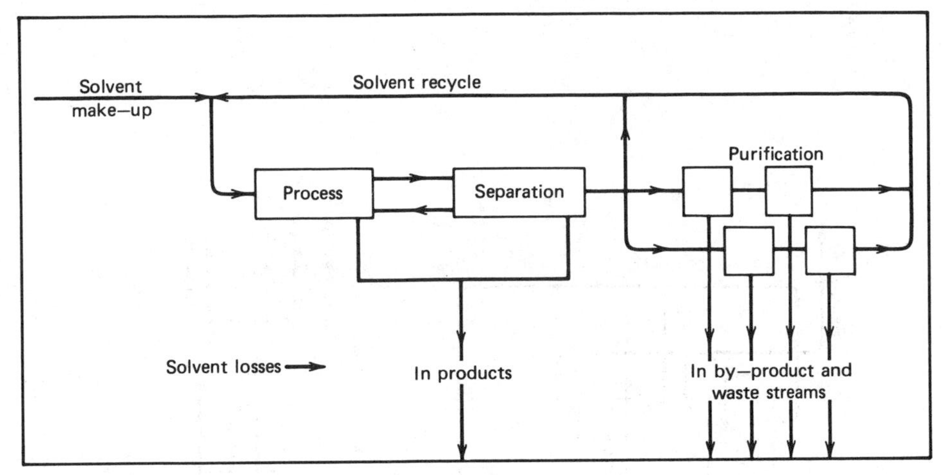

Figure 1. Solvent flow in generalized solvent-recovery system.

Solvent-Recovery Systems

Solvent-recovery operations may differ greatly in design, in scale, and in the way they interface with the main process. This is well illustrated by the series of flow diagrams shown below adapted from actual installations or commercial designs. Adsorption (see Adsorptive separation), eg, plays an important part in Figure 2, absorption (qv) in Figure 3, crystallization (qv) in Figure 4, vaporization and recompression in Figure 5 (see Supercritical fluids; Supplement Volume), and liquid–liquid extraction (qv) in Figure 6. Figure 7 is exclusively a distillation (qv) process, the system in Figure 8 uses vacuum, and those in Figures 5 and 9 operate to a large extent under pressure. A batch process is illustrated in Figure 10, semicontinuous processes in Figures 2 and 9, and truly continuous processes in the other figures. Filtration (qv), drying (qv), refrigeration (qv), decantation, and evaporation (qv) are utilized in varying degrees and modifications in the different applications. Differences in design are often called for even where the application is the same. Thus, Figures 2 through 10 are not standard designs for the industries in question, nor do they necessarily represent best practice for the particular application indicated. For example, the water-absorption system shown in Figure 3 for recovering acetone from cellulose acetate spinning is economical only when a large supply of cold water is available. Where water temperatures are high, an adsorption system similar to that shown in Figure 2 might be suitable for the same application.

The principal similarity which exists among solvent-recovery systems lies in the approach that must be taken in order to ensure a satisfactory design. The same techniques, such as absorption, adsorption, extraction, filtration, distillation, and condensation are available in all cases. The same factors such as volatility, solubility, thermal stability, corrosion, purity requirements, capacity, steam and water conditions, safety, and economics must be taken into account. In general, each unit must be individually designed. Standardization is desirable only where all factors are substantially the same.

Solvent-recovery units may be classified according to the method used to make the initial separation between solvent and product streams as shown in Figure 1. This separation may be made by any one of the following: mechanical separation, such as

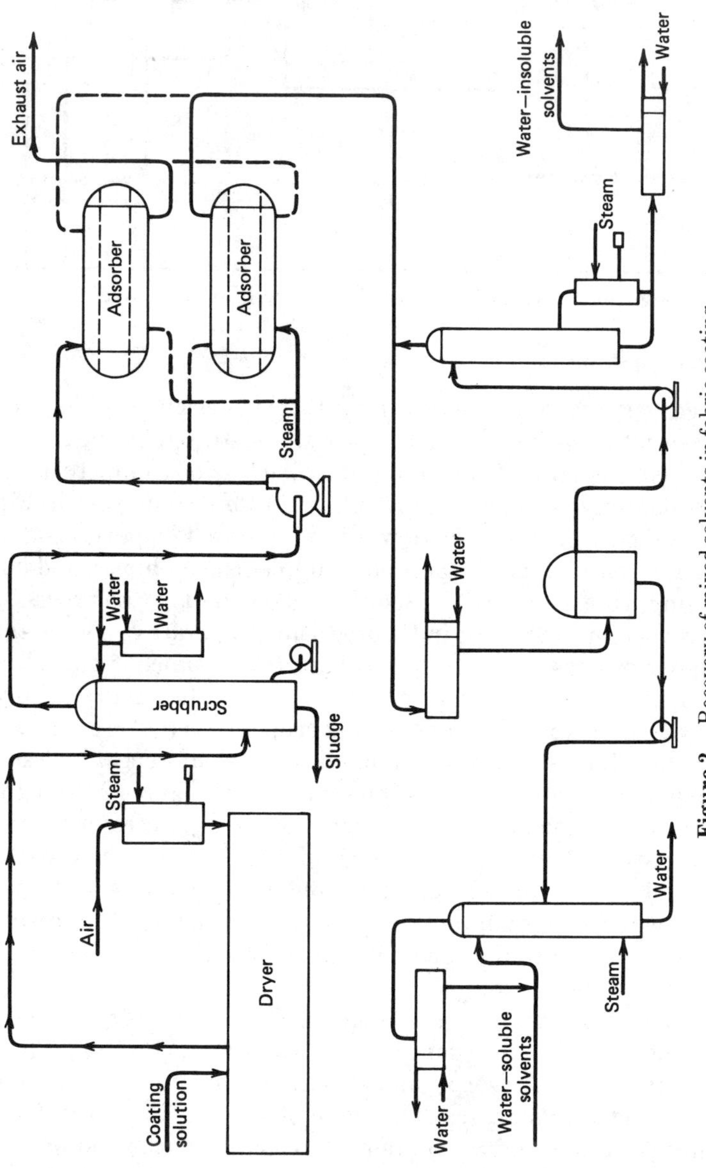

Figure 2. Recovery of mixed solvents in fabric coating.

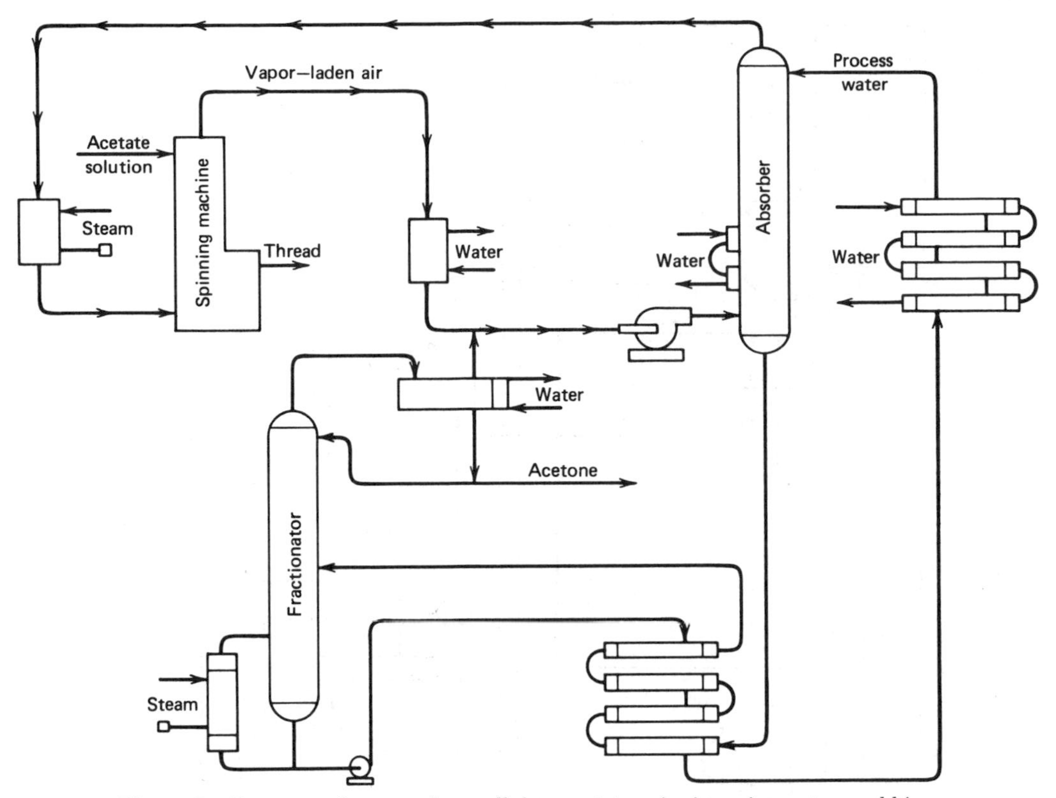

Figure 3. Recovery of acetone from cellulose acetate spinning using water scrubbing.

filtering, settling, draining, decanting, centrifuging, or pressing; extraction, using another liquid to wash out solvent or product; evaporation; fractional distillation; drying the product in the absence of air or gas; or drying in the presence of air or inert gas followed by condensation, absorption, or adsorption.

Although a solvent-recovery system may be classified by the initial step, subsequent steps may be of a different type. Thus in Figure 10, the mechanical separation of garments from solvent is followed by drying in the presence of air. In Figure 2, drying the product in air is followed by adsorption of solvent, evaporation, condensation, decantation, and distillation.

It may also be noted that some flow diagrams show more than one type of solvent-recovery system. Thus, in Figure 9 the separation of the furfural solvent from butadiene and its separation from polymer may be considered different units. The same may be said for separating hexane from solvent-oil mixtures (miscella) and from spent flakes (marc) in Figure 8. In Figure 4 the recoveries of volatile methylene chloride and of a nonvolatile urea solution are accomplished by entirely different techniques.

Solvent-Recovery Techniques

Mechanical Separation. Draining of liquids from solids is a common operation in solvent processes. The solids are usually retained either by stationary or moving screens, perforated plates, baskets, belts, or chains. Sometimes agitation by shaking,

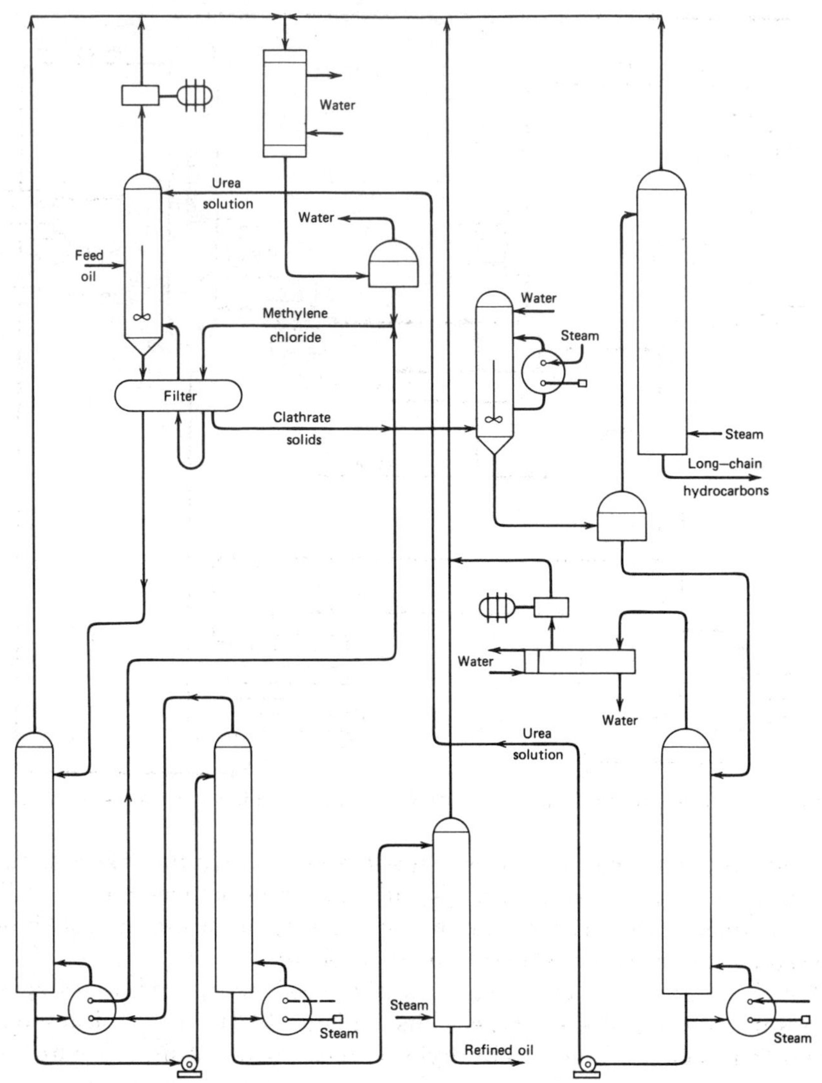

Figure 4. Recovery of diluent and clathrating solution in solvent refining of oil (1–2).

tumbling, or rotation promotes separation. An example of draining is shown in Figure 8, where the conveyor that moves the extracted flakes from the extractor to the dryer allows solvent to drain back into the extractor. Another example is the draining of liquid from fabricated parts in solvent degreasing operations.

Filtration through screens or cloths is indicated where solids are present in small particle sizes. Batch, semicontinuous, or continuous filters using either pressure or vacuum are used for separating solvents from solids. For flammable or noxious solvents, filter closures should be pressure tight or at least fumeproof. Presses or open filters are usually avoided.

Continuous pressure-tight filters are sometimes used, for example, in the solvent refining of lubricating oils. The operation of the filter shown in Figure 4 is typical of such equipment. The solids are first filtered and carried on a moving screen or drum to a section where they are washed with solvent. Then they are moved to sections where

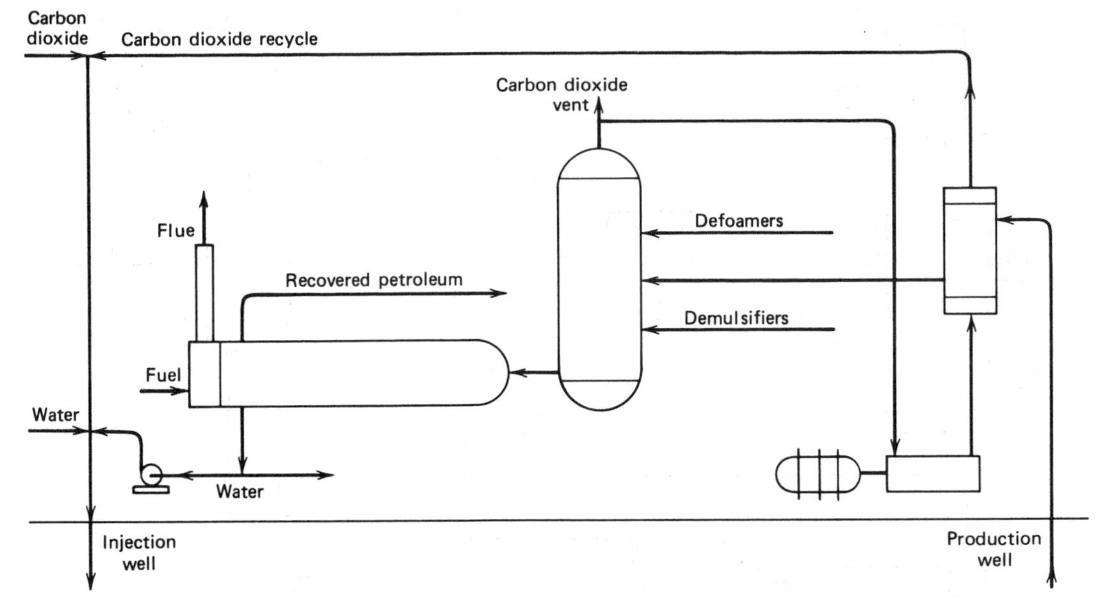

Figure 5. Recovery of supercritical CO_2 solvent in secondary and tertiary petroleum recovery (3–5).

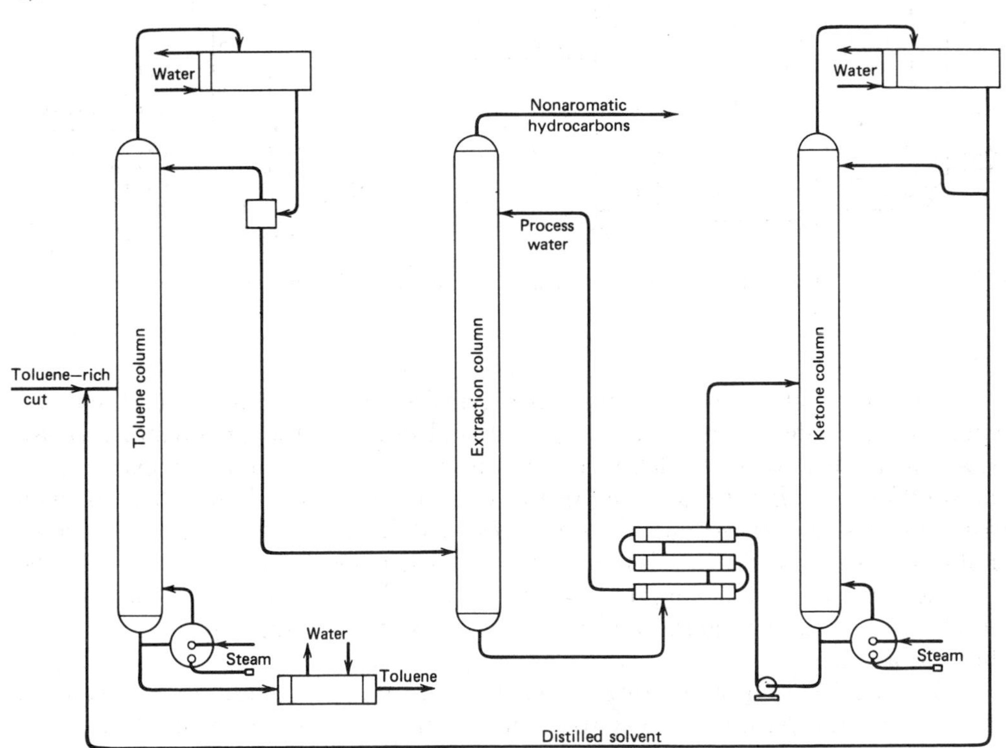

Figure 6. Recovery of methyl ethyl ketone in azeotropic process for producing toluene.

liquid is drawn out and the dry solids dumped into a screw conveyor. The filters are operated under some vacuum in order to conduct any leakage in the rotary joints inward.

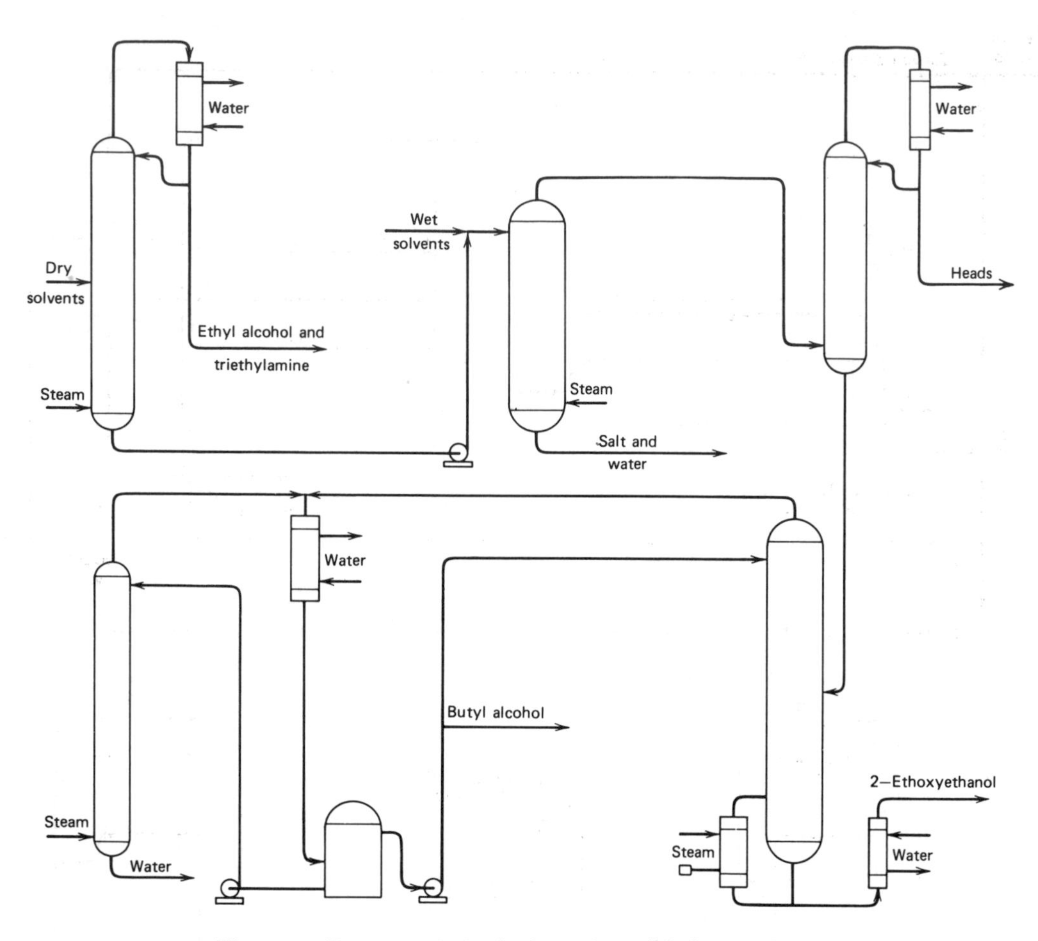

Figure 7. Recovery of mixed solvents in antibiotic manufacture.

Equipment of this type is generally expensive and its use can be justified only where the throughput is relatively large. Batch filtration, on the other hand, may also be expensive because of the labor required for removing the solids. An exception is the separation of dirt, fines, or impurities present only in small quantities. In such cases, disposable filter mediums are often used, and the filter is made sufficiently large that only occasional cleaning is required. Filtration is frequently complicated by the presence of polymer-forming, sticky, or very finely divided materials. Pretreatment with heat or chemicals and the addition of filter aids or surface-active agents may solve these problems.

Settling followed by decantation or drawing off of the separated phases is an obvious and frequently easy way to separate immiscible liquids, solids, and gases. Separating drums of nominal cross-section often give fluid velocities that are far below the free-settling values, and achieve essentially complete separation of the fluid phases with or without the use of baffle, centrifugal-flow, or packed-type entrainment separators. However, solutions of materials occurring in nature or of synthetic polymers, often contain traces of surface-active compounds that produce relatively stable foams,

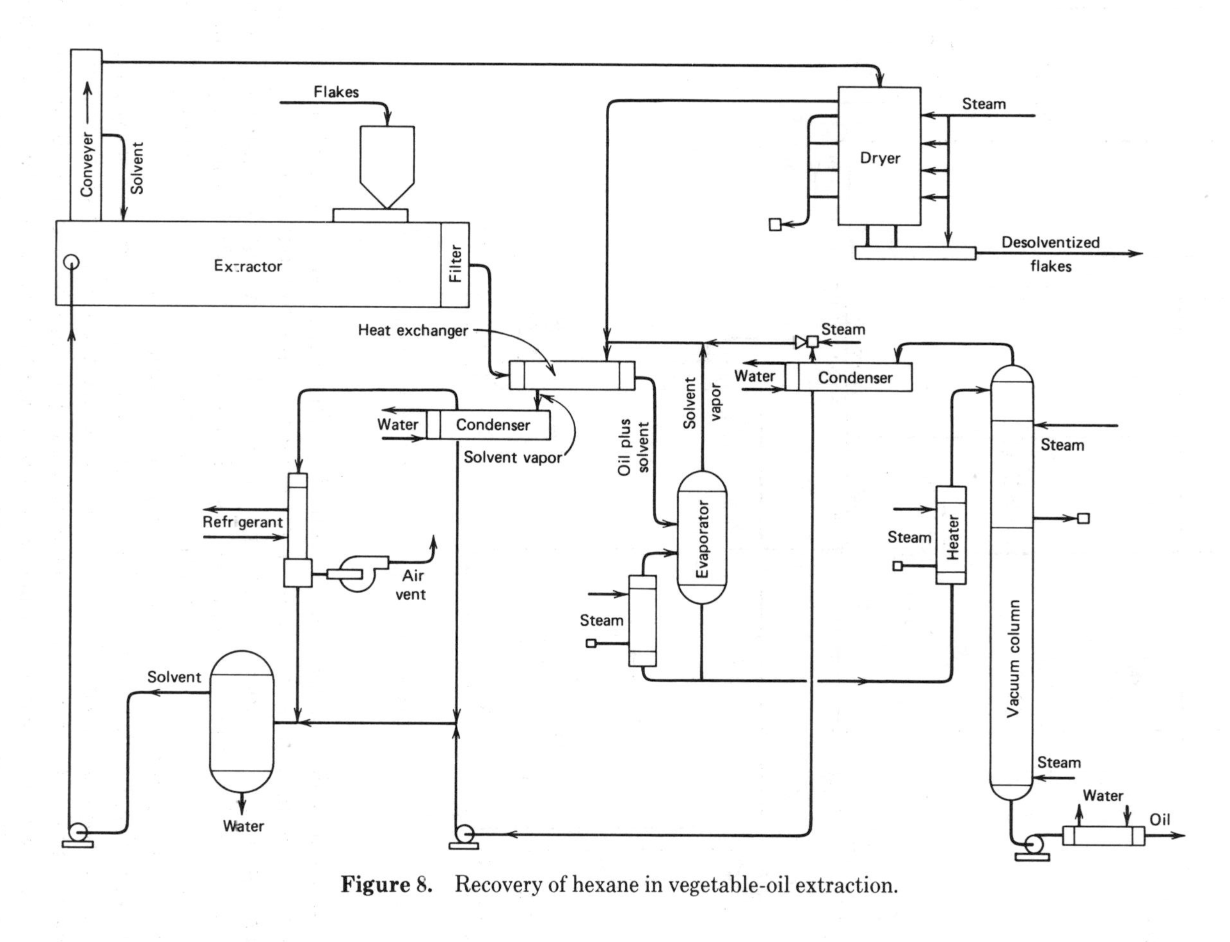

Figure 8. Recovery of hexane in vegetable-oil extraction.

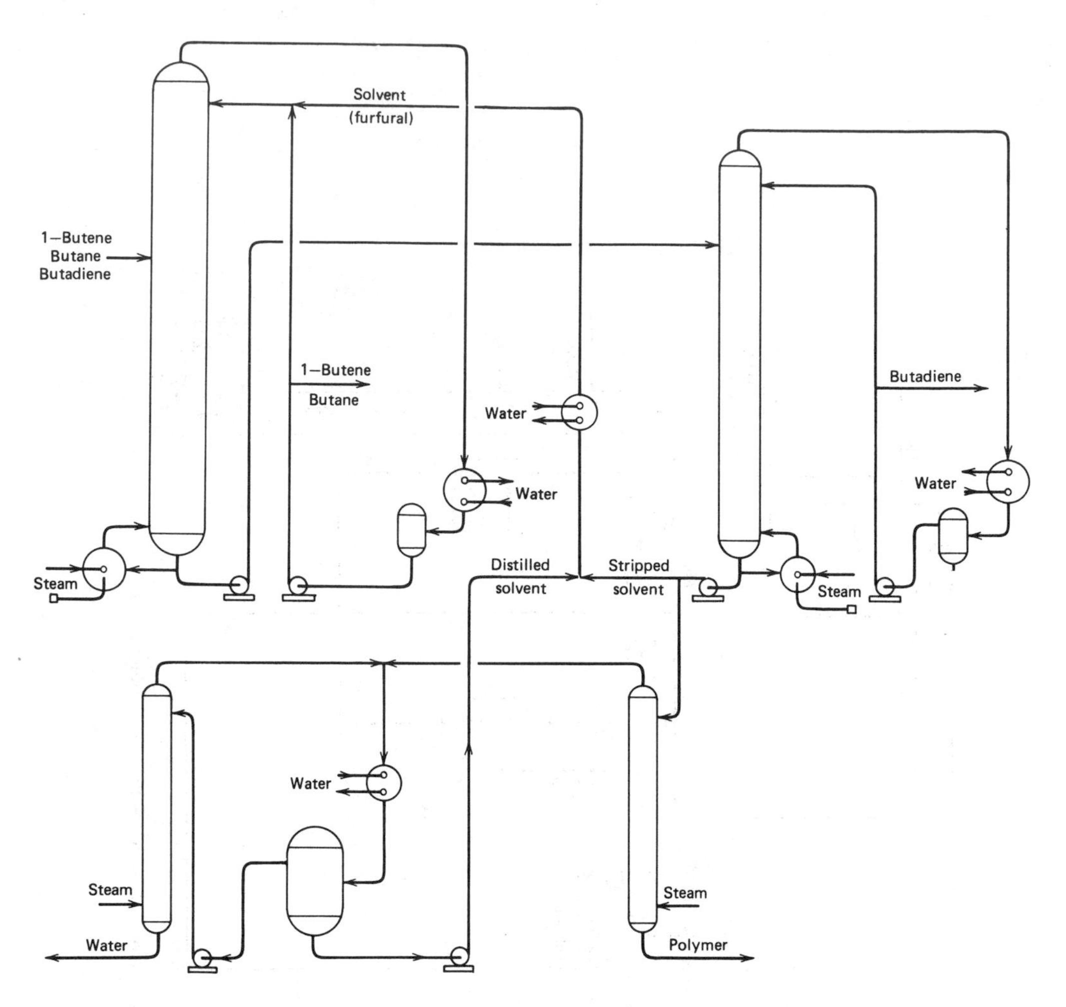

Figure 9. Recovery of furfural in butadiene purification.

emulsions, or solids suspensions. This undesirable surface activity must be counteracted. The appropriate technique may be very specific to a particular solution and much ingenuity and experimentation is often needed to devise the right treatment. The mixture of residual petroleum, carbon dioxide, water, and fine sediment produced from the well in Figure 5 requires the expenditure of a significant amount of heat, energy, and chemicals for the necessary phase separations.

Centrifugal filters, perforate-basket centrifuges, and solid-bowl centrifuges use centrifugal force to increase the efficiency of filtering, draining, and settling operations, respectively (see Centrifugal separation). Examples may be found in the extraction step in dry cleaning, separating crystals from solvent solutions, and the clarification of extracts. Pressing between rolls, inside membranes, or in screw expellers also promotes separation of solvent from fibrous or flaky materials. Magnetic or electrostatic separators may be required in some cases.

The removal of dust, lint, or mist from air streams is often required to prevent the fouling of adsorption beds. Air filters are used except where the gases contain

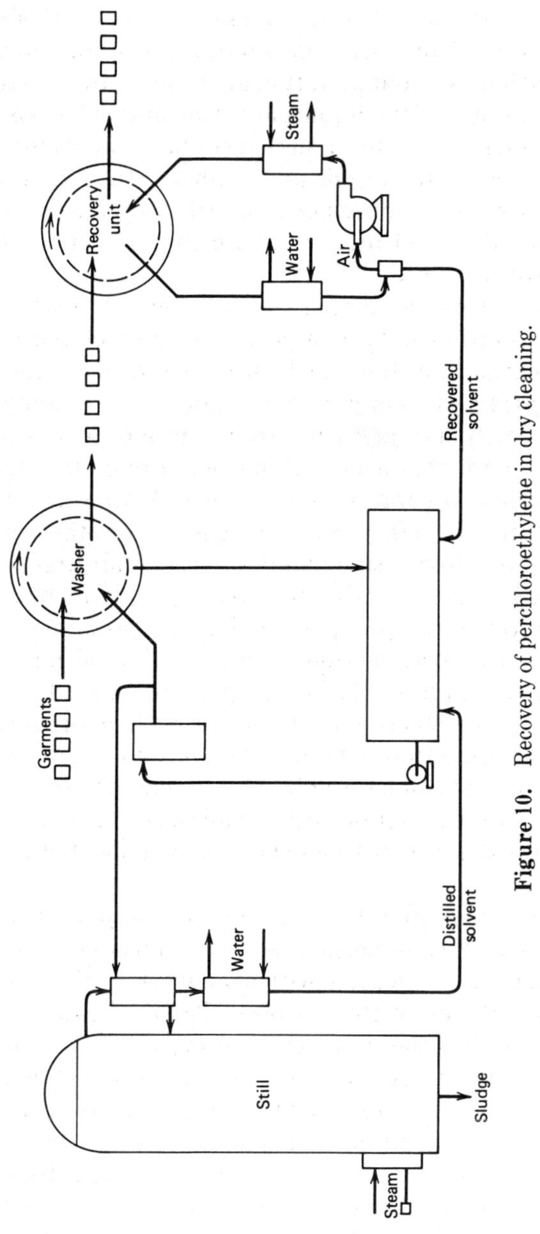

Figure 10. Recovery of perchloroethylene in dry cleaning.

adhesive or varnish-forming materials. In the latter case, entrainment separators or slat-packed towers with water-washed surfaces are employed (see Fig. 2). Such an arrangement may also serve to remove corrosive impurities from air.

Extraction. Solvent and product are sometimes separated by washing with water or another solvent (6–7). In Figure 6, the solvent is washed out of the nonaromatic hydrocarbon product with water. In other cases the product is washed out of the solvent with an aqueous solution. Centrifugal contactors, mixer-settlers and sieve-plate, baffled or packed towers with or without agitation are some of the types of equipment used in liquid–liquid extraction. If the liquids are clean and of low viscosity, extractors of reasonable size give effective countercurrent contact equivalent to ten or more theoretical stages. However, if the liquids are viscous or turbid, or if they tend to form emulsions, very low extraction efficiencies may be expected. Sometimes the solvent must be washed out of a solid product in which case various types of washers or solid–liquid extractors are used.

Extraction is sometimes accompanied by a chemical reaction between a component of the washing solution and the impurity (or product) material being extracted. In other cases, the extraction step may be preceded by a chemical reaction in which the material to be extracted is converted to a more easily extractable compound. For example, in the production of penicillin from dilute fermentation broths, a change in pH from 7.5 to 2.5 converts the penicillin from a preferentially water-soluble compound to a preferentially organic-soluble one (see Antibiotics, β-lactams). By alternately changing pH and extracting from one phase to the other, hundredfold increases in concentration are possible with only minimal expenditures of energy. Similar recovery processes may be expected in the currently evolving biosyntheses.

Chemical reaction techniques are particularly prominent in solvent-recovery systems associated with the extraction of transition-metal compounds from ores (8) (see also Extractive metallurgy). These compounds are made preferentially soluble in organic solvents by the reaction of the cation with an organic complexing agent. Changes in oxidation-reduction potential, pH, or anionic environment converts the compound to a less organophilic form, thus reversing the direction of extraction and regenerating the solvent. Extraction with chemical reaction is performed in the same type of equipment as extraction without chemical reaction, but the design calculations are more complex.

Evaporation. Solvents are often recovered by simple evaporation and condensation. In dry cleaning, for example, grease and dirt must be removed from the solvent (see Fig. 10). Package stills, complete with condenser, feed preheater, and semiautomatic controls are available for this purpose. These units are inexpensive and they are often used in industries other than dry cleaning. For high boiling Stoddard solvent the operation may be under vacuum, whereas for lower-boiling chlorinated solvents it may be at atmospheric pressure. Feed may be either batch or continuous. Heating is usually with indirect steam, but provision for gas or electricity is made in some units. Direct steam may be added for the final removal of solvent from the sludge.

A natural-circulation evaporator is shown in Figure 8. It is followed by a heater which is also an evaporator, but of the once-through type. The latter has a lower boiling point at the inlet and a higher temperature driving force than the former. Other types of evaporators may, of course, be used. Multiple-effect evaporation and its counterpart, vapor-reuse distillation, are growing in popularity because they effectively conserve high cost energy. Evaporation of solvent at high dilution is particularly susceptible to such saving because boiling and condensing temperatures are close together, and solutions are stable over a wide range of temperature and pressure. The double-effect vaporization of methylene chloride is shown in Figure 4. Condensing vapors from the second, high pressure column supply heat to the reboiler of the first column.

Interest is growing in solvents that are high pressure intermediate-density supercritical fluids (3–5). The solvent power of such fluids can be changed by varying the density and less energy is required to separate solute from solvent than with conventional vaporization of liquids. Possible uses include secondary and tertiary petroleum recovery (see Petroleum, chemicals for enhanced recovery). Utilization of CO_2 in such an application is shown in Figure 5. Depending on the pressures and temperatures in the formation, the CO_2 may function at various points as gas, liquid, supercritical fluid, solvent for petroleum fractions, viscosity reducer, driving fluid, or a combination of these. Conversion to a low density phase is accomplished in this case by the reduction in pressure as the production rises to the surface.

When a volatile solvent is separated from a nonvolatile liquid, the product is saturated with solvent at the pressure in the evaporator. The introduction of direct steam into the boiling liquid reduces the solvent partial pressure. However, steam-stripping columns, like the one at the lower right in Figure 9, are much more efficient. The steam not only supplies heat and dilutes the vapor, but its condensate washes the solvent-free polymer out of the column. In the vacuum column shown in Figure 8, condensation of steam and consequent emulsification must be avoided. Indirect heat is supplied to the top of the column to prevent lowering the temperature by solvent vaporization. This piece of equipment is in effect a falling-film evaporator with a steam purge.

Direct steam is commonly used to steam distill solvent from the surfaces or pores of solids. Since the heat for vaporizing the solvent is provided by steam condensation, the separated solid is wet. This moisture can be removed by drying. Drying equipment for evaporating water is generally simpler than that for evaporating solvent. However, the two-step arrangement consumes more energy.

Fractional Distillation. Solvents are commonly separated from products, water, or other solvents by fractional distillation. The mixtures to be separated are usually nonideal and often azeotropic or partially miscible. Where more than one solvent is used, several different types of compounds may be present in the same solution, and such mixtures are very difficult to separate. The feed mixture in Figure 7, for example, contains two alcohols, an amine, an alcohol ether, water, salt, and traces of hydrocarbons.

Such complex mixtures are separated into individual components by a long series of distillation operations. Additional components may be added to serve as entrainers in azeotropic or extractive steps (9) (see Azeotropic and extractive distillation). Sometimes a partial separation of mixed solvents suffices to permit their reuse. In both Figures 2 and 7 additional columns would be needed for complete separations. In Figure 7, a change in just one of the solvents might require a complete revision in the distillation-train arrangement.

Methods have been developed for estimating multicomponent from binary equilibrium and designs of this type can be calculated. Sieve-plate or bubble-cap columns are generally specified, but packed columns and other designs are also used. Operation is normally continuous, except where very small quantities of solvent are to be separated (see Distillation).

Drying in the Absence of Air or Inert Gas. Drying solids by vaporizing the solvent in the absence of circulating air or inert gas has many advantages (see Drying). Explosion hazards are reduced, mechanical circulation of air is eliminated, and the task of recovering the vaporized solvent is greatly simplified. On the other hand, heat transfer to the solid may be more difficult, and a higher temperature is required for vaporization unless reduced pressure is used.

Heat is usually supplied by surfaces heated with steam or other media, or occasionally radiant lamps are used. In Figure 8, a rotary-shelf type is shown. Agitation and flow of solids is obtained by rotary rakes that move the solids inwardly and outwardly across successive steam-heated platens. Steam-jacketed screw conveyors would accomplish the same result in a somewhat different way. Vacuum drum dryers, cylinders, dryers, rotary dryers, or other types of indirect dryers may also be used, provided the solid has suitable properties. Dryers that employ forced circulation of superheated solvent vapors through the solids give good heat transfer and eliminate contact of solids with moving parts or heated surfaces.

At the point where granular solids flow from these dryers, solvent vapors may leak out or air may be drawn in and overload the gas-handling capacity of the condensing system. These problems are alleviated if the pressure in the dryer can be kept close to atmospheric, if mechanical discharge devices are used, or if the void volume can be compressed or displaced with liquid. A purge stream of direct steam flowing back into the dryer through the voids reduces solvent losses without increasing the noncondensable gas load (see Fig. 8). If the dryer must operate at significant pressure or vacuum, the solid can be discharged batchwise or through locks.

Drying with Air or Inert Gas. Drying in the presence of air or a mixture of nitrogen and CO_2 (deoxygenated air) permits vaporization of solvent at lower temperatures than in the absence of air. In addition, hot circulated air provides good heat transfer and does not require contact of product with hot surfaces. The importance of tight closure is reduced since air may be drawn into the dryer without serious consequence. Hot air may be the sole source of heat, or additional conducting or radiating surfaces may be provided inside the apparatus. If combustible solvents are present, a high rate of air circulation keeps their concentration well below the lower explosive limit. Although flammable mixtures still exist near the drying surface, the danger of explosion is greatly reduced. Operation may also be above the upper limit. The use of inert-gas generators to remove oxygen from the air is good safety practice and is justified especially where the air stream is recirculated.

Recovery of solvent from the air may be by adsorption, absorption, or condensation. Since recovery, particularly in the last case, is never complete, the air from the recovery system is often recirculated back to the dryer (see Figs. 3 and 10). Recirculation is not used when return ductwork is expensive, when exhaust air is relatively solvent-free, or when recirculation increases impurities (see Fig. 2). Even with recirculation, a small portion of air is generally discharged through a blower. The resulting suction draws air through the dryer openings and prevents solvent vapors from contaminating the surrounding atmosphere. If dust is present, filtered air may be supplied at the dryer inlet and outlet. A similar arrangement is used for inert gas.

Condensation. Vaporized solvents may be liquefied either in surface condensers or direct-contact condensers. Direct contact with water is largely restricted to solvents with very low water solubilities, but direct contact with cold solvent may be used in any case. Surface condensers are almost exclusively shell-and-tube. With vertical vapor-through-tube condensers, the condensed liquid and vent gas leave at a temperature approaching that of the cooling water. With other types, separate vent condensers are used or additional passes for vent gas provided inside the main condenser. Extended surface designs are advantageous for lean air streams where sensible heat is an appreciable proportion of the total (see Heat-exchange technology).

The air or gas leaving a condenser contains solvent to the extent of its vapor pressure at the condensing temperature. The equilibrium mol fraction in Table 1 gives the volumetric concentration of solvent remaining in the air after condensation at 25°C. Comparison with the lower-explosive-limit column reveals that condensation from circulating air streams operating below the lower explosive limit would not be possible for most common flammable solvents. As a consequence, the use of condensation is restricted to high boiling solvents, nonflammable solvents, systems using inert gas, or systems where relatively small proportions of air are present.

Even in these cases, solvent loss may be high if an appreciable amount of air or gas is discharged to the atmosphere. Refrigeration of the vent gas to 0°C reduces the concentration of solvent to ca 20–30% of that obtained at 25°C (see Fig. 8). Gas compression before cooling and condensation has been much used by the petroleum industry in gasoline and vapor-recovery plants. In solvent recovery, however, compression is not preferred because of the somewhat corrosive nature of many solvent–air mixtures.

The solvent content of the condenser vent-gas stream may also be effectively reduced with small clean-up absorbers or adsorbers. Sometimes the volume of the vent gas can be reduced by degassing the liquid at an earlier point in the process.

Absorption. Solvents may be recovered from air or gas streams more or less completely by scrubbing with a suitable liquid. Such operations constitute absorption. However, although much of the theory of absorption refers to packed or continuous-contact towers, most large absorbers in solvent-recovery plants are of the stagewise type. Bubble-cap, sieve-plate, and baffle-plate columns are used, as are banks of spray

Table 1. Recovery Characteristics of Some Typical Solvents

Solvent	Explosive limits mol fraction in air		Equilibrium mol fraction in air at 25°C	Minimum absorption requirement at 25°C mol water/mol air
	Upper	Lower		
methanol	0.360	0.073	0.16	0.26
ethanol	0.190	0.043	0.075	0.33
propanol	0.135	0.021	0.026	0.37
butanol	0.112	0.014	0.009	0.44
acetone	0.128	0.026	0.29	2.1
methyl ethyl ketone	0.100	0.018	0.13	2.3
ethyl acetate	0.090	0.025	0.12	8
n-butyl acetate	0.076	0.017	0.018	15
ethyl ether	0.480	0.019	0.70	50
benzene	0.071	0.013	0.13	>100
toluene	0.071	0.012	0.037	>100
n-hexane	0.075	0.011	0.20	>100
n-decane	0.054	0.008	0.002	>100
carbon tetrachloride			0.14	>100
trichloroethylene			0.10	>100
tetrachloroethylene			0.025	>100

chambers arranged for countercurrent flow. Baffle-plate designs, spray designs, and randomly packed towers operating well below maximum capacity give low pressure drops and therefore tend to minimize the cost of power for circulating air. However, after considering all factors, the more predictable sieve-plate or bubble-cap columns with moderate pressure-drop designs are usually chosen.

Water is the principal scrubbing liquid used; the amount required depends largely on the type of solvent recovered. This is shown in the fourth column of Table 1, which gives the minimum number of moles of water theoretically needed for complete absorption of the compound from one mole of lean air at 25°C and 101.3 kPa (1 atm). These values would be the same as those shown in the third column, if the scrubbing liquid were not water but one forming an ideal solution with the solvent, that is, one obeying Raoult's law. The ratio of the values in these two columns is the activity coefficient of the compound in dilute aqueous solution. In a homologous series of alcohols the one with the highest vapor pressure is most easily absorbed in water. If an absorber oil were used instead of water, the order would be reversed, and the moles of absorbent required for the hydrocarbon solvents could be taken from the third column of Table 1.

The temperature of absorption is also important in determining the quantity of scrubbing liquid required. With rich air streams the heat of absorption is sometimes removed by indirect cooling either inside the absorption column or externally between stages (see Fig. 3). Refrigerated scrubbing liquids are sometimes used for the same reason (see Absorption).

Adsorption. The predominant method for removing solvent vapors from air streams is adsorption on activated carbon (see Carbon, activated carbon; Adsorptive separation gases) (10–12). As commonly practiced, the air is fed alternately to one of two adsorbing vessels, while the other is being fed with low pressure steam (see Fig. 2).

The steaming operation causes the solvent to be vaporized, and the latent heat is supplied by a portion of the steam that condenses into the bed. With some organic solvents this moisture interferes with subsequent adsorption, but with most it does not. If water can be tolerated, its vaporization into the air stream keeps the temperature down. In any case, if there is a net loss of heat around the cycle the water content rises to a point where it stops adsorption. This water balance is often an important factor in the design of an adsorption unit. Heating, cooling, or special drying periods may have to be included in the cycle of operations.

Although such a system has all the inherent disadvantages of intermittent processes, it offers great advantages for many applications. Unlike water scrubbing, it can be applied to water-insoluble solvents. Unlike condensation, it can reduce the solvent content of the air to a figure as low as desired.

The theoretical steam requirement for carbon adsorption is usually less than for absorption. Since heats of adsorption are high, the vapor pressure of solvent above activated carbon increases more rapidly with temperature than it does above liquid solutions. However, since the steaming operation in adsorption is somewhat inefficient, the actual steam requirement may offer no advantage over a favorable scrubbing application with water or oil. On the other hand, with mixed solvents and with those not ideally suited to absorption, water scrubbing may require excessive quantities of steam for reasons already noted. Low pressure drop may be obtained in adsorption systems by reducing the linear velocity of the air and the thickness of the bed. Beds are usually

30–100 cm thick with uniform pellets ≤3 mm dia. The beds are commonly supported inside horizontal cylinders so as to give a large flow cross section. Special devices must be used for distributing the air.

The success of carbon adsorption depends to a large extent upon proper control of the operating cycle. When the rate of solvent recovery is constant, automatic time-cycle controllers are often used. When flow varies widely, cut points may be determined by bed saturation. This type of operation is particularly well suited to guard against sudden surges in solvent content caused by process disturbances. In large installations with many adsorbers, operations are staggered to eliminate the peaks in steam demand.

Some volatile compounds react chemically on the surface of activated carbon. In the presence of air, organic compounds like aldehydes and ketones are oxidized, and the heat of combustion added to the heat of adsorption may raise the temperature of the active carbon to a point where it burns. With normal well-distributed air flows through the carbon bed, fires are not encountered because the flowing stream cools the carbon. However, with irregular operation or a badly channeled bed, hot spots may develop. Procedures and precautions have been recommended to prevent such an occurrence (13–14). Use of inert gas is an obvious solution.

Solvent vapors can also react on the surface of activated carbon to give undesirable products or to reduce the activity of adsorption. Entrained solids may foul the bed mechanically. Galvanic corrosion of metal parts may be accelerated by the electrical conductivity of the carbon. Sooner or later the pellets deteriorate and must be replaced. Most of the serious problems involved in adsorption processes have been solved by the manufacturers of activated carbon. Maintenance and operating labor has been minimized by sound engineering of process, equipment, and controls. Activated carbons resistant to poisoning and mechanical failure have been produced by careful controls of the manufacturing process.

Although carbon-adsorption operations are usually conducted batchwise, continuous contacting is feasible, especially on a large scale. Both moving-bed and fluidized-bed designs have been used for moving carbon particles from the adsorber to the stripper and back. In one solvent-recovery installation, a series of shallow fluidized beds produces countercurrent contact between the solid and gas phases (see Fluidization).

Adsorption may also be applied in a different way for solvent recovery. Traces of high-molecular-weight compounds, undesirable because of color, odor, or other properties, may be removed by treatment of the liquid with active carbons, clays, or other adsorbents (11). These solids are first slurried with the solvent until adsorption is complete. Then they are filtered from the purified liquid. The operation can be performed continuously, but it is usually done batchwise on a small scale (see Adsorption separation, liquids).

Automatic Control. Instrumentation has been an important factor in the development of satisfactory solvent-recovery units. When these units are only auxiliary to the manufacturing operation, they should be designed to require as little attention as possible. The trend has been toward completely automatic control, including start-up and shut-down periods. A number of fairly large installations require no manual labor other than the operation of a start-and-stop switch. Sequenced with the aid of cycle control, many batch operations are as efficient as continuous operations. A single piece of equipment may thus be used for a whole series of operations. For example, in vegetable-oil extraction, a single heated vessel with automatic control throughout has been used for solvent extraction of oil from flakes, filtration of the marc from the miscella, vacuum vaporization of the solvent from the marc, and discharge of the dried solids (see Instrumentation and control; Vegetable oils).

Economic Aspects

Solvent-recovery decisions, like most engineering and management decisions, are based mainly on economic analysis (see Economic evaluation). Such analysis involves not only careful engineering calculations, but also a prediction of future prices of raw materials, utilities, and products, of interest rates on borrowed capital, tax rates, and of government standards and regulations with regard to pollution and occupational safety. In an inflationary economy and with government regulation in a state of flux, such predictions are highly tenuous and, unless great profitability is expected, new expenditures for solvent recovery may be delayed unless they are required by law.

For processes employing large quantities of solvents, the installation of a solvent-recovery system is a foregone conclusion, since recovery under any circumstances would be profitable. The type of system used may significantly contribute to the total production cost.

For these reasons, the recovery system should be designed early in the development of a new process. It may, eg, prove desirable to evaporate the solvent with steam rather than air since recovery in the latter case is more expensive and less efficient. Savings may be effected if the process can use a different solvent. A nonflammable, less corrosive, or easily recovered solvent may be preferable. Mixed solvents are generally more difficult to recover than single compounds, and individual solvents may vary greatly in their ease of recovery. Methanol, for instance, is easily dehydrated by straight fractional distillation. Ethanol and propanol form constant-boiling mixtures with water, and recovery in the anhydrous form requires azeotropic entrainers. Butanol, on the other hand, is easily dehydrated since the two layers can be decanted and stripped. When purification of the recovered solvent is difficult, a less pure solvent might be acceptable for recycle.

Selection of the optimum design involves minimizing cost of unrecovered solvent, utilities, and labor, interest on borrowed capital and lost return on investment. The cost of unrecovered solvent is of most concern in an inefficient recovery system handling large volumes of solvent over a long period of time. If only 95% of the solvent is recovered, the cost of making up the unrecovered 5% would normally be much larger than any of the other charges. Better engineering design aimed at recoveries >99% would usually not increase the other costs significantly and should be sought in every case.

Even with efficient recovery, make-up solvent is always required. Designers of solvent-recovery equipment often claim that the solvent content of the various waste streams can be held to negligible amounts. Although analyses of these streams usually corroborate such claims, inventory balances general show losses to be much higher than those calculated. Only by a careful study of spillage, venting, gauging practice, chemical instability, etc, can this discrepancy be eliminated.

For large long-lived systems with recoveries approaching 100%, utilities often present the largest cost. Considerable increases in investment and in the complexity of the system have been made to reduce steam costs by as little as $0.01/kg solvent recovered. For example, in the recovery of acetic acid from cellulose acetate manufacture, azeotropic distillation with a volatile entrainer was first used to reduce the steam required for straight fractional distillation of water from acid. This was followed by the use of less volatile entrainers, and then by liquid extraction with low boiling solvents. Steam savings were realized in each successive process, and these were further improved by the use of multiple-effect and vapor-reuse techniques. Now it is evident that extraction with higher boiling solvents gives even lower steam consumption.

By comparison, interest on borrowed capital and loss of return on investment are small charges against a large long-lived solvent-recovery system, as long as the economy is noninflationary. With inflation, these charges increase several-fold, reflecting mainly anticipated reduction in the real value of the dollar. Aside from these considerations, solvent-recovery decisions could be made on the basis of sound engineering rather than on concern for the future of the general economy.

Investment and operating costs increase as the quantity of solvent increases, but not proportionally. There are many fixed costs, and investment, labor, and overhead costs per kilogram recovered can be unreasonably high for small or short-lived solvent-recovery systems. In the past, most small manufacturers have simply discarded spent solvent and other wastes, but with increasing enforcement of environmental regulations this may no longer be possible. A recent EPA publication states that as many as 232 small paint-formulating plants may have to close as a result of a proposed regulation of plant effluents (15). Remedies like piping or trucking effluents to a common treatment installation have been suggested. A small combination solvent-recovery and disposal unit, if developed, might find a sizable market.

Many larger plants are engaged in a number of small-volume solvent applications for which individual recovery and disposal systems would not be economical. Such solvent effluents can be collected and stored, and periodically recovered. However, the mix of waste streams must not include materials that produce health and safety hazards during disposal. Indiscriminate mixing of wastes, followed by sale or payment for disposal to a private contractor does not relieve the waste generator of legal responsibility for whatever environmental harm may result (see Wastes, industrial).

Health and Safety Factors

Safety considerations are of prime importance in the design of most solvent-recovery processes. They may form the basis for choosing a particular solvent and motivate the installation of the recovery system. Safety considerations may determine important operating conditions, or call for certain types of building designs, equipment, controls, or accessories. Although safety is usually thought of in terms of flammable solvents, consideration must also be given to heat and mechanical hazards, and to the noxious properties of solvent vapors (see Plant safety).

Accepted practice with regard to the handling of flammable solvents is contained in UL codes and those of the National Fire Protection Association (16). They form the basis for municipal safety laws and insurance inspection requirements. Subjects covered include location, design, and ventilation of buildings; size, type, and construction of vessels; arrangement of piping, valves, pumps, and vents; distribution of alarm and extinguishing equipment; instruction of personnel; and many others. The codes for dry cleaning, for solvent extraction of fats and oils, and for deposition of coatings by sprays and dip tanks include rather detailed descriptions of the standard equipment items available to these industries. Such standardization is in the interest of safety especially for small installations with operating personnel lacking in scientific and engineering background. At the same time, it is the intent of these codes to permit genuine improvements in design.

The use of most common flammable solvents creates conditions classified as Class I, Group D by the National Electrical Code (17). Motors and other electrical equipment in these locations must be explosion proof and specifically approved for this class of service. Sources of static electricity such as belt drives should be eliminated, and open flames such as welding torches should be prohibited during operation of the unit. Vents should be protected with flame arresters. Leakage of noxious or flammable vapors into operating rooms should be minimized and adequate ventilation provided. Pressure-relief systems with remote catch tanks should be used on equipment wherever there is a possibility of excessive internal pressure.

Mixtures of air and solvent should be kept below the explosive range whenever they are present in significant volumes. Convenient automatic instruments are available for this purpose. It has been recommended that concentrations inside equipment be kept below one half of the lower explosive limit. For concentrations higher, special precautions should be taken to guard against mishaps. Unsafe practices lead, sooner or later, to disaster.

Concentrations of solvent vapors outside equipment and in areas accessible to employees are limited by OSHA standards (18). Allowable concentrations for both nonflammable and flammable vapors under these standards are much lower than those dictated by fire safety. Compliance with these standards may require costly additions to process equipment, control and monitoring instrumentation, and employee training.

Increased interest in the long-range effect on human health of exposure to solvents and other chemicals has led to the following developments: Legislation has been enacted requiring governmental agencies to identify and regulate health risks and requiring industrial concerns to comply with the appropriate regulations in their operating, labeling, and reporting practices (see Industrial hygiene and toxicology). Courts have awarded sizeable damages against chemical manufacturers and users on the basis of exposure to a variety of substances by employees, customers, and others. Finally, some large chemical companies in response to, or in anticipation of, the foregoing developments have adopted policies of "cradle-to-grave stewardship" over the chemicals they use or produce, monitoring both their own and their customers' safety precautions. The effect of these developments on solvent-recovery design will, of course, be an increased preference for innocuous solvents and for systems giving essentially complete recovery.

Applications

Processes for the manufacture of synthetic fibers and sheets, impregnated articles, and related products often use solvents as volatile vehicles for the deposition of polymeric materials. Although many resins are deposited from water suspensions or are formed directly from fused solid or by chemical reaction, deposition from organic solvents represents a significant fraction of the total. In 1980 these industries in the U.S. recovered ca 3×10^6 metric tons of solvent. Recovery of acetone from cellulose acetate spinning and film production was the largest operation in this group. The products in these applications are usually air-dried and the solvent is recovered by absorption or adsorption. Superheated steam has also been used for drying followed by condensation of the vapors. The recovery of solvents vaporized in painting is generally avoided unless the operation is large or continuous. The automotive industry with its very extensive painting operations has long been reluctant to modify these operations in order to recover the solvents. However, under a new cooperative arrangement with the Michigan Department of Natural Resources, the large manufacturers expect to reduce their volatile organic emissions substantially (19).

Solvent extraction of vegetable oils and other solid–liquid extraction processes in the United States recover $>2 \times 10^6$ t/yr of solvent. The main application is extraction of oil from soybeans (qv). Recovery is principally by evaporation from the miscella and drying the marc with indirect heat or steam (16,20). Other oils are also extracted in quantity but the expression process is favored for seeds of higher oil content. The pharmaceutical industry uses solvents extensively for the extraction of both natural and synthetic materials. Field tests have been conducted on the use of solvents to extract petroleum *in situ* from oil sands, but only minor attention was given to the solvent-recovery aspects (5) (see Oil shale; Tar sands). Other interesting processes involving the solvent extraction of metal ores, fossil fuel materials, and other natural resources have reached various stages of development (8,21–22), many with government financing.

Solvent refining of lubricating oils and other solvent operations in the petroleum industry entail the recovery of solvents in quantities far exceeding all other applications. A single installation may circulate $>5 \times 10^8$ t/yr (23). Extractive distillations or gas-treating operations can be of similar magnitudes. If steam stripping of absorber oils is classed as a solvent-recovery operation, the total volume of solvent recovered by this industry is truly immense. Liquid–liquid extraction is also being used extensively in the solvent extraction of nuclear intermediates and of other metallic compounds from ores (8). The organic streams in these metallurgical operations are ordinarily prepared for reuse by chemical treatment, additional extractions, and mechanical separations. However, evaporation and distillation are also used. Liquid–liquid extractions of appreciable size are used in pharmaceutical manufacture.

Washing operations using solvents, eg, dry cleaning, are small in size but large in number. In the U.S. about 3×10^5 metric tons of solvent is consumed each year by such operations. Statistics are not available on the amount of redistillation practiced in this industry but $1–2 \times 10^6$ t/yr seems a reasonable estimate. Improvements in aqueous detergents and the increased popularity of washable fabrics account for the recent decline in dry cleaning (qv).

The use of solvents for solution polymerizations and other chemical reactions is a growing area for solvent recovery. About 10^7 metric tons of cyclohexane are recovered each year in the manufacture of high density polyethylene alone (see Olefin polymers). Recovery is mainly by mechanical separation and redistillation. Polypropylene manufacture and other polymerization processes also include large solvent-recovery operations. Recovery of acetic acid used as a solvent medium for the reaction of acetic anhydride with cellulose now amounts to about 10^6 t/yr in the U.S.

BIBLIOGRAPHY

"Solvent Recovery" in *ECT* 1st ed., Vol. 12, pp. 641–654, by C. M. Cooper, Michigan State College; "Solvent Recovery" in *ECT* 2nd ed., Vol. 18, pp. 549–564, by C. M. Cooper, Michigan State University.

1. H. L. Hoffman and co-workers, *Hydrocarbon Process. Pet. Refiner* **45**(9), 220 (1966).
2. E. J. Fuller in J. J. McKetta, ed., *Encyclopedia of Practice and Design*, Marcel Dekker, Inc., New York, 1979.
3. C. A. Irani and E. W. Funk in N. N. Li, ed., *Recent Developments in Separations Science*, Vol. III-A, CRC Press, Cleveland, Ohio, 1977, pp. 171–193.
4. L. W. Holm and V. A. Josendal, *J. Pet. Technol.* **26,** 1427 (1974).
5. T. B. Reid and H. J. Robinson, *J. Pet. Technol.* **33,** 1723 (1981).
6. R. E. Treybal in R. H. Perry and C. H. Chilton, eds., *Chemical Engineers' Handbook*, 5th ed., McGraw-Hill Book Co., New York, 1973, Sec. 15.
7. L. A. Robbins in P. A. Schweitzer, ed., *Handbook of Separation Techniques for Chemical Engineers*, McGraw-Hill Book Co., New York, 1979.
8. G. M. Ritcey and A. W. Ashbrook, *Solvent Extraction*, Elsevier Scientific Publishing Co., Amsterdam, 1979.
9. B. D. Smith in R. H. Perry and C. H. Chilton, eds., *Chemical Engineers' Handbook*, 5th ed., McGraw-Hill Book Co., New York, 1973, pp. 13-36–13-47.
10. T. Vermeulen, G. Klein, and N. K. Hiester in ref. 9, Sec. 16.
11. P. N. Cheremisinoff and F. Ellerbusch, *Carbon Adsorption Handbook*, Ann Arbor Science Publishers, Inc., Ann Arbor, Mich., 1978.
12. J. W. Drew in P. A. Schweitzer, eds., *Handbook of Separation Techniques for Chemical Engineers*, McGraw-Hill Book Co., New York, 1979.
13. A. A. Naujokas, *CEP Technical Manual, Loss Prevention* **12,** 128 (1979).
14. M. J. Chapman and D. L. Field, *CEP Technical Manual, Loss Prevention* **12,** 136 (1979).
15. J. R. Berlow, *Proposed Effluent Limitations for the Paint Formulating Point Source Category*, EPA 440/1-79/049-b, U.S. Environmental Protection Agency, Washington, D.C., 1979.
16. *National Fire Codes*, Vol. 3, National Fire Protection Association, Boston, Mass., 1978.
17. Ref. 16, Vol. 6.
18. R. J. Lewis and R. L. Tatken, eds., *Registry of Toxic Effects of Chemical Substances*, U.S. Dept. of Health and Human Services, Cincinnati, Ohio, 1980.
19. D. Rector, *SAE Transactions*, No. 790370, V. 88, 1979.
20. D. Swern, ed., *Bailey's Industrial Oil and Fat Products*, Interscience Publishers, a division of John Wiley & Sons, Inc., New York, 1964, pp. 704–711.
21. R. P. Anderson, *Chem. Eng. Prog.* **71,** 72 (Apr. 1975).
22. R. Katzen, R. Frederickson, and B. F. Brush, *Chem. Eng. Prog.* **76,** 62 (Feb. 1980).
23. J. G. Speight, *The Chemistry and Technology of Petroleum*, Marcel Dekker, Inc., New York, 1980.

C. M. COOPER
Michigan State University

SOLVENTS, INDUSTRIAL

The term industrial solvents is generally applied to organic compounds used on an industrial scale to dissolve, suspend, or change the physical properties of other materials. These applications include production of finished products of dissolved or suspended materials for the ink or coatings industries, reaction solvents for the process industry, or cleaning agents for the maintenance or dry-cleaning industries. Although any organic liquid can act as a solvent on an industrial scale, practical consideration limits those that are used. Generally, these compounds are aromatic or aliphatic hydrocarbons, alcohols, aldehydes, ketones, amines, esters, ethers, glycols, glycol ethers, or alkyl or aromatic halides that boil at 75–220°C.

Selection of the proper solvent or development of a blend is made easier when the interaction between solvent and solute is understood. Traditionally, trial and error methods were used to make this selection. However, as more information on solvent–solute interactions has been gathered, there has been a gradual shift from the empirical methods to more fundamental ones. Environmental and safety rules and regulations have required substitution of new solvents for older, less acceptable ones. Selection of these substitutions as well as the development of new products is aided by knowledge of the solvent–solute interaction.

Often, complete solution is not possible or desired, depending upon the materials involved and their application. In these cases, solubility parameters have little usefulness. However, solubility parameters are useful, particularly in the coatings industry, because they characterize systems for experimental evaluation. Solubility parameters, though useful, must be verified by experimental evaluation.

Solubility

Solution of one compound by another is based on the fundamental concept that the process must be spontaneous. This ideal is expressed by the equation for the free energy of mixing:

$$\Delta G = \Delta H - T\Delta S$$

where ΔG is the free energy of mixing, ΔH is the enthalpy of mixing, and ΔS is the entropy of mixing. ΔS can be considered positive for the dissolution process, as it is usually considered in industrial applications. Therefore, the controlling term for the spontaneous process is the enthalpy of mixing. If it is negative or a small positive number, then the process could be considered spontaneous.

Single-Component Solubility Parameter. An expression was developed for the heat of mixing

$$\Delta H_m = \frac{x_1 x_2 V_1 V_2}{x_1 V_1 + x_2 V_2}\left[\frac{a_1^{1/2}}{V_1} - \frac{a_2^{1/2}}{V_2}\right]^2$$

where x_1 and x_2 are mole fractions, a_1 and a_2 are interaction constants, and V_1 and V_2 are volumes (1). Cohesive energy of a mole of liquid mixture could be expressed as

$$\Delta E_m = (x_1 V_1 + x_2 V_2)\left[\left(\frac{\Delta E_1^v}{V_1}\right)^{1/2} - \left(\frac{\Delta E_2^v}{V_2}\right)^{1/2}\right]^2 \phi_1 \phi_2$$

Table 1. Physical Properties of Common Industrial Solvents

Solvent	CAS Registry No.	Common name (trade name)	Empirical formula	KB value[a,b]	Solubility parameter δ, $(J/m^3)^{1/2} \times 10^{-3}$ [c]
Alcohol					
methanol	[*67-56-1*]	methyl alcohol, wood alcohol	CH_4O		7.07
ethanol	[*64-17-5*]	ethyl alcohol, grain alcohol	C_2H_6O		6.20
1-propanol	[*71-23-8*]	*n*-propyl alcohol	C_3H_8O		5.80
2-propanol	[*67-63-0*]	isopropyl alcohol, IPA	C_3H_8O		5.61
1-butanol	[*71-36-3*]	butyl alcohol, *n*-butanol	$C_4H_{10}O$		5.56
2-butanol	[*78-92-2*]	*sec*-butyl alcohol	$C_4H_{10}O$		5.27
2-methyl-1-propanol	[*78-83-1*]	isobutyl alcohol, IBA	$C_4H_{10}O$		5.12
2-methyl-2-propanol	[*75-65-0*]	*tert*-butyl alcohol	$C_4H_{10}O$		5.17
furfuryl alcohol	[*98-00-0*]	2-hydroxymethyl-furan	$C_5H_7O_2$		6.10
tetrahydrofurfuryl alcohol (THFA)	[*97-99-4*]	(2-tetrahydrofur-anyl)methanol	$C_5H_{11}O_2$		
1-pentanol	[*71-41-0*]	amyl alcohol	$C_5H_{12}O$		5.32
3-methyl-1-butanol	[*123-51-3*]	isoamyl alcohol	$C_5H_{12}O$		4.88
allyl alcohol	[*107-18-6*]	2-propenol, vinylcarbinol	C_3H_6O		5.76
cyclohexanol	[*108-93-0*]	cyclohexyl alcohol (Hexalin)	$C_6H_{12}O$		5.56
1-hexanol	[*111-27-3*]	*n*-hexyl alcohol	$C_6H_{14}O$		5.22
4-methyl-2-amyl alcohol	[*54972-97-3*]	4-methyl-2-pent-anol	$C_6H_{14}O$		
2-ethylbutyl alcohol	[*97-95-0*]	*sec*-hexyl alcohol	$C_6H_{14}O$		5.12
hexyl alcohol (mixture)	[*111-27-3*]	commercial hexanol	$C_6H_{14}O$		
benzyl alcohol	[*100-51-6*]	α-hydroxytoluene, phenylcarbinol	C_7H_8O		5.90
2-octanol	[*123-96-6*]	*sec*-capryl alcohol[l]	$C_8H_{18}O$		
2-ethylhexanol	[*104-76-7*]	2-hexyl alcohol	$C_8H_{18}O$		4.63
2-ethyl-4-methyl-1-pentanol	[*106-67-2*]		$C_8H_{18}O$		
Aliphatic hydrocarbons					
pentane	[*109-66-0*]	*n*-pentane	C_5H_{12}	28	3.4
cyclohexane	[*110-82-7*]	hexamethylene, benzene hydride	C_6H_{12}	50	4.0
hexane	[*110-54-3*]	*n*-hexane	C_6H_{14}	30.2	3.6
heptane	[*142-82-5*]	*n*-heptane	C_7H_{14}	30.8	3.6
n-octane	[*111-65-9*]	*n*-octane	C_8H_{18}		3.7
α-pinene	[*80-56-8*]	2-pinene, mixed isomers	$C_{10}H_{16}$		
β-pinene	[*127-91-3*]	2(10)-pinene (nopinene)	$C_{10}H_{16}$		
rubber solvent				33.0	3.6
VM&P (Varnish Makers' and Painters') solvent		petroleum ether, naphtha		39.0	3.7

Water solubility (at 25°C, except where noted otherwise), wt %[d,e]		Azeotrope, wt %/°C	Boiling range at 101.3 kPa (= 1 atm), °C[f]	Vapor pressure[g] (at 25°C except where noted otherwise), kPa[h]	Specific gravity[i] (at 20°C except where noted otherwise)	Refractive index[j] (at 20°C except where noted otherwise)	Freezing point[k], °C
In water	Water in						
∞	∞	none	64–65	16.7	0.79129	1.32840	−98
∞	∞	96/78	75–78	8	0.78939	1.36143	−114
∞	∞	72/87	96–98	2.8	0.80375	1.38556	−126
∞	∞	88/80	81–84	6	0.78545	1.37720	−88
7.5	20.5	57/93	116–118	0.83	0.8097	1.3993	−89
12.5	44.1	73.2/87	97–100	2.4	0.8069	1.3972	−115
10.0	16.9	67/90	107–110	1.3	0.8016	1.3959	−108
∞	∞	88/80	82	5.6	0.7812_{25}	1.3877	26
∞	∞	20/98	170	0.08	1.1285	1.4868	−14
∞	∞		178	0.1	1.0524	1.4520	80
2.2	7.5	46/96	119–138	0.32	0.8151	1.4100	−78
2.7	9.6	50/95	131	0.32	0.8104	1.4071	−117
∞	∞	72/89	97	3.7	0.85511_{15}	1.4135	−129
3.8	11.8	31/98	160–162	0.13	0.9684_{25}	1.46477_{25}	25
0.7	7.4	33/98	157	0.1	0.8196	1.4181	−45
1.6	6.4	57/94	128–134	1.1	0.8080	1.4112	−90
0.6	4.6	41/97	147	0.51	0.8333	1.4224	−114
0.08	8.4	9/100	205	0.013	1.04127_{25}	1.54035	−15
			179		0.8193		−39
0.07	2.6	20/99	184	0.013	0.8332	1.42305	−76
0.0038	0.012	98/35	35–39	68.2	0.62624	1.35748	−129
0.0055	0.01	91/69	79–81	13	0.77855	1.42623	6
0.00095	0.011	94.4/62	64–67	20.1	0.65937	1.37486	−95
0.00029	0.0091	87.1/79.2	91–96	6.1	0.68376	1.38764	−91
66×10^{-6}	0.0095	74.5/90	126	1.9	0.70252	1.39743	−57
			159	0.5	0.8582	1.4658	−64
			166	0.61	0.8667	1.4768_{25}	−62
			41–96	65_{38}	0.691_{15}	1.3850	<−80
			119–137	3.5_{38}	0.735_{15}	1.4233	<−80

Table 1 (*continued*)

Solvent	CAS Registry No.	Common name (trade name)	Empirical formula	KB value[a,b]	Solubility parameter δ, $(J/m^3)^{1/2} \times 10^{-3}$ [c]
high flash VM&P solvent		petroleum ether		33.0	3.6
short-range mineral spirits		(Quick Dry)		33.0	3.7
Rule 66 mineral spirits		(Stoddard solvent)		35.7	3.7
140 solvent		dry-cleaners' solvent		32.0	3.8
low odor 140 solvent				30.7	3.6
mineral seal oil		(535 solvent)		22.0	3.5
Alkyl halides					
methylene chloride	[*75-09-2*]	dichloromethane	CH_2Cl_2	136	4.7
chloroform	[*67-66-3*]	trichloromethane	$CHCl_3$		4.5
carbon tetrachloride	[*56-23-5*]	tetrachloromethane	CCl_4	104	4.2
perchloroethylene	[*127-18-4*]	tetrachloroethylene	C_2Cl_4	90	4.5
1,1,2-trichloro-1,2,2-trifluoroethane (TTE)	[*76-13-1*]	(Halocarbon 113)	$C_2Cl_3F_3$		3.6
trichloroethylene	[*79-01-6*]	ethylene trichloride (Triclene)	C_2HCl_3	130	4.5
1,1,2,2-tetrachloroethane	[*79-34-5*]	acetylene tetrachloride	$C_2H_2Cl_4$		4.7
1,1,1-trichloroethane	[*71-55-6*]	methylchloroform	$C_2H_3Cl_3$	124	4.1
1,1,2-trichloroethane	[*79-00-5*]	vinyl trichloride	$C_2H_3Cl_3$		4.7
1,2-dichloroethane	[*107-06-2*]	ethylene dichloride	$C_2H_4Cl_2$		
1,2-dibromoethane	[*106-93-4*]	ethylene dibromide	$C_2H_4Br_2$		5.07
ethyl chloride	[*75-00-3*]		C_2H_5Cl		4.5
ethyl bromide	[*74-96-4*]		C_2H_5Br		4.7
propylene dichloride	[*78-87-5*]	1,2-dichloropropane	$C_3H_6Cl_2$		4.4
1,2,4-trichlorobenzene	[*120-82-1*]		$C_6H_3Cl_3$		4.8
o-dichlorobenzene	[*95-50-1*]	1,2-dichlorobenzene	$C_6H_4Cl_2$		4.88
chlorobenzene	[*108-90-7*]	phenyl chloride, MCB	C_6H_5Cl	90	4.6
p-chlorotoluene	[*106-43-4*]		C_7H_7Cl		4.3
Amines					
n-propylamine	[*107-10-8*]		C_3H_9N		
isopropylamine	[*75-31-0*]	2-aminopropane	C_3H_9N		
butylamine (mixture)	[*109-73-9*]		$C_4H_{11}N$		4.2
tert-butylamine	[*75-64-9*]		$C_4H_{11}N$		
diethylamine	[*109-89-7*]		$C_4H_{11}N$		3.9
diethylenetriamine	[*111-40-0*]		$C_4H_{13}N_3$		
cyclohexylamine	[*108-91-8*]		$C_6H_{13}N$		
triethylamine	[*121-44-8*]		$C_6H_{15}N$		
diisopropylamine	[*108-18-9*]		$C_6H_{15}N$		
toluidines (mixture)	[*26915-12-8*]	methylaniline	C_7H_9N		
dibutylamine	[*111-92-2*]		$C_8H_{19}N$		4.0
ethanolamine	[*141-43-5*]	amino ethanol	C_2H_7ON		
morpholine	[*110-91-8*]	diethyleneimide oxide	C_4H_9ON	10.8	
ethylaminoethanol	[*110-73-6*]		$C_4H_{11}ON$		

Water solubility (at 25°C, except where noted otherwise), wt %[d,e] In water	Water in	Azeo-trope, wt %/°C	Boiling range at 101.3 kPa (= 1 atm), °C[f]	Vapor pressure[g] (at 25° except where noted otherwise), kPa[h]	Specific gravity[i] (at 20°C except where noted otherwise)	Refractive index[j] (at 20°C except where noted otherwise)	Freezing point[k], °C
			141–160	0.77$_{20}$	0.771$_{15}$		
			157–179	1$_{38}$	0.776$_{15}$	1.4294	<−80
			157–196	0.8$_{38}$	0.787$_{15}$	1.4339	−70
			182–199	0.07$_{38}$	0.786$_{15}$	1.4345	−53
			183–206	0.07$_{38}$	0.783$_{15}$	1.4359	−53
			254–321		0.816$_{15}$		
1.30	0.20	98.5/38	39–43	58.1	1.326	1.42416	−95
0.82	0.072	97.8/56.1	60–62	26	1.47988$_{25}$	1.44293$_{25}$	−63
0.077	0.010	95.5/66	75–76	15.3	1.58439$_{25}$	1.45739$_{25}$	−23
0.015	0.0105	84/87.7	121–123	2.4	1.6063$_{30}$	1.50566	−22
0.017	0.011	99/44.5	46	44	1.56354$_{25}$	1.35572$_{25}$	−36
0.11	0.32	94/73.6	84–86	6.3	1.4514$_{30}$	1.4767	−86
0.287$_{20}$		66/94.3	134–146	0.8	1.60	1.4910$_{25}$	−44
0.13$_{20}$	0.034	92/65	72–86	16.1	1.3376	1.4379	−30
0.45	0.12	83.6/86	114	5$_{35}$	1.4416	1.4711	−35
0.81	0.15	92/72	83–84	11	1.2531	1.4448	−36
0.429	0.071	73.7/91	131	1	2.1791	1.5387	10
0.44			12	217.7	0.9028$_{15}$	1.3738$_{10}$	−136
0.91		99.1/37	38	62.4	1.451	1.4239	−118
0.04	0.27	88/78	97	5$_{19.4}$	1.1593	1.4345	17
<0.1			213	0.13$_{38}$	1.454$_{25}$		
0.026	0.309	33/98	180	0.17	1.30589	1.55145	−17
0.0488	0.0327	71.6/90.2	132	1.57	1.1063	1.52481	−46
			162	1.3$_{44}$	1.0695$_{24.4}$		7.3
∞			48	40.9	0.7173	1.3882	−83
∞			32	76.6	0.6875	1.3742	−95.2
			77		0.74–0.76		−50
∞			44	48.3	0.6958	1.3788	−73
∞			55	31.1	0.7070	1.3854	−50
			207	0.029$_{20}$	0.9586		−39
∞			135	1.2	0.87128$_{15}$	1.45926	−18
5.5	4.6	90/75	90	7.6	0.7276	1.4010	−115
.11	40	90.8/74	84	8	0.7153	1.39236	−96
			199	<0.13	1.004		−16
0.47$_{20}$	6.2$_{20}$	49.5/97	159	0.31	0.7619	1.4177	−62
∞			171	0.048	1.0195$_{15}$	1.4539	10.5
∞			129	1.35$_{25}$	1.0049$_{15}$	1.4542	−3
			161		0.92		

Solvent	CAS Registry No.	Common name (trade name)	Empirical formula	KB value[a,b]	Solubility parameter δ, $(J/m^3)^{1/2} \times 10^{-3}$ [c]
dimethylethanolamine	[*108-01-0*]		$C_4H_{11}ON$		
diethanolamine	[*111-42-2*]		$C_4H_{11}O_2N$		
diisopropanolamine	[*110-97-4*]		$C_6H_{15}O_2N$		
triethanolamine	[*102-71-6*]		$C_6H_{15}O_3N$		
Aromatic hydrocarbons					
benzene	[*71-43-2*]	benzol, coal-tar naphtha	C_6H_6	112	4.5
toluene	[*108-88-3*]	toluol	C_7H_8	105	4.3
ethylbenzene	[*100-41-4*]	ethylbenzol	C_8H_{10}		4.3
xylene (mixture)	[*1330-20-7*]	xylol	C_8H_{10}	98	4.3
o-xylene	[*95-47-6*]	1,2-dimethylbenzene, *o*-xylol[l]	C_8H_{10}	104	4.3
m-xylene	[*108-38-3*]	1,3-dimethylbenzene, *m*-xylol[l]	C_8H_{10}		4.3
p-xylene	[*106-42-3*]	1,4-dimethylbenzene, *p*-xylol[l]	C_8H_{10}	93	4.3
aromatic 100 solvent		(aromatic 10, 1 solvent)		94	4.2
aromatic 150 solvent		(aromatic 15, 2 solvent)		92	4.1
Esters					
ethyl acetate	[*141-78-6*]	acetic ester	$C_4H_8O_2$		4.4
n-propyl acetate	[*109-60-4*]	propyl acetate	$C_5H_{10}O_2$		4.3
isopropyl acetate	[*108-21-4*]	*sec*-propyl acetate	$C_5H_{10}O_2$		4.1
methyl acetoacetate	[*105-45-3*]		$C_5H_8O_3$		
ethyl acetoacetate	[*141-97-9*]		$C_6H_{10}O_3$		
n-butyl acetate	[*123-86-4*]	butyl ethanoate	$C_6H_{12}O_2$		4.1
sec-butyl acetate	[*105-46-4*]	1-methylpropyl acetate	$C_6H_{12}O_2$		4.0
isobutyl acetate	[*110-19-0*]	2-methylpropyl acetate	$C_6H_{12}O_2$		4.0
amyl acetate	[*628-63-7*]	1-pentyl acetate	$C_7H_{14}O_2$		4.1
isoamyl acetate	[*123-92-2*]	3-methyl-1-butyl acetate	$C_7H_{14}O_2$		3.8
4-methyl-2-pentanol, acetate	[*108-84-9*]		$C_8H_{16}O_2$		3.9
isobutyl isobutyrate	[*97-85-8*]		$C_8H_{16}O_2$		3.9
benzyl acetate	[*140-11-4*]		$C_9H_{10}O_2$		
Glycol ethers					
ethylene glycol monomethyl ether	[*109-86-4*]	2-methoxyethanol (methyl Cellosolve)	$C_3H_8O_2$		5.56
ethylene glycol monoethyl ether	[*110-80-5*]	2-ethoxyethanol (Cellosolve solvent)	$C_4H_{10}O_2$		5.12
ethylene glycol monomethyl ether acetate	[*110-49-6*]	2-methoxyethyl acetate (methyl Cellosolve acetate)	$C_5H_{10}O_3$		4.5
diethylene glycol monomethyl ether	[*111-77-3*]	(methyl Carbitol)	$C_5H_{12}O_3$		4.1

Water solubility (at 25°C, except where noted otherwise), wt %[d,e]		Azeotrope, wt %/°C	Boiling range at 101.3 kPa (= 1 atm), °C[f]	Vapor pressure[g] (at 25°C except where noted otherwise), kPa[h]	Specific gravity[i] (at 20°C except where noted otherwise)	Refractive index[j] (at 20°C except where noted otherwise)	Freezing point[k], °C
In water	Water in						
			131		0.8866		
96.4		none	268	3×10^{-5}	1.0899_{30}	1.4747_{30}	28
			83–84		0.722		
∞			335	$<1 \times 10^{-4}$	1.1196_{25}	1.4835_{25}	22
0.18	0.063	91.2/69	80	13	0.87901	1.50112	5.5
0.052	0.033	79.8/85	110	3.7	0.86696	1.49693	−95
0.015	0.043	67/92	136	1.3	0.86702	1.49588	−95
			138–144	0.8–1	0.8702	1.4983	−45
0.018		50.1/94	144	0.88	0.8802	1.50545	−25
0.02	0.04	54.5/93	139	1.1	0.86417	1.49722	−48
0.19		54.9/93	138	1.2	0.86105	1.49582	13.2
			316–342	0.07_{38}	0.8756	1.4998	<−90
			361–405	0.07_{38}	0.8956	1.5107	−35
8.08	2.94	91.5/70.3	77	12	0.90063	1.37239	−84
2.3	2.9	86/82.2	101	4.3	0.89377_{15}	1.38442	−93
2.9	1.8	89.4/17.6	88	8.1	0.8718	1.3773	−73
			172	0.8	1.0747	1.4186	−80
12	4.9	none	181	0.3	1.0213_{25}	1.4192	−39
0.43	1.86	71.3/90.2	126	1.7	0.87636_{25}	1.3827_{30}	−74
0.62_{20}	1.65_{20}	77.5/87	112	3.2	0.8720	1.38941	
0.67_{20}	1.64_{20}	83.4/87.4	118	2.6	0.8745	1.39018	−99
0.17_{20}	1.15_{20}	59/95.2	149	0.8	0.8752	1.4028	<−100
2.0	1.6	63.7/93.6	143	0.6	0.8719	1.4007	−78
0.5		60.6/95.5	146	0.51_{20}	0.8598		−64
			148	0.64	0.8542	$1.3986_{18.2}$	−81
slight		12.5/99.6	215	0.189	1.0515_{25}	1.5232	−52
∞		15.3/100	124	1.3	0.97459	1.4021	−85
∞		28.8/99.4	136	0.71	0.92945	1.4077	<−90
			120		0.919_{25}		−97
∞		48.5/97	145	0.7	1.0049	1.4022	−65
∞		none	194	0.024	1.021	1.4264	−76

Table 1 (*continued*)

Solvent	CAS Registry No.	Common name (trade name)	Empirical formula	KB value[a,b]	Solubility parameter δ, $(J/m^3)^{1/2} \times 10^{-3}$ [c]
ethylene glycol monoethyl ether acetate	[*111-15-9*]	2-ethoxyethyl acetate (Cellosolve acetate)	$C_6H_{12}O_3$		4.2
ethylene glycol dimethyl ether	[*110-71-4*]	(dimethyl Cellosolve)	$C_6H_{14}O_2$		
ethylene glycol monobutyl ether	[*111-76-2*]	2-butoxyethanol (butyl Cellosolve	$C_6H_{14}O_2$		4.6
diethylene glycol monoethyl ether	[*111-90-0*]	(Carbitol)	$C_6H_{14}O_3$		4.98
ethylene glycol monophenyl ether	[*122-99-6*]	(butyl Carbitol)	$C_8H_{10}O_2$		
ethylene glycol monobutyl ether acetate	[*112-07-2*]	(butyl Cellosolve) acetate)	$C_8H_{16}O_3$		4.1
diethylene glycol monoethyl ether acetate	[*112-15-2*]	(Carbitol acetate)	$C_8H_{16}O_4$		4.1
diethylene glycol monobutyl ether	[*112-34-5*]	(butyl ethyl Cellosolve)	$C_8H_{18}O_3$		4.6
propylene glycol monophenyl ether	[*770-35-4*]	(Polysolve PM)	$C_9H_{12}O_2$		4.93
diethylene glycol mono-*sec*-butyl ether acetate	[*124-17-4*]	(butyl Carbitol acetate)	$C_{10}H_{20}O_4$		4.1
Ketones					
acetone	[*67-64-1*]	2-propanone, dimethyl ketone	C_3H_6O		4.8
methyl ethyl ketone (MEK)	[*78-93-3*]	2-butanone	C_4H_8O		4.5
mesityl oxide	[*141-79-7*]	4-methyl-3-penten-2-one	$C_6H_{10}O$		4.4
cyclohexanone	[*108-94-1*]	cyclohexyl ketone	$C_6H_{10}O$		4.8
methyl *n*-butyl ketone (MBK)	[*591-78-67*]	2-hexanone	$C_6H_{12}O$		4.0
methyl isobutyl ketone	[*108-10-1*]		$C_6H_{12}O$		4.1
diacetone alcohol	[*123-42-2*]	diacetone, 4-hydroxy-4-methyl-2-pentanone	$C_6H_{12}O_2$		4.5
methyl amyl ketone	[*110-43-0*]	2-heptanone	C_7H_4O		4.1
methyl isoamyl ketone	[*110-12-3*]		$C_7H_{14}O$		4.1
diisobutyl ketone	[*106-83-8*]		C_8H_8O		3.8
isophorone	[*78-59-1*]	3,5,5-trimethyl-2-cyclohexen-1-one	$C_8H_{14}O$		4.4
Others					
ethylene glycol	[*107-21-1*]		$C_2H_6O_2$		7.12
propylene glycol	[*57-55-6*]		$C_3H_8O_2$		6.15
diethylene glycol	[*111-46-6*]		$C_4H_{10}O_3$		5.90
acetonitrile	[*75-05-8*]	methyl cyanide	C_2H_3N		5.80

Water solubility (at 25°C, except where noted otherwise), wt %[d,e]		Azeotrope, wt %/°C	Boiling range at 101.3 kPa (= 1 atm), °C[f]	Vapor pressure[g] at 25°C except where noted otherwise), kPa[h]	Specific gravity[i] (at 20°C except where noted otherwise)	Refractive index[j] (at 20°C except where noted otherwise)	Freezing point[k], °C
In water	Water in						
22.9	6.5	44.4/97.5	156	0.145	0.9730	1.4023_{25}	−62
			83	6.4_{20}	0.8692		−60
∞		20.8/98.8	170	0.11	0.90075	1.4198	
∞		none	202	0.017	0.9885	1.4273	
2.2			215		1.03		
1.1	1.6		192		0.9424		−64
∞		none	217	0.013	1.0096	1.4213	−25
∞			231	0.003	0.9553		−68
6.5	3.7		247	0.0013	0.981		−32
∞		none	56	24.3	0.78998	1.35868	−95
24_{20}	10_{20}	88.7/73	79	12	0.8049	1.3788	−87
			130	1.3	0.8539		−59
2.3_{20}	8.0_{20}	45/96	156	0.64	0.95099_{15}	1.45097	−32
1.4			127	1.3_{39}	0.830_{0}		−57
1.7	1.9	75.7/87.9	116	2.7	0.8008	1.3957	−84
∞		15.7/99.5	168	0.23	0.9387	1.4235	−44
0.4	1.5		150	1_{38}	0.817	1.4110	
6.5	1.2		144		0.8132		−74
0.05	0.8		166		0.81	1.4230	−47
1.2	4.3		215–220		0.922		
∞		none	197	0.016	1.1135	1.4318	−13
∞		none	187	0.017	1.0362	1.4329	−60
∞		none	244	<0.0013	1.1164	1.4475	−6.5
∞		84.2/76.7	82	11.8	0.7822	1.34411	−44

Table 1 (*continued*)

Solvent	CAS Registry No.	Common name (trade name)	Empirical formula	KB value[a,b]	Solubility parameter δ, $(J/m^3)^{1/2} \times 10^{-3}$ [c]
butyronitrile	[109-74-0]	n-butyl nitrile	C_4H_7N		5.12
N-methyl-2-pyrrolidinone	[872-50-4]	(M-Pyrol)	C_5H_9NO		5.51
nitromethane	[75-52-5]		CH_3O_2N	6.20	
2-nitropropane	[79-46-9]	sec-nitropropane (Ni Par)	$C_3H_7O_2N$		4.8
nitrobenzene	[98-95-3]		$C_6H_5O_2N$		4.88
m-nitrotoluene	[99-08-1]	m-nitrotoluol[l]	$C_7H_7O_2N$		
N,N-dimethylformamide (DMF)	[68-12-2]		C_3H_7ON		5.90
ethyl ether	[60-29-7]		$C_4H_{10}O$		3.6
isopropyl ether	[108-20-3]		$C_6H_{10}O$		
n-butyl ether	[142-96-1]		$C_8H_{18}O$		
diphenyl oxide	[101-84-8]		$C_{12}H_{10}O$		
propylene oxide	[75-56-9]		C_3H_6O		4.5
tetrahydrofuran	[109-99-9]		C_4H_8O		4.4
1,4-dioxane	[123-91-1]	diethylene oxide	$C_4H_8O_2$		4.88
γ-butyrolactone	[96-48-0]		$C_4H_6O_2$		6.15
furfural	[98-01-1]		$C_5H_4O_2$		5.46
pine oil	[8002-09-3]			>500	4.2
turpentine	[8006-64-2]	gum spirits, spirits of turpentine	$C_{10}H_{16}$	56	4.0
cresylic acid (mixture of o-, m-, and p-cresol)	[1319-77-3]		C_7H_8O		
m-cresol	[108-39-4]	3-methylphenol	C_7H_8O		4.98

[a] Kauri-butanol test is used to determine the relative solvency of hydrocarbons and can be used as an indication of their relative aromatic content.
[b] ASTM D 1133.
[c] To convert $(J/m^3)^{1/2}$ to $(cal/cm^3)^{1/2}$, multiply by 2050.
[d] ∞ = infinitely soluble.
[e] ASTM D 890, ASTM D 1364, ASTM D 1744, ASTM D 3401, ASTM E 123, and ASTM E 203.

where ΔE^v is the energy of vaporization and ϕ_1 and ϕ_2 are volume fractions. This can be rewritten as

$$\Delta H_m = V_t \left[\left(\frac{\Delta E_1^v}{V_1} \right)^{1/2} - \left(\frac{\Delta E_2^v}{V_2} \right)^{1/2} \right]^2 \phi_1 \phi_2$$

The term $\Delta E^v/V$, the energy of vaporization per unit volume, can be taken as a measure of the internal pressure. It is often called the solubility parameter δ. In other words

$$\left(\frac{\Delta E^v}{V} \right)^{1/2} = \delta = \frac{a^{1/2}}{V}$$

Table 1 (*continued*)

Water solubility (at 25°C, except where noted otherwise), wt %[d,e] In water	Water in	Azeotrope, wt %/°C	Boiling range at 101.3 kPa (= 1 atm), °C[f]	Vapor pressure[g] (at 25°C except where noted otherwise), kPa[h]	Specific gravity[i] (at 20°C except where noted otherwise)	Refractive index[j] (at 20°C except where noted otherwise)	Freezing point[k], °C
3.3	trace	67.5/88.7	118	2.55	0.7911	1.3838	−112
∞			202		1.027_{25}		−24
11.1	2.09	76.4/23.6	101	4.88	1.13816	1.38118	−28.5
1.71	0.53	70.6/88.5	120	2.4	0.98839	1.39439	−91
0.19_{20}	0.24_{20}	12.0/98.6	211	0.037	1.20824_{15}	1.55457_{15}	5.8
			232	0.13_{50}	1.163_{15}		15.1
∞		none	153	0.49	0.94873	1.43047	−60
6.04	1.47	98.7/34.2	34	71.2	0.71337	1.35243	−116
1.2_{20}	0.57	95.5/62.2	68	19.7	0.7235	1.3681	−86
0.03_{20}	0.19_{20}	67/92.9	142	1.67	0.7684	1.3992	−95
0.39		3.25/99.3	258	0.003	1.066_{30}	1.57625_{30}	27
40.5_{20}	12.2_{20}	none	34	59.3	0.8287	1.36603	−112
∞		93.3/63.4	66	26.3	0.8892	1.40716	−109
∞		82/87.8	101	4.9	1.0336	1.4224	11.8
∞		none	204	1.3	1.1254_{25}	1.4348_{25}	−43
1.02	0.1	34.5/98	162	0.33	1.1598	1.5261	−37
			200–220		0.86	1.481	
			154–170		$0.8454–868_{25}$	1.465	−40
			191–203	0.13_{38-53}	$1.03–1.038_{25}$		11–35
2.5		none	202	0.019	1.0341	1.5414	12

[f] ASTM D 1078-70.
[g] ASTM D 2984, ASTM D 2551, and ASTM D 233.
[h] To convert kPa to mm Hg, multiply by 7.5.
[i] ASTM D 1217, ASTM D 941, ASTM D 1298, ASTM D 2111, and ASTM D 891.
[j] ASTM D 1218.
[k] ASTM D 1015 and ASTM D 1016.
[l] Name unacceptable, no longer in use.

is the expression most commonly used for the solubility parameter and δ has the units of $(J/m^3)^{1/2}$ [$= 4.88 \times 10^{-4}$ $(cal/cm^3)^{1/2}$]. Therefore, the free energy of mixing is given by the expression

$$\Delta G = V_t[\delta_1 - \delta_2]\phi_1\phi_2 + RT[x_1 \ln x_1 + x_2 \ln x_2]$$

and solution should be assured as δ_1 approaches δ_2. According to this theory, two substances should mix when the solubility parameters are equal.

The influence of several important factors has been oversimplified by this theory; nevertheless, it is widely used as a formulating tool. The solubility parameter for a

Table 2. Solvent Characteristics of Common Industrial Solvents

Name	Viscosity neat[a], mPa·s (= cP)$_{°C}$	Surface tension, mN/m (= dyn/cm)$_{°C}$	Blush resistance (at 27°C), % rh	Dilution ratio[b] Toluene	Dilution ratio[b] Aliphatic naphtha	Coefficient of expansion[c], cm^3/°C
Alcohols						
methanol	0.5506_{20}	22.55_{20}		2.2	0.5	0.0012
ethanol	1.078_{25}	22.32_{20}		latent	latent	0.0011
1-propanol	1.722_{20}	23.7_{20}		latent	latent	0.00096
2-propanol	1.765_{30}	20.9_{30}		latent	latent	0.00105
1-butanol	2.271_{30}	23.7_{30}		latent	latent	0.00094
2-butanol	3.180_{30}	22.6_{30}		latent	latent	0.00091
2-methyl-1-propanol	3.91_{25}	22.1_{30}		latent	latent	0.00095
2-methyl-2-propanol	3.316_{30}	20.0_{26}		latent	latent	0.0013
furfuryl alcohol	4.62_{25}	38_{20}		latent	latent	
tetrahydrofurfuryl alcohol	6.24_{25}	37_{25}		latent	latent	0.00052
1-pentanol	3.35_{25}	25.6_{20}		latent	latent	0.00092
3-methyl-1-butanol	2.96_{30}	23.44_{30}		latent	latent	0.00092
allyl alcohol	1.072_{30}	25.68_{20}		latent	latent	
cyclohexanol	41.06_{30}	33.47_{30}		latent	latent	0.00077
1-hexanol	4.592_{25}	24.48_{20}		latent	latent	0.00087
4-methyl-2-amyl alcohol	4.074_{25}	22.63_{25}		latent	latent	0.00101
2-ethylbutyl alcohol	5.892_{25}	24.32_{25}		latent	latent	0.00089
benzyl alcohol	4.650_{30}	39.96_{20}		latent	latent	0.00075
2-ethylhexanol	9.8_{20}			latent	latent	0.00087
Aliphatic hydrocarbons						
pentane	0.235_{20}	15.48_{25}				
cyclohexane	0.980_{20}	24.98_{20}				0.0012
hexane	0.3126_{20}	17.9_{25}		12.1		0.0013
heptane	0.4181_{20}	20.3_{20}		12.2		
n-octane	0.5466_{20}	21.14_{25}				
α-pinene	1.40_{20}	15.8_{30}				
β-pinene	1.70_{20}					
rubber solvent	0.52_{20}	19.0_{20}		14.0		0.0013
VM&P solvent	0.83_{20}	23.3_{20}		14.7		0.0010
short-range mineral spirits	1.14_{20}	24.2_{20}		12.3		0.0010
Rule 66 mineral spirits	1.20_{20}	25.0_{20}		12.7		0.0010
140 solvent	1.63_{20}	25.7_{20}		11.3		0.0009
low odor 140 solvent	1.66_{20}	25.3_{20}		10.7		0.0009
mineral seal oil		27.5_{20}				0.0005
Alkyl halides						
methylene chloride	0.393_{30}	28.12_{20}				0.00137
chloroform	0.514_{30}	27.16_{20}				0.00126
carbon tetrachloride	1.038_{15}	26.15_{25}				0.00127
perchloroethylene	0.798_{30}	31.27_{30}				0.00102
1,1,2-trichloro-1,2,2-trifluoroethane	0.771_{20}	17.75_{20}				
trichloroethylene	0.566_{20}	29.5_{20}				0.00117
1,1,2,2-tetrachloroethane	1.844_{15}	36.04_{20}				0.00099
1,1,1-trichloroethane	0.903_{15}	25.56_{20}				
1,2-dichloroethane	0.887_{15}	32.23_{20}				0.00121
1,2-dibromoethane	1.88_{15}	38.91_{20}				
ethyl chloride	0.292_{20}	20.64_{10}				0.0021
ethyl bromide	0.379_{25}	24.15_{20}				
propylene dichloride		32.9_{20}				0.0011
o-dichlorobenzene	1.324_{25}	26.84_{20}				0.00085
chlorobenzene	0.79_{20}	33.28_{20}				0.00098
Amines						
n-propylamine	0.353_{25}	$22.21_{19.2}$				

Table 2 (*continued*)

Name	Viscosity neat[a], mPa·s (= cPa)°C	Surface tension, mN/m (= dyn/cm)°C	Blush resistance (at 27°C), % rh	Dilution ratio[b] Toluene	Aliphatic naphtha	Coefficient of expansion[c], $cm^3/°C$
isopropylamine	0.36_{25}	19.53_{20}				
diethylamine	$0.3878_{10.2}$	20.63_{15}				
cyclohexylamine	1.662_{20}	31.51_{20}				0.001164
triethylamine	0.363_{25}	20.66_{20}				0.00126
diisopropylamine		20.04_{16}				
dibutylamine	0.95_{20}	24.75_{20}				0.00079
ethanolamine	30.855_{15}	48.89_{20}				
morpholine	2.534_{15}	37.63_{20}				
diethanolamine	380_{30}					
triethanolamine	613_{25}					
Aromatic hydrocarbons						
benzene	0.6487_{20}	28.88_{20}			29.4	0.00138
toluene	0.5866_{20}	28.53_{20}			28.4	0.0011
ethylbenzene	0.6783_{20}	29.04_{20}				
xylene (mixture)		28.7_{20}			27.3	0.001
o-xylene	0.809_{20}	30.03_{20}				
m-xylene	0.617_{20}	28.63_{20}				
p-xylene	0.644_{20}	28.31_{20}				
aromatic 100 solvent		29.7_{20}			27.0	0.0009
aromatic 150 solvent		30.6_{20}			26.3	0.0008
Esters						
ethyl acetate	0.426_{25}	23.75_{20}	44	3.2	1.2	0.00139
n-propyl acetate	0.585_{20}	24.28_{20}	76	3.2	1.4	0.00131
isopropyl acetate	0.569_{20}	22.1_{20}	69	2.7	1.2	
methyl acetoacetate	1.704_{20}					
ethyl acetoacetate	1.5081_{25}	$32.47_{14.8}$				
n-butyl acetate	0.770_{15}	25.09_{20}	83	2.9	1.3	0.00121
sec-butyl acetate		$23.33_{21.1}$	76	2.4	1.2	0.00118
isobutyl acetate	0.697_{20}	23.7_{20}	80	2.7	1.1	0.00126
amyl acetate	0.924_{20}	25.68_{20}	91	2.3	1.3	0.00115
isoamyl acetate	$0.872_{19.9}$	$24.62_{21.1}$				0.00119
4-methyl-2-amyl acetate		22.6_{20}	91	1.7	1.0	0.0011
isobutyl isobutyrate		33.8_{-76}		1.3	0.8	
benzyl acetate	1.399_{45}					
Glycol ethers						
ethylene glycol monomethyl ether	1.72_{20}	$31.82_{14.9}$	42	4.0	0	0.00095
ethylene glycol monoethyl ether	2.05_{20}	28.2_{25}	59	4.9	1.1	0.00097
ethylene glycol monomethyl ether acetate			80	2.3	0.6	0.0011
diethylene glycol mono-methyl ether	3.48_{25}	34.8_{25}	76	2.3	0	0.00086
ethylene glycol monoethyl ether acetate	1.025_{25}	31.8_{25}	94	2.5	0.9	
ethylene glycol monobutyl ether	3.15_{25}	27.4_{25}	96	3.5	2.3	0.00092
diethylene glycol monoethyl ether	3.85_{25}	31.8_{25}	76	1.9	9.2	
ethylene glycol monobutyl ether acetate			96	1.8	1.2	
diethylene glycol monoethyl ether acetate	2.8_{20}		92	2.2	0.6	0.00101
diethylene glycol monobutyl ether				3.9	1.9	

Table 2 (*continued*)

Name	Viscosity neat[a], mPa·s (= cPa)$_{°C}$	Surface tension, mN/m (= dyn/cm)$_{°C}$	Blush resistance (at 27°C), % rh	Dilution ratio[b]		Coefficient of expansion[c], $cm^3/°C$
				Toluene	Aliphatic naphtha	
diethylene glycol monobutyl ether acetate			92	2.2	0.6	
Ketones						
acetone	0.3371_{15}	23.32_{20}	20	4.5	0.7	0.00144
methyl ethyl ketone	0.423_{15}	$23.97_{24.8}$	51	4.3	0.9	0.00076
mesityl oxide			83	4.0	1.1	
cyclohexanone	2.453_{15}	34.5_{20}	92	5.7	1.1	0.00094
methyl n-butyl ketone			80	4.0	1.1	
methyl isobutyl ketone	0.542_{25}	23.64_{20}	78	3.6	1.0	0.00116
diacetone alcohol	2.9_{20}	31.0_{20}	76	3	0.5	0.00099
methyl amyl ketone	0.77_{20}		92	3.9	1.2	0.00114
methyl isoamyl ketone	0.73_{20}		89	4.1	1.2	0.00107
diisobutyl ketone	0.95_{20}	22.5_{20}	95	1.5	0.7	0.00102
isophorone	2.3_{20}		97	6.2	0.3	
Other						
ethylene glycol	26.09_{15}	46.49_{20}				0.00064
propylene glycol	56.0_{20}	72_{25}				0.00072
diethylene glycol	35.7_{20}	48.5_{20}				0.000635
acetonitrile	0.375_{20}	19.1_{20}				0.00137
n-butyronitrile	0.624_{15}	27.33_{20}				
N-methyl-2-pyrrolidinone	1.65_{25}	40.7_{20}				
nitromethane	0.647_{20}	37.48_{20}		1.2		0.00115
2-nitropropane	0.770_{20}	29.87_{20}	82	1.2	0.4	0.00104
nitrobenzene	2.165_{15}	43.35_{20}				
N,N-dimethylformamide	0.9243_{20}	36.76_{20}		7.7	0.2	
ethyl ether	0.242_{20}	17.06_{20}				
isopropyl ether	0.329_{20}	$17.34_{24.5}$				
n-butyl ether	0.741_{15}	23.4_{15}				
diphenyl oxide	2.459_{40}	38.8_{30}				
propylene oxide	0.327_{20}					0.00151
tetrahydrofuran	0.55_{20}	26.4_{25}	50	2.8	1.0	
1,4-dioxane	1.439_{15}	34.45_{15}				
γ-butyrolactone	1.7_{25}					
furfural	2.41_{20}	37.6_{20}				0.00097
pine oil						0.00090
m-cresol	24.6_{15}	38.01_{15}				

[a] ASTM D 83, ASTM D 341, ASTM D 445.
[b] The ratio of hydrocarbon diluent to active solvent required to cause persistent heterogeneity in a solution of cellulose nitrate, ASTM D 1720-62.
[c] ASTM D 1903.

blend is given by

$$\delta_{\text{blend}} = \delta_1\phi_1 + \delta_2\phi_2 + \delta_3\phi_3 \ldots \delta_n\phi_n$$

where ϕ represents the volume fraction.

A list of single-component parameters for common industrial solvents is given in Table 1.

Table 3. Cost and Production Data for Industrial Solvents

Name	1970		1975		1979		1981 cost bulk, $/kg (except where noted)
	Production, 1000 t	Cost, $/kg	Production, 1000 t	Cost, $/kg	Production, 1000 t	Cost, $/kg	
Alcohols							
methanol	2237	0.01	2348	0.03	3342	0.03	0.21[a]
ethanol	887	0.03	648[b]	0.06	639[b]	0.09	0.49[a]
1-propanol	27	0.05	53	0.09	85	0.12	0.19
2-propanol	870	0.03	690	0.05	862	0.08	0.54[a]
1-butanol					347	0.10	0.17
2-butanol							0.17
2-methyl-1-propanol					65	0.07	0.14
2-methyl-2-propanol							0.12
furfuryl alcohol							0.29
tetrahydrofurfuryl alcohol							0.28
1-pentanol							0.15
3-methyl-1-butanol							0.33
allyl alcohol							0.34
cyclohexanol	5	0.06					0.29
1-hexanol	7	0.05			20	0.15	0.23
4-methyl-2-amyl alcohol							0.21
2-ethylbutyl alcohol							0.93
hexyl alcohol (mixture)							0.15
benzyl alcohol		0.14	3	0.37	3	0.40	0.65
2-octanol							0.16
2-ethylhexanol	207	0.04	176	0.09	144	0.10	0.10
Aliphatic hydrocarbons							
pentane	518	0.01	331	0.04	447	0.05	
cyclohexane	835	0.01	7861	0.05	1100	0.10	0.48[a]
hexane	147	0.02	114	0.03	176	0.06	0.27[a]
heptane							0.28[a]
n-octane							0.33[a]
α-pinene							0.11
β-pinene							0.12
VM&P solvent							0.34[a]
short-range mineral spirits							0.47[a]
Rule 66 mineral spirits							0.37[a]
140 solvent							0.34[a]
low odor 140 solvent							0.47[a]
Alkyl halides							
methylene chloride	182	0.04	225	0.07	287	0.09	0.08[a]
chloroform	108	0.03	118	0.07	161	0.09	0.12
carbon tetrachloride	459	0.02	411	0.06	324	0.05	
perchloroethylene	320	0.03	308	0.06	351	0.05	0.10
1,1,2-trichloro-1,2,2-trifluoroethane							0.45
trichloroethylene	277	0.03	132	0.07	145	0.05	0.12
1,1,2,2-tetrachloroethane							0.21
1,1,1-trichloroethane	166	0.05	208	0.08	325	0.07	0.15
1,1,2-trichloroethane							0.16
1,2-dichloroethane	3383	0.01	3618	0.04	5350	0.05	0.06
1,2-dibromoethane	134	0.08	125	0.12			0.17
ethyl chloride			261	0.05	264	0.08	0.10
ethyl bromide							0.33
propylene dichloride	47		38	0.02	31		
o-dichlorobenzene	30	0.05	25	0.13	26	0.14	0.23
chlorobenzene	220	0.03	139	0.12	147	0.12	
p-chlorotoluene							0.32

Table 3 *(continued)*

Name	1970 Production, 1000 t	1970 Cost, $/kg	1975 Production, 1000 t	1975 Cost, $/kg	1979 Production, 1000 t	1979 Cost, $/kg	1981 cost bulk, $/kg (except where noted)
Amines							
n-propylamine							0.34
butylamine (mixture)	8	0.14	10	0.31			
tert-butylamine							0.54
diethylamine	4	0.10	5				0.43
cyclohexylamine							0.44
triethylamine							0.55
diisopropylamine	1	0.10					0.44
toluidines (mixture)							0.33
ethanolamine	39	0.05	37	0.14			
ethylaminoethanol							0.20
dimethylethanolamine							0.45
diethanolamine	42	0.05	39	0.15			
triethanolamine	38	0.06	40	0.15			0.25
Aromatic hydrocarbons							
benzene	3765	0.01	3402	0.04	5579	0.08	0.46[a]
toluene	2730	0.01	2322	0.03	3323	0.07	0.42[a]
ethylbenzene	2190	0.02	2187	0.04	3832	0.07	0.14
xylene (mixture)	1761	0.01	2095	0.03	3187	0.07	0.40[a]
o-xylene							0.12
m-xylene							0.14
Esters							
ethyl acetate	94	0.08	108	0.12	143	0.15	0.19
n-propyl acetate	10	0.05	16	0.10	23	0.13	0.22
isopropyl acetate							0.18
methyl acetoacetate							0.39
ethyl acetoacetate							0.44
n-butyl acetate	34	0.05	36	0.10	63	0.13	0.22
isobutyl acetate							0.20
amyl acetate							0.16
4-methyl-2-amyl acetate							0.27
isobutyl isobutyrate							0.19
benzyl acetate	1	0.19	1	0.41	1	0.49	0.73
Glycol ethers							
ethylene glycol monomethyl ether	44	0.05	41	0.11	46	0.13	0.22
ethylene glycol monoethyl ether	62	0.06	81	0.12	112	0.14	0.22
propylene glycol monophenyl ether							0.15
ethylene glycol monomethyl ether acetate							0.20
diethylene glycol monomethyl ether	6	0.05	7	0.12	9	0.15	0.25
ethylene glycol monoethyl ether acetate							0.24
ethylene glycol monobutyl ether	48	0.07	59	0.12	99	0.15	0.22
diethylene glycol monoethyl ether	16	0.06	15	0.12	15	0.13	0.24
ethylene glycol monobutyl ether acetate							0.28
diethylene glycol monoethyl ether acetate							0.23

Table 3 (*continued*)

Name	1970 Production, 1000 t	1970 Cost, $/kg	1975 Production, 1000 t	1975 Cost, $/kg	1979 Production, 1000 t	1979 Cost, $/kg	1981 cost bulk, $/kg (except where noted)
diethylene glycol monobutyl ether	7	0.08	12	0.14	19	0.16	0.25
diethylene glycol monobutyl ether acetate							0.23
Ketones							
acetone	733	0.02	744	0.05	782	0.07	0.14
methyl ethyl ketone	218	0.04	192	0.08	298	0.10	0.19
mesityl oxide			20	0.11	10	0.13	0.21
cyclohexanone	324	0.05	251	0.16	396	0.17	0.29
methyl *n*-butyl ketone							0.13
methyl isobutyl ketone	90	0.05	68	0.10	86	0.13	0.21
diacetone alcohol		0.06	18	0.10	20	0.15	0.21
methyl amyl ketone							0.13
methyl isoamyl ketone							0.23
diisobutyl ketone							0.24
Other							
ethylene glycol	1378	0.03	1728	0.10	2145	0.10	0.16
propylene glycol	194	0.09	177	0.12	277	0.12	0.18
diethylene glycol	155	0.03	146	0.11	178	0.10	0.14
acetonitrile							0.24
n-butyronitrile							0.24
nitromethane							0.70
2-nitropropane							0.15
nitrobenzene	248		188	0.10	432	0.11	0.15
m-nitrotoluene			140				0.52
N,N-dimethylformamide							0.27
ethyl ether							0.11
isopropyl ether							0.07
n-butyl ether							0.47
diphenyl oxide							0.48
propylene oxide	535	0.04	691		1020		0.19
tetrahydrofuran					54	0.36	0.36
1,4-dioxane			6	0.33			0.43
γ-butyrolactone							0.39
furfural							0.29
pine oil							0.21
turpentine							0.37[a]
cresylic acid (mixture)	1	0.01	2	0.03	1	0.04	0.22[a]
m-cresol							0.44[a]

[a] Values listed are in $/L.
[b] Synthetic products.

Two-Component Solubility Parameter. One of the main problems with the preceding theory is its restriction to nonpolar compounds. Polar molecules or molecules that hydrogen bond have interactions that influence both the enthalpy and entropy terms. The correction for the heats of mixing caused by polar effects has been calculated and an additional term included:

$$\frac{\Delta E^v}{V} = \delta^2 + \omega^2$$

Table 4. Health and Safety Data for Industrial Solvents

Name	PEL[a,b], ppm	IDLH[c], ppm	Flash point[d], °C	Lower explosive limit, vol %	Upper explosive limit, vol %	Evaporation rate (ether = 1)	Autogeneous ignition temperature[e], °C	Vapor density, g/L
Alcohols								
methanol	200	25,000	11, CC	6.7	36	4.8	385	1.11
ethanol			13, CC	3.3	19.0	6.3	423	1.59
1-propanol	200	4,000	25, CC	2.1	13.5	9.0	440	2.07
2-propanol	400	20,000	12, CC	2.0	12.0	7.4	399	2.07
1-butanol	100	8,000	35, CC	1.4	11.2	20.5	365	2.55
2-butanol	150	10,000	31, OC	1.7	9.8	13.0	406	2.55
2-methyl-1-propanol	100	8,000	28, CC	1.2	10.9	17.3	427	2.55
2-methyl-2-propanol	100	8,000	11, CC	2.4	8.0		480	2.55
furfuryl alcohol	50	250	65, CC	1.8	16.3		491	3.37
tetrahydrofurfuryl alcohol			84, CC	1.5	9.7		282	3.5
1-pentanol			33, CC	1.2	10.0	28.0	300	3.04
3-methyl-1-butanol	100	8,000	43, CC	1.2	9.0		350	3.04
allyl alcohol	2	150	21, CC	2.5	18.0		378	2.00
cyclohexanol	50	3,500	68, CC			146.8	300	3.45
1-hexanol			63, CC					3.52
4-methyl-2-amyl alcohol			46, TCC	1.0	5.5	33.1		3.53
2-ethylbutyl alcohol			63, CC					
hexyl alcohol (mixture)								
benzyl alcohol			101, CC				436	3.72
2-octanol			88, CC					4.48
2-ethylhexanol			81, CC			above 150		4.49
2-ethylisohexyl alcohol			77, CC			above 150		
Aliphatic hydrocarbons								
pentane	1,000	5,000	−40, TOC	1.5	7.8		260	2.48
cyclohexane	300	10,000	−17, TCC	1.3	8.4	1.5	245	2.90
hexane	500	5,000	−22, TCC	1.2	7.5	1.3	225	2.97
heptane	500	4,250	4, CC	1.05	6.7	2.3	215	3.45
n-octane	500	3,750	13, CC	1.0	6.5		220	3.86
α-pinene			33, CC				255	4.7
β-pinene								

rubber solvent			−40, CC	1.2	7.5	1.5	316	
VM&P	500	10,000	12, CC	1.2	6.3	8.3	287	4.1
high flash VM&P solvent	500	10,000	32, CC	1.0	6.0	13.0		
short-range mineral spirits	500	10,000	39, CC	1.0	6.2	30.0	278	
Rule 66 mineral spirits	500	10,000	39, CC	0.9	6.1	41.4	276	
140 solvent	500	10,000	61, CC	0.9	6.0	90.6	257	
low odor 140 solvent	500	10,000	61, CC	0.9	6.0	91.4	257	
mineral seal oil	500	10,000	129, CC	0.9	6.0	<2000.0	254	
Alkyl halides								
methylene chloride	500	5,000	none	14.8	22.0	1.4	662	2.93
chloroform	50	1,000	none			1.9	>982	4.12
carbon tetrachloride	10	300	none			1.8	>982	5.32
perchloroethylene	100	500	none	none	none	6.0	none	5.83
1,1,2-trichloro-1,2,2-trifluoroethane	1,000	4,500	none				680	
trichloroethylene	100	1,000	none	8.0	10.5	2.5	420	4.53
1,1,2,2-tetrachloroethane	5	150	none			17.0		4.88
1,1,1-trichloroethane	350	1,000	none	8.0	10.5	1.9	537	4.55
1,1,2-trichloroethane	10	500	none	8.4	13.3		460	4.55
1,2-dichloroethane	50	1,000	13, CC	6.2	15.9	3.0	413	3.35
1,2-dibromoethane	20	400	none					6.48
ethyl chloride	1,000	20,000	−50, TCC	3.8	15.4		519	2.22
ethyl bromide	200	3,500	−20, CC	6.7	11.3		511	3.76
propylene dichloride	75	2,000	16, CC	3.4	14.5	4.1	557	3.9
1,2,4-trichlorobenzene			110, CC					6.26
o-dichlorobenzene	50	1,700	66, CC	2.2	9.2	65.0	648	5.05
chlorobenzene	75	2,400	30, CC	1.3	7.1	10.0	638	3.88
p-chlorotoluene			52, OC					
Amines								
n-propylamine			−37, CC	2.0	10.4		318	
isopropylamine	5	4,000	−26, CC				402	
butylamine (mixture)	5	2,000	−12, OC	1.7	9.8		312	2.52
tert-butyl amine			−9, CC	1.7 at 100°C	8.9 at 100°C		380	2.5
diethylamine	25	2,000	−26, CC	1.8	10.1		312	2.53
diethylenetriamine			102, OC				399	3.48
cyclohexylamine			32, OC				293	3.42
triethylamine	25	2,000	−7, CC	1.2	8.0			3.48
diisopropylamine	5	1,000	−6, CC					3.5

Table 4 (*continued*)

Name	PEL[a,b] ppm	IDLH[c], ppm	Flash point[d], °C	Lower explosive limit, vol %	Upper explosive limit, vol %	Evaporation rate (ether = 1)	Autogeneous ignition temperature[e], °C	Vapor density, g/L
toluidines (mixture)	5	100	85, CC				482	3.69
dibutylamine			52, OC					4.46
ethanolamine	200	6,000	85, CC					2.11
morpholine	20	8,000	38, OC				310	3.00
ethylaminoethanol			71, OC					3.06
dimethylethanolamine			41, OC					3.03
diethanolamine			137, OC				662	3.65
diisopropanolamine			−1, OC					4.59
triethanolamine			179, OC					5.14
Aromatic hydrocarbons								
benzene	1	2,000	−11, TCC	1.3	7.1	2.6	562	2.77
toluene	200	2,000	4, TCC	1.27	7.0	4.6	480	3.14
ethylbenzene	100	2,000	15, TCC	1.0	6.7		432	3.66
xylene (mixture)	100	10,000	28, TCC	1.1	7.0	11.5	529	3.66
o-xylene	100	10,000	32, TCC	1.0	6.0		465	3.66
m-xylene	100	10,000	29,.TCC	1.1	7.0		530	3.66
p-xylene	100	10,000	27, TCC	1.1	7.0		530	3.66
aromatic 100 solvent			44, CC	1.0	6.0	28	462	
aromatic 150 solvent			63, CC	1.0	6.0	94.5	486	
Esters								
ethyl acetate	400	10,000	−4, TCC	2.2	11.0	2.4	427	3.04
n-propyl acetate	200	8,000	14, TCC	2.0	8.0	4.3	450	3.52
isopropyl acetate	250	16,000	4, CC	1.8	7.8	2.5	460	3.52
methyl acetoacetate			77, CC	1.8	8.0		280	4.0
ethyl acetoacetate			84, CC				295	4.48
n-butyl acetate	150	10,000	22, CC	1.7	7.6	9.1	425	
sec-butyl acetate	200	10,000	31, OC	1.7		5.0		4.00
isobutyl acetate	150	7,500	18, CC	1.7	10.5	6.1	423	
amyl acetate	100	4,000	25, TCC	1.1	7.5	16.2	379	4.5
isoamyl acetate	100	3,000	25, CC	1.0 at 99°C	7.5		360	4.49
4-methyl-2-amyl acetate			45, COC		18.7			4.97

isobutyl isobutyrate			49, CC					
benzyl acetate			102, CC				461	5.1
Glycol ethers								
ethylene glycol monomethyl ether	25	4,500	42, CC	2.5	14	18.7	285	2.62
ethylene glycol monoethyl ether	200	6,000	44, 202[f]; CC	1.8	14	24.4	235	3.10
propylene glycol monophenyl ether			38, CC					
ethylene glycol monomethyl ether acetate	25	4,500	56, CC	1.7	8.2	25.2		4.07
diethylene glycol monomethyl ether			93, CC			345.0		4.14
ethylene glycol monoethyl ether acetate	100	2,500	57, 134, TOC	1.7		37.4	379	4.72
ethylene glycol dimethyl ether			40, CC					3.11
ethylene glycol monobutyl ether	50	700	61, CC	1.1	10.6	88.5		
diethylene glycol monoethyl ether			96, OC			518		4.62
ethylene glycol monobutyl ether acetate			88, CC			246		
diethylene glycol monoethyl ether acetate			110, COC			1444		6.07
diethylene glycol monobutyl ether			78, CC			18000	228	5.58
ethylene glycol monophenyl ether								
diethylene glycol monobutyl ether acetate			115, OC			6500	299	
Ketones								
acetone	1000	20,000	−18, TCC	2.6	12.8	1.8	465	2.0
methyl ethyl ketone	200	3,000	−1, TCC	1.8	10.0	2.5	516	2.42
mesityl oxide	25	5,000	31, CC				344	3.38
cyclohexanone	50	5,000	44, TCC	1.1 at 38°C		30.2	420	3.4
methyl *n*-butyl ketone			35, OC	1.22	8.0		533	3.45
methyl isobutyl ketone			23, TCC	1.4	7.5	5.6	459	3.45
diacetone alcohol	50	2,100	64, CC	1.8	6.9	68	603	4.0
methyl amyl ketone	100	4,000	49, OC			2.61	533	3.94
methyl isoamyl ketone			43, OC			17.3		
diisobutyl ketone			60, CC	0.8 at 100°C	6.2 at 100°C	40.5		4.9
isophorone	25	800	96, CC			364		
Others								
ethylene glycol			116, OC	3.2		1550	400	2.14
propylene glycol			99, COC	2.6	12.6	907	371	2.62
diethylene glycol			124, OC			>2000	229	3.66
acetonitrile	40	4,000	6, CC	4.4	16		524	1.42

Table 4 (*continued*)

Name	PEL[a,b], ppm	IDLH[c], ppm	Flash point[d], °C	Lower explosive limit, vol %	Upper explosive limit, vol %	Evaporation rate (ether = 1)	Autogeneous ignition temperature[e], °C	Vapor density, g/L
n-butyronitrile			29, TOC					
N-methyl-2-pyrrolidinone			96, OC	2.18	12.24		346	3.4
nitromethane	100	1,000	44, TOC	7.3		6.5	418	2.11
2-nitropropane	25	2,300	39, TOC	2.6		7.9	428	3.06
nitrobenzene			88, CC	1.8 at 93°C			482	4.25
m-nitrotoluene	5	200	112, CC	2.2				4.72
N,N-dimethylformamide	10	3,500	67, TOC	2.2 at 38°C	15.2 at 38°C		445	2.51
ethyl ether			−45, CC	1.85	36	1.0	160	2.56
isopropyl ether			−28, CC	1.4	7.9		443	3.52
n-butyl ether			25, CC	1.5	7.6		194	4.48
diphenyl oxide			115, CC	0.8	1.5		620	5.86
propylene oxide			−35, CC	2.8	37			2.0
tetrahydrofuran			−14, CC	2.3	11.8		321	2.5
1,4-dioxane	100	200	12, CC	2.0	22.2		180	3.03
γ-butyrolactone			98, OC					3.0
furfural			60, CC	2.1	19.3		316	3.31
pine oil			78, CC			1036		
turpentine	100	1,900	35, CC	0.8		21.6	249	4.84
cresylic acid (mixture)	5	250	81, CC	1.35 at 149°C				3.72
m-cresol	5	250	94, CC	1.1 at 150°C			559	3.72

[a] PEL = permissible exposure limit; values given are for 8-h work-shift time-weighted average levels.

[b] Ref. 10.

[c] IDLH = the concentration that is immediately dangerous to life or health; represents a maximum level from which one could escape within 30 min without any escape-impairing symptoms or any irreversible health effects.

[d] CC = closed-cup, ASTM D 3278; OC = open-cup; TCC = tag closed-cup, ASTM D 56-64; TOC = tag open-cup, ASTM D 1310-63; COC = Cleveland open-cup, ASTM D 92-66.

[e] ASTM D 2155.

[f] Two values reported.

where δ signifies the dispersion forces and ω the dipole forces. The g factor is introduced to account for the effect of the dipole moment on the nearest neighbors and is described as follows (2):

$$g = 1 + z \int \cos \gamma \cdot \exp(-W/kt)d\omega$$

where γ is the angle between dipole moments and W is the potential of the average torque hindering their rotation. Values of g are usually near one for dilute solutions, but at high concentrations they approach the value of the pure polar solute. The effect on the cohesive energy would be small except for highly polar or hydrogen-bonded molecules.

Three-Component Solubility Parameters. Three-component parameters have been developed in order to extend the solubility parameter concept to polar and hydrogen-bonding systems (see Solubility Parameters). Here it is assumed that the cohesive energy, $-E$, per unit volume is given as follows:

$$-\frac{E}{V} = -\frac{E_d}{V} - \frac{E_p}{V} - \frac{E_h}{V}$$

where $-E_d$ is the dispersion interaction, $-E_p$ is the polar interaction, and $-E_h$ is the hydrogen-bonding interaction. This gives the following total solubility parameter:

$$\delta_o^2 = \delta_d^2 + \delta_p^2 + \delta_h^2$$

where δ_o^2 is the total solubility parameter for the three-component system. An extensive list of these three-component solubility parameters and of solubility parameters for commercial polymers has been compiled (3). Also listed are methods for determinations of the solubility parameter from a variety of physical constants.

Common Industrial Solvents

Information concerning the physical properties, solvent characteristics, economic and production data, and health and safety factors of common industrial solvents is listed in Tables 1, 2, 3, and 4, respectively (3–9).

Regulations

There are at least five Federal agencies and many more state and local agencies that regulate the transportation, use, and disposal of industrial solvents (see Regulatory agencies). The purpose of these agencies is to protect the general public, employees, and environment from hazardous and toxic materials. The following Federal agencies and acts should be consulted by solvent users and manufacturers: the Environmental Protection Agency (EPA): Clean Water Act (1972 and 1977), Clean Air Act (1977), Resource Conservation and Recovery Act (1976), Safe Drinking Water Act (1977), Toxic Substance Control Act (TSCA) (1976), and the Comprehensive Environment Response, Compensation, and Liability Act (1980); the Food and Drug Administration (FDA): Federal Food, Drug, and Cosmetic Act (1938–1978); the Occupational Safety and Health Administration (OSHA): Occupational Safety and Health Act (1970); the Consumer Product Safety Commission (CPSC): Consumer Product Safety Act (1972), Federal Hazardous Substance Act (1970), and Poison Prevention Packaging Act (1970); and the Department of Transportation (DOT) (11). In addition to the preceding Federal agencies, state and local fire marshals regulate the storage and handling of flammable liquids. These state and local agencies must be consulted. The hazard and toxic classification of all industrial solvents are given in the literature (see also Table 4). Current listings for the specific compounds should be consulted for an accurate assessment of the potential hazard of any given solvent (see Industrial hygiene and toxicology).

Uses

Cleaning. ***Dry Cleaning.*** Dry-cleaning solvents are normally either perchloroethylene, mineral spirits, or 140 flash naphtha. Limited applications of fluorocarbons, eg, Freon 113, are made, but they are expensive (see Drycleaning and laundering, drycleaning). The impact of environmental regulations on chlorinated and chlorofluorinated hydrocarbons under which they are classified as hazardous may restrict their use as dry-cleaning solvents. Petroleum solvents, although they present a potential fire and explosion hazard and suffer from a decrease in solvency, may become the primary solvent for the dry-cleaning industry.

Cold Cleaning. The use of solvents or blended solvents for cold cleaning has been extensive. These applications include metal degreasing, parts cleaning, maintenance, paint and varnish strippers, and printing-ink solvents. The solvency requirements for cold cleaning are different than those used for coating. In some cases it is not desirable to dissolve the polymeric materials but only to cause them to soften and swell so they can be removed easily.

Vapor Degreasing. Vapor degreasing is a very efficient method of cleaning metal parts where solvent residues would cause production problems. The cold part is suspended in the vapors of the solvent. The refluxing solvent bathes the part in fresh distilled solvent until the part is heated to the vapor temperature.

Coatings. Solvents in paints and varnishes dissolve resins that provide protective coatings and support pigments for color. The solvent evaporates after application leaving the pigment and resin on the surface. Formulation of the solvent system is important because it controls ease and method of application (spraying vs brushing, for example), drying time, and nature of the resin film. New regulations from the EPA and OSHA reflecting both environment and safety requirements have caused many reformulations. This also has resulted in a shift toward water-base rather than solvent-base systems for some applications (see Coatings, industrial; Paint). However, some states have exempted the use of 1,1,1-trichloroethane and methylene chloride.

Staining and Wood Treatment. Treating is required when wood is to be used in exterior applications to provide protection from weathering and to prevent attack from insects and fungi which cause decay. The solvents contain insecticides and fungicides which pass into the wood. Pigmentation and resin can be added to the solvent to produce wood stains as well as protective coatings (see Stains, industrial).

Printing Inks. The preparation of a printing ink is similar to a paint in that a resin and pigment must be blended with a solvent to produce a coating. For solvents used in printing inks, the solvents must evaporate very fast. However, because of the resins used, the solvents are usually petroleum products (see Inks).

Agricultural Products. Many of the insecticides or herbicides (qv) used for pest and weed control are dissolved in solvent systems that contain emulsifiers which give a product that can be diluted with water for application to crops (see Insect control technology).

Reaction Solvents. Many chemical systems produced for the chemical and pharmaceutical industries use solvents as inert reaction media, where the components of the reaction are brought together in solution. Solvents are also used in purification steps for extraction or recrystallization. Other industries, eg, petroleum refining or re-refining industries, use large quantities of solvents for purification steps. Solvents are also used as additives for no-lead gasoline formulations (see Crystallization; Extraction; Gasoline; Hydrocarbons, C_1–C_6; Petroleum).

BIBLIOGRAPHY

"Solvents, Industrial" in *ECT* 1st ed., Vol. 12, pp. 654–686, by Arthur K. Doolittle, Carbide and Carbon Chemicals Company, A Division of Union Carbide and Carbon Corporation; "Solvents, Industrial" in *ECT* 2nd ed., Vol. 18, pp. 564–588, by John W. Wyart, Celanese Chemical Company, and Mark F. Dante, Shell Chemical Company.

1. J. H. Hildebrand and R. L. Scott, *The Solubility of Nonelectrolytes*, 3rd ed., Dover Publications, Inc., New York, 1964.
2. J. G. Kirkwood, *J. Chem. Phys.* **7**, 911 (1939).
3. H. Burrell, *Polymer Handbook*, 2nd ed., Wiley-Interscience, New York, 1975.
4. J. A. Riddick and W. B. Bunger, *Organic Solvents*, 3rd ed., Wiley-Interscience, New York, 1970.
5. N. I. Sax, *Dangerous Properties of Industrial Materials*, 5th ed., Van Nostrand Reinhold Company, New York, 1979.
6. *Chem. Mark. Rep.* (July 20, 1981).
7. *Am. Paint J.* **66**(2), 26 (1981).
8. F. W. Mackinson, R. S. Strieoff, and L. J. Partridge, Jr., eds., *NIOSH/OSHA Pocket Guide to Chemical Hazards*, DHEW (NIOSH) Pub. No. 78-210, Washington, D.C., 1980.
9. *U.S. Tariff Commission TC 1.33.979*, U.S. International Trade Commission, Washington, D.C., 1979
10. 29 CFR 1910.1000, Jan. 1, 1977.
11. *Hazardous Materials Transportation Guide for Hazardous Materials Shipping Papers*, CFR Title 49, Parts 100–199, Subpart C, 1979.

CLYDE F. PARRISH
Indiana State University

SORBIC ACID

Sorbic acid [*110-44-1*] (*trans,trans*-2,4-hexadienoic acid) is a white crystalline solid. It was first isolated in 1859 by hydrolysis of the oil distilled from unripe mountain-ash berries (1). The name was derived on the basis of the scientific name for the rowan tree, which is the parent plant of the mountain ash, *Sorbus aucuparia* Linné. In 1900, Doebner performed the first synthesis (2). Interest in this compound was minimal until its antimicrobial effect was discovered in 1940. Early interest in manufacturing sorbic acid first centered around its use as a tung oil replacement when supplies were curtailed in the United States during World War II. High manufacturing costs prohibited expanded use until its applicability as a food preservative was approved in 1953 (see Food additives). It is widely used in moist foods below pH 6.5 where control of bacteria, molds, and yeasts is essential to obtain safe and economical storage life.

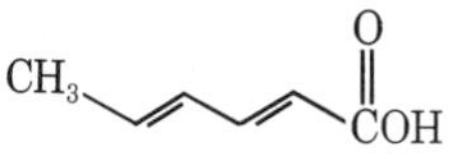

sorbic acid

Physical Properties

The sorbic acid crystal has a well-ordered morphology as a result of its hydrogen bonding and trans,trans structure. Some physical properties are given in Table 1 along with those of the most commercially used salt, potassium sorbate [*24634-61-5*]. Table 2 shows the solubility in various solvents. More extensive data on solubility is given in refs. 3 and 4. The unsaturated conjugated double bonds make sorbic acid dust an explosion hazard, particularly when mixed with free-radical initiators or oxidizing agents. Minimum explosive limits are 0.02 g/L of air for sorbic acid (5).

Chemical Properties

The chemical reactivity of sorbic acid is determined by the carboxyl group and the conjugated double bonds.

Conjugated Double Bonds. Sorbic acid is brominated faster than other olefinic acids (7). Reaction with HCl gives predominately 5-chloro-3-hexenoic acid (8).

Reactions with amines at high temperatures under pressure lead to mixtures of dehydro-2-piperidinones (9):

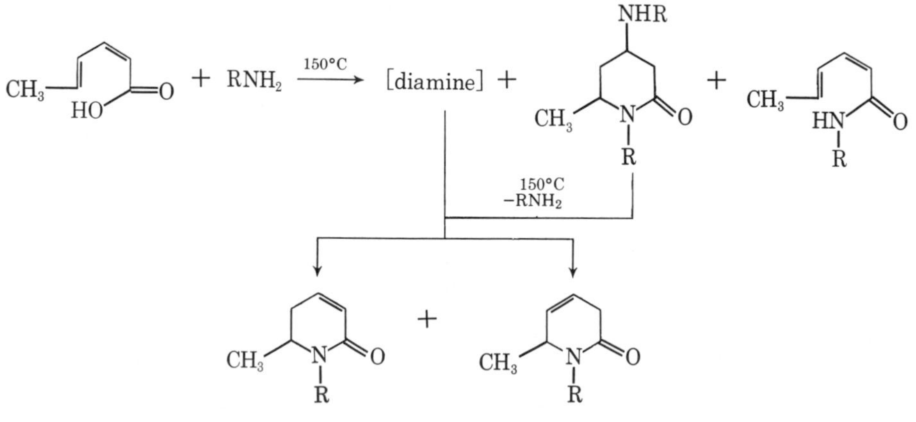

Table 1. Physical Properties of Sorbic Acid and Potassium Sorbate

Properties	Values	
	Sorbic acid	Potassium sorbate
mol wt	112.13	150.22
melting point, °C	134.5	270°C dec
boiling point (at 101.3 kPa (= 1 atm)), °C	228	
density, g/cm^3		
at 19°C	1.204	
at 20°C		1.363
flash point, °C	126–130	
dissociation constant (at 25°C), mol/L	1.73×10^{-5}	
pK_a^{25} (H_2O)	4.76	
pK_a^{25} (50 wt % ethanol)	4.62	
pK_a (0.1 *M* NaCl)	4.51	
dissociation constant of dimer, K^{24} mol/L (CCl_4)	1.96×10^{-4}	
specific heat, J/(g·K)[a]	1.84	
latent heat of fusion, kJ/mol[a]	13.6	
heat of combustion, kJ/mol[a]	3107	
heat of neutralization, kJ/mol[a]	6.07	
vapor pressure, kPa[b]		
at 130°C	1.3	
at 150°C	3.7	
at 170°C	9.3	

[a] To convert J to cal, divide by 4.184.
[b] To convert kPa to mm Hg, multiply by 7.5.

A yellow crystalline complex melting at 198°C is formed from sorbic acid and iron tricarbonyl (10):

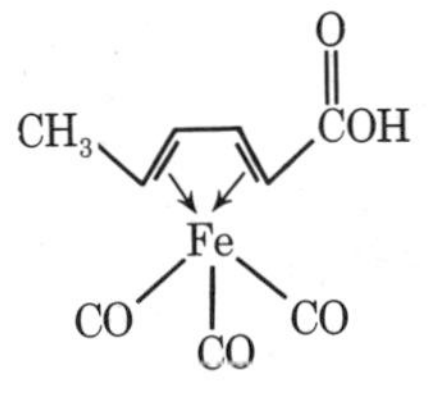

Similar coordination occurs also in the presence of other di- and trivalent metals. Reduction of the double bonds can produce various hexenoic acid mixtures.

Sorbic acid is oxidized rapidly in the presence of molecular oxygen or peroxide compounds. The decomposition products indicate that the double bond farthest from the carboxyl group is oxidized (11). More complete oxidation leads to acetaldehyde, acetic acid, fumaraldehyde, fumaric acid, and polymer products. Sorbic acid undergoes Diels-Alder reactions with many dienophiles and undergoes self-dimerization. Dimerization leads to the eight possible isomeric Diels-Alder structures and all their isomers (12):

Table 2. Solubility of Sorbic Acid and Potassium Sorbate

Solvent	Temperature, °C	Solubility, g/100 g solvent		References
		Sorbic acid	Potassium sorbate	
water	0	0.14		4, 6
	20	0.15	58.2	
	40	0.34		
	60	0.72		
	80	1.6		
	100	3.9		
pH[a]				
4.25	20	0.33		3
6.25	20	3.1		
7.25	20	12.0		
acetic acid (glacial)	23	11.5		6
acetone	20	9.2	0.1	6
butanol	25	11.3		4
carbon tetrachloride	20	1.3	<0.01	
cyclohexane	20	0.28		
ethanol, anhydrous	20	12.9	2.0	4, 6
60 wt %	20	6.4		6
ethyl ether	20	5.0	0.1	
glycerol	20	0.31	0.2	
isopropyl alcohol	20	12.9		4
methanol, anhydrous	20	12.9	16	6
50 wt %	20	1.6		3, 6
corn oil	20	0.7	0.01	3
soybean oil	20	0.52		3
propylene glycol	20	5.5	20	6
sodium chloride, 15 wt %	20	0.038	15	3–4, 6

[a] Controlled pH.

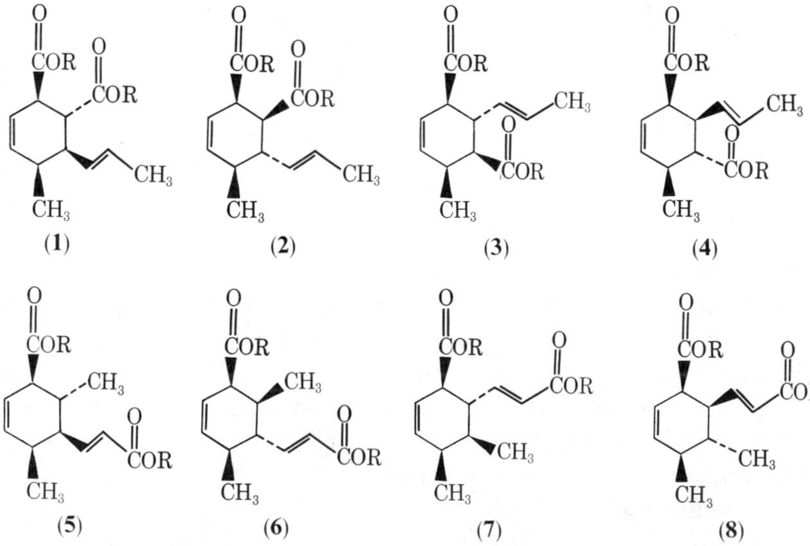

(1) (2) (3) (4)

(5) (6) (7) (8)

Sorbic ester dimers

Polymerization catalyzed by free radicals occurs with sorbic acid. The copolymers formed have high molecular weights with linear structures; thus, the trans form of the residual double bond is preserved (13).

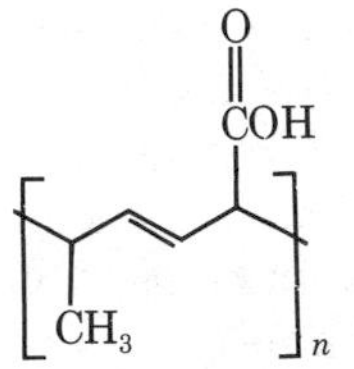

$$\left[\begin{array}{c} \text{(CH}_3\text{)CH–CH=CH–CH(COOH)} \end{array} \right]_n$$

Copolymers with acrylonitrile, butadiene, isoprene, acrylates, piperylene, styrene, and polyethylene have been studied. The high cost of sorbic acid as a monomer has prevented large-scale uses. The ability of sorbic acid to polymerize, particularly on metallic surfaces, has been used to explain its corrosion inhibition for steel, iron, and nickel (14).

Carboxylic Acid Group. Sorbic acid undergoes the normal acid reactions forming salts, esters, amides, and acid chlorides. Industrially, the most important compound is the potassium salt because of its stability and high solubility. Sodium sorbate [*7757-81-5*] (*E,E* form [*42788-83-0*]) is less stable and not commercially available. The calcium salt [*7492-55-9*], which has limited solubility, has some use in packaging materials.

Sorbic acid anhydride [*13390-06-2*] can be prepared by heating the polyester of the 3-hydroxy-4-hexenoic acid with sorboyl chloride [*2614-88-2*] or by reaction of sorbic acid with oxalyl chloride (15–16). Preparation of the esters of sorbic acid must be controlled to prevent oxidation and polymerization. The lower sorbic acid esters have a pleasant odor.

Synthesis and Manufacture

The first synthesis of sorbic acid was from crotonaldehyde and malonic acid in pyridine (2,17–18). The 32% yield can be improved with the use of malonic acid salts (19). One of the first commercial methods involved the reaction of ketene and crotonaldehyde in the presence of BF_3 in ether at 0°C (20–21) (see Ketenes and related substances). A lactone identified as

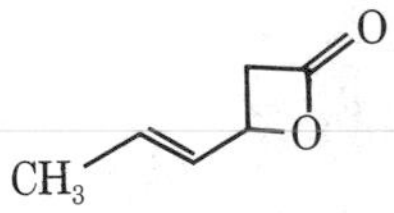

$$CH_3CH{=}CH\text{–}\underset{\text{(β-lactone, C=O, –O–)}}{CH\text{–}CH_2}$$

forms and then reacts with acid, giving a 70% yield. At present, most commercial sorbic acid is produced by a modification of this route. Catalysts comprised of metals, eg, zinc, cadmium, nickel, copper, manganese, and cobalt; metal oxides; or carboxylate salts of bivalent transition metals, eg, zinc isovalerate, produce a condensation adduct with ketene and crotonaldehyde (22–24) which has been identified as

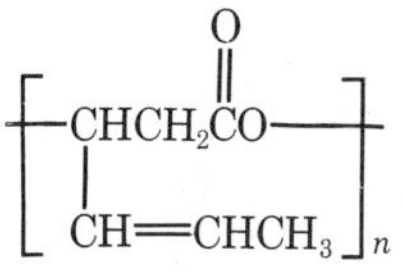

$$\left[\begin{array}{l} \text{–CHCH}_2\overset{\overset{\text{O}}{\|}}{\text{C}}\text{O–} \\ \;\;| \\ \;\text{CH{=}CHCH}_3 \end{array} \right]_n$$

An excess of crotonaldehyde or aliphatic, alicyclic, and aromatic hydrocarbons and their derivatives is used as a solvent to produce compounds of molecular weights of 1000–5000 (25–28). After removal of unreacted components and solvent, the adduct referred to as polyester is decomposed in acidic media or by pyrolysis (29–36). Proper operation of acidic decomposition can give high yields of pure *trans,trans*-2,4-hexadienoic acid, whereas the pyrolysis gives a mixture of isomers which must be converted to the pure trans,trans form. The thermal decomposition is carried out in the presence of alkali or amine catalysts. A simultaneous codistillation of the sorbic acid as it forms and the component used as the solvent can simplify the process scheme. The catalyst remains in the reaction batch. Suitable solvents and entraining agents include most inert liquids that boil preferably at 200–300°C, eg, aliphatic hydrocarbons. When the polyester is split thermally at 170–180°C and the sorbic acid is distilled directly with the solvent, production and purification can be combined in a single process step. The solvent can be reused after removal of the sorbic acid (34). The isomeric mixture can be converted to the thermodynamically more stable trans,trans form in the presence of iodine, alkali, or sulfuric or hydrochloric acid (37–38).

Further purification to meet food-grade specifications is required in the form of carbon treatments and recrystallization from aqueous or other solvent systems. The illustrated flow scheme for sorbic acid production shown in Figure 1 is greatly simplified because of the techniques and amount of equipment required for each step.

The ketene–crotonaldehyde route through polyester with various modifications and improvements is being practiced by Hoechst, Daicel, Ueno, Chisso, Nippon Gohsei, and Monsanto. Differences in their processes consist mostly of the methods of polyester splitting and first-stage purification. Production of the potassium salt can be from finished sorbic acid or from a stream in the sorbic acid production route before the final drying step. More than 100 patents on the process for producing sorbic acid and potassium sorbate from this route are given in the literature.

Union Carbide abandoned the ketene–crotonaldehyde route in 1953 in favor of the oxidation of 2,4-hexadienal made by acetaldehyde condensation. A silver compound was used as the catalyst (39–40) and prevented the peroxidation of the ethylenic bonds. Their plant operated until 1970.

A method of sorbic acid preparation based on the reaction of crotonaldehyde and acetone followed by oxidation of the crotonylidenacetone is of interest in the USSR (41–42):

$$CH_3CH{=}CHCH{=}O + CH_3C({=}O)CH_3 \longrightarrow CH_3CH{=}CHCH{=}CHC({=}O)CH_3$$

$$\downarrow \text{NaOCl}$$

$$CH_3CH{=}CHCH{=}CHC({=}O)CCl_3$$

$$\downarrow \text{NaOH}$$

$$\text{sorbic acid} \xleftarrow{\text{HCl}} CH_3CH{=}CHCH{=}CHC({=}O)ONa$$

Other methods include ring opening of *para*-sorbic acid [*108-54-3*] (δ-lactone of 5-hydroxy-2-hexenoic acid) in hydrochloric acid or in alkaline solutions (43–44); the ring opening of γ-vinyl-γ-butyrolactone in various catalysts (45–46); or isomerization of 2,5-hexadienoic acid esters (47–48). Other methods are described in refs. 6, 49, and 50.

Economic Aspects

Sorbic acid is produced and marketed in the United States in several forms. The soluble potassium sorbate is marketed as a powder or as granules. Sorbic acid is marketed in the dust-free crystalline form for food use or as powder for feed use. The 1981 list price is $6.25/kg. In addition to Monsanto, which is the only U.S. producer, there are four Japanese producers and a producer in the FRG. Worldwide consumption in 1980 was ca $(1.3–1.5) \times 10^4$ metric tons and the nameplate productive capacity was ca $(2.0–2.4) \times 10^4$ t.

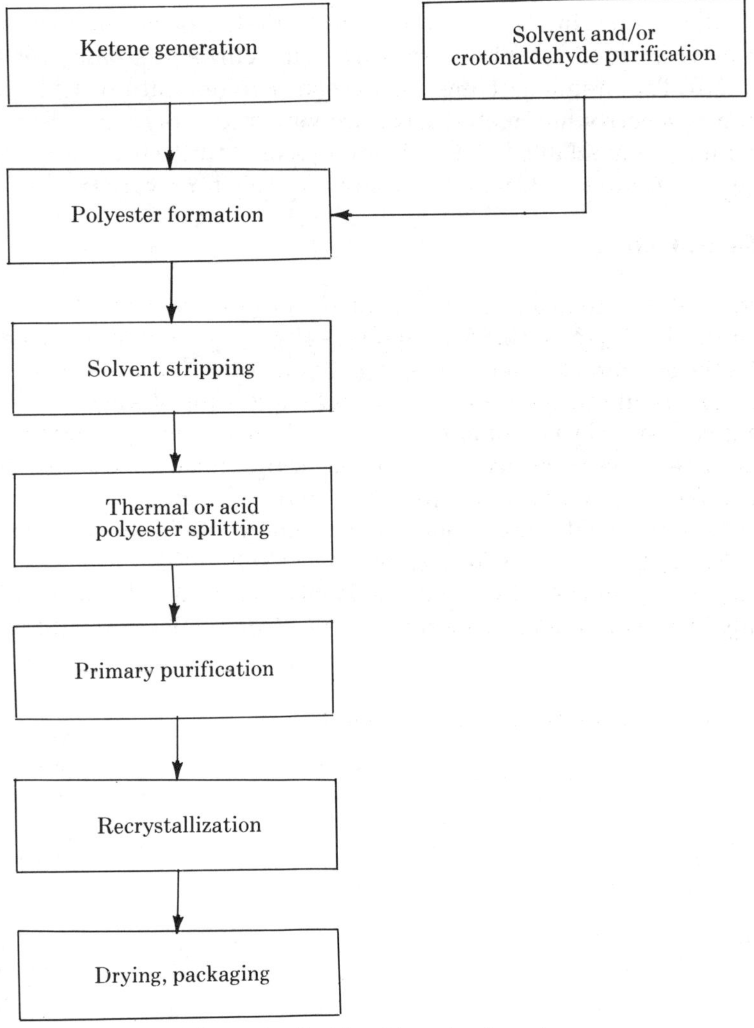

Figure 1. Sorbic acid production scheme.

Purification Specifications

Sorbic acid and its salts are highly refined in order to obtain the necessary purity for use in foods. The quality requirements are defined by the U.S. Food Chemicals Codex (see Table 3) (51). Codistillation or recrystallization from water, alcoholic solutions, or acetone is used to obtain sorbic acid and potassium sorbate of a purity that passes not only the Codex requirements but is sufficient for long-term shortage. Measurement of the peroxide content and heat stability can further determine the presence of low amounts of impurities. The presence of isomers other than the trans,trans form causes instability and affects the melting point.

Analytical Techniques

Sorbic acid is normally assayed spectrophotometrically by absorbance at ca 260 nm (52). When there are interfering compounds in the food substrate, the sample should be extracted with suitable solvents (53–54). The prepared sample can then be subjected to absorbance in a spectrophotometer (55), gas chromatography using thermal conductivity (54,56), or high pressure liquid chromatography with a uv detector (53,57–58). The bromine titrimetric method or thiobarbituric acid method can be used when a spectrophotometer, gas chromatograph, or high pressure liquid chromatograph is not available (59–60). High pressure liquid chromatography is the method of choice because of its simplicity and accuracy for a variety of substrates.

Health and Safety Factors

The extremely low toxicity of sorbic acid enhances its desirability as a food preservative. The oral LD_{50} for sorbic acid in rats is 10 g/kg compared to 5 g/kg for NaCl (61–62). In subacute and chronic toxicity tests in rats, wt 5% sorbic acid in the diet results in no abnormal effects after 90 d or lifetime feeding studies (62–63). A level of 10% in rat diets results in a slight enlargement of the liver, kidneys, and thyroid gland (64). Studies of the long-term toxicity of sorbic acid in mice and in rats indicate no carcinogenic effects at dietary levels up to 10% (64).

In humans, sorbic acid is metabolized to CO_2 and water in the same way as other fatty acids, releasing 27.6 kJ/g (6.6 kcal/g) (65). As a result of the favorable toxicological and physiological aspects of sorbic acid, the WHO has allowed for it the highest acceptable daily intake of all food preservatives, ie, 25 mg/kg body weight (66).

Table 3. Specifications for Sorbic Acid and Potassium Sorbate [a]

Specification	Sorbic acid	Potassium sorbate
melting point, °C	132–135	
water, wt % max	0.5	1.0
purity of titration, wt %	99.0–101.0	98.0–101.0
heavy metals (as Pb), ppm, max	10	10
ash, wt % max	0.2	
arsenic, ppm max	3	3
acidity (as sorbic acid)		passes test
alkalinity (as K_2CO_3)		passes test

[a] Ref. 51.

Uses

Sorbic acid and its potassium salt, collectively called sorbates, are used primarily in a wide range of food and feed products and to a lesser extent in certain cosmetics (qv) (67), pharmaceuticals (qv), and tobacco products (see Pet and livestock feeds). Although the calcium and sodium salts have been used in limited applications, the acid and its potassium salt are used almost exclusively.

Since the first demonstration of antimicrobial activity in the 1940s, sorbates have been shown to have inhibitory activity against a wide spectrum of yeasts, molds, and bacteria, including most food-borne pathogens (68–72). As bacterial inhibitors, sorbates are least effective against lactic acid bacteria. Although this can be a problem for foods that suffer from lactic spoilage, it has proven to be a positive point in cases of yeast suppression during lactic fermentations (73–76). The antimicrobial activity of sorbates has been reviewed (77) (see also Industrial antimicrobial agents).

The inhibitory activity of sorbates is attributed to the undissociated acid molecule. The activity, therefore, depends upon the pH of the substrate; the upper limit for activity is ca pH 6.5 in moist applications and the degree of activity increases as the pH decreases. The upper pH limit can be extended higher in low water-activity systems. Table 4 indicates the effect of pH on the dissociation of sorbic acid. The activity of the sorbates at a higher pH is one distinct advantage over the two other most commonly used food preservatives, ie, benzoic and propionic acids, since the upper pH limits for activity of these compounds are ca pH 4.5 and pH 5.5, respectively. Although the effect of sorbates can be microbiocidal under certain conditions, activity is most often manifested as a microbial growth retardant.

The exact mechanism of inhibition by sorbic acid has not been thoroughly elucidated even though it has been the subject of a great deal of research and a number of hypotheses have been proposed. A number of enzyme systems in fungi and bacteria have been designated as sites of sorbate inhibition (78–84). Sorbic acid has been shown to inhibit the transport of carbohydrates into yeast cells, inhibit oxidative and fermentative assimilation, and uncouple oxidative phosphorylation in a variety of bacteria (83,85–87). The above studies were conducted for various systems including whole cells, cell-free extracts, and isolated enzyme systems. Although all may be valid under specific sets of conditions, no single mechanism proposed seems to account wholly for the inhibitory activity of sorbates. More likely, microbial inhibition by sorbates

Table 4. Percent Undissociated Sorbic Acid at Various pH Levels[a]

pH	Wt %
3.0	98
3.5	95
4.0	86
4.5	65
4.76 (pK_a)	50
5.0	37
5.5	15
6.0	6
6.5	2
7.0	<1

[a] Refs. 77 and 88.

is the result of a combination of events which may differ from one organism to another and from one set of conditions to another.

To date, there has been no evidence to indicate that microorganisms can develop resistance to sorbates, as occurs with antibiotics (qv) and certain other antimicrobial chemicals (see Antimicrobial agents). However, there is variation in the sorbate sensitivity of microorganisms from one genus to another, between different species in the same genus, and even between different strains of the same species. *Saccharomyces bailii* is resistant to sorbates, benzoates, and other short-chain monocarboxylic acids because of an inducible enzyme system which transports these compounds out of the cell (89). Some molds, when present in extremely high numbers, can metabolize sorbates. This has been attributed to typical β-oxidation as occurs with other fatty acids (78). Species of *Penicillium*, particularly *P. roqueforti*, can decarboxylate sorbates to 1,3-pentadiene, resulting in a hydrocarbon odor (90).

Food applications of sorbates expanded rapidly after issuance of the original patents in 1945 (91). The first uses were based on their excellent fungistatic properties and thus involved foods with low pH and/or low water activity, in which cases yeasts and molds are the primary spoilage agents. More recent application research has been directed toward utilizing the bacteriostatic properties of sorbates (see also Food additives).

Sorbates are classified as GRAS in the United States, with no upper limit set for foods that are not covered by Standards of Identity. They are also allowed in over 70 food products having Standards of Identity. Examples of products that often contain sorbates are natural and processed cheeses and other cheese products, salad dressings, bakery products, prepared salads, fermented vegetable products, dried fruits, fruit juices (qv), margarine, wine, fish products, jams, and jellies (92). The use levels in food products are 0.01–0.5 wt % (Table 5). Compared with other antimicrobial preservatives, sorbates can be used in higher concentrations without affecting the flavor of foods. The level of sorbates necessary for preservation of a specific product depends upon a number of factors including the product composition (pH, moisture, other inhibitors,

Table 5. Sorbate Use Levels, wt %

Product	Typical use level	References
natural cheese	0.3[a]	93–95
processed cheese and cheese spreads	0.2[a]	93–95
cottage cheese	0.075	96–98
bakery products		
chemically leavened	0.05–0.3	99–101
yeast-raised	0.03–0.3	102–103
tortillas	0.03–0.5	104
fillings, icings, toppings	0.05–0.1	95
pet foods, semimoist	0.1–0.3	105–106
margarine	0.1	107–109
prepared salads	0.05–0.1	110
salad dressings, pourable	0.05–0.1	95
fruit drinks and juices	0.025–0.075	95, 111
wine	0.01–0.025	112–113
pickles	0.05–0.1	73–78, 95
syrup	0.02–0.1	114

[a] Maximum use level allowed by Standard of Identity.

fat content), the initial contamination level, packaging, and storage temperature. Maximum shelf-life extension with sorbates is achieved when products have low initial levels of microbial contamination and when products are properly handled and stored. Therefore, the preservative cannot be used to mask poor quality product or poor handling practices.

Sorbates can be applied to food by any of several methods including direct addition, dipping in or spraying with an aqueous sorbate solution, dusting with sorbate powder, or impregnating food-packaging materials. The potassium salt is used in applications where high water solubility is desired.

Margarine. Improvements in the technology of making margarine have greatly reduced the spoilage problems associated with this product (see Fats and fatty oils; Vegetable oils). Sorbates and benzoates are both used in margarine, with sorbates being the more effective of the two because of the high product pH and the comparative oil:water distribution coefficients for the two compounds, ie, sorbic acid 3:1, benzoic acid 6:1 (107–108). Thus a greater proportion of the sorbate remains in the water phase where microbial spoilage occurs. Sorbates are generally used at 0.1 wt % and are most often used in those product forms that have higher spoilage potential, eg, low fat or unsalted margarine or product that is packaged in plastic tubs leaving a headspace.

Wines. Sorbic acid is used in table wines to prevent secondary fermentation of residual sugar (see Wine). It is used at 0.01–0.025 wt % in addition to SO_2. The addition of sorbic acid affords protection against recontamination by yeasts for wines that have been heated or filter-sterilized, but at these low levels it does not provide adequate protection against undesirable malolactic or acetic acid bacteria (112–113).

Dairy Products. The dairy industry is the largest commercial user of sorbates with the largest portion used in processed cheeses (see Milk and milk products). Data on the antimicrobial efficacy of sorbates in cheese are published in ref. 93. The most common methods of application include dipping or spraying with potassium sorbate solutions for natural cheeses and direct addition of sorbic acid to processed cheeses and cold pack cheeses. Sorbate impregnated wrapping material can be used for packaged cheese slices and pieces. For cottage cheese, sorbic acid is added to the cream dressing prior to pasteurization to a level of 0.075 wt % in the finished product (96–97). Most cheese products are covered by Standards of Identity and, with the exception of cottage cheese, a maximum use level is set (Table 5).

Seafood. Sorbates are used in a number of ways to extend the shelf life of a variety of seafood products, both fresh and processed. For smoked or dried fish, an instantaneous dip in 5 wt % potassium sorbate or a 10-min dip in 1.0 wt % potassium sorbate prior to drying or smoking inhibits the development of yeast and mold in the finished product (115–116). For fresh fish, sorbates can be incorporated at ca 0.5 wt % into the ice, refrigerated sea water, or ice-water slush in which fish are packed, or applied as a 2.5–5.0 wt % potassium sorbate dip for fillets (117–118). Sorbates inhibit the growth of psychrotrophic spoilage bacteria, but it is important that the treatment be applied while the product is very fresh (119–120). It has also been shown that sorbate can extend the shelf life as well as delay *Clostridium botulinum* type E toxigenesis in fish that has been packaged in modified atmospheres (121–122) (see also Aquaculture).

Fruit and Vegetable Products. Sorbates are applied at 0.05–0.1 wt % as a fungistat for prunes, pickles, relishes, maraschino cherries, olives, and figs (94,123). The same levels also extend the shelf life of prepared salads, eg, potato salad, cole slaw, and tuna salad (119). In fermented vegetables, it is used to retard yeasts during fermentation or in the cover brine to protect the finished product (74–76,94,108).

Sorbates are effective in reducing postharvest losses of fresh citrus fruit, particularly in cases where the spoilage fungi have become resistant to current chemical treatments (124–125). They also prevent discoloration of celery butts (126).

In recent years, sorbates have shown considerable value for treating raisin grapes to prevent losses caused by mold during the drying period. An additional benefit of this treatment is the reduction in time required for drying (127).

Sorbate combined with mild heat has a synergistic effect with regard to microbial destruction; thus, in the presence of 0.025–0.06 wt % sorbate, products such as apple juice, peach and banana slices, fruit salads, and strawberries, could be treated with less severe heat treatments to achieve products with extended shelf life (128–129). Sorbates increase the heat sensitivity of various spoilage fungi under varying conditions of pH and water activity (130–131). A similar synergistic effect has been reported for the combination of sorbate with irradiation (134).

Bakery Products. Sorbates are used in and/or on yeast-raised and chemically leavened bakery products (see Bakery processes and leavening agents). The internal use of sorbates in yeast-raised products at one fourth the amount of calcium/sodium propionate which is normally added provides a shelf life equal to that of propionate without adversely affecting the yeast fermentation. Sorbates added at one tenth the propionate level reduce the mix time by 30% (103). This internal treatment in combination with an external spray of potassium sorbate can provide the same or an increased shelf life of pan breads, hamburger and hot-dog buns, English muffins, "brown'n' serve" rolls, and tortillas. The total sorbate useful in or on these baked goods ranges from 0.03 wt % for pan breads to 0.5 wt % for tortillas; 0.2–0.3 wt % sorbic acid protects chemically leavened yellow and chocolate cakes (99). Fruit-pie fillings and icings can be protected with 0.03–0.1 wt % sorbates.

Meat and Poultry. In 1981, the only sorbate treatment of meat permitted by the USDA was a 2.5 wt % potassium sorbate dip for dry sausages to prevent mold growth. Recently work has been done to support increased sorbate use in meat and poultry; 0.26 wt % potassium sorbate with 0.004 wt % sodium nitrite in curing bacon reduces the nitrosamine formation during frying and provides a safe antibotulinal shelf life (135–136) (see *N*-Nitrosamines). In cooked cured sausages, ie, beef, pork, and chicken frankfurter emulsions, 0.004 wt % nitrite/0.20 wt % sorbic acid results in delayed germination and outgrowth of *Clostridium botulinum* (137–138).

For fresh poultry, a potassium sorbate dip significantly reduces the total number of viable bacteria and doubles the refrigerated shelf life of ice-packed broilers (139). In cooked, uncured, vacuum-packaged turkey and poultry stored at 4°C, 0.2–0.25 wt % potassium sorbate suppresses microbial growth for up to 10 d (140). Country-cured hams have been sprayed with a 10 wt % potassium sorbate solution, resulting in no mold growth for up to 30 d (141). A complete review of sorbate use in meat products has been published (142). In 1981, the FDA accepted for consideration a petition which would affirm sorbates as GRAS for use in meats, fresh poultry, and poultry products (143) (see Meat products).

Pet Foods and Commercial Animal Feeds. For many years, it has been known that very stable, long shelf-life, intermediate moisture pet foods can be prepared through the use of 0.1–0.3 wt % sorbates (see Pet and livestock feeds). In these products, the antimicrobial effectiveness of sorbates is enhanced by combination with moderate heat treatment, adjustment of pH, and reduction of water activity with humectants, eg, propylene glycol, or by adjustment of sugar and salt content. These techniques have been reviewed extensively (105–106).

As the cost of energy has escalated in recent years, the use of high moisture food by-products in commercial animal feeds has also escalated, particularly in beef cattle and dairy rations, as a means of reducing the cost of production. Because of the very broad spectrum of activity of sorbates, they are extremely effective in preservation of wet by-products, eg, brewers' and distillers' grains, beet pulp, citrus pulp, and condensed whey (106).

Sorbic acid is not only suitable for preservation of feedstuffs but also improves the feed utilization and weight gain of chickens. This has proven to be of economic value under practical conditions when sorbic acid is added to the feed at 0.02–0.04 wt % (144–146). Similar effects have been observed for the use of sorbic acid in swine feeds (147).

BIBLIOGRAPHY

"Sorbic Acid" in *ECT* 1st ed., Vol. 6, pp. 272–274, by J. A. Field, Union Carbide and Carbon Corp.; Suppl. 1, pp. 840–849, by A. E. Montagna, Union Carbide Chemicals Company; "Sorbic Acid" in *ECT* 2nd ed., Vol. 18, pp. 589–599, by S. W. Moline, C. E. Colwell, and J. E. Simeral, Union Carbide Corporation.

1. A. W. Hofmann, *Lieb. Ann. Chem. Pharm.* **110,** 129 (1859).
2. O. Doebner, *Ber. Dtsch. Chem. Gas.* **33,** 2140 (1900).
3. Trolle-Lassen Co., *Arch. Pharm. Org. Chem.* **66**(23), 1235 (1959).
4. N. G. Polyanskii and co-workers, *Zh. Prikl. Khim. (Leningrad)* **39,** 2005 (1966).
5. *U.S. Bureau of Mines, Report 7132*, May 1968.
6. E. Lück, *Sorbinsaure*, Band 1, B. Behr's Verlag GmbH, Hamburg, 1969.
7. J. J. Sudborough and J. Thomas, *J. Chem. Soc.* 97, 2450 (1910).
8. C. K. Ingold, G. J. Pritchard, and H. G. Smith, *J. Chem. Soc.*, 79 (1934).
9. R. Fittig, *Lieb. Ann. Chem. Pharm.* **161,** 307 (1872).
10. U.S. Pat. 3,126,401 (March 24, 1964), G. Ecke (to Ethyl Corporation).
11. P. Heinänen, *Ann. Acad. Sci. Fenn. Ser. A* **49**(4), 112 (1938).
12. J. J. Bloomfield, unpublished data, Monsanto Co., St. Louis, Mo., 1981.
13. K. Fugjiwana and co-workers, *Nippon Hoshasen Kobunshi Kenkyu Kyokai Nempo* **4,** 183 (1962).
14. I. N. Putilova, *Tr. Mezhd. Kongr. Korrozii Metallov (Moscow)* **2,** 32 (1966).
15. Ger. Pat. 1,283,832 (Jan. 27, 1965), H. Fernholz and H. J. Schmidt (to Hoechst).
16. R. Adams and L. Ulich, *J. Am. Chem. Soc.* **42,** 599 (1920).
17. O. Doebner, *Ber. Dtsch. Chem. Ges.* **23,** 2372 (1890).
18. *Ibid.*, **35,** 1136 (1902).
19. Pol. Pat. 47,632 (Oct. 14, 1963), I. Nagrodzka and co-workers.
20. U.S. Pat. 2,484,067 (Oct. 11, 1949), A. B. Boese, Jr. (to Union Carbide Corp.).
21. H. J. Hagemeyer, *Ind. Eng. Chem.* **41,** 765 (1949).
22. Jpn. Pat. 39-13,849 (Aug. 5, 1967), H. Nakamura (to Nippon Gohsei).
23. Jpn. Pat. 45-9,368 (April 14, 1970), O. Nakamura (to Nippon Gohsei).
24. Ger. Pat. 1,042,573 (Nov. 6, 1958), H. Fernholz (to Hoechst).
25. U.S. Pat. 3,022,342 (Feb. 20, 1962), H. Fernholz, K. Ruths, and K. Heinmann-Trosien (to Hoechst).
26. U.S. Pat. 3,021,365 (Feb. 13, 1962), H. Fernholz and E. Munolos (to Hoechst).
27. Ger. Pat. 1,150,672 (June 27, 1963), O. Probst (to Hoechst).
28. U.S. Pat. 3,499,029 (March 3, 1970), H. Fernholz and H. Neu (to Hoechst).
29. Jpn. Pat. 44-26,646 (Nov. 7, 1969), I. Nakajima (to Nippon Gohsei).
30. U.S. Pat. 3,759,988 (Sept. 18, 1973), G. Kunstle (to Wacker).
31. Can. Pat. 8,982,73 (April 18, 1972), R. Smith and E. Jeans (to Chemcell).
32. Ger. Pat. 2,203,712 (Aug. 16, 1973), H. Fernholz, H. J. Schmidt, and F. Wunder (to Hoechst).
33. Ger. Pat. 1,153,742 (Sept. 5, 1963), K. Ruths and O. Probst (to Hoechst).
34. Ger. Pat. 1,059,899 (June 25, 1959), H. Fernholz (to Hoechst).
35. U.S. Pat. 3,461,158 (Aug. 12, 1969), L. Hörnig and H. Neu (to Hoechst).

36. N. G. Polyanskii and co-workers, *Tr. Tombovsk Inta Khim. Mashinostr.* **2,** 94 (1968).
37. U.S. Pat. 3,642,885 (Feb. 15, 1972), L. Hörnig and co-workers (to Hoechst).
38. Ger. Pat. 1,281,439 (Oct. 31, 1968), L. Hörnig and O. Probst (to Hoechst).
39. U.S. Pat. 2,887,496 (May 19, 1959), E. Lashley (to Union Carbide).
40. G. F. Woods and co-workers, *J. Am. Chem. Soc.* **77,** 1800 (1955).
41. V. S. Markevich and S. M. Markevich, *Khim. Promst.* (*Moscow*) **12,** 898 (1973).
42. U.S.S.R. Pat. 169,520 (March 17, 1965), S. M. and V. S. Markevich.
43. R. Joly and C. Amiaro, *Bull. Soc. Chem. Fr.*, 139 (1947).
44. U. Eisner, J. Elvidoe, and R. Lindstead, *J. Chem. Soc.*, 1372 (1953).
45. U.S. Pat. 4,022,822 (May 10, 1977), Y. Tsu Jino (to Nippon Gohsei).
46. U.S. Pat. 4,158,741 (June 19, 1979), M. Goi (to Nippon Gohsei).
47. Ital. Pat. 719,380 (Nov. 2, 1966), G. P. Chiusoli, S. Merzoni, and G. Cometti (to Montecatini).
48. G. P. Chiusoli, *Angew. Chem.* **72,** 74 (1960).
49. B. N. Utkin, *Khimiya Sorbinovoi Kislofy*, M., NIITE Khim. (1970).
50. N. G. Polyanskii, *Khim. Promst.* (*Moscow*) **1,** 20 (1963).
51. *Food Chemicals Codex*, 3rd ed., National Academy of Science, National Academy Press, Washington, D.C., 1981.
52. G. Alderton and J. C. Lewis, *Food Res.* **23,** 338 (1958).
53. J. L. Owen, C. M. Calvert, R. L. Simmons, L. Ruff, and E. M. Emery, Monsanto Co. Report No. OA-81-M4, St. Louis, Mo., 1981.
54. A. Graveland, *J. Assoc. Off. Anal. Chem.* **55,** 1024 (1972).
55. J. J. Maxstadt and A. B. Karasz, *J. Assoc. Off. Anal. Chem.* **55,** 7 (1972).
56. D. E. LaCroix and N. P. Wong, *J. Assoc. Off. Anal. Chem.* **54,** 361 (1971).
57. M. C. Bennett and B. R. Petrus, *J. Food Sci.* **42,** 1220 (1977).
58. G. L. Park and D. B. Nelson, *J. Food Sci.* **45,** 1629 (1981).
59. P. Spanyar and A. Sandor, *Z. Lebensm. Unters. Forsch.* **108,** 402 (1958).
60. H. Schmidt, *Dtsch. Lebensm. Rundsch.* **58,** 1 (1962).
61. *Acute Toxicity of Sorbic Acid*, Monsanto Co. Report No. 4-71-0032, St. Louis, Mo., 1971.
62. H. J. Deuel, R. Alfin-Slater, C. S. Weil, and H. F. Smyth, *Food Res.* **19,** 1 (1954).
63. K. Lang, *Arzneim. Forsch.* **10,** 997 (1960).
64. I. F. Gaunt, K. R. Butterworth, J. Hardy, and S. D. Gangoli, *Food Cosmet. Toxicol.* **13,** 31 (1975).
65. E. Lück, *Sorbinsaure Chemie—Biochemie—Mikrobiologie—Technologie*, Band 2, B. Behr's Verlag, Hamburg, 1972, p. 21.
66. E. Lück, *International Flavors and Food Additives*, United Trade Press Ltd., London, 1976, pp. 122–127.
67. R. Woodford and E. Adams, *Am. Perfum. Cosmet.* **85,** 25 (1970).
68. E. S. Beneke and F. W. Fabian, *Food Technol.* **9,** 486 (1955).
69. T. A. Bell, J. L. Etchells, and A. F. Borg, *J. Bacteriol.* **77,** 573 (1959).
70. R. H. Vaughn and L. O. Emard, *Bacteriol. Proc.* **5,** 38 (1951).
71. L. O. Emard and R. H. Vaughn, *J. Bacteriol.* **63,** 487 (1952).
72. G. K. York, dissertation, University of California, Davis, 1960.
73. G. F. Phillips and J. O. Mundt, *Food Technol.* **4,** 291 (1950).
74. R. N. Costilow, W. E. Ferguson, and S. Ray, *Appl. Microbiol.* **3,** 341 (1955).
75. R. N. Costilow, F. M. Coughlin, D. L. Robach, and H. S. Rasheb, *Food Res.* **21,** 27 (1956).
76. R. N. Costilow, F. M. Coughlin, E. K. Robbins, and W. T. Hsu, *Appl. Microbiol.* **5,** 373 (1957).
77. J. N. Sofos and F. F. Busta, *J. Food Prot.* **44,** 614 (1981).
78. D. Melnick, F. H. Luckmann, and C. M. Gooding, *Food Res.* **19,** 44 (1954).
79. J. J. Azukas, R. N. Costilow, and H. L. Sadoff, *J. Bacteriol.* **81,** 189 (1961).
80. G. K. York and R. H. Vaughn, *Bacteriol. Proc.* **55,** 20 (1955).
81. W. Martoadiprawito and J. R. Whitaker, *Biochem. Biophys. Acta* **77,** 536 (1963).
82. J. R. Whitaker, *Food Res.* **24,** 37 (1959).
83. G. K. York and R. H. Vaughn, *J. Bacteriol.* **88,** 411 (1964).
84. J. A. Troller, *Can. J. Microbiol.* **11,** 611 (1965).
85. T. Deak and E. K. Novak, *Yeasts, The Proceedings of the Second Symposium on Yeasts*, Slovac Academy of Sciences, Bratislava, Czech., 1966, p. 533–536.
86. A. G. Man in *Bacteria, a Treatise on Structure and Function*, Vol. 1, Academic Press, New York, 1960.
87. G. K. York and R. H. Vaughn, *Bacteriol. Proc.* **60,** 47 (1960).

88. F. Sauer, *Food Technol.* **31,** 66 (1977).
89. A. D. Warth, *J. Appl. Bacteriol.* **43,** 215 (1970).
90. E. H. Marth, C. M. Capp, L. Hasengahl, J. W. Jackson, and R. V. Hussong, *J. Dairy Sci.* **49,** 1197 (1966).
91. U.S. Pat. 2,379,294 (June 26, 1945), C. M. Gooding (to Best Foods, Inc.).
92. *Sorbic Acid and Potassium Sorbate for Preserving Food Freshness and Market Quality*, IC/FI-13A, Monsanto Co. Technical Bulletin, St. Louis, Mo., 1978.
93. M. D. Bonnor and L. G. Harmon, *J. Dairy Sci.* **40,** 1599 (1957).
94. *Sorbic Acid and Potassium Sorbate*, Monsanto Co. Publication No. IC/FI-13A, St. Louis, Mo., 1977.
95. 21 CFR 133, revised April 1, 1981.
96. *Sorbic Acid and Potassium Sorbate for Use in Cottage Cheese*, Monsanto Co. Publication No. IC/FI-20, St. Louis, Mo., 1977.
97. E. B. Collins and H. H. Moustafa, *J. Dairy Sci.* **52,** 439 (1969).
98. 21 CFR 133.128, revised April 1, 1981.
99. D. Melnick, H. W. Vahlteich, and A. Hackett, *Food Res.* **21,** 133 (1956).
100. *New Easy Way To Extend The Shelf Life of Brown'N' Serve Rolls*, Monsanto Co., St. Louis, Mo., 1981.
101. *Sorbic Acid and Potassium Sorbate for Use in Chemically Leavened Bakery Products*, Monsanto Co. Publication No. IC/FI-14, St. Louis, Mo., 1980.
102. C. S. Hickey, *Baker's Dig.* **54**(4), 20 (1980).
103. *Potassium Sorbate Surface Treatment for Yeast Raised Bakery Products*, Monsanto Co. Publication No. IC/FI-21, 1977.
104. *Good News for Tortillas*, Monsanto Co., St. Louis, Mo., 1981.
105. R. Davis, G. G. Birch, and J. J. Parker, eds., *Intermediate Moisture Foods*, Applied Science Publishers, Ltd., London, 1976.
106. N. W. Desrosier, *The Technology of Food Preservation*, AVI Publishing Co., Westport, Conn., 1970, 365–383.
107. C. J. Doherty, *Technical Service Report*, Union Carbide Corp., 1965.
108. E. Becker and I. Roeder, *Fette Seifen Austrichm.* **49,** 321 (1957).
109. 21 CFR 166.110, revised April 1, 1981.
110. *Sorbic Acid and Potassium Sorbate for Use In Prepared Salads*, Monsanto Co. Publication No. IC/FI-15R, St. Louis Mo., 1978.
111. 21 CFR 146, revised April 1, 1981.
112. G. Wurdig, *Brew. Distill. Int.* **6,** 42 (1976).
113. R. C. Auerbach, *Wines Vines* **40,** 26 (1959).
114. 21 CFR 168, revised April 1, 1981.
115. J. J. Geminder, *Food Technol.* **13,** 459 (1959).
116. J. W. Boyd and H. L. A. Tarr, *Food Technol.* **9,** 411 (1953).
117. J. P. H. Wessels, E. Lamprecht, J. de A. Rodrigues, and C. K. Simmonds, *J. Food Technol.* **7,** 303 (1972).
118. N. Tomlinson, S. E. Geiger, G. A. Gibbard, D. T. Smith, B. A. Southcott, and J. W. Boyd, *Technical Report No. 783*, Canadian Fisheries and Marine Service, Vancouver, 1978.
119. J. M. Debevere and J. P. Voets, *J. Appl. Bacteriol.* **35,** 351 (1972).
120. M. S. Fey, Ph.D. dissertation, Cornell University, Department of Poultry Science, Ithaca, N.Y., 1980.
121. R. C. Lindsay, *J. Food Prot.*, in press.
122. R. C. Lindsay, Department of Food Science, University of Wisconsin, Madison, Wi., personal communication, 1981.
123. *CRC Handbook of Food Additives*, 2nd ed., CRC Press, Boca Raton, Fla., 1972.
124. J. J. Smoot and A. A. McCormack, *Proc. Fla. State Hortic. Soc.* **91,** 119 (1978).
125. *Use of Potassium Sorbate in Protecting Citrus Fruit*, Monsanto Co. Publication No. IC/NC-602, St. Louis, Mo., 1981.
126. C. B. Hall, *Proc. Fla. State Hortic. Soc.* **72,** 280 (1959).
127. *Use of MP-11 for Mold Protection of Field Drying Raisins*, Monsanto Co. Publicaton No. IC/NC-603, St. Louis, Mo., 1981.
128. J. R. Robinson and C. H. Hills, *Food Technol.* **16,** 77 (1962).
129. U.S. Pat. 2,992,114 (July 11, 1961), E. A. Weaver (to U.S. Government).

130. L. R. Beuchat, *J. Food Sci.* **46,** 771 (1981).
131. L. R. Beuchat, *J. Food Prot.* **44,** 450 (1981).
132. L. R. Beuchat, *Appl. Environ. Microbiol.* **41,** 472 (1981).
133. L. R. Beuchat, *J. Food Prot.* **44,** 765 (1981).
134. C. F. Niven, Jr. and W. R. Chesbro, *Antibiot. Annu.* 855 (1956–1957).
135. *Shelf Life Sensory, Cooking, and Physical Characteristics of Bacon Cured with Varying Levels of Sodium Nitrate and Potassium Sorbate*, U.S. Department of Agriculture, July 1979.
136. F. J. Ivey, K. J. Shaver, L. N. Christiansen, and R. B. Thompkin, *J. Food Prot.* **41,** 621 (1978).
137. M. C. Robach, *Appl. Environ. Microbiol.* **38,** 840 (1978).
138. J. N. Sofos, F. F. Busta, and C. E. Allen, *Appl. Environ. Microbiol.* **37,** 1103 (1979).
139. E. C. To and M. C. Robach, *J. Food Technol.* **15,** 543 (1980).
140. M. C. Robach, E. C. To, S. Meydev, and C. F. Cook, *J. Food Sci.* **45,** 638 (1980).
141. J. D. Baldock, P. R. Frank, P. P. Graham, and F. J. Ivey, *J. Food Prot.* **42,** 780 (1979).
142. M. C. Robach and J. N. Sofos, *J. Food Prot.* **45,** 374 (1982).
143. S. A. Miller, *Fed. Regist.* **46**(76), 22808 (April 21, 1981), Petition No. GRASP 1GO272.
144. G. Dust, in E. Lück, eds., *Sorbinsaure*, Band 3, B. Berhr's Verlag GmbH 131, Hamburg, FRG, 1970.
145. B. C. Dilworth, T. C. Chen, E. J. Day, R. D. Miles, A. S. Arafa, R. H. Harms, G. L. Romoser, V. G. DiFate, and K. J. Shaver, *Poult. Sci.* **6,** 1445 (1979).
146. M. Suwathep, C. R. Parkhurst, F. M. McCorkle, and F. J. Ivey, *Poult. Sci.* **7,** 1741 (1981).
147. T. Veum, unpublished data, University of Missouri, Columbia, Mo., 1981.

C. L. KELLER
S. M. BALABAN
C. S. HICKEY
V. G. DIFATE
Monsanto Co.

SORBITE. See Steel.

SORBITOL, $CH_2OH(CHOH)_4CH_2OH$. See Alcohols, polyhydric.

SORGHUMS. See Wheat and other cereal grains.

SOYBEANS AND OTHER OILSEEDS

The four principal oilseed crops grown in the United States are soybeans, cottonseed, peanuts, and sunflowers. Some are consumed directly as foods, and all serve as sources of edible oils. After removal of these oils, the resulting meals are rich in proteins and are used mainly for animal feeds. Food uses of oilseed proteins, however, are increasing. Table 1 shows botanical names, geographic distribution, and main uses.

Soybeans are now the principal oilseed crop in the United States. They are believed to have been domesticated in the eastern half of northern China around the 11th century BC or earlier. They were later introduced and established in Japan and other parts of Asia, brought to Europe, and were introduced in the United States in the late 18th century or early 19th century. Soybeans became established as an oilseed crop in the United States in the late 1920s and attained major commercial importance during World War II. Cotton has a long history; its origins can be traced as far back as 3000 BC through spun cotton yarn found in the Indus valley. It is indigenous to many parts of the world, and its beginnings as an oilseed in the United States are associated with the invention of the cotton gin by Eli Whitney in 1794 (see Cotton). Peanuts probably originated in South America and were later introduced to Africa and Asia. Subsequent cultivation of peanuts in North America was started with plants imported from Africa (see Nuts). Sunflowers are native to North America and likely originated in the southwestern United States. They were introduced to Spain by the early Spanish explorers and then spread eastward to Russia where they became established as an oilseed crop. Sunflowers did not become a significant U.S. oilseed crop until 1967, when

Table 1. Botanical Classification, Area of Production, and Uses of Oilseeds

Common name	Botanical classification		Principal production areas	Uses
	Family	Genus and species		
soybean	Leguminosae (legume)	*Glycine max Merrill*	United States, Brazil, People's Republic of China, Argentina	edible oil, animal feed, food, edible proteins, industrial oil
cottonseed	Malvaceae (mallow)	*Gossypium arboreum*, Sri Lanka cotton; *G. herbaceum*, Levant cotton; *G. barbadense*, Sea Island cotton; *G. hirsutum*, Upland cotton	USSR, People's Republic of China, United States, India	edible oil, animal feed
peanut	Leguminosae (legume)	*Arachis hypogaea*	India, People's Republic of China, United States, Senegal, Sudan	food, edible oil, animal feed
sunflower	Compositae (composite)	*Helianthus annuus*	USSR, United States, Argentina, People's Republic of China, Romania	edible oil, animal feed, food

varieties were developed with high oil content and improved agronomic characteristics, such as increased resistance to diseases and pests.

References to oilseeds and vegetable-oil sources not included in this article are given under General References (see Vegetable oils).

Physical Characteristics

Plants and seeds of the four oilseeds vary in growth habit, size, shape, and other features. Some of their important physical characteristics are summarized below. A common feature of the structures of the four seeds is storage of the bulk of the protein and oil in distinct membrane-bound, subcellular organelles called protein bodies and lipid bodies, as illustrated for soybeans in Figure 1. Although not shown in Figure 1, the protein bodies contain inclusions referred to as globoids ($\geq 0.1\ \mu m$) that are storage sites for phytate and cations such as potassium, magnesium, and calcium (2). During germination, the contents of the storage organelles are mobilized and utilized by the growing seedling.

Soybean. ***Plants.*** Soybeans grow on erect, bushy annual plants, 0.3–2.0 m high with hairy stems and trifoliolate leaves. The flowers are white or purple or combinations of white and purple. Growing season varies with latitude (120–130 d in central Illinois).

Seeds. Seeds are produced in pods, usually containing three almost spherical to oval seeds weighing 0.1–0.2 g. Commercial varieties have a yellow seed coat plus two cotyledons, plumule, and hypocotyl–radicle. Cotyledons contain protein and lipid bodies.

Cottonseed. ***Plant.*** Cotton grows as an annual or perennial herb or shrub, sometimes treelike, 0.25–3 m or more high depending on species, with 3-, 5-, or 7-lobed leaves with white to yellow or purple-red flowers. The herbage is irregularly dotted

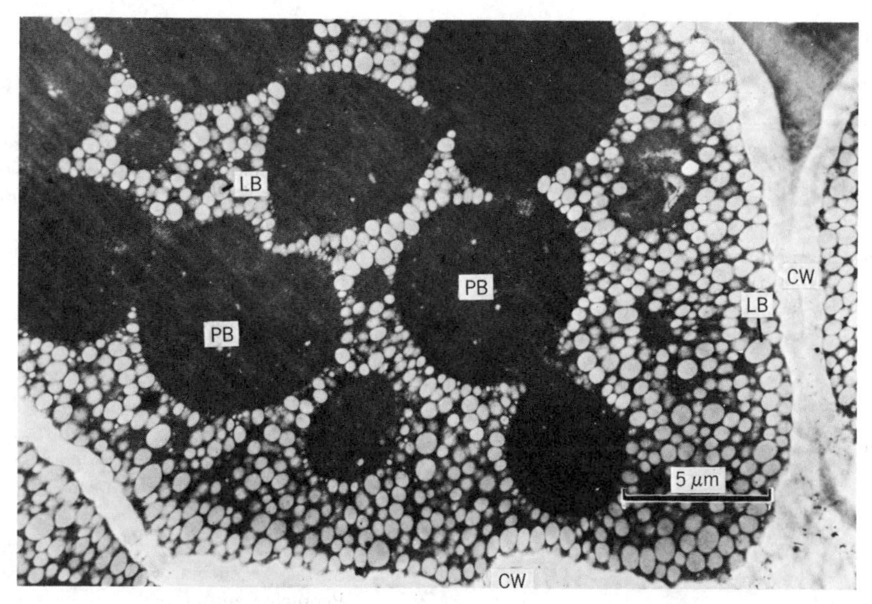

Figure 1. Transmission electron micrograph of a section of a mature, hydrated soybean cotyledon. Protein bodies (PB), lipid bodies (LB), and cell wall (CW) are identified (1).

with pigment glands. The growing season for Upland cotton grown in the United States is 120 d.

Seeds. The seeds are produced in leathery capsules (boll), covered with lint and fuzz fibers. Seeds are ovoid, 8–12 mm long with brown to nearly black seed coats. The interior consists of radicle, hypocotyl, epicotyl, and two cotyledons. Seed tissue contains protein bodies and lipid bodies. Most varieties contain pigment glands (storage site of gossypurpurin and gossypol) 100–400 μm long (3). Glandless varieties of cottonseed are available, but are not yet grown on a significant scale.

Peanut. ***Plant.*** Peanuts grow on annual herbaceous plants, bushy upright (erect) or spreading (runner) types, 25–50 cm high with bijugate leaves, with hairy stems and bright yellow flowers whose peduncles bend after fertilization and push the pods underground where they develop and ripen. Growing season is 120–140 d.

Seeds. The seeds are produced in pods containing two or three seeds. The kernels are almost spherical to roughly cylindrical (0.4–1.1 g each) and consist of a thin coat (testa) containing two cotyledons and the germ. Cotyledons contain protein bodies, lipid bodies, and starch grains.

Sunflower. Sunflowers are grown in two types. The varieties grown for oilseed production are generally black-seeded with thin seed coats that adhere to the kernels. They contain 40–50% oil and ca 20% protein. Nonoilseed varieties, sometimes referred to as confectionary, striped, or large-seeded sunflowers, have striped seed coats and relatively thick hulls that do not adhere to the kernels; they contain 20–30% oil and are usually larger than seeds of oilseed varieties.

Plant. The sunflower is an erect annual, 1–4 m high, with alternating leaves; the heads are $\leq$30 cm wide with orange-yellow rays and dark center disks. The growing season is ca 120 d.

Seeds. The sunflower seed (achene) is four-sided and flattened (ca 9 mm long by 4–8 mm wide) with a black or striped grey and black seed coat (pericarp) enclosing a kernel. The kernel contains protein and lipid bodies.

Chemical Composition

The compositions of the four oilseeds are given in Table 2. All have a high seed-coat or hull content except soybeans. Because of the high hull content, the crude-fiber content of the other seeds is also high; confectionary varieties of sunflower seed may contain up to 28% crude fiber on a dry basis (6). Soybeans differ from the other oilseeds in their high protein and low oil content. All these oilseeds, however, yield high protein meals when they are dehulled and defatted.

Proteins. The proteins found in the four oilseeds are complex mixtures consisting of four characteristic fractions with molecular weights of ca 8,000–700,000, as illustrated by the ultracentrifuge pattern for soybean proteins (Fig. 2). The 7 S and 11 S fractions usually predominate, but in sunflower seeds, the 2 S and 11 S proteins are the main fractions (8). The 7 S and 11 S fractions are considered to be storage proteins and are located in the protein bodies (9). Of the four oilseeds, the proteins of soybeans are the best characterized. The principal portion of soybean 7 S fraction, β-conglycinin, consists of at least seven isomers resulting from various combinations of three subunits (α, α', and β): $\alpha\beta_2$, $\alpha'\beta_2$, $\alpha'\alpha\beta$, $\alpha_2\beta$, $\alpha\alpha'_2$, α_3, and β_3 (10–11). Based on the sizes of the subunits, the β-conglycinin isomers have molecular weights between 126,000 and 171,000 (see Polypeptides; Proteins).

Table 2. Compositions[a] of Oilseeds, Percent[b]

Oilseed	Hulls	Oil	Protein[c]	Ash	Protein in dehulled, defatted meal[d]
soybean	8	20	43	5.0	52
cottonseed					
acid delinted	36	21.6	21.5	4.2	
kernels		36.4	32.5	4.7	63
peanut	20–30				
kernels	2–3.5[e]	50.0	30.3	3.0	57
sunflower					
arrowhead variety, low oil type[f]	47	29.8	18.1		67
armavirec variety, high oil type[f]	31	48.0	16.9		60
kernels, armavirec variety, high oil type[f]		64.7	21.2		

[a] Approximate; moisture-free basis.
[b] Ref. 4, except as noted otherwise.
[c] N × 6.25.
[d] Data vary with efficiency of dehulling and oil extraction, variety of seed, and climatic conditions during growth.
[e] Red skins or testa.
[f] Ref. 5.

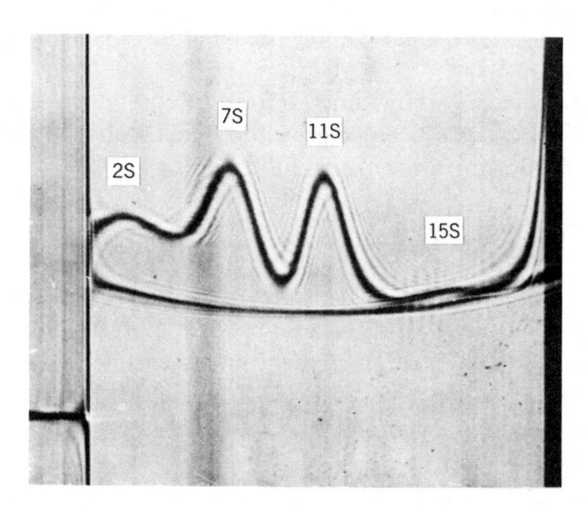

Figure 2. Ultracentrifugal pattern for the water-extractable proteins of defatted soybean meal in pH 7.6, 0.5 ionic strength buffer. Numbers across the top of the pattern are sedimentation coefficients in Svedberg units (S). Molecular weight ranges for the fractions are 2 S: 8,000–50,000; 7 S: 100,000–180,000; 11 S: 300,000–350,000; 15 S: 700,000 (7).

The 11 S molecule, also called glycinin, is more complex than β-conglycinin. It consists of ca 12 polypeptides, half of which have acidic isoelectric points (mol wt ca 37,000–44,000) and half of which have basic isoelectric points (mol wt ca 17,000–22,000); six different acidic and five different basic polypeptides have been separated and identified. Some acidic and basic subunits are linked in nonrandom pairs through disulfide bonds. They are apparently synthesized as single-chain precursors that are subsequently modified by proteolysis to form the acidic and basic polypeptide chains (12–13). Arachin, the 11 S-like protein from peanuts, contains six acidic subunits and six basic subunits per molecule, but they do not appear to be cross-linked by disulfide bonds (14).

In addition to the storage proteins, the oilseeds contain a variety of minor proteins including trypsin inhibitors, hemagglutinins, and enzymes (eg, urease and lipoxygenase in soybeans).

Amino acid compositions of the four oilseeds are given in Table 3 along with the essential amino acid pattern for a high-quality protein that meets human requirements as established by the Food and Nutrition Board of the NRC (see Amino acids). Soybeans meet or exceed the reference pattern except for valine and the sulfur amino acids, which are only slightly low, whereas the other proteins are deficient in four to five amino acids including lysine, threonine, valine, isoleucine, and leucine (see Amino acids).

Lipids. Representative fatty acid compositions of the unprocessed triglyceride oils found in the four oilseeds are given in Table 4 (see Fats and fatty oils). Cottonseed, peanut, and sunflower oils are classified as oleic–linoleic acid oils because of their high content (>50%) of these fatty acids. Although the oleic and linoleic acid content of soybean oil is high, it is distinguished from the others by a content of 4–10% linolenic acid, and hence is called a linolenic acid oil. The distribution of the different fatty acids on the glycerol molecule in the various seed oils is given in ref. 22.

Table 3. Ammonia and Amino Acid Composition of Defatted Oilseed Meals, g/16 g Nitrogen

Amino acid	Cottonseed[a]	Peanut[b]	Soybean[c]	Sunflower[d]	FNB pattern[e]
lysine	4.4	3.4	6.4	3.8	5.1
histidine	2.7	2.4	2.6	2.5	1.7
ammonia	2.1	1.7	1.9	2.2	
arginine	11.6	12.0	7.3	8.9	
aspartic acid	9.2	13.0	11.8	8.7	
threonine	3.0	2.5	3.9	3.2	3.5
serine	4.2	5.2	5.5	3.9	
glutamic acid	21.7	20.6	18.6	21.0	
proline	3.6	5.1	5.5	5.0	
glycine	4.1	6.6	4.3	5.1	
alanine	3.9	3.8	4.3	4.1	
valine	4.5	3.1	4.6	4.8	4.8
cystine	2.6	2.5	1.4	1.8	2.6 (cystine + methionine)
methionine	1.5	1.1	1.1	1.9	
isoleucine	3.1	2.3	4.6	4.0	4.2
leucine	5.8	6.2	7.8	6.1	7.0
tyrosine	3.1	3.6	3.8	2.7	7.3 (tyrosine + phenylalanine)
phenylalanine	5.4	5.0	5.0	4.7	
tryptophan	1.2	1.0	1.4	1.1	1.1

[a] Means for eight glanded seed varieties (15).
[b] Means for 16 varieties (16) except for tryptophan (17).
[c] Means based on 32 hydrolysates except for proline, cystine, and tryptophan (18).
[d] Means for 7 varieties (5).
[e] Essential amino acid pattern (Food and Nutrition Board pattern) for high-quality protein for humans (19). An essential amino acid cannot be synthesized by an organism at a sufficiently rapid rate to meet metabolic needs.

Table 4. Fatty Acid Composition of Unprocessed Oilseed Oils, Percent

Carboxylic acid[a]	Cottonseed[b]	Peanut[c]	Soybean[b]	Sunflower[b]
Saturated fatty acids				
10:0	0.48			
12:0	0.38		0.10	
14:0	0.79		0.16	0.1
16:0	22.0	10.5	10.7	5.81
18:0	2.24	3.2	3.87	4.11
20:0	0.19	1.4	0.22	0.29
22:0		2.1		0.61
24:0		0.7		
Unsaturated fatty acids				
16:1	0.78		0.29	0.10
18:1	18.1	50.3	22.8	20.7
18:2	50.3	30.6	50.8	63.5
18:3	0.40		6.76	0.32
20:1		1.0		0.10

[a] See Carboxylic acids for nomenclature (eg, number of carbons, 10; number of double bonds, 0; etc).
[b] Ref. 20.
[c] Mean value for 1968 crop of 82 peanut genotypes (21).

In addition to the triglycerides, the four oilseeds also contain phosphatides. Soybean phosphatides consist mainly of phosphatidylcholines, phosphatidylethanolamines, phosphatidylinositols, and phosphatidylserines (23). Generally, ca one-half the phosphatides are extracted from the oilseeds with hexane. Total phosphatide contents of cottonseed and peanut kernels are estimated to be 1.5–1.9 and 0.8%, respectively (24).

Sterols are present in concentrations of 0.2–0.4% in oilseed oils; their compositions are given in Table 5.

The sterols exist in the seeds in four forms: free, esterified, nonacylated glucosides, and acylated glucosides. Soybeans contain a total of 0.16% of these sterol forms in the ratio of ca 3:1:2:2 (26) (see Steroids).

Carbohydrates. Oilseeds contain two types of carbohydrates (qv): soluble mono-

Table 5. Composition of Oilseed Sterols[a], Percent

Oil	Campesterol	Stigmasterol	β-Sitosterol	Δ^5-Avenasterol
soybean	20	20	53	3
cottonseed	4	1	93	2
peanut	15	9	64	8
sunflower	8	8	60	4

[a] Ref. 25.

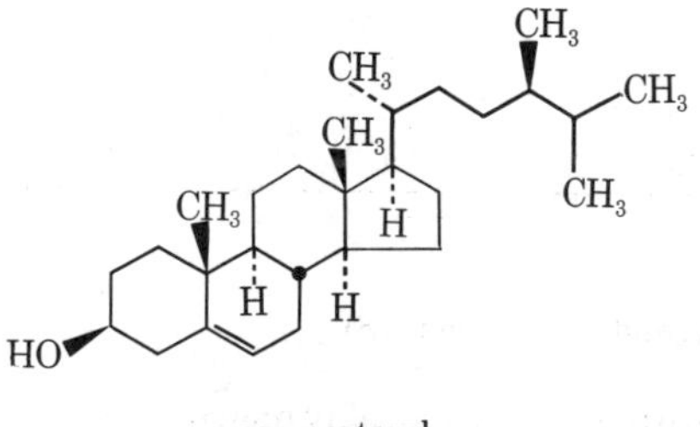

campesterol
(ergost-5-en-3-ol)

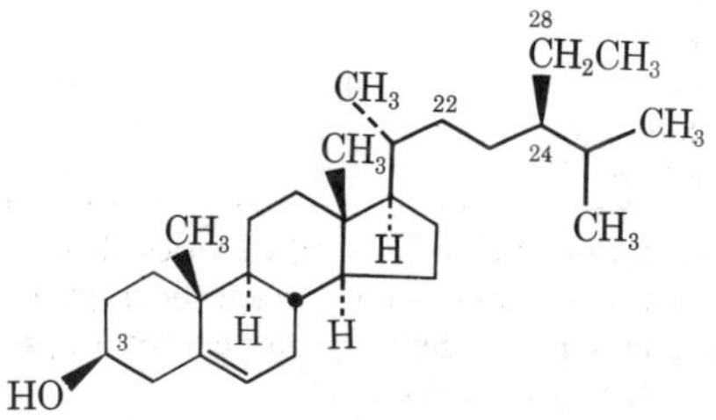

β-sitosterol
(stigmast-5-en-3-ol)
stigmasterol is stigmasta-5,22-dien-3β-ol.
Δ^5-avenasterol is stigmasta-5,24(28)-dien-3β-ol.

and oligosaccharides and insoluble polysaccharides. The contents of soluble sugars and total carbohydrates in the defatted oilseed meals are given in Table 6; values for intact seeds are different because of their oil and seed-coat contents. Sucrose, raffinose, and stachyose are the principal soluble sugars present. The polysaccharide (not including crude-fiber) content is roughly equal to the total carbohydrates minus total soluble sugars, and it ranges from 11 to 19% of the flours.

Minor Constituents. All four oilseeds contain minor components that affect the defatted seeds, especially when used for feed and food. Contents of phytic acid, for example, are given below.

Seed	*Phytic acid, %*	*Ref.*
soybeans	1.0–1.5	28
cottonseed kernels	2.2–3.8	24
peanut kernels	0.8	24

Gossypol and related pigments found in cottonseed affect nutritional properties and the color of the oil and meal. Kernels contain 1.1–1.3% gossypol, whereas hexane-defatted flours from the same varieties contain 1.2–2.0% gossypol (15).

Cottonseed also contains the cyclopropenoid fatty acids, malvalic and sterculic acids, as glycerides that are largely extracted with hexane. Crude oils from 22 commercial cottonseed varieties contained 0.6–1.0% of these acids (29), whereas 11 commercially refined cottonseed oils had values of 0.04–0.42% (30).

Soybeans contain only small amounts of phenolic acids, but these compounds are significant constituents of sunflower seeds. Defatted sunflower meal contains 2.7% chlorogenic acid, 0.38% quinic acid, and 0.2% caffeic acid (31). Chlorogenic acid turns from yellow to green and finally to brown as the pH is raised from 7 to 11. These chromophoric properties seriously limit the use of sunflower protein preparations in foods (32) (see Alkaloids; Coffee).

The isoflavone glucosides genistin, daidzin, and glycetein-7-β-glucoside, plus small amounts of the corresponding aglycones, are constituents of soybeans. Isoflavone contents range from 0.047 to 0.36% (33).

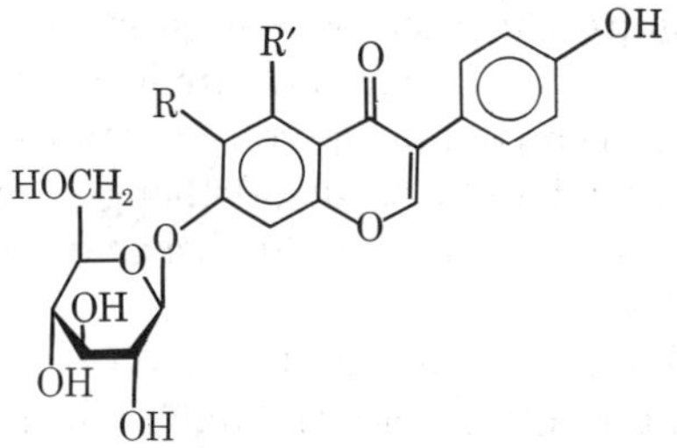

R′ O OH R HOCH$_2$ O O OH OH OH

daidzin, R = R′ = H
genistin, R = H, R′ = OH
glycetein-7β-glucoside, R = OCH_3, R′ = H

Soybeans and peanuts also contain saponins (34). Recent results indicate that saponin content of soybeans is 5.6% and that of defatted soy flour 2.2–2.5% (35); these values are ca tenfold higher than those obtained in earlier investigations.

Harvesting and Storage

Soybeans. The U.S. soybean crop normally is harvested in September or October. Ideal moisture content for harvesting is 13%, and the crop can be successfully stored at this moisture content until the following summer. Soybeans at $\leq$12% moisture can be stored for 2 yr or more with no significant deterioration. Beans with moisture above safe storage limits are dried or placed in aerated bins for gradual moisture reduction.

Table 6. Carbohydrate Contents of Defatted Oilseed Flours[a], Percent

Constituent	Cottonseed		Peanut	Soybean	Sunflower
	Deglanded[b]	Glandless			
Soluble sugars					
glucose	trace	trace	2.12	trace	0.60
sucrose	2.41	2.62	7.70	7.80	2.29
trehalose					0.79
raffinose	7.93	11.95	trace	1.25	3.22
stachyose	0.95	0.68		6.30	
Total	*11.29*	*15.25*	*9.82*	*15.35*	*6.90*
Total carbohydrate[c]	*22.5*	*26.8*	*22.4*	*34.0*	*24.2*

[a] Ref. 27.
[b] Prepared by liquid-cyclone process.
[c] Obtained by difference: 100 − (protein + oil + ash + crude fiber) = nitrogen-free extract.

Soybeans are trucked from the farm to country elevators. From there they are moved by truck or rail to processing plants, subterminal elevators, or terminal elevators. From the terminal elevators, the beans are shipped by rail or barge to export elevators or processors.

Soybeans are stored in concrete silos 6–12 m dia with heights of ≥46 m. The silos are often arranged in multiple rows, and the resulting interstitial silo areas are likewise used for storage. In bulk storage, seasonal temperature changes cause variations in temperature between the different portions of the grain mass (eg, in the winter, soybeans next to the outer walls are colder than those in the center of the silo). Such temperature differences initiate air currents that transfer moisture from warm to cold portions of the seed mass. Thus, bulk soybeans originally at safe moisture concentrations may, after storage, have localized regions of higher moisture that cause growth of microorganisms, which in turn can lead to heating. Under these conditions, the beans turn black and may even ignite. Such seasonal moisture transfer also occurs with other oilseeds. In commercial practice the temperature is carefully monitored and, when it rises, the beans are either remixed or processed. Aflatoxin contamination is not a problem as it is with cottonseed and peanuts. Although fungi invade soybeans stored at high moisture and temperatures, *Aspergillus flavus* does not grow well on soybeans and aflatoxins are negligible.

Cottonseed. In the United States, cotton harvesting begins in late July when the ripened boll bursts and the cotton dries and fluffs. The harvest is usually completed by the end of December. After picking, the cotton is processed in gins to separate the lint from the cottonseed. Moisture, temperature, initial quality, previous history, and length of storage determine how long cottonseed can be held before processing; moisture is the most important factor. For safe storage, moisture should be <10%. Cottonseed is usually stored in Muskogee-type warehouses (low, flat metal buildings with roofs that slope close to the angle of repose of the cottonseed) that are equipped with aeration and temperature-monitoring systems. High temperatures cause rapid deterioration of the seed through formation of free fatty acids, hence the need for ventilating fans and aeration ducts when the temperature rises. At high moisture and temperatures of 28–37°C, cottonseed is also prone to invasion by *Aspergillus flavus*, a fungus that produces aflatoxins (36). Aflatoxins are highly toxic compounds that produce liver cancer in animals (see also Food toxicants, naturally occurring).

Peanuts. When the kernels are fully developed and are taking on a mature color, the plants are dug mechanically, shaken to remove the soil, and inverted into windrows to dry (cure) and mature completely. Ideally, the peanuts are left to cure for several days until the moisture content drops to ca 10%; then they are harvested mechanically. Green harvesting is practiced under adverse weather conditions yielding peanuts with 18–25% moisture; artificial drying reduces the moisture to ca 10%. After the moisture is equilibrated between the kernels and hulls, the former contain 7–8% moisture which is safe for storage.

The cured peanuts are stored or shelled; <10% of the U.S. crop is retailed in the shell. Shelled peanuts should have a moisture content of ca 7% (6.5% is optimum) for safe storage and are best stored under refrigeration. Moisture control is critical to maintain quality. At high moisture, molds grow, including *Aspergillus flavus* which produces aflatoxins. The Agricultural Marketing Service and the FDA limit aflatoxin to 25 ppb in raw peanuts and 20 ppb in processed peanut products.

Sunflowers. In the northern United States, sunflower seeds are harvested in late September and October, ca 120 d after planting. Harvesting is frequently delayed until after a killing frost to speed drying. In rainy fall conditions, seed may contain >20% moisture. Such seed must be dried rapidly to ≤10% moisture. The crop dries easily because air readily passes through beds of the large seed. A moisture content of 9.5% is safe for short-term storage, but ≤7% is recommended for long-term storage without aeration (37). Marketing channels lead from the farm to country elevators to processors and export terminals. Since most of the United States production is in North Dakota, Minnesota, and South Dakota, the bulk of the seeds is shipped overseas from Duluth, Minn.

Processing

Cottonseed. Cottonseed is converted into oil and meal by hydraulic pressing, screw-pressing, prepress–solvent extraction, or direct solvent extraction. In the United States, almost all cottonseed is processed by the last three techniques (see Fig. 3). First, the seed is cleaned to remove sticks, stones, leaves, and other foreign materials. Second, the cotton fibers remaining after ginning are removed mechanically (delinting). Next, the hulls are split and removed, and the separated kernels are passed between smooth rolls to form flakes. The flakes may then be extracted with hexane to remove the oil. Alternatively, some processors remove the oil mechanically by screw-pressing or by a combination of prepressing, which removes 50–85% of the oil, and solvent extraction. Before screw-pressing, and sometimes before solvent extraction, the flakes are cooked to facilitate removal of the oil. Screw-pressing yields a cake containing 2.5–4.0% oil; the cake is ground into a meal, and ground cottonseed hulls are blended back to adjust protein content to trading standards. Meal emerging from solvent extractors is freed of solvent by heating; the resulting meal contains about 1% residual oil (see Extraction, liquid–solid extraction).

Screw-pressed oil is allowed to stand to settle out suspended solids, filtered through a plate filter press, and then pumped to storage. The oil-rich solvent (miscella) from the solvent-extraction process is filtered or clarified, and most of the solvent is removed in a long tube evaporator. Finally, the concentrated oil passes through a stripping column where sparging steam is introduced to remove the residual solvent.

A metric ton of cottonseed yields ca 91 kg linters, 247 kg hulls, 162 kg oil, and 455 kg meal.

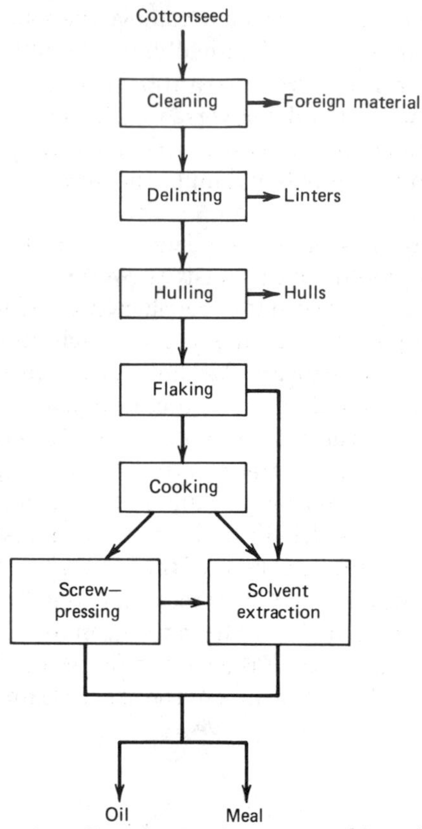

Figure 3. Schematic outline of processing cottonseed into oil and meal.

Peanuts. Only 10–15% of the U.S. peanut crop is converted into oil and meal. Processing is carried out by screw-pressing or prepressing followed by solvent extraction. For screw-pressing, the peanuts are shelled, cooked, and pressed to yield a crude oil plus a cake containing ca 5% residual oil. The cake is ground, and the ground peanut hulls are blended back to adjust protein content. In prepressing–solvent extraction, the cooked meats are screw-pressed at low pressure to remove a portion of the oil and then extracted with hexane to reduce the residual oil to ca 1%. Residual hexane in the meal is recovered by applying jacket or live steam in a so-called desolventizer (equipment for vaporizing and removing solvent). Hexane in the miscella is recovered by evaporation as in the processing of cottonseed.

A metric ton of peanuts yields ca 317 kg oil and 418 kg meal; the remainder is shells and foreign matter.

Soybeans. Virtually all soybeans processed in the U.S. are solvent-extracted (Fig. 4). Beans arriving at the plant are cleaned and dried, if necessary, before storage. When the beans move from storage to processing, they are cleaned further, cracked to loosen the seed coat or hulls, dehulled, and then conditioned to 10–11% moisture. The conditioned meats are flaked and extracted with hexane to remove the oil. Hexane and oil in the miscella are separated by evaporation and the hexane is recovered. Residual hexane in the flakes is removed by steam treatment in a desolventizer–toaster. The moist heat treatment inactivates antinutritional factors (trypsin inhibitors) in the raw flakes and increases protein digestibility (38–39).

A metric ton of soybeans yields ca 180 kg oil and 790 kg meal.

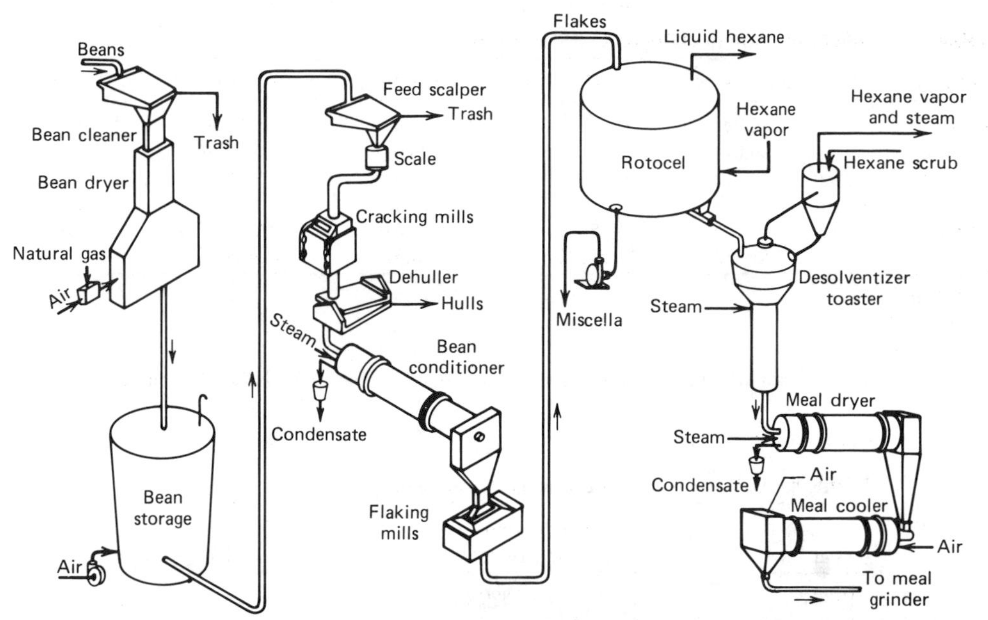

Figure 4. Schematic outline for processing soybeans into oil and meal by solvent extraction. Courtesy of Dravo Corp.

Sunflowers. Like cottonseed and peanuts, sunflower seeds are processed by screw-pressing, solvent extraction, or both. In screw-pressing and prepress–solvent extraction, 50–75% of the hulls are removed before pressing. The cake obtained after pressing is hammer-milled or broken up by passing through corrugated rolls to yield expeller meal. For prepress–solvent extraction, the technique preferred in the United States, the expeller meal is conditioned, flaked, and extracted with hexane. In some new plants the sunflower hulls are separated and burned to generate part of the power used in processing.

When processed without dehulling, a metric ton of sunflower seed yields ca 400 kg oil and 550 kg high fiber meal.

Economic Aspects

Soybeans are the predominant oilseed crop in the world, ca 50% of the total oilseed production, followed by cottonseed, peanuts, and sunflower seed (Table 7). Soybean production in the United States has shown a phenomenal growth since the crop became established in the 1920s and contributes significantly to the agricultural economy (41) (see Table 8).

Average U.S. prices of the four oilseeds and derived products for 1976–1980 are summarized in Table 9.

The United States is the principal producer of soybeans, India is the largest grower of peanuts, and the USSR leads in the production of sunflower seed and shares leadership in cottonseed production with the People's Republic of China (see Table 10).

Soybeans are the most important oilseed in international trade; the United States, Brazil, and Argentina are the main suppliers to the export markets. In 1980–1981,

Table 7. World Production of Oilseeds[a], 1000 metric tons

Oilseed	1977–1978	1978–1979	1979–1980	1980–1981[b]	1981–1982[c]
soybean	72,041	77,406	93,622	80,939	88,752
cottonseed	25,012	23,940	25,025	25,576	27,524
peanut	16,528	17,601	16,944	16,677	18,297
sunflower seed	12,876	12,901	15,434	13,248	14,226
others[d]	20,523	22,780	22,519	23,454	24,629
Total	*146,980*	*154,628*	*173,544*	*159,894*	*173,428*

[a] Ref. 40.
[b] Preliminary.
[c] Forecast.
[d] Includes rapeseed, sesame seed, safflower seed, flaxseed, castor beans, copra, and palm kernel.

Table 8. U.S. Soybean Production and Crop Value

Year	Production, 1000 metric tons	Crop value, $\$10^6$
1930	379	19
1940	2,124	70
1950	8,144	738
1960	15,107	1,185
1970	30,675	3,205
1980	48,772	13,825

Table 9. Average U.S. Prices of Oilseeds and Derived Products, $/t[a]

Product	1976–1977	1977–1978	1978–1979	1979–1980	1980–1981
Soybeans					
seed	267	228	262	240	274
meal	219	181	210	201	239
oil	525	539	603	535	502
Cotton					
seed		78	126	133	138
meal		154	182	181	221
oil		560	697	560	573
Peanuts					
seed	456	474	463	463	502
Sunflowers					
seed	243	224	236	200	238
meal			136	100	125
oil	243		728	560	575

[a] Refs. 42–43.

United States soybean exports of 19.7×10^6 metric tons represented 41% of the domestic crop, 24% of the world crop, and 78% of the total soybeans traded internationally. The European Common Market and Japan are the largest importers of soybeans (44). Japan produces <5% of its soybean needs; it is the largest single soybean customer of the United States, importing ca 4×10^6 t annually (45). The United States

Table 10. World Oilseed, Oil, and Meal Production, 1980–1981[a], 1000 metric tons

Country	Cottonseed			Peanuts			Soybeans			Sunflowers		
	Seed	Oil	Meal	Seed	Oil	Meal	Seed	Oil	Meal	Seed	Oil	Meal
Argentina							3,500	149	683	1,260	425	540
Brazil				310	54	74	15,500	2,590	10,584			
People's Republic of China	5,269	590	1,715	3,600	576	693	7,880			908		
India	2,700	200	1,001	6,000	1,412	2,045						
Pakistan	1,422	148	488									
Romania										650	233	310
Senegal				450	58	68						
USSR	5,500	691	2,272							4,652	1,611	1,711
Sudan				800	122	135						
United States	3,955	602	1,733	1,047	63	84	48,772	5,112	22,055	1,748	298	439
others	6,967	971	1,986	4,524	687	468	5,290	4,750	22,873	3,873	2,080	3,000
World total	*25,542*	*3,202*	*9,195*	*16,731*	*2,972*	*3,567*	*80,942*	*12,711*	*56,195*	*13,091*	*4,647*	*5,515*

[a] Seed figures include northern hemisphere crops harvested in late 1980 plus southern hemisphere crops harvested in early 1981 (44).

also exports processed soybean products, eg, oil and meal; more than half of the U.S. crop is exported as whole beans and processed products.

U.S. exports of the other oilseeds are smaller and follow different patterns (42). Exports account for 1–3% of the cottonseed crop and 50–60% of the processed oil production. Of the peanut crop ca one half is consumed as whole nut products and 20–30% are exported. The bulk of the sunflower-seed crop is exported; in 1980–1981, 1.5×10^6 t (85% of the crop) was shipped overseas, primarily to Europe. Domestic processing plants are under construction, however, and interest in sunflower-seed oil is increasing.

About 94 mills with an annual capacity of 39×10^6 t process soybeans in the United States (46) and operate at an average of ca 80% of capacity.

Although soybeans contribute about one-half the world's production of oilseeds, they supply only one-third the total edible vegetable fats and oils (Table 11) because of their relatively low oil content. Nonetheless, the production of soybean oil exceeds the production of cottonseed, peanut, and sunflower-seed oils.

The bulk of the oil obtained from the four oilseeds is consumed in food products. For many uses, the different oils can be substituted for each other; consequently, the proportions used in a product may depend upon small price differences.

In 1980, consumption of edible fats and oils, in thousand metric tons, was distributed as follows (47): soybean, 4,195; cottonseed, 322; peanut, 68; sunflower, 125; total, 6,985. For 1980, estimates of the largest domestic uses, in thousand metric tons, were as follows (42):

Product	*Soybean*	*Cottonseed*	*Peanut*
salad and cooking oil	1833	209	69
shortening	1207	86	
margarine	749	11	
Total	*3789*	*306*	*69*

Industrial (nonfood) utilization of vegetable oils is much smaller than food usage. Usage of U.S. soybean oil (6% of total production) in 1976 is given in Table 12 (48).

Free substitution of protein meals in feeds is much more restricted than interchange of oils. Because of its good balance of essential amino acids, soybean meal is an indispensible ingredient for efficient feeding of nonruminants, eg, poultry and swine. Soybeans thus provide almost 65% of the world's protein meals (including fish meal)

Table 11. World Production of Vegetable Oils[a], 1000 metric tons

Oil	1977–1978	1978–1979	1979–1980	1980–1981[b]	1981–1982[c]
soybean	10,854	11,696	14,374	12,223	13,464
cottonseed	3,182	3,011	3,193	3,206	3,503
peanut	3,083	3,311	3,068	2,966	3,240
sunflower seed	4,719	4,705	5,581	4,781	5,120
others[d]	13,338	14,547	14,822	16,213	16,472
Total	*35,176*	*37,270*	*41,038*	*39,389*	*41,799*

[a] Ref. 40.
[b] Preliminary.
[c] Forecast.
[d] Includes rapeseed, sesame seed, safflower seed, olive, corn, coconut, palm kernel, palm, and babassu.

Table 12. Nonfood U.S. Use Distribution of Soybean Oil, 1976

Product	1000 metric tons
paint and varnish	38
resins and plastics	37
other drying-oil products	2
fatty acids	12
others (includes soap)	15
foots[a] and losses	126
Total	*230*

[a] Material precipitated in refining fatty oils with alkali; also called soap stock.

(Table 13). Of the 22.1×10^6 t soybean meal produced in the United States in 1980–1981, 16.3×10^6 t was used in feeds and the rest was exported. Poultry consumed 7.5×10^6 t, and swine used 4.8×10^6 t (49) (see Pet and livestock feeds).

The edible-oilseed-protein industry is in its early stages and has not developed as was predicted ca 1970. Data on production of edible peanut flour are unavailable. Production estimates for edible soybean proteins in the United States in 1980 and wholesale prices as of February 1982, were as follows (50):

Product	*1000 metric tons*	*$/kg*
flours and grits	261	0.27
protein concentrates	35	0.90
textured flours	41	0.64
textured concentrates		1.15
protein isolates	33	2.42

Estimates of the amounts of soybeans used for traditional soy foods in the United States in 1981 were as follows (51):

Product	*Metric tons*
soy sauce	13,815
tofu and related products	8,236
soy milk, including products made with soybean protein isolates	7,347
miso	412
tempeh	256

Table 13. World Production Oilseed Protein Meals[a], 1000 metric tons

Protein meal	1977–1978	1978–1979	1979–1980	1980–1981[b]	1981–1982[c]
soybean	49,778	53,604	65,612	56,011	61,614
cottonseed	9,109	8,644	9,136	9,206	10,024
peanut	3,699	3,973	3,682	3,560	3,887
sunflower	4,406	4,418	5,276	4,562	4,892
others[d]	14,120	15,446	15,125	15,559	16,086
Total	*81,112*	*86,085*	*98,831*	*88,898*	*96,503*

[a] Ref. 40.
[b] Preliminary.
[c] Forecast.
[d] Includes rapeseed, sesame seed, safflower seed, linseed, copra, palm kernel, and fish meal.

Nutritional Properties and Antinutritional Factors

Oil. Because of their high linoleic acid contents, unhydrogenated and partially hydrogenated cottonseed, peanut, soybean, and sunflower oils are good sources of this essential fatty acid. Soybean oil is the principal vegetable oil consumed (4.195×10^6 t in 1980), and ca three fourths is partially hydrogenated to impart high temperature stability to cooking oil, extend shelf life, and improve flavor stability and physical and plastic properties. Linoleic and linolenic acid contents of soybean oil are reduced by hydrogenation, but more important from a nutritional viewpoint are the migration of double bonds up and down the carbon chain and the conversion of cis to trans isomers, ie, positional and geometrical isomerization. Although long-term studies with rats and short-term tests with humans have failed to reveal toxic effects on ingestion of partially hydrogenated soybean oil, this complex problem is under active investigation (52).

Heating and oxidation of fats, especially under severe conditions, results in the formation of a variety of compounds including hydrocarbons, cyclic hydrocarbons, alcohols, cyclic dimeric acids, and polymeric fatty acids. Some of these compounds are toxic, but the present consensus is that an oil such as soybean oil is safe and nontoxic when used under normal cooking conditions (52).

Cyclopropenoid fatty acids found in cottonseed oil are biologically active in several animal species. When ingested by the laying hen, these fatty acids are deposited in the egg yolks. In storage, the yolks become rubbery and the whites turn pink (53). In rainbow trout, cyclopropenoid fatty acids act as synergists with aflatoxins and as liver carcinogens (54). The long history of cottonseed oil in the human diet, however, has not revealed any adverse effects; hence, there is presumptive evidence that humans are not affected at past and present levels of ingestion (55). The cyclopropenoid acid content of crude cottonseed oil is lowered by processing; the largest decrease occurs during deodorization (30).

Proteins and Meals. Nutritional properties of the oilseed protein meals and their derived products are determined by the amino acid compositions, content of biologically active proteins, and various nonprotein constituents found in the defatted meals. Phytic acid, which is common to all four meals, is believed to interfere with dietary absorption of minerals such as zinc, calcium, and iron (56) (see Food toxicants, naturally occurring; Mineral nutrients).

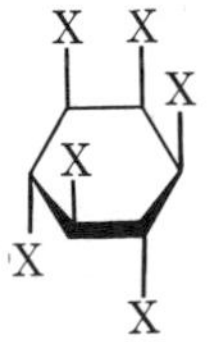

X X X X X X

phytic acid, X = $\mathrm{O\overset{\displaystyle O}{\overset{\|}{P}}(OH)_2}$

Cottonseed. When compared with an ideal amino acid pattern (Table 3), cottonseed proteins are low in lysine, threonine, isoleucine, and leucine. These deficiencies can be minimized by blending cottonseed flour with cereal flours as exemplified by Incaparina, a blend of cottonseed flour, corn flour, sorghum flour, and torula yeast. When cottonseed is treated with moist heat, the ϵ-amino group of lysine and gossypol form a derivative that is biologically unavailable.

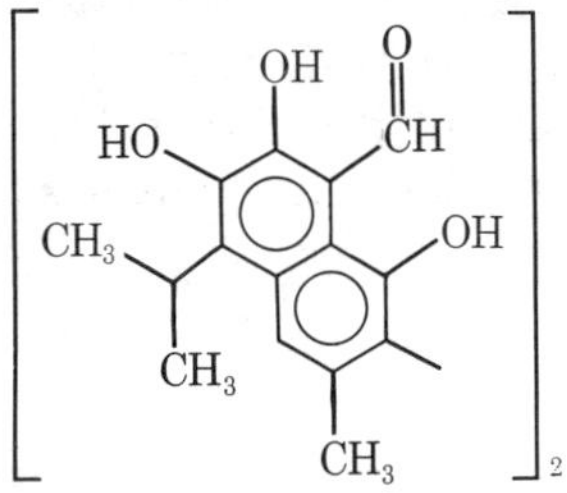

gossypol

Gossypol has other adverse effects; it is toxic to monogastric animals and, when present in rations of laying hens, causes yolk discoloration (3,53). Ruminants are generally resistant to gossypol toxicity, but ingestion by dairy cows of large amounts of free gossypol resulted in the formation of polyphenolic pigment in the blood plasma and liver, hemolysis of red blood cells, and other physiological alterations (57). Cottonseed flour containing gossypol has been fed to humans in a number of studies, but no instances of toxicity have been reported (3). The FDA limits the content of free gossypol (the portion extractable with acetone:water, 70:30 (v/v)) in edible cottonseed flour to 450 ppm. Gossypol content of cottonseed flour can be lowered by liquid (58) or air classification (59) to remove the pigment glands, but neither approach is used commercially. Glandless gossypol-free cottonseed is produced on a limited scale near Lubbock, Texas; the roasted kernels are used in breads and confection items.

The cyclopropenoid fatty acids, malvalic acid and sterculic acid, exist as glycerides in the seed. Extraction with n-hexane reduces their concentration from ca 0.05–0.2% in the seed (15,29) to 0.002–0.008% in cottonseed meal (60). Glandless cottonseed kernels contain the normal concentrations of these acids (15). In rainbow trout, the cyclopropenoid acids cause cancer of the liver either alone or by acting synergistically with aflatoxin B_1. However, similar effects in mammals or humans have not been demonstrated (54).

Peanuts. Like cottonseed proteins, peanut proteins tend to be low in lysine, threonine, isoleucine, and leucine (Table 3). The nutritive value of cornmeal is improved by blending with peanut flour and still more by soy flour (61). Raw peanuts contain ca one-fifth the trypsin inhibitor found in raw soybeans, but this concentration is high enough to cause hypertrophy (enlargement) of the pancreas in rats. Wet-heating inactivates the inhibitor but has little if any effect on the nutritional value of peanut flour (62).

Soybeans. Numerous studies have demonstrated that methionine is the first limiting amino acid in soybean proteins; ie, it is in greatest deficit for meeting the nutritional requirements of a given species. Although it is common practice to add synthetic methionine to broiler feed to compensate for this deficiency, there is growing evidence that this is not necessary when soy proteins are fed to humans, with the possible exception of infants (63). The presence of trypsin inhibitors in soybeans is well-documented, and when ingested, their primary physiological effect is to enlarge the pancreas (39). They are largely inactivated by moist heat, and there are no documented cases where ingestion of soybean proteins by humans has affected the pancreas. Long-term effects of ingestion of soy products by rats are under study (64).

Sunflower Seed. Compared to the ideal protein for humans, sunflower proteins are low in lysine and leucine and are borderline in threonine and isoleucine content (Table 3). Because of their low lysine content, sunflower proteins are not suitable for blending with cereals but can be used with legume and animal proteins (65). Heating before oil extraction improves the nutritive value of the defatted meal (66); heat-labile antinutritional factors may be present, but none has been conclusively identified to date (32).

Oilseed Products and Uses

Oil. Most crude oil obtained from oilseeds is processed further and converted into edible products. Only a small fraction of the total oil from cottonseed, peanuts, soybeans, and sunflower seed is utilized for industrial (nonedible) purposes.

Edible Oil. For edible purposes, oilseed oils are processed into salad and cooking oils, shortenings, and margarines (67). These products are prepared by a series of steps as shown for soybean oil in Figure 5.

Degumming removes the phosphatides and gums, which are refined into commercial lecithin (qv) or returned to the defatted flakes just before the solvent-removal and toasting step. Next, free fatty acids, color bodies, and metallic prooxidants are removed with aqueous alkali. Some processors omit the water-degumming step and remove the phosphatides and free fatty acids with alkali in a single operation. High vacuum steam distillation in the deodorization step removes undesirable flavors to yield a product suited for salad oil. Partial hydrogenation, under conditions where linolenate is selectively hydrogenated, results in an oil with greater stability to oxidation and flavor deterioration. After winterization (cooling and removal of solids that crystallize in the cold), the product is suited for use as salad and cooking oils. Alternatively, soybean oil can be hydrogenated under selective or nonselective conditions to increase its melting point and produce hardened fats. Such a partially hydrogenated soybean oil, by itself or in a blend with other vegetable oils or animal fats, is used for shortening and margarine. Blends of soybean or other oils of varying melting point ranges are utilized to obtain desired physical characteristics, eg, mouth feel and plastic melting ranges, and the least expensive formulation.

Polyunsaturated fatty acids in vegetable oils are subject to oxidative deterioration. Linolenic esters in soybean oil are particularly sensitive to oxidation; even a slight degree of oxidation, commonly referred to as flavor reversion, results in undesirable flavors, eg, beany, grassy, painty, or fishy. Oxidation is controlled by the exclusion of metal contaminants (iron and copper), addition of metal inactivators (citric acid), minimum exposure to air, protection from light, and selective hydrogenation to decrease the linolenate content to ca 3% (67).

Nonfood Uses. Vegetable oils are utilized in a variety of nonedible applications, but only ca 6–7% of U.S. soybean oil production is used for such products. Soybean oil can be converted into alkyd resins (qv) for protective coatings, plasticizers, dimer acids, surfactants (qv), and a number of other products (67).

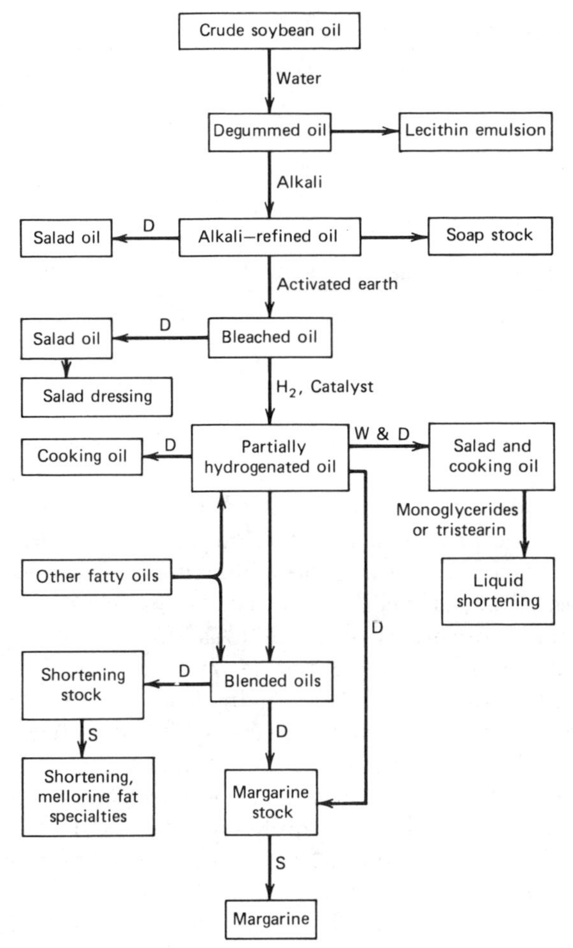

Figure 5. Schematic outline for manufacture of edible soybean oil products. D = deodorization, W = winterization, and S = solidification (68). Courtesy of the American Soybean Association and the American Oil Chemists' Society.

Protein Products. Most of the meal obtained in processing of oilseeds is used as protein supplements in animal feeds. Only in the last two decades have appreciable amounts been converted into products for human consumption, and these have been almost exclusively derived from defatted soybean flakes.

Feeds. Because of their high content of protein, oilseed meals are essential ingredients of poultry and livestock feeds. Proximate compositions are shown in Table 14. Because of its gossypol content, high fiber, and low lysine content, cottonseed meal is used primarily for beef and dairy feeds. Less than one-third the U.S. cottonseed meal supply is used for poultry feeds. Peanut meal likewise is high in fiber but limited in lysine and methionine, and low in tryptophan; hence, its main outlet is in feeds for dairy and beef cattle. Sunflower meal is high in fiber and low in lysine, and thus fed mainly to ruminants (sheep, beef, and dairy cattle). In contrast, dehulled soybean meal is low in crude fiber and high in lysine. Although limiting in methionine, soybean meal is a key ingredient for blending with corn in formulating feeds for nonruminants, eg, poultry and swine. The two proteins complement each other; soy supplies the lysine

Table 14. Proximate Compositions of Oilseed Protein Meals[a], wt %

Meal	Process	Dry matter	Crude protein	Crude fat	Crude fiber	Ash
soybean						
with hulls	solvent	89.6	44.0	0.5	7.0	6.0
dehulled	solvent	89.3	47.5	0.5	3.0	6.0
cottonseed	prepress–solvent	89.9	41.0	0.8	12.7	6.4
peanut, with hulls	solvent	92.3	47.0	1.0	13.0	4.8
sunflower[b]	prepress–solvent	89.4	31.0	2.5	21.8	6.0

[a] Ref. 69, except as noted otherwise. Analytical values on as is basis.
[b] Ref. 70.

and corn the methionine necessary to provide a balanced ration at relatively low cost.

Edible Products. At present, only defatted peanut and soybean flakes are converted into edible-grade products. Peanut flour and grits are manufactured from prepress–solvent-extracted flakes as protein ingredients in processed foods (71).

Defatted soybean flakes for edible purposes are prepared essentially as outlined in Figure 4, except that more attention is paid to sanitation than in processing for feed use, and the solvent is removed in a vapor desolventizer–deodorizer or flash desolventizer to permit preparation of flakes ranging from raw to fully cooked (72). Desolventizer–toasters are used to prepare fully cooked (toasted) flakes to give maximum nutritive value (38). Degree of cooking is determined by estimating the amount of water-soluble protein remaining after a moist heat treatment with the protein dispersibility index (PDI) or the nitrogen solubility index (NSI) (73). A raw, uncooked flake has a PDI or NSI of ca 90, whereas a fully cooked flake has values of 5–15.

Defatted soybean flakes give three classes of products differing in minimum protein content (expressed on a dry basis): flours and grits (50% protein); protein concentrates (70% protein); and protein isolates (90% protein). Typical analyses are shown in Table 15. Flours and grits are made by grinding and sieving flakes. Concentrates are prepared (7) by extracting and removing the soluble sugars from defatted

Table 15. Typical Compositions of Soybean Protein Products and Their Uses[a], wt %

Constituent	Defatted flours and grits[b]	Protein	
		Concentrates[c]	Isolate[d]
protein[e]	56.0	72.0	96.0
fat	1.0	1.0	0.1
fiber	3.5	4.5	0.1
ash	6.0	5.0	3.5
soluble carbohydrates	14.0	2.5	0
insoluble carbohydrates	19.5	15.0	0.3

[a] Analytical values on a moisture-free basis (74).
[b] For baked goods, ground and processed meats, breakfast cereals, diet foods, infant foods, meat substitutes, confections, and milk replacers.
[c] For ground and processed meats and infant foods.
[d] For processed meats, meat substitutes, infant foods, and dairy substitutes.
[e] N × 6.25.

flakes by leaching with dilute acid at pH 4.5; leaching with aqueous ethanol; or moist heat treatment followed by leaching with water. Protein isolates (sodium soy proteinates) are obtained by extracting the soluble proteins with water at pH 8–9, precipitating at pH 4.5, centrifuging the resulting protein curd, washing, redispersing in water (with or without adjusting the pH to ca 7), and finally spray drying (7) (see Fig. 6). Flours, concentrates, and isolates are also specially processed into textured products that are used as meat extenders and substitutes (75). These and other food applications are summarized in Table 15.

Oilseed proteins are used in foods at concentrations of 1–2 to nearly 100%. Because of their high protein contents, textured soy flours and concentrates serve as meat substitutes. At low concentrations, the proteins are added primarily for their functional properties, eg, emulsification, fat absorption, water absorption, texture, dough formation, adhesion, cohesion, elasticity, film formation, and aeration (7,76) (see Food processing).

The use of some oilseed proteins in foods is limited by flavor, color, and flatus effects. Raw soybeans, for example, taste grassy, beany, and bitter. Even after processing, residues of these flavors may limit the amounts of soybean proteins that can be added to a given food (77). The use of cottonseed and sunflower-seed flours is restricted by the color imparted by gossypol and phenolic acids, respectively. Flatus production by defatted soy flours has been attributed to raffinose and stachyose (78). These sugars are removed by processing the defatted flours into concentrates and isolates.

Industrial Products. In the United States, only soybean protein isolates are used for industrial applications. They are available commercially in unhydrolyzed (unmodified), hydrolyzed, and chemically modified forms. The unhydrolyzed types are made by the process outlined in Figure 6; the acid-precipitated curd is washed, dried, and then ground. For the hydrolyzed types, the acid-precipitated curd is suspended

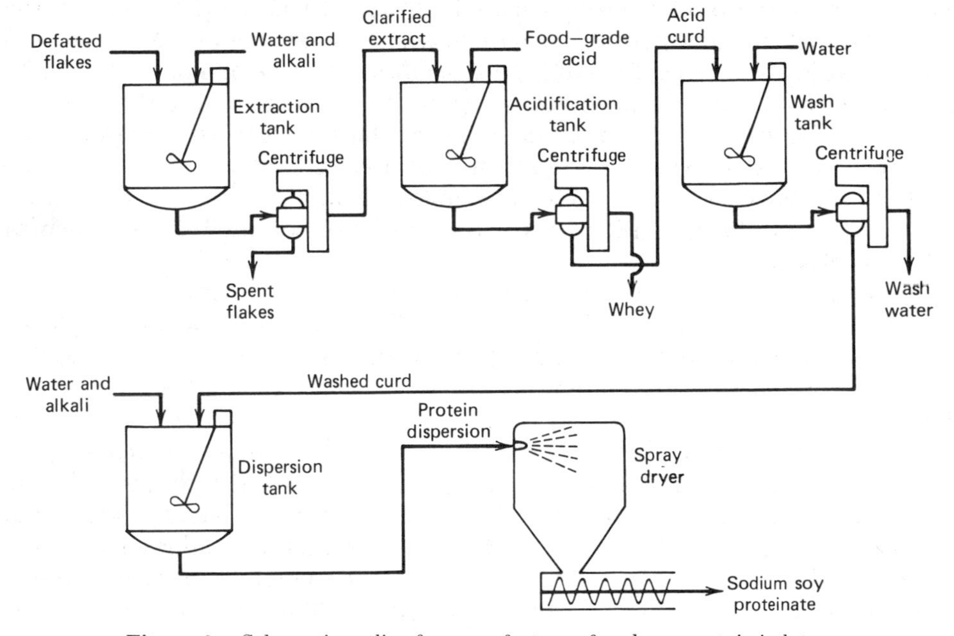

Figure 6. Schematic outline for manufacture of soybean protein isolates.

in alkali and heated to dissociate the proteins into subunits and partially hydrolyze the polypeptide chains. The reaction is terminated by acidifying to a pH of ca 4.5 to precipitate the modified protein which is dried (79). The chemically modified isolates are made by proprietary processes. Isolates are employed primarily as adhesives for clays used in coating of paper and paperboard to render surfaces suitable for printing.

Food Products. Peanuts, soybeans, and sunflower seeds are consumed as such or are processed into edible products.

Peanuts. In the United States, ca half the crop is processed into edible products, mainly peanut butter; other products include peanut candy, salted nuts, and roasted in-the-shell peanuts. Peanut butter is made and consumed primarily in the United States. The peanuts are shelled and dry-roasted, the skins are removed, and the nuts are finely ground. This material is blended with salt and other ingredients that may include hydrogenated fat, dextrose, corn-syrup solids, lecithin, and antioxidants (see Nuts).

Soybeans. Small amounts of soybeans are roasted and salted for snacks. Nut substitutes for baked products and confections are also manufactured from soybeans. Larger amounts are used in oriental foods, some of which are becoming increasingly popular in the United States.

Soy milk. In the traditional Chinese process, soybeans are soaked in water, ground into a slurry, cooked, and filtered to remove the insoluble cell wall and hull fractions. A number of process modifications have been made since the early 1960s, including a heat treatment before or during grinding to inactivate the enzyme lipoxygenase and thus prevent formation of grassy and beany flavors. Markets for the blander products made by the new processes have developed rapidly in Japan since the late 1970s. Some of the products are sweetened and flavored; fermented (yogurt-like) soy milks are also available in Japan (80).

Tofu. This product is prepared by adding a coagulant such as calcium sulfate to soy milk to precipitate the protein and oil into a gelatinous curd. The curd is then separated from the soluble portion (whey), pressed, and washed to yield a market-ready product that is a traditional food in Japan (81). It has become popular in the United States since the late 1970s and was produced in 1981 by more than 150 shops.

Miso. This paste-like food resembles peanut butter in consistency and is made by fermenting cooked soybeans and salt with or without a cereal such as rice or barley (82). It is used as a base for soups and is consumed in Japan, China, Indochina, Indonesia, and the East Indies. In 1981, it was produced by ca 10 small establishments in the United States.

Tempeh. Cooked beans are inoculated with the mold, *Rhizopus oligosporus*, and allowed to ferment for 24 h. The mold mycellium binds the soybean cotyledons together. When sliced and deep-fried in fat, the product is crisp and golden brown (83). Tempeh, a traditional food of Indonesia, was made by more than 30 concerns in the United States in 1981.

Soy sauce. This condiment, well-known to U.S. consumers, is made by fermentation or acid hydrolysis. In the fermentation process, cooked soybeans or defatted soybean meal are mixed with roasted wheat and the mixture blended with a pure culture of *Aspergillus oryzae*. Brine is added and the mixture is allowed to ferment 6–8 months. The product is then filtered and pasteurized (84). In the acid hydrolysis process, defatted soybean flour is refluxed with hydrochloric acid until the proteins are hydrolyzed. The hydrolysate is then filtered, neutralized, and bottled.

Sunflower Seed. Confectionary-type seed is roasted both in the shell and in the dehulled form. Roasted nutmeats are used in candies, cookies, snack items, cakes, and spreads. Small amounts are sold through health food stores; one variety is grown and sold solely as bird feed (32).

BIBLIOGRAPHY

"Soybeans" in *ECT* 1st ed., Vol. 12, pp. 689–701, by J. C. Cowan, Northern Regional Research Center, U.S. Department of Agriculture; "Soybeans" in *ECT* 2nd ed., Vol. 18, pp. 599–614, by J. C. Cowan.

1. K. Saio and T. Watanabe, *Nippon Shokuhin Kogyo Gakkai-Shi* **15,** 290 (1968).
2. J. N. A. Lott in *The Plant Cell*, Vol. 1 of N. E. Tolbert, ed., *The Biochemistry of Plants, A Comprehensive Treatise*, Academic Press, New York, 1980, pp. 589–623.
3. L. C. Berardi and L. A. Goldblatt in I. E. Liener, ed., *Toxic Constituents of Plant Foodstuffs*, 2nd ed., Academic Press, New York, 1980, pp. 183–237.
4. A. K. Smith in A. M. Altschul, ed., *Processed Plant Protein Foodstuffs*, Academic Press, Inc., New York, 1958, pp. 249–276.
5. F. R. Earle, C. H. Van Etten, T. F. Clark, and I. A. Wolff, *J. Am. Oil Chem. Soc.* **45,** 876 (1968).
6. P. J. Wan, G. W. Baker, S. P. Clark, and S. W. Matlock, *Cereal Chem.* **56,** 352 (1979).
7. W. J. Wolf, *J. Agric. Food Chem.* **18,** 969 (1970).
8. E. H. Rahma and M. S. Narasinga Rao, *J. Food Sci.* **44,** 579 (1979).
9. E. Derbyshire, D. J. Wright, and D. Boulter, *Phytochemistry* **15,** 3 (1976).
10. G. E. Sykes and K. R. Gayler, *Arch. Biochem. Biophys.* **210,** 525 (1981).
11. V. H. Thanh and K. Shibazaki, *Biochim. Biophys. Acta* **490,** 370 (1977).
12. P. E. Staswick, M. A. Hermodson, and N. C. Nielsen, *J. Biol. Chem.* **256,** 8752 (1981).
13. N. E. Tumer, V. H. Thanh, and N. C. Nielsen, *J. Biol. Chem.* **256,** 8756 (1981).
14. T. Yamada, S. Aibara, and Y. Morita, *Agric. Biol. Chem.* **43,** 2563 (1979).
15. J. T. Lawhon, C. M. Cater, and K. F. Mattil, *J. Am. Oil Chem. Soc.* **54,** 75 (1977).
16. C. T. Young, G. R. Waller, and R. O. Hammons, *J. Am. Oil Chem. Soc.* **50,** 521 (1973).
17. E. W. Lusas, *J. Am. Oil Chem. Soc.* **56,** 425 (1979).
18. J. F. Cavins, W. F. Kwolek, G. E. Inglett, and J. C. Cowan, *J. Assoc. Off. Anal. Chem.* **55,** 686 (1972).
19. Committee on Dietary Allowances, Food and Nutrition Board, *Recommended Dietary Allowances*, 9th rev. ed., National Academy of Sciences, Washington, D.C., 1980, pp. 39–54.
20. C. A. Brignoli, J. E. Kinsella, and J. L. Weihrauch, *J. Am. Diet. Assoc.* **68,** 224 (1976).
21. R. E. Worthington, R. O. Hammons, and J. R. Allison, *J. Agric. Food Chem.* **20,** 727 (1972).
22. C. Litchfield, *Analysis of Triglycerides*, Academic Press, New York, 1972, 355 pp.
23. C. R. Scholfield, *J. Am. Oil Chem. Soc.* **58,** 889 (1981).
24. W. A. Pons, Jr., M. F. Stansbury, and C. L. Hoffpauir, *J. Assoc. Off. Agric. Chem.* **36,** 492 (1953).
25. T. Itoh, T. Tamura, and T. Matsumoto, *J. Am. Oil Chem. Soc.* **50,** 122 (1973).
26. T. Hirota, S. Goto, M. Katayama, and S. Funahashi, *Agric. Biol. Chem.* **38,** 1539 (1974).
27. G. F. Cegla and K. R. Bell, *J. Am. Oil Chem. Soc.* **54,** 150 (1977).
28. G. M. Lolas, N. Palamidis, and P. Markakis, *Cereal Chem.* **53,** 867 (1976).
29. A. V. Bailey, J. A. Harris, E. L. Skau, and T. Kerr, *J. Am. Oil Chem. Soc.* **43,** 107 (1966).
30. J. A. Harris, F. C. Magne, and E. L. Skau, and T. Kerr, *J. Am. Oil Chem. Soc.* **41,** 309 (1964).
31. C. M. Cater, S. Gheyasuddin, and K. F. Mattil, *Cereal Chem.* **49,** 508 (1972).
32. J. A. Robertson, *CRC Crit. Rev. Food Sci. Nutr.* **6,** 201 (1975).
33. A. C. Eldridge and W. F. Kwolek, *J. Agric. Food Chem.*, in press.
34. Y. Birk in I. E. Liener, ed., *Toxic Constituents of Plant Foodstuffs*, Academic Press, Inc., New York, 1969, pp. 169–210.
35. D. E. Fenwick and D. Oakenfull, *J. Sci. Food Agric.* **32,** 273 (1981).
36. T. A. P. Hamsa and J. C. Ayres, *J. Am. Oil Chem. Soc.* **54,** 219 (1977).
37. E. H. Gustafson, *J. Am. Oil Chem. Soc.* **55,** 751 (1978).
38. E. Sipos and N. H. Witte, *J. Am. Oil Chem. Soc.* **38**(3), 11 (1961).
39. I. E. Liener, *J. Am. Oil Chem. Soc.* **58,** 406 (1981).
40. *Oilseeds and Products, Foreign Agriculture Circular*, FOP 1-82, USDA, Foreign Agricultural Service, Washington, D.C., Jan. 1982.

41. *Soya Bluebook*, American Soybean Association, St. Louis, Mo., 1981, pp. 145, 149.
42. *Fats and Oils Outlook and Situation*, FOS-304, USDA, Economic Research Service, Washington, D.C., July 1981.
43. *Oilseeds and Products, Foreign Agriculture Circular*, FOP 22-81, USDA, Foreign Agricultural Service, Washington, D.C., Dec. 1981.
44. *Oilseeds and Products, Foreign Agriculture Circular*, FOP 7-81, FOP 15-81, FOP 22-81, USDA, Foreign Agricultural Service, Washington, D.C., Apr., Sept., Dec. 1981.
45. *Oilseeds and Products, Foreign Agriculture Circular*, FOP 10-81, USDA, Foreign Agricultural Service, Washington, D.C., May 1981.
46. *Fats and Oils Situation*, FOS-301, USDA, Economics and Statistics Service, Washington, D.C., Oct. 1980.
47. *Fats and Oils Situation*, FOS-302, USDA, Economics and Statistics Service, Washington, D.C., Feb. 1981.
48. *Fats and Oils Situation*, FOS-290, USDA, Economics and Statistics and Cooperative Service, Washington, D.C., Feb. 1978.
49. *Feed Outlook and Situation*, FdS-283, USDA, Economic Research Service, Washington, D.C., Nov. 1981.
50. N. R. Lockmiller, personal communication, A. E. Staley Manufacturing Co., Decatur, Ill., Aug. 1981 and Feb. 1982.
51. R. Leviton and W. Shurtleff, press release, Soycrafters Association of North America, Colrain, Mass., and The Soyfoods Center, Lafayette, Calif., Sept. 1981.
52. E. A. Emken in D. R. Erickson, E. H. Pryde, O. L. Brekke, T. L. Mounts, and R. A. Falb, eds., *Handbook of Soy Oil Processing and Utilization*, American Soybean Association, St. Louis, Mo., and American Oil Chemists' Society, Champaign, Ill., 1980, pp. 439–458.
53. R. A. Phelps, F. S. Shenstone, A. R. Kemmerer, and R. J. Evans, *Poult. Sci.* **44,** 358 (1965).
54. J. D. Hendricks, R. O. Sinnhuber, P. M. Loveland, N. E. Pawlowski, and J. E. Nixon, *Science* **208,** 309 (1980).
55. F. H. Mattson in *Toxicants Occurring Naturally in Foods*, 2nd ed., National Academy of Sciences, Washington, D.C., 1973, pp. 189–209.
56. J. W. Erdman, Jr., *J. Am. Oil Chem. Soc.* **56,** 736 (1979).
57. T. O. Lindsey, G. E. Hawkins, and L. D. Guthrie, *J. Dairy Sci.* **63,** 562 (1980).
58. H. K. Gardner, Jr., R. J. Hron, and H. L. E. Vix, *Cereal Chem.* **53,** 549 (1976).
59. D. W. Freeman, R. S. Kadan, G. M. Ziegler, Jr., and J. J. Spadaro, *Cereal Chem.* **56,** 452 (1979).
60. R. S. Levi, H. G. Reilich, H. J. O'Neill, A. F. Cucullu, and E. L. Skau, *J. Am. Oil Chem. Soc.* **44,** 249 (1967).
61. G. N. Bookwalter, K. Warner, R. A. Anderson, and E. B. Bagley, *J. Food Sci.* **44,** 820 (1979).
62. K. Anantharaman and K. J. Carpenter, *J. Sci. Food Agric.* **20,** 703 (1969).
63. H. L. Wilcke, D. T. Hopkins, and D. H. Waggle, *Soy Protein and Human Nutrition*, Academic Press, Inc., New York, 1979, 406 pp.
64. J. J. Rackis, J. E. McGee, M. R. Gumbmann, and A. N. Booth, *J. Am. Oil Chem. Soc.* **56,** 162 (1979).
65. F. Sosulski, *J. Am. Oil Chem. Soc.* **56,** 438 (1979).
66. H. E. Amos, D. Burdick, and R. W. Seerley, *J. Anim. Sci.* **40,** 90 (1975).
67. Ref. 52, 598 pp.
68. Ref. 52, p. 68.
69. R. D. Allen, *Feedstuffs* **53**(30), 25 (1981).
70. D. H. Kinard, *Feed Manage.* **32**(6), 16 (1981).
71. J. L. Ayres, L. L. Branscomb, and G. M. Rogers, *J. Am. Oil Chem. Soc.* **51,** 133 (1974).
72. K. W. Becker, *J. Am. Oil Chem. Soc.* **55,** 754 (1978).
73. American Oil Chemists' Society, *Official and Tentative Methods*, 3rd ed., Champaign, Ill., Methods Ba 10-65 and Ba 11-65, 1976.
74. F. E. Horan, *J. Am. Oil Chem. Soc.* **51,** 67A (1974).
75. F. E. Horan in *Technology*, Vol. 1A of A. M. Altschul, ed., *New Protein Foods*, Academic Press, Inc., New York, 1974, pp. 366–413.
76. J. E. Kinsella, *J. Am. Oil Chem. Soc.* **56,** 242 (1979).
77. J. J. Rackis, D. J. Sessa, and D. H. Honig, *J. Am. Oil Chem. Soc.* **56,** 262 (1979).
78. J. J. Rackis, *J. Am. Oil Chem. Soc.* **58,** 503 (1981).

79. R. A. Olson and P. T. Hoelderle in R. Strauss, ed., *Protein Binders in Paper and Paperboard Coating*, TAPPI Monograph Series No. 36, 1975, pp. 75–96.
80. W. Shurtleff and A. Aoyagi, *Tofu and Soymilk Production*, Vol. 2 of *The Book of Tofu*, New-Age Foods Study Center, Lafayette, Calif., 1979, 336 pp.
81. W. Shurtleff and A. Aoyagi, *Food for Mankind*, Vol. 1 of *The Book of Tofu*, Autumn Press, Hayama-shi, Kanagawa-ken, Jpn., 1975, 334 pp.
82. W. Shurtleff and A. Aoyagi, *The Book of Miso*, Autumn Press, Brookline, Mass., 1976, 254 pp.
83. W. Shurtleff and A. Aoyagi, *The Book of Tempeh—A Super Soyfood from Indonesia*, Harper & Row Publishers, Inc., New York, 1979, 158 pp.
84. D. Fukushima, *J. Am. Oil Chem. Soc.* **58,** 346 (1981).

General References

A. M. Altschul, *Processed Plant Protein Foodstuffs*, Academic Press Inc., New York, 1958, 955 pp.

A. E. Bailey, *Cottonseed and Cottonseed Products*, Interscience Publishers, Inc., New York, 1948, 936 pp.

B. E. Caldwell, *Soybeans: Improvement, Production, and Uses*, American Society of Agronomy, Inc., Madison, Wisc., 1973, 681 pp.

D. W. Cobia and D. E. Zimmer, eds., *Sunflower Production and Marketing*, Extension Bull. 25 (Revised), North Dakota State University of Agriculture and Applied Science, Fargo, N.D., 1978, 73 pp.

D. R. Erickson, E. H. Pryde, O. L. Brekke, T. L. Mounts and R. A. Falb, *Handbook of Soy Oil Processing and Utilization*, American Soybean Association, St. Louis, Mo., and American Oil Chemists' Society, Champaign, Ill., 1980, 598 pp.

T. Hymowitz and C. A. Newell, *Econ. Bot.* **35,** 272 (1981).

A. G. Norman, *Soybean Physiology, Agronomy, and Utilization*, Academic Press, Inc., New York, 1978, 249 pp.

W. O. Scott and S. R. Aldrich, eds., *Modern Soybean Production*, Farm Quarterly, Cincinnati, Ohio, 1970, 192 pp.

A. K. Smith and S. J. Circle, *Proteins*, Vol. 1 of *Soybeans: Chemistry and Technology*, Avi Publishing Co., Inc., Westport, Conn., 1972, 470 pp.

D. Swern, ed., *Bailey's Industrial Oil and Fat Products*, 4th ed., Vol. 1, John Wiley & Sons, Inc., New York, 1979.

W. J. Wolf and J. C. Cowan, *Soybeans as a Food Source*, rev. ed., CRC Press, Inc., Cleveland, Ohio, 1975, 101 pp.

J. G. Woodruff, ed., *Peanuts: Production, Processing, Products*, 2nd ed., Avi Publishing Co., Inc., Westport, Conn., 1973, 330 pp.

Copra (coconut)

L. V. Curtin in A. M. Altschul, ed., *Processed Plant Protein Foodstuffs*, Academic Press, Inc., New York, 1958, pp. 645–676.

Flaxseed

S. W. Peterson in A. M. Altschul, ed., *Processed Plant Protein Foodstuffs*, Academic Press, Inc., New York, 1958, pp. 593–617.

Palm fruit and palm kernel

J. A. Cornelius, *Prog. Chem. Fats Other Lipids* **15,** 5 (1977).

J. G. Collingwood in A. M. Altschul, ed., *Processed Plant Protein Foodstuffs*, Academic Press, Inc., New York, 1958, pp. 677–701.

Rapeseed

L. A. Appleqvist and R. Ohlson, *Rapeseed*, Elsevier Publishing Co., Amsterdam, The Netherlands, 1972, 391 pp.

Safflower seed

J. A. Kneeland in A. M. Altschul, ed., *Processed Plant Protein Foodstuffs*, Academic Press, Inc., New York, 1958, pp. 619–644.

Sesame seed

E. A. Weiss, *Castor, Sesame, and Safflower*, Barnes & Noble, Inc., New York, 1971, 901 pp.
L. A. Johnson, R. M. Suleiman, and E. W. Lusas, *J. Am. Oil Chem. Soc.* **56,** 463 (1979).

WALTER J. WOLF
U.S. Department of Agriculture

SPACE CHEMISTRY

The abundance of the elements, their present chemical forms, and their distribution within the universe are the result of lengthy and complex evolutionary processes. The formation of elements began with the origin of the universe $(1–2) \times 10^{10}$ yr ago and has continued in the interiors of stars to the present (1). As a result of extensive processing and recycling of material, the relative abundances of the elements do not in general vary greatly from one region of the universe to another. Significant exceptions to this uniformity do occur, however, and a prime example is the Earth. Most atoms in the universe are currently in environments where particle kinetic energy is much too high to permit the existence of stable chemical bonds, but in selected environments stable bonds do exist and chemistry plays an important role. Examples of such regions are planetary bodies, interstellar clouds, atmospheres of cool stars, and the interiors of some white dwarf stars. In some of these environments, chemical processes have strongly fractionated groups of elements sharing common chemical properties, and the relative abundances of the elements can differ greatly from their original cosmic abundances.

Space chemistry has a practical aspect in that astrophysical processes determined the composition and environment of Earth and to some extent its evolution. Processes observed on the other plants provide insight into aspects of terrestrial evolution that cannot be deduced from study of the Earth alone. An eventual utility of space chemistry will be the exploitation of near-Earth bodies, eg, the Moon and asteroids, as sources of raw materials. If large-scale operations are ever to be conducted in space, this use will probably be a necessity.

Origin of the Elements

Nucleosynthesis, the creation of elements, occurs in astrophysical sites where particle energies are high enough for charged particles to penetrate the coulomb barriers of nuclei and produce nuclear reactions. Originally the entire universe was hot enough for such reactions, but this situation was very short-lived. For most of the age of the universe, significant nucleosynthesis has occurred only inside stable or exploding stars and, to a much lesser extent, in interstellar gas by reactions with cosmic rays.

Ninety-nine percent of the atoms in existence were created in a single event, ie, the origin of the universe. This synthesis produced most of the atoms, but essentially none of the elements heavier than lithium. The heavier elements were created within stars from the earlier produced hydrogen and helium. The origin of the universe is referred to as the big bang, and the reality of this event is widely accepted in the astrophysics community. Only 17 years ago it was hotly debated whether the universe really had a beginning or whether it had always been in existence. In 1965 Penzias and Wilson (2) discovered a microwave flux that fills the sky and whose flux and spectra are consistent with an origin from a 2.7 K blackbody radiator. This radiation is believed to be relic radiation produced 10^6 yr after the origin of the universe at the time when most of the matter of the universe decoupled or stopped frequently interacting with the photon flux of the universe. At the time of decoupling, the matter and photon temperatures were 3000 K, and after that time the density of the universe dropped to the point where the time scale for interaction of typical photons and gas became longer than the age of the universe. In the intervening time, the photon's energy spectrum cooled with the expansion of the universe, and after nearly 2×10^{10} yr it now appears as a 2.7 K background flux uniform over the sky. This radiation is strong evidence for the big-bang origin of the universe.

In the initial stages of the big bang, the temperature of the universe was $>10^{13}$ K and the density was $>10^{15}$ g/cm^3. Above 10^{10} K there were no composite nuclei, but as the temperature dropped, helium and lithium formed. After a few tens of minutes, element production associated with the origin of the universe was finished. The primary reactions that occurred during this period and produced most of the atoms in the universe are as follows (3):

$$^{1}H + {}^{1}H \rightarrow {}^{2}H + \gamma$$

$$^{1}H + {}^{2}H \rightarrow {}^{3}He + \gamma$$

$$^{3}He + {}^{3}He \rightarrow {}^{4}He + 2\,{}^{1}H + \gamma$$

$$^{3}He + {}^{4}He \rightarrow {}^{7}Be + \gamma$$

$$^{7}Be + e \rightarrow {}^{7}Li + \nu$$

Beryllium-7 and free neutrons are unstable, so the output from this synthesis was hydrogen, deuterium, the two isotopes of helium, and some lithium-7. These isotopes were produced in the following approximate relative abundances: ^{1}H(1), ^{2}H(10^{-4}), ^{3}He(10^{-7}), ^{4}He(10^{-1}), and ^{7}Li(10^{-9}). Essentially no heavier elements were produced, and without later synthesis in stellar interiors, chemistry would basically be confined to H and Li. Although less than one percent of the original atoms produced in the big bang have undergone subsequent reactions to produce heavier elements, these heavier elements are responsible for the abundant chemistry and diversity in the universe. Processing by stars managed to produce all known stable isotopes. Some isotopes are rare, but there are no unfilled positions in the nuclide chart (ie, the graph of neutron abundance vs proton abundance in nuclei).

The elements up to iron can, and to a certain extent are, produced by fusion reactions in stars. Up to iron the binding energy per amu (atomic mass unit) generally increases with mass so that thermonuclear fusion reactions are exothermic. Most of the visible stars are powered by the fusion of hydrogen to helium in which ca 0.7% of the reacted mass is converted to energy. By laboratory standards many of those reactions are quite slow. In the core of the sun where the temperature is $>10^7$ K and the density is >100 g/cm^3, the lifetime of a proton against fusion is ca 10^{10} yr. In stars of less or equal mass, conversion of hydrogen to helium occurs via the proton–proton chain where the protons interact to form deuterium. Deuterium in turn interacts with a proton to form ^3He, and two ^3He nuclei then collide to form ^4He. In more massive stars with their higher internal temperatures, the dominant reaction scheme is the CNO cycle where carbon, nitrogen, and oxygen act as catalysts in the transformation of four protons into a helium nucleus (see Fusion energy).

Stars spend most of their lifetimes converting hydrogen to helium, but when they eventually exhaust a significant fraction of the hydrogen in their cores, their temperatures rise as high as 3×10^9 K where fusion reactions occur among elements between helium and iron. These reactions are often referred to as carbon, oxygen, neon, or silicon burning. Typical reactions are of the form:

$$^{12}C + {}^4He \rightarrow {}^{16}O + \gamma$$

$$^{16}O + {}^4He \rightarrow {}^{20}Ne + \gamma$$

$$^{16}O + {}^{16}O \rightarrow {}^{32}S + \gamma$$

$$^{16}O + {}^{16}O \rightarrow {}^{24}Mg + 2\,{}^4He$$

$$^{18}O + {}^4He \rightarrow {}^{21}Ne + n$$

Beyond the iron peak elements, elements form primarily by neutron-capture reactions. When the available neutron flux is low, as is the case in evolved but stable stars such as red giants, the dominant process is the *s* (slow) process. In the *s* process, sequential neutron captures build successively more neutron-rich isotopes of a given element until a radioactive isotope is reached. The time between neutron captures is long, and before a new neutron can be interacted with, the isotope undergoes β decay, thereby forming the next element in the periodic table. Neutron capture by the new element then proceeds until another neutron-rich unstable isotope is reached and another β decay forms the next higher atomic number. The *s*-process nucleosynthesis follows the path in the nuclide chart known as the valley of β stability. The abundances of the *s*-process isotopes are proportional to their binding energy or inversely proportional to their neutron cross sections.

In violent stellar events such as supernova explosions, intense neutron fluxes are produced and an *r* (rapid) process neutron capture occurs. Here the mean time between neutron captures is shorter than the lifetimes of radioactive element to β decay. Successive neutron captures on a given element produce extremely neutron-rich and highly unstable nuclei. These nuclei decay by emission of β particles until a stable or long-lived isotope is reached. Some isotopes can be produced only by the *r* process, and in fact the elements beyond bismuth are formed only in this way. Some isotopes are shielded from the *r* process by stable nuclei and can only be produced by the *s* process, but many of the isotopes of heavy elements can be produced by either the *r* or *s* process. Neutron-poor isotopes, which cannot be produced by either the *r* or *s* process, are formed by the *p* (proton) process. The *p*-process isotopes are very rare in comparison with neutron-capture isotopes of the same element.

In terms of nucleosynthesis, the most anomalous elements in the periodic table are lithium, beryllium, and boron. These elements along with deuterium are destroyed in stellar interiors. Some ^7Li was synthesized in the big bang and some is produced in stars, but the other isotopes of these elements are probably not made in stars. The principal source of these exceedingly rare elements may be spallation of heavily interstellar gas atoms owing to collisions with cosmic rays (4).

Abundances of the Elements

Mass-loss processes and catastrophic explosions of stars recycle newly created elements from stars back into interstellar gas and dust where they are eventually incorporated into new stars. This mixing and reworking of material has not only increased the number of heavy elements in the universe over time, but it has also led to a degree of homogenization. The elemental composition of most stars in our galaxy is very close to that of the Sun, and in general, most galaxies have compositions similar to our own. Deviations from solar abundances are commonly found in matter that has not experienced extensive recycling between stellar interiors and interstellar gas and dust. For such stars the ratio of the heavier elements to hydrogen is lower than that in the Sun by a factor of up to 1000. But even in these objects, the relative abundances of the heavy elements are close to solar.

Because the composition of the Sun is believed to be close to the composition of the entire galaxy, the solar abundances are often referred to as cosmic abundances. The abundances in the sun are most directly determined from the strengths of spectral absorption lines. Unfortunately, abundances determined from spectroscopy are not of high accuracy owing to the weakness of many of the lines and to large uncertainties in the oscillator strengths (f values) required to compute abundance from measured line intensities. Many of the abundances referred to as solar are actually determined from meteorites. The carbonaceous-chondrite meteorites appear to have formed from the solar nebula that preceded the solar system in a manner that did not significantly fractionate elements capable of forming stony objects (5). In Figure 1 the abundances of a carbonaceous chondrite are compared with those determined from solar spectroscopy. Within the errors of the solar measurements there is excellent agreement except for the elements that cannot totally be incorporated into a meteoritic body at room temperature. These elements include the noble gases plus hydrogen, carbon, nitrogen, and oxygen. Oxygen is incorporated into the meteorite, but only at a level allowed by stoichiometry. Because the carbonaceous chondrites appear to match solar abundances closely for most elements and because they can be analyzed in the laboratory to high accuracy, most cosmic abundances actually are derived from meteorites. However, carbonaceous meteorites are complex materials with complex histories, and their abundances are only an approximation to true solar abundances.

What is believed to be an accurate estimation of solar abundances is listed in Table 1 (7–8). The abundances of H, C, N, and O are obtained by solar spectroscopy, and the abundances of the noble gases are derived from solar cosmic-ray studies and a variety of nucleosynthetic criteria. The other determinations are from meteorites. The data of Table 1 are plotted in Figure 2. Gaps occur in the figure where the lifetime of an element is short compared with the age of the Sun. The abundances in Figure 2 are the result of ca 10^{10} years of nucleosynthesis and recycling of stars. The shape of the curve is a result both of the various astrophysical environments of the elements and of the basic physics of nuclear interactions. The main trend in the cosmic-abundance curve is a decrease in abundance with increasing atomic number. Abundance drops by a factor of 10^{12} from hydrogen to uranium. Superimposed on this decrease is fine

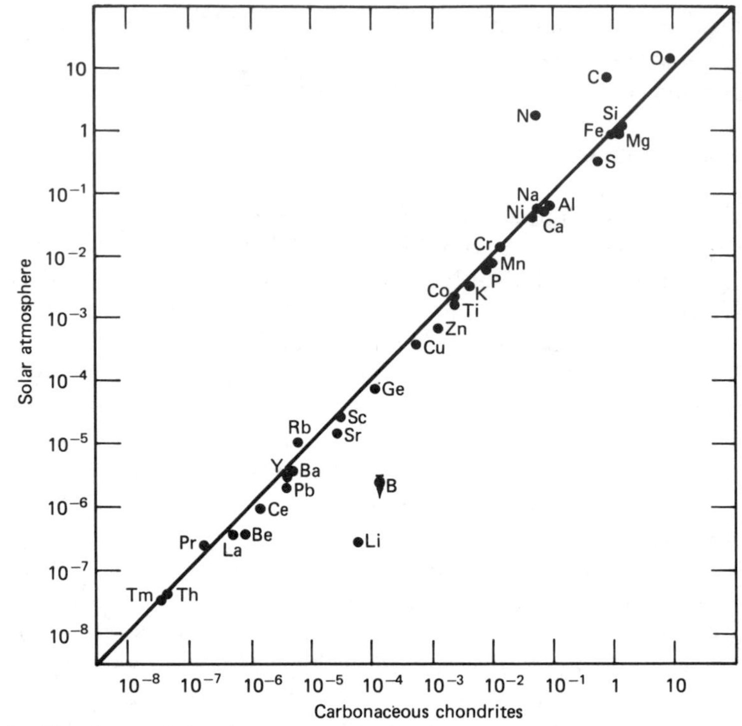

Figure 1. The elemental abundances, normalized to Si, in a Type I carbonaceous-chondrite meteorite compared to abundances determined for the atmosphere of the Sun (6).

structure resulting from differing nuclear binding energies or nuclear cross sections. The zig-zag odd–even effect relates to the relatively high binding energy per mass unit of nuclei having an even number of protons. On the average, an element with an even atomic number is approximately ten times as abundant as its odd-numbered neighbors in the periodic table. Besides the odd–even effect there are other peaks in the abundance curve associated with high binding energy. The most conspicuous is Fe and the iron group elements, but smaller peaks appear where elements have so-called magic numbers of either protons or neutrons associated with unusual binding energy (see Actinides and transactinides). Light elements whose nuclei are multiples of ^4He show these high binding energies. After ^1H and ^4He, the isotopes ^{12}C, ^{16}O, ^{20}Ne, ^{24}Mg, ^{28}Si, ^{32}S, and ^{56}Fe are the most abundant in the universe. Of all elements, iron has the highest binding energy per mass unit and as a result its abundance is a thousand times higher than the curve of Figure 2 which ignores the iron-group elements. The most dramatic deviation from a smooth fit to the abundance curve is for the light elements lithium, beryllium, and boron. Unlike other elements, these are generally destroyed at the temperatures of stellar interiors.

Interstellar Gas and Dust

Interstellar gas and dust is both the raw material from which stars form and the material returned to the interstellar medium when stars lose mass gradually or explode violently. Interstellar material typically is very tenuous; the gas density in the Earth's galaxy averages less than one atom per cubic centimeter. The total abundances in the

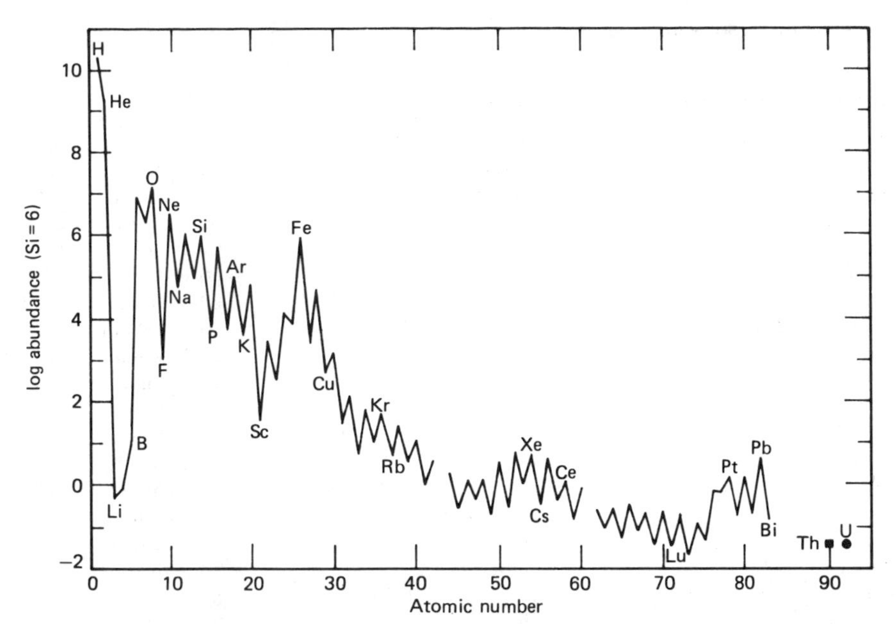

Figure 2. Cosmic abundances plotted from Table 1.

interstellar medium are thought to be close to solar, but the elements are strongly fractionated into either gas or dust depending on the element's ability to form solid grains under interstellar conditions. Refractory elements that can condense from dense clouds of solar-abundance gas at roughly 1400 K are depleted in the gas by factors of ≥1000 (10). More volatile materials including carbon, nitrogen, and oxygen are depleted significantly, but to a lesser extent than the refractory elements. The majority of atoms that can form grains in interstellar conditions do exist in the form of grains. The grains are believed to consist of rather refractory, stony cores of iron, magnesium, silicon, oxygen, and less abundant elements, surrounded by a less stable mantle of hydrogen, carbon, nitrogen, and oxygen. Laboratory studies have shown that the uv, x-ray, and charged-particle fluxes in the interstellar medium should convert the low atomic weight mantles to complex organic molecules (11).

Complex gas molecules usually have lifetimes <100 yr before they are destroyed by photodissociation. A very important exception, however, is the case of molecular clouds where the high dust density acts as a shield blocking external radiation. The gas in these clouds has been extensively studied by radio telescopes, and spectroscopic studies have identified a large array of molecular species (Table 2) (12). Clouds occupy only 2% by volume of interstellar medium, but because the gas they contain is dense, the clouds constitute roughly half of the interstellar mass. Clouds can be classified as either cool dark clouds, which have particle density of 10^3–10^4/cm^3 and temperatures of 5–15 K, or as warm clouds with densities of 10^4–10^7/cm^3 and temperatures up to 70 K. The warm clouds are usually close to and heated by newly formed hot stars. The formation of the molecules is not well understood, but the observed high abundance of HCO^+ and N_2H^+ indicate that ion–neutral reactions play an important role (12). The formation, evolution, and destruction of molecules in the clouds involves atoms, ions, free radicals, molecular ions, neutral molecules, solid surfaces, cosmic rays, shock waves, and uv radiation. The relative importance of shocks, grain surfaces, and other parameters is unknown.

Table 1. Estimates of Solar Abundances[a]

Element	Atomic abundance	Element	Atomic abundance	Element	Atomic abundance
1 hydrogen	2.2×10^{10}	33 arsenic	6.1	63 europium	8.8×10^{-2}
2 helium	1.4×10^{9}	34 selenium	68	64 gadolinium	0.30
3 lithium	49	35 bromine	9.7	65 terbium	5.4×10^{-2}
4 beryllium	0.81	36 krypton		66 dysprosium	0.34
5 boron	9.0[b]	37 rubidium	6.1	67 holmium	8.4×10^{-2}
6 carbon	9.3×10^{6}	38 strontium	26	68 erbium	0.22
7 nitrogen	2.0×10^{6}	39 yttrium	3.8	69 thulium	3.4×10^{-2}
8 oxygen	1.5×10^{7}	40 zirconium	11	70 ytterbium	0.23
9 fluorine	1000	41 niobium	1.0	71 lutetium	3.5×10^{-2}
10 neon	3.4×10^{6}	42 molybdenum	3.9	72 hafnium	0.18
11 sodium	5.8×10^{4}	43 technetium	unstable	73 tantalum	2.0×10^{-2}
12 magnesium	1.0×10^{6}	44 ruthenium	1.9	74 tungsten	0.12
13 aluminum	8.5×10^{4}	45 rhodium	0.30	75 rhenium	5.6×10^{-2}
14 silicon	1.0×10^{6}	46 palladium	1.5	76 osmium	0.69
15 phosphorus	7200	47 silver	0.45	77 iridium	0.64
16 sulfur	4.9×10^{5}	48 cadmium	1.5	78 platinum	1.4
17 chlorine	4400	49 indium	0.19	79 gold	0.21
18 argon	1.1×10^{5}	50 tin	3.5	80 mercury	0.21
19 potassium	3800	51 antimony	0.31	81 thallium	0.19
20 calcium	6.2×10^{4}	52 tellurium	6.3	82 lead	3.9
21 scandium	35	53 iodine	1.0	83 bismuth	0.14
22 titanium	2700	54 xenon	5.4	84–89	unstable elements
23 vanadium	290	55 cesium	0.39	90 thorium	4.3×10^{-2} unstable
24 chromium	1.4×10^{4}	56 barium	4.7	91 protactinium	unstable
25 manganese	9000	57 lanthanum	0.44	92 uranium	2.7×10^{-2} unstable
26 iron	8.7×10^{5}	58 cerium	1.1	93–106	unstable elements
27 cobalt	2300	59 praseodymium	0.16		
28 nickel	5.0×10^{4}	60 neodymium	0.78		
29 copper	470	61 promethium	unstable		
30 zinc	1400	62 samarium	0.23		
31 gallium	38				
32 germanium	120				

[a] Refs. 7–8. [b] Ref. 9.

The Solar System

The solar system consists of the Sun, nine planets, the planetary satellites, and an enormous number of asteroids and comets. These bodies have diverse physical properties influenced in part by composition, mass, distance from the Sun, and mode of origin. The planets are distinguished as two groups, the inner or terrestrial planets and the outer or giant planets.

Terrestrial Planets. The planets Mercury, Venus, Earth, and Mars are Earthlike objects that differ greatly from the outer planets. The outer planets are gaseous objects that consist mainly of low atomic weight elements; the inner planets are solid bodies composed of stony and metallic materials. Roughly 90% of the mass of each terrestrial planet is composed of magnesium, silicon, iron, and oxygen; the cation atomic ratio of the metals is close to their approximate solar ratio of 1:1:1. Oxygen is the most abundant element, but its abundance is less than solar and is determined by the amount required to form stoichiometric ratios with the principal cations. Most of the other elements that can form stony objects and form compounds less volatile than gold probably occur in roughly cosmic proportions. The condensation temperature in the solar nebula was on the order of 900 K. With the exception of sulfur, all of these ele-

Table 2. Known Interstellar Molecules[a]

Formula	Name
Molecules of 2 atoms	
H_2	hydrogen
OH	hydroxyl radical
CH^+	methylidyne ion
CH	methylidyne
CO	carbon monoxide
CN	cyanogen radical
CS	carbon monosulfide
NO	nitric oxide
$C{\equiv}C$	ethynyl radical
SO	sulfur monoxide
NS	nitrogen sulfide
SiO	silicon monoxide
SiS	silicon monosulfide
Molecules of 3 atoms	
HCO^+	formyl ion
N_2H^+	diazenylium ion
H_2O	water
HCN	hydrogen cyanide
HNC	hydrogen isocyanide
CCH	ethynyl radical
HCO	formyl radical
HOC^+	isoformyl
HNO	nitrosyl radical
OCS	carbonyl sulfide
H_2S	hydrogen sulfide
SO_2	sulfur dioxide
Molecules of 4 atoms	
NH_3	ammonia
H_2CO	formaldehyde
$HC{\equiv}CH$	acetylene
C_3N	cyanoethynyl radical
HNCO	isocyanic acid
H_2CS	thioformaldehyde
HCNS	isothiocyanic acid
Molecules of 5 atoms	
CH_4	methane
$HC{\equiv}CC{\equiv}C$	butadiynyl radical
$HC{\equiv}CCN$	cyanoacetylene
$H_2C{=}C{=}O$	ketene
NH_2CN	cyanamide
CH_2NH	methanimine
HCOOH	formic acid
Molecules of 6 atoms	
CH_3CN	acetonitrile
NH_2CHO	formamide
CH_3OH	methanol
CH_3SH	methyl mercaptan
C_2H_4	ethylene
Molecules of 7 atoms	
$CH_3C{\equiv}CH$	methylacetylene
CH_3CHO	acetaldehyde
CH_3NH_2	methylamine
$H_2C{=}CHCN$	acrylonitrile
$HC{\equiv}CC{\equiv}CCN$	cyanobutadiyne
Molecules of 8 atoms	
$HCOOCH_3$	methyl formate
Molecules of 9 atoms	
CH_3CH_2OH	ethanol
CH_3CH_2CN	propionitrile
CH_3OCH_3	dimethyl ether
$HC{\equiv}C(C{\equiv}C)_2CN$	cyanohexatriyne
Molecules of 10 atoms	
none	
Molecules of 11 atoms	
$HC{\equiv}C(C{\equiv}C)_3CN$	cyanooctatetrayne

[a] Ref. 9.

ments are cosmically rare and should not constitute more than a small fraction of the mass of the terrestrial planets. The elements that could not be totally incorporated into dust grains in the environment in which the terrestrial planets formed are highly depleted in the inner planets. These elements include the volatile metals, eg, indium, thallium, and bismuth; the elements that form icy grains, eg, hydrogen, carbon, nitrogen, and oxygen; and the noble gases.

The mean densities of the inner planets (Table 3) provide an important clue to their interior compositions. Even taking into consideration different gravitational compressions by estimating their uncompressed densities, the densities differ considerably, from 5.4 g/cm^3 for Mercury to 3.3 g/cm^3 for Mars (13). Most of the mass of the planets is in iron, magnesium, and silicon. The latter two are totally oxidized; iron, however, can exist in three oxidation states, and the ratio of metallic iron to oxidized iron explains much of the density variation seen in Table 3. The uncompressed densities probably result from the mixing of a metal component with a density of 7.6 g/cm^3 and a silicate component of ca 3.3 g/cm^3 (15). The density of Mercury indicates that >60 wt % of the planet is metallic iron. The density is so high that not only must es-

Table 3. Basic Data of the Terrestrial Planets[a]

Property	Mercury	Venus	Earth	Mars
mean distance from Sun, 10^6 km	57.9	108.2	149.6	227.9
mean distance from Sun, relative, Earth = 1[b]	0.3871	0.7233	1.000	1.5237
orbital period, yr, relative, Earth = 1	0.2409	0.6152	1.000	1.881
orbital eccentricity	0.2056	0.0068	0.0167	0.0934
orbital inclination, degrees, relative, Earth = 0	7.00	3.39	0	1.85
rotational period, h	1406	5832	23.93	24.62
obliquity of rotation axis, degree	ca 0	−2	23.45	23.98
mass, 10^{26} g	3.303	48.70	59.76	6.421
mass, relative, Earth = 1	0.0558	0.8150	1	0.1074
equatorial radius, km	2439	6050	6378	3398
equatorial radius, relative, Earth = 1	0.382	0.949	1	0.532
escape velocity from equator, km/s	4.3	10.3	11.2	2.38
mean density, g/cm^3	5.42	5.25	5.52	3.94
mean "uncompressed" density, g/cm^3	5.4	ca 4.2	ca 4.2	3.3
equatorial surface temperature, K, day/night	720/95	750/749	300/280	250/185
atmospheric surface pressure, relative, Earth = 1	10^{-15}	90	1	0.007

[a] Refs. 13–14.
[b] Earth's mean distance from Sun = AU (astronomical unit).

sentially all of the iron be in the metallic state, but in addition the abundance of iron relative to Mg and Si must be above solar. Suggestions for the anomalously high total iron abundance in Mercury are either that Mg and Si were incompletely condensed into dust grains (16) at the time of planetary accretion or that gas-dynamic drag effects depleted the inner region of the solar nebula in the lower density silicate particles (17). The opposite extreme of highly reduced Mercury is the highly oxidized interior of Mars. The low density implies that only a small fraction of the iron in Mars can be in the metallic state.

The variation of Fe^{2+}–Fe among the terrestrial planets is probably due to the state of iron in the dust grains from which the planets originally accreted. The implication is that iron was primarily in metallic form at distances of Mercury's orbit from the Sun but largely oxidized at the orbit of Mars. This is compatible with a model for formation of the terrestrial planets that involves accretion of grains that formed by equilibrium condensation from a massive gaseous solar nebula. Pressures in this hypothetical nebula were high, ca 10 Pa (10^{-4} atm), and the temperature gradient decreased from ca 1400 K at the orbit of Mercury to ca 400 K at the orbit of Mars. The composition of the dust grains, and hence of the planets that formed from them, would depend on the solids that could exist in chemical equilibrium with the solar-nebula gas at a particular distance from the Sun. Figure 3 illustrates the different minerals that can exist as solid grains in complete equilibrium with the gas at various temperatures (13). This sequence involves both condensation and back reactions with the gas (18). For example, iron condenses as metal at 1400 K and does not exist in silicates at high temperatures. At lower temperatures iron can exist to an increasing extent in silicates, and at 700 K free iron grains react with gaseous H_2S in the nebula to form

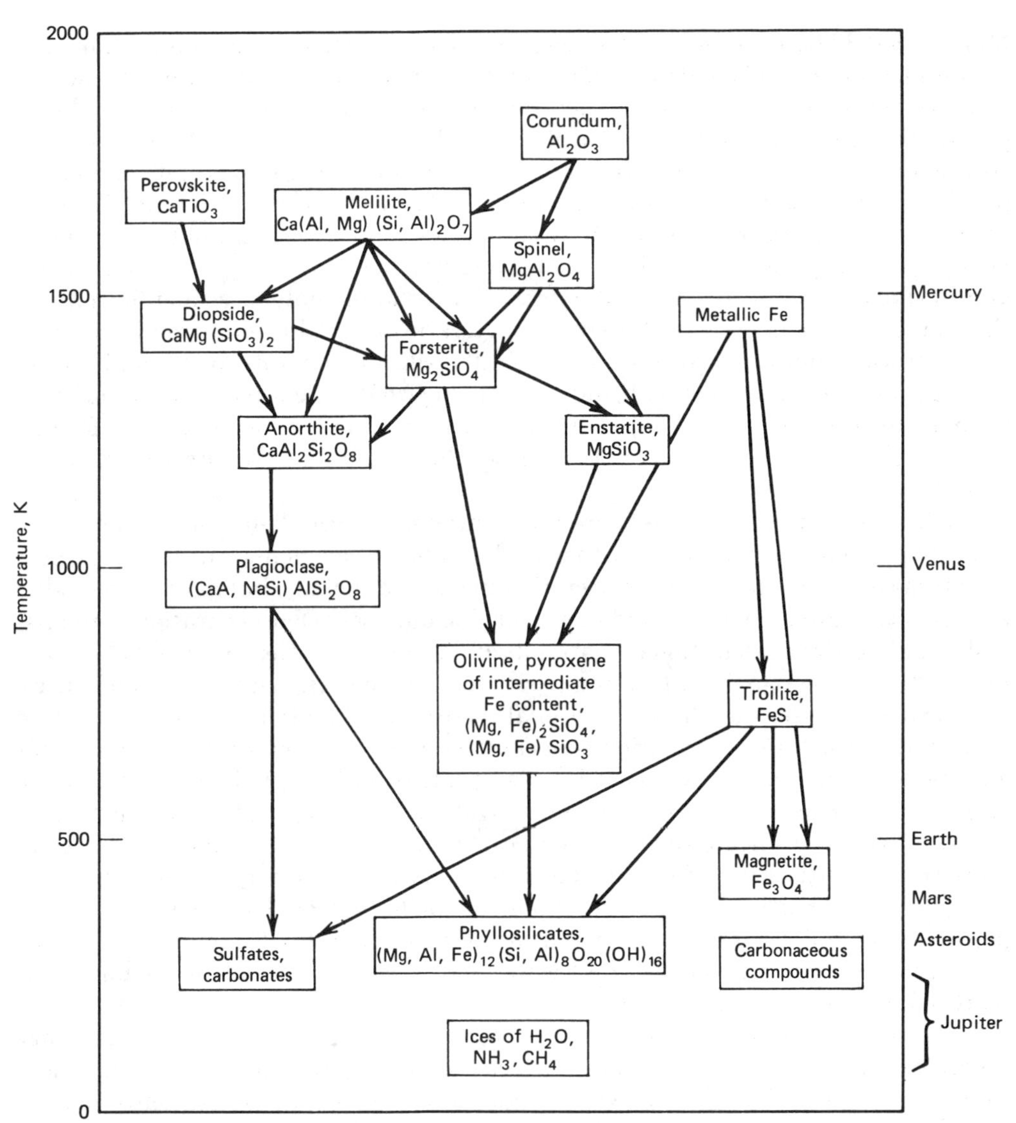

Figure 3. The condensation and reaction sequence for small mineral grains in total chemical equilibrium with a cooling solar composition gas at pressure of ca 10 Pa (ca 10^{-4} atm). The symbols on the right indicate the temperature regimes at which the different planets could have accreted in a hot dense solar nebula (13). Courtesy of Prentice-Hall, Inc.

FeS. Metallic iron remaining to temperatures of 400 K should react with H_2O gas to form magnetite (Fe_3O_4).

The sequence shown in Figure 3 (12) requires chemical equilibrium between solids and gas and does not take into account effects such as supersaturation. For the most obvious application, ie, planet formation, it also requires a massive hot nebula in which preexisting solids have been vaporized. The degree of equilibration and the existence of a massive hot nebula are questionable, but the condensation sequence does explain many of the properties of meteorites and is consistent with the densities of the planets (19). On the right-hand margin of Figure 3 are the estimated temperature regimes where the individual terrestrial planets could have formed in a massive hot nebula.

Mercury would have formed at a temperature where essentially all calcium, aluminum, titanium, and iron had condensed. The condensation of iron metal before magnesian silicates can explain the high iron abundance in Mercury, but it requires nebular pressures higher than 10 Pa (10^{-4} atm) (20). The equilibrium model places the accretion of Venus at a temperature where sulfur exists in the gaseous state and hence predicts sulfur abundance in Venus and Mercury below the cosmic S:Si ratio. The dust grains that formed the Earth would have equilibrated with cooler gas, and sulfur should be in its cosmic proportion. In both Venus and Earth, iron would exist both as metal and in silicates, as contrasted to Mercury, where it would only be as metal. In Mars, the condensation theory would predict a high ratio of Fe^{2+}:Fe and possible incorporation of water due to the accretion of hydrated silicate minerals. Water and carbon are highly depleted on the Earth and presumably what little it has is contamination from a late accretion of stray objects formed farther out in the solar system or formed in place under conditions cooler than those when the bulk of the Earth's mass accreted.

All of the terrestrial planets are differentiated bodies; their compositions vary radially in their interiors. Part of the redistribution of elements within the planets may be a relic of nonhomogenous accretion in the solar nebula, but most of the radial zoning is the result of igneous and metamorphic processes. The separation of immiscible fluids and the separation of materials of different densities produced planetary regions that differ significantly in composition from their original cosmic abundances. The principal effect was the gravitational separation of an immiscible metal phase to form iron cores. Core formation concentrates a planet's metallic iron at its center and depletes the rest of the planet in siderophile (iron-loving) elements. These are largely the transition elements that concentrate in the metallic phase of a melt composed of iron and silicate phases. Elements with strong siderophilic tendencies in terrestrial conditions include nickel, palladium, platinum, iridium, osmium, rhenium, and gold. In the Earth's mantle, siderophiles are highly depleted, but not to the extent predicted by equilibrium between metal and silicate liquids (6). Either the siderophile separation was not efficient in the Earth, or further accretion of solar-abundance material occurred after core formation.

In contrast to the siderophiles that concentrate in planetary cores, planetary processing enriches lithophile elements in the crusts. The lithophile elements are those that form strongly ionic bonds with oxygen or are incompatible elements such as uranium, thorium, and the rare earths that, because of their valence states and large ionic sizes, do not readily substitute for abundant elements in common rock-forming minerals. Lithophile elements include calcium, aluminum, lithium, rubidium, strontium, potassium, scandium, titanium, barium, and vanadium. One aspect of lithophile concentration is that uranium, potassium, and thorium, the internal heat sources of the inner planets, are largely confined to the outer regions of the planets. Table 4 lists the geochemical affinities of the elements.

The surfaces of the terrestrial planets indicate that each has had a unique geologic history. The Earth is the most active, and Mercury is the least active. Mercury's heavily cratered regions suggest that much of Mercury's surface dates back to the era of heavy bombardment, which ended 3.9×10^9 yr ago. The heavy bombardment period apparently saturated the surfaces of all the terrestrial planets with hypervelocity-impact craters and multiringed basins ranging in sizes up to 1400 km. This cratering overprinted all previous terrains. Mercury also contains less heavily cratered plains. These

Table 4. Geochemical Classification of the Elements[a]

Periodic group	Lithophile[b]	Siderophile[b]
IA	Li, Na, K, Rb, Cs	
IIA	Be, Mg, Ca, Sr, Ba, Ra	
IIIB	Sc, Yt, La	
IVB	Ti, Zr, Hf, Th	
VB	V, Nb, Ta	(V)
VIB	Cr, (Mo), W, U	(Cr), Mo
VIIB	Mn	Mn
VIII	Fe, (Co), (Ni)	Fe, Co, Ni, Ru, Rh, Pd, Os, Ir, Pt
IB		(Cu), (Ag), Au
IIB	(Zn), (Cd)	(Zn), (Cd)
IIIA	B, Al, (Ga), (Tl)	Ga
IVA	Si, (Sn), (Pb)	Ge, Sn, Pb
VA	(P), (As), (Sb)	P, As, Sb, Bi
VIA		
VIIA	F, Cl, Br, I	

[a] Ref. 21.
[b] Parentheses indicate subsidiary tendencies.

appear to be igneous flows that occurred after the era of heavy bombardment. The fact that there are no fresh regions on Mercury with extremely low crater densities implies that Mercury, like the Moon, is currently an inert body and has been geologically inactive for most of the age of the solar system. Mercury's active period probably ended $2\text{–}3 \times 10^9$ yr ago as the available thermal energy in the planet was no longer capable of causing surface volcanism. The thrust (compressive) faults of Mercury's crust are indicative of a global shrinkage of the planet owing to cooling or phase changes (22). Most of the existing knowledge of Mercury came from Mariner 10, the only spacecraft mission to the planet (23).

Venus is essentially a twin of the Earth in that its mass, density, and distance from the Sun are very similar. Data on Venus, however, are limited because a dense atmosphere and permanent cloud cover obscure the planet's surface. Direct information on Venus' surface has been obtained by the USSR landers Venera 8, 9, 10, 13, and 14, and radar mapping has been done from the Earth and from the U.S. Pioneer Venus spacecraft. The surface compositions measured by the Venera spacecraft are comparable with basalt for four of the sites and with granite for the other. Surface images from the Venera landers showed landscapes ranging from dense fields of angular boulders to smooth plains (24).

Most of the surface of Venus was mapped to a spatial resolution of 30 km by radar altimetry and radar imaging on the Pioneer Venus mission (25–26). The data show that although the total range in elevations on Venus is similar to that on the Earth, the distribution is quite different. The distribution of elevations on Earth is bimodal with a peak at −5 km (below sea level) corresponding to the ocean floor and a peak at 1 km (above sea level) owing to continental land masses. Elevations on Venus are strongly peaked at a single value. Although the resolution of maps is low, they indicate a diversity of types of terrain. There are highlands, lowlands, ancient heavily cratered regions, features that appear to be shield volcanoes, and rolling plains. Some of the low-lying basins have properties consistent with having been filled by lava. The rolling plains, which may be exposures of the ancient crust of Venus, appear to be granitic on the basis of the Venera results. There is no evidence for the existence on Venus of the ridges and subduction troughs that are prominent features on the Earth as a result from active continental drift. It appears that the dichotomy of ancient continents and young sea-floor material recycled by active plate movement is a process unique to the Earth. Venus has had a complex geological history, but one significantly different from that of the Earth.

Mars is a small planet with only 10% the mass of the Earth. It has been studied by U.S. and USSR flyby spacecraft, by orbiters, and by the two Viking landers. Unlike Mercury and the Moon, Mars has been active geologically through much of the planet's lifetime (27). However, the existence of impact craters on most Martian land forms indicates that most Martian features are ancient compared to the Earth's. Mars has wide topographic diversity. Ancient heavily cratered regions date back to the time of heavy bombardment and have not greatly changed for 3×10^9 yr, but other surface features such as polar ice caps and sand dunes change annually (one Mars year = 1.88 Earth yr). Mars has experienced extensive basaltic volcanism, and it contains the largest known volcano in the solar system, Olympus Mons. This enormous shield volcano is 600 km across and 24 km high with an 80 km summit caldera. Although plate tectonics did not play a role in sculpturing the surface of Mars, there are large-scale features resulting from faulting; the largest is Valles Marineris, a giant canyon 4000 km long and 4 km deep. The most mysterious features on Mars are those formed by water. There are erosional landforms which often are many hundreds of kilometers in extent and consist of channeled lowlands and sculptured highlands shaped by a directed flow. These features may have been formed by transient flooding by liquid water or scouring by glacial ice (28–29). Their existence may imply significant climatic changes on Mars, for at present, neither liquid water nor ice is stable at midlatitudes. Water ice does form on the midwinter polar cap, however. Although water may exist on Mars, any organics would likely be destroyed by photocatalytic reactions, leaving little chance of biological activity there (30).

The compositions of the atmospheres of the terrestrial planets are shown in Table 5. Mercury's atmosphere is merely material picked up from the solar wind and is not

Table 5. Atmospheric Compositions of the Terrestrial Planets[a]

Planet	Relative pressure	Principal gases, %	Other gases, ppm
Mercury	10^{-15}	He, ca 98 H, ca 2	
Venus	90	CO_2, 96 N_2, 3.5	H_2O, ca 100; SO_2, 150; Ar, 70 CO, 40; Ne, 5; HCl, 0.4; HF, 0.01
Earth	1	N_2, 77 O_2, 21 H_2O, 1 Ar, 0.93	CO_2, 330; Ne, 18; He, 70; Kr, 1.1 Xe, 0.087; CH_4, 1.5; H_2, 5; N_2O, 0.3 CO, 0.12; NH_3, 0.01; NO_2, 0.001 SO_2, 0.002; H_2S, 0.0002; O_3, 0.4
Mars	0.007	CO_2, 95 N_2, 2.7 Ar, 1.6	O_2, 1300; CO, 700; H_2O, 300 Ne, 2.5; Kr, 0.3; Xe, 0.08; O_3, 0.1

[a] Ref. 14.

made of materials accreted during the origin of the solar system as is the case for the other inner planets. Of the inner planets with real atmospheres, the Earth is the most unusual. The main elements in the Earth's atmosphere are N_2 and O_2. Carbon dioxide, the principal gas in the atmospheres of the other terrestrial planets, is found only in trace amounts. The Earth's atmosphere is anomalous because its composition is determined by biology and not by chemistry alone. With the exception of argon, the gases in the atmosphere are controlled and rapidly recycled by biological processes (13). The most remarkable aspect of this control is the maintenance of a large oxygen abundance above a planetary surface that would rapidly react with practically all atmospheric oxygen in the absence of life (31). Without biological recycling, oxygen would also be depleted by reactions with N_2 and H_2O to produce nitric acid. The normal composition for the atmosphere of a terrestrial planet is nearly pure CO_2 plus small amounts of N_2 and ^{40}Ar produced by the decay of ^{40}K. Earth originally must have had an atmosphere consisting of H_2O and CO_2, but in time the water formed the oceans and CO_2 formed carbonate rocks. Earth actually has roughly as much CO_2 as the dense 8 MPa (80 bar) atmosphere of Venus, but on Earth the CO_2 is locked up in surface deposits (32). The locking up of H_2O and CO_2 on the Earth's surface is unique in the solar system because the base of the Earth's atmosphere is near the triple point of H_2O where vapor, liquid, and solid water can coexist. On Mars nearly all available water is in hydrated minerals or as permafrost. On Venus the surface temperature of 750 K precludes the existence of water, but H_2O is also strongly missing in the atmosphere. Either water was not incorporated into Venus when it formed, or it was subsequently lost to space. Earth's two closest neighbors in space have atmospheres either nearly 100 times as dense and intensely hot (Venus) or roughly 100 times less dense and contrarily cold (Mars).

Although the atmosphere has remained fairly stable for much of Earth's history, it is possible that it could become more like either the atmospheres of Mars or Venus with appropriate perturbations. Such perturbations might be large-scale deforestation, conversion of large amounts of coal to CO_2, or the gradual but inevitable increase in the brightness of the Sun.

The Outer Planets. The outer planets include Pluto and the giant planets (Jupiter, Saturn, Uranus, and Neptune). Very little is known about Pluto, but it appears to be small, even in comparison with the inner planets, and it clearly is quite different from the other outer planets (32–33). The giant planets are totally different than the inner planets. They are larger; they have lower internal density; they do not have surfaces; they are primarily composed of low atomic weight elements; and they have extensive satellite and ring systems (Table 6). The measured densities of the giant planets gives the best insight into their bulk compositions. In Figure 4 the masses and sizes of the planets are compared with those computed for model planets of different bulk compositions (34). The figure indicates that Jupiter and Saturn have interiors composed primarily of hydrogen. Their densities are compatible with their solar-abundance mix of the elements. Hydrogen and helium, the principal constituents of Jupiter and Saturn, would not have formed solid grains under conditions which could have existed when the solar system formed. Unlike the terrestrial planets, these two planets could not have formed by accretion of solid materials. They could only have formed by direct gravitational collapse of solar-nebula gas. Figure 4 also shows that Uranus and Neptune have a distinctly different composition. They are too dense to be composed of hydrogen and helium in solar proportions. These planets are probably composed of hydrogen,

Table 6. Basic Data for the Outer Planets [a]

Property	Jupiter	Saturn	Uranus	Neptune	Pluto
mean distance from Sun, 10^8 km	7.783	14.27	28.70	44.97	59.00
mean distance from Sun, AU	5.203	9.539	19.18	30.06	39.44
orbital period, yr	11.86	29.46	84.01	164.8	247.7
orbital eccentricity	0.0485	0.0556	0.0472	0.0086	0.256
orbital inclination, degrees	1.30	2.49	0.77	1.77	17.2
rotational period, h	9.841	10.23	15.5	15.8	153.3
obliquity of rotation axis, degrees	3.08	29	97.92	28.8	≥50
mass, 10^{28} g	189.9	56.86	8.66	10.30	ca 0.001
mass, relative, Earth = 1	317.9	95.15	14.54	17.23	ca 0.0005
equatorial radius, 10^3 km	71.90	60.00	26.15	24.75	ca 1.4
equatorial radius, relative, Earth = 1	11.27	9.44	4.10	3.88	ca 0.2
escape velocity, km/s	59.5	35.6	21.22	23.6	ca 1
mean density, g/cm^3	1.314	0.69	ca 1.2	ca 1.7	ca 1

[a] Ref. 14.

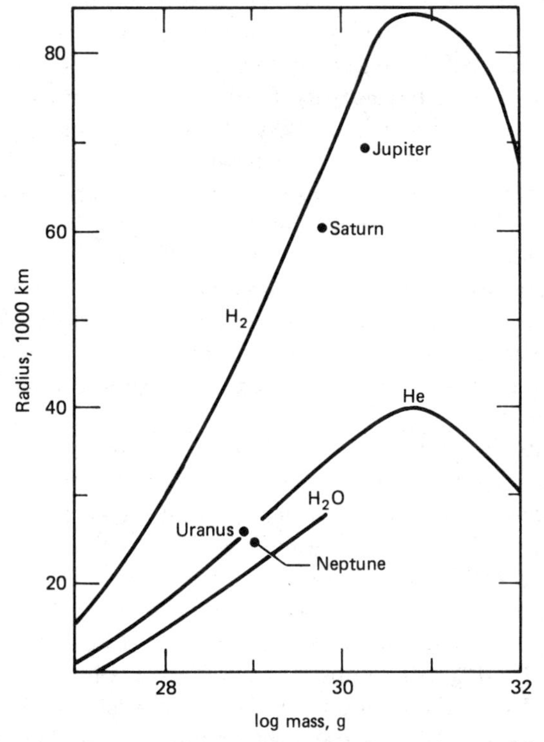

Figure 4. Comparison of the outer planets with model planets computed for different compositions. The plot distinguishes Jupiter and Saturn as H_2, He planets and Uranus as being composed of heavier elements such as those which form ices (H_2, CN, and O) (34).

carbon, nitrogen, and oxygen in ratios determined by the amounts of water, methane, and ammonia ices and clathrates that existed in the outer regions of the solar nebula. The terrestrial planets formed from stony and metallic dust grains, but in the outer reaches of the solar system the temperature was evidently low enough for both dust and ice grains to exist as solids. Uranus and Neptune formed by accretion of these grains but are primarily composed of the icy materials because hydrogen, carbon, nitrogen, and oxygen are more cosmically abundant than the stony elements. The composition of Uranus and Neptune is probably similar to the bulk composition of comets.

The interiors of the giant planets are complex. They are differentiated with cores of stony composition matter surrounded by material of lower atomic weight (34). In Saturn and Jupiter at pressures >300 GPa (>3 Mbar), hydrogen exists in the metallic state and resembles a simple alkali metal (35). In Uranus and Neptune at pressures >100 GPa (>1 Mbar), there is evidence that methane may break down into carbon and hydrogen (36). The stable phase of carbon would be either diamond or metallic carbon. Both Jupiter and Saturn radiate considerably more energy than they absorb from the Sun. Part of this energy may originate in the separation of immiscible hydrogen and helium phases at high pressures (35). The gravitational separation of helium rain falling into the planet's core is a heat source much as the separation of the earth's iron core was a heat source early in its history.

The outer planets, with the possible exception of Pluto, do not have surfaces, but detailed spacecraft observations of their atmospheres and other phenomena were made by Voyager 1 and 2 spacecraft (14,37).

Asteroids and Comets. Asteroids and comets are believed to be surviving relic bodies from the planetary formation process. Asteroids range in size up to 1000 km in diameter and the largest observed comets are estimated to be only 25 km in size. The asteroids, which apparently formed between Mars and Jupiter, seem to be small planetesimals that did not accrete to form a planet. There is no evidence that they are fragments of a planet-sized body that disintegrated. Spectral reflectance curves measured for asteroids indicate that there are many different compositional types (38), some consistent with meteorite types found on the earth. Asteroids are widely believed to be the main, and possibly the sole, source of recovered meteorites.

In Table 7 chemical analyses are given for three meteorite types that are probably representative of three common asteroid types. The table illustrates the variety of objects that formed in the asteroid region of the solar system. Most meteorites that hit the Earth, and also presumably most of the asteroids, are undifferentiated objects whose elemental compositions approximate solar abundances for many elements. Even among these objects, however, there is diversity. They range from the carbonaceous chondrites with no metal and large water and carbon contents to the H chondrites which contain only minimal amounts of water and carbon but large amounts of metallic iron. Although rare, some meteorites and asteroids were strongly differentiated due to melting and element separation early in the history of the solar system. The iron meteorite in Table 7 is the result of an asteroid analogue to planetary-core formation, and other differentiated materials resemble basalts and even material from a core–mantle interface. Actual melting is rare but most meteorites contain evidence of prolonged heating at moderate temperatures. The source of heat that would melt bodies only a few hundred kilometers in diameter has always been a mystery because the heat liberated from uranium, potassium, and thorium, the heat sources of the terrestrial planets, is totally inadequate to melt small bodies like asteroids. The discovery of the existence of ^{26}Al in the early solar system (41) provides a possible solution to this problem (see Radioactivity, natural). Aluminum-26, with a half-life of $<10^6$ yr, is an extinct radioisotope that existed in some meteorites at concentrations high enough to cause melting in asteroid-size bodies (see Radioisotopes). This heating oc-

Table 7. Meteorite Compositions, wt % [a]

	Type I (carbonaceous chondrite)	Type H (ordinary chondrite)	Typical iron meteorite
Fe (metal)	0.00	16.30	90.6
Ni (metal)	0.00	1.74	7.9
Co (metal)	0.00	0.09	0.5
FeS		5.48	
SiO_2	22.56	36.74	
TiO_2	0.07	0.12	
Al_2O_3	1.65	2.04	
Cr_2O_3	0.36	0.55	
FeO	23.70	10.24	
MnO	0.19	0.32	
MgO	15.81	23.44	
CaO	1.22	1.60	
Na_2O	0.74	0.90	
K_2O	0.07	0.09	
P_2O_5	0.28	0.27	0.3
H_2O	19.89	0.15	
C	3.10	0.02	0.04
NiO	1.23		
CoO	0.06		
S	5.49		0.7

[a] Refs. 39–40.

curred only for a few million (10^6) years and could occur only a few half-lives after the formation of ^{26}Al. Aluminum-26 is believed to have been synthesized in a supernova that preceded the formation of the solar system by less than a few million (10^6) years.

A report of enrichment of levorotatory amino acids (qv) in the Murchison meteorite is a finding for which there is no scientific explanation (42).

Comets resemble asteroids in that they too are small early solar-system objects that escaped incorporation into planets. The comets differ, however, in that they must have accumulated much farther out in the solar system in the regions where conditions were cooler and volatile ices could survive as grains. According to the standard model for a comet nucleus, the Whipple model, a comet is a "dirty-snowball" mixture of ice and dust in approximately equal proportions (43). Cometlike objects were presumably the building blocks of Neptune, Uranus, and the icy satellites of the outer planets. Water ice is believed to be the main volatile in most comets, but the telescopic detection of CO_2^+, CO^+, CH, CN, CH_3CN, NH_2, C_2, C_3, and CS in comet spectra indicates that the volatile inventory of comets is complex.

Because comets are unstable near the Sun, they must have formed in cold conditions and remained at great distance from the Sun for most of the age of the solar system. Orbital considerations indicate that most of the solar system's comets are presently >10,000 AU from the Sun. It is estimated that there are about 10^{11} comets orbiting the Sun at this distance in what is known as Oort's cloud of comets. Occasionally a comet is gravitationally perturbed from the cloud into an orbit that passes close to the Sun and into the realm of the terrestrial planets. Near the asteroid belt, solar heating becomes sufficient to sublime gas from a comet's nucleus, usually only a kilometer across, and departing gas and dust produce a visible coma surrounding the nucleus and a tail up to 2×10^8 km long. When they are too far from the Sun to have tails or gas–dust comas, comets are totally undetectable by existing technology; when close to the Sun, their nuclei are obscured. Until Halley's comet is visited in 1986 by USSR, Japanese, and European spacecraft, the structure of cometary nuclei will remain unknown.

The origin of comets is unknown and perhaps even unknowable as their present orbits contain no information whatsoever about their place of origin. The most popular theories for cometary origin is that they either formed near Uranus and Neptune and then were gravitationally ejected into Oort's cloud or that they formed at some intermediate location between Pluto and Oort's cloud. Because they formed so far from the Sun, it is possible or even probable that the dust particles in comets are actually well-preserved samples of the presolar interstellar grains that formed the solar system. If so, the grains are older than the Sun and planets. At least part of the volatiles and organic materials in comets might also be the residue of the low atomic weight mantles on the grains (11). Possible samples of cometary material have been collected in the form of small interplanetary dust particles that entered the Earth's atmosphere without strong heating (44).

Satellites. Satellites are primarily a phenomenon of the outer planets. The only satellites around the inner planets are Phobos and Diemos, the two tiny satellites of Mars, and the Earth's moon. The Moon is a substantial satellite but is anomalous in many ways. It is so large in relation to the earth that many people consider it a sister planet rather than a subordinate satellite. The Moon is the only satellite about which a great deal is known, and yet it is still not understood why and how it came into being. The great scientific returns from the Apollo missions revealed the Moon to be a body surprisingly different than expected. The Moon had an active evolutionary period that included formation of a calcium-, aluminum-, and silicon-rich crust floating on a magma ocean or oceans, a period of huge impact events which resurfaced the moon, and a 10^9-yr period of episodic basaltic volcanism. Quite unlike the Earth, however, the Moon became an inactive body after only 1.5×10^9 yr of existence.

The most remarkable aspect of the Moon is its composition. Its average density, 3.3 g/cm^3, implies that, unlike the Earth and Venus, it cannot have a significant iron core. Regardless of its density, the surface materials on the Moon are highly depleted in the siderophile elements as if the Moon did have a large iron core. The Moon is also depleted in volatiles. In comparison with the Earth, carbon and water are essentially nonexistent on the Moon. What little carbon there is in lunar soils is probably extralunar, brought in by the solar wind and meteorites. The returned samples show no positive evidence of even trace amounts of indigenous water. A case has been made that the bulk composition of the Moon is similar to the composition of the Earth's mantle and that the Moon formed from mantle material ejected from the Earth during its final stages of accretion (6). To provide a depletion of siderophile elements, this would have to have occurred after the formation of the Earth's iron core.

The satellite systems of the outer planets are extensive (Table 8). The regular satellites, those with circular orbits in the equatorial plane of a planet, undoubtedly formed in conjunction with the processes that formed the mother planet. In the Jovian system there appears to have been a miniature solar nebula where heat from the newly formed Jupiter prevented the existence of ice grains near the planet, for the inner Jovian satellites are stony objects and the outer ones icy. This compositional change

Table 8. Data for Principal Satellites[a]

Planet	Satellite	Diameter, km	Orbital period, d	Mean distance from planet, km	Mean density, g/cm^3
Earth	Moon	3,476	72.32	384,000	3.35
Mars	Phobos	22	0.319	9,380	ca 1.9
	Deimos	12	1.262	23,500	ca 2.1
Jupiter	Amalthea	240	0.489	181,300	
	Io	3,632	1.769	412,600	3.55
	Europa	3,126	3.551	670,900	3.04
	Ganymede	5,276	7.155	1,070,000	1.93
	Callisto	4,820	16.69	1,880,000	1.81
	Himalia	170	250.6	11,470,000	
Saturn	Mimas	390	0.9642	188,224	1.2
	Enceladus	500	1.389	240,192	1.1
	Tethys	1,050	1.990	296,563	1.0
	Dione	1,120	2.755	379,074	1.4
	Rhea	1,530	4.529	527,828	1.3
	Titan	5,140	15.94	1,221,432	1.9
	Hyperion	290	21.73	1,502,275	
	Iapetus	1,440	79.24	3,559,400	1.2
	Phoebe	200	406.5	10,583,200	
Uranus	Miranda	320	1.414	130,000	ca 2
	Ariel	860	2.520	192,000	ca 2
	Umbriel	900	4.144	276,000	ca 2
	Titania	1,040	8.706	438,000	ca 2
	Oberon	920	13.463	586,000	ca 2
Neptune	Triton	3,800	5.877	355,000	ca 2
	Nereid	940	359.9	5,562,000	ca 2
Pluto	Charon		6		ca 1

[a] Refs. 14 and 37.

with radial distance mimics that of the planets. In the Saturnian system all the satellites are mixtures of icy materials with a usually minor component of stony materials. The stony fraction of the satellites actually increases with distance from Saturn. It has been suggested that because of the slow formation of the Saturn system, the early formed stony dust grains might have spiraled into Saturn owing to gas drag in the dense inner regions of the gas cloud surrounding the newly formed planet. The difference between the Jovian and Saturnian systems is probably the result of the different masses of the planets and the time scales of their formation.

The only resolved images of the Saturnian and Jovian satellites are those returned by the Voyager 1 and 2 spacecraft (37). The Voyager results revealed an incredible range in physical character of the satellites. Some are stony objects and could be thought of as terrestrial. Most of the satellites, however, contain varying amounts of icy materials and are compositionally different from the terrestrial planets. The icy-composition satellites must have assembled from materials similar in composition to those that formed comets.

Saturn's moon Titan, the second largest satellite in the solar system, is intermediate in size between Mercury and Mars and unique in that it has an atmosphere. Voyager measurements indicate that its surface pressure is ca 150 kPa (ca 1.5 atm). Because of the lower surface gravity, this pressure implies a very massive atmosphere.

The composition of the atmosphere is nitrogen, methane, and argon in descending order of importance. Voyager also detected, in ppm to percent concentrations, ethane, acetylene, ethylene, hydrogen cyanide, and, tentatively, methylacetylene and propane. Titan is unique in the solar system in that it has an atmosphere where organic synthesis can occur and the products then gravitationally settle downward and concentrate on a solid surface. Organic synthesis also occurs in the atmospheres of the giant planets but there the materials eventually settle or are transferred by convection downward into hot atmospheric layers where molecules are destroyed. The surface of Titan is probably covered with organics synthesized in its atmosphere by the interaction of methane, hydrogen, and ammonia with the solar wind, energetic charged particles, and internal energetic events such as electrical discharges. Because of the huge amount of carbon in the atmosphere and in the satellite, the "organic" surface layer on Titan is likely to be quite thick. Because the surface pressure and temperature of Titan is close to the triple point of methane, methane may play a role in Titan meteorology similar to that of water in the Earth's atmosphere. The existence of ponds, lakes, or oceans of methane is a definite possibility. Liquid nitrogen probably also exists on Titan but perhaps only as atmospheric droplets.

Most satellites are heavily cratered objects that have been geologically inactive since the early history of the solar system. Callisto, a large Jovian satellite, contains an enormously high density of impact craters that probably date back to the actual accretion period of the satellite. The Saturnian satellites Tethys and Mimas show craters produced by impacts of energy almost great enough to destroy the satellite. Craters were expected on all satellites. What was not expected was that several satellites would have active histories of other geological processes and that one satellite would be so active that rapid resurfacing has obliterated all impact craters. The satellites were commonly presumed to be cold dead objects because they would not have sufficient internal heat sources to drive surface processes. Some satellites such as Io and Enceladus, however, are internally heated by tidal friction (45). In the case of Io the heating is strong enough to make Io the most geologically active body in the solar system. Io has even fewer impact craters than the Earth. Internal heating drives a unique sulfur volcanism of such intensity that several simultaneous eruptions occur on the body constantly. The surface of Io is renewed on a time scale of only a few million (10^6) years or less. The surface is covered by sulfur flows, sulfur dioxide snow, volcanic ash, and hot sulfur lakes. Active processes are less dramatic on other satellites, but bodies such as Enceladus, Iapetus, Europa, and Ganymede have also been extensively altered by geological processes that obliterate ancient crater terrains.

Near-Earth Extraterrestrial Resources

If large-scale manned operations are ever conducted in space, the utilization of available extraterrestrial materials may become a necessity (46). Economic and environmental considerations place practical limits on the amount of material that can be lifted from the Earth's surface by rockets and placed into orbit. Projects such as large orbiting habitats, space factories, solar-power satellites, and large lunar bases could benefit greatly from use of materials that are already in space. The specific energy required to remove material from selected near-Earth bodies is of course less than that required to lift materials from the Earth into orbit. Various forms of extraterrestrial resources could be used. One form of material would be highly processed products such as GaAs solar cells or PtIr reaction chambers. Such products, made of rare elements, could be refined and processed at the material source, requiring only relatively small amounts of mass to be transported to the use site. Perhaps the most valuable use of extraterrestrial materials, however, would be as bulk building materials. These could consist of only moderately refined materials such as fertile soil for growing plants or construction beams and slabs made of metallic nickel–iron and glass materials that are massive and would require ambitious schemes for transportation to the use site.

The potential suppliers of near-Earth space resources are the Moon and the Apollo (Earth-crossing) asteroids. For lunar operations the prime source would be lunar material, but for low or high Earth orbit either lunar or asteroidal materials could be used. In the case of asteroids, the approach to obtaining large quantities of bulk material would be to move an entire asteroid into Earth orbit. Choosing Apollo asteroids with orbits very similar to the Earth's would allow this maneuver to be done with a total velocity change on the order of only 5 km/s. This cannot be accomplished using external rockets, but it can be done using a fraction of the asteroid as reaction mass ejected from the body at high velocity. This is possible in principle using a series of controlled thermonuclear explosions, but the most popular approach is to use solar-energy-powered catapults that would launch small parcels of asteroidal material at high velocity. The launchers that have been investigated are electric rail guns and electromagnetic mass drivers (47). These devices would launch continuous streams of rather small projectiles at velocities of several kilometers per second. The thrust from this ejection would be low but continuous and over a period of years would be sufficient to place the asteroid into Earth orbit. In this process a significant fraction of the asteroid mass would be consumed in producing projectiles. The Earth capture phase would probably involve gravitational interaction with the Moon as this reduces the total energy requirement. The asteroid propulsion system would consist of a solar-power station, a means of excavating and forming projectiles and the launcher itself.

The future of asteroid mining is a question of economics, not technology. It seems perfectly possible that, with sufficient funding, near-Earth asteroids in the 100-m to several-kilometer size could be moved from solar orbit to Earth orbit (48). These objects could then be used as both habitats and ore bodies for building large-scale space structures. The materials of these bodies would be similar to meteorites (Table 7) and those discussed in the asteroid–comet section. These asteroids can provide enormous quantities of the most important materials for large-scale manned space operations (49). A 1-km carbonaceous chondrite would essentially be an orbiting mountain composed of 10–20 wt % water and nearly 5 wt % hydrocarbons and other organic materials. Ordinary stony meteorite types contain up to 20 wt % of iron–nickel metal. In the 10^{-4} *g* gravity of small asteroids and in the high vacuum of space, complex facilities could be constructed that would be impossible on Earth. With abundant and constant solar energy, unorthodox building techniques such as vapor deposition of evaporated materials and *in situ* foaming of molten materials in approximately weightless conditions could be utilized. In addition, more conventional methods such as the production of metal and glass fibers, sheets, and beams could be used to build farms, factories, and cities.

Even merely as pieces of real estate, the Earth-orbiting asteroids would be of great value. The obvious application, unfortunately, is military. A solid FeNi asteroid only a few hundred meters across could be an orbiting fortress easy to defend and nearly impossible to destroy. Peaceful uses of orbiting real estate are probably limitless, but the economics of this use will probably become attractive only if population, technology, motivation, and forward thinking continue to increase in the future.

Asteroids might also be brought through the atmosphere to Earth's surface, but their greatest value is as a resource of raw materials in orbit. In the future, however, the depletion of critical resources such as elements used in semiconductors, catalysts, and superalloys may make it attractive to bring refined asteroidal materials to the Earth. The advantage of the asteroids is that they are not depleted in elements that have been highly depleted on the surface of the Earth and concentrated in its iron core. For example, a kilometer-sized iron asteroid contains 44.5×10^8 metric tons of nickel, 1.8×10^7 t cobalt, 4.5×10^4 t gallium, 1.8×10^4 t platinum, 5400 t osmium, and 2700 t gold—all in metallic forms. The present market value would be $>\$10^{13}$. How asteroidal materials could be brought through the atmosphere is unclear, but presumably it would involve fabrication of low density airplane-sized bodies that could survive atmospheric entry with minimal ablation. One suggestion has been to produce low density metallic-foam lifting bodies which would land in the ocean and then be towed to appropriate recovery sites.

Lunar resources can be utilized for either lunar or Earth-orbital projects. For Earth-orbital uses, material would have to be launched into lunar-escape trajectories (50). This could be done using mass drivers or rail guns to launch lunar-material packets to the lunar escape velocity of 2 km/s. The launch velocity and aiming would have to be precise in order for the material packets to be received by a "catcher" in space.

The natural application of lunar materials is for construction and operation of a lunar base. A base could range from a simple tunnel habitat similar to a submarine to large highly complex settlements. Mining, processing, and use of lunar materials would be quite different from similar operations on the Earth. The low surface gravity (17% of the Earth's gravity) and hard vacuum would provide new opportunities, although remoteness and lack of natural water and air would cause hardships. Mining on the Moon presumably would be done at sites local to a manned lunar base. The ore would consist of the lunar soil, which covers the Moon to depths of several meters. Because the soil is fine grained, typically <100 μm (-140 mesh), crushing would not be necessary although sieving, disaggregation, and separation by physical properties would be done before beneficiation. As the product of intense meteoroid impact on basalt and on anorthosite, a rock composed largely of plagioclase feldspar, the lunar soil is composed of glass and mineral fragments. The common minerals that might be used in beneficiation schemes are anorthite, $Ca(Al,Si)_4O_8$; pyroxene, $(Mg,Fe)SiO_3$; olivine, $(Mg,Fe)_2SiO_4$; ilmenite, $FeTiO_3$; troilite, FeS; and iron metal which contains appreciable amounts of nickel and cobalt. Much of the separation of these grains could be done by electrostatic and magnetic separators designed to take advantage of the low gravity and high vacuum of the Moon. Schemes such as carbothermic–silicothermic reduction have been studied as a means of producing metallic aluminum, iron, titanium, and calcium from these lunar ores (51–52). The Moon is depleted in volatiles, water in particular, but hydrogen, carbon, nitrogen, and the rare gases required for industrial processes can be recovered from soil. These volatiles are not indigenous to the Moon but they exist in soil because they are implanted from the solar wind. For example, lunar soil typically contains 50 ppm of hydrogen (9) implanted in the outer

0.1 μm of grain surfaces and nearly 0.1 cm^3/g (STP) of helium. Select sizes and compositions of grains, eg, small ilmenite grains, contain up to 0.1 wt % hydrogen. The volatiles can be extracted from lunar soil merely by heating. Water for vital industrial and human uses would have to be produced using hydrogen recovered from grain surfaces and oxygen derived from decomposition of silicates and oxides.

Construction materials for significant lunar structures cannot be launched from the surface of the Earth. They must be made on the Moon largely from lunar materials. All of the primary materials for massive construction exist on the Moon, and there is no apparent reason why semi-autonomous stations could not be built and maintained there. Such projects would of course be very expensive in comparison with installations in existing remote habitats such as nuclear submarines and polar-research stations. Utilization of lunar materials will only occur when and if there is a strong economic or political need.

Chemical Processes in Space

The first chemical experiment conducted in the weightless environment of space took place in the space shuttle Columbia. From a latex mixture heated to a constant 70°C, a 14-h chemical reaction produced tiny polystyrene latex spheres of a size, quality, and uniformity not possible with current earthbound technology. The Columbia system produced monodisperse spheres as large as 5-μm. They will be used as seeds for further polymerization of spheres as large as 20-μm. The spheres could be used as calibration standards and as parts of medical diagnostic applications (53).

BIBLIOGRAPHY

1. S. Weinberg, *The First Three Minutes*, Basic Books, New York, 1978.
2. A. A. Penzias and R. W. Wilson, *Astrophys. J.* **142,** 419 (1965).
3. J. Audouze and S. Vauclair, *An Introduction to Nuclear Astrophysics*, D. Reidel, Boston, Mass., 1980.
4. H. Reeves, *Annu. Rev. Astron. Astrophys.* **12,** 437 (1979).
5. E. Anders, *Geochim. Cosmochim. Acta* **35,** 516 (1971).
6. A. E. Ringwood, *Origin of the Earth and Planets*, Springer-Verlag, Inc., New York, 1979.
7. J. T. Wasson, *Meteorites*, 1982, in press.
8. A. G. W. Cameron, *Center for Astrophysics*, preprint 1357, Harvard University, Cambridge, Mass., 1982.
9. S. Epstein and H. P. Taylor, *Geochim. Cosmichim. Acta* **2**(Suppl. 6), 1771 (1975).
10. L. Spitzer and E. B. Jenkins, *Annu. Rev. Astron. Astrophys.* **13,** 133 (1975).
11. J. M. Greenberg in C. Ponnamperuma, ed., *Comets and the Origin of Life*, D. Reidel, Boston, 1981, pp. 111–127.
12. B. Turner, *J. Mol. Evol.* **15,** 79 (1980).
13. J. Wood, *The Solar System*, Prentice-Hall, Inc., Englewood Cliffs, N.J., 1979.
14. J. K. Beatty, B. O'Leary, and A. Chaikin, *The New Solar System*, Sky Publishing Co., Cambridge, Mass., 1981.
15. H. C. Urey, *The Planets*, Yale University Press, New Haven, Conn., 1952.
16. S. S. Barshay and J. S. Lewis, *Annu. Rev. Astron. Astrophys.* **14,** 81 (1976).
17. S. J. Weidenschilling, *Icarus* **26,** 361 (1975).
18. L. Grossman and J. Larimer, *Rev. Geophys. Space Phys.* **12,** 71 (1974).
19. J. L. Lewis, *Science* **186,** 74, 440 (1974).
20. L. Grossman, *Geochim. Cosmochim. Acta* **36,** 597 (1972).
21. V. Goldschmidt in A. Muir, ed., *Geochemistry*, Clarendon Press, Oxford, UK, 1954.

22. B. Murray, M. C. Malin, and R. Greeley, *Earthlike Planets*, W. H. Freeman & Co., Publishers, San Francisco, Calif., 1981.
23. Special Mariner 10 issue, *J. Geophys. Res.* **80,** 2341 (1975).
24. C. P. Florensky, L. B. Ronca, A. T. Trakhtman, V. P. Volkov, and V. V. Zazetsky, *Geol. Soc. Amer. Bull.* **88,** 1537 (1977).
25. H. Masursky, E. Eliason, P. G. Ford, G. E. McGill, G. H. Pettengill, G. G. Schaber, and G. Schubert, *J. Geophys. Res.* **85,** 8232 (1980).
26. Special Pioneer Venus issue, *J. Geophys. Res.* **85,** 7573 (1980).
27. T. A. Mutch, R. E. Arvidson, J. W. Head, K. L. Jones, and R. S. Saunders, *The Geology of Mars*, Princeton University Press, Princeton, N.J., 1976.
28. B. K. Lucchitta, D. M. Anderson, and H. Shoji, *Nature* **290,** 759 (1981).
29. V. R. Baker, *Science* **202,** 1249 (1978).
30. *Nature* **295,** 43 (1982).
31. L. V. Berkner and L. C. Marshall, *J. Atmos. Sci.* **22,** 225 (1965).
32. A. J. Meadows, *Planet. Space Sci.* **21,** 1467 (1973).
33. *Icarus* **44,** 1 (1980).
34. W. B. Hubbard, *Science* **214,** 145 (1981).
35. D. J. Stevenson, *Phys. Rev. B* **12,** 3999 (1975).
36. M. Ross, *Nature* **292,** 435 (1981).
37. Special Voyager issues, *Science* **204,** 945 (1979); **206,** 925 (1979); **212,** 159 (1981).
38. C. R. Chapman and M. J. Gaffey in T. Gehrels, ed., *Asteroids*, The University of Arizona Press, Tucson, Ariz., 1979, p. 655.
39. V. F. Buchwald, *Handbook of Iron Meteorites*, University of California Press, Berkeley, Calif., 1975.
40. B. Mason, *Handbook of Elemental Abundances in Meteorites*, Gordon and Breach, New York, 1971.
41. C. M. Gray, D. A. Papanastassiou, and G. J. Wasserburg, *Icarus* **20,** 213 (1973).
42. M. H. Engel and B. Nagy, *Nature* **296,** 837 (1982).
43. F. L. Whipple in ref. 11, p. 1.
44. D. E. Brownlee in ref. 11, p. 63.
45. S. J. Peale, *Science* **203,** 892 (1979).
46. G. K. O'Neill, *Phys. Today* **27,** 32 (1974).
47. W. H. Arnold, S. Bowen, K. Fine, D. Kaplan, M. Kolm, H. Kolm, J. Newman, G. O'Neill, and W. R. Snow in *Space Resources and Space Settlements, NASA SP-428*, 1979, p. 87.
48. B. O'Leary, *Science* **197,** 363 (1977).
49. M. J. Gaffey, E. F. Helin, and B. O'Leary in ref. 41, p. 191.
50. D. R. Criswell in ref. 41, p. 207.
51. R. J. Williams, D. S. McKay, D. Giles, and T. E. Bunch in ref. 41, p. 275.
52. D. Bhogeswara, U. V. Choudary, T. E. Erstfeld, R. J. Williams, and Y. A. Cheng in ref. 41, p. 257.
53. *Ind. Chem. News* **3**(7), 10 (1982).

D. E. Brownlee
University of Washington

SPANDEX AND OTHER ELASTOMERIC FIBERS. See Fibers, elastomeric.

SPRAYS

A spray is a liquid-in-gas dispersion in the form of a multitude of drops. In chemical-processing applications, the drops are formed during the process of atomization in which a liquid column or sheet is broken up because of mechanical energy in the liquid or in another fluid that is mixed with the liquid to be atomized, or mechanical energy applied externally through a rotating or vibrating device. The device producing the dispersion is the atomizer. The resultant spray can comprise drops ranging in diameter from submicrometer to several thousand micrometers.

In the home, sprays are produced by shower heads, garden-hose nozzles, and aerosol-can jets (see Aerosols). They are commonly used for applying agricultural chemicals to crops, coating surfaces, dispersing liquid fuels for combustion, spray drying wet solids, gas-liquid mass-transfer operations, and a host of other applications. Natural sprays include waterfall mists, rains, and ocean sprays. Although in this article emphasis is given to sprays that are useful in chemical processing, the principles of spray formation and the properties of the dispersions apply to all types of sprays. The term droplet is used to emphasize small-particle dispersions.

Characteristics

Mechanics of Breakup. Sprays are produced by the breakup of liquid filaments or sheets. A filament issuing from an orifice of diameter d_o is inherently unstable, and in an early analysis it was predicted that the filament would break up into essentially spherical drops with a uniform diameter equal to 1.89 times the orifice diameter (1). Careful studies using high speed photography have shown this prediction to be remarkably valid for low viscosity liquids. For more viscous liquids, a modified diameter ratio appears appropriate (2):

$$\frac{D}{d_o} = 1.89\left[1 + \frac{3We_L^{1/2}}{Re_L}\right]^{1/6} \quad (1)$$

where the two dimensionless groups, the liquid Weber number and the liquid Reynolds number, are defined as follows:

$$We_L = \frac{u_o^2 \rho_L d_o}{\sigma} \quad (2)$$

$$Re_L = \frac{d_o u_o \rho_L}{\mu_L} \quad (3)$$

There are limits to this simplified Rayleigh breakup ratio. At low liquid rates ($We_L < 8$), there is nonuniform dribbling from the orifice (3). At high liquid rates, gas inertia imposes additional sinuous disturbances on the filament, and the liquid tends to atomize at or near the orifice. A criterion for the onset of atomization is (4–5):

$$We_L > 2.8 \times 100 \left[\frac{We_L^{1/2}}{Re_L}\right]^{-0.82} \quad (4)$$

Applications for simple filament breakup are limited to cases where uniform drop size is needed, as in the production of spherical prills from molten salts or earlier-day manufacture of lead shot.

Of greater commercial significance is the breakup of liquid sheets issuing from atomizing devices. These sheets have surface waves that tend to cause rupture at the nodes, producing filaments which, in turn, disintegrate to droplets in a range of sizes. If the liquid is relatively nonviscous, the thickness of the sheet t_s at the point of rupture and the sheet velocity of rupture U_s are related by:

$$U_s = \left[\frac{4\sigma}{\rho_L t_s}\right]^{1/2} \tag{5}$$

Once the filaments are formed, the Rayleigh–Weber criteria for further breakup can be applied. Droplet size is a function of ruptures per unit area and the thickness of the sheet. Thus, droplet diameter can be estimated (3) by:

$$D = 1.89\left[\frac{8t_s^2}{\pi^2\sqrt{3}n_r}\right]^{1/4} \tag{6}$$

where n_r = number of ruptures per unit area. The observed range of values of D indicates ranges of t_s and n_r, and experimental evaluation of these parameters is required. This general approach served to obtain droplet diameter relationships for swirl-type pressure nozzles (6):

$$D = C_1\left[\frac{\sigma t_s l}{U_s^2}\right]^{1/2}(\rho_L/\rho_G)^{1/6} \tag{7}$$

where the constant C_1 is determined experimentally.

The usual criterion for stability is the gas Weber number for the droplet; a value of ca 13 for the critical Weber number has been proposed (7). On this basis:

$$D_{max} = 13\,\frac{\sigma}{U_d^2\rho_G} \tag{8}$$

where U_d is the droplet velocity relative to the gas environment. The concept of maximum stable droplet size has application in many gas-liquid contacting operations.

Droplet Size Distribution. The typical spray pattern consists of a range of droplet sizes. Some knowledge of size distribution is helpful in evaluating process applications of sprays, especially when it is the question of transfer of heat and mass between the dispersed liquid and its surrounding gas. The careful selection and design of the atomizing device is critical in such applications, and droplet size distribution is an important function of fluid properties and atomizing pressure as well as of the geometry of the device itself (see Fluidization; Fluid mechanics).

In a general sense, distribution of droplet sizes may be considered in relation to the normal or Gaussian distribution. A useful function for evaluating spray distribution has the log-normal form:

$$\frac{dn}{dD} = f(D) = \frac{1}{Ds_g\sqrt{2\pi}}\,e^{-(\ln D\, -\, \ln D_g)^2/2\, s_g^2} \tag{9}$$

where n = number of droplets, D_g = number geometric mean droplet size, and s_g = geometric standard deviation

A preponderance of small droplets often causes heavy skewing of the distribution curve, and the volume (or mass) mean equivalent of equation 9 is appropriate:

$$\frac{dV}{dD} = f(D^3) = \frac{1}{Ds_g\sqrt{2\pi}}\,e^{-(\ln D\, -\, \ln D_v)/2\, s_g^2} \tag{10}$$

where V = droplet volume and D_v = geometric volume mean diameter. On either a number or volume basis, measured droplet diameters may be placed in size classifications and represented as a histogram as shown in Figure 1 (8). The equivalent continuous curve (eq. 9) is also shown; the curve is based on differential, instead of discrete, droplet size groupings. The dashed curve of Figure 1 is based on volume, rather than number, fractions. It is skewed to the right because of weighting in proportion to D^3 (eq. 10).

A normal distribution of volumes (or mass) of droplets is represented by a straight line on log-normal probability graph paper. A representative distribution is shown as the upper curve of Figure 2; details of the device producing this distribution are described below. The data of Figure 2 show the expected departure from a normal distribution (9).

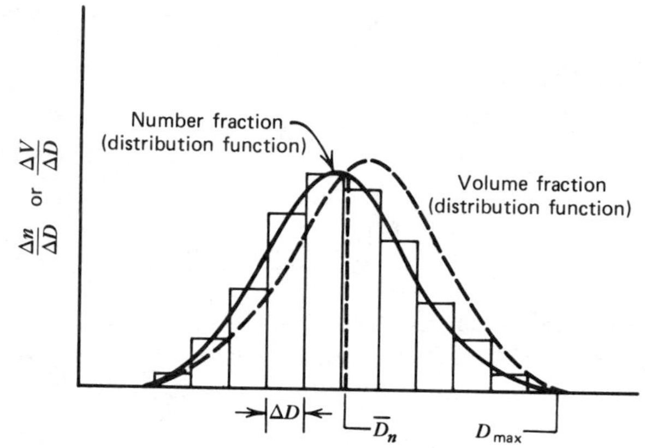

Figure 1. Histogram of droplet number per size class (8).

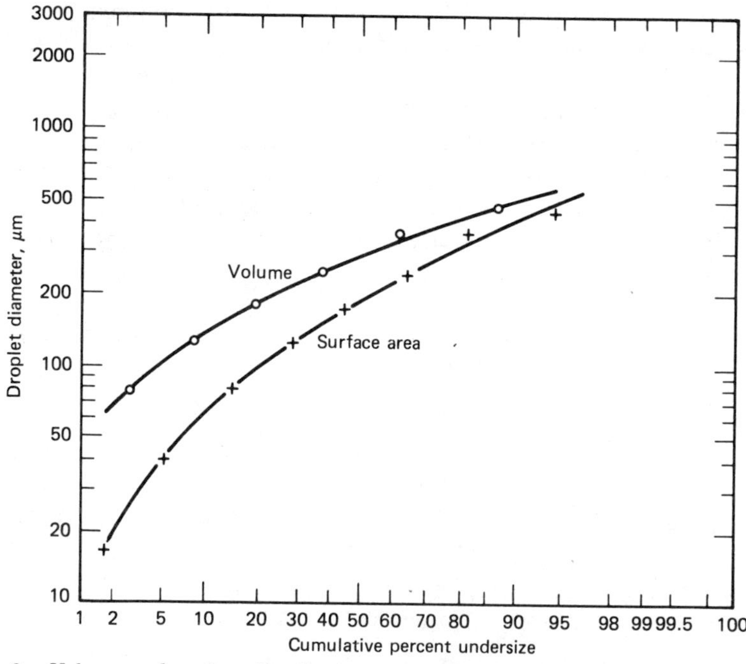

Figure 2. Volume and surface distributions; spray droplets produced by a hollow-cone (swirl-chamber) nozzle; d_o = 1.60 mm; ΔP = 690 kPa (100 psi); flow = 1380 cm^3/s; water into air (9).

Diameters. Several diameters are used for characterizing spray dispersions. Median diameters divide the spray into two equal portions by number, surface area, or volume. The number median $\overline{D}_n$ is the smallest, because of the preponderance of small-size droplets. The volume median $\overline{D}_v$ (or mass median, since liquid density does not change) is easily determined from cumulative distribution plots such as Figure 2; for the example shown, the volume or mass median size is 290 μm. The surface-area median $\overline{D}_s$ is also obtained from a plot such as that shown in Figure 2; the lower curve is based on surface area and the value of $\overline{D}_s$ is 190 μm. The number median for this particular spray pattern is <17 μm.

Mean diameters refer to hypothetical droplet sizes that represent particular composite properties of sprays. The common mean diameters are defined and expressed as follows:

Arithmetic mean. Simple weighted average based on all the individual droplets in the spray.

$$D_a = \frac{\sum_{i=1}^{k} (N_i D_i)}{\sum_{i=1}^{k} N_i} \tag{11}$$

Geometric mean. Simple average based on the logarithm of the diameters of all the individual droplets in the spray.

$$\ln D_g = \frac{\sum_{i=1}^{k} (\ln D_i) N_i}{\sum_{i=1}^{k} N_i} \tag{12}$$

Surface mean. Diameter of a droplet whose surface area, if multiplied by the total number of droplets, equals the surface of all particles in the spray.

$$D_s = \left[\frac{\sum_{i=1}^{k} N_i D_i^2}{\sum_{i=1}^{k} N_i} \right]^{1/2} \tag{13}$$

Volume mean. Diameter of a droplet whose volume, if multiplied by the number of droplets, equals the total volume of the sample.

$$D_v = \left[\frac{\sum_{i=1}^{k} (N_i D_i^3)}{\sum_{i=1}^{k} N_i} \right]^{1/3} \tag{14}$$

Sauter mean. Diameter of a droplet whose ratio of volume to surface area is equal to that of the entire spray.

$$D_{vs} = \frac{\sum_{i=1}^{k} (N_i D_i^3)}{\sum_{i=1}^{k} (N_i D_i^2)} \tag{15}$$

Of these diameters, the Sauter mean has frequent application in mass-transfer analyses which involve a combination of surface area per unit volume and boundary-layer effects for a particular droplet size and shape (see Mass transfer). The Sauter mean diameter is often about 75–85% of the volume median diameter. For the spray of Figure 2, the Sauter mean diameter is 215 μm. Methods for predicting Sauter mean diameters are given below.

Drop Size Measurement. Many techniques have been used for the measurement of drop size distribution in sprays (10–13). These techniques vary greatly in accuracy, cost, time for data reduction, and compatibility with the spray liquid and system geometry. Many reported measurements have been based on droplet collection on a slide coated with wax or similar material that can stop the droplets long enough for counting by manual or automatic means. A less frequently used but related approach is to measure the distribution of crater sizes formed by impaction of droplets on a slide coated with magnesium oxide.

Another approach is to freeze the drops quickly and then use sieving or microscopic techniques to evaluate the size distribution. This method is best suited for molten materials that are solid under ambient conditions. Another approach uses photographic methods where the drops are counted from enlarged reproductions.

More recent methods are based on optical, inertial, and electrical techniques, such as light scattering of the particles in a collimated laser beam (see Lasers), or an interferometric method for the off-axis detection of scattered light (14–15). The latter can be used for very small droplet sizes and provides information on velocities as well as diameters (see Fig. 3) (15).

Atomizers

The geometry of the atomizing device is based on the characteristics of the spray desired. A number of manufacturers make these devices, each offering a broad assortment of designs. The manufacturer can usually supply flow and pressure-drop data, based on water discharging into atmospheric air, and, in some cases, data on droplet size distribution. Accordingly, data available from the manufacturer should always be consulted before any quantitative analysis of spray characteristics for a given application is undertaken.

Centrifugal Pressure Nozzles. ***Swirl Chamber (Hollow Cone).*** For this device, a circular orifice outlet is preceded by a chamber in which the liquid is given a swirl, either by special tangential inlets or by slotted distribution intervals. The device is called a hollow-cone nozzle because of the spray pattern developed. The included spray angle ranges from 30–160°, depending on design. The elements of a typical swirl-chamber nozzle are shown in Figure 4 (16). As pressure drop across the nozzle increases, liquid capacity increases and dispersion becomes finer, as shown in Figure 5.

The swirling liquid creates an air (or gas) core extending from the rear of the chamber through the center of the orifice. The issuing sheet of liquid breaks into droplets that fall around a central gas core; thus the hollow-cone description.

Flow through the nozzle may be correlated by the conventional orifice equation:

$$u_o = K\sqrt{\frac{2\,\Delta P}{\rho_L}} \tag{16}$$

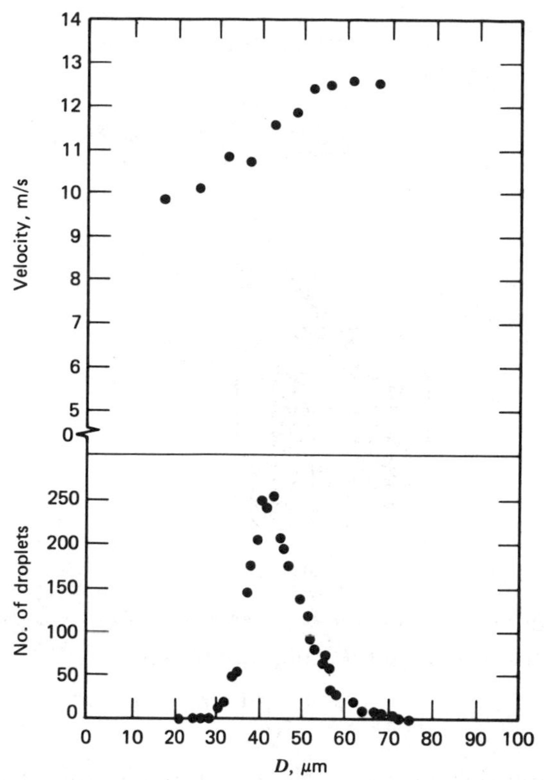

Figure 3. Spray droplet size variations for a sonic-type nozzle dispersing water into air with a nozzle air pressure of 6.89 kPa (1 psi). The sample was obtained 5 cm below the nozzle exit on the centerline of the spray. The volume-surface (Sauter) mean diameter was 49 μm (15).

where u_o is a superficial liquid velocity based on the orifice area. However, the total area is not utilized by liquid, because of the gas core, and the discharge coefficient K can thus be 0.2 or less. If the gas-core diameter d_c is known, the value of K can be estimated with the following equations (17):

$$a = 1 - \left(\frac{d_c}{d_o}\right)^2 = \frac{\text{liquid flow area}}{\text{orifice area}} \tag{17}$$

$$K = \sqrt{a^3/(2-a)} \tag{18}$$

Data for the ratio d_c/d_o are plotted in Figure 6 for several nozzles in which the swirl was imparted by grooved inserts in the whirl chamber (17). Values of K for swirl-chamber nozzles are often in the range of 0.2–0.5.

The actual axial velocity of the issuing liquid is higher than the superficial velocity:

$$u_L = u_o(A_o/A_c) = u_o(d_o^2/(d_o^2 - d_c^2)) = u_o/a \tag{19}$$

The mean diameters of these nozzles can be predicted, but not with high accuracy. Some experimental data are desirable, but if a rough estimate is required, in the absence of data, a correlation for the Sauter mean diameter is useful and reasonably reliable (18):

$$\frac{D_{vs}}{d_o'} = 5.0\left[\frac{d_o'\rho_L u_L'}{\mu_L}\right]^{-0.35}\left[\frac{u_L'\mu_L}{\sigma}\right]^{-0.20} \tag{20}$$

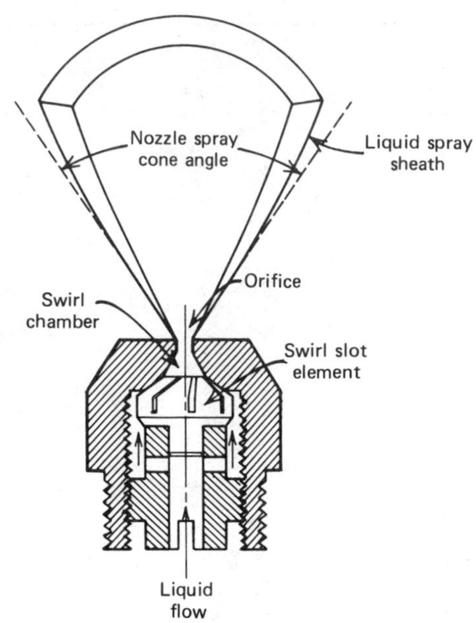

Figure 4. Typical swirl-chamber (hollow-cone) nozzle (16).

with the effective orifice diameter d_o' computed from

$$d_o' = \sqrt{\frac{4\,Q_L}{\pi\,u_L}} \tag{21}$$

The volumetric liquid flow rate Q_L is usually available from the manufacturer and is a function of nozzle pressure.

The maximum size of droplet in the distribution can be estimated (8):

$$\frac{D'_{max}}{\overline{D}_v} = \frac{\overline{D}_v(D_{90} + D_{10}) - 2\,D_{90}D_{10}}{\overline{D}_v^2 - D_{90}D_{10}} \tag{22}$$

It has been suggested that the maximum diameter is ca three times the volume or mass median diameter (19).

Typical data of swirl-chamber hollow-cone nozzles are shown in Figure 7 (20). The mass or volume median diameter is 120 μm and the Sauter mean diameter is 100 μm. By means of equations 17–22, the predicted volume/surface (Sauter) mean diameter is 112 μm, the volume mean about 130 μm, and the maximum droplet size 380 μm. Flow rate, mass median diameter, and Sauter mean diameter change with nozzle pressure, within constraints of throughput, as shown in Figure 7**b**. It would be necessary to change nozzles, for example, if droplet diameters were to be changed while holding flow rate constant.

Droplet size data supplied by nozzle manufacturers are usually based on water dispersed in air. For conversion to liquids other than water, the Mugele equation (eq. 23) suggests the following relationship (18):

$$\frac{(D_{vs})_1}{(D_{vs})_2} = \left(\frac{\rho_{L,2}}{\rho_{L,1}}\right)^{0.35}\left(\frac{\mu_{L,1}}{\mu_{L,2}}\right)^{0.15}\left(\frac{\sigma_1}{\sigma_2}\right)^{0.20} \approx \frac{D_{v,1}}{D_{v,2}} \tag{23}$$

In a similar relationship, the liquid density, viscosity, and surface tension ratios have exponents of 0.3, 0.2, and 0.5, respectively (5); a viscosity ratio exponent of 0.32 has been proposed (21).

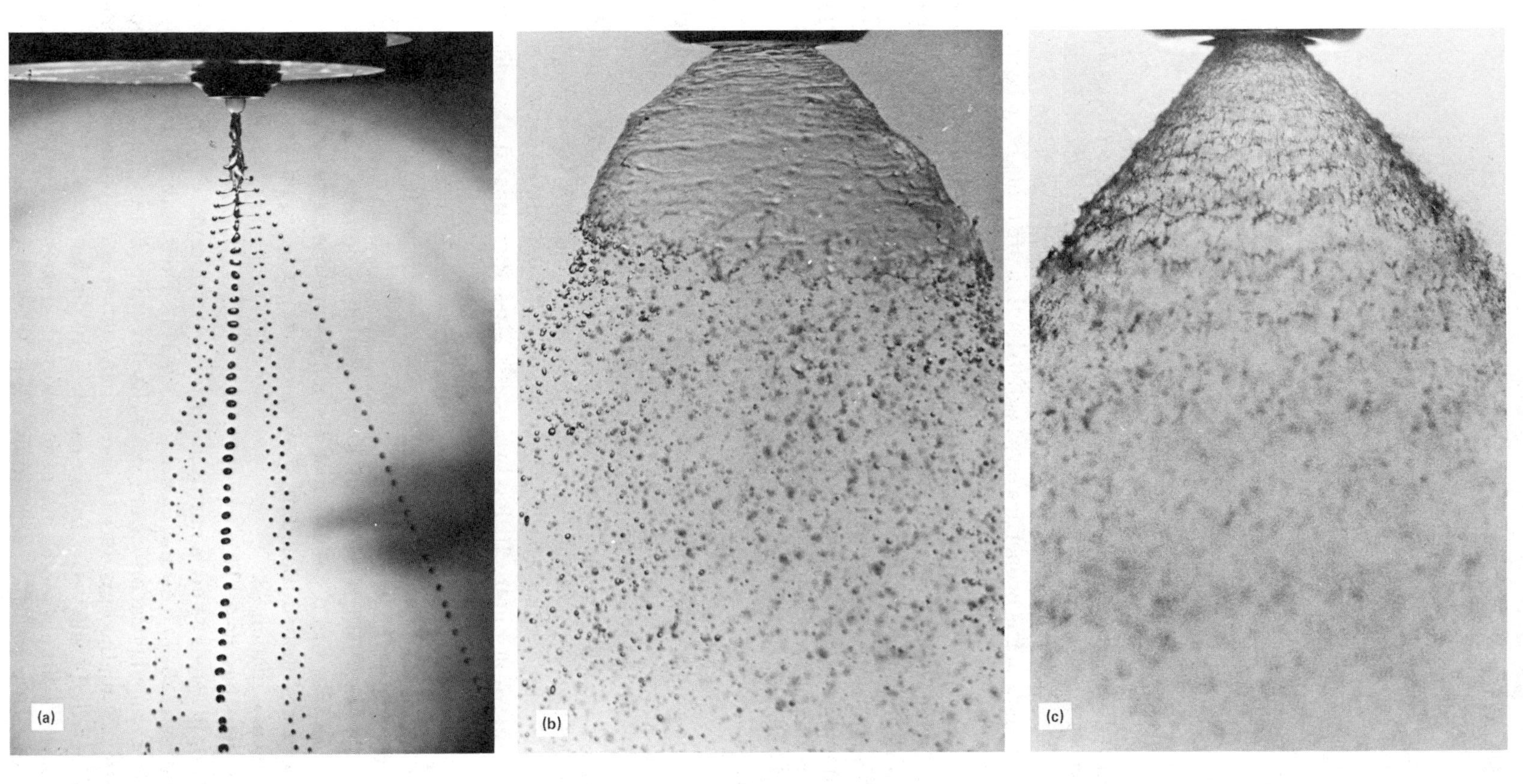

Figure 5. High speed photographs showing progressive development of hollow-cone spray with increase in nozzle pressure. Courtesy Delavan Manufacturing Co.

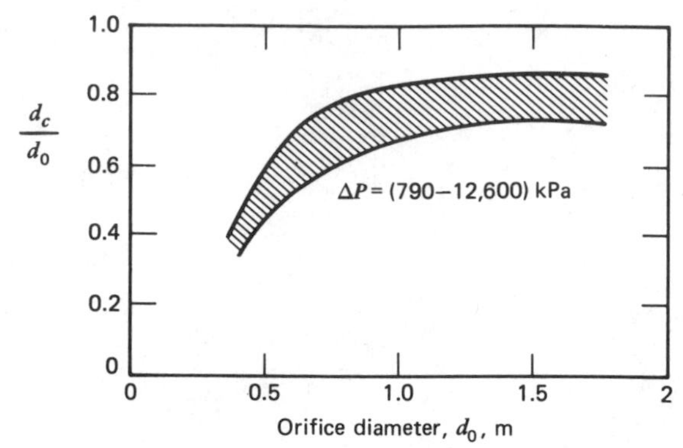

Figure 6. Correlation of air-core diameter d_c with orifice diameter d_o for different grooved-core nozzle combinations (17). To convert kPa to psi, multiply by 0.145.

The effects of gas properties on size distribution are thought to be minor, as indicated by equation 20. Significant increases in surrounding gas density can cause a decrease in average size, but the measurements shown in Figure 8 indicate minima (16); the higher sizes measured at higher pressures are attributed to coalescence caused by entrained gas flowing through the cone sheath.

This entrainment of gas has been the subject of a number of investigations, and aerodynamic studies show how entrainment influences droplet trajectories and coalescence (22–23). Aside from its effect on spray pattern, the gas can cause unwanted losses of volatiles during spray drying (24–25).

Solid-Cone Nozzle. This device is similar in design to the swirl-chamber nozzle except that a special core or axial jet fills the center of the conical pattern. The resulting full-volumetric coverage enhances rates of mass and heat transfer between the spray liquid and gas passing through the cone. A typical solid-cone spray nozzle is shown in Figure 9 (26). The included spray angle ranges from 30–120°.

Measurements of radial variations in spray uniformity show somewhat higher flow toward the center of the cone, with the correspondingly larger droplet sizes. In some cases, a binodal distribution is created (27). Typical droplet size distribution data are shown in Figure 10 for two nozzle sizes operating at the same pressure drop.

Flow through the nozzle may be obtained from manufacturers' data or estimated from the orifice relationship, equation 16. For an approximate average, the discharge coefficient may be taken as 0.7.

Droplet size distribution may be estimated by equations 20 and 21 (18) and equation 22 (8), with $u_c = u_o$, since the solid-cone nozzle has no central gas core. Manufacturers' distribution data may be converted to liquids other than water by variations of equation 23.

Fan-Spray Nozzles. This type of atomizer is used in the coating industry and in special applications where a narrow elliptical pattern is more appropriate than a circular pattern (23,28) (see also Coating processes). The spray is formed either by discharge from an elliptical orifice or from a circular orifice impinging on a curved surface. The deflector method produces a somewhat coarser spray pattern. The cone is full, and its included angle varies from near-straight discharge to >100°.

Relatively little research has been reported on fan-spray nozzles, but the general relationships for conical sprays apply. Data from manufacturers are essential for

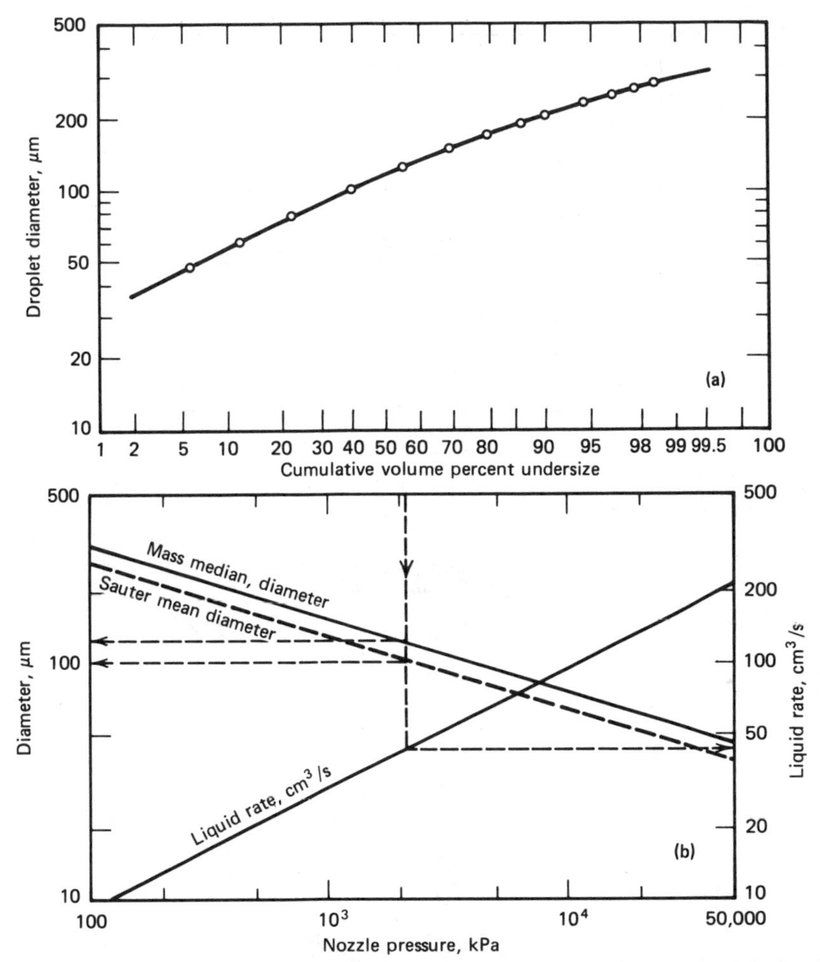

Figure 7. Droplet size distribution (**a**) and flow characteristics (**b**) of a typical swirl-chamber (hollow-cone) nozzle, water into air. Nozzle data: single inlet, swirl chamber, 42 cm^3/s water flow at 2068 kPa (20). To convert kPa to psi, multiply by 0.145.

quantitative treatment of fan-spray nozzle applications. Fan-spray distribution data for water and water-oil dispersions are given in ref. 25.

Two-Fluid Atomizers. In this device, a separate gas flow promotes atomization of the liquid, and by adjustment of gas-to-liquid ratios of flow and pressure, extremely fine dispersions can be produced. The gas and liquid may be mixed within the nozzle before discharge through a single-outlet orifice. Alternatively, gas and liquid may discharge through separate orifices arranged in such a manner that the gas impinges on the liquid at or just outside the orifice. The two-fluid atomizer (pneumatic nozzle) produces a finer spray than simple pressure nozzles, but consumes more energy.

The two-fluid atomizer may be used on a siphoning or aspirating basis in which the liquid is pumped from a pressure lower than that of the discharge. These atomizers may also be used to handle highly viscous materials that would not disperse well from previous nozzles. Typical designs of two-fluid nozzles are shown in Figure 11. Spray angles may be quite wide, approaching 180°, but because of turbulence and gas entrainment the patterns pull into a more narrow plume a short distance from the nozzle.

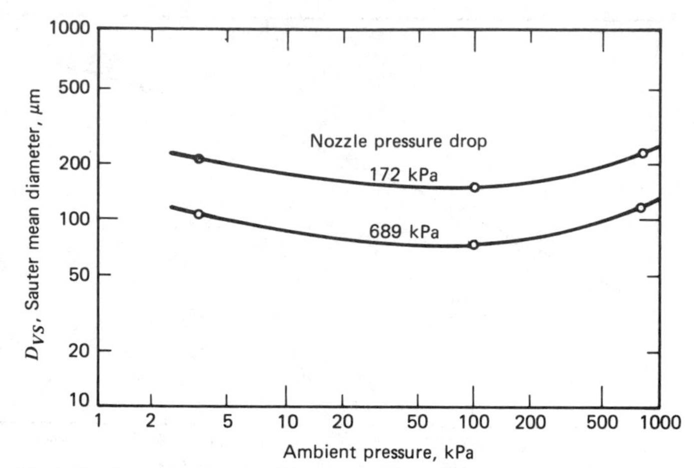

Figure 8. Variation in mean droplet diameter with ambient pressure. Hollow-cone nozzle, air–oil (16). To convert kPa to psi, multiply by 0.145.

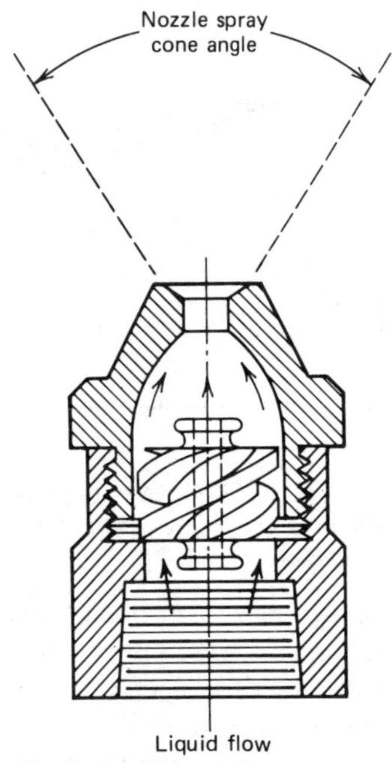

Figure 9. Typical solid- or full-cone spray nozzle (26).

An older, but still useful, empirical correlation predicts mean droplet size in two-fluid nozzle sprays (29). Equation 24 is based on several hundred tests with small air-atomizing nozzles, using several liquids plus a variety of pressure and flow conditions:

$$D_{vs} = \frac{585}{U_r}\left[\frac{\sigma}{\rho_L}\right]^{0.50} + 75.2\left[\frac{\mu_L}{\sqrt{\sigma\rho_L}}\right]^{0.45}\left[\frac{1000}{Q_G/Q_L}\right]^{1.5} \tag{24}$$

This is a dimensional equation, with the following units: D_{vs} = Sauter mean diameter, μm; U_r = relative velocity, gas to liquid, in nozzle, m/s; σ = surface tension, mN/m (= dyn/cm); ρ_L = liquid density, g/cm^3; μ_L = liquid viscosity, mPa·s (= cP); and Q_G,Q_L = volumetric flow rates of gas and liquid.

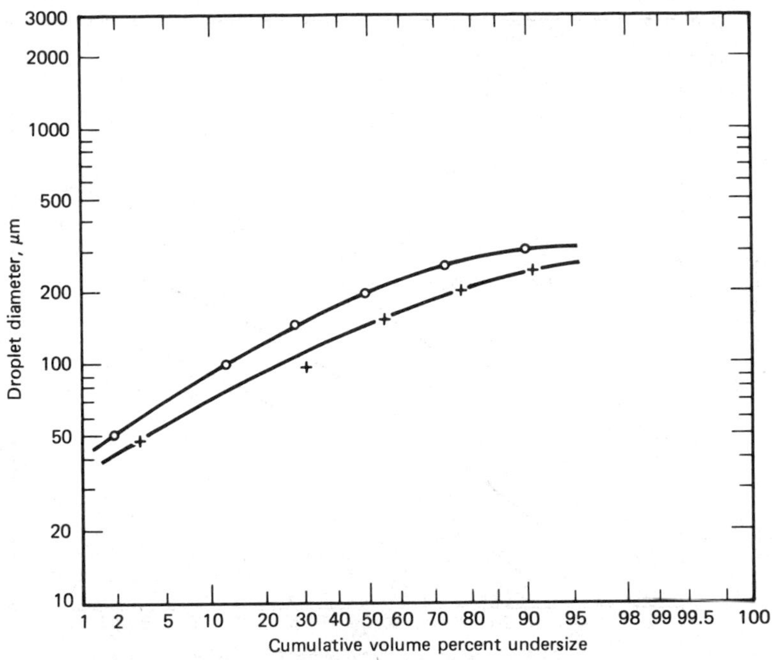

Figure 10. Typical droplet size distribution data for two solid-cone nozzles operating at the same pressure drop. ○ = 3.2 mm orifice, ΔP = 345 kPa (50 psi), flow = 0.0412 cm^3/s; + = 2.0 mm orifice, ΔP = 345 kPa, flow = 0.0114 cm^3/s (26).

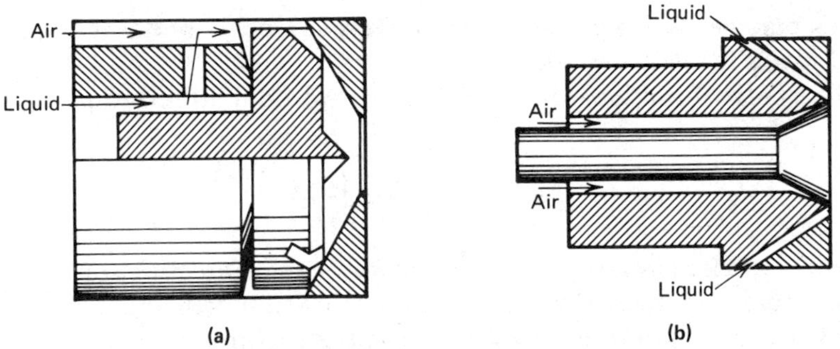

Figure 11. Two-fluid atomizers. (**a**) Internal-mixing type. (**b**) External-mixing type.

Experimental support for the equation covers the following ranges of physical properties: viscosity, 0.3 to 50 mPa·s; density, 0.7 to 1.2 g/cm^3; surface tension, 19 to 73 mN/m; and gas velocity, subsonic.

The Nukiyama-Tanasawa equation (eq. 24) has been applied to many atomization cases, with fairly good success. For example, it gives a reasonable fit to distribution data obtained for liquids atomized by a high temperature gas stream (30). It can be used to represent air–molten wax data at relative velocities of 100 m/s and higher (Fig. 12) (31). It was applied successfully to nitrogen–oil and hydrogen–oil atomizer test data, and even represented data from single-fluid (pressure-atomizing) nozzles (32). The equation should always be used with caution, and highly accurate results should not be expected from it since it makes no allowance for variations in nozzle geometry. The first term in the equation tends to dominate, except at gas/liquid volumetric ratios

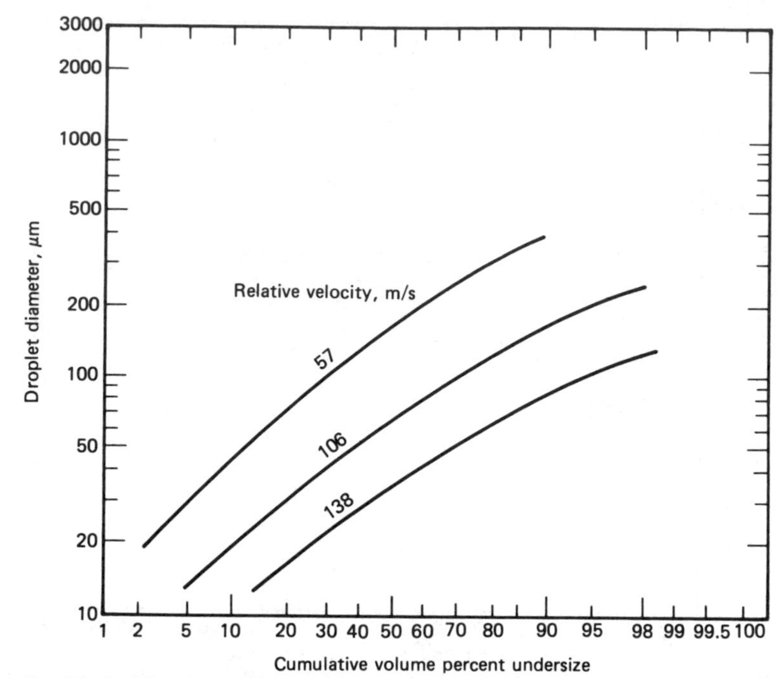

Figure 12. Typical droplet size distribution data for a two-fluid nozzle operating at different relative velocities. System: air–molten wax. Liquid viscosity = 9 mPa·s (= cP); surface tension = 29.5 mN/m (= dyn/cm); liquid density = 0.83 g/cm^3 (17,31).

<5000; this may be too simplistic a functional relationship under the usually higher gas/liquid ratios in two-fluid nozzles.

An alternative functional relationship for two-fluid atomization is given below (21):

$$D_{vs} = \frac{a}{(U_r^2 \rho_G)^\alpha} + b\left(\frac{Q_G \rho_G}{Q_L \rho_L}\right)^\beta \tag{25}$$

where α and β are functions of nozzle design and a and b are functions both of nozzle design and liquid properties. Equations 24 and 25 are similar in form and may be useful in the conversion of data from one set of conditions to another.

General trends of reported data on two-fluid atomizers have been summarized (20,33) and are shown in Figure 13.

Sonic Atomizer. A special type of two-fluid device is the sonic atomizer. Gas (normally air) is accelerated within the device to sonic velocity. The sound waves produced impinge on a plate or annular cavity (resonation chamber) and are reflected to the path of the emerging liquid sheet. The waves exert a chopping action, breaking up the liquid into droplets usually ranging downward in size from 50 μm. The spacing of the plate or cavity can be easily adjusted in order to modify the dispersion. The generation of high intensity sound for atomization is discussed in refs. 34 and 35. Because dispersion characteristics are highly dependent on nozzle geometry, manufacturers should be consulted for details (see Ultrasonics).

Rotary Atomizers. These disk- or cup-shaped devices are rotated by a separate power source and are capable of handling slurries and other materials which might clog the narrow passages of pressure and two-fluid nozzles. The devices may be smooth

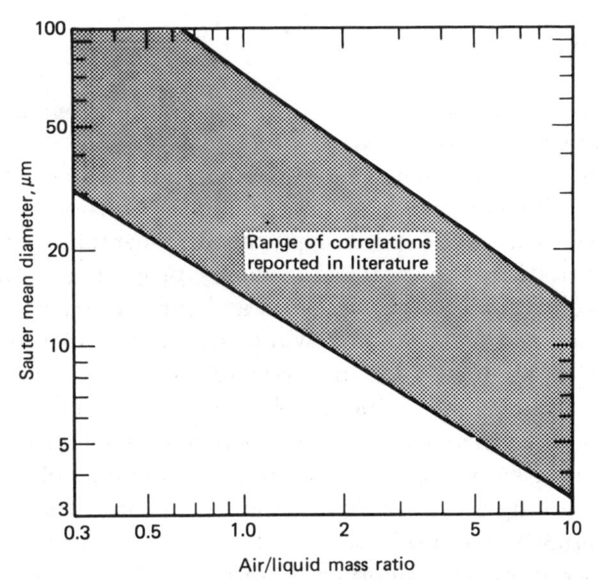

Figure 13. General trend of mean droplet size with air–liquid mass ratio in two-fluid nozzles (20).

or vaned, flat or bowl shaped, and may have multiple tiers. Many are designed with straight or curved slots between flat disks, with the channels so formed guiding the liquid to the periphery. Aside from liquid properties, the main design variables are liquid flow rate and periphery speed plus, of course, device geometry. The application of rotary atomization to spray drying is discussed in detail in ref. 17 (see Drying).

Representative droplet size distribution data, presented in the form of a design chart, are given in Figure 14 (36). The significant effect of peripheral speed should be noted. For a given speed, an increase in feed rate results in a coarser spray.

The mechanism of droplet formation is similar to that for spray development from pressure nozzles; sheets or filaments of liquid formed are unstable and break up shortly after discharge from the atomizer. Specific tests for a given flow rate, system, and geometry are required if a rotary atomizer is to be used for a critical service. In a very general way, the following dimensionless equation appears to represent the key variables for atomizer selection and design (37):

$$\frac{D_{vs}}{r} = 0.4\left(\frac{\Gamma}{\rho_L N r^2}\right)^{0.6}\left(\frac{\mu_L}{\Gamma}\right)^{0.2}\left(\frac{\sigma \rho_L L}{\Gamma^2}\right)^{0.1} \qquad (26)$$

where r = disk radius; Γ = liquid feed rate based on wetted periphery, mass/(time-length); N = rotational speed of disk, revolutions per time unit; L = vane length; ρ_L = liquid density; μ_L = liquid viscosity; and σ = surface tension (see also Dimensional analysis).

Sprays Formed by Flashing Liquid Jets. Although not strictly regarded as an atomizer, an orifice through which a high pressure liquid expands with partial vaporization can produce a spray pattern. If the low pressure zone conditions and space are so designed as to accommodate the two-phase mixture, the droplet size distribution can be measured. This flashing atomization process has been studied on several different nozzle designs and two liquids, and droplet size distributions were measured (38). As might be expected, under flashing conditions, higher pressure drops produced finer sprays.

Chemical Applications

A remarkable number of processes involving production of chemicals and metals, and utilization of chemicals, fuels, and foodstuffs require sprays (20); some applications are given in Table 1. Though a small component in most systems, the atomizer must be chosen carefully. The following factors enter into the proper selection, and might be regarded as a check list when specifying spray equipment: the nature of the process, including characteristics of the sprayed products; properties of the liquid to be atomized, eg, temperature, density, viscosity, and surface tension; percent solids in suspensions, slurries, and pastes; total flow rate; required droplet distribution or mean droplet diameter; liquid or gas pressures available for nozzles, or power required for rotary atomizers; ambient gas temperature, pressure, and flow patterns; conditions that may contribute to wear and corrosion; size and shape of vessel, enclosure, or combustor containing spray, or area to be treated; economics of the spray operation, taking into account long-term performance and operating expenses; and analysis of all pertinent information and selection of atomizer type, spray pattern, number of atomizers, flow per atomizer, and other operating conditions.

Table 1. Spray Applications

Production or processing
spray drying (dairy products, coffee and tea, starch, pharmaceuticals, soaps and detergents, pigments, etc)
spray cooling
spray reactions (absorption, roasting, etc)
atomized suspension technique (effluents, waste liquors, etc)
powdered metals
Treatment
evaporation and aeration
cooling (spray ponds, towers, reactors, etc)
humidification and misting
air and gas washing and scrubbing
industrial washing and cleaning
Coating
surface treatment
spray painting (pneumatic, airless, and electrostatic)
flame spraying
insulation, fibers, and undercoating materials
multicomponent resins (urethanes, epoxies, polyesters, etc)
particle coating and encapsulation
Combustion
oil burners (furnaces and heaters, industrial and marine boilers)
diesel-fuel injection
gas turbines (aircraft, marine, automotive, etc)
rocket-fuel injection
Miscellaneous
medicinal sprays
dispersion of chemical agents
agricultural spraying (insecticides, herbicides, fertilizer solutions, etc)
foam and fog suppression

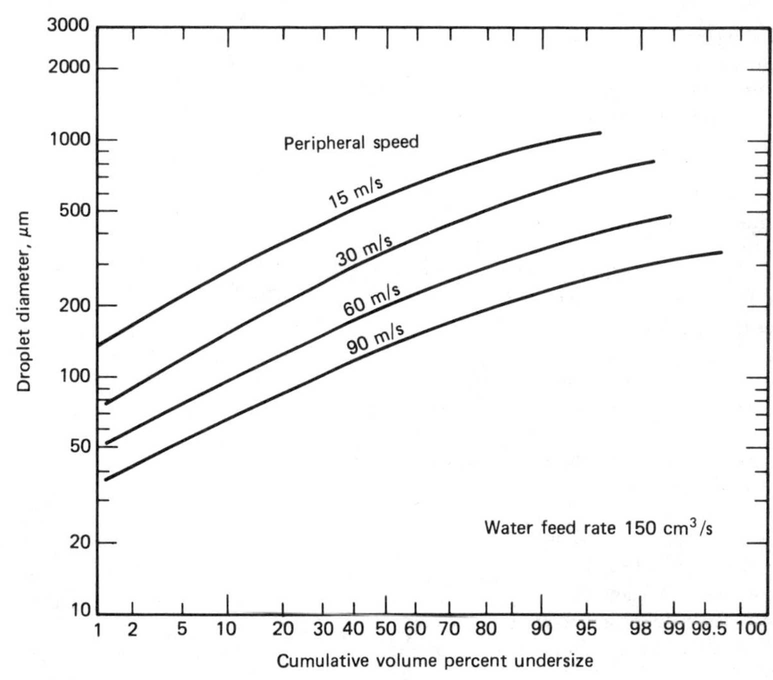

Figure 14. Droplet size distribution performance chart for water fed to a 12.7 cm diameter rotating-disk atomizer. Disk contained 24 vanes, 0.95 cm high (36).

Nomenclature

a	= factor in equation 17; coefficient in equation 25
A_c	= cross-sectional area of gas core
A_o	= cross-sectional area of orifice
b	= coefficient in equation 25
C_1	= constant in equation 7
d_c	= gas-core diameter
d_o	= orifice diameter
d_o'	= effective orifice diameter (eq. 21)
D	= droplet diameter
D_a	= arithmetic mean diameter
D_g	= number geometric mean diameter
D_i	= diameter of size i
D_{max}	= maximum stability stable droplet size (eq. 8)
D'_{max}	= maximum droplet size in distribution
$\overline{D}_n$	= number median diameter
D_s	= surface mean diameter
$\overline{D}_s$	= surface median diameter
D_v	= volume (or mass) mean diameter
$\overline{D}_v$	= volume (or mass) median diameter
D_{vs}	= volume/surface, or Sauter, mean diameter
D_{90}	= 90th percentile droplet diameter, based on cumulative volume
D_{10}	= 10th percentile droplet diameter, based on cumulative volume
i	= number of droplets in size class
k	= limit denoting largest size class sampled
K	= discharge coefficient
l	= distance from orifice

L = vane length
n = number of droplets
n_r = ruptures of film per unit area
N = rotational speed, revolutions per time unit
N_i = number of droplets in size class i
P = pressure
Q_G = volumetric flow rate of gas
Q_L = volumetric flow rate of liquid
r = disk radius
Re = Reynolds number (dimensionless)
s_g = geometric standard deviation
t_s = sheet (film) thickness
u_L = actual liquid velocity
u'_L = relative velocity, liquid to surrounding gas
u_o = superficial liquid velocity
U_d = droplet velocity, relative to surrounding gas
U_r = relative velocity, gas to liquid, in two-fluid nozzle
U_s = velocity of rupture
V = droplet volume
We = Weber number (dimensionless)
α = exponent in equation 25
β = exponent in equation 25
Γ = liquid mass rate, based on wetted periphery
ΔP = pressure drop across atomizer
μ = viscosity
ρ = density
σ = surface tension
π = 3.1416. . . .

Subscripts

G = gas phase
L = liquid phase

BIBLIOGRAPHY

"Sprays" in *ECT* 1st ed., Vol. 12, pp. 703–721, by W. E. Meyer and W. E. Ranz, The Pennsylvania State University; "Sprays" in *ECT* 2nd ed., Vol. 18, pp. 634–654, by R. W. Tate, Delavan Manufacturing Company.

1. Lord Rayleigh, *Proc. London Math. Soc.* **10,** 4 (1878).
2. C. Weber, *Z. Angew. Math. Mech.* **11,** 136 (1931).
3. W. E. Ranz, *On Sprays and Spraying*, Engineering Research Bulletin B-65, College of Engineering and Architecture, The Pennsylvania State University, University Park, Penn., 1956.
4. W. Ohnesorge, *Z. Angew. Math. Mech.* **16,** 355 (1936).
5. D. Steinmeyer in R. H. Perry, and C. H. Chilton, eds., *Chemical Engineers' Handbook*, 5th ed., McGraw-Hill Book Co., New York, 1973, pp. **18**-58–**18**-81.
6. N. Dombrowski and D. L. Wolfsohn, *Trans. Inst. Chem. Eng.* **50,** 259 (1972).
7. J. O. Hinze, *AIChE J.* **1,** 289 (1953).
8. R. A. Mugele and H. D. Evans, *Ind. Eng. Chem.* **43,** 1317 (1951).
9. H. G. Houghton in J. H. Perry, ed., *Chemical Engineers' Handbook*, 3rd ed., McGraw-Hill Book Co., New York, 1950, p. 1170.
10. C. E. Lapple, *Chem. Eng. N.Y.* **70**(11), 149 (May 20, 1968).
11. R. W. Tate, *AIChE J.* **7,** 574 (1961).
12. R. R. Irani and C. F. Callis, *Particle Size Measurement, Interpretation and Application*, John Wiley & Sons, Inc., New York, 1963.
13. D. A. Lundgren and co-eds., *Aerosol Measurement*, University of Florida Press, Gainesville, Fla., 1979.

14. D. W. Roberds, *Appl. Opt.* **16,** 1861 (1977).
15. W. D. Bachalo, C. F. Hess, and C. A. Hartwell, *J. Eng. Power* **102,** 799 (1980).
16. S. M. DeCorso, *Trans. ASME J. Eng. Power* **82,** 10 (1960).
17. W. R. Marshall, *Atomization and Spray Drying*, Chemical Engineering Progress Monograph Series No. 2, American Institute of Chemical Engineers, New York, 1954.
18. R. A. Mugele, *AIChE J.* **6,** 3 (1960).
19. W. Licht, *AIChE J.* **20,** 595 (1974).
20. R. W. Tate, *Chem. Eng.* (*N.Y.*) **67**(15), 157 (July 19, 1965).
21. K. Y. Kim and W. R. Marshall, *AIChE J.* **17,** 575 (1971).
22. P. H. Rothe and J. A. Block, *Int. J. Multiphase Flow* **3,** 263 (1977).
23. F. E. J. Briffa and N. Dombrowski, *AIChE J.* **12,** 708 (1966).
24. T. G. Kieckbusch and C. J. King, *AIChE J.* **26,** 718 (1980).
25. J. A. Zakarian and C. J. King, *Ind. Eng. Chem. Proc. Des. Devel.* **21,** 107 (1982).
26. R. L. Pigford and C. Pyle, *Ind. Eng. Chem.* **43,** 1649 (1951).
27. E. Sada, K. Takahashi, K. Morikawa, and S. Ito, *Can. J. Chem. Eng.* **56,** 455 (1978).
28. C. J. Clarke and N. Dombrowski, *Chem. Eng. Sci.* **26,** 1949 (1971).
29. S. Nukiyama and Y. Tanasawa, *Trans. Soc. Mech. Eng.* (*Jpn.*) **4**(14), 86 (1938).
30. J. J. Laskowski and W. E. Ranz, *AIChE J.* **16,** 802 (1970).
31. R. H. Wetzel, Ph.D. thesis, University of Wisconsin, 1952.
32. H. C. Lewis, D. G. Edwards, M. J. Goglia, R. I. Rice, and L. W. Smith, *Ind. Eng. Chem.* **40,** 67 (1948).
33. J. Gretzinger and W. R. Marshall, *AIChE J.* **7,** 312 (1961).
34. R. L. Wilcox and R. W. Tate, *AIChE J.* **11,** 69 (1965).
35. M. J. Ashley, *Chem. Eng.* (*London*), 368 (June 1974).
36. W. M. Herring and W. R. Marshall, *AIChE J.* **1,** 200 (1955).
37. S. J. Friedman, G. A. Gluckert, and W. R. Marshall, *Chem. Eng. Progr.* **48,** 181 (1952).
38. R. Brown and J. L. York, *AIChE J.* **8,** 149 (1962).

JAMES FAIR
University of Texas

SPUTTERING. See Film deposition techniques; Metallic coatings.

STAINS, INDUSTRIAL

In this article, a stain is defined as a solution or dispersion of colorants in a vehicle designed primarily to be applied to various substrates to impart color effects rather than to form a protective coating. Thus, stains differ from paints and other coatings.

Industrial stains are applied by many methods to a wide variety of materials. Although their principal function is to change the color of the substrate, they are frequently formulated to serve as glazes, impregnants, inks, polishes, drawer coatings, and many other purposes. Stains are used commercially as architectural finishes and for concrete, paper products (paper plates), plastics, textiles, and numerous wood products. As no single type of stain could possibly be used for all these various applications, they vary widely in composition and are usually expressly formulated for a specific use (see also Coatings; Paint; Shellac).

Because of the great variations in compositions, stain prices cover a wide range. The price increases since 1970 are given below.

Product	*Comparison basis*		
	1970	*1975*	*1980*
amber wiping stain	1.0	1.78	1.60
dark walnut non-grain-raising stain	1.0	1.52	1.72
red spirit stain	1.0	1.52	1.94

Stain Types

Water Stains. Water stains are 1–3% solutions of water-soluble dyes. Although almost any water-soluble colorant may be used, most industrial stains are prepared from a limited number of dyes because of cost, solubility, and other factors. Acid dyes are used for high grade furniture applications where maximum lightfastness is required. Basic dyes are used on short-lived objects such as fruit crates, baskets, etc. Water stains can be applied by practically every staining process, although spraying, dipping, infusion, and impregnation are preferred. Water stains raise the grain of wood to such an extent that sanding is required. For this reason, they have been replaced by non-grain-raising stains for most wood applications (1–2) (see also Dyes and dye intermediates).

Non-Grain-Raising Stains. Non-grain-raising stains (NGR) are solutions of the acid dyes or selected spirit-soluble dyes, or both, in a nonaqueous vehicle. These stains are composed of 0.5–6% dyes, 5–10% dye solvent, 20–50% alcohols (usually methanol), and the remainder medium boiling hydrocarbons (especially toluene), or other liquids nonmiscible with water. The dye solvent is usually a high boiling or practically non-volatile liquid that is the last fraction to evaporate, thereby maintaining the dye in solution long enough to stain the wood. The diethyl ether of ethylene glycol, and methyl and ethyl lactate are typical dye solvents. Others are ethylene glycol, diethylene glycol, and the monomethyl, monoethyl, and isopropyl ethers of ethylene glycol and propylene glycol (see also Dye carriers).

These stains give bright, transparent, light-resistance effects. Because of their inherent brilliance, they are seldom used as a single-stain application. In most instances, they are used as the base stain to attain a specified undertone. Then a more opaque stain such as a pigmented wiping stain is applied which tones down the excessive brilliance in the final finish.

Spirit Stains. Spirit stains are generally 0.5–12% solutions of dyes in alcohols. Frequently, they also contain shellac and other alcohol-soluble resins or various chemical mordants. Their grain-raising effect on wood is less than that of water stains, but they do raise wood grain more than non-grain-raising and oil stains. Spirit stains are characterized by quick and deep penetration and rapid drying; the latter feature complicates application to large areas, except with machines (calender staining). The better grades of these stains are colored with the faster-to-light alcohol-soluble, metalized azo dyes, usually incorporated with lacquers (lacquer stains) or dissolved in glycol ether–alcohol mixtures for staining wood.

Basic dyes are used for tinctorially powerful and bright color effects despite their poor fastness to light and water. Their tendency to bleed into water can be greatly reduced by combining them with mordants (tannin). Large quantities of dye–mordant complexes are used in the calender staining of paperboard. When applied without binders, basic dyes tend to crock (rub off), although this tendency is less when they are applied to materials containing natural mordants (for example, kraft paper). For high speed and offset printing inks, 10–20% solutions of the basic dyes or dye–mordant complexes, frequently containing 4% of shellac, are widely used. These inks are commonly called aniline inks or flexographic inks (see Inks). The same dyes in alcohol–toluene–*o*-dichlorobenzene solutions are useful leather stains (shoe stains, brush dyeing) (see Leather).

Many water-soluble acid dyes and a few direct dyes can be incorporated in spirit stains, particularly if first dissolved in a suitable glycol ether. Some oil-soluble dyes can also be incorporated in spirit stains, particularly with alcohol–acetone–toluene solvents. These stains are occasionally useful for special refinishing applications. Spirit-stain dyes tend to bleed into topcoatings, particularly lacquers. Except for those containing oil-soluble dyes, the spirit stains are used to finish the insides of cabinets and drawers and to impart color to a wide variety of materials. They are useful in certain branding, marking, stencil, and duplicator inks, although with the latter the solvent is applied to the copy sheet.

Oil Stains. Oil stains are essentially 1–3% solutions of oil-soluble dyes in aromatic hydrocarbons, such as toluene or coal-tar naphtha, containing 5–10% resins, dying oils, or varnish. Part or all of the aromatic hydrocarbon can be replaced with aliphatic hydrocarbons (petroleum fractions) for certain formulations, depending on the solubility characteristics of the dyes. Most dyes used for such stains tend to migrate into almost any topcoatings, although a first coating of shellac greatly minimizes this migration. These stains penetrate well and impart brilliant color effects. However, the dyes used in this type stain have extremely poor lightfastness and should only be used on inexpensive articles intended for interior use.

Pigmented Stains. Solvent-borne stains are the most widely used stains in the furniture industry today (see Pigments).

Wiping Stains. Wiping stains are composed of inorganic and organic pigments usually ground in an alkyd resin of the linseed or soya type or a drying oil. For maximum adhesion to subsequent seal coats, they are formulated with a pigment-to-binder ratio of 1.5–1.75:1.0. The solvents are most likely a blend of aliphatic and aromatic hydrocarbons (mineral spirits, VM&P naphtha, xylene, and toluene). Many formulations also contain ester- or ketone-type solvents, depending on the customers finishing schedule and the topcoats. A small percentage of ethyl or isopropyl alcohol and water may be present for suspension purposes. These stains usually contain 15–30 wt % solids.

Nonwipe Spray Stains. These stains are basically low solids wiping stains. A higher percentage of lower boiling solvents is used in their formulation. These fast-evaporating solvents allow the stain to be applied in a wet coat for maximum penetration, yet flash off quickly enough to prevent sags or runs. Nonwipe spray stains do not impart the depth of color or enhance the wood grain like a wiping stain. However, they are servicable and their use is growing, because of savings in wiping cloths and labor costs, and today's faster line schedules. Many manufacturers of cabinet frames, headboards, and similar furniture use this type stain exclusively.

Dip Stains. Solvent-borne dip stains differ from nonwipe-spray stains only in solvent balance. This type stain contains just enough high boiling solvent to attain the proper flow or drain-off. Dip stains may or may not be wiped, depending on the configuration of the article being stained.

Glazes. Glazes are essentially slow-drying wiping stains formulated for brushing over a seal coat. They impart depth to a quality wood finish to which a non-grain-raising stain, washcoat, and wiping stain has been applied. They are also applied over colored basecoats for an antiquing effect. The vehicle may contain alkyd resin, drying oil, rosin, rosin esters, acrylic resin (for plastics), or any combination of these materials (see Alkyd resins; Driers and metallic soaps; Terpenoids; Acrylic ester polymers). Retarders or plasticizers in the formulation increase brushability after most of the solvent has evaporated. The plasticizer also serves to prevent blush or incompatibility with seal coats.

Wood Fillers. Wood fillers are used primarily for filling open-pore woods such as oak, ash, walnut, mahogany, etc (see Wood). They are heavy-bodied materials with 80–90 wt % solids. Oak and ash woods require a filler cut of 0.96–1.2 g/L (8–10 lb/gal) of solvent, usually VM&P naphtha. Walnut and mahogany are more readily filled by a cut of 0.72–1.08 g/L (6–9 lb/gal). Wood fillers consist of insoluble inert or extender pigments mixed with an alkyd, drying oil, rosin, rosin ester, or a combination of same. The inert pigments are usually calcium carbonate, sulfate, or silicate, colored with a wide variety of inorganic and organic pigments. Wood fillers are brushed or sprayed over the wood. After a short flash time, the filler is padded in, that is, wiped with pressure by pad or cloth in a circular motion to fill the pore. Subsequent wiping in the direction of the wood grain removes any residue from the surface.

Pigmented Stains, Waterborne. Many waterborne resins and colorants, eg, acrylic solutions, acrylic latex, vinyl acrylic, polyesters, etc, are used in pigmented stains and glazes. They raise the wood grain but not as much as true water stains. Waterborne pigmented stains are still in the development stage for most furniture-type wood applications.

Toners. Transparency is a distinguishing feature of the old-craftsman or traditional finish. With the advent of the modern trend in furniture design, very light finishes have come into vogue. Colors are lighter or of an entirely different hue than the natural color of the wood being finished.

The transparent colors of the traditional finish are known as additive colors. The color of the wood, plus the color of each of the finishing materials, ie, stain, filler, glaze, and shading lacquer, together produce the final color. Therefore, the color of the wood has to be taken into account for a traditional brown finish. The stain should contain blue-black and yellow with a lesser amount of red to allow for the red in the wood. To produce the same color on walnut, the stain might contain only red and yellow, since the blue-black tint is inherent in walnut wood. With this method, the color of the finish has greater intensity than that of the wood. Traditionally, it was common practice to produce a finish lighter than the color of the wood or of a different hue by removing the natural coloring from the wood.

This procedure required bleaching. Bleaches vary considerably in composition, depending on the type of wood being bleached or the degree of lightness desired. Oxalic acid on oak removes greyness and iron streaks, whereas a solution of caustic and hydrogen peroxide whitens the wood (see Bleaching agents).

The modern blond or simulated bleached finishes are achieved with toners. The term toner has been applied to several different materials, but here it is defined as a pigmented lacquer of varying degrees of opacity. The variations in the toners are infinite; the pigments can be opaque, or transparent, or a mixture. The vehicle for the pigments can be any of a number of materials as well as nitrocellulose lacquer.

Toners offer a new coloring aid, namely white. Whites have, as a rule, good opacity and can cover and hide other colors. Thus, the toner offers the means of subtracting color from the wood, and colors are produced lighter than the wood itself without resorting to the expensive, time-consuming bleaching operation.

Toners import varying degrees of opacity, and the effects produced have been accepted by the consumer and have become very popular.

Other Classifications. Products with such diverse features as synthetic dyes, pigments, or other chemical agents capable of changing the color of wood may be classified in many different ways, depending on the special features or characteristics that are of particular interest to the user, supplier, or manufacturer (see Table 1) (3).

Ingredients

Some stains are merely solutions or dispersions of colorants in alcohols, hydrocarbons, water, or similar liquids. However, the professional formulator adds many other ingredients, including additives and modifiers, eg, driers, catalysts, antiskinning agents, antisettling agents, dispersing and wetting agents, viscosity-control agents, antifoaming agents, fungicide and mildew controls, pH-control agents, adhesives, antioxidants, and surface-active agents (surfactants).

Table 1. Classification of Stains

Final coloration	Synthetic pigments	Solvent polarity	Functional groups
in situ pigmentation	cationic in organic solvents	nonpolar, naphtha, white spirit	azo
metal-complex formation	anionic, water-soluble sodium salts	moderately polar, alcohols, esters	nitroso
chemical reaction of phenolic compounds	metal chelates	highly polar, water-soluble dyes, aqueous dispersions	anthraquinone
contact staining	dye salts		phthaleins

Dyes. Acid dyes are sodium sulfonates of azo, anthraquinone, triphenylmethane, azine, xanthene, ketonimine, nitro, and nitroso compounds. Basic dyes are derived from azo, acridine, anthraquinone, methane, oxazine, thiazine, triarylmethane, and xanthane compounds. Spirit dyes are nonionic and are the free acid or free base forms of certain azo, anthraquinone, xanthrene, or triarylmethane compounds (4).

Pigments. Pigments enhance or obscure the surface to which they are applied. In addition, they increase the body of the stain, build up the film for glazing purposes, fill the wood pores, and improve the durability of exterior stains (see Pigments, dispersed pigment concentrates).

Alkyd Resins. Any resin that is a polymer of an ester-type monomer is a polyester resin. In this broad sense, alkyds are polyesters. However, general usage restricts the term alkyd to polyesters that are modified with a triglyceride oil or the acids of such an oil. Moreover, when a resin is designated as polyester, it is commonly understood that alkyds are not included in the term. In this article, the term alkyd resin generally designates the reaction product of a polybasic acid, a polyhydric alcohol, and a monobasic fatty acid or oil (see Alkyd resins).

Oils. Oils and fats are triglycerides; they consist of one molecule glycerol and three molecules of long-chain fatty acids. The former are liquids and the latter are solids at normal temperatures. The various oils differ greatly in drying properties and other characteristics, depending on the type of fatty acids. The most common oils used in wood stains today are linseed, soybean, tung, and oiticica. They are mainly used in wood glazes. Oils have been mostly replaced by alkyd resins because the latter are faster drying and have other more desirable properties (see Fats and fatty oils; Vegetable oils).

Rosin. Rosin is marketed as gum rosin (natural exudation of living pine trees) and wood rosin (produced by steam-and-solvent extraction from tree stumps). It is relatively cheap and is used in some stains, glazes, and fillers, mainly as oil replacements. The brittleness and tackiness imparted by rosin can be alleviated in several ways, and at present a large variety of modified rosins is marketed. Rosin is improved by hydrogenation, esterification (ester gum), and addition of maleic anhydride (maleic modified rosins) (see Gums; Resins, natural; Terpenoids).

Solvents. Hydrocarbons used as stain solvent include petroleum naphthas, toluene, xylene, aromatic petroleum solvents, turpentine, and dipentene. Petroleum naphthas, some aromatic petroleum solvents, and turpentine are mixtures. Aliphatic hydrocarbons or paraffins make up the bulk of petroleum naphthas, usually 85–95%. Aromatic hydrocarbons are much better solvents than aliphatic hydrocarbons (5).

Alcohols differ greatly in their solvent properties from hydrocarbons. For example, the low boiling alcohols, eg, methyl, ethyl, and isopropyl, are miscible with water whereas hydrocarbons are not. Most alcohols dissolve materials such as shellac, Manila gum, natural kauri, Vinsol resin, zein, and certain synthetic resins. Most of these materials are not soluble in hydrocarbons. Conversely, many products soluble in hydrocarbons are not dissolved by alcohols, eg, coumarone resin (see Hydrocarbon resins) and ester gums. Some materials are dissolved by both hydrocarbons and alcohols, eg, rosin.

Esters, ketones, and glycol ethers are also used as stain solvents. Glycol ethers are excellent coupling agents and solvents for many resins and dyes (see Solvents, industrial).

Application Techniques

Stains are applied by brush, spray, dip, flow coat, padding, and tumbling. Because of today's fast finishing lines, some manufacturers construct, sand, stain, and coat their product within the same day. Such procedures require specialty stains adapted for this purpose (see also Coating processes).

Stainers have at their command a very wide variety of stains that enable them to produce color effects quickly, easily, safely, and at minimum labor costs. Successful application requires skill and artistry because of the differences in color, structure, absorbency, and other properties of surfaces, and the application characteristics of the various types of stains. Frequently, the ability to match colors and design color harmony is required. Technology has improved the composition and application methods for stains, but proper use remains an art.

Pretreatment. Formamide mixed with an equal volume of water is the most effective penetration promoter when applied to veneered panels. Possible grain-raising produced by the pretreatment depends on the type of wood used (species); the solids content and film thickness of the lacquer applied; and the standard of quality set by the cabinetmaker.

Impregnation. Frequently, articles to bc stained are first exposed to a vacuum for $1/2$ to 2 h or longer to remove air and moisture from capillaries and cells; a stain is then drawn or pumped into the chamber and allowed to remain in contact, often under pressure overnight, until the desired depth of penetration is obtained. Impregnation is, therefore, a special dip-staining process. It is often used to color pencil slats, logs (railroad ties), walking canes, wood for musical instruments, and densified (colorant-resin-impregnated) wood. Antirot agents are often incorporated.

Spattering. Spattering is a special spraying process in which the air–stain dispersion is not uniform, and consequently the stain is deposited in relatively large drops scattered irregularly over the surface. Very coarse sprays from a gun are used, although experienced operators can produce these effects by merely drawing a stick across a stiff brush wet with the stain. It is a method of obtaining irregularly spotted effects over uniformly colored grounds. Fast-drying stains minimize any tendency for the scattered stain drops to spread, run, or streak.

Differential or Contrast Staining. This technique is used for wood that includes hard and soft tissues, eg, prominent surface markings caused by natural tissues, or prominent annual rings clearly differentiated into alternating zones of dense tissue with small pores and more open tissue with wider vessels. Wood possessing these and similar features is likely to absorb stains unevenly, the lighter and softer tissues generally assuming richer and deeper colors than the more firmly packed, less porous dense tissue. Differential staining is considered to be of esthetic value. It is often applied by mixing different species of wood for veneer inlays of various geometric designs.

In general, waterborne stains are likely to prove more effective for such applications than the solvent-dispersible types.

Flowing. Surfaces are often colored by simply pouring on a liquid stain and letting the excess, unabsorbed stain drain off. It is a convenient means of applying stains to test strips. It is the reverse of dipping, but leaves the top surfaces of the object in the same draining condition as when removed from a dipping bath. Flowing is useful for limiting the staining to one surface or portion thereof, in contrast to the overall staining obtained in regular dipping baths. It is most adaptable to coloring flat and convex surfaces. Usually, the lower portions of the draining surfaces appear more strongly colored than the upper portions because the former receive more stain.

Graining. Graining is a technique requiring great skill. The varnish stain is applied in transparent films of uneven thickness over a light, opaque ground color to simulate the grain effect of wood; special tools are used for this purpose. For a more lightfast but similar effect, a dispersion of insoluble pigments in drying oils (nonpenetrating oil stains) is first applied. Graining tools or an artist's brush are used, and when this application has dried thoroughly, an uncolored or colored transparent-gloss varnish lacquer or other coating is applied. This process resembles hand printing.

Transfers and Decalcomanias. These products impart stainlike effects, even though they are not strictly classifiable as stains. They are usually print effects on thin sheet stocks or film, produced with stains composed of insoluble colorants and adhesives. When wet with suitable liquids and pressed against a surface, the colorants transfer and are glued to the new surface. Although the process differs from inhibition printing, the same effect is obtained.

Stains for Exterior Wood Finishes

Stains are most effective on rough lumber or plywood surfaces, but they are also satisfactory on smooth surfaces. They are available in a variety of colors and are especially popular in brown tones since they give a natural or rustic wood appearance. They are not available in white. They are an excellent finish for weathered wood (6), but are not effective when applied over a solid-color stain or over old paint.

Solid-Colors. Solid-color stains, also called heavy-bodied stains, are opaque finishes that come in a wide range of colors. They have a much higher concentration of pigment than semitransparent penetrating stains. As a result, they obscure the natural wood color and grain. Oil-based solid-color stains tend to form a film much like paint and may peel loose from the substrate. Latex-based solid-color stains are similar to thinned paints and form a film like oil-based solid-color stains (7).

Semitransparent Penetrating Stains. Semitransparent penetrating stains are growing in popularity (8–9). They are moderately pigmented and thus do not totally hide the wood grain. These stains penetrate the wood surface and leave a porous surface; they do not form a surface film like paints. As a result, they do not blister or peel even if moisture gets into the wood. Penetrating stains are alkyd- or oil-based and may contain a fungicide or water-repellent. Latex-based (waterborne) stains do not penetrate the wood surface like oil-based stains (10).

Health and Safety Factors; Environmental Considerations

Safe handling of industrial stains requires adequate ventilation; removal of all sources of possible ignition; avoidance of prolonged exposure to solvents, both by respiratory and skin contact; use of safety glasses and respirators; and good housekeeping, ie, no spilled stain left around for possible contact. Overexposure to solvent vapors may result in irritation of the respiratory tract or acute nervous system depression.

Solvent vapor concentrations are expressed as TLV and represent "the conditions under which nearly all workers can be repeatedly exposed day after day without adverse effect." These values vary between 5 and 1000 ppm. Dermatological problems could include rash, blistering, or inflammation. Immediate cleansing of any exposed area is recommended.

The Federal Lead Toxic law (Sept. 1, 1977), prohibits the use of lead-containing ingredients (pigments, driers, etc) in coatings applied to objects used in the house. Stains on furniture and kitchen cabinets and architectural stains are included.

Environmental Considerations. With advent of governmental regulations limiting solvent emission, there is an increased interest in waterborne stains in the furniture industry. However, these materials are still in the developmental stage. They are used in moderate amounts in Europe, in only minor amounts in the United States.

BIBLIOGRAPHY

"Stains, Industrial" in *ECT* 1st ed., Vol. 12, pp. 722–740, by W. H. Peacock, American Cyanamid Company; "Stains, Industrial" in *ECT* 2nd ed., Vol. 18, pp. 654–671, by W. H. Peacock, American Cyanamid Company.

1. *Pigm. Resin Technol.* **20,** 23 (March 1976).
2. S. Fabriz, *Ind. Lackier. Betr.* **46,** 55 (1978).
3. S. Dombey and B. Stangroom, *J. Oil Colour Chem. Assoc.* **7,** 176 (1980).
4. *Colour Index*, 3rd ed., 1975, and *Supplement*, 1963, Society of Dyers and Colourists, Bradford, UK, and AATCC, U.S., revised 1981.
5. R. W. Tess, *Solvents, Theory and Practice*, 1973.
6. W. Feist, E. Mraz, and J. Black, *For. Prod. J.* **27,** 13 (1977).
7. C. Metzger, *Paint Varn. Prod.*, 35 (June 1973).
8. J. Black, D. Laughnan, and E. Mraz, *Research Note FPL-046*, U.S. Department of Agriculture, 1979.
9. D. Cassens and W. Feist, *Selection and Application of Exterior Finishes for Wood*, Publication no. 135, U.S. Department of Agriculture, Forest Products Laboratory, 1979, pp. 3 and 7.
10. J. Vander Elburg and R. Dooper, *Verfkroniek* **50,** 288 (1977).

General References

W. H. Peacock in Mattiello, ed., *Protective and Decorative Coatings*, Vol. 3, John Wiley & Sons, Inc., New York, p. 769.

Farg. Fernissa **40**(4), 20 (1976).

M. Goehel, *Coating* **8,** 246 (1975).

E. H. Miller, paper presented at the *Proceedings of the BRE Symposium*, Garston, March 12, 1975.

R. Dooper, *Decorat. Contractor* **76,** 9, 19 (1976).

ROBERT S. BAILEY
Lilly Industrial Coatings, Inc.

STARCH

Starch [*9005-25-8*], $(C_6H_{10}O_5)_n$, the principal reserve polysaccharide in plants, constitutes a substantial portion of the human diet. It is the principal component of most seeds, tubers, and roots and is produced commercially from corn, wheat, rice, tapioca, potato, sago, and other sources. Most commercial starch is produced from corn which is comparatively cheap and abundant throughout the world. Wheat, tapioca, and potato starch are produced on a smaller scale and at higher prices.

Nonfood uses of starch have a long history, and Egyptian papyrus cemented with a starchy adhesive has been dated to 3500–4000 BC. Pliny the Elder (130 BC) described the uses of a modified wheat starch in sizing papyrus to produce a smooth surface (see Paper; Papermaking additives). Today, starch is used extensively in the paper, textile, and adhesive industries and has many other applications. Emergence of immobilized-enzyme technology has given rise to the production of sweeteners (qv) from corn starch (ie, corn syrup, high-fructose corn syrup), and work is underway in many laboratories to determine the extent to which cereal grains or their starches can be utilized as sources of chemical feedstocks, replacing nonrenewable petroleum sources (see Chemurgy; Enzymes, immobilized).

Starch is a mixture of linear (amylose) and branched (amylopectin) polymers of α-D-glucopyranosyl units. Natural starch occurs usually as granules composed of both linear and branched starch molecules. However, some starches are composed only of branched molecules, and these are termed waxy starches because of the vitreous sheen of a cut seed surface. Some mutant seed varieties have been produced with starches having up to 85% linear molecules, although most starches have ca 25% linear and 75% branched molecules.

The quasi-crystalline granules that characterize starch ultrastructure are insoluble in water at ambient temperature and relatively resistant to carbohydrases other than α-amylases (see also Carbohydrates; Sugar).

Physical Properties

Starch occurs in plants in the form of granules which may vary in diameter from 2–150 μm (1). Rice starch has the smallest granules and potato starch the largest (2). Microscopic examination of starch granules reveals a distinct cleft called the hilum (1). The hilum is the botanical center of the granule, ie, it is the nucleus around which granule growth occurred. Microscopic examination with polarized light reveals a birefringence which, along with x-ray diffraction, is evidence of semicrystallinity of the granule. Between crossed Nicol prisms of the microscope, a black cross, ie, a cross of isoclines, is observed at the center of the hilum. Cereal starches give an A-type x-ray spectrum, tuber starches a B-type pattern, and a few starches give an intermediate diffraction pattern, the C type (3).

Undamaged starch granules are insoluble in cold water but imbibe water reversibly and swell slightly. The percent increase in granule diameter ranges from 9.1% for corn to 22.7% for waxy corn (4). However, in hot water, a larger irreversible swelling occurs producing gelatinization. Gelatinization takes place over a discrete temperature range that depends on starch type.

Starch	*Range*, °C
potato	59–68
tapioca	58.5–70
corn	62–72
waxy corn	63–72
wheat	58–64

At a certain temperature in the heating process (the lower limit of the gelatinization temperature), the kinetic energy of the system is sufficient to overcome the hydrogen bonding in the interior of the starch granule. The amorphous regions of the granule are solvated first and the granule swells rapidly, eventually to many times its original size. During swelling, and as a consequence of it, some of the linear amylose molecules are leached out of the granule into solution. When a cooked starch paste containing amylose molecules, swollen granules, and granule fragments is cooled, the mixture thickens and, if sufficiently concentrated, may form a gel. The property of forming thick pastes or gels is the basis of many starch uses. The gelatinization temperature is the range between the lower limit, indicating onset of granule swelling, and the upper limit, corresponding to the point where almost all granules are 100% gelatinized. The gelatinization range depends upon the method used to measure it. The most sensitive method follows microscopically the loss of birefringence of a starch slurry heated on a Kofler hot stage. Other chemicals present in the slurry may affect the gelatinization range in a predictable way, and this information may be important to certain industrial applications of starch. Certain chemicals, such as sodium sulfate, sucrose, and dextrose, inhibit gelatinization and increase the gelatinization temperature, probably by competing for available water. Other chemicals, such as sodium nitrate, alkali, and urea, lower the gelatinization temperature range, possibly by disrupting granular intermolecular hydrogen bonds.

The physical properties of starch are altered by mechanical treatment. If in the dry state the granular integrity is disrupted, as by grinding, the starch gelatinizes more readily in cold water. Furthermore, the granule is readily attacked by chemicals and enzymes. In the swollen state, the granule is more fragile, and vigorous agitation of a cooked paste results in rupture of most, if not all, granules. Consequently, the cold paste loses its viscosity and gelling ability.

Specific optical rotation values $[\alpha]_D$ for starches range from 180 to 220° (3); for pure amylose and amylopectin fractions, $[\alpha]_D$ is 200°. Measurements of these values are hampered by the limited solubility of these polymers and the opacity of some of their dispersions. More recently, the structure of amylose has been investigated with infrared spectroscopy (6). Analysis of the spectrum shows it to be consistent with the proposed ground-state conformation of the monomer D-glucopyranosyl units. The nature of intramolecular bonding in amylose has also been investigated with nmr spectroscopy (7).

Chemical Properties

Most common starches contain two different types of D-glucopyranose polymers. Amylose is essentially a linear polymer of α-D-glucopyranosyl units linked (1→4) as shown in Figure 1. This polymer may be separated from the starch by complete gelatinization and vigorous dispersion of the hot starch solution with a complexing agent such as 1-butanol (8) in water. On cooling, an amylose–butanol complex crystallizes

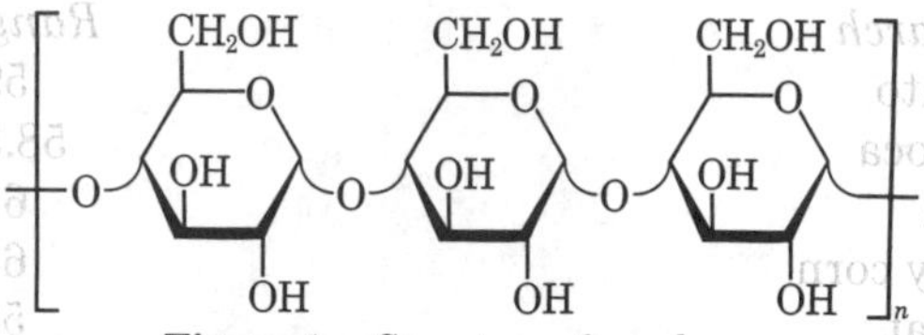

Figure 1. Structure of amylose.

and may be removed by centrifugation. Recrystallization of the amylose–1-butanol complex and removal of the 1-butanol produces highly pure amylose.

Amylopectin is a highly branched polymer of α-D-glucopyranosyl units containing 1→4 links with 1→6 links at branch points (Fig. 2). In an investigation of the amylopectin structure (9), 4.67% tetra-*O*-methyl-D-glucopyranose was obtained from amylopectin and 0.32% from amylose. This methyl ether could only derive from nonreducing end groups, and it was therefore suggested that in amylopectin the average branch length is ca 25 D-glucose units. On the basis of these data, linear amylose should consist of ca 350 D-glucose monomers per average polymer chain.

Amylose behaves as a random coil or double helix in aqueous solution and molecules in a starch dispersion associate via hydrogen bonding to form a precipitate. This tendency to associate, called retrogradation, is important in many food systems and is implicated in staling of baked goods. For instance, corn starch is not employed for starch-thickened frozen cream pies or gravies because of its tendency to retrograde on cooling and freezing. Such products made with regular corn starch have an unacceptable texture and appearance. However, amylopectin (or a modified amylopectin) does not easily retrograde and is the thickener of choice in the food systems that demand good freeze–thaw stability. Varieties of corn, barley, and rice having no amylose are the starches of choice for starch-thickened frozen foods.

Starch is not only heterogeneous in relation to polymer structure but also in relation to polymer molecular weight. Numerous methods have been used to determine the molecular weight of starch polymers, among them end-group analysis (used only for amylose), osmotic-pressure determination, light-scattering methods, and ultracentrifugation techniques (see Analytical methods). For amylose, osmotic-pressure measurements gave a range of 10,000–60,000 (10). With an anerobic isolation technique to prevent possible oxidative degradation, a range of $(1.6\text{–}7.0) \times 10^5$ was obtained (11). With DMSO extraction, an amylose was isolated with a molecular weight of 1.9×10^6 as measured by light-scattering techniques.

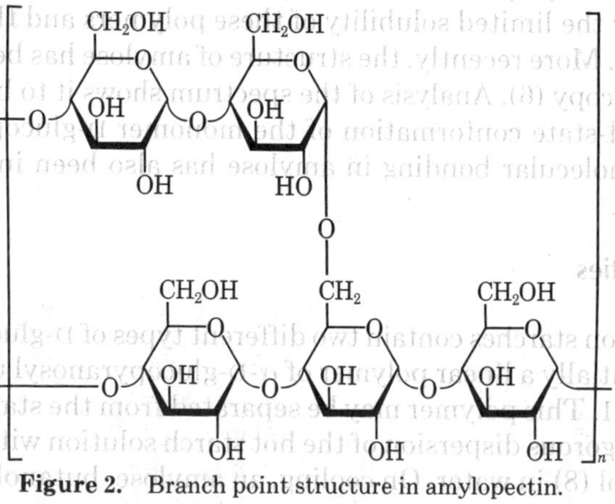

Figure 2. Branch point structure in amylopectin.

Amylopectin is a much larger molecule, ranging in mol wt between 50,000 and 10^6 (10). More recent measurements based on light scattering reveal very large amylopectin molecules in Easter lily starch, mol wt = 250×10^6 (12), and in potato starch, mol wt = 36×10^6 (13). Clearly, these molecules are larger than first believed and are much larger than their linear companion in starch, amylose.

Hydrolysis of starch is an important industrial reaction which is accomplished by acid, enzymes, or both. Starch is hydrolyzed enzymatically with α-amylase; glucoamylase produces commercial D-glucose. The action of acids on starch produced a sugarlike substance (14). The principal degradation product was identified as D-glucose (15). Treatment with acid causes random cleavage of the $\alpha 1 \rightarrow 4$ and $\alpha 1 \rightarrow 6$ linkages in the starch molecules. Other products may also be formed, among them oligosaccharides and acid breakdown products of D-glucose such as 5-hydroxymethylfurfural and levulinic acid. Oligosaccharides are produced either by incomplete breakdown of the starch molecules or by reversion (acid-catalyzed recombination of two or more D-glucose monomers).

α-Amylase hydrolyzes starch to a mixture of D-glucose, maltose, and a limit dextrin obtained from amylopectin. α-Amylase acts by randomly cleaving α-(1→4) glucosidic bonds. However, the 1→6 α-D-bonds of amylopectin are resistant to this enzyme and when a branch point is encountered, hydrolysis stops. The remainder of amylopectin, after the accessible outer branches have been hydrolyzed, is called the α-amylase-limit dextrin. α-Amylase may be of animal origin (salivary and pancreatic amylase), plant origin, or microbial origin. Enzymes of plant and animal origin are necessarily involved in starch conversion *in vivo*, but the principal enzymes used in commercial starch digestion are of microbial origin. These enzymes are used in the distilling, brewing, baking, textile, and paper industries.

β-Amylase occurs in many plants, eg, barley, wheat, rye, soy beans, and potatoes, where it is usually accompanied by some α-amylase. The action pattern of β-amylase on starch molecules is much more precise than that of α-amylase. Beginning at the nonreducing end of amylose or amylopectin, β-amylase removes maltose units successively until, in the case of amylose, the reducing end of the molecule is encountered or, in the case of amylopectin, an α(1→6) branch point is met. The remaining structure from amylopectin, after treatment with β-amylase, is called the β-amylase limit dextrin. Many workers have used β-amylase as a probe of the fine structure of amylopectin (16–19).

Glucoamylase hydrolyzes both amylose and amylopectin, yielding D-glucose as a single product. Thus, this enzyme has the ability to split both 1→4-α-D and 1→6-α-D-glucosidic bonds, although the 1→4 bonds are lysed more rapidly than the 1→6 bonds. Glucoamylase removes single D-glucose units from the nonreducing end(s) of the molecule. Glucoamylases are produced by some species of fungi of the *Aspergillus* and *Rhizopus* groups as well as by certain yeasts and bacteria.

Although acid-catalyzed hydrolysis of starches has been fairly thoroughly investigated, less is known about the reaction of these polymers in alkaline solution. The degrading reaction of starch in alkaline solution proceeds more slowly than the acid lysis and yields derivatives of isosaccharinic acids as its product rather than D-glucose. The mechanism is a beta-elimination-type reaction (20).

Oxidation of the hydroxyl groups of starch, eg, with hypochlorite, gives aldehydes, ketones, or carboxylic acids. Treatment with periodic acid results in the formation of aldehyde groups at *C*-2 and *C*-3 of the D-glucopyranose ring. Other starch oxidants include nitrogen dioxide, chlorine, permanganate, dichromate, and ozone.

Etherification and esterification of the hydroxyl groups give starch derivatives (see below). Derivatives are also obtained through graft polymerization. A free radical, initiated on the starch backbone, reacts with monomers such as vinyl or acrylic derivatives. A number of such copolymers have been prepared and evaluated for extrusion processing (21). A starch–acrylonitrile graft copolymer has received the most attention; it rapidly absorbs many hundred times its weight in water and has been patented (22). Potential applications of this starch-based product include disposable diapers and medical supplies (see Cyanoethylation).

A relatively new application of starch is the encapsulation of chemical pesticides to improve safety in handling and reduce losses due to leaching and other causes (23–24). The pesticide is encapsulated in a cross-linked starch–xanthate. Little or no pesticide is lost by drying. These encapsulated formulations have excellent shelf life in a dry state but, when placed in water or soil, the pesticide is released readily from the matrix (see Insect-control technology).

Starch gives a characteristic blue color with iodine. This color is due to the amylose which forms a helical complex with iodine on the inside of the helix. Amylopectin also forms a complex, but its color is purple to reddish-brown, depending on the source of amylopectin (25). This color reaction is usually obscured by the amylose–iodine blue. This characteristic blue color has been used both as a qualitative and quantitative test for starch in various systems.

Manufacture

Wet-Milling of Corn Starch. Milling of corn, *Zea mays*, provides corn starch, which is extensively used in food and nonfood applications. Although other cereal grains are milled for starch, corn is readily available at a low and steady price. Starch processes have recently undergone radical changes because of the introduction of continuous automation. Factories have been rebuilt and at least eight new plants have been constructed in the United States in the past decade. Corn may be dry-milled using screening and air classification of particle size, but this process does not completely separate oil, protein, starch, and hull (26). Better separation is obtained by wet-milling (27).

To better understand the milling process, it is necessary to examine the structure of the corn kernel (see Fig. 3) (28). The principal parts of the kernel are the tip cap, the main entry for water absorption by the kernel which consists of 0.8 wt % of the kernel, the pericarp or hull (5%), the germ or embryo (11%), and the endosperm (82%). The tip cap and pericarp comprise the fiber fraction in wet-milling, or the bran fraction in dry-milling. The germ is composed mainly of protein and lipids, whereas the endosperm consists of starch granules embedded in proteinaceous cell walls. The principal U.S. corn crop, dent corn, has two distinct regions of endosperm, termed floury and horny. The floury endosperm at the grain center has loosely packed starch in a light protein cellular structure, whereas the horny endosperm at the grain edges contains densely-packed starch granules with a high protein content. Starch granules in the horny endosperm are angular in shape as opposed to the round granules in the floury endosperm. In most dent corn the ratio of floury to horny endosperm is ca 1:2. The horny endosperm requires thorough steeping to ensure complete starch recovery. The germ next to the endosperm is composed of the scutellum, a repository of enzymes required for hydrolyzing the endosperm for the embryo from which the new corn plant

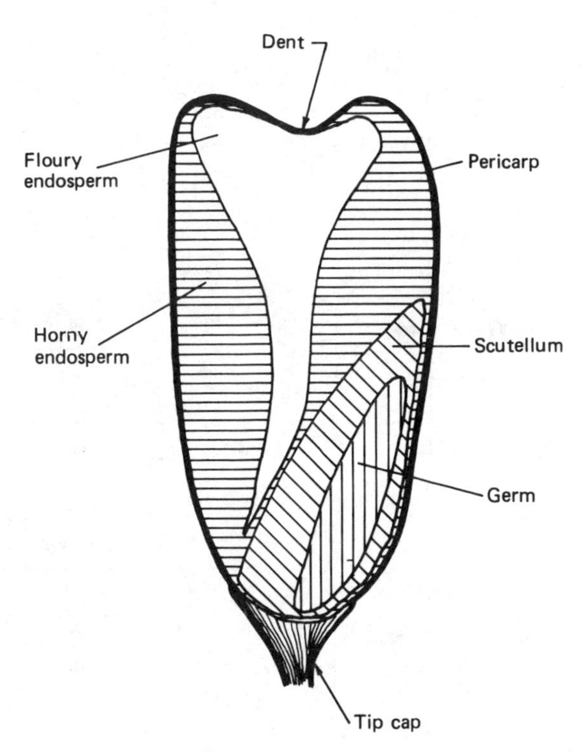

Figure 3. Schematic cross-sectional view of a corn kernel.

germinates. The scutellar epithelium is bound strongly to the endosperm, and effective germ separation requires prolonged steeping. An average composition of corn grain on a dry basis is 71.3% starch, 9.91% protein, and 4.45% fat (29–30). The normal water content is 10–15%.

A flow chart for a typical corn-milling operation is shown in Figure 4. The corn is first cleaned by screening to remove cob, sand, and other foreign material, and then by aspiration to remove the lighter dust and chaff. It is then placed in large vats, called steeps, for the steeping process of softening the kernels for milling. Steeping requires careful control of water flow, temperature (48–52°C), sulfur dioxide concentration (0.1%), and pH at 3–4. Corn introduced to the steeps at a moisture content of 15% attains a moisture concentration of 45% at the end of 30–40 h. This water absorption rate is accelerated by the sulfur dioxide (31) in the steep water and results in a 55–65% increase in kernel volume (32–33,35).

Sulfur dioxide was first added to corn steep water to prevent the growth of putrefactive microorganisms, but it is indispensible in maximizing starch yield. It acts on the nitrogen-containing components of corn which consist of 10% albumin and nonprotein nitrogen, 10% globulin, 38% zein, and 42% glutelin (34). Sulfur dioxide effects softening of the glutelin matrix, followed by dispersion (35). This action allows maximum starch release and recovery, especially from the horny endosperm. Although sulfur dioxide inhibits the growth of microorganisms, after several hours its concentration decreases and lactic acid bacteria resembling *Lactobacillus bulgaricus* start to grow.

The steeped corn is coarsely ground in an attrition mill to break loose the germ. The mill gap during this step must be adjusted to maximize the amount of germ freed

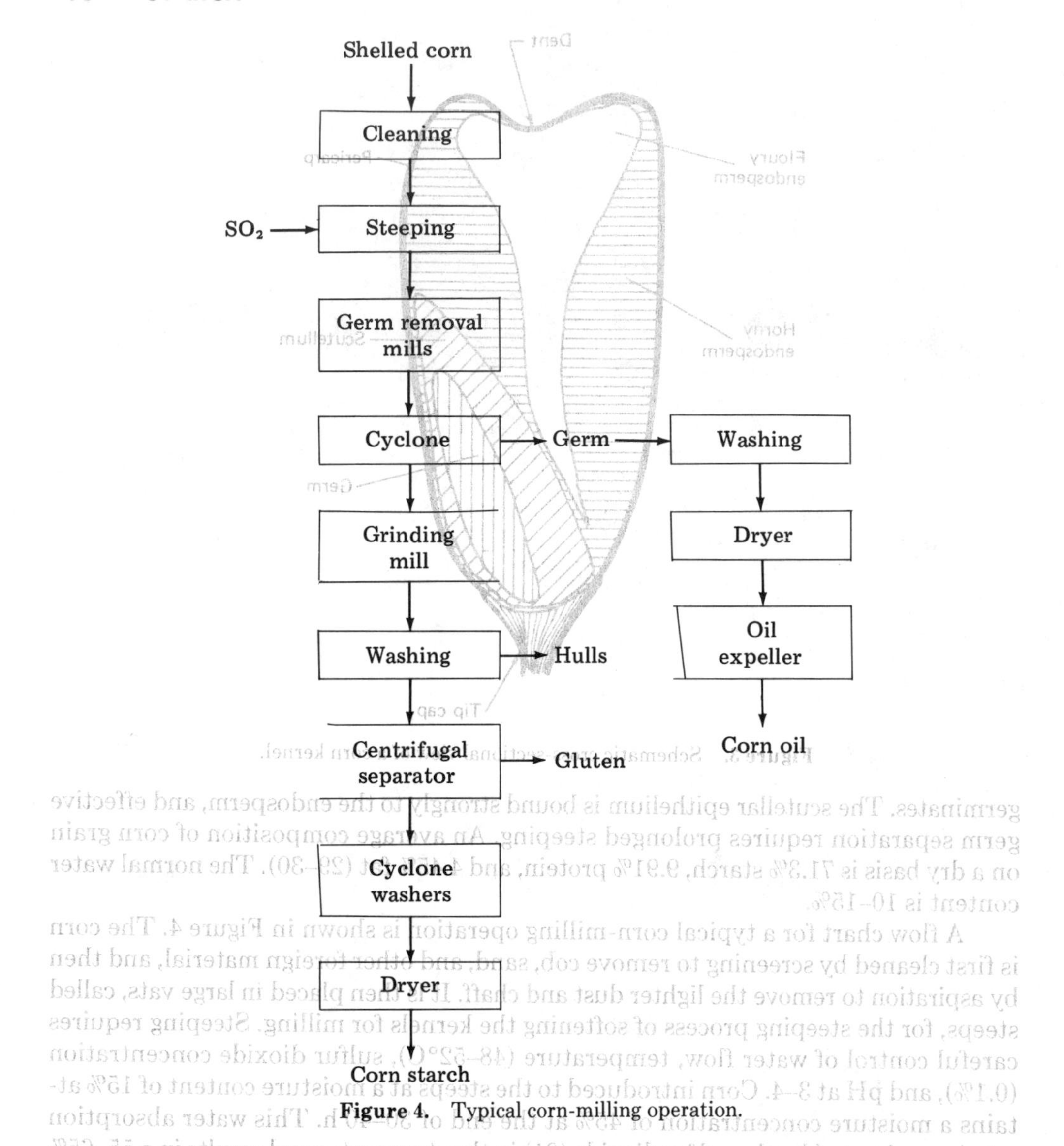

Figure 4. Typical corn-milling operation.

but minimize rupture of the germ, which would cause loss of oil and present problems in the purification step. Germ is removed from the aqueous slurry in a hydroclone (Figure 5), ie, a cyclone separator. Particles separate by density (36–37), the endosperm and fiber leaving in the hydroclone underflow and the germ from the center. The germ fraction is then pumped onto screens, washed several times to remove residual starch, dewatered to 50–55% water content, and processed for corn oil.

The cyclone underflow is milled a second time for complete release of the starch granules. Some starch factories use a Bauer mill, which is a combination attrition–impact mill (38); others favor the Entoleter mill, which is an impact mill only (39). Following the second milling, the kernel suspension contains starch, gluten, and fiber. The fiber used to be removed by screening the suspension through agitated nylon screens, but today the slurry is flowed over fixed concave screens. The fiber is retained

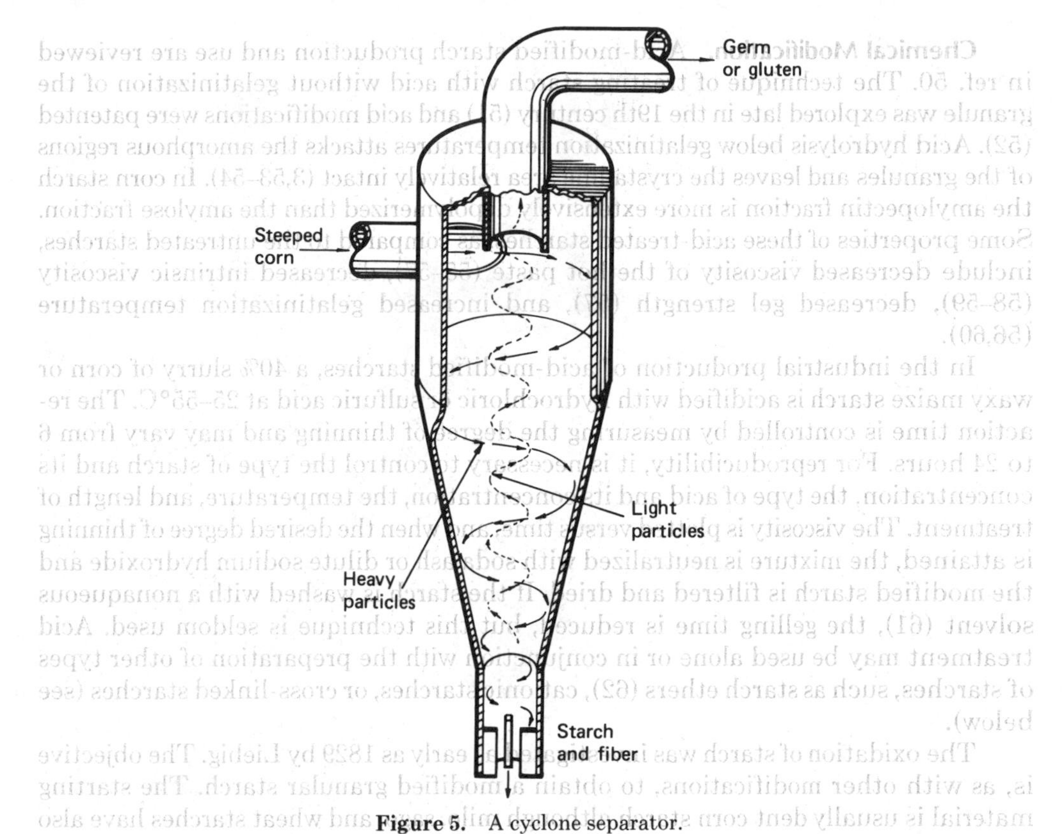

Figure 5. A cyclone separator.

on the screen, while the starch and gluten pass through. Collected fiber is slurried and rescreened to remove residual starch and protein. The fiber is later combined with 21% gluten for feed use.

The starch–gluten suspension, commonly known as mill starch is concentrated by centrifugation to reduce soluble material. The low density of gluten, compared to starch, permits easy separation by centrifugation. Protein content is thus reduced to 1–2%. The starch suspension from the centrifugal separator is diluted and subjected to 8–14 stages of hydroclone washing (40–41). The concentrated starch underflow from this process is once again diluted and passed through a final battery of hydroclones to wash the starch and remove the last traces of protein.

The starch suspension may be processed dry and marketed as unmodified corn starch, modified by chemical or physical means, gelatinized and dried, or hydrolyzed to corn syrup. The wet-milling process requires ca 0.2 m^3 H_2O/100 kg (20 gal/100 lb) corn; this water must be removed before marketing. The corn is usually dewatered by centrifugation, followed by injection into a column of hot air (200–260°C). The starch granules dry very rapidly and are collected in cyclones (42–43). The large amounts of energy consumed in evaporating and drying starch make the wet-milling industry the second most energy-intensive food industry in the United States (44). The principal product of milling, unmodified corn starch, is a white powder with a pale yellow tint. Absolute whiteness requires bleaching. The final product usually has a water content of 11% and may contain 1% protein, ash, lipids, and fiber (5,45–49).

Chemical Modification. Acid-modified starch production and use are reviewed in ref. 50. The technique of treating starch with acid without gelatinization of the granule was explored late in the 19th century (51) and acid modifications were patented (52). Acid hydrolysis below gelatinization temperatures attacks the amorphous regions of the granules and leaves the crystalline area relatively intact (3,53–54). In corn starch the amylopectin fraction is more extensively depolymerized than the amylose fraction. Some properties of these acid-treated starches, as compared to the untreated starches, include decreased viscosity of the hot paste (55–57), decreased intrinsic viscosity (58–59), decreased gel strength (57), and increased gelatinization temperature (56,60).

In the industrial production of acid-modified starches, a 40% slurry of corn or waxy maize starch is acidified with hydrochloric or sulfuric acid at 25–55°C. The reaction time is controlled by measuring the degree of thinning and may vary from 6 to 24 hours. For reproducibility, it is necessary to control the type of starch and its concentration, the type of acid and its concentration, the temperature, and length of treatment. The viscosity is plotted versus time, and when the desired degree of thinning is attained, the mixture is neutralized with soda ash or dilute sodium hydroxide and the modified starch is filtered and dried. If the starch is washed with a nonaqueous solvent (61), the gelling time is reduced, but this technique is seldom used. Acid treatment may be used alone or in conjunction with the preparation of other types of starches, such as starch ethers (62), cationic starches, or cross-linked starches (see below).

The oxidation of starch was investigated as early as 1829 by Liebig. The objective is, as with other modifications, to obtain a modified granular starch. The starting material is usually dent corn starch although milo, sago, and wheat starches have also been employed.

The oxidant commonly used is sodium hypochlorite, prepared from chlorine and aqueous sodium hydroxide.

$$2\,NaOH + Cl_2 \rightarrow NaCl + NaOCl + H_2O$$

This reaction is exothermic, and cooling must be provided during the preparation of the oxidant.

In the manufacture of oxidized starch, a slurry of starch granules is treated with alkaline hypochlorite, neutralized, washed to remove salts, and dried to a moisture content of 10–12%. Temperature, an important variable for oxidation control, is generally in the range of 21–38°C. Structure of the unmodified starch influences the type and extent of oxidation (63–64). As with acid treatment, the principal part of the oxidation occurs in the amorphous region of the starch granules (63–65). Chemical changes (63–64) include formation of carboxyl and carbonyl groups and scission of some D-glucosyl linkages, resulting in a decrease in molecular weight of the carbohydrate polymers. Oxidized starches are bleached white by hypochlorite. Because they are anionic, they absorb cationic pigments such as methylene blue. Oxidation results in lower pasting temperatures, decreased thickening power, and lower paste setback (66).

Changes in starch on heat treatment were investigated in the 1930s (67–68). Later, experiments in starch dextrinization showed that the products of dry heat treatment, ie, the pyrodextrins, were soluble in cold water and did not retrograde from solution (69). In dextrinization, starch is hydrolyzed and low molecular weight dextrins are produced, some of which recombine to form a more highly branched structure. Furthermore, incipient pyrolysis eliminates water from a few D-glucopyranosyl groups.

In the manufacture of pyrodextrins, dried starch is sprayed with an acid, usually hydrochloric but sometimes nitric, and then dried to 1–5% water content. The acidified starch is hydrolyzed and reverted by heating. At a final temperature of 95–120°C, a white dextrin is produced, and at 150–180°C, a canary-yellow dextrin. Without acid, but with a longer reaction time and a final temperature of 170–195°C, the product is a British gum. In the last step, cooling must be rapid to prevent overconversion. The acid may be neutralized at this point if so desired. Chemical changes produced by this acid–heat treatment lead to starches with lower water content, decreased solution viscosity, and greater solubility in hot water compared to the original starch.

Economic Aspects

Commercial starch is mainly corn starch, although some sorghum, wheat, and potato starch is also produced. In 1980, ca 5% of the total corn production, 21×10^6 m^3 (6×10^8 bu), were ground for starch and other products; 1 m^3 corn weighs ca 721 kg and yields 438 kg starch, 26 kg oil, and 142 kg combined gluten and hulls (or 1 bu of 56 lb gives 34 lb starch, 2 lb oil, and 11 lb gluten and hulls).

The corn-refining industry is organized to deal with very large output and in 1977 corn millers shipped 2.49×10^6 metric tons of starch worth ca $\$408.2 \times 10^6$ (70). The development of high fructose corn syrup (HFCS) stimulated the growth of the corn-refining industry. In 1967, HFCS was consumed at a rate of 45 g per capita; in 1979 it was estimated at 6.8 kg per capita, accounting for ca 12% of the total corn-sweetener market (71). The growth rate of corn-derived sweeteners has been phenomenal, increasing from 1.13×10^6 t in 1960, to more than 3.40×10^6 t in 1980 (72). This figure represents >25% of the total sweetener market and is expected to grow in the future. The growth potential for the corn wet-milling industry continues to be good because of predicted increases in HFCS production and demand for corn alcohol (see Sweeteners).

Starch imports account for a small portion of the total U.S. starch industry and consists mainly of tapioca starch from Thailand. The 1977 imports are given below (73).

tapioca and cassava	37,460 t
all others	25,730 t
Total	*63,190 t*

Uses

Nonfood Uses. Unmodified corn starch is used mainly in nonfood applications in the mining, adhesives, and paper industries. Although pregelatinized starch is unmodified chemically, it is modified physically. Pregelatinized starches minimize water losses in oil-well drilling muds, and are used in cold-water-dispersable wallpaper pastes and in papermaking as internal fiber adhesives (see Petroleum, drilling fluids; Papermaking additives).

Modified starches are acid-modified, oxidized, or heat-treated. Acid-modified, or thin-boiling starches, are employed mainly in the textile industry as warp sizes and fabric finishes (see Textiles). In this application they serve to increase yarn strength and abrasion resistance and generally improve weaving efficiency. Thin-boiling starches also find some applications in paper manufacture and laundry-starch preparations.

Oxidized starches, principally those prepared by hypochlorite oxidation, are mainly used in paper coatings and adhesives to improve surface characteristics for printing or writing. Oxidized starches are also employed as textile warp sizes and finishes, in the manufacture of insulation and wallboard, and in laundry spray starch.

Starch pyrodextrins and British gums have the ability, in aqueous solution, to form films capable of bonding like or unlike materials. As such, they have extensive use as adhesives for envelopes, postage stamps, and other products (see Adhesives). These dextrins are used in glass-fiber sizing, to protect the extruded fiber from abrasion, and as binders for metal core castings, water colors, briquettes, and other composite materials.

Various organic chemicals, eg, ethanol, isopropyl alcohol, n-butanol, acetone, 2,3-butylene glycol, glycerol, and fumaric acid, are produced from starch by fermentation (74) (see Fermentation). Other polyhydroxy compounds from starch include D-glucose, sorbitol, methyl α-D-glucopyranoside, and glycerol or glycol D-glucopyranosides.

Food Uses. Unmodified starch is used in food preparations requiring thickening, gelling, or similar properties. Such applications include puddings, salad dressings, pie fillings, and candies. Pre-gelatinized starch is used in a variety of products where thickening is required but cooking is to be avoided, such as instant pudding, pie fillings, and cake frostings.

Acid-modified starches are used for the manufacture of gum candies because they form hot concentrated pastes that gel firmly on cooling. Heat-treated starches are used in food applications to bind and carry flavors and colors. Sweetening agents are made from starch by enzymatic or acid treatment.

Derivatives

Starches, as polyhydroxy compounds, undergo many reactions characteristic of alcohols, including esterification and etherification. Since the D-glucopyranosyl monomers contain, on average, three free hydroxyl groups, the degree of substitution (DS) may be at most three. The more important commercial starch derivatives are very lightly derivatized (DS <0.1). Such modifications may produce distinct changes in colloidal properties and generally produce polymers with properties useful in a wide variety of applications.

Hydroxylalkyl Starch Ethers. Hydroxyethyl starch ethers of low DS (0.05–0.10) have been produced in a number of ways, usually near the end of the wet-milling process on a high solids–starch suspension. The derivative, a modification of ungelatinized starch, is easily filtered and can be produced economically in a fairly pure form (see also Cellulose derivatives).

In the wet-milling process, a 40–50% solids–starch suspension is treated with a Group I or Group II metal hydroxide and ethylene oxide at ca 50°C. A DS of 0.1 is easily obtained, and the product is still conveniently purified by filtration. Higher DS products are obtainable but are more difficult to purify and filter because the derivatized granules swell during washing.

Introduction of hydroxyethyl groups at low DS results in marked modification of physical properties. Among them are reduced gelatinization temperature (75), increased rate of granule swelling (76), and lowered tendency of starch pastes to gel and retrograde.

Low DS hydroxyethyl starches are widely used as paper coatings and sizes to improve sheet strength and stiffness. They are also used as paper-coating color adhesives and to increase internal fiber bonding in paper products. Additionally, hydroxyethyl starches are used as textile warp sizes.

Hydroxypropyl and other hydroxylalkyl starches are used as additives in salad dressing, pie fillings, and other food thickening applications (see Food additives).

Cationic Starches. The two commercial cationic starches are the tertiary and quaternary aminoalkyl ethers. Tertiary aminoalkyl ethers are prepared by treating a basic starch slurry with a tertiary amine containing a β-halogenated alkyl, 3-chloro-2-hydroxypropyl radical, or a 2,3-epoxypropyl group. Under these conditions, starch ethers are readily formed that contain tertiary amine free bases. Treatment with acid gives the cationic form. Amines that have been used in this reaction include 2-dimethylaminoethyl chloride, 2-diethylaminoethyl chloride, and *N*-(2,3-epoxypropyl)diethylamine. Commercial preparation of low DS derivatives requires reaction times of 6–12 h at 40–45°C for complete reaction. The product is filtered, washed, and dried.

Cationic starches exhibit decreased gelatinization temperature and increased hot paste viscosity. Pastes remain clear and fluid even at room temperatures and show no tendency to retrograde. This stability is undoubtedly due to repulsion between the cationic starch molecules in dispersion.

Quaternary ammonium alkyl ethers are prepared in similar fashion: an alkaline starch reacts with a quaternary ammonium salt containing a 3-chloro-2-hydroxypropyl or 2,3-epoxypropyl radical. Alternatively, these derivatives may be prepared by simple quaternization of tertiary aminoalkyl ethers by reaction with reagents such as methyl iodide. Sulfonium (77) and phosphonium (78) starch salts have also been prepared and examined.

Quaternary ammonium starches, like the tertiary ammonium derivatives, show decreased gelatinization temperatures, increased paste clarity and viscosity, and reduced tendency to retrograde. Quaternary ammonium starch salts exhibit cold-water swelling at a DS as low as 0.07. Cationic starches are used on paper mainly for fiber and pigment retention. Their application also improves bursting strength and fold endurance. They have been employed as emulsifiers for water-repellent paper sizes. Because of their relatively high costs, cationic starches are not widely used as textile sizes, although they have been employed in ore refining as flocculating agents (qv).

Starch Phosphates. Starch phosphate monoesters may be prepared by heating a dry mixture of starch and acid salts of ortho-, pyro-, or tripolyphosphoric acid at 50–60°C for 1 h. Degree of substitution is generally low (DS <0.15), but higher DS derivatives can be prepared by increasing the temperature, phosphate salt concentration, and reaction time.

Compared to the unmodified starches, the monophosphate esters have a lower gelatinization temperature and swell in cold water at a DS of 0.07. Like other derivatives, starch phosphates have increased paste viscosity and clarity and decreased set-back or retrogradation. Their properties are similar to those of potato starch, which also contains phosphate groups.

Starch monophosphates are very useful in food applications because of their excellent freeze–thaw stability. As thickeners in frozen gravy and frozen cream-pie preparations, they are superior to other starches. A pregelatinized starch phosphate has been developed (79) which is dispersable in cold water for use in instant dessert powders and icings. Nonfood applications include use as core binders for metal molds, in papermaking to improve fold strength and surface characteristics, as a textile size, in aluminum refining from bauxite ores, and as detergent builder.

In contrast to monophosphates, starch phosphate diesters contain ester cross-links between two or more starch chains. The covalent linkage of the starch chains in the granule produces a starch which, compared to the unmodified starch, swells less and is more resistant to heat, agitation, acid, and rupture of swollen granules.

Starch in aqueous suspension may be cross-linked to form diesters with phosphorus oxychloride, phosphorus pentachloride, and thiophosphoryl chloride (80). Cross-bonded starches may also be produced by the reaction with trimetaphosphates (81), which require more vigorous conditions than phosphorus oxychloride. Typically, a starch slurry and 2% trimetaphosphate salt react at pH 10–11 and 50°C for 1 h.

Starch phosphate diesters show a significant increase in the stability of the swollen granules. Depending on the degree of derivatization, the hot paste viscosity may be greater or less than that of the parent starch. In contrast to starch phosphate monoesters, the pastes of the diesters do not increase in clarity. Starches with high DS have exceptional stability to high temperatures, low pH, and mechanical agitation. If the degree of cross-linking is sufficient, swelling can be completely inhibited, even in boiling water.

Cross-linked starch derivatives are used as thickeners and stabilizers in baby foods, salad dressings, fruit pie filling, and cream-style corn. They are superior to unmodified starches in their ability to keep food in suspension after cooking, increased resistance to gelling and retrogradation, freeze–thaw stability, and lack of syneresis on standing. They are also used to produce high wet-strength paper (80), as ion-exchangers, and metal sequesterants to prevent oxidative rancidity in oils.

Starch Acetates. Starch acetates may be either low DS or high DS. The commercial importance of low DS acetates is based mainly on the stabilization of aqueous polymer sols. Light derivatization of low DS acetate inhibits association of the amylose polymers and the longer outer chains of amylopectin. These properties are important in food applications. More highly derivatized starches (DS 2–3) are useful because of their solubility in organic solvents and ability to form films and fibers.

Low DS starch acetates are produced by treatment of granular starch with acetic acid or preferably acetic anhydride, either alone or in acetic acid, pyridine, or aqueous alkaline solution. Dimethyl sulfoxide has been used as a cosolvent with acetic anhydride to make low DS starch acetates; ketene or vinyl acetate have also been investigated. Commercially, the acetic anhydride-aqueous alkali system is employed at pH 7–11 and room temperature to give a DS of 0.5.

Low DS starch acetates have lower gelatinization temperatures and less tendency to retrograde after pasting and cooling. Gelling may be completely inhibited if the DS is high enough. Low DS starch acetate polymers also form films useful in textile and paper manufacture.

Lightly derivatized starch acetates are used in food because of their clarity and stability. These applications include frozen fruit pies and gravies, baked goods, instant puddings, and pie fillings. Starch acetates are employed in textiles as warp sizes and in paper to improve printability, surface strength, and solvent resistance.

High DS starch acetates are prepared by the methods used for low DS acetates, but with longer reaction time. In general, high DS starch acetates and amylopectin acetates give weak and brittle films and fibers. However, amylose triacetate is useful in forming film and fiber. It is soluble in organic solvents such as acetic acid, pyridine, and chloroform. Films of this high DS acetate, cast from chloroform solution, are pliable, lustrous, transparent, and colorless. These properties are useful in packaging material.

BIBLIOGRAPHY

"Starch" in *ECT* 1st ed., Vol. 12, pp. 764–778, by R. W. Kerr, Corn Products Refining Company; "Starch" in *ECT* 2nd ed., Vol. 18, pp. 672–691, by Stanley M. Parmerter, Corn Products Company.

1. R. W. Kerr in R. W. Kerr, ed., *Chemistry and Industry of Starch*, Academic Press, Inc., New York, 1950, pp. 3–25.
2. O. B. Wurzburg in T. E. Furia, ed., *Handbook of Food Additives*, The Chemical Rubber Company, Boca Raton, Fla., Vol. 1, 1973.
3. D. French in ref. 1, pp. 157–178.
4. N. N. Hellman, T. F. Boesch, and E. H. Melvin, *J. Am. Chem. Soc.* **74,** 348 (1952).
5. T. J. Schoch and E. C. Maywald in R. L. Whistler and E. F. Paschall, eds., *Starch: Chemistry and Technology*, Vol. II, Academic Press, Inc., New York, 1967, pp. 637–647.
6. B. Casu and M. Reggiani, *Stärke* **18,** 218 (1966).
7. M. St. Jacques, P. R. Sundararajan, J. K. Taylor, and R. H. Marchassault, *J. Am. Chem. Soc.* **98,** 4386 (1976).
8. T. J. Schoch, *Cereal Chem.* **18,** 121 (1941).
9. W. Z. Hassid and R. M. McCready, *J. Am. Chem. Soc.* **65,** 1157 (1943).
10. K. H. Meyer in H. Mark and A. V. Tobolsky, eds., *Physical Chemistry of High Polymeric Systems*, 2nd ed., Interscience Publishers, Inc., New York, 1950, p. 456.
11. C. T. Greenwood, *Stärke* **12,** 169 (1960).
12. B. H. Zimm and C. D. Thurmond, *J. Am. Chem. Soc.* **74,** 1111 (1952).
13. L. P. Witnauer, F. R. Senti, and M. D. Stern, *J. Polym. Sci.* **16,** 1 (1955).
14. G. S. C. Kirchoff, *Acad. Imp. Sci., St. Petersbourg, Mem.* **4,** 27 (1811).
15. T. deSaussere, *Bull. Pharm.* **6,** 499 (1814).
16. G. N. Bathgate and D. J. Manners, *Biochem. J.* **101,** 3C (1966).
17. E. Y. C. Lee, C. Mercier, and W. J. Whelan, *Arch. Biochem. Biophys.* **125,** 1028 (1968).
18. C. Mercier, *Stärke* **25,** 78 (1973).
19. J. J. Marshall and W. J. Whelan, *Arch. Biochem. Biophys.* **161,** 234 (1974).
20. J. N. BeMiller in R. L. Whistler and E. F. Paschall, eds., *Starch: Chemistry and Technology*, Vol. I, Academic Press, Inc., New York, 1965, p. 521.
21. E. B. Bagley, G. F. Fanta, R. C. Burr, W. M. Doane, and C. R. Russell, *Polym. Eng. Sci.* **17,** 311 (1977).
22. U.S. Pat. 3,935,099 (Apr. 3, 1974); U.S. Pats. 3,981,100, 3,985,616, and 3,997,484 (Sept. 8, 1975), M. O. Weaver, E. B. Bagley, G. F. Fanta, and W. M. Doane (to United States of America as represented by Secretary of Agriculture).
23. B. S. Shasha, W. M. Doane, and C. R. Russell, *J. Polym. Sci. Polym. Lett. Ed.* **14,** 417 (1976).
24. W. M. Doane, B. S. Shasha, and C. R. Russell, *Am. Chem. Soc. Symp. Ser.* **53,** 74 (1977).
25. D. Grebel, *Mikrokosomos* **57,** 111 (1968).
26. O. L. Brekke in G. E. Inglett, ed., *Corn: Culture, Processing, Products*, AVI Publishing Company, Westport, Conn., 1970, pp. 262–291.
27. S. A. Watson in ref. 5, pp. 1–51.
28. M. J. Wolf, C. L. Buzan, M. M. MacMasters, and C. E. Rist, *Cereal Chem.* **29,** 321 (1952).
29. G. Bianchi, P. Avato, and G. Mariani, *Cereal Chem.* **56,** 491 (1979).
30. G. C. Shove in ref. 26, pp. 60–72.
31. L. T. Fan, H. C. Chen, J. A. Shellenberger, and D. S. Chung, *Cereal Chem.* **42,** 385 (1965).
32. L. T. Fan, P. S. Chu, and J. A. Shellenberger, *Biotechnol. Bioeng.* **4,** 311 (1962).
33. R. A. Anderson, *Cereal Chem.* **39,** 406 (1962).

34. J. S. Wall in Y. Pomeranz, ed., *Advances in Cereal Science and Technology*, Vol. II, American Association of Cereal Chemists, St. Paul, Minn., 1978, pp. 135–219.
35. M. J. Cox, M. M. MacMasters, and G. E. Hilbert, *Cereal Chem.* **21**, 447 (1964).
36. Br. Pat. 701,613 (1953), Anonymous.
37. U.S. Pat. 2,913,122 (Nov. 17, 1959), P. L. Stavenger and D. E. Wuth (to Dorr-Oliver, Inc.).
38. U.S. Pat. 3,040,996 (June 26, 1962), M. E. Ginaven (to The Bauer Brothers Company).
39. U.S. Pat. 3,029,169 (Apr. 10, 1962), D. W. Dowie and D. Martin (to Corn Products Company).
40. U.S. Pat. 2,689,810 (Sept. 21, 1954), H. J. Vegter (to Stamicarbon N.V.).
41. U.S. Pat. 2,778,752 (Jan. 22, 1957), H. J. Vegter (to Stamicarbon N.V.).
42. F. Baunack, *Stärke* **15**, 299 (1963).
43. U.S. Pat. 4,021,927 (May 10, 1977), L. R. Idaszak (to CPC International, Inc.).
44. *Industrial Energy Study of Selected Food Industries for the Federal Energy Office*, Final Report, U.S. Dept. of Commerce, Washington, D.C., 1974.
45. R. L. Sims, *Sugar Azucar*, 50 (Mar. 1978).
46. *Corn Starch*, 3rd ed., Corn Industries Research Foundation, Washington, D.C., 1964.
47. W. R. Morrison in ref. 34, p. 221.
48. T. J. Schoch and E. C. Maywald, *Anal. Chem.* **28**, 382 (1956).
49. W. Bergthaller and G. Tegge, *Stärke* **24**, 348 (1972).
50. P. Shildneck and C. E. Smith in ref. 5, p. 217.
51. C. J. Litner, *J. Prakt. Chem.* **34**, 378 (1886).
52. U.S. Pat. 675,822 (Jan. 12, 1899), C. B. Duryea; U.S. Pat. 696,949 (May 24, 1901), C. B. Duryea.
53. K. H. Meyer and P. Bernfeld, *Helv. Chim. Acta* **23**, 890 (1940).
54. W. C. Mussulman and J. A. Wagoner, *Cereal Chem.* **45**, 162 (1968).
55. G. V. Caesar and E. E. Moore, *Ind. Eng. Chem.* **27**, (1935).
56. W. Gallay and A. C. Bell, *Can. J. Res. Sect. B* **14**, 381 (1936).
57. W. G. Bechtel, *J. Colloid Sci.* **5**, 260 (1950).
58. R. W. Kerr in ref. 1, p. 682.
59. S. Lansky, M. Kooi, and T. J. Schoch, *J. Am. Chem. Soc.* **71**, 4066 (1949).
60. H. W. Leach and T. J. Schoch, *Cereal Chem.* **39**, 318 (1962).
61. U.S. Pat. 3,446,628 (May 27, 1969), T. J. Schoch, D. F. Stella, and H. J. Wolfmeyer (to Corn Products Company).
62. E. T. Hjermstad in ref. 5, p. 425.
63. J. Schmorak, D. Mejzler, and M. Lewin, *Stärke* **14**, 278 (1962).
64. J. Schmorak and M. Lewin, *J. Polymer Sci.* **A1**, 2601 (1963).
65. F. F. Farley and R. M. Hixon, *Ind. Eng. Chem.* **34**, 677 (1942).
66. T. J. Schoch, *Tappi* **35**, 4 (1952).
67. J. R. Katz, *Rec. Trav. Chim.* **53**, 555 (1934).
68. J. R. Katz and A. Weidinger, *Z. Phys. Chem. Abt. A* **184**, 100 (1939).
69. B. Brimhall, *Ind. Eng. Chem.* **36**, 72 (1944).
70. *Census of Manufacturers*, Bureau of the Census, U.S. Department of Commerce, Washington, D.C., 1977.
71. *Sugar and Sweetener Report*, SSR Vol. 4, No. 12, U.S. Department of Agriculture, Washington, D.C., 1979.
72. *Nutritive Sweeteners from Corn*, 2nd ed., Corn Refiners Association, Inc., Washington, D.C., 1979.
73. *U.S. Imports*, FT-246, Bureau of the Census, U.S. Department of Commerce, Washington, D.C., 1977.
74. G. E. Tong, *Chem. Eng. Prog.* **74**, 70 (1978).
75. T. J. Schoch and E. C. Maywald, *Anal. Chem.* **28**, 385 (1956).
76. A. Harsveldt, *Tappi* **45**, 85 (1962).
77. U.S. Pat. 2,989,520 (Apr. 22, 1959), M. W. Rutenberg and J. L. Volpe (to National Starch and Chemical Corporation).
78. U.S. Pat. 3,077,469 (June 28, 1961), A. Aszalos (to National Starch and Chemical Corporation).
79. U.S. Pat. 2,884,346 (Dec. 24, 1957), J. A. Korth (to Corn Products Company).
80. U.S. Pat. 2,328,537 (Aug. 9, 1940), G. F. Felton and H. H. Schopmeyer (to American Maize Products Company).
81. U.S. Pat. 2,938,901 (Nov. 23, 1956), R. W. Kerr and F. C. Cleveland, Jr. (to Corn Products Company).

General References

Refs. 1, 5, and 20 are also general references.

W. Banks and C. T. Greenwood, *Starch and Its Components*, Edinburgh University Press, Edinburgh, 1975.

C. T. Greenwood, *Adv. Carbohydr. Chem.* **22**, 483 (1967).

C. T. Greenwood and E. A. Milne, *Adv. Carbohydr. Chem.* **23**, 282 (1968).

R. V. MacAllister, *Adv. Carbohydr. Chem.* **36**, 15 (1979).

D. J. Manners, *Adv. Carbohydr. Chem.* **17**, 371 (1962).

J. J. Marshall, *Adv. Carbohydr. Chem.* **30**, 257 (1974).

J. A. Radley, ed., *Examination and Analysis of Starch and Starch Products*, Applied Science Publishers, Ltd., London, 1976.

J. A. Radley, ed., *Industrial Uses of Starch and Its Derivatives*, Applied Science Publishers, Ltd., London, 1976.

J. A. Radley, ed., *Starch Production Technology*, Applied Science Publishers, Ltd., London, 1976.

R. L. Whistler, ed., *Methods in Carbohydrate Chemistry*, Vol. 4, Academic Press, Inc., New York, 1964.

ROY L. WHISTLER
JAMES R. DANIEL
Purdue University

STEAM

Steam [*7732-18-5*] is the most important industrially used vapor and, after water, the most common and important fluid used in chemical technology. It can be generated by evaporation of water at subcritical pressures, by heating water above the critical pressure, and by sublimation of ice. Steam is used in electric-power generation, for driving mechanical devices, for distribution of heat, as a reaction medium; a solvent; a cleaning, blanketing or smothering agent; and as a distillation aid. Steam is so widely used because of water's availability and steam's easy generation and distribution, high latent heat, moderate density, nonpolluting properties, and the ease of temperature control it allows in processes and heating applications.

Steam is generated and used saturated (or dry-saturated), wet, or superheated. Saturated steam has no moisture or superheat, wet steam contains moisture, and superheated steam has no moisture and its temperature is above the saturation temperature. The range of applications of steam extends from subatmospheric pressures to 36.1 MPa (356 atm). Temperatures up to 650°C are used and superheating to temperatures as high as 1090°C has been considered. Because steam cycles with higher pressures and temperatures usually have better thermodynamic efficiency, utilizing steam at higher parameters is desirable. The pressure and temperature usable are limited by the mechanical, creep, oxidation, and corrosion properties of metals used to contain the high pressure steam and water.

The thermodynamic and physical properties of steam are well established over the range of pressures and temperatures used today. The chemical properties of steam and of substances in steam, their molecular structures and interactions with the solid surfaces of containments need to be more explored.

Steam is generated from water by boiling, flash evaporation, and throttling from high to low pressure. The phase change occurs along the saturation line with the specific volume of steam larger than that of the boiling water. Thermal energy (heat of evaporation) is absorbed during the process. At the critical and supercritical pressures, the water–steam transition occurs at a constant specific volume, and the heat of evaporation is zero (there is no evaporation).

Subcritical pressure boilers or steam generators range from simple fire-tubed units to water-tubed natural or controlled-circulation drum types, to once-through boilers. For generation of supercritical pressure steam, once-through boilers are always used. The heat comes from the combustion of fossil fuel or waste products, or from nuclear fission, geothermal sources, or process streams. Utilization of the ocean thermal gradient (temperature difference between the surface and deep water), magma, nuclear fusion, and solar energy for steam generation is under development (see also Power generation; Water; Energy management; Fuels; Fusion energy; Geothermal energy; Furnaces; Heat-exchange technology).

Properties of Steam

Properties of steam can be divided into thermodynamic, transport, physical, and chemical properties. In addition, molecular structure and chemical composition of steam are of interest. It was at the start of industrialization in the eighteenth century that thermodynamic relationships were first measured, about 1763 by Watt. A century later in 1859, Rankine published his *Manual of the Steam Engine* which gave a practical thermodynamic basis for the design and performance of steam engines. The first steam table for practical use was based on Regnault's data and began to appear toward the end of the nineteenth century. A thermodynamically consistent set of equations for fitting data was devised in 1900 by H. L. Callendar and was adopted by Mollier and others. Further measurements and developments followed rapidly. The library of the NBS contains six different steam tables published between 1897 and 1915.

Because of the universal use of water and steam, the need for international research cooperation and for property formulations was recognized as early as 1929, when the First International Steam-Table Conference was held in London. Since then, nine international conferences on the properties of steam have been held, and in 1972 the International Association for the Properties of Steam (IAPS) was formed. Today, there are four IAPS Working Groups studying equilibrium properties; transport properties; the chemical thermodynamics of power cycles; and other properties (especially surface and electrical). The IAPS recommended pressure range is 0.1–1000 MPa (1–10,000 bars); and the recommended temperature range is 1–1000°C. Besides the releases and publications by IAPS, description of steam properties and related research can be found in the collection of papers that are deposited with the IAPS Executive Secretary (Office of Standard Reference Data, National Bureau of Standards, Washington, D.C.), and refs. 1–12.

Steam properties in the form of graphs, tables and theoretical and empirical equations are widely used in design and service analysis of steam engines, power systems, and heat-transfer and process equipment.

Molecular Structure and Composition. The molecular structure of steam is not as well known as that of ice and water (qv). During the water–steam phase change, rotation of molecules and vibration of atoms within the water molecules do not change considerably, but the translation movement increases. This accounts for the volume increase when water is evaporated at subcritical pressures. There are indications that even in the steam phase some H_2O molecules are associated into small clusters of two and more molecules (13). Values for the dimerization enthalpy and entropy of water have been determined from recent measurements of the pressure dependence of the thermal conductivity of water vapor at 358–386 K and 13.3–133.3 kPa (100–1000 mm Hg). Upward curvature in experimental thermal conductivity–pressure plots in this range as well as in plots of published data at higher temperature indicates the presence of one or more higher polymers. Upper limits of the equilibrium constants for the formation of the trimer ($n = 3$), tetramer ($n = 4$), pentamer ($n = 5$), and hexamer ($n = 6$) are estimated from the thermal conductivity data to be 1.9×10^{-8}, 5.8×10^{-11}, 3.8×10^{-13}, and 2.8×10^{-15} $kPa^{-(n-1)}$, respectively, where n is the cluster size. The results are consistent with theoretical calculations of the energy of these water clusters. There is evidence, based on quantum-mechanical calculations for water and methanol clusters and on experiments with methanol, that the most significant higher polymer in water vapor is the tetramer. Knowledge of the properties of water clusters is important for studies of the atmosphere and cloud formation. The thermodynamic parameters for these clusters in water vapor may be important in constructing an accurate equation of state for water, and the experimental results can help to substantiate theoretical calculations of cluster energies.

Pure water and steam contain three isotopes of hydrogen (^1H, ^2H, and ^3H) and three isotopes of oxygen (^{16}O, ^{17}O, and ^{18}O). Although tritium and oxygen-17 are present only in extremely minute concentrations (see Deuterium and tritium; Oxygen), there is about 200 ppm of deuterium and 1000 ppm of oxygen-18 in water and steam. These six components can combine in 18 different ways.

Electrical resistance of condensed steam often approachs pure-water values of 18 MΩ·cm, conductivity is typically 0.2 μS/cm or better, and impurity content is less than 1 ppm. Geothermal steam contains much higher concentrations of impurities (up to percentage amounts), reflecting composition of the bedrock. It often has high concentrations of hydrogen sulfide, calcium carbonate and bicarbonate, carbon dioxide, sodium chloride, silica, ammonia, sulfates, potassium, and other impurities. Chemical composition of steam from several geothermal fields is given in refs. 14–16 (see Geothermal energy).

Equilibrium Thermodynamic Properties. The various thermodynamic properties are not independent. If temperature T and pressure p are independent variables, the dependent basic properties are volume, v; internal energy, u; entropy, s; enthalpy (formerly called heat content), $h = u + pv$; and the thermodynamic potentials are the Helmholtz function, $f = u - Ts$, and the Gibbs function, $g = h - Ts = f + pv$.

Thermodynamic Consistency. With given values of these properties and their derivatives, a thermodynamic surface of a substance can be constructed. If a surface is defined, then all other properties follow by consistent thermodynamic relations. For example, if temperature and pressure are chosen as independent variables of a formulation, then expressions for the specific volume, entropy, enthalpy, and all other thermodynamic properties may be derived directly by partial differentiation of the

Gibbs function:

$$g = g(T,p)$$

Volume:

$$v = \left(\frac{\partial g}{\partial p}\right)_T$$

Entropy:

$$s = -\left(\frac{\partial g}{\partial p}\right)_p$$

Enthalpy:

$$h = g + Ts = g - T\left(\frac{\partial g}{\partial T}\right)_p$$

If specific volume v and temperature T are chosen as independent variables, then expressions for pressure, specific entropy and internal energy and all other properties follow by partial differentiation of the Helmholtz function, $f = f(T,v)$. Although Helmholtz and Gibbs functions are not measurable quantities, they are preferred in modern description of thermodynamic properties. They are known as fundamental functions or equations and represent the thermodynamic surface of a substance (see Thermodynamics).

The latest formulations of the thermodynamic equilibrium properties in British units are in ref. 10, in SI units in ref. 11. The pressure–temperature diagram for water and steam (Fig. 1) shows the saturation line, critical point, and ranges for which the properties have been formulated (10). A pressure–temperature diagram covering a wider range (up to 25 GPa or 250 kbars) is in Figure 2. At high pressure, supercritical fluid and water can change to ice even at high temperatures. In power generation, and

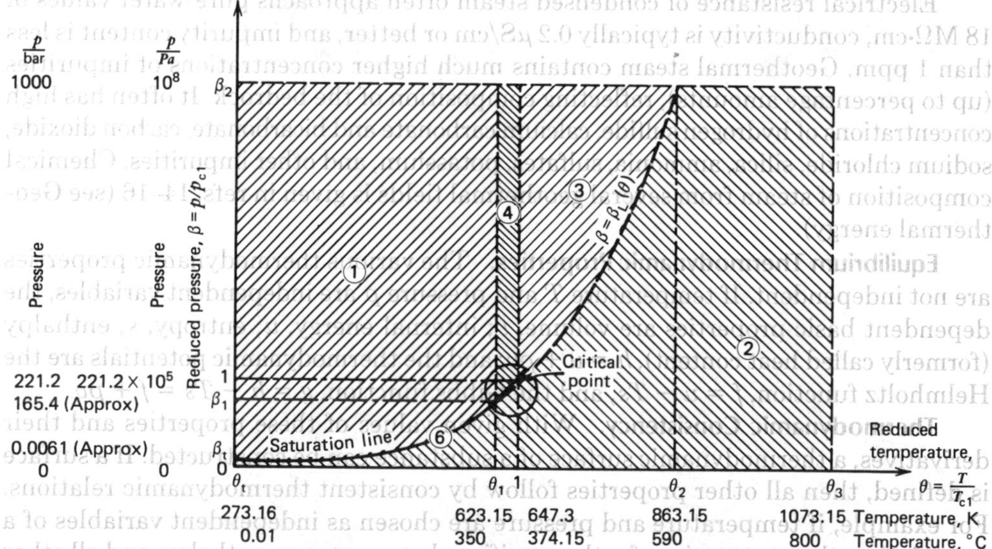

Figure 1. Pressure–temperature diagram with subregions covered by analytical representations (10).

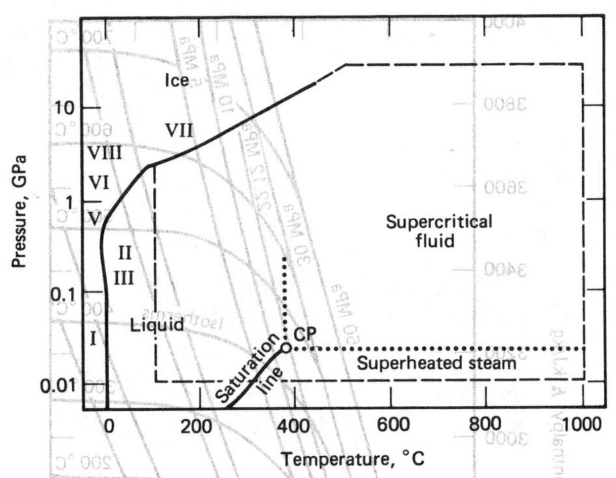

Figure 2. P–T projection of high pressure portion of the phase diagram of water (6). To convert GPa to psi, multiply by 145,000.

in the design of steam cycles, enthalpy–entropy (Mollier) diagrams are often used (see Fig. 3).

The saturation line is the transition between water and steam at subcritical conditions. During slow heating or cooling (equilibrium), the liquid–vapor phase change occurs at the saturation pressure and temperature on the saturation line. During rapid heating and cooling and steam expansion, as in many industrial steam cycles, water can be superheated or steam subcooled when the phase change occurs. The spontaneous onset of condensation during fast steam expansion, such as in a steam turbine, occurs on the so called Wilson line which is at about 4% of the theoretical (equilibrium) moisture level in the wet-steam region. At and above the critical pressure, the phase change occurs without boiling or evaporation and the enthalpy of evaporation is zero. The quality of wet steam is the percentage of steam in a steam–water mixture.

Transport Properties. Dynamic and kinematic viscosity and thermal conductivity have been grouped as the transport properties. The derived parameters, thermal diffusivity ($\lambda/\rho C_p$) and the Prandtl number ($\mu C_p/\lambda$), are two of the useful transport parameters used in engineering calculations. In the above formulas, λ is thermal conductivity, ρ is density, C_p is specific heat at constant pressure, and μ is dynamic viscosity. A plot of dynamic viscosity versus temperature is shown in Figure 4, kinematic viscosity in Figure 5, and an isometric projection of the thermal conductivity surface is shown in Figure 6. Notice the sharp increase of thermal conductivity near the critical point.

Other Physical Properties. Surface tension and the two-phase critical wavelength (Laplace constant) are given in Figure 7. Surface-tension data are important in evaluations of condensation, droplet formation and transport, surface wetting, and emulsions. Surface tension is very sensitive to chemical impurities (1).

Static dielectric constant (permittivity) for electric fields of very low frequency and moderate intensity is shown in Figure 8 (19). Static dielectric constant of saturated and superheated subcritical steam is between 1 and 3. It is higher for high pressure saturated steam.

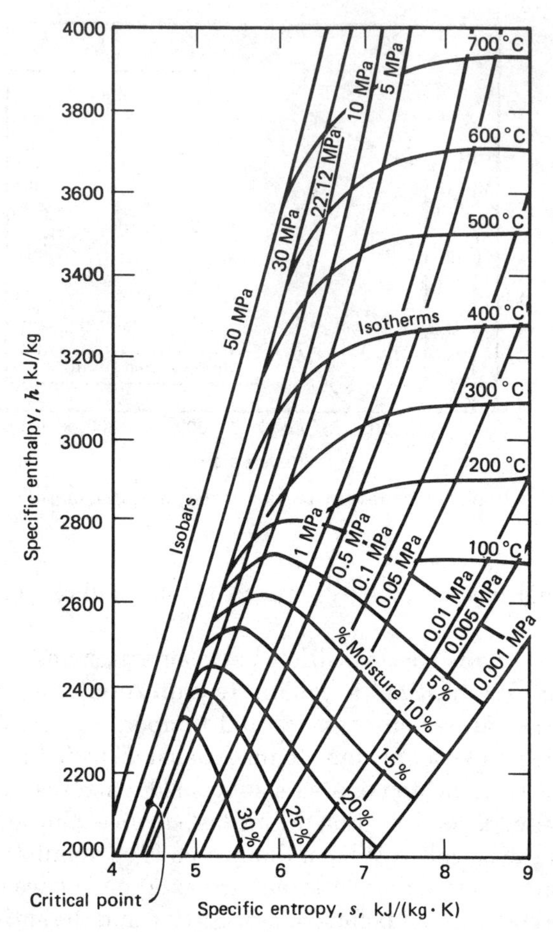

Figure 3. Enthalpy–entropy (Mollier) chart for steam (17). To convert MPa to psi, multiply by 145. To convert kJ/kg to Btu/lb, divide by 2.324; to convert kJ/(kg·K) to Btu/(lb·°F), divide by 4.184.

Chemical Properties. Chemical properties of steam and steam and water with impurities include ionization equilibria, dissociation, solvent and solute properties, and properties governing chemical reactions. Of as much interest as the bulk properties of superheated steam and steam mixtures (one phase) are properties of steam with moisture (two phase), steam over water, steam with suspended solids, and steam–solid interfaces—all under both equilibrium and dynamic conditions. Of these properties, only a fraction has been measured.

Ion Product. Ionization constant (ion product, ion-product constant) is the most important characteristic of an electrolyte, for it has a fundamental influence on acid–base and other related equilibria (see Water).

For pure H_2O or very dilute solutions, the ion-product constant K_w equals $[H_3O^+][OH^-]$. An equation for the ion product valid over a wide range of pressure and temperature has been developed (20):

$$\log K_w^* = A + B/T + C/T^2 + D/T^3 + (E + F/T + G/T^2) \log \rho_w^*$$

where $K_w^* = K_w/(\text{mol/kg})^2$, density $\rho_w^* = \rho_w/(\text{g/cm}^3)$, T (kelvins), and values for the parameters are: $A = -4.098$, $B = -3245.2\,K$, $C = +2.2362 \times 10^5\,K^2$, $D = -3.984 \times 10^7$

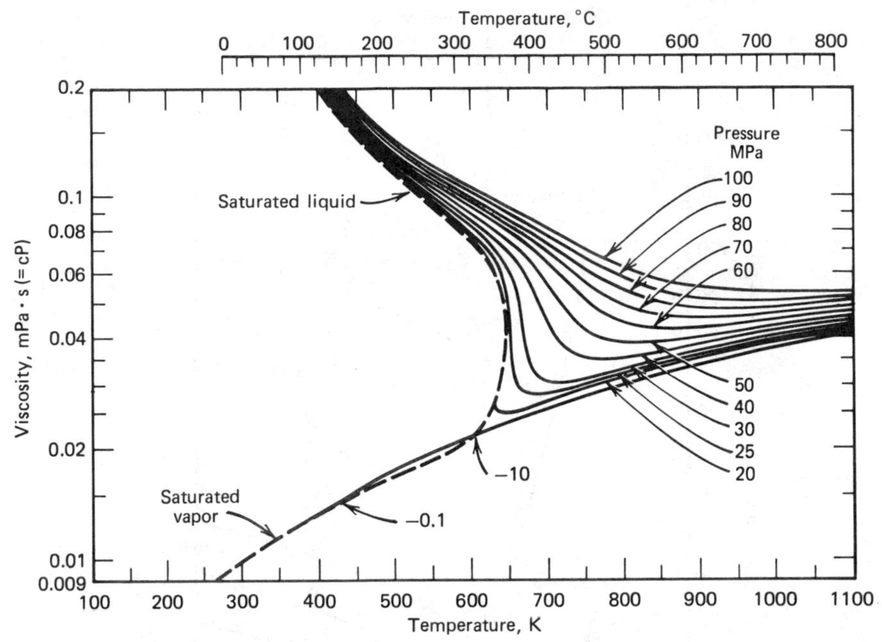

Figure 4. Dynamic viscosity of water and steam from the triple point to 1100 K and pressures to 100 MPa (1000 atm) (18). To convert MPa to psi, multiply by 145.

K^3, $E = +13.957$, $F = -1262.3\ K$, $G = +8.5641 \times 10^5\ K^2$. This equation for K_w^* is represented for density $\rho^* = 1$ g/cm^3 in Figure 9. Numerical values are in Table 1.

Solubility and Deposition. Solubility of chemicals in steam reflects the capacity of steam to transport them. When impurity concentration exceeds solubility, chemicals precipitate and can concentrate by deposition (21–22). Although the solubility of most salts of interest in water is high (typically of the order of percent), their solubility in superheated steam is in the ppb to the ppm range (higher at higher pressures). Because of the low solubility in superheated steam, the concentration of soluble impurities has to be kept very low. Steam also dissolves many metal oxides and is completely miscible with volatile substances and gases. A review of high temperature aqueous solutions, including steam is given in refs. 23–24. Dynamic solubility in rapidly expanding steam may be much higher than the equilibrium solubility.

Investigation of solubility in steam and of vaporous carryover (the distribution of substances between water and vapor) started after 1940 with the advent of higher pressure boilers and problems with steam-turbine deposits. Before that, it was not believed that steam could dissolve inorganic compounds with very low vapor pressure such as salts, hydroxides, and oxides (25–27). Silica solubility has been measured and reported (28). The solubility diagram is shown in Figure 10. Compilations and some new data on the solubility of sodium hydroxide and sodium chloride are in ref. 30; equilibrium solubility of sodium hydroxide is shown in Figure 11. Solubilities of some other compounds, including magnetite and cupric oxide, are in Figure 12. As can be seen in Figures 10 and 11, solubility in superheated steam rapidly decreases as the steam pressure decreases or specific volume increases. Such changes, which occur in turbines, pressure-reducing stations, and other components of steam cycles can result in local precipitation and deposition, which in turn may cause corrosion, clogging, and other malfunctions. The lowest solubility, and therefore lowest tolerance to impurities,

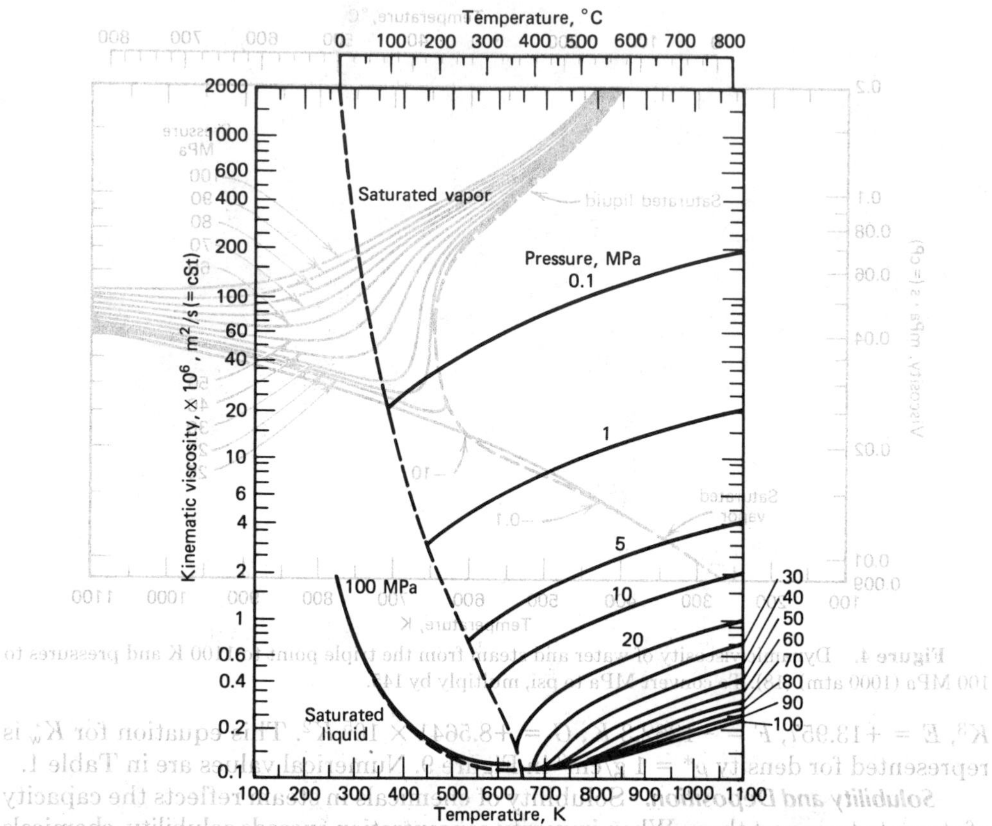

Figure 5. Kinematic viscosity of water and steam (18). To convert MPa to psi, multiply by 145.

is in the lowest pressure areas, just before the saturation line. This is illustrated in Figure 10, where a low pressure turbine, steam-expansion line is added to the silica-solubility diagram. At the last superheated stage (number 24), the solubility is only about 0.02 ppm. The sharp decrease of the solubility of cupric oxide (see Fig. 12) often results in its deposition on a few high pressure turbine blade stages.

Characteristic distribution of deposited species found in utility turbines is in Figure 13. Up to 100 wt % of some chemical compounds has been found in some deposits as indicated by the length (average to maximum) of vertical lines in Figure 13.

Water–Vapor Distribution. Distribution of substances between water and vapor (also called vaporous carryover, distribution ratio constant or coefficient, and partition coefficient) is one factor governing the concentration of low-vapor-pressure substances transferred into steam. It has been reported elsewhere (21,29,31–35) and theoretically treated (31,33–34). Distribution ratios for undissociated species are called the true ratios (K_D), for a partially dissociated species, the apparent ratios (K_D or K_{app}).

For constant concentrations, compositions, and pH and in the absence of chemical reactions, the distribution ratios depend only on the ratio of water and steam densities ρ (see Fig. 14):

$$K = \left(\frac{\rho_w}{\rho_s}\right)^n$$

where n is a constant for each chemical compound and concentration.

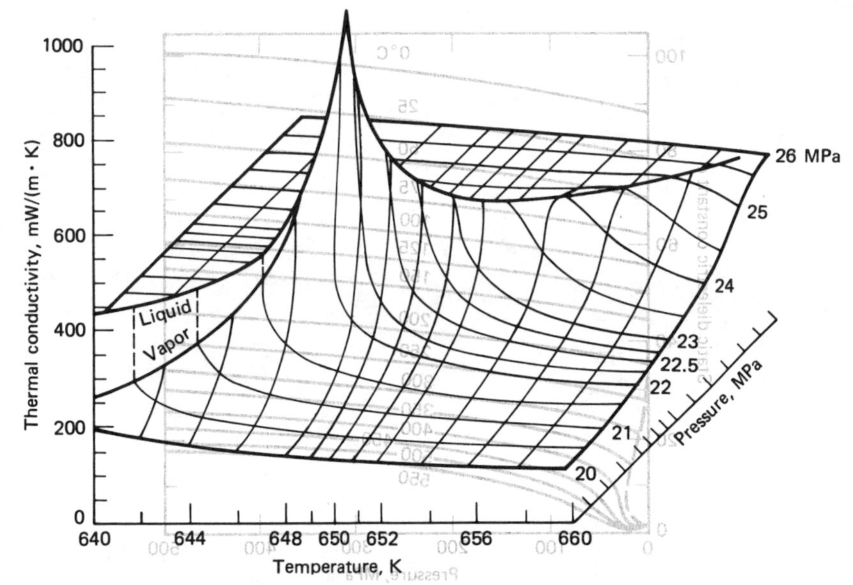

Figure 6. Isometric projection of thermal-conductivity surface. Courtesy R. C. Hendricks of NASA-Lewis. To convert MPa to psi, multiply by 145.

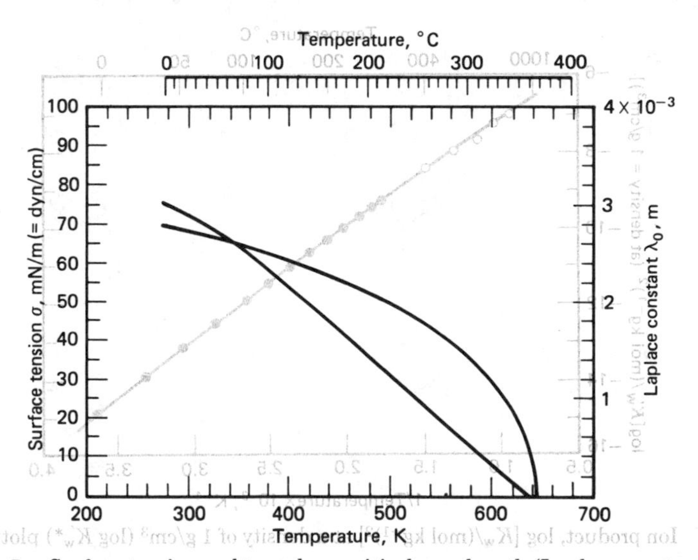

Figure 7. Surface tension and two-phase critical wavelength (Laplace constant) (11).

The compounds ionized in water (strong electrolytes) are hydrated and not so easily transferred into steam as the weakly hydrated molecules of weak electrolytes such as Al_2O_3, B_2O_3, and SiO_2. Amphoteric compounds which can react with both bases and acids (such as Al_2O_3) show maxima in the relationship of the apparent distribution ratio as a function of pH. The pH dependency of the distribution ratio for SiO_2 is shown in Figure 15. Dependence of the apparent molecular and ionic distribution ratios on the concentration of NaCl in boiler water is theoretically treated in ref. 31.

Hydrolysis. Superheated steam can hydrolyze salts and form acids and hydroxide such as in the hydrolysis of NaCl:

$$NaCl(s) + H_2O(g) = NaOH(s) + HCl(g)$$

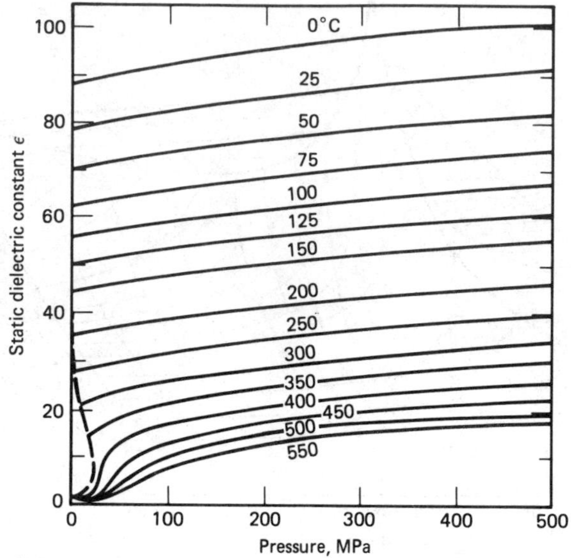

Figure 8. Static dielectric constant as a function of pressure and temperature (19). To convert MPa to psi, multiply by 145.

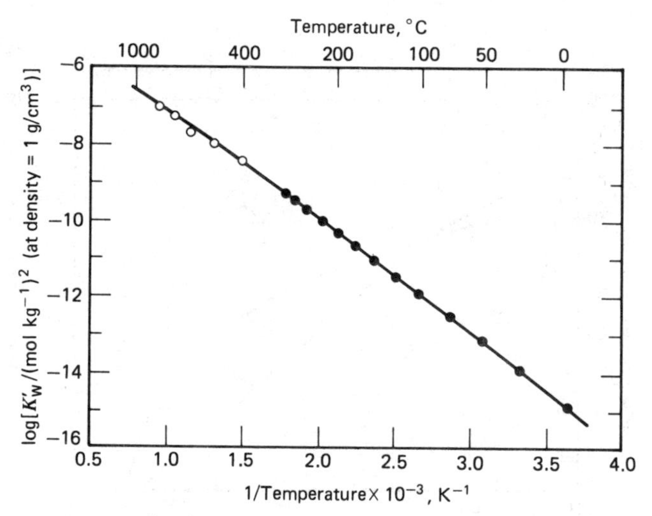

Figure 9. Ion product, log $[K_w/(\text{mol kg}^{-1})^2]$ at a density of 1 g/cm^3 (log $K_w'^*$) plotted against $1/T$, 0–1000°C (20).

Hydrolysis of NaCl has been observed at temperatures as low as 250°C when superheated steam was passed over solid NaCl crystals (33). Hydrolysis of solid sodium chloride in steam is also described in ref. 36, and of calcium chloride in ref. 37. Hydrolysis of ammonium chloride in steam has been observed at temperatures above 200°C.

Steam Generation

Steam below the critical pressure is normally generated in boilers though it may also be generated in evaporators. In the former case, the steam is generated for its

Table 1. The Negative Logarithm (base 10) of the Ion Product of Water Divided by (mol/kg)², $-\log K_w^*$ [a]

Pressure, MPa[b]	Temperature, °C							
	0	25	100	200	300	400	600	1000
saturated vapor	14.938	13.995	12.265	11.289	11.406			
25	14.83	13.90	12.18	11.16	11.14	19.43	23.27	24.93
50	14.71	13.82	12.10	11.05	10.86	11.88	18.30	20.8
75	14.62	13.73	12.03	10.95	10.66	11.17	15.25	18.39
100	14.53	13.66	11.96	10.86	10.50	10.77	13.40	16.72
150	14.34	13.53	11.84	10.71	10.26	10.29	11.59	14.5
200	14.21	13.40	11.72	10.57	10.08	9.98	10.73	12.97
250	14.08	13.28	11.61	10.45	9.91	9.74	10.18	12.02
300	13.97	13.18	11.53	10.34	9.76	9.54	9.78	11.24
350	13.87	13.09	11.44	10.24	9.63	9.37	9.48	10.62
400	13.77	13.00	11.37	10.16	9.52	9.22	9.23	10.13
500	13.60	12.83	11.22	10.00	9.34	8.99	8.85	9.42
600	13.44	12.68	11.09	9.87	9.18	8.80	8.57	8.97
700	13.31	12.55	10.97	9.75	9.04	8.64	8.34	8.64
800	13.18	12.43	10.86	9.64	8.93	8.50	8.13	8.38
900	13.04	12.31	10.77	9.54	8.82	8.37	7.95	8.12
1000	12.91	12.21	10.68	9.45	8.71	8.25	7.78	7.85

[a] Ref. 20.
[b] To convert MPa to atm, divide by 0.1013.

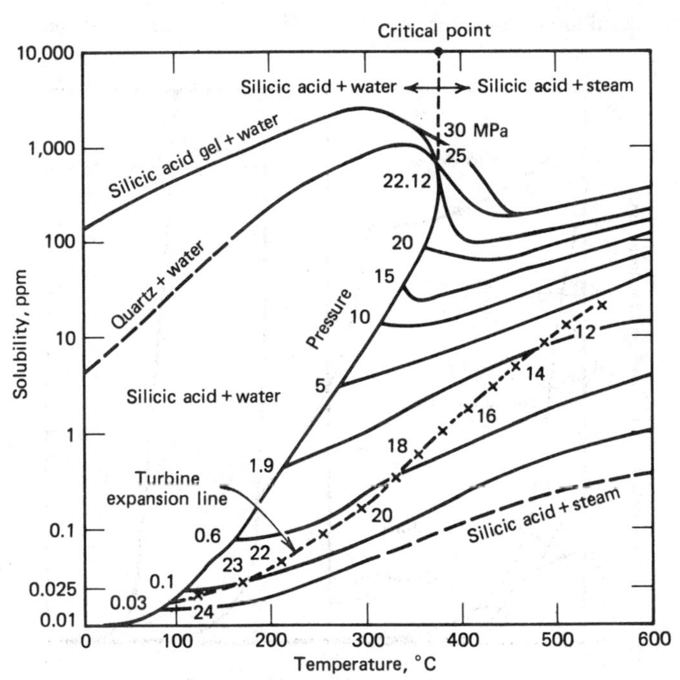

Figure 10. Silica solubility diagram (29). To convert MPa to psi, multiply by 145.

power or heating value; in the latter case, it is generated to produce, through its condensation, pure water from an impure source. Some subcritical steam generation is performed in once-through units where all the feedwater to the unit is transformed into steam in a single passage through the generator. Some very small subcritical steam

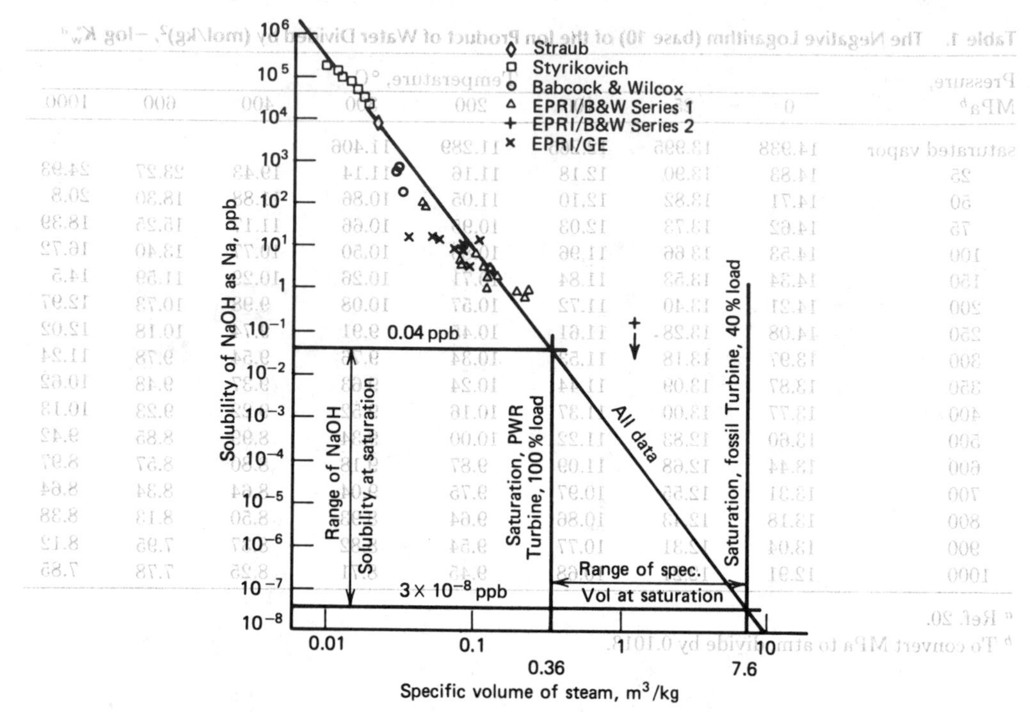

Figure 11. Equilibrium solubility of sodium hydroxide and turbine steam conditions.

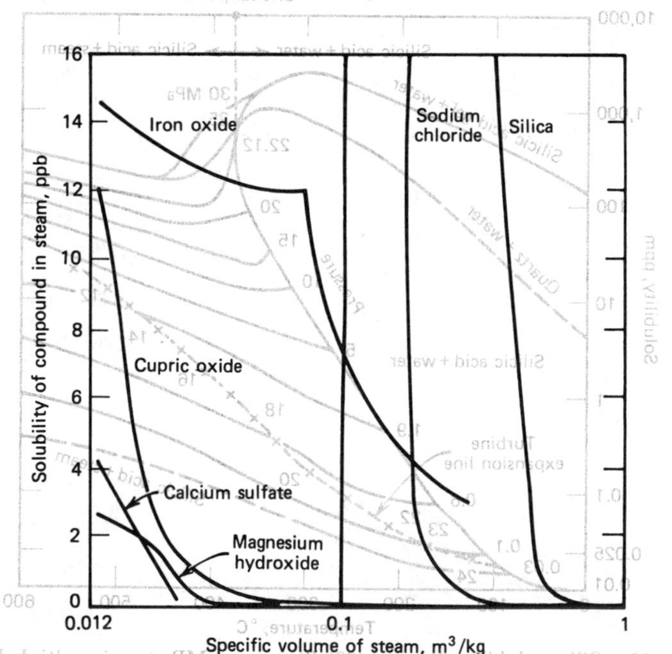

Figure 12. Solubility in superheated steam under typical steam turbine conditions.

generators are of the flash type: the water is simply dropped on a heated surface and completely converted to steam.

The limits of steam production are fixed in practice by the ability of the con-

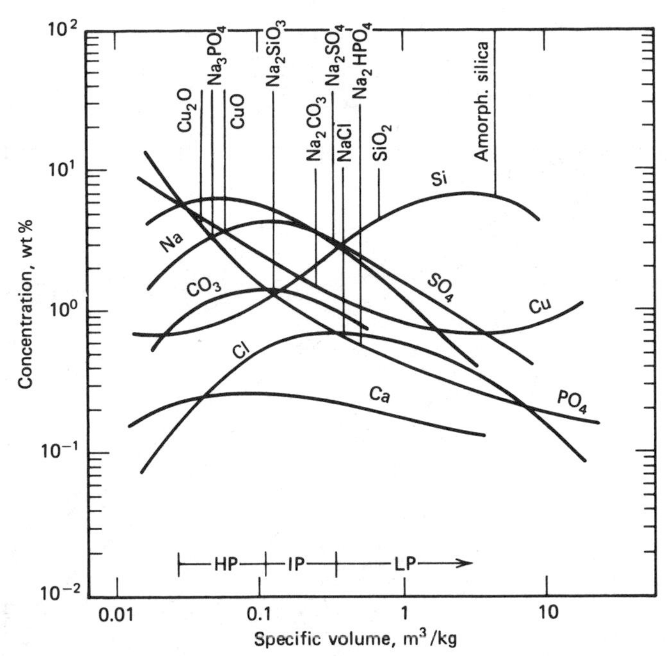

Figure 13. Characteristic turbine deposition versus specific volume of steam. The vertical lines give the average to maximum range of concentrations and the most prevalent chemical compounds.

struction materials to withstand the pressures generated at the temperatures employed. Steam generators have been built for and operated at pressures in excess of 31 MPa (4500 psia) and at temperatures in excess of 700°C, although not necessarily concurrently.

Fuel-Fired Boilers. Boilers are of many types. They may be classified by their physical design, such as stationary or movable; by type of energy source, such as fossil fuel, waste-heat, nuclear, or solar; by type of fossil fuel; by type of heating surface, such as fire-tubed or water-tubed; and, if water-tubed, by whether they are natural or controlled circulation or are once-through generators.

Both process steam generating boilers and central heating boilers normally operate at 0.2–10.4 MPa (15–1500 psig). Most of the fire-tube boilers that are currently in service are of the small package type, completely instrumented and controlled for unattended operation, with capacities up to 25 metric tons per hour available in standard designs. Such units are used primarily in heating service in apartment, office, and industrial buildings. However, more and more building heating systems are employing high temperature, high pressure water as a replacement for steam.

Most modern boilers for process-steam generation and large central heating are of the water-tube type, in which the water is circulated through tubes that are exposed to radiant or convection gas heating to provide the energy necessary to transform the water into steam. Such boilers are practically all of the thermosyphon circulating type. These boilers operate on a recirculating principle: water from an upper steam–water disengaging drum is carried down to the lower end of banks of water-filled tubes in which steam is generated as the mixture rises to the steam-separating drum under the influence of the heat absorbed in the furnace of the boiler. Such natural-circulation boilers represent the greatest number of water-tubed boilers in service and are available

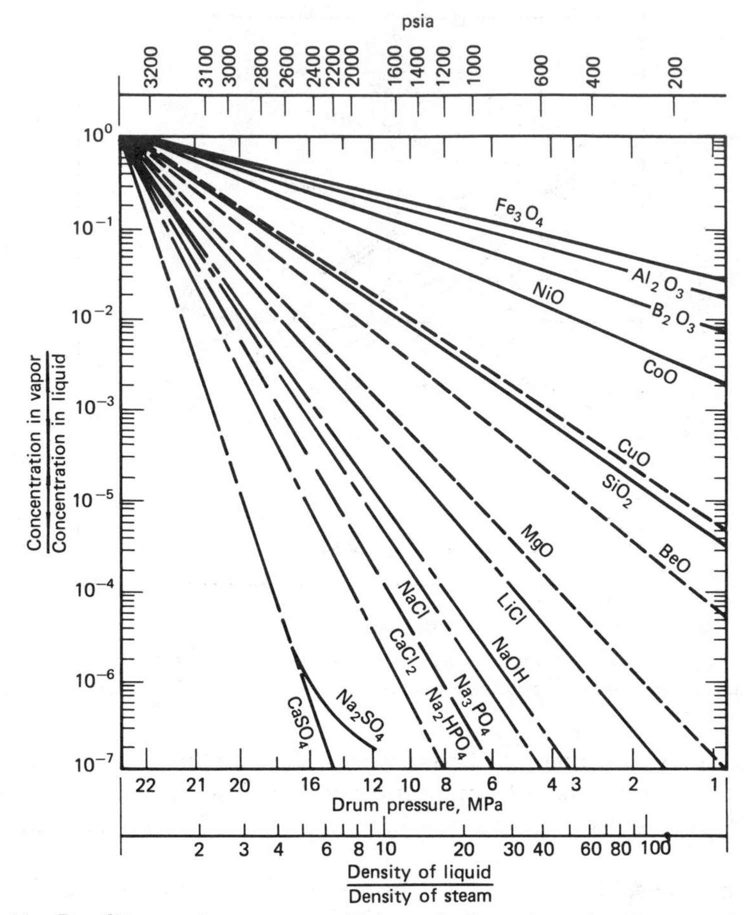

Figure 14. Ray diagram of carry-over coefficients of salts and metal oxide contaminants in boiler water (29).

in many different varieties, depending on fuel to be used, size, and pressure (see Power generation).

Steam generators used for power generation in central power stations can be designed for either subcritical or supercritical pressures; selection is based on the economic evaluation of unit capacity, fuel type, and cost. In general, units up to about 600 MW in size tend to be subcritical, and larger units, supercritical.

In all drum-type boilers, whether natural or controlled circulation, means must be provided within the drum to separate the steam formed in the generating tubes from the remaining boiler water. The required efficiency of this separation depends to a great extent on the intended use of the steam and whether it is to be superheated or not. Where the steam is to be used saturated, as in heating installations without long distribution lines, steam containing 0.25–0.5% moisture can be tolerated without loss of control of the drum water level or slugging of steam traps and development of water hammering in the steam line (see below). Where the steam is to be superheated, the more usual case, and used for heating or in a process, separation of water within the drum must be good enough that the steam entering the superheater contains no more than 1 ppm of dissolved salts. Amounts in excess of this concentration are likely

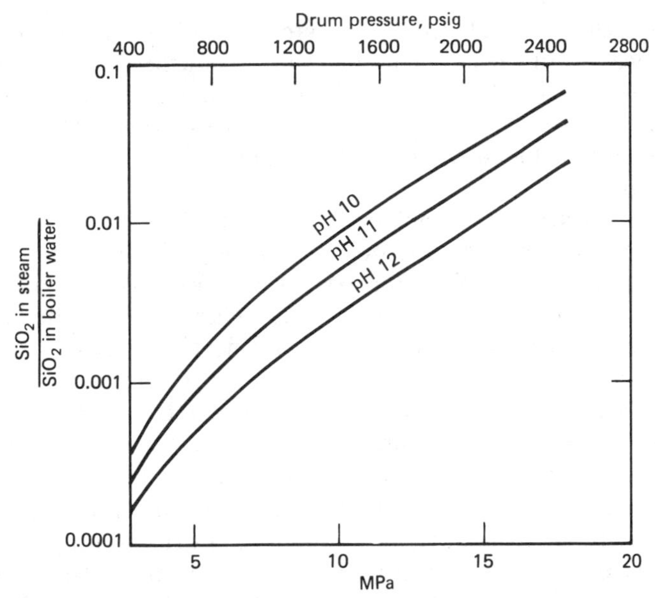

Figure 15. Ratio of silica in steam to silica in boiler water versus boiler drum pressure at selected boiler-water pH (35).

to form deposits on the steam side in the superheater tubes, which can cause excessive elevation of temperature and failure of the tubes. Where the superheated steam is to be used primarily in steam turbines, the quality of the steam produced is even more important. In this latter case, the amount of salts entering the superheater in the saturated steam should not exceed a few parts per billion in order to avoid rapid formation of deposits on the blades in the steam turbine.

The continuing effort to increase the thermal efficiency of fossil-fuel steam power generating plants has led to the development of steam at pressures above the critical point. In such plants, the feedwater to the steam generator is raised to the desired pressure above the critical pressure of water, 22.1 MPa (3206 psia); it is then passed once through the unit to emerge at the desired temperature, eg, 540°C. In the process of passage through the steam generator, there is no boiling in the usual sense since there is never any phase change. The water fed to such supercritical steam generators must be of exceptionally high purity; generally, it should contain less than 50 ppb dissolved and suspended matter (see also Furnaces, fuel-fired; Supercritical fluids, Supplement Volume).

Package Units. Steam generators known as package or shop-assembled units have become a large part of the total number of steam generators under 50 t/h in single boilers installed. These units normally have capacities of from a few metric tons per hour to more than 100 t/h in a single boiler. They are completely assembled by the manufacturer and are mounted on bases that support the entire unit except for the stack and the few auxiliaries. These units have their own forced-air blowers, oil- or gas-burning equipment, and controls for pressure, feedwater flow, and drum level. Such package units cost much less per pound of steam capacity than the corresponding size of field-erected boilers because of the savings in erection cost at the job site and the duplication of design in fabrication.

Combined Cycle. A combination of a steam turbine and one or more large gas turbines with a water-tube heat-recovery steam generator utilizing heat from the exhaust gases is being utilized to increase fuel efficiency in the generation of electric power, often in combination with generation of process steam. Thermodynamic efficiency of these cycles is higher than that of steam cycles, about 40%. Generating capacities are up to 600 MW.

Cogeneration. Large industrial users of steam are increasingly combining generation of electric power by high pressure steam and combined cycles with production of low pressure process steam. Turbine extraction or exhaust steam (high back pressure) is used as the process steam. This has the advantage of the higher efficiency of the high pressure cycle and a source of steam and electricity. Excess electricity is sold (see Power generation).

Nuclear Generation. In nuclear-fission systems, steam is generated from water heated either directly in the reactor core or in steam generators by a circulating primary cycle heat-transfer medium (water, heavy water, sodium, helium, or carbon dioxide) (see Nuclear reactors). Fusion-power reactors are in a conceptual design stage. They may become operational around the year 2000 (see Fusion energy).

Owing to the demands on investment and operation, nuclear systems are built mostly for generation of electric power and are of high capacity, up to 1300 MW_e. Light water pressurized water reactors and boiling-water reactors are the most frequent.

Solar Energy. Many concepts for generation of steam using solar energy (qv) have been proposed (38). There are not yet any commercial steam-generating plants operating, but a potential exists for generation of electric power and industrial steam (see Table 2). Low pressure steam can be generated by ocean thermal energy conversion (OTEC), involving flash-evaporation of seawater in an open Rankine cycle. The temperature difference between the warm surface water and the cold deep water can sustain such operation (38).

Waste-Heat Recovery. Process energy demand can be significantly reduced by a recovery of residual heat and heat generated in the chemical processes. The recovered heat is used in the process itself, for heating feedwater in steam generation, or for generation of steam in recovery or waste-heat boilers. Figure 16 shows a typical unit that takes advantage of energy recovery from chemical processes, many of which are exothermic. The recovery comes from gas-turbine exhaust gases, exothermic chemical

Table 2. Generic Types of Solar Systems [a]

Collector type	Energy transport	Storage	Energy conversion
distributed systems			
fixed V-trough	organic fluid	fixed system considered without storage	organic Rankine
one-axis			
parabolic trough	water or toluene	sensible thermal	central-steam Rankine
variable slats	steam	sensible thermal	central-steam Rankine
two-axis			
parabolic dish	steam	sensible thermal	central-steam Rankine
parabolic dish	electric	advanced battery	small heat engine mounted on dish
central receiver			
heliostat (two-axis)	optical	sensible thermal	central-steam Rankine

[a] Ref. 38.

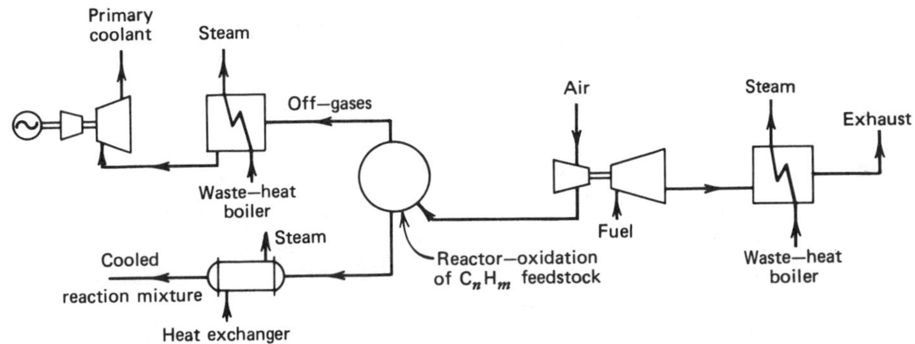

Figure 16. Waste-heat recovery system in a chemical plant (17).

reactions, heat exchange from high temperature off-gases, expansion of off-gases, and incineration of acid residues. Oxidation of hydrocarbon feedstocks gives a high temperature reaction mixture and hot off-gases, and steam is produced from both of these. The heat of reaction is removed from the reaction mixture in a heat exchanger, from which low pressure steam is separated in a steam drum; the heat from the off-gases is removed in waste-heat boilers. The cooled off-gases then pass through a turbo-expander which generates electricity, and the cooled gases, at about −60°C, act as primary coolant in later stages of the process. Air for the hydrocarbon oxidation is supplied by gas-turbine-driven compressors, whose exhaust gases raise high pressure steam in waste-heat boilers. Any acid residues are burned in modified coal-fired boilers to provide more steam for electricity generation in conventional turbines and for other plant uses.

Steam generation in heat-recovery systems can be roughly classified into the following four categories:

1. Removal of heat from vessels in which exothermic chemical reactions are to proceed at essentially isothermal temperature conditions (eg, oxidation of naphthalene or *o*-xylene to phthalic anhydride).
2. Cooling of process stream and/or removal of latent heat of condensation at temperatures capable of generating steam (eg, heat removal from secondary reformer and shift-converter effluents in ammonia synthesis-gas production).
3. Combustion of burnable materials that are subjected to oxidation processes in order to convert them into gases, and in which the combustion gases thus produced are capable of generating steam (eg, carbon removal from catalysts, incineration of waste products, and sulfur oxidation to sulfur dioxide).
4. Combustion of fuel to produce elevated temperatures primarily for physical or chemical processes in which the by-product flue gases thus produced are capable of generating steam (eg, drying, evaporation, calcining, smelting, and primary reforming of methane–steam gas mixtures).

The generation of steam as a means of controlling the temperature at which an exothermic reaction takes place is of considerable interest in the chemical industry. The reaction may require careful temperature control in order to avoid production of undesired by-product materials or, perhaps, in order to avoid short life or physical breakdown of the catalyst. Systems of this sort operating in the temperature ranges of 90–200°C are frequently held in careful balance by the use of steam generation.

Figure 17 shows three methods of heat recovery. The reaction may be controlled by direct removal of heat in the reaction mass (Fig. 17**a**), by indirect control such as circulation of the cooling fluid (Fig. 17**b**), or by circulation of an intermediate indirect-cooling medium (Fig. 17**c**) from which steam is generated.

In some instances, the catalyst chambers or reaction vessels are filled with tubes which may contain the catalyst on the inside or the outside of the tube. Where relatively low temperatures are involved, so that the pressure in the steam generator is correspondingly low, reaction may take place in the tube of a tubular reaction vessel with water and steam circulating outside the tubes in the shell of the vessel, such as a large vertical heat exchanger. The steam generated by the circulating water is taken off at an upper steam drum and unvaporized water recirculated. Such a system is illustrated in Figure 17**b**. In this instance, the circulation is maintained by the thermosyphon (natural circulation) characteristics of the system.

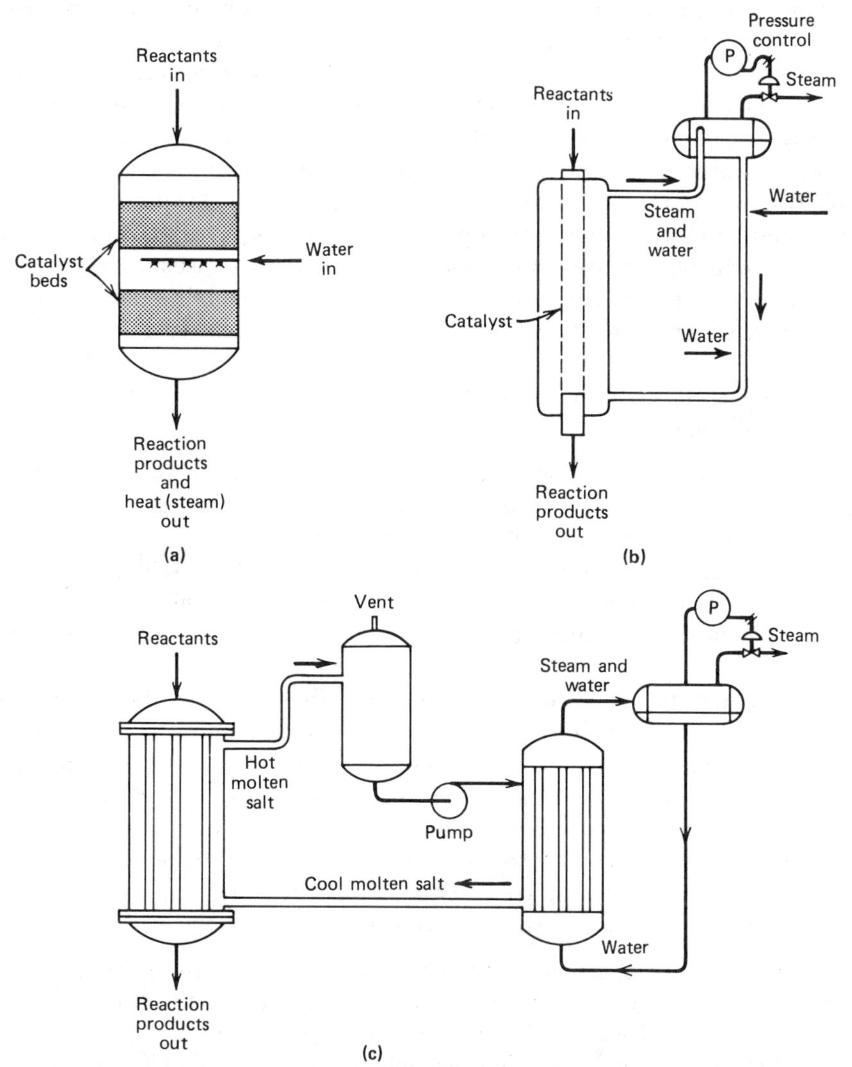

Figure 17. Heat-recovery systems using steam: (**a**) direct quenching with water; (**b**) circulation of water and steam; (**c**) generation of steam by molten salt.

When the temperature of the operation is high, which may result in generation of steam at high pressure, the catalytic reaction may take place on the shell side of the heating unit. In this case, the steam is generated in the tubes at higher pressure. Provided the process side operates at low pressure, steam is also generated under these conditions, but the cost of the apparatus is less because the pressure is controlled inside the smaller surfaces of the tubular vessel. The degree of temperature control necessary on the reaction side may govern the arrangement of the equipment. When only small temperature changes are permitted in the reaction mass, small tubes or small spaces between tubes are required in order to prevent the temperature in the middle or center of the catalyst mass from exceeding what may be a safe temperature at the wall of the reaction tank or tube.

A good example of high pressure steam generation as a by-product of the heat of reaction of a chemical process is a large synthetic-ammonia plant where steam is generated on the inside of bayonet-type tubes inserted in the reaction vessel. Steam at a pressure of 11.7 MPa (1700 psi) generated from the waste-heat unit constitutes the entire source of steam required to generate the necessary mechanical power through a topping turbine driving a compressor. The exhaust steam is used in the reaction process as well as for other drives and for heating service. In this instance, the need for bayonet-type generating tubes brings the requirement of very careful control of the purity of the boiler water circulating within those tubes. Modern ethylene plants are likewise large generators of steam from waste heat from the reaction stream from the ethylene furnaces (see Ethylene).

Another example of heat recovery by generation of steam is in some catalytic cracking operations utilizing the fluid technique. Here the cracking catalyst may be passed through or be in contact with the tubes of a steam evaporator which takes heat from the main mass or reaction to control the temperature.

Geothermal Steam. Geothermal steam comes from wells as steam or is produced by flash evaporation of hot water or brine. Water from which the geothermal steam is generated is either *magmatic water* (released from solidifying magma), *meteoric water* (rain and snow), or a mixture of the two. Depending upon its salt content and application, geothermal fluids may be used directly or through a secondary fluid cycle. Geothermal steam is used in power generation, heating, and in industrial processes. Steam temperature ranges from 185–370°C, salinity from less than 1000 ppm to several hundred thousands parts per million, and content of noncondensable gases is up to 6%. Geothermal steam power stations generating up to 250 MW have been built. Problems with deposition and corrosion must be considered in utilizing this resource, which has a large potential but relatively small utilization so far (see Geothermal energy).

Magma. The feasibility of producing steam by pumping water into the molten rock or inserting a conventional closed heat exchanger was demonstrated in April 1981, at the Kilauea Iki Lava Lake in Hawaii, where the entire 35-m thick, molten zone of the lake was drilled and cored. A demonstration plant could be operational in 10–20 yr. Magma temperature is 850–1100°C, and the magma bodies considered for steam production are within 10 km of the earth's surface. The United States Geological Survey estimates that the energy content of these magma energy sources in the continental United States is 800–8000 times the present U.S. annual energy consumption.

Table 3. Steam Chemistry Limits for Normal Operation[a,b]

Operation[c]	pH	Conductivity, μS/cm		Maximum ppb									
		Specific	Cation	O_2	Na	Cl^-	SiO_2	Cu	Fe	SO_4	PO_4	Free OH	Other
Westinghouse, steam*			0.3	10	5	5	10	2	20				
Westinghouse, PWR–AVT	8.5–9.6 Cu[b]		(2.0)	5	0.25	0	4					0.4	SS (1.0)
General Electric, steam*			0.2		3								
U.S. NRC and General Electric, BWR	5.6–8.6	(1.0)	0.1			0.5							O_2 ca 30 ppm
Allis Chalmers Limit*			0.3 D		Na + K <10		20	3	20				
			0.20 T		Na + K <10		20	3	20				
Allis-Chalmers recommended*			0.1		Na + K <10		10	1	10				
VGB, steam*			0.2		Na + K <10		20	3	20				
VGB, recycle, FW	>9.0			20				10	50				FW CO_2 <1 ppm, oil <0.5 ppm
VGB, supercritical	9.0–9.5		0.2	20			20	3	20				FW CO_2 <1 ppm, oil <0.3 ppm
VGB, BWR	5.6–8.6	(1.0)				0.5	16						
VGB, PWR–AVT	>9.0		0.2 FW	20 FW		2.5	16	3	20				
VGB, PWR–PO4	8.8–9.5	(5.0)	0.2 FW	20 FW		2.5	16	3	20		15		Na/PO_4 <2.6
KWU, steam*			0.2		Na + K <5		20	FW 3	FW 20				

CEGB, steam D*					10		20					
CEGB, drum	9.1–9.3					1.0	20					
CEGB, supercritical	9.0–9.3		0.1	15	5		20	10	Fe + Ni <3			
CEGB, Magnox FW	9.0–9.3		0.1	15	10		20	Cu + Fe + Ni <20				
CEGB, AGR, FW	9.0–9.3		0.1	15	5		20	Cu + Fe + Ni <10				
JIS, recycle	8.5–9.5		0.3	7			20		FWS	3		
JIS, supercritical	9.0–9.5		0.3	7			20	2	10			
USSR, Martynova*	8.9–9.1						15	5	20			
Mitsubishi, PWR	8.8–9.2	(5.0)	(0.3)	5		0.25	2	5	15		0	Ni5, Na/Cl <0.7
Hitachi, supercritical			0.3		0.15		20	2	10			
B&W, PWR–OT	8.8–9.2 Cu		0.5	7	2		20	2	20			TS50, Pb1
	9.3–9.5											
Combustion, D–AVT	8.5–9.0											TS (2)
Combustion, D–PO4	9.4–9.7						20			20		TS (50)
Combustion, supercritical	9.2–9.4			5	10							TS50
Calgon, drum steam*	8.0–9.0	(2.0)					10			0	0	
Averages	*8.55–9.26*		*0.22*	*11.6*	*6.5*	*1.5*	*17.4*	*5.0*	*17.4*	*7.0*	*0.1*	

[a] Ref. 22.

[b] D = drum or recycle boiler; OT = once-through boiler. AVT = all volatile treatment; PO4 = phosphate treatment; Cu = systems with copper; FW = feedwater limits; TS = total solids; SS = suspended solids. Numbers in parenthesis are ppm concentrations or conductivity of boiler water.

[c] Only the lines with asterisk are direct steam limits.

Steam Accumulators. Some plants in the process industries require steam at a very high rate for a short period of time and then at a low rate for a considerable length of time. In a brewery, for example, there may be frequent cyclic demands for a tremendous amount of heat to quickly sterilize a large fermenter. The problem can be solved by the use of accumulators. These are insulated high pressure reservoirs containing water into which steam is charged at high pressure during times of low steam demand. At the time of peak withdrawal from the system, the rated output of the boiler, augmented by the flow of flash steam from the accumulator, is released to the low pressure flow line. For more information, see refs. 39–40.

Steam and Water Chemistry

Purification of Water Used in Steam Generation. To avoid excessive corrosion or erosion, the problems associated with deposition and oxide buildup, and contamination of the products directly in contact with steam, high purity water is generally used for steam generation. Suspended solids and dissolved inorganic and organic impurities are removed by sedimentation (qv), coagulation, flocculation, filtration (qv), precipitation softening, lime (qv) softening, absorption on activated carbon, ion exchange (qv), reverse osmosis (qv), electrodialysis (qv), ultrafiltration (qv), electromagnetic filtration, evaporation (qv), degasification, and membrane separation. Gases are removed by deaeration and oxygen scavenging before the water is used as a makeup or feedwater to boilers. Combinations of these water-purification processes can reduce impurity content to low ppb levels. Typical problems with the water purification are described in refs. 17, 33–34, 41–45 (see also Flocculating agents; Water, reuse; Carbon and artificial graphite; Ion exchange; Filtration; Membrane technology).

Steam and Water Purity Requirements. Generally, the higher the system pressure and heat fluxes, the higher purity water is needed. For the steam purity, wet steam systems can tolerate higher concentration of impurities (under the conditions of no heat exchange) than superheated steam systems. In the superheated steam systems, the lower the pressure of steam at saturation, the lower the solubility of impurities in steam and the lower the tolerance for dissolved impurities. As an example, see Figure 10 for silica, where the solubility at saturation is about 0.02 ppm. This lowest solubility should be the control limit.

In many countries, specifications exist which give required compositions of water and steam for different types of boilers and applications (7,29,46–52). Parameters which should be controlled are, together with average specified values for steam used in electric power generation, shown in Table 3 (22). Other steam applications can tolerate different limits. Recommended boiler water limits for watertube boilers for various pressure ranges are shown in Table 4 (48).

Water Treatment. While harmful impurities are being removed from the water used in steam generation, other chemicals are added to protect the system against corrosion and scale formation. The leading boiler-water chemical treatment programs and their advantages and disadvantages are listed in Table 5. Mixed phosphates are used to maintain proper sodium to phosphate molar ratios in the phosphate treatments. Common water-treatment additives and their applications are listed in Table 6. Principal chemical reactions in boiling water are listed in Table 7. Chemical-injection points around a typical cycle are in Figure 18 (see Water, industrial water treatment).

Table 4. ABMA[a] Recommended Boiler-Water Limits and Associated Steam Purity for Water-tube Boilers[b]

Drum pressure, MPa[c]	Range of total dissolved solids[d] boiler water ppm (max)	Range of total alkalinity[e] boiler water ppm	Suspended solids boiler water ppm (max)	Range of total dissolved solids[e,f,g], steam ppm (max, expected value)
		At steady-state full-load operation *Drum-type boilers*		
0.1–2.2	700–3500	140–170	15	0.2–1.0
2.2–3.2	600–3000	120–600	10	0.2–1.0
3.2–4.2	500–2500	100–500	8	0.2–1.0
4.2–5.3	400–2000	80–400	6	0.2–1.0
5.3–6.3	300–1500	60–300	4	0.2–1.0
6.3–7.0	250–1250	50–250	2	0.2–1.0
7.0–12.5	100	[f]	1	0.1
12.5–16.3	50		na	0.1
16.3–18.0	25		na	0.05
18.0–20.1	15		na	0.05
		Once-through boilers		
≥9.8	0.05	na	na	0.05

[a] American Boiler Manufacturers' Association.
[b] Ref. 48.
[c] To convert MPa to psi, multiply by 145.
[d] Actual values within the range reflect the total dissolved solids (TDS) in the feed water. Higher values are for high solids, lower values are for low solids in the feed water.
[e] Actual values within the range are directly proportional to the actual value of TDS of boiler water. Higher values are for high solids, low values are for low solids in the boiler water.
[f] Dictated by boiler-water treatment.
[g] These values are exclusive of silica.

Relationships between the molar ratio, orthophosphate concentration and pH are shown in Figure 19. In boilers, pH changes due to impurities are compensated by changing the molar ratio and phosphate concentration.

Concentration of Impurities in Steam Systems. The principal impurity concentration mechanisms for steam systems are precipitation and deposition from superheated steam, drying of moisture or water on hot surfaces, precipitation of acid droplets near the steam saturation line, concentration in liquid owing to a temperature gradient, concentration in oxides, heterogenous crystal or film growth on surfaces and suspended particles, and adsorption.

Precipitation and deposition occur when the concentration of an impurity exceeds its solubility in superheated steam. Any system component where the specific volume of steam is rapidly increasing is prone to deposition of dissolved impurities. In boilers and other heat-exchange equipment, the temperature gradient, particularly in stagnant areas such as in crevices, results in impurity concentrations. Impurities also deposit from boiling water films, particularly at high heat fluxes (above 1200 kW/m^2 (0.38×10^6 $Btu/(ft^2{\cdot}h)$) for water on a steel surface) under the so-called departure from nucleate boiling (DNB) conditions above the critical heat flux.

Steam Sampling and Analysis. Sampling of steam is needed because its composition cannot be accurately predicted from the composition of feedwater and boiler water. A representative sample must be withdrawn, condensed, cooled and transported to an analyzer or a grab-sample tap. There is not yet any accurate, verified wet or superheated steam-sampling system available for all needed steam components, ie,

Table 5. Characteristics of the Leading Boiler-Water Chemical-Treatment Programs

Program	Favorable	Unfavorable
conventional phosphate, Na_3PO_4 (+ NaOH)	hardness salts are converted to form readily removed by bottom blowdown; relatively high levels of suspended solids are successfully controlled; acids neutralized; surface passivation by PO_4	high pressure boilers cannot tolerate intentional formation of boiler sludge; required alkalinity levels are too high for operation above 10.4 MPa (1500 psig); oil or organic contamination produces highly adherent deposit; possible "under deposit" corrosion by concentrated NaOH
coordinate phosphate, $Na{:}PO_4 < 3$	caustic corrosion may be eliminated; deposit form makes for easy removal; produces low solids levels and high steam purity; acids neutralized; surface passivation by PO_4	in boilers containing deposits, chemical interaction of iron and phosphate can lead to caustic corrosion; at very low $Na{:}PO_4$ molar ratios (<2.1) corrosion by phosphoric acid is possible
congruent phosphate, $2.6 < Na{:}PO_4 < 2.8$	caustic and acid corrosion eliminated; deposit form makes for easy removal; produces low solids levels and high steam purity; acids neutralized; surface passivation by PO_4	control of the $Na{:}PO_4$ molar ratio may be difficult; continuous feed and blowdown may be required
sodium or lithium hydroxide	acid neutralization	can cause rapid corrosion when concentrated in high quality regions or under deposits; vaporous carryover in steam at higher pressures; difficult to analyze
chelates, ethylenediaminetetraacetic acid (EDTA), nitrilotriacetic acid (NTA)	optimum heat transfer and boiler efficiency obtained under good feedwater-quality conditions; elimination of boiler sludge prevents formation of adherent deposits involving oil/organics	inability to accurately test/control free chelant residual can lead to overfeed and subsequent corrosion (condition is heightened when treatment is applied to deposit-bearing boiler); will not complex iron or copper under normal boiler-water pH conditions; presence of oxygen in boiler water can cause dechelation and excessive corrosion
all-volatile (AVT) (ammonia, morpholine, cyclohexylamine)	"near-zero solids" in boiler water and high purity steam realized under ideal feedwater conditions, with some corrosion protection; no carry-over of solids, no surface concentration; condenser leakage detection by sodium measurement; boiler deposition of corrosion products easy to remove by chemical cleaning	feedwater contamination may exceed inhibiting ability of volatile feed, leading to boiler corrosion; introduction of contaminants into feedwater produces deposits that may be hard to remove; marginal acid neutralization, particularly with ammonia in wet steam regions; interference with condensate polishing; corrosion of copper alloys by NH_4OH and oxygen
neutral, oxygen added ca 200 ppb O_2	no interference of additives with condensate polishing; low corrosion rates of ferritic steels	requires very low concentrations of impurities in water; no corrosion protection in a case of an upset; corrosion of copper alloys; precise control required

Table 6. Common Water-Treatment Additives and Their Applications

Additive	Application
sodium orthophosphates	pH control, hardness precipitation
sodium hydroxide	pH control (acid neutralization)
lithium hydroxide	pH control (acid neutralization)
neutralizing volatile amines	once-through and pressurized-water-reactor (PWR) cycles, high heat flux steam generators; control of preboiler corrosion-product generation and transport; no control of scale formation by feedwater contaminants
ammonia	
morpholine	
cyclohexylamine	
diethanolamine	
hydrazine	oxygen scavenging
sodium sulfite	oxygen scavenging for drum boilers at <10.3 MPa (1500 psi) pressure
chelating agents	chelating Ca and Mg hardness,
ethylenediaminetetraacetic acid (EDTA)	<8.3 MPa (1200 psi)
nitrilotriacetic acid (NTA)	<6.2 MPa (900 psi)
sludge conditioners	dispersion of sludge for easy removal by blowdown in drum boilers, inhibiting of scale formation <6.9 MPa (1000 psi)
synthetic sulfonated polymer	
synthetic carboxylated polymer	
polyacrylic acid	
carboxymethyl cellulose	<4.1 MPa (600 psi)
organophosphonate	
lignin	<2.1 MPa (300 psi)
oxygen or hydrogen peroxide	to improve surface passivation in all ferrous high purity water systems (ca 200 ppb of dissolved oxygen in feedwater)
filming amines	surface protection against condensate corrosion
octadecylamine and some of its salts	
antifoams	to reduce foaming and carry-over in boilers
polyglycols	
polyamides	

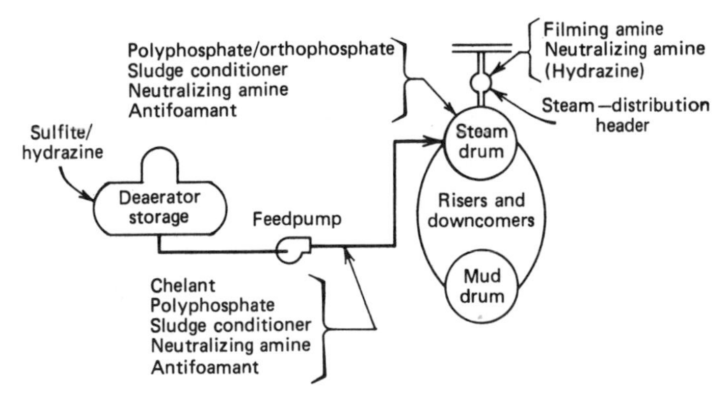

Figure 18. Chemical-injection points are determined by reactions involved and environment. Thus, pressure limits on polymers dictate their addition upstream of boiler (53).

Table 7. Chemical Reactions in Boiler Water

$10\ Ca^{2+}$ hardness + $2\ OH^-$ alkalinity + $6\ PO_4^{3-} \rightarrow Ca_{10}(OH)_2(PO_4)_6$ hydroxyapatite
Mg^{2+} hardness + SiO_2 + $2\ OH^-$ alkalinity $\rightarrow$ ($MgO + SiO_2.H_2O$) serpentine
$2\ Na_2SO_3 + O_2 \rightarrow 2\ Na_2SO_4$ sodium sulfate
$N_2H_4 + O_2 \rightarrow N_2 + H_2O$
$3\ N_2H_4 \rightarrow 4\ NH_3 + N_2$
$NaHCO_3 \rightarrow NaOH + CO_2$

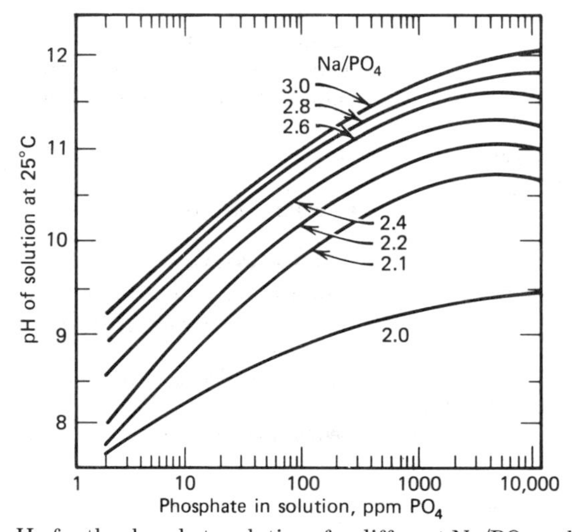

Figure 19. pH of orthophosphate solutions for different Na/PO_4 molar ratios (41).

volatile, nonvolatile, suspended and gaseous impurities (54–55). Some general recommendations can, however, be made: (*1*) The sampling device (nozzle) should be isokinetic, smooth and without corners where impurities can deposit or be trapped. Attemperation (softening) helps. (*2*) It should be immediately followed by a condenser–cooler. The sampling nozzle should be located where the analyzed steam components are distributed homogeneously. (*3*) The sample line should be short, smooth, and down-sloping with a minimum of bends. It should be made of stainless steel or other noncorroding material. (*4*) The Reynolds number of the flow should be >10,000. (*5*) Sample flow should be continuous, constant or changing slowly. In the United States, ASTM steam-sampling systems have been standardized (ASTM D 1066) and another system has been evaluated (55–62).

After condensation of a steam sample, impurities in steam are analyzed by continuous instruments or by different chemical methods. New instruments and methods are now being used as well as standard wet chemical techniques such as the ASTM methods (63). They include liquid ion chromatography, ion chromatography exclusion for organic acids; inductively coupled plasma for metals, ion-sensitive electrodes for sodium and chloride, and infrared spectroscopy or flame ionization for organic carbon. For most steam constituents, a detection limit of 1 ppb is desirable (55). If samples are preconcentrated by evaporation, ion exchange, precipitation, or filtration, parts per trillion (10^{12}) detection is achievable (see Analytical methods). Moisture in steam can be measured directly *in situ* with a laser scattering probe.

Steam-Borne Deposits. Deposits are a concentrated form of impurities carried with steam as the mechanical and vaporous carryover. Soluble impurities precipitate and deposit when their concentration in steam exceeds solubility; insoluble impurities are carried in suspension. More than 120 chemical compounds and elements have been identified in turbine, boiler, and piping deposits, including metal oxides, carbonates, chlorides, sulfates, phosphates, silicates, and some metals. Some chemical compounds, such as NaOH, Na_2CO_3, SiO_2, NaCl, CuO, Fe_3O_4, and Na_2SO_4, have been found at concentrations approaching 100% (see Fig. 13). Origins of these chemicals are: the makeup water, cooling-water inleakage, air inleakage, water-treatment chemicals, corrosion products, erosion products, and paints and preservatives. Some react after deposition, such as acids with alkaline chemicals; some after they are exposed to air (such as NaOH forming Na_2CO_3).

In addition to indicating steam and water-chemistry problems, deposits can cause corrosion, impede heat transfer, change flows and pressures, and cause clogging and malfunctions of valves, seals, etc (21,64–67). With today's improved water purification methods, silica no longer constitutes the threat to turbine capacity and efficiency that it once did. Corrosive impurities are now the prime concern.

Semiquantitative, wet chemical, and x-ray diffraction methods are used in deposit analysis. Atomic absorption, inductively coupled plasma, and liquid-ion chromatography are also useful.

Uses of Steam

Steam is used for generation of electrical and mechanical power, space and process heating, evaporation, steam distillation, oil recovery, chemical and petrochemical processes, cleaning, sterilization, blanketing and in many other applications. It is used by all principal manufacturing industries (68). In the United States, the greater portion of the energy used by the industrial sector is in the form of thermal energy rather than in the form of electrical energy. In 1968, the total industrial use of energy was 2.64×10^{19} J (25×10^{15} Btu or 25 quads) for all purposes, including 2.3×10^{18} J (2.2×10^{15} Btu) as feedstock. The breakdown of this total industrial use of energy is as follows:

process steam	40.6%
electric drive	19.2%
electrolytic process	2.8%
direct process heat	27.8%
feedstock for chemicals	8.8%
other	0.8%

Process steam and direct process heat together accounted for 68.4% of the total industrial use of energy. Table 8 summarizes process-heat requirements for selected industries in the U.S. It shows how much steam is used in proportion to other heat sources (69).

Table 8. Summary of Process-Heat Requirements for Selected Industries in the United States, PJ/yr[a,b]

Standard industrial code (SIC)	Description	Hot water <100°C	Steam		Direct heat/hot air			Rounded totals
			100–177°C	>177°C	<100°C	100–177°C	>177°C	
1211	bituminous coal and lignite						12	12
1477	sulfur mining		46					46
20	food and kindred products	63	290	105	11	100	16	585
22	textile mill products	20	201	4		73	14	312
24	lumber and wood products	5	22	4	112	4	74	221
26	paper and allied products		490				99	589
28	chemicals		1,480			474	260	2,200
2911	petroleum refining		126	401			2,740	3,300
32	stone, clay, and glass	8	21	39		25	1,139	1,230
3312	blast furnaces and steel mills		69				1,804	1,873
3331	primary copper	15			2.5		57.3	76
3334	primary aluminum			40			66.5	108
3711 3712 3713	automobile and truck manufacturing	14	1		22.5	11	1	50
	Rounded totals	*127*	*2,740*	*593*	*148*	*687*	*6,287*	*10,580*
	Percent of total	*1.2*	*25.9*	*5.6*	*1.4*	*6.5*	*59.4*	

[a] To convert PJ (10^{15} J) to 10^{12} Btu, divide by 1.054.
[b] Ref. 69.

Steam Cycles. In a typical steam cycle (see Fig. 20), heat is extracted from steam and converted to work in a turbine or a steam engine (points 1–2); rejected heat is re-

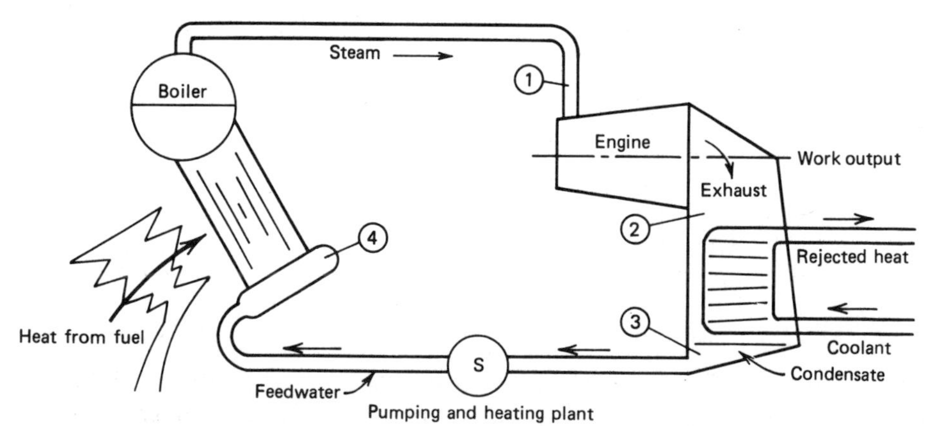

Figure 20. The basic plant of the steam cycle. (Follow the state-points through the diagrams of Figure 21: points 1, 2, 3, and 4 all correspond) (17).

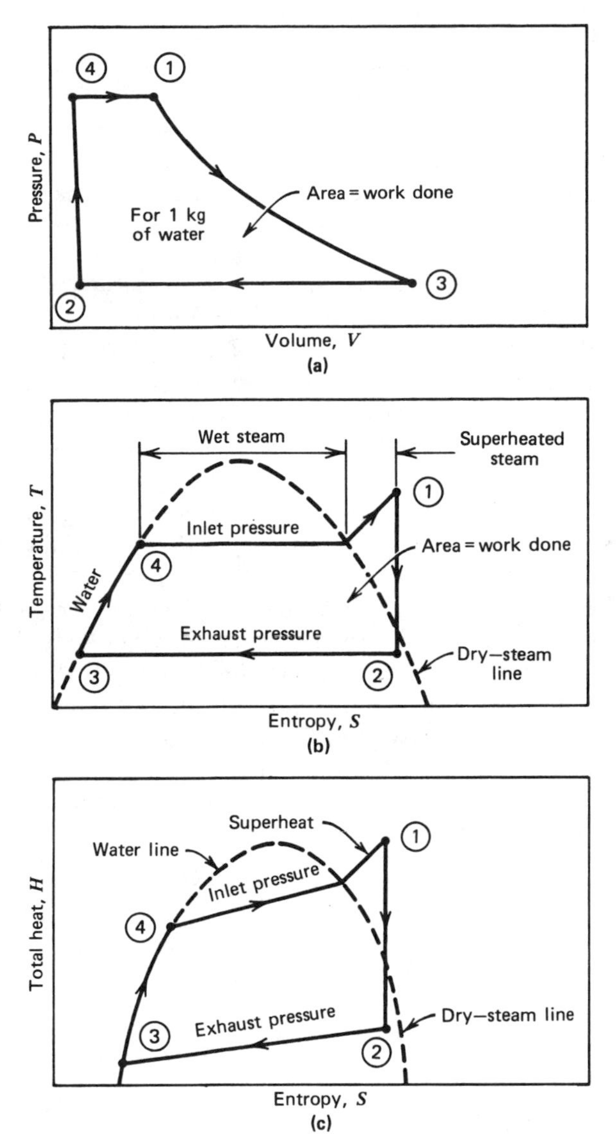

Figure 21. (**a**) Pressure–volume diagram for the steam cycle; (**b**) temperature–entropy diagram; (**c**) total heat–entropy diagram (17).

moved by a condensate coolant (2–3); condensate is preheated in feedwater heaters and compressed to the boiler pressure (3–4); and the water is evaporated and steam often superheated in the boiler by accepting the heat from fuel. The cycle is shown in three different thermodynamic diagrams in Figure 21.

Carnot efficiency η_c is a theoretical maximum efficiency of a perfect cycle (without losses):

$$\eta_c = \frac{T_1 - T_2}{T_1} \times 100\%$$

where T_1 is the absolute temperature at the engine (or turbine) inlet and T_2 is the absolute temperature at the exhaust. Increasing the inlet or lowering the exhaust temperature increases cycle efficiency. Cogeneration and combined cycle systems all serve to increase T_1; using condensing cycles decreases T_2. Energy balance of a real cycle is in Figure 22. Highest losses are the exhaust losses, where heat is removed from the cycle by the condensate cooling water (see Thermodynamics).

A modern supercritical single reheat cycle is in Figure 23. Its temperature–entropy diagram is in Figure 24. Cycle efficiency, which may reach 46%, is increased by feedwater preheating in bleed heaters and by flue gases in the economizer, and by superheating and reheating of steam. Notice the entropy increase in the turbine caused by the turbine losses. There are two electric generators driven by the turbine's cross-compound unit. Typical capacity of this type of unit is 900–1100 MW_e.

Heat-Rate and Thermal-Efficiency Trends. As shown above, the maximum thermal efficiency of a steam cycle is the Carnot efficiency, which, in all practical cases, is less than 100%. Improvements in the overall thermal efficiency of steam power cycles between 1915 and 1980 are shown in Figure 25 (70). The principal factor allowing the efficiency improvement through increases of pressures and temperatures (top of Figure 25) was the development of boiler-tube materials more resistant to creep and oxidation (lower part of Fig. 25).

Steam Turbines. The steam turbine is the simplest and most efficient engine for converting large amounts of heat energy into mechanical work. As the steam is allowed to expand, it acquires high velocity and exerts force on the turbine blades. There are three principles of action employed in steam turbines: impulse; reaction; and shear torque. The impulse principle involves a stationary nozzle and moving blades which absorb the mechanical energy from the steam as it flows over the blades. In the reaction turbine, the nozzles are themselves attached to the shaft. In one case, the motivating force is one of the impulse of a stream against a blade; in the other, one of a reaction

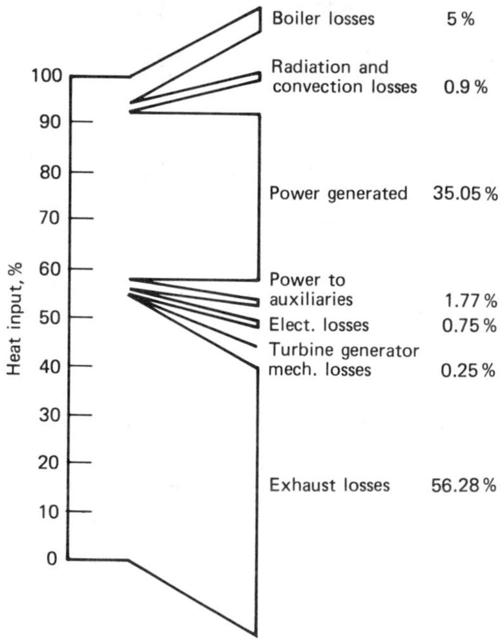

Figure 22. Energy distribution in a simple cycle of a steam power plant using a boiler, turbine, and condenser (17).

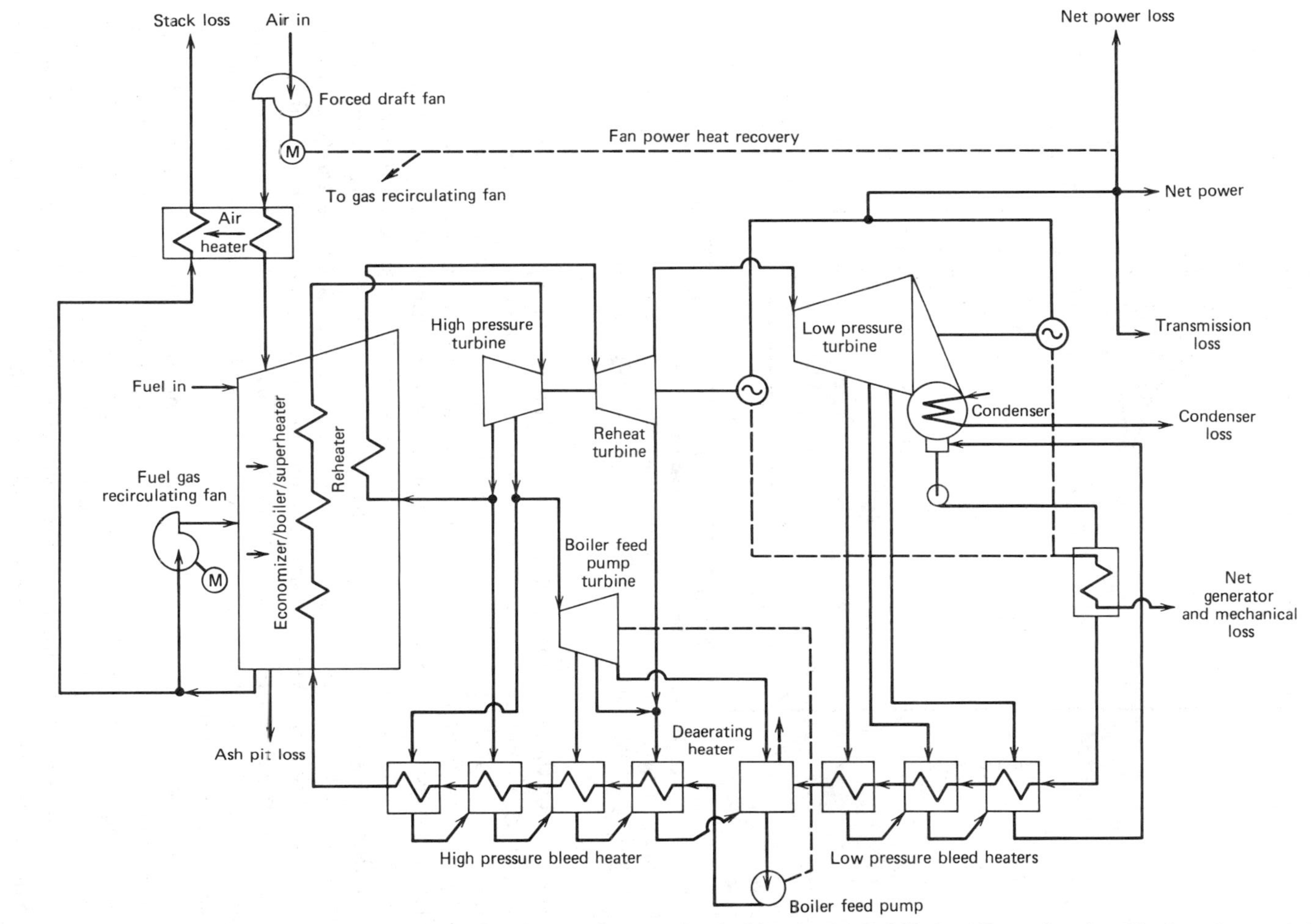

Figure 23. Power-cycle diagram, fossil fuel—single reheat, eight-stage regenerative feed heating—24.2 MPa/540°C superheat/540°C reheat (3515 psia/1000°F/1000°F steam) (7).

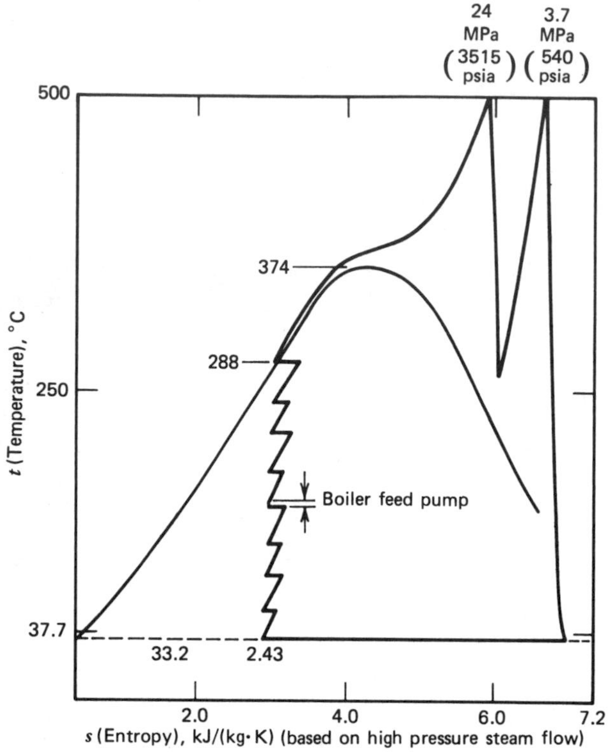

Figure 24. Steam cycle for fossil fuel—temperature–entropy diagram—single reheat, eight-stage regenerative feed heating; same cycle as in Figure 23 (7). To convert kJ/(kg·K) to Btu/(lb·°F), divide by 4.184.

force created by the acceleration of the steam in the moving nozzles. In a shear-torque turbine, drag (friction force) is exerted by high velocity steam (usually wet) on rotating disks; there are no blades. Turbines range in size from a few kilowatts (electric) to 1300 MW_e, from one stage to multiple stage, multiple component (high pressure, intermediate pressure, and up to three low pressure turbines) units. After the shear-torque turbine, the Terry turbine is the simplest (see Fig. 26). For mechanical drives, single- and double-stage turbines are generally used. Most larger modern turbines are multi-stage axial-flow turbines, such as the single cylinder, 30 MW extraction turbine in Figure 27. The largest utility turbines may be five-cylinder, tandem, double-flow supercritical pressure turbine sets. A special type of turbine is the turboexpander (expander, expansion turbine) used to recover power from steam and other hot gases and to reduce steam pressure.

Turbines have few moving parts. The rotor with disks and blades is usually supported by two radial bearings; axial forces are supported by an axial thrust bearing. Babbitted bearings are usually used (see Bearing materials). Where the shaft of the turbine projects through the casing, means are provided for packing against leakage of the steam outward at the high pressure end, and infiltration of air at the low pressure end. Packed stuffing boxes are used on small turbines. Sealing glands, which are built into the turbines and which seal the shaft with steam or water, are used on large turbines. Outward leakage from the glands is minimized by labyrinths.

Governing of steam turbines is accomplished by three methods: throttling at the

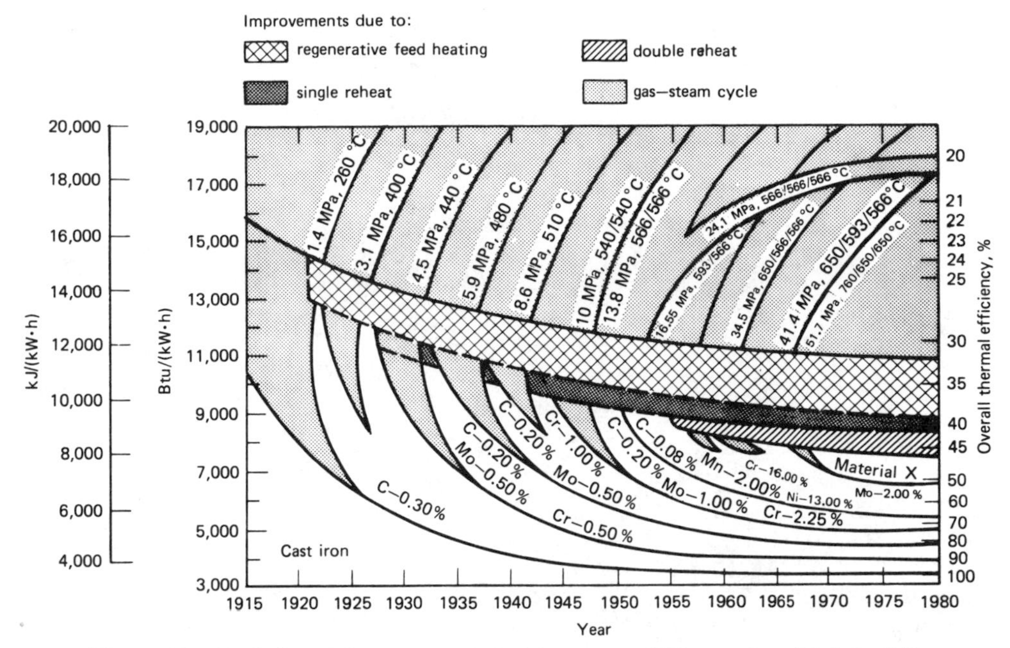

Figure 25. Evolution of the steam cycle (70). To convert MPa to psi, multiply by 145.

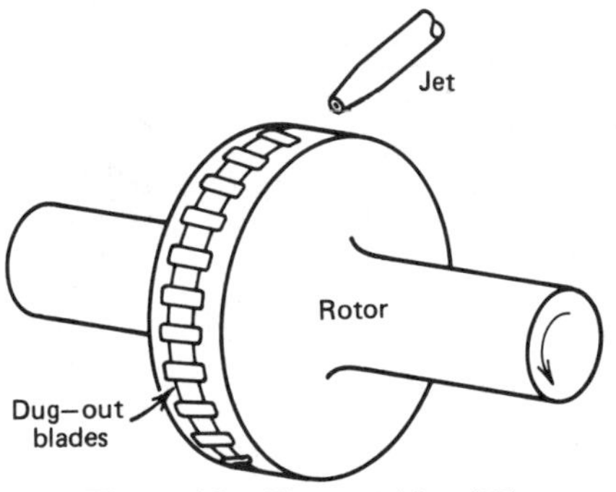

Figure 26. Terry turbine (17).

inlet; varying the number of inlet nozzles in action; and varying the duration of full-pressure puffs (blasts), of which there are several per second. In addition, some turbines are provided with hand-operated by-pass valves which, by admitting high pressure steam to the low pressure stage, enable the turbine to carry more load.

For the synchronous operation of the electric-power generator, utility turbines are operated at constant rpm (1500 or 3000 for 50 Hz systems, 1800 or 3600 for 60 Hz). Turbines for mechanical drives, such as marine turbines, compressor and blower drives, operate at variable speeds. Steam pressure and temperature conditions are shown in Figure 25. Efficiency of steam turbines ranges from about 20% for small low-pressure units to about 46% for supercritical turbines. Maximum steam velocity in turbines exceeds the velocity of sound, maximum steam expansion rate, $1/p\ \partial p/\partial t$, exceeds 3000 s^{-1}. See also refs. 3, 7, 17.

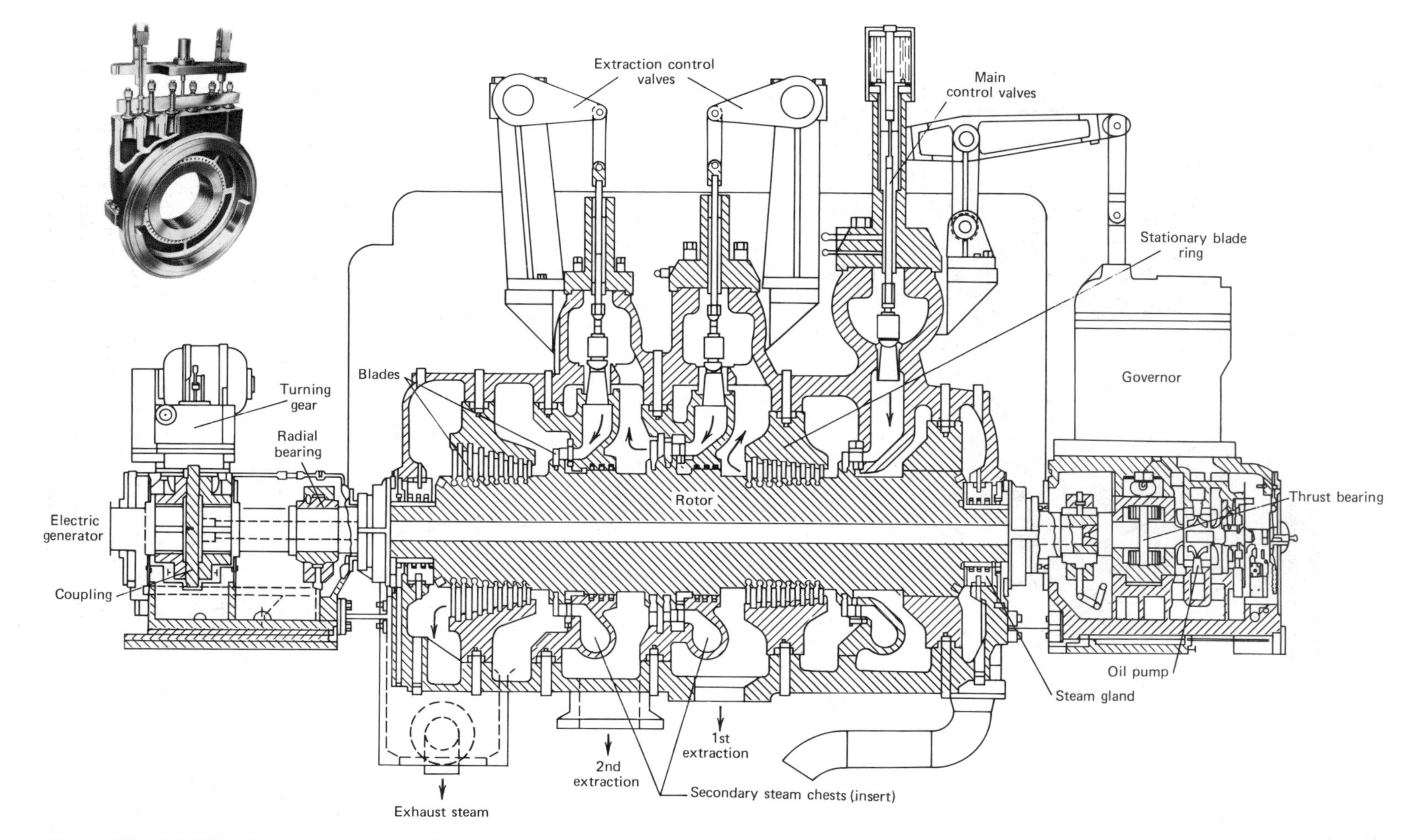

Figure 27. 30 MW double-extraction noncondensing 3600-rpm steam turbine. Inlet pressure 8.6 MPa (1250 psi), inlet temperature 440°C, back pressure 0.17 MPa (25 psi). Automatic extractions at 2.9 MPa (425 psi) and 1.14 MPa (165 psi). Secondary steam chests with extraction control valves (insert) control the extraction flows. Courtesy of Westinghouse Electric Corporation.

Mechanical Drives. Steam turbines are very efficient at high load ratings, and depending upon the steam balance in a plant, they are normally considered as drive units if more than 37 kW (50 hp) is required. Where many small loads are to be handled separately in a plant, it may be preferable to generate electric power by passing the steam through a back-pressure turbine connected to an electric generator; the generated electricity in turn can be fed into the motors through the plant. Steam turbines operate very effectively at high speeds (3,000–10,000 rpm) and thus lend themselves well to large power-output, high speed drives for gas compressors, multi-stage high pressure pumps, such as boiler feedpumps, and other high speed rotating equipment. In such installations, the steam can be fed to the turbine and extracted at various stages of its expansion corresponding to the desired process operating pressures. In ship propulsion, steam-turbine drives with rpm reduction are used in most large commercial and Navy vessels.

Steam Traps. Steam traps remove condensate from wet steam streams as required for safe and efficient operation of most steam-handling equipment. A steam trap is an automatic valve that discharges condensate. Mechanical steam traps use floats or buckets; thermostatic traps sense the difference in temperature between steam and condensate; thermodynamic traps sense forces generated by flashing condensate and steam flowing through orifices; and labyrinth traps allow condensate to flow freely but their orifices are choked by flashing steam. See also ref. 17.

Water Hammer. Water hammer is caused by the condensate collected in a section of pipework or other equipment and driven rapidly by steam. When it hits an obstruction such as a valve, steam trap, or turbine blades, it can cause a shock and considerable damage.

Steam Ejectors. Steam may be used directly to move and pump fluids by means of ejector eductors. These devices depend on the velocity of steam flowing through a nozzle to induce the flow of other fluids without the use of any mechanial moving parts. Efficiency is low if the only desired effect of the steam is to move the fluid, such as air or a noncondensable gas. However, if the steam is also used to heat the fluid so that the entire heating value of the steam is utilized, the overall efficiency may be higher than for any other type of pumping. One of the most common uses of steam ejectors is the production of suction or vacuum.

Steam is a very useful medium for ejecting air or other noncondensable mediums from a process vessel or fractionating system. Probably the biggest single use for multi-stage steam-jet ejectors is in evacuating noncondensable gases and air from steam condensers.

Generation of Electric Power. Steam cycles for generation of electric power use various types of boilers, steam generators, and nuclear reactors, operate at subcritical or supercritical pressures, use makeup and often also condensate-water purification systems and chemical additives for feedwater and boiler-water treatment. They are designed to maximum cycle efficiency and reliability. A fossil once-through supercritical cycle is in Figure 23. Nuclear steam cycles are similar. In 1974, 6253 GW·h of electricity was generated in the United States. Because of energy conservation and other factors, electric-energy growth rate in the United States has been decreasing since then. The average annual growth rate projected for ten years is 3.7% (71). Because of the high cost of oil, many generating plants are converting to coal and more electricity is being generated by industrial users of steam in co-generation. In the United States, 8424 MW of generating capacity (50 units) will be converted from oil to coal in the period 1981–1990 (71).

Out of the total capacity of 558,237 MW (summer) the actual U.S. installed generating capacity in 1980 was divided as follows (71): nuclear, 49,199 MW; hydro, 66,094 MW; pumped storage, 11,319 MW; geothermal, 800 MW; steam-coal, 231,784 MW; steam-oil, 96,115 MW; steam-gas, 57,052 MW; combustion turbine-oil, 34,155 MW; combustion turbine-gas, 6,013 MW; combined cycle-oil, 3,477 MW; combined cycle-gas, 1,908 MW; other, 321 MW. As these figures show, steam is used for generation of most of the electric power in the U.S. To operate all the above steam capacity would require about 10^9 kg of steam to be generated every hour and 10^{13} kg of steam every year. As of September 1, 1981, the number of commercial nuclear power stations, operable, under construction or ordered was as follows (72): World, 535 with generating capacity 408,098 MW_e; United States, 172 with generating capacity 163,509 MW_e. Sizes of these generating units range from kilowatts to 1300 MW electric.

Steam Heating. Wet and saturated steam has a definite pressure for each fixed boiling or condensing temperature which it is desired to maintain. Therefore, the control of the desired temperature for any process-heating requirement may be fixed by choosing the steam pressure. It is customary for steam to be generated under slightly superheated conditions if the steam generator is to be located at a considerable distance from the various users.

When higher temperatures are needed, the higher corresponding pressure becomes important in the design of equipment. The steam pressure rises quite rapidly with temperature and becomes of major concern for heating conditions at temperatures above 180°C. The use of steam for heating is normally limited to pressures <2.2 MPa (300 psig) because the cost of the heat-transfer equipment becomes too high for sizable heating units as higher pressures are required. When temperatures in excess of 180–205°C are required, other heating media are usually preferred (see Heat-exchange technology).

Evaporators. Steam-heating systems have often been installed in a cascade system, as in multiple-effect evaporators (see Evaporation). This arrangement makes possible the recovery of heat at several successive levels merely by reducing the pressure at each of the stages. Condensing all the steam vaporized in one stage by heating and vaporizing the material present in the succeeding stage at lower operating pressure produces good economy of heat. This system of heating operates best in the low pressure range because at higher pressures the equilibrium temperature changes more slowly with the pressure.

Control. When close temperature control is required in order to prevent overheating of material being processed or to assure a high heating density, steam is the medium normally used. A pressure regulator controlling the steam pressure of a heating unit maintains temperatures usually within degrees of the design conditions on the process side (see also Instrumentation and control).

Selected Industrial Processes. ***Evaporation and Distillation.*** Steam is used to supply heat to most evaporation and distillation (qv) processes, such as in sugar-juice processing and alcohol distillation. In evaporation, pure solvent is removed and a low volatility solute is concentrated. Distillation involves relative enrichment of the lower-boiling components of a liquid mixture by partial evaporation. In evaporation of concentrated solutions, there may be substantial boiling-point elevation. For example, the boiling point of 80% NaOH solution at the atmospheric pressure is 226°C.

Steam-Reforming Processes. In the steam-reforming process, light hydrocarbon feedstock such as natural gas, liquefied petroleum gas, naphtha, or in some cases, heavier distillate oils, after being purified of sulfur compounds, react with steam over a nickel-containing catalyst to produce a mixture of hydrogen, methane and carbon oxides. Essentially total decomposition of compounds containing more than one carbon atom per molecule is obtained (see Ammonia; Hydrogen; Petroleum).

Steam reforming in industrial practice falls into two main classes according to the catalyst type and reactor equipment used: hydrogen production by high temperature reforming, generally at above 700°C, and methane manufacture by low temperature reforming at below 550°C. The amount of steam consumed by reaction in the overall process depends upon the choice of product gas. Taking the formula of a petroleum naphtha per atom of carbon as approximately CH_2 (such hydrocarbons typically have about six to eight carbon atoms per molecule), the extreme cases are

$$4\,[CH_2] + 2\,H_2O \rightarrow 3\,CH_4 + CO_2$$

Steam reacted = 0.5 mol/atom carbon (0.64 kg/kg naphtha)

$$4\,[CH_2] + 8\,H_2O \rightarrow 12\,H_2 + 4\,CO_2$$

Steam reacted = 2.0 mol/atom carbon (2.6 kg/kg naphtha)

In practice, total consumption of steam by reaction as indicated by the above equations is never achieved (despite total decomposition of the hydrocarbon) because both the methane-reforming and carbon monoxide-shift reactions approach a thermodynamic equilibrium instead of proceeding to completion:

$$CH_4 + H_2O \rightleftharpoons CO + 3\,H_2$$

$$CO + H_2O \rightleftharpoons CO_2 + H_2$$

Steam is a reactant in both of these reactions, which are reversible over the catalyst. The equilibrium composition of the product gas mixture can be altered by choice of suitable temperature, pressure, and steam/feedstock ratio to produce a gas mixture consisting largely of methane or largely of hydrogen with varying proportions of carbon monoxide, in each case with some carbon dioxide which can subsequently be removed. It is often convenient to use more than one reaction stage to modify the gas mixture produced from the primary gasification step toward the desired composition.

Synthetic Ammonia. Steam–methane reforming is used in the production of synthetic ammonia (qv) where the gas-reforming process produces the hydrogen, which in turn reacts with nitrogen from the air at high pressure and temperature over a suitable catalyst to produce ammonia. In this synthetic-ammonia process, ca 0.75 kg of water as steam is required in the reaction to produce 1 kg of ammonia. A very large synthetic-ammonia plant uses a single-train process with steam-turbine-driven centrifugal compressors; the exhaust steam is used for the foregoing gas-reforming reaction. The high pressure steam thus generated from waste heat is the heart of the energy system for the process which in turn is the key to the process economics (73).

Coal Gasification. Coal-gasification processes involve the reaction of coal at high temperature with steam and air or oxygen to produce a mixture of gases, typically carbon monoxide, carbon dioxide, hydrogen, and methane, as well as hydrogen sulfide from the reactions with the sulfur in the coal. The gases from the gasifier are further upgraded in several steps. By reaction with water, the carbon monoxide in the gas is converted to hydrogen and carbon dioxide. Carbon dioxide and hydrogen sulfide are removed in a purification system and the hydrogen is converted to methane by reaction with carbon monoxide. There are at least seven low-energy and seven high-energy-coal gasification processes, some of which use steam (74–75).

Coal Liquefaction. Coal liquefaction processes involve the following steps: the coal is crushed, dried and pulverized, and a solvent is added to produce a slurry; the slurry is heated, usually in the presence of hydrogen, to dissolve the coal; the extract is cooled to remove hydrogen, hydrocarbon gases, and hydrogen sulfide; and the liquid is flashed at low pressure to separate condensable vapors from the extract. Mineral matter and organic solids are then separated and used to produce hydrogen for the process.

Additional steps include separation of the solvent and major products and desulfurization of the extract. Steam is used to produce the hydrogen used in the dissolution of coal. There are at least six different liquefaction processes.

Enhanced Oil Recovery. Increasing amounts of steam are being used for steam flooding as a tertiary oil recovery technique (see Petroleum, chemicals for enhanced recovery). Steam pumped into the partly depleted oil reservoirs through input wells decreases the viscosity of crude trapped in the porous rock of a reservoir, displaces it, and maintains the pressure needed to push it toward the production well. Steam is also used in hot-water extraction of oil from tar sands (qv) in the caustic conditioning before the separation in a flotation tank (76).

Synthetic Rubber. In the synthetic rubber solution-polymerization process, steam is used for removing solvent and for crumb dewatering (see Elastomers, synthetic).

Steam Distillation and Stripping. Steam distillation is used to separate water-immiscible liquid mixtures of components of varying volatility. The separation is achieved when the more volatile components evaporate under the mass-transfer concentration gradients existing between the liquid interface and the bulk stream flow. The less volatile impurities remain in the residue passing through the still. Steam distillation can be continuous or of a batch type. Steam stripping is a similar process applied to the water-immiscible solids, such as stripping of monomers from polymer

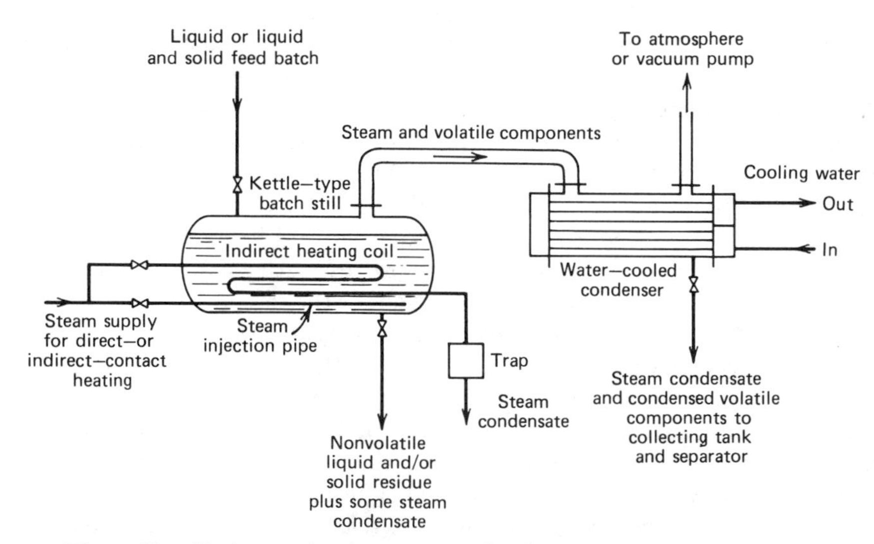

Figure 28. Single-stage batch-type steam distillation or stripping system (17).

fibers or other products. A single-stage batch-type steam distillation or stripping system is shown in Figure 28 (17).

Steam Cleaning. High pressure steam can be used to produce a high velocity jet with some superheating by expansion through a suitable nozzle to atmospheric pressure. The high velocity is effective in removing dirt and loose scale from solid surfaces, and the high temperature encourages the melting or vaporization of oil and grease deposits, thus releasing the solid deposits for mechanical-blast removal. Some condensation of steam on the initially cold surfaces also takes place, which may help to dissolve and release dirt and scale.

Desalination. Low pressure turbine steam is often extracted for low pressure distillation of salt water to produce drinking water. Condensed steam is returned into the cycle (see Water, supply and desalination).

Corrosion in Steam

Use of metals in hot steam is limited by their oxidation rate, mechanical strength and creep resistance (see Corrosion and corrosion inhibitors). Temperature and stress limits and corrosion allowances are specified in national standards and pressure vessel codes (77–78). General corrosion rates in pure steam (see Table 9) are about the same as in high purity deoxygenated water, except for gray iron, nickel, lead and zirconium, which corrode faster in steam. Iron-base alloys, including the austenitic and ferritic stainless steels, are used extensively in contact with steam. They oxidize to form a protective film of the spinel oxide Fe_3O_4 (magnetite) or, in the case of stainless steels, R_3O_4 (where R = iron, chromium, or nickel). Gamma Fe_2O_3 has also been found on ferrous alloys in degassed high temperature water and steam. Its physical properties are very similar to those of Fe_3O_4; it is magnetic and has an almost identical crystal structure.

Only Fe_3O_4 and gamma Fe_2O_3 are considered to be protective films. Both are adherent and good electronic conductors. Alpha Fe_2O_3, which forms in water and steam containing oxygen, is not adherent, is less protective, and is an insulator. FeO, which does not form at temperatures below 570°C, is nonprotective.

A good summary of the behavior of steels in high temperature steam is given in ref. 80. Calculated scale thickness for ten years' exposure of ferritic steels in 593°C, 13.8 MPa (2000 psi) superheated steam is about 0.64 mm for 5 Cr–½ Mo steels, and 1 mm for 2¼ Cr–1 Mo steels. Steam pressure does not seem to have much influence. The steels form duplex layer scales of a uniform thickness. Scales on austenitic steels in the same test also formed two layers but were irregular.

Generally, the higher the alloy content, the thinner the oxide scale. Extensively thick oxide scale can exfoliate and be prone to the "under the scale" concentration of corrodents and corrosion. Exfoliated scale can cause solid particle erosion of the downstream equipment and clogging.

Thick scale on boiler tubes impairs heat transfer and results in an increase of metal temperature.

Where corrosive impurities from steam or water concentrate on metal surfaces, corrosion can be severe (21–22,29,81). General and pitting corrosion, stress-corrosion cracking, corrosion fatigue, corrosion-erosion, caustic gauging, and hydrogen embrittlement and cracking have been observed. Combined corrosion effects and tensile stresses can produce sudden failures of materials. Corrosion is not caused by the steam,

Table 9. Corrosion of Metals and Alloys in Steam[a]

Material	Temperature, °C	Corrosion rate, mm/yr
cast irons		
gray	25–112	<0.5
nickel	25–350	<0.05
silicon	25–300	<0.05
mild steel	25–510	<0.05
austenitic stainless steels		
Alloys		
302, 304, 321, 347,	25–790	<0.05[b]
316, 317	25–350	<0.05
ACI CN 20, 20 Cr–30 Ni	25–350	0.5–1.3
copper	25–250	<0.05
Alloys		
brass 70–80 Cu + Zn, Sn or Pb	25–212	<0.05
brass, 57–93 Cu + Al, Zn or As	25–212	<0.05
cupro–nickel 66–88:11–13	25–300	<0.05[c]
nickel 99	25–425	0.5–1.3
Ni–Cu 66–32	25–350	<0.05
Ni–Cr–Fe T6–16–7	25–815	0.5–1.3[d]
Ni–Mo 62–28 + Fe,V	25–350	<0.05
Ni–Cr–Mo 54–15–16 + Fe,W	25–350	<0.05
aluminum	50–250	0.5–1.3
Alloys	300	>1.3[d]
gold	25–350	<0.05
lead	25–150	0.5–1.3
	300	>1.3
platinum	25–350	<0.05
silver	25–350	<0.05
tantalum	25–250	<0.05
titanium	25–350	0.0[e]
zirconium	25–350	>1.3

[a] Extracted from ref. 79.
[b] Intergranular crack.
[c] No Zn.
[d] Stress cracks.
[e] Aerated steam.

but by the concentrated steam impurities, such as chlorides, caustic, inorganic and organic acids, carbonates, sulfates, H_2S, and their mixtures. Oxygen, copper oxides, and lead and nickel oxide can aggravate the corrosion.

Hydrogen in water and steam can diffuse into steels. Hydrogen damage is a mechanism observed frequently in low alloy steel boiler tubes where corrosion hydrogen diffuses into the metal and reacts with carbon, forming methane bubbles which weaken the material. Hydrogen sulfide is also known to cause very fast stress corrosion of some steels, such as the low alloy steel turbine disks (82). In several cases, the H_2S present was a product of decomposition of sodium sulfite used for oxygen scavenging. Oxygen, hydrogen, hydrazine, some oxides and any other chemicals that can cause a shift of corrosion potential need to be considered in corrosion of steam-cycle components. Transition corrosion-potential regions, before and after the region of passivity, should be given particular attention since they are often associated with a localized corrosion attack such as pitting and stress corrosion.

More than 160 chemical compounds (corrodents, water-treatment additives, corrosion products, process chemicals, etc) have been found in steam and steam-borne deposits (21,65–66).

In relation to the thermodynamic state of steam, accelerated corrosion most frequently occurs within the narrow region of superheat above the saturation line in the so-called salt zone (22,83). In this zone, many salts and acids can be present as concentrated aqueous solutions (up to 30 wt % for sodium chloride). Their ionic conductivity is high enough to cause rapid corrosion. At higher superheats, most salts are dry; in the moisture region, they can be diluted and washed away. Hygroscopic impurities, such as sodium hydroxide, exist as a solution over a wide range of superheat and can, therefore, attack a larger part of a steam system. Corrosion–erosion is usually associated with the moisture region below the saturation line where low pH, high velocity water droplets may attack, particularly, carbon and low alloy steels. Different corrosion regions are represented in the Mollier diagram (entropy–enthalpy) in Figure 29.

To minimize corrosion in steam systems, impurities must be controlled, acid impurities neutralized, oxygen minimized at its sources or scavenged, and metal surfaces kept clean and covered with good protective oxide (such as a thin homogeneous layer of magnetite) and possibly passivated. Any steam and water additives chosen should not decompose under the high temperature, high pressure conditions of a cycle to form harmful products.

Corrosion of steam-cycle components while out of service may be more severe than corrosion during operation, for without a layup protection by dry air and/or vapor-phase inhibitors, metal surfaces with deposited steam-borne corrodents may be exposed to oxygen and humidity.

Corrosion is of concern not only because of the conversion of the load-carrying metals into oxides, but also because of the effects of corrosion products on other system components and processes, such as erosion, corrosion, clogging, sludge deposition, catalyst poisoning, product contamination, and radioactive contamination.

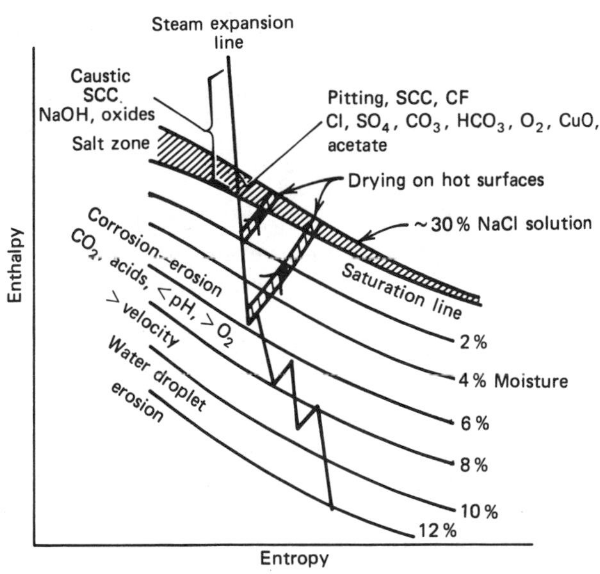

Figure 29. Mollier diagram with regions of active corrosion and erosion mechanisms. SCC = stress corrosion cracking; CF = corrosion fatigue.

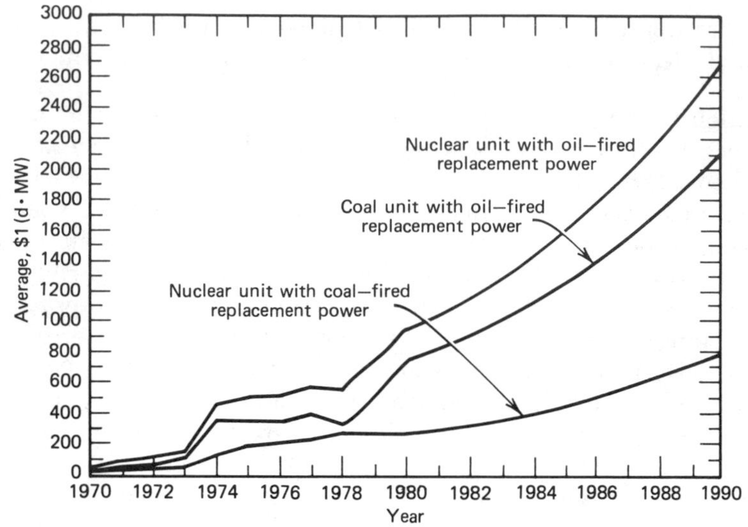

Figure 30. U.S. electric-utility outage cost trends, historical and forecast.

Steam additions increase oxidation of steels in hot-air service by decreasing the plasticity of the protective scale (84).

In steam–hydrocarbon reforming used for the production of hydrogen and carbon monoxide, carburization (an uptake of carbon by steels) can occur and lead to embrittlement of the surface layer of system components.

Corrosion behavior in water and steam of several materials used for nuclear-reactor core applications is summarized in ref. 85.

General Economics

With increasing fuel costs and sizes of industrial and utility installations, the emphasis in economical considerations has been shifting to the high thermal efficiency, reliability, and availability. The investment, operating, maintenance, transmission, insurance, and other costs and depreciation must also be considered, but they are often less important.

Thermal efficiency and heat rate directly influence the fuel cost. Increasing cycle pressure and temperature and using superheat, reheat, and condensing cycles results in a significant increase of efficiency and lower heat rate (see Fig. 25). In the process industry, this can be best achieved by cogeneration; in the electric-utility industry, by combined gas turbine–steam turbine cycles and by further increasing cycle parameters. A study (86) of advanced power cycles indicates that a 1000 MW_e (1.34×10^6 hp) unit can save 150,000 metric tons of coal/yr, or $\$10^7$ at \$65/t, if the cycle parameters can be increased from 24.1 MPa (3500 psi)/540°C inlet/540°C reheat to 34.5 MPa (5000 psi)/650°C inlet/593°C double reheat. The change would result in an average gross plant-efficiency improvement from 39.83 to 42.59% and in a heat-rate improvement from 7.728 to 7.222 kJ (1.847 to 1.726 kcal)/kW·h. The saving would be \$3600 per percent per MW per year.

Reliability and availability affect the cost of replacement steam and power and the cost of aborted or delayed processes and curtailed production. Cost of the replacement electric power is approaching 10^6 dollars a day for a large utility unit, with premium cost up to $100 per MW·h. Average costs of replacement power and their forecasts are in Figure 30. Because of the large impact of reliability, it is often better to invest in more expensive, higher quality equipment and spare parts that can be quickly replaced, eg, heat exchangers and condensers with titanium tubes and spare turbine rotors.

Installation costs and production expenses vary widely. The Utility Cost Study, 1981 (87), gives indications of the escalating construction costs, production expenses, and heat rates. Results from a survey of 239 conventional and 31 nuclear plants are included in the study summarized in Table 10.

Table 10. Summary of the Utility Cost Study, 1981[a]

		Cost, $ per kW installed	Initial operation	Fuel, mills per kW·h	Total production cost, mills per kW·h
conventional	high	940	1981	112.2	198.4
steam plants	low	85	1960	5.4	9.0
combined cycle		143	1974	12.0	13.0
(one station only)					
nuclear plants	high	1571	1985	10.8	53.4
	low	173	1972	3.0	7.0

[a] Ref. 87.

BIBLIOGRAPHY

"Steam" in *ECT* 1st ed., Vol. 12, pp. 778–793, by H. N. La Croix, Foster Wheeler Corporation; "Steam" in *ECT* 2nd ed., Vol. 18, pp. 692–715, by J. K. Rice, Cyrus Wm. Rice and Company.

1. *Proceedings of the International Conferences on Properties of Steam*, International Association for Properties of Steam, 9 conferences held to 1982.
2. *Proceedings of the International Water Conference*, Engineering Society of Western Pennsylvania, Pittsburgh, Pa., 42 conferences held to 1982.
3. A. Stodola, *Steam and Gas Turbines*, Peter Smith, New York, 1945.
4. K. Raznjevic, *Handbook of Thermodynamic Tables and Charts*, Hemisphere Publishing Corp., 1976.
5. B. J. McBride and co-workers, *Thermodynamic Properties to 6000 K for 210 Substances Involving the First 18 Elements*, NASA, 1963.
6. F. Franks, ed., *Water, a Comprehensive Treatise*, Plenum Press, New York, 1972.
7. *Steam—Its Generation and Use*, 39th ed., Babcox and Wilcox Company, New York, 1978.
8. J. Straub and K. Schaffler, *Proceedings of the 9th International Conference on the Properties of Steam*, Munich, FRG, September, 1979, Pergamon Press, Oxford, UK, 1980.
9. *J. Phys. Chem. Ref. Data*, the American Chemical Society and the American Institute of Physics for the National Bureau of Standards.
10. C. A. Mayer and co-workers, *ASME Steam Tables*, 4th ed., American Society of Mechanical Engineers, New York, 1979.
11. *U.K. Steam Tables in SI Units 1970*, Edward Arnold Ltd., London, 1970.
12. K. Keenan, Hill, and Moore, *Steam Tables (International Edition—Metric Units)*, John Wiley & Sons, Inc., New York, 1969.
13. L. A. Curtiss, D. J. Frurip, and M. Blander in ref. 8, p. 521.
14. J. Kestin and co-eds., *Sourcebook on the Production of Electricity from Geothermal Energy*, Brown University, Providence, R.I., 1980.

15. L. A. Casper and T. R. Pinchback, eds., *Geothermal Scaling and Corrosion*, STP 717, American Society for Testing and Materials, Philadelphia, Pa., 1980.
16. *Compilation of Geothermal Information*, Report LBL-3220, Energy and Environment Division, Lawrence Berkeley Laboratory, University of California, Berkeley, sponsored by ERDA, 1977.
17. P. M. Goodall, ed., *The Efficient Use of Steam*, IPC Science and Technology Press, Guilford, Surrey, UK, 1980.
18. R. C. Hendricks, R. B. McClintock, and G. J. Silvestri, *J. Eng. Power* **99,** 664 (1977).
19. M. Uematsu and E. V. Franck, *J. Phys. Chem. Ref. Data* **9,** 1291 (1980).
20. W. L. Marshall and E. V. Franck, *J. Phys. Chem. Ref. Data* **10,** 295 (1981).
21. O. Jonas, *Identification and Behavior of Turbine Steam Impurities*, *Corrosion 77*, National Association of Corrosion Engineers, San Francisco, 1977.
22. O. Jonas, W. T. Lindsay, Jr., and N. A. Evans in ref. 8, p. 595.
23. J. W. Cobble, *Science* **152,** 1979 (1966).
24. P. A. Akolzin and co-workers, *Water Regime for Supercritical Units*, Energia, Moscow, 1972, Chapt. 6.
25. D. W. Rudorff, *Eng. Boiler House Rev.*, 2 (Jan. 1945).
26. O. Fuchs, *Z. Elektrochem.* **47,** 101 (1941).
27. F. F. Straus, *Steam Turbine Blade Deposits*, University of Illinois, Bulletin No. 59, Urbana, Ill., June 1, 1946.
28. H. G. Heitman, *Mitl. Ver. Grosskesselbetr.* **90,** 171 (1964).
29. O. Jonas, *Combustion* **50,** 11 (1978).
30. W. E. Allmon and co-workers, *Proceedings of the 41st International Water Conference*, Engineering Society of Western Pennsylvania, 1980, pp. 138–172.
31. M. A. Styrikhovich, O. I. Martynova, and Z. S. Belova, *Teploenergetika* **12,** 86 (1965).
32. O. I. Martynova, *NACE Conference on High Temperature High Pressure Electrochemistry in Aqueous Solutions*, Reprint Number: C-2, 1973.
33. O. Jonas, *paper presented at the 42nd Annual Meeting International Water Conference*, Pittsburgh, Pa., October 26–28, 1981.
34. P. Cohen, *Water Coolant Technology of Power Reactors*, American Nuclear Society, 1980.
35. *Manual on Industrial Water and Industrial Waste Water*, 2nd ed., ASTM STP 148-L, American Society for Testing and Materials, Philadelphia, Pa., 1967, p. 963.
36. O. I. Martynova and Ju. F. Samojlov, *J. Inorg. Chem.* **2,** 2829 (1957).
37. O. I. Martynova and co-workers, *J. Inorg. Chem.* **5,** 16 (1960).
38. J. K. Kreider and F. Kreith, *Solar Energy Handbook*, McGraw-Hill, New York, 1981.
39. Ref. 17, pp. 389–405.
40. W. Goldstern, *Steam Storage Installations*, 2nd ed., Pergamon Press, Oxford, 1970.
41. *Drew Principles of Industrial Water Treatment*, Drew Chemical Corp., Boonton, N.J., 1979.
42. *Betz Handbook of Industrial Water Conditions*, Betz, Trevose, Pa., 1962.
43. F. N. Kemmer and J. McCallion, ed., *The Nalco Water Handbook*, Nalco Chemical Co., McGraw-Hill, New York, 1979.
44. S. B. Applebaum, *Demineralization by Ion Exchange*, Academic Press, New York, 1968.
45. Degrement, *Water Treatment Handbook*, Société General d'Epuration et d'Addainissement, 1973.
46. British Standard BS1170, Methods for treatment of water for marine boilers.
47. British Standard BS2486, Treatment of water for land boilers.
48. *Boiler Water Limits and Steam Purity Recommendations for Watertube Boilers*, American Boiler Manufacturers Association, 1981.
49. *CEGB, Chemical Control of Boiler Feedwater, Boiler Water, and Saturated Steam for Drum-type and Once-Through Boilers*, Generation Operation Memorandum 72, Issue 4, CEGB, London, UK, 1975.
50. *VGB-Kraftwerks-Technick Mitteilugen* **52,** 167 (1972).
51. Japanese Standards JIS B8223 (Water Quality of the Feedwater and the Boiler Water for Recirculating Boilers) and JIS B8224 (Feedwater Quality for Once-Through Boilers).
52. D. E. Simon II, *Proceedings of the 36th International Water Conference*, Engineering Society of Western Pennsylvania, 1976, p. 65.
53. W. H. Yost, *Power*, 93 (Feb. 1982).
54. R. Svoboda and P. Schmid, *Brown Boveri Rev.* **3-78,** 179 (1978).
55. R. W. Lane, B. Otten, eds., *Power Plant Instrumentation for Measurement of High Purity Water Quality*, ASTM STP 742, American Society for Testing and Materials, Philadelphia, Pa., 1981.

56. R. V. Cobb and E. E. Coulter, *ASTM Proceedings* **61,** 1386 (1961).
57. M. Lawlor and C. Clar, *An Experimental Investigation of Some of the Factors Which Influence the Accuracy of Steam Sampling*, ASME Paper No. 61-WA-266, American Society of Mechanical Engineers, New York, 1961.
58. P. Goldstein and F. B. Simmons, *Proc. Am. Power Conf.* **26,** 720 (1962).
59. T. A. Miskimen, *Results of Steam Sampling Nozzle Tests on Evaporator Vapor*, ASME Paper No. 59-A-301, American Society of Mechanical Engineers, New York, 1959.
60. J. Jackson, *Power Eng.* **64,** 627 (Nov. 1960).
61. *Tentative Method of Sampling Steam*, ASTM D 1066-67T, American Society for Testing and Materials, Philadelphia, Pa., 1967.
62. D. F. Pensenstadler and co-workers, *Program for Steam Purity Monitoring: 1. Instrumentation and Sampling*, ASTM STP 742, American Society for Testing and Materials, Philadelphia, Pa., 1981, pp. 55–70.
63. *1982 Annual Book of ASTM Standards*, Part 31 Water, American Society for Testing and Materials, Philadelphia, Pa., 1982.
64. F. G. Straub, *Trans. ASME* **57,** 447 (1935).
65. O. Jonas, *Proceedings of the 34th International Water Conference*, Engineering Society of Western Pennsylvania, Pittsburgh, Pa., 1973, pp. 73–81.
66. H. Kirsch and S. Pollmann, *Chem. Ing. Tech.* **40,** 897 (1968).
67. O. I. Martynova and B. S. Rogatskin, *Teploenergetika* **17,** 50 (1970).
68. *The Industrial Energy Efficiency Improvement Program*, DOE/CE-0015, U.S. Department of Energy, Dec. 1980.
69. Battelle Columbus Laboratories, *Survey of the Application of Solar Thermal Energy Systems to Industrial Process Heat*, Report No. TID-27348/1, U.S. Energy Research and Development Administration, Jan. 1977, p. 15.
70. S. B. Bennett and R. L. Bannister, *Mech. Eng.*, 19 (Dec. 1981).
71. *Electric Power Supply and Demand 1981–1990*, National Electric Reliability Council, 1981.
72. *Nucl. News* (*Hinsdale, Ill.*) **25**(2), (Feb. 1982).
73. *Chem. Eng.* **74,** 112 (1967).
74. A. L. Hammond, *Energy and the Future*, American Association for the Advancement of Science, 1973.
75. R. Loftness, *Energy Handbook*, Van Nostrand Reinhold Co., New York, 1978.
76. A. R. Allen, *Coping With the Oil Sands*, Great Canadian Oil Sands, Ltd., 1974.
77. British Standards BS749, 1113, 2790.
78. *ASME Boiler and Pressure Vessel Code*, The American Society of Mechanical Engineers, New York, 1977.
79. N. E. Hammer, *Metals Section Corrosion Data Survey*, 5th ed., National Association of Corrosion Engineers, Houston, Texas, 1974, pp. 174–175.
80. G. E. Lien, ed., *Behavior of Superheater Alloys in High Temperature, High Pressure Steam*, The American Society of Mechanical Engineers, New York, 1971.
81. R. J. Lindinger and R. M. Curran, *Experience with Stress Corrosion Cracking in Large Steam Turbines*, *Corrosion 81*, National Association of Corrosion Engineers, Toronto, Ontario, 1981.
82. R. M. Curran and co-workers, *Stress Corrosion Cracking of Steam Turbine Materials*, Southeastern Electric Exchange, General Electric, Clearwater, Fla., April 21–22, 1969.
83. W. T. Lindsay, Jr., *Power Eng.* **83**(5), 68 (May 1979).
84. A. J. Sedricks, *Corrosion of Stainless Steels*, John Wiley & Sons, Inc., New York, 1979, pp. 244–245.
85. W. E. Berry, *Corrosion in Nuclear Applications*, John Wiley & Sons, Inc., New York, 1971.
86. T. Suzuki, *The Development of Coal Firing Power Unit with Ultra High Performance in Japan*, EPRI Fossil Plant Heat Rate Improvement Workshop, Charlotte, N.C., Aug. 1981.
87. *Turbomachinery International*, Utility Cost Studies, Business Journals Inc., Norwalk, Conn., 1981.

OTAKAR JONAS
Westinghouse Electric Corp.

STEARIC ACID, $CH_3(CH_2)_{16}COOH$.
See Carboxylic acids.

STEATITE.
See Talc.

STEEL

Steel is the generic name for a group of ferrous metals composed principally of iron which, because of their abundance, durability, versatility, and low cost, are the most useful metallic materials known. Steel, in the form of bars, plates, sheets, structural shapes, wire pipe and tubing, forgings, and castings is used in the construction of buildings, bridges, railroads, aircraft, ships, automobiles, tools, cutlery, machinery, furniture, household appliances, and many other articles upon which the convenience, comfort, and safety of today's society depend. Steel is an essential material for spacecraft and their supporting facilities and is found in practically every kind of material needed for the national defense.

The abundance of steel is indicated by the fact that, in a typical year, over 700 $\times 10^6$ metric tons of raw steel is produced throughout the world.

The durability and versatility of steel are shown by the wide range of mechanical and physical properties possessed by different kinds of steel. By the proper choice of carbon content and alloying elements, and by suitable heat treatment, steel can be made so soft and ductile that it can be cold drawn into complex shapes such as automobile bodies. Conversely, steel can be made extremely hard to resist wear, or tough enough to withstand enormous loads and shock without deforming or breaking. In addition, some steels are made to resist heat and corrosion by the atmosphere and a wide variety of chemicals.

The usefulness of steel is enhanced by the fact that it is inexpensive: its price ranges from ca 44¢/kg for the common grades to several dollars per kg for special steels such as some tool steels.

The uses of steel are too diverse to be listed completely or to serve as a basis of classification. Inasmuch as grades of steel are produced by more than one process, classification by method of manufacture is not advantageous. The most useful classification is by chemical composition into the large groups of carbon steels, alloy steels, and stainless steels. Within these groups are many subdivisions based on chemical composition, physical or mechanical properties, or uses.

It has been known for many centuries that iron ore, embedded in burning charcoal, can be reduced to metallic iron (1–2). Iron has been made by this method as early as 1200 BC, when iron was known throughout the ancient world. It is not known when the first metal resembling modern carbon steel was made intentionally. Some authorities believe that steel was being made in India around 1000 BC. Remains of swords with steel blades have been found in Luristan, a western province of Iran (formerly Persia) at sites dated at ca 800 BC.

Consisting almost entirely of pure iron, the first man-made iron closely resembled modern wrought iron, which is relatively soft, malleable, ductile, and readily hammer-welded when heated to a sufficiently high temperature. It was used for many purposes, including agricultural implements and various tools (see Iron).

Most steel contains >98% iron. However, it also contains carbon which, if present in sufficient amounts (up to ca 2%), gives a steel a property unmatched by any of the metals available to ancient man. This property is the ability to become extremely hard if cooled very quickly (quenched) from a high enough temperature, as by immersing it in water or some other liquid. The hardness of steel and its ability to take and hold a sharp cutting edge make it an extremely valuable metal for weapons, tools, cutlery, surgical instruments, razors, and other special forms.

It was not until the eighteenth century that carbon was recognized as a chemical element, and it is quite certain that no early metallurgist was aware of the basis of the unique properties of steel as compared with those of wrought iron. Carbon can be alloyed with iron in a number of ways to make steel, and all methods described here have been used at various times in many localities for perhaps 3000 years or more.

For example, low carbon wrought-iron bars were packed in airtight containers together with charcoal or other carbonaceous material. By heating the containers to a red heat and holding them at that temperature for several days, the wrought iron absorbed carbon from the charcoal; this method became known as the cementation process, described below. If the iron is made in primitive furnaces, carbon can be absorbed from the charcoal fuel during and after its reduction. The Romans and others recognized this and built and operated furnaces that produced a steel-like metal instead of wrought iron. In neither of the two foregoing cases did the iron or steel become molten.

In very ancient times in India, a steel called wootz was made by placing very pure iron ore and wood or other carbonaceous material in a tightly sealed pot or crucible which was heated to a very high temperature for a considerable time. Some of the carbon in the crucible reduced the iron ore to metallic iron which absorbed any excess carbon. The resulting iron–carbon alloy was an excellent grade of steel. In a similar way, pieces of low carbon wrought iron were placed in a pot along with a form of carbon and melted to make a fine steel. A variation of this method, in which bars that had been carburized by the cementation process were melted in a sealed pot to make steel of the best quality, became known as the crucible process (see below).

Before the invention of the Bessemer process for steelmaking in 1856, only the cementation and crucible processes were of any industrial importance. Although both these processes had been known in the ancient world, their practice seems to have been abandoned in Europe before the Middle Ages. The cementation process was revived in Belgium around 1600, whereas the crucible process was rediscovered in the UK in 1742. Both processes were practiced in secret for some time after their revival, and little is known of their early history. The cementation process flourished in the UK during the eighteenth and nineteenth centuries and continued to be used to a limited extent into the early part of the twentieth century. At red heat, a low carbon ferrous metal, in contact with carbonaceous material such as charcoal, absorbed carbon which, up to the saturation point of about 1.70%, varied in amount according to the time the metal was in contact with the carbon and the temperature at which the process was conducted. A type of muffle furnace or pot furnace was used and the iron and charcoal were packed in alternate layers.

In the softer grades (average carbon content ca 0.50%), the composition of the bar was uniform throughout. In the harder grades (average carbon content as high as 1.50%), the outside of a bar might have a carbon content of 1.50–2.00%, whereas the center contained 0.85–1.10%. Steels made by this method were called cement steels.

The crucible process gave steels that were not only homogeneous throughout but were free from occluded slag originating in the wrought iron used to make cement steel. Crucible steel was so superior to cement steel for many purposes that the crucible process quickly became the leader for the production of the finest steels. Its drawback was, however, that each crucible held only ca 50 kg steel.

The steelmaking processes discussed above were all eventually supplanted by entirely new methods (3–4). The first of the new techniques was the pneumatic or Bessemer process, introduced in 1856. Shortly thereafter, the regenerative-type furnace, now known as the open-hearth furnace, was developed in the UK. It was adapted to steelmaking and the open-hearth process was the principle method for producing steel throughout the world until 1970.

Since then, the most widely used processes for making liquid steel have been the oxygen steelmaking processes, in which commercially pure oxygen (99.5% pure) is used to refine molten pig iron. In the top-blown basic oxygen process, oxygen is blown down onto the surface of the molten pig iron. In the bottom-blown basic oxygen process, the oxygen is blown upward through the molten pig iron. The direction in which the oxygen is blown has important effects on the final steel composition and on the amount of iron lost in the slag.

Liquid steel is also made by melting steel scrap in an electric-arc furnace. This process is gradually finding more and more application in large-scale production.

Pig iron and iron and steel scrap are the sources of iron for steelmaking in basic-oxygen and open-hearth furnaces; electric furnaces rely on iron and steel scrap. In the basic-oxygen and open-hearth furnaces, the pig iron is used in the molten state as obtained from the blast furnace; in this form, pig iron is referred to as hot metal.

Pig iron consists of the iron combined with numerous other elements, including mostly carbon, manganese, phosphorus, sulfur, and silicon. Depending upon the composition of the raw materials used in the blast furnace, principally iron ore (beneficiated or otherwise), coke, and limestone, and the manner in which the furnace is operated, pig iron may contain 3.0–4.5% carbon, 0.15–2.5% or more of manganese, as much as 0.2% sulfur, 0.025–2.5% phosphorus, and 0.5–4.0% silicon. In refining pig iron before converting it to steel, these elements must either be removed almost entirely or at least reduced drastically in amount. The same is true of these and any other unwanted elements that may be contained in scrap.

Modern steelmaking processes are either acid or basic processes. Carbon, manganese, and silicon are removed with relative ease by either process. The removal of phosphorus and sulfur requires special conditions that can be met only by basic processes, where lime is added to give a basic slag that is capable of forming compounds with phosphorus and sulfur during the refining operations, thereby removing them from the metal. Because of the chemical nature of the slags, each process must be carried out in equipment lined with refractories of the proper chemical composition; otherwise the slags would react with and be neutralized by the lining material and thereby destroy the lining rapidly. Thus, the basic processes are carried out in vessels or furnaces lined with basic refractories such as dolomite or magnesite, whereas acid processes are carried out in equipment lined with acid refractories, such as silica brick, sintered silica sand, or ganister (see Refractories).

Oxidation is employed to convert a molten bath of pig iron and scrap, or scrap alone, into steel. Each steelmaking process has been devised primarily to provide some means by which controlled amounts of oxygen can be supplied to the molten metal undergoing refining. The oxygen combines with the unwanted elements (with the exception of sulfur) and, unavoidably, with some of the iron, to form oxides which either leave the bath as gases or enter the slag. The mechanism by which sulfur is removed does not involve direct reaction with oxygen but depends instead on whether the slag is sufficiently basic and temperatures are high enough. As the purification of the pig iron proceeds, owing to the removal of carbon, the melting point of the bath is raised, and sufficient heat must be supplied to keep the bath molten.

In general, steels with similar chemical compositions have similar mechanical and physical properties, no matter by which process they are made.

For a number of years, much of the steel made in the United States was produced by the basic open-hearth process. Acid open-hearth furnaces have been and are still being used for castings and specialty steels. Most electric furnaces for making steel for ingots and continuous-cast products are of the basic type, whereas acid electric furnaces are used primarily in foundries for producing steel for castings. The basic oxygen process has become the leading steelmaking method in the United States. The production of steel in electric furnaces has now surpassed that in open-hearth furnaces. Bessemer installations have been phased out.

Interest is increasing in making iron and steel directly from ore, without first reducing the ore in the blast furnace to make pig iron which has to be purified in a second step. These processes, generally referred to as direct-reduction processes, are employed where natural gas is readily available for the reduction (see also Iron by direct reduction).

Open-Hearth Process

The open-hearth furnace is a large structure comprised of several parts, all of which are constructed primarily of refractory bricks (5) (see also Furnaces, fuel-fired). The metallic charge of steel scrap and liquid blast-furnace iron or hot metal is placed on the bottom of an elongated tunnel-like furnace (see Fig. 1). A fuel such as heavy fuel oil or tar is heated and injected into the furnace through an end wall. Preheated air burns the fuel, thereby heating the charge until it melts. The excess carbon and silicon in the hot metal are removed by oxidation. The oxygen for this oxidation comes from air above the bath of liquid steel, iron ore added to the charge, and oxygen gas blown down into the liquid steel bath through water-cooled pipes or lances.

The air used to burn the fuel is preheated in large checker chambers which are heated by the combustion products leaving the furnace. This furnace is of the regenerative type in which the furnace is fired from one end for 10–15 min, and then fired from the other end. The bricks in the checker chamber at the outgoing end are heated by the combustion products and the air is heated by the bricks in the checker chamber at the incoming end. When the direction of firing is reversed, the air direction is also reversed. This cycle of firing is repeated throughout the 4-to-12 hour period required to make a batch or heat of steel.

It has been found desirable to group a number of steelmaking furnaces into one building called a shop. An open-hearth shop often has 5–10 open-hearth furnaces arranged in a line to facilitate charging the furnaces (see Fig. 2).

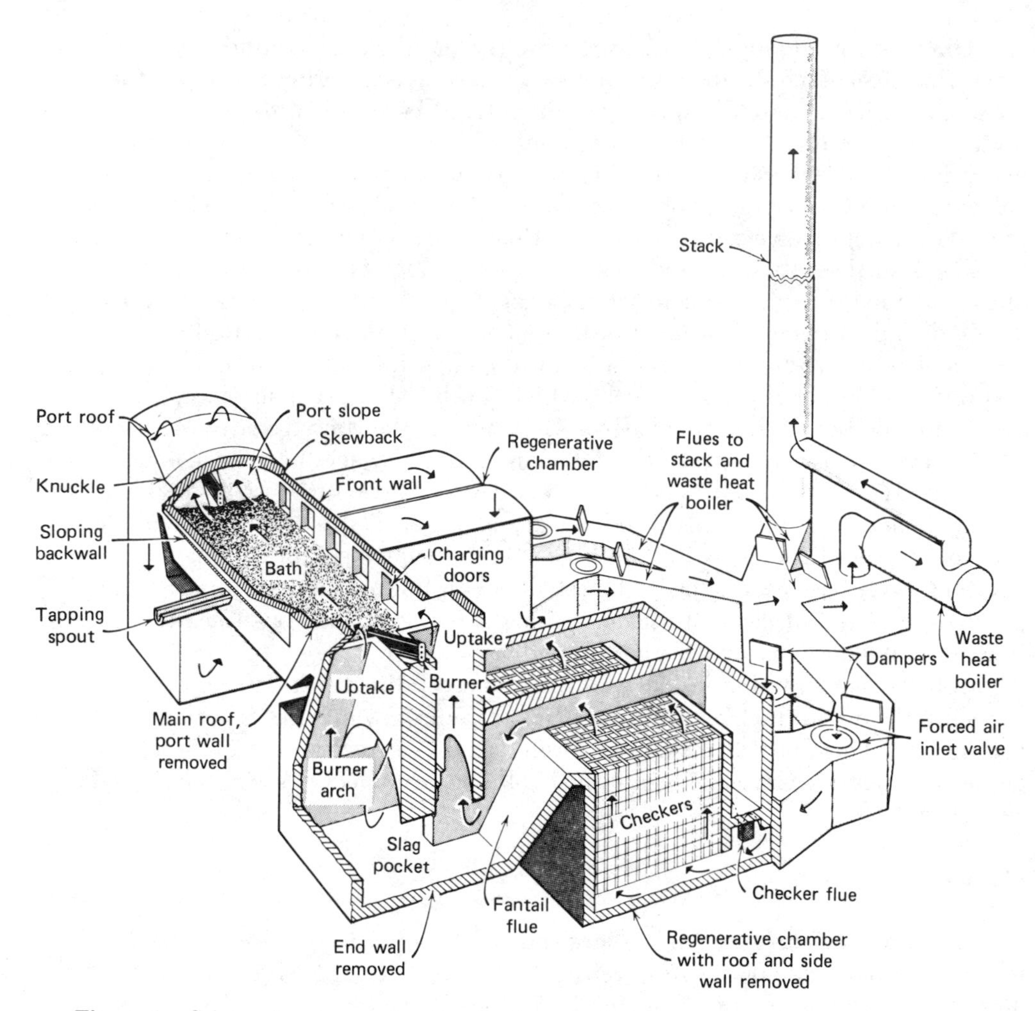

Figure 1. Schematic cut-away drawing showing principal parts of an open-hearth furnace. Heavy curved arrows indicate direction of flow of air, flame, and waste gases when a liquid fuel is fired through a burner in the trench at the right end of the furnace. Direction of arrows reverses when the furnace is "reversed" and fired from the left end (1).

In the United States in 1980, ca 12% of all the raw steel made was produced in open-hearth furnaces (6). Most of the steels formerly made in basic open-hearth furnaces are now made by the basic oxygen steelmaking process because of lower costs.

Electric-Furnace Processes

The principal electric-furnace steelmaking processes are the electric-arc furnace, induction furnace, consumable-electrode melting, and electroslag remelting. The raw material for all these processes is solid steel. As implied by the term steelmaking, in the first two processes steel is made that is different in composition and shape from the starting material which is usually steel scrap or direct-reduced iron (DRI). The steel used for the last two processes closely resembles the desired steel ingot which

Figure 2. Pouring of hot metal into an open-hearth furnace. Part of a charging machine is shown in the foreground. The fronts of three open-hearth furnaces are shown.

is subsequently rolled or forged. The last two processes are employed for very high quality steel for applications with extremely strict requirements.

Electric-Arc Furnace. The electric-arc furnace is the principal electric steelmaking furnace. The carbon arc was discovered by Sir Humphry Davy in 1800, but it had no practical application in steelmaking until Sir William Siemens of open-hearth fame constructed, operated, and patented furnaces operating on both direct- and indirect-arc principles in 1878. At that early date, the availability of electric power was limited and very expensive. Furthermore, carbon electrodes of the quality to carry sufficient current for steel melting had not been developed (see Furnaces, electric).

In indirect-arc heating, the arcs pass between electrodes supported above the metal in the furnace, which thus is heated solely by radiation from the arc. In direct-arc heating, the current must flow through the metal bath in order that the heat developed by the electrical resistance of the metal, slight though it is, is added to that radiated from the arcs. The path of the current in a direct-arc furnace is through one electrode and thence through the arc between the electrode base and the bath, then through the bath and up through an arc between the bath and an adjacent electrode, completing the circuit through this second electrode.

The first successful direct-arc electric furnace, patented by Heroult in France, was placed in operation in 1899. The patent covered single- or multi-phase furnaces with the arcs placed in series through the metal bath. This type of furnace, utilizing three-phase a-c power, has been the most successful for steel production.

The first direct-arc furnace in the United States was a single-phase two-electrode rectangular furnace of four tons capacity at the Halcomb Steel Company, Syracuse, New York, which made its first heat in 1906. A similar but smaller furnace was installed two years later at the Firth-Sterling Steel Company in McKeesport, Pennsylvania. In 1909, a 15-ton three-phase furnace was installed in the South Works of the Illinois Steel Company, in Chicago, Illinois, which was, at that time, the largest electric steelmaking furnace in the world. It was the first round instead of rectangular furnace and operated on 25-cycle power at 2200 V.

In all the foregoing furnaces, steel was made for ingots. The first electric furnace for the production of steel for commercial castings was installed in 1911 by the Treadwell Engineering Company, Easton, Pennsylvania. It was a single-phase two-electrode furnace with a capacity of two tons.

In recent years, the direct-arc principle has been applied to vacuum consumable-electrode furnaces.

The chemical reactions taking place in the electric-arc furnace are similar to those in other steelmaking processes. When the carbon, phosphorus, and sulfur concentrations are decreased and the temperature has been raised to the desired level, the steel is tapped from the furnace into a ladle at the same time that ferromanganese, ferrosilicon, and other addition agents are added to deoxidize the steel and obtain the desired composition. For alloy steel, copper, nickel, and molybdenum can be added at any time without loss by oxidation. Chromium, being easily oxidized, is added as ferrochromium after the steel has been deoxidized. Alloy elements such as aluminum, titanium, zirconium, vanadium, and boron are added in the ladle because they are easily oxidized. As soon as chemical analyses establish that the chemical composition of the bath is correct and the bath temperature is within the desired range, the steel is tapped by tilting the furnace to allow the molten steel to run through a spout into a ladle.

Electric-arc steelmaking furnaces produce liquid steel primarily from steel scrap. Other forms of metallic iron, including pig iron and direct-reduced iron, are often added to the charge. The energy required to reduce iron ore is ca 2200 kW·h/t metallic iron produced. The consumption of this amount of energy is avoided in the electric-furnace process because metallic iron in one form or another is used as the raw material in place of iron ore. As a result, only the energy for melting the metallic iron is required, ie, ca 500 kW·h/t liquid steel produced, including the normal heat losses in the electric furnace (7–10).

In general terms, liquid steel is produced in electric-arc furnaces from a metallic steel charge of about the same composition as the steel to be made. The charge is put into the furnace, melted, and refined moderately; the liquid steel is then tapped out. The thermal energy required for melting the steel scrap is provided by three a-c arcs between three graphite electrodes and the scrap charge. The three arcs are used to balance the electrical load on the three-phase electrical supply system. Carbon, alloy, and stainless steels are made in electric-arc furnaces. In the United States since 1974, electric-arc-furnace steel production has overtaken open-hearth production. Thus, in 1980, ca 28% of the U.S. steel production was made in electric-arc furnaces (6).

Electric-arc furnaces offer the advantages of low construction costs, flexibility in the use of raw materials, and the ability to produce steels over a wide range of compositions (carbon, alloy, and stainless) and operate below full capacity.

These furnaces are lined with either acid or basic refractories. Technical and economic obstacles to the use of select scrap (low in phosphorus and sulfur) and the increasing utilization of alloy steels have greatly decreased the use of acid-lined furnaces. Almost all furnaces used for ingot-steel production and most foundry furnaces used for making steel for castings are now lined with basic refractories.

In the acid process, because of the selected materials in the charge, the use of a single siliceous oxidizing slag predominates. However, under these conditions, alloying elements such as chromium and manganese are rapidly oxidized and lost to the slag. Oxidizable elements can be added to the furnace only after the heat has been deoxidized, just before tapping, or in the ladle after the steel is tapped. Large additions to the ladle have a chilling effect and cannot mix readily with the molten steel. Alloying elements that do not oxidize readily (eg, copper, nickel, and molybdenum) may be added to the furnace at any time.

In the basic process, a slag containing ca 50% CaO is used for steels of the highest quality with low sulfur and phosphorus contents. The furnace is lined with basic refractories because a basic slag rapidly erodes acid refractories.

Typical refractories used in acid and basic linings of electric-arc steelmaking furnaces are shown in Figure 3, and the arrangement of the electrodes for a three-phase furnace, their supporting arms, and the electrical power leads in Figure 4.

Basic Electric-Arc Process. With the power turned off, the solid scrap and other components of the charge are placed in the furnace. Alloying materials that are not easily oxidized can be and usually are charged in the furnace prior to melting down the charge. Excess carbon is used in the melt bath, permitting some carbon to be removed by ore additions or oxygen injection. This excess carbon is removed as carbon monoxide gas, which bubbles out of the liquid steel bath, thereby stirring the bath. This action equalizes composition and temperature of the bath and allows control

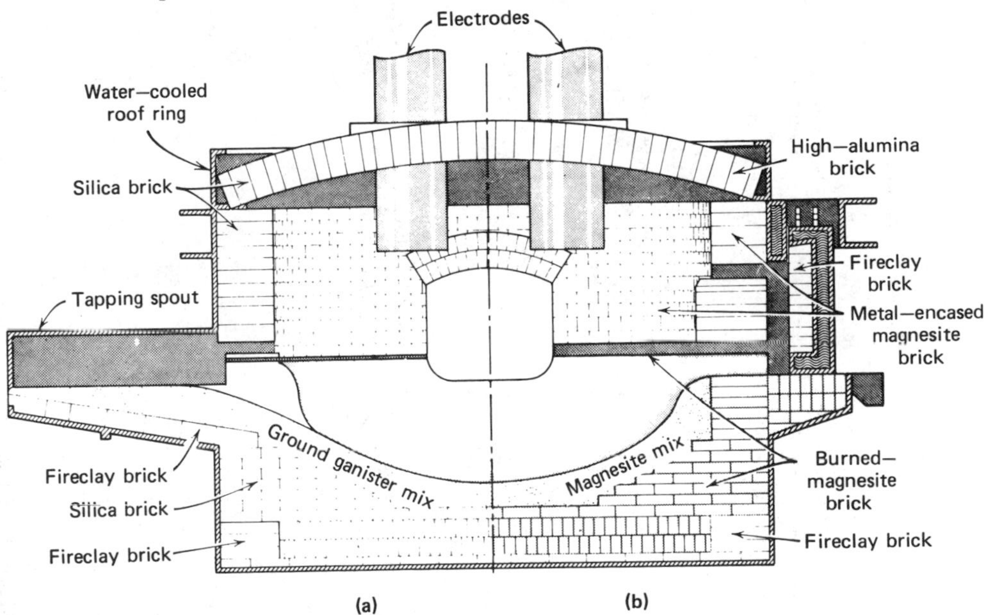

Figure 3. Schematic cross section of a Heroult electric-arc furnace, indicating typical refractories employed in an acid lining (**a**) and a basic lining (**b**). Although only two electrodes are shown, furnaces of this type have three electrodes and operate on three-phase current.

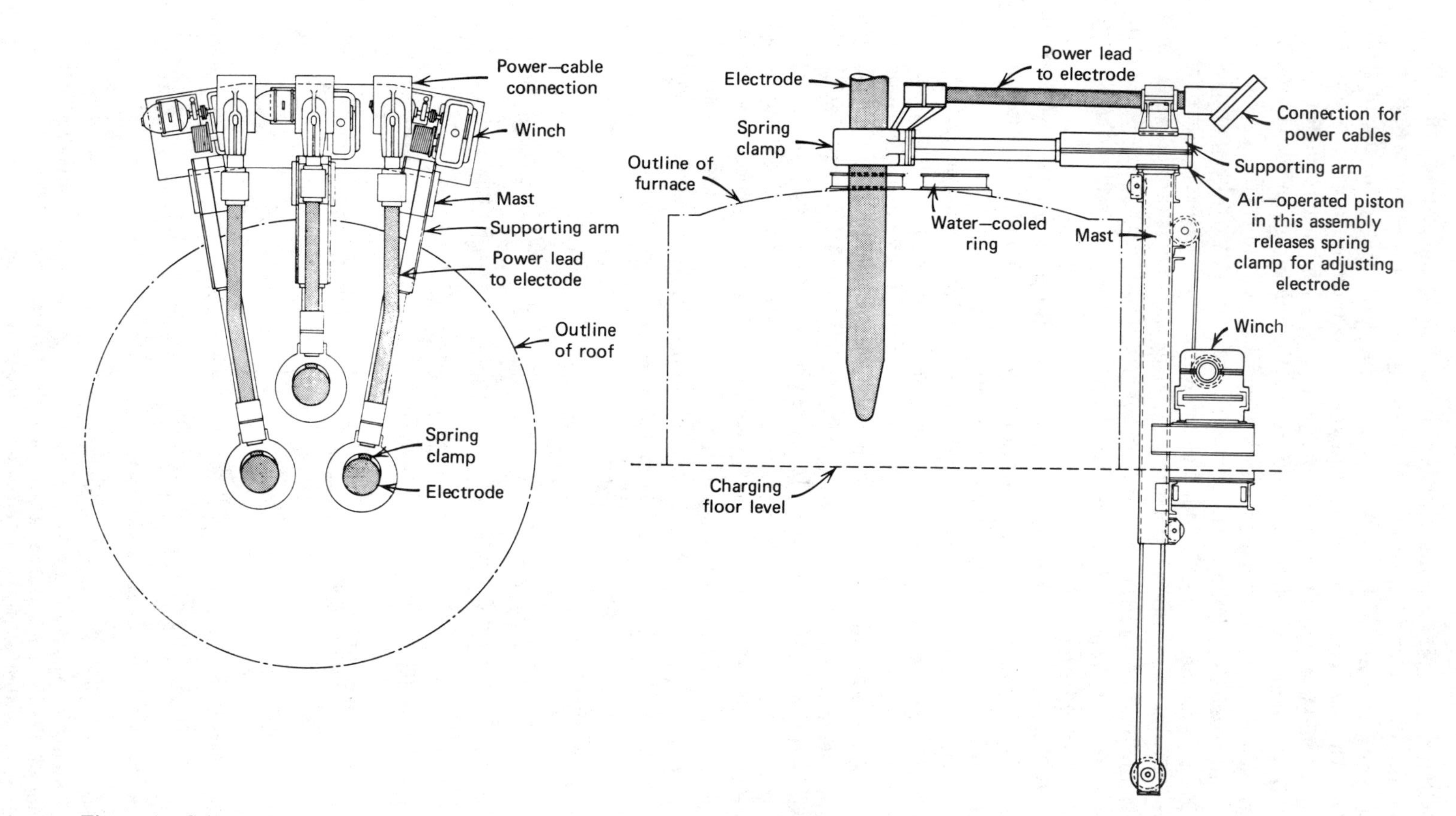

Figure 4. Schematic arrangement of the electrodes, their supporting masts, and the electrical power leads for an electric-arc steelmaking furnace (1).

within narrow specifications. If the metallic charge is too low in carbon, a recarburizer in the form of coke or scrap electrodes is charged with the scrap to achieve a carbon content at melt-down that is 0.15–0.25% higher than the carbon content of the finished steel.

Although iron ore or other forms of iron oxide may be used to lower the carbon content, gaseous oxygen is generally injected into the molten bath.

After charging has been completed, the arcs are struck and the charge is melted as quickly as possible. Since the arcs melt the portion of the charge directly beneath each electrode, the electrodes bore through the solid charge with the melted metal forming a pool on the hearth. From the time this pool forms, the charge is heated from the bottom up by radiation from the pool, heat from the arcs, and the resistance offered to the current by the scrap until the charge is completely melted.

From the time the molten metal begins to form until the entire charge is in solution, oxidation of phosphorus, silicon, manganese, carbon, and other elements occurs in varying degrees. Oxygen for these and other oxidation reactions is obtained from oxygen gas injected into the bath, oxygen in the furnace, calcination of limestone (if used), oxides of alloying elements added, and iron ore or other iron oxides (if charged or added later). As oxidation progresses, the temperature of the bath is raised to remove carbon and increase the fluidity of the bath to permit inclusions to rise through the molten metal into the slag. Carbon and oxygen give carbon monoxide gas which bubbles out of the bath, thereby stirring the bath and making it more nearly homogeneous with respect to composition and temperature. In addition, this action removes some of the hydrogen formed which decreases the tendency for gas bubbles to form in the steel as it freezes and reduces the hydrogen content of the solid steel. Consequently, the steel has less tendency to crack, especially in thick sections.

Acid Electric-Arc Process. Acid electric furnaces operate with partial oxidation, complete oxidation with a single slag, or with silicon reduction and double-slag technique. Partial oxidation is used chiefly to produce steel for inexpensive castings that do not require any tests other than superficial surface inspection. The double-slag process is employed where control of the oxidizing power of the finishing slag is desired. Silicon reduction is employed in Europe but not in the United States. Most U.S. steel foundries employ the complete-oxidation process with a single slag.

Charging and melting procedures in the acid process are similar to those in the basic process, except for the necessity of using scrap with a low content of phosphorus and sulfur which cannot be removed in the acid process. As soon as the charge has melted down, the slag formed by the oxidation of silicon, manganese, and iron is tested to make certain that its iron oxide (FeO) content is sufficiently high. The carbon content of the metal is determined at the same time, and as in the basic process, should be higher than the carbon content desired in the finished steel. The iron oxide content of the slag is adjusted by adding silica sand or iron ore, as required.

After the bath is covered with the proper oxidizing slag and the carbon content of the metal is high enough, the temperature is raised to bring the bath to a boil. The boiling is caused by the reaction of carbon with oxygen in the steel and promotes the manufacture of clean steel, since the hotter steel, being more fluid, and the agitation caused by carbon monoxide bubbling out of the bath both promote the rise of solid oxidized products to the slag. Thus, the number of nonmetallic inclusions that make steel dirty is reduced. The amounts of carbon and oxygen in the bath should be sufficient to maintain boiling for at least 10 min, during which time the carbon content of the metal falls. When the carbon content has reached the desired concentration, silicon and manganese in the form of ferroalloys are added to the bath to kill (deoxidize) the steel and prevent further lowering of the carbon content. The furnace contents should be tapped as soon as these additions have melted completely and are diffused through the bath.

Induction Furnace. The high frequency coreless induction furnace is used in the production of complex, high quality alloys such as tool steels (9). It is used also for remelting scrap from fine steels produced in arc furnaces, melting chrome–nickel alloys and high manganese scrap, and more recently, has been applied to vacuum steelmaking processes.

The induction furnace was first patented in Italy in 1877 as a low frequency furnace. It was first commercially applied, installed, and operated in Sweden. The first installation was made in 1914 by the American Iron and Steel Company in Lebanon, Pennsylvania; however, it was not successful. Other low frequency furnaces have been operated successfully, especially for stainless steel.

The first high frequency coreless induction furnaces were built and installed by the Ajax Electrothermic Corporation. Commercial production with this type was started in Sheffield, UK, in 1921. In the United States, it was first installed by the Heppenstall Forge and Knife Company in Pittsburgh, Pennsylvania, in 1928.

The high frequency induction furnace consists of a refractory crucible surrounded by a water-cooled copper coil through which an ac flows. The rapidly changing magnetic field of the coil at high flux density generates heavy secondary currents in the solid metal of the charge. Resistance of the metal to flow of the induced current generates heat which melts the charge.

Most commonly, the melting procedure is essentially a dead-melt process, that is, the solid constituents of the charge melt smoothly and mix with each other. Little if any refining is attempted in ordinary induction melting, and no chemical reactions occur whose gaseous products agitate the molten bath. The charge is selected to produce the composition desired in the finished steel with a minimum of further additions, except possibly small amounts of ferroalloys as final deoxidizers. For ordinary melting, the high frequency induction furnace is not used extensively. It is, however, employed in the production of complex high quality alloys such as alloy tool steels, and also for remelting scrap from fine steels produced in arc furnaces, as well as for melting chrome–nickel alloys and high manganese scrap. More recently, it has been applied to the vacuum steelmaking processes described below.

Vacuum and Atmosphere Melting. A coreless high frequency induction furnace of the type described above is enclosed in a container or tank which can be either evacuated or filled with a gaseous atmosphere of any desired composition or pressure. Provision is made for additions to the melt, and tilting the furnace to pour its contents into an ingot mold also enclosed in the tank or container without disturbing the vacuum or atmosphere in the tank (Fig. 5).

Although vacuum melting often has been employed as a remelting operation for very pure materials or for making electrodes for the vacuum consumable-electrode furnace described below, it is generally more useful in applications that include refining. Oxygen, nitrogen, and hydrogen are removed from the molten metal in vacuum melting, as well as carbon when alloys of very low carbon content are being produced.

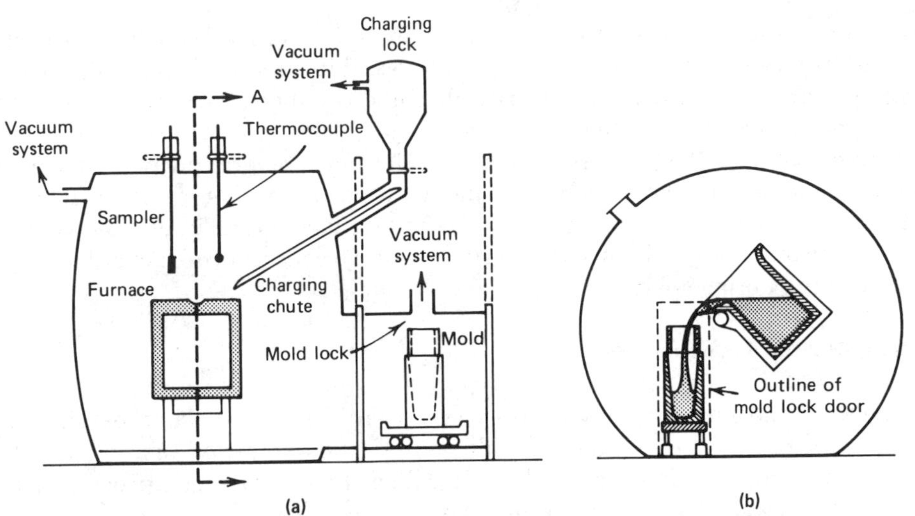

Figure 5. Schematic arrangement of a furnace in a vacuum chamber equipped with charging and mold locks for vacuum induction melting (1). (**a**) Front cross section; (**b**) section AA during pouring.

Consumable-Electrode Melting. This refining process produces special-quality alloy and stainless steels by casting or forging the steel into an electrode that is remelted and cast into an ingot in a vacuum (Fig. 6). These special steels include bearing steels, heat-resistant alloys, ultrahigh-strength missile and aircraft steels, and rotor steels.

A consumable-electrode furnace consists of a tank above ground level that encloses the electrode and a water-cooled copper mold below ground level. After the furnace has been evacuated, power is turned on and an arc is struck between the electrode and a starting block that is placed in the mold before operation begins. Heat from the arc progressively melts the end of the electrode. Melted metal is deposited in a shallow pool of molten metal on the top surface of the ingot being built up in the mold. Rate of descent of the electrode is automatically controlled to maintain the arc. The re-

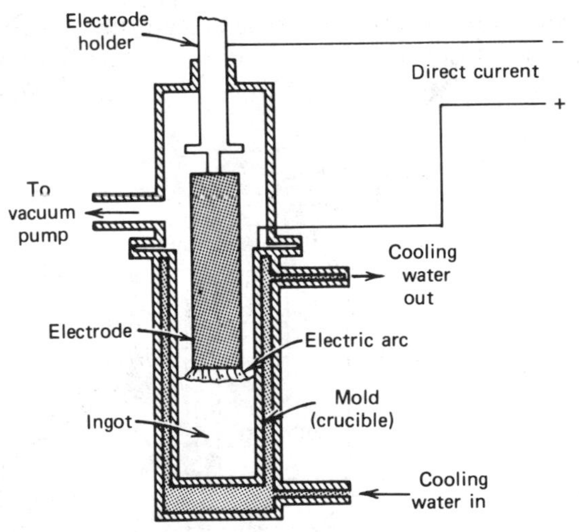

Figure 6. Schematic representation of the principle of design and operation of a consumable-electrode furnace for melting steels in a vacuum (1).

melting operation removes gases (hydrogen, oxygen, and nitrogen) from the steel, improves its cleanliness, produces an ingot that exhibits practically no center porosity or segregation, and improves hot workability and the mechanical properties of the steel at both room and high temperatures.

Electroslag Remelting. This process has the same general purpose as consumable-electrode melting, and a conventional air-melted ingot serves as a consumable electrode; no vacuum is employed. Melting takes place under a layer of slag which removes unwanted impurities. Grain structure and orientation are governed by controlled cooling during solidification.

Oxygen Steelmaking Processes

In oxygen steelmaking, 99.5% pure oxygen gas is mixed with hot metal, causing the oxidation of the excess carbon and silicon in the hot metal and thereby producing steel. In the United States, this process is called the basic oxygen process or BOP (11–23). The first U.S. commercial installation began operation in 1955.

Blowing with oxygen was investigated in the FRG and Switzerland some years before the first commercial steelmaking plants to use this method began operation in Austria in the early 1950s. This operation was designed to employ pig iron produced from local ores that were high in manganese and low in phosphorus. The process spread rapidly throughout the world.

Top-Blown Basic Oxygen Process. The top-blown basic oxygen process is conducted in a cylindrical furnace somewhat similar to a Bessemer converter. This furnace has a dished bottom without holes and a truncated cone-shaped top section wherein the mouth of the vessel is located. The furnace shell is made of steel plates ca 50 mm thick; it is lined with refractory 600–1200 mm thick.

A jet of gaseous oxygen is blown at high velocity onto the surface of a bath of molten pig iron at the bottom of the furnace by a vertical water-cooled retractable pipe or lance inserted through the mouth of the vessel (Fig. 7). The furnace is mounted in a trunnion ring and can be tilted backward or forward.

With the furnace tilted toward the charging floor, which is on a platform above ground level, solid scrap is dumped by an overhead crane into the mouth of the furnace. Scrap can form up to 30% of the charge unless it is preheated, when up to 45% may be used. The crane then moves away from the furnace and another crane carries a transfer

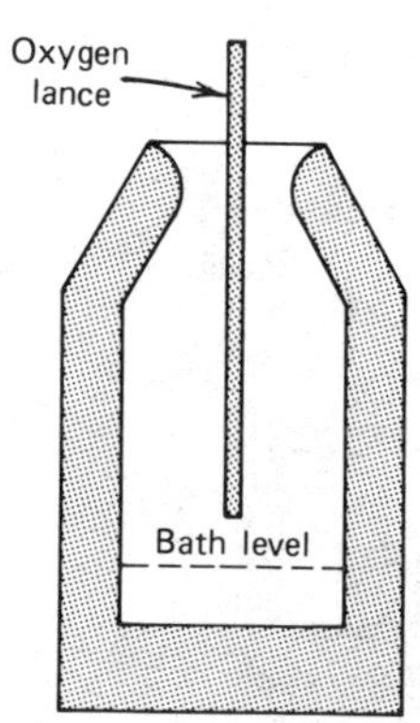

Figure 7. Sketch of a basic oxygen steelmaking furnace, with oxygen lance inserted through the mouth of the furnace (1).

ladle of molten pig iron to the furnace and pours the molten pig iron on top of the scrap (see Fig. 8).

The manganese residue of the blown metal before ladle additions is generally higher than in an open-hearth process and is closely related to the amount of manganese in the furnace charge.

High slag fluidity and good slag–metal contact promote transfer of phosphorus from the metal into the slag, even before the carbon reduction is complete.

Efficiency of sulfur elimination is as good as or better than that in the basic open-hearth process because the bath action is more vigorous, the operating temperatures are higher, and the fuel does not contain sulfur.

Because the basic oxygen process uses a refining agent containing practically no nitrogen, the product has a low nitrogen content. Oxygen residues are comparable to those obtained in open-hearth processes for steels of the same carbon content.

Residual alloying elements such as copper, nickel, or tin are usually considered undesirable. Their main source is purchased scrap. Because of the generally high consumption of hot metal in the basic oxygen process, the residual alloy content is low, since little purchased scrap is needed.

No external heat source is required. In fact, a main problem in the basic oxygen process is one of limiting temperature rather than increasing it. To minimize the

Figure 8. Pouring molten pig iron into a basic oxygen steelmaking furnace prior to the commencement of the production of a heat of steel. The furnace has been tilted to receive the molten metal.

amount of refractories consumed, the final temperature of the molten steel has to be controlled. If this temperature is higher than needed for the subsequent casting, refractory consumption is increased. To control the final temperature, sufficient solid-steel scrap or other cooling agents such as iron ore must be added to the charge.

The furnace is immediately returned to the upright position, the lance is lowered into it to the desired height above the bath, and the flow of oxygen is started. Striking the surface of the liquid bath, the oxygen immediately forms iron oxide, part of which disperses rapidly through the bath. Carbon monoxide generated by the reaction of iron oxide with carbon is evolved, giving rise to a violent circulation that accelerates refining.

Slag-forming fluxes, chiefly burnt lime, fluorspar, and mill scale (iron oxides), are added in controlled amounts from an overhead storage system shortly before or after the oxygen jet is started. These materials, which produce a slag of the proper basicity and fluidity, are added through a chute built into the side of a water-cooled hood positioned over the mouth of the furnace. This hood collects the gases and the dense reddish-brown fume emitted by the furnace during blowing, and conducts them to a cleaning system where the solids are removed from the effluent gas before it is discharged to the atmosphere. When no contaminants are present in the recovered solids, the solids can be added to the mix fed to sintering machines used to agglomerate fine ore, blast-furnace dust, and so on. Otherwise, the solids are disposed of by dumping.

The oxidizing reactions take place so rapidly, that a 300-metric-ton heat, for example, can be processed in ca 30 min. The intimate mixing of oxygen with the molten pig iron permits this rapid refining. Mixing oxygen with the molten pig iron gives a foamy emulsion of oxygen, carbon monoxide, slag, and molten metal which occupies a volume ca five times that of the slag and molten metal. The furnace must be big enough to accommodate this foamy mixture.

The mechanics of carbon elimination are similar to those of the open-hearth processes, ie, oxidation of carbon to carbon monoxide and carbon dioxide. The chemical reactions and results are the same in both cases. The progress of the reaction is plotted in Figure 9.

The general reaction in open-hearth technology, whereby silicon is oxidized to silica and transferred to the slag, also applies to the basic oxygen process. Oxidation of silicon is important mainly because of its thermal effects. Only a trace of silicon remains in the steel at the end of the refining period.

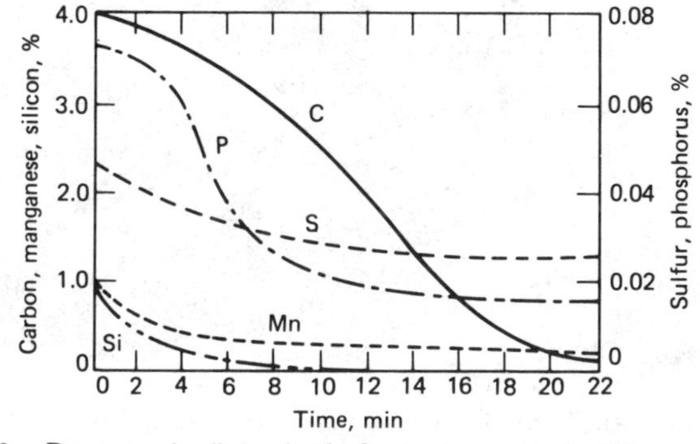

Figure 9. Progress of refining in the basic oxygen steelmaking process (1).

After the oxygen blowing is completed, the lance is withdrawn and the furnace is tilted to a horizontal position. A temperature reading is taken with an immersion-type thermocouple and a sample of steel is withdrawn for chemical analysis. If the steel is too hot, the furnace is returned to the vertical position and cooled by adding scrap or limestone through the chute in the hood. If the steel is too cold, the lance is again lowered and oxygen is blown for a short period.

When the temperature and composition are satisfactory, the furnace is tilted toward its taphole side and the steel is tapped into a waiting ladle; alloying additions are added to the steel in the ladle through a chute (Fig. 10).

After the steel has been tapped, the furnace is inverted by tilting toward its opposite side and the remaining slag is dumped into a slag pot. Then, the furnace is turned to charging position and is ready for the next heat.

Bottom-Blown Basic Oxygen Process. The bottom-blown basic oxygen process is also conducted in a furnace similar to a Bessemer converter. The furnace comprises two parts, the bottom and the barrel. Both the bottom and the barrel consist of an outer steel shell that is about 50 mm thick and lined with refractory that is 600–1200 mm thick. The bottom contains 6–24 tuyeres or double pipes (Fig. 11). Oxygen is blown into the furnace through the center pipes and natural gas or some other hydrocarbon is blown into the furnace through the annular space between the two pipes (24–29).

Considerable heat is generated when the oxygen gas oxidizes the carbon, silicon,

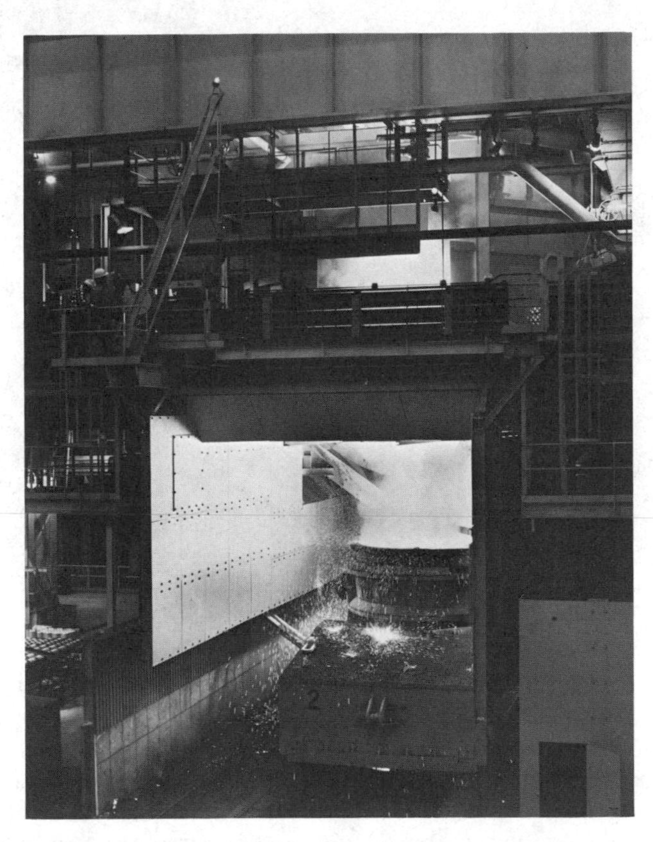

Figure 10. Tapping liquid steel from BOP furnace into a ladle.

Figure 11. Liquid blast-furnace iron (hot metal) is being charged into a 200-metric ton Q-BOP furnace in preparation for making a heat of steel.

and iron in the molten pig iron. If special precautions are not taken, the temperature of the refractory adjacent to the oxygen inlet increases when the oxygen is blown upward through the furnace bottom, as in a Bessemer converter. No commercial refractory has been found that withstands such temperatures without cooling. Surrounding the oxygen jet with a cylindrical stream of a hydrocarbon such as methane imparts the required cooling by thermal decomposition of the hydrocarbon (Fig. 12). The bottom-blown oxygen process is based on this technique.

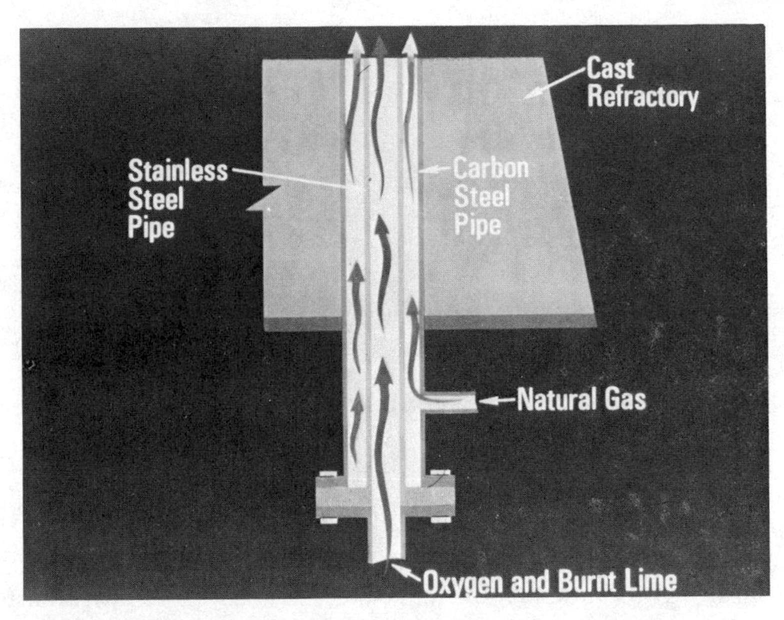

Figure 12. Tuyere for bottom-blown basic oxygen process (5).

In Europe, this process is called the OBM process and it is generally used with high phosphorus (1.5–2.7%) hot metal which does not permit the use of the catch-carbon technique (see below). Instead, all heats of steel are blown with oxygen until the carbon content is decreased to about 0.05%; the phosphorus is then removed by the oxygen, and the steel is tapped into a ladle where carbon and ferroalloys are added to obtain the desired composition.

In the United States and Japan, this process is called the Q-BOP process and is used with low phosphorus (usually <0.2%) hot metal, which permits the use of the catch-carbon practice described below (see Fig. 11).

Both the OBM and the Q-BOP processes are operated in about the same way as the top-blown process. The furnace is charged with steel scrap and hot metal and oxygen is blown into this mixture to produce steel of the desired composition. However, the method of adding lime to the furnace is different. In the bottom-blown processes, powdered lime is added to the oxygen before it is blown into the furnace. Thus, the oxygen is used as a carrier gas to pneumatically transport the powdered lime into the furnace. The powdered lime dissolves rapidly in the slag, permitting increased oxygen-blowing flow rate and thus a decrease in blowing time. Furthermore, slopping is avoided, ie, the sudden ejection of appreciable amounts of slag and metal from the furnace. In the top-blown process, pebble-sized lime is used with pieces 2.5–7.5 cm in diameter.

The chemistry of the bottom-blown processes is similar to that of the top-blown process. However, the slags in the bottom-blown processes contain significantly less iron oxide causing a ca 2% increase in yield, that is, in the amount of liquid steel produced from a given charge of hot metal and scrap. This increase in yield contributes greatly to decreasing the costs. The lower iron oxide content appears to be the result of better mixing of slag and metal.

The decrease in the iron oxide content of the slag also causes an increase in the amount of manganese remaining in the liquid steel at the end of the oxygenation stage ("oxygen blow"). Manganese is desirable in the finished steel and is almost always added as the steel is tapped from the furnace into a ladle. Thus, a smaller amount of manganese is added to bottom-blown steel during tapping, which results in decreased costs.

However, the low iron oxide content of the slag increases the slag viscosity, and liquid steel is entrapped in the slag unless sufficient fluorspar or other slag fluidizer is added to the charge.

Another difference between the top- and bottom-blown oxygen processes is the size of scrap that can be melted. When liquid steel is cast into ingots, occasionally an imperfect ingot is produced which is scrapped instead of being rolled to the desired product. The last ingot made from a heat of steel is almost always too short to be properly rolled. It is called an ingot butt and must be scrapped. In the top-blown process, it is generally impossible to completely melt ingot butts and condemned ingots during the normal oxygenation time of 15–20 min. The liquid steel bath is not agitated enough for sufficient heat to be transferred to the ingot, usually 50–75 cm thick, to melt within 15–20 min. However, in the bottom-blown process, ingot butts and condemned ingots are readily melted during the normal oxygenation time of 12–15 min because increased agitation within the liquid steel bath improves heat transfer to the ingot.

Top- and Bottom-Blown Basic Oxygen Processes. During the 1970s several combinations of top and bottom blowing were developed. In the lance-bubbling-equilibrium (LBE) process, nitrogen or argon is injected through a number of porous refractory plugs installed in the furnace bottom, while oxygen is top-blown into the furnace through a lance. The bottom injection of nitrogen or argon causes more intimate mixing of slag and metal, and hence most of the advantages of the Q-BOP bottom-blowing process are obtained in furnaces designed for top blowing. The LD-KG process is similar to the LBE process but the nitrogen or argon is injected through tuyeres in the furnace bottom rather than through porous refractory plugs (30–32).

Catch-Carbon Technique. In many countries, including the United States, the iron ore used in blast furnaces to make molten pig iron or hot metal contains less than about 0.2% phosphorus. Using such low phosphorus hot metal simplifies the steelmaking operation. Commercial steels are generally specified to contain no more than ca 0.025% phosphorus, primarily to increase the ductility of the steel. In the basic oxygen process, some phosphorus is removed with the carbon. With low phosphorus hot metal, medium carbon (0.20–0.40%) and high carbon (0.45–1.00% or higher) steels are made by the catch-carbon technique, in which the oxygenation stage is stopped when the desired carbon content is reached. In general, sufficient phosphorus is removed to meet specifications.

In Europe, high phosphorus iron ore is used and hot metal containing 1.5–2.0% phosphorus is produced. When this high phosphorus hot metal is used in the basic oxygen process, the catch-carbon technique is excluded. Instead, the hot metal must be blown with oxygen until the carbon content is decreased to ca 0.05%, at which point sufficient phosphorus is removed to meet the steel phosphorus specification.

Charge Control. A wide variety of carbon and alloy steels is made with the basic oxygen process, with carbon contents ranging between 0.02 and 1.00% or higher. For this reason, the amount of carbon to be removed varies over a substantial range, which affects the amounts of oxygen to be used and heat generated. The only source of heat in this process is the reaction of oxygen with silicon, carbon, manganese, iron, and phosphorus from both the hot metal and the steel scrap. Because it is advantageous to oxygenate a heat of steel only once, it is desirable to charge the correct amounts of hot metal, scrap, and oxygen, giving the steel the desired temperature and carbon content at the end of the oxygenation. This aim is accomplished in more than 50% of the heats made by the catch-carbon technique even though the temperature and carbon content limits are as close as ±8°C and ±0.05%, respectively.

Because the final carbon content, the hot-metal composition, and the temperature vary from heat to heat, a different combination of hot metal, scrap, and oxygen is required for each heat. The proper amounts of hot metal, scrap, and oxygen are governed by the heat and materials requirements of the process. The final slag composition and the average composition of the carbon monoxide–carbon dioxide gas mixture that is produced when the oxygen reacts with carbon from the hot metal can be predicted mathematically. A digital-computer program can also be employed for solution of the pertinent equations. A computer program substantially increases the heats made with only one oxygen blow, thereby increasing the productivity of the process (11).

Determination of Chemical Composition

The chemical composition of a given steel is generally specified by the customer within rather narrow limits for each element (other than residual elements). The chemical composition of the steel at various stages of refining is followed closely, since the carbon content dictates the course of refining.

In the early days of steelmaking, experienced melters could estimate the carbon content fairly closely by examining the fracture of a broken sample of the steel that had been taken molten from the furnace and poured into a mold. This empirical method was not good enough as specifications became more rigid and steelmaking processes took less time. Considerable effort was devoted to the development of rapid analytical methods for the determination of carbon, manganese, sulfur, silicon, and phosphorus. Today, analytical instruments are stationed near the furnaces for rapid carbon analysis during the refining operation.

Analytical methods for determining other elements, eg, manganese, silicon, phosphorus, and sulfur, still require that samples be analyzed in a chemical laboratory, sometimes located at a considerable distance from the furnaces.

At the present time, the spectograph is preferred for rapid analysis. Samples are sent to the laboratory by pneumatic tube or other fast means, the samples are analyzed, and the results are transmitted to the furnace area by teletype. A dozen elements or more can be determined quickly and accurately by this method.

Scrap as Raw Material

Scrap consists of the by-products of steel fabrication and wornout, broken, or discarded articles containing iron or steel (33) (see Recycling). It is a principal source of the iron for steelmaking; the other source is iron from blast furnaces, either molten as it comes from the furnace (hot metal) or in solid pig form (see Table 1). Scrap is of great practical value. Every ton of scrap consumed in steelmaking is estimated to displace and conserve for future use 3.5–4 t of natural resources including iron ore, coal, and limestone. On the average, the steel industry consumes ca one-fifth more pig iron (mostly as hot metal) than scrap. According to the American Iron and Steel

Table 1. U.S. Consumption of Scrap, Hot Metal and Pig Iron by Steelmaking Processes and Blast Furnaces in 1980[a]

Process	1,000 Metric tons	
	Scrap	Hot metal and pig iron
open-hearth furnaces	6,705	7,644
basic oxygen process	19,819	71,014
electric furnaces	30,110	690
blast furnaces	3,322	

[a] Ref. 5.

Institute (AISI), the U.S. steel industry consumes an average of about 56×10^6 t iron and steel scrap for the production of 100×10^6 t raw steel. The basic oxygen steelmaking process uses 15–30% scrap (up to 45% with preheating); the open-hearth processes 35–60%. Electric-arc furnaces are charged almost entirely with scrap.

Home Scrap. Unsalable products, which result unavoidably in the course of steelmaking and finishing operations, are termed home scrap or revert scrap. These include pit scrap; ingots too short to roll; rejected ingots; crop ends from slabs, blooms, and billets; shear cuttings from trimming flat-rolled products; pieces damaged in handling and finishing; ends cut from bars, pipe, and tubing, and so on.

In general, ca 30×10^6 tons of home scrap would result from the manufacture of 100×10^6 tons of raw steel and the processing of this steel into finished products. Since, as stated earlier, 55×10^6 tons of scrap is required to produce 100×10^6 tons of raw steel, 25×10^6 tons of purchased scrap would have to be used to supplement 30×10^6 tons of home scrap.

Purchased Scrap. ***Dormant Scrap.*** Such scrap comprises obsolete, worn-out, or broken products of the consuming industries, eg, discarded steel furniture, washing machines, stoves, and other outdated consumer goods; beams, angles, channels, girders, railings, grilles, pipe, etc, arising from the demolition of buildings; farm machinery; broken or damaged industrial machinery; old ships; railroad rails and rolling stock; and junked automobiles. This type of scrap, because of its miscellaneous nature, requires careful sorting and classification to prevent the contamination of steel in the furnace with unwanted chemical elements. The scrap should be small enough to facilitate handling and charging into the furnace. The need for proper classification and preparation of dormant scrap is emphasized by the existence of over 70 different specifications covering various grades of scrap. In addition, the Association of American Railroads has 45 specifications applying to railroad scrap.

Junked automobiles represent enormous resource of steel scrap. Formerly, after stripping, entire bodies were squeezed into compact blocks in huge hydraulic presses. However, the expense of collection and stripping them of copper, lead, chromium, nickel, and other unwanted metals and combustibles gradually reduced the profitability of junked cars. Nevertheless, in recent times, partly for esthetic reasons, a concerted effort has been made to empty the auto graveyards and dispose of abandoned cars. After removal of engines, transmissions, etc, autobodies are fed into huge machines called shredders that cut the body into small pieces. The shredded material is then passed through magnetic separating equipment which discards the nonferrous material to produce a high quality steel scrap (see Magnetic separation). Other methods are available. A recently developed device can crush an engine block into sortable chunks in 42 s.

Prompt Industrial Scrap. Prompt industrial scrap is generated by the steel consumers. It may consist of the unwanted portions of plate or sheet, trimmings resulting from stamping and pressing operations, machine turnings, rejected products scrapped during manufacture, short ends, flashes from forgings, and other types of scrap. Prompt industrial scrap can usually be identified easily as to source and composition, provided proper segregation has been maintained in its handling.

Chemical Composition. Scrap should be free of unknown and unwanted elements referred to as tramp alloys. The increasing use of alloy steels aggravates the tramp-alloy problem, since more and more purchased scrap may be expected to include unidentified alloy steels.

The segregation of home scrap according to its chemical composition is relatively simple. Purchased scrap, especially dormant scrap, presents some problems because a large percentage is of unknown origin and composition. Chemical analysis of selected samples from individual lots is sometimes employed in scrap classification. Spectrographic analysis is faster than chemical analysis; however, both are relatively time-consuming and expensive and require careful selection and preparation of samples. Some less expensive but less accurate tests are commonly used including magnetic tests, spark tests, spot tests, and pellet tests.

Alloying Elements. Scrap can be a valuable source of alloying elements, and full advantage is taken of this source in the production of alloy steels in electric furnaces. However, since most open-hearth production consists of carbon and low alloy steels, alloying elements in scrap are generally a source of trouble.

Tin, copper, nickel, and other elements found in scrap alloy readily with steel and, in many cases, render it unfit for use. Relatively small amounts of these metals can contaminate an entire heat of steel. Tin and copper in certain ranges of composition cause brittleness and bad surface conditions. Nickel and tin contaminate heats and may also leave a residue in the furnace that is absorbed by subsequent heats. Lead is extremely harmful to furnace bottoms and refractories and, if present in sufficient quantities, may cause furnaces to crack.

Addition Agents

In steelmaking, various elements are added to the molten metal to effect deoxidation, control of grain size, improvement of the mechanical, physical, thermal and corrosion properties, and other specific results. Originally, the chemical element to be incorporated into the steel was added to the bath in the form of an alloy that consisted mainly of iron but was rich in the desired element. Such alloys, because of their high iron content, became known as ferroalloys, and were mostly produced in iron blast furnaces. Later, the production of alloys for steelmaking was carried out in electric-reduction and other types of furnaces, and a number of alloys now produced contain very little iron. For this reason, the term addition agent is preferred to describe the materials added to molten steel for altering its composition or properties; ferroalloys are a special class of addition agents.

Included in the ferroalloy class are alloys of iron with aluminum, boron, calcium, chromium, niobium, manganese, molydenum, nitrogen, phosphorus, selenium, silicon, tantalum, titanium, tungsten, vanadium, and zirconium. Some of these chemical elements are available in addition agents that are not ferroalloys, as well as in almost pure form; these include relatively pure metals such as aluminum, calcium, cobalt, copper, manganese, and nickel; oxides of molybdenum, nickel, and tungsten; carbon, nitrogen, and sulfur in various forms; and alloys consisting principally of combinations of two or more of the foregoing elements. Some rare-earth alloys are also used for special purposes, but to a minor extent.

Addition agents may be added to the charge in the furnace, the molten bath near the end of the finishing period, the ladle, or the molds. Timing of the alloy additions depends upon the effect on the temperature of the molten metal, the ease with which specific addition agents go into the solution, the susceptibility of a particular addition agent to oxidation, and the formation and removal of reaction products.

The economical manufacture of alloy steels requires consideration of the relative affinity of the alloying elements for oxygen as compared with the affinity of iron for oxygen. For example, copper, molybdenum, or nickel may be added with the charge or during the working of the heat and are fully recovered. Chromium and manganese, because they are easily oxidized, should be added late in the heat, and all or part of these two may be added in the ladle. Easily oxidized materials such as aluminum, boron, titanium, vanadium, and zirconium are added in the ladle in order to minimize oxidation losses.

Chill Effects. Admixture of an addition agent at ambient temperature with liquid steel lowers the temperature. The addition agent absorbs heat as its temperature is increased and as it melts to dissolve in the liquid steel. The most notable exception to this rule is silicon. When silicon dissolves in liquid steel, more heat is evolved than is required to heat the solid silicon from ambient temperatures to the liquid-steel temperature and to melt the silicon; the liquid-steel temperature thus rises.

Knowledge of these thermal or chilling effects of addition agents permits determination of the most economical combination of addition agents for making each grade of steel. The chilling effects of various addition agents are listed in Table 2.

Table 2. Chill Values of Various Addition Agents[a]

Addition agent	Chemical composition, wt %	Chill value, °C for 1 kg per metric ton steel
aluminum	95 Al–2 Cu–1 Fe–1 Mn–1 Si	−0.12
carbon, graphite	100 C	+0.04
high carbon ferrochrome	68 Cr–25 Fe–5 C–1.5 Si–0.5 Mn	+2.50
cobalt	100 Co	+1.40
medium carbon ferromanganese	81 Mn–16 Fe–1.5 C–1.5 Si	+1.90
high carbon ferromanganese	76 Mn–16 Fe–7 C–1 Si	+2.28
nickel	100 Ni	+1.30
silicomanganese	61 Mn–17 Si–14 Fe–2 C	+1.78
ferrophosphorus	71.5 Fe–24 P–3 Mn–1.5 Si	+2.60
ferrosilicon, %		
10	88 Fe–10 Si–2 C	+1.70
50	50 Si–48 Fe–2 Al	+0.74
75	75 Si–23 Fe–2 Al	−0.52
steel scrap	100 Fe	+1.64
sulfur	100 S	−1.46

[a] Ref. 34.

Ladle Metallurgy

The finished steel, from whatever type of furnace (basic oxygen, open-hearth, or electric) is tapped into ladles. Most ladles hold all the steel produced in one furnace heat. Some slag is allowed to float on the surface of the steel in the ladle to form a protective blanket. Excess slag flows from the ladle through a spout and is either collected in pots or allowed to run onto the floor, where it solidifies and is removed.

A ladle consists of a steel shell, lined with refractory brick, with an off-center opening in its bottom equipped with a nozzle (Fig. 13). A stopper-rod assembly makes it possible to enlarge or close the opening to control the flow of steel through the nozzle.

Ladle metallurgy, the treatment of liquid steel in the ladle, is a rapidly growing field in which several new processes, or new combinations of old processes, are developed every year (35–40) with the following objectives: removal of sulfur, oxygen, hydrogen, and carbon; addition of ferroalloys with very high recoveries; and decrease or increase of liquid-steel temperature to meet temperature specifications for continuous casting.

Argon Treatment. ***Ladle-Without-Cover.*** In argon stirring, argon gas is passed through liquid steel in a ladle in order to mix ferroalloys with the steel, homogenize the steel with respect to chemical composition and temperature, accelerate cooling, and remove oxide and sulfide inclusions. The use of argon in treating liquid steel in the ladle has greatly enhanced the flexibility of steelmaking operations, and significantly improving the surface and internal quality of the steel.

CAS process. CAS is an acronym for composition adjustment by sealed argon bubbling. An enlarged chute or immersion tube is dipped into the surface of the liquid steel in the area through which argon bubbles are emerging and pushing the slag away. The argon is injected into the liquid steel through a porous plug in the ladle bottom. Ferroalloys are added to the liquid steel in the ladle through the chute or immersion tube. The main objective is to decrease ferroalloy consumption by adding ferroalloys to the liquid steel under the protection of the argon gas that fills the immersion tube.

The SAB process (sealed argon bubbling) is very similar to the CAS process, but a synthetic slag of CaO, Al_2O_3, and SiO_2 is placed on the liquid steel inside the immersion tube to make clean steel.

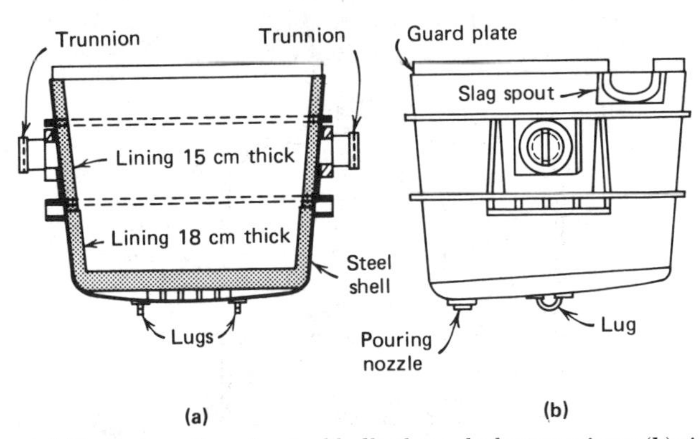

Figure 13. (**a**) Vertical section of a steel ladle through the trunnions; (**b**) side view (1).

AOD process. AOD is an acronym for argon–oxygen decarburization. In this process, liquid steel is tapped from the furnace into a ladle and then poured into the AOD vessel. Either argon–oxygen gas mixtures or pure argon are blown into the steel through tuyeres in or near the vessel bottom. Although the AOD process primarily removes carbon from stainless steel, it also removes carbon and inclusions from carbon and alloy steels.

Ladle-With-Cover. In these processes, argon gas is injected either through a lance inserted downward into the steel or through a porous plug in the ladle bottom. The injection lance is used in the TN (Thyssen Niederrhein) and KIP (Kimitsu injection) processes, both designed primarily to remove sulfur. In these processes, it is very important to prevent the steelmaking slag from accompanying the liquid steel as it is tapped into the ladle because it contains iron oxide which is detrimental to sulfur removal. In the TN process, either CaSi or Mg is injected into the steel; in the KIP process, a mixture of 90% CaO–10% CaF_2 is used. The ladle cover keeps air away, and should fit so closely that the space above the steel becomes filled with argon, thereby preventing oxidation. Both processes produce steel with low sulfur and oxygen contents.

CAB Process. The best known ladle-with-cover process using the porous plug for argon injection is the CAB (capped argon bubbling) process. In this process, steelmaking slag is kept out of the ladle and a 40% CaO–40% SiO_2–20% Al_2O_3 synthetic slag is added. The ladle is covered and argon gas bubbled through the porous plug and through the steel. Ferroalloys are added through a chute in the ladle cover. This process produces cleaner steel and requires less ferroalloy.

Vacuum Processes. The more complex operation of treating liquid steel by exposure to vacuum is accomplished in many processes, including the stream degassing, DH, RH, RH-OB, RH-FR, and VOD processes. Vacuum may be used to decrease the hydrogen content to 2 ppm or less, thereby preventing hydrogen embrittlement in such steel products as rails and rotors for electrical generators. Vacuum may also be used to remove carbon and add ferroalloys under nonoxidizing conditions.

Stream-Degassing Process. Stream degassing is accomplished by placing an empty ladle (or mold) in a tank (see Fig. 14). A bottom-pour ladle containing the molten steel to be degassed is set upon the evacuated tank; the bottom of the ladle and the top of

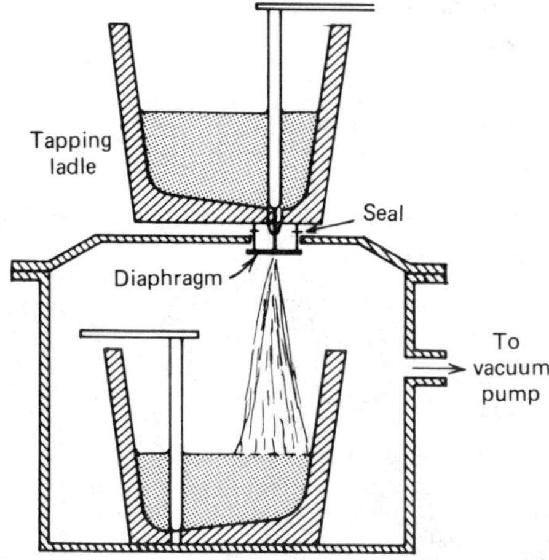

Figure 14. Schematic arrangement of equipment for stream degassing (1).

the tank are equipped with mating seals to exclude air. When the stopper rod of the tapping ladle is raised, molten metal flows through the nozzle, melts a metal diaphragm that seals the opening to the tank, and passes into the ladle (or mold) in the vacuum tank. As the stream of molten metal enters the evacuated space, it breaks up into tiny droplets, exposing an enormous surface to vacuum degassing. After the tank is purged, its internal pressure is raised to 101.3 kPa (1 atm) and the steel removed and teemed (poured) in the usual manner.

DH Process. In the DH (Dortmund-Hoerder) process, a refractory-lined chamber with one hollow leg or pipe extending from the bottom is inserted into the liquid steel in a ladle (see Fig. 15). After the chamber is evacuated, liquid steel is moved back and forth between the ladle and the chamber by moving them together and apart. By repeating this cycle 10–20 times, the liquid steel is exposed to vacuum each time it is drawn into the chamber.

RH Process. In the RH (Ruhrstahl-Heraeus) process, a refractory-lined chamber similar to the DH degasser is used, but with two hollow legs instead of one. Liquid steel moves continuously from the ladle, up through one leg into which argon gas in injected, into the vacuum chamber, down the other leg, and back into the ladle. The injection of argon gas in one leg provides the means to move the liquid steel. In both the DH and RH degassing processes, the vacuum in the vacuum chamber is a good industrial vacuum, <130 Pa (1 mm Hg). Treatment times are usually ca 20 min.

In the RH light-treatment process, a relatively poor industrial vacuum is used to save energy. Starting at a pressure of 101.3 kPa (1 atm), the pressure in the vacuum chamber is gradually reduced to ca 20 kPa (150 mm Hg) in 10 min. During this time, the oxygen dissolved in the liquid steel reacts with carbon to lower the carbon content from 0.08 to 0.05% and the oxygen content from 400 to ca 50 ppm. Aluminum is then added to produce an aluminum-killed steel. Very little Al_2O_3 is formed, because the oxygen content of the steel is low when the aluminum is added. This technique produces a cleaner steel and requires less aluminum. After the aluminum addition, the pressure in the vacuum chamber falls to ca 4 kPa (30 mm Hg) at the end of the treatment. Because the steel is tapped from the BOP furnace at ca 0.08% rather than 0.05% carbon, the loss of iron as iron oxide in the slag is somewhat less and the yield of steel is increased.

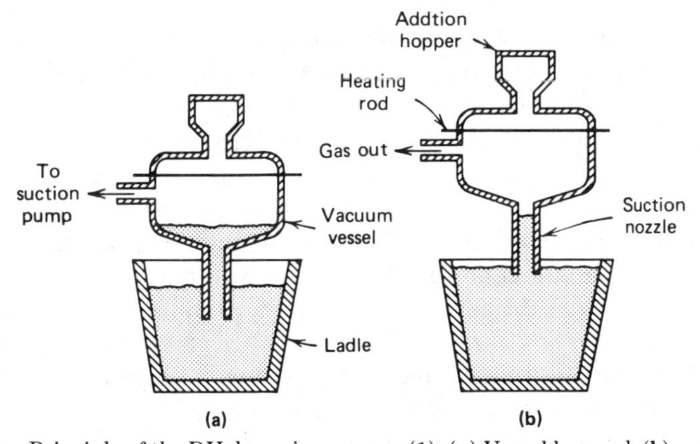

Figure 15. Principle of the DH degassing process (1). (**a**) Vessel lowered; (**b**) vessel raised.

RH-OB Process. The RH-OB (Ruhrstahl-Heraeus oxygen blowing) process co-bines the RH vacuum degassing process with oxygen blowing. The oxygen is injected into the liquid steel in the vacuum chamber through double tuyeres, which are protected from burning by the use of argon and kerosene. With this process, steel can be tapped from the BOP furnace at about 0.10% carbon, further increasing the yield of steel and recovery of ferroalloys. The RH-OB process produces cleaner steel than conventional processing. The pressures in the vacuum chamber in the RH-OB process are similar to those in the RH light-treatment process.

VOD Process. The VOD (vacuum-oxygen-decarburization) process was originally developed for producing stainless steel but has also been used to refine carbon steels. The liquid steel is tapped into a ladle into which argon is injected through a porous plug in the bottom. The ladle is placed in a tank which is then evacuated to ca 2.6 kPa (20 mm Hg), and oxygen blown through a lance onto the steel to lower the carbon content to ca 0.01%.

Electric-Arc Processes. ***LF Process.*** In the LF (ladle–furnace) process, the liquid steel is refined by argon stirring with a synthetic slag and heated by an electric arc; vacuum is not used. This process produces cleaner steel, and removes oxygen and sulfur.

VAD Process. In the VAD (vacuum-arc degassing) process, the liquid steel is refined by heating under vacuum while argon is injected through a porous plug in the ladle bottom. The ladle is placed in a tank which is evacuated. The tank cover has openings through which the three graphite electrodes extend onto the surface of liquid steel. The steel is refined by removal of hydrogen, oxygen, and sulfur.

ASEA-SKF Process. In the ASEA-SKF process, named for two Swedish companies, the liquid steel is tapped into a ladle, and the liquid steel exposed to vacuum while argon is injected through a porous plug. The liquid steel is also stirred by means of an electromagnetic stirring coil located outside of the ladle. The vacuum is then removed and a cover with three graphite electrodes placed over the ladle to heat the steel. This process removes hydrogen, oxygen, and sulfur.

Ingots

The ladle is carried by an overhead crane to a pouring platform where the steel is teemed into a series of molds, where it solidifies to form castings called ingots (41–45).

Until recently, the ingot was the first step in steel processing in the sequence of rolling and other operations required to make finished products. Most steel produced today is still cast into ingots. After removal from the mold by a process called stripping, the ingot is placed with other ingots in a pit-type furnace called a soaking pit where it is brought to the proper temperature for hot working, ie, rolling on primary rolling mills into semifinished products called blooms, slabs, and billets. These products form the starting material for further hot rolling on secondary and finishing mills. Today, some blooms, slabs, and billets are produced by continuous casting (see below).

The size and shape of an ingot depend upon the product to be made and the type of equipment available for hot working. For example, ingots for rolling slabs from which flat-rolled products such as plates and sheets are made range in weight from 10–40 t, with many in the 20-t range. Some forging ingots weigh 300 t.

Ingot molds are made of cast iron. Their cross sections may be square, rectangular, or round. The mold cavity is tapered from top to bottom to facilitate removal of the ingot.

As a mold is being filled with molten steel, the metal next to the mold walls and mold stool (a cast base on which the mold rests) is chilled by contact with the relatively cold surfaces and solidifies to form a shell or skin. Early during solidification, the ingot skin contracts and forms an air gap between itself and the mold walls. This gap reduces the rate at which heat can be transferred to the mold and thence to the atmosphere. As solidification proceeds, the thermal gradients become less steep. For these reasons, the thickness of the ingot skin (frozen zone) increases rapidly at first but much more slowly as solidification proceeds. During solidification, the thickness of the ingot in millimeters is about equal to 25 times the square root of the solidification time in minutes.

Liquid steel readily dissolves oxygen up to a solubility limit of about 0.2%. When carbon is present, as it usually is, this solubility is decreased. The product of the carbon and oxygen contents in percent is equal to ca 0.002 under normal conditions in steel-making furnaces. If the carbon content is 0.10%, the solubility limit of oxygen is 0.002/0.10 or 0.02% oxygen, assuming no silicon or aluminum is present. However, the solubility of oxygen in solid steel is ca 0.005% at the freezing point and decreases further as the temperature decreases. If the liquid steel is saturated with much more than 0.005% oxygen, most of the oxygen is expelled as the steel freezes. If carbon is present in the liquid steel and silicon and aluminum are not, the oxygen escaping from the solution combines with the carbon to form carbon monoxide bubbles. The addition of deoxidizing agents such as silicon and aluminum to the liquid steel reduces the oxygen content. The degree of deoxidization (silicon and aluminum contents) establishes three types of steel: killed, semikilled, and rimmed (41).

Molten steel does not solidify at one definite temperature but over a temperature range. As a result, the gases evolved from still liquid portions may be trapped at solid–liquid interfaces to produce blowholes. In Figure 16, eight typical conditions in commercial ingots are shown, cast in identical bottle-top molds, in relation to the degree of gas evolution. The dotted line indicates the height to which steel was poured originally. Figure 16(**a**) represents a fully killed ingot that evolved no gas because it was completely deoxidized. Its top is slightly concave, and directly below the top is an intermittently bridged shrinkage cavity called pipe. Actually, fully killed steels are almost always poured in big-end-up molds, equipped with refractory hot tops, exothermic inserts, or some other means for maintaining the steel at the top of the ingot molten long enough for the pipe to be kept in a small region at the top of the ingot that can be discarded later with a minimum of waste (Fig. 17).

Figure 16(**b**) shows a typical semikilled ingot in which only a small amount of gas was evolved. Nevertheless, the volume of the resulting blowholes compensated fully for shrinkage during solidification. Ferrostatic pressure (pressure exerted by liquid steel owing to gravity) prevented the formation of blowholes in the lower half of the ingot. The pressure of the gases trapped in the blowholes caused the surface of the ingot to bulge.

Figure 16(**e**) represents a typical capped ingot. It evolved so much gas that the resulting strong upward currents along the sides in the upper half of the ingot swept away the gas bubbles that otherwise would have formed blowholes. Even in the lower half of the ingot, the blowholes could not form until the gas evolution had moderated. As a result, a thick solid skin formed first, followed by the zone containing the blow-

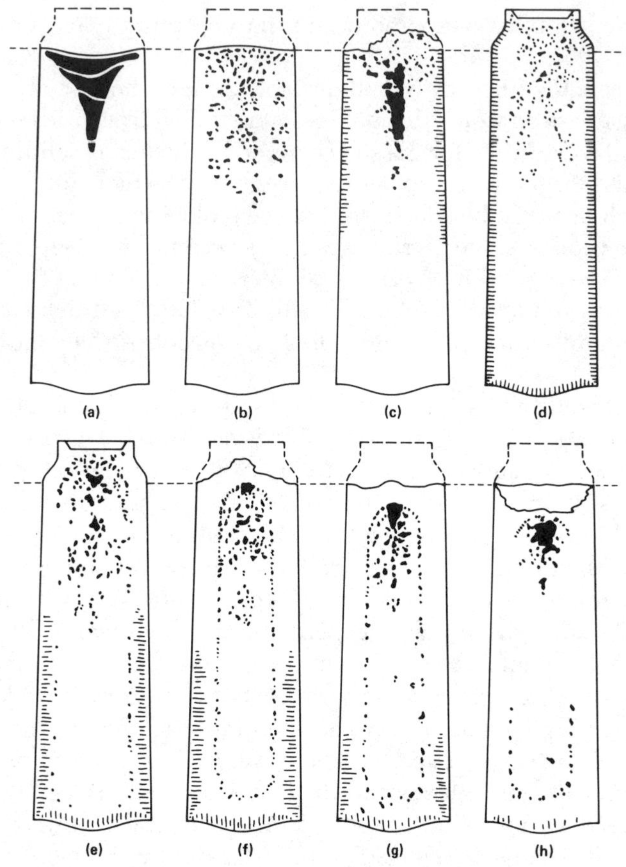

Figure 16. Series of ingot structures.

holes. The increase in the apparent volume of the ingot was sufficient to cause the steel to rise to the cap at the top of the mold.

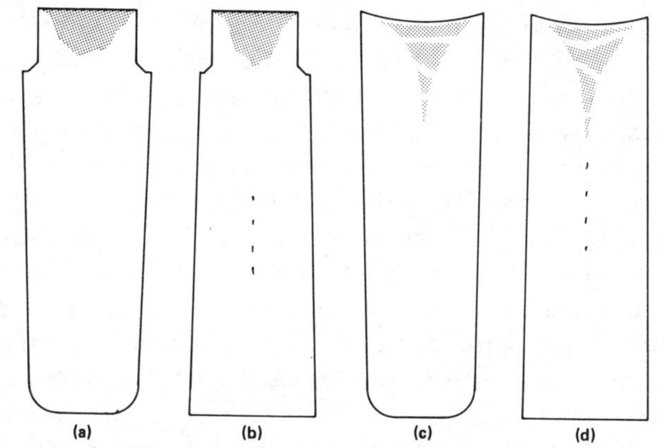

Figure 17. Types of killed ingots. (**a**) Big-end-up, hot-topped; (**b**) big-end-down, hot-topped; (**c**) big-end-up, not hot-topped; (**d**) big-end-down, not hot-topped.

Figures 16(**f**)–(**h**) show rimmed ingots. In the ingot in Figure 16(**f**), the evolution of gas, although greater than in the ingot in Figure 16(**e**), was insufficient to prevent the honeycomb blowholes from exceeding in volume the amount required to offset solidification shrinkage, and the top surface of the ingot rose slightly as it froze in from the sides of the mold.

Figure 16(**g**) represents a typical rimmed ingot in which gas evolution was so strong that the formation of blowholes was confined to the lower quarter of the ingot. The apparent increase in volume caused by blowholes offset the shrinkage that occurred during solidification. As a result, the top of the ingot did not rise or fall appreciably during solidification.

Figure 16(**h**) illustrates a violently rimming ingot, typical of low-metalloid steel. Honeycomb blowholes could not form, and the top surface of the ingot fell markedly during solidification.

From the foregoing, it is seen that, in all except killed steels, the evolution of gas produces cavities of roughly cylindrical shape (skin or honeycomb blowholes) or spherical shape (located deeper in the ingot). Blowholes serve a useful purpose in diminishing or preventing the formation of pipe and increasing the amount of usable steel in an ingot. Blowholes, unless located within ca 10 cm of the top, tend to have interiors free of oxide coating and clean enough to weld easily and become completely closed during rolling. If the blowholes extend to the ingot surface or lie so near to the surface as to become exposed by scaling of the ingot surface during heating for rolling, they can become oxidized and do not weld, giving rise to seams in the rolled product. In properly made ingots, the gas evolution is so controlled that a skin of adequate thickness covers the blowholes closest to the surface.

Segregation. When a liquid consisting of more than one component freezes, the solid is seldom uniform in composition; this is termed segregation. When liquid steel freezes, it generally exhibits positive segregation in which the part that freezes first is highest in iron content and the part that freezes last is lowest. Segregation is a complex phenomenon that involves numerous factors, including the densities of the liquid and solid steel, the effects of the alloying elements present at the freezing point, the mode of freezing (including whether heat is removed in more than one direction) and the extent to which the remaining liquid moves with respect to the solidified portion during freezing. In general, the amount of segregation is more or less proportional to the size of the solid that is formed. Continuous-cast slabs 200 mm thick have much less segregation than ingots 750 mm thick. Those locations in the cast ingot or slab where the content of an element is above average are said to exhibit positive segregation; when the content is below average, it is termed negative segregation.

Some elements tend to segregate more than others; sulfur segregates most. The following elements, in descending order, also segregate, but to a smaller degree: phosphorus, carbon, silicon, and manganese. The tendency for elements to segregate is increased by increasing time for solidification. Therefore, large ingots tend to show more segregation than smaller ones.

Turbulence caused by gas evolution during solidification tends to increase segregation. Therefore, killed are less segregated than semikilled steels, and semikilled are less segregated than capped or rimmed steels. In a rimmed ingot, the first part of the steel to freeze and form the skin is low in carbon, phosphorus, and sulfur; that is, it exhibits negative segregation. The core of the ingot, or the final portion to freeze, exhibits positive segregation. The boundary between the rim and core zones of a rimmed ingot is very sharp, and these zones are so different with respect to chemical composition that they seem to be different steels.

Applications. The thick skin of relatively clean metal on rimmed-steel ingots makes them desirable for rolling products where the surface of the finished products is most important. The higher range of carbon content of most rimmed steel is 0.12–0.15%, the lower is 0.06–0.10%.

Capped steel has a thin rimmed zone that is relatively free from blowholes, and a core zone that is less segregated than that of a rimmed ingot of the same volume. Capped steel is used to advantage when the carbon content is above 0.15%. Steel of this type is used for sheet, strip, skelp, tin plate, wire, and bars. (Skelp is a hot-rolled flat strip used for butt-weld pipe.)

Semikilled steel has wide application in structural shapes, plates, and commercial-grade bars. Its carbon content ranges from 0.15–0.30%, as required for a given product.

Killed steel is generally used when a homogeneous structure is required in the product. Alloy steels, forging steels, and steels for carburizing are of this type, when strength, toughness, and absence of defects is required. In general, all steels with more than 0.30% carbon are killed steels.

Casting

Continuous Casting. Formerly, all steel for hot working was cast into ingots by the methods described above. Although this series of operations is both time-consuming and expensive and results in some loss of steel, in the United States most steel (70–80%) is handled this way.

The possibility of casting molten steel continuously into useful shapes equivalent to conventional semifinished shapes, and thus eliminating the ingot and primary-mill stages of rolled-steel production, led to a long series of attempts by many investigators, using a variety of machine designs. Because of the high melting point, high specific heat, and low thermal conductivity of steel, most attempts failed or were only partially successful in the case of ferrous metals. For nonferrous metals, on the other hand, continuous-casting proved practical. This success eventually led to the solution of the problem.

In the United States, the first continuous-casting machine for steel was developed in the late 1930s and early 1940s. Relatively little progress was then made until ca 1960. By 1980, 94 continuous-casting machines were installed and numerous others are planned. Steelmakers in other countries have adopted continuous casting on a large scale.

Although the continuous casting of steel appears deceptively simple in principle, many difficulties are inherent in the process. For example, when molten steel comes into contact with a water-cooled mold, a thin solid skin forms on the wall (see Fig. 18). However, because of the physical characteristics of steel, and because thermal contraction causes the skin to separate from the mold wall shortly after solidification, the rate of heat abstraction from the casting is so low that molten steel persists within the interior of the section for some distance below the bottom of the mold. The thickness of the skin increases because of the action of the water sprays as the casting moves downward and, eventually, the whole section solidifies (46–51).

The mass of the solid casting is supported as it descends by driven pinch rolls that also control the speed of descent by controlling the rate of withdrawal of the casting from the mold. Any tendency for the casting to adhere to the mold wall may cause the

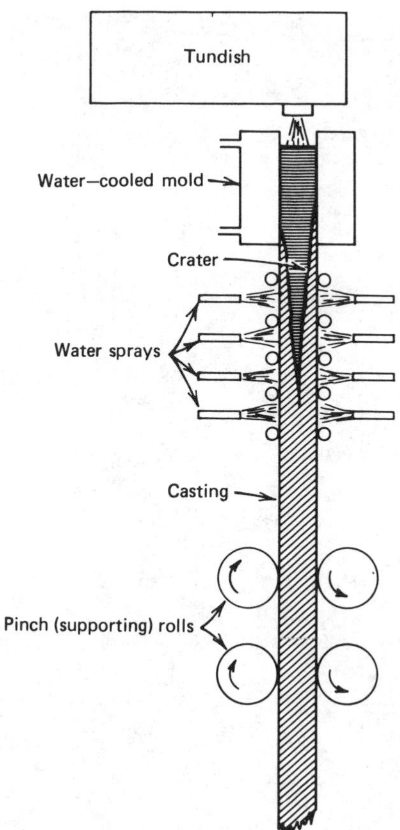

Figure 18. Schematic representationn of the progress of cooling of steel during continuous casting.

skin to rupture. However, molds that move up and down for a predetermined distance at controlled rates during casting have practically eliminated the sticking problem (see Fig. 19). The continuous-cast section issuing from the withdrawal or supporting rolls may be disposed of in several ways, some of which are illustrated in Figure 20.

A number of other designs of continuous-casting machines are being used in the production of semifinished sections.

Continuous casting is now widely accepted as an inexpensive method of casting steel. It gives a higher yield than ingot casting and avoids the cost of rolling ingots into slabs since the slabs are produced directly from liquid steel. Excessive segregation during solidification creates problems of low ductility in regions of high carbon and alloy content. This phenomenon causes cracking during processing. In general, segregation increases with increasing thickness. Continuous-cast slabs are only ca 20–25 cm thick and exhibit much less segregation than slabs rolled from ingots which are 3–5 times as thick. Consequently, there is usually less variation in steel composition from one heat to the next in continuous-cast steel than there is from the top to the bottom in a single ingot.

Numerous attempts to continuously cast rimmed steels have not been successful. At normal rates, the solidifying slab moves downward at ca 125 cm/min. At the same time, the pressure on the steel at the solidification front increases rapidly because of

Figure 19. Continuous casting of square steel blooms. The liquid steel is poured into molds, which are not shown, at the top of the photograph. The solidifying steel blooms move downward and then toward the camera. In the foreground, the blooms are cut to the proper length.

the weight of liquid steel (the ferrostatic head) above it. As a result, the pressure at the solidification front soon becomes high enough to prevent the formation of carbon monoxide bubbles which are needed to obtain the rimming action.

Continuous-cast aluminum-killed steel is similar in composition and properties to ingot-cast aluminum-killed steel. It contains 0.025–0.060% aluminum and has ex-

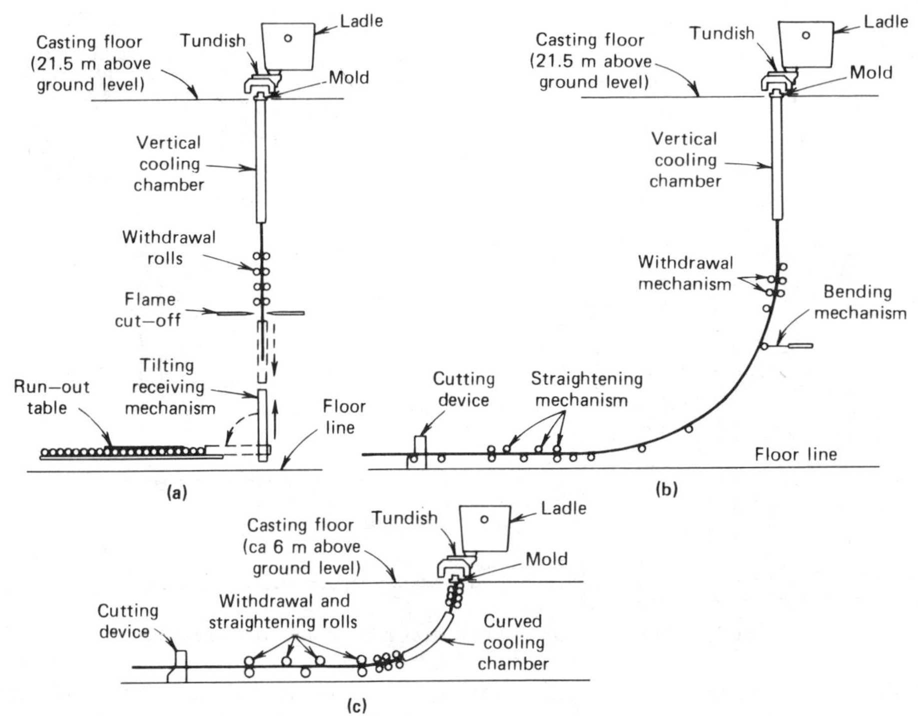

Figure 20. Continuous casting of steel (1).

cellent deep-drawing characteristics. However, special techniques are required, including an immersed nozzle between the tundish and the continuous-casting mold and a flux covering in the mold. An immersed nozzle between the ladle and the tundish is also desirable, as is a synthetic flux or some other type of tundish, and an argon gas cover over the steel.

Bottom-Pressure Casting. In bottom-pressure casting, illustrated in Figure 21, a ladle filled with molten steel is placed in a pressure vessel covered with a lid in which a pouring tube has been inserted that dips down into the molten steel almost to the ladle bottom. A gooseneck tube connects the pouring tube to the lower end of mold which is supported in an inclined position. When air pressure is applied to the vessel, molten metal is forced upward through the pouring tube and gooseneck into the mold. Metal rises in the mold until it reaches the upper end where a riser fills with molten metal to feed the casting as it solidifies to prevent formation of a pipe. When metal appears at the proper level in the riser, a gate at the lower end of the mold is closed to isolate the mold from the gooseneck. After a short interval, the pressure in the vessel is released to drain the metal in the gooseneck back into the ladle. After the casting has solidified, the mold is removed and prepared for the next casting. This method gives better yields than ingot casting and produces slabs and billets with surfaces that require little conditioning (52–53).

Plastic Working of Steel

Plastic working of a metal such as steel is the permanent deformation accomplished by applying mechanical forces to a metal surface. The primary objective is

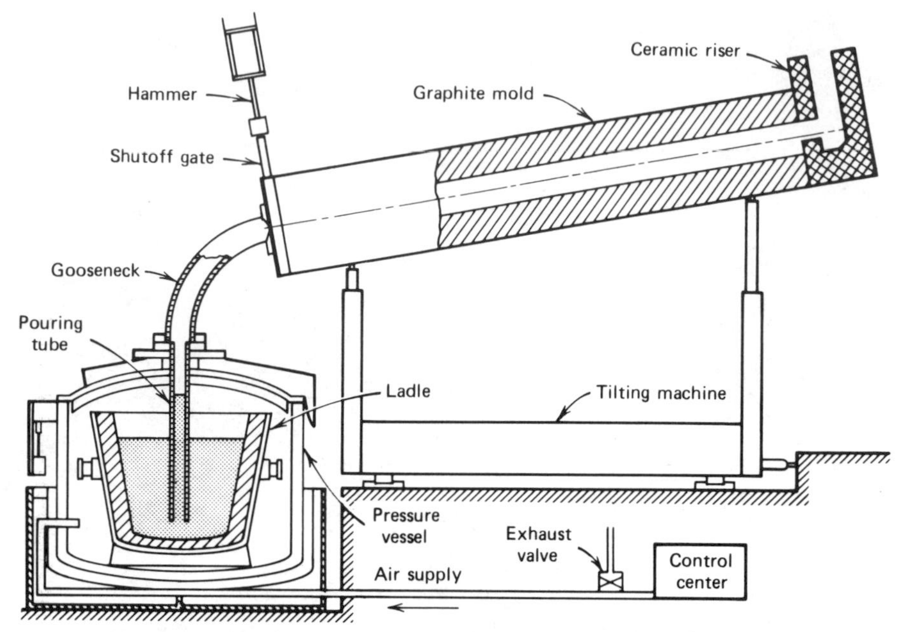

Figure 21. The bottom-pressure casting method as applied to slabs (2).

usually the production of a specific shape or size (mechanical shaping), although in some cases it may be the improvement of certain physical properties and mechanical properties of the metal (mechanical treatment). These two objectives are often attained simultaneously (54–60).

Plastic deformation of steel can be accomplished by hot working or cold working. In hot working, the steel is heated to 1090–1310°C, depending on the grade and the work to be accomplished. The force required to deform the metal is very sensitive to the rate of application; however, after deformation, the basic strength of the steel is essentially unchanged. In cold working, on the other hand, steel is not heated before working, and the force required to cause deformation is relatively insensitive to the application rate and temperature variations, but the basic strength of the steel is permanently increased.

The principal hot-working techniques are hammering, pressing, extrusion, and rolling (see Fig. 22); the first two of these are called forging. Other methods include rotary swaging, hot spinning, hot deep-drawing, roll forging, and die forging. Hot working of steel is generally a shaping process, but can also improve mechanical properties.

Cold working is generally applied to bars, wire, strip, sheet, and tubes. It reduces the cross-sectional area of the piece being worked by cold rolling, cold drawing, or cold extrusion. Cold working imparts improved mechanical properties, better machinability, bright surface, and production of thinner material than hot working can accomplish economically (see Metal treatments).

Metallography and Heat Treatment

The great advantage of steel as an engineering material is its versatility, which arises from the fact that its properties can be controlled and changed by heat treatment

Figure 22. Hot rolling of steel slabs to plate. In the foreground is a steel plate mill. Behind the plate is the next slab to be rolled.

(61–63). Thus, if steel is to be formed into some intricate shape, it can be made very soft and ductile by heat treatment; on the other hand, heat treatment can also impart high strength.

The physical and mechanical properties of steel depend upon its constitution, that is, the nature, distribution, and amounts of its metallographic constituents as distinct from its chemical composition. The amount and distribution of iron and iron carbide determine the properties, although most plain carbon steels also contain manganese, silicon, phosphorus, sulfur, oxygen, and traces of nitrogen, hydrogen, and other chemical elements such as aluminum and copper. These elements may modify, to a certain extent, the main effects of iron and iron carbide, but the influence of iron carbide always predominates. This is true even of medium-alloy steels, which may contain considerable amounts of nickel, chromium, and molybdenum.

The iron in steel is called ferrite. In pure iron–carbon alloys, the ferrite consists of iron with a trace of carbon in solution, but in steels it may also contain alloying elements such as manganese, silicon, or nickel. The atomic arrangement in crystals of the allotropic forms of iron is shown in Figure 23.

Cementite, the term for iron carbide in steel, is the form in which carbon appears in steels. It has the formula Fe_3C, and thus consists of 6.67% carbon and 93.33% iron. Little is known about its properties, except that it is very hard and brittle. As the hardest constituent of plain carbon steel, it scratches glass and feldspar but not quartz. It exhibits ca two-thirds the induction of pure iron in a strong magnetic field.

Austenite is the high temperature phase of steel. Upon cooling, it gives ferrite

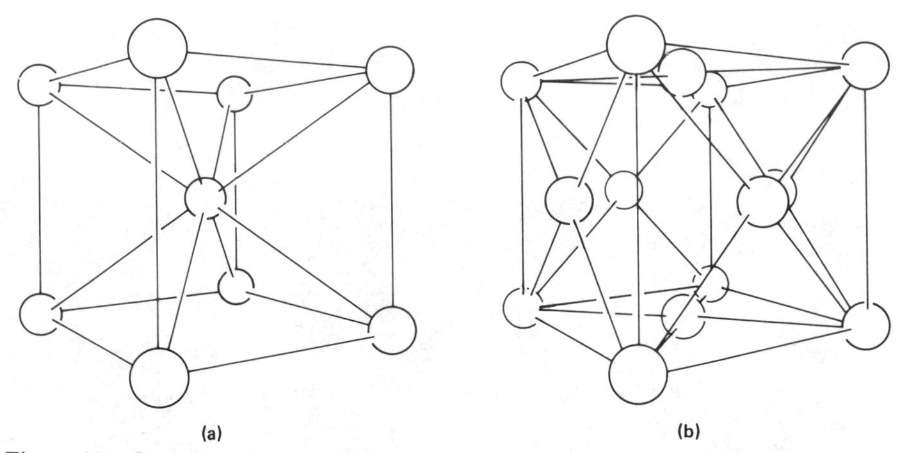

Figure 23. Crystalline structure of allotropic forms of iron. Each white sphere represents an atom of (**a**) α and δ iron in bcc form, and (**b**) γ iron, in fcc (1).

and cementite. Austenite is a homogeneous phase, consisting of a solid solution of carbon in the γ form of iron. It forms when steel is heated above 790°C. The limiting temperatures for its formation vary with composition and are discussed below. The atomic structure of austenite is that of γ iron, fcc; the atomic spacing varies with the carbon content.

When a plain carbon steel of ca 0.80% carbon content is cooled slowly from the temperature range at which austentite is stable, ferrite and cementite precipitate together in a characteristically lamellar structure known as pearlite. It is similar in its characteristics to a eutectic structure but, since it is formed from a solid solution rather than from a liquid phase, it is known as a eutectoid structure. At carbon contents above and below 0.80%, pearlite of ca 0.80% carbon is likewise formed on slow cooling, but excess ferrite or cementite precipitates first, usually as a grain-boundary network, but occasionally also along the cleavage planes of austentite. The excess ferrite or cementite rejected by the cooling austenite is known as a proeutectoid constituent. The carbon content of a slowly cooled steel can be estimated from the relative amounts of pearlite and proeutectoid constituents in the microstructure.

Bainite is a decomposition product of austenite consisting of an aggregate of ferrite and cementite. It forms at temperatures lower than those where very fine pearlite forms and higher than those at which martensite begins to form on cooling. Metallographically, its appearance is feathery if formed in the upper part of the temperature range, or acicular (needlelike) and resembling tempered martensite if formed in the lower part.

Martensite in steel is a metastable phase formed by the transformation of austenite below the temperature called the M_s temperature, where martensite begins to form as austenite is cooled continuously. Martensite is an interstitial supersaturated solid solution of carbon in iron with a body-centered tetragonal lattice. Its microstructure is acicular.

Iron–Iron Carbide Phase Diagram

The iron–iron carbide phase diagram (Fig. 24) furnishes a map showing the ranges of compositions and temperatures in which the various phases such as austenite, ferrite, and cementite are present in slowly cooled steels. This diagram covers the temperature

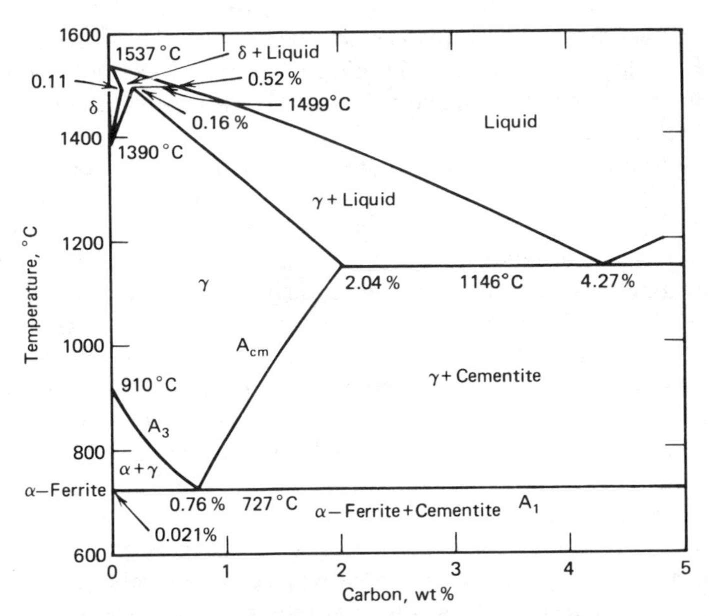

Figure 24. Iron–iron carbide phase diagram (1).

range from 600°C to the melting point of iron, and carbon contents from 0–5%. In steels and cast irons, carbon can be present either as iron carbide (cemenite) or as graphite. Under equilibrium conditions, only graphite is present because iron carbide is unstable with respect to iron and graphite. However, in commercial steels, iron carbide is present instead of graphite. When a steel containing carbon solidifies, the carbon in the steel usually solidifies as iron carbide. Although the iron carbide in a steel can change to graphite and iron when the steel is held at ca 900°C for several days or weeks, iron carbide in steel under normal conditions is quite stable.

The portion of the iron–iron carbide diagram of interest here is that part extending from 0–2.01% carbon. Its application to heat treatment can be illustrated by considering the changes occurring on heating and cooling steels of selected carbon contents.

As already stated, iron occurs in two allotropic forms, α or δ (the latter at a very high temperature) and γ (see Fig. 23). The temperatures at which these phase changes occur are known as the critical temperatures and the boundaries in Figure 24 show how these temperatures are affected by composition. For pure iron, these temperatures are 910°C for the α-γ phase change and 1390°C for the γ-δ phase change.

Changes on Heating and Cooling Pure Iron. The only changes occurring on heating or cooling pure iron are the reversible changes at ca 910°C from bcc α iron to fcc γ iron and from the fcc δ iron to bcc γ iron at ca 1390°C.

Changes on Heating and Cooling Eutectoid Steel. Eutectoid steels are those that contain 0.8% carbon. The diagram shows that at and below 727°C the constituents are α ferrite and cementite. At 600°C, the α-ferrite may dissolve as much as ca 0.007% carbon. Up to 727°C, the solubility of carbon in the ferrite increases until, at this temperature, the ferrite contains about 0.02% carbon. The phase change on heating an 0.8% carbon steel occurs at 727°C which is designated as A_1, as the eutectoid or lower critical temperature. On heating just above this temperature, all ferrite and cementite transform to austenite, and on slow cooling the reverse change occurs.

When a eutectoid steel is slowly cooled from ca 738°C, the ferrite and cementite form in alternate layers of microscopic thickness. Under the microscope at low magnification, this mixture of ferrite and cementite has an appearance similar to that of a pearl and is therefore called pearlite.

Changes on Heating and Cooling Hypoeutectoid Steel. Hypoeutectoid steels are those that contain less carbon than the eutectoid steels. If the steel contains more than 0.02% carbon, the constituents present at and below 727°C are usually ferrite and pearlite; the relative amounts depend upon the carbon content. As the carbon content increases, the amount of ferrite decreases and the amount of pearlite increases.

The first phase change on heating, if the steel contains more than 0.02% carbon, occurs at 727°C. On heating just above this temperature, the pearlite changes to austenite. The excess ferrite, called proeutectoid ferrite, remains unchanged. As the temperature rises further above A_1, the austenite dissolves more and more of the surrounding proeutectoid ferrite, becoming lower and lower in carbon content until all the proeutectoid ferrite is dissolved in the austenite, which now has the same average carbon content as the steel.

On slow cooling the reverse changes occur. Ferrite precipitates, generally at the grain boundaries of the austenite, which becomes progressively richer in carbon. Just above A_1, the austenite is substantially of eutectoid composition, 0.8% carbon.

Changes on Heating and Cooling Hypereutectoid Steels. The behavior on heating and cooling hypereutectoid steels (steels containing >0.80% carbon) is similar to that of hypoeutectoid steels, except that the excess constituent is cementite rather than ferrite. Thus, on heating above A_1, the austenite gradually dissolves the excess cementite until at the A_{cm} temperature the proeutectoid cementite has been completely dissolved and austenite of the same carbon content as the steel is formed. Similarly, on cooling below A_{cm}, cementite precipitates and the carbon content of the austenite approaches the eutectoid composition. On cooling below A_1, this eutectoid austenite changes to pearlite and the room temperature composition is, therefore, pearlite and proeutectoid cementite.

Early iron–carbon equilibrium diagrams indicated a critical temperature at ca 768°C. It has since been found that there is no true phase change at this point. However, between ca 768 and 790°C there is a gradual magnetic change, since ferrite is magnetic below this range and paramagnetic above it. This change, occurring at what formerly was called the A_2 change, is of little or no significance with regard to the heat treatment of steel.

Effect of Alloys on the Equilibrium Diagram. The iron–carbon diagram may, of course, be profoundly altered by alloying elements, and its application should be limited to plain carbon and low alloy steels. The most important effects of the alloying elements are that the number of phases which may be in equilibrium is no longer limited to two as in the iron–carbon diagram; the temperature and composition range, with respect to carbon, over which austenite is stable may be increased or reduced; and the eutectoid temperature and composition may change.

Alloying elements either enlarge the austenite field or reduce it. The former include manganese, nickel, cobalt, copper, carbon, and nitrogen and are referred to as austenite formers.

The elements that decrease the extent of the austenite field include chromium, silicon, molybdenum, tungsten, vanadium, tin, niobium, phosphorus, aluminum, and titanium; they are known as ferrite formers.

Manganese and nickel lower the eutectoid temperature, whereas chromium, tungsten, silicon, molybdenum, and titanium generally raise it. All these elements seem to lower the eutectoid carbon content.

Grain Size. ***Austenite.*** A significant aspect of the behavior of steels on heating is the grain growth that occurs when the austenite, formed on heating above A_3 or A_{cm}, is heated even higher; A_3 is the upper critical temperature and A_{cm} the temperature at which cementite begins to form. The austenite, like any metal composed of a solid solution, consists of polygonal grains. As formed at a temperature just above A_3 of A_{cm}, the size of the individual grains is very small but, as the temperature is increased above the critical temperature, the grain sizes increase. The final austenite grain size depends, therefore, upon the temperature above the critical temperature to which the steel is heated. The grain size of the austenite has a marked influence upon transformation behavior during cooling and upon the grain size of the constituents of the final microstructure. Grain growth may be inhibited by carbides that dissolve slowly or by dispersion of nonmetallic inclusions. Hot-working refines the coarse grain formed by reheating steel to the relatively high temperatures used in forging or rolling and the grain size of hot-worked steel is determined largely by the temperature at which the final stage of the hot-working process is carried out. The general effects of austenite grain size on the properties of heat-treated steel are summarized in Table 3.

Microscopic-Grain-Size Determination. The microscopic grain size of steel is customarily determined from a polished plane section prepared in such a way as to delineate the grain boundaries. The grain size can be estimated by several methods. The results can be expressed as diameter of average grain in millimeters (reciprocal of the square root of the number of grains per mm^2), number of grains per unit area, number of grains per unit volume, or a micrograin-size number obtained by comparing the microstructure of the sample with a series of standard charts.

Fine- and Coarse-Grain Steels. As mentioned previously, austentic-grain growth may be inhibited by undissolved carbides or nonmetallic inclusions. Steels of this type are commonly referred to as fine-grained steels, whereas steels that are free from grain-growth inhibitors are known as coarse-grained steels.

The general pattern of grain coarsening when steel is heated above the critical temperature is as follows: Coarse-grained steel coarsens gradually and consistently as the temperature is increased, whereas fine-grained steel coarsens only slightly, if at all, until a certain temperature known as the coarsening temperature is reached,

Table 3. Trends in Heat-Treated Products

Property	Coarse-grain austenite	Fine-grain austenite
Quenched and tempered products		
hardenability	increasing	decreasing
toughness	decreasing	increasing
distortion	more	less
quench cracking	more	less
internal stress	higher	lower
Annealed or normalized products		
machinability		
rough finish	better	inferior
fine finish	inferior	better

after which abrupt coarsening occurs. Heat treatment can make any type of steel either fine or coarse grained; as a matter of fact, at temperatures above its coarsening temperature, the fine-grained steel usually exhibits a coarser grain size than the coarse-grained steel at the same temperature.

Making steels that remain fine grained above 925°C involves the judicious use of deoxidation with aluminum. The inhibiting agent in such steels is generally conjectured to be a submicroscopic dispersion of aluminum nitride or, perhaps at times, aluminum oxide.

Phase Transformations. ***Austenite.*** At equilibrium, that is with very slow cooling, austenite transforms to pearlite when cooled below the A_1 temperature. When austenite is cooled more rapidly, this transformation is depressed and occurs at a lower temperature. The faster the cooling rate, the lower the temperature at which transformation occurs. Furthermore, the nature of the ferrite–carbide aggregate formed when the austenite transforms varies markedly with the transformation temperature and the properties are found to vary correspondingly. Thus, heat treatment involves a controlled supercooling of austenite, and in order to take full advantage of the wide range of structures and properties that this treatment permits, a knowledge of the transformation behavior of austenite and the properties of the resulting aggregates is essential.

Isothermal Transformation Diagram. The transformation behavior of austenite is best studied by observing the isothermal transformation at a series of temperatures below A_1. The transformation progress is ordinarily followed metallographically in such a way that both the time–temperature relationships and the manner in which the microstructure changes are established. The times at which transformation begins and ends at a given temperature are plotted, and curves depicting the transformation behavior as a function of temperature are obtained by joining these points (Fig. 25). Such a diagram is referred to as an isothermal transformation (IT) diagram, a time-temperature-transformation (TTT) diagram or, an S curve (64).

The IT diagram for a eutectoid carbon steel is shown in Figure 25. In addition to the lines depicting the transformation, the diagram shows microstructures at various stages of transformation and hardness values. Thus, the diagram illustrates the characteristic subcritical austenite transformation behavior, the manner in which microstructure changes with transformation temperature, and the general relationship between these microstructural changes and hardness.

As the diagram indicates, the characteristic isothermal transformation behavior at any temperature above the temperature at which transformation to martensite begins (the M_s temperature) takes place over a period of time, known as the incubation period, in which no transformation occurs, followed by a period of time during which the transformation proceeds until the austenite has been transformed completely. The transformation is relatively slow at the beginning and toward the end, but much more rapid during the intermediate period in which ca 25–75% of the austenite is transformed. Both the incubation period and the time required for completion of the transformation depend on the temperature.

The behavior depicted in this diagram is typical of plain carbon steels, with the shortest incubation period occurring at ca 540°C. Much longer times are required for transformation as the temperature approaches either the Ae_1 or the M_s temperature. This A_1 temperature is lowered slightly during cooling and increased slightly during heating. The 540°C temperature, at which the transformation begins in the shortest

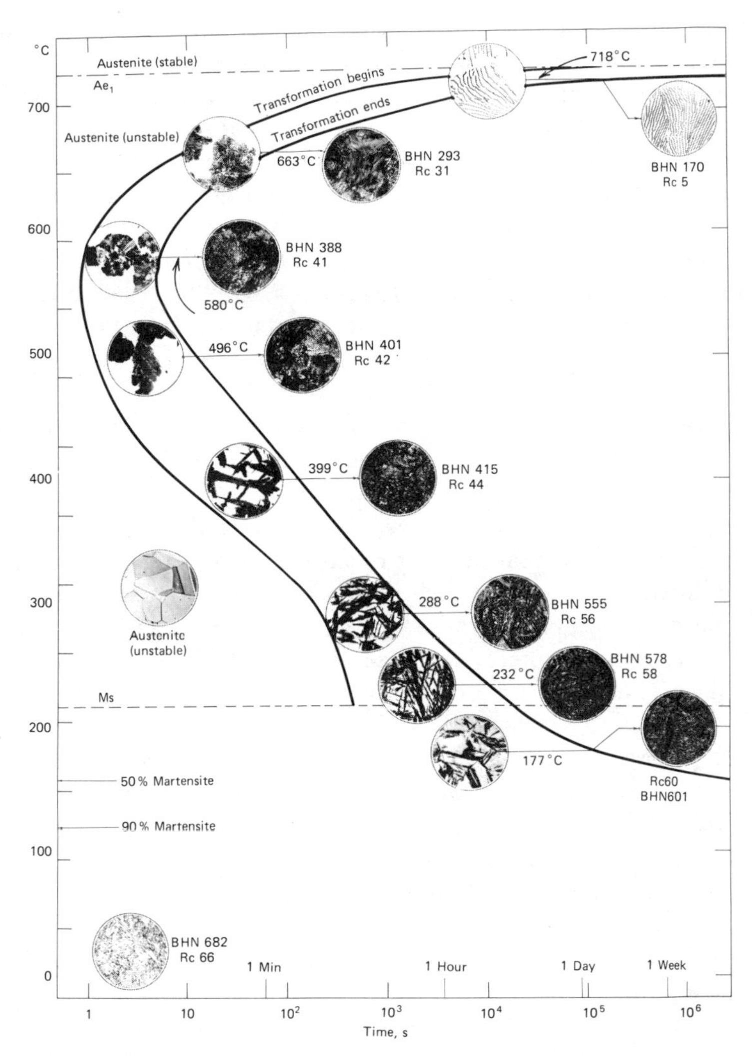

Figure 25. Isothermal transformation diagram for a plain carbon eutectoid steel (1); $Ae_1 = A_1$ temperature at equilibrium; BHN = Brinell hardness number; Rc = Rockwell hardness scale C. C, 0.89%; Mn, 0.29%; austenitized at 885°C; grain size, 4–5; photomicrographs originally ×2500.

time period, is commonly referred to as the nose of the IT diagram. If complete transformation is to occur at temperatures below this nose, the steel must be cooled rapidly enough to prevent transformation at the nose temperature. Microstructures resulting from transformation at these lower temperatures exhibit superior strength and toughness.

Pearlite. In carbon and low alloy steels, transformation over the temperature range of ca 700–540°C gives pearlite microstructures of the characteristic lamellar type. As the transformation temperature falls, the lamellae move closer and the hardness increases.

Bainite. Transformation to bainite occurs over the temperature range of ca 540–230°C. The acicular bainite microstructures differ markedly from the pearlite microstructures. Here again, the hardness increases as the transformation temperature decreases, although the bainite formed at the highest possible temperature is often softer than pearlite formed at a still higher temperature.

Martensite. Transformation to martensite, which in the steel illustrated in Figure 25 begins at ca 230°C, differs from transformation to pearlite or bainite because it is not time-dependent, but occurs almost instantly during cooling. The degree of transformation depends only on the temperature to which it is cooled. Thus, in this steel of Figure 25, transformation to martensite starts on cooling to 230°C (designated as the M_s temperature). The martensite is 50% transformed on cooling to ca 150°C, and the transformation is essentially completed at ca 90°C (designated as the M_f temperature). The microstructure of martensite is acicular. It is the hardest austenite transformation product but brittle; this brittleness can be reduced by tempering, as discussed below.

Phase Properties. ***Pearlite.*** Pearlites are softer than bainites or martensites. However, they are less ductile than the lower temperature bainites and, for a given hardness, far less ductile than tempered martensite. As the transformation temperature decreases within the pearlite range, the interlamellar spacing decreases, and these fine pearlites, formed near the nose of the isothermal diagram, are both harder and more ductile than the coarse pearlites formed at higher temperatures. Thus, although as a class pearlite tends to be soft and not very ductile, its hardness and toughness both increase markedly with decreasing transformation temperatures.

Bainite. In a given steel, bainite microstructures are generally found to be both harder and tougher than pearlite, although less hard than martensite. Bainite properties generally improve as the transformation temperature decreases and lower bainite compares favorably with tempered martensite at the same hardness or exceeds it in toughness. Upper bainite, on the other hand, may be somewhat deficient in toughness as compared with fine pearlite of the same hardness (64).

Martensite. Martensite is the hardest and most brittle microstructure obtainable in a given steel. The hardness of martensite increases with increasing carbon content up to the eutectoid composition. The hardness of martensite, at a given carbon content, varies with the cooling rate.

Although for some applications, particularly those involving wear resistance, the hardness of martensite is desirable in spite of the accompanying brittleness, this microstructure is mainly important as starting material for tempered martensite structures, which have definitely superior properties.

Tempered Martensite. Martensite is tempered by heating to a temperature ranging from 170–700°C for 30 min to several hours. This treatment causes the martensite to transform to ferrite interspersed with small particles of cementite. Higher temperatures and longer tempering periods cause the cementite particles to increase in size and the steel to become more ductile and lose strength. Tempered martensitic structures are, as a class, characterized by toughness at any strength. The diagram of Figure 26 describes, within ±10%, the mechanical properties of tempered martensite, regardless of composition. For example, a steel consisting of tempered martensite, with an ultimate strength of 1035 MPa (150,000 psi), might be expected to exhibit elongation of 16–20%, reduction of area of between 54% and 64%, yield point of 860–980 MPa (125,000–142,000 psi), and Brinell hardness of about 295–320. Because of its high ductility at a given hardness, this is the structure that is preferred.

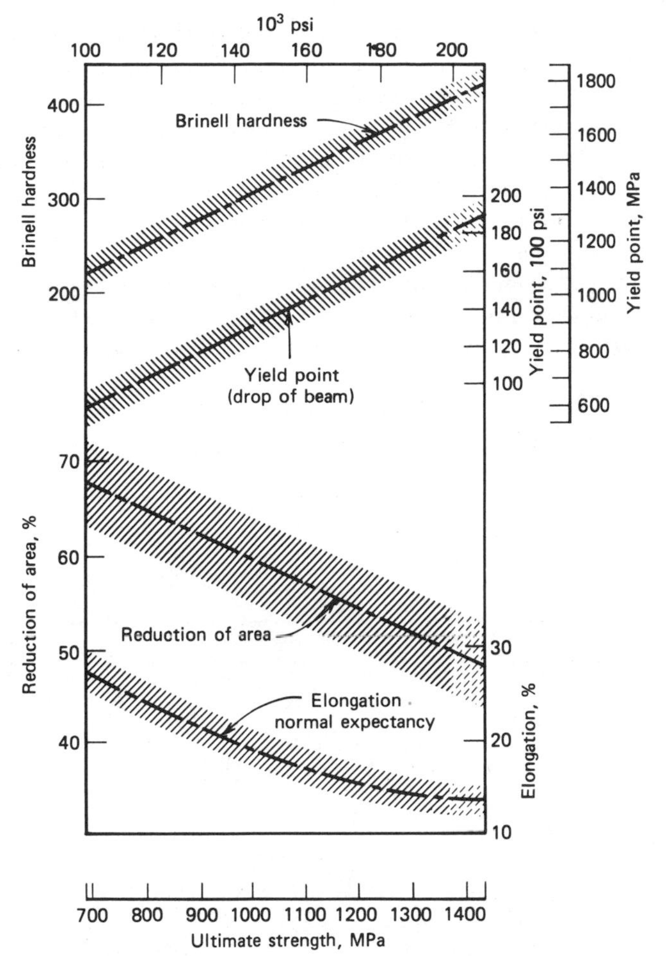

Figure 26. Properties of tempered martensite (1). Fully heat-treated miscellaneous analyses, low alloy steels; 0.30–0.50% carbon.

Transformation Rates. The main factors affecting transformation rates of austenite are composition, grain size, and homogeneity. In general, increasing carbon and alloy content as well as increasing grain size tend to lower transformation rates. These effects are reflected in the isothermal transformation curve for a given steel.

Continuous Cooling. The basic information depicted by an isothermal transformation diagram illustrates the structure formed if the cooling is interrupted and the reaction completed at a given temperature. The information is also useful for interpreting behavior when the cooling proceeds directly without interruption, as in the case of annealing, normalizing, and quenching. In these processes, the residence time at a single temperature is generally insufficient for the reaction to go to completion; instead, the final structure consists of an association of microstructures which were formed individually at successively lower temperatures as the piece cooled. However, the tendency to form several structures is still explained by the isothermal diagram (65–66).

The final microstructure after continuous cooling depends on the times spent at the various transformation-temperature ranges through which a piece is cooled.

The transformation behavior on continuous cooling thus represents an integration of these times by constructing a continuous-cooling diagram at constant rates similar to the isothermal transformation diagram (see Fig. 27). This diagram lies below and to the right of the corresponding isothermal transformation diagram if plotted on the same coordinates; that is, transformation on continuous cooling starts at a lower temperature and after a longer time than the intersection of the cooling curve and the isothermal diagram would predict. This displacement is a function of the cooling rate, and increases with increasing cooling rate.

Several cooling-rate curves have been superimposed on Figure 27. The changes occurring during these cooling cycles illustrate the manner in which diagrams of this nature can be correlated with heat-treating processes and used to predict the resulting microstructure.

Considering first the relatively low cooling rate (<22°C/h), the steel is cooled through the regions in which transformations to ferrite and pearlite occur which

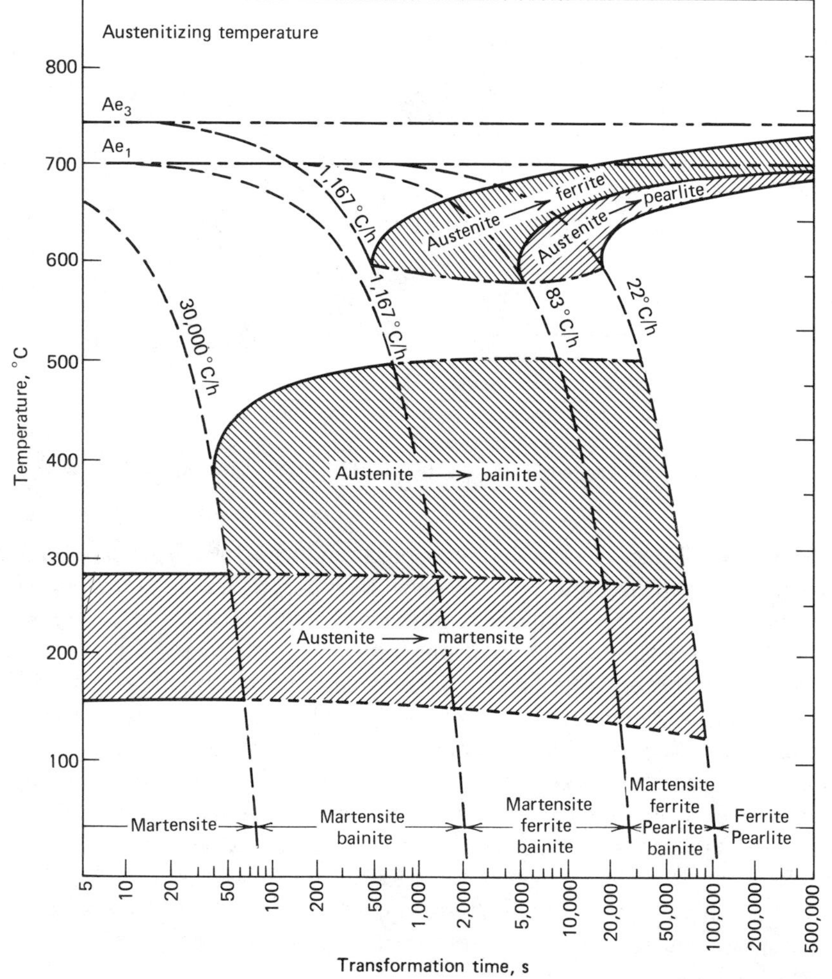

Figure 27. Continuous-cooling transformation diagram for a Type 4340 alloy steel, with superimposed cooling curves illustrating the manner in which transformation behavior during continuous cooling governs final microstructure (1). Ae_3 = critical temperature at equilibrium.

constitute the final microstructure. This cooling rate corresponds to a slow cooling in the furnace such as might be used in annealing.

At a higher cooling rate (22–83°C/h), such as might be obtained on normalizing a large forging, the ferrite, pearlite, bainite, and martensite fields are traversed and the final microstructure contains all these constituents.

At cooling rates of 1,167–30,000°C/h, the microstructure is free of proeutectoid ferrite and consists largely of bainite and a small amount of martensite. A cooling rate of at least 30,000°C/h is necessary to obtain the fully martensitic structure desired as a starting point for tempered martensite.

Thus, the final microstructure, and therefore the properties of the steel, depend upon the transformation behavior of the austenite and the cooling conditions, and can be predicted if these factors are known.

Hardenability

Hardenability refers to the depth of hardening or to the size of a piece that can be hardened under given cooling conditions, and not to the maximum hardness that can be obtained in a given steel (67–68). The maximum hardness depends almost entirely upon the carbon content, whereas the hardenability (depth of hardening) is far more dependent upon the alloy content and grain size of the austenite. Steels whose IT diagrams indicate a long time interval before the start of transformation to pearlite are useful when large sections are to be hardened, since if steel is to transform to bainite or martensite, it must escape any transformation to pearlite. Therefore, the steel must be cooled through the high temperature transformation ranges at a rate rapid enough for transformation not to occur even at the nose of the IT diagram. This rate, which just permits transformation to martensite without earlier transformation at a higher temperature, is known as the critical cooling rate for martensite. It furnishes one method for expressing hardenability; for example, in the steel of Figure 27, the critical cooling rate for martensite is 30,000°C/h or 8.3°C/s.

Although the critical cooling rate can be used to express hardenability, cooling rates ordinarily are not constant but vary during the cooling cycle. Especially when quenching in liquids, the cooling rate of the steel always decreases as the steel temperature approaches that of the cooling medium. It is, therefore, customary to express hardenability in terms of depth of hardening in a standardized quench. The quenching condition used in this method of expression is a hypothetical one in which the surface of the piece is assumed to come instantly to the temperature of the quenching medium. This is known as an ideal quench; the diameter of a round steel bar, which is quenched to the desired microstructure, or corresponding hardness value, at the center in an ideal quench, is known as the ideal diameter for which the symbol D_I is used. The relationships between the cooling rates of the ideal quench and those of other cooling conditions are known. Thus, the hardenability values in terms of ideal diameter are used to predict the size of round or other shape which have the same cooling rate when cooled in actual quenches whose cooling severities are known. The cooling severities (usually referred to as severity of quench) which form the basis for these relationships are called H values. The H value for the ideal quench is infinity; those for some commonly used cooling conditions are given in Table 4.

Hardenability is most conveniently measured by a test in which a steel sample is subjected to a continuous range of cooling rates. In the end-quench or Jominy test,

Table 4. H Values Designating Severity of Quench for Commonly Used Cooling Conditions[a]

Degree of agitation of medium	Quenching medium		
	Oil	Water	Brine
none	0.25–0.30	0.9–1.0	2
mild	0.30–0.35	1.0–1.1	2.0–2.2
moderate	0.35–0.40	1.2–1.3	
good	0.40–0.50	1.4–1.5	
strong	0.50–0.80	1.6–2.0	
violent	0.80–1.1	4.0	5.0

[a] H values are proportional to the heat-extracting capacity of the medium.

a round bar, 25 mm diam and 102 mm long, is heated to the desired austenitizing temperature and quenched in a fixture by a stream of water impinging on only one end. Hardness measurements are made on flats that are ground along the length of the bar after quenching. The results are expressed as a plot of hardness versus distance from the quenched end of the bar. The relationships between the distance from the quenched end and cooling rates in terms of ideal diameter (D_I) are known, and the hardenability can be evaluated in terms of D_I by noting the distance from the quenched end at which the hardness corresponding to the desired microstructure occurs and using this relationship to establish the corresponding cooling rate or D_I value. Published heat-flow tables or charts relate the ideal-diameter value to cooling rates in quenches or cooling conditions whose H values are known. Thus, the ideal-diameter value can be used to establish the size of a piece in which the desired microstructure can be obtained under the quenching conditions of the heat treatment to be used. The hardenability of steel is such an important property that it has become common practice to purchase steels to specified hardenability limits. Such steels are called H steels.

Heat-Treating Processes

In heat-treating processes, steel is usually heated above the A_3 point and then cooled at a rate that results in the microstructure that gives the desired properties (69–70).

Austenitization. The steel is first heated above the temperature at which austenite is formed. The actual austenitizing temperature should be high enough to dissolve the carbides completely and take advantage of the hardening effects of the alloying elements. In some cases, such as tool steels or high-carbon steels, undissolved carbides may be retained for wear resistance. The temperature should not be high enough to produce pronounced grain growth. The piece should be heated long enough for complete solution; for low alloy steels in a normally loaded furnace, 1.8 min/mm of diameter or thickness usually suffices.

Excessive heating rates may create high stresses, resulting in distortion or cracking. Certain types of continuous furnaces, salt baths, and radiant-heating furnaces provide very rapid heating, but preheating of the steel may be necessary to avoid distortion or cracking, and sufficient time must be allowed for uniform heating throughout. Unless special precautions are taken, heating causes scaling or oxidation, and may result in decarburization; controlled-atmosphere furnaces or salt baths minimize these effects.

Quenching. The primary purpose of quenching is to cool rapidly enough to suppress all transformation at temperatures above the M_s temperature. The cooling rate required depends upon the size of the piece and the hardenability of the steel. The preferred quenching media are water, oils, and brine. The temperature gradients set up by quenching create high thermal and transformational stresses which may lead to cracking and distortion; a quenching rate no faster than necessary should be employed to minimize these stresses. Agitation of the cooling medium accelerates cooling and improves uniformity. Cooling should be long enough to permit complete transformation to martensite. Then, in order to minimize cracking from quenching stresses, the article should be transferred immediately to the tempering furnace (Fig. 28).

Tempering. Quenching forms very hard, brittle martensite with high residual stresses. Tempering relieves these stresses and improves ductility, although at some expense of strength and hardness. The operation consists of heating at temperatures below the lower critical temperature (A_1).

Measurements of stress relaxation on tempering indicate that, in a plain carbon steel, residual stresses are significantly lowered by heating to temperatures as low as 150°C, but that temperatures of 480°C and above are required to reduce these stresses to very low values. The times and temperatures required for stress relief depend upon the high temperature yield strength of the steel, since stress relief results from the localized plastic flow that occurs when the steel is heated to a temperature at which yield strength decreases. This phenomenon may be affected markedly by composition, and particularly by alloy additions. The toughness of quenched steel, as measured by the notch impact test, first increases on tempering up to 200°C, then decreases on tempering between 200 and 310°C, and finally increases rapidly on tempering at 425°C and above. This behavior is characteristic and, in general, temperatures of 230–310°C should be avoided.

In order to minimize cracking, tempering should follow quenching immediately. Any appreciable delay may promote cracking.

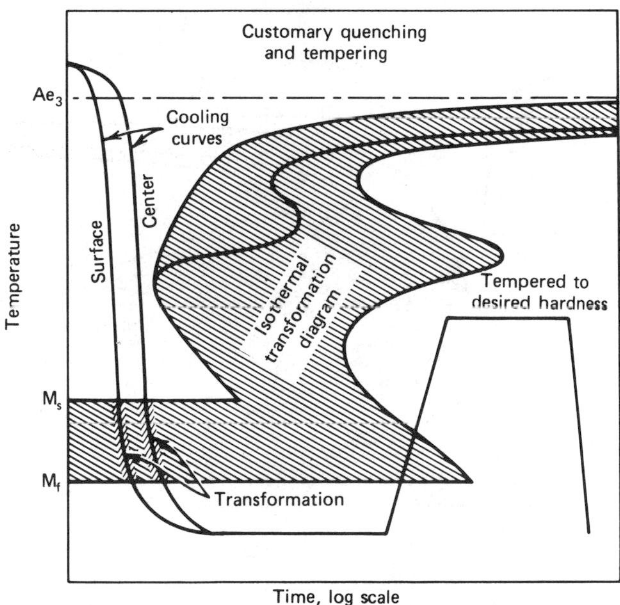

Figure 28. Transformation diagram for quenching and tempering martensite; the product is tempered martensite.

The tempering of martensite results in a contraction, and if the heating is not uniform, stresses result. Similarly, heating too rapidly may be dangerous because of the sharp temperature gradient set up between the surface and the interior. Recirculating-air furnaces can be used to obtain uniform heating. Oil or salt baths are commonly used for low temperature tempering, lead or salt baths at higher temperatures.

Some steels lose toughness on slow cooling from ca 540°C and above, a phenomenon known as temper brittleness; rapid cooling after tempering is desirable in these cases.

Martempering. A modified quenching procedure known as martempering minimizes the high stresses created by the transformation to martensite during the rapid cooling characteristic of ordinary quenching (see Fig. 29). In practice, it is ordinarily carried out by quenching in a molten-salt bath just above the M_s temperature. Transformation to martensite does not begin until the piece reaches the temperature of the salt bath and is removed to cool relatively slowly in air. Since the temperature gradient characteristic of conventional quenching is absent, the stresses produced by the transformation are much lower and a greater freedom from distortion and cracking is obtained. After martempering, the piece may be tempered to the desired strength.

Austempering. As discussed earlier, lower bainite is generally as strong as and somewhat more ductile than tempered martensite. Austempering, which is an isothermal heat treatment that results in lower bainite, offers an alternative heat treatment for obtaining optimum strength and ductility.

In austempering the article is quenched to the desired temperature in the lower bainite region, usually in molten salt, and kept at this temperature until transformation is complete (see Fig. 30). Usually, it is held twice as long as the period indicated by the isothermal transformation diagram. The article may be quenched or air cooled to room

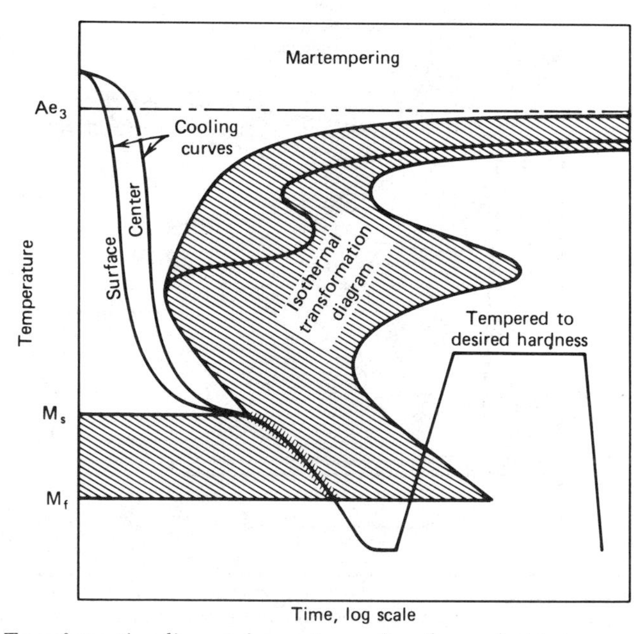

Figure 29. Transformation diagram for martempering; the product is tempered martensite.

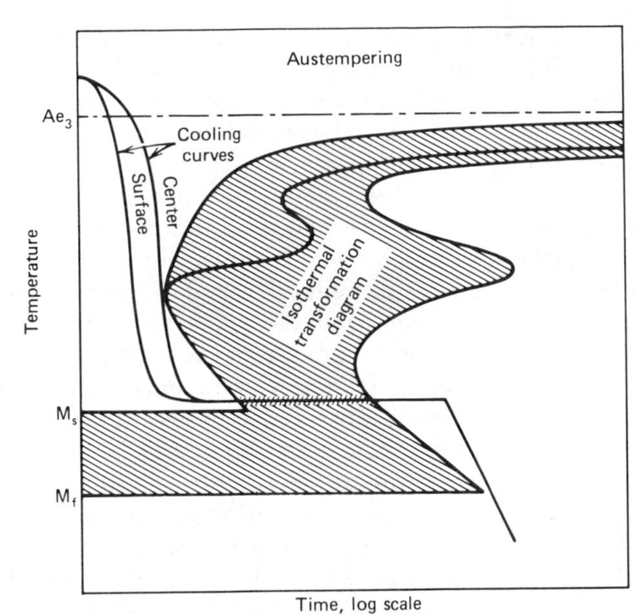

Figure 30. Transformation diagram for austempering; the product is bainite.

temperature after transformation is complete, and may be tempered to lower hardness if desired.

Normalizing. In this operation, steel is heated above its upper critical temperature (A_3) and cooled in air. The purpose of this treatment is to refine the grain and to obtain a carbide size and distribution that is more favorable for carbide solution on subsequent heat treatment than the earlier as-rolled structure.

The as-rolled grain size, depending principally on the finishing temperature in the rolling operation, is subject to wide variations. The coarse grain size resulting from a high finishing temperature can be refined by normalizing to establish a uniform, relatively fine-grained microstructure.

In alloy steels, particularly if they have been slowly cooled after rolling, the carbides in the as-rolled condition tend to be massive and are difficult to dissolve on subsequent austenitization. The carbide size is subject to wide variations, depending upon the rolling and slow cooling. Here again, normalizing tends to establish a more uniform and finer carbide particle size which facilitates subsequent heat treatment.

The usual practice is to normalize at 50–80°C above the upper critical temperature; however, for some alloy steels considerably higher temperatures may be used. Heating may be carried out in any type of furnace that permits uniform heating and good temperature control.

Annealing. Annealing relieves cooling stresses induced by hot- or cold-working and softens the steel to improve its machinability or formability. It may involve only a subcritical heating to relieve stresses, recrystallize cold-worked material, or spheroidize carbides; it may involve heating above the upper critical temperature (A_3) with subsequent transformation to pearlite or directly to a spheroidized structure on cooling.

The most favorable microstructure for machinability in the low or medium carbon steels is coarse pearlite. The customary heat treatment to develop this microstructure is a full annealing, illustrated in Figure 31. It consists of austenitizing at a relatively high temperature to obtain full carbide solution, followed by slow cooling to give transformation exclusively in the high temperature end of the pearlite range. This simple heat treatment is reliable for most steels. It is, however, rather time-consuming since it involves slow cooling over the entire temperature range from the austenitizing temperature to a temperature well below that at which transformation is complete.

Isothermal Annealing. Annealing to coarse pearlite can be carried out isothermally by cooling to the proper temperature for transformation to coarse pearlite and holding until transformation is complete. This method, called isothermal annealing, is illustrated in Figure 32. It may save considerable time over the full-annealing process described previously, since neither the time from the austenitizing temperature to the transformation temperature, nor from the transformation temperature to room temperature, is critical; these may be shortened as desired. If extreme softness of the coarsest pearlite is not necessary, the transformation may be carried out at the nose of the IT curve, where the transformation is completed rapidly and the operation further expedited; the pearlite in this case is much finer and harder.

Isothermal annealing can be conveniently adapted to continuous annealing, usually in specially designed furnaces, when it is commonly referred to as cycle annealing.

Spheroidization Annealing. Coarse pearlite microstructures are too hard for optimum machinability in the higher carbon steels. Such steels are customarily annealed to develop spheroidized microstructures by tempering the as-rolled, slowly cooled, or normalized materials just below the lower critical temperature range. Such an operation is known as subcritical annealing. Full spheroidization may require long holding times at the subcritical temperature and the method may be slow, but it is simple and may be more convenient than annealing above the critical temperature.

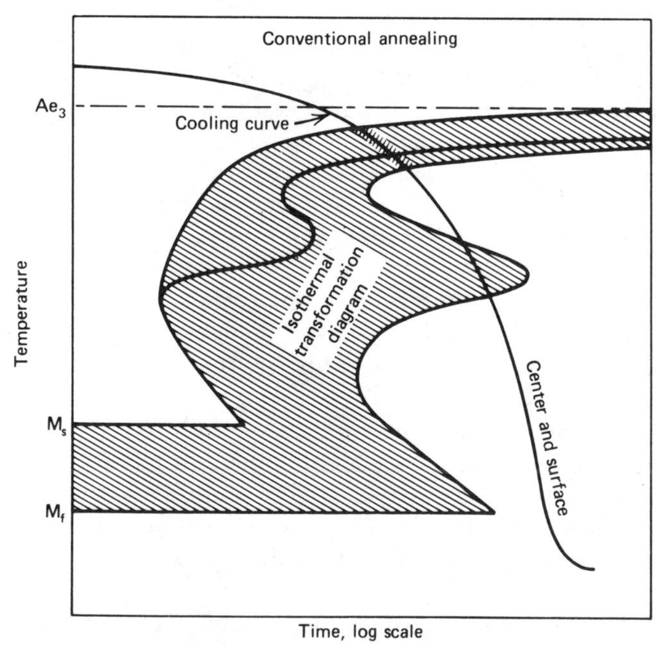

Figure 31. Transformation diagram for full annealing; the product is ferrite and pearlite.

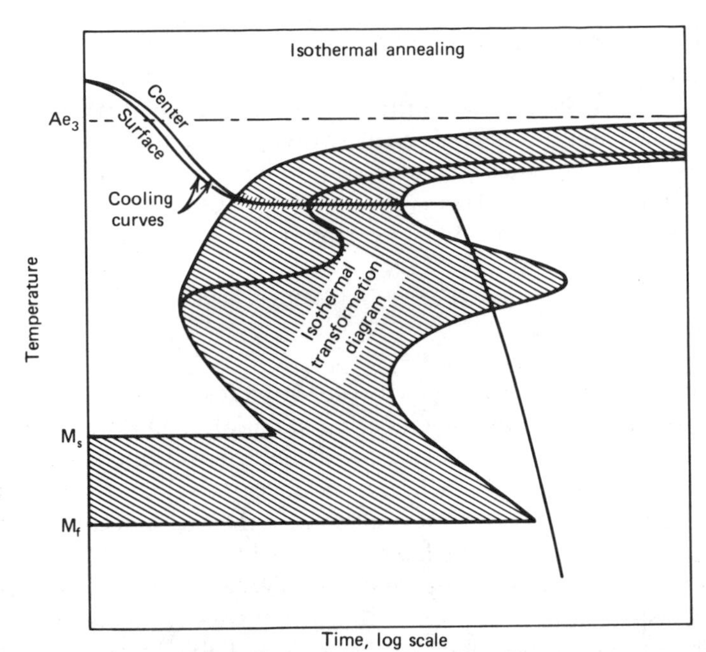

Figure 32. Transformation diagram for isothermal annealing; the product is ferrite and pearlite.

The annealing procedures described above to produce pearlite can, with some modifications, give spheroidized microstructures. If free carbide remains after austenitizing, transformation in the temperature range where coarse pearlite ordinarily would form proceeds to spheroidized rather than pearlitic microstructures. Thus, heat treatment to form spheroidized microstructures can be carried out like heat treatment for pearlite, except for the lower austenitizing temperatures. Spheroidization annealing may thus involve a slow cooling similar to the full-annealing treatment used for pearlite, or it may be a treatment similar to isothermal annealing. An austenitizing temperature not more than 55°C above the lower critical temperature is customarily used for this supercritical annealing.

Process Annealing. Process annealing is the term used for subcritical annealing of cold-worked materials. It customarily involves heating at a temperature high enough to cause recrystallization of the cold-worked material and to soften the steel. The most important example of process annealing is the box-annealing of cold-rolled low carbon sheet steel. The sheets are enclosed in a large box which can be sealed to permit the use of a controlled atmosphere to prevent oxidation. Annealing is usually carried out between 590 and 700°C. The operation usually takes ca 24 h, after which the charge is cooled slowly within the box; the entire process takes ca 40 h.

Carburizing. In carburizing, low carbon steel acquires a high carbon surface layer by heating in contact with carbonaceous materials. On quenching after carburizing, the high carbon skin hardens, whereas the low carbon core remains comparatively soft. The result is a highly wear-resistant exterior over a very tough interior. This material is particularly suitable for gears, camshafts, etc. Carburizing is most commonly carried out by packing the steel in boxes with carbonaceous solids, sealing to exclude the atmosphere, and heating to about 925°C for a period of time depending upon the depth desired; this method is called pack carburizing. Alternatively, the steel may be heated in contact with carburizing gases in which case the process is called gas carburizing;

or, least commonly, in liquid baths of carburizing salts, in which case it is known as liquid carburizing.

Nitriding. The nitrogen case-hardening process, termed nitriding, consists of subjecting machined and (preferably) heat-treated parts to the action of a nitrogenous medium, commonly ammonia gas, under conditions whereby surface hardness is imparted without requiring any further treatment. Wear resistance, retention of hardness at high temperatures, and resistance to certain types of corrosion are also imparted by nitriding.

Carbon Steels

The plain carbon steels represent by far the largest volume produced, with the most diverse applications of any engineering material, including castings, forgings, tubular products, plates, sheet and strip, wire and wire products, structural shapes, bars, and railway materials (rails, wheels, and axles). Carbon steels are made by all modern steelmaking processes and, depending upon their carbon content and intended purpose, may be rimmed, semikilled, or fully killed (71–75).

The American Iron and Steel Institute has published standard composition ranges for plain carbon steels, which in each composition range are assigned an identifying number according to a method of classification (see Table 5). In this system, carbon steels are assigned to one of three series: 10xx (nonresulfurized), 11xx (resulfurized), and 12xx (rephosphorized and resulfurized). The 10xx steels are made with low phosphorus and sulfur contents, 0.04% max and 0.050% max, respectively. Sulfur in amounts as high as 0.33% max may be added to the 11xx and as high as 0.35% max to

Table 5. Standard Numerical Designations of Plain Carbon and Constructional Alloy Steels (AISI–SAE Designations)

Series designation[a]	Types	Series designation[a]	Types
10xx	nonresulfurized carbon-steel grades	47xx	1.05% Ni–0.45% Cr–0.20% Mo
11xx	resulfurized carbon-steel grades	48xx	3.50% Ni–0.25% Mo
12xx	rephosphorized and resulfurized carbon-steel grades	50xx	0.28 or 0.40% Cr
13xx	1.75% Mn	51xx	0.80, 0.90, 0.95, 1.00, or 1.05% Cr
23xx	3.50% Ni	5xxxx	1.00% C–0.50, 1.00, or 1.45% Cr
25xx	5.00% Ni	61xx	0.80 or 0.95% Cr–0.10 or 0.15% V
31xx	1.25% Ni–0.65% Cr	86xx	0.55% Ni–0.50 or 0.65% Cr–0.20% Mo
33xx	3.50% Ni–1.55% Cr	87xx	0.55% Ni–0.50% Cr–0.25% Mo
40xx	0.25% Mo	92xx	0.85% Mn–2.00% Si
41xx	0.50 or 0.95% Cr–0.12 or 0.20% Mo	93xx	3.25% Ni–1.20% Cr–0.12% Mo
43xx	1.80% Ni–0.50 or 0.80% Cr–0.25% Mo	98xx	1.00% Ni–0.80% Cr–0.25% Mo
46xx	1.55 or 1.80% Ni–0.20 or 0.25% Mo		

[a] The first figure indicates the class to which the steel belongs; 1xxx indicates a carbon steel, 2xxx a nickel steel, and 3xxx a nickel–chromium steel. In the case of alloy steels, the second figure generally indicates the approximate percentage of the principal alloying element. Usually, the last two or three figures (represented in the table by x) indicate the average carbon content in points or hundredths of 1 wt %. Thus, a nickel steel containing ca 3.5% nickel and 0.30% carbon would be designated as 2330.

the 12xx steels to improve machinability. In addition, phosphorus up to 0.12% max may be added to the 12xx steels to increase stiffness.

In identifying a particular steel, the letters x are replaced by two digits representing average carbon content; for example, an AISI No. 1040 steel would have an average carbon content of 0.40%, with a tolerance of ±0.03%, giving a range of 0.37–0.44% carbon.

Properties. The properties of plain carbon steels are governed principally by carbon content and microstructure. The fact that properties can be controlled by heat treatment has been discussed under Metallography and Heat Treatment. Most plain carbon steels, however, are used without heat treatment.

The properties of plain carbon steels may be modified by residual elements other than the carbon, manganese, silicon, phosphorus, and sulfur that are always present, as well as gases, especially oxygen, nitrogen, and hydrogen, and their reaction products. These incidental elements are usually acquired from scrap, deoxidizers, or the furnace atmosphere. The gas content depends mostly upon melting, deoxidizing, and pouring procedures; consequently, the properties of plain carbon steels depend heavily upon the manufacturing techniques.

The average mechanical properties of as-rolled 2.5-cm bars of carbon steels as a function of carbon contents are shown in Figure 33. This diagram is an illustration

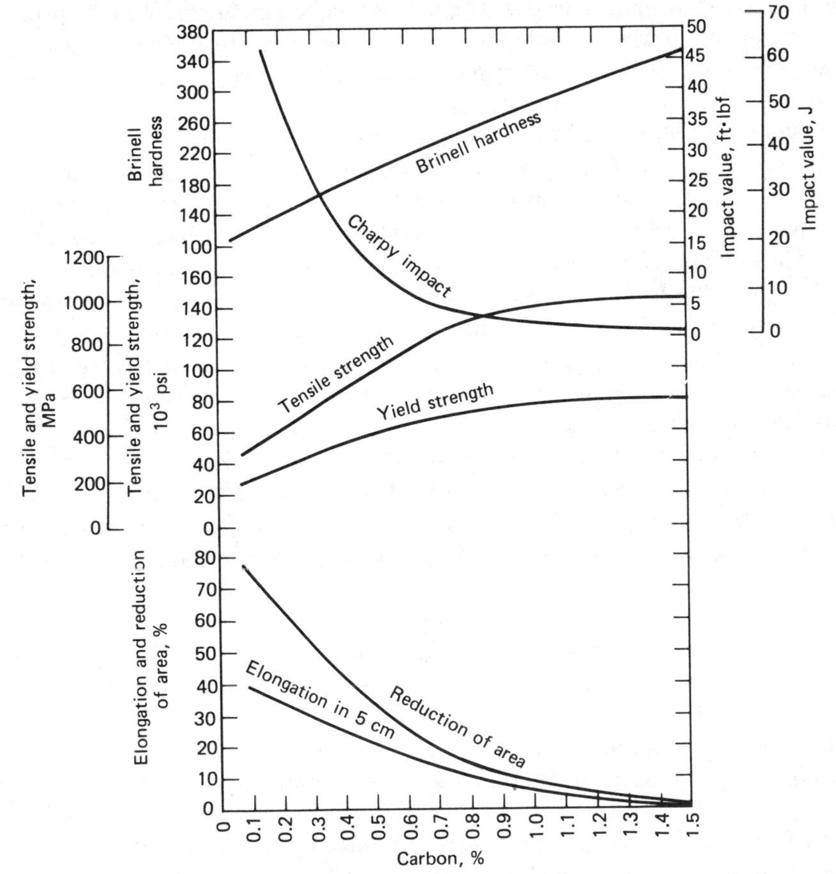

Figure 33. Variations in average mechanical properties of as-rolled 2.5-cm bars of plain carbon steels, as a function of carbon content (1).

of the effect of carbon content when microstructure and grain size are held approximately constant.

Microstructure and Grain Size. The carbon steels with relatively low hardenability are predominantly pearlitic in the cast, rolled, or forged state. The constituents of the hypoeutectoid steels are, therefore, ferrite and pearlite, and of the hypereutectoid steels, cementite and pearlite. As discussed earlier, the properties of such pearlitic steels depend primarily upon the interlamellar spacing of the pearlite and the grain size. Both hardness and ductility increase as the interlamellar spacing or the pearlite-transformation temperature decreases, whereas the ductility increases with decreasing grain size. The austenite-transformation behavior in carbon steel is determined almost entirely by carbon and manganese content; the effects of phosphorus and sulfur are almost negligible; and the silicon content is normally so low as to have no influence. The carbon content is ordinarily chosen in accordance with the strength desired, and the manganese content selected to produce suitable microstructure and properties at that carbon level under the given cooling conditions.

Microstructure of Cast Steels. Cast steel is generally coarse-grained, since austenite forms at high temperature and the pearlite is usually coarse, in as much as cooling through the critical range is slow, particularly if the casting is cooled in the mold. In hypoeutectoid steels, ferrite ordinarily precipitates at the original austenite grain boundaries during cooling. In hypereutectoid steels, cementite is similarly precipitated. Such mixtures of ferrite or cementite and coarse pearlite have poor strength and ductility properties, and heat treatment is usually necessary to obtain suitable microstructures and properties in cast steels.

Hot Working. Many carbon steels are used in the form of as-rolled finished sections. The microstructure and properties of these sections are determined largely by composition, rolling procedures, and cooling conditions. The rolling or hot working of these sections is ordinarily carried out in the temperature range in which the steel is austenitic, with four principal effects: Considerable homogenization occurs during the heating for rolling, tending to eliminate dendrite segregation present in the ingot; the dendritic structure is broken up during rolling; recrystallization takes place during rolling, with final austenitic grain size determined by the temperature at which the last passes are made (the finishing temperature); and dendrites and inclusions are reoriented, with markedly improved ductility, in the rolling direction.

Thus, homogeneity and grain size of the austenite is largely determined by the rolling technique. However, the recrystallization characteristics of the austenite and, therefore, the austenite grain size characteristic at a given finishing temperature, may be affected markedly by the steelmaking technique, particularly with regard to deoxidation.

The distribution of the ferrite or cementite and the nature of the pearlite are determined by the cooling rate after rolling. Since the usual practice is air cooling, the final microstructure and the properties of as-rolled sections depend primarily on composition and section size.

Cold Working. The manufacture of wire, sheet, strip and tubular products often includes cold working, with effects that may be eliminated by annealing; however, some products, particularly wire, are used in the cold-worked condition. The most pronounced effects of cold working are increased strength and hardness and decreased ductility. The effect of cold working on the tensile strength of plain carbon steel is shown in Figure 34.

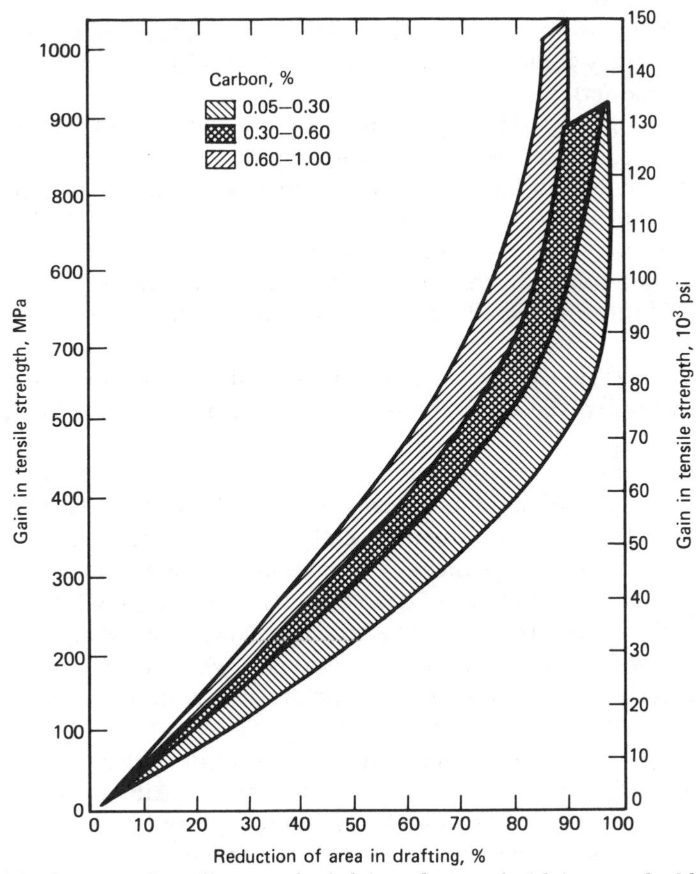

Figure 34. Increase of tensile strength of plain carbon steel with increased cold working.

Upon reheating cold-worked steel to the recrystallization temperature (400°C) or above, depending upon composition, extent of cold work, and other variables, the original microstructure and properties may be restored.

Heat Treatment. Although most wrought (rolled or forged) carbon steels are used without a final heat treatment, it may be employed to improve the microstructure and properties for specific applications.

Annealing is applied when better machinability or formability is required than would be obtained with the as-rolled microstructure. A complete annealing is generally employed to form coarse pearlite, although a subcritical annealing or spheroidizing treatment is occasionally used. Process annealing for optimum formability is universal with cold-rolled strip and sheet and cold-worked tubing.

The grain size of as-rolled products depends largely upon the finishing temperature but is difficult to control. A final normalizing treatment from a relatively low temperature may establish a fine, uniform grain size for applications in which ductility and toughness are critical.

Quenching and tempering of plain carbon steels are being more frequently applied. In one type of treatment, the steel is heat treated to produce tempered martensite, but because of relatively low hardenability, the operation is limited to section sizes of not more than 10–13 mm. In the other type, large sections of plain carbon steels are quenched and tempered to produce fine pearlite microstructures with much better strength and ductility than those of the coarse pearlite microstructures in as-rolled or normalized products.

Thin sections of carbon steels (≤5 mm) are particularly suitable for the production of parts requiring toughness at high hardness by austempering.

Residual Elements. In addition to the carbon, manganese, phosphorus, sulfur, and silicon which are always present, carbon steels may contain small amounts of gases, such as hydrogen, oxygen, or nitrogen, introduced during the steelmaking process; nickel, copper, molybdenum, chromium, and tin, which may be present in the scrap; and aluminum, titanium, vanadium, or zirconium, which may be introduced during deoxidation.

Oxygen and nitrogen cause the phenomenon called aging, manifested as a spontaneous increase in hardness at room temperature and believed to be a precipitation effect.

An embrittling effect, the mechanism of which is not completely understood, is caused by a hydrogen content of more than ca 3 ppm. As discussed earlier, the content of hydrogen and other gases can be reduced by vacuum degassing.

Alloying elements such as nickel, chromium, molybdenum, and copper, which may be introduced with scrap, do, of course, increase the hardenability although only slightly since the concentrations are ordinarily low. However, the heat-treating characteristics may change, and for applications in which ductility is important, as in low carbon steels for deep drawing, the increased hardness imparted by these elements may be harmful.

Tin, even in low amounts, is harmful in steels for deep drawing; for most applications, however, the effect of tin in the quantities ordinarily present is negligible.

Aluminum is generally desirable as a grain refiner and tends to reduce the susceptibility of carbon steel to aging associated with strain. Unfortunately, it tends to promote graphitization and is, therefore, undesirable in steels used at high temperatures. The other elements that may be introduced as deoxidizers, such as titanium, vanadium, or zirconium, are ordinarily present in such small amounts as to be ineffective.

Dual-Phase Sheet Steels

Dual-phase steels derive their name from their unique microstructure of a mixture of ferrite and martensite phases. This microstructure is developed in hot- and cold-rolled sheet by using a combination of steel composition and heat treatment that changes an initial microstructure of ferrite and pearlite (or iron carbide) to ferrite and martensite (76–82).

Normally, high strength hot-rolled sheets are manufactured by hot rolling and cooling on a hot-strip mill, which produces a microstructure of ferrite and pearlite. On heating to ca 750–850°C, a microstructure of ferrite and austenite is produced, and by cooling at an appropriate rate (which depends on steel composition or hardenability), the austenite is transformed to a very hard martensite phase contained within the soft, ductile ferrite matrix. The final ferrite–martensite microstructure, which may contain 5–30% martensite (increasing amounts increase the strength), may be considered as a composite; the strength may therefore be estimated, according to a simple law of mixtures, from the strengths and volume fractions of the individual phases.

The properties of a steel (0.11 C, 1.6 Mn, 0.60 Si, 0.04 V) in the hot-rolled (ferrite and pearlite) and in the heat-treated (ferrite–martensite double-phase) state are given below (to convert MPa to psi, multiply by 145):

	Yield strength, MPa	*Tensile strength, MPa*	*Total elongation, %*
hot rolled	480	4275	24
dual phase	345	4516	32

Although the tensile (ultimate) strength of the steel is little affected by heat treating, the yield strength is substantially reduced and the ductility markedly improved. The low yield strength allows for the easy initiation of plastic deformation during press forming of dual-phase sheet material. However, dual-phase steels have the unique capacity to strain harder rapidly so that after a few percent deformation (3–5%) the yield strength exceeds 550 MPa (80,000 psi).

Dual-phase steels have found application in automotive bumpers and wheels where high ductility is required to form the complex shapes. The development of very high strength of ca 550 MPa (80,000 psi) allows thinner, lighter weight sheet to be used, instead of steels having strengths of only 200–350 MPa (30,000–50,000 psi). However the heat-treating step increases production costs.

Alloy Steels

As a class, alloy steels may be defined as steels having enhanced properties owing to the presence of one or more special elements or larger proportions of elements (such as silicon and manganese) than are ordinarily present in carbon steel. Steels containing alloying elements are classified into high strength low alloy (HSLA) steels; AISI alloy steels; alloy tool steels; stainless steels; heat-resistant steels; and electrical steels (silicon steels). In addition, there are numerous steels, some with proprietary compositions, with exceptional properties developed to meet unusually severe requirements. The relatively small production of such steels does not reflect their engineering importance (83–84) (See also High temperature alloys; Tool materials).

Functions of Alloying Elements. In the broadest sense, alloy steels may contain up to ca 50% of alloying elements which directly enhance properties. Examples are the increased corrosion resistance of high chromium steels, the enhanced electric properties of silicon steels, the improved strength of the high strength low alloy steels, and the improved hardenability and tempering characteristics of the AISI alloy steels.

Thermomechanical Treatment. The conventional method of producing high strength steels has been to add alloy elements such as Cr, Ni, and Mo to the liquid steel. The resulting alloy steels are often heat treated after rolling to develop the desired strength without excessive loss of toughness (resistance to cracking upon impact). In the 1970s, a less expensive method was developed to produce high strength low alloy (HSLA) steels with improved toughness and yield strength ranging from 400–600 MPa (60,000–85,000 psi). In this thermomechanical treatment, the working of the steel is controlled while its temperature is changing and it is being hot rolled between 1300 and 750°C to its final thickness (85–89). The HSLA steels that are commonly strengthened by thermomechanical treatment, also called controlled rolling, generally contain 0.05–0.20% carbon, 0.40–1.60% manganese, 0.05–0.50% silicon, plus 0.01–0.30%

of one or more of the following elements: aluminum, molybdenum, niobium, titanium, and vanadium. Thermomechanical treatment usually involves a substantial degree of rolling, such as a 50–75% decrease in thickness in the last rolling passes; temperature maintained between 750 and 950°C; a controlled rate of cooling after hot rolling. This procedure gives a very fine steel grain size and imparts strength and toughness. Steels so treated are increasingly used in automobiles and oil and gas pipelines.

High Strength Low Alloy Steels (HSLA). These steels are categorized according to mechanical properties, particularly the yield point; for example, within certain thickness limits they have yield points ranging from 310–450 MPa (45,000–65,000 psi) as compared with 225–250 MPa (33,000–36,000 psi) for structural carbon steel. This classification is in contrast to the usual classification into plain-carbon or structural-carbon steels, alloy steels, and stainless steels on the basis of alloying elements.

The superior mechanical properties of HSLA steels are obtained by the addition of alloying elements (other than carbon), singly and in combination. Each steel must meet similar minimum mechanical requirements. They are available for structural use as sheets, strips, bars, plates, and in various other shapes. They are not to be considered as special-purpose steels or requiring heat treatment.

To be of commercial interest, HSLA steels must offer economic advantages. They should be much stronger and often tougher than structural carbon steel. In addition, they must have sufficient ductility, formability, and weldability to be fabricated by customary techniques. Improved resistance to corrosion is often required. The abrasion resistance of these steels is somewhat higher than that of structural carbon steel containing 0.15–0.20% carbon. Superior mechanical properties permits the use of HSLA steels in structures with a higher unit working stress; this generally permits reduced section thickness with corresponding decrease in weight. Thus, HSLA steels may be substituted for structural carbon steel without change in section, resulting in a stronger and more durable structure without weight increase.

AISI Alloy Steels. The American Iron and Steel Institute defines alloy steels as follows: "By common custom steel is considered to be alloy steel when the maximum of the range given for the content of alloying elements exceeds one or more of the following limits: manganese, 1.65%; silicon, 0.60%; copper, 0.60%; or in which a definite range or a definite minimum quantity of any of the following elements is specified or required within the limits of the recognized field of constructional alloy steels: aluminum, boron, chromium up to 3.99%, cobalt, columbium (niobium), molybdenum, nickel, titanium, tungsten, vanadium, zirconium, or any other alloying element added to obtain a desired alloying effect" (90). Steels that contain 4.00% or more of chromium are included by convention among the special types of alloy steels known as stainless steels discussed below (91–97).

Steels that fall within the AISI definition have been standardized and classified jointly by AISI and SAE as shown in Table 5. They represent by far the largest alloy-steels production and are generally known as AISI alloy steels. They are also commonly referred to as constructional alloy steels.

The effect of the alloying elements on AISI steels is indirect since alloying elements control microstructure through their effect on hardenability. They permit the attainment of desirable microstructures and properties over a much wider range of sizes and sections than is possible with carbon steels.

Alloy Tool Steels. Alloy tool steels are classified roughly into three groups: Low alloy tool steels, to which alloying elements have been added to impart hardenability higher than that of plain carbon tool steels; accordingly, they may be hardened in heavier sections or with less drastic quenches to minimize distortion; intermediate alloy tool steels usually contain elements such as tungsten, molybdenum, or vanadium, which form hard, wear-resistant carbides; high speed tool steels contain large amounts of carbide-forming elements, that serve not only to furnish wear-resisting carbides, but also promote the phenomenon known as secondary hardening and thereby increase resistance to softening at elevated temperatures (see Carbides).

Stainless Steels. Stainless steels are more resistant to rusting and staining than plain carbon and low alloy steels (98–106). This superior corrosion resistance is due to the addition of chromium. Although other elements, such as copper, aluminum, silicon, nickel, and molybdenum, also increase corrosion resistance, they are limited in their usefulness.

No single nation can claim credit for the development of the stainless steels; Germany, the UK, and the United States share alike in their development. In the UK in 1912, during the search for steel that would resist fouling in gun barrels, a corrosion-resistant composition was reported of 12.8% chromium and 0.24% carbon. It was suggested that this composition be used for cutlery. In fact, the composition of AISI Type 420 steel (12–14% chromium, 0.15% carbon) is similar to that of the first corrosion-resistant steel.

The higher chromium–iron alloys were developed in the United States from the early 20th century on, when the effect of chromium on oxidation resistance at 1090°C was first noticed. Oxidation resistance increased markedly as the chromium content was raised above 20%. Even now and with steels containing appreciable quantities of nickel, 20% chromium seems to be the minimum amount necessary for oxidation resistance at 1090°C.

The austenitic iron–chromium–nickel alloys were developed in Germany around 1910 in a search for materials for use in pyrometer tubes. Further work led to the versatile 18% chromium–8% nickel steels, so-called 18–8, which are widely used today.

The chromium content seems to be the controlling factor and its effect may be enhanced by additions of molybdenum, nickel, and other elements. The mechanical properties of the stainless steels, like those of the plain carbon and lower alloy steels, are functions of structure and composition. Thus, austenitic steels possess the best impact properties at low temperatures and the highest strength at high temperatures, whereas martensitic steels are the hardest at room temperature. Thus, stainless steels, which are available in a variety of structures, exhibit a range of mechanical properties which, combined with their excellent corrosion resistance, makes these steels highly versatile from the standpoint of design.

The standard AISI and SAE types are identified in Table 6.

Martensitic Stainless Steels. Martensitic stainless steels are iron–chromium alloys that are hardenable by heat treatment. They include Types 403, 410, 414, 416, 420, 431, 440A, 440B, 440C, 501, and 502 (see Table 6). The most widely used is Type 410, containing 11.50–13.50% chromium and <0.15% carbon. In the annealed condition, this grade may be drawn or formed. It is an air-hardening steel, affording a wide range of properties by heat treatment. In sheet or strip form, Type 410 is used extensively in the petroleum industry for ballast trays and liners. It is also used for parts of furnaces operating below 650°C, and for blades and buckets in steam turbines.

Table 6. Standard Stainless and Heat-Resisting Steel Products

AISI type number	SAE type[a] number	Chemical composition, %			
		Carbon	Chromium	Nickel	Other
201	30201	0.15 max	16.00–18.00	3.50–5.50	Mn 5.50–7.50[b] P 0.06 max[c] N 0.25 max
202	30202	0.15 max	17.00–19.00	4.00–6.00	Mn 7.50–10.00 P 0.06 max N 0.25 max
301	30301	0.15 max	16.00–18.00	6.00–8.00	
302	30302	0.15 max	17.00–19.00	8.00–10.00	
302B	30302B	0.15 max	17.00–19.00	8.00–10.00	Si 2.00–3.00[d] P 0.20 max S 0.15 min[e]
303	30303	0.15 max	17.00–19.00	8.00–10.00	Mo 0.60 max
303Se	30303Se	0.15 max	17.00–19.00	8.00–10.00	P 0.20 max S 0.06 max Se 0.15 min
303SeA		0.08 max	17.25–18.75	11.50–13.00	Se 0.15–0.35
304	30304	0.08 max	18.00–20.00	8.00–10.00	
304L		0.030 max	18.00–20.00	8.00–10.00	
305	30305	0.12 max	17.00–19.00	10.00–13.00	
307		0.07–0.15	19.50–21.50	9.00–10.50	Mo residual only
308	30308	0.08 max	19.00–21.00	10.00–12.00	
308 Mod		0.07–0.15	19.50–21.50	9.00–10.50	Mo residual only
309	30309	0.20 max	22.00–24.00	12.00–15.00	
309S	30309S	0.08 max	22.00–24.00	12.00–15.00	
309SCb		0.08 max	22.00–24.00	12.00–15.00	NbTa min, 10 times carbon Ta 0.10 max
309SCbTa		0.08 max	22.00–24.00	12.00–15.00	NbTa min, 10 times carbon
310	30310	0.25 max	24.00–26.00	19.00–22.00	
314	30314	0.25 max	23.00–26.00	19.00–22.00	
316	30316	0.08 max	16.00–18.00	10.00–14.00	Mo 2.00–3.00
316L	30316L	0.030 max	16.00–18.00	10.00–14.00	Mo 2.00–3.00
317	30317	0.08 max	18.00–20.00	11.00–15.00	Mo 3.00–4.00
318		0.10 max	16.00–18.00	10.00–14.00	Mo 2.00–3.00 NbTa min, 10 times carbon
D319		0.07 max	17.50–19.50	11.00–15.00	Mn 2.00 max Si 1.00 max Mo 2.25–3.00
321	30321	0.08 max	17.00–19.00	9.00–12.00	Ti min, 5 times carbon
330		0.25 max	14.00–16.00	33.00–36.00	
347	30347	0.08 max	17.00–19.00	9.00–13.00	NbTa min, 10 times carbon
348	30348	0.08 max	17.00–19.00	9.00–13.00	NbTa min, 10 times carbon Ta 0.10 max Co 0.20 max
403	51403	0.15 max	11.50–13.00		
405	51405	0.08 max	11.50–14.50		Al 0.10–0.30

AISI type number	SAE type[a] number	Chemical composition, %			
		Carbon	Chromium	Nickel	Other
410	51410	0.15 max	11.50–13.50		
410Mo		0.15 max	11.50–13.50		Mo 0.40–0.60
414	51414	0.15 max	11.50–13.50	1.25–2.50	
416	51416	0.15 max	12.00–14.00		P 0.06 max S 0.15 min Mo 0.60 max
410Se	51410Se	0.15 max	12.00–14.00		P 0.06 max S 0.06 max Se 0.15 min
420	51420	>0.15	12.00–14.00		
420F	51420F	>0.15	12.00–14.00		S[f]
430	51430	0.12 max	14.00–18.00		
430F	51430F	0.12 max	14.00–18.00		P 0.06 max S 0.15 min Mo 0.60 max
430Ti		0.10 max	16.00–18.00		Ti 0.30–0.70
431	51431	0.20 max	15.00–17.00	1.25–2.50	
434A		0.05–0.10	15.00–17.00		Cu 0.75–1.10
442	51442	0.25 max	18.00–23.00		
446	51446	0.20 max	23.00–27.00		N 0.25 max
501	51501	>0.10	4.00–6.00		Mo 0.40–0.65
502	51502	0.10 max	4.00–6.00		Mo 0.40–0.65

[a] SAE chemical composition (ladle) ranges may differ slightly in certain elements from AISI limits.
[b] Manganese: All steels of AISI Type 300 series—2.00% max. All steels of AISI Type 400 and 500 series—1.00% max except 416, 416Se, 430F, and 430Se (1.25% max) and Type 446 (1.50% max).
[c] Phosphorus: All steels of AISI Type 200 series—0.060% max. All steels of AISI Type 300 series—0.045% max except Types 303 and 303Se (0.20% max). All steels of AISI Type 400 and 500 series—0.040% max except Types 416, 416Se, 430F, and 430FSe (0.060% max).
[d] Silicon: All steels of AISI Type 200, 300, 400, and 500 series—1.00% max except where otherwise indicated.
[e] Sulfur: All steels of AISI Type 200, 300, 400 and 500 series—0.30% max except Types 303, 416, and 430F (0.15% min) and Types 303Se, 416Se, and 430FSe (0.060% max).
[f] No restriction.

Type 420, with ca 0.35% carbon and a resultant increased hardness, is used for cutlery. In bar form, it is used for valves, valve stems, valve seats, and shafting where corrosion and wear resistance are needed. Type 440 may be employed for surgical instruments, especially those requiring a durable cutting edge. The necessary hardness for different applications can be obtained by selecting grade A, B, or C, with increasing carbon content in that order.

Other martensitic grades are Types 501 and 502, the former with >0.10% and the latter <0.10% carbon; both contain 4.6% chromium. These grades are also air-hardening, but do not have the corrosion resistance of the 12% chromium grades. Types 501 and 502 have wide application in the petroleum industry for hot lines, bubble towers, valves, and plates.

Ferrite Stainless Steels. These steels are iron–chromium alloys that are largely ferritic and not hardenable by heat treatment (ignoring the 475°C embrittlement). They include Types 405, 430, 430F, and 446 (see Table 6).

The most common ferritic grade is Type 430, containing 0.12% carbon or less and

14–18% chromium. Because of its higher chromium content, the corrosion resistance of Type 430 is superior to that of the martensitic grades. Furthermore, Type 430 may be drawn, formed, and with proper techniques, welded. It is widely used for automotive and architectural trim. It is employed in equipment for the manufacture and handling of nitric acid to which it is resistant. Type 430 does not have high creep strength but is suitable for some types of service up to 815°C and thus has application in combustion chambers for domestic heating furnaces.

The high chromium content of Type 446 (23–27% chromium) imparts excellent heat resistance, although its high temperature strength is only slightly better than that of carbon steel. Type 446 is used in sheet or strip form up to 1150°C. This grade does not have the good drawing characteristics of Type 430, but it may be formed. Accordingly, it is widely used for furnace parts such as muffles, burner sleeves, and annealing baskets. Its resistance to nitric and other oxidizing acids makes it suitable for chemical-processing equipment.

Austenitic Stainless Steels. These steels are iron–chromium–nickel alloys not hardenable by heat treatment and predominantly austenitic. They include Types 301, 302, 302B, 303, 304, 304L, 305, 308, 309, 310, 314, 316, 316L, 317, 321, and 347. In some recently developed austenitic stainless steels, all or part of the nickel is replaced by manganese and nitrogen in proper amounts, as in one proprietary steel and Types 201 and 202 (see Table 6).

The most widely used austenitic stainless steel is Type 302, known as 18–8; it has excellent corrosion resistance and, because of its austenitic structure, excellent ductility. It may be deep drawn or strongly formed. It can be readily welded, but carbide precipitation must be avoided in and near the weld by cooling rapidly enough after welding. Where carbide precipitation presents problems, Types 321, 347, or 304L may be used. The applications of Type 302 are wide and varied, including kitchen equipment and utensils; dairy installations; transportation equipment; and oil-, chemical-, paper-, and food-processing machinery.

The low nickel content of Type 301 causes it to harden faster than Type 302 because of reduced austenite stability. Accordingly, although Type 301 can be drawn successfully, its drawing properties are not as good as those of Type 302. For the same reason, Type 301 can be cold-rolled to very high strength.

Type 301, because of its lower carbon content, is not as prone as Type 302 to give carbide precipitation problems in welding. In addition, its somewhat higher chromium content makes it slightly more resistant to corrosion. It is used to withstand severe corrosive conditions in the paper, chemical, and other industries.

The austenitic stainless steels are widely used for high temperature service.

Types 321 and 347, with additions of titanium and niobium, respectively, are used in welding applications and high temperature service under corrosive conditions. Type 304L may be used as an alternative for Types 321 and 347 in welding and stress-relieving applications below 426°C (see Welding).

The addition of 2–4% molybdenum to the basic 18–8 composition produces Types 316 and 317 with improved corrosion resistance. These grades are employed in the textile, paper, and chemical industries where strong sulfates, chlorides, and phosphates and reducing acids such as sulfuric, sulfurous, acetic, and hydrochloric acids are used in such concentrations that the use of corrosion-resistant alloys is mandatory. Types 316 and 317 have the highest creep and rupture strengths of any commercial stainless steels.

The austenitic stainless steels most resistant to oxidation are Types 309 and 310. Because of their high chromium and nickel contents, these steels resist scaling at temperatures up to 1090 and 1150°C and, consequently, are used for furnace parts and heat exchangers. They are somewhat harder and not as ductile as the 18–8 types, but they may be drawn and formed. They can be welded readily and have increasing use in the manufacture of jet-propulsion motors and industrial-furnace equipment (see High temperature alloys).

For applications requiring good machinability, Type 303 containing sulfur or selenium may be used.

High-Temperature Service, Heat-Resisting Steels. The term high temperature service comprises many types of operations in many industries. Conventional high temperature equipment includes steam boilers and turbines, gas turbines, cracking stills, tar stills, hydrogenation vessels, heat-treating furnaces, and fittings for diesel and other internal-combustion engines. Numerous steels are available from which to select the one best suited for each of the foregoing applications. Where unusual conditions occur, modification of the chemical composition may adapt an existing steel grade to service conditions. In some cases, however, entirely new alloy combinations must be developed to meet service requirements. For example, the aircraft and missile industries have encountered design problems of increased complexity, requiring metals of great high temperature strength for both power plants and structures, and new steels are constantly under development to meet these requirements (107–108).

A number of steels suitable for high temperature service are given in Table 7.

The design of load-bearing structures for service at room temperature is generally based on the yield strength or for some applications on the tensile strength. The metal behaves essentially in an elastic manner, that is, the structure undergoes an elastic deformation immediately upon load application and no further deformation occurs with time; when the load is removed, the structure returns to its original dimensions.

At high temperature, the behavior is different. A structure designed according to the principles employed for room-temperature service continues to deform with time after load application, even though the design data may have been based on tension tests at the temperature of interest. This deformation with time is called creep,

Table 7. Alloy Composition of High Temperature Steels

Ferritic steels	Austenitic steels	AISI type
0.5% Mo	18% Cr–8% Ni	304
0.5% Cr–0.5% Mo	18% Cr–8% Ni with Mo	316
1% Cr–0.5% Mo	18% Cr–8% Ni with Ti	321
2% Cr–0.5% Mo	18% Cr–8% Ni with Nb	347
2.25% Cr–1% Mo	25% Cr–12% Ni	309
3% Cr–0.5% Mo–1.5% Si	25% Cr–20% Ni	310
5% Cr–0.5% Mo–1.5% Si		
5% Cr–0.5% Mo, with Nb added		
5% Cr–0.5% Mo, with Ti added		
9% Cr–1% Mo		
12% Cr		410
17% Cr		430
27% Cr		446

since at the design stresses at which it first was recognized occurred at a relatively low rate.

In spite of the fact that plain carbon steel has lower resistance to creep than high temperature alloy steels, it is widely used in such applications up to 540°C, where rapid oxidation commences and a chromium-bearing steel must be employed. Low alloy steels containing small amounts of chromium and molybdenum have higher creep strengths than carbon steel and are employed where materials of higher strength are needed. Above ca 540°C, the amount of chromium required to impart oxidation resistance increases rapidly. The 2% chromium steels containing molybdenum are useful up to ca 620°C, whereas 10–14%-chromium steels may be employed up to ca 700–760°C. Above this temperature, the austenitic 18–8 steels are commonly used; their oxidation resistance is considered adequate up to ca 815°C. For service between 815 and 1090°C, steels containing 25% chromium and 20% nickel, or 27% chromium are used.

The behavior of steels at high temperature is quite complex, and only a few design considerations have been mentioned here (see High temperature alloys).

Quenched and Tempered Low Carbon Constructional Alloy Steels. A class of quenched and tempered low carbon constructional alloy steels has been very extensively used in a wide variety of applications such as pressure vessels, mining and earth-moving equipment, and in large steel structures (109–111).

As a general class, these steels are referred to as low carbon martensites to differentiate them from constructional alloy steels of higher carbon content, such as AISI alloy steels, that develop high carbon martensite upon quenching. They are characterized by a relatively high strength, with minimum yield strengths of 690 MPa (100,000 psi), toughness down to −45°C, and weldability with joints showing full joint efficiency when welded with low hydrogen electrodes. They are most commonly used in the form of plates, but also sheet products, bars, structural shapes, forgings, or semifinished products.

Several steel-producing companies manufacture such steels under various trade names; their compositions are proprietary.

Maraging Steels. A group of high nickel martensitic steels called maraging steels contain so little carbon that they are referred to as carbon-free iron–nickel martensites (112–113).

Iron–carbon martensite is hard and brittle in the as-quenched condition and becomes softer and more ductile when tempered. Carbon-free iron–nickel martensite, on the other hand, is relatively soft and ductile and becomes hard, strong, and tough when subjected to an aging treatment at 480°C.

The first iron–nickel martensitic alloys contained ca 0.01% carbon, 20 or 25% nickel, and 1.5–2.5% aluminum and titanium. Later an 18% nickel steel containing cobalt, molybdenum, and titanium was developed, and still more recently a series of 12% nickel steels containing chromium and molybdenum came on the market.

By adjusting the content of cobalt, molybdenum, and titanium, the 18% nickel steel can attain yield strengths of 1,380–2,070 MPa (200,000–300,000 psi) after the aging treatment. Similarly, yield strengths of 12% nickel steel in the range of 1,035–1,380 MPa (150,000–200,000 psi) can be developed by adjusting its composition.

Silicon-Steel Electrical Sheets. The silicon steels are characterized by relatively high permeability, high electrical resistance, and low hysteresis loss when used in magnetic circuits (see Magnetic materials). First patented in the UK around 1900, the silicon steels permitted the development of more powerful electrical equipment and have furthered the rapid growth of the electrical power industry. Steels containing 0.5–5% silicon are produced in sheet form for the laminated magnetic cores of electrical equipment and are referred to as electrical sheets (114–116).

The grain-oriented steels, containing ca 3.25% silicon, are used in the highest efficiency distribution and power transformers and in large turbine generators. They are processed in a special way to give them directional properties related to orientation of the crystals making up the structure of the steel in a preferred direction.

The nonoriented steels are subdivided into low silicon steels, containing ca 0.5–1.5% silicon, used mainly in rotors and stators of motors and generators. Steels containing ca 1% silicon are used for reactors, relays, and small intermittent-duty transformers.

Intermediate-silicon steels (2.5–3.5% Si) are used in motors and generators of average to high efficiency and in small- to medium-size intermittent-duty transformers, reactors, and motors.

High silicon steels (ca 3.75–5.00% Si) are used in power transformers and high efficiency motors, generators, and transformers, and in communications equipment.

Economic Aspects

The production of steel is of great importance in most countries because modern civilization depends heavily on steel, the raw material for many industries. As a result, most countries have an active steel industry.

Production. World production of raw steel (ingots and continuous-cast products) is shown in Table 8 (117–121); U.S. production since 1940 is given in Table 9. The leading producers in 1979 were the USSR, United States, Japan, the FRG, and the People's Republic of China. Between 1969 and 1979, world steel production has increased at an annual rate of 3%, and the world population increased at an annual rate of 1.8%.

Prices. The U.S. finished-steel composite prices are shown in Table 10. This composite price is a weighted price for steel bars, structural shapes, plates, wire, rails, black pipe, and hot- and cold-rolled sheet and strip. From 1960–1979, this price increased at an average annual rate of 7.3%.

Products. The various steel products manufactured by the U.S. steel industry and the productions are shown in Table 11. Production is led by hot- and cold-rolled sheets, followed by hot-rolled bars and plates. The distribution of U.S. shipments to various industries is given in Table 12. The largest steel consumers are the automotive and construction industries.

U.S. exports and imports of steel products are shown in Table 13. In 1980, imports exceeded exports by over 10^7 t; the largest category was sheets and strip. In 1980, imports accounted for about 17% of the steel products consumed in the United States, whereas in 1960 imports accounted for only 5%.

Health and Safety Factors

The hazards associated with steelmaking are many and varied, but have been sharply reduced by industry-wide efforts.

Table 8. World Raw-Steel Production, 1000 Metric Tons

	1969	1979[a]
Argentina	1,690	3,200
Australia	7,032	8,200
Austria	3,926	4,900
Belgium	12,832	13,100
Brazil	4,925	13,900
Bulgaria	1,515	2,500
Canada	9,351	16,000
People's Republic of China	16,000	32,000
Czechoslovakia	10,802	15,500
France	22,510	23,600
FRG	45,316	46,800
GDR	5,140	7,000
Hungary	3,032	3,800
India	6,557	10,100
Italy	16,428	24,000
Japan	82,166	112,000
People's Republic of Korea[a]	2,000	5,100
Republic of Korea	373	7,600
Luxembourg	5,521	5,000
Mexico	3,467	7,000
Netherlands	4,712	5,800
Poland	11,251	19,500
Romania	5,540	12,000
South Africa	4,625	8,900
Spain	5,982	12,100
Sweden	5,322	4,700
United Kingdom	26,896	22,000
USSR	110,315	153,000
United States	128,151	124,000
Yugoslavia	2,220	3,500
others	9,038	21,500
Total	*574,635*	*748,300*

[a] Estimates (122).

Table 9. U.S. Raw-Steel Production, 1000 Metric Tons[a]

Year	BOP	Electric furnace	Open hearth	Bessemer	Total
1940	0	1,542	55,858	3,365	60,765
1945	0	3,136	65,263	3,905	72,304
1950	0	5,478	78,256	4,114	87,848
1955	279	7,303	95,580	3,012	106,174
1960	3,035	7,601	78,352	1,079	90,067
1965	20,755	12,523	85,450	532	119,260
1970	57,452	18,291	43,565	0	119,308
1975	65,137	20,575	20,104	0	105,816
1980	61,339	28,273	11,842	0	101,454

[a] Ref. 5.

Table 10. U.S. Finished-Steel Composite Prices[a]

Year	Cents/kg
1960	13.66
1965	14.04
1970	16.87
1975	28.88
1979	44.06

[a] Ref. 123.

Accidents. Studies over the past ten years show that over 60% of serious accidents in steel mills involved machinery, cranes, railroad equipment, and motor vehicles.

In furnace areas and hot-rolling mills, contact with a hot or molten metal, or a hot piece of equipment can result in a serious burn. If molten metal or slag contacts water, an explosion may occur because of the sudden generation of steam. Gases used and generated in steelmaking processes require special attention.

Severe injuries may be inflicted by rollers or from shearing, cropping, and trimming machines. It is essential that all machinery be equipped with guarding devices to minimize the number of pinching points.

Personal protective equipment is essential for prevention of many injuries, and includes hard hats, safety shoes, and eye shields. Gaiters, arm shields, and flame retardant suits should be worn when necessary.

Statistics indicate that accident frequencies for the steel industry are well below those for all industries in the United States. Steel companies with well designed and effective safety programs compare favorably in safety with other industries such as the chemicals, communications, and petroleum industries.

Health Hazards. Steelmaking, a complex series of operations, presents many industrial hygiene and occupational health hazards. Radiant heat may be extremely high around furnaces, hot-rolling mills, and soaking pits, leading to excessive fluid loss through perspiration, and ultimately to dehydration and collapse. Potential exposure to carbon monoxide is a serious concern around steel-mill furnaces and ovens. For example, as both a by-product of and a fuel for blast furnace operations, carbon monoxide presents a constant threat to workers in these areas. Since it is odorless and colorless, carbon monoxide has the potential to asphyxiate a worker without warning. Furnaces and ovens are often relined with a ceramic material containing silica. The risk of developing silicosis is thus present, depending upon the amount of free silica in the air. Exposures to sulfuric and hydrochloric acid mists may occur when steel is pickled to remove iron oxide scale formed during hot rolling. Excessive exposures to these acids can cause irritation of eyes and respiratory passageways. Exposure to lead fumes may occur during the production of leaded steel, during wire patenting, terneplate manufacturing, and babbitting. Lead exposure must be carefully monitored and controlled. Noise is a problem in most areas of steelmaking. Rolling mills; structural, bar, plate, and hot-strip mills; and blast furnaces are some of the potential sources of high noise levels. The noise may be constant, intermittent, or impulsive. Although worker exposures may vary, noise exposures of sufficient intensity, duration, and frequency can lead to hearing loss (see Noise pollution).

The steel industry has recognized these and other potential health hazards and is resolving exposure problems by good work practices and engineering controls. For

Table 11. Net Shipments of U.S. Steel Products, 1000 Metric Tons[a]

Steel products	1960	1970	1980
ingots and steel castings	307	833	404
blooms, slabs, billets, sheet bars	1,286	4,328	1,982
skelp	118	1	22
wire rods	833	1,525	2,438
structural shapes (heavy)	4,387	5,049	4,410
steel piling	384	449	314
plates	5,562	7,316	7,330
rails			
standard (over 30 kg/m)	606	779	984
all other	42	38	48
joint bars	24	23	15
tile plates	113	160	186
track spikes	41	69	77
wheels (rolled and forged)	221	226	144
axles	101	147	175
bars			
hot rolled (includes light shapes)	6,274	7,355	6,270
reinforcing	2,009	4,437	4,249
cold finishing	1,257	1,352	1,438
tool steel	79	80	72
pipe and tubing			
standard	1,949	2,317	1,608
oil country goods	1,086	1,186	3,277
line	2,440	1,900	1,082
mechanical	692	955	1,051
pressure	246	215	135
structural		455	516
stainless		29	37
wire			
drawn	2,213	2,143	1,263
nails and staples	291	272	157
barbed and twisted	43	89	64
woven wire fence	94	109	111
bale ties and baling wire	57	108	83
black plate	523	553	457
tin plate	4,958	5,176	3,779
tin-free steel		777	887
all other tin-mill products		65	55
sheets			
hot rolled	7,249	11,175	10,991
cold rolled	13,124	12,927	12,077
sheets and strip			
galvanized	2,773	4,359	4,757
all other metallic coated	237	500	655
electrical	515	663	545
strip			
hot rolled	1,209	1,173	607
cold rolled	1,203	1,045	844
Total steel products	*64,544*	*82,358*	*75,786*

[a] All grades, including carbon, alloy, and stainless steels (124).

Table 12. Distribution of Net Shipments of U.S. Steel Products to Consuming Industries, 1000 Metric Tons[a]

Market classification	1960	1970	1980
steel for converting and processing	2,656	3,123	3,735
forging[b]	763	951	846
bolts, nuts, rivets, and screws	973	912	535
steel service centers and distributors	11,322	14,538	14,671
construction, including maintenance	8,767	8,086	7,931
contractor's products	3,268	4,028	2,856
automotive			
vehicles, parts, etc	12,877	12,710	10,755
forgings	377	421	244
rail transportation			
freight cars, passenger cars, and locomotives	1,599	1,819	1,709
rails and all other	691	992	1,153
shipbuilding and marine equipment	564	779	1,090
aircraft and aerospace	71	51	48
oil and gas industry	367	3,220	4,872
mining, quarrying, and lumbering	261	451	412
agricultural			
agricultural machinery	694	734	773
all other	216	288	352
machinery, industrial equipment and tools	3,591	4,689	4,121
electrical equipment	1,885	2,444	2,214
appliances, utensils and cutlery	1,597	1,960	1,565
other domestic and commercial equipment	1,777	1,613	1,543
containers, packaging and shipping materials			
cans and closures	4,516	5,660	3,777
barrels, drums, and shipping pails	764	747	658
all other	554	647	601
ordnance and other military	150	1,109	161
export (reporting companies only)	2,325	5,430	2,354
nonclassified shipments	1,923	4,791	7,094
Total shipments	*64,548*	*82,193*	*76,070*

[a] All grades including carbon, alloy, and stainless steels (125).
[b] Not classified below.

Table 13. U.S. Imports and Exports of Steel Products, 1000 Metric Tons[a]

Products	1960 Imports	1960 Exports	1970 Imports	1970 Exports	1980 Imports	1980 Exports
steel-mill products						
ingots, blooms, billets, slabs, etc	58	108	155	2,885	141	827
wire rods	370	9	958	148	752	193
structural shapes and piling	288	268	1,076	175	1,659	139
plates	192	83	879	164	1,869	188
rails and accessories	9	121	65	89	376	122
bars and tool steel	762	77	1,237	294	780	394
pipe and tubing	435	177	1,748	294	3,426	427
wire and wire products	496	26	804	37	649	49
tin-mill products	36	622	304	383	342	804
sheets and strip	395	1,208	4,899	1,937	4,063	576
Total steel-mill products	*3,041*	*2,709*	*12,125*	*6,406*	*14,057*	*3,719*
other steel products	199	224	574	361	922	509
Total steel products	*3,240*	*2,933*	*12,699*	*6,767*	*14,979*	*4,228*

[a] Ref. 125.

example, blast furnaces are routinely equipped with carbon monoxide monitors and alarms, and self-contained breathing apparatus are available. Heat stress is reduced through the use of protective radiation shields, air conditioning of control stations, pulpits, and crane cabs, and protective clothing. Ventilation control and air cleaning reduce exposure to dust, fumes, and vapors; personal respiratory protection is also provided. Noise is reduced by modification of equipment, mufflers for air exhaust and intakes, and sound-proof enclosures and hearing protectors (see Insulation, acoustic).

In the areas of safety and health, there is a continuous exchange of information between steel companies. The Safety and Health Committees of the American Iron and Steel Institute, as well as its various technical committees, provide an active forum for exchanging information to make the steel mill a safer and healthier place to work.

BIBLIOGRAPHY

"Steel" in *ECT* 1st ed., Vol. 12, pp. 793–843, by Walter Carroll, Republic Steel Corporation; "Steel" in *ECT* 2nd ed., Vol. 18, pp. 715–805, by Harold E. McGannon, United States Steel Corporation.

1. H. E. McGannon, ed., *The Making, Shaping, and Treating of Steel*, 9th ed., U.S. Steel Corporation, Pittsburgh, Pa., 1971.
2. R. F. Tylecote, *A History of Metallurgy*, Metals Society Publication 182, Metals Society, London, 1976.
3. J. L. Bray, *Ferrous Production Metallurgy*, John Wiley & Sons, Inc., New York, 1954.
4. W. H. Dennis, *Foundations of Iron and Steel Metallurgy*, Elsevier, New York, 1967.
5. G. Derge, ed., *Basic Open Hearth Steelmaking*, 3rd ed., AIME, Warrendale, Pa., 1964.
6. *Annual Statistical Report 1980*, American Iron and Steel Institute, Washington, D.C., 1981.
7. C. E. Sims, ed., *Electric Furnace Steelmaking*, Vols. 1 and 2, AIME, Warrendale, Pa., 1962.
8. *Electric Furnace Proceedings*, issued yearly, Iron and Steel Society-AIME, Warrendale, Pa.
9. W. E. Schwabe, *Iron Steel Eng.* **46,** 132 (Sept. 1969).
10. O. K. Hill and C. G. Robinson, *Iron Steel Eng.* **56,** 33 (July 1979).
11. J. M. Gaines, ed., *BOF Steelmaking*, Vols. 1 and 2, Iron and Steelmaking Society-AIME, Warrendale, Pa., 1982.
12. *Basic Oxygen Steelmaking*, Metals Society Publication 197, Metals Society, London, 1979.
13. A. Jackson, *Oxygen Steelmaking for Steelmakers*, George Newnes, Ltd., London, 1969.
14. *Steelmaking Proceedings*, issued yearly, Iron and Steelmaking Society-AIME, Warrendale, Pa.
15. R. D. Pehlke, *Unit Process of Extractive Metallurgy*, American Elsevier, New York, 1973.
16. *Oxygen Steelmaking*, Association of Iron and Steel Engineers, Pittsburgh, Pa., 1966.
17. R. D. Pehlke, *Iron and Steelmaker* **7,** 15 (Dec. 1980); **8,** 18 (Jan. 1981).
18. J. K. Tien and J. F. Elliott, eds., *Metallurgical Treatises*, The Metallurgical Society-AIME, Warrendale, Pa., 1981.
19. *Proceedings of the Third International Iron and Steel Congress*, American Society for Metals, Metals Park, Ohio, 1979.
20. E. T. Turkdogan, *Physical Chemistry of High Temperature Technology*, Academic Press, New York, 1980.
21. W. K. Lu, ed., *The Role of Slag in Basic Oxygen Steelmaking Processes*, McMaster University, Hamilton, Ontario, Canada, 1976.
22. *Developments in Metallurgical Control in Basic Oxygen Steelmaking*, British Steel Corp., Redcar, UK, 1979.
23. G. C. Carter, ed., *Applications of Phase Diagrams in Metallurgy and Ceramics*, Vols. 1 and 2, U.S. National Bureau of Standards Special Publication 496, Washington, D.C., 1978.
24. Fr. Pat. 1,450,718 (July 18, 1966), G. Savard and R. Lee (to Air Liquide).

25. K. Brotzmann, *Technik und Forschung* **47,** 718 (1968); British Iron and Steel Industry translation 7255.
26. H. N. Hubbard and W. T. Lankford, Jr., *Iron Steel Eng.* **50,** 37 (Oct. 1973).
27. M. J. Papinchak and T. M. Weaver, in *Proceedings of the Third International Iron and Steel Congress*, American Society for Metals, Metals Park, Ohio, 1979.
28. G. Denier, *Steelmaking Proceedings* **63,** 131 (1980).
29. M. Saigusa, J. Nagai, F. Sudo, H. Bada, and S. Yamada, *Ironmaking Steelmaking* **7,** 242 (1980).
30. T. Ueda, M. Taga, Y. Tozaki, and T. Hirata, *Iron Steelmaker* **8,** 50 (Aug. 1981).
31. F. Schleimer, R. Henrion, F. Goldert, G. Danier, and J. C. Grosjean, *Iron Steel Eng.* **58,** 34 (Dec. 1981).
32. H. Jacobs, B. Ceschin, and P. Dauby, *Iron Steel Eng.* **58,** 39 (Dec. 1981).
33. *Handbook Including Specifications for Iron and Steel Scrap*, Institute of Scrap Iron and Steel, Inc., Washington, D.C., 1979.
34. R. J. King and W. R. Chilcott, *Physical Chemistry of Production or Use of Alloy Additives*, The Metallurgical Society-AIME, Warrendale, Pa., 1974, p. 69.
35. *Secondary Steelmaking*, Metals Society Publication 190, Metals Society, London, 1978.
36. J. S. Kirkaldy, *Ladle Treatment of Carbon Steel*, McMaster University, Hamilton, Ontario, Canada, 1979.
37. *Scaninject II*, Mefos, Lulea, Sweden, 1980.
38. W. G. Wilson and A. McLean, *Desulfurization of Iron and Steel and Sulfide Shape Control*, ISS-AIME, Warrendale, Pa., 1980.
39. L. S. Darken and R. W. Gurry, *Physical Chemistry of Metals*, McGraw-Hill, New York, 1953.
40. O. Kubaschewski and C. B. Alcock, *Metallurgical Thermochemistry*, Pergamon Press, New York, 1979.
41. E. T. Turkdogan, *J. Iron Steel Inst. London* **210,** 21 (1972).
42. G. J. Roe and B. L. Bromfitt, *Steelmaking Proceedings* **63,** 288 (1980).
43. D. P. Helliwell, *Steelmaking Proceedings* **64,** 288 (1981).
44. M. A. Orehoski and W. M. Keenan, *Steelmaking Proceedings* **64,** 172 (1981).
45. J. P. Barrett, T. W. Brown, and W. J. Stolnacker, *Steelmaking Proceedings* **64,** 178 (1981).
46. *Continuous Casting of Steel*, Metals Society Publication 184, Metals Society, London, 1977.
47. *Solidification and Casting of Metals*, Metals Society Publication 192, Metals Society, London, 1979.
48. *A Study of the Continuous Casting of Steel*, International Iron and Steel Institute, Brussels, 1977.
49. *Continuous Casting of Steel*, Iron and Steel Society-AIME, Warrendale, Pa., 1981.
50. R. D. Pehlke, *Unit Processes of Extractive Metallurgy*, American Elsevier, New York, 1973.
51. F. Gallucci and E. S. Szekeres, *Iron Steelmaker* **7,** 23 (Oct. 1980).
52. L. W. Jeffreys, *Iron Steel Eng.* **55,** 62 (April 1978).
53. W. J. Link, *Iron Steel Eng.* **55,** 29 (July 1978).
54. *Flat Rolling*, Metals Society Publication 260, Metals Society, London, 1979.
55. C. M. Sellars and G. J. Davies, eds., *Hot Working and Forming Processes*, Metals Society Publication 264, Metals Society, London, 1980.
56. F. A. A. Crane, *Mechanical Working of Metals*, Macmillan, New York, 1964.
57. W. L. Roberts, *Cold Rolling of Steel*, Marcel Dekker, New York, 1978.
58. J. E. Jenson, *Forging Industry Handbook*, Forging Industry Association, Cleveland, Ohio, 1966.
59. "100 Years of Metalworking," *Iron Age Magazine*, Radnor, Pa., June 1955.
60. R. Serjeantson, ed., *Metallurgical Plantmakers of the World*, Metal Bulletin, Inc., New York, 1981.
61. G. L. Kehl, *The Principles of Metallographic Laboratory Practice*, McGraw-Hill, New York, 1949.
62. *Applications of Modern Metallographic Techniques*, STP 480, American Society for Testing Materials, Philadelphia, Pa., 1970.
63. W. C. Leslie, *The Physical Metallurgy of Steels*, Hemisphere Publishing, McGraw-Hill, New York, 1981.
64. E. C. Bain and H. W. Paxton, *Alloying Elements in Steel*, American Society for Metals, Metals Park, Ohio, 1961.
65. *Heat Treatment '79*, Metals Society Publication 261, Metals Society, London, 1980.
66. K. E. Thelning, *Steel and its Heat Treatment: Bofors Handbook*, Butterworths, Boston, Mass., 1975.

67. D. V. Doane and J. S. Kirkaldy, *Hardenability Concepts with Applications to Steel*, The Metallurgical Society-AIME, Warrendale, Pa., 1978.
68. C. A. Siebert, D. V. Doane, and D. H. Breen, *The Hardenability of Steels*, American Society for Metals, Metals Park, Ohio, 1977.
69. G. Krauss, *Principles of Heat Treatment of Steel*, American Society for Metals, Metals Park, Ohio, 1980.
70. M. Atkins, *Atlas of Continuous Cooling Transformation Diagrams for Engineering Steels*, British Steel Corp., Sheffield, UK, 1978.
71. J. S. Blair, *The Profitable Way: Carbon Sheet Steel Specifying and Purchasing Handbook*, General Electric Technology Marketing, Schenectady, N.Y., 1978.
72. J. S. Blair, *The Profitable Way: Carbon Strip Steel Specifying and Purchasing Handbook*, General Electric Technology Marketing, Schenectady, N.Y., 1978.
73. J. D. Jevons, *The Metallurgy of Deep Drawing and Pressing*, John Wiley & Sons, Inc., New York, 1942.
74. *Low Carbon Structural Steels for The Eighties*, Institution of Metallurgists, London, 1977.
75. J. S. Blair, *The Profitable Way: Carbon Plate Steel Specifying and Purchasing Handbook*, General Electric Technology Marketing, Schenectady, N.Y., 1978.
76. R. A. Kot and J. W. Morris, eds., *Structure and Properties of Highly Formable Dual-Phase Steels*, The Metallurgical Society-AIME, Warrendale, Pa., 1979.
77. A. T. Davenport, ed., *Formable HSLA and Dual-Phase Steels*, The Metallurgical Society-AIME, Warrendale, Pa., 1979.
78. A. B. Rothwell and J. M. Gray, eds., *Welding of HSLA (Microalloyed) Structural Steels*, American Society for Metals, Metals Park, Ohio, 1978.
79. R. A. Kot and B. L. Bramfitt, *Fundamentals of Dual-Phase Steels*, The Metallurgical Society-AIME, Warrendale, Pa., 1981.
80. P. E. Repas, *Iron Steelmaker* **7,** 12 (Aug. 1980).
81. M. D. Baughman, K. L. Fetters, G. Perrault, Jr., and K. Toda, *Iron Steel Eng.* **56,** 52 (Aug. 1979).
82. A. P. Coldren and G. T. Eldis, *J. Met.* **32,** 41 (March 1980).
83. B. P. Bardes, ed., *Metals Handbook*, 9th ed., Vol. 1, American Society for Metals, Metals Park, Ohio, 1978, p. 127.
84. *Nickel Alloy Steels Data Book*, International Nickel Co., Inc., New York, 1967.
85. *Micro Alloying 75: Proceedings of an International Symposium on High-Strength Low-Alloy Steels*, Union Carbide Corp., New York, 1977.
86. F. B. Pickering, *Physical Metallurgy and the Design of Steels*, Applied Science Publishers, London, 1978.
87. G. R. Speich and D. S. Dabkowski, in J. B. Ballance, ed., *The Hot Deformation of Austenite*, The Metallurgical Society-AIME, Warrendale, Pa., 1977.
88. *Iron Age* **214,** MP9 (Dec. 9, 1974).
89. A. T. Davenport, D. R. DiMicco, and D. W. Dickinson, *J. Met.* **32,** 28 (March 1980).
90. *Steel Products Manual; Strip Steel*, American Iron and Steel Institute, Washington, D.C., 1978.
91. *Alloy Cross Index*, Mechanical Properties Data Center, Battelle's Columbus Laboratories, Columbus, Ohio, 1981.
92. P. M. Unterweiser, *Worldwide Guide to Equivalent Irons and Steels*, American Society for Metals, Metals Park, Ohio, 1979.
93. *Unified Numbering System for Metals and Alloys*, Society of Automotive Engineers, Warrendale, Pa., 1977.
94. *Handbook of Comparative World Steel Standards*, International Technical Information Institute, Tokyo, Japan, 1980.
95. R. B. Ross, *Metallic Materials Specification Handbook*, E. and F. N. Spon Ltd., New York, 1980.
96. C. W. Wegst, *Key to Steel (Stahlschluessel)*, Verlag Stahlschluessel Wegst KG, Marbach/Neckar, Federal Republic of Germany, 1974.
97. M. J. Wahll and R. F. Frontani, *Handbook of Soviet Alloy Compositions*, Metals and Ceramics Information Center, Battelle's Columbus Laboratories, Columbus, Ohio, 1976.
98. *Source Book on Stainless Steels*, American Society for Metals, Metals Park, Ohio, 1976.
99. K. G. Brickner and co-workers, *Selection of Stainless Steels*, American Society for Metals, Metals Park, Ohio, 1968.
100. D. Peckner and I. M. Bernstein, *Handbook of Stainless Steels*, McGraw-Hill, New York, 1977.

101. W. F. Simmons and R. B. Gunia, *Compilation and Index of Trade Names, Specifications, and Producers of Stainless Alloys and Superalloys*, Data Series, DS45A, American Society for Testing Materials, Philadelphia, Pa., 1972.
102. G. E. Rowan and co-workers, *Forming of Stainless Steels*, American Society for Metals, Metals Park, Ohio, 1968.
103. R. A. Lula, ed., *Toughness of Ferritic Stainless Steels*, STP 706, American Society for Testing Materials, Philadelphia, Pa., 1980.
104. C. R. Brinkman and H. W. Garvin, eds., *Properties of Austenitic Stainless Steels and Their Weld Metals*, STP 679, American Society for Testing Materials, Philadelphia, Pa., 1978.
105. J. J. Demo, ed., *Structure, Constitution, and General Characteristics of Wrought Ferritic Stainless Steels*, STP 619, American Society for Testing Materials, Philadelphia, Pa., 1977.
106. F. B. Pickering, ed., *The Metallurgical Evolution of Stainless Steels*, American Society for Metals, Metals Park, Ohio, 1979.
107. E. F. Bradley, ed., *Source Book on Materials for Elevated-Temperature Application*, American Society for Metals, Metals Park, Ohio, 1979.
108. G. V. Smith, ed., *Ductility and Toughness Considerations in Elevated Temperature Service*, American Society of Mechanical Engineers, New York, 1978.
109. R. L. Brockenbrough and B. G. Johnston, *USS Steel Design Manual*, ADUSS 27-3400-03, U.S. Steel Corp., Pittsburgh, Pa., 1974.
110. J. H. Gross, *Trans. ASME; J. Pressure Vessel Technology* **96,** 9 (Feb. 1974).
111. *Annual Book of ASTM Standards, Part 4-Steel*, American Society for Testing Materials, Philadelphia, Pa., 1982, p. 465.
112. S. Floreen and G. R. Speich, *Trans. ASM* **57,** 714 (1964).
113. *Third Maraging Steel Project Review*, AD 425 299, Air Force Systems Command Technical Documentary Report No. RTD-TDR-63-4048, Nov. 1963, available from National Technical Information Service of the United States Department of Commerce, Springfield, Va.
114. A. E. DeBarr, *Soft Magnetic Materials Used in Industry*, Reinhold, New York, 1953.
115. F. Brailsford, *Magnetic Materials*, John Wiley & Sons, New York, 1960.
116. R. M. Bozorth, *Ferromagnetism*, Van Nostrand, New York, 1951.
117. R. Cordero, ed., *Steel Traders of the World*, Metal Bulletin, Inc., New York, 1980.
118. R. Packard, *Metal Bulletin Handbook*, Metal Bulletin, Inc., New York, 1981.
119. R. Cordero and R. Serjeantson, eds., *Iron and Steel Works of the World*, Metal Bulletin, Inc., New York, 1978.
120. *Directory of Iron and Steel Works of the United States and Canada*, American Iron and Steel Institute, Washington, D.C., 1980.
121. *Directory: Iron and Steel Plants*, Association of Iron and Steel Engineers, Pittsburgh, Pa., 1981.
122. *Iron Age* **223,** 87 (Jan. 7, 1980).
123. *Iron Age* **186,** 109 (July 7, 1960); **196,** 104 (July 8, 1965); **206,** 90 (July 9, 1970); **216,** 104 (July 7, 1975); **222,** 62 (July 9, 1979).
124. *Annual Statistical Report*, American Iron and Steel Institute, Washington, D.C., 1960, p. 86; 1970, p. 24; 1980, p. 30.
125. *Annual Statistical Report*, American Iron and Steel Institute, Washington, D.C., 1960, p. 94; 1970, p. 26; 1980, p. 32.

General References

The Handbook of Steel Pipe, American Iron and Steel Institute, Washington, D.C., 1980.
A. B. Dove, ed., *Steel Wire Handbook*, Vols. 1–3, The Wire Association, Stamford Conn., 1965.
Steel Products Manual, Secs. 1–16, American Iron and Steel Institute, Washington, D.C, 1974–1978.
Useful Information on the Design of Plate Structures, American Iron and Steel Institute, Washington, D.C., 1979.

ROBERT J. KING
United States Steel Corporation

STERILIZATION TECHNIQUES

The term sterilization is defined as the total absence of living organisms; a more accurate definition is the following: free from living microorganisms with a probability previously agreed to be acceptable.

The nature of the process of killing microorganisms is similar in its kinetics to other first-order reactions. However, because of the uniqueness of the desired outcome; namely, the total destruction of every microorganism, it requires certain special considerations.

Sterilization technology is of primary importance in industries as diverse as food processing and space exploration. Generally, however, it is more readily associated with the health-care profession and industry. Most industries employing sterilization technology are regulated by some Federal agency. The introduction of foods, pharmaceuticals, and medical devices into interstate commerce is regulated by the FDA. The registration of chemical sterilants is regulated by the EPA.

The U.S. market in hospital sterilizing equipment is ca $\$70 \times 10^6$. Manufacturers of various types of equipment are given in Table 1.

Table 1. Manufacturers of Steam, Ethylene Oxide Sterilizers, and Dry Heat Sterilizers

Manufacturer	Location
Amsco/American Sterilizer Co.	Erie, Pa.
H. W. Anderson Products, Inc.	Oyster Bay, N.Y.
Bard International CR	Durham, UK
Baumer Equipment Medico Hospitalar S/A	Sao Paulo, Brazil
Be Venue Laboratories, Inc.	Bedford, Ohio
British Sterilizer Co., Ltd.	Essex, UK
Britains Hospital Supplies, Ltd.	Cheddleton, UK
Castle Co., division of Sybron Corp.	Rochester, N.Y.
Consolidated Stills & Sterilizers	Boston, Mass.
Danspital, Ltd., Turn-Key Hospitals	Roedovre, Denmark
Dent & Hellyer, Ltd.	Andover, UK
Downs Surgical, Ltd.	Surrey, UK
Drayton Castle, Ltd.	Middlesex, UK
Electrolux Wascator	Alingas, Sweden
Environmental Tectonics Corp.	Southampton, Pa.
Harsanyi Labor Mim	Budapest, Hungary
Intermed	Stafford, UK
Labotal, Ltd.	Jerusalem, Israel
O.C.R.A.S. Zambelli S.A.S.	Torino, Italy
Sakura Finetechnical Co., Ltd.	Tokyo, Japan
Scardi Construzioni Sanitario	Milano, Italy
Surgical Equipment Supplies, Ltd.	London, UK
Minnesota Mining and Manufacturing Co.	St. Paul, Minn.
Vacudyne Altair	Chicago Heights, Ill.
Vernitron Medical Products	Carlstadt, N.J.
Webecke & Co.	Bad Schwartau, FRG
Radiation sterilizers:	
Atomic Energy of Canada, Ltd.	Ottawa, Canada

A distinction must be made between sterilization and certain other processes which are called sterilization by popular misconception. These may not have the potential or medical capability for the total destruction of microbial life, but can render an object microbiologically safe for certain applications. In most instances, a judgement on the suitability of a sterilization or a substitute process can only be made by a microbiologist. Well-known exceptions are the common practice of boiling infants' feeding bottles under ambient atmospheric pressure for 5–15 min with the aim to sterilize them, or the practice of soaking objects in 70% alcohol. Boiling at atmospheric pressure does not result in sterilization, and sterilization in alcohol requires seven days of soaking (1).

Methods and procedures less rigorous than sterilization are often used with the result that the object might be safe for certain applications. Such procedures include disinfection, sanitization, and the use of antiseptics and bacteriostats (see also Disinfectants and antiseptics; Food processing; Industrial antimicrobial agents).

It is necessary to establish a criterion for microbial death when considering a process resulting in the death of microbial cells. With respect to the individual cell, the irreversible cessation of all vital functions such as growth and reproduction is a most suitable criterion. On a practical level, it is necessary to establish test criteria that permit a conclusion without having to observe individual microbial cells. The failure to reproduce in a suitable medium after incubation at optimum conditions for some acceptable time period is traditionally accepted as satisfactory proof of microbial death and, consequently, sterility. The application of such a testing method is, for practical purposes, however, not considered possible. The cultured article cannot be retrieved for subsequent use and the size of many items totally precludes practical culturing techniques. In order to design acceptable test procedures, the kinetics and thermodynamics of the sterilization process have to be studied.

Kinetics

An overwhelming body of evidence, starting with the earliest investigations (2), supports the contention that the rate of destruction of microorganisms is logarithmic, ie, first order with respect to the concentration of microorganisms. The process can be described by the expression:

$$\frac{N_o}{N_t} = e^{-kt} \tag{1}$$

in which N_t = the number of organisms alive at time t, N_o = the initial number of organisms, and k = the kinetic rate constant. It can be seen that N_t approaches zero as t approaches infinity. Absolute sterility, accordingly, is impossible to attain. As a practical matter, total kill is assumed to have taken place when an appropriate attempt at culturing results in no growth.

First-order kinetics yield linear plots on semilogarithmic graphs when plotted against time, and it has been found convenient to express the rate of kill in terms of a decimal reduction rate or D value. The D value represents the time of exposure (at given conditions for a given microorganism) required for a tenfold decrease in the viable population. This principle has found great utility in the health-care industry as well as in food processing (3). The practical significance of the D-value concept is that it simplifies the design of sterilization cycles with any desired degree of effectiveness.

A sufficiently large D reduction for any process results in a negative log N_t which, in a practical sense, represents the probability of survival of the last remaining microorganism. In the health sciences, a 6 D reduction from the last survivor or a 10^{-6} concentration of microorganisms is generally regarded as an acceptable criterium for sterility, although exceptions in the direction of higher or lower values do exist, depending on the type of sterilization process used, or the purpose of the sterile product.

Thermodynamics

The Arrhenius rate theory, an empirical derivation, holds for the sterilization process:

$$k = \mathrm{A}e^{-\Delta E/RT} \tag{2}$$

where k = kinetic constant, R = gas-law constant, T = absolute temperature, and ΔE = the activation energy.

The Eyring equation for the theory of absolute reaction rates is more accurate:

$$k = \frac{k_{\mathrm{B}}T}{h} e^{(\Delta S/R)(-\Delta H/RT)} \tag{3}$$

where k_{B} = Boltzman's constant, h = Plank's constant, and ΔS and ΔH are the standard entropy and enthalpy changes, respectively. Determinations of ΔE or ΔH and ΔS usually yield values of 167–335 J/mol (40–80 cal/mol). Such values are often characteristic of protein denaturation and microbial death may involve irreversible denaturation of some or even all of the cell's proteins.

The relationship between the D value and k can be derived by considering the meaning of D:

$$D = \frac{t_2 - t_1}{\log_{10} N_1 - \log_{10} N_2} \tag{4}$$

where $N_1 = 10\, N_2$

Substituting into equation 1:

$$D = \frac{2.3}{k} \tag{5}$$

Of great practical value is the derivation for the effect of temperature, the z value. It is defined as the temperature change necessary to effect a tenfold reduction (1 D)

$$z = \frac{T_2 - T_1}{\log_{10} D_1 - \log_{10} D_2} \tag{6}$$

A convenient multiple of D is the thermal death time F_o. It is the exposure time required for less than 1×10^{-6} probability of survival (4–5). The relationship between F_o and z becomes:

$$z = \frac{T_2 - T_1}{\log_{10} F_o^1 - \log_{10} F_o^2} \tag{7}$$

The comparison of the effectiveness of sterilization cycles at different temperatures becomes possible. For example, for steam sterilization:

$$F_o^{121} = \frac{t}{\operatorname{antilog} \dfrac{121 - t}{z}} \tag{8}$$

F_o^{121}, the thermal death time at 121°C is accepted to be 12 min (6) and the z value for *Bacillus stearothermophilus*, the most commonly used test organism is 10°C. Substituting into equation 8, one obtains an equivalent thermal death time t of 0.9 min at 132°C or 0.5 min at 135°C.

Testing and Monitoring

Direct testing for sterility by culturing is a destructive test method, and the product is rendered useless for food or medical purposes. Indirect testing methods usually rely on a statistically valid sampling pattern for a product. In the case of sterilization, where the desired outcome has to demonstrate a $<10^{-6}$ probability of failure, even a sample size of 500,000 cultures could only provide a 50% chance of detecting a failure in the process. Accordingly, product monitoring for sterilization is of very limited value. Product monitoring is only utilized if no other information is available for the particular process cycle for a given lot of products, eg, a referee test. Since sterilization is a highly reproducible and well-understood process, it has been found that process monitoring is far more suitable for purposes of sterility assurance. Process monitoring can be accomplished by measurement of individual parameters, each known to be critical for success, or by methods that are capable of integrating all critical conditions into a single display which can be observed or measured.

Biological Monitoring. Biological indicators integrate the process into a single display, ie, death, the desired outcome of the process itself. Biological indicators are preparations of specific microorganisms particularly resistant to the sterilization process they are intended to monitor. The organisms are inoculated onto filter paper strips or disks. It is also possible to prepare indicators with other organisms, or to inoculate product or packaging samples selected to resemble as closely as possible the actual product being sterilized.

When designing industrial sterilization cycles, the "bioburden" or "bioload" is determined first. The bioload is the number of organisms present on or in an entire chamber load. It has been found to be highly reproducible from lot to lot for mass-produced items. An appropriate number of biological indicators is added to the load, reflecting the bioload and the desired degree of safety. If the bioload were small, the number of organisms on the biological indicator would have to be 10^6. All of them must be killed during the process to indicate that the process produced the 10^{-6} concentration considered necessary.

The carriers containing the biological indicators are retrieved following exposure, transferred aseptically into sterile culture media, and incubated for the required length of time, usually 7 to 14 days. Some unexposed indicators are also incubated to prove that the spores were viable. If no growth is observed while the viability control displays the required growth, the conclusion is made that the sterilization cycle was successful. In order for the biological indicator concept to work, it is necessary to place the indicators into that portion of the sterilizer load considered the most difficult for the sterilant to reach. The lot of products is quarantined to prevent dissemination in case of incomplete sterilization.

Hospital and health-care institutions face a different problem. The sterilizer loads are diverse and generally prepared manually. Therefore, the bioburden varies and it is impossible to determine it for each load. Some prior assumptions are made about the bioburden when designing hospital cycles, and their design includes a sufficient degree of additional safety factors. Unfortunately, such extended cycles are difficult to monitor with biological indicators alone, because the manufacturers' recommended exposures considerably exceed the resistance requirements of biological indicators. An additional problem is the inability of hospitals to quarantine sterilized supplies, both for lack of an adequate number of instruments and storage space. For hospital sterilization, biological indicators are prepared in such a way that by a combination of the number of organisms and their resistance they demonstrate generally the adequacy of a given type sterilization cycle.

Spores of *Bacillus stearothermophilus* are frequently used for steam sterilization because of their high resistance to this type of sterilization. *Bacillus subtilis* is used for dry-heat and ethylene oxide sterilization; for radiation, *Bacillus pumilus* is used. The viability of any spore preparation is known to change with time, and such preparations have limited shelf lives. It is always advisable to test the viability of a given lot of spore preparations by culturing an unexposed sample from the same lot alongside the exposed samples. The labeling of commercially prepared products usually includes information on the number of organisms present, the *D* values, performance characteristics, instructions for culturing, and the expiration date.

Typical performance characteristics for some of the most widely used biological indicators are given in Table 2.

According to the *D* values shown in Table 2, a 6 *D* reduction with the 1.5 min requirement for steam at 121°C results in a 9 min cycle. Yet a minimum 12-min exposure is recommended, and with the safety factor, an 18-min steam contact is required (see Table 3) for hospital cycles at 121°C.

Culturing is time-consuming, and quarantining of packages sterilized in the hospital is not feasible. Under such circumstances, it is not possible to obtain information about a specific sterilizer load before the contents of that load are used. However, a general picture of the sterilization activity of the hospital can be gained.

Nevertheless, it is recommended by authoritative sources such as the Joint Commission on Accreditation of Hospitals (JCAH) (8), the Sterilization Standards Committee of the Association for the Advancement of Medical Instrumentation (AAMI) (9), and the Association of Operating Room Nurses (AORN) (10) that biological testing of all sterilizers be conducted at least once a week, but preferably every

Table 2. Typical Performance Characteristics of Some Biological Indicators

Culture	Sterilization process	Approximate *D*-value
Bacillus subtilis spores	ethylene oxide at 50% rh and 54°C	
	600 mg/L	3 min
	1200 mg/L	1.7 min
Bacillus stearothermophilus spores	saturated steam at 121°C	1.5 min
Bacillus pumilus spores	gamma radiation	
	wet preparations	2×10^{-6} Gy[b]
	dry preparations	1.5×10^{-6} Gy[b]
Bacillus subtilis spores	dry-heat at 121–170°C	60–1 min
Clostridium sporogenes spores	saturated steam at 112°C	3.5–0.7 min

[a] Ref. 6.

[b] To convert Gy to rad, multiply by 100.

Table 3. Minimum Exposure Periods for Sterilization of Hospital Supplies[a]

Load	Temperature, °C	Holding time, min	Air removal method	Heat-up time, min	Safety factor	Exposure time, min
hard goods						
unwrapped	121–123	12	gravity	1	2	15
	133–135	2	gravity[b]	<1	0.5	3
	133–135	2	pulsing	<1	1	4
	133–135	2	prevacuum	1	1	4
wrapped	121–123	12	gravity	5	3	20
	133–135	2	gravity	7	1	10
	133–135	2	pulsing	<1	1	4
	133–135	2	prevacuum	1	1	4
fabrics, packs	121–123	12	gravity	12	6	30
	133–135	2	pulsing	<1	1	1
	133–135	2	prevacuum	1	1	4
	141–142	0.5	pulsing	<1	0.5	2
liquids	121–123	12	gravity	[c]	[c]	[c]

[a] Ref. 7. Courtesy of Charles C Thomas Publisher.
[b] High speed.
[c] Depending upon liquid volume and container.

day for steam sterilizers, and every cycle for ethylene oxide sterilizers. These tests are carried out by placing the biological indicators in test packs of specific constructions, and placing the test packs in otherwise normally loaded sterilizers.

Since hospital sterilizer loads vary in composition, the challenge presented to the test organism could vary considerably, depending on the type and contents of packages they are placed in. The benefits of a standardized test-pack construction and test protocol are obvious, and such recommendation is made by AAMI for steam sterilizers (11). At this date, there is no concensus on the construction of a test pack for ethylene oxide sterilization.

Monitoring by Electromechanical Instrumentation. According to basic engineering principles, no process can be conducted safely and effectively unless instantaneous information is available about its conditions. All sterilizers are equipped with gauges, sensors, and timers for the measurement of the various critical process parameters. Some sterilizers are equipped with computerized control to eliminate the possibility of human error. However, electromechanical instrumentation is subject to random breakdowns or drifts from calibrated settings and even with regular preventive maintenance procedures, sterilization failures can occur because of instrumentation error.

An inherent problem is the location of the sensors. It is not possible to locate the sensors inside the packages which are to be sterilized. Electromechanical instrumentation is, therefore, capable of providing information only on the conditions the packages are exposed to, but cannot detect failures as the result of inadequate sterilization conditions inside the packages. Such instrumentation is considered a necessary, but not sufficient, means of monitoring the sterilization process.

Chemical Monitoring. Chemical indicators are devices employing chemical reactions or physical processes. They are designed in such a way as to permit observation of changes in a physical condition, such as color or shape, and to monitor one or more process parameters.

In many respects, chemical indicators combine the functions of biological indicators and electromechanical instrumentation. The sensors can be located inside the packages, and the results are observable immediately when the package is opened for use. Most chemical indicator types available today, however, are not capable of fully integrating all the critical process parameters. They are, therefore, not accepted as a guarantee of sterility, although some can indicate if conditions were adequate for sterilization.

A particular advantage of chemical indicators is the manufacturer's ability to formulate a relatively large, homogeneous batch of reagent mixture that can be deposited on inexpensive substrates, such as paper, by high speed printing techniques, resulting in relatively low unit costs. Since hospital sterilization cycles are standardized, they can benefit considerably from the use of chemical indicators, because hospital sterilization systems cannot employ quarantines.

Industrial sterilization cycles tend to vary considerably, not only from manufacturer to manufacturer, but often from product type to product type, depending on the bioburden present on a given load. Chemical indicators have, until recently, been used only to differentiate between sterilized and nonsterilized packages. Recent developments resulted in the availability of chemical dosimeters of sufficient accuracy to permit their application either as total monitors or as critical detectors of specific parameters.

Dry-Heat Sterilization

Dry-heat sterilization is generally conducted at 160–170°C for ≥2 h. Specific exposures are dictated by the bioburden concentration and the temperature tolerance of the products under sterilization. At considerably higher temperatures, the required exposure times are much shorter. The effectiveness of any cycle type must be tested. Biological indicators can be utilized both to demonstrate the effectiveness of a cycle and for routine monitoring. The organism recommended is *Bacillus subtilis.* For dry-heat sterilization, forced-air type ovens are usually specified for better temperature distribution; temperature-recording devices are recommended.

It is an axiom of sterilization technology that appropriate conditions must be established throughout the material to be sterilized. The time-temperature conditions are critical when considering any sterilization method using heat.

Chemical indicators for dry-heat sterilization are available either in the form of pellets enclosed in glass ampuls, or in the form of paper strips containing a heat-sensitive ink. The former displays its end point by melting, the latter by a color change (see Chromogenic materials).

Steam Sterilization

Steam sterilization specifically means sterilization by moist heat. The process cannot be considered adequate without assurance that complete steam penetration takes place to all parts and surfaces of the load to be sterilized (see Fig. 1). Steam sterilization at 100°C and atmospheric pressure is not considered effective. The process is invariably carried out under higher pressure in autoclaves using saturated steam. The temperature can be as low as 115°C, but is usually 121°C or higher (see Steam).

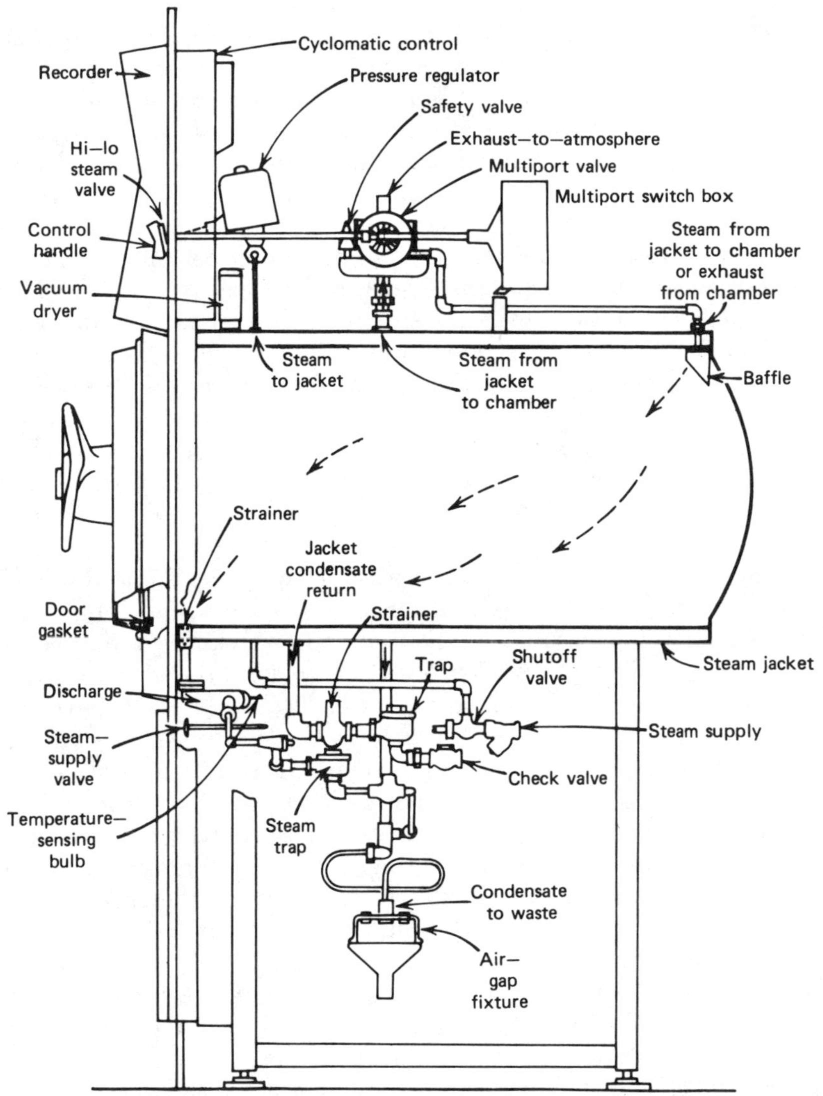

Figure 1. Longitudinal cross-section of steam sterilizer (1).

Great care is needed in the design of autoclaves and sterilization cycles because of the requirement for the presence of moisture. The autoclave must be loaded to allow complete steam penetration to occur in all parts of the load before timing of the sterilization cycle commences. The time required for complete penetration, the so-called heat-up time, as well as exposure time, the so-called holding time, varies with different construction of autoclaves, and different types of loads and packaging materials. The volume of each container has a considerable effect on the heat-up time whenever fluids are sterilized. Thermocouples led into the chamber past the door gasket or by other means are often employed to determine heat-up times and peak temperatures. The pressure is relieved at the end of each fluid sterilization cycle. Either vented containers must be used or specific provisions be made to allow for the safe removal of nonvented, hermetically sealed containers from the autoclave.

The elimination of air from the chamber and complete steam penetration of the load may be accomplished by gravity displacement or prevacuum techniques.

The gravity-displacement type autoclave relies on the relative nonmiscibility of steam and air to allow the steam that enters to rise to the top of the chamber and fill it. The air is pushed out through the steam-discharge line located at the bottom of the chamber. Gravity-displacement autoclaves are utilized for the sterilization of liquids, fabric loads, fabric-wrapped instruments (usually at 121°C), and for unwrapped medical instruments at 134°C (flash or emergency sterilization).

The prevacuum technique, as its name implies, eliminates air by creating a vacuum. This procedure facilitates steam penetration and permits the use of higher pressures. Consequently, higher temperatures can be used (134–141°C) which results in shorter cycle times. Prevacuum cycles employ either a vacuum pump/steam (or air) ejector combination to reduce air residuals in the chamber or rely on the pulse-vacuum technique of alternating steam injection and evacuation until the air residuals have been removed. Pulse-vacuum techniques are generally more economical; vacuum pumps or vacuum-pump–condenser combinations may be employed. The vacuum pumps used in these systems are water-seal or water-ring types, because of the problems created by mixing oil and steam. Prevacuum cycles are used for fabric loads and wrapped or unwrapped instruments.

In prevacuum autoclaves, problems are created by the removal of air and the air-insulation systems. As the result of continued routine use, it is only a matter of time before leaks develop in the door gaskets and valve packings, or the vacuum pump breaks down. A specific test was developed to evaluate the ability of prevacuum sterilizers to eliminate air from their chambers, and to prevent its leaking back during the prevacuum phase. It is called the Bowie & Dick Test (12). This test utilizes a pack of specific construction (or its proven equivalent) placed in the empty chamber and exposed to specific test conditions (9,12). The pack contains a chemical indicator sheet. A correctly functioning sterilizer produces a uniform color change. A nonuniform color change indicates the presence of air which requires the attention of a qualified mechanic. The daily testing of all prevacuum sterilizers is recommended (1,9–10).

The critical parameters of steam sterilization are temperature, time, air elimination, steam quality, and the absence of superheating. Temperature and time are interrelated, as shown in equation 8. The success of steam sterilization is dependent on direct steam contact which can be prevented by the presence of air in the chamber. The ability of steam to heat a surface to a given temperature is considerably reduced by the presence of air. Air elimination, therefore, is regarded as an absolute parameter. If the required amount of air has not been eliminated from the chamber and the load, no combination of time and temperature results in complete sterilization.

The term steam quality refers to the amount of dry steam present relative to liquid water (in the form of droplets). The steam delivered from the boiler usually contains some water. Excessive amounts can result in drying problems following exposure, and >3% water or <97% quality steam is considered unacceptable. Excessive amounts of water deposits dissolved boiler chemicals onto the load to be sterilized. Boiler chemicals are used to prevent corrosion in the lines. Inappropriate boiler chemicals, also called boiler amines, may introduce toxicity problems (see Corrosion and corrosion inhibitors).

Superheated steam results when steam is heated to a temperature higher than that which would produce saturated steam. The equilibrium between liquid and vapor is destroyed, and the steam behaves as a gas. It loses its ability to condense into moisture when in contact with the cooler surface of the article to be sterilized. This process resembles dry-heat sterilization more than steam sterilization and, under ordinary time–temperature conditions for steam sterilization, does not produce sterility.

The selection of an appropriate steam-sterilization cycle must be made after a careful study of the nature of the articles to be sterilized, the type and number of organisms present, type and size of each package, type of packaging material used, etc.

Cycle-development studies may be conducted with full autoclave loads that produce fractional kill. The results are plotted as D value of the load versus time of exposure, and the logarithmic plot is extrapolated to the desired degree of safety.

Since hospital loads are not uniform, certain assumptions were made by the manufacturers of sterilizers in arriving at recommended cycles. These recommendations include a safety factor, as well as allowance for heat-up time. Table 4 gives typical exposure recommendations for various types of articles.

Biological indicators for steam sterilization utilize *Bacillus stearothermophilus*. In monitoring industrial cycles, a sufficient number of preparations each with a known degree of resistance are added to the load and retrieved after exposure and cultured. In monitoring hospital cycles, two indicator preparations (often referred to as spore strips) of known resistance are placed inside the test pack.

Electromechanical monitors for steam sterilization include temperature–time recording charts and pressure–vacuum sensing gauges. Most recording charts are also capable of displaying pressure–vacuum values. The temperature sensor is invariably located in the chamber-drain line, which is considered to be the coolest area since air would exit from the chamber via the drain line. Unfortunately, there is no way of locating the sensors inside the packages being sterilized except under specialized test conditions.

Chemical indicators for steam sterilization can be classified into four categories as discussed below.

Some indicators integrate the time–temperature of exposure. Some of these operate throughout the temperature range utilized for steam sterilization (121–141°C), others function in specific time–temperature cycles. Some are capable of monitoring the safety factor in the exposure, others only minimal conditions. All are capable of indicating air entrapment or incomplete steam penetration. The results are indicated by a color change. Certain types change from a specific initial color through a series of intermediate shades to a specific final color. An incomplete color change is an indication of incomplete processing. Other types function by having a color column advancing along the length of the test strip, in a manner reminiscent of paper chromatography. Inadequate processing is indicated by an advance that falls short of a predesignated finish line.

Indicators can determine whether a specific temperature has been achieved. Since the entrapment of large amounts of air can result in the lowering of steam temperatures, these indicators react to some critical defect in sterilization conditions. Temperature detectors usually consist of a pellet sealed in a glass ampul. The attainment of correct temperature is indicated when the pellet melts and loses its original shape. The pellets are also capable of changing color when melted, which aids in the determination of the correct end point. The chemical composition of the pellet is such that it melts within the desired sterilization temperature range. Obviously, for each type of cycle utilizing a different temperature, a different type must be used.

Table 4. Exposure Periods for Sterilization of Specific Articles[a]

Article	Minimum time required, min sterilizer type, °C			
	Gravity, 121	Gravity, 134	Prevacuum, 134	Pulsing, 134
ampuls, spinal, heat-stable, in test tubes (on sides)	15			
brushes, synthetic fibers, in dispensers or individually wrapped	30	15	4	4
dressings				
wrapped in paper or muslin	30	15	4	4
loosely packed, in canisters, on sides	30	15	4	4
glassware, empty, inverted	15	3	4	4
inhalational (heat-stable) therapy equipment	30			
instruments, metal				
any number in perforated tray, unwrapped	15	3	4	4
in lightly covered or padded tray	20	10	4	4
unwrapped, combined with sutures, tubing, or other porous materials	20	10	4	4
combined with other porous materials in lightly covered or padded tray	30	15	4	4
instruments, wrapped in muslin, four thicknesses, for storage	30	15	4	4
linen, packs[b]	30		4	4
needles, hollow, individually packed				
in glass tubes, lumen moist, on sides	30	15	4	4
in paper wrappers, lumen moist	30	15	4	4
rubber gloves, wrapped in muslin or paper	20		4	4
rubber catheters, drains, tubing, lumen moist				
unwrapped	20	10	4	4
individually packaged in muslin or paper	30	15	4	4
rubber or plastic (heat-stable) sheeting interleaved with layer of muslin	30		4	4
treatment trays, wrapped in muslin or paper	30		4	4
solutions, Square-Pak bottles				
75–250 mL	20			
500–1000 mL	30			
1500–2000 mL	40			
suture				
silk, cotton or nylon, wrapped in paper or muslin	30	15	4	4
wire, on metal reel, wrapped	30	15	4	4
syringes, disassembled, individually packaged in muslin or paper	30	15	4	4
utensils, on edge				
unwrapped	15	3	4	4
wrapped in muslin or paper	20	10	4	4

[a] Ref. 12. Courtesy of Charles C Thomas Publisher.
[b] 30 × 30 × 50 cm, 5.4 kg.

Indicators can determine if uniform steam penetration has been achieved during a Bowie & Dick-type test. They are produced in the form of sheets (23 × 30 cm) and they are capable of uniform color change over their entire surface when exposed to pure saturated steam under test conditions. Nonuniform color development is an indication of failure of the test.

Indicators can be utilized to distinguish packages that have been processed from those that have not been processed. These are external indicators that do not have the capability to detect critical shortcomings in cycle parameters since they are not located inside the packages. A well-known example of this type is autoclave tape, which is also used to hold together packages wrapped in muslin or other kinds of wrap-type packaging materials.

Gas Sterilization

Certain articles, particularly those used in the medical industry and in space exploration, require sterilization. However, they cannot withstand the temperatures and moisture of steam sterilization or exposure to radiation. Gaseous sterilants that function at relatively low temperatures offer an attractive alternative. Although many chemical compounds or elements can be considered sterilants, an obvious practical requirement is that the gas selected should allow safe handling, and that its residues should volatilize relatively quickly if absorbed by components of the article sterilized. Ethylene oxide (qv) satisfies most of these requirements, and is the most frequent choice (1,6,14–16) (see Fig. 2). Since it is highly flammable, it must be used in a carefully controlled manner, and is either dispensed from a single-use cartridge or diluted with inert gases until no longer flammable. The most frequently used diluents are fluorocarbon gases or carbon dioxide.

It is necessary to determine the bioburden and make cycle-verification studies when this form of sterilization is used, as it is with other sterilization methods. The manufacturer of hospital sterilization equipment provides cycle recommendations based on the expected bioburden and the consideration of an appropriate safety factor. In ethylene oxide sterilization, it is necessary to determine if residues of the sterilant are absorbed by the sterilized article, and to examine the possible formation of other potentially toxic materials as a result of reaction with ethylene oxide.

The critical parameters of ethylene oxide sterilization are temperature, time, gas concentration, and relative humidity. The critical role of humidity has been demonstrated by a number of studies (11,17–18). Temperature, time, and gas concentration requirements are dependent not only on the bioburden, but also on the type of hardware and gas mixture used. If cycle development is not possible, as in the case of hospital sterilization, the manufacturer's recommendations should be followed.

Provisions must be made for allowing residues of the sterilant absorbed by the product to dissipate by storage for a long time under ambient conditions, or by using aeration cabinets with forced air circulation at elevated temperatures. The amount of absorbed sterilant should be determined before releasing the sterilized articles. If such studies are not feasible, the recommendations of the manufacturer of the articles sterilized or of the aeration equipment should be obtained. The permissible residue concentrations are 10–250 ppm, depending on the size of the article and on its intended use.

Biological monitoring of ethylene oxide sterilization is conducted with spores

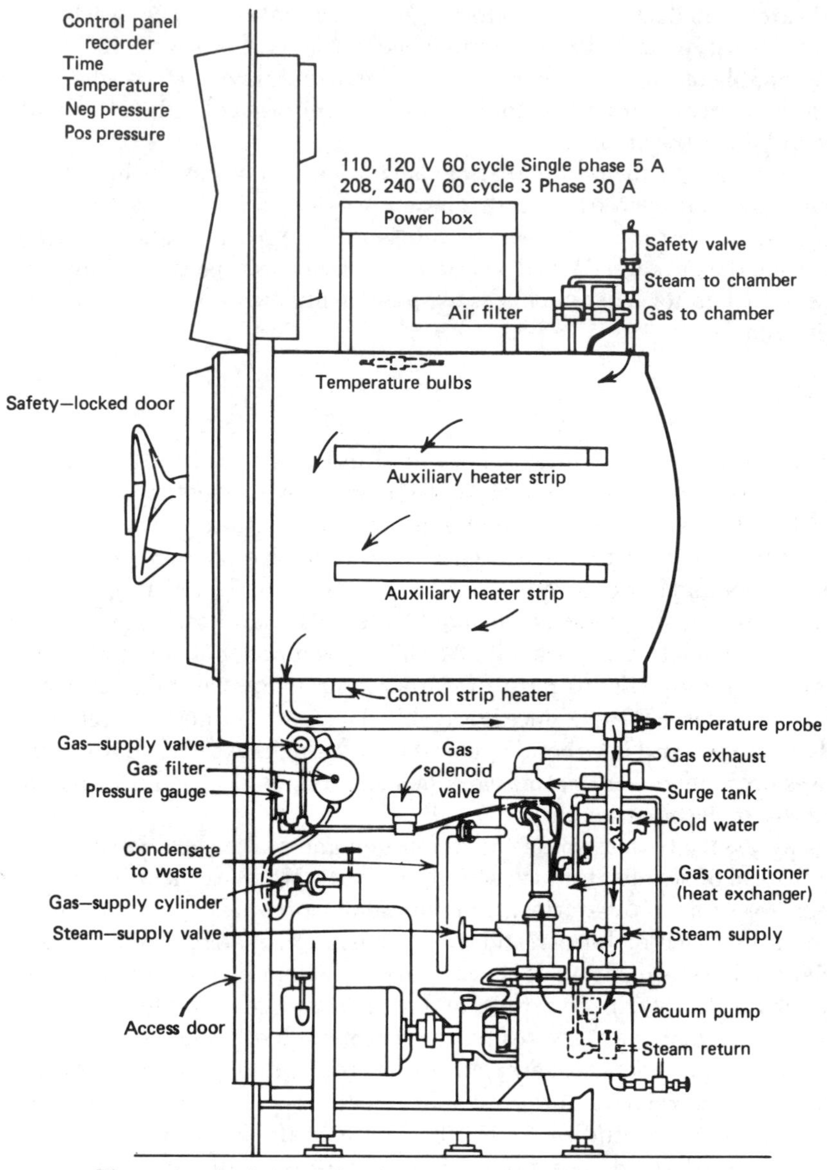

Figure 2. A schematic drawing of an ethylene oxide sterilizer (1).

of *Bacillus subtilis* organism. See Table 2 for required performance characteristics.

Chemical indicators for ethylene oxide sterilization are usually of the color-change type and are capable of indicating the presence of ethylene oxide under some minimal conditions of temperature, time, and gas concentration. A few types are also capable of indicating the absence of moisture. External processing indicators similar to autoclave tape, but with ethylene oxide-responsive chemicals are also available.

General recommendations for instrumentation include monitoring gas concentration, temperature/time, and the moisture content of the chamber. Hospital sterilizers are not usually equipped with instrumentation providing direct display of gas concentration and moisture content. They rely instead on a specific sequence of steps performed automatically or by the operator of the sterilizer.

Ionizing Radiation

Radiation sterilization, as currently practiced, employs electron accelerators or radioisotopes (qv). Electrons have relatively low penetration ability, and the use of accelerators requires careful control. Gamma-radiation sterilization usually employs ^{60}Co and occasionally ^{137}Cs as the radioisotope source. A very wide range of packaging materials can be used because gamma rays possess a considerably greater penetrating ability. However, they must not be degraded to the point where the quality of the aseptic barrier is compromised.

Since materials tend to undergo chemical changes when exposed to gamma radiation, it is generally recommended that the effect of the radiation on any material be determined before gamma-ray sterilization is attempted.

Since exposure requirements are measured in terms of the delivered dose of radiation, the procedure becomes time independent. Bioburden determinations should be carried out to establish the effective dose. A dose of 2.5×10^4 Gy (2.5 megarad) is selected for many articles, although lower doses would probably be adequate. Radiation exposure can be monitored with biological indicators utilizing *Bacillus pumilus* as the test organism. Counters and electronic measuring devices are also used.

Chemical dosimeters based on ferrous sulfate, ferrous cupric sulfate, or ceric sulfate are generally used. Color-change process indicators are also used, but these cannot measure the radiation dose.

Filtration and Liquid Sterilization, Other Sterilants

This process depends on the physical retardation of microorganisms from a fluid by a filter membrane or similarly effective medium. The effectiveness of this process is also influenced by the bioburden (6). The obvious requirements are that the filtrate be sterile, relatively free of particles, and not lose its effectiveness or safety.

Formalin, a solution of methanal (formaldehyde (qv)) in water, has sterilizing properties (19), as does glutaraldehyde (20). The sterilizing action of these liquids is dependent upon complete contact with the surface of the article to be sterilized, which can be prevented by grease, smudges, fingerprints, or other impurities. No sterilant can penetrate a greasy layer, and careful precleaning is required before processing. The time of exposure is also critical, and the recommendations of the manufacturer must be followed exactly. Although some sterilants can be used at higher temperatures, their stability can be affected adversely. Changes in the pH of the solutions may have similar adverse effects, and it is strongly recommended that the instructions of the sterilant manufacturer be followed faithfully.

Liquid sterilants are known to corrode metal parts of articles and instruments that are to be sterilized. Articles composed exclusively of glass or certain types of corrosion-resistant metal alloys can be safely processed, but the instructions of the manufacturer should be consulted. Since the degree of corrosion is related to length of exposure, many articles are merely disinfected in a shorter exposure time. Disinfection may be suitable for certain applications, but judgement about the safety of this choice must be made by a qualified microbiologist.

Other chemicals have also been used in the past as liquid sterilants. A solution based on hydrogen peroxide (qv) was offered commercially but has been withdrawn from the market by its manufacturer (Lehn and Fink, Division of Sterling Drug, Inc.).

Since liquid sterilants or their residues can be harmful to living tissues, it is always necessary to rinse articles with sterile water or saline solution following sterilization.

Processing in liquid sterilants results in wet products which require highly specialized packaging. Therefore, liquid sterilization should only be considered if the sterilized article is to be used almost immediately.

Liquid sterilization is an extremely useful method for articles that cannot withstand the conditions of steam sterilization, but problems associated with its use limit its application.

Other Sterilants. It is possible to conceive of other chemical agents or physical conditions that destroy microbial cells. Specific processes may be developed to suit certain types of products or processes, but these will have limited general application. Ultraviolet light has sterilizing properties, but cannot penetrate many materials, whereas others are destroyed by exposure to a sterilizing dose. Some chemicals, such as 3-hydroxy-2-oxetanone (α-hydroxy-β-propiolactone), have been used in the past but were found to be carcinogenic. If it is necessary to utilize a hitherto untested sterilant, its effectiveness should be determined according to some acceptable test standard such as those listed in the Official Methods of the Association of Official Analytical Chemists (21). The manufacture and sale of chemical sterilants and disinfectants is regulated in the United States by EPA.

Sterilization Packaging

In rare instances, it is possible to sterilize an article at the time and place of its use. However, sterilization generally takes place at another location and before the article is used. The main purpose of packaging is to protect the sterility of the contents. When an article is placed in its protective container and subsequently sterilized, the process is called terminal sterilization. When it is sterilized first and then placed in a presterilized container, the process is called sterile filling.

A sterile-filling operation requires an environment that excludes or diminishes the possibility of recontamination before the sterile product is sealed. Completely sealed units, such as glove boxes, are suitable. Specially constructed clean hoods or rooms can be used for sterile filling, which utilize laminar air flow with highly filtered air. Personnel working in such environments have to be garbed in special protective garments and wear masks, hair covers, etc. The packaging materials (qv) used in sterile filling can be of types that provide a hermetically sealed environment, such as glass, metal cans, or metal foil.

Packaging material used for terminal sterilization must permit full sterilant penetration as well as provide a microbial barrier. Consideration must also be given to the conditions to which the sterile package will be exposed until used, such as storage, transportation, frequency of handling, etc.

Storage time by itself is not expected to affect the maintenance of sterility. However, longer storage time may increase the incidence of potentially harmful conditions. Frequent handling, wetness, and possible deterioration of the packaging material are typical examples of conditions that may compromise sterility and limit the shelf life of a package. The package contents may have a specific shelf life.

Most industrially prepared, presterilized packages contain inserts with statements that sterility is guaranteed only if the package is not opened or damaged. The wide choice of packaging materials and methods available for industrial processes allows the selection of packaging materials, package designs, and processes that provide maximum protection. Indeed, with appropriate packaging, sterilization methods, and materials, sterility can be protected for an indefinite length of time.

Hospital sterilization is more limited in the availability of sterilization methods and of packaging materials, which permits eventual microbial invasion, particularly when articles are wrapped in protective fabrics such as muslin (140-thread-count cotton) or newer synthetic, single-use nonwoven wrappers. Therefore, the expected shelf life of hospital-wrapped and sterilized articles is considered to be ca 21–30 d, if a so-called double-wrapping technique is used. Double wrapping requires two successive wraps, each with layer or layers of an approved packaging material.

Another type of hospital packaging system employs the so-called peel-open packages. These are constructed by heat sealing two webs of packaging material around the edges. One layer is usually a plastic film of composite construction, the other is a surgical-grade kraft paper designed to give an effective microbial barrier. Shelf life is extended to a time that is determined by need rather than sterility protection.

Whichever packaging method is used, provisions must be made for the opening of the package and the retrieval of the sterilized article in a manner that does not compromise its sterility.

Related Techniques

Procedures that are less thorough than sterilization may be used for the preparation of foods and medical supplies. Some of these processes are capable of rendering an object microbiologically safe for a given purpose when they are employed with proper safeguards. The cooking of food may result in the reduction of spoilage because some potentially harmful organisms are destroyed, even though the temperatures seldom exceed 100°C. Contrary to popular belief, boiling baby formula is not a sterilization procedure, but suffices in most cases.

Pasteurization is the heating of certain fluids, frequently milk or dairy products. It destroys potentially harmful organisms such as mycobacteria, *M. tuberculosis*, *M. bovis*, or *M. avium*. Since pasteurization is carried out at 62°C for 30 min or at 72°C for 15 s, it is not a sterilization procedure (see Milk and milk products). Disinfection destroys pathogenic organisms. This procedure can render an object safe for use, and has to be judged by a person with appropriate training in microbiology. Disinfectants include solutions of hypochlorites, tinctures of iodine or iodophores, quaternary ammonium salts (qv), ethyl alcohol, formaldehyde, and glutaraldehyde (see Aldehydes).

Tyndalization, or fractional sterilization, is a procedure no longer considered to be acceptable. Spores of vegetative organisms are the most difficult entities to destroy. In this procedure, rather than destroying them as spores, they are prompted to germinate and then destroyed by boiling water.

Bacteriostasis is the process of preventing the growth and reproduction of microorganisms; however, when the bacteriostat is removed or its power is exhausted, the organisms can resume growth. Bis(2-hydroxy-3,5,6-trichlorophenyl)methane, (hexachlorophene) is a bacteriostatic agent that belongs in this category.

Sanitization is a cleaning procedure that reduces microbial contaminants on certain surfaces to safe or relatively safe levels, as defined by the EPA or public health authorities. The article is usually cleaned with hot water and various detergents. Sanitization can be safe for a product in contact with intact skin or for food utensils, but it is not considered safe for articles inserted in the body. Effective sanitization is a requirement in the reprocessing of reusable medical supplies before packaging and sterilization. It is also a requirement in the maintenance of utensils and containers used for food preparation.

Decontamination is a procedure to render safe for handling, disposal, or the subsequent processing of an article that may contain a large amount of potentially infectious organisms. Decontamination and sterilization are similar procedures, except that in the former case the bioburden is higher. In both cases, all organisms present are destroyed. However, decontamination is not expected to result in the 10^{-6} probability of microbial survival, as in sterilization, because of the higher bioburden. Decontamination may include sanitization and disinfection steps, but it most frequently involves sterilization procedures with exposure times 2–4 times longer than usual. Incineration is a frequent choice for decontamination of single-use articles (see Incinerators).

Germicides are agents capable of killing some forms of microorganisms.

Sterilization in the Food-Processing Industry

The concept of heat processing of foods in hermetically sealed containers was introduced in 1810 (22). The role of microorganisms was still unknown at the time, and only the so-called agents of putrefaction were eliminated.

The problem of microbial contamination of foods is twofold: foods may act as nutrients for, and carriers of, pathogenic organisms, and foods may be spoiled by the action of certain organisms (see Food processing). Generally, four specific types of organisms are considered to be food poisons, although it is possible for a number of different organisms present in foods to cause disease. Salmonella and *Clostridium perfringens* require the ingestion of the organisms in large numbers. *Staphylococcus* and *Clostridium botulinum* produce toxins and the organisms need not be ingested for the symptoms to occur. *Salmonella* is relatively easy to destroy by boiling, and so is *Staphylococcus*, although the toxin is more heat resistant. The *Clostridia* are spore-forming organisms. The germinated cells can easily be destroyed, but the spores are heat resistant.

Food spoilage can be caused by enzymatic action or microorganisms. Certain preservation techniques, such as freezing, may retard spoilage by preventing the multiplication of microorganisms or the catalytic action of enzymes. Freezing, however, does not kill the organisms and should not be regarded as a sterilization method.

The most widely used sterilization method in the food industry is moist heat. The heat is usually supplied by high pressure steam, but since most foods already contain moisture the role of the steam is to heat the food to the required temperature. The cooking and sterilization processes can frequently be combined into one. The food may be sealed into impervious containers of glass, metal, or plastic film and undergo terminal sterilization, or it may be presterilized in batches or in a continuous operation and then filled into a presterilized container. The latter process is called sterile filling.

The effect of pH has been well known for some time. Acidic foods, such as fruits, tend to retard microbial growth and resist certain types of contamination. For this reason, the standards adopted industrywide have been based on the processing of foods with low acidity. In the United States, the FDA has regulatory responsibility over the preparation, sterilization, and distribution of foods.

The sterilizers or retorts used to process canned or prepackaged foods must be designed in such a way as to assure uniform temperature distribution throughout. Adequate venting permits complete air removal; the air vent is located at the opposite end from the steam inlet (23). The retorts may be horizontal or vertical in design.

The destruction of microorganisms in foods follows the same kinetic relationship as described earlier. The process is strongly influenced by the nature of the food, the size of the container, and the temperature. Industrywide standards have been established by the National Canners Association (23).

With the large volume of food production, continuous processing offers economic advantages. The continuous sterilization of prefilled cans may be accomplished with relative ease because some cans are capable of withstanding the high pressures generated in order to attain the required sterilization temperatures. Such a process is, in reality, not truly continuous, since the prefilled containers represent a discontinuous phase. Time delays in heating full large-volume cans to sterilization temperatures place a limitation on the can sizes. Presterilization followed by aseptic filling solves this problem. The containers are presterilized by heat, steam, radiation, etc. The food may be processed in large discontinuous batches, or in modern continuous retorts. Sterilization conditions are checked by appropriate instrumentation.

A truly continuous cooking-sterilization process employs back-pressure valves to allow the attainment of pressures that result in the required temperatures.

There are four types of food sterilization processes: terminal sterilization in prefilled containers in a batchwise process; terminal sterilization in prefilled containers of appropriate design heated to the required temperatures in a continuous process; aseptic filling following batchwise cooking in an appropriate retort; and aseptic filling in a continuous cooking system equipped with appropriate valves to allow the necessary pressures for attainment of the required sterilization temperatures.

BIBLIOGRAPHY

"Sterilization" in *ECT* 1st ed., Vol. 12, pp. 896–916, by E. L. Gaden, Jr., and E. J. Henley, Columbia University; "Sterilization" in *ECT* 2nd ed., Vol. 18, pp. 805–829, by J. E. Doyle, Castle Company.

1. J. J. Perkins, *Principles and Methods of Sterilization in the Health Sciences*, 2nd ed., Charles C Thomas, Springfield, Ill., 1969.
2. H. Chick, *J. Hyg.* **8,** 92 (1908).
3. L. I. Katzin, L. A. Sandholser, and M. E. Strong, *J. Bacteriol.* **45,** 265 (1943).
4. G. B. Phillips and W. Miller, *Industrial Sterilization*, Duke University Press, Durham, N.C., 1973, pp. 239–282.
5. G. B. Phillips and W. Miller, *Validation of Steam Sterilization*, In. Technical Monograph #1, Parenteral Drug Association, 1978.
6. *The United States Pharmacopeia XX* (*USP XX–NF XV*), The United States Pharmacopeial Convention, Inc., Rockville, Md., 1980, pp. 1037–1040.
7. Ref. 1, p. 163.
8. *Accreditation Manual for Hospitals*, Joint Commission on Accreditation of Hospitals, Chicago, Ill., 1980.
9. *Steam Sterilization and Sterility Assurance*, Association for Advancement of Medical Instrumentation, Arlington, Va., Jan. 1980.

10. *Assoc. Oper. Room Nurses J.* **32,** 222 (Aug. 1980).
11. R. R. Ernst, *Industrial Sterilization: Ethylene Oxide Gaseous Sterilization for Industrial Applications*, Duke University Press, Durham, N.C., 1973.
12. Ref. 1, p. 165.
13. J. H. Bowie, J. C. Kelsey, and G. R. Thomson, *Lancet*, 586 (March 16, 1963).
14. U.S. Pat. 2,075,845 (1937), P. M. Gross and J. F. Dickson.
15. C. R. Phillips and S. Kaye, *Am. J. Hyg.* **50,** 270 (1949).
16. C. R. Phillips, *Am. J. Hyg.* **50,** 280 (1949).
17. S. Kaye and C. R. Phillips, *Am. J. Hyg.* **50,** 296 (1949).
18. R. R. Ernst and J. E. Doyle, *Biotechnol. Bioeng.* **10,** 1 (1963).
19. A. Heyl, *Surgical Business*, 42 (Jan. 1972).
20. U.S. Pat. 3,016,328 (Jan. 9, 1962), R. E. Pepper and E. R. Lieberman (to Ethicon).
21. W. Horwitz, *Official Methods of Analysis of Association of Official Analytical Chemists*, AOAC, Washington, D.C.
22. *Chem. Week*, 40 (March 10, 1982).
23. *Processes for Low Acid Foods*, National Canners Association, Washington, D.C., latest edition.

General References

S. S. Block, *Disinfection, Sterilization, and Preservation*, 2nd ed., Lea and Febinger, Philadelphia, Pa., 1977.

G. Sykes, *Disinfection and Sterilization*, 2nd ed., Spon, London, 1965.

S. Turco and R. E. King, *Sterile Dosage Forms, Their Preparation and Clinical Application*, Lea and Febinger, Philadelphia, Pa., 1974.

N. A. M. Eskin, H. M. Henderson, and R. S. Townsend, *Biochemistry of Foods*, Academic Press, New York, 1971.

National Conference of Spacecraft Sterilization, California Institute of Technology, U.S. National Aeronautics and Space Administration Technical Information Division, 1966.

Canned Food, Principles of Thermal Processing, Food Processing Institute, Washington, D.C.

THOMAS A. AUGURT
Propper Manufacturing Co., Inc.

STEROIDS

Steroids are ubiquitous members of a large class of marine and terrestrial organic compounds that have the perhydro-1,2-cyclopentenophenanthrene ring system (**1**) as a common structural feature (1–6).

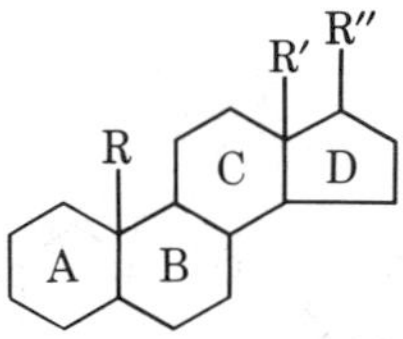

(**1**)

The vast diversity of the natural and synthetic members of this class depends primarily upon variations in the side chains R, R′, and R″, as well as on nuclear substitution, degree of unsaturation, and stereochemical relationships at the ring junctions. Included under the designation steroid are the naturally occurring and synthetic substances, eg, marine, myco-, phyto-, and zoosterols; adrenocortical hormones; steroidal plant and animal alkaloids; antibiotics; antihypertensive drugs; bile acids; cardiac aglycones; contraceptive drugs (qv); insect hormones; insect moulting hormones; sapogenins; male and female sex hormones; toad poisons; and vitamin D (see Antibiotics; Cardiovascular agents; Hormones, adrenal-cortical hormones; Hypnotics, sedatives, anticonvulsants; Insect-control technology; Silk; Vitamins, vitamin D).

The term steroid is derived from the isolation of the initial solid crystalline secondary alcohols from the nonsaponifiable lipid extracts of plants and animals called sterols. Study of steroids has led to significant advances in synthetic and mechanistic chemistry, as well as in biochemistry, biology, and the medical sciences. Pharmaceutical research has developed a wide variety of novel steroid-based medicines with sales of ca $\$10^{9}$.

Steroid research in the 19th century involved isolation of the sterols and bile acids from natural sources but little structural investigation. DeFourcroy is generally credited with the discovery of cholesterol (**2**) in 1789, and Gavelin described an impure preparation of cholic acid (**3**) from the saponification of ox bile (7). Early in the 20th century, lithocholic acid (**4**) was isolated from the same source (8). The chemical and physical similarities between compounds (**2**) and (**4**) and their cooccurrence in ox bile suggested that the bile acids were steroids, an inference which was proven by degradative studies (9).

The structural elucidation of cholesterol was begun in 1903 (10–11). X-ray crystallographic data demonstrated that the steroid nucleus is a thin lath-shaped structure (12). These data led to the proposal that 1,2-cyclopentenophenanthrene is the correct structure for cholesterol (**2**) (13–14). This structure accommodated all the experimental evidence and was confirmed by the identification of Diels hydrocarbon (**5**) and by the total synthesis of equilenin (**6**) (15–17).

During the final stages of the structural determination of cholesterol, the sex hormones estrone, progesterone, estradiol, and testosterone, were isolated and their structures determined (17). Also in the middle 1930s, investigations of steroid pro-

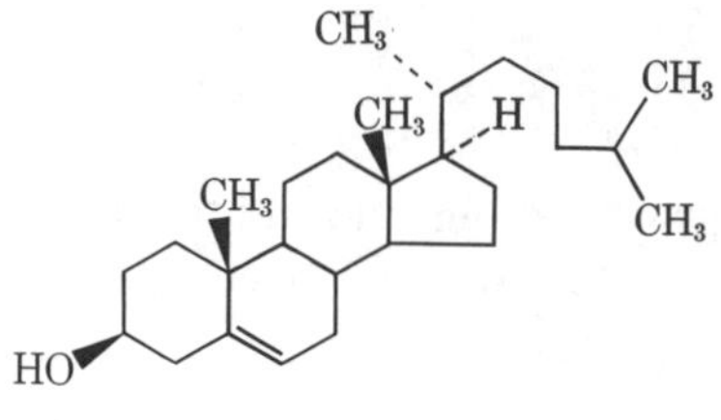

(2) 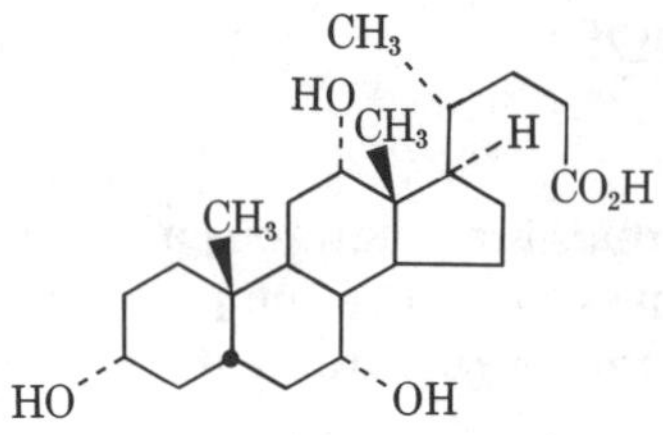 (3)

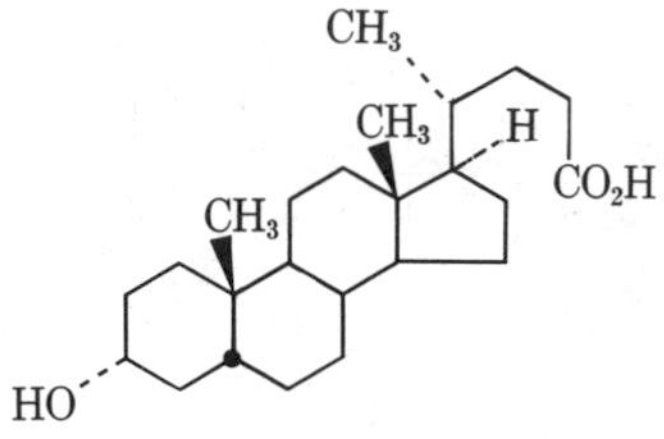

(4)

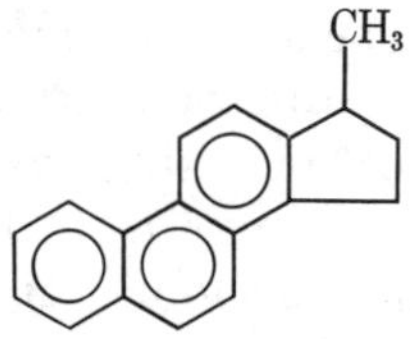

(5) 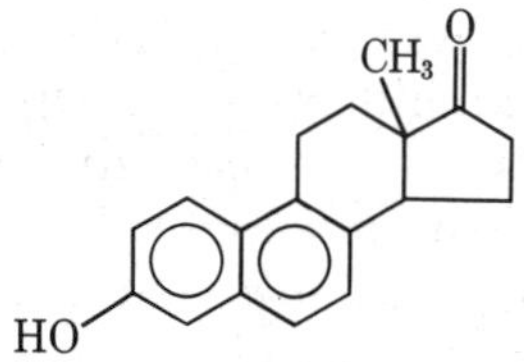 (6)

duction in the cortex of the adrenal glands began. Among the many steroids isolated were cortisol, which controls glucose metabolism, and aldosterone, which controls water homeostasis. The commercial steroid industry began in the 1940s with the preparation of pregnenolone from the sapogenin, diosgenin, in Mexican yams (18). The discovery in 1948 that the adrenal cortex steroid cortisone dramatically alleviates the symptoms of arthritis led to intensive research on the anti-inflammatory properties of these types of steroids. Great effort went into the development of partial and total syntheses for the commercial preparation of cortisone (19). The search for alternative methods of producing cortisone and the discovery of much more potent analogues greatly stimulated academic and industrial interest in steroid research.

Although by 1950 very many steroid reactions had been documented, their reaction mechanisms were not very well-understood. The key role of conformational analysis in the understanding of the properties and reactions of alicyclic compounds and, in particular, steroids was demonstrated (20–22). The demonstration that steroids not only have a rigid skeleton but also have a chemistry which is logically predictable resulted in the enthusiastic adoption of steroid research by organic and physical-organic chemists.

The discovery of the reduction of phenolic ethers to dihydro derivatives in 1950 led to the development of 19-norsteroids derived from estrone methyl ether (23). The preparation of orally active progestins based on 19-norsteroids led to their commercial introduction in 1957 for the regulation of menstrual disorders and, subsequently in 1960, for oral contraception.

Continuing investigations in the 1960s disclosed that the insect moulting hormone, ecdysone, was a steroid and led to the development of new methods of insect control (24). Similar inquiries into the nature of the arrow poison derived from a Central American tree frog revealed it to be a very novel steroidal alkaloid (25). Besides its lethality, interest in this steroid stems from its persistent depolarizing effects on neuronal transmission.

Beginning in the early 1970s, interest in steroid sources intensified because of problems with supplies of Mexican diosgenin. This led to new commercial fermentation methods for the preparation of commercially important steroids from cholesterol and sitosterol. At the same time, studies began on the control of cholesterol biosynthesis on a molecular level through selective enzymic inhibition. The fascinating interconversions among the vitamin D-like compounds were also described, and the active form of vitamin D was determined and its biochemical effects were delineated.

Nomenclature

The naming of steroids according to rules approved by the International Union of Pure and Applied Chemistry in 1972 are based on IUPAC 1957 rules as subsequently revised (26–28). Tentative rules for the symbolic shorthand system designation of steroids have also been proposed (29). Compounds are systematically named as derivatives of the parent hydrocarbons. Substituents attached to the plane of the steroid ring system from above are called β and designated by a solid line; those attached below the plane of the steroid are α and are shown by a dotted line; those substituents of unknown configuration are indicated with a wavy line and designated ξ. Although the nomenclature rules of references 26–28 are generally followed, many if not most of the more important steroids are designated by trivial names.

Biosynthesis

In its simplest terms, the equation for the formulation of cholesterol from acetyl coenzyme A can be written:

$$18\ CH_3CO—S—CoA + 10\ H^+ + \tfrac{1}{2}\ O_2 \rightarrow C_{27}H_{46}O + 9\ CO_2 + 18\ CoA—SH\ - 10\ e$$

This equation scarcely does justice to the actual step-by-step formation of cholesterol which involves ten enzymes from acetyl-CoA to squalene and at least 19 enzymic steps from squalene to cholesterol (30). Acetate is converted to acetyl-CoA, which self-condenses to form acetoacetyl-CoA, which then undergoes an aldol reaction with another molecule of acetyl-CoA to form β-hydroxy-β-methylglutaryl-CoA (31–32). This is then reduced to mevalonic acid which is decarboxylated and phosphorylated to isopentenyl pyrophosphate. An isomerase catalyzes the formation of γ,γ-dimethylallyl pyrophosphate from isopentenyl pyrophosphate, and under nonequilibrium conditions condensation with one another to form *trans*-geranyl pyrophosphate. Addition of a second molecule of isopentenyl pyrophosphate to *trans*-geranyl pyrophosphate forms farnesyl pyrophosphate, which intermolecularly forms a cyclopropane ring to generate presqualene pyrophosphate (33). Ring opening of the cyclopropane then leads to squalene, which is epoxidized at its 2,3-position. This last phenomenon has been

gonane [*4732-76-7*] 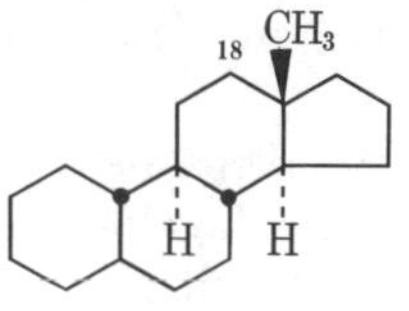 estrane [*24749-37-9*] 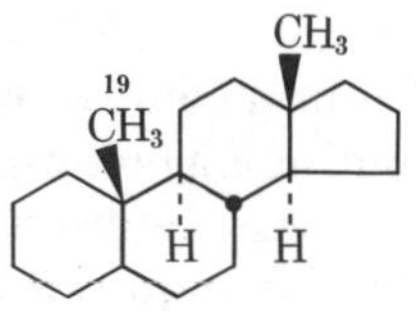 androstane [*24887-75-0*]

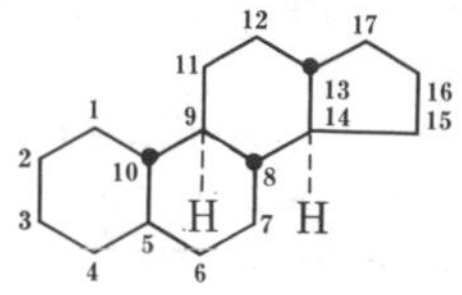

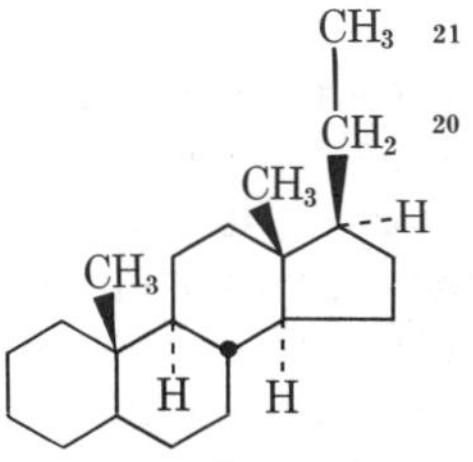

pregnane [*24909-91-9*] 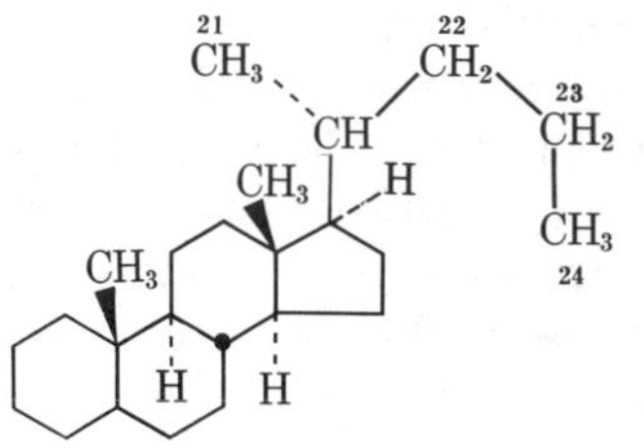 cholane [*548-98-1*]

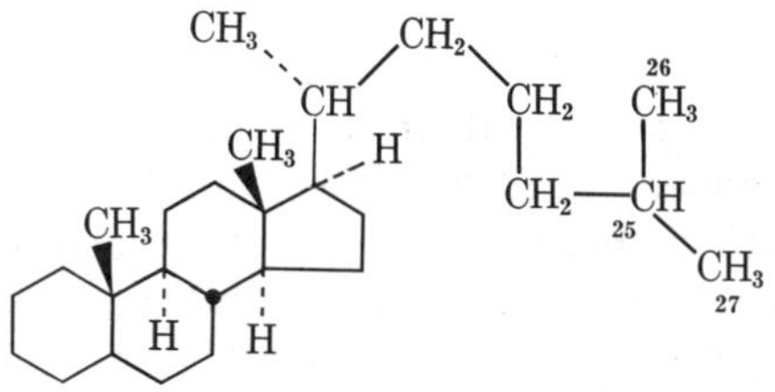

cholestane [*14982-53-7*]

duplicated in the laboratory (34). Opening the oxirane ring in the monocyclic precursor then forms the tetracyclic lanosterol (**7**). This is an amazing transformation considering that the highly flexible polyene must attain the folded conformation in order to form lanosterol (35). Although this cyclization is under enzyme control, similar observations of polycycle formation from acyclic precursors have been made (36). The next step in the conversion of lanosterol to cholesterol requires the saturation of the *C*-24 double bond and the oxidative elimination of the three methyl groups at *C*-4 and *C*-14 (37). Removal of the 14α-methyl group is accomplished by hydroxylation, subsequent oxidation to the aldehyde, and then 15α-hydroxylation. Elimination forms the 8,14-diene system and formic acid (38). The reduction of Δ^{14} probably involves trans addition of a hydride ion from NADPH at the 14α-carbon and of a solvent proton (39). After reduction of the 14-double bond, the remaining double bond isomerizes to *C*-7. After reduction and isomerization, the methyl groups at *C*-4 are removed by selective hydroxylation of the 4α-methyl group and subsequent oxidation to the carboxylic acid. The 3β-hydroxyl group is next oxidized to the ketone which induces decarboxylation and epimerization of the remaining 4β-methyl group to the 4α-configuration. Reduction of the 3-ketone regenerates the 3β-alcohol. The process is then repeated to remove the remaining methyl group (39). The resulting Δ^7-steroid, ie, lathosterol (**8**), is then dehydrogenated to the 5,7-diene, 7-dehydrocholesterol (**9**), followed by hydrogenation of the Δ^7-double bond to form cholesterol (39). The enzymic conversions from acetate through the formation of farnesyl pyrophosphate, with the exception of HMG (hydroxymethylglutaryl)-CoA reductase, all occur in the cytoplasm of the cell. In subsequent steps leading to cholesterol, both enzymes and substrates are microsomally bonded (40–41).

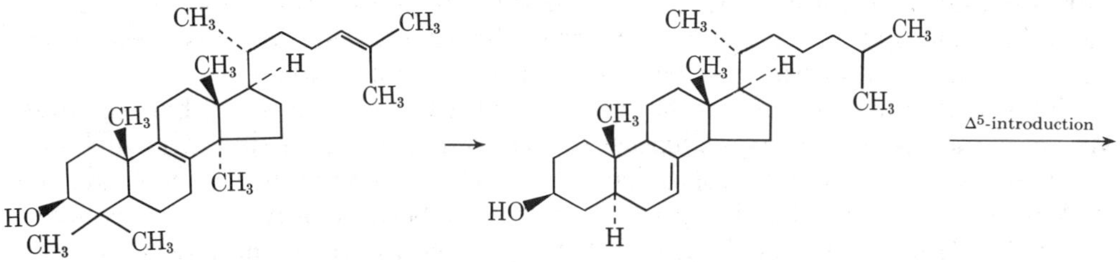

(**7**) 3β-lanosterol → (**8**) lathosterol $\xrightarrow{\Delta^5\text{-introduction}}$ (**9**) 7-dehydrocholesterol $\xrightarrow{\Delta^7\text{-reduction}}$ cholesterol (**2**)

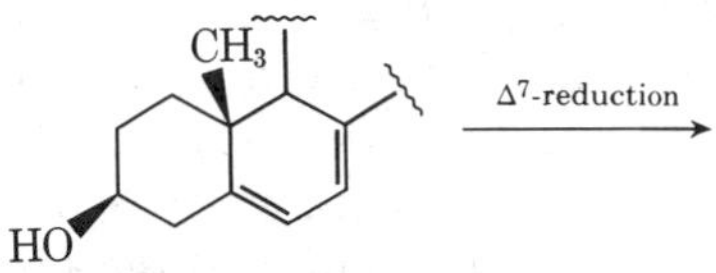

In the mammalian systems, cholesterol in the prime progenitor of the steroids (42). A very simplified relationship showing the main biosynthetic pathways leading to steroids and steroid hormones is shown below.

cholesterol → progesterone → mineralocorticoids, glucocorticoids

cholesterol ↓ bile acids; progesterone ↓ testosterone

bile acids → testosterone → estradiol

Structural Types

The structural features in the main types of naturally occurring steroids are diverse because of the ubiquitous presence of steroids in terrestrial and marine plants and animals, and because of their differing biological functions.

Sterols. The natural sterols are crystalline C_{26}–C_{30} alcohols containing an aliphatic side chain at *C*-17; they are widely distributed and occur both free and in combination as esters and glycosides (43). Although cholesterol is the main sterol in man and the higher animals, the lower animals and plants show an enormous variety of oxidation and unsaturation patterns as well as side-chain diversity.

Cholesterol (**2**) is present in all mammalian tissue either free or esterified and serves an important biochemical function in membranes (44). An average adult man weighs ca 70 kg, of which 145 g is cholesterol. Of this, 32 g (22%) occurs in brain and nervous tissue. The rest of the remaining cholesterol is not distributed evenly: connective tissue and body fluids account for 22%; muscle, 21%; and skin, 11% (45). Cholesterol functions as a metabolic precursor for the bile acids and mammalian steroid hormones as well as for the cardenolides, ecdysones, and many other steroid types (46–47).

In the 1930s, cholesterol was a precursor for androgen synthesis but was supplanted by sapogenins (48). However, advances in microbial fermentation by the Japanese, who have large supplies of cholesterol as a by-product of wool defatting, have again made cholesterol a commercial source of steroid hormones (49). A series of oxygenated cholesterol derivatives has been developed as inhibitors of sterol synthesis (50). Lanosterol (7) is the best known of the trimethyl sterols. It is a tetracyclic triterpene and occurs with other tripenes and cholesterol in wool fat.

The D vitamins are steroids that are part of the hormonal system that regulates calcium homeostasis. Rickets is a disease caused by faulty calcium deposition in bone. It was observed that patients responded favorably to sunlight or ingestion of cod liver oil. Attempts to isolate the vitamin were ineffective because of its very low concentration (51). However, it was discovered that superficially purified cholesterol contains precursors which, upon irradiation, are converted into the vitamin. Provitamin D_2 was identified as ergosterol (**10**), which is a characteristic sterol of yeast and other fungi, and provitamin D_3 was identified as 7-dehydrocholesterol (**9**). Irradiation yields vitamin D_2 (**11**) and D_3 (**12**), respectively (52). The thermal and photochemical reactions of the 5,7-diene system are illustrated in Figure 1. Diene (**9**) undergoes a photochemical conrotatory ring opening to yield 9,10-secocholesta-5(10),6,8-triene (**13**) (previtamin D_1), which undergoes a thermal sigmatropic 1,7-hydrogen shift to yield a cisoid conformer of vitamin D (**12**) (53). However, previtamin D (**13**) can be photochemically excited to form lumisterol (**14**) by conrotatory ring closure and tachysterol (**15**) by isomerization about the central triene double bond. At higher temperatures, triene (**13**) undergoes a disrotatory thermal ring closure to give pyrocalciferol (**16**), which upon irradiation is transformed into photopyrocalciferol (**17**). Thermally, (**17**) reverts to pyrocalciferol (**16**) (54) (see Vitamins, vitamin D).

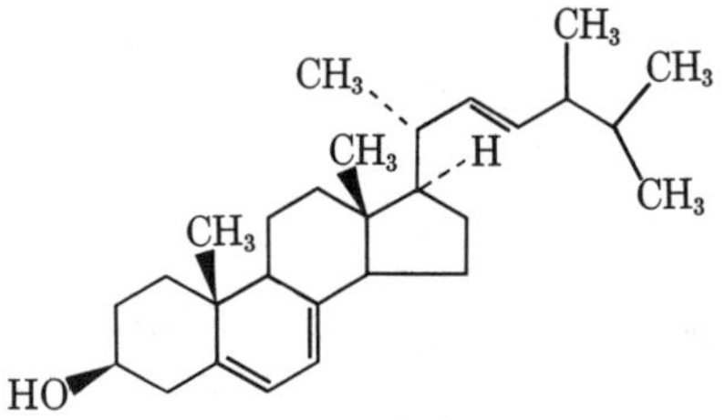

(**10**)

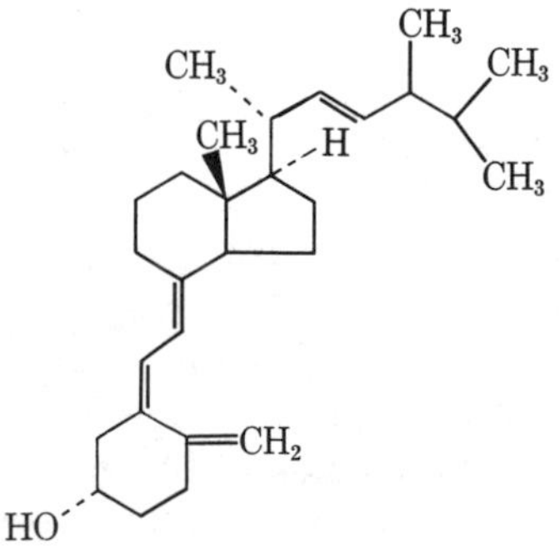

(**11**)

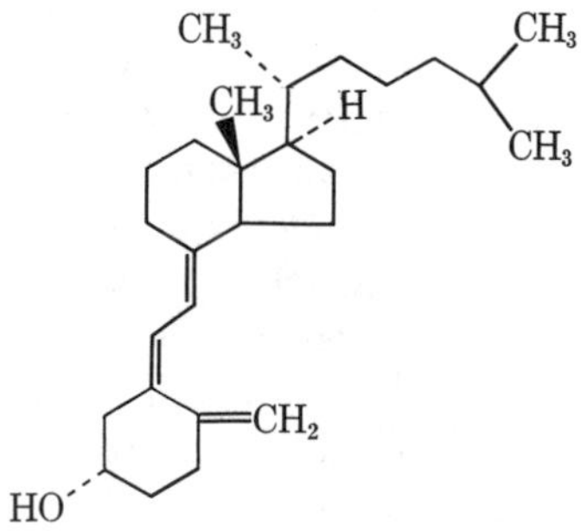

(**12**)

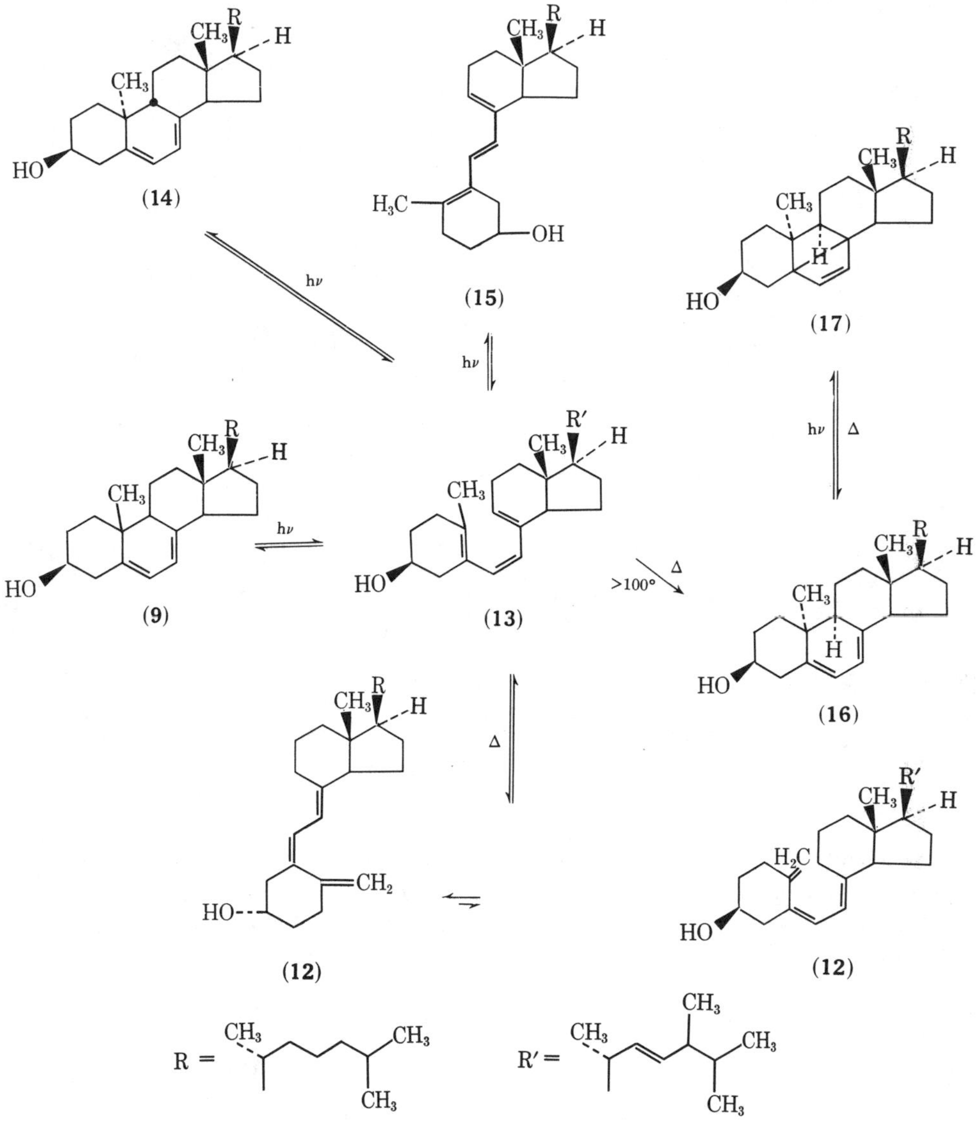

Figure 1. The photochemical and thermal reactions of vitamin D (**13**).

The plant sterols are exemplified by compounds, eg, stigmasterol (**18**) and sitosterol (**19**). Stigmasterol (**18**) can be isolated from the calabar bean, but it occurs chiefly in soybean oil as an approximately 1:4 mixture with sitosterol (**19**) (55–56). Sterol (**18**) serves as starting material for the preparation of the sex hormone progesterone (**21**) and corticoids through oxidation to the enone, ozonolysis, and oxidation of the subsequently formed enamine (**20**) with air (57). The by-product, ie, sitosterol, became an important source for steroid hormones with the discovery of microbiological methods for side-chain cleavage (58).

In higher plants, cyclization of (3*S*)-squalene 2,3-epoxide occurs with formation of the 9,19-cyclopropane containing steroid cycloartenol (**22**). This compound has a key role in sterol biosynthesis that is analogous to that of lanosterol in animals and

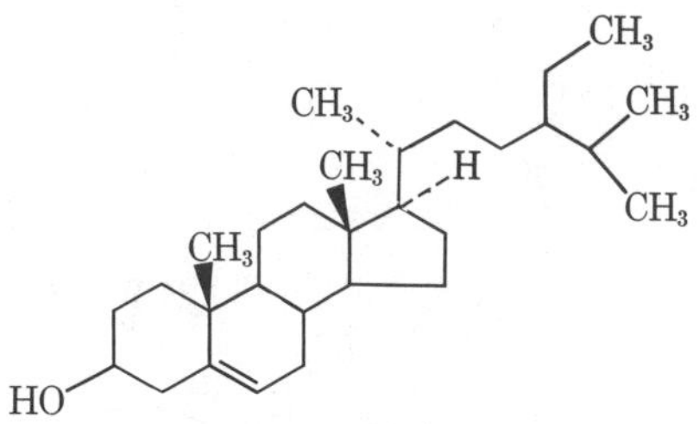

(19)

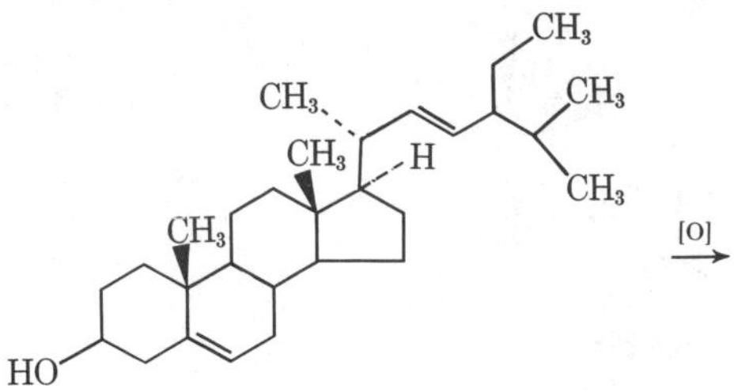

(18)

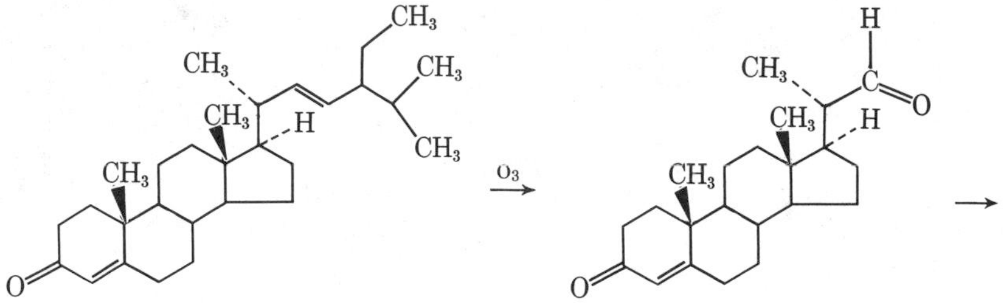

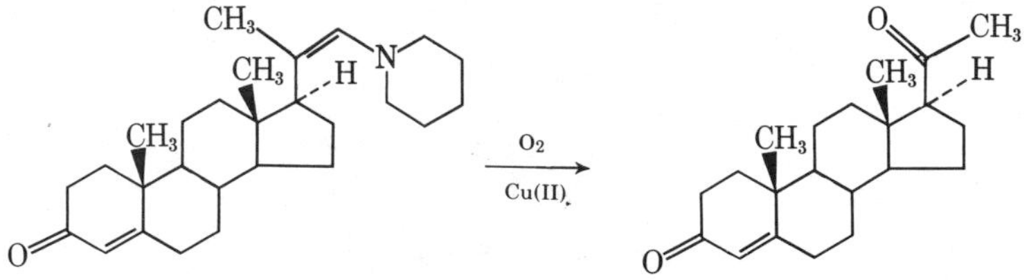

(20)

(21)

fungi, with the cyclopropane methylene group becoming an angular methyl group (59). These cyclopropane steroids are widely distributed in the plant kingdom. The triterpene abietospirane (**23**) comprises 14% of the bark of the white fir (*Abies alba*) (60).

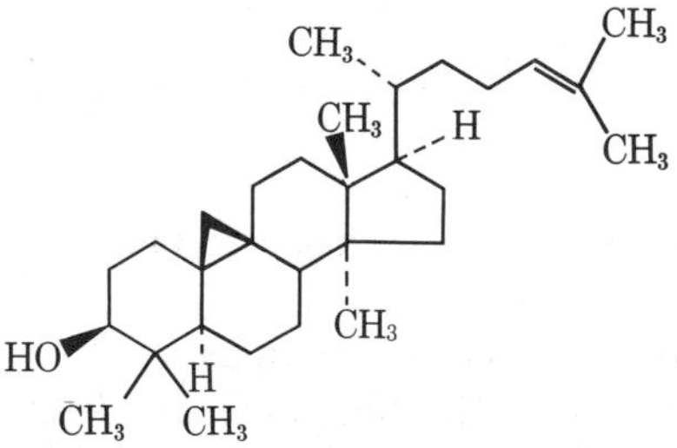

(22)

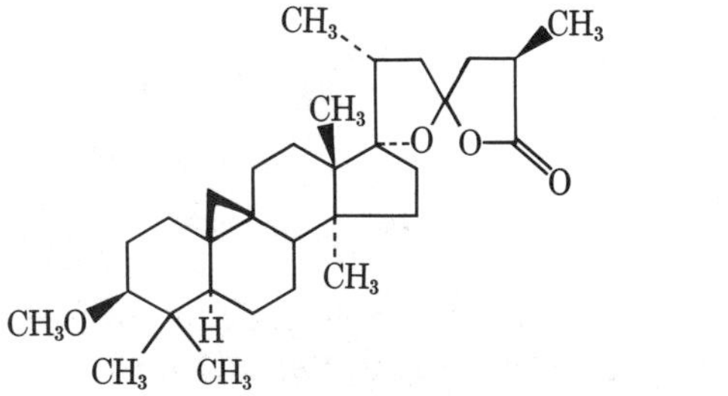

(23)

Marine sterols generally have a carbon range of C_{26}–C_{30}, although very minor or trace levels of sterols with conventional cholesterol nuclei but with biosynthetically, unusually short hydrocarbon side chains (total carbon range of C_{19}–C_{25}) are present in extracts of a wide variety of marine organisms (61–62). The carbon variation occurs almost exclusively in the side chain. The structural diversity other than carbon content which exists among marine sterols is also almost entirely in the side chain. The C_{26} sterols, eg, asterosterol (**24**), are a widespread class of marine origin and are probably formed from a C_{27} steroid, eg, occelasterol (**25**), by biological *C*-27 demethylation. The C_{27} steroids are also formed by demethylation of a C_{28} precursor, eg, 24-methylenecholesterol (**26**). Biological alkylation at *C*-27 by *S*-adenosylhomocysteine leads to the C_{29} aplysterol (**27**), and further alkylation forms the C_{30} verongulasterol (**28**) (63).

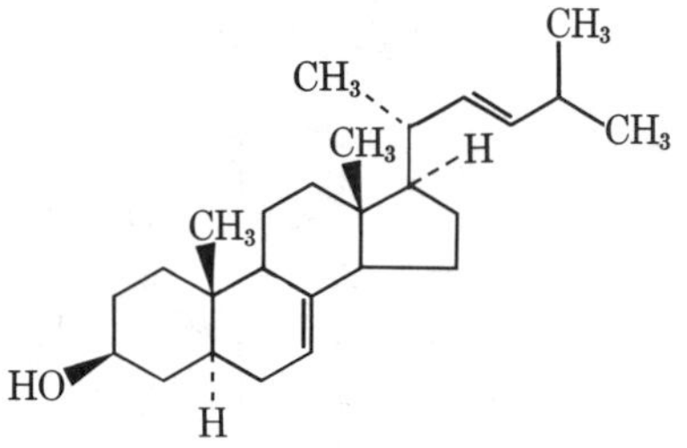

(**24**)

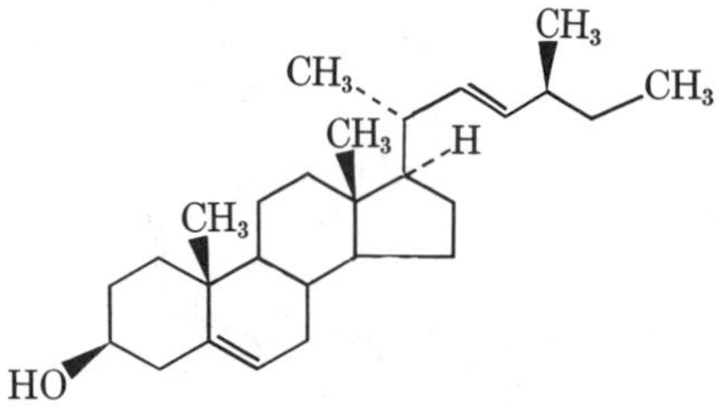

(**25**)

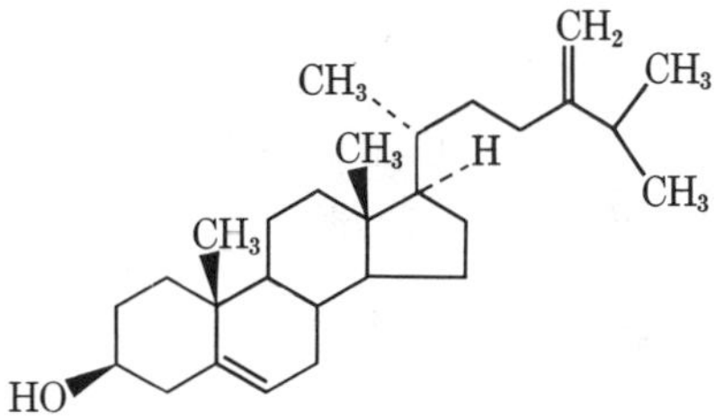

(**26**)

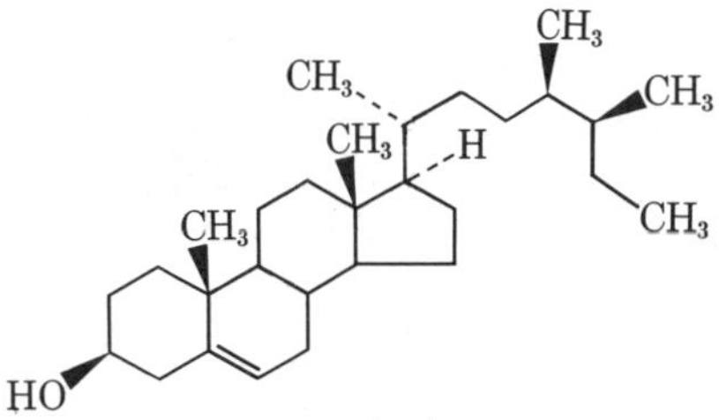

(**27**)

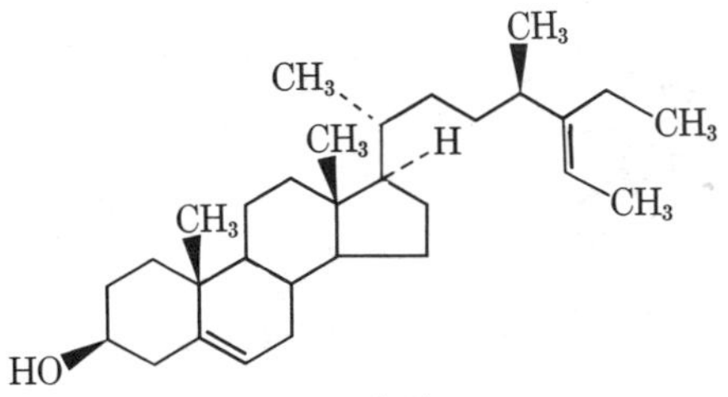

(**28**)

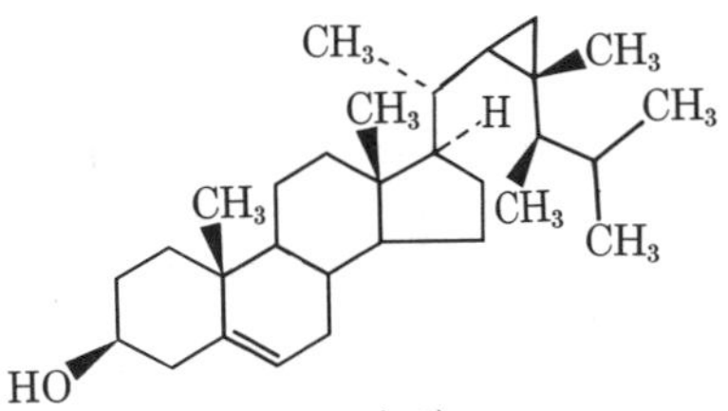

(**29**)

The steroid gorgosterol (**29**) has a cyclopropane ring in the side chain, which seems characteristic of a class of marine steroids (64). Other novel steroids isolated from marine sources are calysterol (**30**), which contains a cyclopropene ring, and the first naturally occuring allene (**31**) (65–66). A totally different type of steroid found in the coral *Telesto rüsei* is the 18-oxygenated pregnatriene (**32**) (67). The 18-substitution

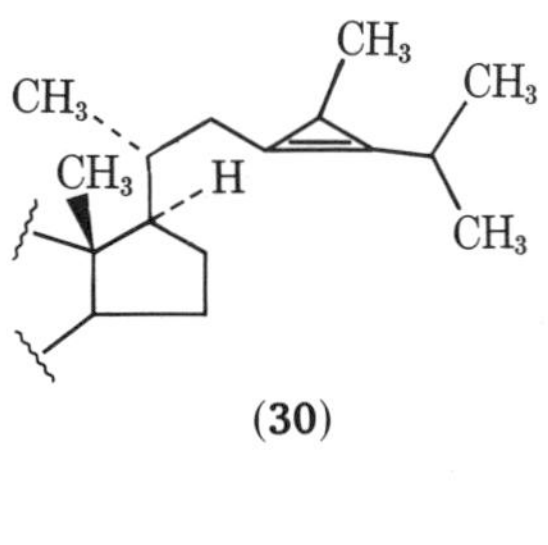

(**30**)

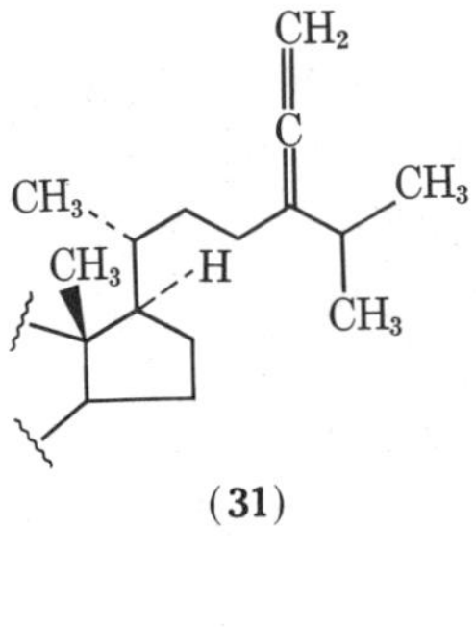

(**31**)

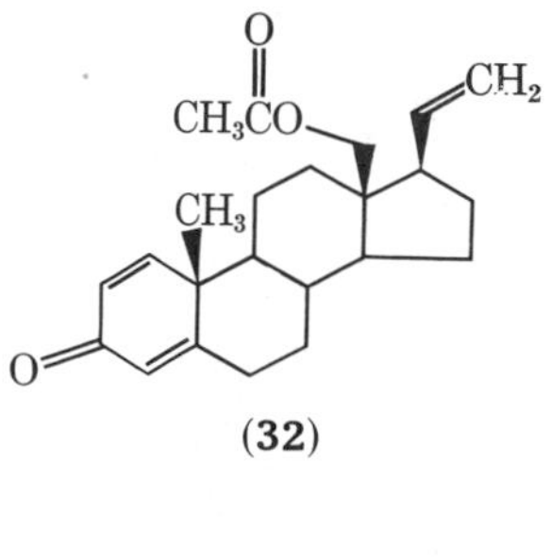

(**32**)

is characteristic of intermediates in the biosynthesis of aldosterone in the human adrenal glands.

The steroidal antibiotics of the fusidane series, ie, fusidic acid (**33**), helvolic acid (**34**), and cephalosporin C (**35**), are fungal metabolites with antibiotic properties. They are of great use in combatting staphylococcus infections.

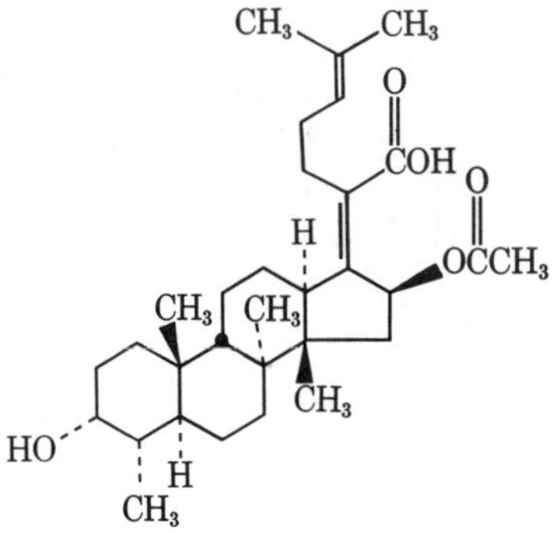

(**33**)

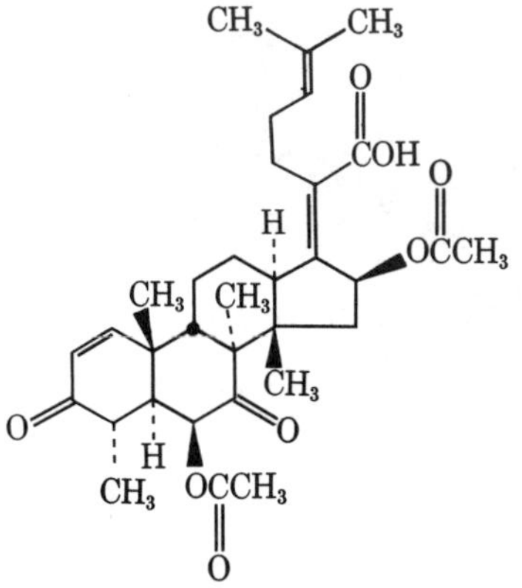

(**34**)

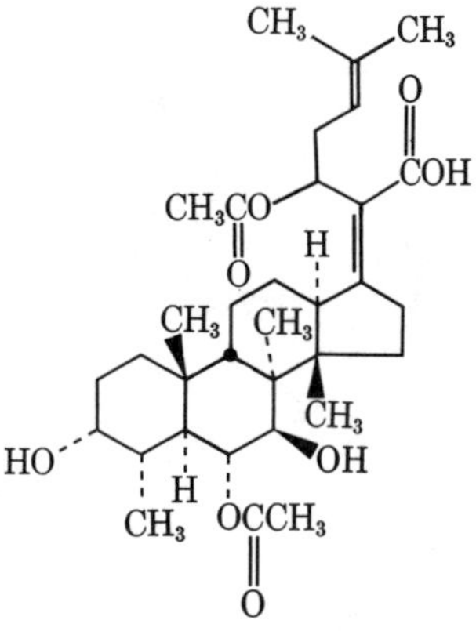

(**35**)

One of the remarkable features of these steroids is the unusual trans-syn-trans-anti-trans stereochemistry which forces ring B into the boat configuration (68). The cucurbitacins, exemplified by the diosphenol elaterin (**36**), have been studied for their cytotoxic and antitumor activity as well as their natural bactericidal action (69).

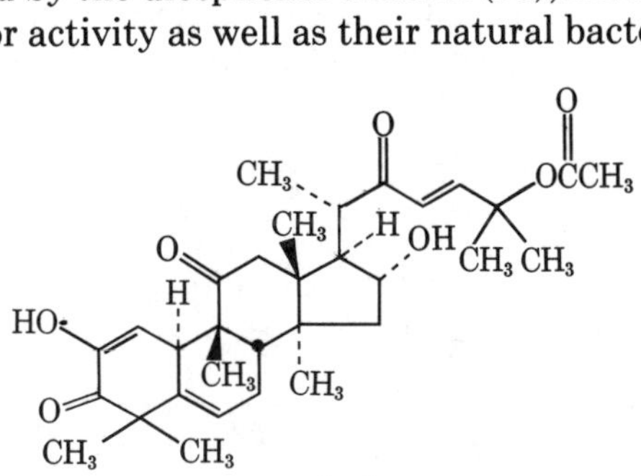

(**36**)

Bile Acids. Bile, a golden brown to greenish fluid, is a product of the liver and is stored in the gall bladder, and its principal effect is upon fat absorption. The active constituents of bile are the bile salts, which are either glycine or taurine conjugates of a polyhydroxic steroidal acid. The principal bile acids of mammals are the hydroxy derivatives of 5β-cholan-24-oic acid. Ox bile is the most important commercial source of these acids and contains primarily cholic acid (**3**) with lesser amounts of deoxycholic acid (**37**), which lacks the 7α-hydroxyl group of structure (**3**) (70). Deoxycholic acid was important as a starting material for the commercial production of steroid hormones, but its importance has lessened considerably with the development of alternative methods based on diosgenin, stigmasterol, and sitosterol.

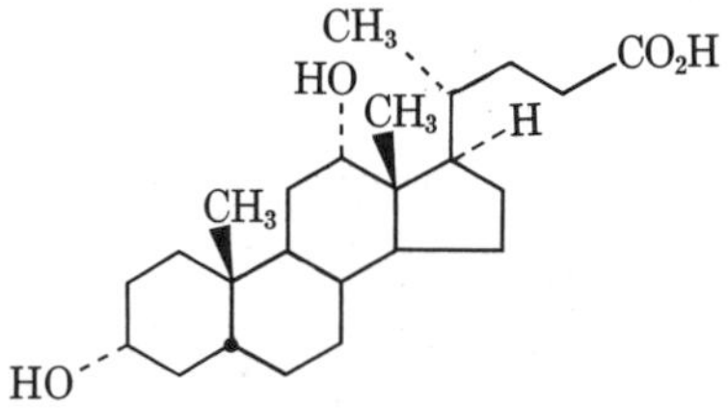

CH_3 HO CO_2H CH_3 H CH_3 HO

(**37**)

The bile acids are biosynthesized in the liver from cholesterol (**2**) (71). In mammals, the 3β-hydroxyl group of compound (**2**) is epimerized to the 3α position and the configuration at *C*-5 is inverted from the AB-trans stereochemistry in cholestanol to the AB-cis characteristic of the cholane system. In the biosynthesis of cholic acid (**3**), all nuclear changes appear to precede oxidation of the side chain. The first and rate-determining step is 7α-hydroxylation of cholesterol, followed consecutively by 12α-hydroxylation, reduction of the double bond, and epimerization of the 3β-alcohol. The cleavage and oxidation of the side chain occurs through formation of the *C*-27-carboxylic acid and then β-hydroxylation at *C*-24. Beta-scission, similar to the β-oxidation of long-chain fatty acids, then yields cholic acid (**3**) and propionic acid as its coenzyme-A ester (72).

Saponins and Sapogenins. The steroidal saponins are plant glycosides that give soapy solutions and the crude plant extracts have been used as detergents. They cause hemolysis and have been used as fish poisons but are not toxic to mammals, probably because they are not absorbed from the gut. The saponins form molecular complexes with several classes of compounds; notable are the very insoluble complexes formed between digitonin [*11024-24-1*], cholesterol, and related alcohols of cholesterol. The saponins are not easy to isolate in a pure state and until recently have not been studied. The complete structures of several have been elucidated and, with few exceptions, the sugar is linked to the steroid through its 3β-hydroxyl group. Thus digitonin consists of the aglycone digitogenin (**38**) and two glycose, two galactose, and one xylose sugar moieties. Interest is primarily in sapogenin aglycones, particularly diosgenin (**39**) and hecogenin (**40**), which are commercially useful as raw materials for the steroid industry (73). The sapogenins are *C*-27 cholestane derivatives and are oxygenated at carbon atoms 16, 22, and 26; in most cases this takes the form of 16,22- and 22,26-oxide bridges in the characteristic spiroacetal grouping. Diosgenin (**39**) has been developed as a source for the 20-oxopregnanes and by further conversions into the other sex hormones (74). Hecogenin (**40**), which used to be plentifully available from sisal processing, was

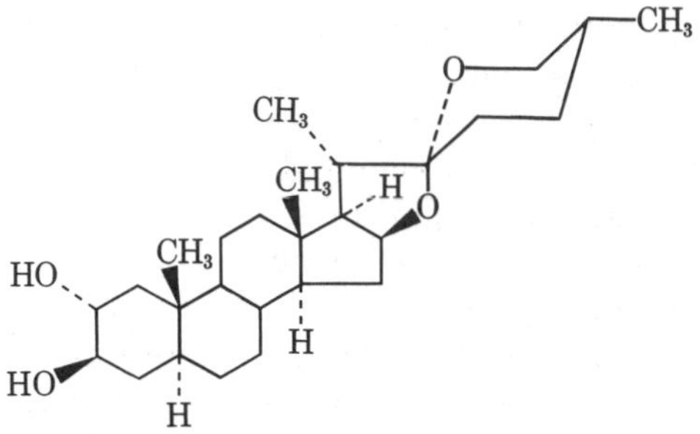

(38)

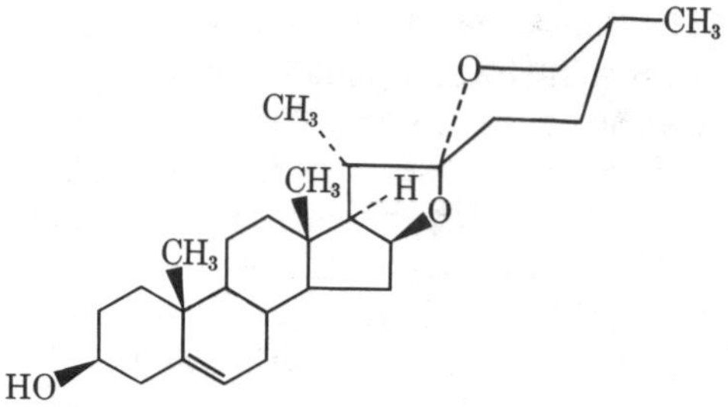

(39)

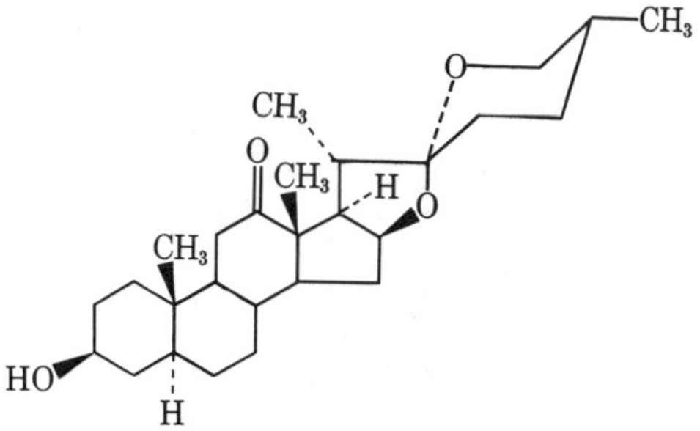

(40)

important for 11-oxygenated corticoid synthesis because of its 12-keto group (75–76).

Steroidal Alkaloids. ***Plant Alkaloids.*** The steroidal plant alkaloids are very widespread in nature. Despite their complexity, these alkaloids may be divided into two main groups: normal steroidal alkaloids where the alkaloid has a normal steroid skeleton with a nitrogen in the side chain, attached to the rings, or as part of an additional ring; and modified steroidal alkaloids where the steroid nucleus is modified and with a nitrogen in the side chain and usually in a ring (77).

The normal steroidal alkaloids are divided into five distinct groups, all of which are derived from 5α-pregnane and differ in the number and position of the amino group(s). Funtumine (**41**), an example of a 3-amino-5α-pregnane, is obtained from a *Holarrhena* species and has been considered as a possible raw material for steroid synthesis (ca 15–20 g/kg of leaves). The 20-amino-5α-pregnanes, eg, terminaline (**42**), occur much less frequently in nature as do the other normal structural types. The 3,20-diamino-5α-pregnane alkaloid holarrhidine (**43**) occurs in the small Indian shrub *Holarrhena antidysenterica.* This shrub is used in Indian folk medicine as a remedy for amoebic dysentery. The alkaloid is of interest because of the unusual 18-hydroxyl group, which characterized the alkaloid as a possible starting material for the synthesis of aldosterone and related 18-oxygenated pregnanes. The related 20-amino-pregnene pachystermine A (**44**) is the only known example of a steroid containing the β-lactam ring. Conessine (**45**), which is a 3-aminoconanine, is the most abundant of the Kurchi alkaloids and was the starting material for the commercial synthesis of 18-hydroxyprogesterone and 18-hydroxydesoxycorticosterone (78). However, this synthesis has been superseded by a sequence starting from pregnenolone (79). The last group of normal alkaloids consists of bases related to 20-piperidyl-5α-pregnane. These are frequently referred to as the solanum, veratrum, or fritillaria alkaloids, and are formed in plants of the *Solanaceae* and *Liliaceae* families. These are conveniently subdivided into three groups: simple bases, solanidanes, and spirosolanes. Solacongestidine (**46**) is the principal alkaloid of *Solanum congestiflorum* and is an example of the first type. The spirosolanes, which occur in potatoes and tomatoes, are of commercial interest as steroid starting materials. For instance, both solasidine (**47**) and its 5α-dihydro derivative, tomatidine, have been converted into 16-dehydropregnenolone derivatives by Marker's process for the degradation of diosgenin (80). The third type, ie, the solanidanes, is characterized by an indolizidine ring system fused across the steroidal 16,17-positions, as observed in solanidine (**48**) (81–82).

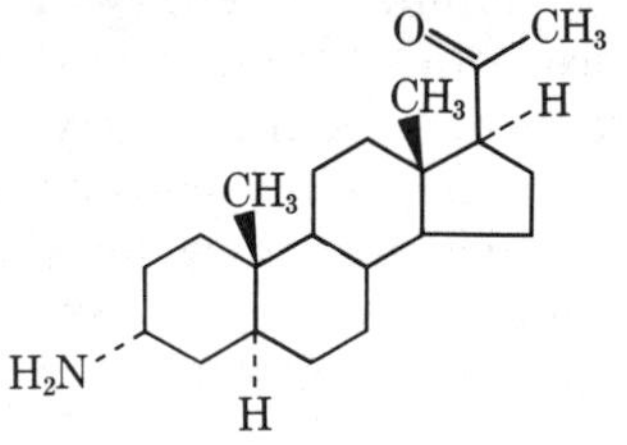

(**41**)

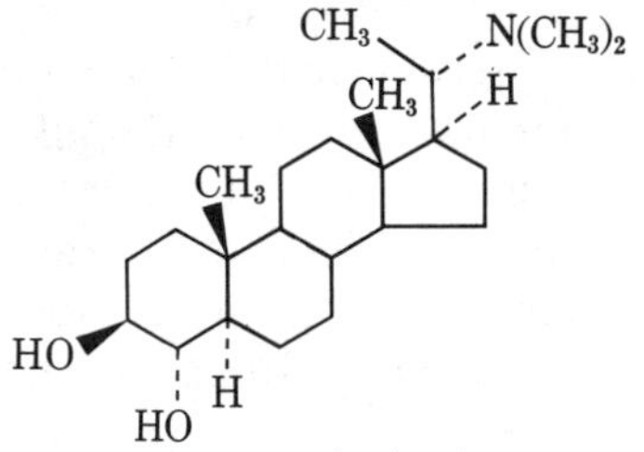

(**42**)

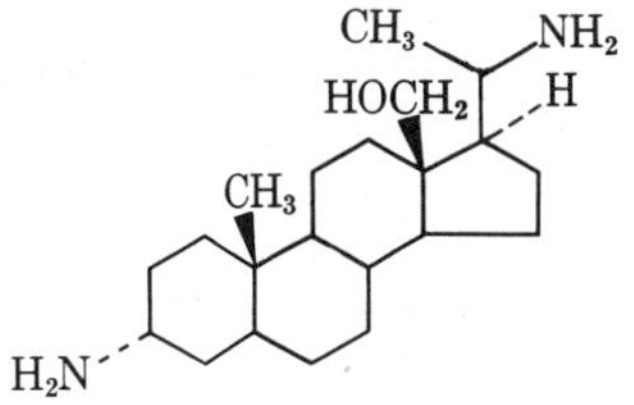

(**43**)

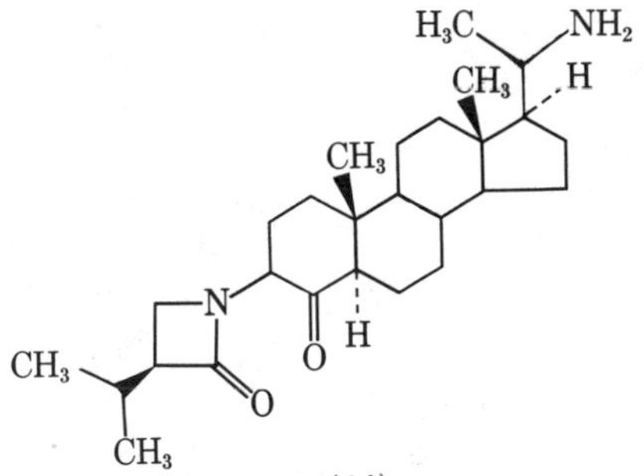

(**44**)

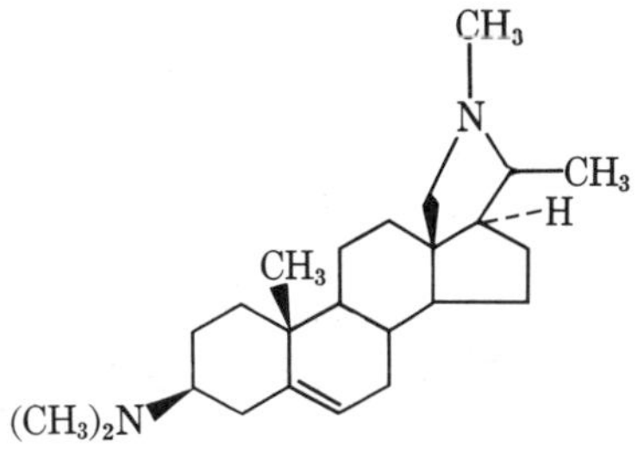

(**45**)

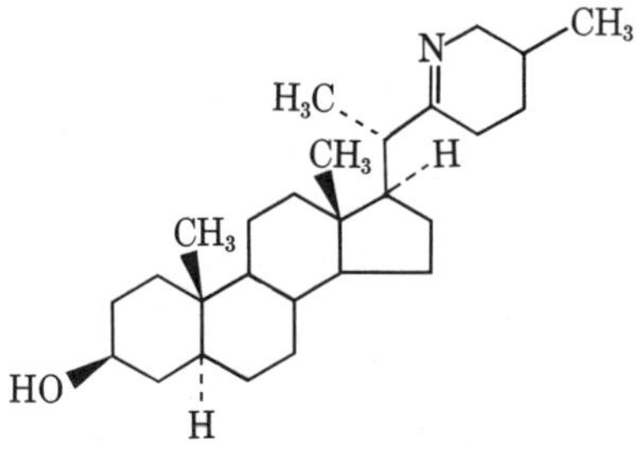

(**46**)

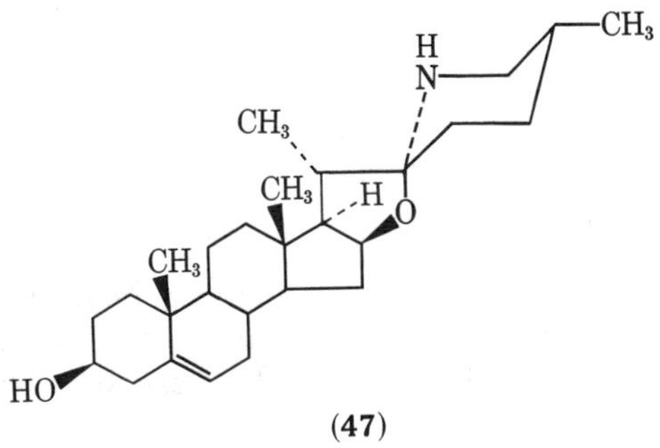

(**47**)

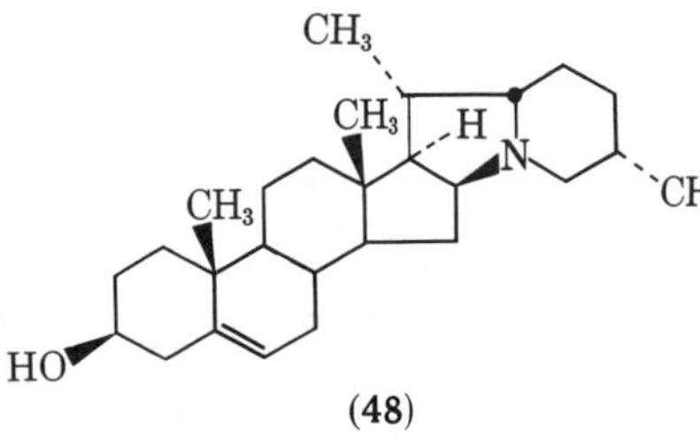

(**48**)

The modified plant steroidal alkaloids are of two types: the C-nor-D-homo steroids, which have a modified ring system where ring C has been contracted to a five-membered ring and ring D has been expanded to a six-membered ring; and the buxus alkaloids, which usually have a cyclopropane ring fused into a normal steroid nucleus at positions 9 and 10. The C-nor-D-homo alkaloids are subdivided into three types: the alkamines, ester alkaloids, and fritellaria alkaloids. Plants of the lily family contain these alkaloids and have been used since the middle ages for witchcraft, febrifuges, cardiotonica, emetics, and poisons. The alkamines, eg, veratramine (**49**), possess 2 or 3 oxygen atoms, whereas the ester alkaloids contain 7–9 oxygen atoms which are esterified with various organic acids, eg, acetic, angelic, veratric, vanillic, etc. The ester alkaloid protoverine (**50**) has hypotensive and vasodilatory properties. The fritellaria alkaloids are structurally intermediate between the alkamines and the highly oxygenated ester alkaloids. They occur in the Chinese folk drug pei-mu, which consists of extracts of the dried bulbs of *Fritillaria roylei* (*Liliaceae*) and is used as a general medical panacea. The active ingredient is the alkaloid verticine (**51**) (77,81–82). Synthetic routes to these complicated structures have been summarized (83).

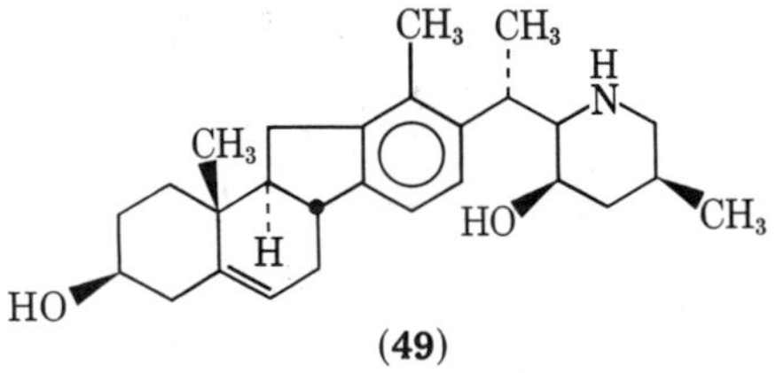

(**49**)

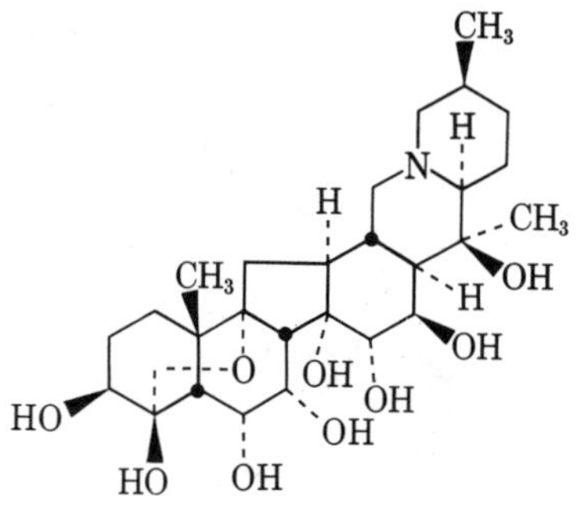

(**50**)

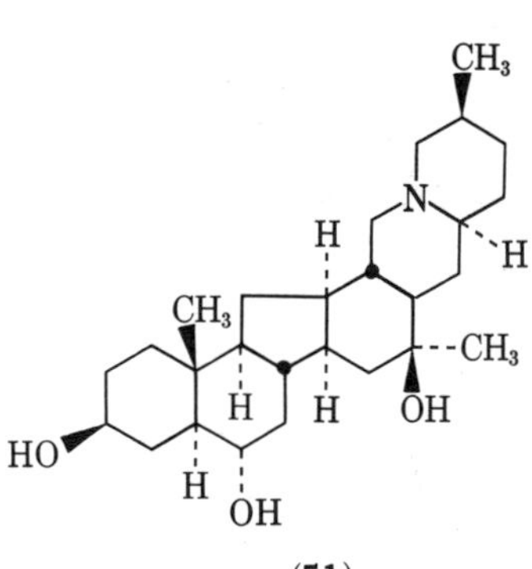

(**51**)

The buxus alkaloids occur in the dogwood tree and concoctions from it have been used as folk remedies for various disorders including venereal disease, tuberculosis, cancer, and malaria. A wide variety of closely related alkaloids have been isolated; the commonest of which is cyclobuxine D (**52**) (77,81).

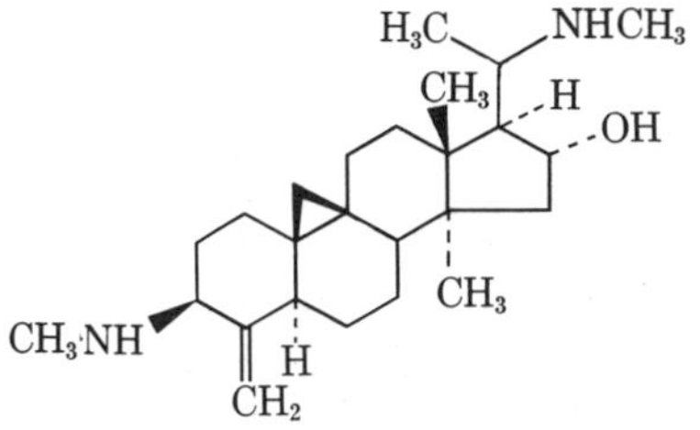

H3C NHCH3 CH3 H OH CH3 CH3·NH H CH2

(52)

Animal Alkaloids. Salamanders are venomous and the venom is secreted in skin glands. Structural studies, including x rays, indicate structure (**53**) for the main alkaloid samandarine. The alkaloid has a novel oxazolidine ring system and a cholic acid configuration. Cycloneosamandione (**54**) is the other main structural type in salamanders. It has the same basic C—N as samandarine (**53**) with the essential difference being an β-aldehydic group at position 10, although a 10α-configuration has originally been assigned (84). Poisoning with samandarine causes convulsions followed by respiratory paralysis; the steroid is toxic to all higher animals. The salamanders die from their own venom if it reaches their circulatory system. The LD_{50} varies among species, but in rabbits it is 1 mpk (milligrams per kilogram) (85).

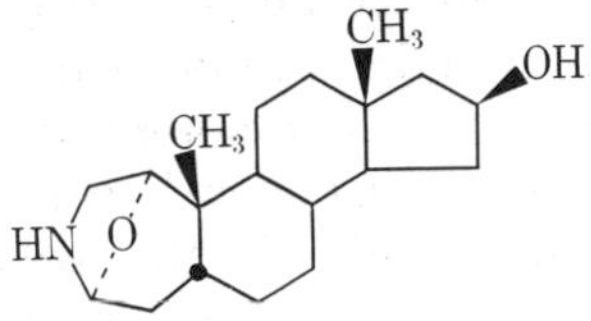

CH3 OH CH3 HN O

(53)

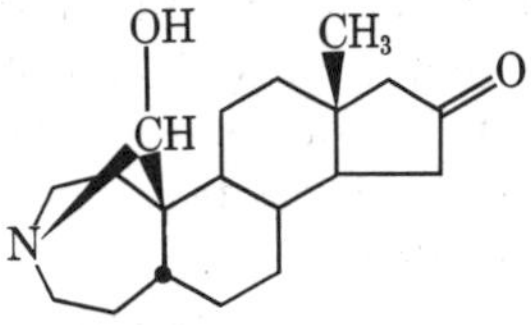

OH CH3 O CH N

(54)

Synthetic approaches have either involved secosteroids or a biogenetic approach (86–87). Basic hydrogen peroxide epoxidation of the trienone (**55**) yields 1α,2α-epoxide (**56**) which, upon hydrogenation, converts to the 1α-hydroxy-5β-derivative (**57**). Beckmann rearrangement of the oxime of (**57**) yields the lactam (**58**), which is reduced with lithium to yield the salamander alkaloid (**59**). The hydroxyl group in (**59**) is transposed to *C*-16 to yield samandarine (87).

Batrachotoxin. For centuries, the skin secretions of the small, vividly colored frog *Phyllobates amotaenia* have been used by Indians in the Colombian rain forest to prepare poisoned darts. The active constituents were isolated and shown to be 20α esters of the steroidal alkaloid batrachotoxin A (**60**) with various pyrrole carboxylic acids (88). Batrachotoxin, the 20α ester of (**60**) with 2,4-dimethylpyrrole-3-carboxylic acid, is among the most lethal substances known with an LD_{50} of 2 μg/kg subcutaneously in mice; strychnine, for comparison, has an LD_{50} of 500 μg/kg. Batrachotoxin is also used for studies of neuromuscular transmission, because it causes depolarization by an irreversible increase in membrane permeability to sodium ions (25). Because of the difficulty of obtaining supplies from natural sources, a synthesis was devised and is illustrated in Figure 2 (89). The lactone (**61**) is reduced to the diol and the ketal is hydrolyzed to form the enone, which is epoxidized to protect this double bond during further synthetic transformations. Selective acetylation of the *C*-18 hydroxyl followed by oxidation gives the progesterone derivative (**62**). Cis-hydroxylation of the

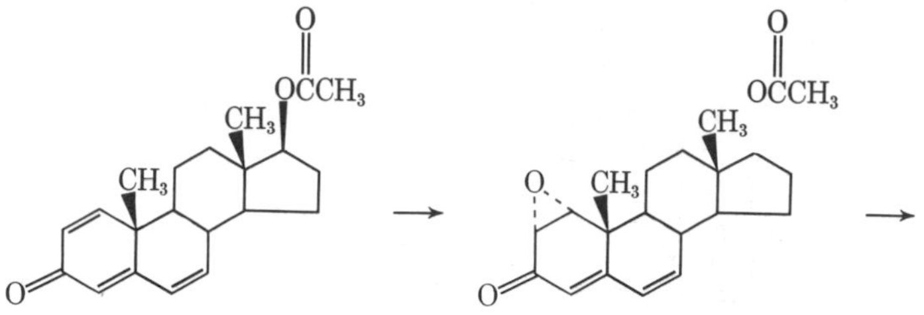

(55) → (56) →

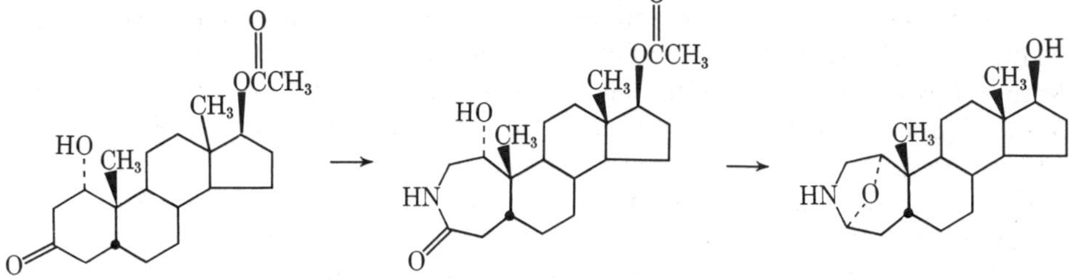

(57) → (58) → (59)

$\Delta^{9,11}$-double bond with osmium tetroxide, acetylation of the 11α-hydroxyl, and subsequent regeneration of 3-keto-4-ene by reductive elimination with zinc yields the enone (**63**). A Δ^6-double bond can be introduced via a high potential quinone, stereospecific epoxide formation, and reductive opening of the epoxide with chromium(II) yields the 7α-alcohol (**64**). Catalytic hydrogenation and methanol treatment yields the ketal (**65**). After acetylation of the 7α-alcohol, the $\Delta^{14,16}$-dienic system is introduced by successive bromination–debromination and peracid yields the epoxide (**66**). Catalytic hydrogenation gives the 14β-alcohol which, with the *C*-18 alcohol, is protected as a ketal to allow hydride reduction of the 20-ketone to the 20α-alcohol. Acetylation and deketalization gives the diol (**67**). The 18-alcohol in (**67**) is oxidized to the aldehyde by dimethylsulfoxide and acetic anhydride, and the 18-methylamino is generated by hydride reduction of the Schiff base. Chloroacetylation forms the amide, which is cyclized with base to generate the seven-membered ring in (**68**). The three acetate groups in lactam (**68**) are hydrolyzed and the less hindered hydroxyls at *C*-11 and *C*-20 are reacetylated. The Δ^7-double bond is introduced by dehydration of the 7α-hydroxyl to form structure (**69**). Hydride reduction and acidic workup generates batrachotoxinin A (**60**). The overall yield for the 36 steps is 0.12%.

Steroidal Lactones. ***Cardenolides.*** Digitalis and other glycosides have a powerful action on the myocardium and are invaluable in the treatment of congestive heart failure. Historically, many plant extracts containing cardiac glycosides have been used in various areas of the world as arrow and ordeal poisons. Squill from *Urginea* (*Scilla*) *maritima* is mentioned as a medicine in the Egyptian Ebers Papyrus (ca 1500 BC). The Romans used it as a diuretic, emetic, and rat poison. Digitalis was mentioned in 1250 in the writing of Welsh doctors. The use of digitalis in 1785 for the treatment of edema is described in ref. 90. Structural elucidation of the active principle indicates the presence of a mono-, di-, tri- or tetrasaccharide, a steroid nucleus, and a lactone attached to the steroid at *C*-17. The aglycone with the simplest chemical structure is digitoxigenin (**70**) although more complicated structures, eg, ouabagenin (**71**), are

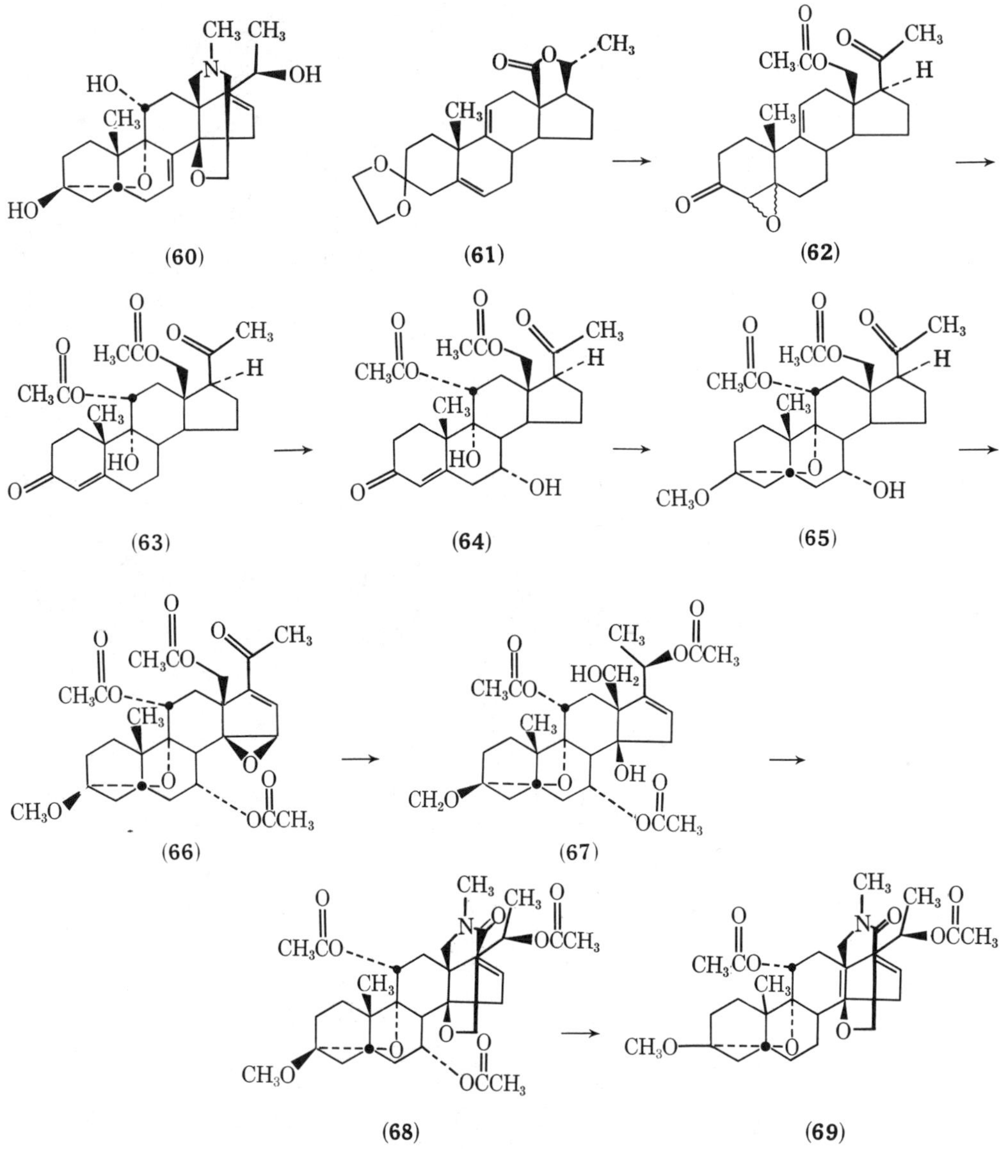

(60) (61) (62) (63) (64) (65) (66) (67) (68) (69)

Figure 2. The synthesis of the South American tree frog arrow poison, batrachotoxin (**60**). The synthesis of batrachotoxin, perhaps the most structurally complicated steroid known, from a commonly available simple steroid starting material, vividly illustrates the power of modern organic synthesis.

more common. Noteworthy are the AB and CD cis ring junctions and the presence of the 14β-hydroxyl group, although the AB trans junction occurs in uzarigenin and a C-5 double bond in xysmalogenin (91).

Synthetic approaches to the construction of cardenolides have centered on the construction of the lactone ring and introduction of the 14β-hydroxyl group (92). The hydroxyl group is usually introduced by elimination of a 15α-hydroxyl group, bromohydrin formation, and bromine removal. Methods for lactone construction are shown in Figure 3 (93–95).

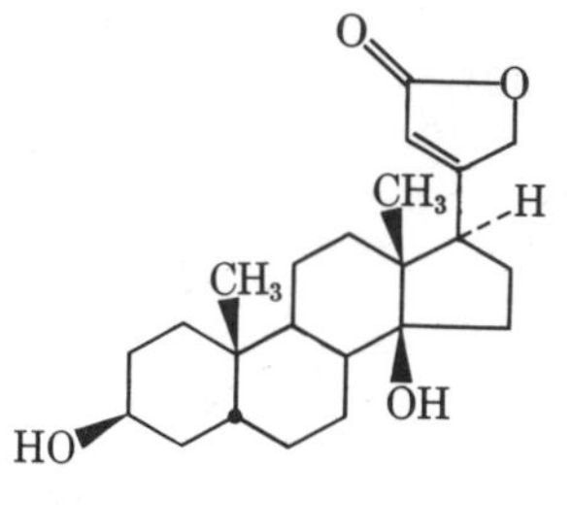

(70) 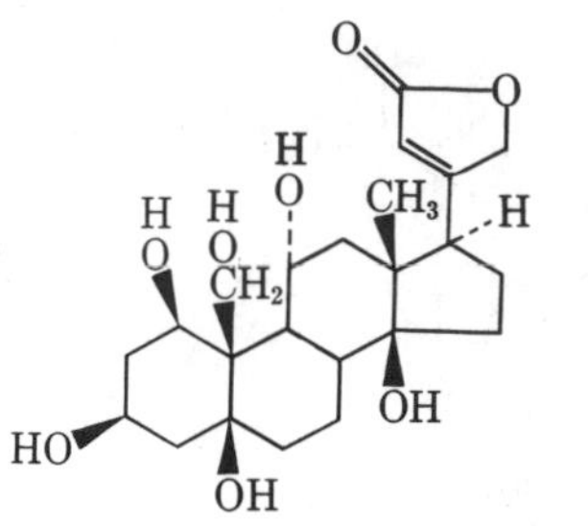 (71)

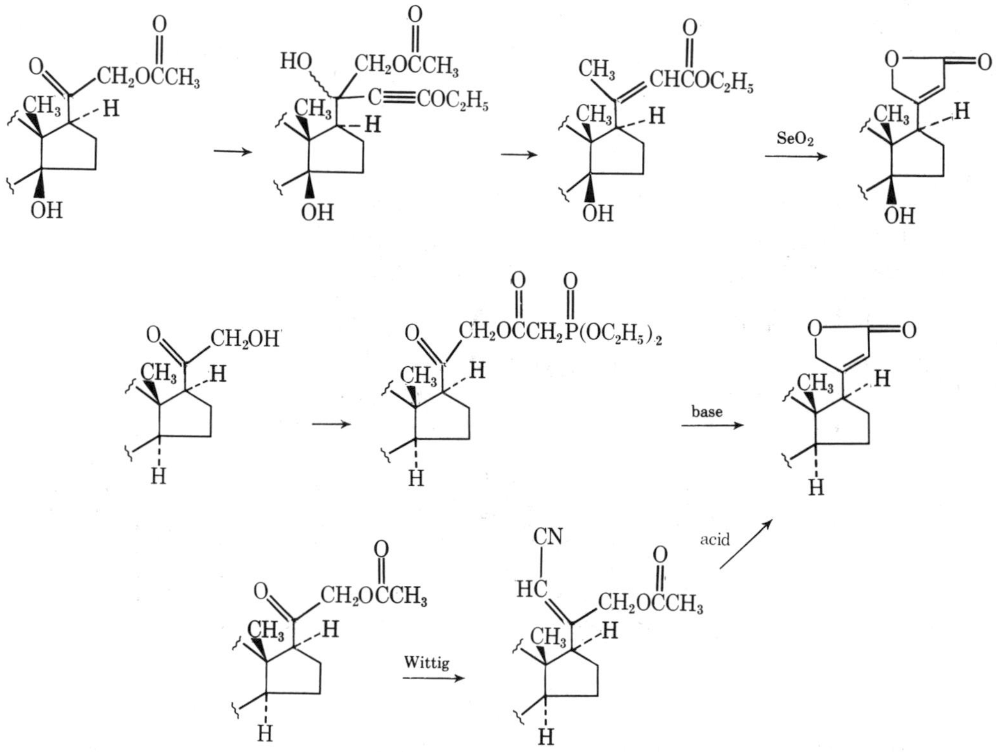

Figure 3. Synthetic methods for the preparation of the cardenolide lactone ring.

Bufadienolides. Except for a 5-substituted 2-pyrone at the 17β-position, the bufadienolides resemble the cardenolides in respect to functional groups, stereochemistry, and biological activity. They occur in several plant families and in the venom of certain Asian toads (96). Scillarenin (**72**) has been isolated from the red squill and bufalin (**73**) can be obtained from the dried venom of the Chinese toad *Bufo bufo gargarizans* (97). The synthesis of these molecules, which is more difficult than the corresponding cardenolides, has been reviewed (96,98).

Holothurigenins. Sea cucumbers (holothurians) constitute one of the five classes of animals in the phylum Echinodermata. The others are sea urchins, brittlestars, sea

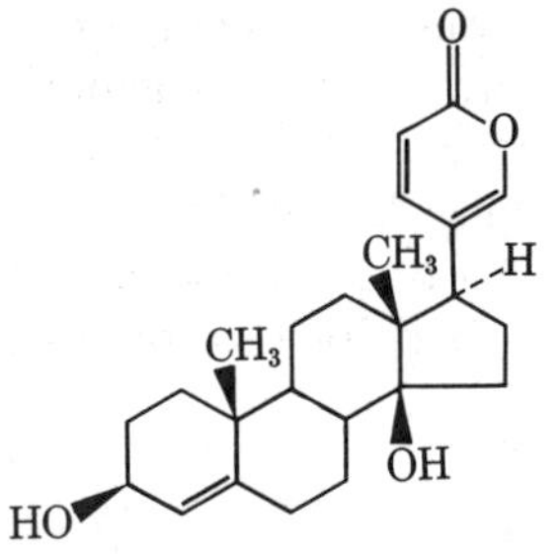

(72) 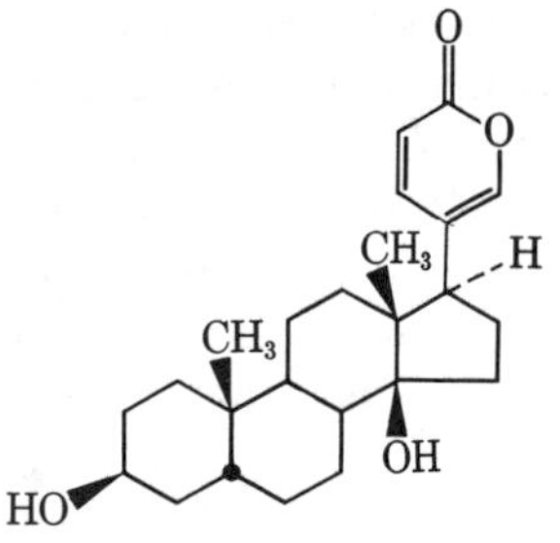 (73)

stars, and sea lilies. The sea cucumbers secrete saponins which are toxic to fish and have hemolytic and oncolytic properties (61,99). Acidic hydrolysis of the saponins gives holothurigenins, eg, seychellogenin (**74**), which is probably an artifact formed by dehydration of either a 12α (**75**) or 12β (**76**) alcohol (100–101). The synthesis of (**74**) has been reported (102). The nitrite ester (**78**) has been prepared from the lanostane alcohol (**77**) with nitrosyl chloride and irradiation resulting in the formation of the oxime (**79**). Oxidation and acetate cleavage with base forms seychellogenin (**74**).

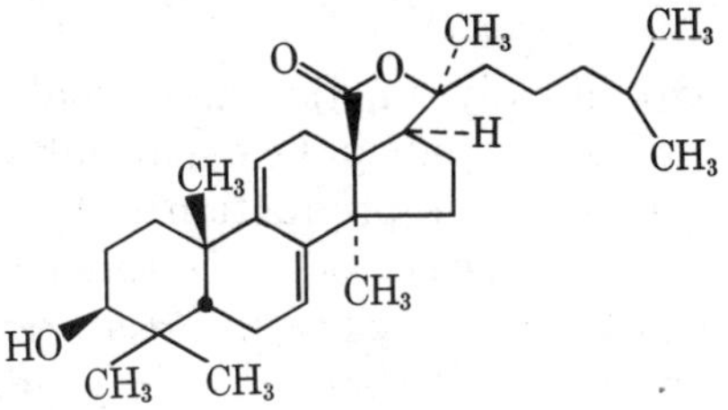

(74) 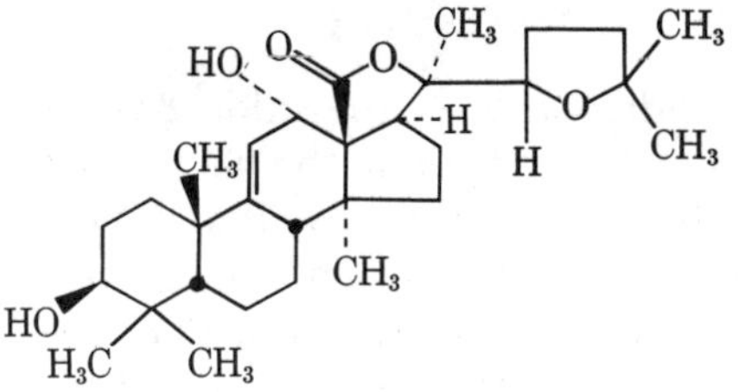 (75)

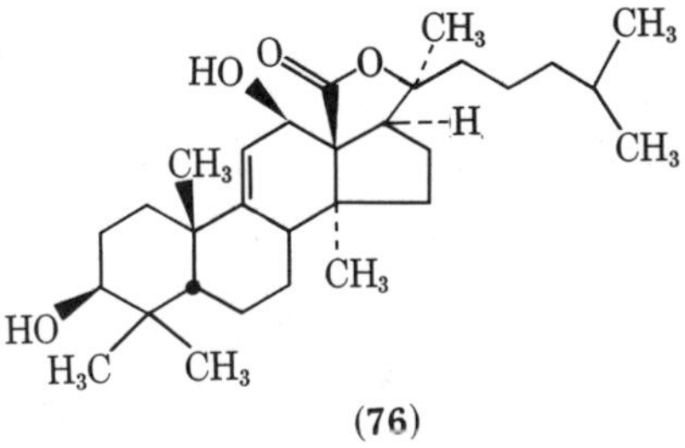

(76)

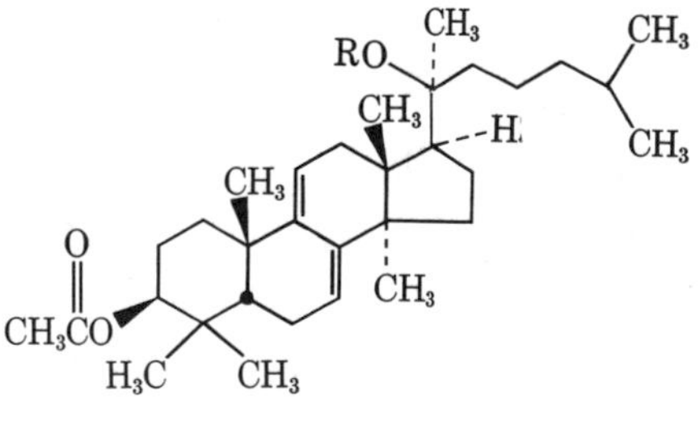

(77) R = H
(78) R = NO

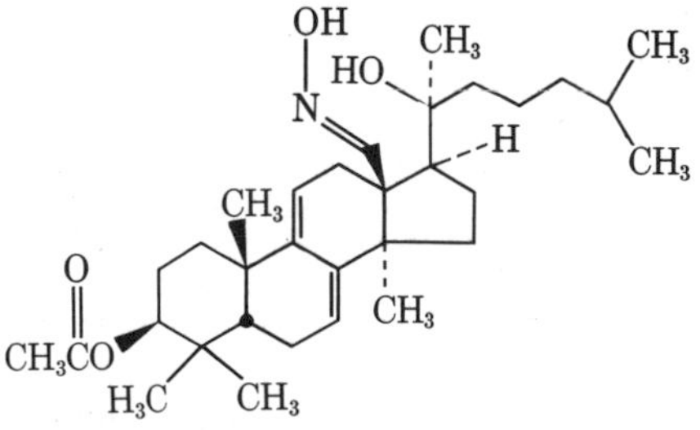

(79)

Withanolides. The withanolides constitute a series of C_{28} lactones of the ergostane type with a pyran ring in the side chain and they form a single biogenetic group which occurs in plants of the family *Solanaceae* (103). Interest has centered on these compounds because of their reported anticancer activity. An unusual structural characteristic of most of these steroids is the presence of a 1-oxo-2-ene chromophore, as in withaferin A (**80**), and the very unusual presence of a chlorine atom as part of a chlorohydrin in Taborosalactone C (**81**) (104).

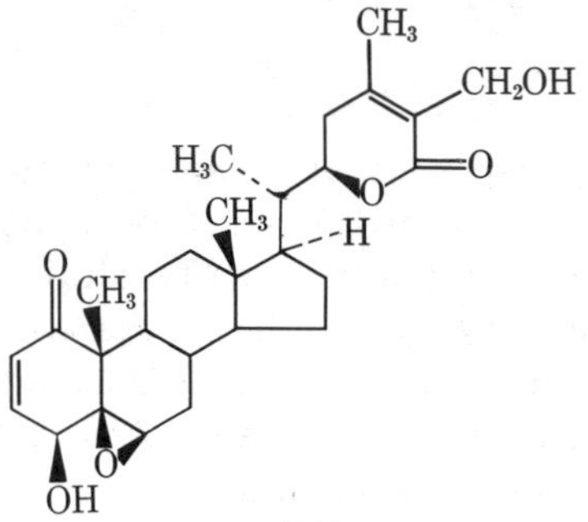

(**80**) 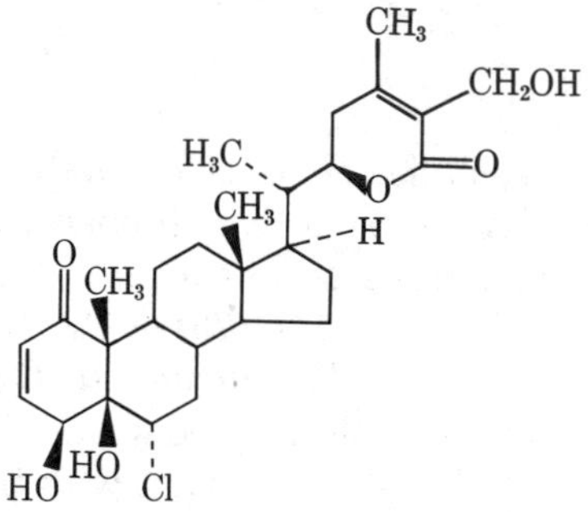 (**81**)

Antheridiol. Antheridiol (**82**) was the first specifically functioning steroid sex hormone determined in the plant kingdom and, when expressed by the female mycelia of the aquatic fungus *Achyla bisexualis*, initiates sexual reproduction (105). Its structural determination was rapidly followed by its synthesis (106–107). Aldol condensation of aldehyde (**83**) with a 3,4-dimethylpent-2-enoate yields the unsaturated lactone (**84**) which can be opened to the dienic acid (**85**). Perbenzoic acid yields a mixture of lactones which, after resolution and elaboration of the 7-ketone, yields antheridiol (**82**) (Fig. 4) (107–108).

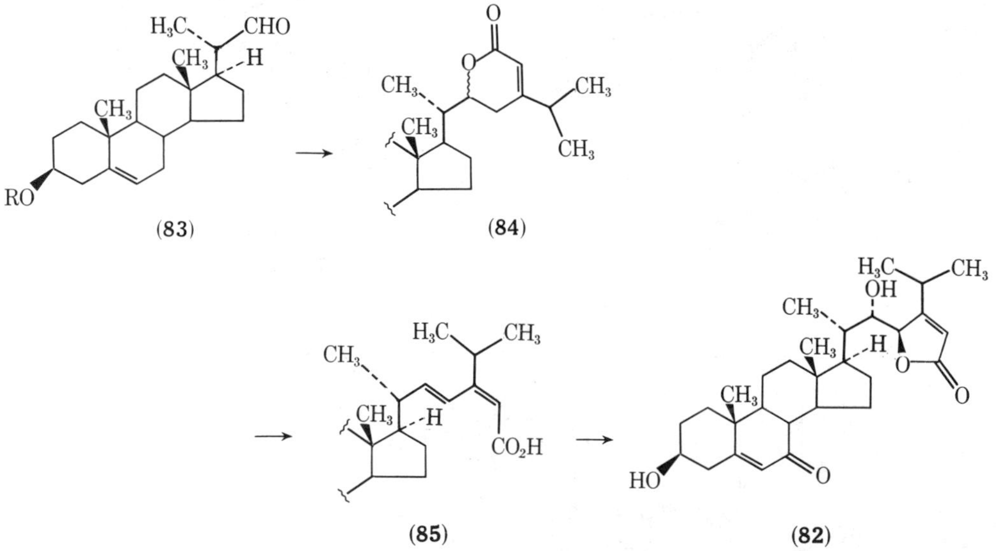

(**83**) (**84**)

(**85**) (**82**)

Figure 4. The synthesis of antheridiol (82), the sex hormone of the aquatic fungus, *Achyla bisexualis*.

Insect Moulting Hormones. The moulting processes which are essential to the postembryonic growth and development of insects are under hormonal control. Every stage of moulting from larva through pupa to adult requires the presence of a metamorphosis hormone and larval growth is mediated by juvenile hormone methyl *trans, trans,cis*-10-epoxy-7-ethyl-3,11-dimethyl-2,6-tridecadienoate. The moulting hormone was long considered to be secreted from the prothoracic glands; however, experiments demonstrate hormone biosynthesis outside this gland (109). These substances and insect biochemistry in general are of interest since they may lead to new and selective means of insect control. In 1954, 25 mg of the crystalline moulting hormone of the silkworm, *Bombyx mori*, was isolated from 500 kg of silkworm pupae (110). Chemical studies and eventually an x-ray analysis of the hormone α-ecdysone (**86**) established its steroidal nature (111). The closely related β-ecdysone (**87**) containing a 20*R*-hydroxyl group also occurs in the silkworm (see Silk).

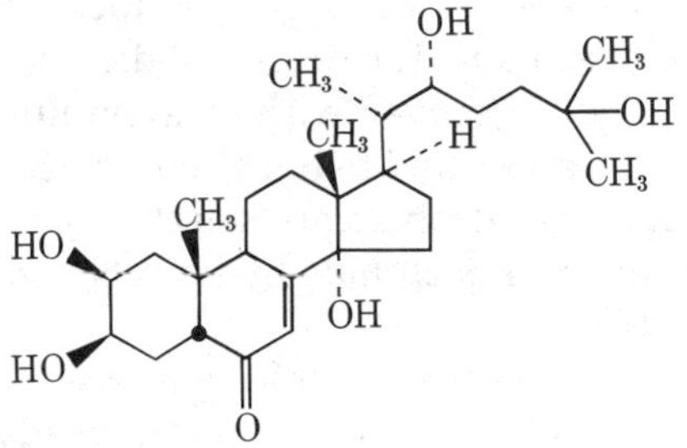

α-ecdysone (**86**)

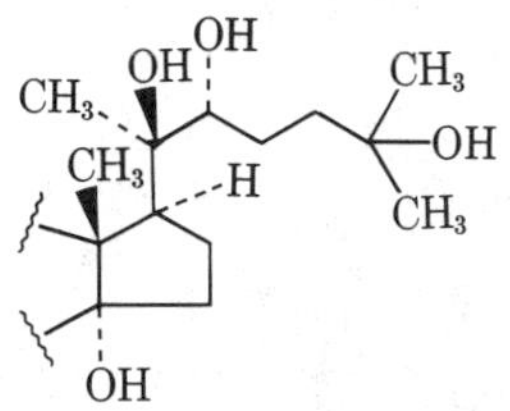

β-ecdysone (**87**)

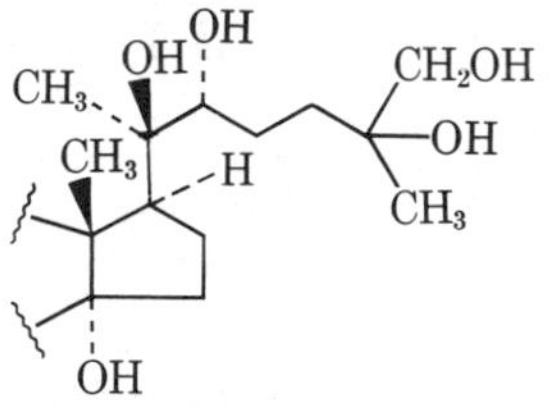

26-hydroxy-β-ecdysone (**88**)

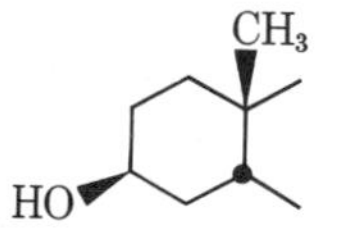

2-deoxy-β-ecdysone (**89**)

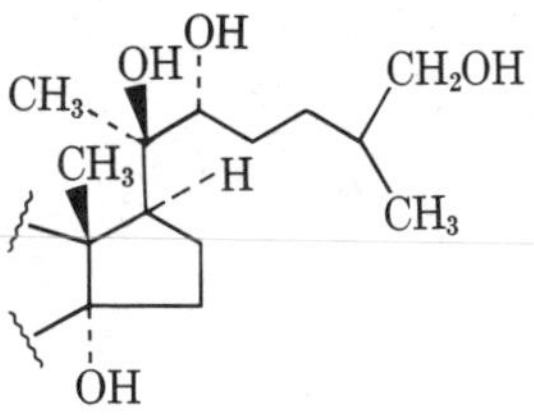

inokosterone (**90**)

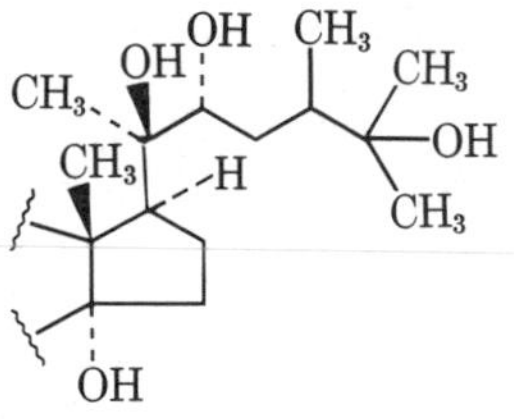

makisterone A (**91**)

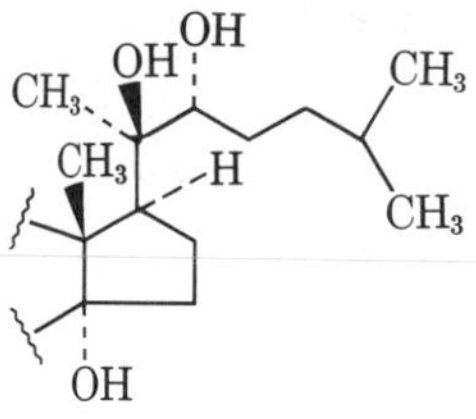

ponasterone A (**92**)

A variety of these moulting hormones or zooecdysones have been subsequently isolated from insects and crustaceans. The tobacco hornworm contains 26-hydroxy-β-ecdysone (**88**) and the crayfish 2-deoxy-β-ecdysone (**89**). From the crab, both inokosterone (**90**) and makisterone A (**91**) have been isolated (112). In 1966, shortly after the structural elucidation of ecdysone, a group of steroids with moulting activity, ie, phytoecdysones, were isolated from plants (113). They are widely distributed and usually occur in large quantities. For example, one kilogram of dry leaves from *Podocarpus nakaii* yields

1–2 g of ponasterone A (**92**) (114). As of 1979, 47 different phytoecdysones had been discovered and all of the zooecdysones, except (**88**), have also been extracted from plants (115).

The straightforward partial synthesis of ecdysone (**86**) was reported by two different groups. The Syntex route started from bisnorcholenic acid methyl ester (**93**) (Fig. 5). The diol, derived from compound (**93**) by epoxidation and hydrolysis, is oxidized to the ketol (**94**) by positive bromine. The double bond in structure (**95**) is introduced by selective tosylation of the 3β-hydroxyl group in structure (**94**) and its subsequent elimination. Cis-hydroxylation of (**94**) is by the Prevost reagent to give the bis-acetate (**96**). The double bond in structure (**95**) is introduced by the usual bromination–elimination sequence, and after protection of the ketol hydroxyl, the 14α-alcohol is formed by selenium dioxide oxidation. Reductive elimination of the ketol acetate in compound (**97**) gives the kinetic 5α-product which, upon basic treatment, yields the natural 5β-configuration. Acetonide formation then yields compound (**99**). The side chain is introduced selectively by reducing the 6-ketone with hydride. Carbons 23–26 are introduced using the reagent indicated in step 12 in Figure 5. The sulphoxyl group is reductively removed with an aluminum amalgam to give the derivative (**100**). The 22-ketone is reduced to the alcohol with the proper stereochemistry, and then the allylic epimeric 6-alcohols are selectively oxidized to the ketone with manganese dioxide to yield, after acetonide hydrolysis, ecdysone (**86**). The yield from compound (**99**) is 12% (116).

A later, more economical synthesis, which superceded the Schering-Hoffmann-LaRoche original synthesis (117), was based on the ready availability of ergosterol (**11**) (118).

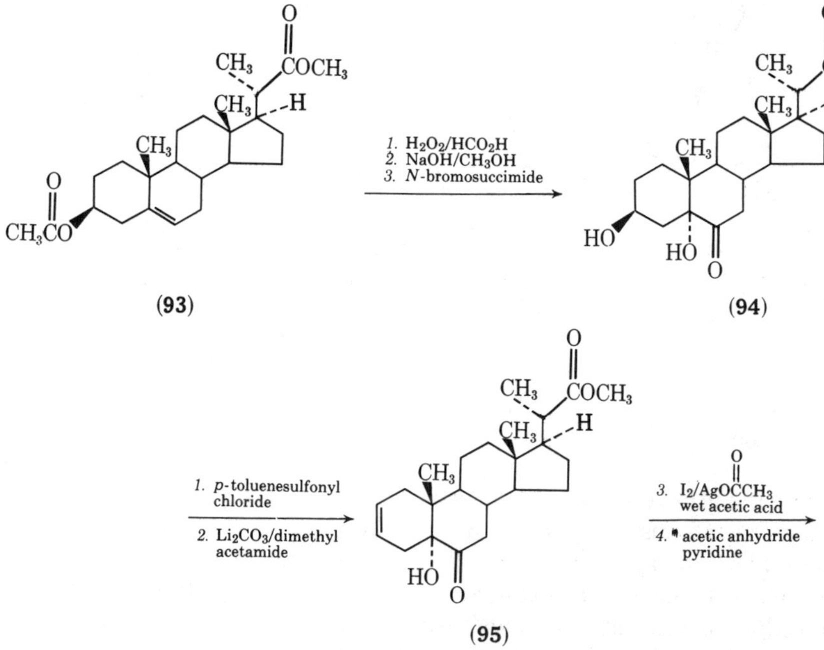

(**93**) → 1. H_2O_2/HCO_2H 2. $NaOH/CH_3OH$ 3. *N*-bromosuccinimide → (**94**)

1. *p*-toluenesulfonyl chloride 2. Li_2CO_3/dimethyl acetamide → (**95**) → 3. $I_2/AgOCCH_3$ (O) wet acetic acid 4. acetic anhydride pyridine →

Figure 5. The syntex route to ecdysone (**86**).

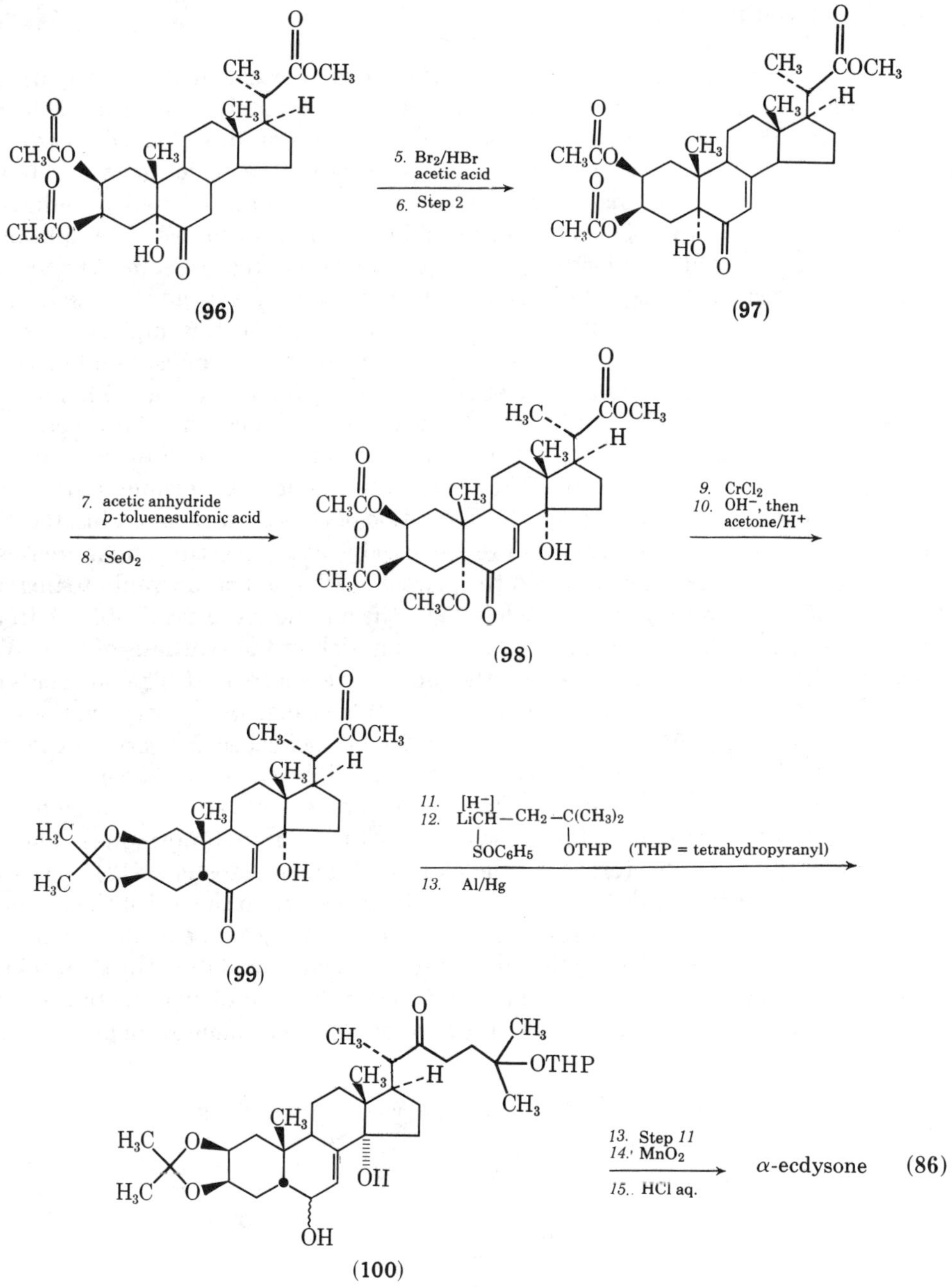

(96)

5. Br_2/HBr acetic acid
6. Step 2

(97)

7. acetic anhydride p-toluenesulfonic acid
8. SeO_2

(98)

9. $CrCl_2$
10. OH^-, then acetone/H^+

(99)

11. [H^-]
12. $LiCH(SOC_6H_5)-CH_2-C(CH_3)_2(OTHP)$ (THP = tetrahydropyranyl)
13. Al/Hg

(100)

13. Step 11
14. MnO_2
15. HCl aq.

α-ecdysone (86)

Figure 5. (*continued*)

In the intervening period, ecdysone syntheses based on stigmasterol (**18**) and diosgenin (**39**) have been reported (119–120). An elegant method for the direct introduction of the 2,7-dien-6-one grouping from the 3β-tosylate-5-ene has significantly shortened the synthetic pathways to ecdysones and its analogues (121).

Manufacture and Synthesis

The choice of starting materials may either be of petrochemical (total synthesis) or animal or vegetable (partial synthesis) origin. Approximately two-thirds of the raw material for chemical synthesis of the steroid hormones produced has depended on diosgenin (**39**) obtained from the plant *Dioscorea compositae*, which grows in hot and humid mountainous areas, mainly in Mexico, but also in Central America, India, and China. From the middle 1950s to the early 1960s, it is likely that well over 50% of all steroids manufactured worldwide originated from Mexican diosgenin. As shown in Figure 6, the great advantage of diosgenin is its versatility. Cleavage of the side chain in diosgenin leads to 16-dehydropregnenolone and the progestins, and by microbial hydroxylation to the corticoids. Conversion to androstenedione and subsequent steps forms estrogens, contraceptives, diuretics (spironolactone), and the male anabolic and androgenic hormones. In addition, for over two decades, the *Dioscorea* plant supply, and hence diosgenin, was plentiful and cheap. The Mexican diosgenin situation changed perceptibly in the early 1970's when the Mexican government started to control, and eventually to nationalize, the collection of *Dioscorea* plants. This, together with the decrease in diosgenin content (6% to 4%) by overharvesting and increases in transportation and labor costs, caused the price of a kilogram of diosgenin to increase from ten dollars to over one hundred dollars by the middle seventies (Table 1). In the United States, stigmasterol (**18**) serves as raw material for the synthesis of corticoids, whereas in Europe, hecogenin (**40**) and the bile acids are also used. The bile acids are used, primarily in France, for corticoid synthesis because of the well worked-out technology and plant investment and because the bile acids became very inexpensive after microbial 11-hydroxylation was commercialized. However, what was needed in 1975 was to find a readily available, inexpensive, renewable steroid source for the androstenedione intermediate for the synthesis of most steroid hormones. The two most commonly available sterols are cholesterol (**2**) and β-sitosterol (**19**). The 1980 annual world resources of these sterols is ca 100,000 t each. In the use of these sterols as sources of steroid hormones, the critical step is the cleavage of the nonfunctionalized side chain. This has been accomplished by microbiological oxidation. In 1976, a large plant for the production of androstadienedione from cholesterol by fermentation was established in Hasaki in Japan. At about the same time, a fermentation plant for the

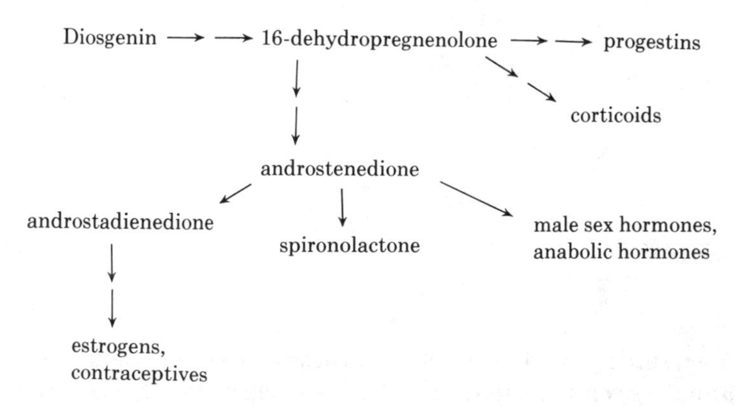

Figure 6. The central role of 16-dehydropregnenolone in the manufacture of steroid products. This critical intermediate was prepared commercially from diosgenin (**39**) obtained from Mexican yams.

Table 1. Diosgenin Price Fluctuations

Year	Price of diosgenin, $/kg
1973	20
1974	25
1975	35–56
1976 (January)	126
1977	60–100
1978	55
1979	25
1980	25
1981	25

large-scale conversion of sitosterol to androstenedione and androstadienedione went into operation in Harbor Beach, Mich. It is likely that the future world requirements for steroid hormones are going to be met by fermentation and the diosgenin source will become of less importance unless, as in China and Mexico, political considerations dictate otherwise (122–123).

The production of steroid hormones is a huge worldwide business with 1980 sales of ca $\$4 \times 10^9$. The steroids include male and female hormones, contraceptives, diuretics, anticancer agents, and anabolic agents. The growth of the world market for steroids is shown in Table 2 (122). The amounts of corticoids produced are expected to increase significantly because of their availability without prescription in the United States. The amounts required for contraceptives are somewhat misleading since the human dose is small, ie, <250 mg/yr, so that 1 t of final product is equivalent to 5×10^6 woman-years (123). Spironolactone (**101**) is a specific antagonist of aldosterone (**102**) and, as such, functions as a diuretic (qv). Its dosage, which is large compared to all the other steroid hormones, explains the large volume requirements.

Table 2. World Market for Steroids, Metric Tons[a,b]

	1963	1968	1973	1975	1977	1979	1980
corticoids	275	629	844	956	1096	1224	1290
sex hormones	86	116	131	156	169	183	190
contraceptives	91	195	200	220	252	280	295
spironolactone	48	60	165	199	218	273	320

[a] Ref. 46.
[b] The amounts are based on metric tons of diosgenin.

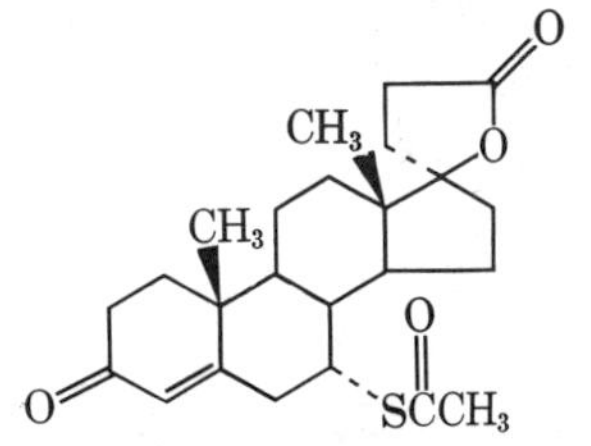

CH_3, CH_3, O, O, O, O, $SCCH_3$

(**101**)

OH, O, CH_2OH, H, O, CH_3, O

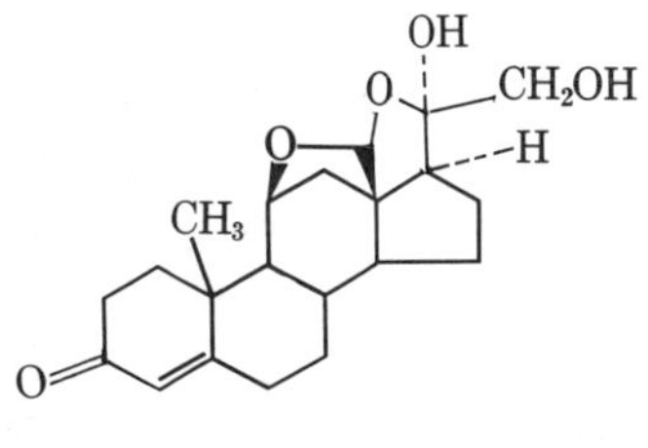

(**102**)

The multitude of natural products containing the steroid skeleton has always presented scientists with the challenge of seeking economically feasible ways of converting these abundant natural sources into medicinally valuable compounds. Initial efforts involved low yielding oxidative processes to convert abundant cholesterol and bile acids into androstanes and pregnanes. Methods were developed for the construction of the more elaborate pregnane side chain from the more readily available androstanes. This approach was superseded by the conversion of diosgenin into pregnanes and then into the corticoids. Recently the shift away from diosgenin to microbially produced androstanes has resulted in the development of newer methods for introducing the pregnane side chain into these molecules. At the same time, the partial synthesis of steroids from natural precursors was effected, and intensive efforts were being conducted on the total synthesis of the tetracyclic steroidal nucleus. These efforts have been successful; eg, 18-methyl steroids and oral contraceptive constituents are produced in this manner.

Partial Synthesis. The main method for steroid synthesis for over two decades was the degradation of diosgenin (**39**) to 16-dehydropregnenolone acetate (**105**) (Fig. 7)

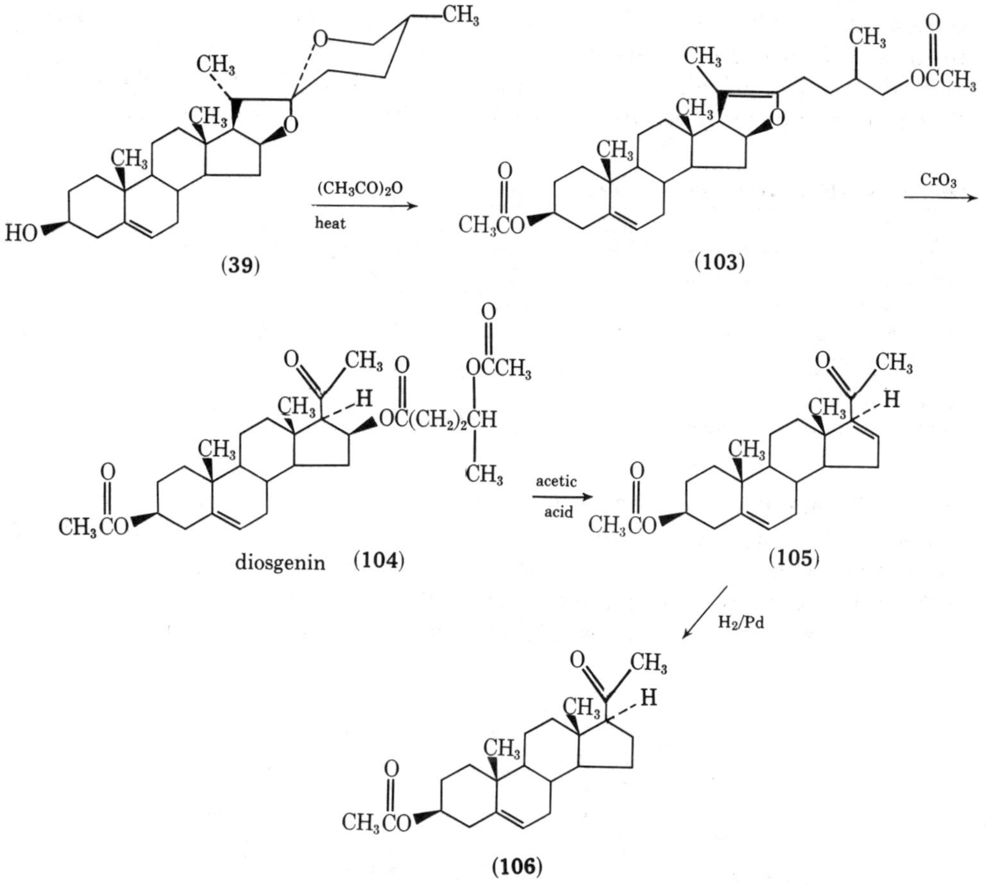

(39) $\xrightarrow[\text{heat}]{(CH_3CO)_2O}$ (103) $\xrightarrow{CrO_3}$

diosgenin (104) $\xrightarrow{\text{acetic acid}}$ (105) $\xrightarrow{H_2/Pd}$

(106)
16-dehydropregnenolone acetate

Figure 7. The commercial synthesis of 16-dehydropregnenolone acetate (**105**) from diosgenin.

(124). When diosgenin is heated to 200°C with acetic anhydride, ring F is cleaved with the introduction of the 20,22-double bond to form the enol ether (**103**). Chromium trioxide oxidation converts this into the keto ester diacetate (**104**) which, upon refluxing in acetic acid, causes elimination to 16-dehydropregnenolone acetate (**105**). Commercially, this conversion is accomplished in over 80% yield. Selective hydrogenation of (**105**) from the alpha face yields pregnenolone acetate (**106**) which is readily converted into progesterone (**21**), but conversion to the androgens and estrogens requires elimination of the side chain at *C*-17. The simplest, most practical, and most widely used method for accomplishing this conversion on a commercial scale is the Beckmann rearrangement of 16-en-20-oximinopregnanes (Fig. 8) (105). Treatment of compound (**105**) with hydroxylamine forms the oxime (**107**) which, after rearrangement with phosphorus oxychloride, yields the enamide (**108**). Because enamides are unstable to acid, they are readily hydrolyzed to 17-ketones (**109**) (126). The same processes are readily applicable to the steroidal alkaloids solasidine (**47**) and its 5α-dihydro derivative tomatidine [*77-59-8*] (46).

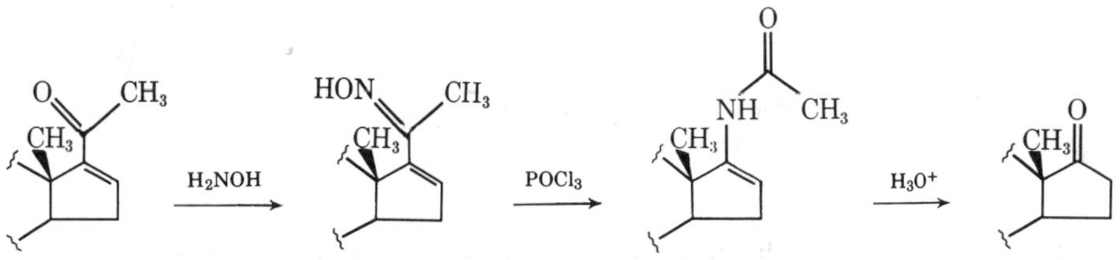

Figure 8. Degradation of the 17-acetyl side chain yields the commercially important 17-keto steroids.

Among the many approaches to the synthesis of corticoids, the route starting from the abundant sapogenin hecogenin (**40**) is still a commercially important process in parts of Europe. In this synthesis (Fig. 9), compound (**40**) is brominated in benzene to yield the 11α,23α-dibromide (**110**). Conversion to the bromine-free ketol diacetate (**111**) is by a two-phase system with sodium hydroxide, acetylation, and debromination and results in an 80% yield. An efficient (80%) reductive elimination of the ketol acetoxy group involves the use of calcium in liquid ammonia to generate 11-ketotigogenin (**112**), which is convertible into cortisone by well-developed processes (127–128). The overall yield of >55% of compound (**112**) from (**40**) made possible the commercial production of cortisone from hecogenin.

The main commercial method for making cortisone (**113**) in the United States starts from stigmasterol (**18**), which is converted chemically in high yield into progesterone (**21**) (Fig. 10). Hydroxylation of progesterone with the microorganism *Rhizophus nigricans* yields 11α-hydroxyprogesterone (**114**) in >90% yield (129). Oxidation of the alcohol (**114**) forms the 11-ketone (**115**), which is condensed with ethyl oxalate to generate the sodium salt (**116**). This salt is brominated to form the 21,21-dibromo-21-carbethoxy derivative (**117**), which is not isolated but immediately subjected to the Favorskii rearrangement to give the ester (**118**). After conversion of the enone to the dienamine (**119**), the ester and the 11-ketone are reduced to the

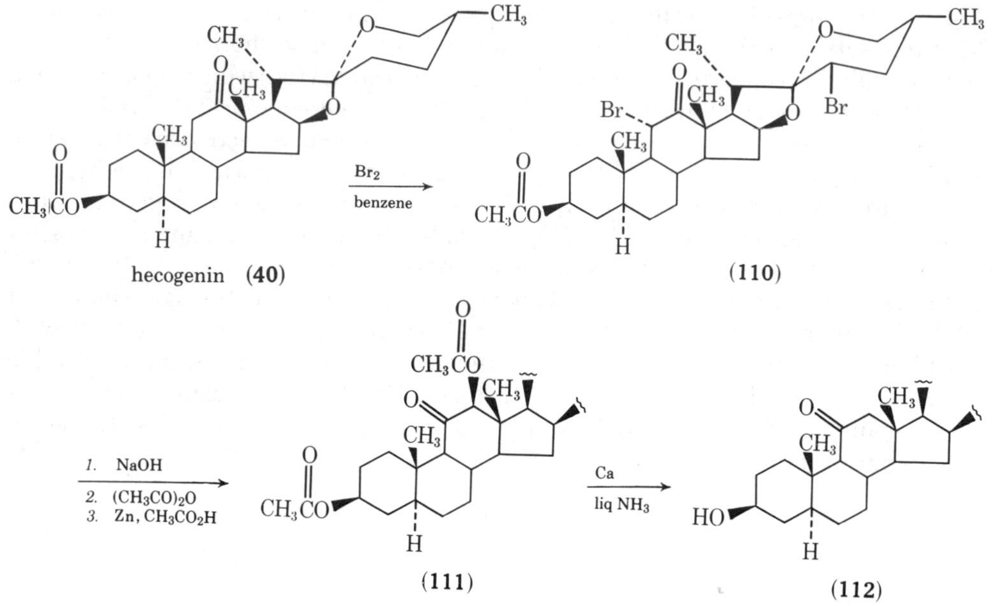

Figure 9. A synthesis of corticosteroids from the plant sapogenin, hecogenin (**40**). This method is used commercially in Europe.

11β,21-diol (**120**) by hydride reagents. Osmium tetroxide-*N*-methylmorpholine-*N*-oxide oxidation then generates hydrocortisone (**113**) (130).

The increasing use of 17-keto steroids as starting materials occurred because of a breakthrough in the microbial degradation of the two commonest steroids, cholesterol and β-sitosterol (131). A simplified degradation pathway of cholesterol by microbes is shown in Figure 11. The homoallylic alcohol in cholesterol (**2**) is oxidized to the enone (**121**) and dehydrogenated to the dienone (**122**) before side-chain cleavage occurs to form the carboxylic acid (**123**). This in turn is cleaved to androstadiendione (**124**). The dienone (**124**) can be reversibly hydrogenated to androstenedione or 9α-hydroxylated to (**126**). Compound (**126**) rapidly undergoes a reverse aldol reaction to form the aromatic secosteroid (**127**) (132).

A Japanese group, starting in 1960, screened 1589 microorganisms for their ability to degrade cholesterol (133). Strains that were potent metabolizers of cholesterol were treated with various enzyme inhibitors to cause accumulation of an intermediate. Metabolism of cholesterol by a mycobacterium species was inhibited at the 9α-hydroxylation step by α,α-dipyridyl so that compound (**124**) was isolated in quantitative yield (134). This process has been commercialized. A similar commercial process was independently developed in the United States, where uv-mutants of a mycobacterium species have been used to produce androstenedione (**125**), androstadienedione (**124**), or an ester of (**123**) from β-sitosterol (**19**) (135).

Because of the shift to 17-keto steroids as starting materials, new methods have been developed for introducing the dihydroxyacetone side chain (Fig. 12). Microbial hydroxylation of androstenedione (**125**) forms the 9α-alcohol (**128**). Dehydration and ethynylation gives (**129**), which is sulfenylated at −70°C to form the ester (**130**). Warming to −40°C causes (**130**) to undergo a [2,3] sigmatropic shift to form the allene

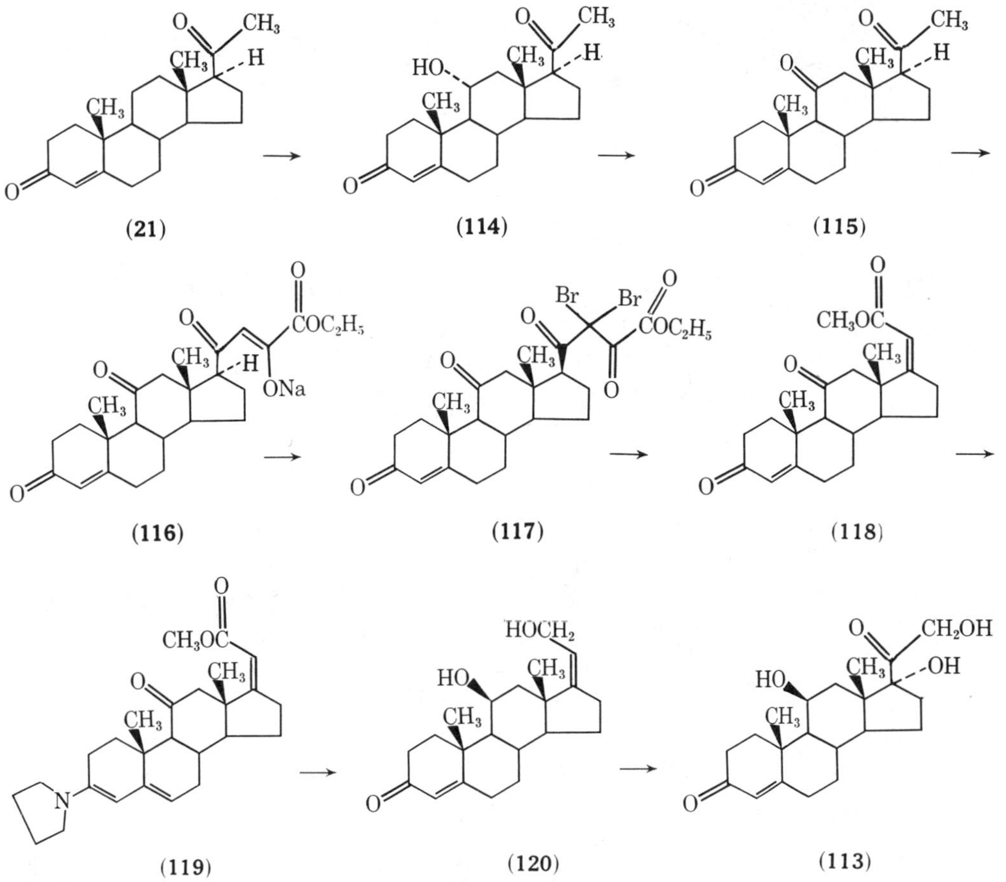

(21) → (114) → (115) →

(116) → (117) → (118) →

(119) → (120) → (113)

Figure 10. The commercial method for the production of corticoid steroids in the United States.

(**131**). Based-catalyzed addition of methanol generates the methoxy sulfoxide (**132**), which is in equilibrium via a [2,3] shift with (**133**). The sulfenyl ester is selectively cleaved with phosphite to form the enol ether (**134**). Enol ethers are readily converted to the dihydroxyacetone side chain (136).

The development and broad acceptance of oral contraceptives in the 1960s led to a great demand for 19-norsteroids, which are the chief components of the pills. As before, attention was turned to easily accessible steroids as precursors of these molecules. The problem was removal of the angular methyl group at *C*-10. The original method was pyrolysis of 1,4-androstradienone (**124**) as a mineral-oil solution at 525°C to produce estrone (**135**) in 15–20% yield (137). Subsequent improvements in technique on an industrial scale have given yields of >70% of estrone (see Fig. 13) (138). An alternative industrial process is the reductive elimination of the angular methyl group of the 17-ketal (**137**) (139).

When (**136**) is treated with the lithium biphenyl radical anion in the absence of ammonia or any other proton donor, the resultant steroidal radical anion expels the *C*-10 angular methyl group as methyl lithium at >35°C with concomittant aromatization of the A-ring. Diphenylmethane is present during the reaction to scavenge the

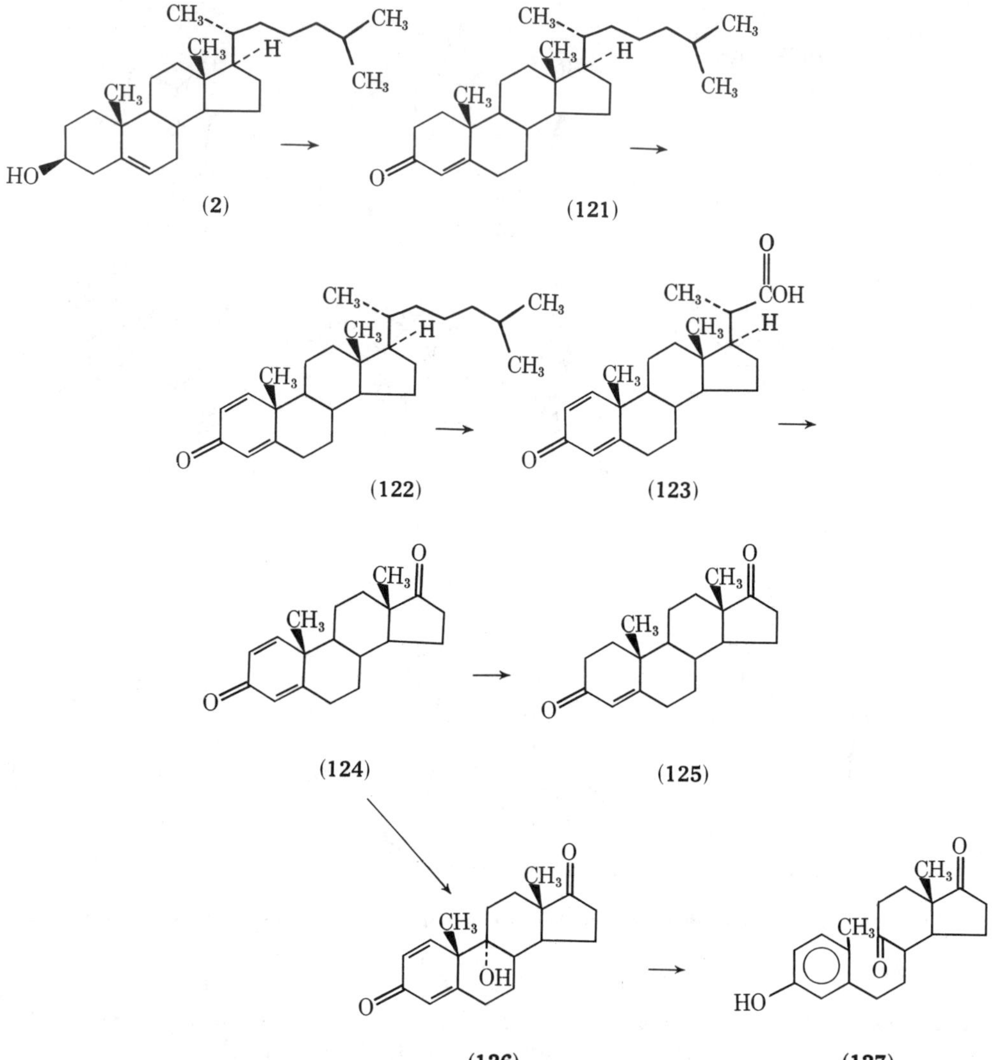

(2) → (121) → (122) → (123) → (124) → (125)

(124) → (126) → (127)

Figure 11. Microbial fermentation of steroids. Bacterial removal of the steroid side chain, forming androstane derivatives, is now a principal commercial method in Japan and the United States for preparing starting materials for pharmaceutical manufacture.

methyl lithium. After methylation of the phenolic group, either estrone or its 17-ethylene ketal can be converted to 19-nor-A-ring-enones by the Birch reduction (Fig. 14) (140). In this reaction, the aromatic ring is reduced by formation of its radical anion by a dissolving metal in ammonia. In the presence of a proton donor, the radical anion is protonated and then a second electron is added to the resultant radical to generate an anion. After protonation, the 1,4-dihydroestrone (**138**) can be isolated. Controlled hydrolysis of the enol ether (**138**) yields the deconjugated enone (**140**), whereas more vigorous hydrolysis of compound (**140**) yields 19-norandrostenedione (**139**) (141).

Intramolecular functionalization of the nonactivated *C*-10 methyl group has also

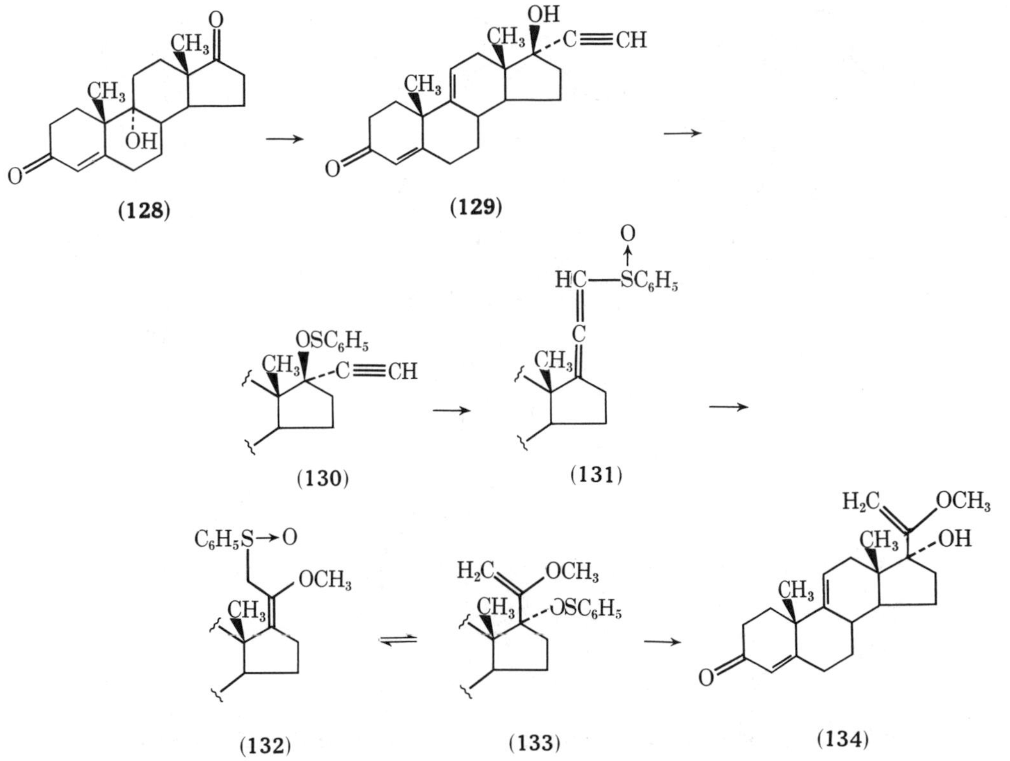

(128) (129)

(130) (131)

(132) (133) (134)

Figure 12. A novel chemical method for introducing the corticoid side chain in 17-ketosteroids.

been used to prepare 19-norsteroids (142). Conversion of compound **(109)** to its 5α-chloro, bromo, or acetoxy, 6β-hydrin **(141)** allows functionalization of the angular methyl group by 6β,19-ether formation **(146)** with lead tetraacetate and iodine (Fig.

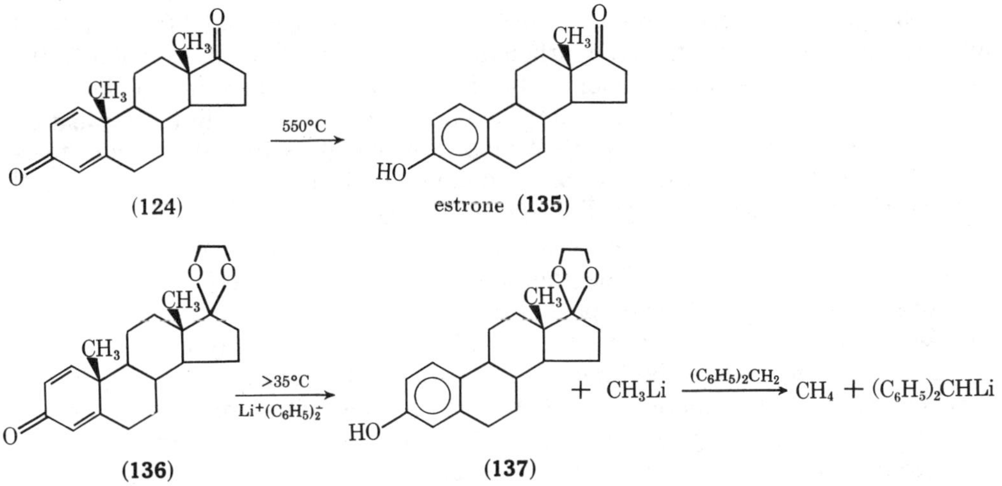

(124) $\xrightarrow{550°C}$ estrone **(135)**

(136) $\xrightarrow[Li^+(C_6H_5)_2^-]{>35°C}$ **(137)**

$$+ \ CH_3Li \xrightarrow{(C_6H_5)_2CH_2} CH_4 + (C_6H_5)_2CHLi$$

Figure 13. Industrial processes for the reductive aromatization of androstanes to estrane derivatives.

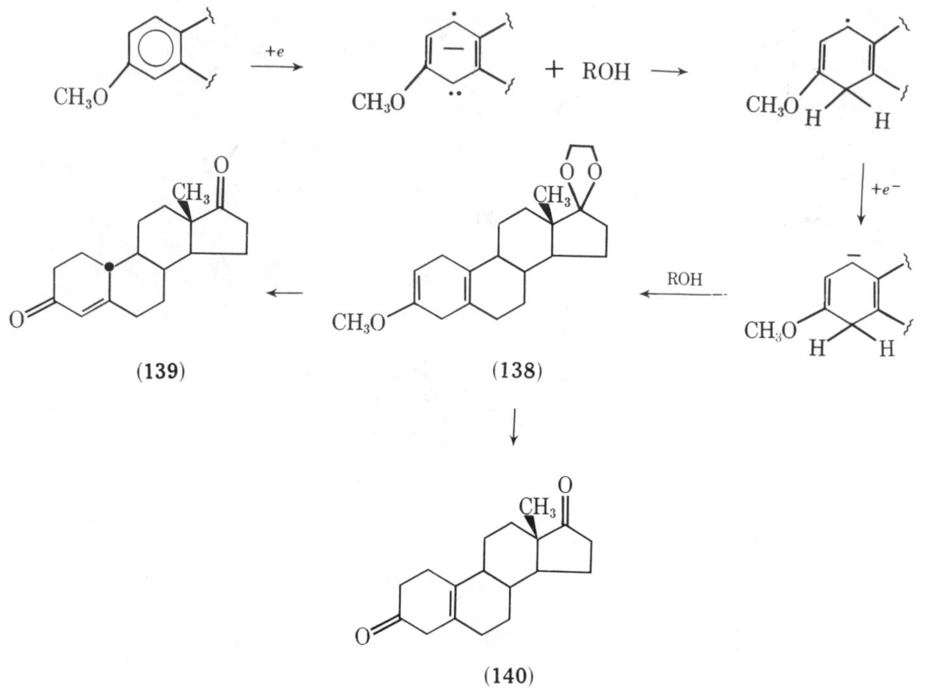

Figure 14. The mechanism of the Birch reaction for the preparation of 19-norsteroids from estranes.

15) (143–145). Reaction of the 6β-alcohol with lead tetraacetate and iodine forms the unstable hypoiodite (**142**), which undergoes homolytic cleavage to the oxygen radical (**143**). This radical abstracts a hydrogen from the proximate angular methyl group to form the carbon radical (**144**), which is oxidized to the carbonium ion which traps iodine (**145**). The iodohydrin (**145**) cyclizes to the 6β,19-ether (**146**). Alternatively, in the absence of iodine, the oxygen radical can be directly generated from lead(IV) and, after hydrogen abstraction, the carbon radical is oxidized to carbonium ion, which is trapped by the 6β-alcohol to yield (**146**). Ester cleavage and oxidation yields the 3-ketone (**147**) which is converted into the enone (**148**) by base-catalyzed elimination of the 5α-substituent. Reductive ring opening with zinc gives 19-hydroxyandrostenedione (**149**). Low temperature chromic acid oxidation forms the 10-carboxylic acid (**150**) which, when warmed in pyridine, decarboxylates to the 3-keto-5(10)-en-19-nor steroid (**151**). This reaction sequence is of general applicability and has been used on highly functionalized steroids. The photolysis of nitrite esters has also been used to functionalize the *C*-10 methyl group. Derivative (**141**) is converted into its nitrite ester (**152**) with nitrosyl chloride in pyridine. Irradiation causes homolysis of the nitrite ester NO bond, generating the same oxygen radical (**143**) formed in the hypoiodite reaction. Hydrogen abstraction gives (**144**), which traps NO to form the 19-nitroso compound which rearranges to the oxime (**153**). Oxidation of compound (**153**) forms the lactone (**154**), which can then enter the reaction sequence at compound (**146**) to form 19-norsteroids (146). The yield of oxime from compound (**152**) is 36–74% (147).

With the well-worked out methodology to prepare 6β,19-ethers and 19-hydroxy steroids, fermentation routes have been developed leading to estrone (**135**) and the enone ether (**148**) (Fig. 16). The 19-hydroxy-cholest-4-en-3-one (**155**) is converted by a *Nocardia* species directly into estrone (**135**) in 30% yield (148), whereas a *Mycobacterium* species converts this same derivative into 19-hydroxyandrostenedione (**149**), which is the prime intermediate in 19-norsteroid synthesis, in >70% yield (149). The 6β,19-oxidocholest-4-en-3-one (**156**) is metabolized into (**149**) in 57% yield and its 9α-hydroxy derivative (**159**) in 20% yield by a *Nocardia* species (150).

The introduction of an 18-methyl group in progestins strongly enhances biological activity and, although these compounds are made commercially by total synthesis, several partial syntheses for their preparation have been described (Fig. 17). The lactone (**158**) prepared by the hypoiodite reaction (151) reacts with only one equivalent of methyl Grignard reagent to form the 13-nor-13-acetyl steroid (**159**), which can be converted by standard methods into 18-methylpregnenolone (**160**) (152). An alternative method for introducing an 18-methyl group is the hypoiodite reaction with the cyanohydrin (**161**). The initially formed 13-methyl radical radical (**162**) is trapped by the nitrile to form the bridged radical (**163**), which rearranges to the 18-cyano steroid (**164**), which is converted into the 13-nor-13-vinyl steroid (**165**) (153–154). A variant of this type of reaction involves building the progesterone side chain from the 17-ketone (**166**) by means of a stabilized phosphonate to generate the α,β-unsaturated nitrile

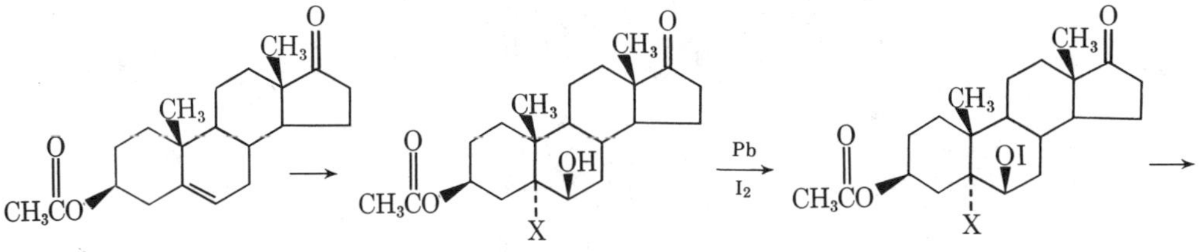

(**109**)

(**141a**) X=Br
(**141b**) X=Cl
(**141c**) X=OCCH$_3$ (O above C)

Pb / I$_2$

(**142a**) X=Br
(**142b**) X=Cl
(**142c**) X=OCCH$_3$ (O above C)

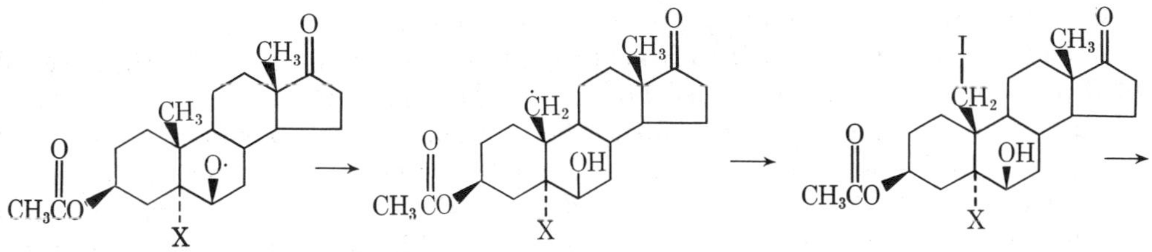

(**143**)

(**144**)

(**145a**) X=Br
(**145b**) X=Cl
(**145c**) X=OCCH$_3$ (O above C)

Figure 15. Alternative methods for the removal of the angular methyl group at C10.

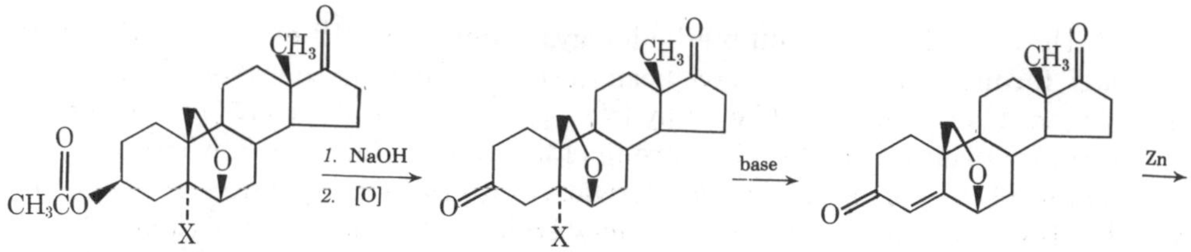

(146 a) X=Br
(146 b) X=Cl
(146 c) X=OCCH$_3$ (O over the C)

1. NaOH
2. [O]

(147 a) X=Br
(147 b) X=Cl
(147 c) X=OCCH$_3$ (O over the C)

base

(148)

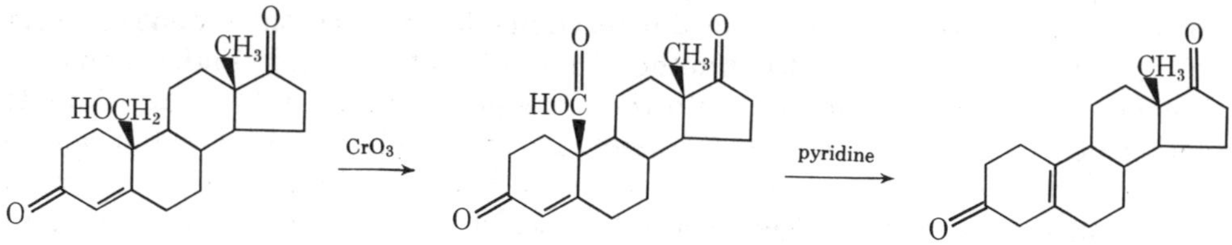

Zn

(149)

CrO_3

(150)

pyridine

(151)

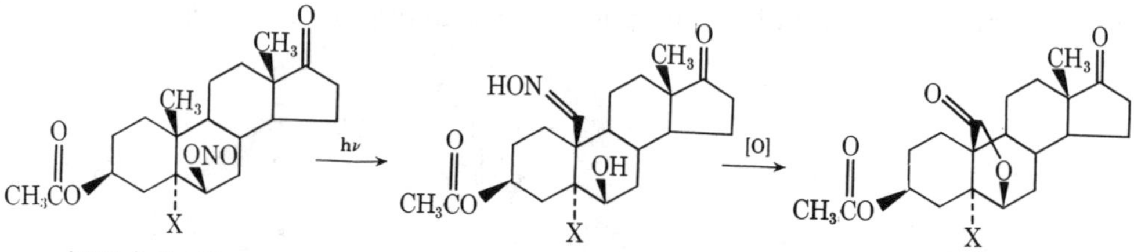

(152 a) X=Br
(152 b) X=Cl
(152 c) X=OCCH$_3$ (O over the C)

hν

(153 a) X=Br
(153 b) X=Cl
(153 c) X=OCCH$_3$ (O over the C)

[O]

(154 a) X=Br
(154 b) X=Cl
(154 c) X=OCCH$_3$ (O over the C)

Figure 15. *(continued)*

(167). Stereospecific reduction and anion formation followed by oxygenation and trapping by acetyl chloride yields the 20-α-peracetoxynitrile **(168)** which, upon photolysis, gives the 13-nitrile **(169)** (155). This reaction has been used to prepare 18-substituted derivatives of desoxycorticosterone (156).

Aldosterone **(102)**, obtained from adrenal extracts, is necessary for the survival of fluid homeostasis of the human body. Two classical approaches to the synthesis of this hormone were reported prior to the development of the commercially used nitrite photolysis described in Figure 18 (157–158). In Figure 18, commercially available corticosterone acetate **(170)** is converted into its nitrite ester **(171)** with nitrosyl chloride in pyridine at 0°C. Irradiation of compound **(171)** in an inert solvent generates the oxime **(172)** (see structures **(141–145)** for the mechanism), which is oxidatively deoximated with nitrous acid to yield aldosterone **(102)**. An alternative synthesis starts from 11β-hydroxy progesterone **(173)** which, when irradiated, forms the epimeric cyclobutanols **(174)** (159) which are dehydrated to the exocyclic olefin **(175)**. Epoxidation **(176)** and base-catalyzed ring opening generates the allylic alcohol **(177)**, which is bis-hydroxylated **(178)** by osmium tetraoxide and cleaved by periodate to aldosterone **(102)** (160).

Because of the interest in vitamin D and its metabolites, the switch in steroidal starting materials from diosgenin to androstenedione, and the need for synthetic comparison samples for sterols obtained from natural sources, much work has been done on new methods for the construction of various *C*-17 side chains (161) (see Fig. 19). Dehydroepiandrosterone, protected as its 3-tetrahydropyranyl ether (THP) (**179**), is condensed with a phosphonate in a Wittig-Horner reaction to form the α,β-unsaturated *E*-ester in 89% yield (162). Selective reduction gives the 17,20-dihydroester (**181**) which, after generation of the enolate, is alkylated with the optically active iodide derived from levulinic acid to give the stereospecifically formed derivative (**182**) (163). Reduction of the ester at *C*-20 to the methyl group is accomplished by hydride reduction to (**183**), tosylate formation with *p*-toluenesulfonyl chloride (**184**), and a second hydride reduction (**185**). The introduction of the 7-double bond is accomplished by sequential bromination–dehydrobromination with 1,3-dibromo-5,5-dimethyl hydantoin and *sec*-collidine. Subsequent hydrolysis of both protecting groups with mild acid frees the three alcohol groups (**186**). Irradiation of (**186**) with a mercury arc and subsequent thermolysis at 80°C gives the natural 25(*S*),26-dihydroxycholecalciferol (**187**) (164–165).

Total Synthesis. Investigations in the field of steroid total synthesis began almost immediately after the precise formula for cholesterol was established early in the 1930s. Most 19-norsteroids are produced in this way (166). Carbon-14 dating techniques have been used to determine whether a steroid has been prepared by partial synthesis from vegetable origins or by total synthesis from petroleum sources (123).

The earliest attempts at total synthesis focused on the A-ring aromatic steroids because of their lesser degree of stereochemical complexity. The AB-aromatic steroid equilenin (**188**) has only two asymmetric centers and estrone (**135**) has four, whereas cholesterol (**2**) has eight asymmetric centers. Equilenin is the simplest representative

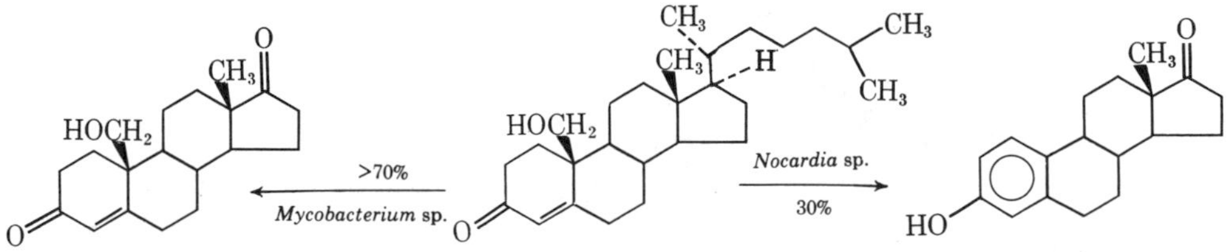

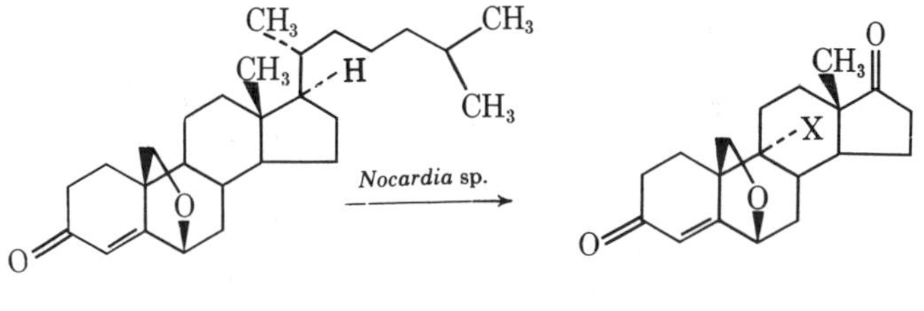

Figure 16. Fermentation routes to 19-norsteroids.

of the natural steroids and was synthesized from a dye-stuff intermediate in 20 steps with an overall yield of 2.7% (16). Its synthesis was reported in 1939 (Fig. 20). The

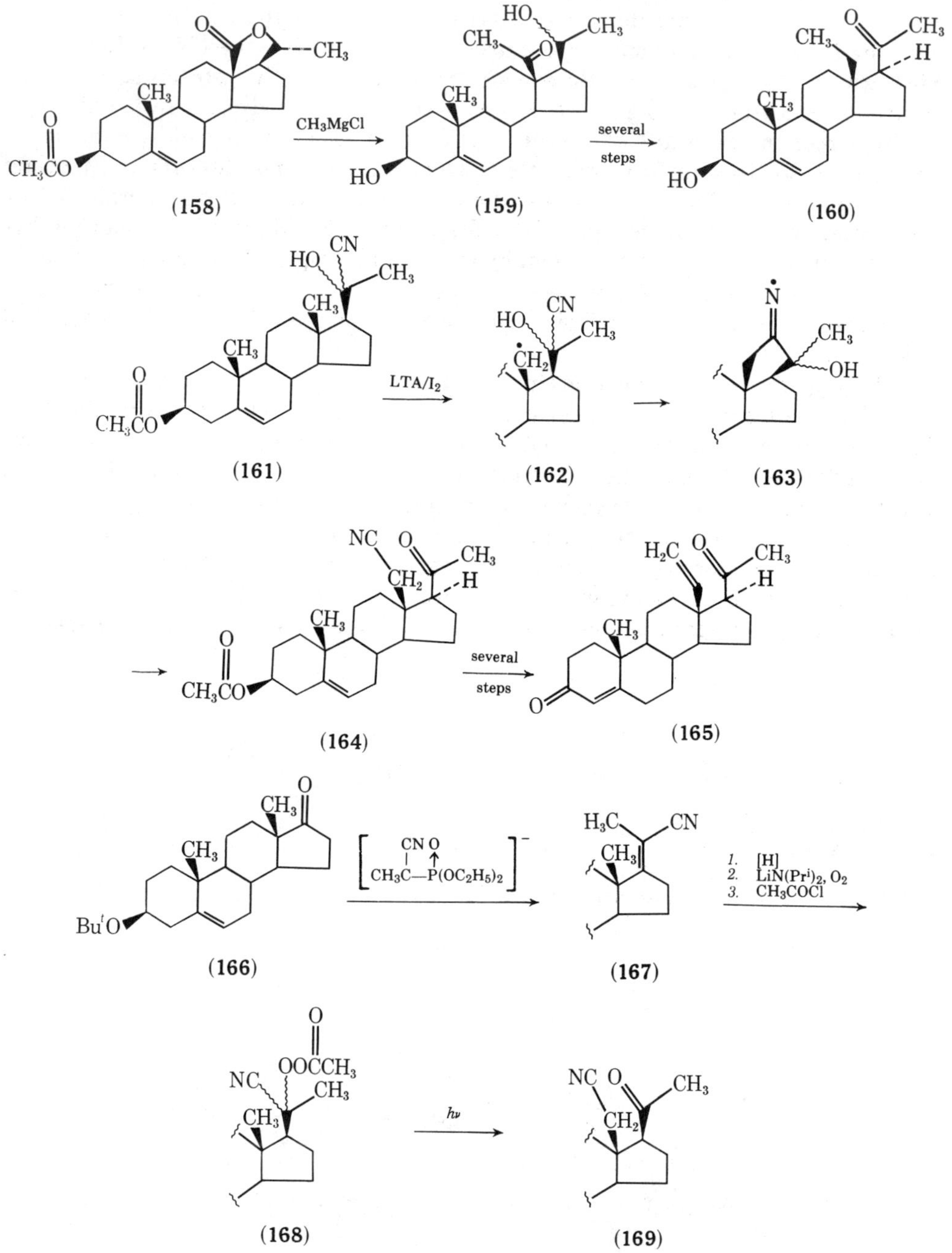

Figure 17. Methods for the introduction of a 18-methyl group. The presence of a 13-ethyl group in progestins leads to a strong enhancement of biological activity.

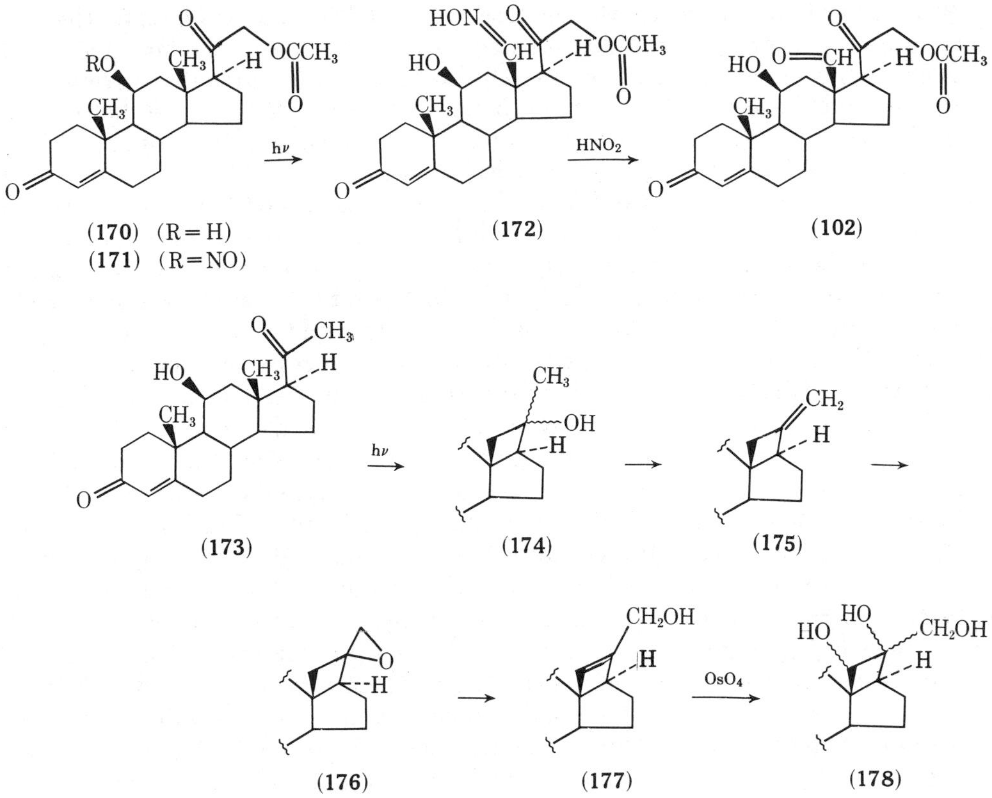

(170) (R=H)
(171) (R=NO)

(172)

(102)

(173)

(174)

(175)

(176)

(177)

(178)

Figure 18. Syntheses of aldosterone.

naphthylamine (**189**) is converted to the naphthylbutyric acid (**190**), which after conversion to its acid chloride and intramolecular Friedel-Crafts reaction, is converted to the tricycle (**190**). Carbalkoxylation and introduction of the eventual *C*-13 angular methyl group yields compound (**192**), which undergoes a Reformatsky reaction, aldol dehydration, and saponification to (**193**). After reduction of the ester double bond (**194**), an Arndt-Eistert homologation gives compound (**195**), and Dieckmann cyclization and ether hydrolysis yields equilenin (**188**) which is resolved as its 1-menthoxyacetic ester. Subsequent studies by another group have increased the overall efficiency of equilenin syntheses (167). Estrone (**135**) was the second natural steroid to be synthesized and was prepared from the benzene derivative (**196**) via (**197**) in 0.1% yield in 18 steps. The route from structure (**197**) was essentially the same as shown in Figure 20 for equilenin (168).

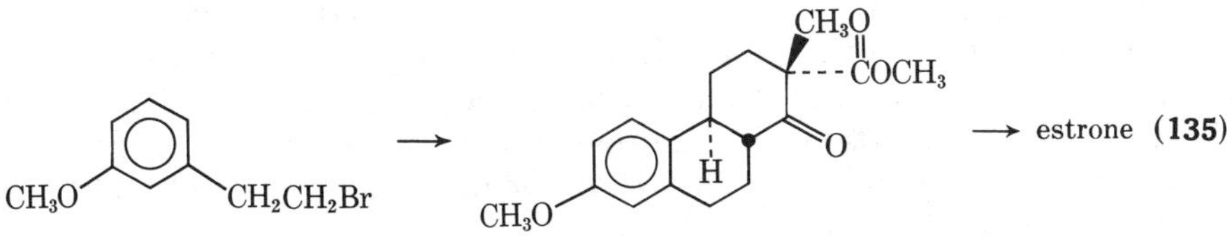

→ estrone (**135**)

(**196**)

(**197**)

A synthesis of estrone derivatives based on the Diels-Alder reaction for the construction of the tetracycle was investigated, but the problem of the introduction of angular methyl group was unresolved (169). Two decades later, the dimethylquinone (**198**) was added thermally to the naphthalene derivative (**199**) to give the D-homo steroid (**200**) with an angular methyl group at *C*-14 instead of *C*-13 (see Fig. 21). The catalyzed Diels-Alder reaction gave the desired regioisomer (**201**) in 70% yield with a cis CD ring junction. Base-catalyzed epimerization gave the natural configuration and selective deoxygenation gave compound (**203**). Catalytic hydrogenation of the 9(11) double bond gave the natural BC trans configuration. Hydride reduction of the enone carbonyl gave the allylic alcohol (**205**) which, after acetylation, was reductively removed forming the olefin. This was bis-hydroxylated and cleaved to the aldehyde ketone (**206**) which cyclized with acid to the aromatic progesterone derivative (**207**). Beckmann rearrangement (Fig. 8) then furnished racemic estrone (**135**) (170).

The approach to estrone originally reported in 1959 and modified in 1960 remains one of the most versatile, efficient, and direct methods for the synthesis of A-ring aromatic steroids (see Fig. 22) (171–173). In this reaction, vinyl Grignard is added to the tetralone (**208**) to produce (**209**). The condensation of vinyl carbinol (**209**) with methyl cyclopentanedione (**210**) is catalyzed by the acidic 1,3-dione in compound (**210**) to yield the tricyclic compound (**211**) (174). Acid cyclization of compound (**211**) forms the estrone derivative (**212**). Reduction of the ketone to the alcohol and hydrogenation of the double bonds gives estradiol methyl ether (**214**) via (**213**). The use of the thiouronium salt (**210**) in place of the vinyl carbinol (**209**) raises the yield of the condensation to >90% (174). The condensation of an asymmetric oxime with the prochiral dione in (**211**) leads to preferential formation of one enantiomer (**216**) which, when cyclized and once the oxime is removed, yields the steroid with the natural optically active configuration of structure (**214**) (175).

An interesting new approach to A-ring aromatic steroids involves ring opening for a benzocyclobutene to an *ortho*-quinodimethane and a subsequent intramolecular Diels-Alder reaction (176). The optically active benzocyclobutene (**217**) is synthesized and then undergoes ring opening at 180°C, forming quinodimethane (**218**), which immediately cyclizes to the estradiol derivative (**219**) (Fig. 23). The optical purity of the estradiol thus formed is 97% (177). An alternative approach (see structures (**220–222**)) is based on the strong affinity of silicon for fluoride ion to generate quinodimethane at room temperature and thus to avoid the high temperatures necessary for benzocyclobutene ring opening (178).

Although the totally synthetic approaches to aromatic steroids are useful, with the Torgov approach having potential commercial significance, the main economic interest has been in the preparation of saturated 19-norsteroids for use in oral contraceptives. Most of the 19-norandrostanes which are used for oral contraceptives are produced by total synthesis, and the 19-nor-18-methyl-steroids have always been produced this way.

The first total syntheses of a nonaromatic steroid were reported in 1951 (179–180). The Woodward approach was based on the Diels-Alder reaction previously described in the section on partial synthesis. The Robinson synthesis of epiandrosterone (**223**) suffered from the disadvantages inherent in all the early syntheses: lack of stereospecificity in many of the steps, the addition and removal of protecting groups, and low yields. The overall yield for epiandrosterone (**223**) from 1,6-naphthalenediol is $7 \times 10^{-5}\%$ for the 40 steps involved.

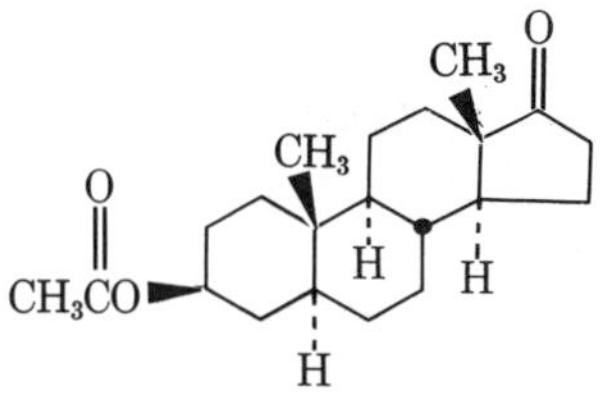

(223)

The tremendous commercial interest in corticoid steroids in the early 1950s led to two total syntheses. One involved a modification of the Woodward synthesis (181).

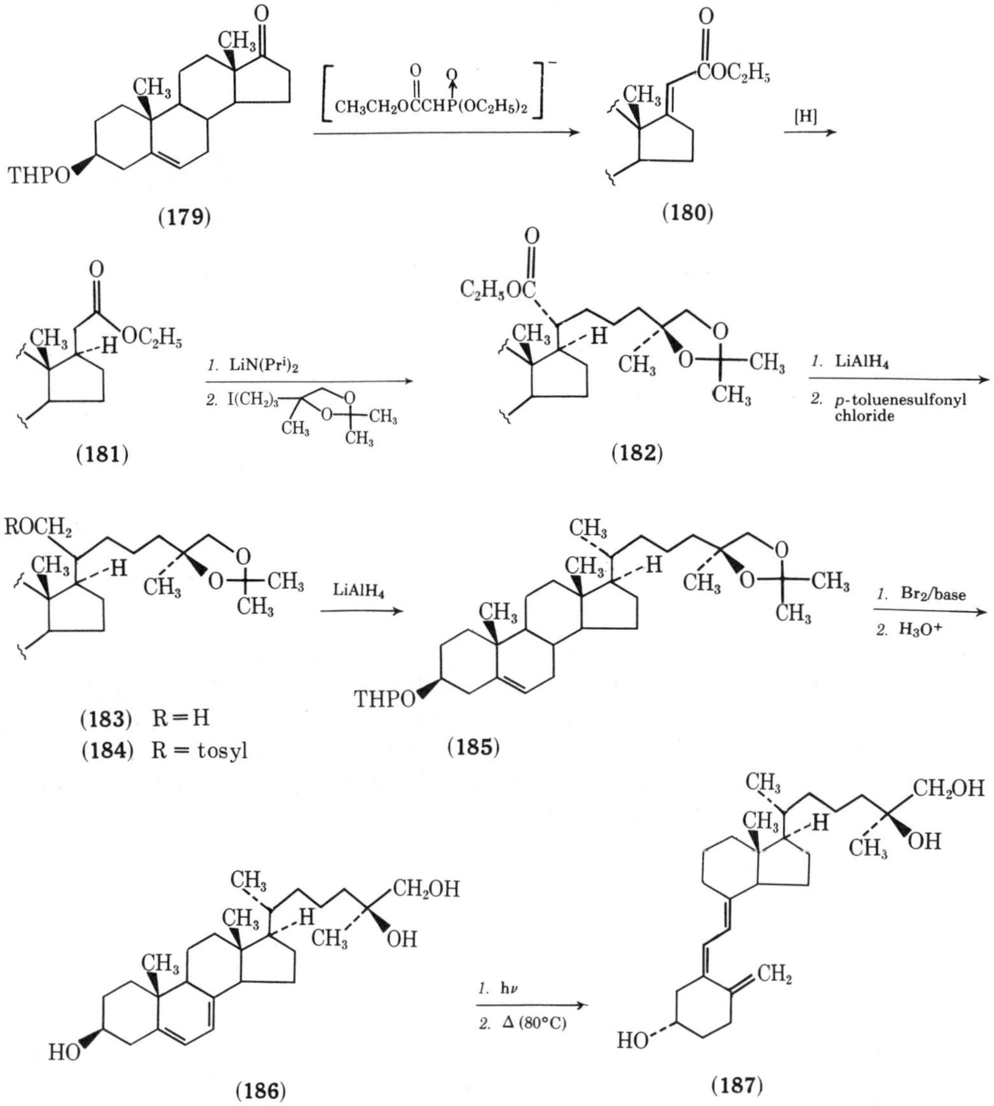

(179) (180) (181) (182) (183) R=H (184) R = tosyl (185) (186) (187)

Figure 19. The introduction of the vitamin D side chain into commercially available 17-ketosteroids. Research of this type reflects the switch from pregnane-based starting materials, obtained from diosgenin, to fermentation derived 17-ketoandrostanes.

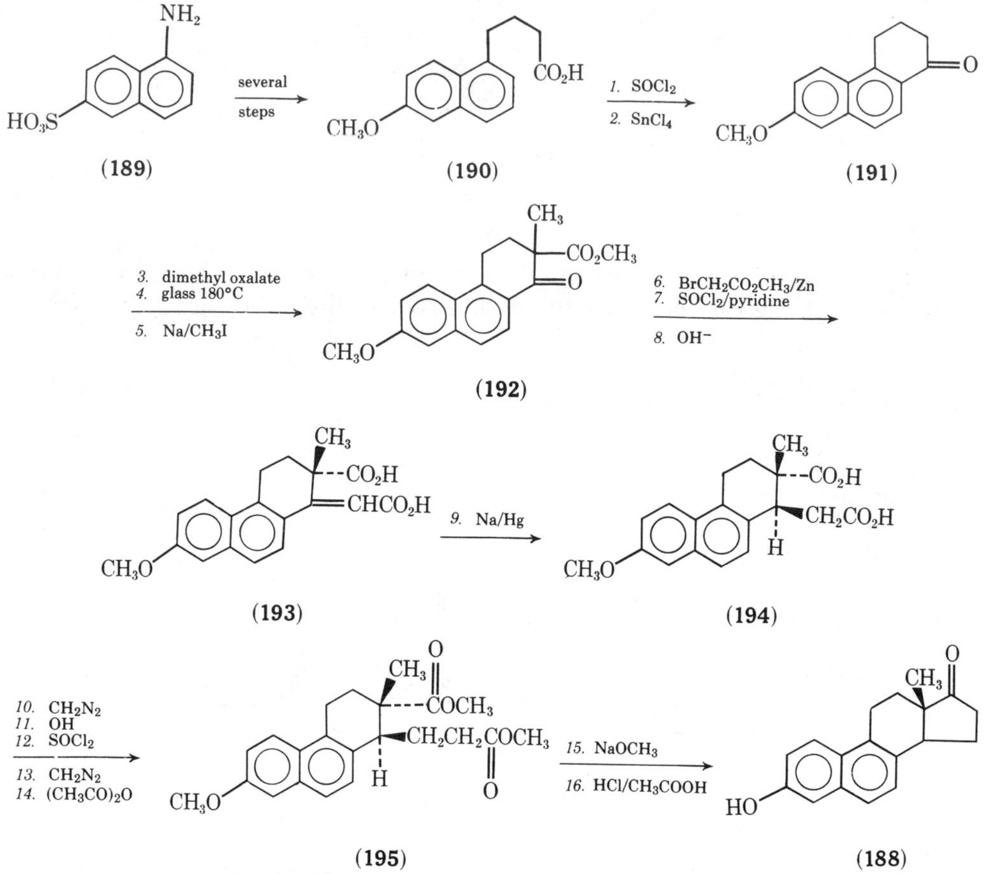

(189) several steps (190) 1. $SOCl_2$ 2. $SnCl_4$ (191)

3. dimethyl oxalate
4. glass 180°C
5. Na/CH_3I

(192)

6. $BrCH_2CO_2CH_3/Zn$
7. $SOCl_2$/pyridine
8. OH^-

(193) 9. Na/Hg (194)

10. CH_2N_2
11. OH
12. $SOCl_2$
13. CH_2N_2
14. $(CH_3CO)_2O$

(195)

15. $NaOCH_3$
16. HCl/CH_3COOH

(188)

Figure 20. The classical approach to the total synthesis of the estrane steroid, equilenin (188).

The other synthesis is characterized by a high degree of stereospecificity in each step and the optical resolution of compound **(224)** (182). The introduction of the dihydroxyacetone side chain was accomplished by sequential alkylation of compound **(225)** with methyl and methallyl iodides, which produces the correct stereochemistry about the incipient C–D ring junction and subsequent *C*-17 hydroxylation. The overall yield of the 22-step process from **(225)** was 1.3%.

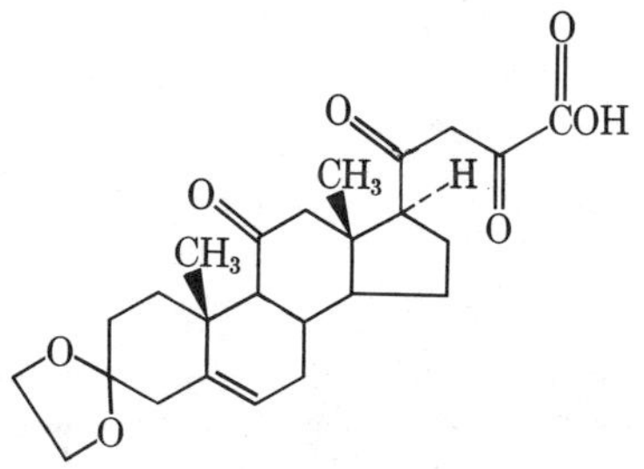

(224) 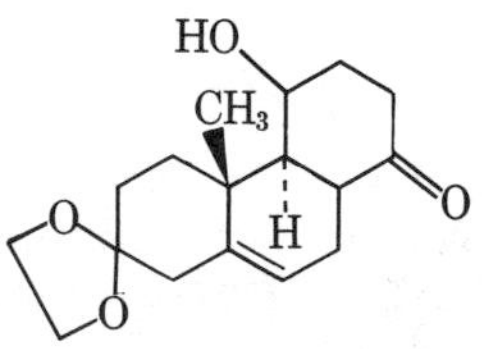 **(225)** 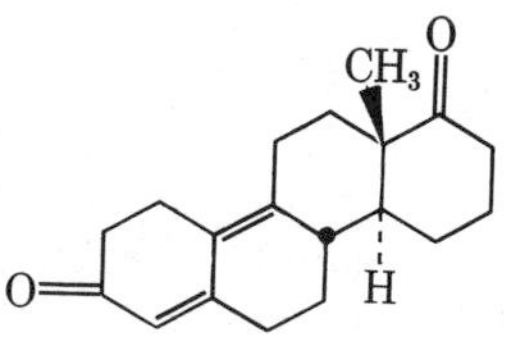 **(226)**

An indication of the effort that went into the development of total synthesis is exemplified by a 1970s synthesis of a D-homo estrane derivative (183). Alkylation with 6-vinyl-2-picoline, ie, a 1,5-diketone equivalent, yields the alkylated enone in high yield. Subsequent manipulations adjust the oxidation level, and Birch reduction results in the formation of the dihydropyridine, which is opened to a 1,5-diketone and then undergoes intramolecular condensation. Aldol condensation yields the homo estrane derivative (**226**). Subsequent studies involving vinylpyridines led to the syntheses of optically active estranes (184).

A synthesis of 19-nortestosterone (**229**), which is an important starting material for oral contraceptives, on an industrial scale is described in ref. 185 (see Contraceptive drugs). In this process, 6-methoxytetralone is converted by isoxazole formation, methylation, and Stobbe condensation to the tricyclic compound. After reduction to the alcohol and ester saponification, the free acid is optically resolved with chloramphenicol (186); however, microbiological resolution at a later stage has been described (187). After decarboxylation, hydrogenation, and esterification, the resultant ester, 2,3,3a,4,5,9b-hexahydro-7-methoxy-3a-methyl-1*H*-benz[*e*]inden-3-ol benzoate, is subjected to Birch reduction, followed by hydrolysis and alkylation with 1,3-dichloro-2-butene to yield the secosteroid (**227**). Hydrolysis of the vinyl chloride in (**227**) forms the ketone (**228**) which, after an intramolecular aldol condensation and dehydration, is hydrolyzed to 19-nortestosterone (**229**).

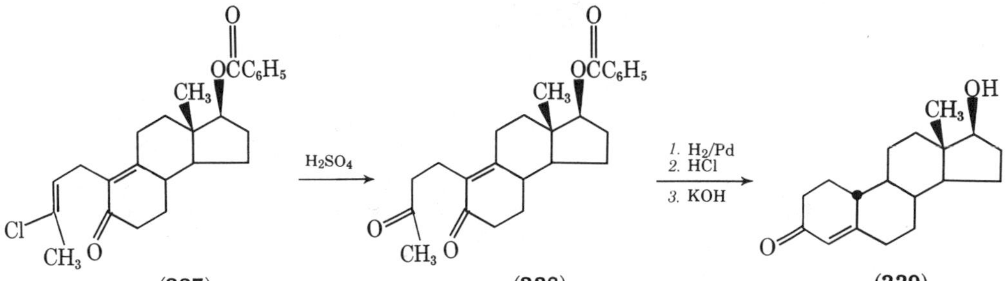

(**227**) $\xrightarrow{H_2SO_4}$ (**228**) $\xrightarrow{1.\ H_2/Pd;\ 2.\ HCl;\ 3.\ KOH}$ (**229**)

A potentially commercially important total synthesis of 19-norsteroids starts from 2-methyl-1,3-cyclopentanedione (**210**), which is the Torgov intermediate, and is depicted in Figure 24 (188). Condensation of (**210**) with methyl vinylketone gives the triketone (**230**). An asymmetric aldol condensation of compound (**230**) followed by acidic dehydration gives the hydrindendione (**231**) in 97% optical purity; the latter's absolute configuration is such that elaboration of the remainder of the steroid molecule leads to a natural series (189). After reduction of the cyclopentane ketone and protection of compound (**232**), condensation with methyl magnesium carbonate gives the acid (**233**), which is in equilibrium with the dienol (**234**). Hydrogenation of this mixture then yields the thermodynamically less stable *trans*-hydrindan (**235**), presumably through selective hydrogenation of compound (**234**) (190). Condensation of (**235**) with formaldehyde gives the exocyclic α-methylene ketone (**236**), which undergoes Michael addition with the β-keto ester (**237**) to the polyketide (**238**) (191). Intramolecular aldol condensation, followed by dehydration yields enone (**239**) which, after hydrogenation of the double bond and deprotection, is cyclized to 19-nortestosterone (**229**). The overall yield of chemically and optically pure compound (**229**) from (**230**) is 27%; thus,

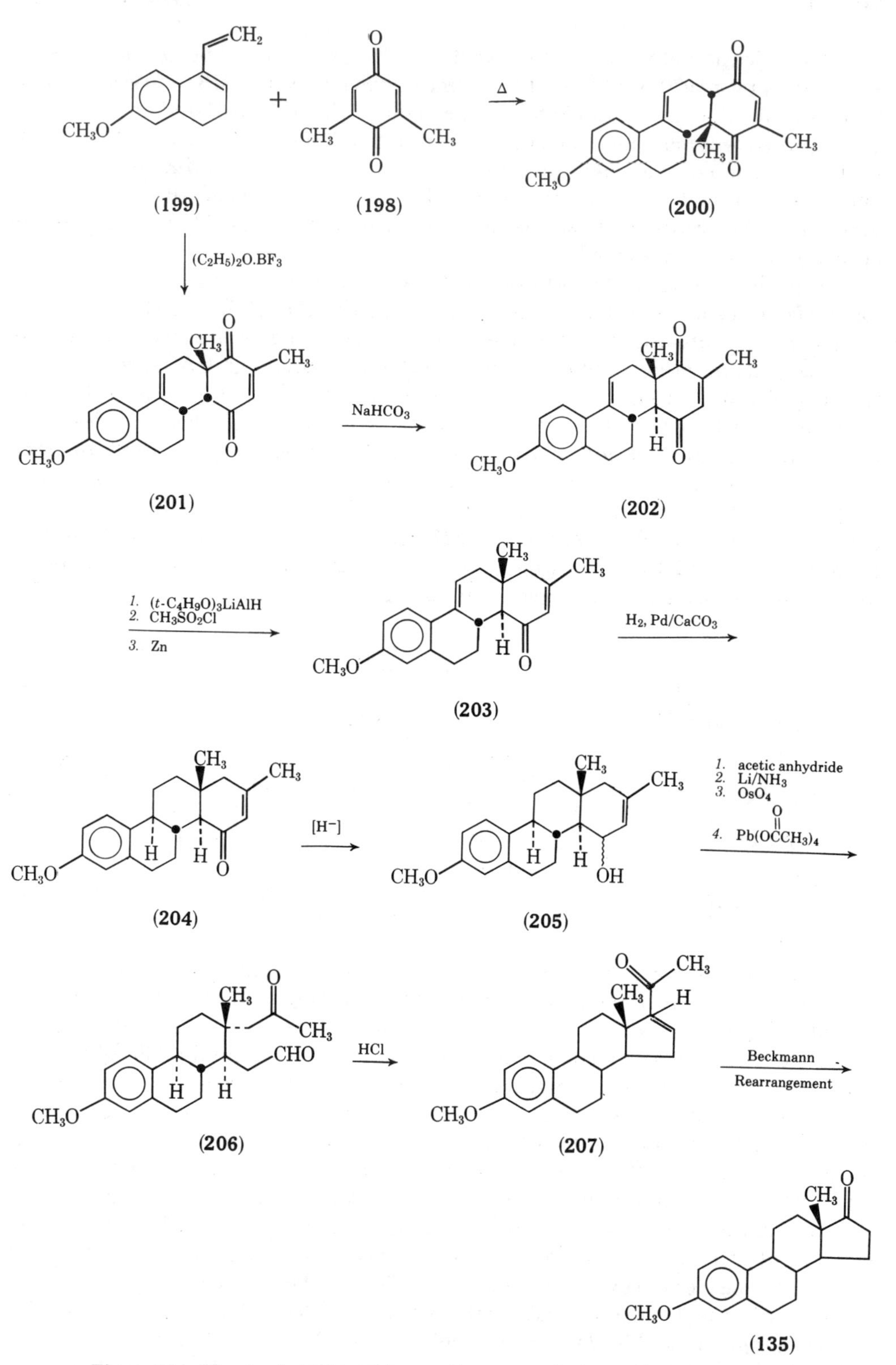

Figure 21. The classical Diels-Alder reaction approach to steroid total synthesis.

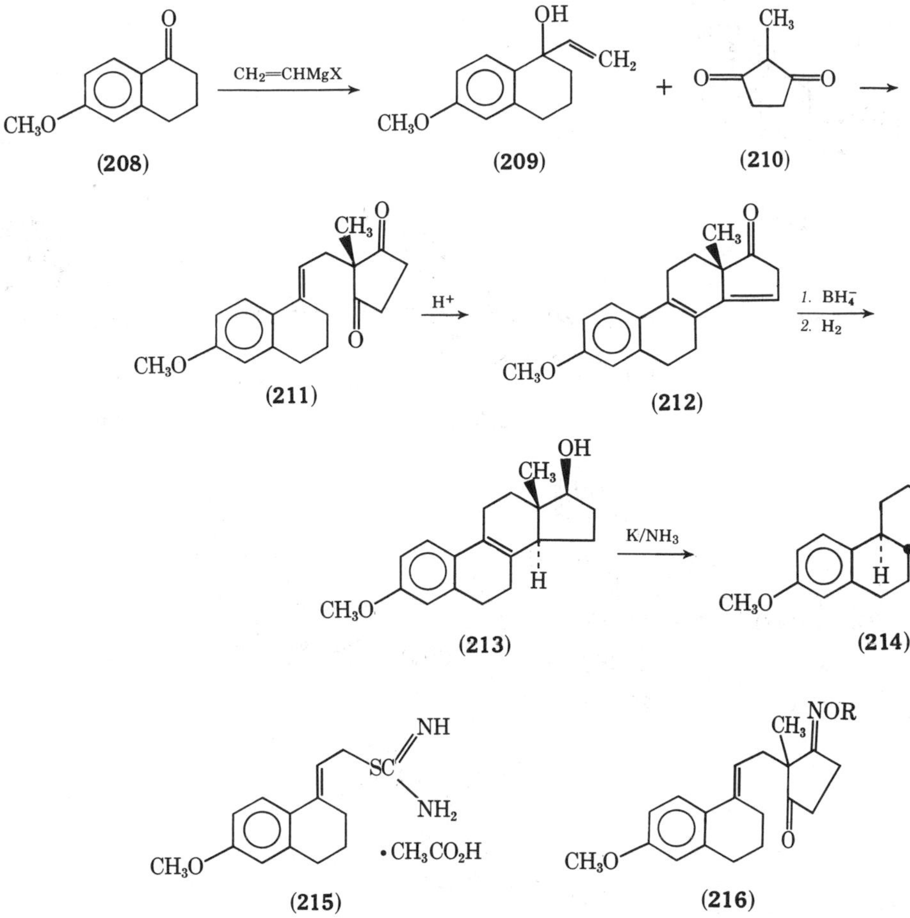

Figure 22. The Torgov synthesis of A-ring aromatic steroids. This method is one of the most versatile, efficient, and direct methods for preparing estrane derivatives.

this is a potentially economically favorable commercial process. A similar series of reactions based on the route **(230)** → **(236)** leads to estradiol methyl ether derivatives (192).

In the 19-norsteroids, replacement of the *C*-13 methyl group with an ethyl group results in enhanced biological potency. For instance, norgestrel **(240)** is a potent progestin used for fertility control. Because the 13-ethyl group is not present in naturally occurring steroids and compound **(240)** cannot be produced economically by partial synthesis, total synthesis is used. The original synthesis is based on the Torgov approach outlined in Figure 25 (193). Condensation of the Michael acceptor **(241)** with ethylcyclopentanedione **(242)** directly forms the 13-ethyl steroid **(243)**. After reduction of the diene and 17-ketone, compound **(244)** undergoes Birch reduction to compound **(245)** which, after oxidation and ethynylation, is hydrolyzed to norgestrel **(240)**. Because the steroids prepared in this way are optically inactive, optical resolution is accomplished with intermediate **(244)** to furnish optically active D-norgestrel (194).

An impressive synthesis of compound **(240)** starts from the optically active hydrindene **(247)** (see Fig. 24), which is condensed with formaldehyde and benzenesul-

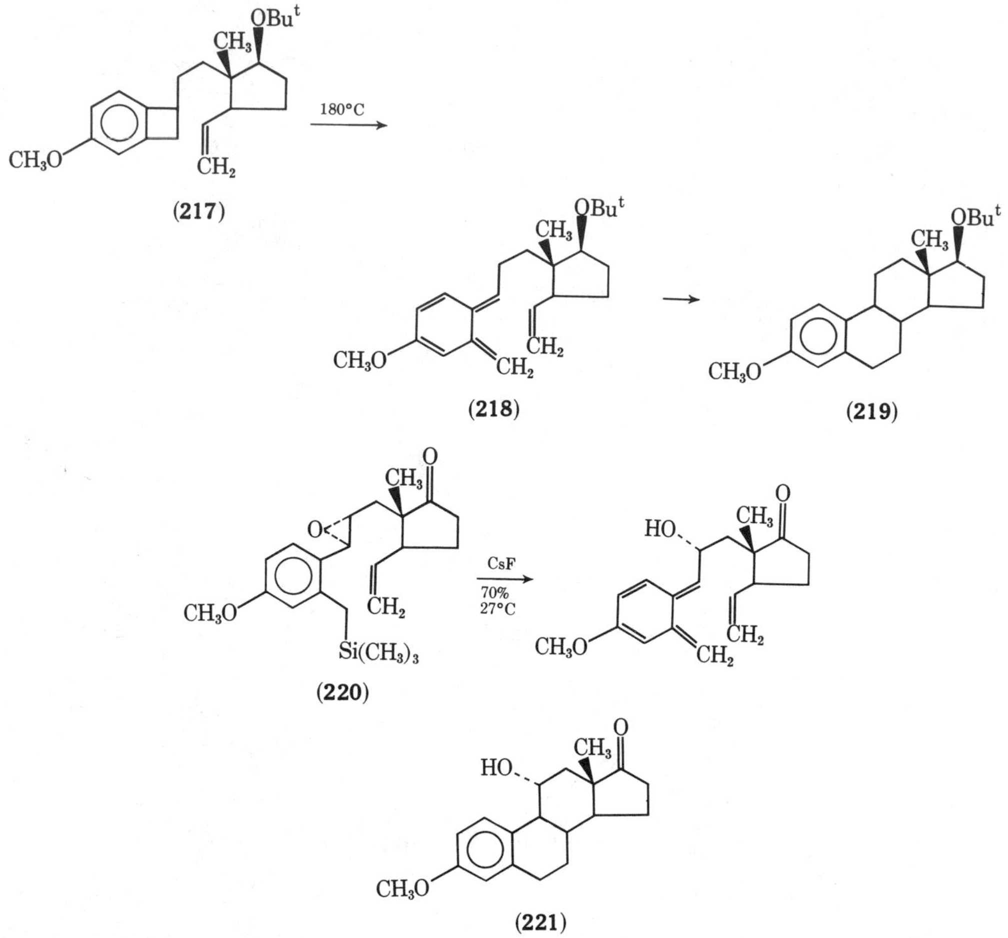

(217) $\xrightarrow{180°C}$ (218) → (219)

(220) $\xrightarrow[70\%,\ 27°C]{CsF}$ → (221)

Figure 23. Modern approaches to the total synthesis of aromatic steroids.

finic acid to generate compound **(248)** in 85% yield. Reduction gives the thermodynamically less stable **(249)**, which is analogous to **(234)** in Figure 24. Condensation with the β-keto ester **(237)** leads to **(250)**, which is the 13-ethyl analogue of the intermediate **(238)** used for the total synthesis of 19-norsteroids. The synthesis of norgestrel then follows the synthetic sequence shown in Figure 24 (195).

The strategy for the total synthesis of steroidal natural products previously discussed has involved, for the most part, step-by-step annelations, ie, the rings are formed one at a time. The biomimetic approach differs in that the synthesis involves the production of a number of rings in a single step by ring closure of an acyclic chain having 1,5-placed trans-olefinic bonds. This process is analogous to the biogenetic conversion of squalene to lanosterol (196). The direct formation of the steroid nucleus by a biomimetic cyclization is shown in Figure 26 (197). In a convergent synthesis, the ketal of the commercially available Hagemann's ester **(251)** is converted to the phosphorane **(260)** in a series of conventional steps. Optical resolution of the acid **(256)** eventually allows the isolation of the optically pure steroid **(265)**. The Wittig reagent

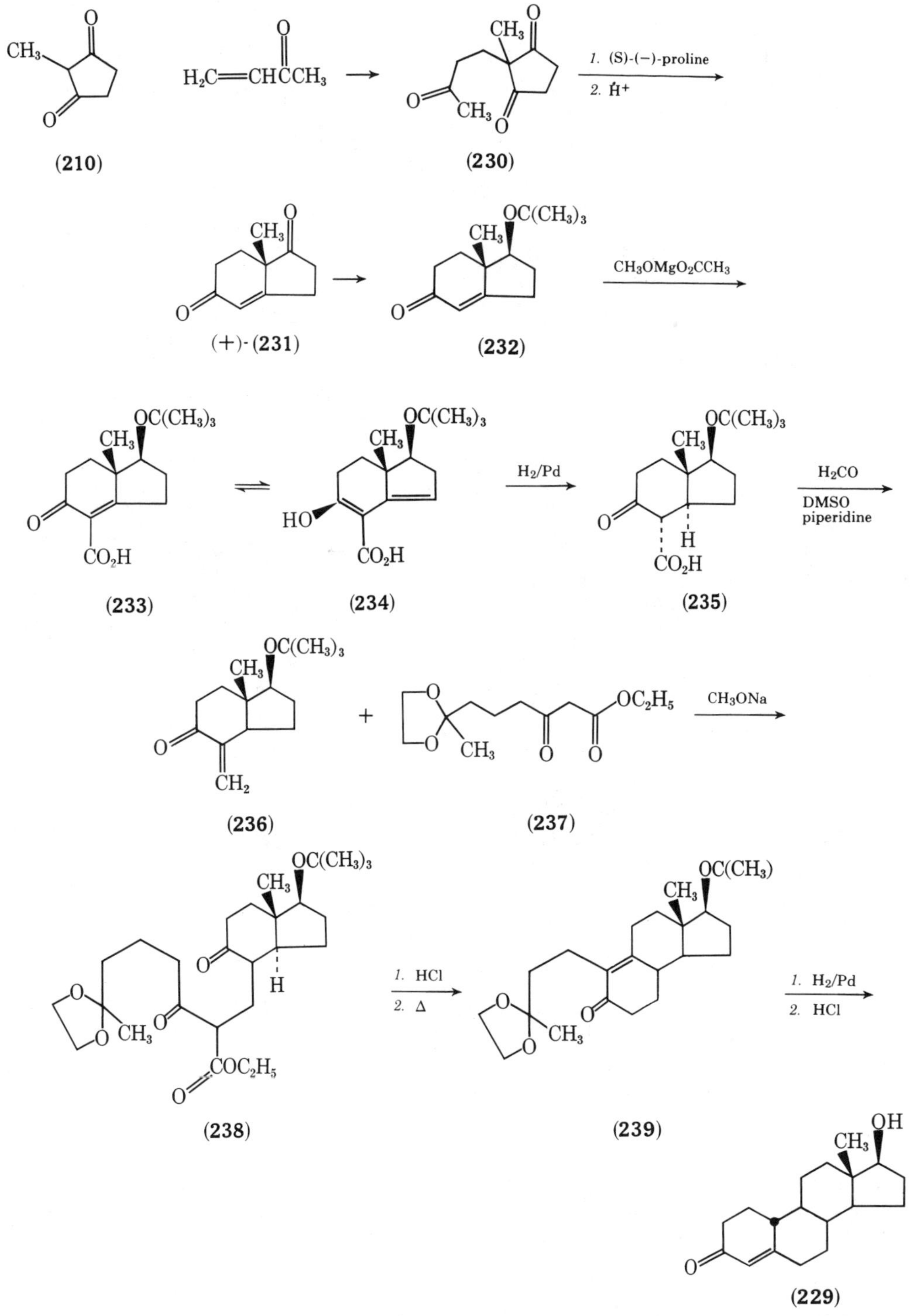

(210)

$H_2C{=}CHCCH_3$ (with C=O) →

(230)

1. (S)-(−)-proline
2. H^+

(+)-(231) → (232)

$CH_3OMgO_2CCH_3$

(233) ⇌ (234)

H_2/Pd

(235)

H_2CO
DMSO
piperidine

(236) + (237)

CH_3ONa

(238)

1. HCl
2. Δ

(239)

1. H_2/Pd
2. HCl

(229)

Figure 24. A potential commercial synthesis of optically active 19-norsteroids.

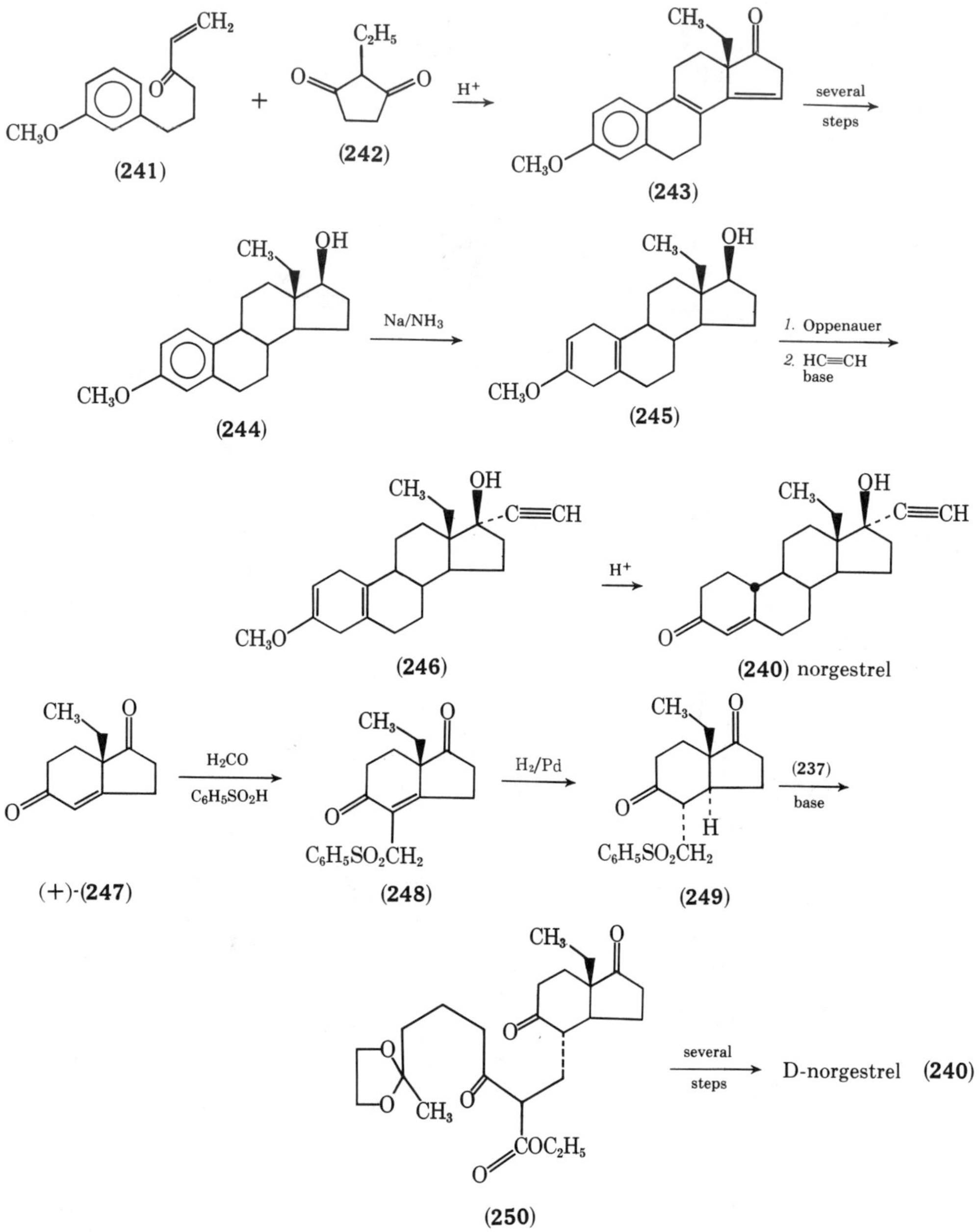

Figure 25. The synthesis of norgestrel (**240**). The presence of an 18-methyl group and the absence of the angular 10-methyl group in this oral contraceptive component requires a total synthesis because of a lack of suitable naturally occurring precursors.

derived from (**260**) is then condensed with the enyne aldehyde (**261**) to form the polyunsaturated monocyclic material (**262**). After dethioketalization and reduction of the enone carbonyl, compound (**264**) is obtained as a mixture of *C*-2 epimers (steroid numbering) with >94% enantiomeric purity with respect to *C*-5. Cyclization in tri-

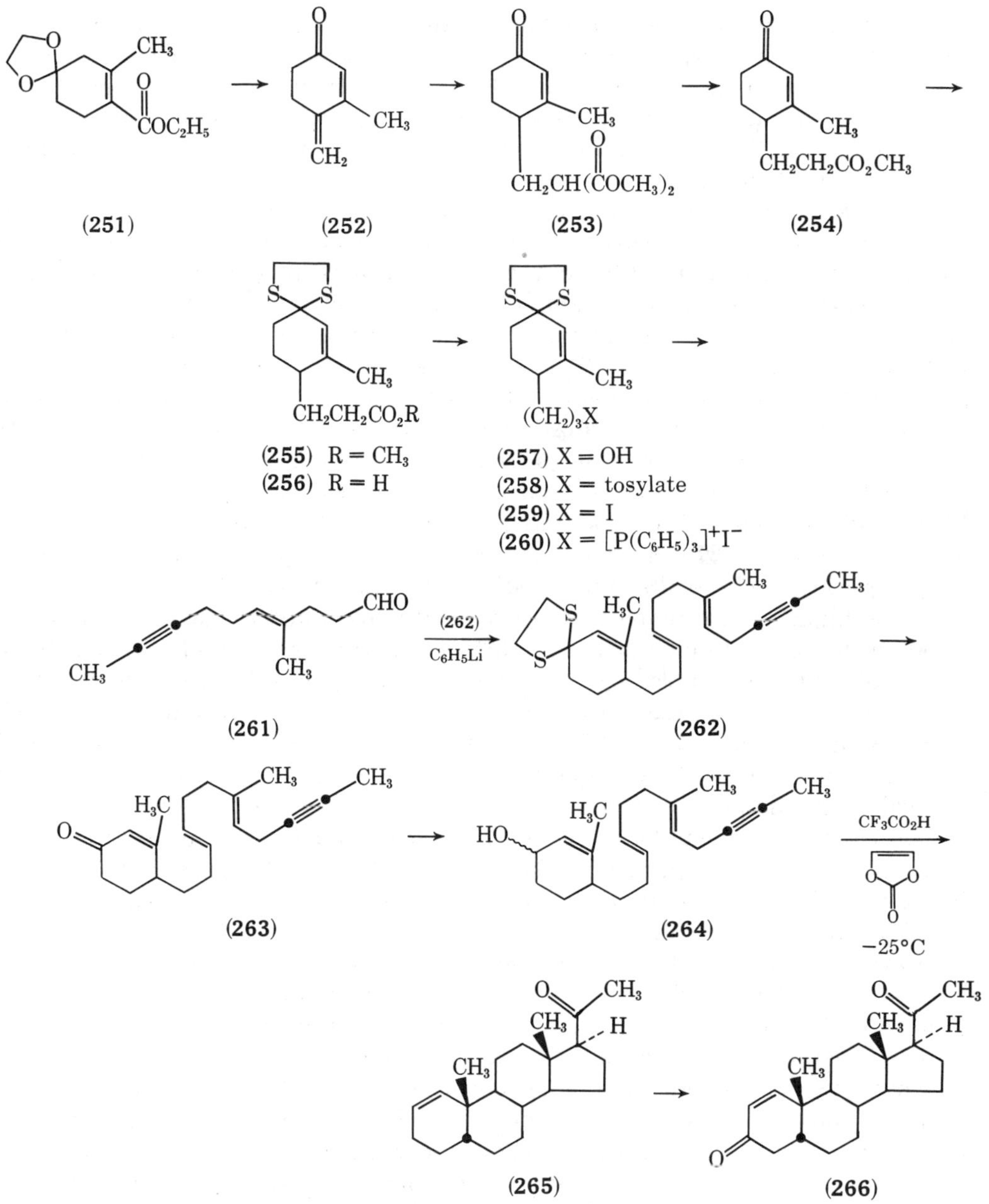

(251) (252) (253) (254)

(255) R = CH_3
(256) R = H

(257) X = OH
(258) X = tosylate
(259) X = I
(260) X = $[P(C_6H_5)_3]^+I^-$

(261) $\xrightarrow[C_6H_5Li]{(262)}$ (262)

(263) → (264) $\xrightarrow[-25°C]{CF_3CO_2H}$

(265) → (266)

Figure 26. The biomimetic approach to steroid total synthesis. This method mimics the *in vivo* cyclization of squalene to cholesterol.

fluoroacetic acid containing ethylene carbonate at −25°C yields the pregnane steroid **(265)** in 65% yield. The ethylene carbonate traps the vinyl cation generated from the acetylenic bond during cyclization and thus introduces the *C*-17-acetyl side chain. Subsequent oxidation of **(265)** with *tert*-butyl chromate yields the progesterone derivative **(266)**. This approach is being developed in Europe as a potential commercial synthesis.

Modified Steroids. Chemical modifications of the naturally occurring compounds have led to new drugs with increased or advantageously modified activities. With the discovery of each new class of biologically active compounds, efforts went into the synthesis of agonists and antagonists based on the parent structures. The emphasis here is placed on the more significant chemical modifications and on those that have general applicability or lead to novel structures.

Steroid Heterocycles. Steroids containing a heterocyclic ring attached to the tetracyclic nucleus have been investigated following the discovery that fusion of a pyrazole ring to the 2,3-positions of the steroid nucleus enhances biological activity in androstane derivatives. For example, stanazolol (**269**) is 10 times as potent an anabolic agent as 17α-testosterone but has only twice the androgenic activity with no lasting adverse side effects (198). The synthesis, which is outlined in Figure 27, is generally applicable to a wide variety of substituted pyrazoles (199). In the corticoid series, a 2′-phenylpyrazole ring substantially enhances the antiinflammatory activity of the parent steroid. The pyrazole (**270**) has antiinflammatory properties 2000 times more potent than that of cortisol; however, moving the phenyl ring to the 1′-position (**271**) results in a sharp drop in activity (200). Because of its potent mineralocorticoid activity, compound (**270**) has not been used clinically. Replacement of the benzene ring by pyridyl (**272**) results in retained activity and reduced side effects (201).

The highly potent activity of the pyrazoles instigated the syntheses of other heterocyclic fused steroids including thiazoles, pyridines, pyrimidines, pteridines, oxadiazoles, pyrroles, indoles, triazoles, and isoxazoles. One of the most potent of these heterocycles is androisoxazole (**273**), which is prepared by the reaction of hydroxylamine with compound (**268**); (**273**) exhibits an oral anabolic:androgenic ratio of 40 (202). However, the 17α-ethynyl derivative (**274**), ie, danazol, has been of most interest clinically. It is devoid of estrogenic side effects and has been evaluated as an oral contraceptive in women (203). In men, (**274**) also functions as an antifertility agent by reducing sperm concentration (204). The cyanoketone (**276**) produced from (**274**) by base-catalyzed fragmentation is similarly active, whereas the 4α,5α-epoxide (**276**), ie, trilostane, is being evaluated as a treatment of Cushings syndrome. The fusion of an oxazoline ring across the 16,17-positions of prednisolone [*50-24-8*] is accomplished by opening the epoxide ring in (**277**) to form the 17α-azide (**278**) (see Fig. 28). Acetylation and then reduction are followed by transacetylation and cyclization to the oxazoline (**279**), which is converted into oxazocort (**280**) (205). In comparison with the 17α-hydroxy derivatives, the mineralocorticoid activity in (**280**) is reduced (206). Oxazocort has no significant effect upon calcium balance, thus making it of particular value for the treatment of postmenopausal rheumatoid arthritis.

A spiro lactone is introduced at *C*-17 as a modification of the cardenolide structure. However, the 3-oxo-4-ene lactone (**281**) blocks the mineralocorticoid action of aldosterone (**102**) (207). The ensuing structural modification results in the more potent and orally active spironolactone (**101**), which is a clinically important aldosterone antagonist functioning as a potassium-sparing diuretic. The synthesis of compound (**101**) is outlined in Figure 29 (208). Ethynylation of compound (**109**) with acetylene and base yields compound (**282**), which is carboxylated with carbon dioxide and base. Catalytic reduction gives the cardenolide analogue (**284**) after acidification. A second reduction yields the spirolactone (**285**), which can be oxidized to the enone (**281**) but in practice was directly oxidized to the dienone (**286**). Addition of thiolacetic acid yields spironolactone (**101**). The initial synthesis was based on diosgenin as a starting ma-

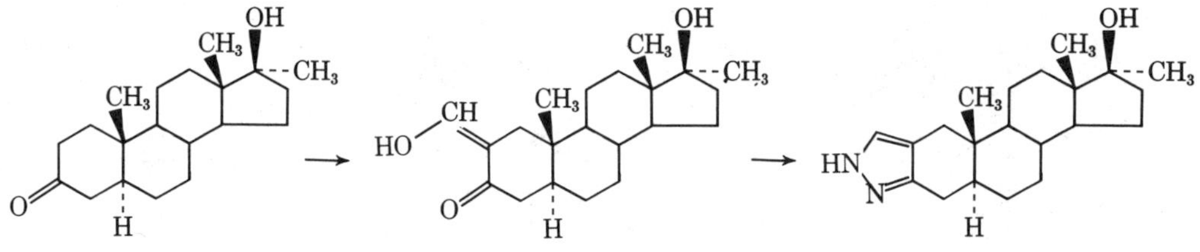

(267) (268) (269)

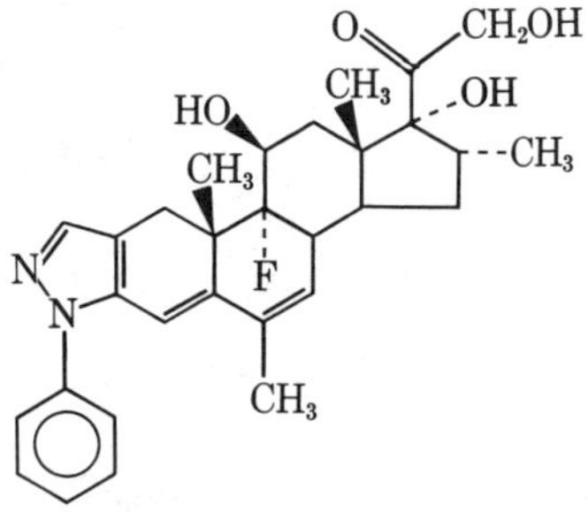

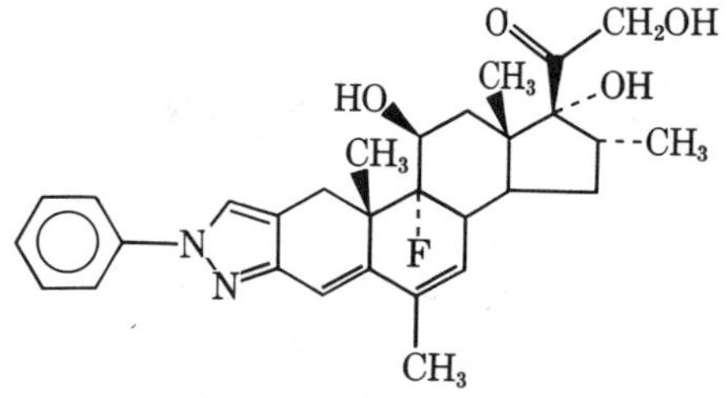

(270) (271)

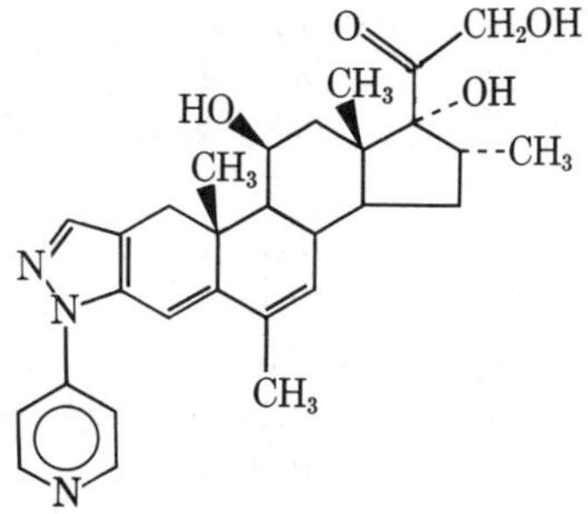

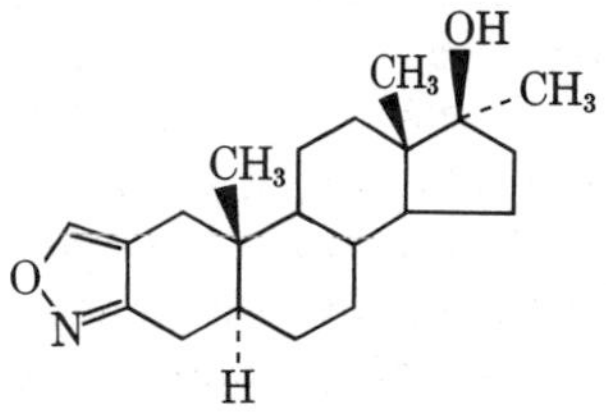

(272) (273)

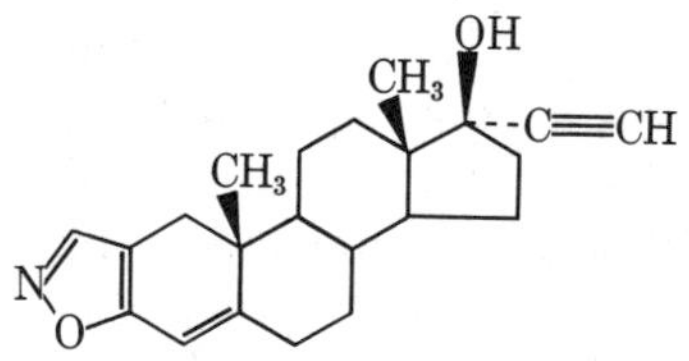

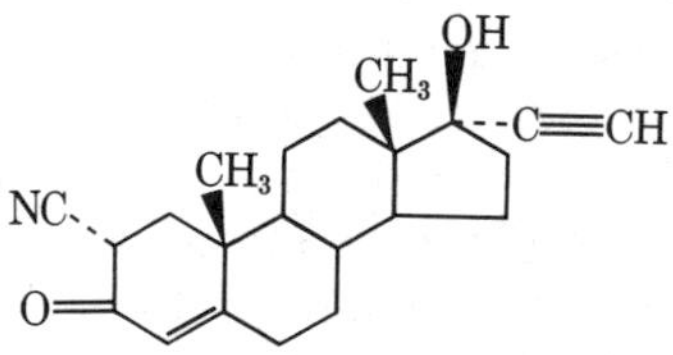

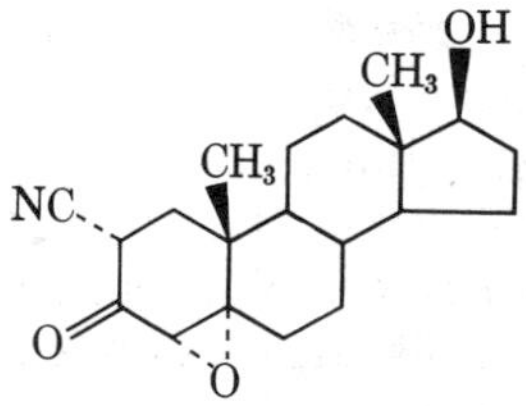

(274) (275) (276)

Figure 27. Steroids with fused heterocyclic rings.

terial, but since the middle 1970s, androstenedione (**125**), which is derived from sitosterol by fermentation, has been used. Essentially the same reaction sequence is used, with the enone in (**125**) being protected as the pyrrolidine dienamine (**287**). Despite the great amount of research that has gone into spironolactones, very few examples of derivatives of (**101**) have been determined to be as biologically active or more active. Among these are the furan derivatives (**288**) and the 7α-ester (**289**) and its 6β-hydroxy metabolite (**290**) (209–211).

Pancuronium (**291**) is a member of a series of bis-quaternary ammonium steroids

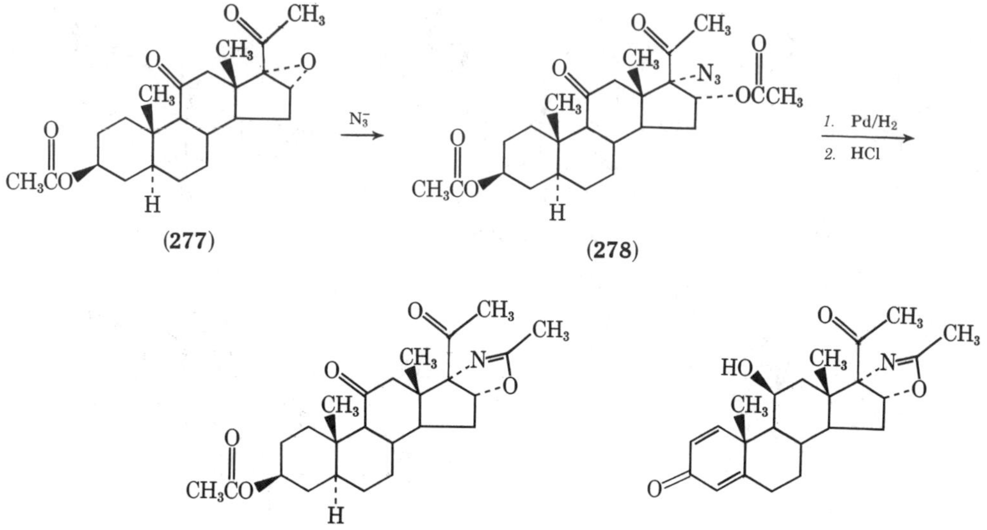

(277) $\xrightarrow{N_3^-}$ (278) $\xrightarrow{1.\ Pd/H_2;\ 2.\ HCl}$

(279) oxazocort (280)

Figure 28. The preparation of oxazocort (**280**).

which cause neuromuscular block (212). Extensive pharmacological and clinical studies have shown that it is five times as potent as *d*-tubocurarine [*57-95-4*] as a competitive neuromuscular blocking agent, with minimal cardiovascular and little or no histamine-releasing or hormonal actions (213).

Steroids Containing a Heteroatom. A special class of modified steroids is represented by steroidal systems containing atoms other than carbon, eg, usually nitrogen, oxygen, or sulfur, and less commonly phosphorus or selenium, at various positions in the cyclopentanophenanthrene skeleton. These are usually classified as heterocyclic steroids (214). Although enormous efforts have been expended in this field, the medicinal chemical success has been very modest and research in this field has dwindled over the last decade.

Partial synthesis. In a synthesis of a series of azacholesterols, one or two of the carbon atoms in the side chain were replaced by nitrogen, which displays interesting hypocholesterolemic activity (215). The compound 22,25-diazacholesterol (**292**) acts by blocking the biosynthetic conversion of desmosterol [*313-04-2*] into cholesterol. When tested in man, compound (**292**) was effective in lowering cholesterol and serum triglyceride levels but accumulated desmosterol and caused reversible myotonia and so was discontinued. It subsequently was used as a birth-control pill for pigeons because it causes a thinning of the egg shell.

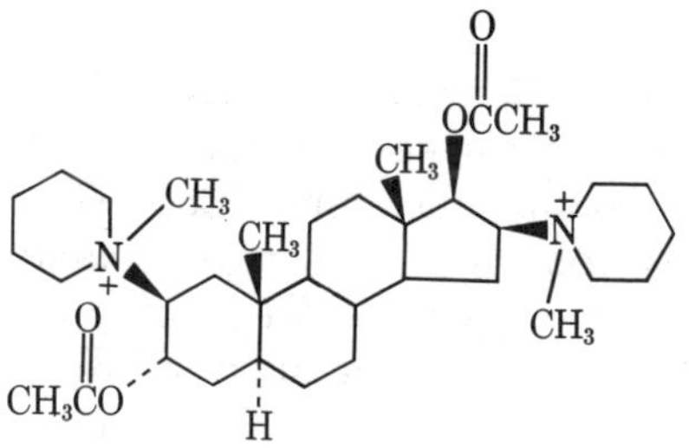

(291) 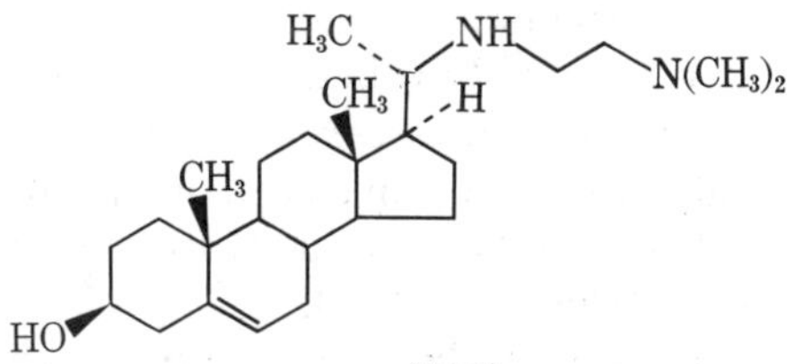 (292)

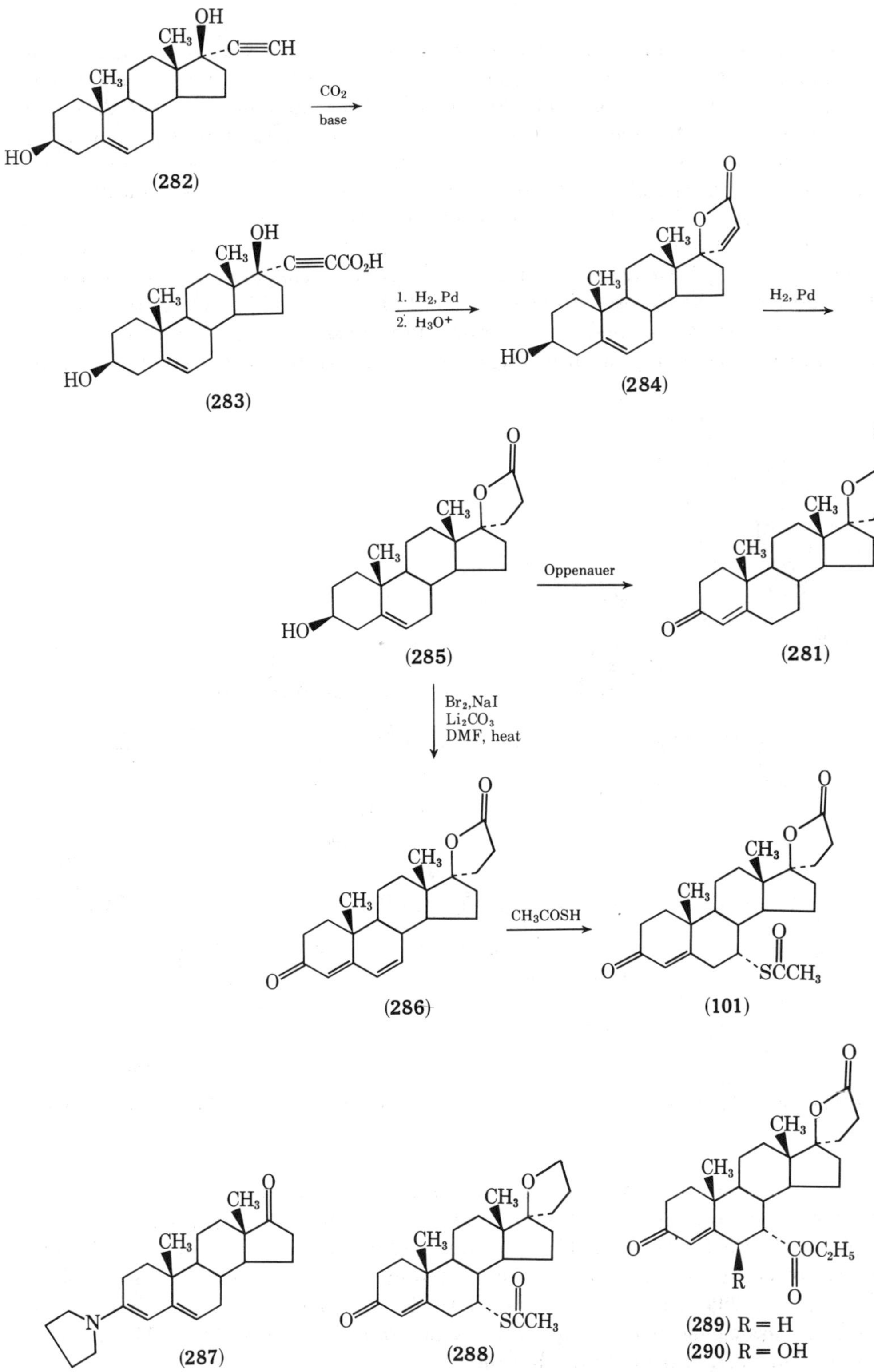

Figure 29. The commercial synthesis of spironolactone (**101**).

One of the advantages of introducing a heteroatom into the A-ring of a steroid is the blocking of the metabolic aromatization of ring A to estrogens. It also makes possible the separation of desired biological activities from undesired ones. Although the 4-oxadihydrotestosterone derivatives have no demonstrable biological activity, the isomeric 2-oxa steroids are potent anabolic agents with reduced androgenic activity. Oxandrolone (**296**) is a clinically useful anabolic agent and is almost totally devoid of androgenic side effects. Its commercial synthesis is outlined in Figure 30 (216). The ozonide derived from the androstane derivative (**293**) is rearranged by heat to form the aldehyde anhydride (**294**), which is solvolyzed in methanol to the aldehyde ester (**295**). Reduction with borohydride yields oxandrolone (**296**) in 80–90% yield. A variety of other 2-oxatestosterone and -progesterone derivatives have also been prepared and retain many of the biological properties associated with the parent molecules (216).

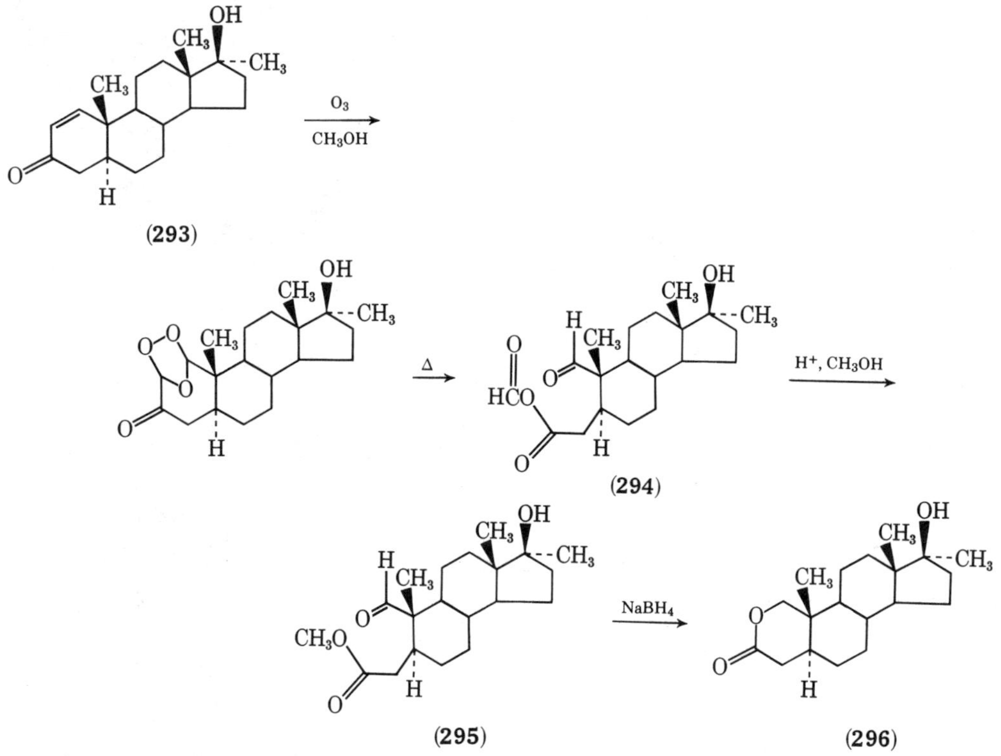

OH CH$_3$ CH$_3$ CH$_3$ O H (**293**) $\xrightarrow[CH_3OH]{O_3}$

O O O O CH$_3$ CH$_3$ OH CH$_3$ O H $\xrightarrow{\Delta}$ O HCO O H CH$_3$ CH$_3$ OH CH$_3$ O H (**294**) $\xrightarrow{H^+, CH_3OH}$

CH$_3$O O O H CH$_3$ CH$_3$ OH CH$_3$ H (**295**) $\xrightarrow{NaBH_4}$ O O CH$_3$ CH$_3$ OH CH$_3$ H (**296**)

Figure 30. The partial synthesis of the heterocyclic steroid, oxandrolone (**296**), a potent anabolic steroid.

The introduction of nitrogen into the steroid nucleus by partial synthesis is illustrated by the conversion of the bridged acetate (**297**) into 2-azaestradiol methyl ether (**304**) in Figure 31 (217). The diosphenol (**298**) forms from compound (**297**) by base-catalyzed oxygenation. Ozonization followed by loss of carbon dioxide yields the aldehyde acid (**299**), which is converted to the lactam (**300**). The 6β,19-ether is eliminated by classical methods to yield the 2-azaestradiol derivative (**303**). Methylation gives the 2-azaestradiol-methyl ether (**304**).

The chemical synthesis of the AB-aromatic-17-phosphasteroid (**341**) has been

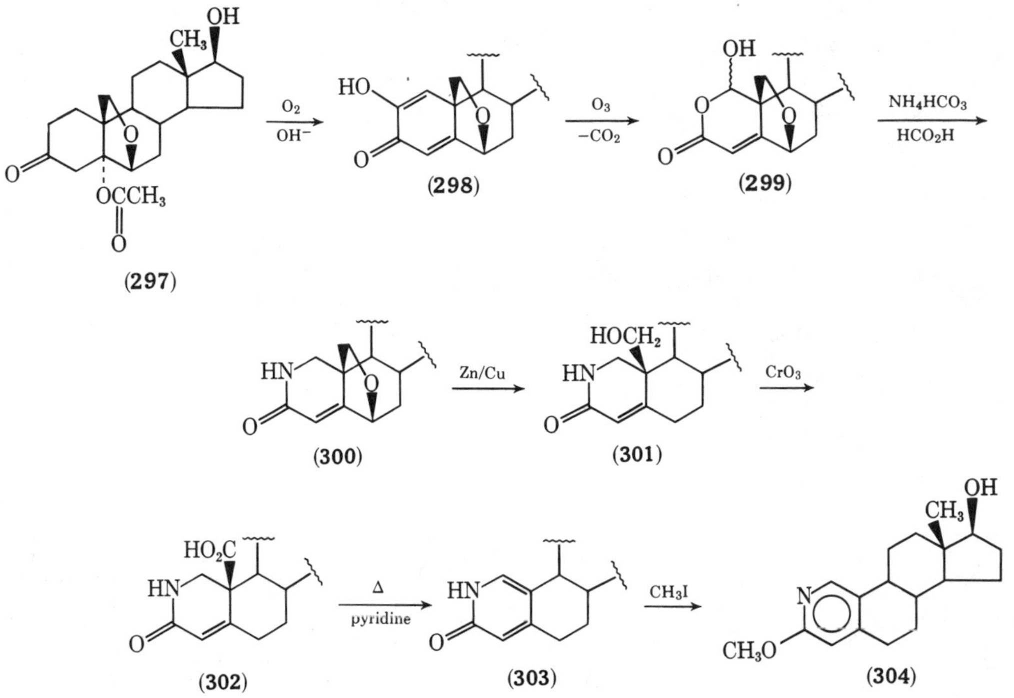

(297) (298) (299) (300) (301) (302) (303) (304)

Figure 31. The partial synthesis of 2-azaestradiol methyl ether (**304**).

reported (218). Testololactone (**306**) is obtained by microbial fermentation of either progesterone or testosterone and is used for treatment of human mammary cancer (219).

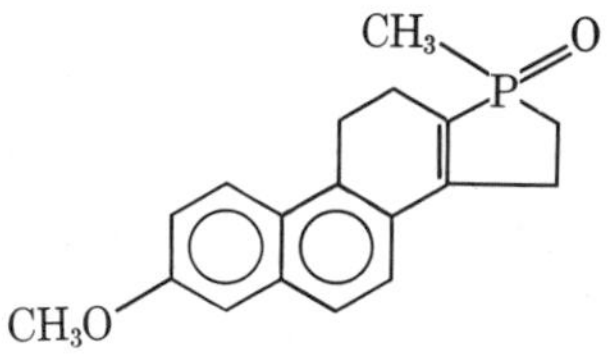

(305) 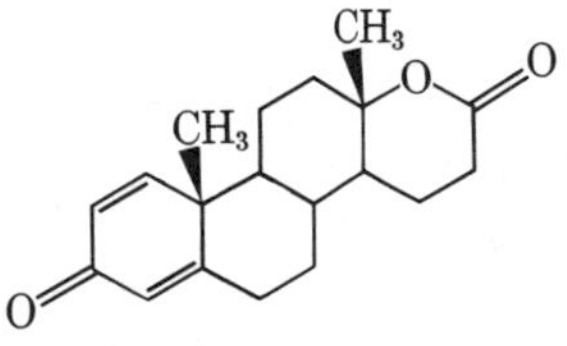 (306)

Total Synthesis. The same type of synthetic approaches used for construction of the carbocyclic skeleton are for the most part amenable to the formation of heterocyclic steroids. The α-tetralone used in the Torgov synthesis of A-ring aromatic steroids can be replaced by aza, oxa, and thio α-tetralones (220). The synthesis of a 6-aza-19-nor steroid by this method is illustrated in Figure 32. Vinylation of the dihydroquinolone (**307**) yields the aza analogue of the Torgov tetralone intermediate. Reaction with methylcyclopentandione gives the intermediate (**308**), which undergoes cyclization and ketalization to compound (**309**) upon heating in ethylene glycol with acid. Subsequent Birch reduction and hydrolysis of the enol ether with concomitant deketalization yields a dione, which is reduced with hydride to N-methyl 6-aza-

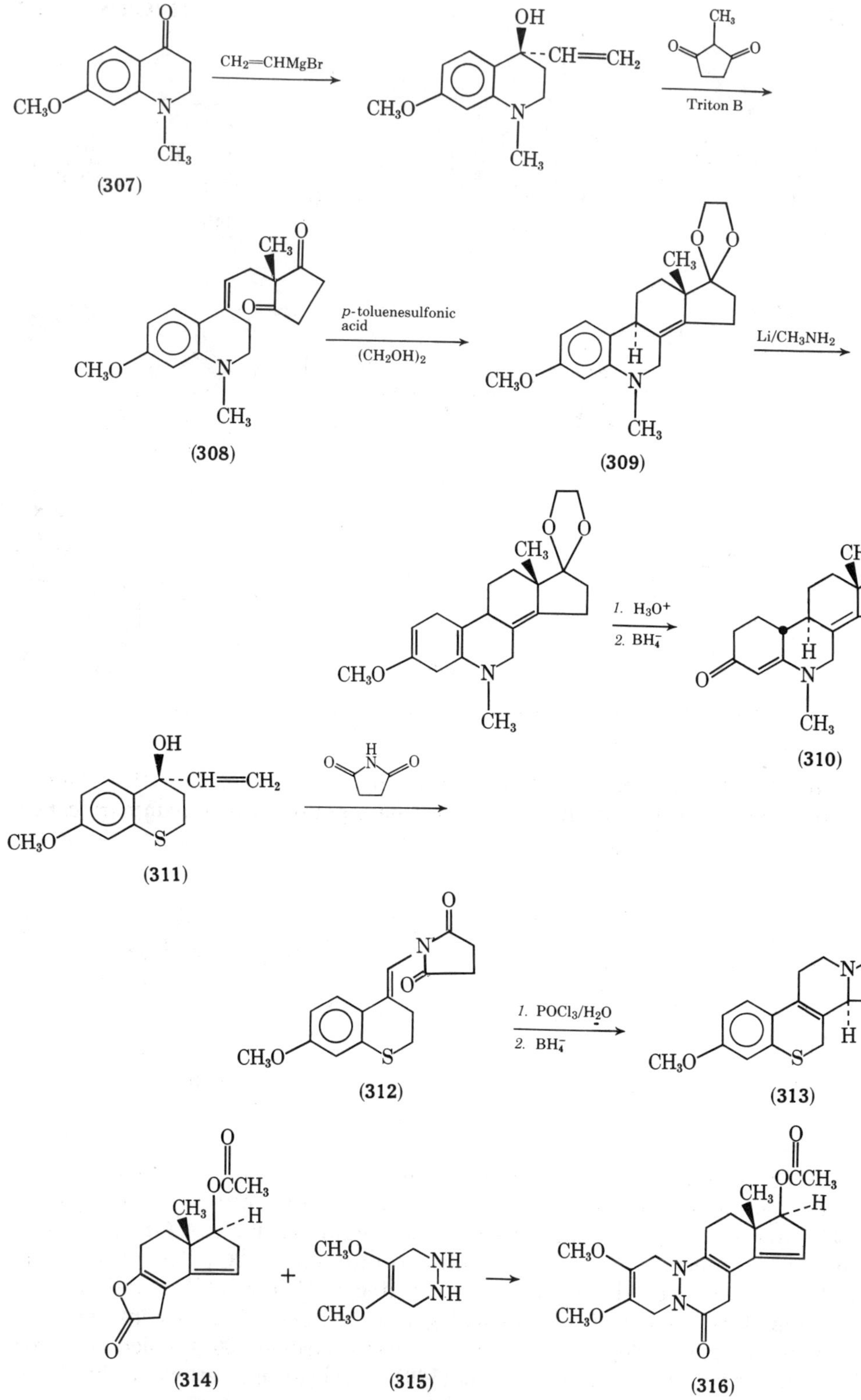

$CH_2{=}CHMgBr$

(307)

Triton B

p-toluenesulfonic acid

$(CH_2OH)_2$

(308)

(309)

Li/CH_3NH_2

1. H_3O^+

2. BH_4^-

(310)

(311)

(312)

1. $POCl_3/H_2O$

2. BH_4^-

(313)

(314) + (315) → (316)

Figure 32. The total synthesis of a heterocyclic steroid based on the Torgov reaction.

8(14)-dehydro-19-nortestosterone (**310**) (221). The alkylcyclopentanedione component in the Torgov reaction is also subject to replacement with a heterocycle, the most commonly used being succinimide. In order to accommodate the altered chemical characteristics of the components, the reaction is run in molten succinimide containing some potassium succinimide. For example, reaction of the thiatetralone (**311**) in a succinimide melt gives the intermediate (**312**). Cyclodehydration of the secosteroid (**312**) is carried out with a mixture of phosphorus oxychloride and water; the latter serves as a proton source for the acid-catalyzed isomerization of the 9(11)-double bond to the 8(9)-position prior to cyclization. The unstable acyl imminium salt is directly reduced to the 6-thia-13-azaestrone 3-methyl ether derivative (**313**) (222). Vicinal diazasteroids (**316**) are available by condensation of the appropriate precursor (**314**) with a dehydropyrazine (**315**) (223).

A Diels-Alder approach to the formation of a 13-azaestrone derivative involves the diene (**317**) and the acyl imine dienophile in a catalyzed reaction to form the tricycle, which upon urethane hydrolysis yields the 13-azaestrone derivative (**318**) (224).

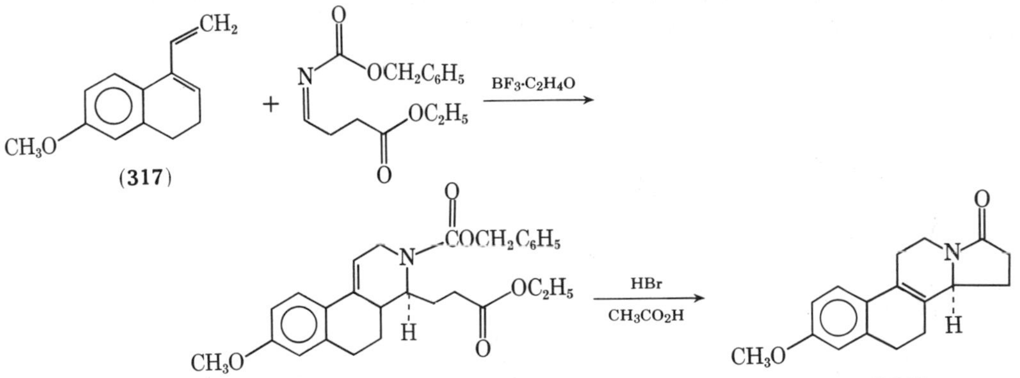

CH_2 $OCH_2C_6H_5$ $BF_3{\cdot}C_2H_4O$ OC_2H_5 CH_3O (**317**) $COCH_2C_6H_5$ OC_2H_5 HBr CH_3CO_2H H CH_3O (**318**)

Cyclopropyl Steroids. Cyclopropyl steroids were for many years a laboratory curiosity and were chiefly used for protecting the 3β-hydroxy-5-ene grouping during reactions involving other portions of the molecule (225). Solvolysis of the 3β-tosylate in methanol yields the methoxycyclopropane, termed an iso-steroid (226).

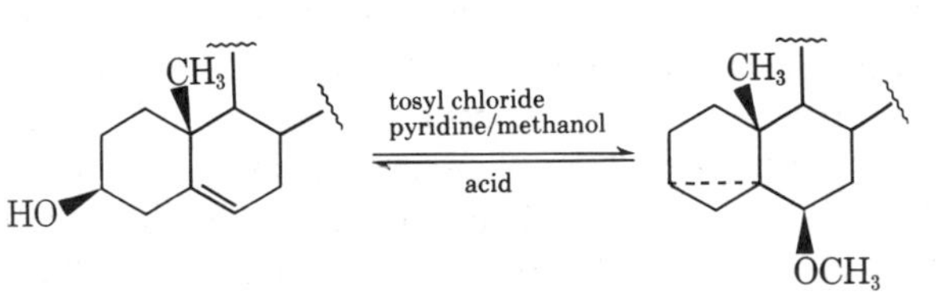

CH_3 HO tosyl chloride pyridine/methanol acid CH_3 OCH_3

Subsequently, a novel synthesis of a cyclopropane ring system involves the photochemical cyclization of the conjugated diene system in compound (**319**) into a bicyclo [1.1.0] bridged ring system, which contains two cyclopropanes and solvolysis of one of the cyclopropane rings to give the ether (**320**) (227).

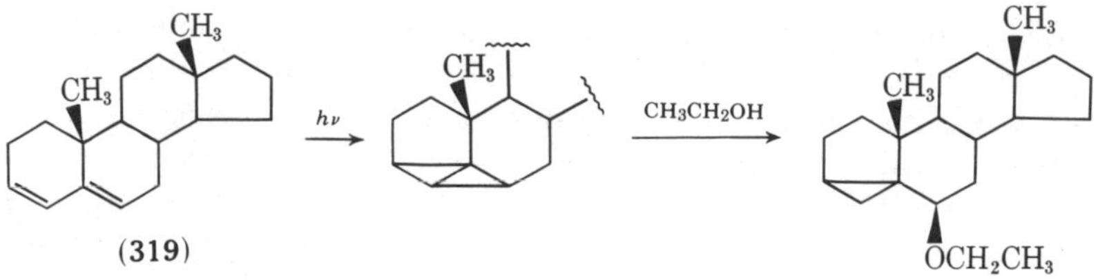

CH3 CH3 (319) $h\nu$ CH3 CH_3CH_2OH CH3 CH3 OCH2CH3 (320)

This reaction is restricted to the unsubstituted diene; if the dienol ether is used, the reaction takes another course. However, a class of cyclopropyl steroids, which have gained importance because of their biological effects, has a cyclopropane ring fused to one of the cyclopentanophenanthrene rings. The 2α,3α-cyclopropane androstane derivative (**321**), where the cyclopropane ring has effectively replaced the 3-keto group, is a potent androgen (228). Initially, the preparation of cyclopropyl-conjugated ketones was accomplished by the addition of diazomethane to an enone double bond to form a pyrazoline (**322**), which eliminates nitrogen either thermally or photochemically to generate the three-membered ring (**323**) (229).

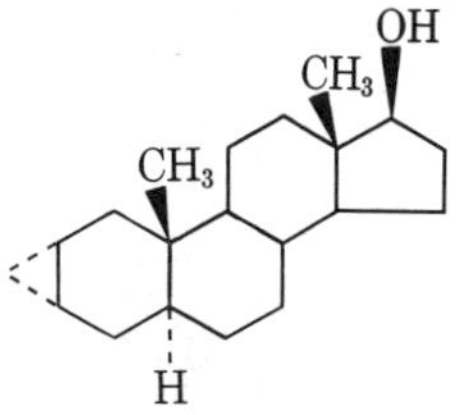

OH CH3 CH3 H (321) CH3 O CH_2N_2 N N CH3 O (322) Δ or $h\nu$ $-N_2$ CH3 O (323)

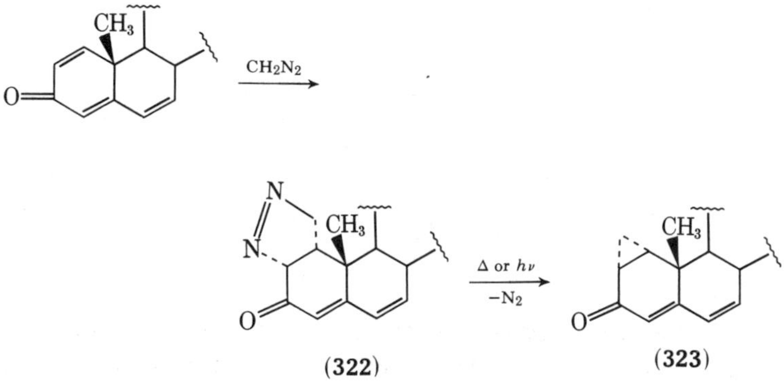

A safer, more general way of introducing the 1α,2α-cyclopropane is the conjugate addition of dimethyloxosulfonium methylide to a 3-oxo-1-ene steroid (229). Intensive investigations resulted in the synthesis of the progesterone derivative (**324**), cyproterone acetate, which is an exceedingly potent progestin and antiandrogen and which suppresses libido (230). It has been suggested as a chemical method for control of sexual offenders (231). The synthesis of (**324**) is outlined in Figure 33. The 17α-hydroxyprogesterone (**325**) is converted into the 1α,2α-cyclopropane (**326**) by the methylide reagent. After acetylation of the 17α-hydroxyl group, the 6α,7α-epoxide is introduced (**327**). Then hydrogen chloride opens the epoxide to the chlorohydrin which is dehydrated to introduce the 6-chloro substituent while the cyclopropane ring is simultaneously opened to yield (**328**). Mild base then closes the cyclopropane ring to form cyproterone acetate (**324**).

To introduce a β-oriented cyclopropane in the 1,2-position, the Simmons-Smith reaction, which involves an electrophilic organozinc reagent, is used (232). Reduction of the 3-oxo-1-ene system with hydride gives the 3β-hydroxy-1-ene (**329**), which upon treatment with zinc and methylene iodide gives the 1β,2β-cyclopropane (**330**). Oxi-

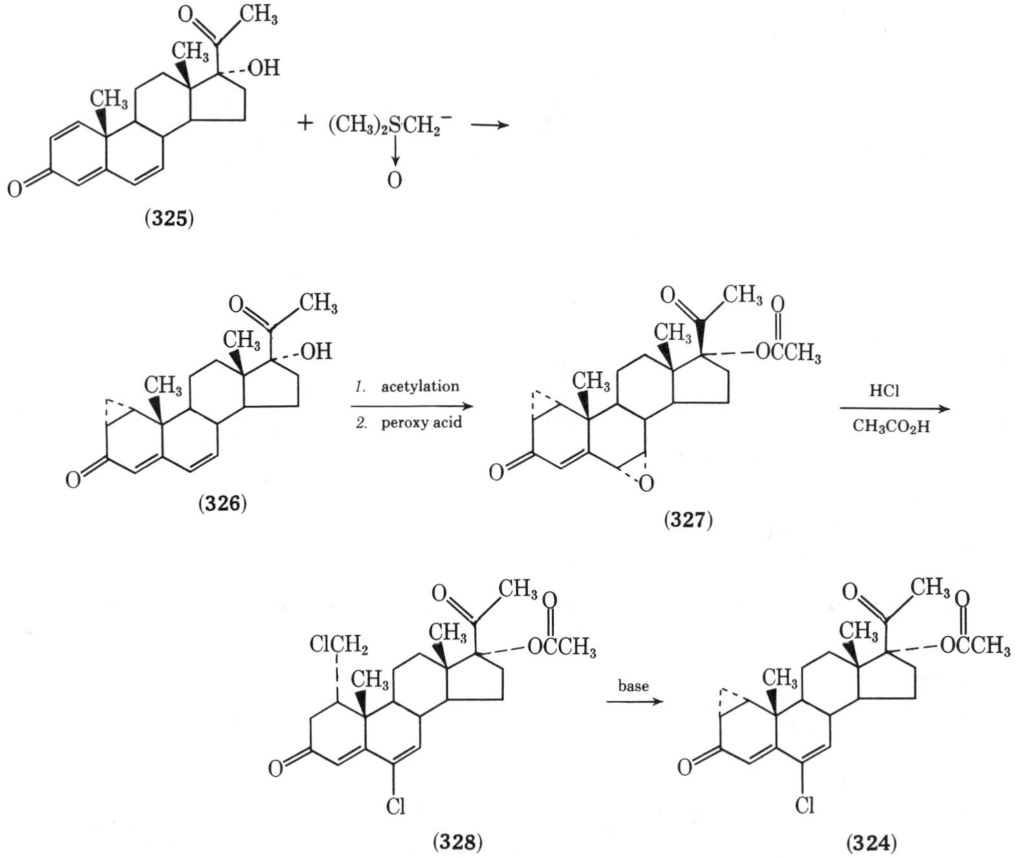

(**325**) + $(CH_3)_2S(\rightarrow O)CH_2^-$ →

(**326**) 1. acetylation 2. peroxy acid → (**327**) HCl, CH_3CO_2H →

(**328**) base → (**324**)

Figure 33. Synthesis of the potent antiandrogenic steroid, cyproterone acetate (**324**). This synthesis illustrates the introduction, opening, and reclosure of the cyclopropane ring, which is important for the enhancement of biological activity.

dation of the hydroxyl group then forms the cyclopropylenone (**331**) (229). This same reaction can stereospecifically form 4,5-methano derivatives (233). Cyclopropanation of the 3β-hydroxy-4-ene cholestene (**332**) yields exclusively the 4β,5β-cyclopropane (**333**), and the 3α-epimer (**334**) stereospecifically forms the 3α-hydroxy-4α,5α-cyclopropane (**335**) (see Fig. 34).

The introduction of 6,7-cyclopropane into a 3-oxo-4,6-diene (**336**) is accomplished via the methylide reagent. However, in this case, a mixture of 6α,7α- (**337**) and 6β,7β- (**338**) epimers are formed with the beta isomer predominating (Fig. 35) (234–235). The ratio of epimers is a function of reaction conditions and substituents remote from the reaction site (236). The 6β,7β-cyclopropylprogesterone derivative (**339**) and the

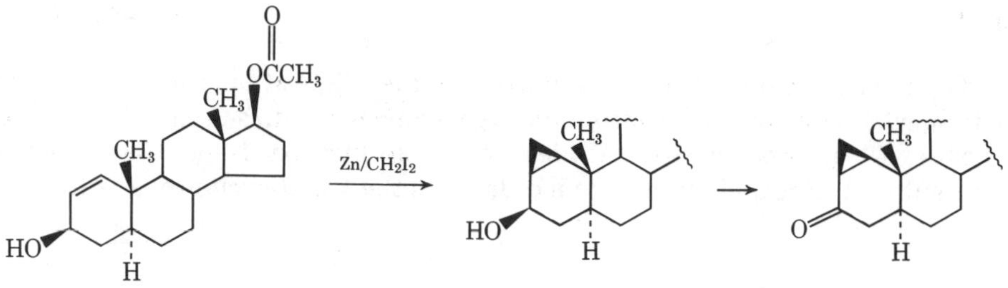

O
OCCH3
CH3
CH3
Zn/CH2I2
CH3
CH3
HO
H
HO
H
O
H

(329) (330) (331)

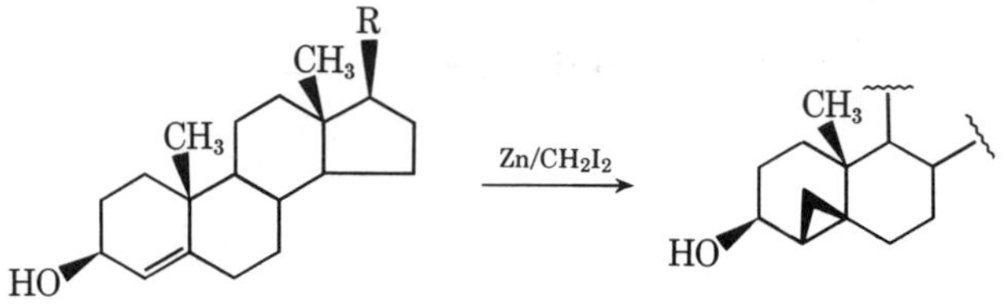

R
CH3
CH3
Zn/CH2I2
CH3
HO
HO

(332) R = cholesteryl side chain (333)

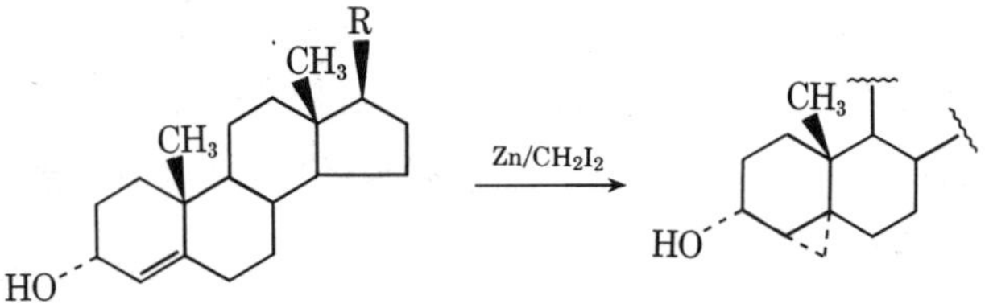

R
CH3
CH3
Zn/CH2I2
CH3
HO
HO

(334) R = cholesteryl side chain (335)

Figure 34. The Simmons-Smith method for the introduction of cyclopropane rings.

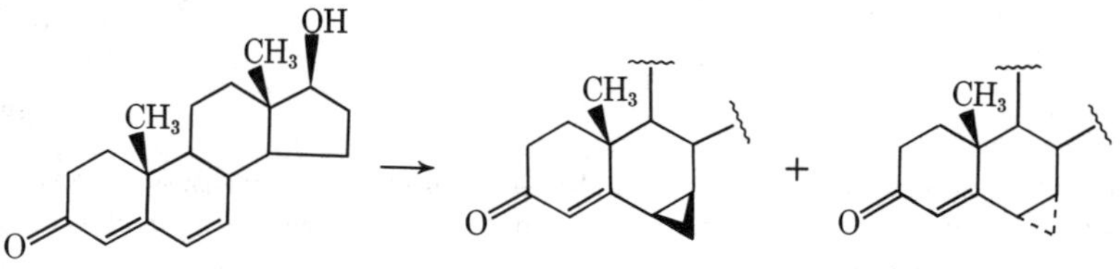

OH
CH3
CH3
CH3
CH3
+
O
O
O

(336) (338) major (337) minor

O
CH3
CH3
CH3
O
O
O
CH3
CH3
O
O

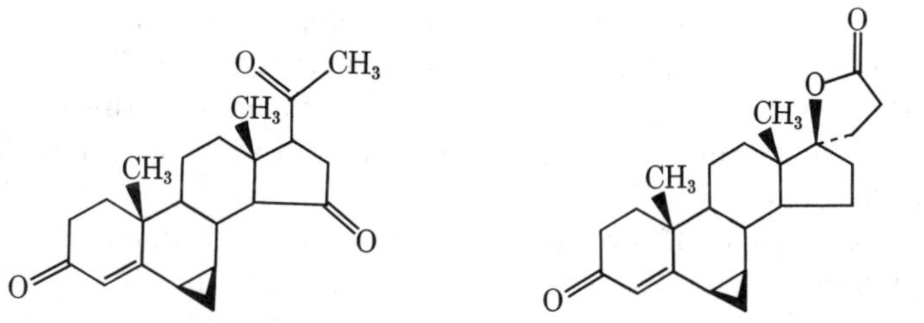

(339) (340)

Figure 35. 1,6-Addition of dimethyloxosulfonium methylide generates γ,Δ-cyclopropylenones.

$6\beta,7\beta$-cyclopropyl-spironolactone (**340**) have been prepared in this way and are potent antimineralocorticoids (236–238).

The 15,16-methylene steroids have also been prepared and their biological properties investigated. The $15\beta,16\beta$-epimer (**341**) is readily accessible from the 17β-hydroxy-15-ene (**341**) by the Simmons-Smith reaction (239). However, $15\alpha,16\alpha$-methylenetestosterone (**343**), which is orally active and has good androgenic activity, is synthesized from the 17α-hydroxyprogesterone derivative with subsequent side-chain cleavage and reduction to produce the 17β-hydroxyl group (Fig. 36) (240). The $16\alpha,17\alpha$-cyclopropanes have been prepared by the action of diethylzinc and methylene iodide on the enol ethers of several 17-keto steroids (241). A novel route to A-nor-5(10)-cyclopropyl steroids (**345**) is through the irradiation of the 3-oxo-5(10)-chromophore (**344**) (242).

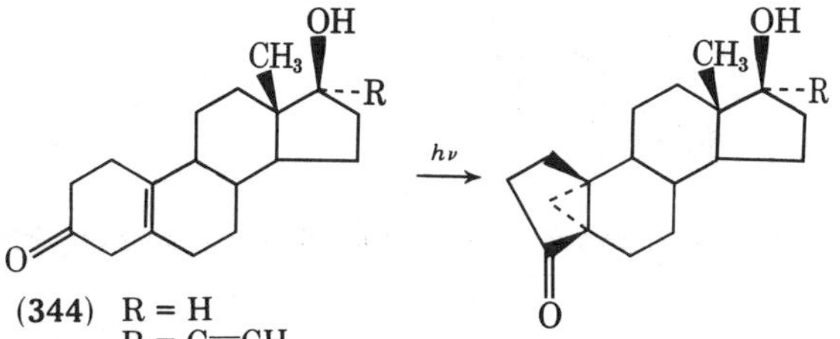

(**344**) R = H
R = C≡CH

(**345**)

The enhancement of biological activity engendered by the introduction of a cyclopropyl ring led to the synthesis of various carbocyclic-annelated steroids. Cyclobutane rings are readily formed by the photocycloaddition of enones to olefins (243). Addition of the 3-oxo-4-ene steroid (**346**) to ethylene gives a mixture of the *cis*-$4\alpha,5\alpha$-cyclobutane (**347**) and the very strained *trans*-$4\alpha,5\beta$-cyclobutane (**348**), which can be isomerized with base to the more stable *cis*-$4\beta,5\beta$-compound (**349**) (244). Similar results are observed with 3-oxo-1-ene steroids (243–245). The hormonal activities of these compounds are, however, substantially less than their unsaturated counterparts (246). The 6,7-cyclobutanes (**350**) are synthesized from the 3-oxo-4,6-diene maleic anhydride product, whereas ordinary olefins furnish only the *trans*-$4\alpha,5\beta$-cyclobutanes (242–248). Cyclohexene adducts are prepared by the photocycloaddition of 6-dehydrotestosterone acetate (**351**) to butadiene. The main product is the trans-$4\alpha,5\beta$-adduct (**352**), which can be epimerized to the more stable cis-$4\beta,5\beta$-compound (**353**) with lesser amounts of cis-adducts (**354**) and (**355**). These cyclohexene derivatives are accompanied by significant amounts of compound (**356**), which contains a linear array of three cyclobutane rings (249).

Retrosteroids. Steroids having an inverted configuration at *C*-9 and *C*-10 are retrosteroids. These compounds were originally prepared through a series of reactions based on studies of vitamin D. Ergosterol (**10**) was converted photochemically into its 10α-epimer (**15**). Oxidation and selective reduction then yielded the enone (**357**), which was converted by standard methods, eg, (**18**) to (**19**), into retroprogesterone (**358**) (250). Conversion to the conjugated dienone (**359**) affords the potent, orally active progestin dydrogesterone. In contrast to progesterone, compound (**359**) is nonthermogenic. A wide variety of retrosteroids in both the sex hormone and corticoid series have been synthesized and evaluated biologically (251).

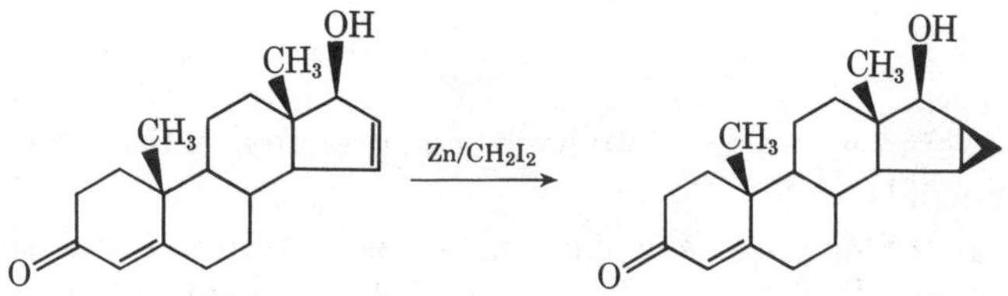

OH
CH_3
CH_3
O
(341)

$\xrightarrow{Zn/CH_2I_2}$

OH
CH_3
CH_3
O
(342)

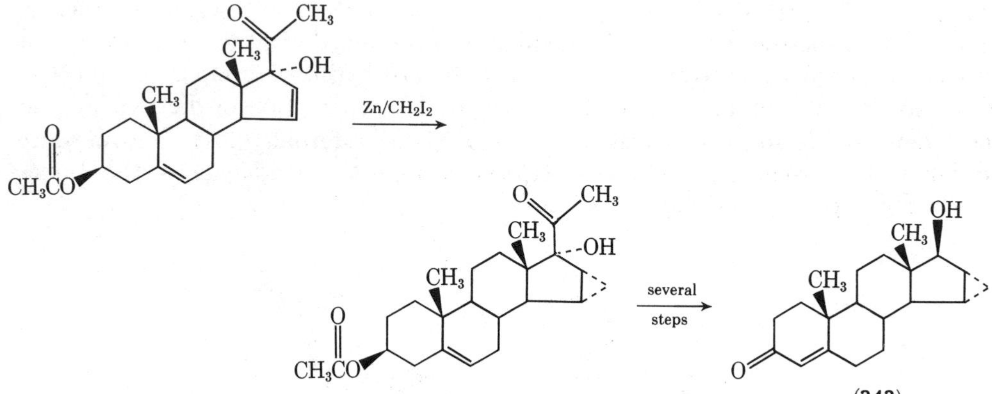

O CH_3
CH_3 --OH
CH_3
O
CH_3CO

$\xrightarrow{Zn/CH_2I_2}$

O CH_3
CH_3 --OH
CH_3
O
CH_3CO

$\xrightarrow{\text{several steps}}$

OH
CH_3
CH_3
O
(343)

Figure 36. The preparation of 15,16-methylene steroids.

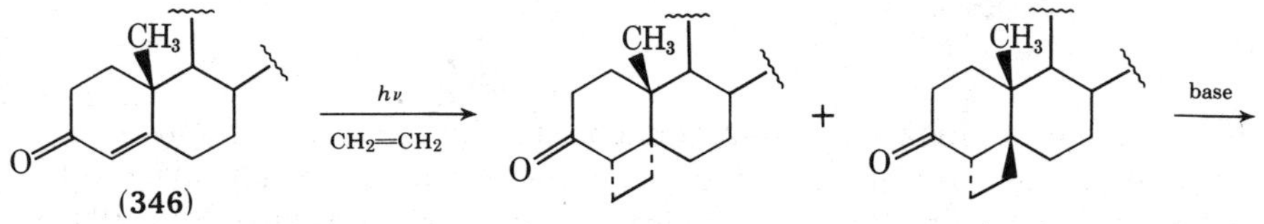

CH_3
O
(346)

$\xrightarrow[CH_2{=}CH_2]{h\nu}$

CH_3
O
(347)

+

CH_3
O
(348)

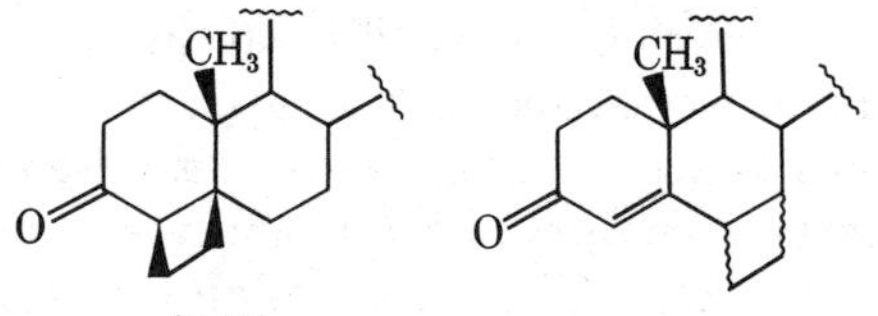

$\xrightarrow{\text{base}}$

CH_3
O
(349)

CH_3
O
(350)

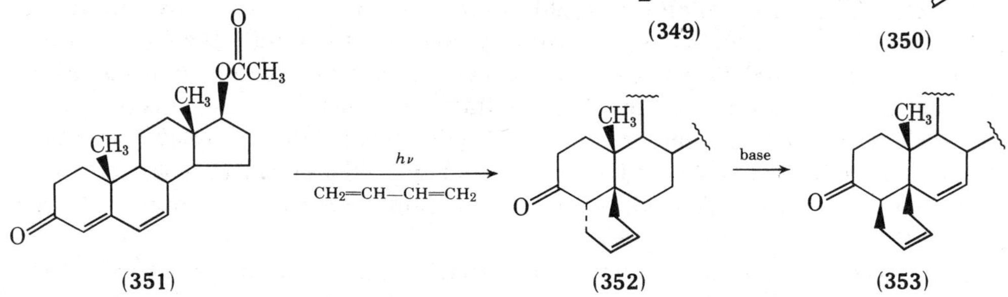

O
$OCCH_3$
CH_3
CH_3
O
(351)

$\xrightarrow[CH_2{=}CH{-}CH{=}CH_2]{h\nu}$

CH_3
O
(352)

$\xrightarrow{\text{base}}$

CH_3
O
(353)

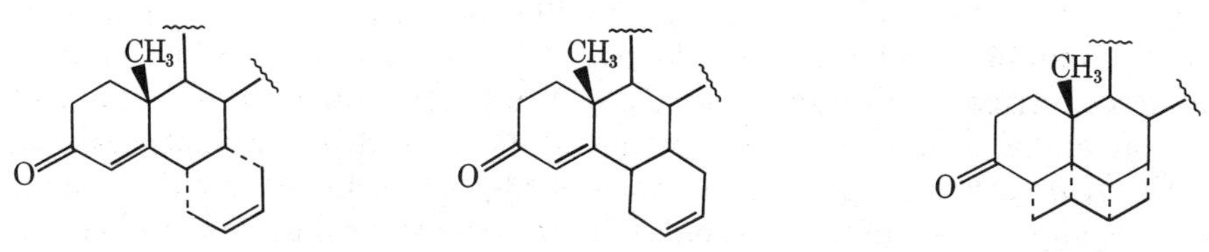

CH_3
O
(354)

CH_3
O
(355)

CH_3
O
(356)

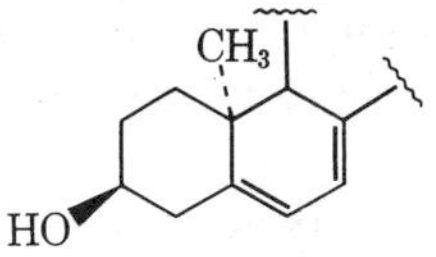

(15) 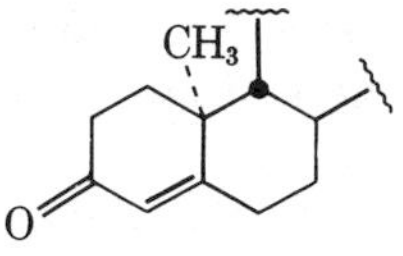 (357) 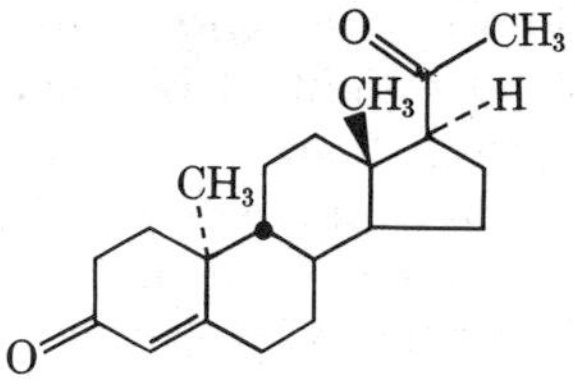 (358)

Δ^6

(359)

In order to avoid the synthetic limitations of the photochemical method, an alternative route to retrosteroids was developed, starting from the steroid fermentation product 11α-hydroxyprogesterone **(114)**. Cleavage of compound **(114)** gives the secosteroid **(360)** which undergoes base-catalyzed fragmentation, elimination, and conjugation to the enone **(361)**. After reduction of the *C*-20 ketone and enone double bond, methyl vinyl ketone annelation, and reoxidation of the *C*-20 alcohol, retroprogesterone **(358)** is formed (252). An alternative thermal method starts with the reduction of a 3-oxo-5(10)-ene steroid to the 3α-hydroxy compound **(362)**. Simmons-Smith methylenation followed by oxidation gives the cyclopropylenone **(363)**, which undergoes base-catalyzed ring opening to give 10α-testosterone **(364)**, in which only the *C*-10 methyl substituent has been epimerized (253).

(114) 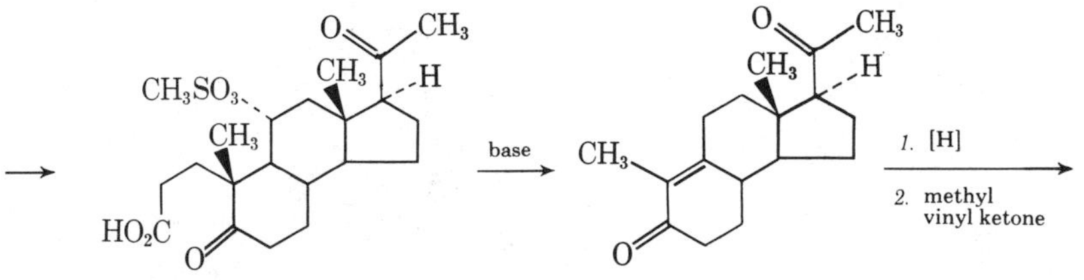 → (360) —base→ (361) —1. [H]; 2. methyl vinyl ketone→ (358)

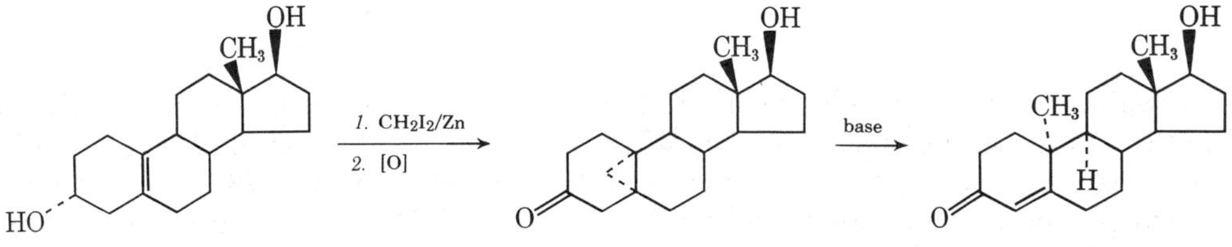

(362) —1. CH_2I_2/Zn; 2. [O]→ (363) —base→ (364)

Homo-, Nor-, and Secosteroids. Homo-, nor-, and secosteroids are steroids that have been modified by expansion or contraction or simultaneous expansion and contraction of one or more of the carbocyclic rings in the tetracyclic nucleus. The driving force for the preparation of these modified steroids has been the enhancement or separation of biological activities (254).

The ring expansion of steroids to their homoderivatives is accomplished by a variety of methods (254). Among the more important is the Tiffeneau Demjanov rearrangement, whereby a ketone is converted to a cyanohydrin, reduced to the amino alcohol, diazotized, and solvolyzed to the homoketone. The 3-oxocholestane derivative (**365**) is converted into its cyanohydrin (**366**) with hydrogen cyanide. After acetylation, reduction with lithium aluminum hydride yields the oxazolidine (**367**) in 68% yield. Diazotization and solvolysis in aqueous acetic acid gives *A*-homo-5α-cholestan-3-one (**368**) in 28% yield (Fig. 37) (255). Isomeric mixtures may result from these rearrangements. The proposed long-range effects of *C*-17-substituents, based upon the supposition that the two isomers are separable, are incorrect. Most of the data in the literature concerning these *A*-homo-3-ketones refer to mixtures (256). Because of problems associated with the reduction, alternative methods for the production of the amino alcohol have been developed. Hecogenin (**40**) is converted into

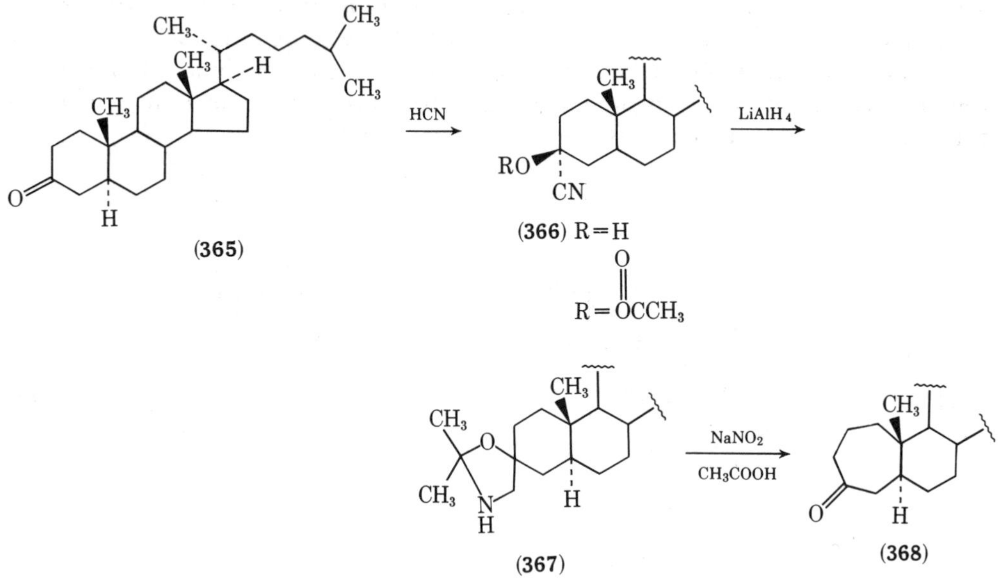

Figure 37. The synthesis of *A*-homosteroids.

the epimeric spiro epoxides (**369**), and nucleophilic ring opening with azide gives a mixture of azido alcohols (**370**) and (**371**), which are readily separated. Reduction of the β-isomer (**370**) with Raney nickel yields the amino alcohol (**372**), which upon diazotization and rearrangement yields the ketone (**373**), which is deoxygenated to C-homohecogenin (**374**) (257). Conversion of compound (**374**) yields C-homodihy-

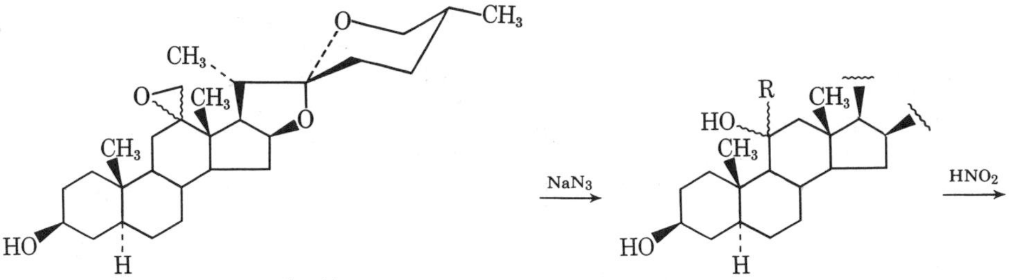

(369) $\xrightarrow{NaN_3}$ (370) $R = \beta\text{-}CH_2N_3$; (371) $R = \alpha\text{-}CH_2N_3$; (372) $R = \beta\text{-}CH_2NH_2$ $\xrightarrow{HNO_2}$

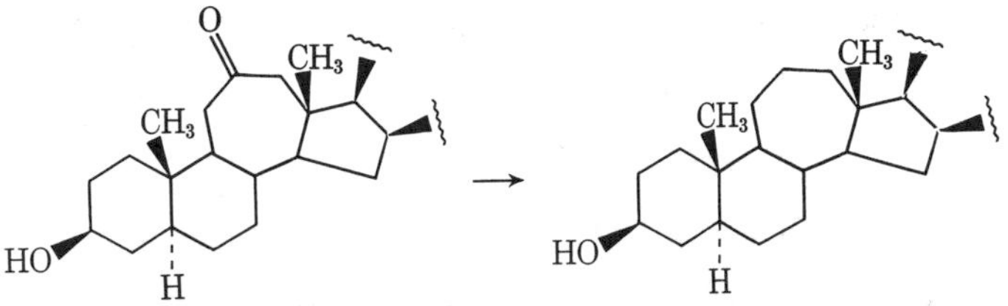

(373) → (374)

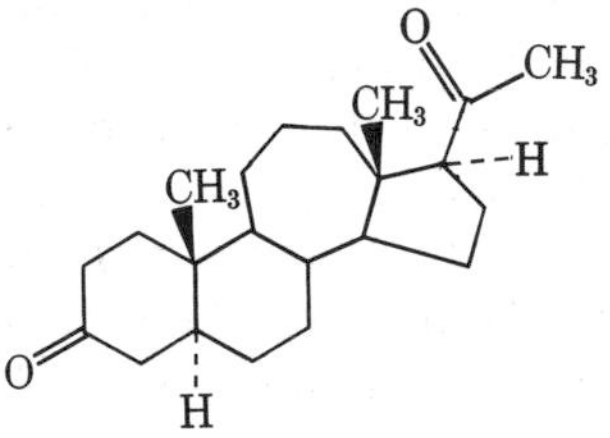

(375) 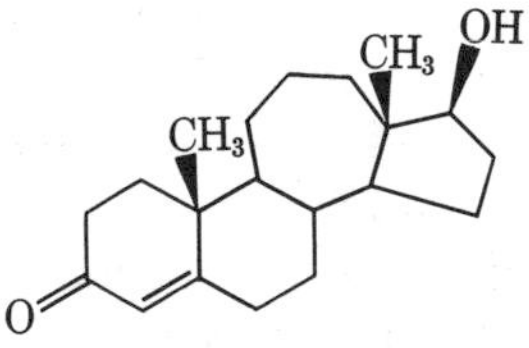 (376) 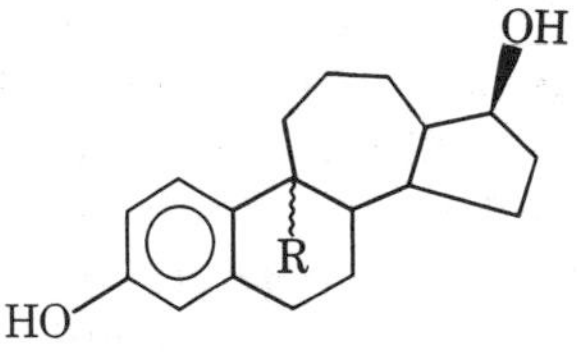 (377) $R = \alpha\text{-H}$; (378) $R = \beta\text{-H}$

droprogesterone (**375**), C-homotestosterone (**376**), and C-homoestradiol (**377**). The estradiol derivative (**377**) is a strong, orally active, estrogen, whereas the others are pharmacologically inactive. Surprisingly, the 9β-C-homoestradiol derivative (**378**), prepared by total synthesis, also has weak estrogenic activity (258).

Ring expansion to homosteroids can also be accomplished with diazomethane on saturated ketones. A valuable extension of this reaction is the homologation of α,β-unsaturated ketones with diazomethane in the presence of boron trifluoride (259). The main product is formed by the insertion of a methylene group between the carbonyl group and the unsaturated α-carbon to give a β,γ-unsaturated ketone. Cholestenone (**121**), which does not react with diazomethane under normal conditions, reacts rapidly with nitrogen evolution in the presence of a small amount of boron trifluoride to yield the *A*-homounsaturated ketone (**379**) in yields of up to 40%.

(121) $\xrightarrow[\text{BF}_3 \; 40\%]{\text{CH}_2\text{N}_2}$ (379)

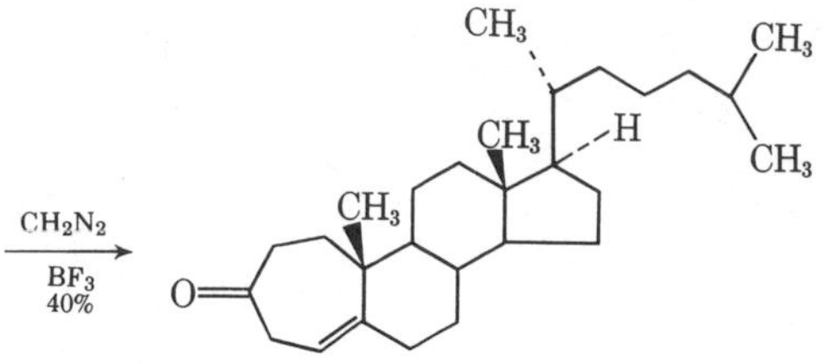

(138) $\xrightarrow{\text{Br}_2\text{C:}}$ $\xrightarrow[\text{2. H}_3\text{O}^+]{\text{1. AgClO}_4}$ (380)

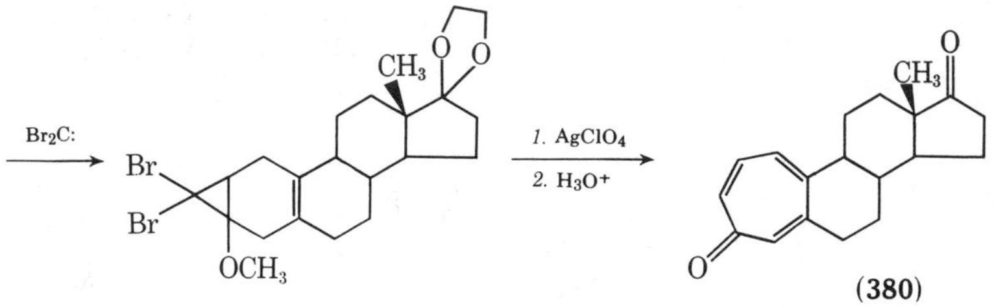

A series of A-ring aromatic tropolones have been prepared by dibromocarbene addition to the Birch reduction product (**138**) to give the cyclopropyl ether. Solvolysis and hydrogen bromide elimination gives the *A*-homosteroid (**380**) (260). D-Homosteroids can be conveniently prepared by the benzylic acid rearrangement of 17-hydroxy-20-ketones (255). The stereochemistry of the resultant D-homohydroxy ketones can be controlled by the choice of reagents (261). For example, chromatography of 17β-hydroxy-20-ketopregnane (**381**) on alumina affords the rearrangement product (**382**) in quantitative yield, whereas methanolic potassium hydroxide gives the isomeric ketol (**383**) in nearly quantitative yield. A novel annulene (**387**) has been prepared from the 19-hydroxy compound (**384**) and (2-chloro-1,1,2-trifluoroethyl)diethylamine to give the 5β,19-cyclodienone (**385**). The cyclopropane ring is readily opened by acetic acid, and subsequent dehydrogenation yields 1,6-methano-[10]annulene (**386**) (262).

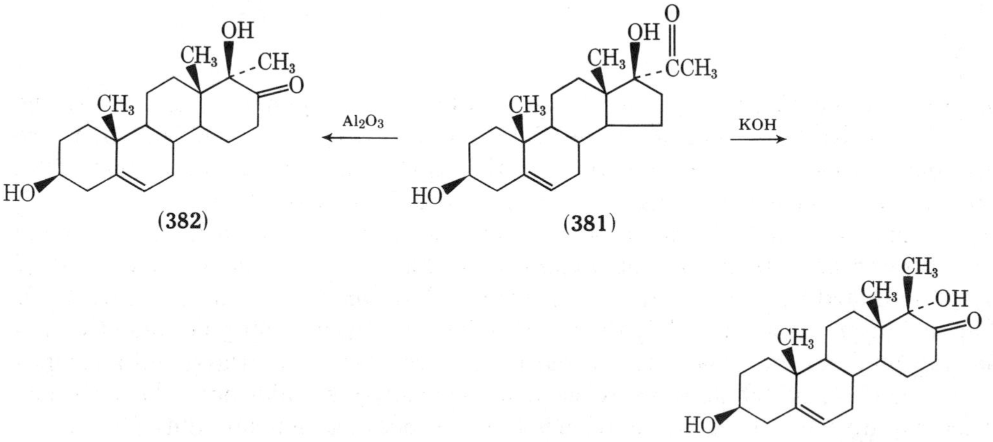

(382) $\xleftarrow{\text{Al}_2\text{O}_3}$ (381) $\xrightarrow{\text{KOH}}$ (383)

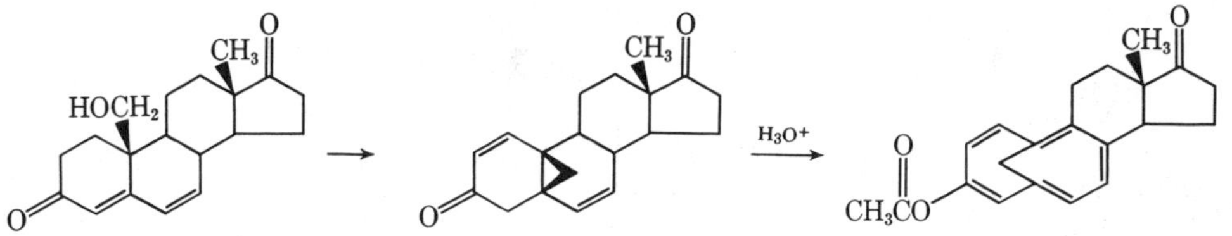

(384) (385) (386)

Formation of the A-norsteroids can be effected by a retropinacol reaction of triterpenoid 4,4-dimethyl-3β-ols (263). 3β-Hydroxy-4,4-dimethyl steroids are usually rearranged with phosphorus pentachloride in inert solvents. Under these conditions, 4,4-dimethylcholestanol (**387**) yields the isopropylidene-A-norsteroid (**388**) in 85% yield. Cleavage of the isopropylidene group by ozone affords the *A*-norketone, isolated as the A/B-trans (5α) isomer (**389**), which is readily equilibrated to a mixture in which the more stable A/B-cis (5β) isomer (**390**) predominates. The ready availability of 2,2- and 4,4-dimethyl-3β-hydroxysteroids enhances the utility of the retropinacol rearrangement as a method for preparing *A*-norsteroids. Biologically, the most interesting *A*-norsteroids are the androstane Anordin (**394**) and the estrane Dinordin (**395**) (264–265). Anordin (**394**) inhibits implantation and is used as an oral contraceptive in China, where it is the so-called vacation pill. Its synthesis starts with chromium trioxide oxidation of the androstane derivative (**391**) to the secosteroid (**392**) which is converted to the *A*-norsteroid (**393**) with sodium acetate in acetic anhydride. Ethynylation with acetylene and potassium hydroxide proceeds stereospecifically at *C*-17 and stereoselectivity at *C*-2 giving predominantly the 2α-epimer (**394**) in a ratio of 95:5 (266). Dinordin (**395**) has been prepared similarly, as well as by total synthesis (265,267).

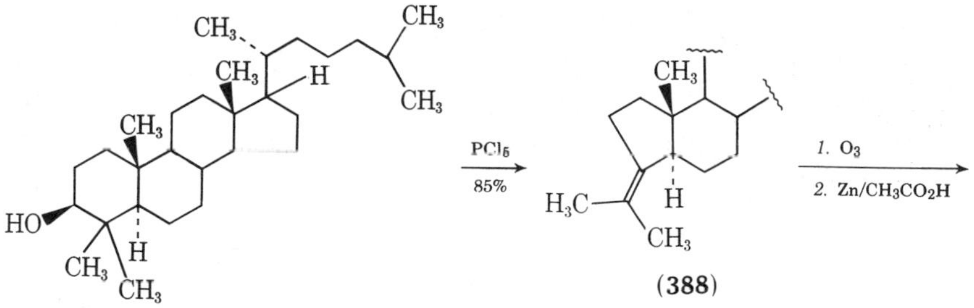

(387) (388) 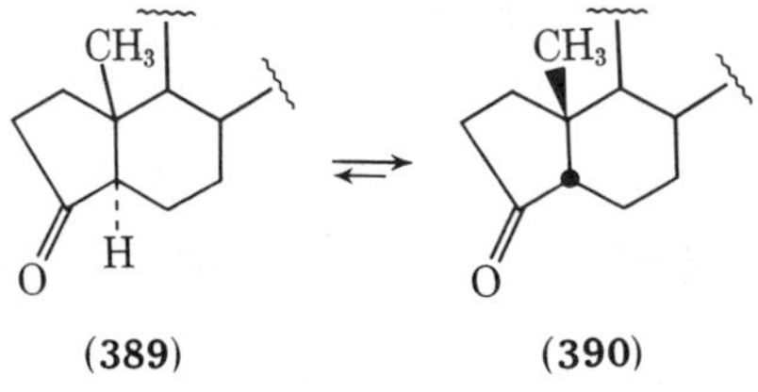 (389) (390)

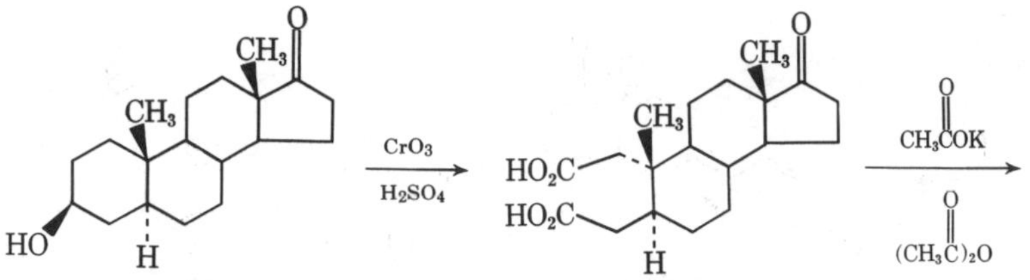

(391) $\xrightarrow[H_2SO_4]{CrO_3}$ (392) $\xrightarrow[(CH_3\overset{O}{\overset{\|}{C}})_2O]{CH_3\overset{O}{\overset{\|}{C}}OK}$

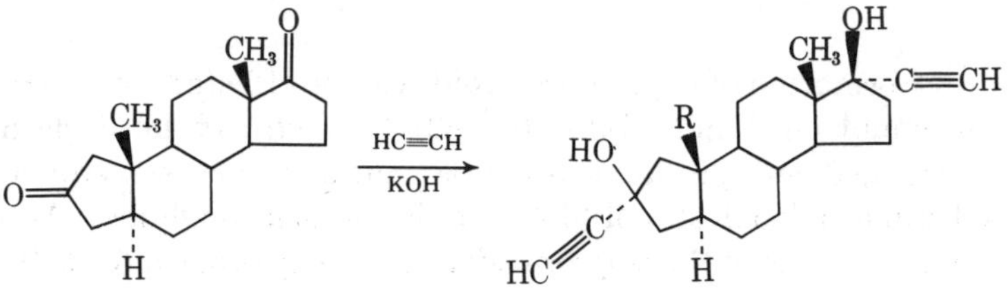

(393) $\xrightarrow[KOH]{HC{\equiv}CH}$ (394) R = H
(395) R = CH_3

B-Norsteroids are generally prepared by chromium trioxide oxidation of 3β-acetoxy-5-ene steroids (268). The steroid (**109**) is oxidized to the enol lactone (**396**), which after hydrolysis to the keto acid undergoes base-catalyzed cyclization to the lactone (**397**), which decarboxylates to give the *B*-norsteroid (**398**).

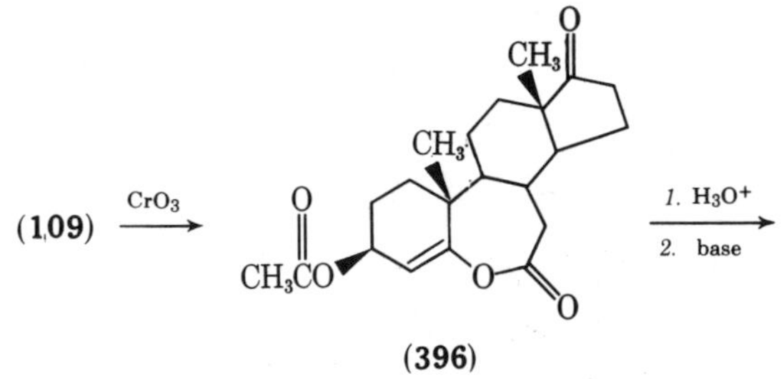

(**109**) $\xrightarrow{CrO_3}$ (**396**) $\xrightarrow[2.\ base]{1.\ H_3O^+}$

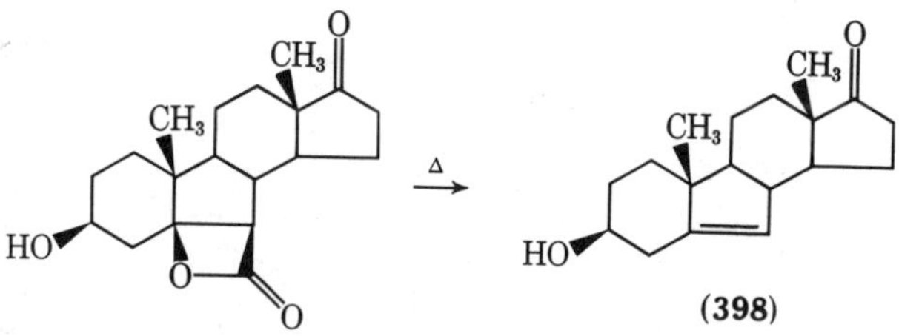

(**397**) $\xrightarrow{\Delta}$ (**398**)

Although the B-nor derivatives of most biologically interesting steroids have been synthesized, their biological activity has been for the most part disappointing. Similarly D-norsteroids are not biologically significant, but were prepared by photochemical ring contraction of a diazoketone (263).

Simultaneous ring expansion and contraction reactions have been extensively used in the preparation of modified steroids (269). A general method for the preparation of *A*-nor-*B*-homosteroids is the photolysis of α,β-epoxy-ketones (269). The reaction normally occurs by initial fission of the $C\alpha$—O bond of the epoxide followed by a 1,2-alkyl or hydrogen shift. The stereochemical course in these reactions is specific; the configuration at *C*-10 is retained and that at *C*-5 is inverted. An alternative method is the fragmentation of the bromoketone (**399**) to give the 6-oxo-5(10)-ene. After the tosylhydrazine is prepared, peracid treatment yields the acetylenic compound (**400**). Compound (**400**) is transformed into the triketone by catalytic hydrogenation and oxidation; subsequent intramolecular condensation with base gives the *A*-nor-*B*-homoestrane (**401**), which has enhanced androgenic and anabolic properties (270). Easier access to these compounds is through ozonization of a 5(10)-double bond followed by intramolecular cyclization (271).

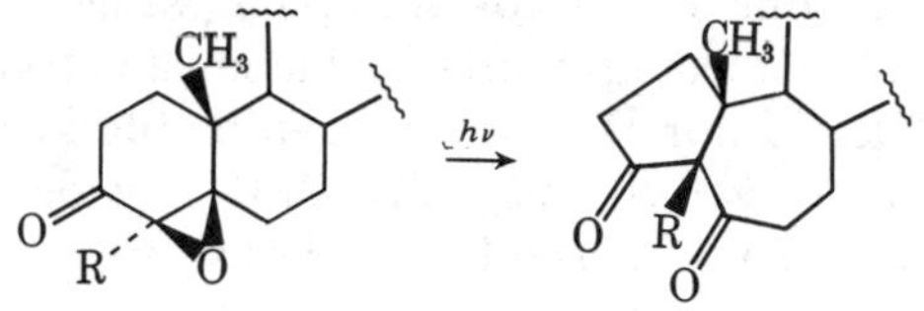

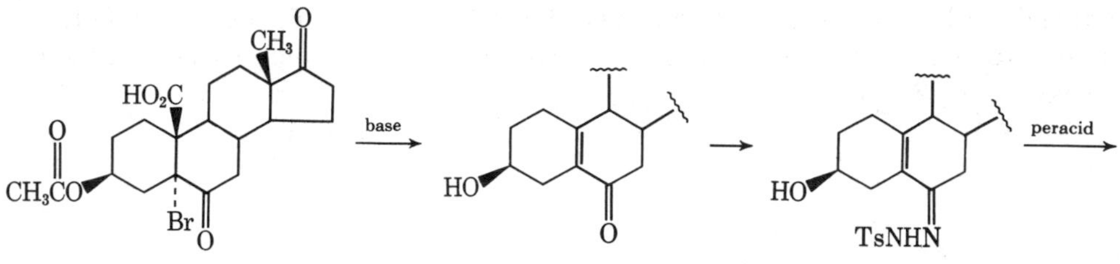

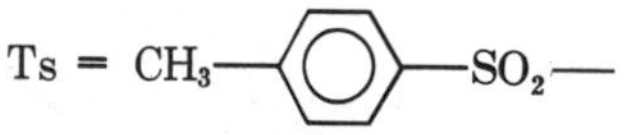

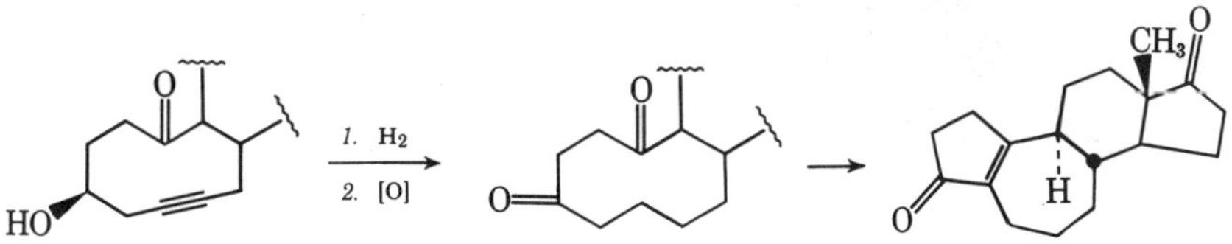

The alternative *A*-homo-*B*-norsteroids (**403**) have been prepared by pinacol rearrangement of the vicinal glycol monotosylate (**402**). The equilibrium mixture of compound (**403**) consists of four parts of the AB-cis steroid to one part of the AB-trans steroid (272).

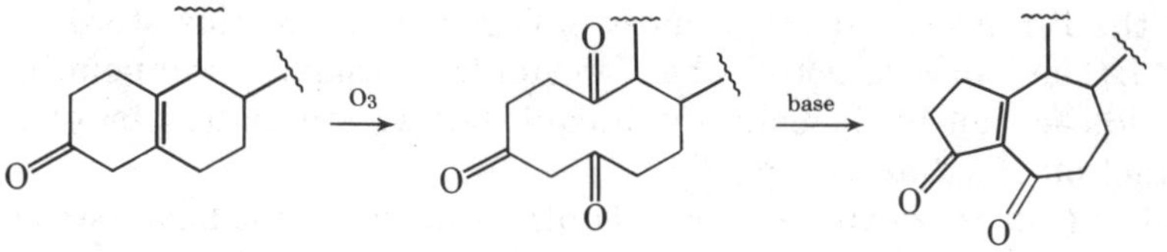

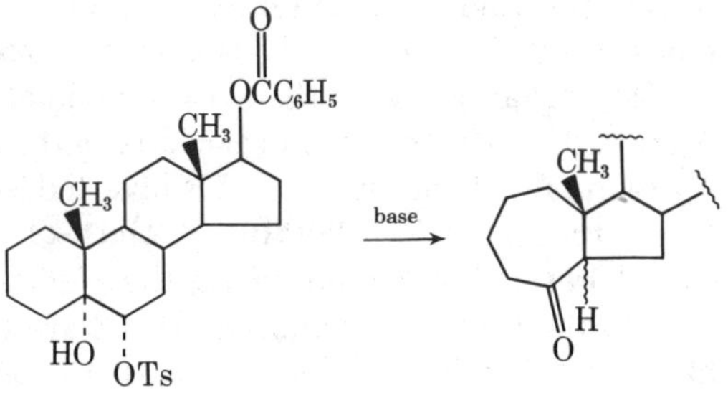

(402) (403)

Most C-ring contractions of steroids are accompanied by simultaneous D-ring expansion. Hecogenin (**40**), for example, after reduction and conversion to its mesylate, undergoes solvolysis to the C-nor-D-homosapogenin (**404**), which is degraded to the C-nor-D-homopregnene derivative (**405**) (273–274). A wide variety of etiojervanes has been prepared by partial as well as total synthesis (275). Among these, compound (**406**) has shown potent antimineralocorticoid activity (276). C-Nor-D-homosteroids produced by hecogenin degradation have the CD-cis configuration; the CD-trans isomers are synthesized by degradation of the steroidal alkaloid jervine and its derivatives (277) (see Alkaloids).

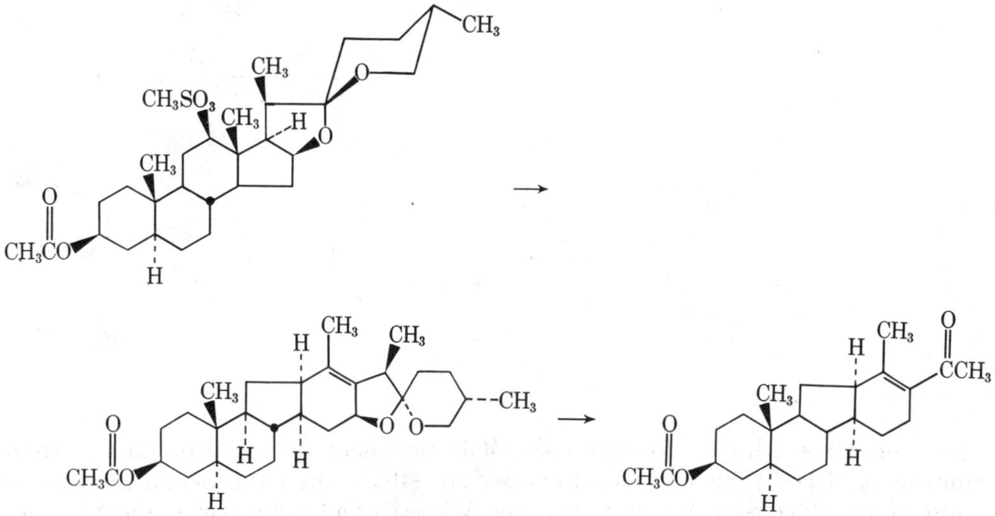

(404) (405)

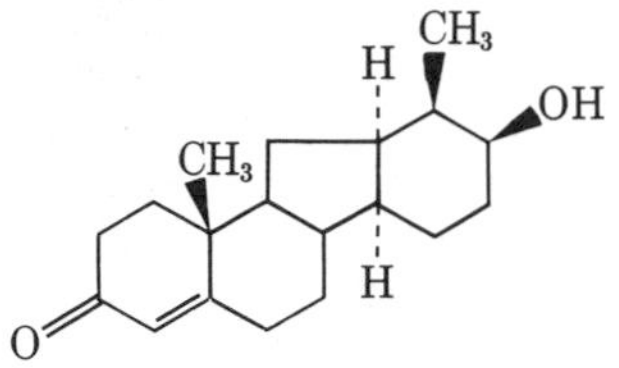

(406)

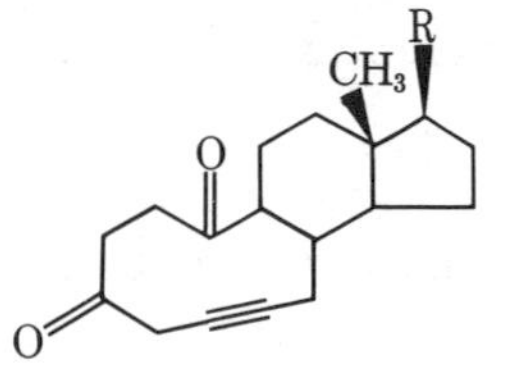

(407) R = (=O)
(408) R = β-$COCH_3$

Many secosteroids have been prepared as intermediates in synthetic sequences. However, secoandrostane (**407**) and secopregnane (**408**) have been prepared as specific irreversible inhibitors of the enzyme 3-oxo-5-ene steroid isomerase (278).

Analytical Methods

Ultraviolet and visible spectroscopy allows the detection and identification of chromophores within a molecule. Compilations of virtually all steroid chromophores are available in refs. 279–280. Physical constants for a large number of steroids are given in ref. 28.

Optical rotatory dispersion and the complementary circular dichroism have been (**295**). Reduction with borohydride yields oxandrolone (**296**) in 80–90% yield. A variety extremely useful in solving stereochemical and configurational problems with steroids containing a chromophore or in those steroids where a chromophore can be attached to a nonchromophoric functional group (282). An empirical analysis of the circular dichroism of ketones has allowed the estimation of the numerical contributions of the rings and substituents to the intensity of the absorption (283).

Proton nuclear magnetic resonance has allowed the direct observation of hydrogen atoms in their steroidal molecular environment (284). Because many of the protons in a steroid have similar molecular shifts, determination of stereochemical and other relationships has been difficult. This has been alleviated by the availability of high magnetic field-strength instruments, which spread the signals over a larger range, and the development of shift reagents which coordinate with polar groups in the steroid and thus induce changes in the chemical shifts of adjacent protons (285). It is possible to obtain routine nmr spectra on trace amounts of material by means of pulse and Fourier transform nmr (286). A more fundamental advance has resulted because of the commercial availability of ^{13}C nmr spectrometers to observe individual carbon atoms in their molecular environment (287). This technique has also been very useful in biosynthesis experiments, in which ^{13}C enriched precursors are used.

Much of the early work on the development of mass spectrometry was based on steroids (288–289). Computer programs are available to aid in the analysis of steroid fragmentation patterns (290).

Extensive x-ray crystal-structure determinations have been done on a wide variety of steroids and these have been collected and listed (291–292). A description of new advances in physical methods are included each year in the *Specialist Report on Terpenoids and Steroids* (6).

See Table 3 for a list of steroids referred to in this article.

Table 3. List of Steroids Referred to in the Text

Steroid	Structure No.	CAS Registry No.
cholesterol	**(2)**	[*57-88-5*]
cholic acid	**(3)**	[*81-25-4*]
lithocholic acid	**(4)**	[*434-13-9*]
16,17-dihydro-17-methyl-15*H*-cyclopenta[*a*]phenanthrene	**(5)**	[*549-38-2*]
equilenin	**(6)**	[*517-09-9*]
3β-lanosterol	**(7)**	[*79-63-0*]
lathosterol	**(8)**	[*80-99-9*]
7-dehydrocholesterol	**(9)**	[*434-16-2*]
ergosterol	**(10)**	[*57-87-4*]
vitamin D_2	**(11)**	[*50-14-6*]
vitamin D_3	**(12)**	[*67-97-0*]
9,10-secocholesta-5(10),6,8-triene	**(13)**	[*50524-96-4*]
lumisterol	**(14)**	[*474-69-1*]
tachysterol	**(15)**	[*115-61-7*]
pyrocalciferol	**(16)**	[*128-27-8*]
photopyrocalciferol	**(17)**	[*41411-05-6*]
stigmasterol	**(18)**	[*83-48-7*]
β-sitosterol	**(19)**	[*83-46-5*]
progesterone	**(21)**	[*57-83-0*]
cycloartenol	**(22)**	[*469-38-5*]
abietospirane	**(23)**	[*71648-15-2*]
asterosterol	**(24)**	[*30674-32-9*]
occelasterol	**(25)**	[*54278-89-6*]
24-methylenecholesterol	**(26)**	[*474-63-5*]
aplysterol	**(27)**	[*38636-49-6*]
verongulasterol	**(28)**	[*70284-74-1*]
gorgosterol	**(29)**	[*29782-65-8*]
calysterol	**(30)**	[*57331-04-1*]
stigmasta-5,24(28),28-trien-3-ol	**(31)**	[*56525-73-6*]
18-(acetyloxy)-pregna-1,4,20-trien-3-one	**(32)**	[*74055-42-8*
fusidic acid	**(33)**	[*6990-06-3*]
helvolic acid	**(34)**	[*29400-42-8*]
cephalosporin C	**(35)**	[*61-24-5*]
diosphenol elaterin	**(36)**	[*18444-66-1*]
deoxycholic acid	**(37)**	[*83-44-3*]
digitogenin	**(38)**	[*511-34-2*]
diosgenin	**(39)**	[*512-04-9*]
hecogenin	**(40)**	[*467-55-0*]
funtumine	**(41)**	[*474-45-3*]
terminaline	**(42)**	[*15112-49-9*]
holarrhidine	**(43)**	[*82182-51-2*]
pachystermine A	**(44)**	[*6156-99-6*]
conessine	**(45)**	[*546-06-5*]
solacongestidine	**(46)**	[*984-82-7*]
solasidine	**(47)**	[*126-17-0*]
solanidine	**(48)**	[*80-78-4*]
veratramine	**(49)**	[*60-70-8*]
protoverine	**(50)**	[*76-45-9*]
verticine	**(51)**	[*23496-41-5*]
cyclobuxine D	**(52)**	[*29784-83-6*]
samandarine	**(53)**	[*467-51-6*]
cycloneosamandione	**(54)**	[*631-72-1*]
17β-(acetyloxy)-androsta-1,4,6-trien-3-one	**(55)**	[*42224-78-2*]
17β-acetyloxy-1α,2α-epoxy-androsta-4,6-dien-3-one	**(56)**	[*20796-94-5*]
17β-(acetyloxy)-1α-hydroxy-5β-androstan-3-one	**(57)**	[*42224-80-6*]
17β-(acetyloxy)-1α-hydroxy-3-aza-5β-*A*-homoandrostan-4-one	**(58)**	[*42224-81-7*]

Table 3 (*continued*)

Steroid	Structure No.	CAS Registry No.
1α,4α-epoxy-3-aza-*A*-homoandrostan-17-ol	**(59)**	*[25484-32-6]*
batrachotoxinin A	**(60)**	*[19457-37-5]*
3,3-[1,2-ethandiylbis(oxyl)]-20-hydroxypregna-5,9(11)-dien-18-oic acid, γ-lactone	**(61)**	*[20918-33-6]*
18-(acetyloxy)-4,5-epoxypregn-9(11)-ene-3,20-dione	**(62)**	*[35596-42-0]*
11α,18-bis(acetyloxy)-9-hydroxy-pregn-4-ene-3,20-dione	**(63)**	*[35596-40-8]*
11α,18-bis(acetyloxy)-7α,9-dihydroxy-pregn-4-ene-3,20-dione	**(64)**	*[35596-37-3]*
11α,18-bis(acetyloxy)-3α,9-epoxy-7α-hydroxy-3α-methoxypregnan-20-one	**(65)**	*[35596-36-3]*
7α,11α-18-tris(acetyloxy)-3α,9:14β,15β-diepoxy-3α-methoxy-5β-pregn-16-en-20-one	**(66)**	*[35875-68-4]*
3α,9-epoxy-3α-methoxy-5β-pregn-16-en-7α,11α,14β,18,20*S*-pentol, triacetate	**(67)**	*[38853-50-8]*
7-(acetyloxy)-7α,8β-dehydro-O^3-methyl-23-oxo-batrachotoxinin A, diacetate ester	**(68)**	*[39024-77-6]*
O^3-methyl-23-oxo-batrachotoxinin A, diacetate (ester)	**(69)**	*[39023-96-6]*
digitoxigenin	**(70)**	*[143-62-4]*
ouabagenin	**(71)**	*[508-52-1]*
scillarenin	**(72)**	*[465-22-5]*
bufalin	**(73)**	*[465-21-4]*
seychellogenin	**(74)**	*[24041-68-7]*
22,25-epoxy-3β,12α,17α,20-tetrahydroxy-lanost-9(11)-en-18-oic acid, gamma-lactone	**(75)**	*[82468-28-8]*
3β,12β,20-trihydroxylanost-9(11)-en-18-oic acid, gamma-lactone	**(76)**	*[67797-17-5]*
lanosta-7,9(11)dien-3β,20-diol	**(77)**	*[66877-85-8]*
3-acetate-20-nitrite-lanost-7,9(11)-diene-3β,20-diol	**(78)**	*[66877-86-9]*
3β-(acetyloxy)-20-hydroxy-lanosta-7,9(11)dien-18-al,18-oxime	**(79)**	*[66877-87-0]*
withaferin A	**(80)**	*[5119-48-2]*
taborosalactone C	**(81)**	*[52329-20-1]*
antheridiol	**(82)**	*[22263-79-2]*
3β-hydroxy-pregn-5-ene-20-carboxaldehyde	**(83)**	*[53906-49-3]*
3β,22*S*-dihydroxy-stigmasta-5,24(28)-dien-29-oic acid, delta-lactone	**(84)**	*[32212-70-7]*
3β-hydroxy-stigmasta-5,22*E*,24*Z*(28)-trien-29-oic acid	**(85)**	*[53421-11-7]*
α-ecdysone	**(86)**	*[3604-87-3]*
β-ecdysone	**(87)**	*[5289-74-7]*
26-hydroxy-β-ecdysone	**(88)**	*[19458-46-9]*
2-deoxy-β-ecdysone	**(89)**	*[17942-08-4]*
inokosterone	**(90)**	*[15130-85-5]*
makisterone A	**(91)**	*[20137-14-8]*
ponasterone A	**(92)**	*[13408-56-5]*
methyl 3β-(acetyloxy)-pregn-5-ene-20*S*-carboxylate	**(93)**	*[6556-99-6]*
methyl 3β,5α-dihydroxy-6-oxo-pregnane-20-carboxylate	**(94)**	*[5241-15-6]*
methyl 5α-hydroxy-6-oxo-pregn-2-ene-20-carboxylate	**(95)**	*[5372-51-0]*
methyl 2,3-bis(acetyloxy)-5α-hydroxy-6-oxo-pregnane-20-carboxylate	**(96)**	*[5241-16-7]*
methyl 2,3-bis(acetyloxy)-5α-hydroxy-6-oxo-pregn-7-ene-20-carboxylate	**(97)**	*[5372-52-1]*
methyl 2β,3β,5α-tris(acetyloxy)14-hydroxy-6-oxo-pregn-7-ene-20-carboxylate	**(98)**	*[5288-94-8]*
methyl 14α-hydroxy-2,3-[(1-methylethylidene)bis(oxyl)]-6-oxo-5β-pregn-7-ene-20-carboxalate	**(99)**	*[18220-32-1]*
6,14-dihydroxy-2,3-[(1-methylethylidene)bis(oxyl)]-25-[tetrahydro-2*H*-pyran-2-yl)oxy]-5β-cholest-7-en-22-one	**(100)**	*[18220-33-2]*
spironolactone	**(101)**	*[52-01-7]*
aldosterone	**(102)**	*[52-39-1]*
(25*R*)-furosta-5,20(22)-diene-3,26-diol diacetate	**(103)**	*[2309-38-8]*
3β-(acetyloxy)-16β-[[5-(acetyloxy)-4-methyl-1-oxopentyl]oxy]-pregn-5-en-20-one	**(104)**	*[58400-99-0]*

Table 3 (*continued*)

Steroid	Structure No.	CAS Registry No.
3β-(acetyloxy)-pregna-5,16-dien-20-one	**(105)**	[*979-02-2*]
3β-(acetyloxy)-pregn-5-en-20-one	**(106)**	[*1778-02-5*]
3β-hydroxy-pregn-5-en-20*E*-one oxime	**(107)**	[*60562-58-5*]
N-[(3β-acetyloxy)androsta-5,16-dien-17-yl]acetamide	**(108)**	[*65732-71-0*]
3β-(acetyloxy)-androst-5-en-17-one	**(109)**	[*853-23-6*]
11α-23α-dibromohecogenin, acetate	**(110)**	[*5130-60-9*]
(25*R*)-3β,12β-bis(acetyloxy)-5α-spirostan-11-one	**(111)**	[*82182-52-3*]
(25*R*)-3β-hydroxy-5α-spirostan-11-one	**(112)**	[*4802-74-8*]
hydrocortisone	**(113)**	[*50-23-7*]
11α-hydroxy-pregn-4-ene-3,20-dione	**(114)**	[*80-75-1*]
pregn-4-ene-3,11-20-trione	**(115)**	[*516-15-4*]
ethyl 22,22-dibromo-3,11,20,23-tetraoxo-21-norchol-4-en-24-oate	**(117)**	[*82182-53-4*]
methyl 3,11-dioxo-pregna-4,17(20)-dien-21-oate	**(118)**	[*31056-07-2*]
methyl 11-oxo-3-(1-pyrrolidinyl)-pregna-3,5,17(20)-trien-21-oate	**(119)**	[*82182-54-5*]
11,21-dihydroxy-pregna-4,17(20)-dien-3-one	**(120)**	[*3494-53-9*]
cholest-4-en-3-one	**(121)**	[*601-57-0*]
cholesta-1,4-dien-3-one	**(122)**	[*566-91-6*]
3-oxo-pregna-1,4-diene-20-carboxylic acid	**(123)**	[*20248-18-4*]
androsta-1,4-diene-3,17-dione	**(124)**	[*897-06-3*]
androst-4-en-3,17-dione	**(125)**	[*63-05-8*]
9α-hydroxyandrosta-1,4-diene-3,17-dione	**(126)**	[*82182-55-6*]
3β-hydroxy-9,10-secoandrosta-1,3,5(10)triene-9,17-dione	**(127)**	[*2394-69-6*]
9α-hydroxyandrost-6-ene-3,17-dione	**(128)**	[*82182-56-7*]
17-hydroxy-pregna-4,9(11)-dien-20-yn-3-one	**(129)**	[*63973-95-5*]
17-[(phenylthio)oxy]-pregna-4,9(11)-dien-20-yn-3-one	**(130)**	[*70518-79-5*]
21-(phenylsulfinyl)-pregna-4,9(11),17(20),20-tetraen-3-one	**(131)**	[*63998-65-2*]
20-methoxy-21-(phenylsufinyl)-pregna-4,9(11),17(20)-trien-3-one	**(132)**	[*63973-96-6*]
20-methoxy-17-[(phenylthio)oxy]-pregna-4,9(11),20-trien-3-one	**(133)**	[*70497-04-0*]
17-hydroxy-20-methoxy-pregna-4,9(11),20-trien-3-one	**(134)**	[*63973-97-7*]
estrone	**(135)**	[*53-16-7*]
androsta-1,4-diene-3,17-dione,cyclic 17-(1,2-ethanediyl acetal)	**(136)**	[*2398-63-2*]
3-hydroxy-estra-1,3,5(10)-trien-17-one, cyclic 1,2-ethanediyl acetal	**(137)**	[*900-83-4*]
3-methoxy-estra-2,5(10)-dien-17-one, cyclic 1,2-ethanediyl acetal	**(138)**	[*1238-30-8*]
estr-4-ene-3,17-dione	**(139)**	[*734-32-7*]
estr-5(10)-ene-3,17-dione	**(140)**	[*3962-66-1*]
3β(acetyloxy)-5α-bromo-6β-hydroxyandrostan-17-one	**(141a)**	[*4229-69-0*]
3β(acetyloxy)-5α-chloro-6β-hydroxyandrostan-17-one	**(141b)**	[*6557-16-0*]
3β-5α-bis(acetyloxy)-6β-hydroxyandrostan-17-one	**(141c)**	[*29246-51-3*]
3β-(acetyloxy)-5α-bromo-6β,19-epoxyandrostan-17-one	**(146a)**	[*2685-64-5*]
3β-(acetyloxy)-5α-chloro-6β,19-epoxyandrostan-17-one	**(146b)**	[*2659-00-9*]
3β,5α-bis(acetyloxy)-6β,19-epoxyandrostan-17-one	**(146c)**	[*807-09-0*]
5α-bromo-6β,19-epoxyandrostane-3,17-dione	**(147a)**	[*24870-00-6*]
5α-chloro-6β,19-epoxyandrostane-3,17-dione	**(147b)**	[*82182-57-8*]
5α-(acetyloxy)-6β,19-epoxyandrostane-3,17-dione	**(147c)**	[*82182-58-9*]
6β,19-epoxyandrost-4-en-3,17-dione	**(148)**	[*6563-83-3*]
19-hydroxyandrost-4-ene-3,17-dione	**(149)**	[*510-64-5*]
3,17-dioxoandrost-4-en-19-oic acid	**(150)**	[*4757-95-3*]
estr-5(10)-en-3,17-dione	**(151)**	[*3962-66-1*]
3β-(acetyloxy)-5α-bromo-6-nitrosooxyandrostan-17-one	**(152a)**	[*82182-59-0*]
3β-(acetyloxy)-5α-chloro-6-nitrosooxyandrostan-17-one	**(152b)**	[*82182-60-3*]
3β,5α-bis(acetyloxy)-6-nitrosooxyandrostan-17-one	**(152c)**	[*82182-61-4*]
3β-(acetyloxy)-5α-bromo-6-hydroxy-17-oxoandrostan-19-al,19-oxime	**(153a)**	[*82182-62-5*]
3β-(acetyloxy)-5α-chloro-6-hydroxy-17-oxoandrostan-19-al,19-oxime	**(153b)**	[*82182-63-6*]
3β,5α-bis(acetyloxy)-6-hydroxy-17-oxoandrostan-19-al,19-oxime	**(153c)**	[*82182-34-1*]
5-bromo-3β,6β-dihydroxy-17-oxo-5α-androstan-19-oic acid, gamma-lactone acetate	**(154a)**	[*13309-29-0*]

Table 3 (*continued*)

Steroid	Structure No.	CAS Registry No.
5-chloro-3β,6β-dihydroxy-17-oxo-5α-androstan-19-oic acid, gamma-lactone acetate	(**154b**)	[*2353-03-9*]
3β,5α-bis(acetyloxy)-6β-hydroxy-17-oxo-5α-androstan-19-oic acid, gamma-lactone	(**154c**)	[*82182-35-2*]
19-hydroxycholest-4-en-3-one	(**155**)	[*13200-77-6*]
6β,19-epoxycholest-4-en-3-one	(**156**)	[*749-00-8*]
6β,19-epoxy-9α-hydroxyandrost-4-ene-3,17-dione	(**157**)	[*74220-48-7*]
3β-(acetyloxy)-(20*R*)-hydroxypregn-5-en-18-oic acid, gamma-lactone	(**158**)	[*3020-10-8*]
1-[3β,20*R*-dihydroxy-18-norpregn-5-en-13-yl]ethanone	(**159**)	[*6182-52-1*]
13-ethyl-3β-hydroxy-18-norpregn-5-en-20-one	(**160**)	[*24357-75-3*]
3β-(acetyloxy)-20-hydroxypregn-5-ene-20-carbonitrile	(**161**)	[*22641-61-8*]
3β-(acetyloxy)-20-oxopregn-5-ene-20-carbonitrile	(**164**)	[*6570-63-4*]
13-ethenyl-18-norpregn-4-ene-3,20-dione	(**165**)	[*23971-17-7*]
3β-*t*-butoxyandrost-5-en-17-one	(**166**)	[*1167-53-9*]
3β-*t*-butoxypregna-5,17(20)-diene-20-carbonitrile	(**167**)	[*60727-73-3*]
20-(acetyldioxy)-3β-*t*-butoxypregn-5-ene-20-carbonitrile	(**168**)	[*62623-48-7*]
3β-*t*-butoxy-20-oxopregn-5-en-18-carbonitrile	(**169**)	[*6570-53-2*]
corticosterone acetate	(**170**)	[*1173-26-8*]
21-(acetyloxy)-11-nitrosooxypregn-4-ene-3,20-dione	(**171**)	[*74220-48-7*]
aldosterone oxime	(**172**)	[*74220-49-8*]
11β-hydroxyprogesterone	(**173**)	[*600-57-7*]
11β,20-dihydroxy-20-methyl-13,21-cyclo-18-norpregn-4-en-3-one	(**174**)	[*82190-26-9*]
11β-hydroxy-20-methylene-13,21-cyclo-18-norpregn-4-en-3-one	(**175**)	[*76807-21-1*]
11β-hydroxyspiro(13,21-cyclo-18-norpregn-4-ene-20,2′-oxiran)-3-one	(**176**)	[*82182-36-3*]
11β-hydroxy-20-[(acetyloxy)methyl]-13,21-cyclo-18-norpregn-4,20-dien-3-one	(**177**)	[*76807-29-9*]
11β,20,21-trihydroxy-20[(acetyloxy)methyl]-13,21-cyclo-18-norpregn-4-en-3-one	(**178**)	[*76807-31-3*]
3β-[(tetrahydro-2*H*-pyran-2-yl)oxy]androst-5-en-17-one	(**179**)	[*19637-35-5*]
3β-[(tetrahydro-2*H*-pyran-2-yl)oxy]pregna-5,17(20)-dien-21-oic acid, ethyl ester	(**180**)	[*64604-58-6*]
3β-[(tetrahydro-2*H*-pyran-2-yl)oxy]pregna-5-en-21-oic acid, ethyl ester	(**181**)	[*69375-37-7*]
25,26-[(isopropylidene)bis(oxy)]-3-[(tetrahydro-2*H*-pyran-2-yl)oxy]-cholest-5-en-21-oic acid, ethyl ester	(**182**)	[*82182-37-4*]
25,26-[(isopropylidene)bis(oxy)]-3-[(tetrahydro-2*H*-pyran-2-yl)oxy]-cholest-5-en-21-ol	(**183**)	[*82182-38-5*]
25,26-[isopropylidenebis(oxy)]-3-[(tetrahydro-2*H*-pyran-2-yl)oxy]-cholest-5-en-21-ol, tosylate	(**184**)	[*82182-39-6*]
25,26-[(isopropylidene)bis(oxy)]-3-[(tetrahydro-2*H*-pyran-2-yl)oxy]-cholest-5-ene	(**185**)	[*77518-21-9*]
cholesta-5,7-diene-3β,25,26-triol	(**186**)	[*53990-60-6*]
(25*S*),26-dihydroxyvitamin D_3	(**187**)	[*55700-58-8*]
equilenin	(**188**)	[*517-09-9*]
17-methoxy-2-methyl-D-homo-18-norandrosta-2,7,13,15,17-pentaene-1,4-dione(5β,9β)	(**200**)	[*82227-90-5*]
3β-methoxy-17-methyl-D-homoestra-1,3,5(10),9(11),16-pentaene-15,17a-dione (14β)	(**201**)	[*38397-71-6*]
3β-methoxy-17-methyl-D-homoestra-1,3,5(10),9(11),16-pentaene-15,17a-dione (14α)	(**202**)	[*38397-73-8*]
3β-methoxy-17-methyl-D-homoestra-1,3,5(10),9(11),16-pentaen-15-one	(**203**)	[*38397-76-1*]
3β-methoxy-17-methyl-D-homoestra-1,3,5(10),16-tetraen-15-one	(**204**)	[*38397-77-2*]
1,2,3,4,4a,9,10,10a-octahydro-7-methoxy-2-methyl-2-(2-oxopropyl)-1-phenanthreneacetaldehyde	(**206**)	[*38397-82-9*]
3β-methoxy-19-norpregna-1,3,5(10),16-tetraen-20-one	(**207**)	[*21321-91-5*]
3-methoxyestra-1,3,5(10),8,14-pentaen-17-one	(**212**)	[*1456-50-4*]
3-methoxyestra-1,3,5(10),8-tetraen-17-ol	(**213**)	[*6733-79-5*]

Table 3 (*continued*)

Steroid	Structure No.	CAS Registry No.
3-methoxyestra-1,3,5(10)-trien-17-ol	**(214)**	*[17337-41-6]*
17β-(1,1-dimethylethoxy)-3-methoxyestra-1,3,5(10)-triene	**(219)**	*[53053-36-4]*
11α-hydroxy-3β-methoxyestra-1,3,5(10)-trien-17-one	**(221)**	*[5210-12-8]*
epiandrostane	**(223)**	*[481-29-8]*
3,11,20,23-tetraoxo-21-norchol-5-en-24-oic acid, cyclic 3-(1,2-ethanediyl acetal)	**(224)**	*[82182-44-3]*
octahydrohydroxy-4′a-methylspiro[1,3-dioxolane-2,2′(5′*H*)-phenanthrenone]	**(225)**	*[2907-08-6]*
estra-4,9-diene-3,17-dione	**(226)**	*[5173-46-6]*
19-nortestosterone	**(229)**	*[434-22-0]*
17β-(1,1-dimethylethoxy)-4,5-secoestr-9-ene-3,5-dione-cyclic 3-(1,2-ethanediyl acetal)	**(239)**	*[24157-07-01]*
norgestrel	**(240)**	*[797-63-7]*
13-ethyl-3-methoxy-gona-1,3,5(10),8,14-pentaen-17-one	**(243)**	*[2436-53-5]*
13-ethyl-3-methoxy-gona-1,3,5(10)-trien-17-ol	**(244)**	*[3625-82-9]*
pregn-1-en-20-one	**(265)**	*[57968-14-6]*
pregn-1-en-3,20-dione	**(266)**	*[57968-15-7]*
17β-hydroxy-2-(hydroxymethylene)-17α-methyl-5α-androstan-3-one	**(268)**	*[434-07-1]*
stanazolol	**(269)**	*[10418-03-8]*
9α-fluoro-11,17,21-trihydroxy-6,16-dimethyl-2′-phenyl-2′*H*-pregna-2,4,6-trieno[3,2-*c*]pyrazol-20-one	**(270)**	*[6793-14-2]*
9α-fluoro-11,17,21-trihydroxy-6,16-dimethyl-1′-phenyl-1′*H*-pregna-4,6-dieno[3,2-*c*]pyrazol-20-one	**(271)**	*[82182-47-6]*
11,17,21-trihydroxy-6,16-dimethyl-2′-(4-pyridinyl)-2′*H*-pregna-2,4,6-trieno[3,2-*c*]pyrazol-20-one	**(272)**	*[52123-46-3]*
androisoxazole	**(273)**	*[360-66-7]*
danazol	**(274)**	*[17230-88-5]*
17-hydroxy-3-oxo-pregn-4-en-20-yne-2-carbonitrile	**(275)**	*[13648-12-9]*
trilostane	**(276)**	*[13647-35-3]*
3β-(acetyloxy)-16α,17-epoxy,5α-pregnane-11,20-dione	**(277)**	*[909-98-8]*
17-azido-3β,16α-bis(acetyloxy)-5α-pregnane-11,20-dione	**(278)**	*[5167-90-8]*
3β-(acetyloxy)-2′-methyl-5′β-*H*-5α-pregnano[17,16-d]oxazole-11,20-dione	**(279)**	*[5070-96-2]*
oxazocort	**(280)**	*[14927-19-6]*
17-hydroxy-3-oxo-17α-pregn-4-ene-21-carboxylic acid, gamma-lactone	**(281)**	*[976-70-5]*
17α-pregn-5-en-20-yne-3β,17-diol	**(282)**	*[3604-60-2]*
3β,17-dihydroxy-17α-pregn-5-en-20-yne-21-carboxylic acid	**(283)**	*[3460-93-3]*
3β,17-dihydroxy-17α-pregna-5,20-diene-21-carboxylic acid, gamma-lactone	**(284)**	*[2844-75-9]*
3β,17-dihydroxy-17α-pregn-5-ene-21-carboxylic acid, gamma-lactone	**(285)**	*[13934-61-7]*
17-hydroxy-3-oxo-17α-pregna-4,6-diene-21-carboxylic acid	**(286)**	*[976-71-6]*
3-(1-pyrollidinyl)-androsta-3,5-dien-17-one	**(287)**	*[905-30-6]*
4′,5′-dihydro-7α-mercapto-spiro[androst-4-ene-17,2′(3′*H*)-furan]-3-one, acetate	**(288)**	*[6673-97-8]*
ethyl-17α-hydroxy-3-oxo-pregn-4-ene-7,21-dicarboxylate, gamma-lactone	**(289)**	*[41020-77-3]*
methyl 6β,17α-dihydroxy-3-oxo-pregn-4-ene-7,21-dicarboxylate, gamma-lactone	**(290)**	*[61265-52-9]*
pancuronium	**(291)**	*[15500-66-0]*
22,25-diazacholesterol	**(292)**	*[24887-57-8]*
17β-hydroxy-17-methyl-5α-androst-1-en-3-one	**(293)**	*[65-04-3]*
oxandrolone	**(296)**	*[53-39-4]*
5α-(acetyloxy)-6,19-epoxy-17-hydroxy-androstan-3-one	**(297)**	*[36334-59-5]*
6,19-epoxy-2,17-dihydroxy-androsta-1,4-diene-3-one	**(298)**	*[36334-60-8]*
6β,19-epoxy-1,17β-dihydroxy-2-oxaandrost-4-en-3-one	**(299)**	*[37147-40-3]*
6,19-epoxy-17-hydroxy-2-azaandrost-4-en-3-one	**(300)**	*[57178-19-5]*

Table 3 (*continued*)

Steroid	Structure No.	CAS Registry No.
17,19-dihydroxy-2-azaandrost-4-en-3-one	**(301)**	*[37147-43-6]*
17β-hydroxy-3-oxo-2-oxaandrost-4-en-19 oic acid	**(302)**	*[83487-92-7]*
17-hydroxy-2-azaestra-1(10),4-dien-3-one	**(303)**	*[37147-47-0]*
3-methoxy-2-azaestra-1,3,5(10)-trien-17-ol	**(304)**	*[57178-23-1]*
2,3,10,11-tetrahydro-7-methoxy-1-methyl-1-oxide-1*H*-naphtho[2,1-*e*]-phosphindole	**(305)**	*[63347-72-8]*
testololactone	**(306)**	*[4416-57-3]*
3-methoxy-6-methyl-6-azaestra-1,3,5(10),8(14)-tetraen-17-one 1,2-ethanediyl acetal	**(309)**	*[19526-31-9]*
17β-hydroxy-6-methyl-6-azaestra-4,8(14)-dien-3-one	**(310)**	*[7680-51-5]*
3,3a,10,11-tetrahydro-7-methoxy-4*H*-[1]benzothiopyrano[4,3-*g*]indolizin-1(2*H*)-one	**(313)**	*[82182-30-7]*
8-(acetyloxy)-1,4,6,7,7a,8,9,11-octahydro-2,3-dimethoxy-7a-methyl-12*H*-cyclopenta[*f*]pyridazinol[1,2-*a*]cinnolin-12-one	**(316)**	*[33606-24-5]*
5,9,10,10a,11,12-hexahydro-2-methoxy-benzo[*f*]pyrrolo[2,1-*a*]isoquinolin-8(6*H*)-one	**(318)**	*[34241-89-9]*
androsta-3,5-diene	**(319)**	*[53875-08-4]*
6-ethoxy-3,5-cycloandrostane	**(320)**	*[2574-55-2]*
2,3-dihydro-3′*H*-cycloprop[2,3]-5α-androst-2-en-17-β-ol	**(321)**	*[17780-46-0]*
1,2-dihydro-17β-hydroxy-5′*H*-androsta-1,4,6-trieno[2,1-*c*]pyrazol-3-one	**(322)**	*[82182-31-8]*
1β,2β-dihydro-17β-hydroxy-3′*H*-cycloprop[1,2]androsta-1,4,6-trien-3-one	**(323)**	*[41982-56-3]*
cyproterone acetate	**(324)**	*[2098-66-0]*
17β-hydroxypregna-1,4,6-triene-3,20-dione	**(325)**	*[39789-83-8]*
1α,2α-dihydro-17-hydroxy-3′*H*-cyclopropa[1,2]pregna-1,4,6-triene-3,20-dione	**(326)**	*[34554-31-9]*
17-(acetyloxy)-6,7-epoxy-1α,2α-dihydro-3′*H*-cyclopropa[1,2]pregna-1,4-diene-3,20-dione	**(327)**	*[40007-74-7]*
17-(acetyloxy)-6-chloro-1α-(chloromethyl)pregna-4,6-diene-3,20-dione	**(328)**	*[17183-98-1]*
17-(acetyloxy)-5-α-androst-1-en-3-ol	**(329)**	*[51505-46-5]*
1,2-dihydro-cycloprop[1,2]androst-1-en-3-ol	**(330)**	*[4834-07-5]*
17-(acetyloxy)-1,2-dihydro-3′*H*-cycloprop[1,2]androst-1-en-3-one	**(331)**	*[4817-19-0]*
cholest-4-en-3β-ol	**(332)**	*[517-10-2]*
3′,4-dihydro-cyclopropa[4,5]cholestan-3-ol	**(333)**	*[13903-59-8]*
cholesta-4,6-dien-3α-ol	**(334)**	*[51505-50-1]*
3′,4-dihydrocyclopropa[4,5]cholestan-3-ol	**(335)**	*[75684-12-7]*
17β-hydroxy-androsta-4,6-dien-3-one	**(336)**	*[2484-30-2]*
17-(acetyloxy)-6α,7α,dihydro-3′*H*-cycloprop[6,7]androsta-4,6-dien-3-one	**(337)**	*[14610-47-0]*
17-(acetyloxy)-6β,7β-dihydro-3′*H*-cycloprop[6,7]androsta-4,6-dien-3-one	**(338)**	*[6218-31-1]*
6,7-dihydro-3′H-cyclopropa[6,7]pregna-4,6-diene-3,15,20-trione	**(339)**	*[53154-89-5]*
6,7-dihydro-17-hydroxy-3-oxo-3′*H*-cyclopropa[6,7]pregna-4,6-diene-21-carboxylic acid, gamma lactone	**(340)**	*[49848-04-6]*
17-hydroxyandrosta-4,15-dien-3-one	**(341)**	*[82182-32-9]*
15β,16β-dihydro-17β-hydroxy-3′*H*-cycloprop[15,16]androsta-4,15-dien-3-one	**(342)**	*[38022-92-3]*
15α,16α-dihydro-17β-hydroxy-3′*H*-cycloprop[15,16]androsta-4,15-dien-3-one	**(343)**	*[25495-19-6]*
17β-hydroxyestr-5(10)-en-3-one	**(344a)**	*[42028-18-2]*
norethynodrel	**(344b)**	*[68-23-5]*
17-hydroxy-5,19-cyclo-A-norpregn-20-yn-3-one	**(345)**	*[53357-40-7]*
17β-hydroxyandrost-4-en-3-one	**(346)**	*[58-22-0]*
3′,4′-dihydro-17β-hydroxy-1′α-*H*-cyclobut[4,5]-5β-androstan-3-one	**(347)**	*[30598-78-8]*
3′,4′-dihydro-17β-hydroxy-1′β-*H*-cyclobut[4,5]-5α-androstan-3-one	**(348)**	*[82432-21-1]*
3′,4′-dihydro-17β-hydroxy-1′β-*H*-cyclobut[4,5]-5β-androstan-3-one	**(349)**	*[82432-22-2]*

Table 3 (*continued*)

Steroid	Structure No.	CAS Registry No.
6,7-dihydro-17-hydroxycyclobut[6,7]androsta-4,6-dien-3-one	**(350)**	[*26332-51-4*]
17β-(acetyloxy)-androsta-4,6-dien-3-one	**(351)**	[*2352-19-4*]
17-(acetyloxy)-3′,6′-dihydro-1′*H*-benz[4α,5β]androst-6-en-3-one	**(352)**	[*38391-35-4*]
17-(acetyloxy)-3′,6′-dihydro-1′*H*-benz[4β,5β]androst-6-en-3-one	**(353)**	[*38391-34-3*]
17-(acetyloxy)-3′,6,6′,7-tetrahydrobenz[6α,7α]androsta-4,6-dien-3-one	**(354)**	[*38391-37-6*]
17-(acetyloxy)-3′,6,6′,7-tetrahydrobenz[6β,7β]androsta-4,6-dien-3-one	**(355)**	[*38391-36-5*]
3-(acetyloxy)hexadecahydro-3a,5b-dimethyl-9,10:11,12-dimethanocyclobuta[*k*]cyclopenta[*a*]phenanthren-8(9*H*)-one	**(356)**	[*38391-33-2*]
9β,10α-ergost-4-en-3-one	**(357)**	[*82228-17-9*]
retroprogesterone	**(358)**	[*2755-10-4*]
dydrogesterone	**(359)**	[*152-62-5*]
11α-hydroxy-5,20-dioxo-3,5-seco-A-norpregnan-3-oic acid methanesulfonate	**(360)**	[*10110-54-0*]
3-acetyl-1,2,3,3a,4,5,8,9,9a,9b-decahydro-3a,6-dimethyl-7*H*-benz[*e*]-inden-7-one	**(361)**	[*10072-88-5*]
estr-5(10)-ene-3,17-diol	**(362)**	[*13864-49-8*]
17β-hydroxycycloprop[5α,10α]estran-3-one	**(363)**	[*4050-22-0*]
10α-testosterone	**(364)**	[*571-41-5*]
5α-cholestan-3-one	**(365)**	[*566-88-1*]
3-hydroxy-5α-cholestane-3-carbonitrile	**(366)**	[*19804-83-2*]
2′,2′-dimethyl-5α-spiro[cholestane-3α,5′-oxazolidine]	**(367)**	[*26435-83-6*]
A-homo-5α-cholestan-3-one	**(368)**	[*13914-51-7*]
(25*R*)-5α-spiro[oxirane-2,12′α-spirostan]-3′β-ol	**(369)**	[*72203-79-3*]
(25*R*)-11α-(azidomethyl)-5α-spirostan-3β,11β-diol	**(370)**	[*83487-90-5*]
(25*R*)-11β-(azidomethyl)-5α-spirostan-3β,11α-diol	**(371)**	[*83487-89-2*]
(25*R*)-11β-(aminomethyl)-5α-spirostan-3β,11α-diol	**(372)**	[*83487-91-6*]
3β-hydroxy-*C*-homo-5α-spirostan-12-one	**(373)**	[*72166-68-8*]
C-homo-5α-spirostan-3-ol	**(374)**	[*72166-69-9*]
C-homoprogesterone	**(375)**	[*82432-23-3*]
C-homotestosterone	**(376)**	[*82182-33-0*]
C-homoestra-1,3,5(10)-triene-3,17-diol	**(377)**	[*72203-78-2*]
9β-*C*-homoestra-1,3,5(10)-triene-3,17-diol	**(378)**	[*65329-29-5*]
A-homocholest-4a-en-3-one	**(379)**	[*51355-04-5*]
A-homoestra-1(10),2,4a-triene-4,17-dione	**(380)**	[*1637-96-3*]
3,17-dihydroxypregn-5-en-20-one	**(381)**	[*387-79-1*]
3β,17aβ-dihydroxy-17a-methyl-*D*-homoandrost-5-en-17-one	**(382)**	[*2460-10-8*]
3β,17aα-dihydroxy-17a-methyl-*D*-homoandrost-5-en-17-one	**(383)**	[*3754-65-2*]
androsta-4,6-diene-3,17-dione	**(384)**	[*633-34-1*]
5,19-cycloandrosta-1,6-diene-3,17-dione	**(385)**	[*38838-72-1*]
3-(acetyloxy)-*AB*(10a)-homoestra-1(10),2,4,6,8-pentaen-17-one	**(386)**	[*38838-87-8*]
lanostan-3β-ol	**(387)**	[*4581-87-7*]
3-(1-methylethylidene)-*A*-norcholestane	**(388)**	[*39932-77-9*]
A-nor-5α-cholestan-3-one	**(389)**	[*6908-02-7*]
A-nor-5β-cholestan-3-one	**(390)**	[*6908-01-6*]
epiandrosterone	**(391)**	[*481-29-8*]
dodecahydro-3a,6-dimethyl-3-oxo-1*H*-benz[*e*]indene-6,7-diacetic acid	**(392)**	[*1165-38-4*]
A-nor-5α-androstane-2,17-dione	**(393)**	[*1032-12-8*]
2-ethynyl-*A*,19-dinorpregn-20-yne-2β,17-diol	**(394)**	[*64675-07-6*]
2-ethynyl-*A*-nor-5α,17α-pregnane-2β,17-diol	**(395)**	[*21501-28-0*]
3-(acetyloxy)-*B*-homo-6-oxaandrost-4-ene-7,17-dione	**(396)**	[*6688-89-7*]
3-(acetyloxy)-5-hydroxy-17-oxo-*B*-norandrostane-6-carboxylic acid, β-lactone	**(397)**	[*39991-58-7*]
3β-hydroxy-*B*-norandrost-5-en-17-one	**(398)**	[*24808-54-6*]
3β-(acetyloxy)-5α-bromo-6,17-dioxoandrostan-19-oic acid	**(399)**	[*41033-59-4*]
11,12-didehydro-3a,5,5a,7,8,9,10,13,13a,13b-decahydro-9-hydroxy-3a-methyl-1*H*-cyclodec[*e*]indene-3,6(2*H*,4*H*)-dione	**(400)**	[*41033-64-1*]

Table 3 (*continued*)

Steroid	Structure No.	CAS Registry No.
B-homo-*A*-norestr-5(10)-ene-3,17-dione	**(401)**	[*23760-16-9*]
5α-androstane-5,6-α,17β-triol-17-benzoate-6-*p*-toluenesulfonate	**(402)**	[*19667-05-1*]
17β-hydroxy-*A*-homonor-5β-androstan-4-one, benzoate	**(403)**	[*19667-06-2*]
17a-methyl-*D*(17a)-homo-*C*,18-dinor-5α-spirost-17-en-3β-ol, acetate	**(404)**	[*84024-86-2*]
3-(acetyloxy)-17a-methyl-*D*-homo-*C*,18-dinor-5α-pregn-17-en-20-one	**(405)**	[*59274-76-9*]
17-hydroxy-17a-methyl-*D*-homo-*C*,18-dinorandrost-4-en-3-one	**(406)**	[*3818-35-7*]
10,11-didehydro-3a,4,5,5a,7,8,10,13,13a,13b-decahydro-3a-methyl-1*H*-cyclodec[*e*]indene-3,6,9(2*H*)-trione	**(407)**	[*60426-56-4*]
3-acetyl-11,12-didehydro-2,3,3a,4,5,5a,7,8,10,13,13a,13b-dodecahydro-3a-methyl-1*H*-cyclodec[*e*]indene-6,9-dione	**(408)**	[*55512-68-0*]

BIBLIOGRAPHY

"Sterols and Steroids" in *ECT* 1st ed., Vol. 12, pp. 917–947, by R. B. Turner, The Rice Institute, and L. I. Conrad, American Cholesterol Products, Inc.; "Steroids" in Suppl. Vol. 1, pp. 848–888 by G. Anner and A. Wettstein, Ciba, Ltd.; "Steroids" in *ECT* 2nd ed., Vol. 18, pp. 830–896, by D. Taub and T. B. Windholz, Merck, Sharp & Dohme.

1. L. F. Fieser and M. Fieser, *Steroids*, Reinhold Publishing Corporation, New York, 1959.
2. *Rodd's Chemistry of Carbon Compounds*, Vol. 2, Parts D and E, Elsevier Publishing Company, London, 1970.
3. *Elsevier's "Encyclopedia of Organic Chemistry,"* Vol. 14 and supplements, Elsevier Publishing Company, Amsterdam, 1940–1965.
4. W. F. Johns, ed., *MTP International Review of Science*, Vol. 8, Butterworths, London, Series 1, 1973; Series 2, 1976.
5. J. Fried and J. A. Edwards, eds., *Organic Reactions in Steroid Chemistry*, Vols. 1 and 2, Van Nostrand Rheinhold Company, New York, 1972.
6. *Terpenoids and Steroids*, The Chemical Society, London. This annual series has summarized chemical work in steroids since 1971.
7. M. de Fourcroy, *Ann. Chim.*, 242 (1789).
8. H. Fischer, *Z. Physiol. Chem.* **73,** 234 (1911).
9. H. Wieland, E. Dane, and E. Scholz, *Z. Physiol. Chem.* **211,** 261 (1932); A. Windaus, *Ber.* **42,** 3770 (1909).
10. A. Windaus, *Ber.* **36,** 3752 (1903); O. Diels and E. Abderhalden, *Ber.* **36,** 3177 (1903).
11. H. Wieland, *Z. Physiol. Chem.* **80,** 287 (1912).
12. J. D. Bernal, *Nature* **129,** 277 (1932); *Chem. and Ind.* **51,** 466 (1932).
13. O. Rosenheim and H. King, *Nature* **130,** 315 (1932); *Chem. and Ind.* **51,** 954 (1932); **52,** 299 (1933).
14. H. Wieland and E. Dane, *Z. Physiol. Chem.* **210,** 268 (1932).
15. O. Diels, W. Gädke, and P. Körding, *Ann.* **459,** 1 (1927).
16. W. E. Bachmann, W. Cole, and A. L. Wilds, *J. Amer. Chem. Soc.* **61,** 974 (1939); **62,** 824 (1940).
17. Ref. 1, p. 445.
18. P. A. Lehmann, A. Bolvar, and R. Quintero, *J. Chem. Ed.* **50,** 195 (1973).
19. Ref. 1, p. 634.
20. D. H. R. Barton, *Experientia* **6,** 316 (1950).
21. H. Sachse, *Ber. Dtsch. Chem. Ges.* **23,** 1363 (1890).
22. O. Hassel, *Tidsskr. Kjemi Bergves. Metall.* **3,** 32 (1943).
23. A. J. Birch, *J. Chem. Soc.*, 367 (1950).
24. P. Karlson, *Dev. Endocrinol.* **7,** 1 (1980).
25. E. X. Alburquerque, J. W. Daly, and B. Witkop, *Science* **172,** 995 (1971).
26. IUPAC-IUB Committee on the Nomenclature of Organic Chemistry, *Pure Appl. Chem.* **31,** 283 (1972); *ibid.*, *Biochem. J.* **127,** 613 (1972).
27. *Ibid.*, *J. Am. Chem. Soc.* **82,** 5577 (1960).
28. *Ibid.*, *Steroids* **13,** 277 (1969); P. E. Verkade and D. Hoffmann-Ostenhof, *J. Org. Chem.* **34,** 1517 (1969); IUPAC-IUB Committee on the Nomenclature of Organic Chemistry, *Arch. Biochem. Biophys.* **147,** 4 (1971).

29. H. Selye, S. Szabo, P. Kourounakis, and Y. Tache, *Adv. Steroid Biochem. Pharmacol.* **3,** 1 (1972).
30. J. R. Sabine, *Cholesterol*, Marcel Dekker, Inc., New York, 1977, pp. 79–103.
31. R. O. Brady and S. Gurin, *J. Biol. Chem.* **189,** 371 (1951).
32. J. L. Rabinowitz and S. Gurin, *J. Biol. Chem.* **208,** 307 (1954).
33. H. C. Rilling and W. W. Epstein, *J. Am. Chem. Soc.* **91,** 1041 (1969).
34. E. E. van Tamelen, J. D. Willett, R. B. Clayton, and K. E. Lord, *J. Am. Chem. Soc.* **88,** 4752 (1966); E. J. Corey, W. E. Russey, and P. R. Ortiz de Montellano, *J. Am. Chem. Soc.* **88,** 4750 (1966).
35. R. B. Woodward and K. Bloch, *J. Am. Chem. Soc.* **75,** 2023 (1953); A. Eschenmoser, L. Ruzicka, O. Jeger, and D. Arigoni, *Helv. Chim. Acta* **38,** 1890 (1955); W. G. Dauben, T. W. Hutten, and G. A. Boswell, *J. Am. Chem. Soc.* **81,** 403 (1959); J. W. Cornforth, R. H. Cornforth, A. Pelter, M. G. Horning, and G. Popjak, *Proc. Chem. Soc. (London)*, 112 (1958); R. K. Mandgal, T. T. Tchen, and K. Bloch, *J. Am. Chem. Soc.* **80,** 2589 (1958).
36. E. E. van Tamelen, *Acc. Chem. Res.* **8,** 152 (1975); W. S. Johnson, *Bioorg. Chem.* **5,** 51 (1976).
37. B. Yagen, J. S. O'Grodnick, E. Caspi, and C. Tamm, *J. Chem. Soc. Perkin Trans. 1*, 1994 (1974).
38. K. Alexander, M. Akhtar, R. B. Boar, J. F. McGhie, and D. H. R. Barton, *J. Chem. Soc. Chem. Commun.*, 1479 (1971); G. F. Gibbons and K. A. Mitropoulos, *Biochim. Biophys. Acta* **380,** 270 (1975).
39. E. Caspi, J. P. Moreau, and P. J. Ramm, *J. Am. Chem. Soc.* **96,** 8107 (1974); L. J. Goad, *Biochem. Soc. Symp.* **29,** 45 (1970).
40. E. D. Beytia and J. W. Porter, *Ann. Rev. Biochem.* **45,** 113 (1976); K. Bloch, *Science* **150,** 19 (1965).
41. G. Popjak, *Ann. Rev. Biochem.* **27,** 533 (1958); J. W. Cornfurth, *Tetrahedron* **30,** 1515 (1974); G. J. Schroepfer, Jr., *Ann. Rev. Biochem.* **50,** 585 (1981); **51,** in press (1982).
42. J. S. Baran, *Intra Sci. Chem. Rep.* **3,** 51 (1969); R. E. Counsell and R. Brueggemeier in M. E. Wolff, ed., *Burger's Medicinal Chemistry*, 4th ed., John Wiley & Sons, Inc., New York, 1979, Chapt. 28, p. 873; R. Deghenghi and M. L. Givner in Chapt. 29, p. 917.
43. Ref. 2, p. 73.
44. J. S. Baran, *Ann. Rep. Med. Chem.* **10,** 317 (1975).
45. B. Khan, G. E. Cox, and K. Asdel, *Arch. Pathol.* **76,** 369 (1963); R. P. Cook, ed., *Cholesterol: Chemistry, Biochemistry and Pathology*, Academic Press, Inc., New York, 1958.
46. R. H. Ode, Y. Kamano, and G. R. Pettit in ref. 4, Series 1, 1973, p. 152.
47. H. H. Sauer, R. D. Bennett, and E. Heftmann, *Phytochemistry* **7,** 2027 (1968); N. J. de Souza, E. L. Ghisalberti, H. H. Reese, and T. W. Goodwin, *Biochem. J.* **114,** 395 (1969); H. Hikino, T. Tohama, and T. Takemoto, *Chem. Pharm. Bull.* **17,** 415 (1969).
48. Ref. 1, p. 504.
49. M. Nagasawa, M. Bae, G. Tamura, and K. Arima, *Agric. Biol. Chem.* **33,** 1636 (1969).
50. A. A. Kandutsch, H. W. Chen, and H.-J. Heiniger, *Science* **201,** 498 (1978).
51. H. Jones and G. H. Rasmussen, *Fortschr. Chem. Org. Naturst.* **39,** 63 (1980).
52. A. Windaus, O. Linsert, A. Luttringhaus, and G. Weidlich, *Justus Liebigs Ann. Chem.* **492,** 226 (1932); A. Windaus, Fr. Schenk, and F. von Werder, *Z. Physiol. Chem.* **241,** 100 (1936).
53. R. B. Woodward and R. Hoffmann, *The Conservation of Orbital Symmetry*, Verlag Chemie, Weinheim, FRG, 1971.
54. H. J. C. Jacobs and E. Havinga, *Adv. Photochem.* **11,** 305 (1979).
55. A. Windaus and A. Hauth, *Ber. Dtsch. Chem. Ges.* **39,** 4378 (1906).
56. J. M. Whitmarsch, *Biochem. J.* **90,** 23 (1964).
57. M. E. Herr and F. W. Heyl, *J. Am. Chem. Soc.* **74,** 3627 (1952).
58. C. K. A. Martin, *Adv. Appl. Microbiol.* **22,** 28 (1977).
59. R. B. Boar and C. R. Romer, *Phytochemistry* **14,** 1143 (1975).
60. W. Steglich, M. Klaar, L. Zechlin, and H. J. Hecht, *Angew. Chem. Int. Ed. Engl.* **18,** 698 (1979).
61. E. Premuzic, *Fortschr. Chem. Org. Naturst.* **29,** 417 (1971).
62. R. M. K. Carlson, S. Popov, I. Massey, C. Delseth, E. Ayangolu, T. H. Varkony, and C. Djerassi, *Bioorg. Chem.* **7,** 453 (1978).
63. C. Djerassi, N. Theobald, W. C. M. C. Kokke, C. S. Pak, and R. M. K. Carlson, *Pure Appl. Chem.* **51,** 1815 (1979); L. J. Goad, *Pure Appl. Chem.* **51,** 837 (1981); C. Djerassi, *Pure Appl. Chem.* **51,** 873 (1981).
64. R. L. Hale, J. LeClerq, B. Turch, C. Djerassi, R. A. Gross, Jr., A. J. Weinheimer, K. Gupta, and P. J. Scheuer, *J. Am. Chem. Soc.* **92,** 2179 (1970); R. L. Hale, N. C. Ling, and C. Djerassi, *J. Am. Chem. Soc.* **92,** 5281 (1970).
65. E. Fattorusso, S. Magno, L. Mayol, C. Santacroce, and D. Sica, *Tetrahedron* **31,** 1715 (1975).

66. N. Theobald, J. N. Shoolery, C. Djerassi, T. R. Erdman, and P. J. Scheuer, *J. Am. Chem. Soc.* **100,** 5574 (1978).
67. R. A. Ross and P. J. Scheuer, *Tetrahedron Lett.*, 4701 (1979).
68. Ref. 4, Series 1, p. 198; Series 2, p. 195.
69. D. Lavie and E. Glotter, *Fortschr. Chem. Org. Naturst.* **29,** 307 (1971).
70. P. P. Nair and D. Kritchevsky, eds., *The Bile Acids*, Vols. 1 and 2, Plenum Publishing Corporation, New York, 1971, 1973.
71. K. Bloch, B. N. Berg, and D. Rittenberg, *J. Biol. Chem.* **149,** 511 (1943).
72. H. Danielsson and J. Sjovall, *Ann. Rev. Biochem.* **74,** 233 (1975).
73. J. Elks in ref. 2, Part E, Chapt. 18.
74. R. E. Marker and E. Rohrmann, *J. Am. Chem. Soc.* **61,** 2072 (1939).
75. J. Elks, G. H. Phillipps, T. Walker, and L. J. Wyman, *J. Chem. Soc.*, 4330 (1956).
76. J. H. Chapman, J. Elks, G. H. Phillipps, and L. J. Wyman, *J. Chem. Soc.*, 4344 (1956).
77. A. R. Pinder in *Rodd's Chemistry of the Carbon Compounds*, Vol. 4, Part G, Elsevier Publishing Company, London, 1978, Chapt. 35, p. 381.
78. R. Pappo, *J. Am. Chem. Soc.* **81,** 1010 (1959).
79. D. N. Kirk and M. S. Rajagopalan, *J. Chem. Soc. Perkin Trans. 1*, 1860 (1975).
80. K. Schreiber, *Pure Appl. Chem.* **21,** 131 (1970).
81. V. Černý and F. Šorm, *Alkaloids* **9,** 305 (1967); J. Tomko and Z. Votický, *Alkaloids* **14,** 1 (1973).
82. Ref. 1, Chapt. 22, p. 847.
83. J. P. Kutney, *Bioorg. Chem.* **6,** 371 (1977); E. Brown and M. Ragult, *Tetrahedron* **35,** 911 (1979).
84. K. Oka and S. Hara, *J. Am. Chem. Soc.* **99,** 3859 (1977).
85. G. Habermehl, *Alkaloids* **9,** 427 (1967).
86. R. B. Rao and L. Weiler, *Tetrahedron Lett.*, 4971 (1973); Y. Shimizu, *J. Org. Chem.* **41,** 1930 (1976).
87. M. Benn and R. Shaw, *Can. J. Chem.* **52,** 2936 (1974).
88. T. Tokuyama, J. Daly, and B. Witkop, *J. Am. Chem. Soc.* **91,** 3931 (1969).
89. R. Imhof, E. Gössinger, W. Graf, L. Berner-Fenz, H. Berner, R. Schaufelberger, and H. Wehrli, *Helv. Chim. Acta* **56,** 139 (1973).
90. B. F. Hoffman and J. T. Bigger, Jr., in A. G. Gilman, L. S. Goodman, and A. Gilman, eds., *The Pharmaceutical Basis of Therapeutics*, 6th ed., MacMillan, Inc., New York, 1980, Chapt. 30, p. 729.
91. R. Tschesche, *Ber. Dtsch. Chem. Ges.* **68,** 7 (1935); E. Yoshii and K. Ozaki, *Chem. Pharm. Bull. (Japan)* **20,** 1585 (1972).
92. R. Deghenghi, *Pure Appl. Chem.* **21,** 153 (1970); R. Thomas, J. Boutagy, and A. Gelbart, *J. Pharm. Sci.* **63,** 1649 (1974); M. B. Gorovits and N. K. Abubakirov, *Khim. Prir. Soed.* **14,** 283 (1978); *Chem. Nat. Cmpd.* **14,** 233 (1978).
93. N. Danieli, Y. Mazur, and F. Sondheimer, *Tetrahedron* **22,** 3189 (1966); R. Deghenghi, A. Philipp, and R. Gaudry, *Tetrahedron Lett.*, 2045 (1963).
94. H.-G. Lehmann and R. Wiechert, *Angew. Chem. Int. Ed. Engl.* **7,** 300 (1968).
95. G. R. Lenz and J. A. Schulz, *J. Org. Chem.* **43,** 2334 (1978); G. R. Pettit, C. L. Herald, and J. P. Yardley, *J. Org. Chem.* **35,** 1389 (1970).
96. R. H. Ode, G. R. Pettit, and Y. Kamano in ref. 4, Series 2, p. 159.
97. Ref. 2, p. 408.
98. R. H. Ode, Y. Kamano, and G. R. Pettit in ref. 4, Series 1, p. 165.
99. J. S. Grossert, *Chem. Soc. Rev.* **1,** 1 (1972).
100. I. Kitagawa, T. Nishino, T. Matsuro, H. Akutsu, and Y. Kyogoku, *Tetrahedron Lett.*, 985 (1978); A. Clastres, A. Ahond, C. Poupat, P. Potier, and A. Intes, *Experientia* **34,** 973 (1978).
101. W. L. Tau, C. Djerassi, J. Fayos, and J. Clardy, *J. Org. Chem.* **40,** 466 (1975).
102. G. Habermehl and K. H. Seib, *Naturwissenschaften* **65,** 155 (1978).
103. A. B. Kundu, A. Mukherjee, and A. K. Dey, *J. Sci. Ind. Res.* **35,** 616 (1976).
104. R. N. Tursunova, V. A. Maslennikova, and N. K. Abubakirov, *Khim. Prir. Soedin.* **13,** 147 (1977); *Chem. Nat. Cmpd.* **13,** 131 (1977); A. V. Kamernitskii, I. G. Reshetova, and V. A. Krivoruchko, *Khim. Prir. Soedin.* **13,** 156 (1977); *Chem. Nat. Cmpd.* **13,** 138 (1977).
105. A. W. Barksdale, *Science* **166,** 831 (1969).
106. G. P. Arsenault, K. Biemann, A. W. Barksdale, and T. C. McMorris, *J. Am. Chem. Soc.* **90,** 5635 (1968).
107. J. A. Edwards, J. S. Mills, J. Sundeen, and J. H. Fried, *J. Am. Chem. Soc.* **91,** 1248 (1969).
108. J. A. Edwards, J. Sundeen, W. Salmund, T. Iwadare, and J. H. Fried, *Tetrahedron Lett.*, 791 (1972).

109. K. Nakanishi, *Pure Appl. Chem.* **25,** 167 (1971); H. Hikino and Y. Hikino, *Fortschr. Chem. Org. Naturst.* **28,** 256 (1970); C. E. Berkoff, *Q. Rev. Chem. Soc.* **23,** 372 (1969).
110. A. Butanandt and P. Karlson, *Z. Naturforsch.* **9b,** 389 (1954).
111. R. Huber and W. Hoppe, *Chem. Ber.* **98,** 2403 (1965).
112. K. Nakanishi in K. Nakanishi, T. Goto, S. Itô, S. Natori, and S. Nozoe, eds., *Natural Products Chemistry*, Vol. 1, Academic Press, Inc., New York, 1974, p. 527.
113. K. Nakanishi, M. Koreeda, S. Sasaki, M. L. Chang, and H. Y. Hsu, *Chem. Commun.*, 915 (1966).
114. D. A. Shooley, G. Weiss, and K. Nakanishi, *Steroids* **19,** 377 (1972).
115. A. Prakash and S. Ghosal, *J. Sci. Ind. Res.* (*India*) **38,** 632 (1979).
116. J. B. Siddall, A. D. Cross, and J. H. Fried, *J. Am. Chem. Soc.* **88,** 862 (1966).
117. U. Kerb, G. Schultz, P. Hocks, R. Wiechert, A. Furlenmeier, A. Fürst, A. Langemann, and G. Waldvogel, *Helv. Chim. Acta* **49,** 1601 (1966).
118. A. Furlenmeier, A. Furst, A. Langemann, G. Waldvogel, P. Hocks, U. Kerb, and R. Wiechert, *Helv. Chim. Acta* **50,** 2387 (1967).
119. H. Mori, K. Shibata, K. Tsuneda, and M. Sawai, *Chem. Pharm. Bull. Jpn.* **16,** 563, 2416 (1968).
120. E. Lee, Y.-T. Liu, P. H. Solomon, and K. Nakanishi, *J. Am. Chem. Soc.* **98,** 1634 (1976).
121. D. H. R. Barton, P. G. Feakins, J. P. Poyser, and P. G. Sammes, *J. Chem. Soc. C*, 1584 (1970).
122. D. Onken and D. Onken, *Pharmazie* **35,** 193 (1980).
123. C. Djerassi, *Proc. R. Soc. London Ser. B* **195,** 175 (1976).
124. R. E. Marker and E. Rohrmann, *J. Am. Chem. Soc.* **61,** 3592 (1939); **62,** 518 (1940); R. E. Marker, R. B. Wagner, P. R. Ulshafer, E. L. Wittbecker, D. P. J. Goldsmith, and C. H. Ruof, *J. Am. Chem. Soc.* **69,** 2167 (1947).
125. E. P. Oliveto in ref. 5, Vol. 2, Chapt. 11, p. 127.
126. G. R. Lenz, *Synthesis*, 487 (1978).
127. T. F. Gallagher and W. P. Long, *J. Biol. Chem.* **162,** 495, 521 (1946); J. Elks, G. H. Phillips, T. Walker, and L. J. Wyman, *J. Chem. Soc.*, 4330 (1956); J. H. Chapman, J. Elks, G. H. Phillips, and L. J. Wyman, *J. Chem. Soc.*, 4344 (1956).
128. G. Rosenkranz and F. Sondheimer, *Fortschr. Chem. Org. Naturst.* **10,** 274 (1953).
129. D. H. Peterson, H. C. Murray, S. H. Epstein, L. M. Reinecke, A. Weintraub, P. D. Meister, and H. M. Leigh, *J. Am. Chem. Soc.* **74,** 5933 (1952); S. H. Epstein, P. D. Meister, D. H. Peterson, H. C. Murray, H. M. Leigh, D. A. Lyttle, L. M. Reinecke, and A. Weintraub, *J. Am. Chem. Soc.* **75,** 408 (1953).
130. F. M. Heyl and M. E. Herr, *J. Am. Chem. Soc.* **72,** 2617 (1950); **79,** 3627 (1952); J. A. Hogg, P. F. Beal, A. H. Nathan, F. H. Lincoln, W. P. Schneider, B. J. Mayerlein, A. R. Hanze, and R. W. Jackson, *J. Am. Chem. Soc.* **77,** 4436, 4438 (1955); V. Van Rheenen, R. C. Kelly, and D. Y. Cha, *Tetrahedron Lett.*, 1973 (1976).
131. L. L. Smith in ref. 6, **4,** 394 (1974).
132. M. Nagasawa, H. Hashiba, N. Watanabe, M. Bae, G. Tamura, and K. Arima, *Agric. Biol. Bull.* **34,** 801 (1970).
133. K. Arima, M. Nagasawa, M. Bae, and G. Tamura, *Agric. Biol. Bull.* **33,** 1636 (1969).
134. *Ibid.*, p. 1644.
135. W. J. Marsheck, S. Kraychy, and R. D. Muir, *Appl. Microbiol.* **23,** 72 (1972).
136. V. Van Rheenan and K. P. Shepherd, *J. Org. Chem.* **44,** 1582 (1979).
137. H. H. Imhoffen, *Angew. Chem.* **53,** 471 (1940); E. B. Hershberg, M. Rubin, and E. Schwenk, *J. Org. Chem.* **15,** 292 (1950).
138. U.S. Pat. 3,994,938 (Nov. 30, 1976), K. Wakabayashi, Y. Chigira, and K. Fukuda (to Mitsubishi Chemical Industries).
139. H. L. Dryden, Jr., G. M. Webber, and J. J. Wieczorek, *J. Am. Chem. Soc.* **86,** 742 (1964).
140. A. J. Birch and S. M. Mukherji, *J. Chem. Soc.*, 2531 (1949); A. L. Wilds and N. A. Nelson, *J. Am. Chem. Soc.* **75,** 5360, 5366 (1953).
141. H. L. Dryden, Jr., in ref. 5, Vol. 1, Chapt. 1, p. 1.
142. K. Heusler and J. Kalvoda in ref. 5, Vol. 2, Chapt. 12, p. 237; J. Kalvoda and K. Heusler, *Synthesis*, 501 (1971).
143. H. Ueberwasser, K. Heusler, J. Kalvoda, Ch. Meystre, P. Wieland, G. Anner, and A. Wettstein, *Helv. Chim. Acta* **46,** 344 (1963).
144. A. Bowers, R. Villoti, J. A. Edwards, E. Denot, and O. Halpern, *J. Am. Chem. Soc.* **84,** 3204 (1962).
145. U.S. Pat. 3,176,014 (March 30, 1965), R. Pappo and L. Nysted (to G. D. Searle).
146. R. H. Hesse, *Adv. Free Radical Chem.* **3,** 83 (1969).
147. K. Heusler and J. Kalvoda in ref. 5, Vol. 2, p. 254.

148. C. J. Sih and K. C. Wang, *J. Am. Chem. Soc.* **87,** 1387 (1965).
149. C. J. Sih, *Prix Roussel Address*, Paris, France, April 24, 1980.
150. C. J. Sih, S. S. Lee, Y. Y. Tsong, K. C. Wang, and F. N. Chang, *J. Am. Chem. Soc.* **81,** 2765 (1965).
151. K. Heusler, P. Wieland, and Ch. Meystre, *Org. Synth.* **5,** 692 (1973).
152. G. V. Baddeley, H. Carpio, and J. A. Edwards, *J. Org. Chem.* **31,** 1026 (1966); V. Cerný, A. Kasal, and F. Šorm, *Coll. Czech. Chem. Commun.* **31,** 1752 (1966); G. R. Lenz, *J. Chem. Soc. Chem. Commun.*, 241 (1978).
153. J. Kalvoda, Ch. Meystre, and G. Anner, *Helv. Chim. Acta* **49,** 424 (1965); J. Kalvoda, *Helv. Chim. Acta* **51,** 267 (1968); J. Kalvoda, *J. Chem. Soc. Chem. Commun.*, 1002 (1970); J. Kalvoda and L. Botta, *Helv. Chim. Acta* **55,** 356 (1972).
154. J. Kalvoda, J. Grob, and G. Anner, *Helv. Chim. Acta* **60,** 1579 (1977).
155. D. S. Watt, *J. Am. Chem. Soc.* **98,** 272 (1976); R. W. Freeksen, W. E. Pabst, M. L. Raggio, S. A. Sherman, R. R. Wroble, and D. S. Watt, *J. Am. Chem. Soc.* **99,** 1536 (1977).
156. G. Neef, U. Eder, G. Haffer, G. Sauer, and R. Wiechert, *Chem. Ber.* **113,** 1106 (1980).
157. J. Schmidlin, G. Anner, J. R. Billeter, and A. Wettstein, *Experientia* **11,** 365 (1955); K. Heusler, H. Ueberwasser, P. Wieland, J. R. Billeter, J. Schmidlin, G. Anner, and A. Wettstein, *Helv. Chim. Acta* **42,** 1586 (1959); A. Lardon, O. Schindler, and T. Reichstein, *Helv. Chim. Acta* **40,** 666 (1957); W. J. Van der Burg, D. A. van Dorp, O. Schindler, C. M. Siegman, and S. A. Szpilfogel, *Rec. Trav. Chim.* **77,** 171 (1958); W. S. Johnson, J. C. Collins, R. Pappo, and M. B. Rubin, *J. Am. Chem. Soc.* **80,** 2585 (1958).
158. D. H. R. Barton, G. M. Beaton, G. M. Geller, and M. M. Pechet, *J. Am. Chem. Soc.* **82,** 2640, 2641 (1960); **83,** 750, 1771, 2400 (1961); D. H. R. Barton, N. K. Basu, M. J. Day, R. H. Hesse, M. M. Pechet, and A. N. Starratt, *J. Chem. Soc. Perkin Trans. 1*, 2243 (1975).
159. P. Buchschacher, M. Cereghetti, H. Wehrli, K. Schaffner, and O. Jeger, *Helv. Chim. Acta* **42,** 2122 (1959); N. C. Yang and D. D. H. Yang, *Tetrahedron Lett.*, 10 (1960).
160. M. Miyano, *J. Org. Chem.* **46,** 1846 (1981).
161. D. M. Piatak and J. Wicha, *Chem. Rev.* **78,** 199 (1978).
162. J. Wicha, K. Bal, and S. Piekut, *Synth. Comm.* **7,** 215 (1977).
163. J. Wicha and K. Bal, *J. Chem. Soc. Chem. Commun.*, 968 (1975); *J. Chem. Soc. Perkin Trans. 1*, 1282 (1978).
164. J. J. Partridge, S.-J. Shiuey, N. K. Chadha, E. G. Baggiolini, J. F. Blount and M. R. Uskoković, *J. Am. Chem. Soc.* **103,** 1253 (1981).
165. B. Lythgoe, *Chem. Soc. Rev.* **9,** 449 (1980).
166. A. A. Akhrem and Y. A. Titov, *Total Steroid Synthesis*, Plenum Publishing Corporation, New York, 1970; R. T. Blickenstaff, A. C. Ghosh and G. C. Wolf, *Total Synthesis of Steroids*, Academic Press, Inc., New York, 1974.
167. W. S. Johnson, J. W. Peterson, and C. D. Gutsche, *J. Am. Chem. Soc.* **67,** 2274 (1945); **69,** 2942 (1947); W. S. Johnson and V. L. Stromberg, *J. Am. Chem. Soc.* **72,** 505 (1950).
168. G. Anner and H. Miescher, *Helv. Chim. Acta* **31,** 2173 (1948); **32,** 1957 (1949); **33,** 1379 (1950).
169. W. S. Johnson and R. G. Christiansen, *J. Am. Chem. Soc.* **73,** 5511 (1951); W. S. Johnson, R. G. Christiansen, and R. E. Ireland, *J. Am. Chem. Soc.* **79,** 1995 (1957).
170. R. A. Dickinson, R. Kubela, G. A. McAlpine, Z. Stojanac, and Z. Valenta, *Can. J. Chem.* **50,** 2377 (1972).
171. S. N. Ananchenko and I. V. Torgov, *Dokl. Akad. Nauk SSR* **127,** 553 (1959).
172. G. A. Hughes and H. Smith, *Chem. Ind.* (*London*), 1022 (1960).
173. J. Weil-Raynal, *Bull. Soc. Chim. France*, 4561 (1969); S. R. Ramadas, S. Padmanabhan, and N. S. Chandrakumar, *J. Sci. Ind. Res.* **39,** 275 (1980).
174. C. H. Kuo, D. Taub, and N. L. Wendler, *Angew. Chem.* **77,** 1142 (1965); *J. Org. Chem.* **33,** 3126 (1970).
175. R. Pappo, R. B. Garland, C. J. Jung, and R. T. Nicholson, *Tetrahedron Lett.*, 1827 (1973).
176. T. Kametani and H. Nemoto, *Tetrahedron* **37,** 3 (1981).
177. T. Kametani, H. Matsumoto, H. Nemoto, and K. Fukumoto, *J. Am. Chem. Soc.* **100,** 6218 (1978); T. Kametani, H. Nemoto, H. Ishikawa, K. Shiroyama, and K. Fukumoto, *J. Am. Chem. Soc.* **98,** 3378 (1976); T. Kametani, H. Nemoto, H. Ishikawa, K. Shiroyama, H. Matsumoto, and K. Fukumoto, *J. Am. Chem. Soc.* **99,** 3461 (1977).
178. S. Djuric, T. Sarkar, and P. Magnus, *J. Am. Chem. Soc.* **102,** 6885 (1980).
179. H. M. E. Cardwell, J. W. Cornforth, S. R. Duff, H. Holtermann, and R. Robinson, *Chem. Ind.* (*London*), 389 (1951); *J. Chem. Soc.*, 361 (1953).

180. R. B. Woodward, F. Sondheimer, D. Taub, K. Heusler, and W. McLamore, *J. Am. Chem. Soc.* **73,** 2403 (1951); **74,** 4223 (1952).
181. L. B. Barkley, W. S. Knowles, H. Raffelson, and Q. E. Thompson, *J. Am. Chem. Soc.* **78,** 4111 (1956); L. B. Barkley, M. W. Farrar, W. S. Knowles, H. Raffelson, and Q. E. Thompson, *J. Am. Chem. Soc.* **75,** 4110 (1953); A. J. Speziale, J. A. Stephens, and Q. E. Thompson, *J. Am. Chem. Soc.* **76,** 5011, 5014 (1954).
182. L. H. Sarett, G. E. Arth, R. M. Lukes, R. E. Beyler, G. I. Poos, W. F. Johns, and J. M. Constantin, *J. Am. Chem. Soc.* **74,** 4974 (1952); G. E. Arth, G. I. Poos, and L. H. Sarett, *J. Am. Chem. Soc.* **77,** 3834 (1955).
183. S. Danishefsky and A. Nagel, *J. Chem. Soc. Chem. Comm.*, 373 (1972).
184. S. Danishefsky and P. Cain, *J. Am. Chem. Soc.* **97,** 5282 (1975).
185. L. Velluz, G. Nominé, J. Mathieu, E. Toromanoff, D. Bertin, J. Tessier, and A. Pierdot, *Compt. Rend. Acad. Sci. Paris* **250C,** 1084 (1960); L. Velluz, J. Mathieu, and G. Nominé, *Tetrahedron, Suppl. 8, Part II*, 495 (1966); L. J. Chinn and H. L. Dryden, Jr., *J. Org. Chem.* **26,** 3904 (1961).
186. L. Velluz, G. Amiard, and R. Heymes, *Bull. Soc. Chim. France*, 904 (1953); 1015 (1954).
187. P. Bellet and G. Nominé, *Compt. Rend. Acad. Sci. Paris* **263C,** 88 (1966).
188. Z. G. Hajos and D. R. Parrish, *J. Org. Chem.* **38,** 3244 (1973).
189. *Ibid.*, **39,** 1612, 1615 (1974).
190. *Ibid.*, **38,** 3239 (1973).
191. N. Cohen, B. L. Banner, W. F. Eickel, D. R. Parrish, G. Saucy, J.-M. Cassal, W. Meier, and A. Furst, *J. Org. Chem.* **40,** 681 (1975).
192. R. A. Micheli, Z. G. Hajos, N. Cohen, D. R. Parrish, L. A. Portland, W. Sciamanna, M. A. Scott, and P. A. Wehrli in ref. 191, p. 675; N. Cohen, B. L. Banner, W. F. Eicher, D. R. Parrish and G. Saucy in ref. 191, p. 681; N. Cohen, *Acc. Chem. Res.* **9,** 412 (1976).
193. H. Smith and co-workers, *J. Chem. Soc.*, 4472 (1964).
194. G. C. Buzby, Jr., D. Hartley, G. A. Hughes, H. Smith, B. W. Gadsby, and A. B. Jansen, *J. Med. Chem.* **10,** 199 (1967).
195. G. Sauer, U. Eder, G. Haffer, G. Neef, and R. Wiechert, *Angew. Chem. Int. Ed. Engl.* **14,** 417 (1975).
196. W. S. Johnson, *Angew. Chem. Int. Ed. Engl.* **15,** 9 (1976); W. S. Johnson, *Bioorg. Chem.* **5,** 51 (1976).
197. W. S. Johnson, B. E. McCarry, R. L. Markezich, and S. G. Boots, *J. Am. Chem. Soc.* **102,** 352 (1980).
198. G. O. Potts, A. L. Beyler, and D. F. Burnham, *Proc. Soc. Exp. Biol. Med.* **103,** 383 (1960).
199. R. O. Clinton, A. J. Manson, F. W. Stonner, A. L. Beyler, G. O. Potts, and A. Arnold, *J. Am. Chem. Soc.* **81,** 1513 (1959).
200. R. Hirschmann, N. G. Steinberg, P. Buchschacher, J. H. Fried, G. J. Kent, M. Tishler, and S. L. Steelman, *J. Am. Chem. Soc.* **85,** 120 (1963).
201. J. Hannah, K. Kelly, A. A. Patchett, S. L. Steelman, and E. R. Morgan, *J. Med. Chem.* **18,** 168 (1975); R. Hirshman, N. G. Steinberg, E. F. Schoenewaldt, W. J. Paleveda, and M. Tishler, *J. Med. Chem.* **7,** 352 (1964).
202. A. J. Manson, F. W. Stonner, H. C. Neumann, R. G. Christiansen, R. L. Clarke, J. H. Ackerman, D. F. Page, J. W. Dean, D. K. Phillips, G. O. Potts, A. Arnold, A. L. Beyler, and R. O. Clinton, *J. Med. Chem.* **6,** 1 (1963).
203. N. H. Lauersen and K. H. Wilson, *Obstet. Gynecol.* **50,** 91 (1977); *Fertil Steril.* **28,** 289 (1977).
204. R. D. Skoglund and C. A. Paulsen, *Contraception* **7,** 357 (1973).
205. G. Nathanson, G. Winters, and E. Testa, *J. Med. Chem.* **10,** 799 (1967); G. Nathanson, G. Winters, and V. Aresi, *Steroids* **13,** 383 (1969).
206. G. Nathanson, C. R. Pasqualucci, P. Radaelli, P. Schiatti, D. Selva, and G. Winters, *Steroids* **13,** 365 (1969); A. Canigga, M. Marchetti, C. Gennari, A. Vattino, and F. B. Nicolis, *Int. J. Clin. Biopharm.* **15,** 126 (1977).
207. J. A. Cella and C. M. Kagawa, *J. Am. Chem. Soc.* **79,** 4808 (1957).
208. J. A. Cella, E. A. Brown, and R. R. Burtner, *J. Org. Chem.* **24,** 743, (1959); E. A. Brown, R. D. Muir, and J. A. Cella, *J. Org. Chem.* **25,** 96 (1960).
209. G. E. Arth, H. Schwam, L. H. Sarett, and M. Glitzer, *J. Med. Chem.* **6,** 617 (1963).
210. R. M. Weier and L. M. Hofmann, *J. Med. Chem.* **18,** 817 (1975).
211. *Ibid.*, **20,** 1304 (1977).

212. W. L. M. Baird and A. M. Raid, *Br. J. Anaesth.* **39,** 775 (1967); W. R. Buckett, C. E. B. Marjoribanks, P. A. Marwick, and M. B. Morton, *Br. J. Pharmacol. Chemother.* **32,** 671 (1968).
213. T. M. Speight and G. S. Avery, *Drugs* **4,** 163 (1972).
214. H. O. Huisman in ref. 4, Series 1, p. 235; H. O. Huisman and W. N. Speckamp in ref. 4, Series 2, p. 207.
215. J. S. Baran, *IntraSci. Chem. Rep.* **3,** 43 (1969).
216. R. Pappo in ref. 215, p. 105; R. Pappo and C. Jung, *Tetrahedron Lett.*, 365 (1962).
217. R. J. Chorvat and R. Pappo, *J. Org. Chem.* **41,** 2864 (1976).
218. C. Symmes, Jr. and L. D. Quin, *J. Org. Chem.* **44,** 1048 (1979).
219. J. Fried and R. W. Thoma, *J. Am. Chem. Soc.* **75,** 5764 (1953).
220. H. O. Huisman, *Bull. Soc. Chim. France*, 13 (1968).
221. W. N. Speckamp, J. A. van Velthuysen, U. K. Pandit, and H. O. Huisman, *Tetrahedron* **24,** 5881 (1968); W. N. Speckamp, J. A. van Velthuysen, M. A. Douw, U. K. Pandit, and H. O. Huisman, *Tetrahedron* **24,** 5893 (1968).
222. J. C. Hubert, W. N. Speckamp, and H. O. Huisman, *Tetrahedron Lett.*, 4493 (1972); J. C. Hubert, W. Steege, W. N. Speckamp, H. O. Huisman, *Synth. Commun.* **1,** 103 (1971).
223. H. Evers, U. K. Pandit, and H. O. Huisman quoted in H. O. Huisman, *Angew. Chem. Int. Ed. Engl.* **10,** 450 (1971).
224. W. N. Speckamp, A. J. de Gee, and H. O. Huisman quoted in ref. 223.
225. H. Laurent and R. Wiechert in ref. 5, Vol. 2, p. 53.
226. D. N. Kirk and M. P. Hartshorn, *Steroid Reaction Mechanisms*, Elsevier Publishing Company, Amsterdam, 1968, p. 236.
227. W. G. Dauben and J. A. Ross, *J. Am. Chem. Soc.* **31,** 6521 (1959).
228. M. Wolff, W. Ho, and M. Honjoh, *J. Med. Chem.* **9,** 682 (1966).
229. R. Wiechert, O. Engelfried, U. Kerb, H. Laurent, H. Miller, and G. Schulz, *Chem Ber.* **99,** 1118 (1966); E. L. Shapiro, T. L. Popper, L. Weber, R. Neri, and H. L. Herzog, *J. Med. Chem.* **12,** 631 (1969).
230. R. Wiechert and F. Neumann, *Arzneimittelforsch.* **15,** 244 (1965); H. Laurent, G. Schulz, and R. Wiechert, *Chem. Ber.* **102,** 2570 (1969).
231. R. Wiechert, *Angew. Chem. Int. Ed. Engl.* **5,** 321 (1970).
232. H. E. Simmons, T. L. Cairns, S. A. Vladuchick, and C. M. Hoiness, *Org. React.* **20,** 1 (1973).
233. W. G. Dauben, P. Laug, and G. H. Berezin, *J. Org. Chem.* **31,** 3869 (1966).
234. N. H. Dyson, J. A. Edwards, and J. H. Fried, *Tetrahedron Lett.*, 1841 (1966).
235. G. Tarzia, N. H. Dyson, I. T. Harrison, J. A. Edwards, and J. H. Fried, *Steroids* **9,** 387 (1967); J. Pfister, H. Wehrli, and K. Schaffner, *Helv. Chim. Acta* **50,** 166 (1967).
236. G. E. Arth, G. F. Reynolds, G. H. Rasmusson, A. Chen, and A. A. Patchett, *Tetrahedron Lett.*, 291 (1974).
237. L. J. Chinn and B. N. Desai, *J. Med. Chem.* **18,** 268 (1975).
238. L. M. Hofmann, L. J. Chinn, H. A. Pedrera, M. I. Krupnick, and O. D. Suleymanov, *J. Pharmacol. Exptl. Therap.* **194,** 450 (1975); L. J. Chinn, K. W. Salamon, and B. N. Desai, *J. Med. Chem.* **24,** 1103 (1981).
239. O. Schmidt, K. Prezewowsky, G. Schulz, and R. Wiechert, *Chem. Ber.* **101,** 939 (1968).
240. R. Wiechert, D. Bittler, and G.-A. Hoyer, *Chem. Ber.* **106,** 888 (1973); G.-A. Hoyer, G. Cleve and R. Wiechert, *Chem. Ber.* **107,** 128 (1974).
241. W. F. Johns and K. W. Salamon, *J. Org. Chem.* **36,** 1952 (1971).
242. J. R. Williams and H. Ziffer, *Tetrahedron* **24,** 6725 (1968); U.S. Pat. 3,819,687 (June 25, 1974), G. R. Lenz (to G. D. Searle).
243. J. A. Waters, Y. Kondo, and B. Witkop, *J. Pharm. Sci.* **61,** 321 (1972); G. R. Lenz, *Rev. Chem. Intermed.* **4,** 369 (1981).
244. M. B. Rubin, T. Maymon, and D. Glover, *Israel J. Chem.* **8,** 717 (1970); G. R. Lenz, *Tetrahedron* **28,** 2195 (1972).
245. P. Boyle, J. A. Edwards, and J. H. Fried, *J. Org. Chem.* **35,** 2560 (1970).
246. P. Boyle and co-workers, *Proc. Int. Symp. Drug Res.*, 206 (1967).
247. P. H. Nelson, J. W. Murphy, J. A. Edwards, and J. H. Fried, *J. Am. Chem. Soc.* **90,** 1307 (1968).
248. G. R. Lenz, *Tetrahedron* **28,** 2211 (1972).
249. G. R. Lenz, *J. Org. Chem.* **44,** 4299 (1979).
250. P. Westerhof and E. H. Reerink, *Rec. Trav. Chim.* **79,** 771 (1960); J. H. Hartog and P. Westerhof, *Rec. Trav. Chim.* **84,** 918 (1965); P. Westerhof and J. H. Hartog, *Rec. Trav. Chim.* **86,** 235 (1967).
251. H. Els, G. Englert, M. Müller, and A. Fürst, *Helv. Chim. Acta* **48,** 989 (1965); J. Hartog, S. J. Halkes, L. Morsink, A. M. DeWachter, and J. L. M. A. Schlatmann, *J. Steroid Biochem.* **6,** 597 (1975).

252. M. Usković, J. Iacobelli, R. Philiou, and T. Williams, *J. Am. Chem. Soc.* **86,** 4538 (1966).
253. R. Ginsig and A. Cross, *J. Am. Chem. Soc.* **87,** 4629 (1965); E. Farkas, J. M. Owen, M. Debono, R. M. Molloy, and M. M. Marsh, *Tetrahedron Lett.*, 1023 (1966).
254. G. A. Boswell, Jr., in ref. 5, Vol. 2, Chapt. 14, p. 354.
255. N. A. Nelson and R. N. Schut, *J. Am. Chem. Soc.* **81,** 6486 (1959).
256. J. B. Jones and P. Price, *Tetrahedron* **29,** 1941 (1973).
257. G. Haffer, U. Eder, G. Neef, G. Sauer, and R. Wiechert, *Justus Liebigs Ann. Chem.*, 425 (1981).
258. *Ibid.*, *Chem. Ber.* **110,** 3377 (1977).
259. W. S. Johnson, M. Neeman, S. P. Birkeland, and N. A. Fedoruk, *J. Am. Chem. Soc.* **84,** 989 (1962).
260. A. J. Birch and G. S. R. Subba Rao, *Tetrahedron Suppl. 7*, 391 (1966).
261. R. B. Turner, *J. Am. Chem. Soc.* **75,** 3484 (1953).
262. P. H. Bentley, M. Todd, W. McCrae, M. L. Maddox, and J. A. Edwards, *Tetrahedron* **28,** 1411 (1972).
263. R. M. Scribner in ref. 5, Vol. 2, Chapt. 15, p. 408.
264. Shanghai Institute of Medicinal Industries, *Hua Hsueh Hsueh Pao* **34,** 301 (1976).
265. P. Crabbe, H. Fillion, Y. Letourneaux, E. Diczfalusy, A. Aedo, J. W. Goldzieher, A. A. Shaikh, and V. D. Castracane, *Steroids* **33,** 85 (1979).
266. Shanghai Institute of Pharmaceutical Industrial Research, *Yao Hsueh Hsueh Pao* **14,** 502 (1979).
267. U. Eder, G. Cleve, G. Haffer, G. Neef, G. Sauer, R. Wiechert, A. Furst, and W. Meier, *Chem. Ber.* **113,** 2249 (1980).
268. S. R. Ramadas, P. K. Sujeeth, T. R. Kasturi, and F. M. Abraham, *J. Sci. Ind. Res.* **35,** 571 (1976).
269. K. Schaffner in ref. 5, Vol. 2, Chapt. 13, p. 288.
270. S. V. Sunthankar and S. D. Mehendale, *Tetrahedron Lett.*, 2481 (1972).
271. L. Velluz, G. Muller, J. Mathieu, and A. Poittevin, *Compt. Rend.* **252,** 4084 (1961).
272. M. Nussim and Y. Mazur, *Tetrahedron* **24,** 5337 (1968).
273. R. Hirschmann, C. S. Snoddy, Jr., C. F. Hiskey, and N. L. Wendler, *J. Am. Chem. Soc.* **76,** 4013 (1954); J. Elks, G. H. Phillips, D. A. H. Taylor, and L. J. Wyman, *J. Chem. Soc.*, 1739 (1954).
274. W. F. Johns, *J. Org. Chem.* **29,** 2545 (1964).
275. W. F. Johns and K. W. Salamon, *J. Org. Chem.* **44,** 958 (1979).
276. W. F. Johns, *J. Org. Chem.* **36,** 711 (1971); W. F. Johns and L. M. Hofmann, *J. Med. Chem.* **16,** 568 (1973).
277. J. Fried and A. Klingsberg, *J. Am. Chem. Soc.* **75,** 4929 (1953); S. M. Kupchan and S. D. Levine, *J. Am. Chem. Soc.* **86,** 701 (1964).
278. F. H. Batzold and C. H. Robinson, *J. Am. Chem. Soc.* **97,** 2576 (1975); D. F. Covey and C. H. Robinson, *J. Am. Chem. Soc.* **98,** 5038 (1976); F. H. Batzold and C. H. Robinson, *J. Org. Chem.* **41,** 313 (1976).
279. L. Dorfman, *Chem. Rev.* **53,** 47 (1953).
280. A. I. Scott, *Interpretation of the Ultraviolet Spectra of Natural Products*, MacMillan Company, New York, 1964.
281. W. Neudert, H. Röpke, and J. B. Leane, *Atlas of Steroid Structure*, Springer Verlag, New York, 1965.
282. P. Crabbé, *Optical Rotatory Dispersion and Circular Dichroism in Organic Chemistry*, Holden-Day, Inc., San Francisco, Calif., 1965; P. Crabbé, *An Introduction to the Chrioptical Methods in Chemistry*, Mexico City, Mexico, 1971.
283. D. N. Kirk and W. Klyne, *J. Chem. Soc. Perkin Trans. 1*, 1076 (1979); 762 (1976); D. N. Kirk, *J. Chem. Soc. Perkin Trans. 1*, 2171 (1976); 2122 (1977).
284. N. S. Bhacca and D. H. Williams, *Applications of NMR Spectroscopy in Organic Chemistry*, Holden-Day, Inc., San Francisco, Calif., 1964.
285. R. E. Sievers, ed., *Nuclear Magnetic Resonance Shift Reagents*, Academic Press, Inc., New York, 1973.
286. T. C. Farrar and E. D. Becker, *Pulse and Fourier Transform NMR*, Academic Press, Inc., New York, 1971; M. L. Martin, G. J. Martin, and J.-J. Delpuech, *Practical NMR Spectroscopy*, Heyden & Son, Inc., Philadelphia, Pa., 1980.
287. E. Breitmaier and W. Voelter, *C13 NMR Spectroscopy*, Verlag Chemie, Weinheim, FGR, 1974; F. W. Wehrli and T. Wirthlin, *Interpretation of Carbon-13 NMR Spectra*, Heyden & Sons, Inc., Philadelphia, Pa., 1976; J. W. Blunt and J. B. Stothers, *Org. Magn. Reson.* **9,** 439 (1977).

288. H. Budzikiewicz, C. Djerassi, and D. H. Williams, *Mass Spectrometry of Organic Compounds*, Holden-Day, Inc., San Francisco, Calif., 1965; Z. V. Zaretskii, *Mass Spectrometry of Steroids*, John Wiley & Sons, Inc., New York, 1976.
289. W. J. A. Van den Heuvel, J. L. Smith, G. Albers-Schönberg, B. Plazonnet, and P. Bélanger in E. Heftmann, ed., *Modern Methods of Steroid Analysis*, Academic Press, Inc., New York, 1973, Chapt. 7.
290. N. A. B. Gray, A. Buchs, D. H. Smith, and C. Djerassi, *Helv. Chim. Acta* **64,** 458 (1981), and previous articles in this series.
291. W. L. Duax and D. A. Norton, *Atlas of Steroid Structure*, IFI/Plenum Data, New York, 1975.
292. Ref. 6, Vol. 4, 1974, p. 531.

General References

Refs. 1–6 are also general references.

W. R. Nes and M. L. McKean, *The Biochemistry of Steroids and Other Isopentenoids*, University Park Press, Baltimore, Md., 1977.

A. S. Mackenzie, S. C. Russell, G. Eglington, and J. R. Maxwell, *Science* **217,** 491 (1982).

Biosynthesis (London), Royal Chemical Society, annual publication since 1973.

Alkaloids (London), Royal Chemical Society, annual publication since 1970.

R. F. Witzmann, *Steroids: Keys to Life*, Van Nostrand Reinhold, New York, 1981, for a mixed audience of drug scientists and the lay public.

GEORGE R. LENZ
Searle Laboratories

STILBENE DERIVATIVES. See Brighteners, fluorescent; Stilbene dyes.

STILBENE DYES

In many cases, stilbene dyes are mixtures of indeterminate structure resulting from the condensation of 5-nitro-2-toluenesulfonic acid (**2**) with itself or with other aromatic compounds. The chemical moiety 2,2′-stilbenedisulfonic acid [*28097-15-6*] (**1**), either as the free acid or in salt form, is common to all stilbene dyes:

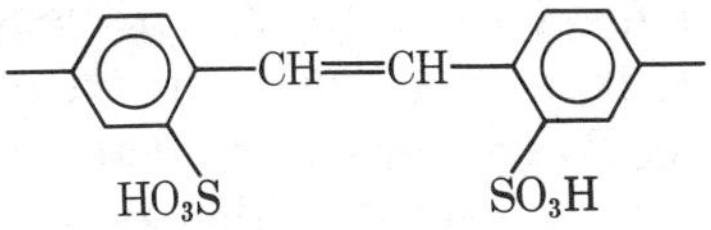

—C₆H₃(SO₃H)—CH=CH—C₆H₃(SO₃H)—

HO_3S SO_3H

(**1**)

The initial self-condensation products of 5-nitro-2-toluenesulfonic acid (**2**) are 4,4′-dinitroso-2,2′-stilbenedisulfonic acid [*58058-72-3*] (**3**), 4,4′-dinitro-2,2′-stilbenedisulfonic acid [*128-42-7*] (**4**), or 4,4′-dinitrodibenzyl-2,2′-disulfonic acid [*6268-17-3*] (**5**); these can act as intermediates for secondary condensations with arylamines or.

with aminoazo compounds.

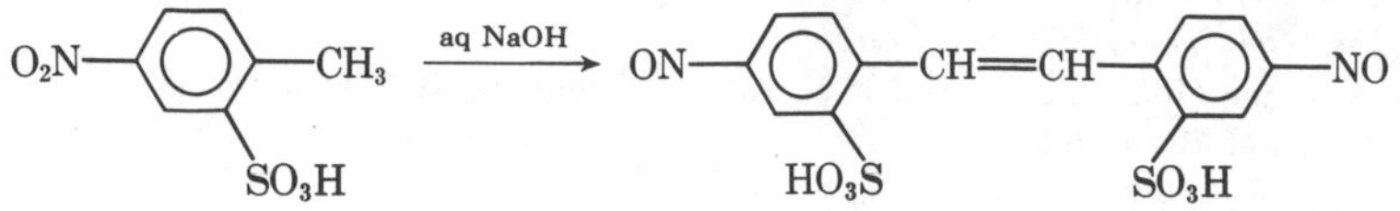

O_2N–C₆H₃(SO_3H)–CH_3 (2) $\xrightarrow{\text{aq NaOH}}$ ON–C₆H₃(SO_3H)–CH=CH–C₆H₃(SO_3H)–NO (3)

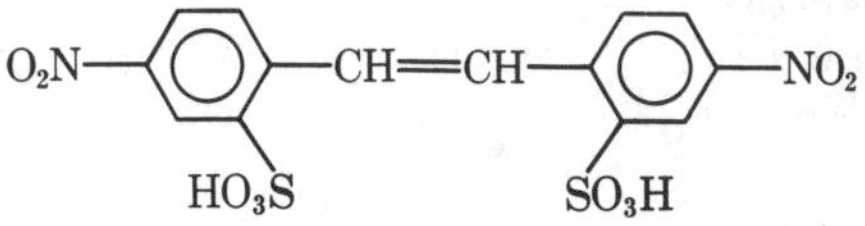

O_2N–C₆H₃(SO_3H)–CH=CH–C₆H₃(SO_3H)–NO_2 (4)

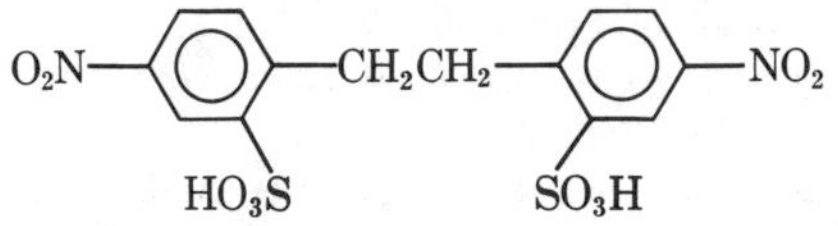

O_2N–C₆H₃(SO_3H)–CH_2CH_2–C₆H₃(SO_3H)–NO_2 (5)

In the *Colour Index*, stilbene dyes are classified according to the complexity of these condensation reactions:

CI Number	*Remarks*
40000–40006	Self-condensation products of 5-nitro-2-toluenesulfonic acid or its derivatives; further products after oxidation or reduction.
40015–40030	Condensation of 5-nitro-2-toluenesulfonic acid and its derivatives with phenols, naphthols, or aminophenols.
40045–40070	Condensation of 5-nitro-2-toluenesulfonic acid and its derivatives with aromatic amines.
40205–40295	Condensation of 5-nitro-2-toluenesulfonic acid and its derivatives with aminoazo compounds.
40500–40510	Azostilbene dyes formed by diazotization of a condensation product containing primary amino groups and coupling with azo dye coupling components.

Because the proportions and concentrations of the reactants and the reaction conditions, eg, time and temperature, vary in these condensation reactions, the exact chemical constitution cannot be given. Therefore, the structure given in the *Colour Index* characterizes a group of related products having similar dyeing and fastness properties.

Dyes prepared by reducing 4,4′-dinitro-2,2′-stilbenedisulfonic acid (**4**) to the diamino compound, tetrazotizing, and coupling are treated as normal azo dyes (qv) with CI numbers 24860–24910. These may be considered as undergoing stoichiometric reactions with resulting exact chemical composition. In general, stilbene dyes are direct dyes used for the dyeing of cellulosic fibers. Some are used extensively in the dyeing of paper and leather. Their lightfastness is excellent. Colors range from lemon-yellows through blacks.

Starting Materials and Manufacture

5-Nitro-2-Toluenesulfonic Acid. The earliest stilbene dyes were reported in 1883 and were manufactured from the intermolecular condensation of 5-nitro-2-toluenesulfonic acid (**2**) (1). When (**2**) is heated with aqueous sodium hydroxide, a variety of yellow dyes is obtained according to the concentration of the caustic, the temperature, and the time of reaction. These are collectively grouped as Direct Yellow 11 [*1325-37-7*] (CI 40000). Economically, they are by far the most important of the stilbene dyes. Although each manufacturer may vary the time, temperature, and concentration, a general procedure is to heat (**2**) at 70–80°C in aqueous sodium hydroxide with an initial concentration of 4.5 wt %; this concentration is increased to 17.5 wt % over 5 h. Although such procedures result in mixtures of undetermined chemical constitution, it is generally agreed that the first step in the self-condensation of (**2**) is the formation of 4,4′-dinitrosostilbenedisulfonic acid (**3**). Continued heating in aqueous caustic yields what is formulated as the nitro-azoxy-tetrakis-stilbene structure [*1325-37-7*] (**6**) (2).

(**2**) —aq NaOH→ (**3**) —aq NaOH→ (**6**)

Such representation as a linear, polymeric condensation product is favored over earlier proposals of closed cyclic structures which are stereochemically impossible (3–4). Structure (**6**) is known as Sun Yellow or Curcumine S. In 1981, 12 dyestuff companies sold the product.

If a reducing agent is added either in the initial condensation of (**2**) or as an aftertreatment of an aqueous solution of (**4**), additional yellow and orange dyes are produced. If glycerol, glucose, or sodium sulfide is used as the reducing agent, Direct Orange 15 [*1325-35-5*] (CI 40002/3) is produced. This is considered to be a mixture of (**6**) and its reduction product (represented as (**7**)), which is called Diamine Orange D.

(**2**) —glycerol, glucose or Na_2S→ (**6**) + (**7**)

It is still widely used in the dyeing of paper with reported U.S. sales in 1979 of more than 10^6 \$. The terminal free amino group can react further with acetic anhydride to produce Direct Orange 22 [*1325-40-2*] (**8**) (CI 40004, Diamine Orange GR).

(**7**) $\xrightarrow{(CH_3CO)_2O}$

SO_3Na SO_3Na SO_3Na SO_3Na

N–⟨⟩–CH=CH–⟨⟩–N=N–⟨⟩–CH=CH–⟨⟩–$NHCOCH_3$

‖

N–⟨⟩–CH=CH–⟨⟩–N=N–⟨⟩–CH=CH–⟨⟩–$NHCOCH_3$

SO_3Na SO_3Na SO_3Na SO_3Na

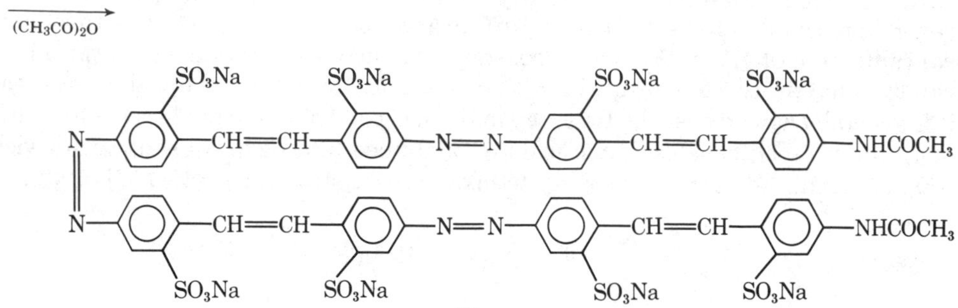

(**8**)

If formaldehyde is used as the reducing agent of (**2**) or (**4**), Direct Yellow 6 [*1325-38-8*] (CI 40001, Stilbene Yellow 3GX) is the product. Direct Yellow 6 is used primarily for paper dyeing.

These same dyes can be produced by alkaline oxidation of (**2**) with sodium hypochlorite or oxygen with isolation of the intermediate 4,4′-dinitro-2,2′-stilbenedisulfonic acid (**4**). Reaction conditions involve heating of (**2**) with 3 wt % sodium hydroxide to 50–55°C, gradual addition of 5 wt % sodium hypochlorite–5 wt % NaOH until a slight excess of chlorine remains, agitating a few hours at 50–55°C, and salting out. Yields of 80–90% of (**4**) are obtained.

(**2**) $\xrightarrow[NaOH]{NaOCl\ or\ O_2}$ O_2N–⟨⟩–CH=CH–⟨⟩–NO_2

HO_3S SO_3H

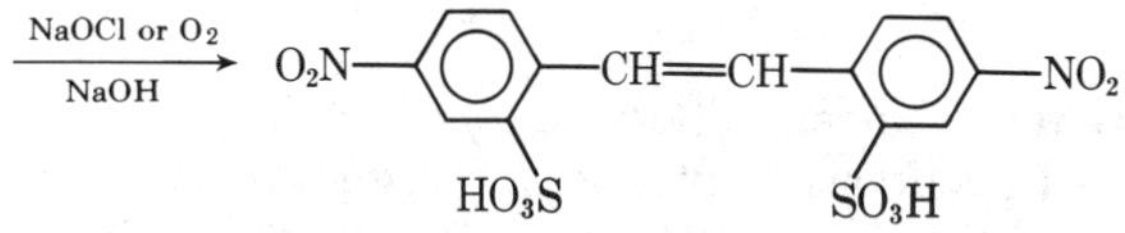

(**4**)

Subsequent reduction with glucose for Direct Orange 15 or with formaldehyde for Direct Yellow 6 proceeds as before. Dyes made by this route are purer and have better fastness properties, but they are more expensive.

If (**6**) is oxidized with nitric acid, sodium hypochlorite, or chromic acid, Mikado Yellow G [*6272-71-5*] (Stilbene Yellow 8G) (**9**) is produced:

SO_3Na SO_3Na SO_3Na SO_3Na

(**6**) $\xrightarrow{oxidize}$ O_2N–⟨⟩–CH=CH–⟨⟩–N=N–⟨⟩–CH=CH–⟨⟩–NO_2

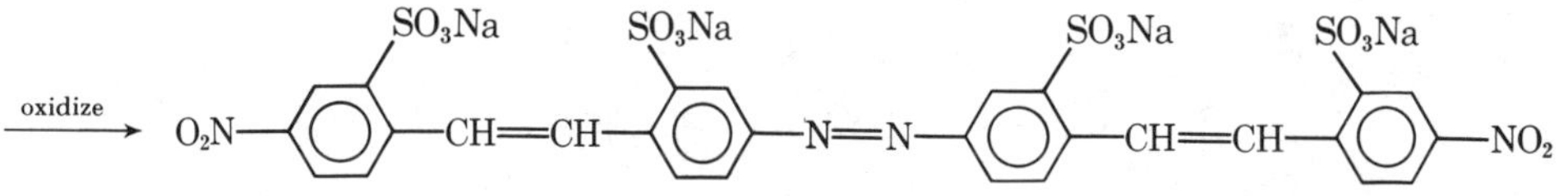

(**9**)

Improved processes for the self-condensation of 5-nitro-2-toluenesulfonic acid are being developed. For example, a recent one-pot process involving a high proportion of the sodium salt of 5-nitro-2-toluenesulfonic acid (**2**), as opposed to the free acid form, yields Direct Yellow 11 without intermediate isolation; it can be spray-dried directly from the reaction vessel (5). Another process eliminates the isolation of the intermediate condensation products, primarily 4,4′-dinitro-2,2′-stilbene disulfonic acid (**4**), by treating the condensation mass directly with 9–13 wt % sodium sulfide (6).

Condensation of Amines with 5-Nitro-2-Toluenesulfonic Acid and Its Derivatives. The condensation of 5-nitro-2-toluenesulfonic acid (**2**) or 4,4′-dinitrostilbene-2,2′-disulfonic acid (**4**) with primary amines leads to a wide range of dyes, primarily in orange-to-brown shades. It is probable that when (**2**) is the starting material, such condensations proceed through the dinitrosostilbene intermediate (**3**). Such condensations with primary amines yield azo dyes and do not involve the diazotization/coupling reaction sequence.

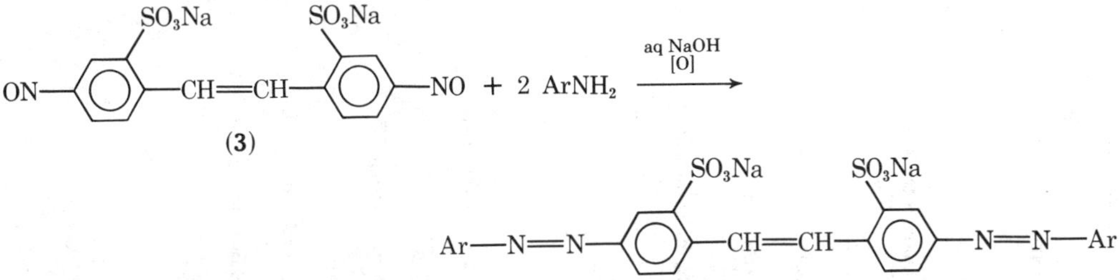

$$ON-C_6H_3(SO_3Na)-CH{=}CH-C_6H_3(SO_3Na)-NO + 2\ ArNH_2 \xrightarrow[{[O]}]{aq\ NaOH}$$

(**3**)

$$Ar-N{=}N-C_6H_3(SO_3Na)-CH{=}CH-C_6H_3(SO_3Na)-N{=}N-Ar$$

These condensation reactions yield mixtures and dyes of indeterminate structures. Manufacturing conditions vary as to concentrations, time, and temperatures. For example, heating (**2**) with aqueous sodium hydroxide for several hours at 27–63°C, then adding 4-aminoazobenzene-4′-sulfonic acid [*104-23-4*] and heating at 103–104°C for 12 h in an autoclave yields Sirius Supra Orange 5G (**10**) [*32829-81-5*] (7).

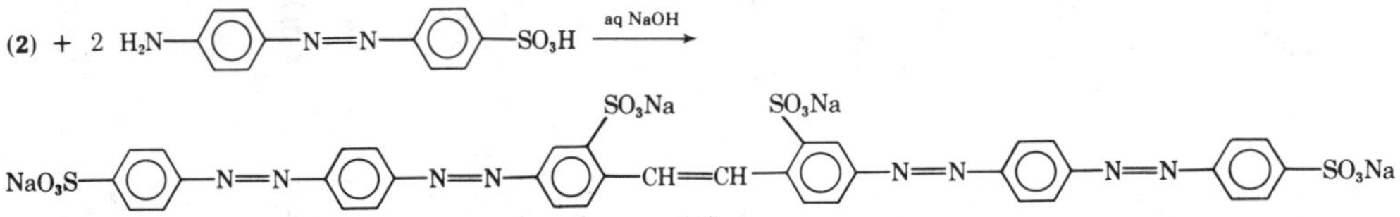

$$(\mathbf{2}) + 2\ H_2N-C_6H_4-N{=}N-C_6H_4-SO_3H \xrightarrow{aq\ NaOH}$$

$$NaO_3S-C_6H_4-N{=}N-C_6H_4-N{=}N-C_6H_3(SO_3Na)-CH{=}CH-C_6H_3(SO_3Na)-N{=}N-C_6H_4-N{=}N-C_6H_4-SO_3Na$$

(**10**)

The great variety of primary amines that have been used in this condensation reaction to yield complex azo dyes without involving a diazotization reaction are listed in Table 1.

These dyes may be considered as intermediates when a free amino group is present. For example, when *p*-phenylenediamine is the condensation moiety, Direct Brown R is obtained; if the free amino groups are tetrazotized and coupled to β-naphthol, Bordeaux shades are obtained; if α-naphthylamine is the coupler, blacks are obtained. If Chicago Orange G is tetrazotized and coupled to salicylic acid, Chicago Orange 3G [*82338-71-4*] (**11**) is obtained:

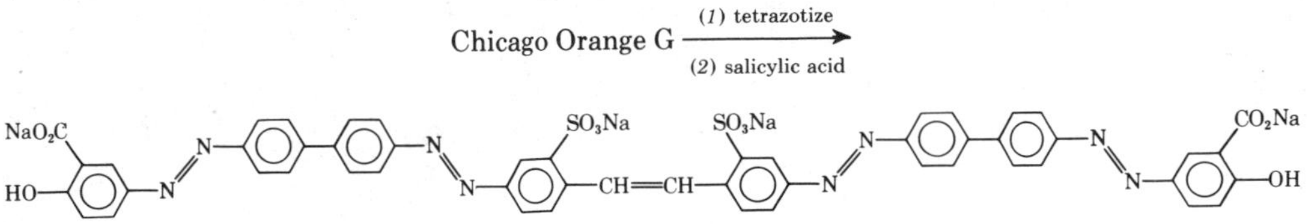

$$\text{Chicago Orange G} \xrightarrow[(2)\ \text{salicylic acid}]{(1)\ \text{tetrazotize}}$$

$$HO(NaO_2C)C_6H_3-N{=}N-C_6H_4-C_6H_4-N{=}N-C_6H_3(SO_3Na)-CH{=}CH-C_6H_3(SO_3Na)-N{=}N-C_6H_4-C_6H_4-N{=}N-C_6H_3(CO_2Na)-OH$$

(**11**)

Table 1. Condensation of 5-Nitro-2-Toluenesulfonic Acid and its Derivatives with Aromatic Amines

Principle product:

$$Ar-N{=}N-C_6H_3(SO_3Na)-CH{=}CH-C_6H_3(SO_3Na)-N{=}N-Ar$$

Aromatic amine condensed, Ar—NH_2	CI name	CAS Registry No.	CI number	Common or trade name	Ref. and remarks
$H_2N-C_6H_4-C_6H_4-NH_2$		[1325-52-6]	40070	Chicago Orange G	U.S. Pat. 601, 063
$C_6H_5-NH_2$				Diphenyl Citronine G	
$H_2N-C_6H_4-NH_2$		[6449-85-0]		Direct Brown R	
$NaO_3S-C_6H_4-N{=}N-C_6H_4-NH_2$		[32829-81-5]		Sirius Supra Orange 5G	
$NaO_3S-C_6H_4-N{=}N-C_6H_2(OCH_3)(CH_3)-NH_2$	Direct Orange 37	[1325-61-7]	40265	Sirius Supra Orange 3R	8; aftertreated with Na_2S
$CH_3O-C_6H_3(NaO_3S)-N{=}N-C_6H_2(CH_3)_2-NH_2$	Direct Orange 41	[1325-58-2]	40235	Sirius Supra Orange RRL	8; aftertreated with glucose
$NaO_3S-C_6H_4-N{=}N-C_6H_3(CH_3)-NH_2$	Direct Orange 34	[12222-37-6]	40220	Sirius Supra Orange 7G	9
$NaO_3S-C_6H_4-N{=}N-C_6H_4-NH_2$	Direct Orange 34/39	[1325-54-8]	40215	Sirius Supra Orange 7GL	10; aftertreated with glucose
$NaOOC-C_6H_4-N{=}N-C_6H_4-NH_2$	Direct Orange 71	[1325-51-5]	40205	Sirius Supra Orange GR	11; aftertreated with glucose
$C_6H_4(SO_3Na)-N{=}N-C_6H_4-NH_2$	Direct Orange 61	[1325-53-7]	40210		
$C_6H_4(SO_3Na)-N{=}N-C_6H_2(CH_3)_2-NH_2$	Direct Orange 35	[1325-56-0]	40225		

CH_3; NaO_3S–⟨⟩–N=N–⟨⟩–NH_2; CH_3	Direct Orange 36	[*1325-57-1*]	40230		12
NaOOC, CH_3; HO–⟨⟩–N=N–⟨⟩–NH_2; NaO_3S, CH_3		[*1325-59-3*]	40240	Benzo Fast Chrome Orange R	13
⟨⟩–N=N–⟨⟩–NH_2; SO_3Na, OCH_3	Direct Orange 70	[*1325-60-6*]	40245		
OCH_3; ⟨⟩–N=N–⟨⟩–NH_2; SO_3Na, CH_3	Direct Orange 37	[*1325-61-7*]	40260		
OCH_3; CH_3O–⟨⟩–N=N–⟨⟩–NH_2; SO_3Na, CH_3	Direct Red 76	[*1325-63-9*]	40270	Sirius Supra Scarlet 2G	13; aftertreated with glucose
NaOOC, OCH_3; HO–⟨⟩–N=N–⟨⟩–NH_2; NaO_3S, CH_3	Direct Red 187	[*1325-64-0*]	40275	Benzo Fast Chrome Red G	13; aftertreated with dichromate
NaO_3S–⟨⟩–N=N–⟨⟩–NH_2	Direct Brown 78 Direct Red 111 Direct Red 112	[*1325-65-1*] [*1325-65-1*] [*12262-21-4*]	40290	Sirius Supra Brown 3R	14
N=N–⟨⟩–NH_2; NaO_3S, SO_3Na		[*1325-67-3*]	40295	Diamine Fast Brown GB	15
SO_3Na; CH_3; S, N; S, N; NH_2		[*1325-47-9*]	40050	Diphenyl Fast Yellow	16

Reactions other than tetrazotizations and couplings can be performed. For example, ethylation of Arnica Yellow [*3051-11-4*] (CI 40025) (**12**), which forms when (**2**) is condensed with *p*-aminophenol, yields Direct Yellow 19 [*1325-45-7*] (CI 40030) whose main chemical constituent is Direct Yellow 12 [*2870-32-8*] (**13**) (CI 24895).

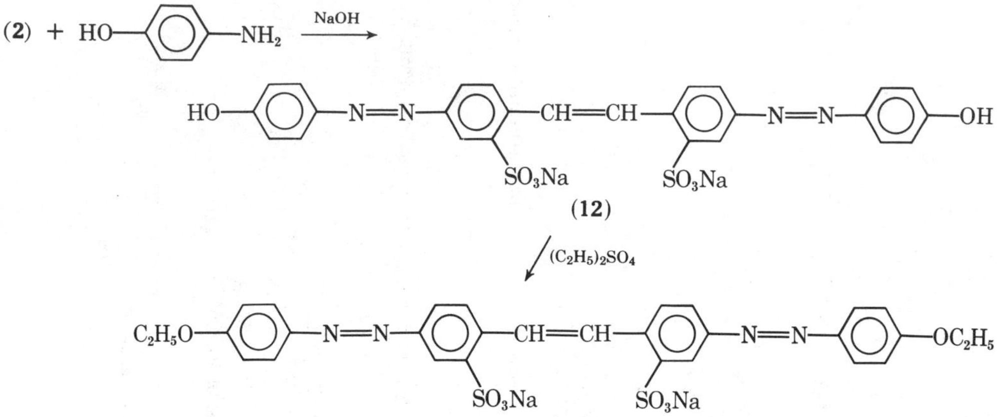

(**2**) + HO–C₆H₄–NH₂ $\xrightarrow{NaOH}$

HO–C₆H₄–N=N–C₆H₃(SO_3Na)–CH=CH–C₆H₃(SO_3Na)–N=N–C₆H₄–OH

(**12**)

$\downarrow (C_2H_5)_2SO_4$

C_2H_5O–C₆H₄–N=N–C₆H₃(SO_3Na)–CH=CH–C₆H₃(SO_3Na)–N=N–C₆H₄–OC_2H_5

(**13**)

Selected metallized dyes can also be produced by this condensation reaction. If *o*-oxy-*o*′-methoxyaminoazo compounds are condensed with 5-nitro-2-toluenesulfonic acid under caustic alkaline conditions with subsequent metallization, dyes, especially in the olive, brown, and gray shades, are obtained. These have excellent lightfastness (17). For example, if (**2**) is condensed with the product of 2-amino-1-phenol-4-sulfonic acid [*98-37-3*] coupled to 2,5-dimethoxyaniline, then metallized with copper, the gray dye (**14**) is obtained:

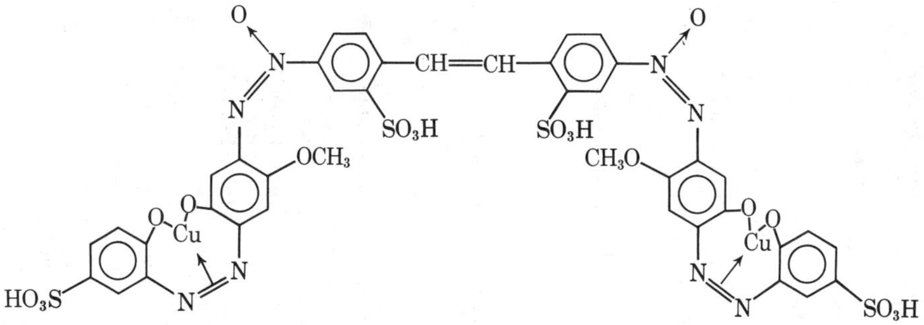

(**14**)

This reaction is possible only with *o*-methoxyaminoazo compounds; *o,o*′-dihydroxyaminoazo compounds decompose under the caustic alkaline condensation conditions.

4,4′-Diamino-2,2′-Stilbenedisulfonic Acid. The dyes described thus far have been obtained primarily from caustic alkaline condensation reactions, which lead to mixtures and products of indeterminate structure. However, if 4,4′-dinitro-2,2′-stilbenedisulfonic acid (**4**) is reduced either by a Béchamp iron reduction or catalytically, the corresponding 4,4′-diamino-2,2′-stilbenedisulfonic acid [*81-11-8*] (**15**) may be tetrazotized and coupled to a variety of aromatic systems to yield azo dyes, which can be characterized with more exact structural assurity:

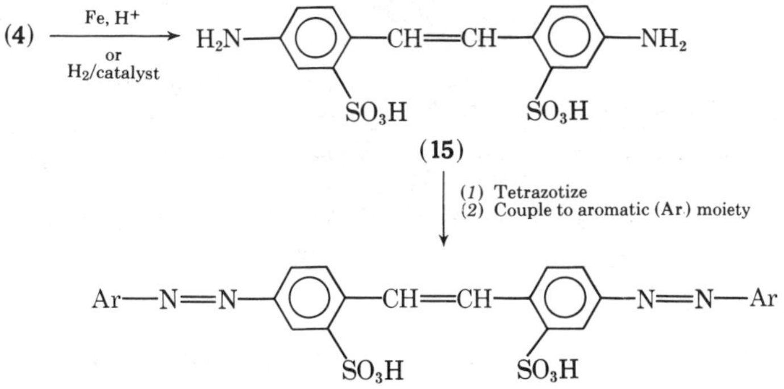

(4) —Fe, H⁺ or H₂/catalyst→ H₂N—C₆H₃(SO₃H)—CH═CH—C₆H₃(SO₃H)—NH₂

(**15**)

(*1*) Tetrazotize
(*2*) Couple to aromatic (Ar) moiety

Ar—N═N—C₆H₃(SO₃H)—CH═CH—C₆H₃(SO₃H)—N═N—Ar

Couplers that have been used in this route are listed in Table 2. Some of these dyes react further to yield more complex structures. For example, when the free hydroxyls on Direct Yellow 4 are ethylated, the resultant diethyl product is Direct Yellow 12 (CI 24895). When made by this route, the product is called Chrysophenine, which is the same structure as (**13**) made by the condensation route. Because of its alkaline sensitivity, the free phenol (**12**) has been used for many years as a pH indicator and the active constituent in Brilliant Yellow paper. Both Brilliant Yellow and Chrysophenine are important dyes economically.

If the free hydroxyls on Direct Yellow 4 are esterified with *p*-toluenesulfonyl chloride, the recently reported Acid Yellow 228 [*82323-97-5*] (CI 24896) (**16**) is obtained (21):

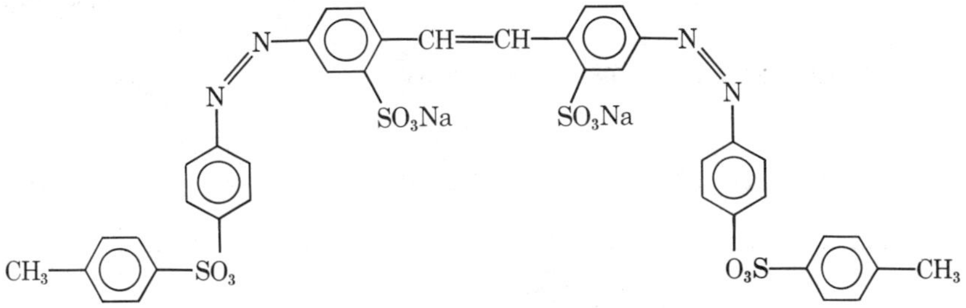

CH_3—C₆H₄—SO_3—C₆H₄—N═N—C₆H₃(SO_3Na)—CH═CH—C₆H₃(SO_3Na)—N═N—C₆H₄—O_3S—C₆H₄—CH_3

(**16**)

Recently this route has been used to obtain dyes with improved water solubility. When (**15**) is tetrazotized and coupled to two moles of *N*,*N*-dialkyl-substituted aniline-type couplers, disazo dyes ranging from yellows to violets with excellent water solubility and good fastness properties on polyamide fibers are obtained. For example, if *N*-β-cyanoethyl-*N*-β-hydroxyethylaniline is the coupler, the yellowish-red dye [*54940-87-3*] (**17**) is obtained (22):

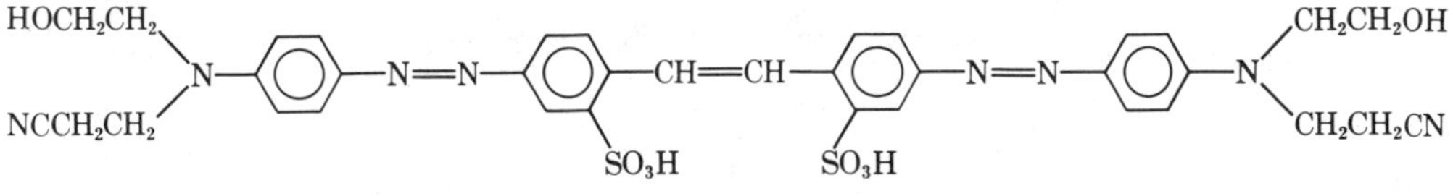

($HOCH_2CH_2$)($NCCH_2CH_2$)N—C₆H₄—N═N—C₆H₃(SO_3H)—CH═CH—C₆H₃(SO_3H)—N═N—C₆H₄—N(CH_2CH_2OH)(CH_2CH_2CN)

(**17**)

Table 2. Dyes Derived from Tetrazotion of 4,4′-Diamino-2,2′-Stilbenedisulfonic Acid

Ar—N=N—⟨⟩—CH=CH—⟨⟩—N=N—Ar

SO_3Na $NaSO_3$

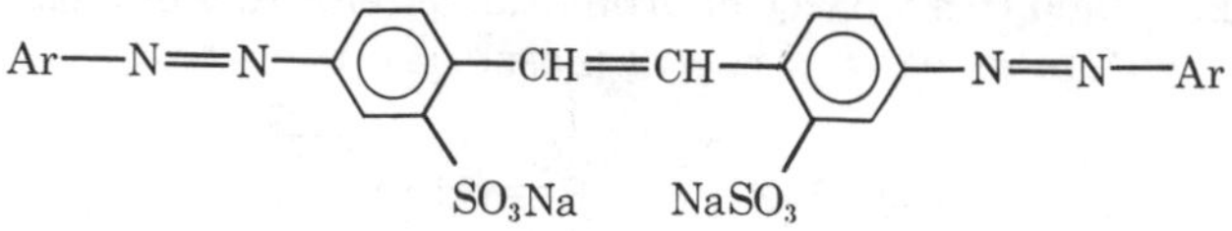

Coupler, Ar	CI dye name	CAS Registry No.	CI number	Common or trade name	Remarks
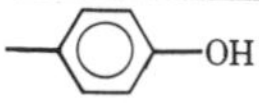 —⟨⟩—OH	Direct Yellow 4	[*91-34-9*, *3051-11-4*]	24890	Brilliant Yellow	18
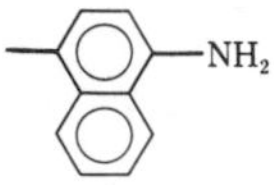 NH_2		[*6459-74-1*]	24860	Hessian Bordeaux	may be diazotized and developed on fiber
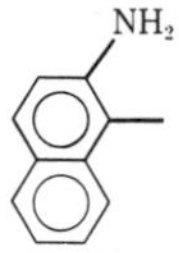 NH_2		[*6459-75-2*]	24865	Hessian Purple N	U.S. Pat. 350,230
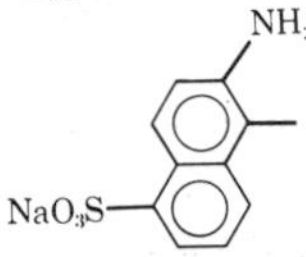 NH_2, NaO_3S		[*6459-76-3*]	24870	Hessian Purple D	U.S. Pat. 350,230
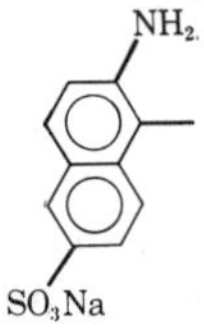 NH_2, SO_3Na		[*6459-77-4*]	24875	Brilliant Hessian Purple	U.S. Pat. 360,553
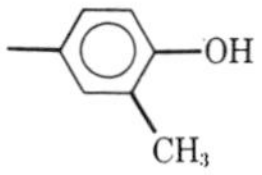 OH, CH_3					intermediate for CI 24900 FIAT 764
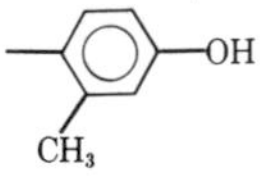 OH, CH_3	Acid Yellow 183	[*50814-29-4*]	24855		19
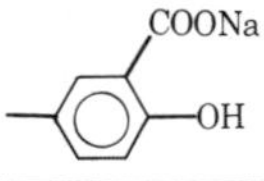 COONa, OH	Direct Yellow 72	[*6459-82-1*]	24910		20; is afterchromed

Very complex and highly conjugated dye systems can be produced by this method. For example, dyes, eg, (**18**) [*57959-13-4*] which yield reddish-black to black shades on paper, have recently been reported by coupling 4,4′-diamino-2,2′-stilbenedisulfonic acid (**15**) to 2-amino-8-hydroxynaphthaline-6-sulfonic acid which is then coupled to resorcinol (23):

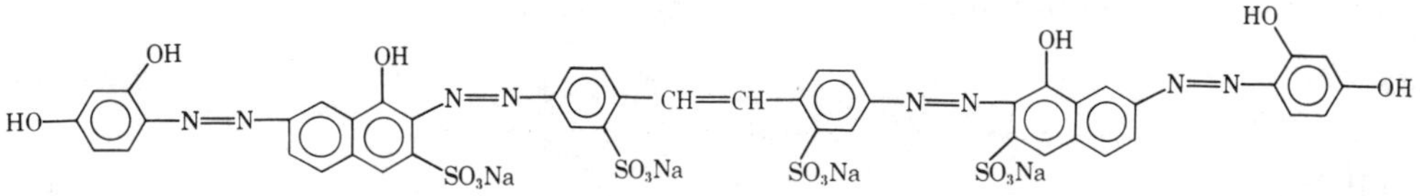

HO—⟨⟩(OH)—N=N—(OH)(naphthalene)(SO_3Na)—N=N—⟨⟩(SO_3Na)—CH=CH—⟨⟩(SO_3Na)—N=N—(naphthalene)(OH)(SO_3Na)—N=N—⟨⟩(HO)—OH

(**18**)

A series of dyes that produce dark blue shades on paper have been made by incorporating the 2-substituted-4-phenyl-3,1-thiazole moiety (24). For example, if (**15**)

is tetrazotized and coupled to 2-amino-8-hydroxynaphthalene-6-sulfonic acid [*90-51-7*] and further tetrazotized and coupled to 2-(4′-sulfophenyl)amino-4-phenylthiazole, a dark blue paper dye [*75447-03-9*] (**19**) with excellent fastness properties is obtained:

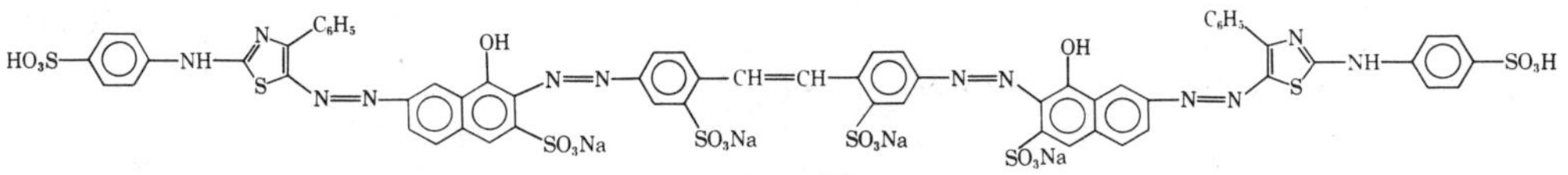

(19)

In contrast to the symmetrical products described, unsymmetrical molecules can be obtained from the reaction of two different couplers to tetrazotized (**15**). For example, Hessian Purple B [*6459-78-5*] (CI 24880) (**20**) is produced when one mole each of Broenner's acid and 7-amino-2-naphthalene sulfonic acid is coupled to one mole of tetrazotized (**15**):

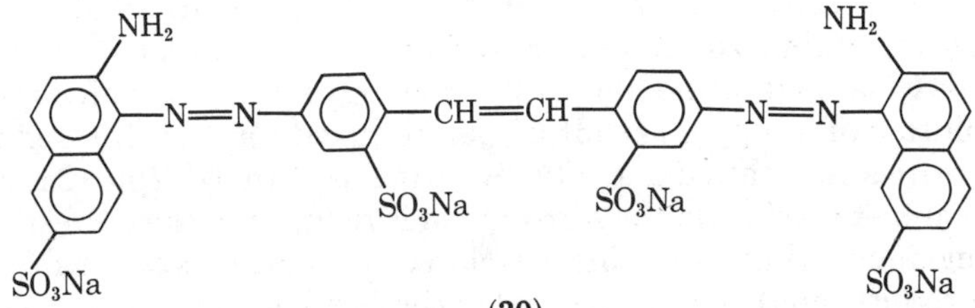

(20)

This route leads to formations of mixtures in the production of unsymmetrical dyes, since the favored product is governed by the faster reactivity rate of the strongest coupler.

4-Nitro-4′-Aminostilbene-2,2′-Disulfonic Acid. When 4,4′-dinitro-2,2′-stilbenedisulfonic acid (**4**) is reduced with sodium sulfhydrate under carefully controlled conditions, 4-nitro-4′-aminostilbene-2,2′-disulfonic acid [*119-72-2*] (**21**) is obtained.

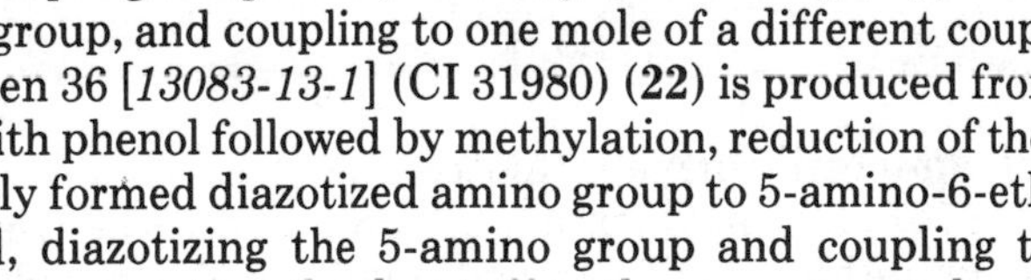

(21)

Unsymmetrical products can be prepared by diazotizing the 4′-amino group, coupling to one mole of a coupling component, reducing the 4-nitro group, diazotizing the newly formed 4-amino group, and coupling to one mole of a different coupling component. Thus, Direct Green 36 [*13083-13-1*] (CI 31980) (**22**) is produced from the reaction of diazotized (**21**) with phenol followed by methylation, reduction of the nitro group and coupling the newly formed diazotized amino group to 5-amino-6-ethoxy-2-naphthalenesulfonic acid, diazotizing the 5-amino group and coupling to *N*-*m*-carboxyphenyl-J-acid, and converting the dyestuff to the copper complex (25):

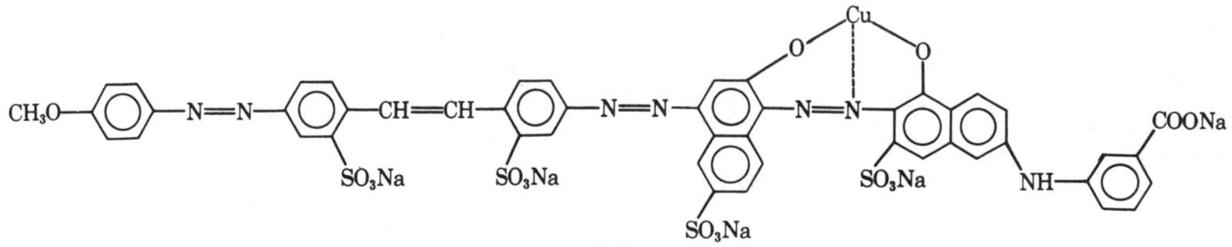

(22)

A similar series of reactions is involved in the production of Direct Green 23 [*13102-26-6*] (CI 31985) (**23**); here *N*-acetyl-S-acid is the final coupling component instead of *N*-*m*-carboxyphenyl-J-acid (**23**) (26):

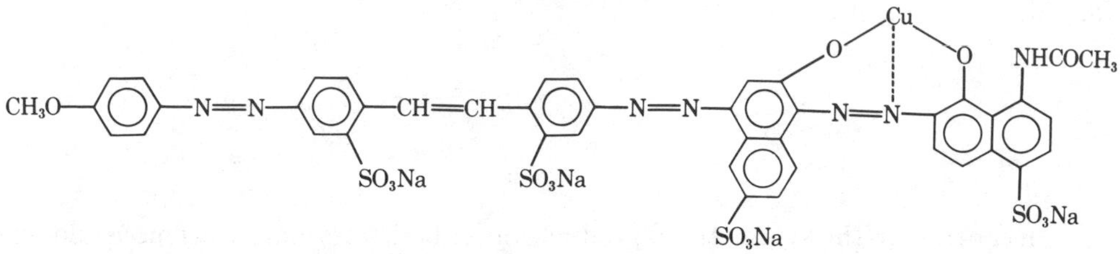

(**23**)

A variation on this reaction involves the formation of aminostilbenenaphthotriazines. 4-Amino-4′-nitro-2,2′-stilbenedisulfonic acid (**21**) can be diazotized and coupled to an equivalent amount of an aminonaphthalenesulfonic acid under weakly alkaline conditions. The reaction medium is heated to 70°C and made alkaline with soda ash, and the resulting aminoazo product [*82198-99-0*] (**24**) is isolated by salting out and filtering. This isolated intermediate is slurried in hot water to which is added an ammoniacal solution of a copper salt. It is then refluxed for 24 h, filtered to remove insoluble impurities, and salted out. The resulting naphthotriazine (**25**) is dissolved in warm, dilute mineral acid and the nitro group is reduced to the corresponding amine. The resulting aminostilbenetriazine is isolated by neutralization, salted out, and filtered. This product can then be diazotized in the usual manner and coupled to a variety of aromatic systems (27):

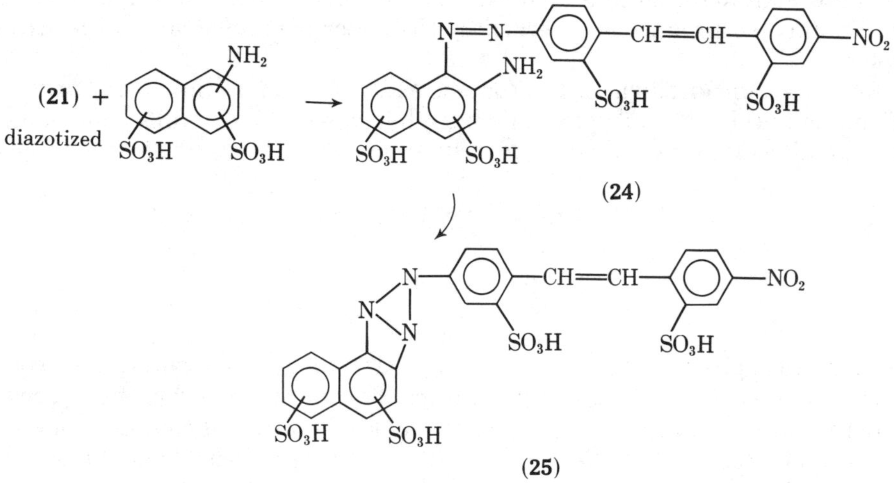

(**24**)

(**25**)

Examples of aminostilbenenaphthiotriazine dyes include Direct Green 34 [*2894-10-2*] (CI 27970) (**26**), in which naphthionic acid is the triazine base with 5-amino-6-ethoxy-2-naphthalenesulfonic acid coupled to *N*-phenyl-J-acid as the aromatic coupler system (28):

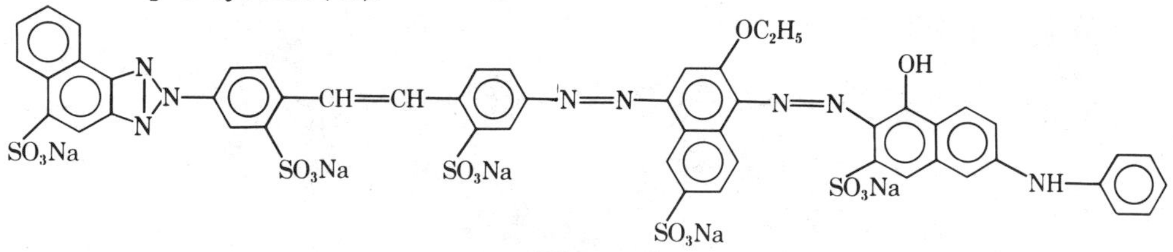

(**26**)

Recent patents claim such stilbenenaphthiotriazines as excellent paper dyes with bright shades, especially for dyeing alum or rosin-sized paper. Examples, eg, (**27**), include couplers having diketone configurations (27):

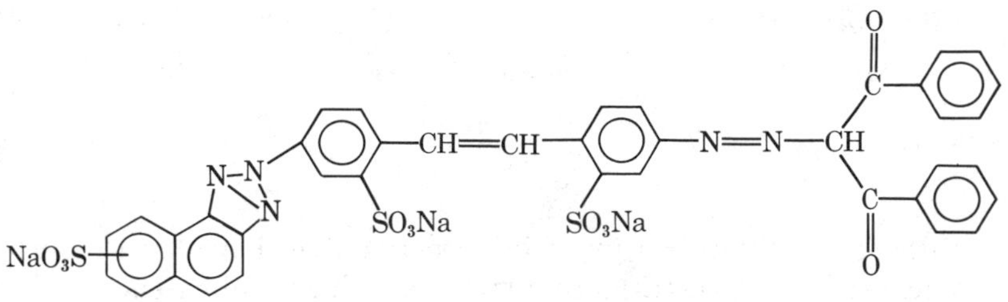

(**27**)

Unreduced 4-nitro-4′-amino-substituted stilbene-2,2′-disulfonic acids are excellent dyes, particularly because of their superior water solubility. For example, the simple dye [*73281-07-9*] (**28**) made by coupling diazotized 4-amino-4′-nitrostilbene-2,2′-disulfonic acid (**21**) to *p*-cresol yields a level yellow shade with excellent fastness properties (29).

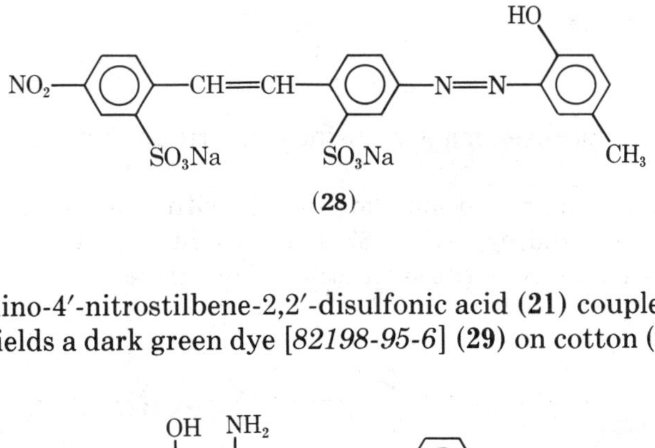

(**28**)

Diazotized 4-amino-4′-nitrostilbene-2,2′-disulfonic acid (**21**) coupled to H-acid and *p*-nitroaniline yields a dark green dye [*82198-95-6*] (**29**) on cotton (30):

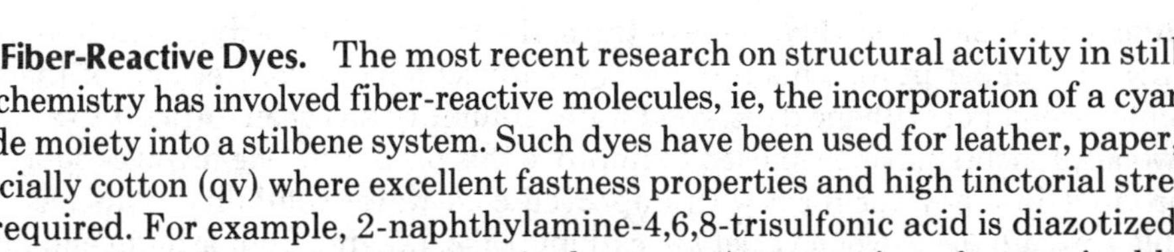

(**29**)

Fiber-Reactive Dyes. The most recent research on structural activity in stilbene dye chemistry has involved fiber-reactive molecules, ie, the incorporation of a cyanuric halide moiety into a stilbene system. Such dyes have been used for leather, paper, and especially cotton (qv) where excellent fastness properties and high tinctorial strength are required. For example, 2-naphthylamine-4,6,8-trisulfonic acid is diazotized and coupled to 2-methoxy-5-methyl-1-aminobenzene. A suspension of cyanuric chloride is added to a slightly acidic solution of this intermediate; the resultant mixture is stirred for 15 h at 15–20°C. After completion of the acylation, 4,4′-diaminostilbene-2,2′-disulfonic acid (**15**) is added and stirred at 50°C for 4 h. The product is isolated by salting out with NaCl; a golden yellow dye [*82209-40-3*] (**30**) with good lightfastness and wetfastness properties is obtained (31):

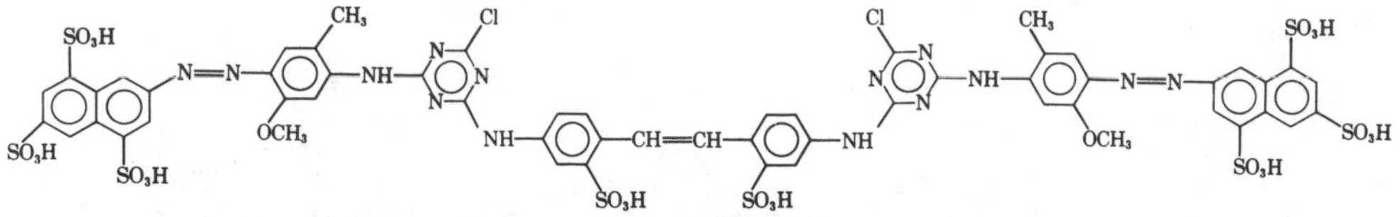

(**30**)

Many stilbene azo fluorotriazinyl fiber-reactive dyes have been reported. For example, condensation of 2,4,6-trifluoro-*s*-triazine with aminobenzene-2,5-disulfonic acid and further condensation with 4-(4″-aminophenylazo)-4′-aminostilbene-2,2′-disulfonic acid gives [*65740-48-9*] (**31**), which is a dye with excellent fastness properties and is used to dye or print cotton bright yellow (32):

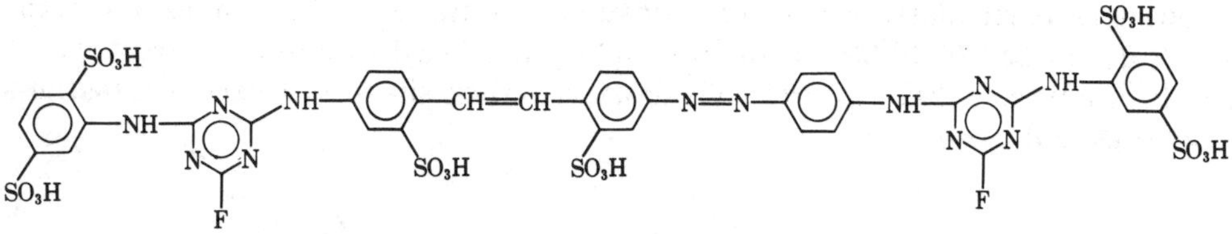

(**31**)

This same structure has been made by using cyanuric chloride as the fiber-reactive hook (33).

Dyes containing one cyanuric halide moiety with a different major structural feature have been reported; eg, yellow (**32**) has both the cyanuric triazine moiety and the naphthotriazine feature in the same stilbene dye molecule (34):

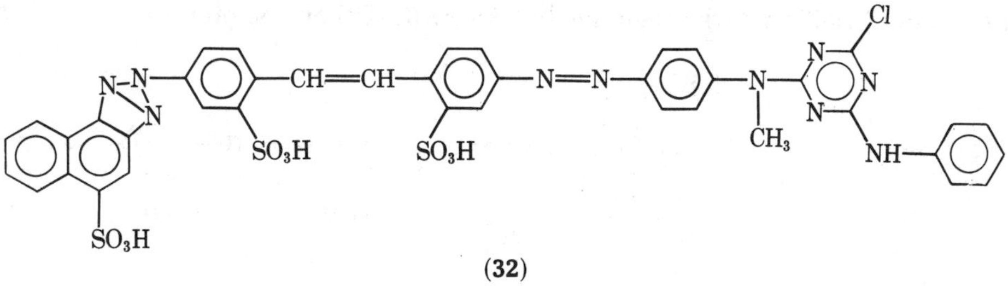

(**32**)

Improved Water Solubility. One of the main uses of stilbene dyes is in the dyeing of paper and paper pulp. This is done in an aqueous medium. Because of the high molecular weight of stilbene dyes, good water solubility has been difficult to achieve, but two approaches have been used to improve this property. One involves supplying the product in a powder form with improved cold-water solubility. This has been possible through the use of a variety of cationic salts, ranging from lithium and mixed lithium–sodium salts to complex tetra-substituted alkyl ammonium cations. For example, condensation yellows, for which (**33**) [*82198-97-8*] is the starting material, have vastly improved cold-water solubilities (35). Likewise, lithium salts, prepared from the reaction of 5-nitro-2-toluenesulfonic acid (2) or 4,4′-dinitro-2,2′-stilbenedisulfonic acid (4) with lithium hydroxide monohydrate, yield stilbene intermediates which, when

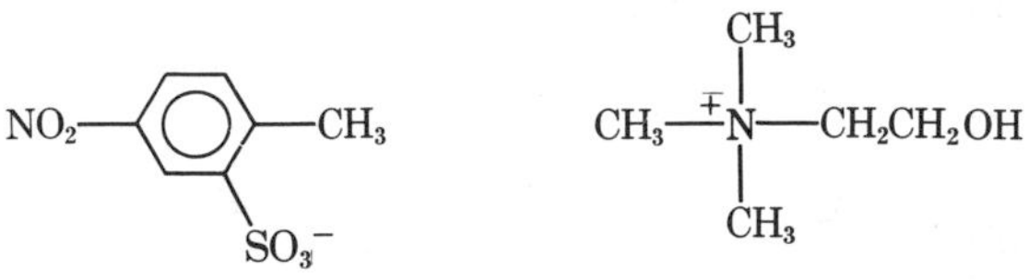

$$NO_2\text{—}C_6H_3(SO_3^-)\text{—}CH_3 \qquad CH_3\text{—}\overset{+}{N}(CH_3)_2\text{—}CH_2CH_2OH$$

(33)

further condensed in the usual manner, yield dyes with improved cold-water solubility (36).

The other method is to supply the paper industry with a liquefied dyestuff product. Recent products include solutions which range from lithium salts exclusively to aqueous mixtures with additives, eg, 5–20 wt % urea, which influence solubilization (37–38).

Economic Aspects

It is difficult to assess the economic impact of stilbene colors on the dyestuff industry. Figures are reported by the U.S. Tariff Commission only when three or more manufacturers produce the same product. For the most recent three years where such data are available (1977–1979), only five stilbene colors have been reported. However, production of these five colors collectively averaged 2.226 metric tons per year, with a value of $8943/yr (see Table 3). Production in 1979 increased 97% from that in 1967.

Health and Safety Factors

As a class, stilbene dyes have not exhibited health or safety properties warranting special precautions; however, standard chemical labeling instructions are required. Of the stilbene dyes of economic importance, the following examples are indicative of the toxicological properties of the class.

Direct Yellow 4. LD_{50} (oral), rat: >5 g/kg; eye irritation; rabbit: mildly irritating; primary dermal irritation, rabbit: nonirritating; acute dermal toxicity, rabbit: 2 g/kg; acute inhalation toxicity, rat: 1100 mg/m^3.

Direct Yellow 11. LD_{50} (oral), rat: >2 g/kg; skin irritation, rabbit: nonirritating; eye irritation, rabbit: slightly irritating.

Direct Yellow 12. LD_{50} (oral), rat: >4.2 g/kg; eye irritation, rabbit: nonirritating; skin irritation, rabbit: slightly irritating.

Direct Orange 15. LD_{50} (oral), rat: >5 g/kg; eye irritation, rabbit: mildly irritating; skin irritation, rabbit: mildly irritating; acute dermal toxicity: >2 g/kg.

Direct Red 111. LD_{50} (oral), rat: >5 g/kg; skin irritation, rabbit: slightly irritating; eye irritation, rabbit: moderately irritating.

The intermediates involved in the preparation of these stilbene dyes should be handled with greater caution. Although little toxicological data is available for the

Table 3. U.S. Production and Sales Data for Stilbene Dyes: 1975–1979[a]

Dyestuff	CAS Registry No.	CI No.	1967 Production, t	1975 Production, t	1975 Sales, 10^3 \$	1975 No. of U.S. producers	1976 Production, t	1976 Sales, 10^3 \$	1976 No. of U.S. producers	1977 Production, t	1977 Sales, 10^3 \$	1977 No. of U.S. producers	1978 Production, t	1978 Sales, 10^3 \$	1978 No. of U.S. producers	1979 Production, t	1979 Sales, 10^3 \$	1979 No. of U.S. producers
Direct Yellow 4	[*91-34-9*, *3051-11-4*]	24890	207	190	896	8	254	1086	6	464	1976	7	325	1776	6	274	1812	7
Direct Yellow 6	[*1324-38-8*]	40001	233	99	536	5	111	647	5	120	1030	5	127	975	6	105	825	4
Direct Yellow 11	[*1325-37-7*]	40000	487	1424	3643	7	1176	3018	7	1319	5477	6	1530	4280	7	1513	4266	6
Direct Yellow 12	[*2870-32-8*]	24895		19	201	7	24	320	5			3	33[b]		3			3
Direct Orange 15	[*1325-35-5*]	40002	103	176	558	5	212	694	3	237	872	5	250	1071	5	213	1098	5
Direct Orange 34	[*1325-54-8*]	40215		25	169	5	28	195	4			2			2	1.5[b]		2
Direct Orange 37	[*1325-62-8*]	40265		4	20	4			2									
Direct Orange 39	[*1325-54-8*]	40215	98	50	313	6	56	367	4	68	450	5	68	450	5	66	471	4
Direct Orange 60	[*12262-20-3*]						5[b]											
Direct Red 76	[*1325-63-9*]	40270				1						1			2			
Direct Red 111	[*1325-65-1*]	40290	[*1325-65-1*]	2.0[b]			6[b]			1.5[b]			0.4[b]					
Direct Red 112	[*12262-21-4*]												0.5[b]					

[a] Ref. 39.
[b] Imported.

basic building block of these dyes, ie, 5-nitro-2-toluenesulfonic acid (**2**), the data should be considered similar to that for the unnitrated analogue *o*-toluenesulfonic acid. This has the highest toxic rating for acute local and acute systemic hazards (40). It may cause death or permanent injury after very short exposure to small quantities by ingestion or by inhalation. The other basic intermediates also may cause moderate-to-severe skin and eye irritation (41), eg, 4,4′-dinitro-2,2′-stilbenedisulfonic acid (**4**): LD_{50} (oral), rat: 12600 mg/kg; skin irritation, rabbit: moderate; eye irritation, rabbit: severe; 4-amino-4′-nitro-2,2′-stilbenedisulfonic acid (**21**): LD_{50} (oral), rat: 14 g/kg; eye irritation, rabbit: severe.

BIBLIOGRAPHY

"Stilbene Dyes" in *ECT* 1st ed., Vol. 12, pp. 949–955, by A. F. Plue, General Aniline and Film Corporation; "Stilbene Dyes" in *ECT* 2nd ed., Vol. 19, pp. 1–13, by H. R. Schwander, J. R. Geigy AG and G. S. Dominguez, Geigy Chemical Corporation.

1. *BIOS Report 1548*, British Intelligence Objectives Subcommittees, UK, 1946, p. 169.
2. *BIOS Doc. No. 2351/2247/1*, British Intelligence Objectives Subcommittees, UK, 1946.
3. A. Green, *J. Chem. Soc.*, 1427 (1904); 1610 (1906); 2076 (1907); 1721 (1908).
4. A. Knight, *J. Soc. Dyers Colour.* **66,** 410 (1950).
5. Br. Pat. 1,511,824 (May 24, 1978), F. Puchner and H. Nickel (to Bayer).
6. U.S. Pat. 3,557,079 (Jan. 19, 1971), J. J. Doody (to Allied Chemical Corporation).
7. *BIOS Report 1548*, British Intelligence Objectives Subcommittees, 1946.
8. Ref. 1, p. 126
9. Ref. 1, p. 124
10. *BIOS Report 1547*, British Intelligence Objectives Subcommittees, England, 1946, pp. 123–125.
11. Ref. 1, p. 122.
12. U.S. Pat. 903,284 (Nov. 10, 1909), A. Gressly (to Cassella Color Co.).
13. *BIOS Report 961*, British Intelligence Objectives Subcommittees, England, 1946, p. 91.
14. Ref. 1, p. 14.
15. Ref. 1, p. 189.
16. Br. Pat. 18,990 (1897) (to J. R. Geigy).
17. U.S. Pat. 2,385,862 (Oct. 2, 1945), E. Keller (to J. R. Geigy); U.S. Pat. 2,468,204 (April 26, 1949), E. Keller (to J. R. Geigy); U.S. Pat. 2,467,262 (April 12, 1949), A. H. Knight (to Imperial Chemical Industries).
18. Ref. 1, p. 147.
19. *Colour Index, 3rd ed. Revision, Number 21*, The Society of Dyers and Colourists, Bradford, Yorkshire, UK, Oct. 1976, p. 601.
20. Ref. 1. p. 100.
21. *Colour Index, 3rd Ed. Revision, Additions and Amendments, Number 25*, The Society of Dyers and Colourists, Bradford, Yorkshire, UK, Oct. 1977, p. 681.
22. Br. Pat. 1,375,787 (Nov. 27, 1974), D. C. Wilson, J. Schofield, J. Teale, and S. Partington (to Yorkshire Chemicals, Ltd.).
23. Ger. Off. 2,504,868 (Nov. 13, 1975), Z. Allan (to Sandoz).
24. Ger. Off. 2,901,654 (July 31, 1980), H. Eilingsfeld, G. Hansen, G. Seybold, and G. Zeidler (to Badishe Aniline. Soda-Fabrik).
25. Ref. 1, pp. 121, 179.
26. Ref. 1, pp. 121, 137.
27. Br. Pat. 1,194,150 (June 10, 1970), W. E. Wallace (to GAF Corporation).
28. *FIAT Report 764*, Field Information Agencies Technical Rev. Ger. Sci., Washington, D.C., 1948.
29. Br. Pub. Pat. Appl. 2,026,008 (June 6, 1979), J. Hildreth, A. Tschopp, and D. Evans (to CIBA-GEIGY).
30. Ger. Off. 2,910,458 (Sept. 25, 1980), H. Nickel (to Bayer).
31. Br. Pub. Pat. Appl. 2,029,436 (June 25, 1979), H. Kaack (to Badishe Aniline. Soda-Fabrik).
32. Br. Pat. 1,549,134 (July 25, 1979), (to CIBA-GEIGY).
33. Ger. Pat. 1,544,447 (Sept. 25, 1975), K. Seitz (to CIBA-GEIGY).

34. U.S. Pat. 3,117,958 (June 14, 1964), R. Starn and W. Gumprecht (to E. I. du Pont de Nemours & Co., Inc.).
35. U.S. Pat. 3,953,419 (April 27, 1976), R. Pedrazzi (to Sandoz); U.S. Pat. 4,126,608 (Nov. 21, 1978), R. Pedrazzi (to Sandoz).
36. U.S. Pat. 3,905,949 (Sept. 16, 1975), M. A. Perkins and H. K. Urion (to E. I. du Pont de Nemours & Co., Inc.).
37. Br. Pat. 1,409,326 (Oct. 8, 1975), A. A. Komander and F. R. Johnson (to Imperial Chemical Industries).
38. U.S. Pat. 4,118,182 (Oct. 3, 1978), S. B. Smith (to E. I. du Pont de Nemours & Co., Inc.).
39. *Synthetic Organic Chemicals: United States Production and Sales*, IFC Publication, U.S. International Trade Commission, U.S. Government Printing Office, Washington, D.C., 1975–1979; *Imports of Benzenoid Chemicals and Products*, USITC Publication, U.S. International Trade Commission, U.S. Government Printing Office, Washington, D.C., 1975–1979.
40. N. I. Sax, *Dangerous Properties of Industrial Materials*, 4th ed., Van Nostrand Reinhold Company, New York, 1975, p. 1175.
41. *Registry of Toxic Effects of Chemical Substances*, 1979 ed., Vol. 2, U.S. Department of Health and Human Services, National Institute for Occupational Safety and Health, Cincinnati, Ohio, 1980, p. 552.

General References

Colour Index, 3rd ed., Vol. 4, The Society of Dyers and Colourists, Bradford, Yorkshire, UK, 1971, pp. 4212–4214, 4365–4373; Vol. 6, 1975, p. 6400.

K. Venkataraman, *The Chemistry of Synthetic Dyes*, Vol. 1, Academic Press, Inc., New York, 1952, pp. 628–635.

RUSSELL E. FARRIS
Sandoz Colors & Chemicals, Inc.

STIMULANTS

Therapeutic agents that stimulate the central nervous system (CNS) include both natural and synthetic compounds and are characterized according to structure, locus of action, and pharmacologic or clinical effects. Therapeutically, they are used to elevate low levels of physiologic activity to normal. However, many stimulants are capable of exciting the CNS beyond the normal range, and a few have been occasionally used for this purpose, eg, in chemical shock therapy.

There is some disagreement about the classes of compounds that should be categorized as CNS stimulants. However, for the purposes of this discussion, the term includes various classes of analeptics, psychostimulants, sympathomimetics, monoamine oxidase inhibitors, and tricyclic and nontricyclic antidepressants.

Analeptics

Analeptics or restoratives are a diverse group of compounds that are capable of restoring depressed medullary and cerebral functions. Although many act principally on the brain stem, all can stimulate the entire CNS even to the point of causing generalized convulsions. Because many older analeptics lack specificity of action and have poor therapeutic indexes, ie, the therapeutic dose compared with the toxic or convulsant dose, they have not been universally accepted as useful therapeutic agents. Recently, though, synthetic analeptics with greater margins of safety have been developed. These compounds are used with increasing frequency for the treatment of barbiturate intoxication, facilitation of arousal after general anesthesia, and treatment of pulmonary dysfunction.

Pharmacologic methods for evaluating analeptic activity have been extensively reviewed (1–2). Measurements of particular importance are the dose required to stimulate respiration, the convulsant dose, the lethal dose, and the dose required to antagonize the depressant effects of various sedative–hypnotics. Further evaluation usually includes studies on the site and mechanism of action.

Many naturally occurring analeptics have been known for centuries. Two of the best known and most thoroughly studied are strychnine (**1**) and picrotoxin [*124-87-8*], a 1:1 combination of picrotoxinin (**2**) and picrotin (**3**). As therapeutic agents, they are obsolete, but they continue to be of historical and experimental interest.

Strychnine was introduced into European medical practice in the early 16th century, notwithstanding the fact that it was first used as a rat poison. Interestingly, it is still used as a rodenticide today (see Poisons, economic). The compound and the related alkaloid brucine are isolated from the seeds of the Indian tree *Strychnos nux-vomica* L. The total synthesis of the complex molecule represents one of the great achievements of organic chemistry (3–4). Strychnine stimulates the entire CNS with preference for the spinal cord. Its effects are thought to result from antagonism of an inhibitory transmitter, possibly glycine (see Alkaloids).

Picrotoxin, unlike strychnine and most other analeptics, is nonnitrogenous. The bitter principle is extracted from the Asian shrub, *Anamirta cocculus L.* Only picrotoxinin has analeptic properties. Both picrotoxinin and picrotin have been synthesized by a multistep process starting with (−)-carvone (5). In earlier studies, their structures and absolute configurations were established (6). The mechanism of picrotoxin's powerful stimulant action has not been firmly established, although some evidence suggests that the compound blocks the action of gamma-aminobutyric acid (GABA), a possible presynaptic inhibitory transmitter (see Neuroregulators). At one time picrotoxin was used in the treatment of overdoses of CNS depressants. However its poor therapeutic index and relatively long latency period preceding a therapeutic response ultimately led to its disuse.

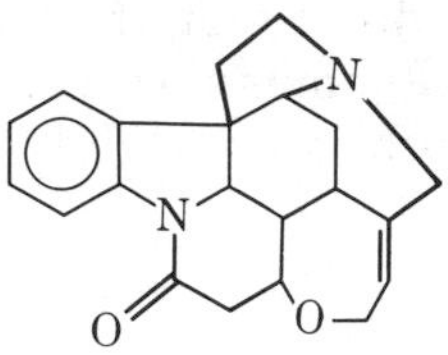

(1) strychnine [*57-24-9*]

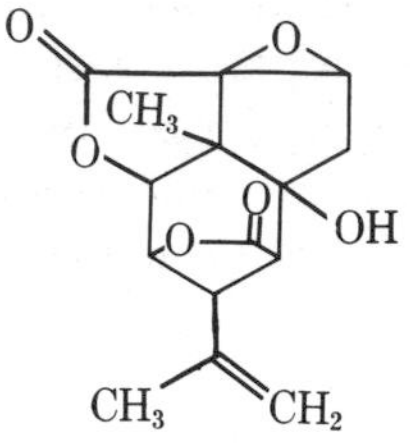

(2) picrotoxinin [*17617-45-7*]

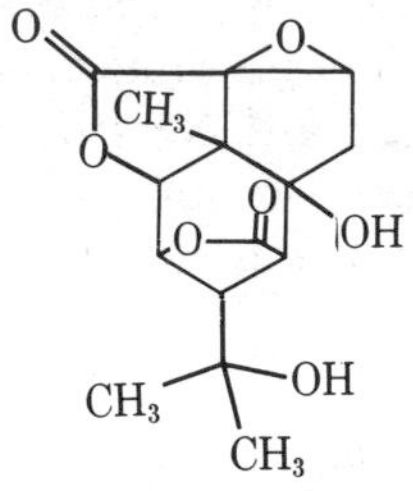

(3) picrotin [*21416-53-5*]

Pentylenetetrazole (**4**) was one of the first totally synthetic analeptics and is prepared by the reaction of cyclohexanone with hydrazoic acid (7). The compound was first introduced in the United States in 1927 as a treatment for barbiturate poisoning and was later used to a limited extent for chemical shock therapy. Today it is used almost exclusively to enhance mental and physical activity in elderly patients and as a diagnostic aid, ie, as an EEG activator, in epilepsy. The compound is also an important laboratory tool for screening anticonvulsant drugs (see Hypnotics, sedatives, anticonvulsants). Although pentylenetetrazole has been extensively modified, none of its derivatives has achieved clinical significance.

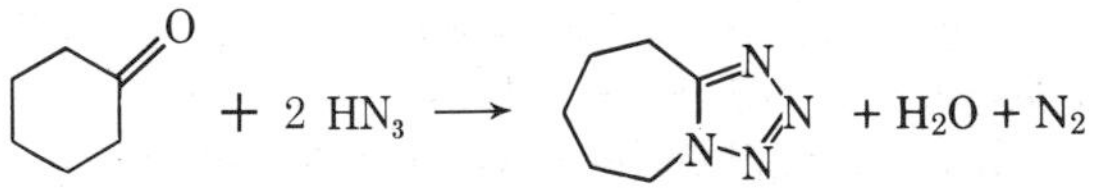

$$+ \ 2\ HN_3 \longrightarrow \quad + H_2O + N_2$$

(4) pentylenetetrazole [*54-95-5*]

Flurothyl (**5**) is an analeptic with strong convulsant properties that has been used for chemical shock therapy (8). The compound is unique in that it is a volatile fluorinated ether and its structure resembles those of the general anesthetics (qv). Comparisons of flurothyl therapy with electroconvulsive therapy indicate that both procedures are comparable with respect to safety, efficacy, and side effects (9). Currently, chemical shock therapy is rarely used.

$CF_3CH_2OCH_2CF_3$

(5) flurothyl [*333-36-8*]

Ideally, analeptics used to enhance depressed respiration should selectively stimulate respiratory centers without affecting other motor centers in the CNS. Although this objective has not been completely achieved, several synthetic analeptics are sufficiently selective to be clinically useful. Currently doxapram (**8**) is by far the leading respiratory stimulant marketed in the United States. In a controlled clinical trial, the compound proved superior to 11 other analeptics with regard to efficacy and side effects (10). Doxapram is prepared by a unique rearrangement of the pyrrolidine (**6**) to the pyrrolidinone (**7**), followed by alkylation with morpholine (11).

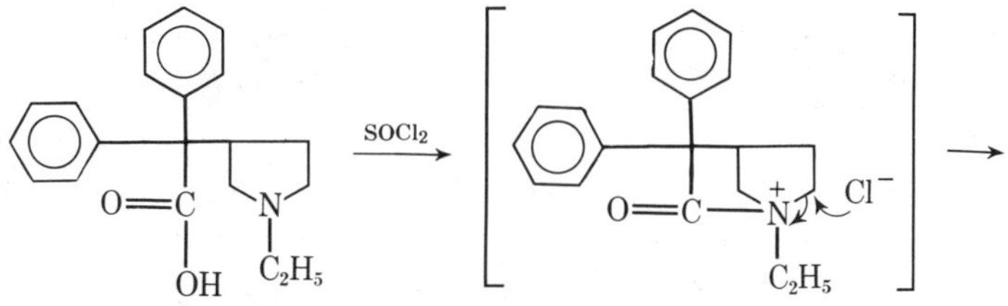

(**6**) [*3471-97-4*] $\xrightarrow{SOCl_2}$ [*82045-62-3*] $\rightarrow$

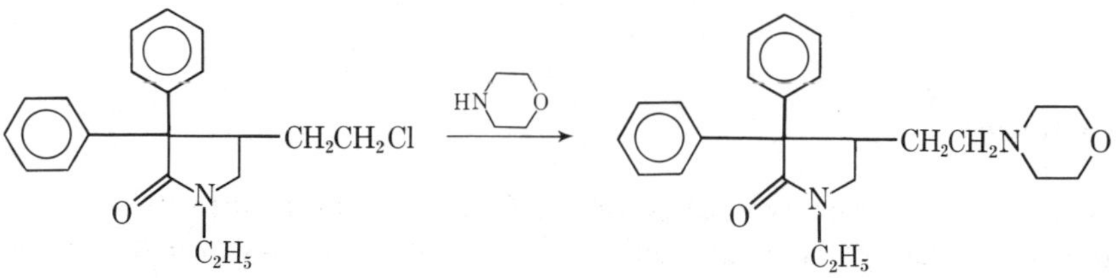

(**7**) [*3192-64-1*] $\rightarrow$ (**8**) doxapram [*309-29-5*]

Nikethamide (**9**) and ethamivan (**10**), which is no longer marketed in the United States, are respiratory stimulants of the arylcarboxamide type. Both are diethylamides and are prepared from the corresponding carboxylic acids by standard methods (12–13).

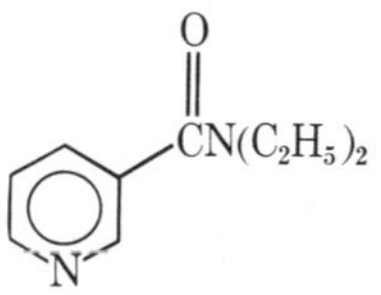

(**9**) nikethamide [*59-26-7*]

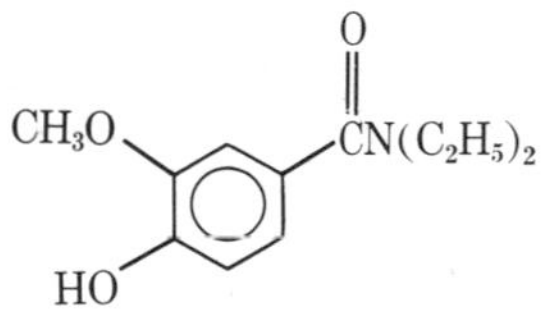

(**10**) ethamivan [*304-84-7*]

Other analeptics marketed as respiratory stimulants outside the United States include almitrine (**11**), amiphenazole (**12**), bemegride (**13**), diethadione (**14**), dimefline (**15**), and fominoben (**16**) (14–19).

Health and Safety Factors. Clinical side effects and LD_{50} values of most commercially available analeptics have been summarized (1). Overdoses produce symptoms of extreme CNS excitation including restlessness, hyperexcitability, skeletal muscle hyperactivity and, in some cases, convulsions.

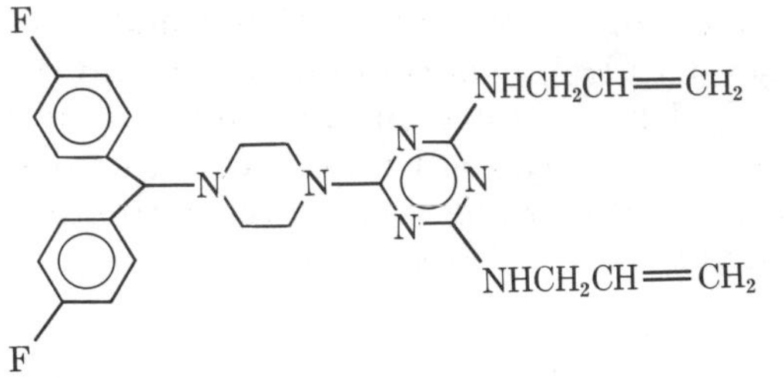

(11) almitrine
[27469-53-0]

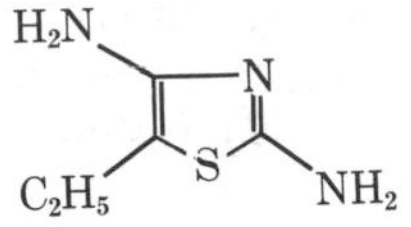

(12) amiphenazole
[490-55-1]

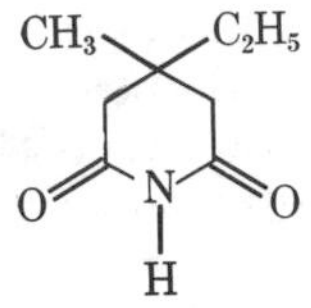

(13) bemegride
[64-65-3]

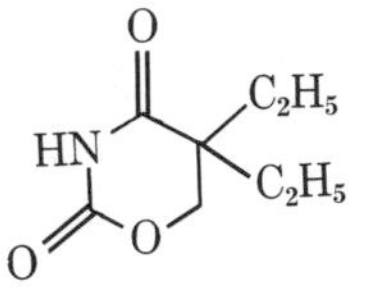

(14) diethadione
[702-54-5]

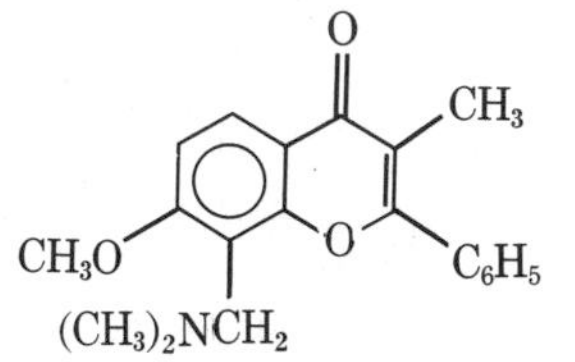

(15) dimefline
[1165-48-6]

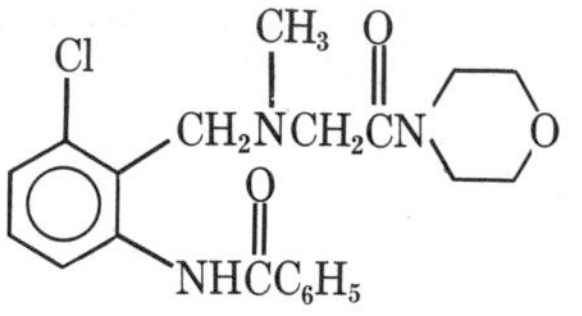

(16) fominoben
[18053-31-1]

Psychostimulants

Compounds with relatively specific cerebral-stimulant properties are classified as psychostimulants or psychoanaleptics. Two such compounds are deanol, which is marketed in the United States as the acetamidobenzoate salt (**17**), and pemoline (**18**). Both have limited use primarily in the treatment of children with attention-deficit disorder and hyperkinetic syndrome. Deanol's mechanism of action is thought to be associated with relief of a cholinergic deficit in the CNS. The compound penetrates the blood-brain barrier and acts as a precursor of acetylcholine. The mechanism of pemoline's stimulant action is unknown. However the compound has weak sympathomimetic properties that may contribute to its clinical effects. Pemoline was first prepared in 1913 (20).

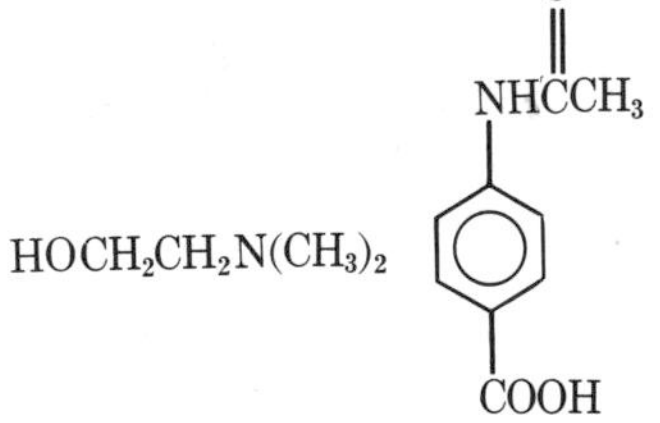

(17) deanol acetamidobenzoate
[3635-74-3]; deanol [108-01-0]

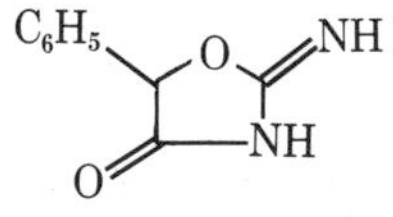

(18) pemoline
[2152-34-3]

Caffeine (**19**) is a mild psychostimulant that has been called the most widely used psychoactive substance on earth (21–22). Annual U.S. consumption has been estimated to be 15,000 metric tons (23). Coffee (qv), tea (qv), and cola-flavored beverages are the main dietary sources of the compound. Average caffeine contents per cup of brewed coffee, instant coffee, tea (bagged), and cola beverages (0.710 L or 12 oz) are 110 mg, 66 mg, 46 mg, and 47 mg, respectively (24) (see also Carbonated beverages). Undoubtedly the popularity of the beverages is, at least to some extent, related to their stimulant effects.

Caffeine has occasionally been used clinically to treat respiratory depression associated with overdoses of CNS-depressant drugs. However, more effective drugs are used currently for this purpose. The compound is marketed primarily in over-the-counter combination products, especially analgesic and antihistamine preparations. At modest doses (100–200 mg), caffeine stimulates the higher centers of the CNS to enhance wakefulness, increase mental activity, and reduce fatigue. Higher doses are required to stimulate respiratory centers. Recent evidence suggests that the stimulant properties of caffeine are associated with its ability to antagonize the effects of the inhibitory transmitter, adenosine (**20**) (21–22).

Caffeine, which is a xanthine alkaloid, is obtained in pure form from tea waste, from the manufacture of caffeine-free coffee, and by total synthesis (25–26).

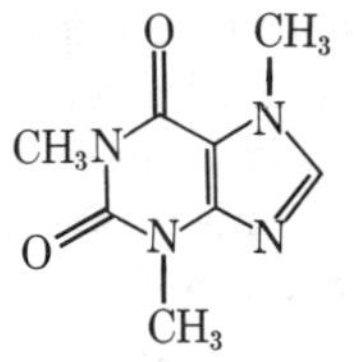

(**19**) caffeine
[*58-08-2*]

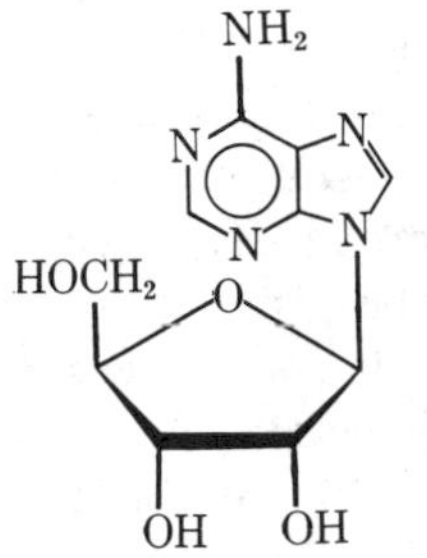

(**20**) adenosine
[*58-61-7*]

Sympathomimetics. Sympathomimetics are a group of mostly synthetic compounds that resemble the neurotransmitters epinephrine and norepinephrine in their actions and to some extent chemically (see Epinephrine and norepinephrine). As a class, they have wide-ranging pharmacologic effects including, in some cases, profound CNS excitatory actions. Sympathomimetics that have selective central effects are used primarily as anorexigenic agents and are prescribed as an adjunct to dietary restraint. Appetite-suppressing agents (qv) have been comprehensively reviewed. Other conditions treated with centrally active sympathomimetics are narcolepsy, depressant-drug poisoning, mild reactive depression (but not severe depression), hyperkinetic syndrome, parkinsonism, and epilepsy as adjunct therapy. Sympathomimetics prescribed for these conditions, apart from those used only as anorectics, are amphetamine (**21**), dextroamphetamine [*51-64-9*] (the + isomer), methamphetamine (**22**), and methylphenidate (**23**) (27–30). All of these compounds are thought to release norepinephrine from storage sites in the sympathetic nerves as their primary mechanism of action.

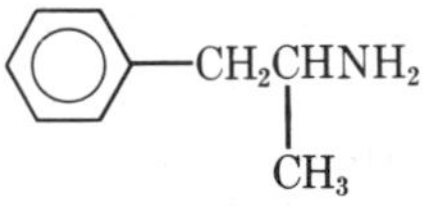

(**21**) amphetamine
[*300-62-9*]

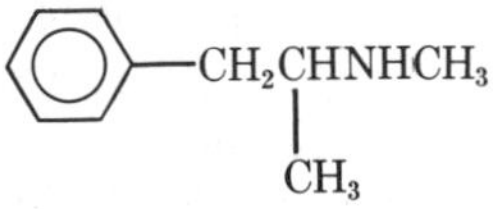

(**22**) methamphetamine
[*537-46-2*]

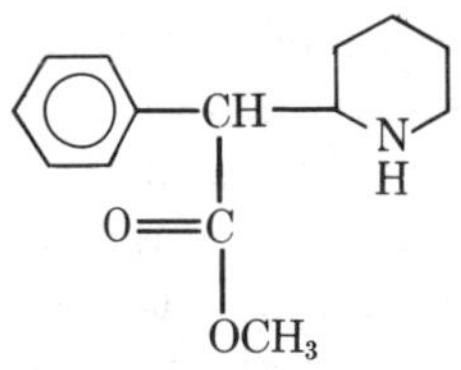

(**23**) methylphenidate
[*113-45-1*]

Side Effects and Abuse Potential. Sympathomimetics are one of the most abused classes of drugs marketed in the United States. Continued use of the compounds to treat obesity or to resist sleep often leads to habituation and the development of tolerance. Toxic symptoms resulting from increasing doses include extreme nervousness, insomnia, visual and auditory hallucinations, and psychosis. Because of the potential for abuse, the manufacture, distribution, and use of sympathomimetics are strictly controlled in the United States by the Drug Enforcement Agency (DEA).

Antidepressants

Depression. Depression is one of many moods that characterize normal daily life. Occasionally, however, the symptoms become severe enough to suggest a serious disorder. It has been estimated that 2–15% of adults in the United States experience significant depressive symptoms in any given year, among whom are some 400,000 who receive treatment (31). Approximately 75% of all hospitalized psychiatric patients suffer from depression (32).

Depressive disorders, often called affective disorders, are divided into several categories. The most common is reactive depression (also called secondary neurotic depression) and is usually precipitated by a serious adverse experience, eg, the loss of a family member, the loss of a job, or a serious illness. Patients with reactive depression frequently recover completely without therapy. A more serious type of depression is endogenous or primary depression. This form is probably of genetic origin and may express itself in one of two ways. The first is a manic-depressive disorder (bipolar) that is characterized by alternating episodes of mania and depression. The second (unipolar) is characterized by depressive episodes only.

Mechanism. Over the last 20 years evidence has accumulated that associates endogenous depression with abnormally low levels of the neurotransmitter norepinephrine, and possibly serotonin in some patients, at central nerve synapses. The reason for this is unclear, but some researchers have speculated that the abnormality may be a compensatory effect resulting from a pathologic hypersensitivity of postsynaptic β-adrenergic or serotonergic receptors in the brain (33). Another theory, based on recent biochemical studies, is that hypersensitive presynaptic α_2-adrenergic receptors may inhibit the release of norepinephrine by way of the negative feedback mechanism (34). Efforts to unify the various clinical and experimental concepts have resulted in the following proposed mechanism for depression (34):

> environmental/genetic factors → hypersensitive α_2 receptors → decreased norepinephrine release → postsynaptic β-adrenoceptors sensitized → clinical symptoms of depression

Treatment. Until the early part of the 20th century, treatment for depression was primarily symptomatic and supportive. The most seriously depressed patients, especially those with suicidal tendencies, were isolated for their own protection; others were moved to more favorable and supportive environments to facilitate recovery. The development and use of psychotherapy after the turn of the century was the next advance in treatment, followed by insulin shock therapy, pentylenetetrazole shock therapy, and electroconvulsive therapy (ECT) in the 1930s. Of these, only psychotherapy and ECT are used to any appreciable extent. The greatest advance in treatment to date began in the 1950s with the discovery and development of two classes of psychotherapeutic drugs, the monoamine oxidase (MAO) inhibitors and the tricyclic antidepressants. Both classes of compounds can elevate the mood of patients with endogenous depression, albiet by different mechanisms. Monoamine oxidase inhibitors inactivate the enzyme MAO, which is responsible for oxidative deamination of norepinephrine and serotonin, thus prolonging their lifetime at the nerve synapses; tricyclic antidepressants inhibit the reuptake of the compounds into nerve terminals. Since the introduction of these two classes of compounds, there has been a significant decline in the number of patients hospitalized for depression.

Monoamine Oxidase Inhibitors. Iproniazid (**24**), which is the prototype MAO inhibitor, was originally introduced in 1951 as a treatment for tuberculosis. Soon thereafter, surprising mood-elevating effects were noted in many of the patients. In 1952 it was shown that the compound had potent MAO inhibitory activity, thus providing a biochemical basis for its effects (35). Later the use of iproniazid for the treatment of depression was investigated and, as a result, in 1957 the compound was introduced into psychiatry (36–37). Since 1957 many different types of MAO inhibitors have been prepared and studied clinically, but relatively few have reached or remained on the market because of serious side effects. The use of iproniazid has been abandoned because of hepatotoxicity.

Currently, there are three MAO inhibitors marketed in the United States for the treatment of depression: isocarboxazid (**25**), phenelzine (**26**), and tranylcypromine (**27**). Isocarboxazid is an acylhydrazine structurally related to iproniazid and is the result of extensive modifications to improve its therapeutic properties. Phenelzine and tranylcypromine may be considered unique analogues of phenylethylamine and amphetamine, respectively, although their pharmacologic effects differ from those of the sympathomimetics.

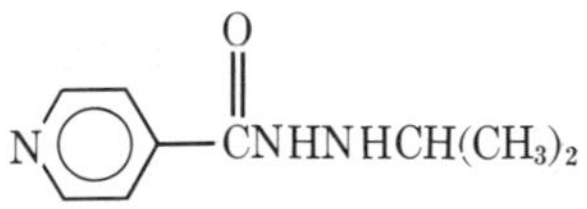

(**24**) iproniazid
[*54-92-2*]

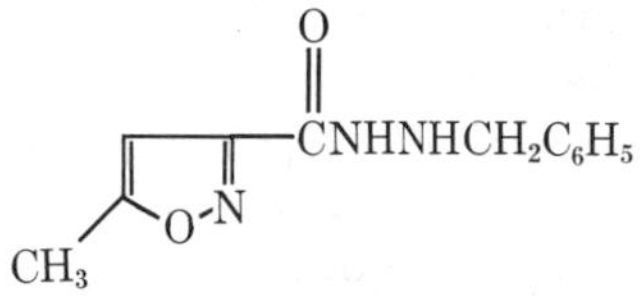

(**25**) isocarboxazid
[*59-63-2*]

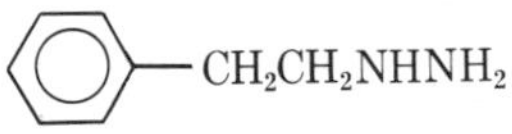

(**26**) phenelzine
[*51-71-8*]

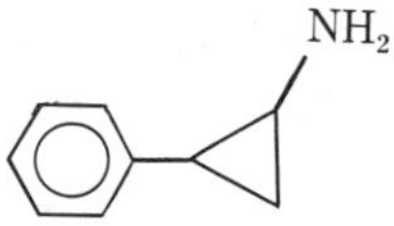

(**27**) *trans*-tranylcypromine
(±) [*155-09-9*]

The relative MAO inhibitory potencies of isocarboxazid, phenelzine, and tranylcypromine, compared with iproniazid, are listed in Table 1, along with other chemical and marketing data.

Nialamide (**28**) and mebanazine (**29**) are two MAO inhibitors marketed in Europe, with structural similarities to iproniazid and phenelzine, respectively. Both compounds are prepared by standard methods (43–44).

(**28**) nialamide
[*51-12-7*]

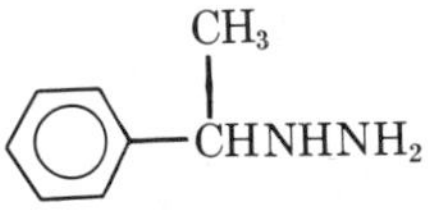

(**29**) mebanazine
[*65-64-5*]

Table 1. Biological, Chemical, and Marketing Data for Important MAO Inhibitors

Generic name	Trade name	Manufacturer	Year of introduction	Relative MAO inhibitory potency[a]	Preparation refs.
iproniazid	Marsilid		1957	1	39
isocarboxazid	Marplan	Roche	1959	3.1	40
phenelzine	Nardil	Parke-Davis	1959	18	41
tranylcypromine	Parnate	Smith, Kline and French	1961	45	42

[a] Rat brain MAO, tyramine as substrate; ref. 38.

The fact that there have been few controlled comparative clinical studies of MAO inhibitors makes it difficult to rate their relative therapeutic value. However, it is generally agreed that tranylcypromine is probably the most effective (45).

Today, the use of MAO inhibitors for the treatment of depression is severely restricted because of potential side effects, the most serious of which is hypertensive crisis. Other side effects include excessive central stimulation, orthostatic hypotension and, in rare cases, hepatotoxicity. Presumably, many of these effects are caused by tyramine-induced release of large quantities of norepinephrine at nerve terminals. Tyramine, a normal constituent of many foods, is normally inactivated by MAO in the liver. To minimize the possibility of side effects, patients taking MAO inhibitors must scrupulously avoid all foods that may be high in tyramine content, eg, cheese and beer. It has recently been shown that at least two types of MAO exist, type A and type B (46). All MAO inhibitors marketed in the United States inhibit both types. However, several selective inhibitors being developed may offer therapeutic advantages and fewer side effects (47).

Tricyclic Antidepressants. Imipramine (**30**), which was the first tricyclic antidepressant to be developed, is one of many useful psychoactive compounds derived from systematic molecular modifications of the antihistamine promethazine (**31**). In this case, the sulfur atom of promethazine was replaced with an ethylene bridge and the dimethylamino group was moved to the terminal carbon of the side chain. The actual synthesis of imipramine is typical of compounds in this class (48).

CH_2Cl, NO_2 → NO_2 NO_2 → NH_2 NH_2 →

N–H → N–$CH_2CH_2CH_2N(CH_3)_2$

(**30**) imipramine
[*50-49-7*]

S, N–$CH_2CHN(CH_3)_2$, CH_3

(**31**) promethazine
[*60-87-7*]

In initial pharmacologic studies, imipramine gave no indication of its ultimate therapeutic potential. The compound displayed weak antihistaminic activity, but it was selected for clinical trials primarily on the basis of its sedative properties. Surprisingly, the expected calming effect on agitated psychotic patients was not seen.

Instead the compound proved very effective in patients suffering from depression. Further clinical studies indicated that imipramine and other members of the tricyclic family should be the treatments of choice for patients with endogenous depression (49) (see Psychopharmacological agents).

Early studies on the mechanism of action showed that imipramine potentiates the effects of catecholamines. This discovery along with other evidence led to the hypothesis that the compound exerts its antidepressant effects by elevating norepinephrine levels at central adrenergic synapses. Subsequent studies have shown that the compound is a potent inhibitor of norepinephrine reuptake, and to a lesser extent serotonin, thus providing biochemical support for the proposed mechanism of action and a basis for predicting clinical efficacy of other compounds.

Current pharmacologic tests for imipraminelike activity include reversal of reserpine- or tetrabenazine-induced ptosis, potentiation of amphetamine-induced increases in spontaneous motor activity or hyperthermia, and potentiation of norepinephrine-induced contractions of the cat nictitating membrane (50–52). The behavioral despair model of Porsolt is also used, along with various *in vitro* and *ex vivo* biochemical tests (53–55).

Following the successful introduction of imipramine, many related compounds were prepared and clinically evaluated for antidepressant effects. Amitriptyline (**32**) was the next compound to be marketed. Structurally, it is closely related to imipramine with a C=CH group replacing the N—CH_2 fragment. Both imipramine and amitriptyline have comparable pharmacologic and clinical effects and are particularly useful in the treatment of depressed patients with psychomotor agitation. Trimipramine (**33**) is also similar to imipramine with a branched side chain and similar activity, although it is slightly less potent.

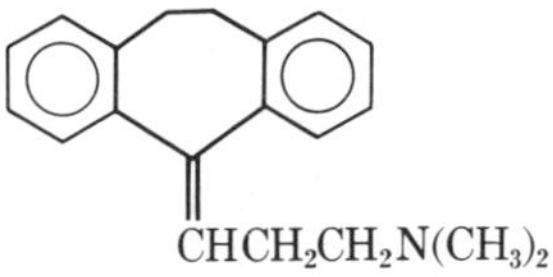

$CHCH_2CH_2N(CH_3)_2$

(**32**) amitriptyline
[*50-48-6*]

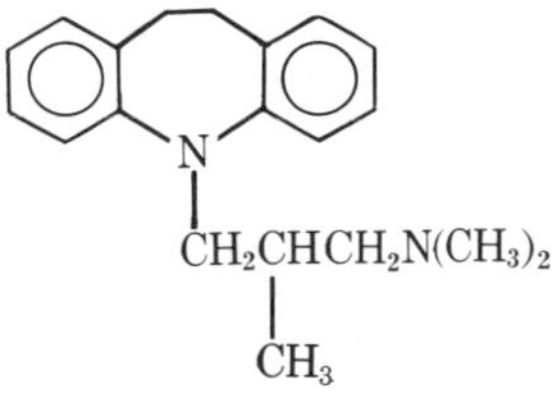

N
$CH_2CHCH_2N(CH_3)_2$
CH_3

(**33**) trimipramine
[*739-71-9*]

Desipramine (**34**) and nortriptyline (**35**) are demethylated derivatives and principal metabolites of imipramine and amitriptyline, respectively. Both compounds produce less sedation and stronger psychomotor effects than their tertiary amine counterparts, probably because tricyclics containing secondary amine groups generally show greater selectivity for inhibiting the reuptake of norepinephrine compared with serotonin. Protriptyline (**36**) is a structural isomer of nortriptyline and another important secondary amine that displays a similar clinical profile.

Doxepin (**37**), unlike other commercially available tricyclics, has an oxygen atom in the bridge between the two aromatic rings. It is marketed as a cis-trans mixture (1:5) of isomers, both of which are active. This close relative of amitriptyline has both sedative and anxiolytic properties associated with its antidepressant profile.

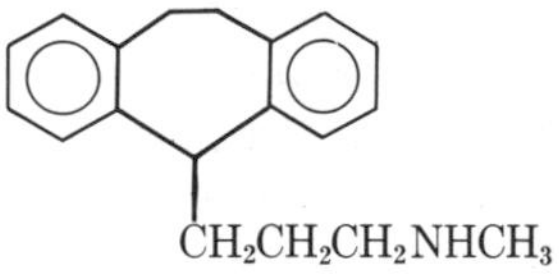

$CH_2CH_2CH_2NHCH_3$

(**34**) desipramine
[*50-47-5*]

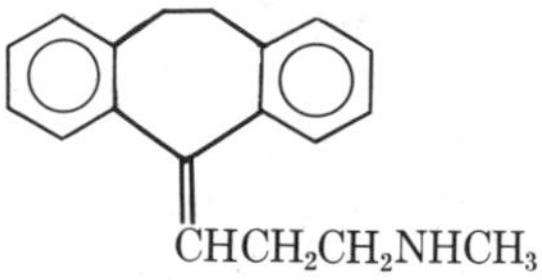

$CHCH_2CH_2NHCH_3$

(**35**) nortriptyline
[*72-69-5*]

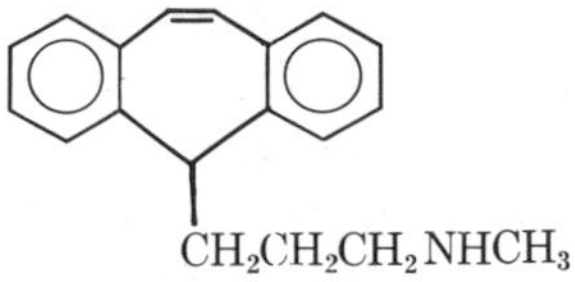

$CH_2CH_2CH_2NHCH_3$

(**36**) protriptyline
[*438-60-8*]

The most recent addition to the U.S. antidepressant market is amoxapine (**38**). It is the desmethyl derivative of the antipsychotic drug loxapine. Though still a tricyclic, its structure has much less resemblance to imipramine than the earlier analogues. Amoxapine may have a slightly faster onset of action than other tricyclics.

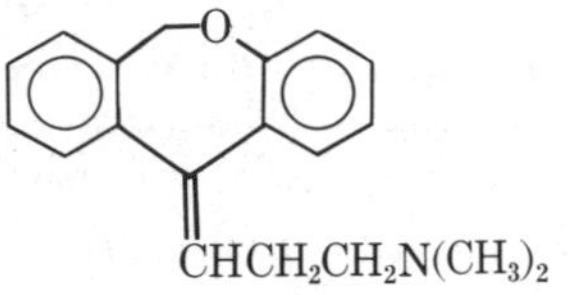

(**37**) doxepin
[*1668-19-5*]

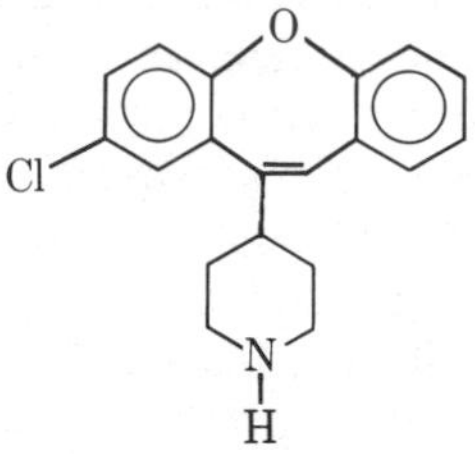

(**38**) amoxapine
[*140-28-44-5*]

Representative brands of tricyclic antidepressants marketed in the United States, and related chemical, biological, and marketing data are listed in Table 2.

Other important tricyclic antidepressants marketed outside the United States are noxiptilin (**39**), butriptyline (**40**) (an isostere of trimipramine), clomipramine (**41**) (the 3-chloro derivative of imipramine), dothiepin (**42**) (the sulfur isostere of doxepine), and dibenzepin (**43**) (66–70). The latter compound is noteworthy because it contains a two-atom bridge, rather than the usual three, between the tricyclic nucleus and the basic amino group.

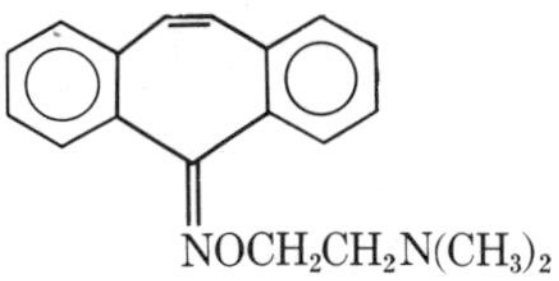

(**39**) noxiptilin
[*3362-45-6*]

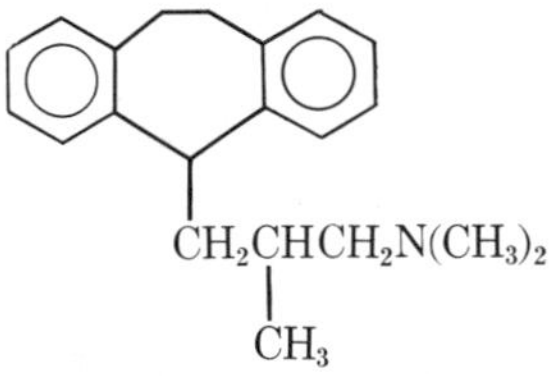

(**40**) butriptyline
[*35941-65-2*]

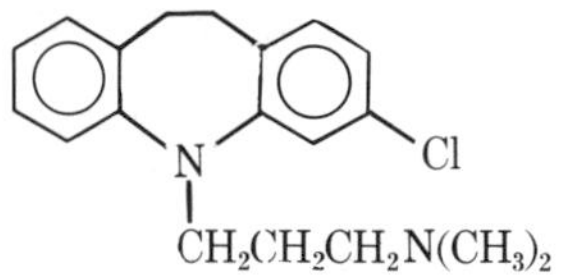

(**41**) clomipramine
[*303-49-1*]

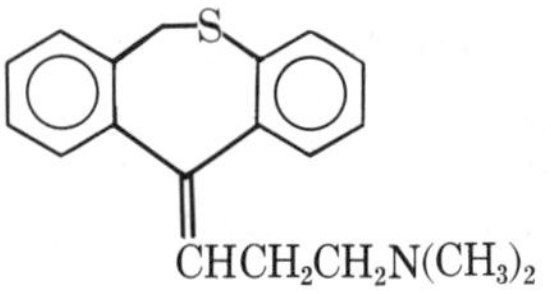

(**42**) dothiepin
[*113-53-1*]

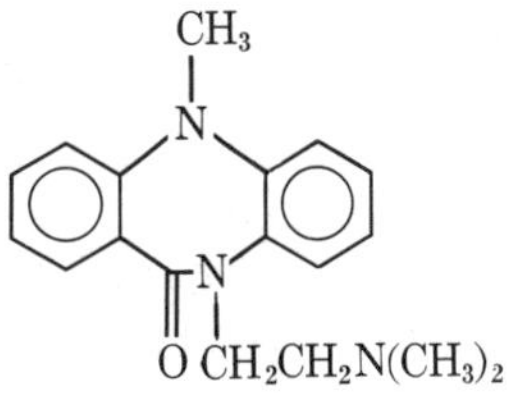

(**43**) dibenzepin
[*4498-32-2*]

Side Effects and Toxicity. Although tricyclic antidepressants have proved much safer to use than MAO inhibitors, undesirable and occasionally serious side effects can occur. The most common side effects, such as blurred vision, dry mouth, constipation, and urinary retention, are associated with the anticholinergic properties of the compounds. More serious, however, are effects on the cardiovascular system that include orthostatic hypotension, tachycardia, arrhythmias, and in rare cases, sudden

Table 2. Biological, Chemical, and Marketing Data for U.S. Marketed Tricyclic Antidepressants

Generic name	Trade name	Manufacturer	Year of introduction	Norepinephrine uptake inhibition, IC_{50}, 10^{-7} M [a]	Serotonin uptake inhibition, IC_{50}, 10^{-7} M [b]	Preparation refs.
imipramine	Tofranil	Geigy	1959	0.75	3.2	48
amitriptyline	Elavil	Merck, Sharp and Dohme	1961	1.3	3.0	59
desipramine	Norpramine	Merrell-National	1964	0.014	18	60
nortriptyline	Aventyl	Lilly	1969	0.29	15	61
trimipramine	Surmontil	Ives	1979	77	260	62
protriptyline	Vivactil	Merck, Sharp and Dohme	1967	0.008	13	63
doxepin	Sinequan	Roerig	1969	5.5	39	64
amoxapine	Asendin	Lederle	1980			65

[a] In mouse atria (56).

[b] In thrombocytes (57–58).

death. Because depressed patients are more prone to suicide than others and because overdoses of tricyclic antidepressants are extremely dangerous, antidepressant poisonings have become a serious problem in recent years. LD_{50} data for most of the antidepressants described herein have been reported (71–72).

Structure-Activity Relationships. Since the development of imipramine, medicinal chemists have sought to correlate structural features and physical properties of antidepressant drugs with pharmacologic effects. Unfortunately, the slow development of suitable animal models delayed useful studies of this type for a number of years. Recent correlations of molecular geometry and conformations of tricyclic antidepressants with norepinephrine reuptake inhibition indicate that two of the most important factors for activity are the relative orientation of the aromatic rings of the tricyclic nucleus and the distance between the basic nitrogen and one of the aromatic rings (73). Maximum activity is usually achieved when the angle between the two aromatic rings is large. Compounds whose rings approach planarity or whose rings are not bridged are generally less active. Aromatic substitution usually decreases activity and frequently affects norepinephrine/serotonin receptor selectivity. A comprehensive review describes recent work on structure–activity correlation and several theories regarding requirements for antidepressant effects (74) (see Pharmacodynamics).

Antidepressants Being Developed. At present there are more than thirty antidepressants undergoing clinical trials worldwide, some of which have already been introduced in overseas markets. Many of the newer compounds, eg, bupropion **(44)**, trazodone **(45)**, pridefine **(46)**, nomifensine **(47)**, viloxazine **(48)**, and zimelidine **(49)**, have unique, nontricyclic structures and in some cases atypical antidepressant profiles (75–80).

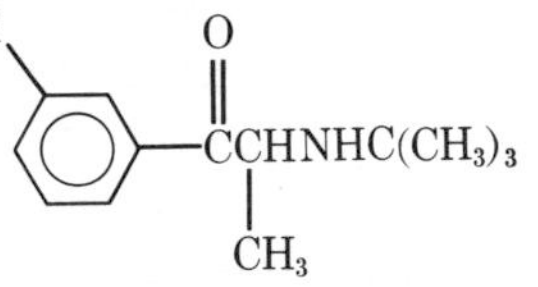

(**44**) bupropion
[*34911-55-2*]

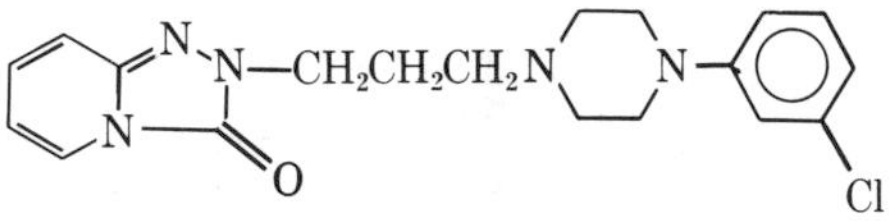

(**45**) trazodone
[*19794-93-5*]

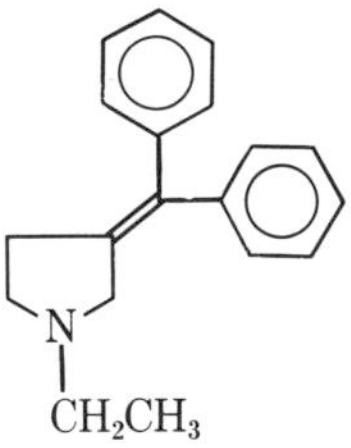

(**46**) pridefine
[*5370-41-2*]

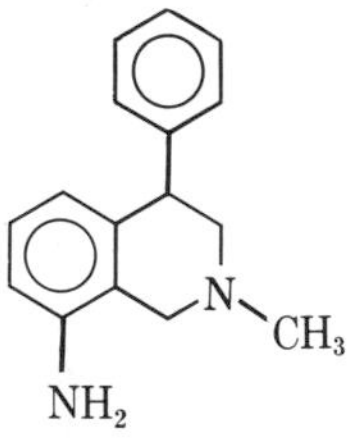

(**47**) nomifensine
[*24526-64-5*]

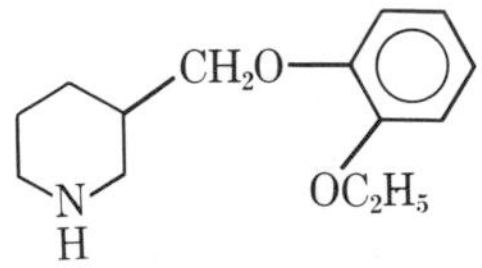

(**48**) viloxazine
[*46817-91-8*]

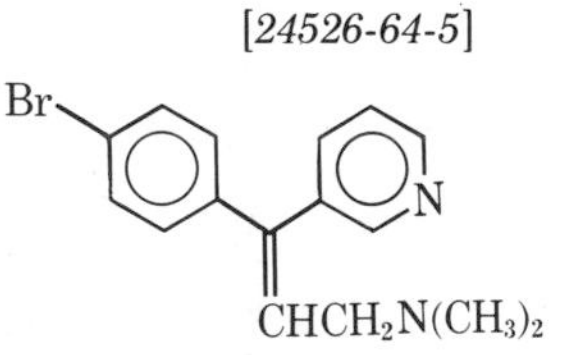

(**49**) zimelidine
[*56775-88-3*]

Maprotiline (**50**) and nisoxetine (**51**) are clinically effective antidepressants that are claimed to be specific inhibitors of norepinephrine reuptake (81–82). Other active compounds, eg, fluoxetine (**52**) (a close relative of nisoxetine) and fluvoxamine (**53**) preferentially inhibit serotonin reuptake (83–84). Compounds with such selectivity may ultimately prove to be beneficial to particular subgroups of depressed patients.

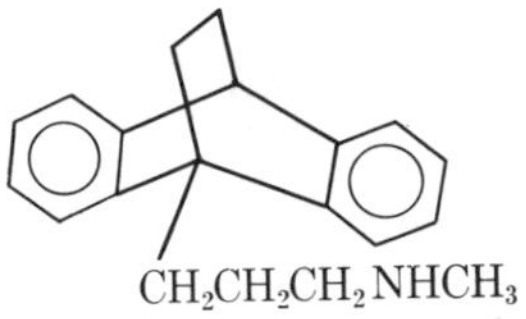

(**50**) maprotiline
[*10262-69-8*]

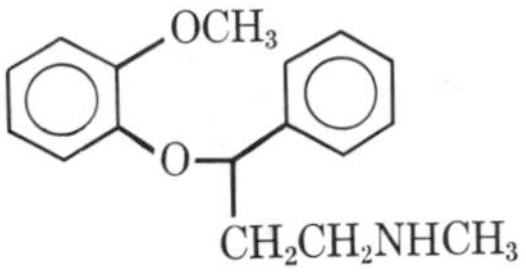

(**51**) nisoxetine
[*53179-07-0*]

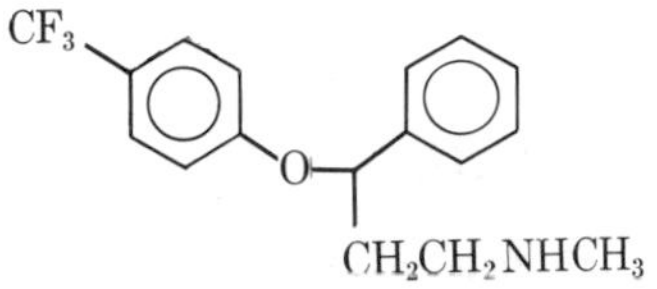

(**52**) fluoxetine
[*54910-89-3*]

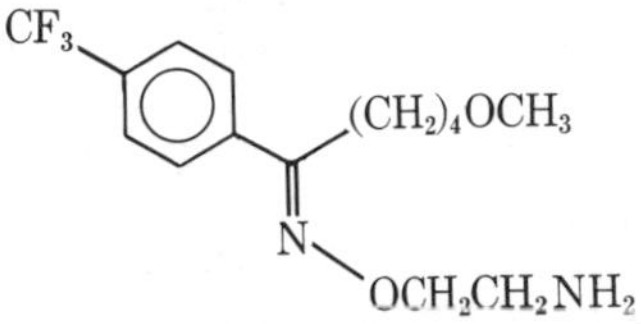

(**53**) fluvoxamine
[*54739-18-3*]

Mianserin (**54**) and iprindole (**55**) are two new antidepressants that apparently act by a unique mechanism (55,85–86). Neither compound at therapeutic doses blocks the reuptake of norepinephrine or serotonin and they have no MAO inhibitory properties. Recent biochemical studies, however, show that mianserin and iprindole, as well as various tricyclic antidepressants and MAO inhibitors and electroconvulsive therapy, substantially reduce the responsiveness (by reducing the number) of postsynaptic beta-receptors over a one- to six-week period. These results correlate well with the slow onset of clinical antidepressant effects characteristic of all antidepressant drugs, and support the hypersensitive beta-receptor theory of depression. It has been postulated that beta-adrenergic receptor desensitization, frequently referred to as down regulation, may be an effect common to all antidepressants though brought about by several different mechanisms (55). The role of serotonin in depression still remains unclear (87).

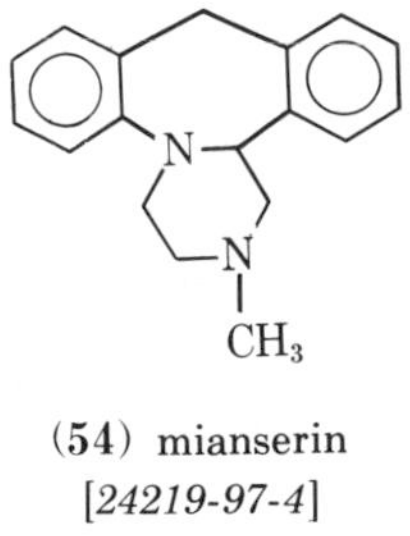

(**54**) mianserin
[*24219-97-4*]

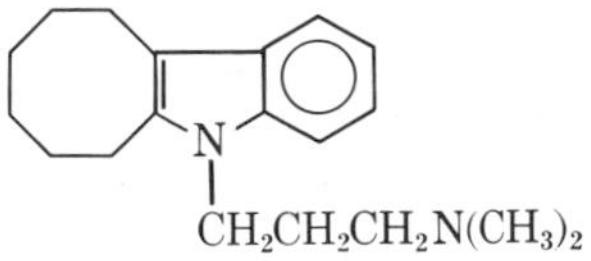

(**55**) iprindole
[*5560-72-5*]

Economic Aspects

According to Department of Commerce surveys, dollar values of manufacturers shipments of CNS stimulants, which include respiratory and cerebral stimulants but not antidepressants or anorexiants rose sharply from $1,484,000 in 1969 to $9,378,000 in 1979 (88–89). During the same period, sales of all antidepressants increased from $49,383,000 to $96,000,000. With the exception of caffeine, all of the compounds described herein are marketed primarily as ethical preparations.

BIBLIOGRAPHY

1. S. C. Wang and J. W. Ward in J. G. Widdecombe, ed., *International Enyclopedia of Pharmacology and Therapeutics*, Vol. 104, Pergamon, Oxford, UK, 1981, p. 85.
2. F. Hahn, *Pharmacol. Rev.* **12,** 447 (1960).
3. R. B. Woodward and co-workers, *J. Am. Chem. Soc.* **76,** 4749 (1954).
4. R. B. Woodward and co-workers, *Tetrahedron* **19,** 247 (1963).
5. E. J. Corey and H. L. Pearce, *J. Am. Chem. Soc.* **101,** 5841 (1979).
6. L. A. Porter, *Chem. Rev.* **67,** 441 (1967).
7. K. F. Schmidt, *Ber.* **57,** 704 (1924).
8. U.S. Pat. 3,363,006 (Jan. 9, 1968), J. F. Olin (to Pennsalt Chemicals Corp.).
9. B. Laurell, ed., *Acta Psychiatr. Scand.* **213,** 5 (1970).
10. A. P. Winnie, *Acta Anaesthesiol. Scand.* **51,** 1 (1973).
11. C. D. Lunsford and co-workers, *J. Med. Chem.* **7,** 302 (1964).
12. P. Oxley and co-workers, *J. Chem. Soc.* 763 (1946).
13. K. Kratzl and E. Krasnicka, *Monatsh. Chemie* **83,** 18 (1952).
14. U.S. Pat. 3,647,794 (Mar. 7, 1972), G. Regnier and co-workers (to Science Union et Cie.-Société Française de Recherche Medicale).
15. W. Davies, J. A. MacLaren, and R. L. Wilkinson, *J. Chem. Soc.* 3491 (1950).
16. F. B. Thole and J. F. Thorpe, *J. Chem. Soc.* **99,** 422 (1911).
17. E. Testa and co-workers, *J. Org. Chem.* **24,** 1928 (1959).
18. P. da Re, L. Verlicchi, and I. Setnikar, *Arzneim. Forsch.* **10,** 800 (1960).
19. G. Kruger and co-workers, *Arzneim. Forsch.* **23,** 290 (1973).
20. W. Traube and R. Ascher, *Chem. Ber.* **46,** 2077 (1913).
21. *Science* **211,** 1408 (1981).
22. S. H. Snyder and co-workers, *Proc. Nat. Acad. Sci.* **78,** 3260 (1981).
23. D. M. Graham, *Nutr. Rev.* **36,** 97 (1978).
24. M. L. Bunker and M. McWilliams, *J. Am. Diet. Assoc.* **74,** 28 (1979).
25. U.S. Pat. 2,817,588 (Dec. 24, 1957), W. E. Barch (to Standard Brands).
26. W. Traube, *Chem. Ber.* **33,** 3052 (1900).
27. W. H. Hartung and J. C. Munch, *J. Am. Chem. Soc.* **53,** 1875 (1931).
28. U.S. Pat. 2,276,508 (Mar. 17, 1942), F. P. Nabenhauer (to Smith, Kline & French Laboratories).
29. A. Ogata, *J. Pharm. Soc. Japan* **451,** 751 (1919).
30. U.S. Pat. 2,507,631 (May 16, 1950), M. Hartmann and L. Panizzon (to Ciba Pharmaceutical Products).
31. L. E. Hollister, *Ann. Intern. Med.* **89,** 78 (1978).
32. *Science* **212,** 432 (1981).
33. A. J. Mandell, D. S. Segaland, and R. Kuczenski in A. J. Friedhoff, ed., *Catecholamines and Behavior*, Vol. 2, Plenum Publishing Corp., New York, 1975, p. 197.
34. B. E. Leonard, *Neuropharmacology* **19,** 1175 (1980).
35. E. A. Zeller and co-workers, *Experientia* **8,** 349 (1952).
36. G. E. Crane, *Psychiatr. Res. Rep.* **8,** 142 (1957).
37. N. S. Kline, *J. Clin. Exp. Psychopathol.* **19,** 72 (1958).
38. D. R. Maxwell, W. R. Gray, and E. M. Taylor, *Br. J. Pharmacol.* **17,** 310 (1961).
39. H. H. Fox and T. Gibbas, *J. Org. Chem.* **18,** 994 (1953).
40. T. S. Gardner, E. Wenis, and J. Lee, *J. Med. Chem.* **2,** 133 (1970).
41. J. H. Biel and co-workers, *J. Am. Chem. Soc.* **81,** 2805 (1959).

42. A. Burger and W. L. Yost, *J. Am. Chem. Soc.* **70,** 2198 (1948).
43. U.S. Pat. 3,040,061 (June 10, 1962), B. M. Bloom and R. E. Carnham (to Pfizer).
44. C. G. Overberger and A. V. DiGiulio, *J. Am. Chem. Soc.* **80,** 6562 (1958).
45. L. S. Goodman and A. Gilman, *The Pharmaceutical Basis of Therapeutics*, 5th ed., McMillan Publishing Co., Inc., New York, 1975, p. 187.
46. H. Y. T. Yang and N. H. Neff, *J. Pharmacol. Exp. Ther.* **189,** 733 (1974).
47. F. Quitkin, *Arch. Gen. Psychiatry* **36,** 749 (1979).
48. W. Schindler and F. Haeflinger, *Helv. Chim. Acta* **37,** 472 (1954).
49. R. Kuhn, *Am. J. Psychiatry* **115,** 459 (1958).
50. A. Cowan and B. A. Whittle, *Br. J. Pharmacol.* **44,** 353 (1972).
51. A. Weissman, B. K. Koe, and S. S. Tenen, *J. Pharmacol. Exp. Ther.* **151,** 339 (1966).
52. C. A. Stone in S. Garattini and M. N. G. Dukes, eds., *Antidepressant Drugs*, Exerpta Medica, Amsterdam, 1967, p. 158.
53. R. D. Porsolt, A. Bertin, and M. Jalfre, *Arch. Int. Pharmacodyn.* **229,** 327 (1977).
54. B. K. Koe in S. Fielding and H. Lal, eds., *Antidepressants*, Futura Publishing Co., Mt. Kisco, N.Y., 1975, p. 143.
55. F. Sulser, J. Vetulani, and P. L. Mobley, *Biochem. Pharmacol.* **27,** 257 (1978).
56. J. Hyttel, *Psychopharmacology* **51,** 225 (1977).
57. G. V. R. Born and R. E. Gillson, *J. Physiol.* **146,** 472 (1977).
58. R. S. Stacey, *Br. J. Pharmacol.* **16,** 284 (1961).
59. U.S. Pat. 3,205,264 (Sept. 7, 1965), E. W. Tristram and R. J. Tull (to Merck & Co.).
60. Brit. Pat. 908,788 (Oct. 24, 1962), (to J. R. Geigy AG).
61. R. D. Hoffsommer, D. Traub, and N. L. Wendler, *J. Org. Chem.* **27,** 4134 (1962).
62. R. M. Jacob and M. Messer, *Compt. Rend.* **252,** 2117 (1961).
63. U.S. Pat. 3,244,748 (Apr. 5, 1966), M. Tishler, J. M. Chemerda, and J. Kollonitsch (to Merck & Co.).
64. K. Stach and F. Bickelhaupt, *Monatsh. Chem.* **93,** 896 (1962).
65. Fr. Pat. 1,508,536 (Jan. 5, 1968), C. F. Howell, R. A. Hardy, and N. Q. Quinones (to American Cyanamid Co.).
66. Brit. Pat. 1,045,911 (Oct. 9, 1966), T. I. Wrigley and P. R. Leeming (to Pfizer Ltd.).
67. U.S. Pat. 3,409,640 (Nov. 5, 1968), F. J. Villani (to Schering Corp.).
68. P. N. Craig and co-workers, *J. Org. Chem.* **26,** 135 (1961).
69. M. Protiva and co-workers, *Experientia* **18,** 326 (1962).
70. Brit. Pat. 961,106 (June 17, 1964), (to A. Wander AG).
71. M. Windholz, ed., *The Merck Index*, 9th ed., Merck & Co., Rahway, N.J., 1976.
72. E. J. Fairchild, ed., *Registry of Toxic Effects of Chemical Substances*, Vol. II, Department of Health, Education, and Welfare, Washington, D.C., 1977.
73. A. I. Salama, J. R. Insalaco, and R. A. Maxwell, *J. Pharmacol. Exp. Ther.* **178,** 474 (1971).
74. F. J. Zeelen in F. Hoffmeister and G. Stille, eds., *Psychotropic Agents*, Part I, Springer-Verlag, New York, 1980, p. 352.
75. Ger. Pat. 2,064,934 (Oct. 21, 1971), D. A. Yeowell (to Wellcome Foundation Ltd.).
76. U.S. Pat. 3,381,009 (Apr. 30, 1968), G. Palazzo and B. Silvestrini (to Aziende Chimiche Riunite Angelini Francesco a Roma, Italy).
77. U.S. Pat. 3,458,635 (July 29, 1969), C. D. Lunsford, G. C. Helsley, and J. A. Richman (to A. H. Robins Co.).
78. Brit. Pat. 1,164,192 (Sept. 17, 1969), (to Farbwerke Hocchst AG).
79. U.S. Pat. 3,712,890 (Jan. 23, 1973), S. A. Lee (to Imperial Chemical Industries, Ltd.).
80. *Drugs of The Future* **3,** 71 (1978).
81. M. Wilhelm and P. Schmidt, *Helv. Chim. Acta.* **52,** 1385 (1969).
82. *Drugs of The Future* **2,** 51 (1977).
83. Ger. Pat. 2,500,110 (July 17, 1975), B. B. Molloy and K. K. Schmiegel (to Eli Lilly Co.).
84. *Drugs of The Future* **3,** 288 (1978).
85. U.S. Pat. 3,534,041 (Oct. 13, 1970), W. J. Vanderburg and J. Delobelle (to Organon Inc.).
86. L. M. Rice, E. Hertz, and M. E. Freed, *J. Med. Chem.* **7,** 313 (1964).
87. A. S. Horn, *Postgrad. Med. J.* **56,** 9 (1980).
88. *Pharmaceutical Preparations, Except Biologicals*, 1969, Current Industrial Reports, Series: Ma-28G(69)-1, U.S. Bureau of the Census, Department of Commerce, Washington, D.C., 1971.
89. Ref. 88, MA-28G(79)-1, 1980.

WILLIAM J. WELSTEAD, JR.
A. H. Robins Company

STOICHIOMETRY, INDUSTRIAL. See Simulation and process design.

STONEWARE. See Ceramics.

STORAX. See Resins, natural.

STOUT. See Beer.

STREPTOMYCIN AND RELATED ANTIBIOTICS. See Antibiotics.

STRONTIUM AND STRONTIUM COMPOUNDS

Strontium

Strontium [*7440-24-6*] is in group IIA of the periodic table between calcium and barium. These three elements are called alkaline-earth metals because their oxides fall between the hydroxides of alkali metals, ie, sodium and potassium, and the oxides of earth metals, ie, magnesium, aluminum, and iron, in chemical properties. Strontium was identified in the 1790s (1). The metal was first produced in 1808 in the form of a mercury amalgam. A few grams of the metal was produced in 1860–1861 by electrolysis of strontium chloride [*10476-85-4*].

Strontium forms 0.02–0.03% of the earth's crust and is present in igneous rocks in amounts averaging 375 ppm. It is the fifth most abundant metallic ion in seawater, occurring in quantities of ca 14 grams per metric ton. Strontium rarely forms independent minerals in igneous rocks but usually occurs as a minor constituent of the rock-forming minerals. Independent strontium minerals do develop in or near sedimentary rocks mainly associated with beds or lenses of gypsum, anhydrite, or rock salt; in veins associated with limestone and dolomite; disseminated in shales, marls, and sandstones; or as a gangue mineral in lead–zinc deposits.

Properties. Strontium is a hard white metal with physical properties as shown in Table 1. Strontium-90 [*10098-97-2*] is a radioisotope that is a product of nuclear fission. Its half-life is 29 yr; it releases no γ radiation but emits β radiation of 0.54 MeV. It has caused concern because it may be a product of fallout of nuclear explosions and, when ingested, may accumulate in animal bone tissues.

The chemical properties of strontium are intermediate between those of calcium and barium; it is more reactive than calcium and less reactive than barium. Strontium is bivalent and reacts with H_2 to form SrH_2 [*13598-33-9*] at reasonable speed at 300–400°C. It reacts with H_2O, O_2, N_2, F, S, and halogens to produce compounds corresponding to its valence (+2). Strontium and the alkaline-earth metals are less active than the alkali metals, but all are strong reducing agents. At elevated temperatures, it reacts with CO_2 according to the reaction (3):

$$5\,Sr + 2\,CO_2 \rightarrow \underset{[12071\text{-}29\text{-}3]}{SrC_2} + \underset{[1314\text{-}11\text{-}0]}{4\,SrO}$$

Metallic strontium dissolves in liquid ammonia. The metal and its salts impart a brilliant red color to flames.

Production. Metallic strontium was first successfully produced by the electrolysis of fused strontium chloride. Although many attempts were made to develop this process, the deposited metal has a tendency to migrate into the fused electrolyte and the method was not satisfactory. A more effective early method was that described in ref. 4. Strontium oxide is reduced thermally with aluminum according to the reaction:

$$3\,SrO + 2\,Al \rightarrow 3\,Sr + Al_2O_3$$

The reaction is conducted in a vacuum and the gaseous metal is condensed in a cooler part of the apparatus.

Uses. There are no commercial uses of strontium metal. The properties of strontium alloys have been investigated by the U.S. Bureau of Mines and others, but little use has been made of the information (5–7). Strontium modifies the properties

Table 1. Physical Properties of Strontium[a]

Property	Value
atomic weight	87.62
melting range, °C	768–791
boiling range, °C	1350–1387
density, g/cm^3	2.6
crystal system	face-centered cubic (fcc)
lattice constant	6.05
latent heat of fusion, kJ/kg[b]	104.7
electrical resistivity, $\mu\Omega$/cm	22.76
stable isotopes, % abundance	
^{84}Sr [*15758-49-3*]	0.56
^{86}Sr [*13982-14-4*]	9.86
^{87}Sr [*13982-64-4*]	7.02
^{88}Sr [*14119-10-9*]	82.56

[a] Refs. 1–3.

[b] To convert J to cal, divide by 4.184.

of low aluminum, silicon-casting alloys and has been demonstrated to improve the machinability of gray-iron castings. Strontium has been used to remove the last traces of gas from vacuum tubes.

Strontium Compounds

Strontium has a valence of +2 and forms compounds that resemble the compounds of the other alkaline-earth metals (see Barium compounds; Calcium compounds). Although many strontium compounds are known, there are only a few that have commercial importance and, of these, strontium carbonate [*1633-05-2*] and strontium nitrate [*10042-76-9*] are made in the largest quantities. The mineral celestite [*7759-02-6*], $SrSO_4$, is the raw material from which strontium carbonate, $SrCO_3$, is made. Strontium nitrate and other strontium compounds are made from strontium carbonate. There are a number of minor uses of strontium compounds in lubricants, pigments, and in the preparation of high purity electrolytic zinc (see Lubrication and lubricants; Pigments, inorganic).

The U.S. demand for strontium compounds in 1979 was ca 20,000 metric tons on a strontium-content basis, which was ca 40% of the total world consumption. Electrical components, chiefly in the form of television picture-tube glass, consumed ca 63% of the strontium used, pyrotechnic materials ca 16%, and electrical materials (ferrites) ca 5% of the total (5) (see Pyrotechnics; Ferrites).

Occurrence. The principal strontium mineral is celestite, which is naturally occurring strontium sulfate. Celestite and celestine both describe this mineral; however, celestite is the form most widely used in English-speaking countries. Celestite has a theoretical strontium oxide content of 56.4 wt %, a hardness of 3–3.5 on Mohs scale, and a specific gravity of 3.96. It is usually white or blueish white and has an orthorhombic crystal form.

Deposits of celestite in Gloucestershire, the United Kingdom, represented the main source of the world supply from 1884 to 1941 and provided up to 90% of the world strontium supply (8). During World War II, shipments to the U.S. and Western Europe from the UK were disrupted, and new celestite deposits in Mexico and Spain were developed. Mexico has become the main supplier for the U.S. market and UK production has dropped to a low level.

There are celestite deposits in the United States in Texas, in Skagit and Whitcom Counties in Washington, in Maricopa County and near Gila Bend in Arizona, and in southern and central California in San Diego and San Bernardino counties. Commercial production is not possible at these deposits because of the relatively low quality of the ore. Canada has celestite deposits in Cape Breton, Nova Scotia, that were developed on a large scale in 1971 by Kaiser Aluminum and Chemical Corp. There are substantial deposits of celestite in Granada, Spain, and deposits are known in Turkey, Iran, Pakistan, Algeria, and Argentina, but the size of these deposits is not well defined. Celestite occurs widely in Mexico but only a few deposits of salable chemical-grade product are known. The first production started in 1941 in San Luis Potosi, but later has been mainly in the state of Coahuila which borders Texas and has rail transportation to the U.S. market. Mexican celestite is white and high in barium content but low in iron.

Strontianite is the naturally occurring form of strontium carbonate. It has a theoretical strontium oxide content of 70.2%, but no economically workable deposits are known. There are some naturally occurring strontium–barium and strontium–calcium isomorphs, but none are of economic importance.

The scale of celestite mining has never been big enough to justify highly efficient mining methods or extensive exploration. Hand digging of the mineral is not uncommon and the ore is usually sorted by hand. Celestite used in the United States has an average strontium sulfate content of 90 wt % with a range of 88–93 wt %.

Almost all celestite is used to make strontium carbonate; however, some ground celestite has been used as an extender and filler in paints and rubber (see Fillers). It has been used as a weighting agent in drilling muds, but it has lost this market to barite, which has a higher specific gravity and is less expensive. The production of celestite was dependent for many years on the demand for pyrotechnics. During World War II and the Korean War, production was high because signaling devices and tracer ammunition were being manufactured at a high level. The use of strontium carbonate in glass for color-television tubes introduced a new large volume use that has changed the historical pattern of use.

Production. There are three main producers of strontium compounds in the United States. FMC Corporation manufactures strontium carbonate and strontium nitrate at Modesto, California, and Chemical Products Corporation manufactures strontium carbonate only at Cartersville, Georgia. Both manufacturers use the black ash process and Mexican celestite as a raw material. Barium carbonate is also made by this process and often in the same equipment. In addition, Barium and Chemicals Inc. produces a broad line of the minor strontium compounds at its plant in Steubenville, Ohio. Outside the United States, strontium carbonate and strontium nitrate are produced in the Federal Republic of Germany, Italy, and Japan.

Economic Aspects. All celestite ore, from which strontium carbonate is made, is imported, largely from Mexico. The quantity and value of imported celestite is reported annually by the U.S. Bureau of Mines (9). Substantial quantities of strontium carbonate and strontium nitrate are imported, principally from the FRG and Italy. During 1981, the U.S. Department of Commerce found that strontium nitrate from Italy was being sold at less than fair value in an antidumping investigation.

Strontium carbonate is sold in two grades: granular and powder. Granular grade, also referred to as glass grade, is made by calcining the powder grade. The powder grade, also known as chemical or ferrite grade, has an average particle size of 1 μm.

The current price of glass-grade strontium carbonate in bulk quantities is $0.695/kg.

Health and Safety Factors. The strontium ion has a low order of toxicity, and strontium compounds are remarkably free of toxic hazards. Chemically, strontium is similar to calcium, and strontium salts, like calcium salts, are not easily absorbed by the intestinal tract. Strontium carbonate has no commonly recognized hazardous properties. Strontium nitrate is regulated as an oxidizer that promotes rapid burning of combustible materials, and it should not be stored in areas of potential fire hazards.

Strontium Acetate. Strontium acetate [*543-94-2*], $Sr(CH_3CO_2)_2$, is a white crystalline salt with a specific gravity of 2.1, and it is soluble to the extent of 36.4 g in 100 mL of water at 97°C. It crystallizes as the tetrahydrate or as the pentahydrate below 9.5°C. When heated, strontium acetate decomposes.

Strontium Carbonate. Strontium carbonate, $SrCO_3$, occurs naturally as strontianite in orthorhombic crystal and as isomorphs with aragonite, $CaCO_3$, and witherite, $BaCO_3$. There are deposits in the United States in Schoharie County, New York; in Westphalia, FRG; and smaller deposits in many other areas, but none are economically workable. Strontianite has a specific gravity of 3.7 and a Mohs hardness of 3.5 and it is colorless, gray, or reddish in color.

Strontium carbonate is a colorless or white crystalline solid having a rhombic structure below 926°C and a hexagonal structure above this temperature. It has a specific gravity of 3.70 and a melting point of 1497°C at 6 MPa (60 atm), and it decomposes to the oxide on heating at 1340°C. It is insoluble in water but reacts with acids, and it is soluble in solutions of ammonium salts.

In the current commercial production of strontium carbonate, celestite ore is crushed, ground, and stored in bins before it is fed to rotary kilns. As the ground ore is being conveyed to the kilns, it is mixed with ground coke. In the kilns, the celestite is reduced to strontium sulfide [*1314-96-1*], which is known as black ash, according to the reaction:

$$SrSO_4(s) + 2\ C(s) \rightarrow SrS(s) + 2\ CO_2(g)$$

The product stream from the kilns is collected in storage bins. Black ash from the bins is fine-ground in a ball mill and fed to a leacher circuit, which is a system of stirred tanks, where it is dissolved in water and the muds are separated by countercurrent decantation. The solution from the decantation is passed through filter presses; the muds are washed, centrifuged, and discarded. The filtered product, a saturated solution containing 12–13 wt % strontium sulfide, is sent to an agitation tank where soda ash is added to cause precipitation of strontium carbonate crystals:

$$SrS(l) + Na_2CO_3(s) \rightarrow SrCO_3(s) + Na_2S(l)$$

After precipitation is complete, the slurry is pumped to vacuum drum filters where a nearly complete liquid–solids separation is accomplished. The liquid is dilute sodium sulfide solution, which is concentrated by evaporation to a flaked 60 wt % sodium sulfide product. The filter cake is a 60 wt % strontium carbonate solid which is fed to a carbonate dryer. After drying, the strontium carbonate product is cooled, ground, and screened for packaging.

Strontium carbonate also precipitates from strontium sulfide solution with carbon dioxide. Hydrogen sulfide is generated as a by-product of this reaction and reacts with sodium hydroxide to produce sodium hydrosulfide, which is sold.

Another commercial process was used in Canada by Kaiser Aluminum and Chemical Corp. at a plant in Sydney, Nova Scotia, which was shut down in 1976 because of celestite ore quality and process problems. Celestite concentrate reacted with soda ash in solution to produce an impure strontium carbonate and a by-product solution containing sodium sulfate. The impure strontium carbonate was calcined at a high temperature to produce strontium oxide, which was quenched with water and leached to extract strontium values as soluble strontium hydroxide. The solution was clarified to remove insoluble gangue material and allowed to react with kiln gas containing carbon dioxide to form strontium carbonate. The strontium carbonate was dried or sintered to produce a chemical-grade or a glass-grade product, respectively. By-product sodium sulfate solution was evaporated and crystallized to produce salt cake (10).

Strontium carbonate can also be made by mixing a finely ground, high grade celestite ore with hydrochloric acid to convert the calcium carbonate and iron oxides into water-soluble chlorides. After washing, strontium sulfate is converted to the carbonate by mixing the resultant slurry with a boiling soda ash solution. The strontium carbonate product is washed, centrifuged, and dried.

The largest use of strontium carbonate is in the manufacture of glass faceplates for color-television tubes. It is present in glass at ca 12–14 wt % on a strontium oxide basis and functions as an x-ray absorber. Strontium carbonate is an effective x-ray barrier because strontium has a large atomic radius, and its presence is required in the relatively high voltage television sets used in the United States and Japan. The lower voltage Western European television faceplates contain the less expensive barium carbonate as an x-ray absorber. Strontium carbonate, when added to special glasses, glass frits, and ceramic glazes, increases the firing range, lowers acid solubility, and reduces pin-holing; strontium carbonate also has low toxicity. It is used in the production of high purity, low lead electrolytic zinc by a process patented by ASARCO, Inc. (11) and it is used in Australia, South Africa, the United States and Japan. At a use level of 4.4–7.7 kg of strontium carbonate per ton of metal produced, it removes lead from the cathode zinc.

Strontium Chromate. Strontium chromate [*7789-06-2*], $SrCrO_4$, is made by precipitation of a water-soluble chromate solution with a strontium salt or chromic acid with a strontium hydroxide solution. It has a specific gravity of 3.84 and is used as a low toxicity, yellow pigment and as an anticorrosive primer for zinc, magnesium, aluminum, and alloys used in aircraft manufacture (12) (see Corrosion and corrosion inhibitors).

Strontium Hexaferrite. Strontium hexaferrite [*12023-91-5*], $SrO.6Fe_2O_3$, is made by combining powdered ferric oxide (Fe_2O_3) and strontium carbonate ($SrCO_3$) and calcining the mixture at ca 1000°C in a rotary kiln (13). The material is crushed, mixed with a binder, and pressed or extruded into finished shapes and sintered at ca 1200°C. The sintered ferrite shapes are used as magnets in small electric motors, particularly in fractional horsepower motors for automobile windshield wipers and window risers. Flexible magnets for use as refrigerator door gaskets are made by blending ferrite powder with a polymer, melting the polymer, and extruding the mixture into the required shape.

Strontium Halides. Strontium halides are made by the reactions of strontium carbonate with the appropriate mineral acids. They are used primarily in medicines as replacements for other bromides and iodides.

Strontium bromide [*10476-81-0*], $SrBr_2$, forms white needlelike crystals, which are very soluble in water (222.5 g in 100 mL water at 100°C) and soluble in alcohol. The anhydrous salt has a specific gravity of 4.216 and a melting point of 643°C.

Strontium chloride [*10476-85-4*], $SrCl_2$, is similar to calcium chloride but is less soluble in water (100.8 g in 100 mL water at 100°C). The anhydrous salt forms colorless cubic crystals with a specific gravity of 3.052 and a melting point of 873°C. Strontium chloride is used in toothpaste formulations (see Dentifrices).

Strontium fluoride [*7783-48-4*], SrF_2, forms colorless cubic crystals or a white powder with a specific gravity of 4.24 and a melting point of 1190°C. It is insoluble in water but is soluble in hot hydrochloric acid.

Strontium iodide [*10476-86-5*], SrI_2, forms colorless crystals which decompose in moist air. It is very soluble in water (383 g in 100 ml water at 100°C) and the anhydrous salt has a specific gravity of 4.549 and a melting point of 402°C.

Strontium Nitrate. Strontium nitrate, $Sr(NO_3)_2$, in the anhydrous form is a colorless crystalline powder with a melting point of 570–645°C and a specific gravity of 2.986. It also exists as a tetrahydrate [*13470-05-8*] which has a density of 2.2 and a melting point of 100°C. The anhydrous salt is the commercially produced product. Strontium nitrate is made by the reaction of milled strontium carbonate with nitric acid. The nitrate slurry is filtered, crystallized, and centrifuged before drying in a rotary dryer. The final product is screened and bagged for shipment. Strontium nitrate is classified as an oxidizer by the U.S. Department of Transportation and may react rapidly enough with easily oxidizable substances to cause ignition. The anhydrous material is normally stable but starts to decompose at ca 500°C with the evolution of nitrogen oxides.

The principal use of strontium nitrate is in the manufacture of pyrotechnics (qv) as it imparts a characteristic, brilliant crimson color to a flame. Railroad fusees and distress or rescue signaling devices are the main uses for strontium nitrate. It is also used to make red tracer bullets for the military.

Strontium Oxide, Hydroxide, and Peroxide. Strontium oxide, SrO, is a white powder that has a specific gravity of 4.7 and a melting point of 2430°C. It is made by heating strontium carbonate with carbon in an electric furnace, or by heating celestite with carbon and treating the sulfide formed with caustic soda and then calcining the product (14). It reacts with water to form strontium hydroxide [*18480-07-4*] and is used as the source of strontium peroxide [*1314-18-7*].

Strontium hydroxide, $Sr(OH)_2$, resembles slaked lime but is more soluble in water (21.83 g per 100 g of water at 100°C). It is a white deliquescent solid with a specific gravity of 3.62 and a melting point of 375°C. Strontium soaps are made by combining strontium hydroxide with soap stocks, eg, lard, tallow, or peanut oil. The strontium soaps are used to make strontium greases, which are lubricants that adhere to metallic surfaces at high loads and are water-resistant, chemically and physically stable, and resistant to thermal breakdown over a wide temperature range (15).

Strontium peroxide, SrO_2, is a white powder with a specific gravity of 4.56 that decomposes in water. It is made by the reaction of hydrogen peroxide with strontium oxide and is used primarily in pyrotechnics and medicines.

Strontium Sulfate. Strontium sulfate, $SrSO_4$, occurs as celestite deposits in beds or veins in sediments or sedimentary rocks. The United States depends on imports from Mexico to supply domestic demand; Mexico shipped 34,100 t of celestite ore to the United States in 1980. Celestite has a specific gravity of ca 3.97 and a Mohs hardness of 3.0–3.5; it is colorless to yellow and often pale blue. Strontium sulfate forms colorless or white rhombic crystals with a specific gravity of 3.96 and an index of refraction of 1.622–1.631. It decomposes at 1580°C and has a solubility of 0.0113 g per 100 mL of water at 0°C.

Strontium Titanate. Strontium titanate [*12060-59-2*], $SrTiO_3$, is a ceramic dielectric material that is insoluble in water and has a specific gravity of 4.81. It is made from strontium carbonate and is used in the form of 0.5-mm thick disks as electrical capacitors in television sets, radios, and computers.

BIBLIOGRAPHY

"Strontium" treated in *ECT* 1st ed., under "Alkaline Earth Metals," Vol. 1, p. 463, by C. L. Mantell, Consulting Chemical Engineer; "Strontium Compounds" in *ECT* 1st ed., Vol. 13, pp. 113–118, by Louis

Preisman, Barium Reduction Corporation, and Desmond M. C. Reilly, Food Machinery and Chemical Corporation; "Strontium" in *ECT* 2nd ed., Vol. 19, pp. 48–49, by L. M. Pidgeon, University of Toronto; "Strontium Compounds" in *ECT* 2nd ed., Vol. 19, pp. 50–54, by Louis Preisman, Pittsburgh Plate Glass Company.

1. J. W. Mellor, *A Comprehensive Treatise on Inorganic and Theoretical Chemistry*, Vol. III, John Wiley & Sons, Inc., New York, 1922.
2. *Metals Handbook*, 9th ed., Vol. 2, American Society for Metals, Metals Park, Ohio, 1979.
3. *Gmelins Handbuch der Anorganischen Chemie*, System Number 29, Verlag Chemie, Weinheim/Bergstr., FRG, 1964.
4. M. Guntz and M. Galliot, *Compt. Rend.* **151,** 813 (1910).
5. *Strontium, A Chapter from Mineral Facts and Problems, 1980 Edition*, Preprint from Bulletin 671, U.S. Bureau of Mines, 1981.
6. H. W. King, "Solid Solutions and Intermetallic Phases Containing Compounds," in *International Conference on Strontium Containing Compounds*, Nova Scotia Technical College, Halifax, Nova Scotia, 1973.
7. H. W. King, "Applications of Strontium Containing Alloys," in ref. 6.
8. *Celestite*, Mineral Dossier No. 6, Mineral Resources Consultative Committee, HMSO, London, 1973.
9. *Mineral Commodity Summaries, 1981*, U.S. Bureau of Mines, U.S. Department of the Interior.
10. D. L. Stein, "Extraction of Strontium Values from Celestite at the Kaiser Plant in Nova Scotia," in ref. 6.
11. U.S. Pat. 2,539,681 (Jan. 30, 1951), R. P. Yeck and Y. E. Lebedeff (to American Smelting and Refining Co.).
12. T. J. Gray and R. J. Rontil, "Strontium Based Pigments and Glazes," in ref. 6.
13. T. J. Gray and R. J. Routil, "Strontium Hexaferrite," in ref. 6.
14. G. D. Parkes, ed., *Mellor's Modern Inorganic Chemistry*, John Wiley & Sons, Inc., New York, 1967.
15. T. J. Gray and T. Betancourt, "Strontium Soaps and Greases," in ref. 6.

ANDREW F. ZELLER
FMC Corporation

STRUCTURE–ACTIVITY RELATIONSHIPS. See Pharmacodynamics.

STYRENE

Styrene [*100-42-5*] (phenylethylene, vinylbenzene, styrol, cinnamene), $C_6H_5CH{=}CH_2$, is the common name for the simplest and by far the most important member of a series of unsaturated aromatic monomers. Styrene is used extensively for the manufacture of plastics, including crystalline polystyrene, rubber-modified impact polystyrene, acrylonitrile–butadiene–styrene terpolymer (ABS), styrene–acrylonitrile copolymer (SAN) and styrene–butadiene rubber (SBR) (see Acrylonitrile polymers; Elastomers, synthetic; Styrene plastics).

Commercial manufacture of the monomer began on a small scale shortly before World War II. Since that time, production of the monomer in the United States has grown enormously, reaching 1.95×10^6 metric tons annually in 1970 and 3.2×10^6 t/yr in 1980. Several factors have contributed to this increase: styrene, bp 145°C, is a liquid that can be handled easily and safely, the activation of the vinyl group by the benzene ring makes styrene easy to polymerize and copolymerize under a variety of conditions, polystyrene is one of the least expensive thermoplastics on a cost-per-cubic-centimeter basis, and polystyrene is one of the easiest thermoplastic materials to mold and extrude and it can be used for a large variety of applications.

Styrene was first isolated in the nineteenth century from the distillation of storax, a natural balsam. Although styrene was known to polymerize, no commercial applications were attempted for many years because the polymers were brittle and cracked easily. The simultaneous development of a process for the manufacture of styrene by the dehydrogenation of ethylbenzene by The Dow Chemical Company and Badische Anilin-und-Soda Fabrik A.G. (BASF) represented the first real breakthrough in styrene technology. In 1937, both of these companies were manufacturing a high purity monomer which could be polymerized to a stable, clear, colorless plastic. During World War II, styrene became important in the manufacture of synthetic rubber and large-scale plants were built. Later, peacetime uses of styrene-based plastics, especially crystalline and high impact polystyrene, have accounted for its continuing rapid growth.

As the technology has matured, many new producers have entered the market. Because styrene has become a large commodity chemical, new entrants have been primarily oil companies and even national governments rather than chemical companies. Ethylbenzene dehydrogenation remains the dominant manufacturing technique, although coproduction of styrene with propylene oxide (qv) accounts for a notable percentage of world production.

Properties

The physical properties of styrene monomer are given in Table 1 (1). Polymerization or copolymerization is the only reaction of styrene that is of commercial significance. Virtually all of the monomer manufactured is consumed in these processes. Styrene monomer can be polymerized by all the common methods used in plastics technology (see Polymerization mechanisms and processes). Techniques of mass, suspension, solution, and emulsion polymerization have been used for the manufacture of polystyrene and styrene copolymers, but processes relating to the first two methods account for most of the polymers being manufactured. Mass polymerization processes

Table 1. Physical Properties of Styrene Monomer[a]

Property	Value					
boiling point (at 101.3 kPa = 1 atm), °C	145.0					
freezing point, °C	−30.6					
flash point (fire point), °C						
Tag open-cup	34.4 (34.4)					
Cleveland open-cup	31.1 (34.4)					
autoignition temperature, °C	490.0					
explosive limits in air, %	1.1–6.1					
vapor pressure, kPa, Antoine equation[b]	$\log_{10} P = 6.08201 - \frac{1445.58}{209.43 + t°C}$					
critical pressure (P_c), MPa[c]	3.81					
critical temperature (t_c), °C	369.0					
critical volume (V_c), cm^3/g	3.55					
refractive index, n_D^{20}	1.5467					
	at 0°C	20°C	60°C	100°C	140°C	150°C
viscosity, mPa·s (= cP)	1.040	0.763	0.470	0.326	0.243	
surface tension, mN/m (= dyn/cm)	31.80	30.86	29.01	27.15	25.30	
density, g/cm^3	0.9237	0.9059	0.8702	0.8346		0.7900
specific heat (liquid), J/(g·K)[d]	1.636	1.690	1.811	1.983	2.238	
specific heat (vapor) at 25°C, C_p, J/(g·K)[d]	1.179					
latent heat of vaporization (ΔH_v), J/g[d]						
at 25°C	428.44					
at 145°C	354.34					
heat of combustion (gas at constant pressure) at 25°C, ΔH_c, kJ/mol[d]	4262.78					
heat of formation (liquid) at 25°C, ΔH_f, kJ/mol[d]	147.36					
heat of polymerization, kJ/mol[d]	74.48					
Q value	1.0					
e value	−0.8					
volumetric shrinkage upon polymerization, vol %	17.0					
cubical coefficient of expansion, per °C						
at 20°C	9.710×10^{-4}					
at 40°C	9.902×10^{-4}					
solubility (at 25°C), wt % monomer in H_2O	0.032					
H_2O in monomer	0.070					
solvent compatibility						
acetone	infinitely soluble					
carbon tetrachloride	infinitely soluble					
benzene	infinitely soluble					
ether	infinitely soluble					
n-heptane	infinitely soluble					
ethanol	infinitely soluble					

[a] Ref. 1.
[b] To convert from $\log_{10} P_{kPa}$ to $\log_{10} P_{mmHg}$, add 0.8751 to the constant.
[c] To convert MPa to psi, multiply by 145.
[d] To convert J to cal, divide by 4.184.

are gradually replacing other techniques because of their significant advantage in energy consumption. A free-radical polymerization of the monomer initiated thermally or with catalysts is generally used (2–7).

In addition to polymerization, styrene also undergoes other reactions of a typical unsaturated compound. Many of these are listed in references 2 and 8.

Manufacture

Many different techniques have been investigated for the manufacture of styrene monomer. Of these, the following methods have been used or seriously considered for commercial production: (1) dehydrogenation of ethylbenzene; (2) oxidation of ethylbenzene to ethylbenzene hydroperoxide, which reacts with propylene to give α-phenylethanol and propylene oxide, after which the alcohol is dehydrated to styrene; (3) oxidative conversion of ethylbenzene to α-phenylethanol via acetophenone and subsequent dehydration of the alcohol; (4) side-chain chlorination of ethylbenzene followed by dehydrochlorination; (5) side-chain chlorination of ethylbenzene, hydrolysis to the corresponding alcohols, followed by dehydration; and (6) pyrolysis of petroleum and recovery from various petroleum processes. The first two methods are the only commercially utilized routes to styrene: dehydrogenation of ethylbenzene accounts for over 90% of the total world production. Method 3 was practiced commercially by Union Carbide Corporation but was later replaced by a dehydrogenation process. Methods 4 and 5, involving chlorine, have generally suffered from the high cost of the raw materials and from the chlorinated contaminants in the monomer. Manufacture of styrene directly from petroleum streams (method 6) is difficult and costly.

The two commercially important routes to styrene are based on ethylbenzene produced by alkylation of benzene with ethylene (see Alkylation). Research programs aimed at replacing expensive benzene and ethylene feedstocks with less costly alternatives have been carried out by a number of companies. None of these processes are expected to be in commercial operation in 1982. The most important new routes include oxidative coupling of toluene to a stilbene intermediate followed by disproportionation of stilbene with ethylene to give two moles of styrene and alkylation of toluene with methanol or synthesis gas (CO/H_2) to form styrene and ethylbenzene (9–11).

Ethylbenzene. ***Alkylation.*** Ethylbenzene was first made commercially from ethylene and benzene by the classical Friedel-Crafts reaction (qv) in an aluminum chloride catalyst complex. Several companies developed variations of this method, including The Dow Chemical Company, BASF, Shell Chemical Company, Monsanto Company, Société Chimique des Charbonnages, and Union Carbide Corporation in cooperation with The Badger Company, Inc. These liquid-phase processes generally involve ethyl chloride or occasionally hydrogen chloride as a catalyst promoter. The aluminum chloride, ethyl chloride, and recycled alkylated benzenes are combined to form a separate catalyst-complex phase, ie, the red oil described in the early literature. A comprehensive discussion of the mechanism of the reaction is given in ref. 12. This complex phase is heavier than the hydrocarbon phase and can be separated readily from it and recycled.

A widely used version is the Union Carbide/Badger process illustrated in Figure 1. In it and in similar processes, the alkylation reaction is carried out in a brick- or glass-lined reactor operating at low pressure and temperature. The reaction mixture of catalyst complex, benzene, and recycled higher alkylated benzenes is agitated to disperse the catalyst-complex phase. Ethylene is sparged into the reaction mixture and essentially complete conversion of ethylene is obtained. The reaction occurs at close to thermodynamic equilibrium conditions. Figure 2 is an illustration of the typical operating range in commercial plants superimposed on an equilibrium diagram, which

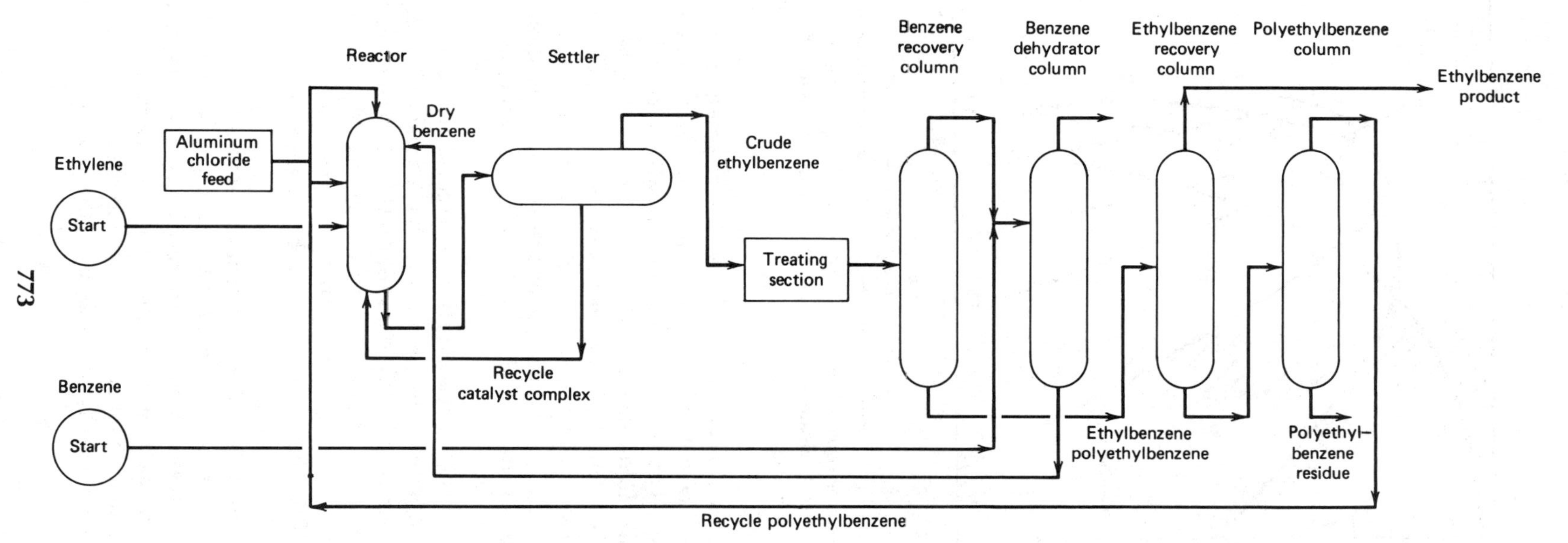

Figure 1. Union Carbide/Badger ethylbenzene process. Courtesy of *Hydrocarbon Processing*.

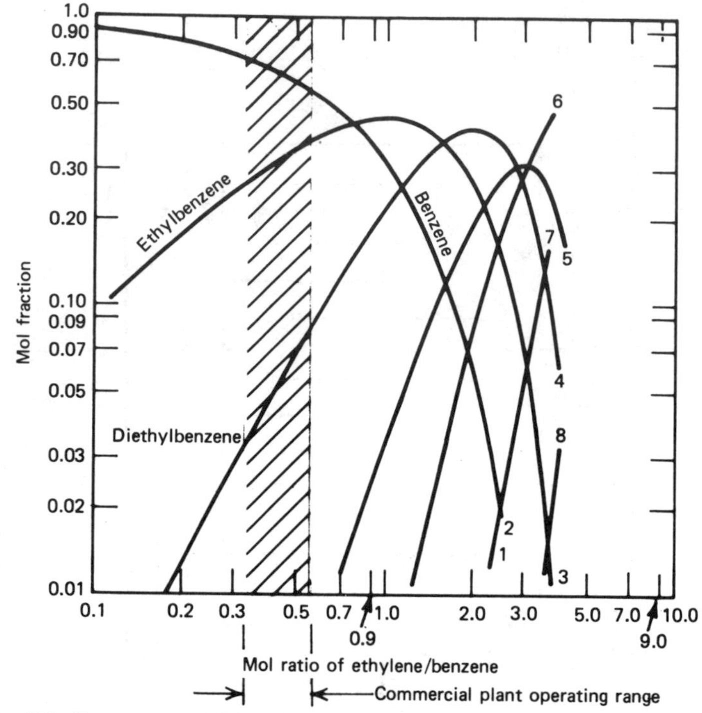

Figure 2. Ethylbenzene equilibrium. Curves represent theoretical thermodynamic equilibrium. 2, Benzene; 3, ethylbenzene; 4, diethylbenzene; 5, triethylbenzene; 6, tetraethylbenzene; 7, pentaethylbenzene; 8, hexaethylbenzene (13).

is a plot of the mole fraction of each species against the molar ratio of ethylene combined with benzene. The optimum operating point is that at which the yield to ethylbenzene is maximized and the recycle of benzene and more highly substituted ethylbenzene is minimized. In general, as the ratio of ethylene to benzene is increased, more of the diethyl-, triethyl-, and higher polyethylbenzenes is produced. This in turn leads to a higher level of residue production and, hence, lower yields. Because the reaction occurs at close to thermodynamic equilibrium conditions, a single reactor can be used to alkylate benzene and to transalkylate polyethylbenzenes to ethylbenzene.

In the typical process, the reactor effluent is cooled and passes to a settler where the organic phase is decanted from the catalyst complex and the catalyst-complex layer is recycled to the reactor. The organic phase is washed with water and caustic to remove traces of catalyst. The waste aqueous phase from the treating steps is neutralized, and a wet aluminum hydroxide sludge is recovered for disposal as a Class I landfill material as defined by the EPA; alternatively, it can be calcined for recovery of aluminum oxide.

Crude ethylbenzene is sent after treatment to a series of three distillation columns for recovery of product ethylbenzene. First, unreacted benzene is recovered as an overhead product. It is then combined with fresh benzene for dehydration and recycle to the reactor. The bottoms from the benzene recovery column flow to the ethylbenzene recovery column, where the product is separated from heavier components. The final column is used to strip polyethylbenzenes from the residue for recycle. The residue is generally used as a fuel. With this type of process, the heavy key component in the ethylbenzene product column is diethylbenzene. It must be held to a low level in the product, since it would readily dehydrogenate to divinylbenzene in the styrene unit. Divinylbenzene would polymerize with styrene in the product recovery area to form a very highly cross-linked popcorn polymer, which is extremely difficult to remove.

In 1974, Monsanto brought on stream an improved liquid-phase aluminum chloride alkylation process; the flow scheme is illustrated in Figure 3. It was determined that the aluminum chloride content in the reaction medium could be reduced to the solubility limit if the process were operated at higher temperatures. Thus, alkylation occurs in the absence of a separate heavy catalyst-complex phase.

Elimination of the catalyst-complex phase results in higher selectivities and higher overall yields, because the catalyst-complex phase preferentially absorbs the more highly alkylated ethylbenzenes which complex most readily with aluminum chloride and ethyl chloride. In this highly reactive environment, undesirable by-products are more likely to form. The higher operating temperature permits recovery of the heat of reaction as steam at a pressure usable elsewhere in the process. The process is suitable for use with low concentration ethylene feeds.

Since the inventory of aluminum chloride is relatively small, the reaction system is sensitive to poisoning from water or other reactive compounds; it also recovers quickly since fresh catalyst is fed continuously. A separate transalkylation reactor is used to minimize the formation of higher polyethylbenzenes, which cannot be transalkylated effectively.

The flow scheme is basically the same as that for the more traditional aluminum chloride processes. It differs in that only dry benzene, rather than benzene and recycled higher alkylated benzenes, plus ethylene, catalyst, and promoter are fed continuously to the alkylation reactor. Effluent from the reactor is mixed with recycle polyethylbenzenes from distillation for transalkylation in a second reactor. The entire transalkylation reactor effluent is then subjected to water and caustic washing for removal of dissolved catalyst. After washing, the crude ethylbenzene is subjected to the usual product recovery sequence in which benzene and recyclable polyethylbenzenes are separated from ethylbenzene in a series of three distillation columns. The organic residue is used as fuel, and aluminum chloride wastes are recovered as in the other processes.

Vapor-phase alkylation at moderate-to-high pressure has been practiced commercially with various solid acid catalysts since 1942. During World War II, the United States government financed the construction of a plant with a fixed-bed catalyst of alumina deposited on silica gel (14–16). Temperatures of over 300°C and pressures of over 6000 kPa (870 psi) were required to carry out the alkylation reaction. The mild alumina-catalyst activity was not sufficient to allow transalkylation even at these operating conditions, which are more severe than those used in the liquid-phase process. Operation at a high ratio of benzene to ethylene was required to minimize formation of higher alkylated ethylbenzenes. In addition, a small dealkylation unit with aluminum chloride catalyst was required to obtain high overall yields comparable to yields obtained in the liquid-phase process.

Solid phosphoric acid catalysts developed by Universal Oil Products (UOP) were used commercially at El Paso Natural Gas Company's Odessa, Texas, plant to produce ethylbenzene (17). The reaction was carried out in the liquid phase with dilute ethylene feedstock. Ethylene conversions of 90–98% were obtained. Earlier phosphoric acid-

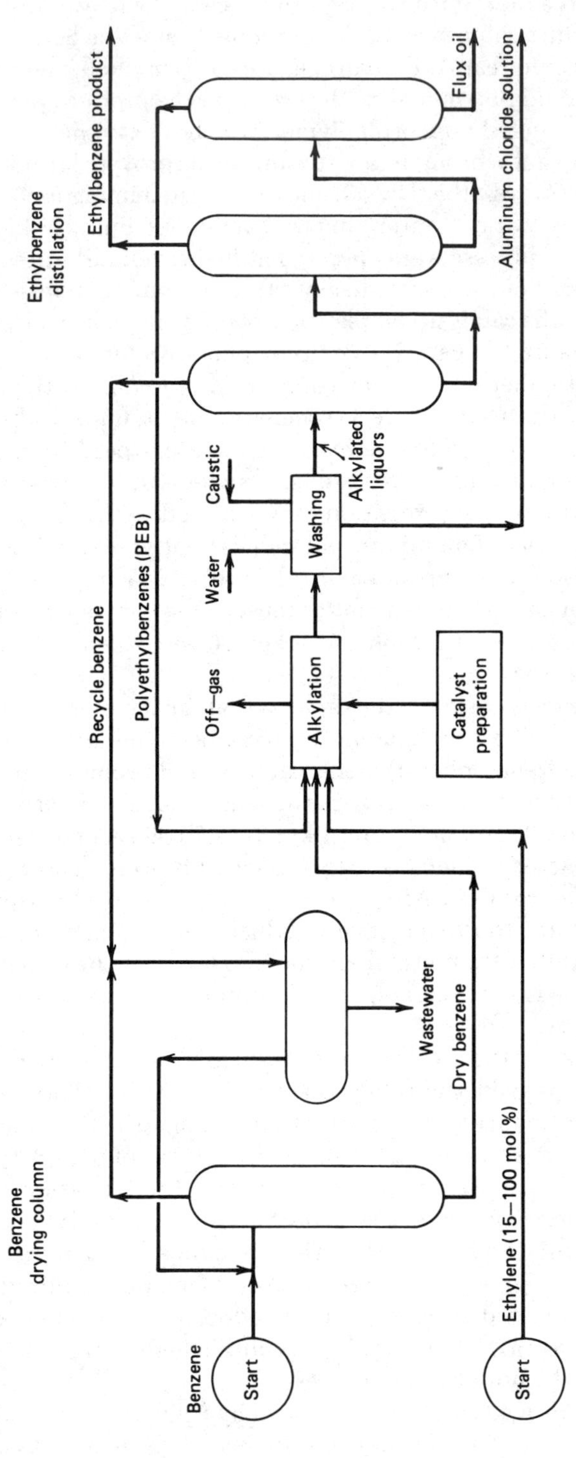

Figure 3. Monsanto ethylbenzene process. Courtesy of *Hydrocarbon Processing*.

based catalysts were limited largely to bench-scale or pilot-plant testing and did not become important commercial factors.

The Mobil/Badger ethylbenzene process was developed during the 1970s and was based on a synthetic zeolite catalyst, ZSM-5, developed by Mobil Oil Corporation (18–21). Although a number of zeolitic or molecular-sieve-type catalysts have been suggested for benzene alkylations with ethylene, most were characterized by very rapid buildup of coke and, consequently, short on-stream time (see Molecular sieves). The Mobil catalyst represents a breakthrough in zeolite catalysis in that it combines high catalytic activity with relatively good resistance to coke formation. The first commercial application of the technology was the conversion in 1976 of the Alkar unit at Cosden's Big Spring, Texas, refinery to a Mobil/Badger unit operating on dilute ethylene feedstock. Nearly quantitative yields of ethylbenzene were obtained from feed gas containing as low as 15 vol % ethylene. In 1980, a plant capable of producing over 4.7×10^5 t/yr of ethylbenzene was brought on stream by American Hoechst Corporation in Bayport, Texas. Following this first application, a number of other units were licensed. Total worldwide capacity based on the Mobil/Badger process is anticipated to be more than 3×10^6 t/yr by 1985.

There are several reasons for the rapid acceptance of this new technology. Most important is the new catalyst system, which is noncorrosive and truly heterogeneous. All of the high alloy materials, brick linings, and catalyst-handling facilities associated with liquid-phase processes are eliminated. Aqueous wastes and, therefore, waste-treatment facilities are eliminated. The catalyst is basically silica–alumina and therefore is inert to the environment. Moreover, since the reaction is carried out at over 400°C in the vapor phase, the reaction heat can be recovered as useful medium-pressure steam.

As illustrated in Figure 4, the Mobil/Badger ethylbenzene process consists of two main sections: reaction and distillation. In the reaction section, fresh and recycled benzene are preheated, vaporized, combined with an alkyl-aromatics recycle stream and fresh ethylene, and then fed to a single reactor containing fixed beds of ZSM-5 catalyst where alkylation of benzene with ethylene occurs in the vapor phase at high temperature and moderate pressure. The high activity of the zeolite catalyst allows simultaneous transalkylation of more highly substituted ethylbenzenes in the same reactor. Two reactors are included in the reaction system; one is on-stream while the other is being regenerated. Reactor effluent vapor is passed to a prefractionator for recovery of unreacted benzene. Condensed prefractionator distillate is recycled to the reactor, and uncondensed light components, which are vented from the system, can be used as fuel gas. A small regeneration heater is provided for the catalyst-decoking cycle.

The distillation section is very much like those in liquid-phase processes. It consists of three columns: benzene recovery, ethylbenzene recovery, and polyethylbenzene recovery columns. In the first, benzene is recovered and recycled to the reactor. Ethylbenzene is then separated from higher boiling components in the second column and the bottoms stream from it are further distilled in the polyethylbenzene recovery column for recycle. The small amount of residue removed from this column can be used as fuel.

The Alkar process, which was developed by UOP, was commercialized in 1960 at Cosden Oil & Chemical Company (22–23). It is based on a solid-acid catalyst, ie, alumina activated with boron trifluoride. The Alkar process is designed to operate

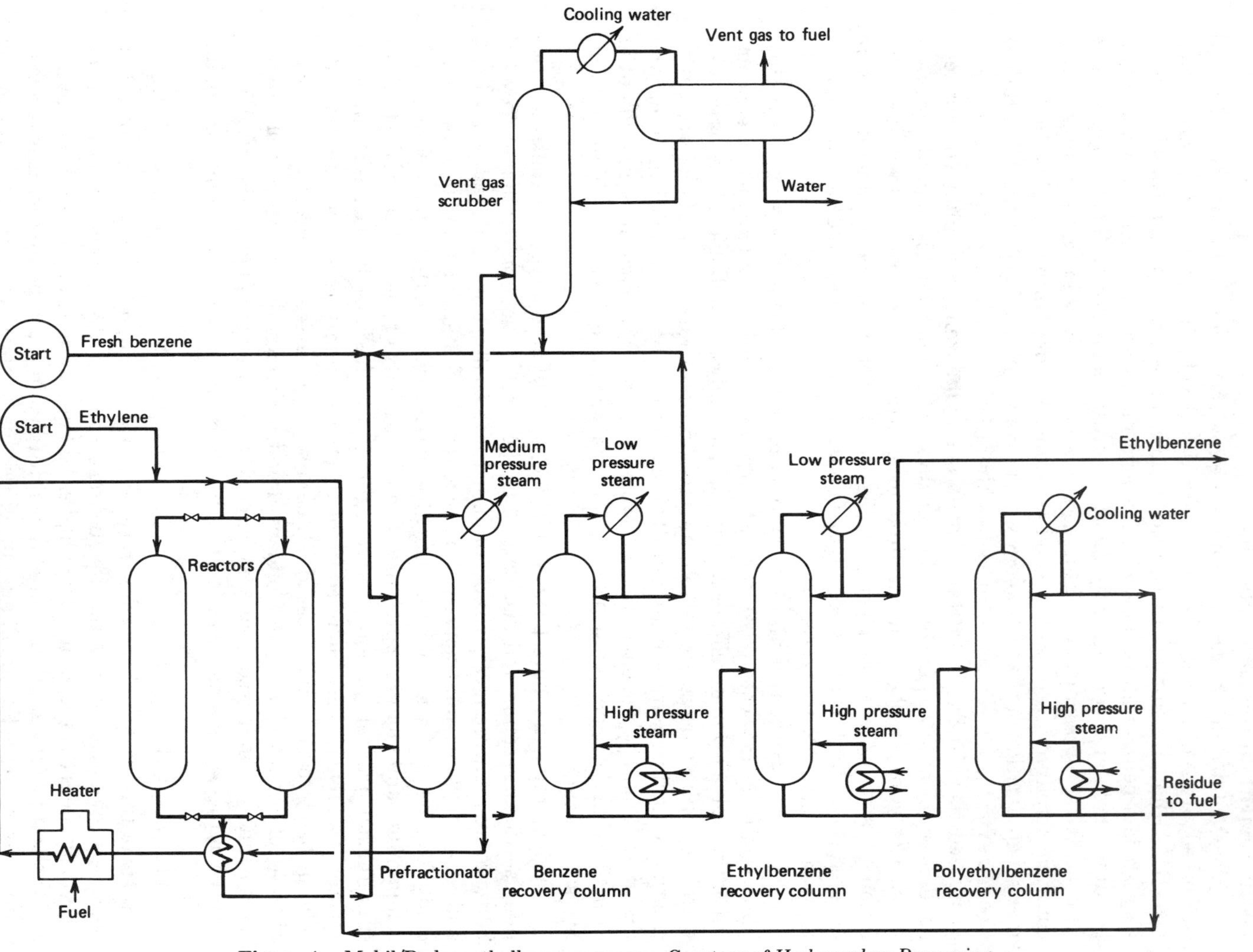

Figure 4. Mobil/Badger ethylbenzene process. Courtesy of *Hydrocarbon Processing*.

with gas streams containing low ethylene concentrations. These include refinery fuel gas streams associated with fluid-bed cracking units and residual oil cokers as well as coke-oven gas streams. Gas streams with ethylene contents as low as 8–10 wt % are suitable feedstocks for the Alkar process. A simplified process flow diagram is shown in Figure 5.

A relatively small amount of boron trifluoride catalyst, which is sensitive to poisoning, is present in the reactor. Feeds must be dry and free of polar compounds, eg, hydrogen sulfide. Like the homogeneous process developed by Monsanto, a separate transalkylation reactor is necessary to achieve high yields. The combined effluent from the two reactors is sent to a flash drum where, in the case of dilute ethylene feed, nonreactive gases are removed. Crude ethylbenzene is then subjected to distillation for recovery of unreacted benzene, ethylbenzene, and polyethylbenzenes. Residue is removed as a purge from the polyethylbenzene stream or as a bottoms product from a third column (see Xylenes and ethylbenzene).

Superfractionation. Superfractionation of mixed C_8 aromatics was first practiced by Cosden Oil & Chemical Company in 1957 at Big Spring, Texas, and was based on a design jointly developed with The Badger Company, Inc. Recovery of ethylbenzene from mixed C_8 aromatics is a process with high capital costs, because the relative volatility difference between ethylbenzene and the less volatile *para*-xylene is very small. Over 300 distillation trays and a high reflux ratio are required to effect the separation. During the 1950s, the process was economically competitive with the relatively small alkylation units in operation at that time. Several commercial superfractionation units were built in the United States, Europe, and Japan during the 1950s and 1960s. However, the supply of mixed xylenes has been growing at a much slower pace than has the demand for styrene, so styrene producers have utilized alkylation processes for an increasing percentage of new capacity. These units have much better economies of scale than superfractionators and, moreover, alkylation efficiencies have been much improved. It is unlikely that significant new superfractionation capacity will be constructed in the future.

Styrene. On a commercial scale, direct dehydrogenation of ethylbenzene accounts for the manufacture of ca 90% of the world capacity of styrene. Peroxidation of ethylbenzene and then reaction with propylene followed by dehydrogenation of α-phenyethanol to styrene accounts for the remaining 10%.

Oxidation Processes. One of the most notable oxidation processes is Halcon International's process to produce styrene and propylene oxide. In this technique, ethylbenzene is oxidized to ethylbenzene hydroperoxide as follows:

$$C_6H_5CH_2CH_3 + O_2 \rightarrow C_6H_5CH(OOH)CH_3$$

The reaction takes place in the liquid phase with air bubbling through the liquid, and no catalyst is required (24–30). However, since hydroperoxides are unstable compounds, exposure to high temperature must be minimized to reduce the rate of decomposition. Fewer by-products are formed from decomposition if the reaction temperature is gradually reduced during the course of the reaction, ie, from 135–160°C during the first half of the reaction to 125–155°C during the second half.

The reaction is more selective toward the production of by-product acids, acetophenone, and α-phenylethanol when it is carried out at constant temperature than when the temperature is gradually reduced. In practice, the temperature is reduced by means of a series of reactors, each of which is maintained at a progressively lower

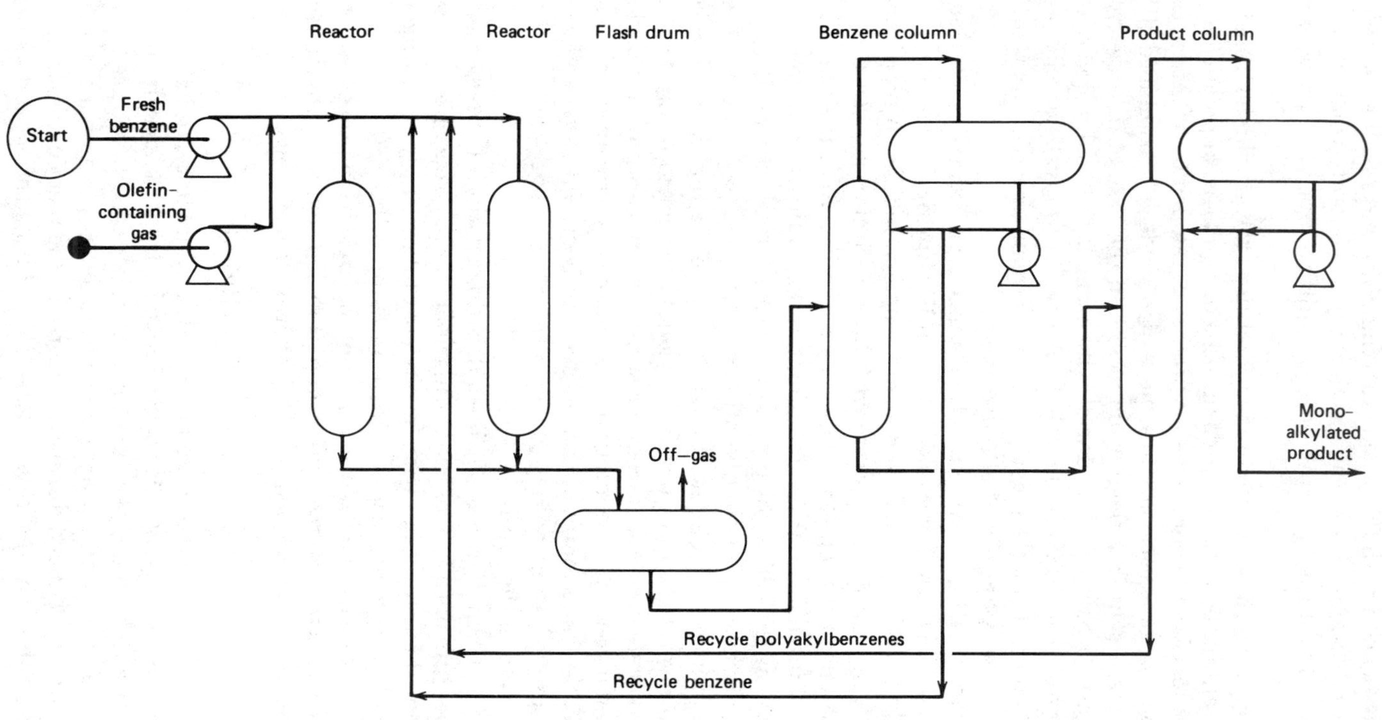

Figure 5. Alkar ethylbenzene process. Courtesy of *Hydrocarbon Processing.*

temperature. The pressure required for the reaction is not critical; 800–1500 kPa (120–220 psi) is sufficient to maintain the reactants in a liquid phase.

The hydroperoxide product reacts with propylene to form propylene oxide and α-phenylethanol (30–40):

$$C_6H_5CH(OOH)CH_3 + CH_3CH{=}CH_2 \rightarrow C_6H_5CH(OH)CH_3 + CH_3\overset{O}{\overbrace{CHCH_2}}$$

Catalysts for this reaction are compounds of metals, eg, molybdenum, tungsten, and vanadium. The epoxidation reaction generally proceeds at 100–130°C in the liquid phase under self-generated pressures. The conversion of ethylbenzene hydroperoxide is nearly complete, and the selectivity of the reaction in producing propylene oxide is greater than 70%. The alcohol can be dehydrated to styrene or it can be reduced to ethylbenzene for recycle if styrene is not desired (see Epoxidation; also Cumene).

Halcon formed a joint venture company, Oxirane Corporation, with Atlantic Richfield to exploit this process (41–42). In 1980, Oxirane was taken over by Atlantic Richfield Chemical Company. The process was first commercialized in 1973 by Montoro, a Spanish joint venture between Oxirane and Enpetrol. The plant had an annual capacity of 7.5×10^4 t of styrene and 3×10^4 t of propylene oxide. It was expanded ca 6 years later to 10^5 t of styrene and 4×10^4 t of propylene oxide. Another application of this technology was by the Japanese joint venture, Nippon Oxirane, between Oxirane, Sumitomo Chemical, and Showa Denka. It has a capacity of 2.25×10^5 t/yr of styrene and 9×10^4 t/yr of propylene oxide. The most recent Oxirane plant in Channelview, Texas, came on stream in 1977 with a stated capacity of close to 4.55×10^5 t/yr of styrene and 1.8×10^5 t/yr of propylene oxide. Styrene–propylene oxide technology was also developed independently by the Shell group, and a plant for 3.3×10^5 t/yr of styrene and 1.25×10^5 t/yr of propylene oxide was started up in 1979 by Shell at Moerdijk, The Netherlands. In the early 1950s, Union Carbide operated a plant based on another oxidation process which had the dual purpose of producing styrene or acetophenone (43). Union Carbide switched to the dehydrogenation of ethylbenzene upon completion of a styrene plant at Seadrift, Texas, in the 1960s.

Dehydrogenation. Several licensed processes are available for the conversion of ethylbenzene to styrene. Although all such processes share in common the catalytic dehydrogenation of ethylbenzene to styrene in the presence of steam, there are two distinct approaches to the reaction section design: the adiabatic process and the isothermal process. As the dehydrogenation reaction is endothermic, requiring ca 1.26–1.33 MJ/kg (540–570 Btu/lb) of ethylbenzene converted at 25°C, heat must be supplied. In the adiabatic process, steam superheated to 800–950°C is mixed with preheated ethylbenzene feed prior to exposure to the catalyst. The isothermal process takes place in a tubular reactor and reaction heat is provided by indirect heat exchange between the process fluid and a suitable heat-transfer medium, eg, flue gas (BASF technology).

For a given dehydrogenation catalyst, the catalyst life, the molar conversion of ethylbenzene (moles of ethylbenzene converted across the reactor divided by moles of ethylbenzene entering the reactor), and the molar selectivity of ethylbenzene to styrene (moles of styrene produced across the reactor divided by moles of ethylbenzene converted across the reactor) are affected by reactor operating pressure, molar steam-to-hydrocarbon ratio, reactor operating temperature, and reactor liquid hourly space velocity (LHSV, volume of liquid hydrocarbon feed per hour divided by total volume of dehydrogenation catalyst). For all catalysts, there is a compromise between activity, ie, ethylbenzene conversion, and styrene selectivity.

Because the dehydrogenation reaction produces two moles of product, ie, styrene and hydrogen, for every mole of reactant, ie, ethylbenzene, the desired reaction course can be enhanced by adding steam to the reactor system to reduce the styrene partial pressure in the reactor and/or by reducing the reactor operating pressure. Both actions favor the forward reaction and thereby increase the molar conversion of ethylbenzene. As conversion is increased by these or other refinements and approaches equilibrium, the rate of the main reaction slows down but the rates of undesired side reactions do not.

In the early 1970s, adiabatic operating pressures were generally above 138 kPa (20 psi), with molar steam-to-hydrocarbon ratios above 14:1. For the isothermal reactor designs, where indirect heating is used to satisfy the endotherm, molar steam-to-hydrocarbon ratios of 6–8:1 were possible. Reactor operating pressures were similar to those for the adiabatic process. As petroleum-derived feedstocks have become significantly more expensive since 1973, adiabatic reactor operating pressures have been reduced below 138 kPa (20 psi) with molar steam-to-hydrocarbon ratios reduced well below 14:1.

Molar conversion of ethylbenzene is also a function of temperature; higher temperatures yield higher conversions in nearly a linear fashion. Therefore, an increase in conversion can be obtained by increasing the reactor temperature. There is, however, an upper limit to reactor temperature; when temperatures are increased above 610°C, thermal cracking of ethylbenzene and styrene occurs. Because in the isothermal process, indirect heating in a tubular reactor provides the heat of reaction, the reactor temperature can be maintained at 580–610°C across the catalyst bed. If this temperature range is maintained, thermal cracking reactions are minimized. Superheated dilution steam is the sole source of reaction heat for the adiabatic reactor. To accommodate the range of molar steam-to-hydrocarbon ratios, the feed mixture is generally introduced at 610–660°C. This results in some thermal cracking of the hydrocarbons. However, by optimizing reactor geometry, including inlet and outlet piping, residence time for thermal cracking reactions can be minimized. Decreasing residence time in the adiabatic process reduces the potential selectivity advantages of the isothermal process.

As for liquid hourly space velocity (LHSV), molar conversion of ethylbenzene is an inverse function of this parameter since higher velocities mean lower residence time. This in turn reduces molar conversion. To balance the effects of rising fuel costs against corresponding feedstock costs, adiabatic reactors are designed for LHSVs of 0.4–0.6 m^3/h ethylbenzene per cubic meter of catalyst. Within this range, a 60–70% molar conversion of ethylbenzene and a 90–95% molar selectivity of ethylbenzene to styrene can be achieved with commercially available dehydrogenation catalysts. It is believed that similar results are obtained with the isothermal process.

The main difference between the isothermal and adiabatic processes is in the way the endothermic reaction heat is supplied. In principle, the isothermal reactor is designed like a shell-and-tube heat exchanger: a fixed bed of dehydrogenation catalyst and reactant gas is on the tube side and a suitable heat-transfer medium is on the shell side. Catalyst is supported on a reactant-gas distribution grid below the bottom tube sheet. The diameters of the self-contained reactor tubes are 10–20 cm and the lengths are 2.4–3.7 m.

Use of the isothermal process on a commercial scale has been limited. Although the process offers advantages over the adiabatic route to dehydrogenation, ie, reduced dilution-steam requirements and reduced thermal cracking of ethylbenzene and styrene, for utilization in plants with capacity of (3–4) × 10^5 t/yr styrene equivalent, several large reactors probably are required to operate with limited pressure drop in the tubular-reactor catalyst beds. Capital costs therefore would be high.

The most important factor in establishing the economics of the dehydrogenation process is the catalyst; ie, the performance of the catalyst determines the ethylbenzene conversion/styrene selectivity relationship, the molar steam-to-hydrocarbon ratio, the appropriate LHSV, and the achievable run life. All commercial catalysts are formulated around an iron oxide (Fe_2O_3) base. Inherent to this ferric compound is the reduction to lower oxides at the dehydrogenation reaction temperatures. Since the presence of lower oxides leads to increased coke formation/deposition on the catalyst and subsequent deactivation, stabilizers and coke retardants are essential to the formulation. The most widely used additives are chromia (Cr_2O_3) as the stabilizer and potassium oxide (as K_2CO_3) as the coke retardant. In addition, steam, in sufficient quantity, retards coke deposition as a consequence of the water-gas-shift reaction: coke and steam react to form carbon dioxide and hydrogen. Thus, the catalyst is self-regenerative, thereby allowing such residues to build up to an equilibrium level.

To suit specific operating requirements, each of the two principal dehydrogenation catalyst manufacturers, ie, United Catalysts Inc. and Shell Chemical Co., offers several catalyst formulations. However, in light of constantly rising raw material and fuel costs, future development by all manufacturers will be based on high selectivity at reduced steam-to-hydrocarbon ratios and reduced pressure drop.

Although all adiabatic processes for ethylbenzene dehydrogenation share common features, there are differences in detail between licensed technologies. A flow sheet for a typical adiabatic dehydrogenation unit is presented in Figure 6. Recycled ethylbenzene from the styrene distillation train is added to fresh ethylbenzene, mixed with steam, heated by the reactor effluent, and pumped to the dehydrogenation reactor section. Before entering the reactor, the resulting vapor mixture of steam and ethylbenzene is raised to the desired reaction temperature by mixing with dilution steam that has been superheated to 800–950°C in a direct-fired heater.

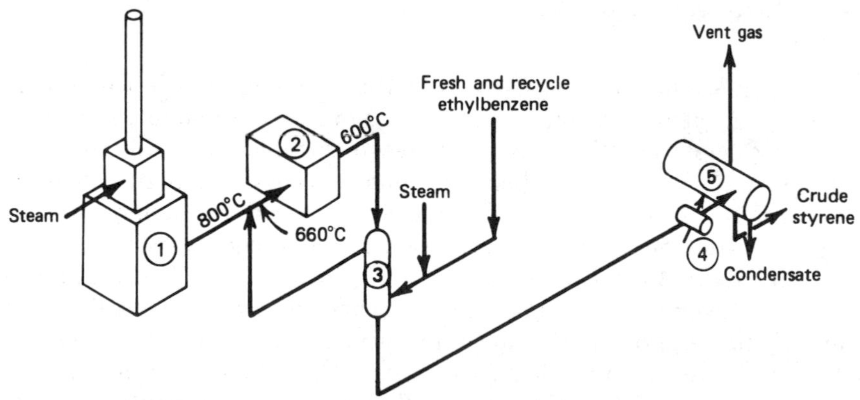

Figure 6. Manufacture of styrene by the adiabatic dehydrogenation of ethylbenzene. 1, Steam superheater; 2, reactor section; 3, feed/effluent exchanger; 4, condenser; 5, settling drum. Courtesy of The Badger Company, Inc.

Normally, the dehydrogenation reactor is operated under vacuum. For maximum selectivity at low steam-to-hydrocarbon ratios, the entire system must be designed for low pressure drop. The reactor is a chrome steel vessel containing a bed of 3-mm-diameter catalyst extrudate. The catalyst is contained in the annular space between two perforated internal cylinders; reactants flow radially outward through the bed and are collected at the reactor wall.

As a consequence of ever-changing raw material and fuel costs, present designs include multiple adiabatic reactors in series with interstage reheating of reactor effluents. Although this design requires additional capital investment, the cost may be offset by the savings in raw material and energy. With a single adiabatic reactor, a conversion of ca 40% can be achieved; with two reactors, conversion can be raised to 65–70%; with the addition of a third reactor, conversion approaches thermodynamic equilibrium, ie, 80–90%. However, for a given dehydrogenation catalyst, as molar conversion of ethylbenzene increases, selectivity of ethylbenzene to styrene decreases.

As shown in Figure 6, vapor effluent leaves the final reactor stage at ca 575–600°C and is heat-exchanged with ethylbenzene feed. Further heat removal and condensation are accomplished by heat exchange with air, cooling water, or refrigeration coolant, or a combination of these methods. In the settling drum, process water, liquid hydrocarbon (crude styrene), and vapor (vent gas) are separated. Compositions for each of the three streams are given in Table 2.

Table 2. Final-Reactor Effluent Composition[a]

Stream	Composition
vapor (vent gas), vol %	
H_2	90–95
CO_2	5
C_1 and C_2	5
liquid hydrocarbon (crude styrene), wt %	
styrene and heavier hydrocarbons	60–65
ethylbenzene	30–35
benzene and toluene	5
process water	water saturated with aromatics

[a] Courtesy of The Badger Company, Inc.

One difference between licensed adiabatic technologies is in the configuration of the dehydrogenation reactor; eg, instead of a radial-flow reactor, a fixed-bed axial reactor may be employed. With this design, reactants typically flow downward through the bed at space velocities and operating conditions essentially the same as for the radial-flow reactor. The inherent disadvantage to the fixed-bed design is the likelihood of increased pressure drop across the catalyst bed; in that bed depth is generally greater than that for a radial-flow reactor. Another difference in technologies is in the reactor feed heat-exchange section. Instead of only steam passing through a fired heater, all of the ethylbenzene and vaporization/dilution steam is combined and the mixture is raised to reaction temperature in a fired heater. Although similar in concept, a third process involves heating the steam and ethylbenzene in separate coils of a fired heater. As temperatures well in excess of 850°C may be encountered with both processes, the potential for thermal cracking of ethylbenzene is increased. Therefore, residence time in the heaters must be minimized.

Purification. To obtain the desired product, crude styrene effluent from the reaction section must be fractionated and undesirable by-products removed. To minimize polymerization of vinylaromatic compounds, low temperature vacuum distillation of the crude styrene is necessary. Further reduction in polymerization is obtained by employing inhibitors in the purification train. As recently as five years ago, sulfur-based inhibitors were used extensively as polymerization retardants. Although reasonably satisfactory results were achieved, the bottoms product leaving the purification train was contaminated with sulfur, which made the stream virtually worthless. In effect, to alleviate polymerization problems associated with styrene purification, an undesirable waste stream was produced. Thus, nonsulfur inhibitors were commercialized. Although generally more expensive than sulfur-based inhibitors, their cost is more than offset by a reduction in styrene losses to the distillation-train bottoms product and by the increased value of the residue stream.

The main nonsulfur inhibitors are nitrogen-substituted aromatics, eg, nitrophenols. Patents describing commercial use of such nonsulfur inhibitors are held by Cosden, Monsanto, and Gulf (44–46). Since these compounds have significantly higher boiling points than styrene, loss to the distillation-train bottoms product results. To retard polymerization of styrene distillate, inhibitors are also needed for storage and shipping. The most commonly used inhibitor for this service is 4-*tert*-butylcatechol (TBC). Depending on storage time and temperature, the quantity of TBC required could be 10–50 ppm (see Antioxidants).

A widely used purification train configuration is that of the Cosden/Badger styrene technology. As illustrated in Figure 7, the styrene-monomer recovery section consists of three vacuum-distillation columns. The liquid hydrocarbon phase from the settler is fractionated in the benzene–toluene column into a small overhead stream consisting primarily of benzene and toluene and a bottoms stream of styrene and unreacted ethylbenzene. The benzene–toluene overhead cut is usually sent to a benzene–toluene splitter to recover benzene for subsequent recycle to the ethylbenzene alkylation reactor. The bottoms stream from the benzene–toluene column is pumped to the ethylbenzene recycle column, where ethylbenzene containing up to 5 wt % styrene is taken overhead and recycled to the reactor feed. The separation of ethylbenzene and styrene is difficult because their boiling points are close. High efficiency, low pressure drop distillation trays are used. However, extensive effort is being made to commercialize the use of packing for this service. Such an approach offers significant reductions in pressure drop and, therefore, polymer formation without increased capital cost.

The ethylbenzene recycle-column bottoms stream, which contains styrene, polymer, high boiling by-products, inhibitor, and up to 1000 ppm ethylbenzene, is pumped to the styrene finishing column. Here, the crude styrene is separated into an overhead product of finished styrene and a bottoms stream containing styrene, polymer, high boilers, and inhibitor. The bottoms from the styrene finishing column are pumped to the continuous bottoms-residue finishing system for recovery of additional styrene. Styrene is stripped from the polymer, high boilers, and inhibitor by circulating it through a reboiler; the resulting residue is used as fuel.

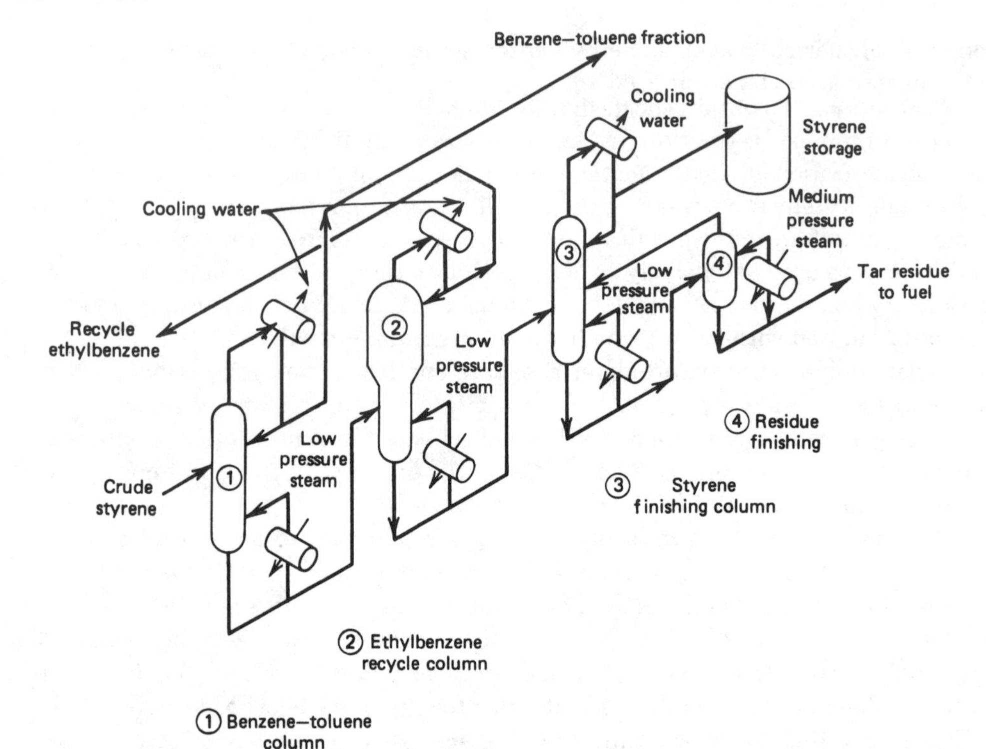

Figure 7. Purification of styrene in the dehydrogenation reactor effluent in the Cosden/Badger styrene process. 1, Benzene-toluene column; 2, ethylbenzene recycle column; 3, styrene finishing column; 4, residue finishing. Courtesy of The Badger Company, Inc.

Economic Aspects

Capacity and Production. The styrene capacity of the United States has grown steadily since production began in the 1930s. From 1960 to 1975, capacity growth averaged ca 10%/yr but slowed to an average annual rate of 4.3% from 1975 to 1980. Capacity rebounded in 1980 to be consistent with the 20-yr average annual growth rate of 8.9% (see Table 3).

Use of styrene averaged ca 91% from 1960 to 1973. This contrasts with the 1974–1977 use which averaged 74%. Major additions to capacity combined with the

Table 3. U.S. Styrene Capacity and Production, 1960–1980[a]

	Annual capacity, thousands of metric tons	Annual production, thousands of metric tons	Average annual increase by five-year period, % Capacity	Production	Utilization %
1960	845	791			94
1965	1417	1295	10.9	10.4	92
1970	2132	1967	8.5	8.7	92
1975	3392	2120	9.7	1.5	62
1980	4653	3129	4.3	8.1	67

[a] Ref. 47.

Table 4. Principal U.S. Styrene Manufacturers, 1981[a]

Company	Nameplate capacity, 10^3 t/yr
American Hoechst Corp., Baton Rouge, La.	290
American Hoechst Corp., Bayport, Texas	408
Amoco Chemical, Texas City, Texas	272
ARCO Chemical Corp., Channelview, Texas	454
ARCO Chemical Corp., Kobuta, Pa.	100
Cos-Mar, Carville, La.	590
Dow, Freeport, Texas	680
Dow, Midland, Mich.	45
El Paso Products, Odessa, Texas	115
Gulf, Donaldsonville, La.	272
Monsanto, Texas City, Texas	680
Sun, Corpus Christi, Texas[b]	36
USS Chemicals, Houston, Texas	54
Total	*3,996*

[a] Ref. 48.
[b] Sun's plant in Corpus Christi, Texas, was acquired by Koch Industries in 1981.

Table 5. Worldwide Styrene Capacity, 1975–1980[b]

	Capacity, 10^3 t/yr		Average annual increase, %
	1975	1980	
North America	3,568	5,279	8.2
South America	165	374	17.8
Western Europe	2,829	2,525	−2.3
Eastern Europe	699	1,442	15.6
Mid-East	25	25	0.0
Asia, Far East	1,511	1,697	2.4
Oceania	28	116	32.9
Africa	18	18	0.0
Total	*8,843*	*11,476*	*5.4*

[a] Ref. 49.

limiting economic conditions of 1974–1975 resulted in a 38% excess capacity in 1975.

Principal United States styrene manufacturers and their capacities are listed in Table 4.

Worldwide styrene capacity by geographic region is summarized in Table 5.

Consumption. Styrene consumption in the United States has increased from ca 900 metric tons in 1940 to nearly 2.6×10^6 t in 1980, as shown in Table 6 (50). The 1940–1960 average annual growth rate was 39.7%, whereas the 1960–1980 rate was 6.6%. Rates for 1960–1970 and 1970–1980 were 9.1%/yr and 4.2%/yr, respectively.

In 1970, total styrene consumption dropped 1% from 1969, with the greatest reductions occurring in the use of styrene in ABS (7.4%) and SBR (5.6%) (see Table 7). Total styrene consumption decreased in 1974 by 1% and in 1975 by 23.9%. The largest decreases occurred in ABS (14%), SBL (14%), and SBR (13%). Total consumption did not exceed the pre-1974 level until 1976.

Table 6. U.S. Styrene Consumption, 1940–1980

	Styrene consumption, 10^3 t/yr	Average annual growth rate by period, %
1940	0.9	
1955	428	50.8
1960	683	9.8
1965	1132	10.6
1970	1773	9.4
1975	1941	1.8
1980	2529	5.4

Table 7. U.S. Styrene Consumption, 1970–1980

Styrene use	10^3 t/yr 1970	10^3 t/yr 1980	Percent of all styrene 1970	Percent of all styrene 1980	Average annual growth rate, %
polystyrene (PS)	1073	1596	60	63	4.1
acrylonitrile–butadiene–styrene (ABS)	134	224	8	9	5.3
styrene–acrylonitrile (SAN)	13	35	1	1	10.4
styrene–butadiene latices (SBL)	104	166	6	7	4.8
styrene copolymers containing more than 50 wt % styrene	88	63	5	3	−3.3
styrene–butadiene elastomers (SBR)	253	236	14	9	−0.7
unsaturated polyester resins (UPE)	97	161	5	6	5.2
polymers containing less than 50 wt % styrene	11	48	1	2	15.9
Total	*1773*	*2529*	*100*	*100*	

Total styrene consumption decreased in 1980 from 1979 by ca 13% to 2.529 × 10^6 t (Table 8).

Price. Styrene prices in the United States declined steadily as a result of economies realized by large-scale production, reaching a low of $0.150/kg in 1973 (see Table 9) (51). During this period, styrene showed little correlation with the price of benzene; however, from 1974 to 1980 when a series of oil-price increases by the producing nations and the removal of domestic oil-price controls forced a consequent rise in benzene price, the price of styrene increased accordingly. The price of styrene will continue to be controlled through ethylbenzene by the price of benzene and, therefore, by the price of crude oil. Approximately 49% of all benzene is consumed in the production of ethylbenzene, and nearly all ethylbenzene is used for styrene production. Benzene capacity far exceeds demand (there is ca 75% utilization), and this condition is expected to continue well into the 1980s. As benzene supply is limited by feedstock availability and price is controlled by the cost of crude, any attempt to forecast the price of styrene must entail a prediction of the actions of the main oil-exporting countries.

Table 8. U.S. Styrene Consumption, 1979 and 1980

Styrene use	10^3 t/yr		Percent of all styrene		Average annual growth rate, %
	1979	1980	1979	1980	
PS	1794	1596	62	63	−11
ABS	293	224	10	9	−24
SAN	39	35	1	1	−10
SBL	197	166	7	7	−16
styrene copolymers containing more than 50 wt % styrene	75	63	3	3	−16
SBR	246	236	8	9	−4
UPE	199	161	7	6	−19
polymers containing less than 50 wt % styrene	54	48	2	2	−11
Total	*2897*	*2529*	*100*	*100*	

Table 9. Average U.S. Styrene Prices, 1955–1980[a]

Year	Price, $/kg	Average annual price change by five-year period, %
1955	0.397	
1960	0.276	−7.0
1965	0.176	−8.6
1970	0.176	0.0
1975	0.485	10.0
1980	0.788	25.5

[a] Ref. 51.

Specifications

Specifications for a typical polymerization-grade styrene monomer product are shown in Table 10. No formal agreement has been reached on styrene monomer specifications by the industry. Through the mid-1970s, plants were designed to produce

Table 10. Specifications for Typical Polymerization-Grade Styrene Monomer Product

Assay	ASTM test method
purity, wt % 99.6	D 3962
color, APHA <10; 10 max (Pt–Co scale)	D 1209
polymer, ppm by wt <10	D 2121
C_8, ppm by wt 400–800	D 3962
C_9, ppm by wt 500–1000	D 3962
aldehydes (benzaldehyde), ppm by wt <50	D 2119
peroxides (hydrogen peroxide), ppm by wt <30	D 2340
inhibitor (TBC), ppm by wt 10–50	D 2120
chlorides (chlorine), ppm by wt <10	

a standard-grade styrene monomer product of 99.6–99.8 wt %. Newer plants are being designed to produce a better grade of product with purities of >99.9 wt %. In part, this phenomenon may be attributed to anticipated tightening of government standards for styrene monomer. This is especially true of monomer used in suspension polymerization facilities to produce food-grade polystyrene.

Analysis

Most ASTM test methods included in Table 10 are based on chemical and physical tests on the finished product. However, for the plants producing high purity products, chromatographic analyses have superseded the more conventional wet methods during manufacturing. Some facilities use continuous, on-line chromatographs to control reaction conditions or distillation-tower operating parameters automatically.

Styrene purity is best determined by gas chromatography, since the results are more accurate than those obtained from freezing-point measurements or titration of the double bond. Color is determined with a spectrophotometer and is generally measured in terms of APHA or the platinum–cobalt scale. 4-*tert*-Butylcatechol (TBC) content in the product is determined spectrophotometrically; when styrene is extracted with aqueous sodium hydroxide, a red quinone forms. The polymer content of styrene monomer is determined by measurement of the degree of turbidity produced by the addition of dry methanol to the styrene monomer. By use of a photometer, polymer content is determined in amounts to 15 mg/kg. Chloride content is determined by diluting the sample with toluene and then allowing the mixture to react with biphenylsodium reagent. The excess sodium reagent is then destroyed with water and nitric acid. The chloride ion in the aqueous layer is determined spectrophotometrically by the ferric thiocyanate procedure. Aldehydes are determined by adding product sample to an alcoholic solution of hydroxylamine hydrochloride; presence of hydrochloric acid is an indication of aldehyde. Peroxide concentration is determined by mixing and boiling a sample of styrene monomer product with a solution of isopropyl alcohol, acetic acid, and sodium iodide. Iodine liberation indicates the presence of peroxides.

Health and Safety Factors

Styrene is mildly toxic and flammable, and it can polymerize violently under specific conditions. However, none of the hazards associated with styrene is severe, and it is considered a relatively safe organic chemical when handled according to appropriate safeguards. Styrene has an odor threshold of 0.05–0.15 ppmv (52). Both the liquid and vapor irritate the eyes and the respiratory system, and high vapor concentrations result in depression of the central nervous system. Irritation to the eye and respiratory tract occurs at 400–500 ppmv but does not result in permanent injury (53–54). Concentrations of up to 2500 ppmv can be tolerated by test animals for one hour without serious systemic effects. Exposure for 30–60 min to concentrations of 10^4 ppmv may be fatal (54).

The ACGIH has adopted an 8-hour time-weighted average exposure limit (TWA) of 50 ppmv and a 15-min short-term exposure limit (STEL) of 100 ppmv for worker exposure to styrene vapor (55). Testing by the National Cancer Institute has led to a determination that the carcinogenicity of styrene is equivocal (56).

Styrene is low in oral toxicity (57). Contact with the eyes is painful but results in transient damage (54,57). Short term contact with the skin (5 min or less) should cause no irritation; however, prolonged contact (one hour or more) may cause irritation, swelling, and blistering (57). Repeated contact results in deoiling of the skin with consequent cracking and inflammation (53). Styrene, as it is commonly stored and transported, contains 4-*tert*-butylcatechol, which is a skin sensitizer. Very small quantities of TBC can produce a rash in sensitive individuals (57).

Styrene monomer is flammable and can form explosive mixtures with air at ambient conditions. Maintaining styrene at room temperature or below is an effective means of avoiding fire hazard (see Fig. 8). Other precautions for the handling of flammable hydrocarbons should also be observed.

The polymerization of styrene is an exothermic reaction and proceeds slowly at room temperature but more rapidly at elevated temperatures. Thus, there is a potential for a runaway polymerization reaction which would result in an accelerating evolution of styrene vapor that could cause a fire or rupture a confining vessel. The polymerization reaction can be prevented by maintaining a TBC inhibitor concentration of at least 10 ppm and by avoiding storage at high temperature.

Effective inhibition of polymerization by TBC occurs only in the presence of some dissolved oxygen, and so storage in an atmosphere-permeable tank is preferred (54). Where inert-gas blanketing of the stored material is to be done, periodic air addition is recommended to maintain the presence of dissolved oxygen (57). In areas where average daily temperatures above 27°C are common, refrigeration of the stored monomer is recommended (58). Equipment for storing or handling of styrene should not be made of rubber or copper-containing materials, because both cause discoloration of the monomer. Also, copper in the monomer interferes with polymerization.

Uses

Styrene use worldwide is dominated by polystyrene demand. In the United States, styrene consumption can be identified by the product resin type and can be subse-

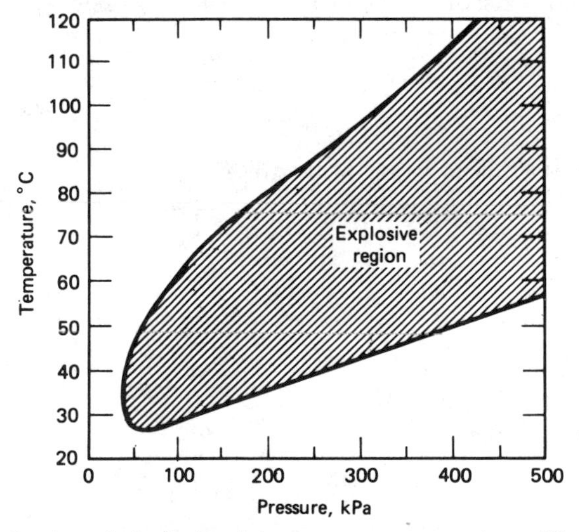

Figure 8. Approximate explosive limits of styrene monomer vapor in equilibrium with liquid styrene in air. Courtesy of the Dow Chemical Co. To convert kPa to mm Hg, multiply by 7.5.

quently characterized by the resulting consumer product. Steady increases in polystyrene and ABS and a significant decrease in SBR usage have occurred (see Table 11) (50,59).

Typical styrene products and their applications are:

Polystyrene. Packaging and disposable serviceware associated with the food packaging and fast-food service markets; housewares, furniture, appliances, television cabinets, recreational goods, and various related consumer products; electrical and electronic equipment represented by audio and visual cassettes; and commercial and industrial molding products including medical, dental, and related laboratory articles.

ABS. Drain, waste, and vent pipes used for industrial and municipal sewer mains and connector pipes; interior and exterior automobile parts, including instrument panels, trim, ductwork, knobs, consoles, dashboards, and headlight and taillight housings; appliance parts, including refrigerator compartments and door liners, air-conditioner and fan housings, and numerous parts for small household items; casings for business machines, eg, word-processing equipment and computer consoles and various smaller components; recreational goods, including vehicle parts for snowmobiles, campers, and boats and accessory equipment; and miscellaneous uses, eg, luggage, toys, and hobby materials.

SB latexes. Paper coatings on publication-grade papers and on board or label stock; carpet backcoating; and miscellaneous applications, including adhesives for floor tiles, cement additives, latex paint, and similar uses.

SBR. Primarily tires and related products; auto industry wares, eg, radiator and heater hoses, belts, and similar parts; and other uses including appliance parts, wire insulation, belts and seals, latex applications, and miscellaneous rubber goods.

UPE (glass-reinforced). Building panels; marine products and household consumer goods; parts for automobiles, trucks, buses, and mass-transit vehicles; and miscellaneous uses, including gel coat resin and electrical components.

UPE (nonreinforced). Casting resins used in the production of synthetic marble and various liners and seals; other applications including resins for putty and a variety of coatings and adhesives.

Table 11. World Styrene Use Distribution as % of All Styrene

	1955	1960	1965	1970	1975	1980
PS	44	46	55	61	62	63
ABS		2 estd	5 estd	7 estd	8	9
SAN		1 estd	1 estd	1 estd	2	1
SBL	8	8	6	6	7	7
styrene copolymers containing more than 50 wt % styrene	6	4	5	5	2	3
SBR	40	35	22	14	11	9
UPE	2	4	6	5	7	6
polymers containing less than 50 wt % styrene	0	0	0	1	1	2

Derivatives

There are three other important styrenic compounds that are commercially significant. These are divinylbenzene, vinyltoluene, and α-methylstyrene. Production of these three chemicals is small and they are considered specialty products. A fourth compound, *p*-methylstyrene, is being developed and may grow from a specialty material to a serious competitor of styrene. *p*-Methylstyrene may have as much as 20% of the styrene market by 1990 (60). In general, these compounds have been restricted to specialty applications because of their narrow ranges of useful properties and, more significantly, because of their high processing costs. The more highly substituted vinyl-aromatic compounds tend to be more reactive than styrene. This leads to increased by-product formation, consequent yield losses, and more complicated product recovery schemes. These compounds also have lower volatilities, which lead to either higher distillation temperatures and increased yield losses because of polymer formation or lower operating pressures and higher investment and operating costs.

Vinyltoluene. Vinyltoluene has been produced commercially in the United States since the late 1940s. The high production costs result from generally lower yields in the toluene alkylation step as well as higher energy costs in the dehydrogenation and product recovery steps. Until recently, the world market of ca 4.5×10^4 t/yr was served almost exclusively by The Dow Chemical Company. The production methods used by Dow are described in ref. 61. The composition of crude reactor effluent in toluene alkylation is given in ref. 58. Typical physical properties and chemical analysis of vinyltoluene are given in Tables 12 and 13, respectively. Toxicological properties of vinyltoluene are similar to those of styrene, although the lower volatility of vinyltoluene makes it easier to maintain workplace concentrations that comply with OSHA standards for styrene.

In the late 1970s, Cosden Oil & Chemical Company entered the vinyltoluene market with a vapor-phase production facility in Big Spring, Texas. In entering the market previously dominated by Dow, Cosden chose to produce a product similar to the typical Dow product, ie, a mixture of 60 wt % *m*- and 40 wt % *p*-vinyltoluene.

Although the vinyltoluene market has never achieved the size or growth of the styrene market, a number of specialty applications have been developed. As a copolymer with styrene, it has value in increasing the operating temperature range of products, eg, paints, coatings, and varnishes. The presence of the methyl group on the aromatic ring increases its solubility in aliphatic solvents. The higher reactivity reduces drying times. Unsaturated polyester resins are used in electrical parts because of their good insulating properties. In general, the market growth for vinyltoluene has largely been restricted by its price relative to those of alternative materials.

***p*-Methylstyrene.** *p*-Methylstyrene (PMS) is a recently introduced monomer available in developmental quantities. It contains ca 97 wt % *p*-vinyltoluene and only 3 wt % *m*-vinyltoluene. The physical properties of the two products are quite similar, with the exception of freezing point, which is lower for the vinyltoluene mixture. However, the polymer properties and potential product application of PMS offer several advantages. The key to making PMS a commercially useful product was the development by Mobil of a para-selective alkylation catalyst (62). This catalyst was derived from the same family as the ZSM-5 catalyst used in the xylene isomerization and ethylbenzene processes. Alkylation by liquid-phase aluminum chloride catalyst is not selective to the para isomer but produces a mixture of *m*-, *p*-, and *o*-ethyltol-

Table 12. Typical Physical Properties of Vinyltoluene[a]

Property	Value
molecular weight	118.17
refractive index, n_D^{20}	1.5422
viscosity, mPa·s (= cP), 20°C	0.837
surface tension, mN/m (= dyn/cm), 20°C	31.66
density, g/cm^3, 20°C	0.8973
boiling point (at 101.3 kPa (= 1 atm)), °C	172
freezing point, °C	−77
flash point (Cleveland open-cup), °C	60
fire point (Cleveland open-cup), °C	68.3
autoignition temperature, °C	575
explosive limits, % in air	1.9–6.1
vapor pressure, kPa[b]	
20°C	0.15
60°C	1.76
160°C	74.66
critical pressure (P_c), MPa[c]	4.19
critical temperature (t_c), °C	382
critical volume (V_c), mL/g	3.33
critical density (d_c), g/mL	0.30
specific heat of vapor (c_p) at 25°C, J/(g·K)[d]	1.2284
latent heat of vaporization (ΔH_v), J/g[d]	
at 25°C	426.10
at boiling point	349.24
heat of combustion (ΔH_c)	
gas at constant pressure at 25°C, kJ/mol[d]	4816.54
heat of formation (ΔH_f)	
liquid at 25°C, kJ/mol[d]	115.48
heat of polymerization, kJ/mol[d]	+66.9 ± 0.2
Q value	0.95
e value	−0.89
shrinkage upon polymerization, vol %	12.6
cubical coefficient of expansion at 20°C	9.361×10^{-4}
solubility, wt % in H_2O at 25°C	0.0089
H_2O in monomer at 25°C	0.47
solvent compatibility	infinitely soluble in acetone; carbon tetrachloride; benzene; diethyl ether; n-heptane; and ethanol

[a] Ref. 1.
[b] To convert kPa to mm Hg, multiply by 7.5.
[c] To convert MPa to psi, multiply by 145.
[d] To convert J to cal, divide by 4.184.

uenes. Separation of the para isomer from the meta isomer by distillation or other means would be prohibitively expensive.

Mobil has undertaken an extensive program to develop polymers based on PMS, which are analogues of styrene polymers and copolymers. In general, the PMS analogues of crystal and high impact polystyrene offer improved high temperature properties (Vicat softening points increased 10–15°C and densities decreased 3–4% as compared to similar polystyrenes). Since polymer products are usually sold on a

Table 13. Typical Chemical Analysis of Vinyltoluene[a]

	Vinyltoluene
purity, wt %	99.6
polymer, ppm	none
phenyl acetylene, ppm	58
aldehydes as CHO, ppm	10
peroxides as H_2O_2, ppm	5
chlorides as Cl, ppm	5
TBC, ppm	12
m-vinyltoluene, wt %	60
p-vinyltoluene, wt %	40

[a] Ref. 1.

volumetric rather than a weight basis, these improved properties result in potential cost savings. Similar physical property advantages are obtained in copolymers with acrylonitrile, acrylonitrile–butadiene, and butadiene. The use of PMS in reinforced polyesters offers potential for high temperature applications and reduced volatility of the monomer in curing operations.

Presumably, the para isomer polymerizes into a regular polymer similar to styrene. The added methyl group causes reduction of polymer densities and an increase in the glass-transition temperature of the various polymers. Poly(*p*-methylstyrene) and its various copolymers have been tested in injection-molding machines in styrene and styrene copolymer applications and are compatible with and can substitute for existing commercial products. Total production of PMS is forecast to be 2×10^4 t/yr by the mid-1980s.

α-Methylstyrene. α-Methylstyrene (AMS) is a specialty monomer and is primarily produced as a by-product of cumene–phenol operations (see Cumene; Phenol). Alkylation of benzene with propylene followed by dehydrogenation of isopropylbenzene to α-methylstyrene was practiced commercially by Dow until 1977. Phenol capacity and, therefore, potential by-product AMS capacity has grown in absolute terms at a faster rate than has the AMS market. Phenol producers have some elasticity in their production capability in that they can recover additional AMS that is recycled and hydrogenated to cumene in existing plants. By-product material has created an oversupply which has reduced profitability and driven direct producers out of the market. Estimated 1980 plant capacities of United States producers of by-product AMS are listed in Table 14. In addition, Amoco Chemicals Corporation is reported to have capacity for 18,000 t/yr of AMS from dehydrogenation; all of it is used captively (63). Although a number of alkylation catalysts, eg, aluminum chloride and boron trifluoride on alumina, have been used as alkylation catalysts, cumene is produced primarily by alkylation of benzene with propylene with a phosphoric acid-based catalyst.

After oxidation of cumene to cumene hydroperoxide, phenol and acetone are produced in a cleavage reaction in an acidic medium As much as 2 wt % of the cumene is converted to by-product AMS in this cleavage step. It is generally recovered and hydrogenated to cumene for recycle to the oxidation section of the plant.

Typical physical properties and typical chemical analysis of AMS are provided in Tables 15 and 16, respectively. It is marketed as a minimum 99.3 wt % pure monomer

Table 14. Producers and 1980 Plant Capacities for α-Methylstyrene [a]

Producer	Capacity, 10^3 t/yr
Allied, Frankford, Pa.	11,000
Clark, Blue Island, Ill.	2,300
Georgia-Pacific, Plaquemine, La.	3,600
Getty, El Dorado, Kan.	900
Union Carbide, Bound Brook, N.J.	4,100
USS, Haverhill, Ohio	17,000
Total	*38,900*

[a] Ref. 63.

inhibited with *tert*-butylcatechol. α-Methylstyrene is a much less reactive monomer than is styrene. It is slow to polymerize and its use is limited to a variety of specialty applications. These include in order of importance: ABS resins, coatings, polyester resins, and hot-melt adhesives. As a copolymer in ABS and polystyrene, it increases the heat-distortion temperature of the resultant product. In coatings and resins, it moderates reaction rates and improves clarity of products. Total demand for AMS has increased from less than 10^4 t/yr in 1965 to over 4×10^4 t/yr in 1981.

Because of its lower vapor pressure and reduced reactivity, storage and handling of the monomer is easier than styrene monomer. Its toxicity is similar to that of styrene and the same precautions should be taken in limiting workplace concentrations to a maximum of 100 ppm. As with other styrenic monomers, the odor can be detected at concentrations below 10 ppm. At higher concentrations, the vapor causes eye and nasal irritation.

Divinylbenzene. Divinylbenzene (DVB), $C_6H_4(CH{=}CH_2)_2$, is a specialty monomer which is of unique value because of its bifunctionality. The presence of two reactive vinyl groups accounts for its strong propensity for cross-linking with itself and other monomers. This cross-linking tendency results in a very hard, brittle homopolymer of DVB with little commercial value. Its real commercial value is that in a copolymer with styrene. Small quantities of DVB markedly modify linear styrene polymers (64–66). The cross-linking results in resins with reduced solubility in most solvents, increased heat-distortion temperatures, increased surface hardness, and improved impact and tensile strengths. Since the amount of DVB included in the copolymer is small, the appearance and optical and electrical properties are much like those of styrene homopolymer.

Dow Chemical and American Hoechst are the two manufacturers of DVB in the United States. Diethylbenzene is recovered as a slipstream from an ethylbenzene alkylation unit. This slipstream contains *o*-, *m*-, and *p*-diethylbenzene isomers in relative concentrations of ca 1:6:3. After removal of other impurities, eg, propyl- and butylbenzenes, the isomer mixture is used as a feedstock for divinylbenzene manufacture. Dehydrogenation of diethylbenzene is carried out in a manner analogous to ethylbenzene dehydrogenation. Temperatures of greater than 600°C and superheated dilution steam are used over typical iron oxide dehydrogenation catalysts. Under these conditions, a mixture of light by-products, ie, benzene, toluene, xylenes, ethylbenzene, styrene, diethylbenzene, vinyltoluene, and ethyltoluene, are produced. *o*-Diethylbenzene is converted to naphthalene which must be removed because of its objectionable odor as well as its adverse effect on copolymer properties. Unreacted di-

Table 15. Typical Physical Properties of α-Methylstyrene[a]

Property	Value
molecular weight	118.18
refractive index, n_D^{20}	1.53864
viscosity, mPa·s (= cP), 20°C	0.940
surface tension, mN/m (= dyn/cm), 20°C	32.40
density, g/cm^3, 20°C	0.9106
boiling point, °C	165
freezing point, °C	−23.2
flash point (Cleveland open-cup), °C	57.8
fire point (Cleveland open-cup), °C	57.8
explosive limits, % in air	0.7–3.4
vapor pressure, kPa[b]	
20°C	0.253
60°C	2.400
100°C	13.066
160°C	88.660
critical pressure (P_c), MPa[c]	4.36
critical temperature (t_c), °C	384
critical volume (V_c), mL/g	3.26
critical density (d_c), g/mL	0.29
specific heat of liquid, J/(g·K)[d]	
at 40°C	2.0460
at 100°C	2.1757
specific heat of vapor at 25°C, J/(g·K)[d]	1.2357
latent heat of vaporization (ΔH_v), J/g[d]	
at 25°C	404.55
at boiling point	326.35
heat of combustion (ΔH_c), gas at constant pressure at 25°C, kJ/mol[d]	4863.73
heat of formation (ΔH_f), liquid at 25°C, kJ/mol[d]	112.97
heat of polymerization, kJ/mol[d]	39.75
Q value	0.76
e value	−1.17
cubical coefficient of expansion at 20°C	9.774×10^{-4}
solubility, in H_2O at 25°C	0.056
H_2O in monomer at 25°C	0.010
solvent compatibility	infinitely soluble in acetone; carbon tetrachloride; benzene; diethyl. ether; n-heptane; and ethanol

[a] Ref. 1.
[b] To convert kPa to mm Hg, multiply by 7.5.
[c] To convert MPa to psi, multiply by 145.
[d] To convert J to cal, divide by 4.184.

ethylbenzene and partially converted ethylvinylbenzene are recycled to the dehydrogenation reactor. The product is recovered as a mixture of divinylbenzene and ethylvinylbenzene. The physical properties of the two commercial grades of the DVB–ethylvinylbenzene mixtures are provided in Table 17; the corresponding chemical analyses of these two grades are provided in Table 18. Dow is the largest producer of DVB mixtures and has established the standard commercial grades; however, new producers have sought to match those grades. In addition to DVB-55, American Hoechst offers DVB-27 and DVB-80. The -27 and -80 designates the approximate content of divinylbenzene.

Table 16. Typical Chemical Analysis of α-Methylstyrene[a]

	α-Methylstyrene
purity, wt %	99.3
polymer, ppm	none
aldehydes as CHO, ppm	10
peroxides as H_2O_2, ppm	3
TBC, ppm	15
α-methylstyrene, wt %	0.5
isopropylbenzene, wt %	0.2

[a] Ref. 1.

Table 17. Typical Physical Properties of Divinylbenzene[a]

	Divinylbenzene	
Property	DVB-22	DVB-55
molecular weight	130.08	130.18
refractive index, n_D^{25}	1.5326	1.5585
viscosity (at 25°C), mPa·s (= cP)	0.883	1.007
surface tension (at 25°C), mN/m (= dyn/cm)	30.55	32.10
density, g/cm³, 20°C	0.8979	0.9162
boiling point, °C	180[b]	195[b]
freezing point, °C		−45
flash point (Cleveland open-cup), °C	57	74
fire point (Cleveland open-cup), °C	57	74
explosive limits, % in air	1.1–6.2	≥1.1[c]
critical pressure (P_c), MPa[d]	2.45[b]	2.45[b]
critical temperature (t_c), °C	348[b]	369[b]
latent heat of vaporization (ΔH_v) at boiling point, J/g[e]	320.49	350.62
solubility, % in H_2O at 25°C	0.0065	0.0052
H_2O in monomer at 25°C	0.051	0.054
solvent compatibility		
acetone	infinitely soluble	infinitely soluble
carbon tetrachloride	infinitely soluble	infinitely soluble
benzene	infinitely soluble	infinitely soluble
ethanol	infinitely soluble	infinitely soluble

[a] Ref. 1.
[b] Calculated.
[c] Could not be measured at 130°C.
[d] To convert MPa to psi, multiply by 145.
[e] To convert J to cal, divide by 4.184.

Product finishing occurs in a series of three columns. Light by-products are removed overhead in the first column with recycleable ethylvinylbenzene and unreacted diethylbenzene. The lights are taken overhead in the second column and they are separated from the recyclables. Product DVB mixtures are taken as an overhead product in the third column, and naphthalene and tar are rejected as a bottoms product. The high reactivity of the unsaturated components requires very low operating pressures and temperatures as well as the use of in-process polymerization inhibitors. The final product is inhibited with TBC and sulfur, as shown in Table 18.

The world market for DVB as 100 wt % DVB was 3000–4000 metric tons in 1981.

Table 18. Typical Chemical Analysis of Divinylbenzene

	DVB-22[a]	DVB-55[a]	DVB-80[b]
polymer, ppm	100	100	
aldehydes as CHO, ppm	40	40	
peroxides as H_2O_2, ppm	5	5	
sulfur as S, ppm	20	230	240
TBC, ppm	1000	1000	1200–1500
total unsaturation (as ethylvinylbenzene), wt %	83.3	149.4	177.3
m-divinylbenzene, wt %	17.1	36.4	60.3
p-divinylbenzene, wt %	8.2	18.6	21.6
total divinylbenzene, wt %	25.3	55.0	81.9
m-ethylvinylbenzene, wt %	23.1	25	6.7
p-ethylvinylbenzene, wt %	10	13	6.8

[a] Ref. 1.
[b] Ref. 67.

By far the largest use was as a copolymer with styrene in ion-exchange resins. These resins are used primarily in home water softeners as well as for municipal and industrial water conditioning (see Ion exchange). Production of these resins is estimated to increase 5–10%/yr. Other markets show little growth, although research at the University of Akron indicates future increased use of thermoplastic elastomers containing DVB as the heart of a radial or star-configuration block copolymer.

Toxicological problems are similar to those encountered with styrene. As with other styrene monomers, the odor can be detected by humans at levels below dangerous concentration, ie, 10–60 ppm. Above 100 ppm, eye and nasal irritation may occur. Vapor toxicity in animals is negligible at repeated exposure to concentrations of 200 ppm or less.

BIBLIOGRAPHY

"Styrene" in *ECT* 1st ed., Vol. 13, pp. 119–146, by A. L. Ward and W. J. Roberts, Pennsylvania Industrial Chemical Corporation; "Styrene" in *ECT* 2nd ed., Vol. 19, pp. 55–85, by K. E. Coulter, H. Kehde, and B. F. Hiscock, The Dow Chemical Company.

1. *Styrene-Type Monomers*, Technical Bulletin No. 170-151B-3M-366, The Dow Chemical Company, Midland, Mich. (no date on brochure).
2. R. H. Boundy and R. F. Boyer, eds., *Styrene, Its Polymers, Copolymers, and Derivatives*, Reinhold Publishing Corp., New York, 1952.
3. H. Ohlinger, *Polystyrol*, Springer-Verlag, West Berlin, 1955.
4. J. Elly, R. N. Haward, and W. Simpson, *J. Appl. Chem.* **1**, 347 (1951).
5. C. H. Basdekis, in W. M. Smith, ed., *Manufacture of Plastics*, Reinhold Publishing Corp., New York, 1964.
6. H. Fikentscher, H. Gerrens, and H. Schuller, *Angew. Chem.* **72**, 856 (1960).
7. C. E. Schildknecht, *Polymer Processes*, Interscience Publishers, Inc., New York, 1956, Chapt. 4.
8. K. F. Coulter, H. Kehde, and B. F. Hiscock, in E. C. Leonard, ed., *Vinyl Monomers*, Interscience Publishers, a division of John Wiley & Sons, Inc., New York, 1969.
9. S. S. Hupp and H. E. Swift, *Ind. Eng. Chem. Prod. Res. Dev.* **18**(2), (1979), pp. 117–122.
10. U.S. Pat. 3,965,206 (June 22, 1976), P. D. Montgomery, P. N. Moore, and W. R. Knox (to Monsanto Co.).
11. U.S. Pat. 4,115,424 (Sept. 19, 1978), M. L. Unland and G. E. Barker (to Monsanto Co.).

12. G. A. Olah, ed., *Friedel-Crafts and Related Reactions*, Interscience Publishers, a division of John Wiley & Sons, Inc., New York, 1964.
13. D. F. Stull, E. F. Westrum, Jr., and G. C. Sinke, *The Chemical Thermodynamics of Organic Compounds*, John Wiley & Sons, Inc., New York, 1969.
14. F. R. Garner and R. L. Iverson, *Oil Gas J.* **63**(43), 86 (1965).
15. A. L. Foster, *Pet. Engr.* **25**(11), C-3 (1953).
16. P. W. Sherwood, *Pet. Refiner* **32**(1), 97 (1953).
17. E. K. Jones, *Oil Gas J.* **58**(9), 80 (1960).
18. U.S. Pat. 3,751,506 (Aug. 7, 1973), G. T. Burress (to Mobil Oil Corp.).
19. U.S. Pat. 4,107,224 (Aug. 15, 1978), F. G. Dwyer (to Mobil Oil Corp.).
20. H. W. Grote and C. F. Gerald, *Chem. Eng. Prog.* **56**(1), 60 (1960).
21. H. W. Grote, *Oil Gas J.* **56**(13), 73 (1958).
22. F. G. Dwyer, P. J. Lewis, and F. M. Schneider, *Chem. Eng.* **83,** 90 (Jan. 5, 1976).
23. P. J. Lewis and F. G. Dwyer, *Oil Gas J.* **75**(40), 55 (1977).
24. Br. Pats. 1,122,732 and 1,122,731 (Aug. 7, 1978), C. Y. Choo (to Halcon International, Inc.).
25. Br. Pat. 1,128,150 (Sept. 25, 1968), C. Y. Choo and R. L. Golden (to Halcon International, Inc.).
26. U.S. Pat. 3,459,810 (Aug. 5, 1969), C. Y. Choo and R. L. Golden (to Halcon International, Inc.).
27. Ger. Pat. 2,631,016 (July 10, 1975), M. Becker (to Halcon International Inc.).
28. U.S. Pat. 3,987,115 (Oct. 19, 1976), J. G. Zajacek and F. J. Hilbert (to Atlantic Richfield Co.).
29. U.S. Pat. 4,066,706 (Jan. 3, 1978), J. P. Schmidt (to Halcon International, Ltd.).
30. S. Afr. Pat. 66 05,917 (April 1, 1968), H. B. Pell and E. I. Korchak (to Halcon International, Inc.).
31. Fr. Pat. 1,548,198 (Nov. 29, 1968), T. W. Stein, H. Gilman, and R. L. Robeck (to Halcon International, Inc.).
32. Ger. Pat. 2,165,027 (July 6, 1973), J. J. Coyle (to Shell Oil Corp.).
33. Br. Pat. 1,345,900 (Feb. 6, 1974), (to ARCO).
34. U.S. Pat. 3,829,392 (Aug. 13, 1974), H. P. Wulff (to Shell).
35. U.S. Pat. 3,849,451 (Nov. 19, 1974), T. W. Stein, H. Gilman, and R. L. Bobeck (to Halcon International, Inc.).
36. U.S. Pat. 3,873,578 (March 25, 1975), C. S. Bell and H. P. Wulff (to Shell Oil Corp.).
37. Neth. Pat. 75,13,859 (April 29, 1976), (to Halcon International, Inc.).
38. Neth. Pat. 75,14,538 (March 31, 1976), (to Halcon International, Inc.).
39. U.S. Pat. 4,059,598 (Nov. 22, 1977), J. J. Coyle (to Shell Oil Corp.).
40. Ger. Pat. 1,939,791 (Feb. 26, 1970), M. Becker and S. Khoobiar (to Halcon International, Inc.).
41. *Chem. Week* **99**(6), 19 (1966).
42. *Chem. Week* **99**(5), 49 (1966).
43. H. J. Sanders, H. F. Keag, and H. S. McCullough, *Ind. Eng. Chem.* **45**(1), 2 (1953).
44. U.S. Pats. 4,105,506 (Aug. 31, 1978), 4,252,615 (Feb. 24, 1981), and 4,272,344 (June 9, 1981), J. M. Watson (to Cosden Technology, Inc.).
45. U.S. Pat. 4,033,829 (July 5, 1977), T. D. Higgins, Jr. and R. A. Newsom (to Monsanto Co.).
46. U.S. Pat. 4,040,911 (Aug. 9, 1977), J. D. Bacha, C. M. Selwitz (to Gulf Research and Development Co.).
47. *Chemical Economics Handbook*, SRI International, Menlo Park, Calif., Styrene Marketing Research Report, April 1980, p. 694.3052-C; *Chemical Economies Handbook, Manual of Current Indicators*, SRI International, April 1981, p. 267.
48. *Chem. Mark. Rep.* **219,** 9 (Jan. 12, 1981).
49. Ref. 47, p. 694.3053-L; The Process Engineering Department, The Badger Company, Inc., Cambridge, Mass.
50. S. A. Cogswell, personal communication, SRI International, Menlo Park, Calif., Aug. 31, 1981.
51. Ref. 47, p. 694.3053-D; *Benzene Marketing Research Report, Chemical Economics Handbook*, SRI International, Menlo Park, Calif., 1979, p. 300.7003W.
52. F. A. Fazzalari, ed., *Compilation of Odor and Taste Threshold Values Data*, ASTM, Philadelphia, Pa., 1977.
53. N. I. Sax, *Dangerous Properties of Industrial Materials*, Reinhold Publishing Co., New York, 1957.
54. *Chemical Safety Data Sheet* SD-37, Manufacturing Chemists Association, Washington, D.C., 1971.
55. *Threshold Limit Values for Chemical Substances in Workroom Air*, American Conference of Governmental Industrial Hygienists, Cincinnati, Ohio, 1981.
56. Written communication between S. d'Arazien, Public Information Office, U.S. Department of Health

and Human Services, and A. M. Sobkowicz, Badger America, Inc., Cambridge, Massachusetts, April 1, 1982.
57. *Storage and Handling of Styrene-Type Monomers*, Form No. 115-575-79 Dow Chemical USA, Organic Chemicals Dept., Midland, Mich., 1979.
58. P. G. Shelley and E. J. Sills, *Chem. Eng. Prog.* **65**(4), 29 (1969).
59. Ref. 47, p. 694.3052-O-P.
60. *Chem. Eng. Prog.* **65**(4), (1969); *Chem. Week* **42,** (Feb. 17. 1982).
61. U.S. Pat. 2,763,702 (Sept. 18, 1956), J. L. Amos and K. E. Coulter (to The Dow Chemical Co.).
62. U.S. Pat. 4,100,217 (July 11, 1978), L. B. Young (to Mobil Oil Corp.).
63. *Chem. Mark. Rep.* **218,** 2 (Aug. 18, 1980).
64. H. Staudinger, *Trans. Faraday Soc.* **32,** 323 (1936).
65. H. Staudinger, *Ber.* **67,** 1164 (1934); *Z. Phys. Chem. Abt. A* **171,** 129 (1934), *Ber.* **68,** 1618 (1935).
66. U.S. Pat. 2,089,444 (Aug. 10, 1973), H. Staudinger and W. Heuer (to I.G. Farbenind A.G.).
67. Private communication between R. Partos, American Hoechst Corporation, Leominister, Mass., and P. J. Lewis, The Badger Company, Inc., Cambridge, Mass., Aug. 26, 1981.

P. J. Lewis
C. Hagopian
P. Koch
The Badger Company, Inc.

STYRENE–BUTADIENE SOLUTION COPOLYMERS. See Elastomers, synthetic.

STYRENE PLASTICS

Polystyrene, the parent of the styrene plastics family, is a high molecular weight linear polymer, which for commercial uses, consists of ca 1000 styrene (qv) units (1–2). Its chemical formula $\text{-}[CH(C_6H_5)CH_2]_n\text{-}$, where $n \sim 1000$, tells little of its properties. The commercially useful form of polystyrene is amorphous and hence possesses high transparency; the phenyl group raises the glass-transition temperature to near 100°C. Therefore, under ambient conditions, the polymer is stiff and clear, whereas above the glass-transition temperature it becomes a viscous liquid which can be easily fabricated with little or no degradation by extrusion or injection-molding techniques. It is this ease with which polystyrene can be converted into useful articles that accounts for the very high volume used in world commerce. Even though crude oil is the source of the polymer, the energy savings accrued during fabrication and use compared to alternative materials more than offsets the short life of many polystyrene articles.

The monomeric precursor, styrene, is a worldwide commodity chemical. Polystyrene, although not a commodity in the same sense since there are many different grades, is produced on a very large scale, usually in continuous polymerization plants. Generally, the key problems associated with manufacture of the polymer are removal of the heat of polymerization and pumping highly viscous solutions; conversion of the monomer to the polymer is energetically very favorable and occurs spontaneously on heating without the addition of initiators. Because it is a continuous polymerization process, material-handling problems are minimized during manufacture. By almost any standard, the polymer produced is highly pure and usually is greater than 99 wt % polystyrene; for particular applications, however, processing aids are often deliberately added to the polymer. Methods for improving the toughness, solvent resistance, and upper use temperature have been developed. Butadiene-based rubbers increase impact resistance, and copolymerization of styrene with acrylonitrile produces heat-resistant and solvent-resistant plastics (see Acrylonitrile polymers). Uses for these plastics are extensive. Packaging applications, eg, disposable tumblers, television cabinets, meat and food trays, and egg cartons, are the largest area of use for styrene plastics. Rigid foam insulation in various forms is being used increasingly in the construction industry, and modified styrene plastics are replacing steel or aluminum parts in automobiles; both applications result in energy savings beyond the initial investment in crude oil. The cost of achieving a given property, eg, impact strength, is among the lowest for styrene plastics as compared to any competitive material.

Properties

The general mechanical properties of these polymers are given in Table 1. Considerable differences in performance can be achieved by using the various styrene plastics. Within each group, additional variation is expected. In choosing an appropriate resin for a given application, other properties and polymer behavior during fabrication must be considered. These factors depend on the combination of inherent polymer properties, the fabrication technique, and the devices, eg, a mold, used for obtaining the final object. Accordingly, consideration must be given to such factors as the surface appearance of the part and the development of anisotropy, and its effect on mechanical strength, ie, long-term resistance of the molding to external strain. Toxicological aspects and applications of styrene plastics are described in ref. 3.

Physical. For molecular weights above 5000, many properties are approximately independent of chain length. Some of the more important values are listed in Table 2; a more extensive compilation is given in ref. 4.

Stress-Strain Properties. The strain energy, derived from the area under the stress-strain curve, is considered to indicate the level of toughness of a polymer. High impact polystyrene (HIPS) has a higher strain energy than an ABS plastic, as shown in Figure 1. Based on different impact-testing techniques, ABS materials are generally tougher than HIPS materials (23–24). The failure of the stress-strain curve to reflect this ductility can be related to the fact that ABS polymers tend to show only localized flow or necking tendency at low rates of extension and, therefore, fail at low elongation. High impact polystyrene extends uniformly during such tests, and the test specimen whitens over all of its length and extends well beyond the yield elongation. At higher testing speeds, ABS polymers also deform more uniformly and give high elongations (24).

Tensile strengths of styrene polymers vary with temperature. Increased temperature lowers the strength. However, tensile modulus in the temperature region of

Table 1. Mechanical Properties of Injected-Molded Specimens of Main Classes of Styrene-Based Plastics[a]

Property	Polystyrene (PS) [9003-53-6]	Poly(styrene-co-acrylonitrile) (SAN)[b] [9003-54-7]	Glass-filled PS[c]	Medium high impact polystyrene	HIPS [9003-53-6]	Medium acrylonitrile–butadiene–styrene terpolymer (ABS)		Standard ABS [9003-56-9]	Super ABS
						Type 1	Type 2		
specific gravity	1.05	1.08	1.20	1.05	1.05	1.05	1.05	1.04	1.04
Vicat softening point, °C	96	107	103	103	95	99	108	103	108
tensile yield, MPa[d]	42.0	68.9	131	39.6	29.6	31.0	53.8	41.4	34.5
elongation, rupture, %	1.8	3.5	1.5	15	58	55	10	20	60
modulus, MPa[d]	3170	3790	7580	2690	2140	2620	2620	2070	1790
impact strength (notched Izod), J/m[e]	21	21	80	96	134	193	187	267	428
dart-drop impact strength	very low	very low	medium high	low	medium high	medium high	high	very high	very high
relative ease of fabrication	excellent	excellent	poor	excellent	excellent	excellent	good	good	medium good

[a] Ref. 2.
[b] 24 wt % acrylonitrile.
[c] 20% glass fibers.
[d] To convert MPa to psi, multiply by 145.
[e] To convert J/m to ft·lbf/in., divide by 53.38.

Table 2. Physical Constants of Polystyrene

Property	Value	Remarks	Refs.
coefficient of thermal expansion, per °C			
linear	$6–8 \times 10^{-5}$	$<T_g$ (unoriented)	5
volume	$1.7–2.1 \times 10^{-4}$	$<T_g$	6
	$5.1–6.0 \times 10^{-4}$	$>T_g$	
compressive modulus MPa[a]	3,000	(unoriented)	7
density, g/cm^3			
amorphous	1.04–1.065		8
crystalline	1.111		9
crystalline	1.12		8, 9
d/dT, g/(cm^3·°C)	-2.65×10^{-4}	$<T_g$	10
d/dT, g/(cm^3·°C)	-6.05×10^{-4}	$>T_g$	10
dielectric constant, e			
amorphous	2.49–2.55	at 1 kHz (curve flat to 1 GHz)	7
crystalline	2.61	at 1 kHz (curve flat to 1 GHz)	11
dissipation factor			
amorphous	15×10^{-4}	at 1 kHz (curve flat to 1 GHz)	7
crystalline	3×10^{-4}	at 1 kHz (curve flat to 1 GHz)	11
glass-transition temperature (T_g), °C	$T_g = 100–120{,}000/(\text{mol wt} + 400)$		10, 12–14
heat capacity (C_p), kJ/(kg·K)[b]			
at 0°C	1.185 (1.139)[c]		7, 15
at 50°C	1.256 (1.394)[c]		7, 15
at 100°C	1.838 (1.821)[c]		7, 15
at 150°C	2.01		7
at 200°C	2.10		7
at 250°C	2.18		7
dC_p/dT, kJ/(kg·°C^2) at 50°C	4.04×10^{-3}		7
heat of combustion, kJ/mol[b,d]	-4.33×10^{3}		16–17
heat of fusion, kJ/mol[b,d]	8.37 ± 0.08		18
crystalline	9.00		19
heat of polymerization, kJ/mol[b,d]	−67.4	to solid polymer	19
	−69.9	to solid polymer	16
heat of solution, kJ/mol[b,d]	−3.60	in monomer	16
melting point T_m, °C	240 (250)[c]		17, 20
melt viscosity–mol wt relationship	$\log \eta_T = 3.4 \log \text{mol wt} - k$		
atactic			21
T, °C	217		
mol wt	≥38,000		
k	13.04		
isotactic			22
T, °C	281		
mol wt	100,000–600,000		
k	14.42		
refractive index, n_D	1.59–1.60	(at λ wavelength of 589.3 nM)	7
dn_D/dT, per °C	-1.42×10^{-4}		

Table 2 (*continued*)

Property	Value	Remarks	Refs.
thermal conductivity, W/(m·K)[e]			7
at 0°C	0.105		
at 100°C	0.128		
at 200°C	0.13		
at 300°C	0.14		

[a] To convert MPa to psi, multiply by 145.
[b] To convert J to cal, divide by 4.184.
[c] Number in parentheses refers to second reference.
[d] Per mole of monomer unit.
[e] To convert W/(m·K) to (Btu·in.)/(h·ft²·°F), divide by 0.1441.

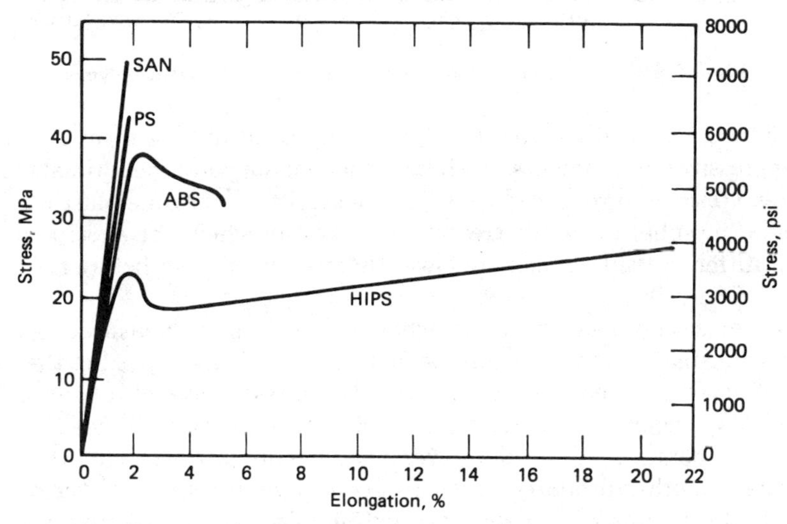

Figure 1. Stress-strain curves for styrene-based plastics.

most tests (−40 to 50°C) is affected only slightly. The elongations of PS and styrene copolymers do not vary much with temperature (−40 to 50°C), but the elongation of rubber-modified polymers first increases with increasing temperature and ultimately decreases at high temperatures.

The molecular orientation of the polymer in a fabricated specimen can significantly alter the stress-strain data as compared with the data obtained for an isotropic specimen, eg, one obtained by compression molding. For example, tensile strengths as high as 120 MPa (18,000 psi) have been reported for PS films and fibers (25). Polystyrene tensile strengths below 14 MPa (2000 psi) have been obtained in the direction perpendicular to the flow.

Creep, Stress Relaxation, and Fatigue. The long-term engineering tests on plastics in intended use environments and temperatures are required for predicting the overall performance of a polymer in a given application. Creep tests involve the measurement of deformation as a function of time at a constant stress or load. For styrene-based plastics, many such studies have been carried out (26–27). Creep curves for styrene and its copolymers at room temperatures show low elongation with only small variation with stress, whereas the rubber-modified polymers exhibit a low elongation region, followed by crazing and increasing elongation, usually to ca 20%, before failure (Fig. 2).

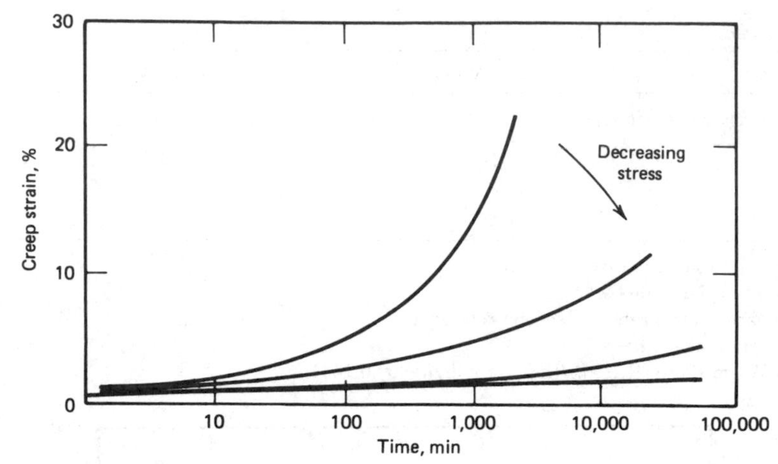

Figure 2. Typical creep behavior for rubber-modified styrene polymers.

Creep tests are ideally suited for the measurement of long-term polymer properties in aggressive environments. Both the time to failure and the ultimate elongation in such creep tests tend to be reduced. Another test to determine plastic behavior in a corrosive atmosphere is a prestressed creep test in which the specimens are prestressed at different loads, which are lower than the creep load, before the final creep test (28).

Stress-relaxation measurements, where stress decay is measured as a function of time at a constant strain, have also been used extensively to predict the long-term behavior of styrene-based plastics (26,29). These tests have also been adapted to measurements in aggressive atmospheres (30). Stress-relaxation measurements are further used to obtain modulus data over a wide temperature range (31).

Fatigue is another property which is of considerable interest to the design engineer. Cyclic deflections of a predetermined amplitude, short of giving immediate failure, are applied to the specimen, and the number of cycles to failure is recorded. In addition to mechanically induced periodic stresses, fatigue failure can be studied when developing cyclic stresses by fluctuating the temperature.

Fatigue in polymers has been reviewed (32). Detailed theory and practice of fatigue testing are covered. Fatigue tests are carried out for two main reasons: to learn the inherent fatigue resistance of the material and to study the relationship between specimen design and fatigue failure. Fatigue tests are carried out both in air and in aggressive environments (33).

Impact Strength. Polystyrene and styrene copolymers are brittle polymers under normal use conditions. A high speed blow at temperatures below T_g causes catastrophic failure without significant deformation, crazing, or yielding in the polymer. Rubber-modified styrene polymers, however, are significantly more impact-resistant. These polymers are characterized by significant whitening of the specimen during the test as a result of craze formation, separation of the rubber phase from the matrix polymer, and cracks (34–35). The mechanism by which the dispersed rubber particles cause such increased toughness continues to be debated (36–38). Craze initiation and termination are controlled by the particles. Under tensile stress, crazes are initiated near the equators of the particles, ie, at the point of maximum stress, and propagate outwards. The highly oriented polystyrene fibrils with the surrounding voids constitute the craze (39). Because of the molecular orientation, such material is load-bearing and ductile. The large number of particles leads to a large number of small crazes, and on a microscopic scale the plastic is ductile instead of brittle. Creation of the craze matter absorbs energy and, so long as the stress does not cause the crazes to become true cracks, then the plastic can recover after the stress is removed. Particle-size sensitivity, which is discussed under Copolymerization in terms of impact resistance, probably results from the ineffectiveness of very small particles in stopping craze growth. They may also be inefficient in producing the necessary stress level around the particle to initiate crazing. Although similar considerations apply to some ABS plastics, the observation of stress whitening or crazing and necking indicates shear yielding (40).

A brittle fracture of a styrene polymer can be brought about by producing uniaxially oriented moldings. Measuring the strength in these moldings across the flow direction or by biaxial loading, such as in a dart-drop test, show the embrittlement of an otherwise tough polymer. Injection moldings of HIPS produced at low temperatures have been shown to be particularly prone to brittle fracture, whereas the ABS polymers in general are less subject to fabrication-induced anisotropy and the consequent embrittlement (41). However, it has been shown that tough moldings of PS can be obtained through the introduction of balanced, multiaxial orientation (42). One way this orientation can be achieved is by molding objects with rotational symmetry at low molding temperatures and by rotating one half of the mold during and after filling, and until the polymer in the mold is cool enough to resist molecular relaxation. In addition to enhanced toughness, craze resistance is improved in such moldings.

Embrittlement of otherwise tough rubber-modified styrene polymers occurs through aging of these polymers (43). The effects of outdoor aging of rubber-modified thermoplastics have been studied (44). Outdoor aging was simulated by laminating a brittle film onto a virgin rubber-modified polymer molding. These experiments showed that aging reduces the energy of crack initiation, so that the final impact strength is determined by the inherent crack-propagation energy of the rubber-modified polymer.

Polystyrene and Copolymers. *General-Purpose Polystyrene.* Polystyrene (PS) is a high molecular weight, ($\overline{M}_w$ = 2–3 × 10^5) crystal-clear thermoplastic that is hard, rigid, and free of odor and taste. Its ease of heat fabrication, thermal stability, low specific gravity, and low cost results in moldings, extrusions, and films of very low unit cost. In addition, PS materials have excellent thermal and electrical properties which make them useful as low cost insulating materials (see Insulation, electric; Insulation, thermal).

Commercial polystyrenes are normally rather pure polymers. The amount of styrene, ethylbenzene, and styrene dimers and trimers is minimized by effective devolatilization or by the use of catalysts for the mass and suspension processes, respectively. Polystyrenes with low overall volatiles content have high heat-deformation temperatures. The very low content of monomer and other solvents, eg, ethylbenzene, in PSs is useful in the packaging of food. The negligible level of extraction of organic materials from PS is of crucial importance in this application.

When additional lubricants, eg, mineral oil, butyl stearate, etc, are added to PS, easy-flow materials are produced. Improved flow is usually achieved at the cost of lowering the heat-deformation temperature. Stiff-flow PS has a high molecular weight and a low volatile level and is useful for extrusion applications. Typical levels of residuals in polystyrene grades are listed in Table 3. Differences in molecular weight distribution are illustrated in Figure 3.

Specialty Polystyrenes. Specialty polystyrenes are available for commercial and scientific use. Although these polymers are essentially pure PS, their molecular structures and/or additives are so adjusted as to make them useful in special applications.

Standard polystyrenes are carefully prepared and characterized PS materials available from the National Bureau of Standards and the Pressure Chemical Company. Materials of extremely narrow and relatively broad molecular weight distribution are available. Pressure Chemical Company has narrow-distribution polystyrenes with molecular weights of $600–2 \times 10^6$.

Monodisperse polystyrene latices are produced by The Dow Chemical Company. Latex particle sizes are 90–20,000 nm. Initially, these latices were primarily used for calibration in electron microscopy and in the mechanism studies of emulsion polymerization. More recently, however, diagnostic medical use has been predominant as has been calibration for scientific measurements, eg, light scattering, ultracentrifuge, filter pore size.

Isotactic polystyrene can be obtained by the polymerization of styrene with stereospecific catalysts of the Ziegler-Natta type (21). As a result of the regular isotactic structure, it can be crystallized and has a threefold helix-chain conformation. The

Table 3. Residuals in Typical Polystyrene, wt %

	Grade	
	Extrusion	Injection-molding
styrene	0.04	0.1
ethylbenzene	0.02	0.1
styrene dimer	0.04	0.1
styrene trimer	0.25	0.8

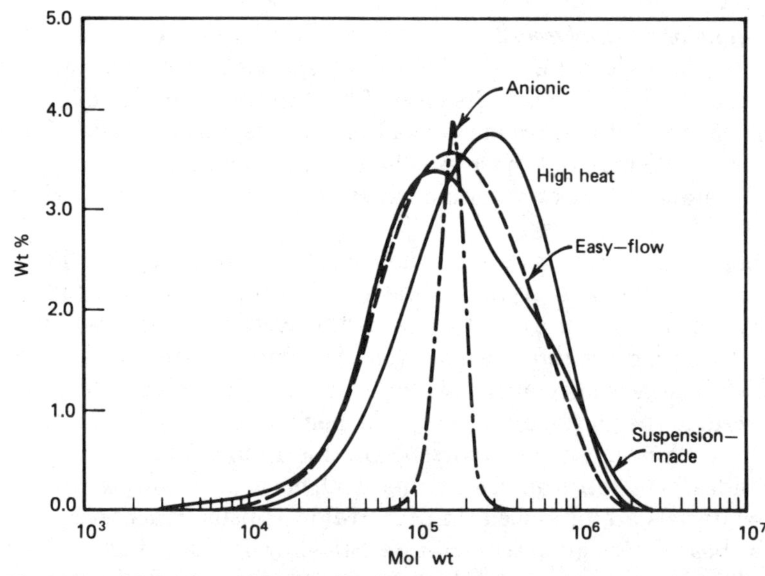

Figure 3. Molecular weight distribution curves for representative polystyrenes.

isotactic polymer exists in the amorphous or crystalline state. Samples quenched from the melt are amorphous but become crystalline if annealed for some time at a temperature slightly below the crystalline melting point. The rate of crystallization is relatively slow compared with other crystallizable polymers, eg, polyethylene or polypropylene. In the amorphous state, the properties of isotactic polystyrene are very similar to those of conventional atactic polystyrene.

Crystalline polystyrene has a high melting temperature indicating a first-order transition temperature of ca 240°C. It is insoluble in most common polystyrene solvents and, as a result of the spherulitic structure of the crystalline phase, is opaque. The density of the 100% crystalline polymer is calculated to be 1.12 g/cm^3 from x-ray data. Although highly isotactic polystyrene has been prepared, only partially crystalline polymers have been obtained and generally with less than 50% relative crystallinity. The lack of commercial interest in isotactic polystyrene may result in part from its low degree of crystallinity and its slow crystallization rate. Syndiotactic polystyrene is unknown.

Stabilized polystyrenes are materials with low volatiles content and added light stabilizers, uv screening agents, antioxidants, and synergistic agents. Early stabilization systems for PS included alkanolamines (qv) and methyl salicylate (22). In recent years, improved stabilizing systems have been developed; these involve a uv-radiation absorber, eg, Tinuvin P (CIBA-GEIGY Corp.) with a phenolic antioxidant. Iron as a contaminant, even at a very low concentration, can cause color formation during fabrication, however this color formation can be appreciably retarded by using tridecyl phosphite as a costabilizer with the uv-radiation absorber and the antioxidant (9). Rubber-modified styrene polymers are heat-stabilized with nonstaining rubber antioxidants, eg, Irganox 1076 (CIBA-GEIGY Corp.). Moldings for lighting fixtures are the main use for these polystyrenes.

Polymers containing flame retardants (qv) have been developed. The addition of flame retardants does not make a polymer noncombustible, but rather increases its resistance to ignition and reduces the rate of burning with minor fire sources. Although the primary commercial developments are in the area of PS foams (see Uses), many other applications have been suggested (see Foamed plastics). Both inorganic (hydrated aluminum oxide, antimony oxide) and organic (alkyl and aryl phosphates) additives have been used (45). Synergistic effects between halogen compounds and free-radical initiators have been reported (46). Several new halogenated compounds and corrosion inhibitors are effective additives (47) (see Corrosion and corrosion inhibitors). The polymer manufacturer's recommendations with regard to maximum fabrication temperature should be carefully observed to avoid discoloration of the molded part or corrosion of the mold or the machine.

Antistatic polystyrenes have been developed in terms of additives or coatings to minimize primarily dust-collecting problems in storage (see Antistatic agents). Large lists of commercial antistatic additives have been published (45). For styrene-based polymers, alkyl and/or aryl amines, amides, quaternary ammonium compounds (qv), anionics, etc, are used (see Additives).

Styrene Copolymers. Acrylonitrile, butadiene, α-methylstyrene, methyl methacrylate, and maleic anhydride have been copolymerized with styrene to yield commercially significant copolymers. Acrylonitrile copolymers with styrene (SAN) have been available for some time (48). Most of these polymers have been prepared at the crossover composition, which is ca 24 wt % acrylonitrile (see Copolymerization).

These copolymers are transparent and, in comparison to PS, are more solvent- and craze-resistant and relatively tough. They also constitute the rigid matrix phase of the ABS engineering plastics (qv). Copolymers are available from continuous-stirred low conversion processes, where the composition drift does not cause loss of transparency. Copolymers with over 30 wt % acrylonitrile are available and have good barrier properties.

Butadiene copolymers are mainly prepared to yield rubbers (see Elastomers, synthetic, styrene–butadiene rubber). Many commercially significant latex paints are based on styrene-butadiene (weight ratio usually 60:40 with high conversion) copolymers (see Coatings; Paint). Most of the block copolymers prepared by anionic catalysts, eg, butyllithium, also are elastomers. However, some of these block copolymers are thermoplastic rubbers, which behave like cross-linked rubbers at room temperature but show regular thermoplastic flow at elevated temperatures (49–50). Diblock (S–B) and triblock (S–B–S) copolymers are commercially available. Typically, they are blended with polystyrene to achieve a desirable property, eg, improved clarity/flexibility (50) (see Polyblends). These block copolymers represent a class of new and interesting polymeric materials (51–52). Of particular interest are their morphologies (53–56), solution properties (57–58), and mechanical behavior (59–60).

Methyl methacrylate copolymers with styrene are clear materials which, when properly stabilized, are similar in their light stability to poly(methyl methacrylate). About 60 wt % methyl methacrylate is needed in the polymer for markedly improved light stability over that of polystyrene.

Maleic anhydride copolymers with styrene tend to have alternating structures. Accordingly, equimolar copolymers are normally produced, corresponding to 48 wt % maleic anhydride. However, by means of continuous-stirred, low conversion processes, copolymers with low maleic anhydride contents can be produced (61). Depending on their molecular weights, these can be used as chemically reactive resins, eg, epoxy systems, coating resins, etc, for PS–foam nucleation, or as high heat-deformation molding materials (62).

Polymers of Styrene Derivatives. Many styrene derivatives have been synthesized and the corresponding polymers and copolymers prepared (63). Glass-transition temperatures T_g for a series of substituted styrene polymers are shown in Table 4. The highest T_g is that of poly(α-methylstyrene), which can be prepared by anionic

Table 4. Glass-Transition Temperatures of Substituted Polystyrene [a]

Polymer	CAS Registry Number	T_g, °C
polystyrene	[*9003-53-6*]	100
poly(*o*-methylstyrene)	[*25087-21-2*]	136
poly(*m*-methylstyrene)	[*25037-62-1*]	97
poly(*p*-methylstyrene)	[*24936-41-2*]	106
poly(2,4-dimethylstyrene)	[*25990-16-3*]	112
poly(2,5-dimethylstyrene)	[*34031-72-6*]	143
poly(*p*-*tert*-butylstyrene)	[*26009-55-2*]	130
poly(*p*-chlorostyrene)	[*24991-47-7*]	110
poly(α-methylstyrene)	[*25014-31-7*]	170

[a] Ref. 64.

polymerization. Because it has a low ceiling temperature (61°C), depolymerization can occur during fabrication with the formed monomer acting as a plasticizer and lowering the heat distortion to 110–125°C (65). The polymer is difficult to fabricate because of its high melt viscosity and it is more brittle than PS, but it can be toughened with rubber.

Some polymers from styrene derivatives seem to meet specific market demands and to have the potential to become commercially significant materials. For example, monomeric chlorostyrene is useful in glass-reinforced polyester recipes since it polymerizes several times as fast as styrene (63). Poly(sodium styrenesulfonate) [*9003-59-2*] is a versatile water-soluble polymer and is used in water-pollution control and as a general flocculant (66–67) (see Water, industrial water treatment; Flocculating agents). Poly(vinylbenzylammonium chloride) [*70504-37-9*] has been useful as an electroconductive resin (68) (see Polymers, conducting).

In addition to the polymerization of styrene derivatives, PS can be chemically modified to produce more reactive species (69).

Rubber-Modified Polystyrene. Rubber is incorporated into polystyrene primarily to impart toughness. The resulting materials are commonly called high impact polystyrenes (HIPS) and are available in many different varieties. In rubber-modified polystyrenes, the rubber is dispersed in the polystyrene matrix in the form of discrete particles. The mechanism of rubber-particle formation and rubber reinforcement and several general reviews of HIPS and other heterogeneous polymers have been published (34–35,70–74).

The photomicrographs in Figure 4 are representative of the morphology of available HIPS materials (75–76). If the particles are much larger than 5–10 μm, poor surface appearance of moldings, extrusions, and vacuum-formed parts are usually noted. Although most commercial HIPS contains ca 3–10 wt % polybutadiene or styrene–butadiene copolymer rubber, the presence of PS occlusions within the rubber particles gives rise to a 10–40% volume fraction of the reinforcing rubber phase (35,77). Accordingly, a significant portion of the PS matrix is filled with rubber particles. Techniques have been published for evaluating the morphology of HIPS (76,78–79).

For effective toughening of otherwise brittle polystyrene with rubbers, the following generalizations can be made: For good impact strength over a wide temperature range, the glass-transition temperature of the rubber must be below −50°C, as measured, eg, by torsion pendulum at 1 Hz (= cycles/s). The use of butadiene rubbers is particularly effective when the rubber is present during the polymerization of styrene. Grafting of some styrene to rubber takes place, and occlusion of polystyrene extends the volume fraction of the dispersed, reinforcing rubber phase. The rubber phase in the final product is cross-linked to some degree for the most effective reinforcement. Since the rubber phase exists in the form of discrete rubber particles, the degree of cross-linking does not significantly influence the melt flow, which is that of a linear, ie, uncross-linked, thermoplastic polymer. A variation in the degree of cross-linking may be needed to optimize product properties for different applications. Depending on the process and rubber concentration used, there is some latitude in the size of the rubber particles that results in a good balance of properties. This range may extend from <1 to >5 μm.

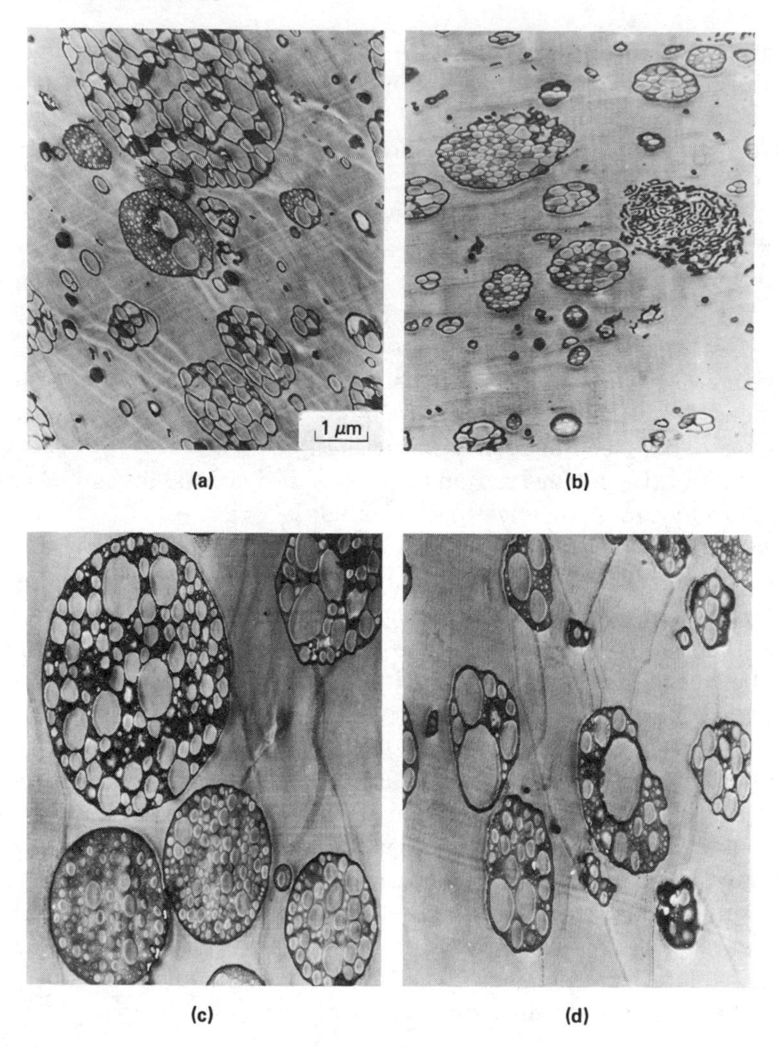

Figure 4. Electron photomicrographs of commercial rubber-modified polystyrene. (**a**) Dow Styron 484; (**b**) Monsanto Lustrex 3350; (**c**) Shell 333; (**d**) Cosden 825. Scale in (**a**) applies to all photomicrographs.

Rubber-Modified Copolymers. *Acrylonitrile–Butadiene–Styrene Polymers.* Acrylonitrile–butadiene–styrene (ABS) polymers have become important commercial products since the mid-1950s. The development and properties of ABS polymers are discussed in detail in ref. 80. ABS polymers, like rubber-modified polystyrene, are two-phase systems in which the elastomer component is dispersed in the rigid styrene–acrylonitrile (SAN) copolymer matrix. The four electron photomicrographs in Figure 5 show examples of the morphology of commercially available ABS polymers. The differences in structure of the dispersed phases are primarily a result of differences in production processes, types of rubber used, and variation in rubber concentrations.

Because of the possible changes in the nature and concentration of the rubber phase, a wide range of ABS polymers is available. Generally, they are rigid (modulus at room temperature, 1.8–2.6 GPa ((2.6–3.8) $\times$ 10^5 psi) and have excellent notched

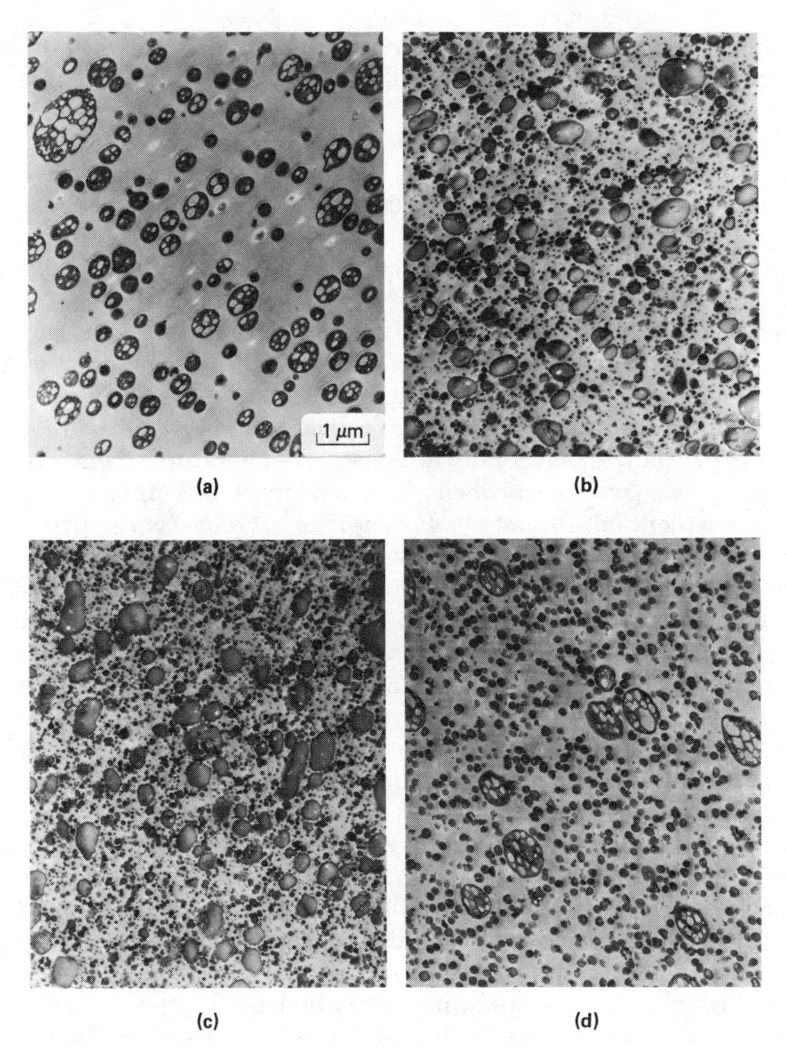

Figure 5. Electron photomicrographs of commercial ABS resins. (**a**) Dow ABS 340; (**b**) Borg-Warner Cycolac T-1000; (**c**) United States Steel Kralastic 606ED, ACFXS53972; (**d**) Monsanto Lustrex I-448. Scale in (**a**) applies to all photomicrographs.

impact strength at room temperature (ca 135–400 J/m ((2.5–7.5) ft·lbf/in.) and at lower temperatures, eg, at −40°C (50–140 J/m ((0.94–2.6) ft·lbf/in.). This combination of stiffness, impact strength, and solvent resistance makes ABS polymers particularly suitable for demanding applications. Another important attribute of several ABS polymers is their minimum tendency to orient or develop mechanical anisotropy during molding (41,81). Accordingly, uniform tough moldings are obtained. In addition, ABS polymers exhibit good ease of fabrication and produce moldings and extrusions with excellent gloss, which can be decorated by many techniques, eg, lacquer painting, vacuum metallizing, and electroplating (82–85). In the case of electroplating, the strength of the molded piece is significantly improved (81). When an appropriate decorative coating or a laminated film is applied, ABS polymers can be used outdoors (86).

Not only are ABS polymers useful engineering plastics, but some of the high rubber compositions are excellent impact modifiers for poly(vinyl chloride) (PVC). Styrene–acrylonitrile–grafted butadiene rubbers have been used as modifiers for PVC since 1957 (87).

New Rubber-Modified Styrene Copolymers. Rubber modification of styrene copolymers other than HIPS and ABS has been useful for specialty purposes. Transparency has been achieved with the use of methyl methacrylate as a comonomer; styrene–methyl methacrylate copolymers have been successfully modified with rubber (88). Improved weatherability is achieved by modifying styrene–acrylonitrile copolymers with saturated, aging-resistant elastomers (89).

Glass-Reinforced Styrene Polymers. Glass reinforcement of PS and SAN markedly improves their mechanical properties. The strength, stiffness, and fracture toughness are generally at least doubled. Creep and relaxation rates are significantly reduced and creep rupture times are increased. The coefficient of thermal expansion is reduced by more than one half, and generally response to temperature changes is minimized (90). Normally, ca 20 wt % glass fibers, eg, 6 mm long, 0.009 mm dia, E glass, can be used to achieve these improvements. Polystyrene, styrene–acrylonitrile copolymer, HIPS, and ABS have been used with glass reinforcement.

Four approaches are currently available for producing glass-reinforced parts: use of preblended, reinforced molding compound; blending of reinforced concentrates with virgin resin; a direct process, in which the glass is cut and weighed automatically and blended with the polymer at the molding machine; and general inplant compounding (81). The choice of any of these processes depends primarily on the size of the operation and the corresponding economics. The use of concentrates, eg, 80 wt % glass, 20 wt % polystyrene, for the subsequent blending to ca 20 wt % glass in the final product has many attractive features and seems to be appropriate for a medium-size operation, whereas the direct process is emphasized in very large-scale operations.

Chemical. ***Degradation.*** Like almost all synthetic polymers, styrene plastics are susceptible to degradation by heat, oxidation, uv radiation, high energy radiation, and shear; although in normal use only uv radiation imposes any real limit on the general usefulness of these plastics (91–92). Thus, it is generally recommended that the use of styrene plastics in outdoor applications be avoided.

Degradation of polystyrene by heat *in vacuo*, ie, generation of volatiles and reduction in molecular weight, is detectable only above 250°C and is significant above 300°C; however, the latter is outside the range of commercial usage. Volatiles generated typically are ca 40 wt % styrene, the remainder being dimer (2,4-diphenyl-1-butene) and trimer (2,4,6-triphenyl-1-hexene) with some higher molecular weight hydrocarbons, benzene, and toluene. The extent of chain scission depends on the source of the polystyrene. A free-radical initiated polymer shows a rapid initial reduction in chain length, which is attributed to backbone irregularities, whereas an anionically prepared polymer is more stable in this respect (93). Poly(α-methylstyrene) unzips to monomer exclusively. Figure 6 is a comparison of the thermal stability of several copolymers. Thermal-oxidative degradation of polystyrene occurs much faster, leading to additional volatile components consisting of aldehydes and ketones, yellowing of the polymer with a very dramatic drop in molecular weight, and some cross-linking. Rates and yields are highly oxygen- and temperature-sensitive. Figure 7 shows the magnitude of oxidative attack on polystyrene and the extent to which an antioxidant can protect the polymer.

Even though polystyrene does not absorb radiation above 300 nm, terrestrial

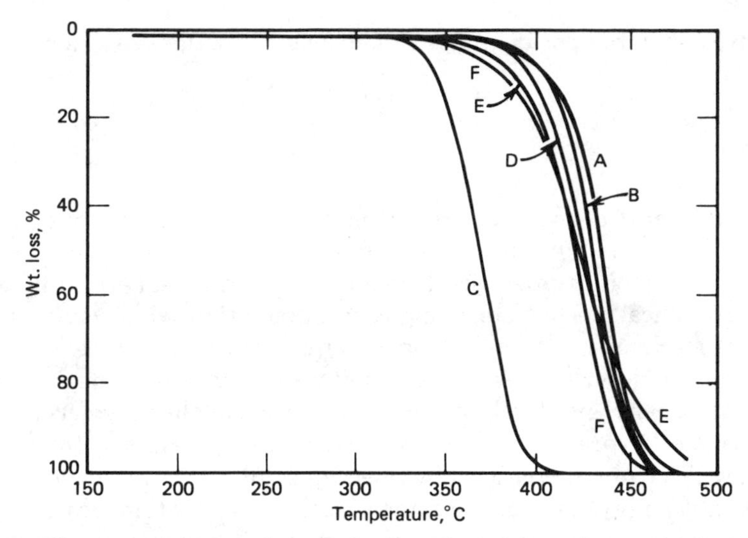

Figure 6. Thermogravimetric analysis of polymers and copolymers of styrene in nitrogen at 10°C/min. A, Polystyrene; B, poly(vinyltoluene); C, poly(α-methylstyrene); D, poly(styrene-*co*-acrylonitrile), 71.5% styrene; E, poly(styrene-*co*-butadiene), 80% styrene; F, poly(styrene-*co*-α-methylstyrene), 75% styrene.

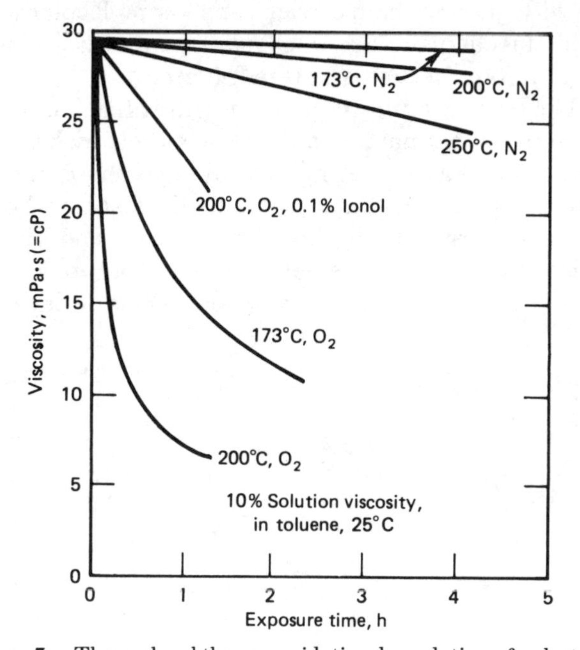

Figure 7. Thermal and thermo-oxidative degradation of polystyrene.

sunlight causes rapid yellowing and embrittlement (94). The chromophore appears to be a π-complex of oxygen (perhaps in the singlet state) with the aromatic ring. Photolysis at 254 nm results in hydrogen generation, cross-linking, embrittlement, and discoloration (95–96). Polybutadiene-modified polymers degrade even more rapidly because of the reactivity of the rubber (97). Ionizing radiation has little or no effect on polystyrene until very large doses are given, in excess of 10^2 Gy (10^6 rad), whereupon some cross-linking occurs (98). Rubber-modified polymers are much more sensitive, since polybutadiene degrades and no longer functions as a reinforcing agent (98).

Polymerization

Styrene and most derivatives are among the few monomers that can be polymerized by all four distinct mechanisms, ie, free-radical, anionic, cationic, and Ziegler mechanisms. These include processes dependent on electromagnetic radiation, which is usually a free-radical mechanism, or high energy radiation, which is either a cationic or free-radical mechanism depending on water content of the system. Free-radical and probably cationic mechanisms yield polymers with a high degree of random placement of the phenyl group relative to the backbone, ie, the polymers are classified as amorphous. Ziegler-prepared polystyrene is a mixture of amorphous and isotactic polystyrene, which can be separated by selective solubility. Anionically made polystyrene is usually amorphous, but in some caases, eg, at low temperatures, isotactic polystyrene has been prepared. Curiously, syndiotactic polystyrene is unknown.

All commercially important styrenic plastics are amorphous, since the vast excess of these polymers is made by a free-radical mechanism. Styrene-butadiene block copolymers are made with anionic chain carriers, and low molecular weight polystyrene is made by a cationic mechanism (99). Analytical standards are available for polystyrene prepared by all four mechanisms (see Initiators).

Free Radical. The styrene family of monomers are almost unique in their ability to undergo spontaneous or thermal polymerization merely by heating to ca 100°C or above. Styrene in essence acts as its own initiator. Most styrene derivatives do the same, with significant rate differences, including some 2,6-disubstituted styrene derivatives and heterocyclic styrenes, eg, vinylpyridine, vinylfuran, and vinylthiophene. The mechanism by which this spontaneous polymerization occurs has been studied extensively (100–101). Styrene undergoes a very slow Diels-Alder dimerization reaction:

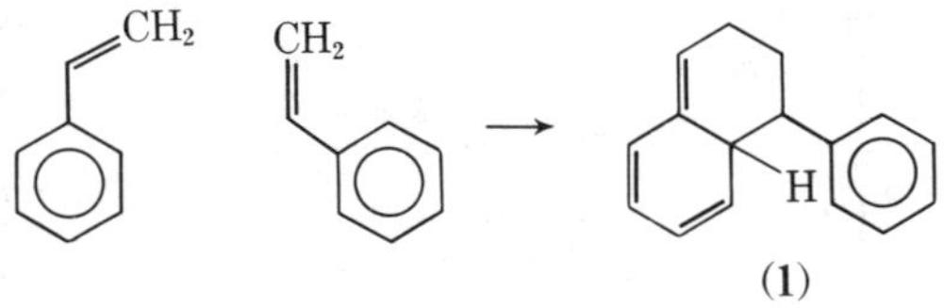

CH_2 CH_2 →

H

(1)

Of the two stereoisomers of (1)

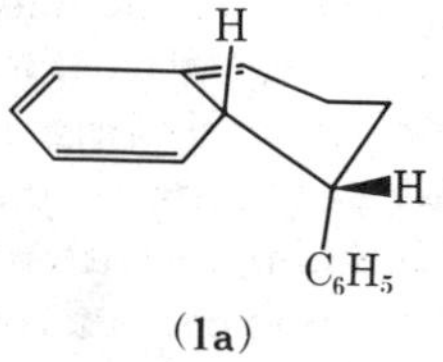

H

H

C_6H_5

(1a)

(axial phenyl)

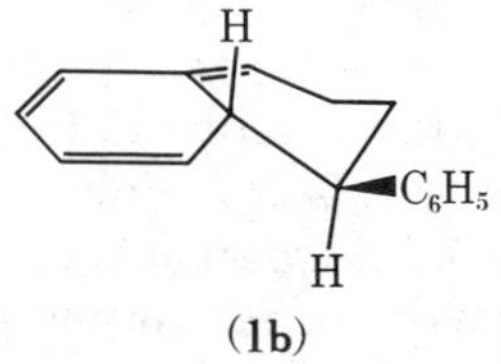

H

C_6H_5

H

(1b)

(equatorial phenyl)

only (**1a**) reacts further in the initiation scheme.

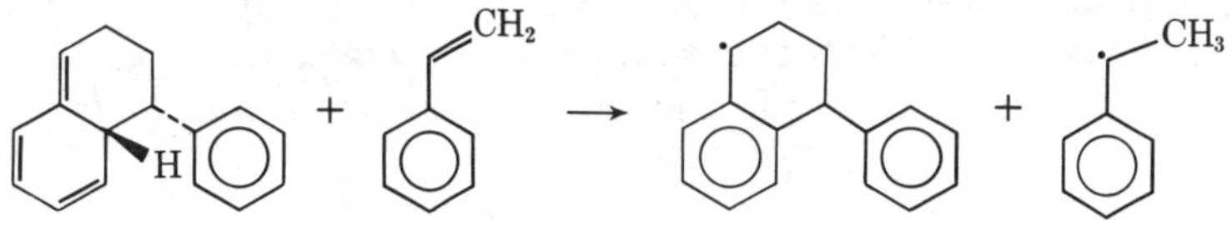

(**1a**)

These two monoradicals then initiate chain growth. Isomer (**1b**) is presumably consumed by reaction with growing chains (101). The Diels-Alder dimer, which has not been isolated or synthesized, contains a hydrogen atom which is both doubly allylic and tertiary and, hence, very labile. In addition to being abstractable by styrene, it is responsible for most of the chain-termination steps by chain transfer (101). The main supportive evidence for this proposal has been summarized (100–101).

Each free radical grows very rapidly by repetitive addition of styrene to the chain in a head-to-tail fashion, until a chain-transfer reaction occurs with the Diels-Alder dimer producing a dead, ie, no longer active, polystyrene chain and a new propagation site. Periodically, a pair of radicals is destroyed in a disproportionation reaction. Attempts have been made to quantify this model with significant success (102–103).

The initial rate of the thermal polymerization of styrene is given by:

$$\text{Initial rate, \%/h} = 3.55 \times 10^{11}\, e^{-(9612/T)}$$

$$T = \text{absolute temperature, K}$$

and the weight-average molecular weight is

$$\overline{M}_w = 6.13 \times 10^6\, e^{-0.0224t}$$

$$t = \text{temperature, °C}$$

with the number average being one half of the weight average. To a good approximation, molecular weight is independent of conversion of styrene to polystyrene for thermal polymerization. These two relationships are useful at 100–180°C, temperatures at which commercial plants operate.

An important consequence of the thermal initiation reaction is the formation of styrene dimers and trimers (100,104). These amount to ca 1 wt % of the polymer and are comprised mainly of:

1-phenyl-4-(1-phenylethyl)tetralin

cis- and *trans*-diphenylcyclobutane

Two other oligomers are always present in polystyrene, namely 2,4-diphenyl-1-butene and 2,4,6-triphenyl-1-hexene, which arise either from a backbiting reaction during chain propagation or from thermal degradation of the polymer (100).

Instead of relying on the spontaneous polymerization reaction, it is often advantageous to induce the polymerization by adding one or more free-radical initiators. A wide range of such materials is available. They differ mainly in the temperature at which each generate free radicals at a useful rate; this often implies half-lives of 1–10 h at the reaction temperature (see Initiators). All function similarly; ie, for peroxidic types:

$$\left(C_6H_5\overset{\overset{\displaystyle O}{\|}}{C}O\right)_2 \longrightarrow 2\ C_6H_5\overset{\overset{\displaystyle O}{\|}}{C}O\cdot$$

benzoyl peroxide

and for azo types:

$$(CH_3)_3CN{=}NC(CH_3)_3 \rightarrow 2\ (CH_3)_3C\cdot + N_2$$

The substituents around the —O—O— or —N=N— groups govern the half-life temperature.

Generally, initiation of the chains is:

$$\text{Initiation, } I \rightarrow 2\ R_x^{\cdot}$$

$$R_x^{\cdot} + S \rightarrow R_1^{\cdot}$$

where S represents styrene. Then, the radical propagates:

$$R_n^{\cdot} + S \xrightarrow[kp]{} R_{n+1}^{\cdot}$$

and termination can occur by chain transfer, which maintains the radical concentration:

$$R_n^{\cdot} + XH \rightarrow PS_n + X\cdot$$

(where XH is a chain-transfer agent and X· is presumed to reinitiate a chain),

$$X\cdot + S \rightarrow R_1^{\cdot}$$

or termination occurs, which destroys a pair of radicals either by disproportionation

$$R_n^{\cdot} + R_m^{\cdot} \rightarrow P_nS + P_mS$$

or by combination

$$R_n^{\cdot} + R_m^{\cdot} \rightarrow P_{n+m}S$$

Below 80°C, combination of radicals is the primary termination mechanism (105). Above 80°C, both disproportionation and chain transfer with the Diels-Alder dimer are increasingly important. The gel or Tromsdorff effects, as manifested by a period of accelerating rate concomitant with increasing molecular weight, is apparent at below 80°C in styrene polymerization; although subtle changes during the polymerization at higher temperature may be attributed to variation of the specific rate constants with viscosity (105–106).

Based on the above general polymerization scheme, it is evident that the rate of polymerization R_p is

$$R_p = k_p[S][R^{\cdot}]$$

where

$$[\mathrm{R}\cdot] = \Sigma\,[\mathrm{R}_n\cdot]$$

Thus, the time to grow one chain is

$$t = \left(\frac{\overline{M}_n}{104\ N}\right) \Big/ \left(k_p \mathrm{S} \frac{1}{N}\right)$$

where $\overline{M}_n$ and k_p are the number-average molecular weight and propagation rate constant at a given temperature, respectively, and N is Avogadro's Number. The number 104 is the molecular weight of styrene and makes possible conversion of polymer molecular weight into degree of polymerization. At 60°C, $\overline{M}_n$ = ca 2.5×10^6, k_p = 176 L/(mol·s), and [S] = 8.4 mol/L. Therefore, the time to grow one chain is 16 s. At 140°C, the time is ca 0.1 s. By comparison, typical reaction times are of the order of hours, meaning that to achieve a desirable molecular weight, the polymerization is run free-radical-starved. Further, once a polymer radical is terminated, it does not become reactivated.

Chain-transfer agents are occasionally added to styrene to reduce the molecular weight of the polymer, although for many applications this is unnecessary. Polymerization temperature alone is sufficient to achieve molecular weight control. Diluents, ie, materials with little or no chain-transfer activity, are sometimes used to reduce engineering problems. Some typical chain-transfer agents for styrene polymerization are listed in Table 5. The chain-transfer agents of commercial significance are α-methylstyrene dimer, terpinolene, dodecane-1-thiol, and 1,1-dimethyldecane-1-thiol. Chain transfer to styrene monomer has been reported, but recent work strongly suggests that this reaction is negligible and transfer with the Diels-Alder dimer is the actual inherent transfer reaction (101,107). Chain transfer to polystyrene has received much attention with little or no indication that such a reaction occurs (108). If it did occur at the benzylic site, then the polymer would be branched.

In some cases, inhibition of polymerization can be regarded as a special type of chain transfer. This is of importance in commercial-scale operations involving styrene storage for extended periods. The majority of inhibitors are of the phenolic/quinone family. All of these species function as inhibitors only in the presence of oxygen, implying either

$$\mathrm{R}\cdot + \mathrm{O}_2 \xrightarrow{\text{fast}} \mathrm{ROO}\cdot$$

Table 5. Chain-Transfer Constants *C* in Styrene Polymerization

Compound	*C*	*T*, °C
benzene	2×10^{-5}	100
toluene	5×10^{-5}	100
isopropylbenzene	2×10^{-4}	100
t-butylbenzene	6×10^{-5}	100
n-butyl chloride	4×10^{-5}	100
n-butyl bromide	4×10^{-5}	100
n-butyl iodide	6×10^{-4}	100
α-methylstyrene dimer	0.3	120
1-dodecanethiol	13	130
1,1-dimethyl-1-decanethiol	1.1	120
1-hexanethiol	15	100

then

$$ROO\cdot + \text{inhibitor} \rightarrow \text{stable species}$$

or

$$\text{inhibitor} + O_2 \rightarrow \text{activated state}$$

then

$$\text{active inhibitor} + R\cdot \rightarrow \text{stable species}$$

In either case, the free radical formed is stable and unable to reinitiate a chain. 4-*tert*-Butylcatechol (TBC) at 12–50 ppm is the almost universally used inhibitor for protecting styrene. At ambient conditions and with a continuous supply of air, TBC has a half-life of 6–10 wk (109). The requirement of oxygen causes complex side reactions, resulting in significant oxidation of the monomer, which causes yellow coloration especially in the vapor phase. An inert gas blanket reduces this problem and flammability hazards, but precautions must be taken to ensure an adequate level of dissolved oxygen in the liquid phase. Another family of inhibitors is that characterized by the N–O bond; these do not seem to require oxygen for effectiveness. They include a variety of nitrophenol compounds, hydroxylamine derivatives, and nitrogen oxides (110–111). Nitric oxide is particularly useful. Presumably, their function depends on the unpaired electron associated with the N—O bond.

Other miscellaneous compounds that have been used as inhibitors are sulfur and some sulfur compounds, picryl hydrazyl derivatives, carbon black, and some soluble transition-metal salts (112). Both inhibition and acceleration have been reported for styrene polymerized in the presence of oxygen. The complexity of this system has been clearly demonstrated (113). The key reaction is the alternating copolymerization of styrene with oxygen to produce a polyperoxide, which above 100°C decomposes to initiating alkoxy radicals. Therefore, depending on the temperature, oxygen can inhibit or accelerate the rate of polymerization.

Ionic. Instead of a neutral unpaired electron, styrene polymerization can proceed with great facility through a positively charged species (cationic polymerization) or a negatively charged species (anionic polymerization). The polymerization reaction is more sensitive to impurities than the free-radical system, and pretreatment of the monomer is usually required. Two general mechanisms are known for anionic initiation: (a) direct attack of a base B: on the monomer and (b) electron transfer from an active donor molecule:

$$\text{(a)}\quad B{:} + CH_2{=}\underset{\displaystyle C_6H_5}{\underset{|}{CH}} \longrightarrow BCH_2{-}\underset{\displaystyle C_6H_5}{\underset{|}{\bar{C}H}}$$

$$\text{(b)}\quad e + CH_2{=}\underset{\displaystyle C_6H_5}{\underset{|}{CH}} \longrightarrow \cdot CH_2{-}\underset{\displaystyle C_6H_5}{\underset{|}{\bar{C}H}}$$

n-Butyllithium is a widely used initiator, acting as in case (a) above. In solution, it exists as six-membered aggregates, and a key step in the initiation sequence is dissociation yielding at least one isolated molecule. Dispersions of alkali metals initiate anionic polymerization by an electron-transfer process. In some cases, free-radical polymerization occurs with anionic polymerization (114–115). Since metal dispersions, being heterogeneous initiator systems, have only a portion of the atoms available at any instant and the surface is highly susceptible to contamination, anionic polymerization initiation is less efficient than with the use of soluble initiators. Stable alkali-metal complexes, eg, sodium naphthalene, which form a radical anion and complexes of alkali metals in liquid ammonia, are much more efficient initiators than the alkali metals (116).

$$\text{(naphthalene radical anion)}^{-}\ \ Na^{+}$$

If the initiation reaction is very much faster than the propagation reaction, then all chains start to grow at the same time, and since there is no inherent termination step, the statistical distribution of chain lengths is very narrow. The average molecular weight is calculated from the mole ratio of monomer-to-initiator sites. Chain (∿) termination is accomplished by adding proton donors, eg, water or alcohols:

$$\sim^{-} + HOR \rightarrow \sim H + RO^{-}$$

or addition with, for example, carbon dioxide:

$$\sim^{-} + CO_2 \rightarrow \sim COO^{-}$$

The term living polymers was coined for the stable, unterminated macroanions (116). Addition of a second monomer to polystyryl anion results in the formation of a block polymer with no detectable free polystyrene. This technique is of considerable importance in the commercial preparation of styrene–butadiene block copolymers, which are used either alone or blended with polystyrene as thermoplastics. Unlike free-radical polymerization, the nature of the solvent plays a crucial role in anionic polymerization.

Cationic polymerization of styrene can be initiated either by strong acids, eg, perchloric acid, or by Friedel-Crafts reagents with a proton-donating activator, eg, boron trifluoride or aluminum trichloride with a trace of a protonic acid or water (117). The solvent again plays an important role, and chain-transfer reactions are very common where the reactants are polymer, monomer, solvent, and counterion. As a result, high molecular weights are more difficult to achieve and molecular weight distributions are often comparable to those obtained from free-radical polymerizations. Commercial use of cationic styrene polymerization is reported only where low molecular weight polymers are desired (99).

Stereospecific polymerization of styrene to produce crystalline, isotactic polystyrene can be achieved either by low temperature anionic polymerization in a poor solvent or by a Ziegler-type initiation system of triethylaluminum and titanium tetrachloride (118–122). The resulting polymer is usually contaminated with atactic (amorphous) polystyrene, which can easily be separated from the isotactic material by extraction with 2-butanone. No commercial application of isotactic polystyrene is known; its high melting point (240°C) makes fabrication difficult, and if crystallization is allowed to occur opacity results.

Copolymerization

Copolymerization makes possible dramatic improvements in one or more properties. There is a vast amount of literature on styrene-containing copolymers (123). The reactivity ratios for the monomer pairs are listed in Table 6. Styrene–acrylonitrile (SAN) copolymers are large-volume thermoplastics with improved mechanical properties and heat and solvent resistance with some loss of color stability (see Acrylonitrile polymers). The disparity of reactivity ratios shown in Table 6 implies an unequal insertion rate of the monomers into the copolymer; Figure 8 illustrates manifestations of this effect. Most SAN copolymers are manufactured at or near the crossover point (A in Fig. 8) to avoid significant drift in composition (see Fig. 9), which results in haze (124–125). Reactors have been designed to overcome composition drift problems on a commercial scale (126). All SAN copolymers produced in the United States are made in a mass or solution continuous process without addition of an initiator. The greater tendency of these copolymers to develop yellow coloration requires more attention to techniques of polymerization and devolatilization (127–130).

Copolymers with butadiene, ie, those containing at least 60 wt % butadiene, are an important family of rubbers. In addition to synthetic rubber, these compositions have extensive uses as paper coatings, water-based paints, and carpet backing. Because of unfavorable reaction kinetics in a mass system, these copolymers are made in an emulsion polymerization system, which favors chain propagation and disfavors termination (131). The result is economically acceptable rates with desirable chain

Table 6. Free-Radical Copolymerization Reactivity Ratios; Styrene = Monomer 1

Monomer 2	r_1	r_2	T, °C
acrylonitrile	0.4	0.04	60
butadiene	0.5	1.40	50
methyl methacrylate	0.54	0.49	60
m-divinylbenzene	0.65	0.60	60

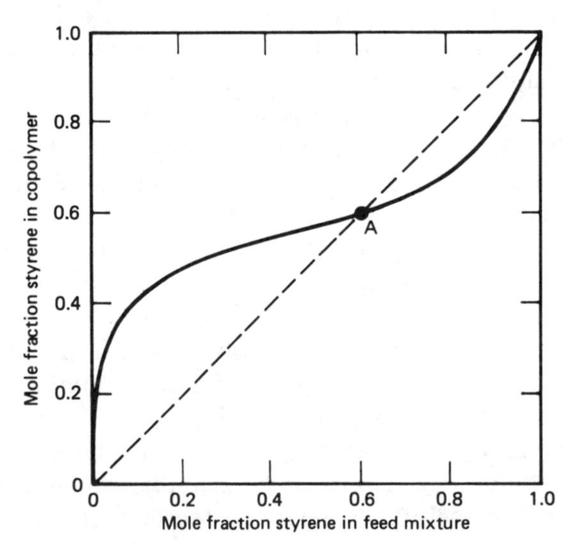

Figure 8. Relationship between feed composition and copolymer composition of styrene–acrylonitrile copolymerization.

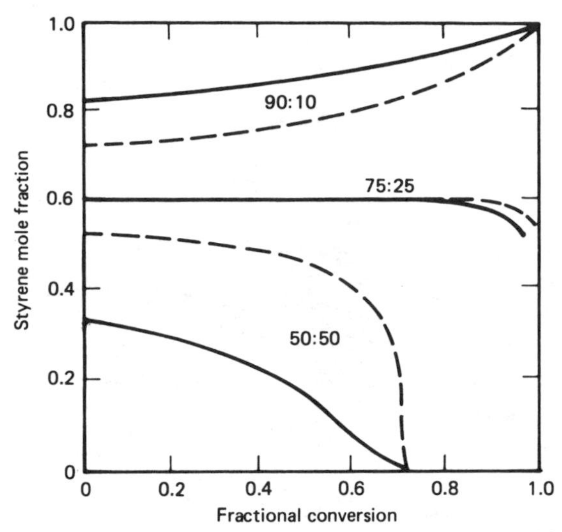

Figure 9. Drift in monomer composition (—) and copolymer composition (- - -) with conversion for three initial monomer mixtures; ratios are based on wt %.

lengths. Usually such processes are run batchwise in order to achieve satisfactory particle-size distribution.

Methyl methacrylate–styrene copolymers are used where improved resistance to outdoor weathering and light stability are required; ca 60 wt % methyl methacrylate is needed. Processes similar to those used for SAN copolymers are used, with care being taken to avoid excessive degradation which results from the ease of unzipping of methyl methacrylate polymers (132).

Divinylbenzene copolymers with styrene are produced extensively as supports for the active sites of ion-exchange resins and in biochemical synthesis. About 1–10 wt % divinylbenzene is used, depending on the required rigidity of the cross-linked gel, and the polymerization is carried out as a suspension of the monomer-phase droplets in water, usually as a batch process. Several studies have been reported recently on the reaction kinetics (133–134).

Polymerization of styrene in the presence of polybutadiene rubber yields a much tougher material than polystyrene. Although the material is usually opaque and has a somewhat lower modulus, the gain in impact resistance has placed this plastic in very wide commercial usage. The polymerization is complex because the rubber has sufficient reactivity, ie, at the allylic hydrogen and 1:2 addition units, to graft and cross-link and the polymerization system is comprised of two phases. The graft copolymer of polystyrene on polybutadiene acts as an emulsifier, stabilizing the dispersion against coalescence. Events taking place during this polymerization can be followed with the ternary phase diagram shown in Figure 10. For a feed mixture made up of, say, 8 wt % rubber in styrene (point A) on an increment of conversion to polystyrene (to D), phase separation occurs. Small droplets of styrene/polystyrene, which are stabilized by the graft copolymer, are dispersed in the styrene/polybutadiene solution. The composition of the two phases is given by points B and C for the polystyrene-rich phase and polybutadiene-rich phase, respectively. The phase-volume ratio, ie, rubber phase/PS phase, is given by the ratio DB/DC. As the reaction proceeds along the line AE, the phase composition and phase-volume ratio can be read from the tie lines. Larger droplets of PS solution form and the small, original ones remain. When the

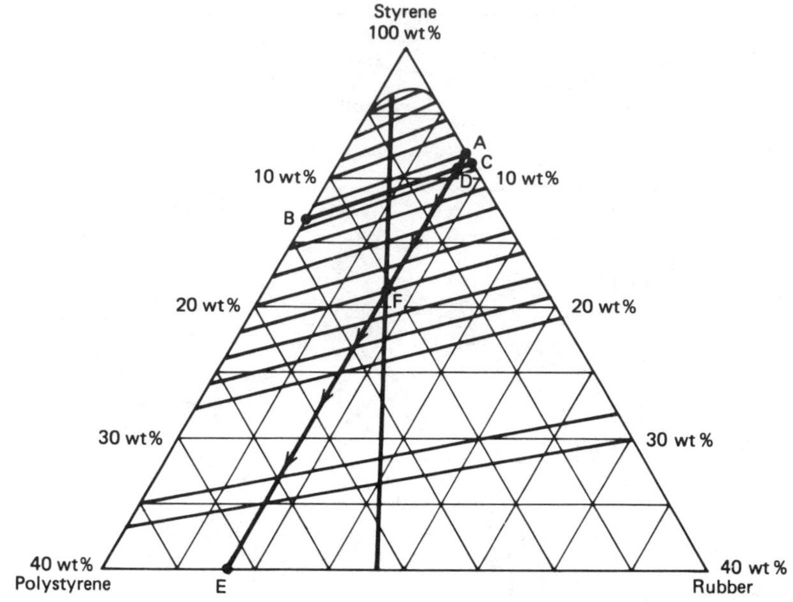

Figure 10. Ternary phase diagram for the system styrene–polystyrene–polybutadiene rubber.

phase-volume ratio is about unity (point F), providing there is sufficient shearing agitation, the larger polystyrene-containing droplets, which have mostly coalesced into large pools, become the continuous phase and polybutadiene-containing droplets are dispersed therein (135–136). However, the latter contain the small, original polystyrene particles as occlusions. The particle size and size distribution are largely controlled by the applied shear rate during and after phase inversion, the viscosity of the continuous phase, the viscosity ratio of the two phases, and the interfacial tension between the phases (55, 135). The viscosity parameters depend on phase compositions, polymer molecular weights, and temperatures; the interfacial tension is largely controlled by the amount and structure of the graft copolymer available at the particle surface. Phase-contrast micrographs of this phase-inversion sequence, wherein the polystyrene phase is dark, are shown in Figure 11. If shearing agitation is insufficient, then phase inversion does not occur and the product obtained is an internetwork of cross-linked rubber with polystyrene and its properties are much inferior. The morphology of the dispersed rubber phase remains much as formed after phase inversion, unless there is a step change in phase concentration by, eg, blending streams. Rubber-phase volume, including the occlusions, largely controls performance; for example, Table 7 illustrates Izod impact strength dependence (137).

Once the desired particle size and morphology with sufficient graft to provide adequate adhesion between the rubber particle and the polystyrene phase have been achieved, the integrity of the particle must be protected against deformation or destruction during fabrication by cross-linking of the polybutadiene chains. These chemical changes taking place on the rubber begin with polymerization. Either anionic or Ziegler-polymerized polybutadiene is used at 5–10 wt % styrene. If peroxide initiators are used, about one half of the rubber is grafted by the time phase inversion occurs; less is grafted in the absence of peroxides (138).

1,2 vinyl adduct   trans 1,4   cis 1,4

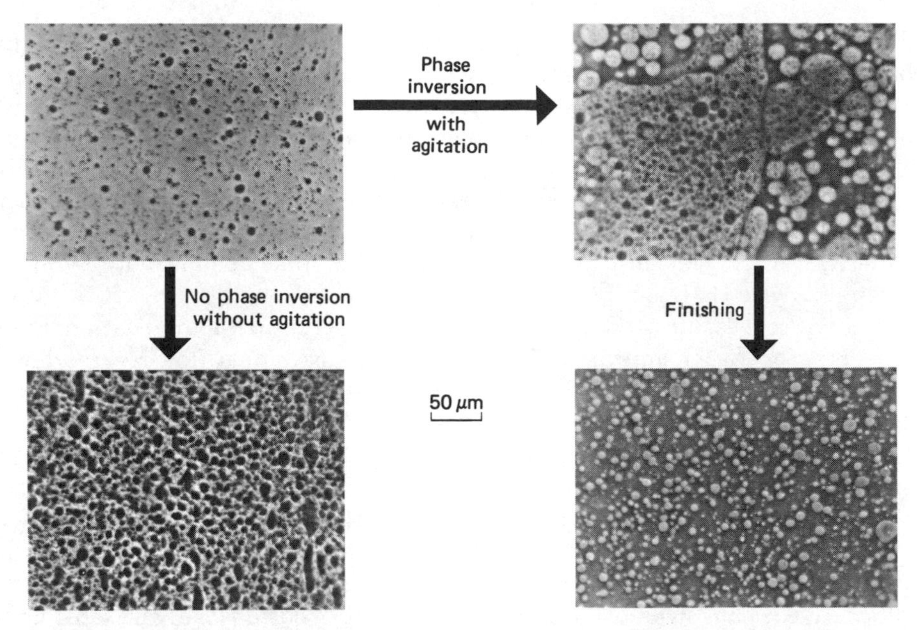

Figure 11. Phase-contrast photomicrographs showing particle formation via phase inversion.

Table 7. Dependence of Izod Impact Strength on Rubber Particle Size

Rubber particle weight average dia, μm	Izod impact strength[a], J/m[b]
0.6	48
0.8	53
1.0	64
1.5	75
2.0	91
2.5	93
3.0	99
3.5	100

[a] ASTM D 256-56.
[b] To convert J/m to ft·lbf/in., divide by 53.38.

Hydrogen abstraction by alkoxy radicals at any of the allylic sites yields a site for styrene polymerization, resulting in a graft (138–139). Model studies based on styrene-butadiene block copolymers show a strong influence of graft structure on particle morphology (55–56). The double bonds in polybutadiene are much less reactive than those in styrene but as conversion exceeds ca 80% and temperature approaches or exceeds 200°C, copolymerization through the polybutadiene (1,2 units) chain seems to occur and leads to cross-linking and, hence, gel formation (140).

Commercial Processes

There are two problems in the manufacture of polystyrene: removal of the heat of polymerization (ca 700 kJ/kg (300 Btu/lb)) of styrene polymerized and simultaneously handling a partially converted polymer syrup with a viscosity of ca 10^8 mPa·s

(= cP). The latter problem strongly aggravates the former. A wide variety of solutions to these problems has been reported. For the four mechanisms described earlier, ie, free radical, anionic, cationic, and Ziegler, several processes can be used. Table 8 summarizes the processes which have been used to implement each mechanism for liquid-phase systems. Free-radical polymerization of styrenic systems primarily in solution is of principal commercial interest. Details of suspension processes, which are declining in importance, are given in refs. 141 and 142. Emulsion processes are described in ref. 143. The historical development of styrene polymerization processes has been reviewed (141,144–145).

Two types of reactors are used for continuous solution polymerization: the linear-flow reactor (LFR), approximating in the ideal case a plug-flow reactor, and the continuous-stirred tank reactor (CSTR) which ideally is isotropic in composition and temperature (see Reactor technology). The linear-flow reactors usually involve conductive heat transfer to many tubes through which a heat-transfer fluid flows. Multiple temperature zones can easily be achieved in a single reactor; agitation is provided by a rotating shaft with arms down the central axis of the tubular vessel. Reactors of this type operate for long periods and handle very high viscosity partial polymers with reliable heat-transfer coefficients. Continuous-stirred tank reactors often make use of evaporative cooling in addition to conduction to the cooled jacket and sensible heat to achieve very good temperature control. The limitation of the reactor occurs at high viscosity, when the rate of vapor disengagement is exceeded by the rate of vapor generation and the ability to mix in the condensed vapor and fresh feed becomes too slow. Typically, this limit occurs at 60–70 wt % solids. Figures 12 and 13 illustrate typical processes based on these two types of reactors. Most general-purpose, ie, unmodified, polystyrene is manufactured in the United States in such facilities. The LFR process (Fig. 12) typically uses up to 20 wt % ethylbenzene mainly to reduce viscosity. The three reactors may each be subdivided into three temperature-control zones, which allows for flexible control of the molecular weight distribution of the polymer as well as optimum output per reactor volume. Temperatures are 100–180°C. At ca 80 wt % solids, the partial polymer is heated from 180°C to 240°C before being devolatilized upon entering a vacuum tank. The recovered solvent and monomer are continuously recycled to the feed. The molten polymer leaves the vacuum tank and is stranded, cooled, and chopped into pellets for shipping. The CSTR process (Fig. 13) differs in that it is typically a single reactor and, therefore, involves a single temperature zone, although more can be added, and operates at ca 60 wt % solids, requiring a large polymer heater to achieve adequate devolatilization (146–147). Less solvent is required.

Table 8. Process vs Mechanism for Styrene Polymerization

Process	Mechanism			
	Free radical	Anionic	Cationic	Zeigler
mass/bulk	used for all styrene plastics	yes	yes	
solution	used for all styrene plastics	styrene–butadiene block copolymers	yes	yes
suspension	used for all styrene plastics			
precipitation	yes			
emulsion	used for styrene–butadiene latices and ABS plastics			

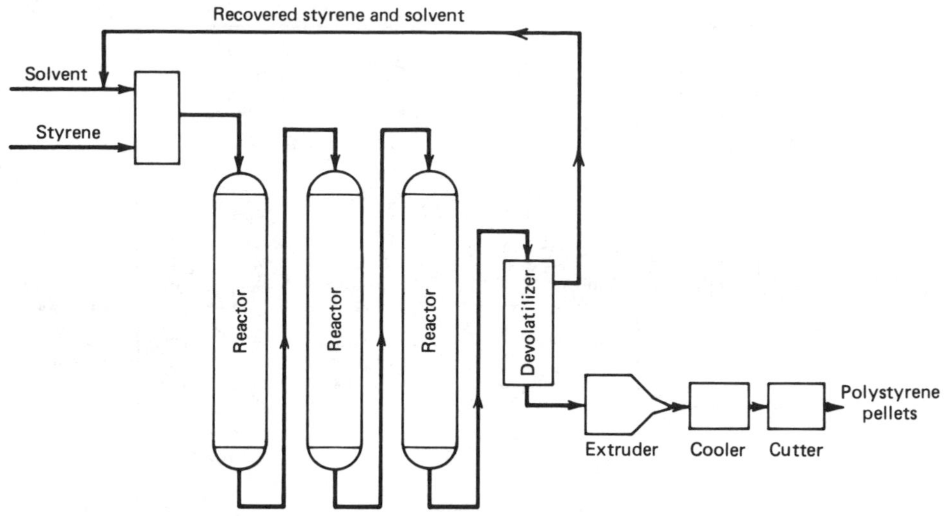

Figure 12. Typical LFR polymerization process.

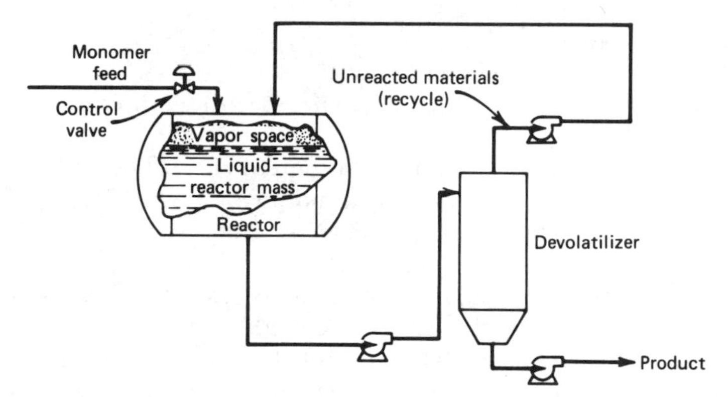

Figure 13. Typical CSTR polymerization process.

In both cases, removal of high boiling oligomers, formed mainly from thermal initiation from the recycle stream, is required; typically, a simple, on-stream distillation process is adequate (148). Rubber-modified polystyrene manufacture places several additional demands: dissolving ca 5–10 wt % polybutadiene in styrene, often a 10–20 h process; shear conditions to achieve phase inversion and the desired particle size for a given product; control of phase compositions to produce the desired particle morphology; and sufficient time and temperature at the end of the process to achieve the necessary cross-linking, gel formation, and swelling index. Graft copolymer formation occurs throughout the polymerization, and sufficient amount of graft copolymer for phase stability is necessary at phase inversion. In the case of a LFR system (Fig. 12), only addition of a rubber dissolver is needed to meet the above requirements. For a CSTR-based system, multiple reactors in series are required to avoid rapid shifts in phase compositions on mixing of one reactor effluent with the next reaction contents with loss of desirable morphology (149–150). Figure 14 shows a multizone CSTR reactor system. Phase inversion and particle sizing is accomplished in the first-stage reactor with the effluent entering a single-tank, baffled reactor operating as five stages with a common vapor space. An alternative is three CSTR reactors in series, followed

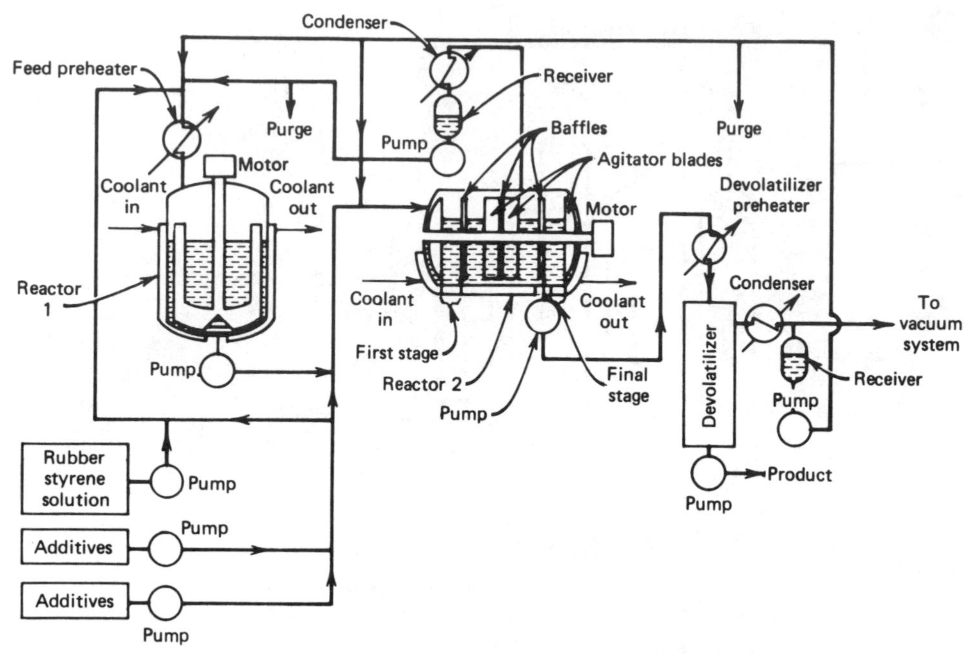

Figure 14. Typical staged CSTR process for rubber-modified polystyrene.

by an adiabatic reactor to achieve high solids content (149). The multiple stages are necessary to achieve favorable rubber-particle morphology with CSTRs; LFR systems inherently avoid sudden phase-composition changes, unless severe internal backmixing is allowed to occur.

Generally, phase inversion and particle sizing occurs in the first reactor. Grafting rate is highest initially, and the amount formed is adequate to stabilize the particles. The shear field in the reactor is critical to control of particle size as well as to meet the minimum shear rate for phase inversion, without which the product consists of a continuous cross-linked rubber network with markedly inferior properties (135–136). After the particles are formed and sized, only gentle agitation is applied to aid heat transfer without disturbing the particle structure. Conditions that produce the desired polystyrene-phase molecular weight are used. Prior to devolatilization, the partial polymer is heated to ca 200°C to cross-link the rubber chain in order to protect the rubber particles from the very high shear fields during fabrication. Stranding, cooling, and chopping into pellets follows.

Although the process is declining in importance, a few manufacturers of rubber-modified polystyrene use a mass-suspension method (151). This is usually a batch process in which phase inversion and particle sizing are accomplished in a mass polymerization (CSTR) reactor before being dispersed in water with the appropriate surfactants, peroxides, etc to finish the polymerization. Although suspension polymerization is an inexpensive way of avoiding the high viscosity, heat-transfer, and devolatilization problems associated with mass processes, contaminated water, extra materials, and loss of the economic advantages of continuous operation weigh against use of mass-suspension processes where the highest efficiency is required. Acrylonitrile–butadiene–styrene plastics are sometimes manufactured by a similar process, although the favored method involves preforming submicrometer, cross-linked polybutadiene particles in an emulsion process, followed by grafting with styrene–acrylonitrile to the desired rubber content or blending in preformed SAN copolymer. This technique achieves the smaller particles, and low molecular weight, rigid phase that is desirable for optimum gloss and impact resistance in many ABS applications.

Polybutadiene rubbers are used almost exclusively as reinforcing agents for styrenic plastics. The very low glass-transition temperature (−80°C or lower) gives good low temperature properties, and the allylic hydrogen atom and weakly active double bonds can provide the desired degree of grafting and cross-linking. Other rubbers have been used: acrylates, ethylene-propylene-diene rubbers, polyisoprene, and polyethylene have been reported, but with limited commercial success because of their low chemical reactivity, not withstanding their acceptable glass-transition temperature (152).

Fabrication. ***Injection Molding.*** There are two basic types of injection-molding machines in use: The reciprocating screw and the screw preplasticator (153). Their simple design, uniform melt temperature, and excellent mixing characteristics make them the preferred choice for injection molding. Machines with shot capacities up to 25 kg for solid injection-molded parts and 65 kg for structural foam parts are available (154–155). Large solid moldings include automotive dash panels, television cabinets, and furniture components. One-piece structural foam parts weighing 35 kg or more are molded for increased rigidity, strength, and part weight reduction (156) (see Plastics processing).

The injection-molding process is basically the forcing of melted polymer into a relatively cool mold where it freezes and is removed in a minimum time. The shape of a molding is defined by the cavity of the mold. Quick entry of the material into the mold followed by quick setup results in a significant amount of orientation in the molded part. The polymer molecules and, in the case of heterogeneous rubber-modified polymers, the rubber particles tend to be highly oriented at the surface of the molding. Orientation at the center of the molding tends to be significantly less because of the relaxation of the molten polymer.

The anisotropy that develops during the molding operation is detrimental to the performance of the fabricated part in several ways. First, highly oriented moldings, which form particularly if low melt temperatures are used, exhibit good gloss, have an abnormally narrow use temperature range causing early warping and, perhaps most importantly, tend to be brittle even though the material is inherently capable of producing tough parts (157). However, this development of polymer orientation during molding can also be used to advantage, as in the case of rotational orientation during the molding operation (158).

The achievement of isotropic moldings is also important when the molded part is to be decorated, ie, painted, metallized, etc. Highly oriented parts that have a high frozen-in internal stress memory tend to give rise to rough or distorted surfaces as a result of the relaxing effect of the solvents and/or heat. For electroplating ABS moldings, it is particularly important to obtain isotropic moldings. Isotropy can be achieved by the use of a high melt temperature, a slow fill speed, a low injection pressure, and a high mold temperature (84–85,159–160).

Injection molding of styrene-based plastics is usually carried out at 200–300°C. For ABS polymers, the upper limit may be somewhat less, since these polymers tend to yellow somewhat if too high a temperature and/or too long a residence time are imposed.

To obtain satisfactory moldings with good surface appearance, contamination, including that by moisture, must be avoided. For good molding practice, particularly with the more polar styrene copolymers, drying must be part of the molding operation. A maximum of 0.1 wt % moisture can be tolerated before surface imperfections appear.

For achieving appropriate economics, injection-molding operations are highly automated and require few operating personnel (153). Loading of the hopper is usually done by an air-conveying system; the pieces are automatically ejected, and the rejects and sprues are ground and reused with the virgin polymer. Also, hot probes or manifold dies are used to eliminate sprues and runners.

Extrusion. Extrusion of styrene polymers is one of the most convenient and least expensive fabrication methods, particularly for obtaining sheet, pipe, irregular profiles, and films. Relatively small extruders, eg, 11.5 cm diameter, can produce well over 675 kg/h of polymer sheet (161). Extrusion is also the method for plasticizing the polymer in screw injection-molding machines and is used to develop the parison for blow molding. Extrusion of plastics is also one of the most economical methods of fabrication, since it is a continuous method involving relatively inexpensive equipment. The extrusion process has been studied in great detail (162–163). Single-screw extruders work extremely well with styrene-based plastics. Machines are available with L/D (length-to-diameter) ratios of 36:1 or more. Some of the longer L/D extruders are used with as many as three vent zones for removal of volatiles, often eliminating the necessity for predrying as is practiced with hygroscopic materials, eg, SAN and ABS. Where venting is inadequate, these polymers must be predried to a maximum moisture content of 0.03–0.05 wt % to obtain high quality sheet (see Film and sheeting materials).

Many rubber-modified styrene plastics are fabricated into sheet by extrusion primarily for subsequent thermoforming operations. Much consideration has been given to the problem of achieving good surface quality in extruded sheet (164–165). Excellent surface gloss and sheet uniformity can be obtained with styrene-based polymers.

Considerable work has been done on mathematic models of the extrusion process, with particular emphasis on screw design. Good results are claimed for extrusion of styrene-based resins using these mathematical methods (163,166).

With the advent of low cost computers, closed-loop control of the extrusion system has become commonplace. More uniform gauge control at higher output rates is achievable with many commercial systems (167–168).

Lamination of polymer films, both styrene-based and other polymer types, to styrene-based materials can be carried out during the extrusion process for protection or decorative purposes. For example, an acrylic film can be laminated to ABS sheet during extrusion for protection in outdoor applications. Multiple extrusion of styrene-based plastics with one or more other plastics has grown rapidly in the last 10 yr. Some sheet coextrusion products based on polystyrene and available in the United States are listed in Table 9.

Thermoforming and Orientation. Thermoforming of HIPS and ABS extruded sheet is of considerable importance in several industries. In the refrigeration industry, large parts are obtained by vacuum-forming extruded sheet. Vacuum forming of HIPS sheet for refrigerator-door liners was one of the most significant early developments promoting the rapid growth of the whole family of rubber-modified polystyrene. When

Table 9. Sheet Coextrusion Products Based on Polystyrene

Polymers[a]	Products	Remarks
GPPS–HIPS	dairy containers, dinnerware, display, signs, etc	attractive high gloss appearance; largest volume coextruded plastic in the United States
HIPS color 1–scrap–HIPS color 2	cold drink cups, packaging	appearance
PE–AD–HIPS	rigid containers	combines improved environmental stress-crack resistance, moisture barrier, and ease of heat sealing of PE with the ease of thermoforming of HIPS
HIPS–AD–Saran–AD–HIPS	rigid containers for long shelf life at room temperature	high barrier properties to oxygen and moisture permeation is gained as a result of the high barrier properties of Saran resins; degree of barrier is dependent on Saran thickness
HIPS–scrap–AD–Saran–AD–scrap–HIPS	rigid containers	valuable when scrap generation is appreciable

[a] GPPS = general-purpose polystyrene; HIPS = high impact polystyrene; PE = polyethylene; AD = adhesion layer, eg, ethylene–vinyl acetate copolymer, etc; Saran = vinyl–vinylidene chloride copolymer, trademark of Dow Chemical; scrap = scrap material which is added between container layers in order to add strength to the container; colors 1 and 2 refer to two different colored polystyrene layers, eg, a blue outer layer for a cup and a white inner layer.

a thermoplastic polymer film or sheet is heated above its glass-transition temperature, it can be formed or stretched. Under controlled conditions, new shapes can be controlled; also, various amounts of orientation can be imparted to the polymer film or sheet for altering its mechanical behavior.

Thermoforming is usually accomplished by heating a plastic sheet above its softening point and forcing it against a mold by applying vacuum, air, or mechanical pressure. On cooling, the contour of the mold is reproduced in detail. In order to obtain the best reproduction of the mold surface, carefully determined conditions for the plastic-sheet temperature, ie, heating time and mold temperature, must be maintained.

Several modifications of thermoforming plastic sheet have been developed. In addition to straight vacuum forming, these are vacuum snapback forming, drape forming, and plug-assist-pressure-and-vacuum forming. Some combinations of these techniques are also practiced. Such modifications are usually necessary to achieve more uniform wall thickness in the finished deep-draw sections. Vacuum forming can also be continuous by using the sheet as it is extruded. An example of this technology is practiced with several high speed European lines in operation in the United States. Precise temperature conditioning allows carefully controlled levels or orientation in the finished part (169).

Thermoforming is perhaps the process with the lowest unit cost. Examples of thermoformed articles are refrigerator-door and food-container liners, containers for dairy products, luggage, etc. Some of the largest formed parts are camper/trailer covers and liners for refrigerated-railroad-car doors (170).

Orientation of styrene-based copolymers is usually carried out at temperatures just above T_g. Biaxially oriented films and sheet are of particular interest. Such orientation increases tensile properties, flexibility, toughness, and shrinkability. Polystyrene produces particularly clear and sparkling film after being oriented biaxially for envelope windows, decoration tapes, etc. Oriented films and sheet of styrene-based polymers are made by the bubble process and by the flat-sheet or tentering process. Fibers and films can be produced by uniaxial orientation (171) (see Film and sheeting materials).

Blow Molding. Blow molding is a multistep fabrication process for manufacturing hollow symmetrical objects. The granules are melted and a parison is obtained by extrusion or by injection molding. The parison is then enclosed by the mold, and pressure or vacuum is applied to force the material to assume the contour of the mold. After sufficient cooling, the object is ejected.

Styrene-based plastics are used somewhat in blow molding but not as much as linear polyethylene and PVC. Rubber-modified polystyrene and ABS are used in specialty bottles, containers, and furniture parts. ABS is also used as one of the impact modifiers for PVC. Clear, tough bottles with good barrier properties are blow-molded from these formulations.

Polystyrene or copolymers are used extensively in injection blow molding. Tough and craze-resistant polystyrene containers have been made by multiaxially oriented injection-molded parisons (172). This process permits the design of blow-molded objects with a high degree of controlled orientation, independent of blow ratio or shape.

Additives. Processing aids, eg, plasticizers (qv) and mold-release agents (see Abherents), are often added to polystyrene. Even though polystyrene is an inherently stable polymer, other compounds are sometimes added to give extra protection for a particular application. Rubber-modified polymers containing unreacted allylic groups are very susceptible to oxidation and require carefully considered antioxidant packages for optimum long-term performance. Ziegler-initiated polybutadiene rubbers are especially sensitive in this respect, since they often contain organocobalt residues from the catalyst complex. For food-contact applications, the additives must be FDA-approved. Important additives used in styrene plastics are listed in Table 10.

Economic Aspects

Styrene monomer is a worldwide commodity product. It is manufactured in many countries throughout the world and is freely shipped by commercial tankers between countries as well as by truck and rail between producers in several nations.

Polystyrene is not a single well-defined product, but rather a product family with as many as 30 members. In a technologically advanced country like the United States, older formulations are either being improved constantly or completely displaced by new types of polymers. This changing technology is exported worldwide, although overseas plants, especially in lesser developed countries, will probably never require the broad product line marketed in the United States. Some plants produce general-purpose polystyrene only, whereas others produce the entire product line of general-purpose polystyrene, high impact polystyrene, styrene–acrylonitrile, and ABS. The estimated 1980 consumption pattern of styrene in the United States is given in Table 11.

Table 10. Additives Used in Styrene Plastics

Type	Compounds	Examples	Amount, wt %	Comments
plasticizers	mineral oil		<4	all cause loss in heat-distortion temperature
	phthalate esters			
	adipate esters			
mold-release agents	stearic acid		0.1	many mold-release agents cause yellowing or haze
	metal stearate			
	organic stearate			
	silicones			
	amide waxes		3	
antioxidants	alkylated phenols	Irganox 1076	1	synergism sometimes achieved with multiple use
	organic phosphite	Polygard[a]	1	
	thioesters	dilauryl thiodipropionate	1	
antistatic agents	quaternary ammonium compounds	Armostat[b]	2	
uv stabilizers	benzotriazoles	Tinuvin P	0.25	
	benzophenones			
ignition-suppression agents	hydrated aluminum oxide, antimony oxide, alkyl and aryl phosphates			

[a] Polygard, trademark of Uniroyal Incorporated.
[b] Armostat, trademark of Noury Chemical Corporation.

Table 11. Estimated Consumption of Styrene in the United States in 1980

Uses	Consumption, thousand metric tons
polystyrene	
general-purpose	334
impact	761
copolymers	
ABS	250
SAN	35
other	25
S–B latex	160
polyesters	130
SBR rubber	227
miscellaneous	78
exports	545
Total	*2545*

Characterization

Four modes of characterization are of interest: chemical analyses, ie, qualitative and quantitative analyses of all components; mechanical characterization, ie, tensile and impact testing; morphology of the rubber phase; and rheology at a range of shear rates. Other properties measured are stress-crack resistance, heat-distortion temperatures, flammability, creep, and others depending on the particular application (173).

Since plastics are almost invariably modified with one or more additives, there are three components of chemical analysis: the high molecular weight portion, ie, the polymer; the additives, ie, plasticizer and mold-release agent; and the residuals remaining from the polymerization process. The relevant characterization methods with references are listed in Table 12. Figure 3 illustrates typical molecular weight distribution curves from size-exclusion chromatography, and Figure 15 shows a capillary gas chromatogram for determining residuals. Mechanical testing is carried out according to the methods listed in Table 13. The Rheometrics variable-rate impact tester is a new and valuable instrument (174).

Rubber particle-size distribution is usually measured with a Coulter counter or directly from electron photomicrographs (76,175); the latter also gives details of particle morphology. Rheological studies are made with a modified tensile tester with capillary rheometer (ASTM D 1238-79) or with the powerful Rheometrics mechanical spectrometer (176). For product specifications, a simple melt-flow test is used (ASTM D 1238 condition G) with a measurement of the heat-distortion temperature (ASTM D 1525). These two tests, with solution viscosity as a measure of molecular weight, have been used historically to characterize polystyrene. Only recently have molecular weight distribution and residuals analyses begun to replace them.

Health and Safety Factors

In pellet form, styrene-based plastics have a very low degree of toxicity (181). Under normal conditions of handling and use, they should pose no unusual problems from ingestion, inhalation, or eye or skin contact. Heating of these polymers usually results in the release of some vapors. The vapors contain some styrene (TLV = 100 ppm (1974); human detection, 5–10 ppm) and, in the case of styrene–acrylonitrile copolymers, some acrylonitrile (TLV = 2 ppm (1978)) as well as very low levels of

Table 12. Chemical Characterization Techniques

High mol wt component			Additives			Residuals		
Variable	Method[a]	Ref.	Variable	Method[a]	Ref.	Variable	Method[a]	Ref.
mol wt	solution viscosity	ASTM 703-44T ASTM D 2857	mineral oil	gc/lc	177	styrene	gc	178
mol wt distribution	sec	ASTM 3536-76 ASTM 3593-80	phthalate esters	gc	177	solvents	gc	178
copolymer composition	ir	177	stearates	ir	177	oligomers	gc	178
						volatile content	mainly styrene and solvent	179
rubber content	ir	177	other	generally lc	177	methanol-soluble		ASTM 703-44T
graft yield	solubility	180						

[a] sec = size-exclusion chromatography (4); gc = gas chromatography; lc = liquid chromatography; ir = infrared analysis.

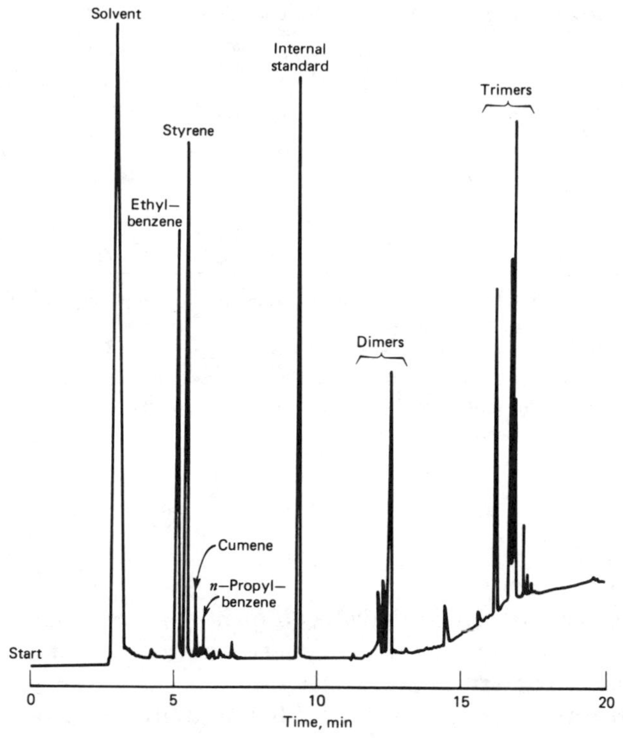

Figure 15. Capillary gas chromatogram showing residuals in polystyrene.

Table 13. Mechanical Tests on Polystyrene

Test	ASTM method
stress-strain	
tensile	D 638
compression	D 695
flexural	D 790
impact	
Izod	D 256
dart	D 3029
hardness	D 785
creep	D 2990

degraded and oxidized hydrocarbons and additives. The warning properties of styrene and any oxidized organics are such that the vapors normally can be detected by humans at very low levels, and hazardous concentrations are usually irritating and obnoxious. Nevertheless, adequate ventilation should be provided.

Styrene polymers burn under the right conditions of heat and oxygen supply. This can occur even when they are modified by addition of ignition-suppression chemicals. Once ignited, these polymers may burn rapidly and produce a dense black smoke. Combustion products from any burning organic material should be considered toxic. Ignition temperatures of several styrenic polymers are listed in Table 14. Fires can be extinguished by conventional means, ie, water and water fog.

Table 14. Ignition Temperatures and Burning Rates of Styrene-Based Polymers[a]

Polymer	Minimum flash-ignition temperature, °C	Minimum self-ignition temperature, °C	Burning rate[b], cm/s
polystyrene, general-purpose	345–360	488–496	<0.06
polystyrene, rubber-modified	385–399	435–474	<0.06
styrene–acrylonitrile copolymer	366	454	<0.06
acrylonitrile–butadiene–styrene	404	>404	<0.06

[a] Ignition temperature data for general-purpose polystyrene and styrene–acrylonitrile copolymer were obtained from ref. 182. Burning rates for all polymers are obtained from Underwriter's Laboratories (UL 94). The rest of the data in Table 15 were generated at Dow Chemical using ASTM D 1929. Test specimens for the burning rate data were 1.27 × 15.24 × 0.318 cm. Descriptions of burning rate and other flammability characteristics developed from small-scale laboratory testing do not reflect hazards presented by these or any other materials under actual fire conditions.
[b] In a horizontal position.

Uses

Many uses of styrene plastics have been mentioned in terms of properties or processing characteristics, eg, coextended sheet materials based on polystyrene in Table 9. Other uses are described below.

Polystyrene Foams. The early history of foamed polystyrene is given in ref. 183 and the theory of plastic foams is discussed in ref. 184. Foamable polystyrene beads were developed in the 1950s by BASF under the trademark of Styropor (185–187). These beads, made by suspension polymerization in the presence of blowing agents such as pentane or hexane, or by postpressurization with the same blowing agents, have had an almost explosive growth, with 200,000 metric tons used in 1980. Some typical physical properties of polystyrene foams are listed in Table 15 (see Foamed plastics).

The following are commercially significant foamed polystyrenes. Extruded planks and boards in the density range of 29 kg/m^3 and largely flame retardant are used for

Table 15. Characteristic Properties of Some Polystyrene Foams

Property	Styrofoam[a] extruded	Bead, board-molded[b]	Foam sheet	ASTM
density, kg/m^3 [c]	35	32	96	
compressive strength, kPa[d]	310	207–276	290	D 1621
tensile strength, kPa[d]	517	310–379	2070–3450	D 1623
flexural strength, MPa[e]	1138	379–517		D 790
thermal conductivity, W/(m·K)[f]	0.030	0.035	0.035	C 177
heat distortion, °C	80	85	85	

[a] Registered trademark of The Dow Chemical Company, ref. 188.
[b] Refs. 189–190.
[c] To convert kg/m^3 to lb/ft^3, multiply by 0.0624.
[d] To convert kPa to psi, multiply by 0.145.
[e] To convert MPa to psi, multiply by 145.
[f] To convert W/(m·K) to (Btu·in.)/(h·ft^2·°F), multiply by 6.933.

low temperature thermal insulation, buoyancy, floral display, novelty, packaging, and construction purposes. Foamed boards and shapes from foaming-in-place (FIP) beads (density 16–32 kg/m^3) are used for packaging, buoyancy, insulation, and numerous other applications. Batch molding of boards and shapes as well as automated molding of continuous planks are used. Extruded foamed polystyrene sheet 1–7 mm thick with densities of 64–160 kg/m^3 are used largely in packaging applications. Extremely fine cell size is required for texture and strength. Special nucleating agents are employed for this purpose, eg, polymers of styrene and maleic anhydride, citric acid–sodium bicarbonate mixtures, and talc (188). High density extruded planks (density 35–64 kg/m^3), are used for heavy-duty structural applications. Styrene copolymer foams containing, eg, acrylonitrile for gasoline resistance, are made either by extrusion or from beads. High density (480–800 kg/m^3) injection-molded objects are made from mixtures of FIP beads and polystyrene granules or more commonly HIPS or ABS with a chemical blowing agent, eg, azodicarbonamide. These moldings have a high density skin of essentially polystyrene and a foamed core. Foamable ABS systems, eg, laminates with ABS skins and a heat-foamable core, are used for structural applications, as in car body parts (189). Of growing commercial significance is coextruded, foam-core ABS pipe (190–191). Reducing the density of the foam core by up to one half and using a chemical blowing agent results in about a 20% overall pipe weight and raw material usage reduction. Syntactic foams are combinations of foamable beads, eg, expandable polystyrene or styrene–acrylonitrile copolymers. These are mixed with a resin, usually thermosetting, which has a large exotherm during curing, eg, epoxy or phenolic resins (192). The mixture is then placed in a mold and the exotherm from the resin cure causes the expandable particles to foam and forces the resin to the surface of the mold. A typical example is expandable polystyrene in a flexible polyurethane-foam matrix used in cushioning applications (193). Sandwich panels are made either by foaming beads between the skin materials or by adhering skins to planks cut to precise dimensions. Foamed polystyrene beads are admixed with concrete for light-weight masonry structures (194).

Extruded Rigid Foam. In addition to low temperature thermal insulation, foamed polystyrenes are used for insulation against ambient temperatures in the form of perimeter insulation and insulation under floors and in walls and roofs. The upside-down roof system, in which a foamed plastic such as Styrofoam plastic foam is applied above the tar-paper vapor seal, thereby protecting the tar paper from extreme thermal stresses which cause cracking, has been patented (195). The foam is covered with gravel or some other wear-resistant topping (see Roofing materials).

In addition to such thermal-insulation uses in buildings, there is a potentially tremendous new area in the form of highway underlayment to prevent frost damage. Damage to roadways, roadbeds, and airfields because of frost action is a costly and aggravating problem. Conventional treatments to prevent such damage are expensive and unreliable. During the 1960s, an improved and more economical solution to the problem was developed (196). It is based on the use of thermal insulation to reduce the heat loss in the frost-susceptible subgrade soil so that no freezing occurs. Many miles of insulated pavements have been built in the United States, Canada, Europe, and Japan. The concept has proved valid, and the performance of extruded polystyrene foam has been completely satisfactory. It is expected that the use of this concept will continue to grow as natural road-building materials continue to become more scarce and expensive and as the weight and speed of land and airborne vehicles continue to increase.

A 2.54 cm Styrofoam plastic foam with thermal conductivity of ca 0.03 W/(m·K) ((0.21 Btu·in.)/(ft·h·°F)) is equivalent to 61 cm of gravel (see Table 15). Any synthetic foam having compressive strength sufficiently high and thermal conductivity sufficiently low is effective. However, the resistance of polystyrene-type foams to water, frost damage, and microorganisms in the soil makes them especially desirable. An interesting and important application of this concept was the use of Styrofoam in the construction of the Alaska Pipeline. In this case, the foam was used to protect the permafrost.

Rigid Foam from Foaming-in-Place (FIP) Beads. In 1954, total U.S. production of general-purpose polystyrene was 91,000–136,000 metric tons with a negligible amount used in foams. By 1980, more than 1.95×10^5 t of expandable polystyrene was sold in the United States (180). Expandable polystyrene (EPS) has been in wide use in packaging since its introduction. Applications range from pallet shipping containers for computer terminals to material-handling stacking trays to packages for fresh fruits and vegetables. Expandable polystyrene serves as a cushioning material, a package insert for blocking and bracing, a flotation item, or as insulation.

The basic resin for EPS is in the form of beads that are expanded to a desired density before molding. Densities for packaging parts typically are 20–40 kg/m^3. Once expanded, the beads are fused in a steam-heated mold to form a specific shape. Most parts are molded of standard white resins, although several pastel colors are available.

Converters who manufacture EPS parts and components for packaging are called shape molders. Others who mold large billets, eg, measuring $0.6 \times 1.2 \times 2.4$ m, are called block molders. The package user can obtain EPS parts without the benefits of a mold; parts can be fabricated from billets by hot-wire cutting. Whether molded or fabricated, EPS packages and their components are typically designed by careful consideration of the compression and cushioning properties of expanded polystyrene (197).

A large number of factors that influence the foaming of foamable polystyrene beads, eg, the molecular weight of the polymer, polymer type, blowing-agent type and content, and bead size, have been analyzed (198). It is suggested that at least part of the pentane blowing agent is present in microvoids in the glassy polystyrene. Thus, the density of beads containing 5.7 wt % n-pentane is 1080 kg/m^3, compared to a value of 1050 kg/m^3 for pure polystyrene beads and compared with a calculated density of 1020 kg/m^3 for a simple mixture in which n-pentane is dissolved in the polystyrene. If all of the pentane were in voids, the calculated density would be 1120 kg/m^3. About 60% of the pentane is held in voids, with the balance presumably dissolved. n-Pentane present in voids has a reduced vapor pressure, which depends on the effective microcapillary diameter. Dissolved n-pentane in polystyrene will also have a reduced vapor pressure, by an amount depending on the thermodynamic interaction between solvent and polymer.

An unspecified experimental styrene copolymer, possibly with acrylonitrile, shows a greatly reduced tendency to lose blowing agent on aging at 23°C (198). FIP beads from chlorostyrene likewise have a greatly enhanced ability to retain blowing agent (194) as indicated in Table 16. The higher heat distortion of this polymer requires steam pressures of 210–340 kPa (30–50 psi) for blowing.

Various other factors that influence the foaming of polystyrene, especially cross-linking, are described (194). In ref. 194, foaming is discussed as a viscoelastic process in which there is competition between the blowing pressure, which increases

Table 16. Loss of Isopentane from 420 μm Foaming-in-Place Chlorostyrene Beads in Open Storage, wt %[a]

Temperature, °C	After 20 h Polystyrene	After 20 h Poly(chlorostyrene)	After 300 h Polystyrene	After 300 h Poly(chlorostyrene)
25	18	1.3	40	10
50	38	12	72	33
80	48	24	80	54

[a] Initial concentration of isopentane = 6 wt %.

with temperature, and the thinning and rupture of cells walls with consequent collapse of the foam. The presence of cross-links reduces the tendency for cell-wall rupture. Figure 16 shows foam volumes attained as a function of temperature for various amounts of divinylbenzene. The effects of temperature and cross-linking on the kinetics of foaming is also discussed in ref. 194.

Another important factor affecting the foaming of polystyrene beads is the differential rate of diffusion of n-pentane compared to steam and air. It is estimated that the quantity of n-pentane normally used is sufficient to produce a foaming volume of thirtyfold and hence a foam density of no more than 32 kg/m^3. However, the steam and/or air which diffuses into the beads contributes to further expansion, so that foaming volumes of 60-fold or more are commonly achieved. This aspect of diffusion has been discussed in some detail (199). In cycle foaming in air, beads are allowed to equilibrate with air after each heat-foaming step through a series of stages. Foaming volumes of 200:1 have been achieved in this manner, with densities down to 3.2–4.8 kg/m^3.

The largest type of expandable polystyrene is in the form of spherical beads. However, The Dow Chemical Company pioneered an elongated shape called Pelaspan-Pac which expands into a wormlike shape ideally suited as a loose fill for pack-

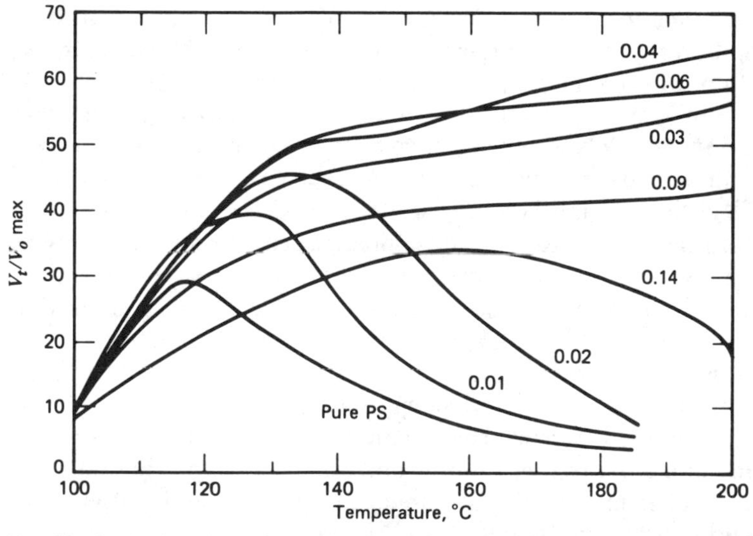

Figure 16. Maximum foaming volume of styrene–divinylbenzene copolymers containing 8.8 wt % CO_2 as a function of divinylbenzene content and temperature. Numeral beside curve indicates wt % divinylbenzene. V_t/V_o = ratio of final volume to initial volume at temperature t. Adapted from ref. 197.

aging. Items packed in spherical beads settle to the bottom of the container during shipping. The irregular elongated form prevents this settling.

Lightweight concrete is made from prefoamed EPS beads, portland cement, and organic binders. Precast shapes are being used to provide structural strength, thermal insulation, and sound deadening.

Polystyrene as a Raw Material for Rigid Thermoplastic Foams. The low cost of PS and its high thermal stability, which provides good recycle efficiency, as well as the possibility of using low cost blowing agents, are important to the use of PS in rigid thermoplastic foams. The heat-distortion point of ca 80°C (in contrast to 65°C for PVC foam) permits its use in most construction applications. The low sensitivity of the polymer to moisture is another important factor.

Because of all of the above and possibly for other reasons, polystyrene has been the dominant material used in rigid thermoplastic foams. Poly(vinyl chloride) is economical and has good physical properties and is flame retarding, but it has marginal heat-distortion properties for some uses and does not readily lend itself to the continuous extrusion process used for polystyrene foam. Thus far, no FIP beads system based on PVC has been perfected. This could well be a result of its high vapor barrier compared to polystyrene. Polyethylene is too permeable to retain a blowing agent and hence is ruled out as a FIP bead material, but it is commercially significant as extruded foam.

Flame-Retardants. The growing use of polystyrene foams in the construction industry has provided impetus for continuing research on flame-retardant additives (see Flame retardants). Organic bromine compounds, eg, hexabromocyclododecane or pentabromomonochlorocyclohexane, have long been used for this purpose in extruded polystyrene foam. Combustion of thermoplastics is reviewed in ref. 200. Use of injection-molded structural foams, based on high impact polystyrene and ABS and containing flame-retardant additives, has grown rapidly since the late 1970s. A typical additive system to impart the required flame-retardant properties is bis(pentabromophenyl) oxide and a synergist, eg, antimony trioxide.

The most common uses of these high density foams are cathode-ray-tube housings, printers, small computers, and point-of-sales terminals. The estimated annual production volume of styrenic polymer-based structural foam by 1982 is 27,000 t (201). The latest technology in structural foam molding is discussed in ref. 202.

Foamed Sheet. Polystyrene foamed sheet is used for foamed trays, egg cartons, disposable dinnerware, and packaging. Foam sheet manufacturing techniques, manufacturers' logistics, and markets are described in ref. 203. Choice of the correct blowing agent for foam sheet is critical in determining ultimate density and physical properties. Further, the choice of blowing agent may be dictated by safety considerations, eg, flammability of hydrocarbons, or environmental concerns, eg, chlorofluorocarbons. These two categories of blowing agents are discussed in ref. 204.

Oriented Polystyrene Film and Sheet. Approximately 45,000 t of polystyrene is being used in the form of film and sheet, both biaxially oriented. Film is <0.08 mm thick, whereas sheet is >0.08 mm thick. The material is predominantly general-purpose polystyrene supplemented with only modest amounts of styrene-acrylonitrile copolymer and rubber-modified polystyrene. This usage does not include foamed sheet (see Film and sheeting materials).

Oriented polystyrene film, in addition to having the lowest cost of any of the rigid plastic materials, offers a high degree of optical clarity, high surface gloss, and excellent dimensional stability (particularly with regard to relative humidity) and is heat-shrinkable. It is not a barrier film; in fact, one of its largest uses, that of packaging field-fresh produce, depends upon its being highly permeable to oxygen and water vapor (see Table 17).

A review of polystyrene film as part of a larger study on films of all types is given in refs. 206 and 207. Other relevant reviews are given in refs. 208–209.

It appears that the first work with biaxially oriented film took place in the FRG, the product mostly being used as a wrapping material for high voltage cables where its high dielectric strength of 20 MV/m at 60 Hz is utilized. There is a modest use in this area both in the FRG and in the United States. Approximately 27,000 metric tons of oriented PS was produced in 1979; it is expected that production will increase 2%/yr to 31,000 t in 1985 (210). In Europe, the film market appears to be dominated by PVC.

Although polystyrene is normally considered a rather brittle material, biaxial orientation imparts some extremely desirable properties, particularly in regard to an increase in elongation. Thus, the 1.5–2% of elongation normally associated with unoriented polystyrene can become as high as 10%, depending on the exact conditions of preparation (211–212).

There have been four principal methods of preparing biaxially oriented polystyrene film. The oldest one was that employed in the FRG; it involved a parabolic or horseshoe-shaped stretcher over which an extruded tube of polystyrene was drawn. Perhaps the second method to gain use was the tenter-frame method, in which the edges of a polystyrene sheet are gripped continuously and pulled sidewise as the sheet is drawn lengthwise in the machine direction by takeup rolls. An eight-sided or pancake stretcher had shown a fair amount of promise but has been abandoned (211–212). The bubble process is used commercially. The high softening point, coupled with the tendency of oriented film to shrink above 85°C, makes heat sealing difficult. Solvent, adhesive, and ultrasonic sealing are used in most applications. Scratch resistance, impact strength, and crease resistance are low. Although the bubble process superficially resembles the common bubble process for making polyethylene, polypropylene, and Saran film, there are a number of substantive differences, and the polystyrene process is considerably more difficult to achieve. In the case of polyethylene, for example, crystallinity stabilizes the blown bubble, whereas with a true thermoplastic, eg, polystyrene, one must stabilize the blown film by cooling below the glass-transition temperature. This involves extremely careful control of temperature in the blowing process.

As a general rule of thumb, there is an economic breakeven point at ca 0.08 mm,

Table 17. Gas and Water-Vapor Permeabilities of Polystyrene Films[a]

Composition	Gas permeability, $(cm^3 \cdot cm)/(10^{10}\ cm^2 \cdot s \cdot kPa)$[b]		
	O_2	CO_2	H_2O
unmodified polystyrene	3.50	14.0	1600
72:28 styrene–acrylonitrile	0.09	4.62	2900

[a] Ref. 205.

[b] To convert kPa to mm Hg, multiply by 7.5.

which coincides with the defined difference between film and sheet. Film is made more economically by the bubble method and sheet by the tenter-frame method. The exact thickness for breakeven depends on technological improvements, which can be made in both processes, in the degree of control which is used in regulating them and in quality requirements.

Polystyrene film contains no plasticizers, absorbs negligible moisture, and exhibits exceptionally good dimensional stability. It does not become brittle with age nor distort when exposed to low or high humidity. Excellent machinability, ie, ability to pass through packaging machinery at high speeds and to be cut and sealed, etc, clarity, and stability are central factors in the use of oriented polystyrene in window envelopes. These same characteristics have also led to applications of film for window cartons. An antifog film 0.025–0.032 mm thick has been used extensively for the windows of cartons for bacon. Substantial amounts of film are also employed as sheet protectors and as inserts in wallets. A thickness of 0.060–0.130 mm is customary for these applications. Polystyrene film can be printed by means of flexographic, rotogravure, and silk-screen methods. A lamination of reverse-printed film and paper has been used for an attractive, high sparkle package for hand soap. The film is an excellent base for metallizing because of the almost complete absence of volatiles. Oriented polystyrene film is also widely used for lamination to polystyrene foamed sheet for preprinted decoration and/or property enhancement (213).

For polystyrene sheet, the use responsible for its rapid growth is in meat trays, where the merchandising value is important, ie, the transparency of the package, as well as its resistance to moisture and dimensional stability. A smaller but growing use is as a photographic-film base where dimensional stability at all humidities and low gel count are both important. The thermoplastic nature of polystyrene and the memory effects built into it through biaxial orientation make it especially suitable for packaging applications. It must be processed under pressure rather than by vacuum-forming equipment to avoid the shrinkage which would otherwise result from the high level of orientation.

Important uses of polystyrene are as windows in envelopes, etc, as mentioned above, and as produce overwrap. High optical clarity is required for mechanized reading of characters through envelope windows, a significant technological trend which is certain to enhance the use of polystyrene film. In regard to produce overwrap, the transmission characteristics of polystyrene for water vapor and oxygen coincide more or less with the metabolic requirements of the produce being packaged, eg, lettuce, so that in addition to giving mechanical protection to something like a head of lettuce, the produce is able to metabolize in a normal manner (214). Low cost and high optical clarity are important. In addition, the feel of polystyrene film gives an impression of crispness as far as the product is concerned. Polystyrene film also has potential use as a type of synthetic paper (209) (see Pulp, synthetic). General-purpose polystyrene must be treated so that it is opaque and this can take the form of either pigmentation, mechanical abrasion, or a surface treatment, eg, a chemical etch. Impact polystyrene is reasonably opaque. Polystyrene film can be given a suitable coating, eg, a latex–clay coating, of the kind that is used on ordinary paper for the purpose of achieving a good printable surface. General discussions of gas permeabilities in plastic films are given in refs. 23, 25–26, and 215–216 (see Barrier polymers). Experimental and theoretical studies on polystyrene have also been reported (217–218).

BIBLIOGRAPHY

"Styrene Resins and Plastics" in *ECT* 1st ed., Vol. 13, pp. 146–179, by J. A. Struthers, R. F. Boyer, and W. C. Goggin, The Dow Chemical Company; "Styrene Plastics" in *ECT* 2nd ed., Vol. 19, pp. 85–134, by H. Keskkula, A. E. Platt, and R. F. Boyer, The Dow Chemical Company.

1. R. H. Boundy, R. F. Boyer, and S. Stoesser, *Styrene, Its Polymers, Copolymers and Derivatives*, American Chemical Society, Monograph No. 115, Reinhold Publishing Corporation, New York, 1952.
2. R. F. Boyer, H. Keskkula, and A. E. Platt in N. M. Bikales, ed., *Encyclopedia of Polymer Science and Technology*, Vol. 13, John Wiley & Sons, New York, 1970, pp. 128–447.
3. C. A. Brighton, G. Pritchard, and G. A. Skinner, *Styrene Polymers: Technology and Environment Aspects*, Applied Science Publishers Ltd., London, 1979.
4. J. F. Rudd, in J. Brandrup and E. H. Immergut, eds., *Polymer Handbook*, 2nd ed., John Wiley & Sons, Inc., 1975, p. V-59–62.
5. Ref. 1, p. 478.
6. R. S. Spencer and G. D. Gilmore, *J. Appl. Polym. Sci.* **20,** 502 (1949).
7. *Plastics Design Data*, Dow Technical Chemical Publication, Form No. 153-5290-65, The Dow Chemical Company, Midland, Mich., 1965.
8. G. Natta, *J. Polym. Sci.* **16,** 143 (1955).
9. R. L. Miller and L. E. Nielsen, *J. Polym. Sci.* **55,** 643 (1961).
10. W. Patnode and W. J. Scheiber, *J. Am. Chem. Soc.* **61,** 3449 (1939).
11. F. L. Saunders and R. C. Mildner, The Dow Chemical Company, unpublished data, 1960.
12. F. P. Reding and co-workers, *J. Polym. Sci.* **57,** 483 (1962).
13. T. G. Fox and P. J. Flory, *J. Appl. Phys.* **21,** 581 (1950).
14. S. Krause and Z-H. Lu, *J. Polym. Sci. Polym. Phys. Ed.* **19,** 1925 (1981).
15. K. Ueberreiter and E. Otto-Laupenmuhlen, *Z. Naturforsch. Teil A* **8,** 664 (1953).
16. D. E. Roberts and co-workers, *J. Polym. Sci.* **2,** 420 (1947).
17. J. W. Breitenbach and J. Derkosch, *Monatsh. Chem.* **81,** 698 (1950).
18. R. Dedeurwaerder and J. F. M. Oth, *J. Chim. Phys.* **56,** 940 (1959).
19. L. K. J. Tong and W. O. Kenyon, *J. Am. Chem. Soc.* **69,** 1402 (1947).
20. C. Y. Liang and S. Krimm, *J. Polym. Sci.* **27,** 241 (1958).
21. G. Natta and F. Danusso, *Stereoregular Polymers and Stereospecific Polymerization*, Pergamon Press, Inc., New York, 1967.
22. U.S. Pat. 2,287,188 (June 23, 1942), L. A. Matheson and R. F. Boyer (to The Dow Chemical Company).
23. H. Keskkula, G. M. Simpson, and F. L. Dicken, *Soc. Plast. Eng. Preprints Ann. Tech. Conf.* **12,** XV-2 (1966).
24. F. J. Furno, R. S. Webb, and N. P. Cook, *Prod. Eng. (N.Y.)* **35,** 87 (Aug. 17, 1964); V. E. Malpass, *Soc. Plast. Eng. Preprints Ann. Tech. Conf.* **13,** 618 (1969).
25. W. E. Brown, *International Plastics Congress, 1966*, N.V. t'Raedthuys, Utrecht, Netherlands, 1967.
26. R. L. Bergen and W. E. Wolstenholme, *SPE J.* **16,** 1235 (1960).
27. G. B. Jackson and J. L. McMillan, *SPE J.* **19,** 203 (1963).
28. S. G. Turley and H. Keskkula, *Polym. Eng. Sci.* **7,** 1 (1967).
29. W. E. Brown, *Performance of Plastics in Building*, No. 1004, Building Research Institute, Inc., Washington, D.C. (1963).
30. R. McFedries, Jr., *Plast. World* **24,** 34 (Oct. 1963).
31. E. Scalco, T. W. Huseby, and L. L. Blyler, Jr., *J. Appl. Polym. Sci.* **12,** 1343 (1968).
32. E. H. Andrews, in J. V. Schmitz and W. E. Brown, eds., *Testing of Polymers*, Vol. 4, Interscience Publishers, a division of John Wiley & Sons, Inc., New York, 1969, p. 237.
33. W. H. Haslett, Jr. and L. A. Cohen, *SPE J.* **20,** 246 (1964).
34. C. B. Bucknall and R. R. Smith, *Polymer* **6,** 437 (1965).
35. J. A. Schmitt, *J. Appl. Polym. Sci.* **12,** 533 (1968).
36. W. Retting, *Angew. Makromol. Chem.* **58/59,** 133 (1977).
37. H. Keskkula, in K. Solc, ed., *Polymer Compatability and Incompatability*, MMI Press, Midland, Michigan, 1981.
38. C. B. Bucknall, *Toughened Plastics*, Applied Science Publishers, London, 1977, Chapts. 7 and 10.

39. R. P. Kambour, *J. Polym. Sci.* **D7,** 1 (1973).
40. Ref. 38, p. 193.
41. H. Keskkula, G. M. Simpson, and F. L. Dicken, *Soc. Plast. Eng. Preprints Annu. Tech. Conf.* **12,** XV-2 (1966).
42. K. J. Cleereman, *SPE J.* **23,** 43 (Oct. 1967); **25,** 55 (Jan. 1969).
43. L. C. Stirik, *Ann. N.Y. Acad. Sci.* **279,** 78 (1976).
44. C. B. Bucknall and D. G. Street, *J. Appl. Polym. Sci.* **2,** 289 (1959).
45. *Modern Plastics Encyclopedia for 1968–69*, Vol. 45, 1968, p. 43.
46. J. Eichhorn, *J. Appl. Polym. Sci.* **8,** 2497 (1964).
47. U.S. Pat. 3,324,076 (June 6, 1967), M. E. Elder, R. T. Dickerson, and W. F. Tousignant (to The Dow Chemical Company).
48. G. P. Ziemba, in *Encyclopedia of Polymer Science and Technology*, Vol. 1, Interscience Publishers, a division of John Wiley & Sons, Inc., New York, 1964, pp. 425–435.
49. U.S. Pat. 3,265,765 (Aug. 9, 1966), G. Holden and R. Milkovich (to Shell Oil Co.).
50. U.S. Pat. 4,267,284 (May 12, 1981), A. G. Kitchen (to Phillips Petroleum Co.).
51. G. E. Molau, *J. Polym. Sci. Part A* **3,** 1267, 4235 (1965).
52. J. Moacanin, G. Holden, and N. W. Tschoegl, eds., *J. Polym. Sci. Polym. Symp.* **26,** (1969).
53. G. Reiss and co-workers, *J. Macromol. Sci. Phys.* **B17,** 355 (1980).
54. G. Reiss, *Polym. Sci. Technol.*, 327, 337 (1977).
55. A. Echte, *Angew. Makromol. Chem.* **58/59,** 175 (1977).
56. *Ibid.*, **90,** 95 (1980).
57. L-K. Bi and L. J. Fetters, *Macromolecules* **9,** 732 (1976).
58. M. R. Ambler, *Chromatogr. Sci. Ser.* **19,** 29 (1981).
59. J. T. Bailey, E. T. Bishop, W. R. Hendricks, G. Holden, and N. R. Legge, *Rubber Age (N.Y.)* **98,** 69 (Oct. 1966).
60. H. L. Hsieh and co-workers, *CHEMTECH* **11,** 626 (Oct. 1981).
61. U.S. Pat. 3,336,267 (Aug. 15, 1967), R. L. Zimmerman and W. E. O'Connor (to The Dow Chemical Company).
62. U.S. Pat. 3,231,524 (Jan. 25, 1966), D. W. Simpson (to The Dow Chemical Company).
63. Ref. 1, p. 736.
64. W. A. Lee and G. J. Knight, in J. Brandrup and E. H. Immergut, eds., *Polymer Handbook*, 2nd ed., Interscience Publishers, a division of John Wiley & Sons, New York, p. III–139, 1975.
65. H. W. McCormick, *J. Polym. Sci.* **25,** 488 (1957).
66. U.S. Pat. 3,206,445 (Sept. 14, 1965), H. Volk (to The Dow Chemical Company).
67. U.S. Pat. 3,340,238 (Sept. 5, 1967), W. E. Smith and H. Volk (to The Dow Chemical Company).
68. U.S. Pat. 3,011,918 (Dec. 5, 1961), L. H. Silvernail and M. W. Zembal (to The Dow Chemical Company).
69. G. D. Jones, in E. M. Fettes, ed., *Chemical Reaction of Polymers*, Interscience Publishers, a division of John Wiley & Sons, Inc., New York, 1964, p. 273.
70. G. E. Molau and H. Keskkula, *J. Polym. Sci. A1* **4,** 1595 (1966).
71. S. Strella, in E. Baer, ed., *Engineering Design for Plastics*, Reinhold Publishing Corp., New York, 1964, pp. 795–814.
72. R. F. Boyer, *Polym. Eng. Sci.* **8,** 161 (1968).
73. C. B. Bucknall, *Toughened Plastics*, Applied Science Publishers, London, 1977.
74. H. Willersinn, *Makromol. Chem.* **101,** 297 (1966).
75. H. Keskkula, S. G. Turley, and R. F. Boyer, *J. Appl. Polym. Sci.* **15,** 351 (1971).
76. K. Kato, *J. Electron Microsc.* **14,** 220 (1965); *Polym. Eng. Sci.* **7,** 38 (1967).
77. D. A. Walker, in ref. 45, p. 334.
78. H. Keskkula and P. A. Traylor, *J. Appl. Polym. Sci.* **11,** 2361 (1967).
79. R. J. Williams and R. W. A. Hudson, *Polymer* **8,** 643 (1967).
80. C. H. Basdekis, *ABS Plastics*, Reinhold Publishing Corp., New York, 1964.
81. *Br. Plast.* **38,** 708 (Dec. 1965).
82. G. M. Kraynak, *Soc. Plast. Eng. Annu. Tech. Conf.* **13,** 896 (1967).
83. *Vacuum Metallizing Cycolac*, Technical Report P 135, Borg-Warner Corp., Chicago, Ill., 1980.
84. E. N. Hildreth, *Modern Plastics Encyclopedia*, Vol. 43, 1966, p. 991.
85. K. Stoeckhert, *Kunststoffe* **55,** 857 (1965).
86. *Mod. Plast.* **45,** 84 (Aug. 1968).
87. U.S. Pat. 2,802,809 (Aug. 13, 1957), R. A. Hayes (to The Firestone Tire and Rubber Co.).
88. L. A. Landers and W. C. Meisenhelder, *SPE J.* **20,** 621 (July 1964).

89. E. Zahn, *Appl. Polym. Symp.* **11,** 209 (1969).
90. W. E. Brown, J. D. Striebel, and D. C. Fuccella, *Automotive Engineering Congress*, paper 680059, Jan. 8, 1968.
91. C. H. Bamford and C. F. M. Tipper, eds., *Chemical Kinetics*, Vol. 14, Elsevier Scientific Publishing Co., New York, 1975.
92. N. Grassie, ed., *Developments in Polymer Degradation*, Vols. 1 and 2, Applied Science Publishers, London, 1977.
93. G. G. Cameron, J. M. Meyer, and I. T. McWalter, *Macromolecules* **11,** 696 (1978).
94. P. J. Burchill and G. A. George, *J. Polym. Sci. Polym. Lett. Ed.* **12,** 497 (1974).
95. R. F. Fox, *Prog. Polym. Sci.* **1,** 47 (1967).
96. N. Grassie and N. A. Weir, *J. Appl. Polym. Sci.* **9,** 975 (1965).
97. A. Ghaffe, A. Scott, and G. Scott, *Eur. Polym. J.* **11,** 271 (1975).
98. Ref. 2, p. 236; Ref. 2, pp. 156–206.
99. U.S. Pat. 4,087,599 (May 2, 1978), J. M. Roe and D. B. Priddy (to The Dow Chemical Company).
100. W. A. Pryor and L. D. Lasswell, in G. H. Williams, ed., *Advances in Free-Radical Chemistry*, Elek Science, London, 1975.
101. O. F. Olaj, H. F. Kauffmann, and J. W. Breitenback, *Makromol. Chem.* **178,** 2707 (1977).
102. A. Hui and A. Hamielec, *J. Polym. Sci.* **C25,** 167 (1968).
103. W. A. Pryor and J. H. Coco, *Macromolecules* **3,** 500 (1970).
104. D. J. Stein and H. Mastof, *Angew. Makromol. Chem.* **2,** 39 (1968).
105. J. C. Bevington, H. W. Melville, and R. P. Taylor, *J. Polym. Sci.* **12,** 449 (1954).
106. A. V. Tobolsky, C. E. Rogers, and R. D. Brickman, *J. Am. Chem. Soc.* **82,** 1277 (1960).
107. Ref. 4, pp. II–61.
108. Ref. 4, pp. II–63.
109. *Form No. 115-575-79*, Dow Chemical Company, Midland, Mich., 1979.
110. U.S. Pat. 4,177,110 (Dec. 4, 1979), J. M. Watson (to Cosden).
111. U.S. Pat. 4,086,147 (April 25, 1978), J. M. Watson (to Cosden).
112. M. H. George in G. E. Ham, ed., *Vinyl Polymerization*, Marcel Dekker Inc., New York, 1967, pp. 186–188.
113. A. A. Miller and F. R. Mayo, *J. Am. Chem. Soc.* **78,** 1017 (1956).
114. K. F. O'Driscoll, R. J. Boudreau, and A. V. Tobolsky, *J. Polym. Sci.* **31,** 115 (1958).
115. K. F. O'Driscoll and A. V. Tobolsky, *J. Polym. Sci.* **31,** 123 (1958).
116. M. Szwarc, *Carbanions, Living Polymers and Electron Transfer Processes*, Interscience Publishers, a division of John Wiley & Sons, New York, 1968.
117. P. H. Plesch, ed., *The Chemistry of Cationic Polymerization*, The Macmillan Co., New York, 1963.
118. J. L. R. Williams, T. M. Loakso, and W. J. Dulmage, *J. Org. Chem.* **25,** 638 (1958).
119. R. J. Kern, *Nature* **187,** 410 (1960).
120. R. J. Kern, H. G. Hurst, and W. R. Richard, *J. Polym. Sci.* **45,** 195 (1960).
121. F. Danusso, G. Natta, and I. Pasquon, *Collect. Czech. Chem. Commun.* **22,** 191 (1957).
122. F. Danusso and D. Sianesi, *Chim. Ind.* (*Milan*) **40,** 450 (1958).
123. Ref. 4, pp. II-303 to II-333.
124. G. E. Molau, *J. Polym. Sci.* **B3,** 1007 (1965).
125. H. J. Karam, in K. Solc, ed., *Polymer Compatibility and Incompatibility*, MMI Press, Midland, Mich., 1981.
126. A. W. Hansen and R. L. Zimmermann, *Ind. Eng. Chem.* **49,** 1803 (1957).
127. Ger. Offen. 2,138,176 (July 30, 1971) (Badishe Anilin u. Soda-Fabrik).
128. U.S. Pat. 4,243,781 (June 6, 1981), R. W. Kent, Jr. (to The Dow Chemical Co.).
129. U.S. Pat. 4,268,652 (May 19, 1981), R. W. Kent, Jr. (to The Dow Chemical Co.).
130. U.S. Pat. 4,206,293 (June 3, 1980), R. L. Kruse (to Monsanto Co.).
131. F. A. Miller, in R. R. Meyers and J. S. Long, eds., *Treatise on Coatings*, Vol. 1, Part 2, Marcel Dekker Inc., New York, 1968, pp. 1–57.
132. Ref. 91, p. 53.
133. G. D. Patterson and co-workers, *Macromolecules* **14,** 86 (1981).
134. R. Okasha, G. Hild, and P. Rempp, *Eur. Polym. J.* **15,** 975 (1979).
135. G. F. Freeguard, *Br. Polym. J.* **6,** 205 (1974).
136. U.S. Pat. 2,694,692 (Nov. 16, 1954), J. L. Amos, J. L. McCurdy, and O. R. McIntire (to The Dow Chemical Co.).
137. U.S. Pat. 4,214,056 (July 22, 1980), R. E. Lavengood (to Monsanto Company).

138. A. Brydon, G. M. Burnett, and G. G. Cameron, *J. Polym. Sci. Chem. Ed.* **11,** 3255 (1973); **12,** 1011 (1974); **18,** 2143 (1980).
139. T. Kotaka, *Makromol Chem.* **177,** 159 (1976).
140. D. J. Stein, G. Fahrbach, and H. Adler, *Adv. Chem. Ser.* **142,** 148 (1975).
141. R. H. M. Simon and D. C. Chappelear, in J. N. Henderson and T. C. Bouton, eds., *Polymerization Reactors and Processes*, ACS Symposium Series 104, American Chemical Society, Washington, D.C., 1979, pp. 71–112.
142. R. B. Bishop, *Practical Polymerization for Polystyrene*, Cahners Books, a division of Chaners Publishing Co., Inc., Boston, Mass., 1971.
143. D. C. Blackley, *Emulsion Polymerization—Theory and Practice*, Applied Science Publishers Ltd., London, 1975.
144. J. L. Amos, *Polym. Eng. Sci.* **14,** 1 (1974).
145. R. F. Boyer, *J. Macromol. Sci. Chem.* **A15,** 1411 (1981).
146. U.S. Pat. 3,884,766 (May 20, 1975), W. G. Bir and J. Novack (to Monsanto Company).
147. U.S. Pat. 3,966,538 (June 29, 1976), C. G. Hagberg (to Monsanto Company).
148. U.S. Pat. 3,719,720 (March 6, 1973), W. G. Bir and L. C. Tsang (to Monsanto Company).
149. U.S. Pat. 4,011,284 (March 8, 1977), G. Gawne and C. Ouwerkerk (to Shell Oil Co.).
150. U.S. Pat. 3,903,202 (Sept. 2, 1975), D. E. Carter and R. H. M. Simon (to Monsanto Co.).
151. U.S. Pat. 4,098,847 (July 4, 1978), J. L. Stevenson and R. A. Fuller (to Cosden Technology, Inc.).
152. E. Martuscelli, R. Palumbo, and M. Kryszewski, *Polymer Blends*, Plenum Press, New York, 1980.
153. J. Lignon, *Modern Plastics Encyclopedia*, Vol. 57, p. 317, 1980–1981.
154. Cincinnati Milacron, *Mod. Plast.* **57**(10A), 332 (1980).
155. *Hoover Universal Sales Literature Machine*, HV-450, Manchester, Mich., undated.
156. *Closeup*, *Plastics Design Forum*, 22 (May/June 1980).
157. H. Keskkula and J. W. Norton, Jr., *J. Appl. Polym. Sci.* **2,** 289 (1959).
158. K. J. Cleereman, *SPE J.* **23,** 43 (Oct. 1967); **25,** 55 (Jan. 1969).
159. P. A. M. Ellis, *Plast. Inst. Trans. J.* **35,** 537 (1967).
160. H. R. Jacobi, *Screw Extrusion of Plastics*, Iliffe Books Ltd., London, 1963.
161. G. A. Kruder, *SPE J.* **28,** 56 (Oct. 1972).
162. E. C. Bernhardt, *Processing of Thermoplastic Materials*, R. E. Krieger Publishing Company, Huntington, N.Y., 1974.
163. Z. Tadmor and I. Klein, *Engineering Principles of Plasticating Extrusion*, Van Nostrand Reinhold, Co., New York, 1970.
164. J. Fredos, ed., *Plastic Engineering Handbook*, Van Nostrand Reinhold Co., New York, 1976.
165. E. P. Weaver, *Polym. Eng. Sci.* **6,** 172 (1966).
166. C. I. Chung, *Plast. Eng.* **33,** 34 (Feb. 1977).
167. R. W. Brand and R. L. Keiks, *Plast. Technol.*, 37 (Feb. 1972).
168. N. Fountas, *Plast. World* **37,** 40 (Nov. 1979).
169. U.S. Pat. 4,039,609 (August 2, 1977), A. W. Thiel, H. Hell (to Bellaplast GmbH).
170. *Cycolac Brand ABS Polymers Sales Bulletin*, Marbon Chemical Division of Borg-Warner Corp., Chicago, Ill., 1980.
171. W. R. R. Park and J. Conrad, *Biaxial Orientation: Encyclopedia of Polymer Science and Technology*, Vol. 2, Interscience Publishers, a division of John Wiley & Sons, Inc., New York, 1965, p. 339.
172. K. J. Cleereman, W. J. Schrenk, and L. S. Thomas, *SPE J.* **24,** 27 (1968).
173. R. A. Bubeck, C. B. Arends, E. L. Hall, and J. B. Vander Sande, *Polym. Eng. Sci.* **21,** 624 (1981).
174. J. Starita, *Plast. World* **35,** 58 (April 1977).
175. D. E. James, *Polym. Eng. Sci.* **8,** 241 (1968).
176. T. Shimada, P. L. Horng, and R. S. Porter, *J. Rheology* **24,** 78 (1980).
177. Either standard procedures or methods supplied by instrument manufacturers apply.
178. D. Simpson, *Br. Plast.* **41,** 78 (May 1968); P. Shapras and G. C. Claver, *Anal. Chem.* **36,** 2282 (1964); L. Rohrschneider, *Z. Anal. Chem.* **255,** 345 (1977).
179. Ref. 1, p. 316.
180. L. D. Moore, W. W. Moyer, and W. J. Frazer, *Appl. Polym. Sci. Symp.* **7,** 67 (1968).
181. *Product Stewardship—Styrene Plastics*, Form. No. 304-109-1280, The Dow Chemical Company, Midland, Mich., 1980.
182. C. Hilado, ed., *Flammability Handbook for Plastics*, 2nd ed., Technomics Publishing Co., Inc., Westport, Conn., 1974.

183. R. N. Kennedy, in R. J. Bender, ed., *Handbook of Foamed Plastics*, Lake Publishing Co., Libertyville, Ill., 1965.
184. K. C. Frisch and J. H. Saunders, *Plastic Foams*, Vol. 1, Part 1, Marcel Dekker, Inc., New York, 1972.
185. U.S. Pat. 2,681,321 (June 15, 1954), F. Stasny and R. Gaeth (to Badische Anilin u. Soda-Fabrik).
186. U.S. Pat. 2,744,291 (May 8, 1956), F. Stasny and K. Buchholtz (to Badische Anilin u. Soda-Fabrik).
187. U.S. Pat. 2,787,809 (April 9, 1957), F. Stasny (to Badishe Anilin u. Soda-Fabrik).
188. U.S. Pat. 3,231,524 (Jan. 25, 1966), D. W. Simpson (to The Dow Chemical Co.); U.S. Pat. 3,089,875 (April 12, 1960), C. H. Pottenger (to Koppers Co., Inc.); U.S. Pat. 3,093,599 (Aug. 16, 1960), H. Muller-Tamm (to Badische Anilin u. Soda-Fabrik).
189. D. C. Wollard, *paper presented by the 12th Annual SPI Conference*, Washington, D.C., Oct. 16–18, 1967.
190. U.S. Pat. 4,249,875 (Feb. 10, 1981), E. Hart and R. Rutledge (to Cosden Technology, Inc.).
191. F. R. Bush and G. C. Rollefson, *paper presented at the 38th SPE-ANTEC*, New York, 1980.
192. *Low Temperature Systems*, Form No. 179-2086-77, The Dow Chemical Co., Midland, Mich., 1977.
193. J. B. Brooks and L. G. Rey, *J. Cell. Plast.* **9,** 232 (1973).
194. L. C. Rubens, *J. Cell. Plast.* **1,** 3 (1965).
195. U.S. Pat. 3,411,256 (Nov. 19, 1968), J. B. Best (to The Dow Chemical Co.).
196. U.S. Pat. 3,250,188 (May 10, 1966), G. A. Leonards (to The Dow Chemical Co.).
197. R. L. Chatman, *Package Eng.* **26,** 56 (July 1981).
198. A. R. Ingram and H. A. Wright, *Mod. Plast.* **41,** 152 (Nov. 1963).
199. S. J. Skinner, S. Baxter, and P. J. Grey, *Plast. Inst. Trans. J.* **32,** 180 (1964).
200. W. Kuryla and A. Papa, *Flame Retardancy of Polymeric Materials*, Vols. 1 and 2, Marcel Dekker, Inc., New York, 1973.
201. Business Communications Co., Inc., Stamford, Conn., 1978.
202. Papers, *SPI Proceedings from Ninth Annual Structural Foam Conference*, Palm Springs, Calif., March 24–26, 1981.
203. R. Martino, *Mod. Plast.* **55,** 34 (Aug. 1978).
204. J. G. Burt, *J. Cell. Plast.* **15,** 158 (1979).
205. V. T. Stannett, *Plast. World* **36,** 74 (Sept. 1978).
206. O. Sweeting, ed., *Science and Technology of Polymer Films*, Vols. 1 and 2, Interscience Publishers, a division of John Wiley & Sons, Inc., New York, 1968–1969.
207. J. Pinsky, ref. 206, Vol. 2.
208. W. R. R. Park, ed., *Plastics Film Technology*, Reinhold Publishing Corp., New York, 1969.
209. C. Saikaishi, *Jpn. Chem. Q.* **1,** 49 (1968).
210. Business Communications Co., Inc., Stamford, Conn., 1980.
211. W. R. R. Park and J. Conrad in N. M. Bikales, ed., *Encyclopedia of Polymer Science and Technology*, Vol. 2, Interscience Publishers, a division of John Wiley & Sons, Inc., 1965, pp. 339–373.
212. H. J. Karam, ref. 206, Vol. 1, pp. 227–253.
213. *Lamination of Trycite Plastic Films to Polystyrene Foam Sheet*, Bulletin No. 500-898-79, The Dow Chemical Co., Midland, Mich., 1979.
214. C. R. Scott, F. J. Butt, and J. Eichhorn, *Mod. Packag.* **38,** 135 (1965).
215. H. Yasuda, H. G. Clark, and V. Stannett in N. M. Bikales, ed., *Encyclopedia of Polymer Science and Technology*, Vol. 9, Interscience Publishers, a division of John Wiley & Sons, Inc., 1968, pp. 794–807.
216. C. E. Rogers in E. Baer, ed., *Engineering Design for Plastics*, Reinhold Publishing Corp., New York, 1964, pp. 609–688.
217. G. V. Schulz and H. Gerrens, *Z. Physik. Chem.* **1,** 182 (1956).
218. W. R. Vieth, P. H. Tam, and A. S. Michaels, *J. Colloid Interface Sci.* **22,** 360 (1966).

A. E. PLATT
T. C. WALLACE
The Dow Chemical Company

SUBERIC ACID, $HOOC(CH_2)_6COOH$. See Dicarboxylic acids.

SUCCINIC ACID AND SUCCINIC ANHYDRIDE

Succinic acid [*110-15-6*] (butanedioic acid), $C_4H_6O_4$, is a constituent of almost all plant and animal tissues. It has also been found in meteorites (1). Succinic anhydride [*108-30-5*] (3,4-dihydro-2,5-furandione) was first prepared by dehydration of the acid, but the direct hydrogenation of maleic anhydride gives nearly quantitative yields.

Uses of succinic acid range from scientific applications such as radiation dosimetry and standard buffer solutions to applications in agriculture, food, medicine, plastics, cosmetics, textiles, plating, and waste-gas scrubbing.

Succinic acid was first obtained as the distillate from amber (Latin, *succinum*) for which it was named. It occurs in beer, meat, molasses, eggs, peat, coal, fruits, honey, and urine. It is formed by the chemical and biochemical oxidation of fats, alcoholic fermentation (qv) of sugar, and in numerous catalyzed oxidation processes. It is a by-product in the manufacture of adipic acid (qv).

Although first prepared by Agricola in 1550, commercial manufacture in the United States began in 1929 when National Aniline started an electrolytic reduction process of maleic anhydride using lead cathodes and a sulfuric acid electrolyte. Succinic anhydride manufacture was started in 1943 using an azeotropic distillation process to dehydrate crystalline succinic acid. Since 1946, National Aniline and then its successor, the Buffalo Color Corporation, have produced succinic anhydride by catalytic hydrogenation of maleic anhydride. The acid is prepared by the hydration of distilled anhydride.

Physical Properties

The acid occurs both as colorless triclinic prisms (α form) and as monoclinic prisms (β form). The latter are triboluminescent and are stable up to 137°C. Both forms dissolve in water, alcohol, diethyl ether, glacial acetic acid, anhydrous glycerol, acetone, and various aqueous mixtures of the last two solvents. Succinic acid sublimes when heated below its melting point; it is purified by sublimation when heated above its melting point (2). In the latter case, the anhydride is formed. A solvent azeotrope, eg, *o*-dichlorobenzene, dehydrates succinic acid at ca 175–220°C (3).

Succinic acid is absorbed from aqueous solution by charcoal or anion-exchange resins (4).

Succinic anhydride forms rhombic pyramidal or bipyramidal crystals. It is relatively insoluble in water and ether, but soluble in boiling chloroform, ethyl acetate, and alcohol. Physical properties of the acid and the anhydride are given in Table 1.

Chemical Properties

Succinic acid undergoes most of the reactions characteristic of dicarboxylic acids (see also Dicarboxylic acids). In addition, the active methylene groups are responsible for many interesting reactions.

Table 1. Physical Properties of Succinic Acid and Succinic Anhydride

Property	Succinic acid	Succinic anhydride	Ref.
mp, °C	188.1	119.6	5
bp, °C	dehydrates at mp	261	
sublimation pt at 267 Pa[a], °C	156–157	90	
specific gravity	1.552–1.577	1.572	
solubility, g/100 g soln			
in water at 0°C	2.88		
at 100°C	121		
96% alcohol, at 15°C	9.99		
ether, at 15°C	1.25		
methylene chloride at bp	insoluble	6.6	3
chloroform at bp	insoluble	3.7	
dissociation constant at 25°C			
$K_1 \times 10^{-5}$	6.52–6.65		
$K_2 \times 10^{-6}$	2.2–2.7		
pK_1 in absolute ethyl alcohol	9.58		6
pK_2 in absolute ethyl alcohol	12.11		6
heat of combustion, J/mol[b]	1491.18 ± 0.19	1554.4 ± 7.7	6
heat of formation at 298.15 K, J/mol[b]	940.35 ± 0.54		6
heat capacity at 298.15 K, J/(mol·K)[b]	152.9		7
heat of solution in 60 pts water, J/mol[b]	27,313		
dielectric constant at 3–97°C, 5 kHz	2.29–2.90		

[a] To convert Pa to μm Hg, multiply by 7.5.
[b] To convert J to cal, divide by 4.184.

On heating, loss of water causes succinic acid to form an internal anhydride with a stable ring structure. Further heating gives the dilactone of γ-ketopimelic acid (8). This reaction can occur with explosive violence when molten succinic anhydride is distilled in the presence of alkali ions. It does not necessarily terminate in a bimolecular reaction.

2 (succinic anhydride) $\xrightarrow{Na^+}$ (dilactone) + $CO_2\uparrow$

Any equipment used for handling succinic anhydride must be scrupulously cleaned to remove traces of alkali if this reaction is to be avoided.

Halogenation. When heated with bromine in a sealed vessel at 100°C, succinic acid yields meso-2,3-dibromosuccinic acid almost quantitatively. The reaction takes place in the presence or absence of water, but with excess water the production of brominated hydrocarbons reduces the yield of the dibromo acid.

The anhydride, when heated with an equivalent of bromine, gives mainly the monobromo derivative (bp, 130–133°C); two moles bromine give *dl*-2,3-dibromosuccinic anhydride (mp, 118–119°C).

Diethyl succinate treated with excess chlorine in sunlight yields small needles of bis(pentachloroethyl) tetrachlorosuccinate,

$$\mathrm{Cl_3CCl_2CO\overset{O}{\overset{\|}{C}}CCl_2CCl_2\overset{O}{\overset{\|}{C}}OCCl_2CCl_3}$$

Succinic acid reacts with an equal molecular amount of phosphorus pentachloride or thionyl chloride to yield succinyl chloride.

The anhydride and thionyl chloride at 110°C over a DMF catalyst give a 90% yield of succinyl chloride (9); phosgene reacts similarly (10).

Condensation with Aldehyde and Ketones. In the presence of a catalyst, eg, thorium sulfate, succinic anhydride and succinic esters react in the vapor phase with excess formaldehyde to yield citraconic anhydride (11). In liquid carbon dioxide with excess formaldehyde, a copolymer of formaldehyde and succinic anhydride is formed (12).

Succinic esters condense with aldehydes and ketones in the presence of bases (eg, sodium alkoxide, piperidine) to form half esters of alkylidenesuccinic acids. This reaction, known as the Stobbe condensation, is specific for succinic esters and substituted succinic esters (13–14). First an aldol or ketol is formed which loses alcohol to give a γ-lactone ester, a so-called paraconate. Rearrangement gives the alkylidenesuccinic acid half ester. The condensation between acetone and diethyl succinate is typical:

$$\mathrm{(CH_3)_2C{=}O + C_2H_5O\overset{O}{\overset{\|}{C}}CH_2CH_2\overset{O}{\overset{\|}{C}}OC_2H_5 \xrightarrow{NaOC_2H_5} (CH_3)_2C(OH)CH(\overset{O}{\overset{\|}{C}}OC_2H_5)CH_2\overset{O}{\overset{\|}{C}}OC_2H_5 \xrightarrow{-C_2H_5OH}}$$

$$\text{ethyl 2,2-dimethylparaconate (}\mathrm{CO_2C_2H_5}\text{ lactone)} \longrightarrow \mathrm{(CH_3)_2C{=}C(\overset{O}{\overset{\|}{C}}OC_2H_5)CH_2\overset{O}{\overset{\|}{C}}OH}$$

ethyl tetrahydro-2,2-dimethyl-5-oxo-3-furancarboxylate (ethyl 2,2-dimethylparaconate) — 1-ethyl 2-isopropylidenesuccinate

Dialkylidenesuccinic acids and anhydrides are formed in a similar manner from aromatic aldehydes:

$$\mathrm{2\ C_6H_5CH{=}CHCHO + NaO\overset{O}{\overset{\|}{C}}CH_2CH_2\overset{O}{\overset{\|}{C}}ONa \xrightarrow[(2)\ HCl]{(1)\ (CH_3\overset{O}{\overset{\|}{C}})_2O,\ CH_3\overset{O}{\overset{\|}{C}}ONa}}\ \text{(}\mathrm{C_6H_5CH{=}CHCH{=}}\text{)}_2\text{-succinic anhydride}$$

cinnamaldehyde — 2,3-dicinnamylidenesuccinic anhydride

Diketones. Although the Stobbe condensation in general takes place in preference to the Claisen reaction, ketones containing reactive methyl or methylene groups form diketones with succinates in the presence of sodium hydride (15) with neither reaction predominating. For example, *tert*-butyl succinate and acetophenone give a 19–34% yield of 1,8-diphenyl-1,3,6,8-octanetetraone in addition to a mixture of the half-esters of β-methyl-β-phenylitaconic acid and its isomers:

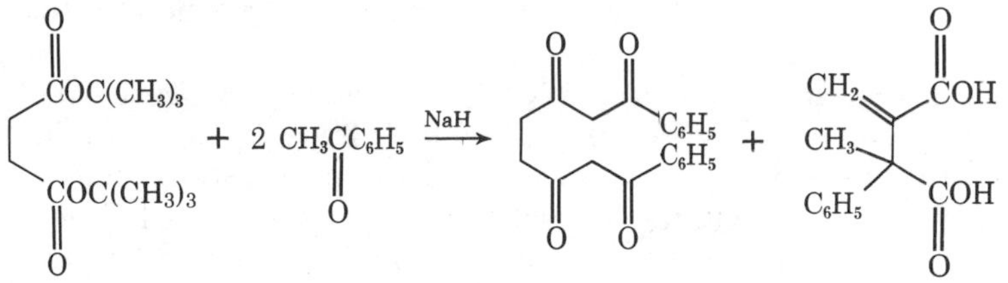

Friedel-Crafts Reactions. Under Friedel-Crafts conditions, succinic anhydride forms alkylbenzoylpropionic acids with alkylbenzenes. In the acylation of indane, the yield is 97% (16):

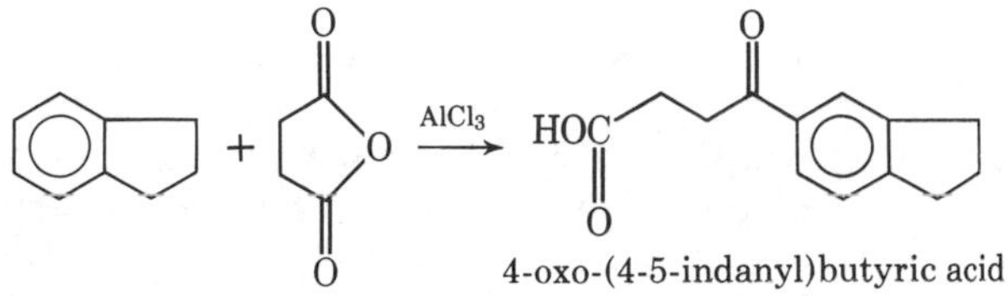

4-oxo-(4-5-indanyl)butyric acid

Reference 16 gives tables and an excellent discussion of succinic anhydride condensations (see Friedel-Crafts reactions). The recent synthesis of 3-(4-biphenylcarbonyl)propionic acid, an inhibitor of blood platelet aggregation (17), is based on Friedel-Crafts acylation:

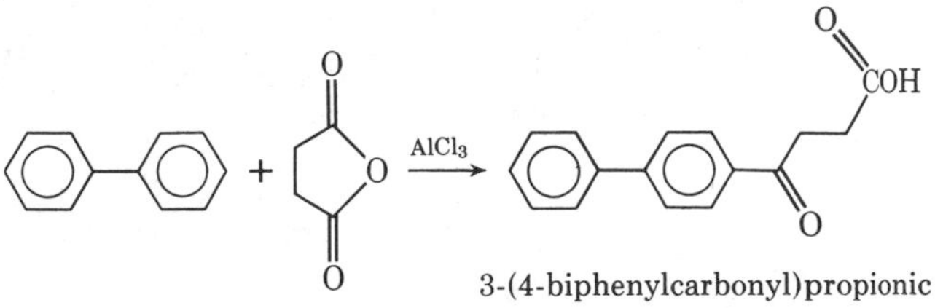

3-(4-biphenylcarbonyl)propionic acid

Esterification. Succinic acid and its anhydride are readily esterified by the usual methods. Esterification is the basis of numerous processes for the recovery and isolation of succinic acid and its esters from the waste streams of adipic acid processes (18–21).

Monomethyl succinate (mp, 58°C) and dimethyl succinate (mp, 19°C) are prepared from acetylene, methanol, and carbon monoxide (22). Dimethyl succinate is also obtained from carbon monoxide, ethylene, methanol, and oxygen in the presence of a palladium catalyst (23). A mixture of mono and diesters is prepared in a flow system by passing succinic acid and methanol or ethanol over activated alumina at room temperature (24).

Esterifications with starch (25–28), cellulose (29), polyols (30–32), and other alcohols (33–34) are responsible for many of the uses of succinic acid and anhydride.

Reactions with Amino Compounds. Succinimide (mp, 126°C) can be prepared from succinic acid or its anhydride with ammonia (35), urea (36), adipamide (37), or an isocyanate. In the commercial process a concentrated aqueous solution of diammonium succinate is heated until water and ammonia are no longer evolved. The molten product is fractionated, and the fraction boiling at 285–290°C is collected.

$$\text{(CH}_2\text{COONH}_4)_2 \xrightarrow{\Delta} \text{succinimide (NH)} + 2\,H_2O + NH_3$$

Succinimides are halogenated with hypobromous or hypochlorous acid. Iodination of succinimide silver salt gives *N*-iodosuccinimide.

N-Bromosuccinimide (mp, 176–177°C) is formed by the addition of bromine to a cold, aqueous solution of succinimide in sodium hydroxide. *N*-Bromosuccinimide is a bromination and oxidation agent in the synthesis of cortisones and other hormones (see also Hormones; Steroids). Its unusual action causes selective halogen substitution at methylene groups adjacent to unsaturated bonds without adding to the ethylene group. By varying the reaction conditions, it can be used as a reagent for selective substitution, including substitution of side chains attached to an aromatic nucleus, controlled dehydrogenation, nuclear addition to aromatic compounds, bromination of bromohydrins, and selective oxidation of alcohol groups to aldehydes or ketones. Olefins, carbonyls, aromatic and heterocyclic compounds, and even saturated compounds have been used in this reaction.

N-Chlorosuccinimide (mp, 150–151°C) is a powerful germicide and deodorant. It has been widely used as disinfectant for drinking water and, to a limited extent, as a halogenating agent similar to *N*-bromosuccinimide. The crystals are orthorhombic with a distinctive, chlorinelike odor (see Chloramine and bromamine).

Ethylenediamine heated with two moles succinic anhydride gives *N*,*N*′-ethylenedisuccinimide:

$$H_2NCH_2CH_2NH_2 + 2\,\text{(succinic anhydride)} \xrightarrow[\Delta]{} \text{(succinimido)}NCH_2CH_2N\text{(succinimido)} + 2\,H_2O$$

Succinimide derivatives are used in pharmaceuticals as anticonvulsants (see Hypnotics, sedatives, and anticonvulsants).

Succinic acid reacts with urea in aqueous solution to give a 2:1 compound (mp, 141°C) (38–39). Its sparing solubility in water forms the basis for recovery of succinic acid from adipic acid manufacture (40–41). Succinamic acid,

$$NH_2\overset{O}{\overset{\|}{C}}CH_2CH_2\overset{O}{\overset{\|}{C}}OH$$

is formed by the reaction of succinic anhydride with ammonia or by partial hydrolysis of succinimide. This reaction is reversed by heating with a dehydrating agent.

Succinamide, $(CH_2CONH_2)_2$, (mp, 268–270°C) is obtained from succinyl chloride with ammonia or by the partial hydrolysis of succinonitrile. Warming succinimides with a primary amine gives *N*-alkylsuccinamides:

$$\text{succinimide (NH)} + 4\,RNH_2 \xrightarrow[\Delta]{} (CH_2CONR_2)_2 + 3\,NH_3$$

Succinic anhydride is aminated when heated with aromatic amines in a eutectic melt (42).

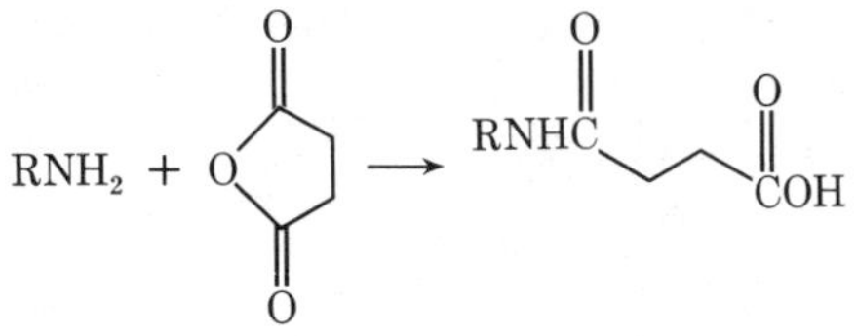

$$RNH_2 + \text{(succinic anhydride)} \longrightarrow RNHC(=O)CH_2CH_2C(=O)OH$$

Reactions with Sulfur Compounds. Diethyl or diphenyl succinate react with potassium hydrogen sulfide to form dipotassium dithiosuccinate. Acidification gives thiosuccinic anhydride with the evolution of hydrogen sulfide:

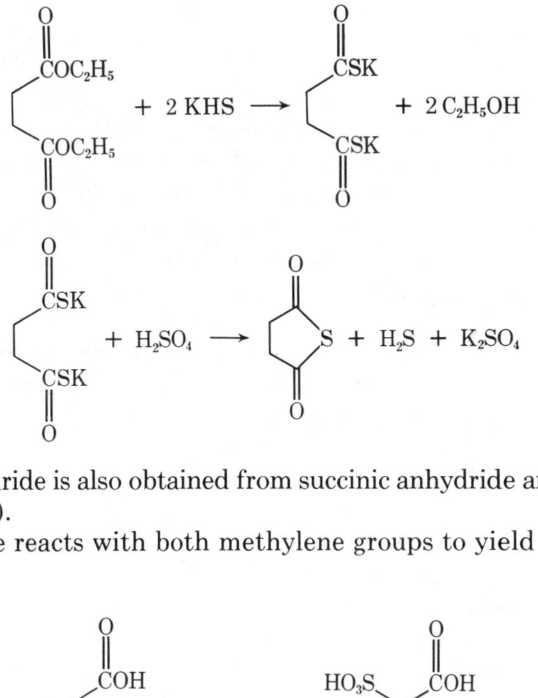

$$C_2H_5OC(=O)CH_2CH_2C(=O)OC_2H_5 + 2\ KHS \longrightarrow KSC(=O)CH_2CH_2C(=O)SK + 2\ C_2H_5OH$$

$$KSC(=O)CH_2CH_2C(=O)SK + H_2SO_4 \longrightarrow \text{(thiosuccinic anhydride)} + H_2S + K_2SO_4$$

This succinic anhydride is also obtained from succinic anhydride and hydrogen sulfide under pressure (43).

Sulfur trioxide reacts with both methylene groups to yield 2,3-disulfosuccinic acid:

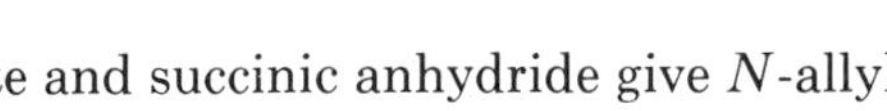

$$HOC(=O)CH_2CH_2C(=O)OH + 2\ SO_3 \longrightarrow HOC(=O)CH(SO_3H)CH(SO_3H)C(=O)OH$$

Allyl isothiocyanate and succinic anhydride give N-allylsuccinimide by loss of carbon oxysulfide:

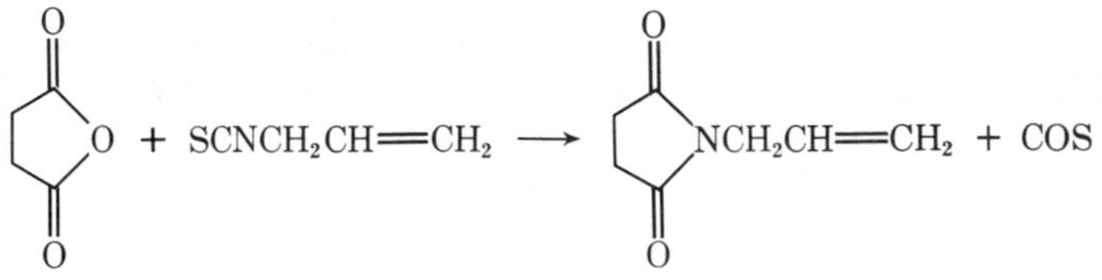

$$\text{(succinic anhydride)} + SCNCH_2CH{=}CH_2 \longrightarrow \text{(succinimide)}NCH_2CH{=}CH_2 + COS$$

Hydration. Succinic anhydride is hydrated practically instantaneously in boiling water. This high reaction rate, as compared to glutaric anhydride, has been attributed to conformational strain (44).

Oxidation. The products of the reaction of succinic acid with hydrogen peroxide depend upon the conditions. Products include peroxysuccinic acid, $(CH_2COOOH)_2$, oxosuccinic acid (oxalacetic acid), malonic acid, or a mixture of acetaldehyde and malonic and malic acids. Succinic anhydride and H_2O_2 in DMF or *N*-methyl-2-pyrrolidinone give monoperoxysuccinic acid,

$$HO\overset{O}{\overset{\|}{C}}CH_2CH_2\overset{O}{\overset{\|}{C}}OOH$$

(mp, 107°C) (45).

Potassium permanganate oxidizes succinic acid to oxalic acid or a mixture of malic and tartaric acids. Sodium perchlorate oxidation yields 3-hydroxypropionic acid.

Reduction. Catalytic hydrogenation of succinic acid or succinic anhydride yields 1,4-butanediol (46), γ-butyrolactone (47), tetrahydrofuran (48), or mixtures of these compounds, depending upon the catalyst and reaction conditions (49). Hydrogenation in a 1,4-butanediol solvent gives γ-hydroxybutyric acid (50).

Degradation. Heating of succinic acid and its salts yields γ-ketopimelic dilactone; cyclohexane-1,4-dione; ethane; a mixture of propionic acid, acetic acid, acrylic acid, and acrolein; oxalic acid; cyclopentanone; and furan. High melting hydrocarbons are formed by heating succinic acid in the presence of zinc at 350–400°C. The electrolysis of succinic acid gives ethylene and acetylene.

Manufacture

Numerous processes are available for the production of succinic acid and its anhydride. The choice depends upon the raw materials available, labor costs, energy considerations, and technology.

Annual world production of adipic acid is in the order of 1.8×10^6 metric tons. Large amounts of by-product succinic acid are obtained, and various techniques are available for its recovery, including separation as the urea adduct (41,51), formation of succinimide (41), extraction (52–56), selective crystallization (57–62), distillation to recover succinic anhydride (63–68), and esterification followed by distillation. The ester mixture may be used as such or separated by fractionation (18–21,69–71). The methyl ester can be converted to the anhydride by a catalytic process (72) in addition to saponification techniques for reclaiming the acid.

Dehydration of succinic acid to its anhydride is followed by fractionation or extraction (3,73–76).

Oxidation of paraffins C_{12}–C_{32} gives mixtures of acids from which succinic acid can be separated (77–78). Oxidation of naphthenic acid with nitric acid in the presence of a vanadium catalyst gives a mixture containing mostly succinic acid (79). Cycloolefin oxidation with ozone (80) or ozone followed by hydrogen peroxide (81) gives better selectivities for succinic acid. The manufacture of succinic acid from furfural by hydrogen peroxide oxidation is claimed in a USSR patent (82). Succinate salts are obtained by the oxidation of butane-1,4-diol in sodium hydroxide solution (83), and of γ-hydroxybutyric acid in the presence of a palladium catalyst with alkaline-earth hydroxides (84).

A Japanese patent claims a nearly quantitative yield of disodium succinate from autoclaving a mixture of acrylonitrile, NaCN, and water in the presence of a palladium catalyst (85).

A U.S. patent (86) describes the reaction of acrylic acid with carbon monoxide in the presence of oleum [$SO_3(H_2SO_4)$]. The product is hydrated with ice, washed free of sulfuric acid, and dried to give an 85% yield of succinic acid (mp, 184–185°C).

Other routes starting with carbon monoxide include the reaction of acetylene with carbon monoxide (87) and the oxidative carbonylation of ethylene in the presence of methanol and a catalyst to give methyl esters of succinic acid (88). Succinic acid yields as high as 75–80% calculated on acetylene are claimed for the first process.

A U.S. patent describes a vapor-phase, catalytic process in which 2-butene is oxidized to give a mixture of 45% fumaric acid and 55% succinic acid in the condensate (89). This product can be dissolved in water and hydrogenated to give succinic acid exclusively. The space–time yield is low in the single example cited in the patent, but improvements in catalyst and operating conditions could lead to a most attractive process. *n*-Butane feedstock might give better yields.

The preparation of succinic acid by fermentation has been studied extensively in Japan (90–93). Yields of succinic acid from glucose as high as 51% based on glucose charged have been reported. Normal paraffins (94), sucrose (95), acetone (96), and isopropyl alcohol (96) are among the feedstocks utilized. Separation methods include lead salt precipitation (97) and ion exchange (98). Although the examples cited in the literature may not be commercially acceptable at this time, the use of a cheap, renewable raw material such as sucrose, or an inexpensive petroleum feedstock could make this route attractive to a producer employing fermentation technology.

The catalytic hydrogenation of maleic acid or its anhydride continues to be a preferred method for the synthesis of succinic acid and its anhydride (see Fig. 1). Palladium on carbon (99), palladium on $CaCO_3$ (100), and carbon-carrier catalysts containing rhodium and ruthenium (100) have been cited as catalysts for the hydrogenation of maleic acid. Nickel on kieselguhr (101), palladium on alumina (102), and calcined nickel aluminate (103) have been mentioned as catalysts for the hydrogenation of maleic anhydride. Reactors similar to those described for phenol hydrogenation can be used (104).

Electrolytic methods for the reduction of maleic acid to succinic are described in Japanese patents (105–106).

Succinic anhydride is made from the acid by removing the water by azeotropic distillation (3), or by using a hydrophilic solvent that condenses at a temperature above that of water vapor (107). In either case, the anhydride is purified by fractionation (see Azeotropic and extractive distillation).

Manufacture of the acid from the anhydride is much simpler. Sales-grade anhydride is dissolved in boiling water. The solution is cooled, and the crystals formed are separated and dried.

The price development of succinic anhydride in tank-car quantities is given

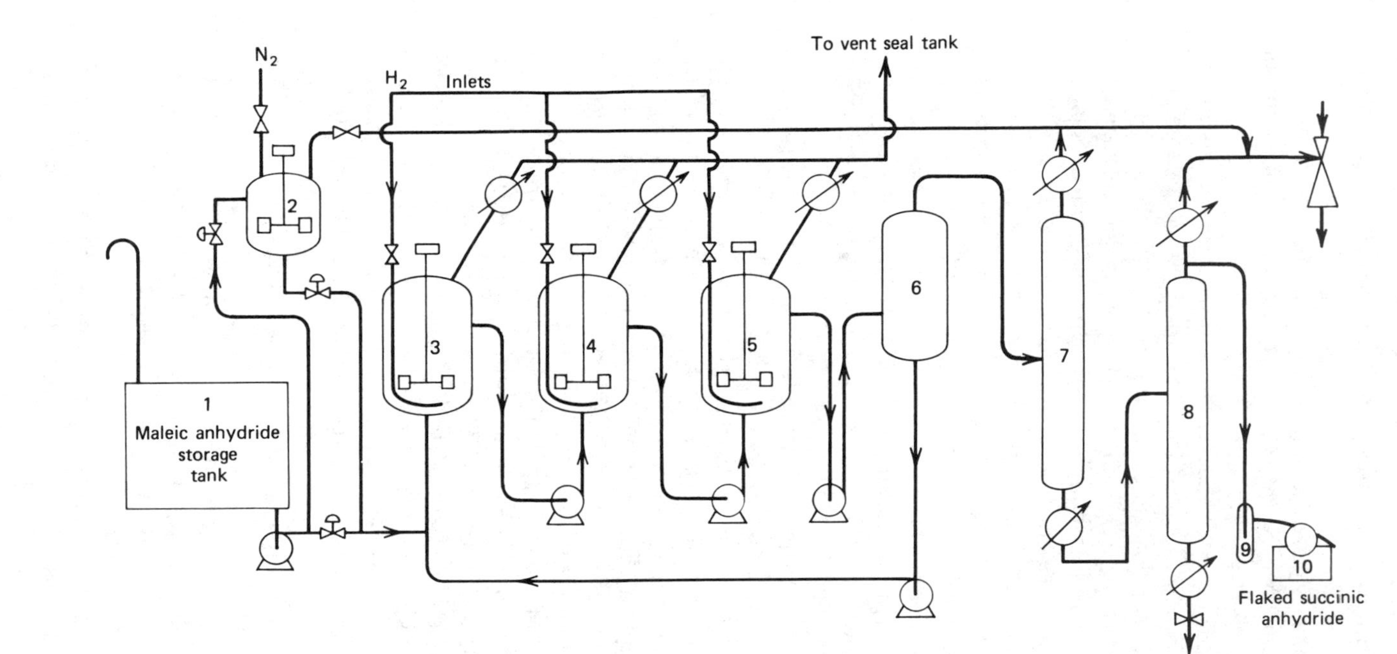

Figure 1. Continuous process for maleic anhydride hydrogenation. This flow chart is based on use of 5% Pd on alumina catalyst received at plant as 50%, water-wet mud. Hydrogenator pressure is 310–930 kPa (30–120 psig) at 120–160°C.

Item no.	*Equipment name*	*Services required*
1	maleic anhydride storage tank	140 kPa (5 psig) steam
2	catalyst dryer and feed tank	310 kPa (30 psig) steam, electricity, nitrogen, vacuum
3	first-stage hydrogenator	930 kPa (120 psig) steam, electricity, nitrogen, hydrogen, cooling water
4	second-stage hydrogenator	930 kPa (120 psig) steam, electricity, nitrogen, hydrogen, cooling water
5	third-stage hydrogenator	930 kPa (120 psig) steam, electricity, nitrogen, hydrogen, cooling water
6	continuous catalyst separator	510 kPa (60 psig) steam, electricity, cooling water, nitrogen
7	lights column	1.82 MPa (250 psig) steam, vacuum, 140 kPa (5 psig) steam, cooling water
8	sales column	2.5 MPa (350 psig) steam, vacuum, 140 kPa (5 psig) steam, cooling water
9	barometric seal	310 kPa (30 psig) steam
10	flaker	310 kPa (30 psig) steam, electricity, cooling water

below:

Year	*$/kg*
1970	0.99
1975	1.98
1980	2.64

Specifications. Commercial specifications of succinic acid and succinic anhydride are given in Table 2.

Analytical Methods

Analytical methods for succinic acid and its anhydride are largely dependent upon the material being analyzed. Simple titration in conjunction with mixed melting points is often sufficient for pure material. Gas chromatography has been recommended for the analysis of plant tissues and fruit juices (110). Methylation with BF_3-methanol followed by glc has been recommended for citrus-tissue analysis (111). Thermal aqueous liquid chromatography has been recommended for determining succinic acid in a mixed acid stream in a manufacturing plant (112). An enzymic method has been developed for the analysis of succinic acid in wine (qv) (113). Mass spectrometry (114) and a combination of gas chromatography with mass spectrometry (115) are used for the analysis of succinic acid in human urine. Thin-layer chromatography is used for the separation of succinic acid from other acids (116–117). Different liquid chromatographic techniques for the analysis of succinic acid in mixture with other acids are compared in ref. 118.

Health and Safety Factors

The GRAS status of succinic acid has been affirmed by the FDA as a flavor enhancer, as a pH control agent in condiments and relishes, and for use in meat products

Table 2. Specifications of Succinic Acid[a] and Succinic Anhydride[b]

Property	Succinic acid	Succinic anhydride
mp, °C, min	187.0	118.3[c]
physical appearance	small white crystals	white flakes
total acidity, %, min	99.0[d]	99.5[e]
unsaturated compounds[f], %, max		0.5
ash, %, max	0.1	
chlorides, %, max		0.15
iron (as Fe), ppm, max	10	
lead (as Pb), ppm, max	5	
sulfates (as SO_4), %, max	0.05	0.04
heavy metals, ppm, max	10	20
turbidity, 5% aq soln, nephelos units[g], max	20	10
fineness[h], %, min	100	

[a] Ref. 108.
[b] Ref. 109.
[c] Solidification pt.
[d] As succinic acid.
[e] As succinic anhydride.
[f] As maleic anhydride.
[g] Coleman Instruments Corporation.
[h] A 2-mm (10-mesh) screen.

Table 3. Uses of Succinic Acid

Use	Ref.
Scientific	
calorimetric standard	129–130
component of standard buffer solutions	131–133
reactor radiation dose rate measurement	134–137
Medicinal	
antidote for toxic substances	138–141
succinylated gelatin as blood plasma substitute	142
succinic acid derivatives in pharmaceuticals	143–153
Agricultural	
seed treatment to enhance germination, growth, and yield	154–155
root stimulator for grapes	156–157
plant-growth accelerator	158–159
soil chelating agent	160
crop-dust antistat	161
cut flower preservative	162
component of barley seed antismut treatment	163
plant-growth regulator	164
Food	
preservation of chicken against *Salmonella*	165–166
dog-food preservative	167
sodium succinate used as flavor enhancer	168
table salt substitute component	169
edible surfactants from esters	170
bread-softening agent	171–173
microencapsulation of flavoring oils	174
protein gelatination	175–177
catalyst for food seasoning preparation	178
Plating	
chemical metal plating	179–180
electroplating baths for Ni, Ag, Cr, Fe, and Au	181–185
anodizing aluminum	186–187
Corrosion inhibition	
succinate ion solutions	188–191
oil derived by succinylation of reaction product of fatty acids with alkanolamines	192
succinate plus succinimide	193–194
Soldering	195–197
Gas scrubbing	
sulfide scrubbing	198–199
sulfur dioxide removal from combustion gases	200–201
Textile	
size preparation	25
shrink-proofing wool	202
modification of polyester resin to improve dyeability	203
modification of caprolactam to improve viscosity stability and fire resistance	204–205
modification of aramid fibers when incorporating a uv absorber	205
preparation of antistats and oil repellents	206–207
Photography	
development accelerators	208
antifoggants	209
photographic stripping layers	210
Cosmetics	
depilatory component	211
acidic ingredient in toothpaste and denture cleaner	212–213

Table 3 (*continued*)

Use	Ref.
Plastics and resins	
color stabilizer for PVC and acrylonitrile resins	214–215
cross-linking agent for acrylic films	216
ingredient of polymers used for water-based paints	30, 217
component of polymers used in electrical insulating tape	30, 217
incorporated in resins used for adhesive bonding	218
component of radiation curable resins	219
cross-link promoter for phenol–formaldehyde resins	220
Detergents and emulsifiers	
starch esters as paint dispersants	26–27
reagent for preparation of nonphosphate detergent builder	221
ester surfactant	222
reagent to prepare emulsifier for vinyl polymerizations	223
Catalysts	
anticlumping agent in tetrafluoroethylene polymerization	224
mechanical stability enhancer for synthetic rubber	225
rubber protector in brake oil	226
dehydration catalyst for monocarboxylic acids to anhydrides	227
Miscellaneous	
esters used in synthetic lubricants	228
absorption refrigeration (ester use)	229
gasoline additive (ester use)	230
miticide (ester use)	33
crystal habit modifier for gypsum	

(119). There is no evidence that succinic acid poses a hazard at concentrations now in use (120). However, caution is advised when dealing with hot vapors given off from succinic acid or anhydride, which may induce coughing (121). Succinic acid is a culture medium for many bacterial strains (122–125), and a sustained concentration of succinic acid in nasal passages caused by breathing hot vapors is to be avoided. A carbon respirator is recommended when handling hot succinic acid or anhydride.

Molten succinic anhydride, being both a dehydrating agent and an acid, causes severe burns on the skin. The molten anhydride should be handled with gloves.

Wastewater from succinic acid processes is suitable to biological degradation by activated sludge (126). Polymeric sorbents (127) and ferric chloride treatment processes (128) can also be used for wastes containing succinic acid.

Uses

A survey of uses of succinic acid and its anhydride is given in Table 3.

BIBLIOGRAPHY

"Succinic Acid and Succinic Anhydride" in *ECT* 1st ed., Vol. 13, pp. 180–191, by W. Howlett Gardner and Lawrence H. Flett, National Aniline Division, Allied Chemical & Dye Corporation; "Succinic Acid and Succinic Anhydride" in *ECT* 2nd ed., Vol. 19, pp. 134–150, by Paul Turi, Sandoz, Inc.

1. J. G. Lawless and co-workers, *Nature* **251**(5470), 40 (1974).
2. U.S. Pat. 1,929,381 (Oct. 3, 1933), A. O. Jaeger and F. P. Fiedler (to Selden Co.).
3. U.S. Pat. 3,014,070 (Dec. 19, 1961), H. Chafetz (to Texaco, Inc.).
4. A. K. Kadyrov and co-workers, *Zh. Prikl. Khim.* **42**(2), 417 (1969).

5. Jpn. Pat. 69 05,857 (Mar. 12, 1969), Y. Inoue and co-workers (to Takeda Chem. Inds., Ltd.).
6. C. E. Vanderzee and co-workers, *J. Chem. Thermodyn.* **4**(4), 533 (1972).
7. C. E. Vanderzee and E. F. Westrum, Jr., *J. Chem. Thermodyn.* **2**(5), 81 (1970).
8. U.S. Pat. 2,302,321 (Nov. 17, 1942), H. Hopf and H. Griesshaber (to Alien Property Custodian).
9. Ya. A. Serednitskii and co-workers, *Visn. L'viv Polstekh. Inst.* **58,** 21 (1971).
10. U.S. Pat. 3,810,940 (May 14, 1974), C. F. Hauser (to Union Carbide Corp.).
11. Ger. Offen. 2,352,468 (May 30, 1974), R. Berg and co-workers (to Pfizer, Inc.).
12. H. Yokota and co-workers, *Bull. Chem. Soc. Jpn.* **42**(8), 2343 (1969).
13. H. Stobbe, *Ber.* **26,** 2312 (1893).
14. W. S. Johnson and G. H. Daub in Vol. 6 of R. Adams and co-eds., *Organic Reactions*, John Wiley & Sons, Inc., New York, 1951, pp. 2–73.
15. G. H. Daub and W. S. Johnson, *J. Am. Chem. Soc.* **72,** 501 (1950).
16. G. Peto in G. A. Olah, ed., *Friedel-Crafts and Related Reactions*, Vol. III, Pt. 1, Interscience Publishers, a division of John Wiley & Sons, Inc., New York, 1964, pp. 550–663.
17. U.S. Pat. 3,966,960 (June 29, 1976), L. Ellenbogen and co-workers (to American Cyanamid Co.).
18. U.S. Pat. 3,726,888 (Apr. 10, 1973), J. L. Hatten and co-workers (to El Paso Products Co.).
19. E. Siimen, Y. Kann, and M. Merendi, *Tr. Tallin Politekh. Inst. Ser. A* **238,** 87 (1966); *Chem. Abstr.* **67,** 43381a (1967).
20. U.S. Pat. 4,105,856 (Aug. 8, 1978), C. A. Newton (to El Paso Prod. Co.).
21. U.S. Pat. 3,991,100 (Nov. 9, 1976), S. Hochberg (to E. I. du Pont de Nemours & Co.).
22. P. Pino and A. Miglierina, *J. Am. Chem. Soc.* **74,** 5551 (1952).
23. Jpn. Kokai 78 40709 (Apr. 13, 1978), S. Umemura and co-workers (to Ube Ind., Ltd.).
24. H. Kojima and co-workers, *Kogyo Kagaku Zasshi* **74**(10), 2207 (1971).
25. Ger. Offen. 2,048,350 (May 27, 1971), F. J. Germino and co-workers (to CPC International, Inc.).
26. U.S. Pat. 4,061,610 (Dec. 6, 1977), R. C. Glowsky and co-workers (to Sherwin-Williams Co.).
27. U.S. Pat. 4,061,611 (Dec. 6, 1977), R. C. Glowsky and co-workers (to Sherwin-Williams Co.).
28. U.S. Pat. 4,231,803 (Nov. 4, 1980), E. M. Bovier and co-workers (to Anheuser-Busch, Inc.).
29. J. A. Cuculo, *Text. Res. J.* **41**(5), 375 (1971).
30. U.S. Pat. 3,329,635 (July 4, 1967), T. J. Miranda (to O'Brien Corp.).
31. K. Geckeler and E. Bayer, *Polym. Bull.* **3**(6–7), 347 (1980).
32. Ger. Offen. 2,012,526 (Oct. 7, 1971), D. Stoye and co-workers (to Chemische Werke Huels A.G.).
33. U.S. Pat. 3,948,961 (April 6, 1976), C. A. Henrick and co-workers (to Zoecon Corp.).
34. U.S. Pat. 3,846,479 (Nov. 5, 1974), J. D. Zech (to ICI Americas, Inc.).
35. Ger. Offen. 2,313,386 (Sept. 19, 1974), E. Fuerst and co-workers (to BASF A.G.).
36. Israeli Pat. 38,852 (Nov. 29, 1974), A. Stern and co-workers (to Makhteshim Chem. Works Ltd.).
37. U.S. Pat. 3,818,081 (June 18, 1974), E. G. Adamek (to DuPont of Canada, Ltd.).
38. L. Semenova and co-workers, *Dokl Akad Nauk Uzb. SSR* **30**(8), 23 (1973).
39. C. H. Walker, *Sep. Sci.* **2**(3), 399 (1967).
40. Jpn. Kokai 77 118,414 (Oct. 4, 1977), J. Nishikido and co-workers (to Asahi Chemical Industries Co. Ltd.).
41. Jpn. Kokai 77 78,818 (July 2, 1977), J. Nishikido and co-workers (to Asahi Chemical Industries Co. Ltd.).
42. H. Singh and co-workers, *Indian J. Chem.* **14B**(10), 809 (1976).
43. T. Takido and co-workers, *Yuki Gosei Kagaku Kyokai Shi* **31**(10), 826 (1973).
44. L. Eberson and co-workers, *Acta Chem. Scand.* **26**(1), 239 (1972).
45. Brit. Pat. 1,139,507 (Jan. 8, 1969), E. G. E. Hawkins (to Distillers Co. Ltd.).
46. Ger. Offen. 2,715,667 (Oct. 12, 1978), D. Freudenberger and co-workers (to Hoechst A.G.).
47. Ger. Offen. 2,553,761 (June 2, 1977), D. Freudenberger and co-workers (to Hoechst A.G.).
48. Ger. Offen. 1,939,882 (Aug. 10, 1968), M. Yamaguchi and co-workers (to Mitsubishi Chemical Industries Co., Ltd.).
49. Ger. Offen. 2,133,768 (Jan. 20, 1972), J. Kanetaka and S. Mori (to Mitsubishi Petrochemical Co., Ltd.).
50. Jpn. Kokai 73 15,821 (Feb. 28, 1973), Y. Fujita and co-workers (to Teijin, Ltd.).
51. Jpn. Kokai 78 79,816 (July 14, 1978), J. Nishikido and co-workers (to Asahi Chemical Industries Co., Ltd.).
52. U.S. Pat. 2,840,607 (June 6, 1958), E. C. Attane, Jr., and co-workers (to Union Oil Co.).
53. U.S. Pat. 2,870,203 (Jan. 20, 1959), A. D. Cyphers, Jr., and A. A. Gruber (to E. I. du Pont de Nemours & Co.).

54. U.S. Pat. 3,329,712 (July 4, 1967), D. E. Danley and G. L. Whitesell (to Monsanto Co.).
55. Brit. Pat. 1,216,844 (Dec. 23, 1970), W. Bowyer and co-workers (to Imperial Chemical Industries Ltd.).
56. Czech. Pat. 139,128 (Nov. 15, 1970), M. Polievka and co-workers.
57. U.S. Pat. 3,338,959 (Aug. 29, 1967), C. T. Sciance and L. S. Scott (to E. I. du Pont de Nemours & Co.).
58. Brit. Pat. 1,366,933 (Sept. 18, 1974), J. E. Lambert (to Imperial Chemical Industries Ltd.).
59. Pol. Pat. 51,539 (July 20, 1966), L. Dworakowski and co-workers (to Spoldzielnia Pracy Chemikow "Argon").
60. Pol. Pat. 63,909 (Nov. 20, 1971), S. Ciborowski (to Instytut Chemii Przemyslowej).
61. USSR Pat. 405,861 (Nov. 5, 1973), I. Ya. Lubyanitskii and co-workers.
62. Jpn. Kokai 77 19,618 (Feb. 15, 1977), BP Chemicals Ltd.
63. U.S. Pat. 3,290,369 (Dec. 6, 1966), J. H. Bonfield and co-workers (to Allied Chemical Corp.).
64. Fr. Pat. 1,532,005 (July 5, 1968), (to Badische Anilin-und Soda-Fabrik A.G.).
65. Ger. Offen. 1,938,103 (Jan. 28, 1971), A. Wegerich (to B.A.S.F. A.G.).
66. Ger. Offen. 2,309,423 (Aug. 29, 1974), H. Heumann and co-workers (to Veba-Chemie A.G.).
67. U.S. Pat. 4,014,903 (Aug. 11, 1975), W. P. Moore (to Allied Chem. Corp.).
68. Brit. Pat. 1,470,169 (Apr. 14, 1977), B. B. Barlow (to Imperial Chemical Industries, Ltd.).
69. Ger. Offen. 2,256,834 (May 24, 1973), S. Ogden and co-workers (to Briggs & Townsend Ltd.).
70. U.S. Pat. 3,810,937 (May 14, 1974), V. P. Kuceski (to C. P. Hall Co.).
71. Ger. Offen. 2,715,293 (Oct. 27, 1977), D. P. Cadogen (to Imperial Chemical Industries, Ltd.).
72. Jpn. Kokai 74 101,324 (Sept. 25, 1974), T. Shimizui and co-workers (to Denki Kagaku K. K.).
73. Eur. Pat. Appl. 2,598 (Dec. 19, 1977), B. Baker (to Imperial Chemical Industries, Ltd.).
74. U.S. Pat. 3,180,878 (Apr. 27, 1965), C. R. Campbell and co-workers (to Monsanto Co.).
75. U.S. Pat. 2,961,462 (Nov. 22, 1960), H. Chafetz (to Texaco Inc.).
76. U.S. Pat. 3,036,126 (May 22, 1962), H. Chafetz (to Texaco, Inc.).
77. U.S. Pat. 3,993,676 (Nov. 23, 1976), A. Onopchenko and co-workers (to Gulf Research and Development Co.).
78. U.S. Pat. 3,388,157 (June 11, 1968), N. Barona (to Ethyl Corp.).
79. Jpn. Kokai 78 37,613 (Apr. 6, 1978), H. Nishino (to Asahi Rika Co., Ltd.).
80. Jpn. Pat. 72 50,085 (Dec. 15, 1972), Y. Araki (to Dainippon Ink and Chemicals, Inc.).
81. U.S. Pat. 3,979,450 (Sept. 7, 1976), J. L. Moskovich and co-workers.
82. USSR Pat. 288,747 (Feb. 15, 1976), E. P. Gendrikov and co-workers.
83. Brit. Pat. 1,129,544 (Oct. 9, 1968), N. R. Ray (to Imperial Chemical Industries, Ltd.).
84. Jpn. Kokai 77 151,117 (Dec. 15, 1977), M. Kawamata and co-workers (to Mitsui Toatsu Chemicals, Inc.).
85. Jpn. Kokai 75 116,414 (Sept. 11, 1975), F. Matsuda and co-workers (to Mitsui Toatsu Chemicals, Inc.).
86. U.S. Pat. 3,341,578 (Sept. 12, 1967), J. F. Vitcha and co-workers (to Air Reduction Co., Inc.).
87. U.S. Pat. 2,851,456 (Sept. 9, 1958), G. Natta and co-workers (to Lonza Elec. & Chem. Works, Ltd.).
88. U.S. Pat. 3,530,168 (Sept. 22, 1970), G. Biale (to Union Oil Co. of Calif.).
89. U.S. Pat. 3,923,881 (Dec. 2, 1975), J. H. Murib and C. E. Frank (to National Distillers & Chem. Corp.).
90. Y. Sasaki and co-workers, *Hahko Kogahu Zasshi* **48**(12), 776 (1970).
91. *Ibid.*, (12), 782 (1970).
92. S. Takao and co-workers, *Hahko Kogahu Zasshi* **51**(1), 19 (1973).
93. K. Hotta and co-workers, *Hahko Kogahu Zasshi* **51**(1), 26 (1973).
94. Jpn. Kokai 73 26,982 (April 9, 1973), T. Furukawa (to Mitsui Petrochem. Ind., Ltd.).
95. Jpn. Kokai 75 63,190 (May 29, 1975), S. Yumimoto and co-workers (to Ajinomoto Co., Inc.).
96. Jpn. Kokai 75 142,782 (Nov. 17, 1975), A. Sato and co-workers (to Agency of Industrial Sciences and Technology).
97. Jpn. Pat. 70 38517 (Dec. 5, 1970), T. Kobayashi and co-workers.
98. Jpn. Kokai 73 58191 (Aug. 15, 1973), K. Yamada and co-workers (to Idemitsu Kosan Co., Ltd.).
99. Jpn. Pat. 69 29,246 (Nov. 28, 1969), R. Uehara and co-workers (to Nikko Physico-Chemical Industry Co., Ltd.).
100. Ger. Pat. 1,259,869 (Feb. 1, 1968), W. Schmitz (to Firma Carl Still).
101. U.S. Pat. 2,245,404 (June 10, 1941), M. A. Kise and co-workers (to The Solvay Process Co.).

102. Jpn. Pat. 73 07,609 (Mar. 7, 1973), J. Kanetaka and co-workers (to Mitsubishi Petrochemical Co., Ltd.).
103. Ger. Offen. 2,455,617 (May 26, 1976), F. J. Broecker and co-workers (to BASF A.G.).
104. U.S. Pat. 2,794,056 (May 28, 1957), L. Winstrom (to Allied Chemical & Dye Corp.).
105. Jpn. Kokai 80 69,281 (May 24, 1980), (to Agency of Industrial Sciences and Technology).
106. Jpn. Kokai 80 50,471 (Apr. 12, 1980), T. Sada and co-workers (to Tokuyama Soda Co., Ltd.).
107. U.S. Pat. 3,957,830 (May 18, 1976), W. Mesch and A. Wittwer (to BASF A.G.).
108. *Technical Data Bulletin D55*, Allied Chemical Corp., New York, 1966.
109. *Technical Data Bulletin AA5*, Allied Chemical Corp., Plastics Division, Morristown, N.J., Oct. 1965.
110. D. K. Stumpf and R. H. Burris, *Anal. Biochem.* **95**(1), 311 (1979).
111. A. Sasson and co-workers, *J. Agric. Food Chem.* **24**(3), 652 (1976).
112. C. L. Guillemen and co-workers, *J. High Resolut. Chromatogr. Chromatogr. Commun.* **4**(6), 280 (1981).
113. A. Joyeaux and co-workers, *Ann. Falsif. Expert. Chim.* **72**(776), 317 (1979).
114. D. Issachar and co-workers, *Biomed. Mass Spectrom.* **6**(2), 47 (1979).
115. S. Lewis and co-workers, *Anal. Chem.* **51**(8), 1275 (1979).
116. V. Gaberc-Porekar and H. Socic, *J. Chromatogr.* **178**(1), 307 (1979).
117. Z. L. Ochotorena, *Res. J. West Mindanao State Univ., Univ. Res. Cent.* **3**(1), 18 (1979).
118. C. Gonnet and co-workers, *Analysis* **7**(7–8), 370 (1979).
119. FDA, *Fed. Regist.* **44**(68), 20656 (Apr. 6, 1979).
120. FDA, *Fed. Regist.* **43**(24), 4635 (Feb. 3, 1978).
121. T. Pyriadi and co-workers, *J. Chem. Educ.* **51**(7), 474 (1974).
122. A. P. Wood and D. P. Kelly, *Arch. Microbiol.* **113**(3), 257 (1977).
123. M. Jones and H. K. King, *Biochem. J.* **108**(1), 11P (1968).
124. H. Nakayama and co-workers, *Bitamin* **39**(1), 39 (1969).
125. N. I. Kalabukov and co-workers, *Probl. Osobo Opasnykh Infek.* **6,** 42 (1973); *Chem. Abstr.* **81,** 102635a (1974).
126. G. W. Malaney and co-workers, *Water Pollut. Control Fed. Pt. 2* **41**(2), R-18 (1969).
127. M. Wojaczynska and co-workers, *Metody Fizykochem Oczszczania Wody Sciekow, Mater. Konf.*, 2nd Nauk Miedzynar Konf., Vol. 1, Paper No. 15, 1979; *Chem. Abstr.* **92,** 46842d (1980).
128. Jpn. Kokai 80 20,652 (Feb. 14, 1980), T. Ogasa and co-workers (to Sumitomo Metal Mining Co., Ltd.).
129. IUPAC Physical Chemistry Division, *Pure Appl. Chem.* **40**(3), 399 (1974).
130. B. N. Oleinik and co-workers, *Vses. Konf. Kalorim* [*Rasshir Tezisy Dokl.*], 7th, **1,** 122 (1977); *Chem. Abstr.* **91,** 217921q (1979).
131. F. Strafelda and co-workers, *Sb. Vys. Sk. Chem.-Technol. Praze, Anal. Chem.* **2,** 219 (1967); *Chem. Abstr.* **72,** 106723q (1970).
132. Yu. N. Levchenko and co-workers, *Tr. Metrol. Inst. SSSR* **161,** 19 (1975).
133. M. Alfenaar and C. L. de Ligny, *Rec. Trav. Chim. Pays-Bas* **86**(12), 1185090 (1967).
134. B. Bartonicek and co-workers, *Radiochem. Radioanal. Lett.* **22**(5), 331 (1975).
135. B. Bartonicek and co-workers, *Int. J. Radiat. Phys. Chem.* **6**(4), 271 (1974).
136. O. Gal and I. Dragnic, *Nucl. Res. Cent. "Democritus" Greece Rep.* **74–75,** 44 (1974).
137. B. Bartonicek, *Nukleon* (3) 22-5 (1979).
138. Fr. Demande 2,019,261 (July 3, 1970), J. M. Melon and co-workers (to Laboratories Sauba S.A.).
139. V. I. Malyuk and co-workers, *Ter. Deistvie Yantarnoi Kisloty*, 127 (1976); *Chem. Abstr.* **89,** 100184u (1978).
140. N. A. Serbinovskaya and co-workers, *Ter. Deistvie Yantarnoi Kisloty*, 179 (1976); *Chem. Abstr.* **89,** 100185v (1978).
141. V. I. Shishov and co-workers, *Ter. Deistvie Yantarnoi Kisloty*, 184 (1976); *Chem. Abstr.* **89,** 85619u (1978).
142. H. R. Stoll and co-workers, *Bibl. Haematol.* **33,** 81 (1969).
143. U.S. Pat. 3,527,798 (Sept. 8, 1970), T. Hara.
144. P. T. Callomon and G. F. Raiziss, *J. Pharmacol.* **79,** 200 (1943).
145. U.S. Pat. 2,533,033 (Dec. 5, 1950), M. L. Moore (to Sharp & Dohme, Inc.).
146. Ger. Offen. 2,441,592 (Apr. 17, 1975), J. M. Bastian (to Sandoz-Patent-G.m.b.H.).
147. U.S. Pat. 4,000,186 (Dec. 28, 1976), A. E. Vanstone (to Biorex Labs., Ltd.).
148. U.S. Pat. 2,414,722 (Jan. 21, 1947), B. C. Cornwell.

149. U.S. Pat. 2,440,218 (Apr. 20, 1948), F. Bergel and A. Cohen (to Hoffmann-La Roche, Inc.).
150. Brit. Pat. 1,114,150 (May 15, 1968), (to Eisai Co., Ltd.).
151. U.S. Pat. 2,786,835 (Mar. 26, 1957), E. R. Pinson, Jr. and co-workers (to Chas. Pfizer & Co., Inc.).
152. B. R. Baker and G. H. Carlson, *J. Am. Chem. Soc.* **64,** 2657 (1942).
153. Hung. Pat. 154,759 (May 29, 1968), J. Rakoczi and J. Torok (to Egyesult Gyogyszer es Tapszergyar).
154. N. A. Drozdov, *Zemledelie*, (5), 55 (1974); *Chem. Abstr.* **81,** 100655h (1974).
155. A. V. Blagoveshchenskii, *Bvull. Mosk. Obshchest. Ispyl. Prir. Otd. Biol.* **72**(5), 34 (1967).
156. R. A. Kiseleva, *Agrokhimiya*, (5), 133 (1976).
157. E. I. Khrenovskii and co-workers, *Sadovod. Vinograd. Vinodel. Mold.* **31**(3), 38 (1976); *Chem. Abstr.* **85,** 88440g (1976).
158. V. Rakitska, *Nauch Tr. Vissh Selskostop, Inst. Sofia, Agron. Fak., Ser. Rastenievud.* **21,** 125 (1969); *Chem. Abstr.* **76,** 69056j (1972).
159. R. A. Kiseleva, *Vinogradarstvo*, 106 (1973); *Chem. Abstr.* **83,** 127322a (1975).
160. D. Bhattacharya and B. Das, *J. Indian Chem. Soc.* **50**(8), 531 (1973).
161. Jpn. Kokai 80 57,501 (Apr. 28, 1980), (to Sankyo Co., Ltd.).
162. Ger. Offen. 2,513,696 (Oct. 23, 1975), H. Kutschera and co-workers (to Lonza A.G.).
163. F. N. Maryutin, *Tr. Khar'k. S-kh. Inst.* **182,** 107 (1973); *Chem. Abstr.* **83,** 142817j (1975).
164. U.S. Pat. 3,506,433 (Apr. 14, 1970), W. W. Abramitis and co-workers (to Armour Ind. Chem. Co.).
165. N. A. Cox and co-workers, *J. Food Sci.* **39**(5), 985 (1974).
166. B. J. Juven and co-workers, *J. Milk Food Technol.* **37**(5), 237 (1974).
167. U.S. Pat. 4,156,706 (June 19, 1979), T. J. Ernst and co-workers (to Ralston Purina Co.).
168. J. Velisek and co-workers, *Nahrung* **22**(8), 735 (1978).
169. Fr. Pat. 1,539,359 (Sept. 13, 1968), (to Kyowa Fermentation Industry Co., Ltd.).
170. U.S. Pat. 3,846,479 (Nov. 5, 1974), J. D. Zech (to ICI Americas, Inc.).
171. U.S. Pat. 3,953,616 (Apr. 27, 1956), P. D. Thomas (to Pfizer Inc.).
172. Pol. Pat. 92,860 (Dec. 15, 1977), R. Majchrzak and co-workers (to Akademia Rolnicza Warszawa).
173. L. N. Kazanskaya and co-workers, *Khlebopeh. Konditershoya Prom.* **11**(11), 4 (1967); *Chem. Abstr.* **69,** 113411g (1968).
174. *Fed. Regist.* **33**(232), 17752 (Nov. 28, 1968).
175. I. M. Klotz, *Methods Enzymol.* **11,** 576 (1967).
176. Y. R. Choi and co-workers, *J. Food Sci.* **46**(3), 954 (1981).
177. U.S. Pat. Appl. 415,469 (Nov. 13, 1973), H. S. Groninger, Jr., and R. Miller (to U.S. Dept. of Commerce).
178. Jpn. Pat. 71 15,070 (Oct. 26, 1966), A. Nishi and co-workers (to Japan Monopoly Corp.).
179. K. Holbrook and P. J. Twist, *Plating* **56**(5), 523 (1969).
180. USSR Pat. 452,631 (Dec. 5, 1974), K. I. Bubnov and co-workers.
181. Ger. (East) Pat. 63,254 (Aug. 5, 1968), H. U. Galgon and H. Penz.
182. V. Skucas and co-workers, *Gal'vanicheskie Khim. Pokrytiya Dragotsennymi Redk. Met. Mater. Semin.*, 18 (1978); *Chem. Abstr.* **93,** 139920p (1980).
183. U.S. Pat. 3,311,548 (Mar. 28, 1967), H. Brown and E. A. Romanowski (to Udylite Corp.).
184. USSR Pat. 823,471 (Apr. 23, 1981), N. T. Kudryavtsev and co-workers.
185. U.S. Pat. 3,551,305 (Dec. 29, 1970), I. M. Dalton and D. Jones.
186. Jpn. Kokai 79 56,039 (May 4, 1979), K. Hiraishi (to Hiraishi, Takanori).
187. Ger. Offen. 2,528,634 (Jan. 15, 1976), S. Kimura.
188. N. K. Zagoruiko and co-workers, *Zh. Prikl. Khim.* **44**(1), 76 (1971).
189. Jpn. Kokai 76 13,338 (Feb. 2, 1976), H. Yamaguchi and co-workers (to Otsuka Chem. Drugs Co., Ltd.).
190. Jpn. Kokai 77 00,737 (Jan. 6, 1977), T. Handa and Y. Ohtsu (to Asahi Denka Kogyo K. K.).
191. Jpn. Pat. 74 20,466 (May 24, 1974), K. Noji and co-workers (to Sumitomo Metal Ind., Ltd.).
192. Ger. Pat. 1,521,710 (Mar. 4, 1971), H. Keil (to Chemische Werke Huels A.G.).
193. U.S. Pat. 3,382,081 (May 7, 1968), P. R. Cutter and co-workers (to Diamond Shamrock Corp.).
194. U.S. Pat. 3,755,003 (Aug. 28, 1973), B. E. Palm and co-workers (to Diamond Shamrock Corp.).
195. U.S. Pat. 4,014,715 (Mar. 29, 1977), J. M. Preston (to General Electric Co.).
196. Jpn. Pat. 74 26,826 (July 12, 1974), J. Tanaka and co-workers (to Senju Metal Industry Co., Ltd.).
197. USSR Pat. 433,000 (June 25, 1974), B. F. Malyshev and co-workers.
198. U.S. Pat. 3,642,431 (Feb. 15, 1972), S. Suzuki and co-workers (to Chevron Research Co.).

199. U.S. Pat. 4,029,744 (June 14, 1977), N. N. Li and R. P. Cahn (to Exxon Research & Engineering Co.).
200. Ger. Offen. 2,814,644 (Oct. 19, 1978), E. Asango (to Chivoda Chemical Engineering and Construction Co., Ltd.).
201. Fr. Demande 2,443,273 (July 4, 1980), A. Kobayashi and co-workers (to Kureha Chem. Industry Co., Ltd.).
202. U.S. Pat. 3,867,095 (Feb. 18, 1975), N. H. Koenig and M. Friedman (to United States Dept. of Agriculture).
203. U.S. Pat. 3,872,183 (Mar. 18, 1975), M. H. Keck (to Goodyear Tire & Rubber Co.).
204. Jpn. Pat. 74 38114 (Oct. 15, 1974), A. Yamamoto (to Teijin Ltd.).
205. Jpn. Kokai 78 98,417 (Aug. 28, 1978), T. Motoi and T. Kubo (to Kanebo, Ltd.).
206. Ger. Offen. 2,027,613 (Dec. 16, 1971), J. Thewis and co-workers (to B.A.S.F. A.G.).
207. U.S. Pat. 4,250,300 (Feb. 10, 1981), M. Iwase and co-workers (to Daikin Kogyo Co., Ltd.).
208. Fr. Demande 2,174,110 (Nov. 16, 1973), L. M. Minsk and co-workers (to Eastman Kodak Co.).
209. U.S. Pat. 3,598,601 (Aug. 10, 1971), D. J. Beavers (to Eastman Kodak Co.).
210. Eur. Pat. Appl. 27,430 (Oct. 12, 1979), P. J. Wright (to Ciba-Geigy A.G.).
211. Jpn. Kokai 77 102,439 (Aug. 27, 1977), M. Tanaka and M. Nakamura (to Shiseido Co., Ltd.).
212. U.S. Pat. 4,165,366 (Aug. 21, 1979), J. R. Mellberg (to Colgate-Palmolive Co.).
213. Ger. Offen. 2,025,338 (Dec. 23, 1970), J. B. Barth (to Colgate-Palmolive Co.).
214. Jpn. Kokai 76 45,150 (Apr. 17, 1976), H. Kaku and co-workers (to Sakai Chem. Ind. Co., Ltd.).
215. U.S. Pat. 3,539,524 (Nov. 10, 1970), W. K. Wilkinson (to E. I. du Pont de Nemours & Co.).
216. Jpn. Pat. 74 11,261 (Mar. 15, 1974), T. Kimura and co-workers (to Mitsubishi Rayon Co., Ltd.).
217. Jpn. Kokai 76 41,098 (Apr. 6, 1976), M. Hayashizaki and co-workers (to Hitachi Chemical Co., Ltd.).
218. U.S. Pat. 4,180,442 (Dec. 25, 1979), N. R. Byrd (to McDonnell Douglas Corp.).
219. U.S. Pat. 4,134,809 (Jan. 16, 1979), J. G. Pacifici and co-workers (to Eastman Kodak Co.).
220. Brit. UK Pat. Appl. 2,053,247 (Feb. 4, 1981), J. G. Robinson and S. A. Brian (to Coal Industry (Patents) Ltd.).
221. Brit. Pat. 1,415,797 (Nov. 26, 1975), V. Lamberti (to Unilever Ltd.).
222. Brit. Pat. 1,465,700 (Feb. 23, 1977), R. I. Hancock (to Imperial Chemical Industries, Ltd.).
223. H. Matsuda and T. Saheki, *J. Macromol. Sci. Chem.* **A16**(6), 1065 (1981).
224. U.S. Pat. 4,189,551 (Feb. 19, 1980), V. Gangal (to E. I. du Pont de Nemours & Co.).
225. Jpn. Kokai 76 80,348 (July 13, 1976), A. Saito and co-workers (to Asahi Chemical Industry Co., Ltd.).
226. Jpn. Kokai 80 129,494 (Oct. 7, 1980), Sanyo Chemical Industries, Ltd.
227. U.S. Pat. 3,513,180 (May 19, 1970), D. M. Fenton (to Union Oil Co. of Calif.).
228. USSR Pat. 810,775 (Mar. 7, 1981), G. N. Dorofeenko and co-workers (to Rostov State University).
229. U. Sellerio and co-workers, *Calore* **40**(4), 249 (1969).
230. Jpn. Kokai 77 126,405 (Oct. 24, 1977), (to Ethyl Corp.).

LEON O. WINSTROM
Consultant

SUGAR

PROPERTIES OF SUCROSE

Sucrose [*57-50-1*] (α-D-glucopyranosyl-β-D-fructofuranoside), $C_{12}H_{22}O_{11}$, formula weight 342.30, is a disaccharide composed of D-glucosyl and D-fructosyl moieties (Fig. 1). Sucrose occurs almost universally in all components of practically every existing phanerogam. As an item of commerce, crystalline sucrose represents the highest volume organic compound produced worldwide in practically pure state (>99.5%). The only possible exception to this statement is ethylene. Table 1 indicates recent volumes of production and consumption of crystalline sucrose.

Universally recognized throughout history as a source of sweetness, sucrose is used widely in food processing as a preservative, bulking agent, flavor enhancer, and texturizer (2). In baking, sugar functions as yeast food, and it contributes to crust characteristics and product stability.

Sucrose is extracted commercially from sugarcane (*Saccharum officinarum L.*), sugar beet (*Beta vulgaris*), and to a lesser extent from sorghum (*Sorghum vulgare*) and from sugar maple (*Acer saccharum*). The latter two sources normally provide sucrose-containing syrups, rather than the more commonly encountered crystalline sucrose of commerce (see Syrups).

The cultivation of sugarcane is restricted to tropical and semitropical regions of the earth; the sugar beet is more suited to temperate zones, as are sorghum and maple. Beets provide the source of sugar for Europe, central Canada, and the central and western United States, and cane supplies the sugar requirements of almost all the rest of the world. Pressure of historic events has often had profound effects on the cultivation of various sugar crops (see Sugar, cane sugar).

Sucrose was first synthesized enzymatically in the laboratory from potassium D-glucosyl-1-phosphate and D-fructose (3). The first chemical synthesis was accomplished by reaction of 3,4,6-tri-*O*-acetyl-1,2-anhydro-α-D-glucopyranose with 1,3,4,6-tetra-*O*-acetyl-D-fructofuranose (4). Because of low yields and other practical difficulties, it appears unlikely that this or other chemical or biochemical syntheses will be cheap enough to soon supersede the production of sucrose by traditionally inexpensive extraction processes. Provided that low cost immobilized enzymes became available, it would be possible to synthesize sucrose from a mixture of glucose and fructose which can be obtained in turn from corn starch. Economics would play an important role in such a development, however (see Enzymes, immobilized).

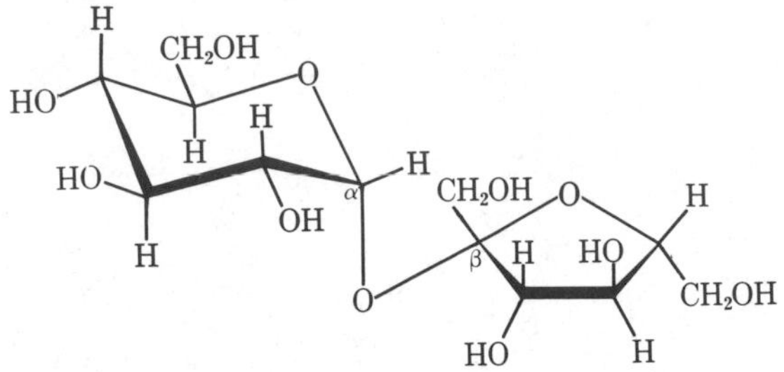

Figure 1. Structure of sucrose.

Table 1. Worldwide Production and Consumption (10^6 Metric Tons) of Refined Sucrose[a]

	Production			Consumption
Crop year	Cane	Beet	Total	Total
1974	50.0	28.5	78.5	77.0
1975	49.9	31.7	81.6	79.2
1976	53.5	32.8	86.3	81.9
1977	57.5	35.0	92.5	86.1
1978	56.4	34.6	91.0	89.4
1979	51.5	33.1	84.6	90.0
1980[b]	55.5	31.6	87.1	89.5

[a] Ref. 1.
[b] Preliminary.

Physical Properties of Sucrose

Crystalline Sucrose. Sucrose crystallizes as monoclinic, hemimorphic (sphenoidal) crystals, with a ratio of axes = 1.2595:1:0.8782, 103°30′. Crystals have a prismatic habit which is strongly influenced by impurities. The presence of raffinose or dextran impurities, for example, causes sucrose to crystallize in long needle-like shapes. A hemiheptahydrate modification, $C_{12}H_{22}O_{11}.3\frac{1}{2}H_2O$, has been examined and a hemipentahydrate is known (5).

Structural aspects of crystalline sucrose have been extensively studied by x-ray, neutron diffraction, and nmr methods. Seven of the hydroxyl groups in the anhydrous crystal are hydrogen bonded, and there are two intramolecular hydrogen bonds (O–2- - -H—O–1′ and O–5- - -H—O–6′).

The melting point of sucrose has been quoted within a range of 160–200°C, with 188°C generally accepted (6). The wide variability of melting points found in the literature indicates the large influence of trace impurities. The effects of these impurities on crystal structure have been extensively investigated (7). Slowly cooling melted sucrose produces an amorphous, glassy state, which crystallizes slowly on standing (8–9).

The density of sucrose is 1.5879 g/cm^3. The linear expansion coefficient ranges from 0.0028–0.0050%, depending on the axis. Characteristic ir absorption bands occur at 1010, 990, 940, 920, 870, 850 cm^{-1} (sharp) and at 680, 580 cm^{-1} (broad). The specific heat of crystalline sucrose is 415.98 J/mol (99.42 cal/mol) at 20°C. The dipole moment is 2.8×10^{-29} C·m (8.3 D).

Sucrose crystals are triboluminescent, due either to gaseous discharge on newly formed surfaces, or to a piezoelectric effect (10–11). The dielectric constant of sucrose is 3.50–3.85, and varies with axis orientation. In commercial bulk granulated sugar, it is about 2.2. The enthalpy of formation is 360 J/(mol·K) [86 cal/(mol·K)] and the enthalpy of combustion is −5.65 MJ/mol (-1.35×10^6 cal/mol). Surface energy is 0.224 pN/m (2.24×10^{-10} dyn/cm). The water sorptive capacity is 0.003 g H_2O/g solid, for a 0.32–0.23 mm (48–65 mesh) fraction of crystalline sucrose (12). Many other physical properties of bulk granulated sucrose as produced commercially are collected in ref. 13.

Sucrose Solutions. Sucrose is very soluble in water, 2.07 g/g H_2O at 25°C, and is readily soluble in aqueous protic solvents such as methanol and ethanol. It is practically insoluble in anhydrous ethanol, ether, chloroform, and anhydrous glycerol. It is moderately soluble in such organic solvents as dimethylformamide, pyridine, and dimethyl sulfoxide. The solubility of sucrose in water, especially in the presence of water-soluble impurities, is of great commercial interest to sugar manufacturers, and extensive solubility and refractive-index tables have been prepared (14–15). The crystallization of sucrose from aqueous solutions has been extensively studied, especially in regard to the requirements for refined sugar (7,16–17).

In aqueous solutions sucrose can be associated with up to ca four molecules of water. Hydration of sucrose compared with other sugars is sucrose > glucose > fructose > maltose. Because of the variable degree of hydration, sucrose behavior in solution is generally treated as nonideal at other than dilute concentrations.

The surface tension of sucrose solutions, in the range of 1–10% concentration, is described by $Y = 73.636 + 0.099\,C - 0.096\,T$ where Y is surface tension in mN/m (= dyn/cm), C is percent solute, and T is temperature, °C (18).

An important commercial property of sucrose in solution is its polarization. Sucrose is purchased, and various stages of manufacturing processes are monitored, by polarization measurements. The specific rotation of pure sucrose ($[\alpha]_D^{20}$) is +66.5, and decreases slightly with increasing concentration (14–15).

Although generally considered to be neutral in solution, acid dissociation constants have been measured for sucrose: $pK_1 = 12.7$, $pK_2 = 13.1$ at 25°C. Color in sucrose solutions is due to the presence of mother liquor in association with the crystal; it may be present as inclusions or as an external surface film. Although sucrose is colorless, and the lack of visible color in solution is used as a simple indication of purity, many of the impurities exhibit fluorescence.

Sweetness of sucrose solutions as compared organoleptically to other sugars has been thoroughly investigated (19–20). The sensation of sweetness is affected by several factors including total acidity, pH, viscosity, temperature, and the presence of other constituents. Alone, in aqueous solution, the threshold concentration for detection of sucrose is about 0.1–0.5%. The sweetness of sucrose is compared to that of some other common sugars in Table 2.

Table 2. Relative Sweetness of Some Common Sugars[a]

Sugar	Relative sweetness
fructose	114
sucrose	100
glucose	69
galactose	63
maltose	46
lactose	39

[a] Ref. 21.

Sucrose Derivatives

Many derivatives have been investigated with a view towards utilization of sucrose as a chemical feedstock. The multifunctionality of the sucrose molecule generally frustrates efforts to prepare high yields of single compounds of sucrose. A wide variety of sucrose esters has been prepared from both low and high molecular weight fatty acids. They have limited use as baking additives, emulsifiers (qv), viscosity modifiers in foods, and are used in detergents (qv), confectionary ingredients, and in drug and cosmetic formulations. Sucrose ethers have many similar properties and uses. Sucrose formulated detergents have the advantage of biodegradability (22). Various reviews have covered the chemistry and use of a multitude of sugar derivatives in a variety of applications (23–25).

Reactions of Sucrose

Metal Salts. In addition to the formation of esters and ethers, sucrose reacts with alkali- and alkaline-earth metal salts to form complexes of varying composition (26). Formation of sucrose complexes with calcium forms the basis for the Steffen process for commercially separating sucrose from beet molasses. Similar processes based on strontium and barium have been employed.

Hydrolysis. The glycosidic linkage is relatively stable in dilute alkali and in neutral solution, with maximum stability occurring at about pH 9. Under mild acid catalysis, sucrose is easily hydrolyzed, and inversion occurs (change in the sign of polarization from +66.5 to a negative value, owing to the larger negative rotation of fructose). Much sucrose is sold in liquid form as "invert" (a roughly equimolar mixture of glucose and fructose) or "medium invert", where about 50% of the sucrose has been hydrolyzed. The hydrolysis of sucrose has been carried out in the presence of a variety of catalysts: mineral acids, ion-exchange resins, and enzymes. Owing to economics, hydrochloric acid is the most common catalyst. Under conditions of acid hydrolysis, a small amount of the fructose formed is converted to D-fructose dianhydrides (27).

A small amount of inversion is detectable under alkaline conditions; the monosaccharides degrade further by way of condensation and disproportionation pathways (27).

Hydrogenation. Sucrose is hydrogenated with Raney nickel to a mixture of sorbitol and mannitol; under more drastic conditions, glycerol and propylene glycol are produced.

Oxidation. Partial or complete oxidation of sucrose with nitric acid or with various metal oxide catalysts produces many products and fragments. Mild oxidation forms oxalic and tartaric acids, as well as a variety of acidic materials designated collectively as saccharic acids.

Thermal Degradation. In general, degradation of dry sucrose at temperatures of 90–200°C begins with cleavage of the glycosidic bond followed by condensation and water formation. At temperatures of 170–210°C, this reaction is referred to as caramelization, and the mixture of products formed is commercially useful as "caramel". At higher temperatures carbon–carbon cleavage occurs (200–300°C), followed by small-molecule formation (300–500°C) and formation of materials not characteristic of the starting material (500–800°C). All of the degradation reactions are sensitive to impurities in the sugar crystal (6). Thermal degradation in solution produces various compounds such as 5-hydroxymethylfurfural and colored condensation products (28). Ionizing radiation causes degradation of sucrose in a somewhat similar manner to thermal processes, both in the crystal and in solution (27,29–30).

Safety and Health Factors

Physically, sucrose is generally considered to be safe to handle, although some hazard may exist when much dust is present (31). From the standpoint of health, extensive studies have concluded that no health hazard exists when sucrose is consumed at current levels (32–33), although a synergistic contribution to the formation of dental caries may exist (34) (see also Sugar, cane sugar).

BIBLIOGRAPHY

"Sucrose" under "Sugars (Commercial)" in *ECT* 1st ed., Vol. 13, pp. 247–251, by J. L. Hickson, Sugar Research Foundation, Inc.; "Sugar (Properties of Sucrose)" in *ECT* 2nd ed., pp. 151–155, by R. A. McGinnis, Spreckels Sugar Co.

1. *Sugar and Sweetener Outlook and Situation*, United States Department of Agriculture, Washington, D.C., Feb. 1981.
2. R. J. Wicker, *Chem. Ind.* **41,** 1708 (1966).
3. W. Z. Hassid, M. Doudoroff, and H. A. Barker, *J. Am. Chem. Soc.* **66,** 1416 (1944).
4. R. V. Lemieux and G. Huber, *J. Am. Chem. Soc.* **75,** 4118 (1953).
5. F. T. Jones and F. E. Young, *Anal. Chem.* **26,** 421 (1954).
6. L. Poncini, *La Sucrerie Belge* **100,** 221 (1981).
7. H. E. C. Powers, *Sugar Technol. Rev.* **1,** 85 (1969–1970).
8. B. Makower and W. B. Dye, *J. Agric. Food Chem.* **4,** 72 (1956).
9. K. J. Palmer, W. B. Dye, and D. Black, *J. Agric. Food Chem.* **4,** 77 (1956).
10. H. Kurten and H. Rumpf, *Chem. Ing. Tech.* **38,** 331 (1966).
11. B. P. Chandra, *J. Phys. D* **10,** 1531 (1977).
12. D. S. Smith, C. H. Mannheim, and S. G. Gilbert, *J. Food Sci.* **46,** 1051 (1981).
13. H. M. Pancoast and W. R. Junk, *Handbook of Sugars*, The AVI Publishing Co., Inc., Westport, Conn., 1980.
14. R. S. Norrish, *Selected Tables of Physical Properties of Sugar Solutions*, Scientific and Technical Surveys No. 51, The British Food Manufacturing Industries Research Association, Leatherhead, Surrey: Randalls Road, England, 1967.
15. F. J. Bates and Associates, *Polarimetry, Saccharimetry, and the Sugars*, U.S. National Bureau of Standards, Circular 440, United States Government Printing Office, Washington, D.C., 1942.
16. B. M. Smythe, *Sugar Technol. Rev.* **1,** 191 (1971).
17. A. VanHook, *Sugar Technol. Rev.* **1,** 232 (1971).
18. M. K. Supran, J. C. Acton, A. J. Howell, and R. L. Saffle, *J. Milk Food Technol.* **34,** 548 (1971).
19. R. S. Shallenberger and G. G. Birch, *Sugar Chemistry*, AVI Publishing Co., Inc., Westport, Conn., 1975.
20. R. M. Pangborn, *J. Food Sci.* **28,** 726 (1963).
21. M. A. Amerine, R. M. Pangborn, and E. B. Roessler, *Principles of Sensory Evaluation of Food*, Academic Press, New York, 1965, p. 95.
22. V. Kollonitch, *Sucrose Chemicals*, International Sugar Research Foundation, Inc., Bethesda, Md., 1970.
23. R. Khan and A. J. Forage, *Sugar Technol. Rev.* **7,** 175 (1979–1980).
24. J. L. Hickson, ed., *Sucrochemistry*, ACS Symposium Series No. 41, American Chemical Society, Washington, D.C., 1977.
25. R. Khan in R. S. Tipson and D. Horton, eds., *Advances in Carbohydrate Chemistry and Biochemistry*, Vol. 33, Academic Press, New York, 1976, pp. 235–294.
26. J. A. Rendleman, Jr. in M. L. Wolfrom, and R. T. Tipson, eds., *Advances in Carbohydrate Chemistry*, Vol. 21, Academic Press, New York, 1966, pp. 209–271.

27. W. Mauch, *Sugar Technol. Rev.* **1,** 239 (1971).
28. F. H. C. Kelly and D. W. Brown, *Sugar Technol. Rev.* **6,** 1 (1978–1979).
29. G. Lofroth, *Int. J. Radiat. Phys. Chem.* **4,** 277 (1972).
30. W. W. Binkley, M. E. Altenburg, and M. L. Wolfrom, *Sugar J.* **34,** 25 (1972).
31. N. I. Sax, *Dangerous Properties of Industrial Materials*, 4th ed., Van Nostrand Reinhold Co., New York, 1975.
32. *Evaluation of the Health Aspects of Sucrose*, FASEB, Food and Drug Administration, Washington, D.C., Bureau of Foods, 1976.
33. E. L. Bierman, *Am. J. Clin. Nutr.* **32,** 2712 (1979).
34. F. Q. Nuttall and M. C. Gannon, *Diabetes Care* **4,** 304 (1981).

General References

R. A. McGinnis, ed., *Beet Sugar Technology, 2nd ed.*, Beet Sugar Development Foundation, Fort Collins, Colo., 1971 (3rd ed., 1982). General coverage with emphasis on manufacturing.

G. P. Meade and J. C. P. Chen, *Cane Sugar Handbook, 10th ed.*, John Wiley & Sons, Inc., New York, 1977. General coverage with emphasis on manufacturing.

G. Vavrinecz, *Atlas of Sugar Crystals*, Verlag Dr. Albert Bartens, Berlin, 1960. Detailed treatment of crystalline state.

W. Mauch and E. Farhoudi, *Sugar Technol. Rev.* **7,** 87 (1979–1980). Discussion of physical properties and composition of commercial white granulated sugar.

F. Schneider, ed., *Sugar Analysis*, ICUMSA, c/o British Sugar Corporation Ltd., Peterborough, UK, 1979. Official analytical methods of the International Commission for Uniform Methods of Sugar Analysis (ICUMSA) relating to commercial sugar.

R. M. SEQUEIRA
Amstar Corporation

SUGAR ANALYSIS

Sugar analysis includes the analysis for sucrose and other sugars as well as other components. References 1–7 are detailed and comprehensive in their coverage of sugar analysis. The International Commission for Uniform Methods of Sugar Analysis (ICUMSA) convenes every four years and their Proceedings (1) reflect the acceptance of methods by the sugar industry. Reference 2 is a compilation of the 1979 ICUMSA methods. The AOAC revises the *Official Methods of Analysis* (3) every five years. The Corn Refiners Association (CRA) established a set of methods (4) as used in the corn sugar industry; ICUMSA, CRA, and AOAC do not always agree on methods but their differences are eventually resolved (see also Analytical methods).

Physical Methods

Polarization. The concentration of a pure sugar solution is determined by measurements of polarization (optical rotation), refractive index, and density. In addition, optical rotation measurements of solutions of known concentration are a means of identifying individual sugars.

Polarization is the measure of the optical rotation of the plane of polarized light as it passes through a solution. The observed or direct rotations, α, is expressed as specific rotation $[\alpha]$:

$$[\alpha] = \alpha/lc$$

where α is the observed rotation in circular degrees, l is the length of the tube containing the solution in dm, and c is the concentration in g/mL. The specific rotation depends upon the temperature and the wavelength of the light. Both must be specified in stating specific rotation. Molecular rotation is the specific rotation multiplied by molecular weight. Specific rotation is a characteristic of each sugar and hence is a means of identification. Tables of specific rotation for different sugars are found in ref. 7.

The saccharimeter is a polarimeter modified for the sugar industry. Its scale reads directly in percent sucrose (sugar degrees) for the standard concentration of exactly 26 g solids per 100 cm^3 and a standard tube length of exactly 2 dm. The 100% sucrose point is $34.616 \pm 0.001°$ at 20°C using white light with a dichromate filter or sodium light of wavelength 589.44 nm. At the mercury arc wavelength of 546.23 nm, the angle is 40.765° (1–2). However, polarimeter scales are not inherently stable or accurate and must be calibrated frequently. For calibration, a quartz control plate is used that has been calibrated by the manufacturer or by one of the national standards laboratories. The quartz plates are very stable and are marked with the equivalent sucrose rotation; ICUMSA is considered the authority on calibration of saccharimeters (1–2).

Saccharimeters used in trade and factory control have the advantage of reading directly in sucrose concentration. This reading is called the direct polarization or, simply, polarization, abbreviated pol. The instrument is calibrated for pure sucrose and is accurate only for solutions containing no other optically active substance, which is almost never the case. In trade transactions involving the monetary value of a cargo of sugar, it is tacitly assumed that the plus errors are exactly balanced by minus errors, and therefore the pol reflects the sugar content. Although this is not correct, the pol can be quickly, easily, accurately, and reproducibly measured and is always the basis for settlement (3–6). Strict adherence to the ICUMSA method is required (1–2). Disagreements are referred to the New York Sugar Trade Laboratory (8) which is the official referee and by definition always correct. Older saccharimeters were manually adjusted to the balance point and could be read to a precision of 0.05% sucrose. Modern instruments have a photoelectric end point and are precise to better than 0.01%.

Mutarotation is a complicating phenomenon in pol measurements. Reducing sugars can exist in different tautomeric forms with different rotations. These forms are stable in the solid state but labile in solution. Therefore, the pol of these solutions depends upon their age. Equilibrium is established in several hours. This phenomenon is of great importance in measuring invert syrups, and is of some importance in evaluating molasses and poor-quality raw sugars containing invert sugar.

Clarification is required for solutions that are so colored or turbid that light cannot pass through them. A minimum amount of clarifying agent, typically lead subacetate, is added to coagulate the dispersed solid to permit filtration. Clarification is a source of error in measuring the pol. This error, called the lead error, amounts to about 0.1%, and has created much controversy.

Double Polarization. The direct polarization, as described above, can give the correct value for sucrose only when no other optically active substances are present. The correct value of sucrose is measured by double polarization or the Clerget method (1–2,6–7). First, the direct polarization is determined; then, the sucrose is inverted and a second polarization measured. The rotation of substances other than sucrose remains constant and the change in polarization is due to the inversion of sucrose. The sucrose is calculated by dividing the change in polarization by the Clerget constant. This constant represents the algebraic difference between the rotations of sucrose and glucose plus fructose under the conditions of the method used. In theory, the Clerget method is simple, but in practice it requires strict adherence to the correct technique. The Jackson and Gillis method (1–3,5–7) uses acid inversion. Inversion with invertase is also used. The constant depends upon the details of the process used.

Density. Density measurement is widely used in the sugar industry to determine the sugar concentration of syrups, liquors, juices, and molasses. The instrument is the standard hydrometer, also called a spindle. Hydrometers graduated in sucrose concentration (percent sucrose by weight) are called Brix hydrometers or Brix spindles. The readings obtained are called spindle Brix. Hydrometers graduated in °Baumé are used for molasses and in the corn sugar industry. The relation between °Baumé and density d, in g/cm^3, is

$$°\text{Baumé} = 145\,(1 - 1/d)$$

Brix spindles are calibrated for pure sucrose. However, the density of other sugar solutions is not very different. In fact, the densities of solutions of all components normally found in sugar are not very different. Therefore, the spindle Brix is considered a measure of total dissolved substances.

The apparent density is the quantity generally used and tabulated. The true density is corrected for weight in vacuum by a factor of 1.00026, which, for practical purposes, amounts to a difference of ca 0.04 spindle Brix. If the apparent density is measured and apparent density tables are used, there is no error.

A density meter has recently been developed that measures the density ten times more accurately than a spindle. It measures the resonant vibration frequency of a glass U-tube that depends upon the mass and hence the density of the fluid inside. This density meter is used in evaluating syrups for the soft-drink industry.

Specific gravity is defined as the density relative to the density of water at the same temperature. Tables of densities and specific gravities of sucrose and other sugars are given in refs. 3–4, 6–7.

Refractive Index. Refractive index is another measurement of sugar concentration in solution. Many refractometers have Brix scales. The readings obtained are called refractometer Brix, or refractometer dry substance (RDS). The calibration is for pure sucrose, but the refractometer is even better than the hydrometer for measuring total dissolved substances. The refractometer is a rugged instrument that holds its calibration very well and is quick and easy to use. The accuracy of the refractometer scale has recently been improved slightly (1). However, the new values differ generally by less than 0.1% from the old scale. Tables of refractive indexes for sucrose and other sugars are given in refs. 1–4, 6–7. Values for sugar mixtures are found in ref. 4.

Isotope Dilution. Isotope dilution is used as a reference method to determine true sucrose (9–10). It is, however, much too complex and expensive to be used on a routine basis. A measured amount of ^{14}C-labeled radioactive sucrose is added to the sample. A specimen of sucrose is recovered, purified, and its radioactivity measured. The quantity of sucrose in the original material is then calculated.

Chemical Methods for the Determination of Reducing Sugars

The reducing action of aldose and ketose sugars is the basis for a large number of methods that have been devised for the quantitative and qualitative determination of aldoses and ketoses. In general, the reagents comprise copper acetate, potassium ferricyanide, tungstates, phenolic reagents in strong acid solution, and alkaline copper solutions stabilized by tartrate, citrate, or carbonate ions. The principal problem encountered with these methods arises from the fact that the reduction is not quite stoichiometric, but varies with experimental conditions; therefore, the chemical equivalence factor is different for each sugar and each method.

The alkaline copper tartrate reagent, devised by Fehling and modified by Soxhlet, is used in a number of methods. The soluble copper tartrate complex ion present in this reagent is deep blue in color (Fehling's solution). Cuprous ions resulting from reduction of the copper do not form a complex with the tartrate but are precipitated from the solution as cuprous oxide.

More than 80 methods using this reagent have been devised. They differ in temperature and time of heating, amount of reagent, and method for the determination of the reduced copper.

Lane and Eynon Method. In this method, the hot reagent is titrated with sugar solution until the copper is completely reduced (2–7). This method is more rapid and more accurate than other procedures, and has, therefore, been generally adopted by the industry. However, it demands skill, and may present difficulties for the occasional user.

For the Soxhlet modification of Fehling's solution, the following reagents are used: 34.64 g pure $Cu_2SO_4.5H_2O$ in water, diluted to 500 mL; 173 g Rochelle salt (sodium–potassium tartrate) plus 50 g carbonate-free sodium hydroxide in water, diluted to 500 mL; and 1% aqueous methylene blue.

In the titration, 10 or 25 mL of mixed Soxhlet's reagent is placed in a 300–400 mL flask. The sugar solution is placed in a 50 mL buret with its outlet offset so that neither the stopcock nor the sugar solution are heated by the steam evolved during the reaction. Almost all of the sugar solution required to completely reduce the copper is added to the cold reagent (the remainder must be between 0.5 and 1.0 mL). The flask containing the reaction mixture is boiled for exactly 2 min, and then 3–5 drops methylene blue solution are added and the titration is completed in exactly 1 min more. The intense blue color of the methylene blue indicator disappears almost instantly with the addition of excess reducing sugar. Duplicate determination should agree within 0.1 mL.

A preliminary titration establishes the amount of sugar solution to add initially to the cold copper reagent so that the final titration is between 0.5 and 1.0 mL. It may have to be repeated several times until the titer is within limits and the time of boiling is exactly 3 min. For calculation, a factor is needed:

$$\text{mg sugar in 100 mL solution} = (\text{factor} \times 100)/\text{titer}$$

The factors for several sugars are given in Table 1; for more extensive tables, see refs. 2–7.

Constant-volume modification (1–2,6) adds sufficient water so that the volume at the end point is always the same. This procedure eliminates the last variable in the Lane and Eynon method and dispenses with the use of tables.

Munson and Walker Method. This method is convenient for an occasional sugar analysis (1–3,6–7). The reduced cuprous oxide is separated by filtration and determined gravimetrically. Factor tables are needed for the calculations.

Ofner Method. This method is used for up to 10% invert sugar in sucrose. It is performed at a pH of 10.4, attained with sodium carbonate instead of sodium hydroxide, and buffered with phosphate. The reduced cuprous oxide is treated with excess standardized iodine, which is then back-titrated with thiosulfate using starch as an indicator.

Knight and Allen Method. Very small amounts of reducing sugars in sucrose are determined by this method (1–2). After the sugar reacts with the copper reagent, the excess unreduced copper is determined with EDTA.

Emmerich Method. Very small amounts of reducing sugars in white sugar are determined by this method (1–2). It is based on the formation of a colored complex of 3,6-dinitrophthalic acid and invert sugar. The absorption is read at 450 nm.

Enzymatic Methods. The enzyme glucose oxidase is highly specific for the oxidation of glucose. A coproduct is hydrogen peroxide which can be determined electrochemically. By using a second enzyme such as invertase, other sugars are converted to glucose and the method extended to other sugars. With immobilized enzymes a very

Table 1. Factors for Use with the Lane and Eynon Method of Sugar Determination[a,b]

Titer, mL	Invert	Glucose	Fructose	Maltose	Lactose
25	124.0	120.5	127.9	194.5	169.9
30	124.3	120.8	128.1	192.8	168.8

[a] Ref. 2.
[b] For 25 mL of reagent.

convenient and specific determination can be made (see Enzymes, immobilized). Automatic instruments are available for quick and continuous analysis.

Inversion. Sucrose by inversion (3,6) is obtained by measuring reducing sugars before and after inversion. The sucrose content is 0.95 times the increase in reducing sugars. This method works well for low purity syrups and molasses. The Clerget method is better for high purity materials.

Micromethods

Micromethods of sugar analysis are used in medical and biological laboratories, in carbohydrate research work, and by the sugar industry when small quantities of sugar are to be determined.

These methods fall into three main categories, depending upon the type of reagent: reduction of ferricyanide to ferrocyanide; reduction of cupric sulfate to cuprous oxide; and the development of color with phenols in strong sulfuric acid solution. As a rule, the copper reagents are more selective than are the ferricyanide reagents, and the phenolic reagents respond to a wide range of carbohydrates in general. Ferricyanide has the advantage of not being oxidized by air. The degree of reduction of either copper or ferricyanide is determined by titration or colorimetrically; the latter method is preferred for small amounts. The colorimetric ferricyanide method is used for blood sugar in clinical laboratories and by the sugar industry for monitoring wastewater for sugar losses.

Copper reagents are preferred for biological work because they are more specific (3). The Somogyi (3) methods are used extensively. With phosphate-buffered reagents, amylases stay in solution. However, the carbonate-buffered reagent is more stable and more widely used. The quantity of copper consumed is determined iodometrically. In the Folin and Wu (3) method, tungstate is added as a color reagent.

Phenolic compounds in strong acid give a color reaction with furfural derivatives that results from the destruction of carbohydrates by the strong acid (11). Various sugars give different colors with this highly sensitive method (11); the detection limit is 1 ppm.

Chromatographic Methods

Qualitative determination of the sugars and sugar derivatives is ideally suited to chromatographic methods (see Analytical methods). Paper chromatography gives a very good separation and, with highly specific sprays, the individual spots can be unequivocally identified. Because a separation time of 12–24 h is required, paper chromatography is being replaced by other more rapid methods, such as thin-layer chromatography, for which the usual separation time is 1–2 h. In addition, a wider selection of substrates permits better separations.

Paper electrophoresis uses a different principle of separation (12). Whereas in paper and thin-layer chromatography an advancing solvent front moves the sample, electrophoresis uses a voltage gradient. Therefore, only charged species are moved, which would seem to rule out sugars. However, the hydroxyl groups on sugars readily form complexes with borate and other ions. A charge is thus imparted and electrophoresis can be applied. By using a voltage gradient of 100 V/cm, good separations are made in 20 min. However, the paper must be cooled by water.

All these methods give good results for qualitative work but lack precision for quantitative determinations. The sugars are detected only by sprays and appear as spots. Even with densitometers, the content of a spot cannot be determined within <10%. Since the samples are very small, recovery for further testing is not possible.

Gas chromatography gives better results but, like all chromatographic methods, precision is not very high, at best ca 1% (1–2). With polarization, precisions of 0.05–0.01% are attained. Polarization measures very precisely the sum of all the sugars, whereas gas chromatography separates and identifies the sugars and measures them individually.

For gas chromatography, the sample is vaporized, and although sugars are not very volatile, their methyl esters and silyl derivatives are. Complete conversion is essential, otherwise several peaks are obtained. Single sugars often give multiple peaks because of the ability of the method to separate anomeric forms. Thus, glucose gives two peaks. Other columns separate five peaks for fructose, corresponding to the five forms in which fructose can exist. In many applications, lower resolution is more convenient.

Liquid chromatography is the most recent instrumental method for sugar analysis and promises eventually to be the method of choice (1). It offers the advantage that no derivatives need be formed and the solvent can be water. It is fast and the separated sugars can be collected. The chromatographic precision is still only 1%, but the individual sugars are separated and determined without interference. Automatic sample injectors give good reproducibility and precision (see also Biomedical automated instrumentation).

Other Compounds

As understood in the sugar industry, purity denotes sucrose content as a percentage of total solids. It is calculated as pol/Brix. Because of the inaccuracies of pol and different ways of measuring Brix, different purities are obtained and are distinguished by such names as apparent, true, or gravity purity. Despite its ambiguity, purity is a good control measure in sugar manufacture and every factory has its own system. The purity of cane juice is generally about 80%, of raw sugar about 98%, and of molasses about 35 to 50% (see also Syrups).

Moisture. In fairly pure syrups and liquors, moisture is determined as the difference between 100 and Brix which is measured by spindle or refractometer (6).

In molasses, low purity syrups, and plant juices, the moisture content is not often needed. Of interest is the dry substance, frequently determined by drying. However, if the period of drying is too long or the temperature too high, the sugar begins to decompose. Optimum drying conditions are controversial, and many different procedures are available. Conditions vary from 1 or 2 h at 105–110°C to vacuum at 70°C. Better drying is obtained from dilute solutions spread thin on sand, kieselguhr, or pumice.

For dry sugars containing up to 0.5% moisture the best method is the Karl Fischer (6).

Ash and Inorganic Constituents. Ash is determined by incineration. The residue is called carbonate ash. Bubbling and swelling of the incinerating sugar mass can be avoided by adding a small amount of concentrated sulfuric acid. In this case, the residue is strongly heated to drive off sulfur trioxide. The ash components are converted to sulfates and the ash is called sulfate ash; it is reported as such. For control, the ash is approximated with an electrical conductived measurement, which is simple, quick and reported as conductivity ash. Standard analytical procedures are used to determine the inorganic constituents, such as K, Ca, sulfate, chloride, and smaller amounts of Na, Mg, Fe, carbonate, phosphate, and silicate (6).

Color. In sugar, lack of color is used to indicate the general degree of refinement and a standard method is prescribed. The standard conditions for colorimetry are 50 Brix concentration and pH 7. The light transmission T is measured at 420 nm (1). Sugar color is expressed as:

$$a^* = -\log T/bc$$

where b is the cell depth, cm; c is the concentration of sugar, g/mL; a^* is the attenuation index. In practice, the number given by the above equation is multiplied by 1000 to give sugar color units. Raw sugars have a color of 1000 or more. Very light-colored syrups have a color of perhaps 100. Refined granulated sugar has a color <35.

BIBLIOGRAPHY

"Sugar Analysis" in *ECT* 1st ed., Vol. 13, pp. 192–203, by E. J. McDonald, National Bureau of Standards; "Sugar Analysis" in *ECT* 2nd ed., Vol. 19, pp. 155–166, by E. J. McDonald, U.S. Department of Agriculture.

1. *ICUMSA*, 17th session, International Commission for Uniform Methods of Sugar Analysis, Peterborough, England, 1978.
2. F. Schneider, *Sugar Analysis ICUMSA Methods*, International Commission for Uniform Methods of Sugar Analysis, 1979.
3. *Official Methods of Analysis of the AOAC*, 13th ed., Association of Official Analytical Chemists, Washington, D.C., 1980.
4. *Standard Analytical Methods*, 6th ed., Corn Refiners Association, Washington, D.C., 1980.
5. R. W. Plews, *Analytical Methods Used in Sugar Refining*, Elsevier, London, 1970.
6. G. P. Meade and J. C. P. Chen, *Cane Sugar Handbook*, 10th ed., John Wiley & Sons, Inc., New York, 1977.
7. F. J. Bates and co-workers, *Polarimetry, Saccharimetry and the Sugars*, C440, U.S. Government Printing Office, Washington, D.C., 1942.
8. New York Sugar Trade Laboratory, 300 Terminal Ave. West, Clark, N.J. 07066.
9. M. J. Sibley, F. G. Eis, and R. A. McGinnis, *Anal. Chem.* **37**, 1701 (1965).
10. H. Horning and H. Hirschmuller, *Z. Zuckerind. Boehm.* **9**, 499 (1959).
11. M. Dubois and co-workers, *Anal. Chem.* **28**, 350 (1956).
12. D. Gross, *J. Chromatogr.* **5**, 194 (1961).

FRANK G. CARPENTER
United States Department of Agriculture

CANE SUGAR

Cane sugar is the name given to sucrose, a disaccharide produced from the sugarcane plant and from the sugar beet (see Sugar, beet sugar). The refined sugars from the two sources are practically indistinguishable and command the same price in competitive markets. However, since they come from different plants, the trace constituents are different and can be used to distinguish the two sugars. One effect of the difference is the odor in the package head space, from which experienced sugar workers can identify the source. Also, since the two plants use different photosynthetic pathways, the carbon isotope ratio in the sugars can be used to distinguish sucrose from the two sources (1). The other major sweetener is corn sugar (glucose), which is a monosaccharide derived from corn starch (see Carbohydrates; Syrups). It is available either as a solid or as a syrup (corn syrup). Corn sugar is essentially glucose (dextrose); corn syrup contains mostly glucose but with several percent of residual, incompletely hydrolyzed higher saccharides. Since the commercialization of enzymatic isomerization in 1968, fructose-containing corn syrup has been available. This product, called high fructose corn syrup (HFCS), is essentially equivalent to invert syrup made from sucrose. It can be substituted for invert syrup and even for sucrose in many food products. Other minor sweeteners are maple sugar, sorghum sugar, and palm sugar, all principally sucrose, and honey, whose composition depends upon the source plant.

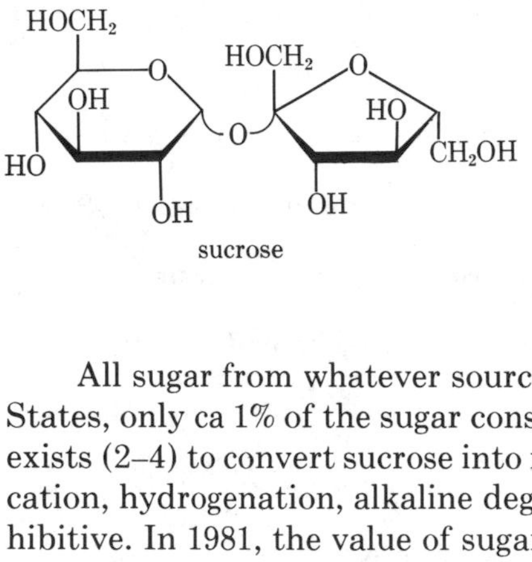

sucrose

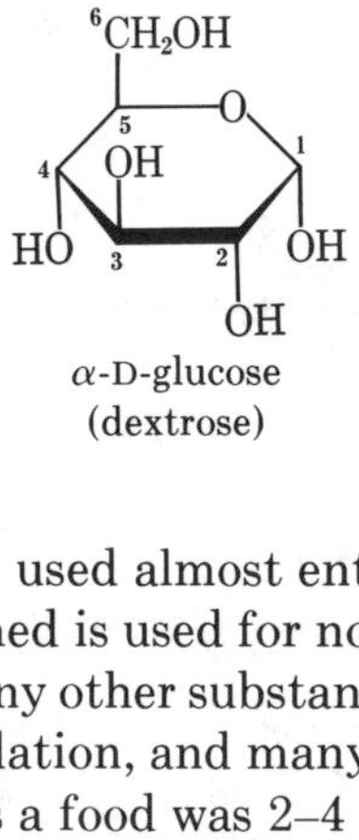

α-D-glucose (dextrose)

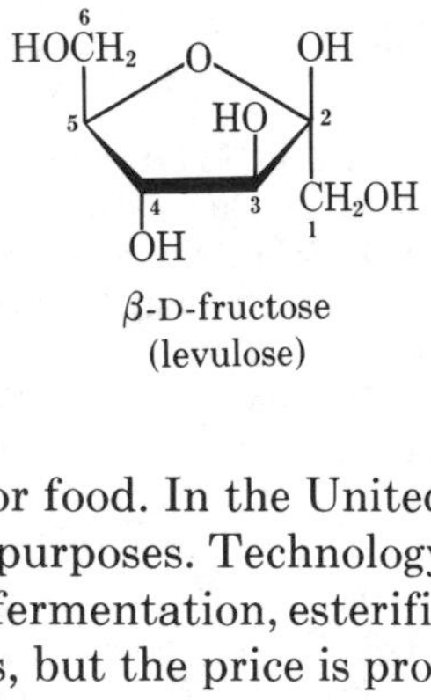

β-D-fructose (levulose)

All sugar from whatever source is used almost entirely for food. In the United States, only ca 1% of the sugar consumed is used for nonfood purposes. Technology exists (2–4) to convert sucrose into many other substances by fermentation, esterification, hydrogenation, alkaline degradation, and many others, but the price is prohibitive. In 1981, the value of sugar as a food was 2–4 times its value as a chemical feedstock. As a food, sugar is all energy [kJ (food calories)], and even brown and raw sugar contain virtually no protein, minerals or vitamins.

Sugar is one of the purest of all substances produced in large volume. Its analysis is approximately: sucrose 99.90%; invert sugars 0.01%; ash (inorganic material) 0.03%; moisture 0.03%; other organic material (mostly polysaccharides) 0.03%. Even raw sugars are >98% sucrose. World production in 1981 was ca 10^8 metric tons.

The *per capita* consumption of sugar is an indicator of degree of economic advancement of a country. Apparently, human nature is such that one of the first uses of income above the subsistence level is to satisfy the sweet tooth. The consumption of sugar in all forms, beet, cane, and corn, in Western Europe and North America is approximately the same, ca 60 kilograms per person each year, and holding steady. However, eastern countries consume much less sugar, and the poorer third-world countries, much less. In some isolated and remote areas, the figure is less than one kilogram per person-year. The world average is about 20 kilograms per person each year and increasing. In the United States, there is a shift toward more use of corn sugar and less of cane and beet sugar because of price, but this applies only in areas where starch is in great surplus.

In the production scheme for cane sugar, the cane cannot be stored for more than a few hours after it is cut because microbiological action immediately begins to degrade the sucrose. This means that the sugar mills must be located in the cane fields. The raw sugar produced in the mills is the item of international commerce. Able to be stored for years, it is handled as a raw material—shipped at the lowest rates directly in the holds of ships or in dump trucks or railroad cars and pushed around by bulldozers. Because it is not intended to be eaten directly, it is not handled as food. The raw sugar is shipped to the sugar refineries which are located in population centers. There it is refined to a food product, packaged, and shipped a short distance to the market. In a few places, there is a refinery near or even within a raw-sugar mill. However, the sugar still goes through the raw stage.

There is, however, another category of cane sugar which is not truly refined. It is called direct-consumption, or plantation white, or merely white sugar. White sugar, which should not be confused with refined sugar, is made directly from the cane without going through the raw-sugar stage. It is a little off-white, frequently has a lingering molasses aroma, and is not as pure as truly refined sugar. It is, of course, perfectly sanitary and quite edible, but it does not keep as well as refined sugar and is usually sold only locally at a reduced price. There is the beginning of a trend, however, to improve the quality of the direct-consumption sugars to make them comparable to refined sugar and thus by-pass the refinery.

Still another class of cane sugar is called noncentrifugal sugar or whole sugar. It is made the way sugar was undoubtedly originally made in prehistoric times, by boiling down whole cane juice without the elimination of any impurities. The whole mixture solidifies upon cooling and is broken into pieces. It is light to dark brown in color, with a consistency between rock-hard and friable. Much of this sugar is made in India where it is called *gur*. In other places, this sugar is known as *panela*, *areado*, *pile*, *piloncillo*, *papelon*, *chancaca*, *muscovado*, *duong-cat*, *panocha*, *shakkar*, and *jaggery*.

The principal by-product of cane sugar production is molasses. About 10–15% of the sugar in the cane ends up in molasses. Molasses is produced both in the raw-sugar manufacture and also in refining (see Syrups). The blackstrap or final molasses is about 35–40% sucrose and slightly more than 50% total sugars. In the United States, blackstrap is used almost entirely for cattle feed. In some areas, it is fermented and distilled to rum or industrial alcohol. The molasses used for human consumption is of a much higher grade, and contains much more sucrose.

History. Historians disagree about whether sugarcane is native to India or New Guinea (5–9). They do agree that ancient people liked sugarcane and carried it with them in their migrations. It spread throughout much of the South Pacific area, except for Australia, by about 1000 BC. By 400 BC, it was known in Persia, Arabia, and Egypt. The word sugar does not appear in the Bible, but the phrase "sweeter than syrup or honey from the honeycomb" (Psalms 19:11) appears, indicating that although sugarcane was possibly known in the Holy Land in biblical times, only syrups could be obtained from it. In the 7th to 10th centuries AD, the Arabs spread sugarcane throughout their region of influence in the Mediterranean. About the same time, it was carried eastward across the Pacific Ocean to Fiji, Tonga, Samoa, Tahiti, Easter Island, and Hawaii. By the 12th century, sugarcane had reached Europe, and Venice was the center of sugar trade and refining. Marco Polo reported advanced sugar refining in China toward the end of the 13th century. Columbus brought sugarcane to the new world on his second voyage. It spread throughout the Western Hemisphere in the next 200 years and by about 1750 sugarcane had been introduced throughout the world.

The process for extracting juice from the cane is also very old. In antiquity, the canes were undoubtedly sucked or chewed for their sweet taste. Also, in the ancient past in various places the canes were cut and crushed by heavy weights, ground with circular stones or by a heavy roller running on a flat surface, pounded in a mortar with a pestle, or soaked in water to better extract the sweet juice (7). The term grinding survives to this day as the name of the process for extracting the cane juice, even though the process no longer involves a true grinding. Parallel rolls, which are used today, were first used in 1449 in Sicily.

The ancient process for obtaining sugar consisted of boiling the juice until solids formed as the syrup cooled. The product looked like gravel and the Sanskrit word for sugar, *shakkara*, has that alternative meaning. Pliny, who traveled widely in the Roman Empire, wrote in 77 AD that sugar was "white and granular." He noted that Indian sugar was more esteemed than Arabian, and that both were used in medicine. By the fourth century AD, the Egyptians were using lime as a purifying agent, and carrying out recrystallization which is still the main step in refining.

Until quite recently, sugar was strictly a luxury item. Queen Elizabeth I is credited with putting it on the table in the now familiar sugarbowl, but it was so expensive that it was used only on the tables of royalty. Sugar production reached large volume at a reasonable price only by the 18th century.

The development of the sugar industry from the 16th century onward is closely associated with slavery which supplied the large amount of labor used at the time. The low cost of labor and the high price for sugar made many fortunes. The abolition of slavery at various times in different countries between 1761 and 1865 profoundly affected the sugar industry. Upon the freeing of the slaves, sugar production fell drastically in many producing areas.

The first use of steam power as a replacement for the animal or human power that drove the cane mills occurred in Jamaica in 1768. This first attempt worked only a short time, but steam drive was used successfully a few years later in Cuba. Steam drive for the mills soon spread throughout the world. The use of steam instead of direct firing was soon applied for evaporating the cane juice.

Probably the most essential piece of equipment in the modern process is the vacuum pan invented by Howard (UK) in 1813. This accomplishes the evaporation of water at a low temperature and lessens the thermal destruction of sucrose. The bone-char process for decolorization dates from 1820. The other essential piece of equipment is the centrifuge which was developed by Weston in 1852 and applied to sugar in 1867 in Hawaii. This machine reduces the time for draining the molasses from the sugar crystals from weeks to minutes by applying a force equal to 1000 G.

The manufacture of sugar was early understood to be an energy-intensive process. Cuba was essentially deforested to obtain the wood that fueled the evaporation of water from the cane juice. When the forests were gone, the bagasse (qv) burner was developed to use the otherwise waste dry cane pulp, bagasse, for fuel. It is to the everlasting credit of the cane-sugar industry that the greatest energy saver of all time was developed in this industry: the multiple-effect evaporator invented by Norbert Rillieux of Louisiana. The 1846 patents of Rillieux describe every detail of the process (10). This system is now used universally by every industry that has to evaporate water.

The principal analytical methods were developed in the mid-19th century: the polariscope by Ventzke in 1842, the Brix hydrometer in 1854, Fehling's method for reducing sugars, and Clerget's method for sucrose in 1846.

Sugar loaves were for centuries the traditional form in which sugar reached the market. These were formed by pouring the mixture of crystals and syrup into a mold. The molds were kept in the hot room to facilitate the draining of the syrup either through the porous mold or through a hole in the bottom. With a little cooling or drying, the crystals stuck together, forming a convenient marketable loaf of sugar. The sugar loaves required no packaging and were broken up by the user as needed. Only a very small amount of sugar now reaches the market in this form.

The sugar business, like many others, has changed structure in the last 100 years from many small units to few large units. In 1861, the Havemeyer and Elder refinery in Brooklyn was the largest in the world and processed 32 metric tons of sugar daily. In 1981, there were several refineries processing >4000 t daily, and the smallest refineries process 500 t each day. In 1861, there were 1200 sugar mills in Louisiana. In 1981, there were 25. This consolidation has been repeated throughout the world. The cane and beet farms have also become larger and fewer.

Cultivation

Sugarcane is a tropical grass. It is killed by freezing, has thick stems about 2 cm in diameter, and grows very tall—3–5 or even 7 m. Sugarcane only reluctantly forms flowers under ideal conditions. In many areas where cane grows well it never flowers. When it does, the flowers are in the form of tassels containing many tiny blossoms. Each tassel forms many thousands of tiny seeds, called fuzz. Experimental stations for cross breeding have been set up in those areas where the climate is right for cane to flower. The station at Canal Point, Florida, has a collection of 30,000 varieties of cane for cross breeding and makes a million (10^6) crosses each year. Since the varieties are all hybrids, seeds will not breed true but each seed is a new variety. The varieties are designated by the initials of the experiment station that developed them followed by a number. The best known varieties are those of POJ (Proefstation Ost Java), CP (Canal Point), NCo (Natal selected from canes bred in Coimbatore, India), and Q (Queensland). One variety that today is cultivated all over the world is NCo 310. Selection criteria for new varieties include agronomic characteristics, sugar and fiber content, harvestability, disease and insect resistance, cold and salt tolerance, and many others. A continuing breeding program is essential because all varieties suffer a gradual yield decline. In this baffling phenomenon, the yield from a variety falls off over a period of years. It is interesting to note that some of the original work on tissue culture was done on sugarcane in Hawaii (11).

The principal commercial cane until the 1920s was the so-called noble cane, *Saccharum officinarium*, with thick stalks and relatively soft rind. However, these canes succumbed to the mosaic virus, which causes a mottling of the leaves and greatly reduced plant vigor. As a result of extensive cross-breeding programs, the experimental stations succeeded in developing multicross hybrids that were resistant to mosaic disease. These new varieties all have a much tougher rind and more fiber and more sugar than the noble canes. The most famous is the miracle cane, POJ 2878, which was the answer to the threat of mosaic disease.

Commercially, cane is grown vegetatively by planting stalks of cane called seed cane or setts. Sugarcane has a bud at each node which is at the base of each leaf. There is also a ring of root primordia. The pieces planted contain one or two buds or may be a whole stalk. When planted, the roots develop and the bud sprouts to form a new plant or stool of the same variety as the cane that was planted.

Upon harvesting, the stalks are cut off even with the ground and the next day more sprouts start from the buds just below ground level. Thus, a field of sugarcane once planted is self perpetuating. This is the source of one of the economic woes of the whole sugar industry. A high price of sugar results from the planting of many cane fields in different parts of the world. Once started, they continue producing year after year whether the price of sugar is high or low. The cane from the first crop is called plant cane and is always best. Subsequent crops are called stubble or ratoon crops. The ratoon crops have many problems. Gaps appear where the harvester dug too low and pulled the whole stool out by the roots. In other areas, the stalk density becomes too great, resulting in too many spindly stalks. Diseases multiply and weeds get thicker than the cane. In many areas, only two or three ratoon crops are taken off before the whole field is plowed out and started over. However, in some tropical areas where conditions are ideal for growing sugarcane, such as Cuba, fields have remained in production for as long as 20 yr.

Good agricultural practice requires the usual close attention to soils, fertilizer, weeds, pests, disease, and water supply. In some cane-growing areas, the land must be drained; in others, irrigated. A common occurrence is for the cane to be knocked down by a thunderstorm or hurricane. The growing tip then turns upward and the stalk grows curved. After several storms the cane field becomes badly tangled, or lodged. Some varieties are more susceptible to lodging, which causes difficult harvesting.

Sugarcane has the very fortunate characteristic that it stores sugar in the stalks at all times. Thus, there is no narrowly defined ripe or mature stage. The sugar content of the stalks increases somewhat with cool weather and decreases again with warm weather. Typically, a cane that had 15% sucrose at its optimum, might have 12% several months before and after the optimum. The harvesting season can therefore be very long. It is nevertheless advantageous to harvest the cane at its highest sucrose content, so considerable attention is paid to the maturity or ripeness of the cane, indicated by a higher sucrose content in the juice in the whole cane stalk. The growing tip has lower sucrose than the rest of the stalk. Tasseling stops growth temporarily and matures the cane. The ripeness can also be controlled to some extent by fertilizer and water supply. Chemical ripeners, which alter the growth of the cane to bring the sucrose content up, can be sprayed on the fields. Because different varieties mature earlier or later in the season, the cane harvest can last six months or more. The shortest harvest season is in Louisiana where freezes occur almost every year and the cane does not start growing again until warm days in spring. In the fall, with the cane only eight months old, harvest is delayed as long as possible so that the cane can grow another day or two and produce a little more sugar. On the other hand, the harvest must be finished by the first freeze, for although frost kills only the growing tip, causing sucrose content to rise, a severe freeze splits open the stalk, and microbial action begins on the next warm day. As a result, the grinding season in Louisiana is compressed to only two months of frenzied activity. At the other extreme, Hawaii normally grinds for more than eleven months in a deliberate and methodical manner. There, the cane is normally grown for three or four years. This results in very heavy yields of badly tangled cane. However, the yield per year of growth is the same for the rest of the world (Table 1).

Raw Sugar Manufacture

At the sugar factory, the cane is piled as a reserve supply in the cane yard so that the factory, which runs 24 h/d, will always have cane to grind. The delivery of the cane to the factory depends upon time of day, weather, and other factors. Very closely controlled operations never have more than a few hours worth of cane in the cane yard, but more generally, the cane yard is fairly full toward evening and nearly empty the next morning.

The cane is moved from the cane yard or directly from the transport to one of the cane tables. Feed chains on the tables move the cane across the tables to the main cane carrier which runs at constant speed carrying the cane into the factory. The operator manipulates the speed of the various tables to keep the main carrier evenly filled.

In order to remove as much dirt and trash as possible, the cane is washed on the main carrier with as much water as is available. This includes recirculated wash water and all of the condenser water. Of the order of 1–2% of the sugar in the cane is washed out and lost in the washing, but it is considered advantageous to wash. In areas where there are rocks in the cane, it is floated through the so-called mud bath to help separate the rocks.

A flow diagram of a raw-sugar factory is shown in Figure 1.

The average yields of sugarcane given in Table 1 are for the whole cane-growing area. Within a cane-growing area, individual fields range from twice this figure to less than half of it. The sugar recovered is nominally 10 wt % of the cane, with some variation from region to region. Sugarcane has the distinction of producing the heaviest yield of all crops, both in weight of biomass and in weight of useful product per unit area of land.

Extraction of the Juice. The juice is extracted from the cane either by milling, in which the cane is pressed between heavy rolls, or by diffusion, in which the sugar is leached out with water (13). In either case, the cane is prepared by breaking into pieces measuring a few cm. In the usual system, magnets first remove tramp iron, and the cane then passes through two sets of rotating knives. The first set, called cane knives,

Table 1. Yield of Sugarcane [a]

Region	t/(ha·yr)
Louisiana	60
Florida	90
Hawaii	100
Philippines	65
Puerto Rico	75
Australia	100
Iran	110

[a] Ref. 12.

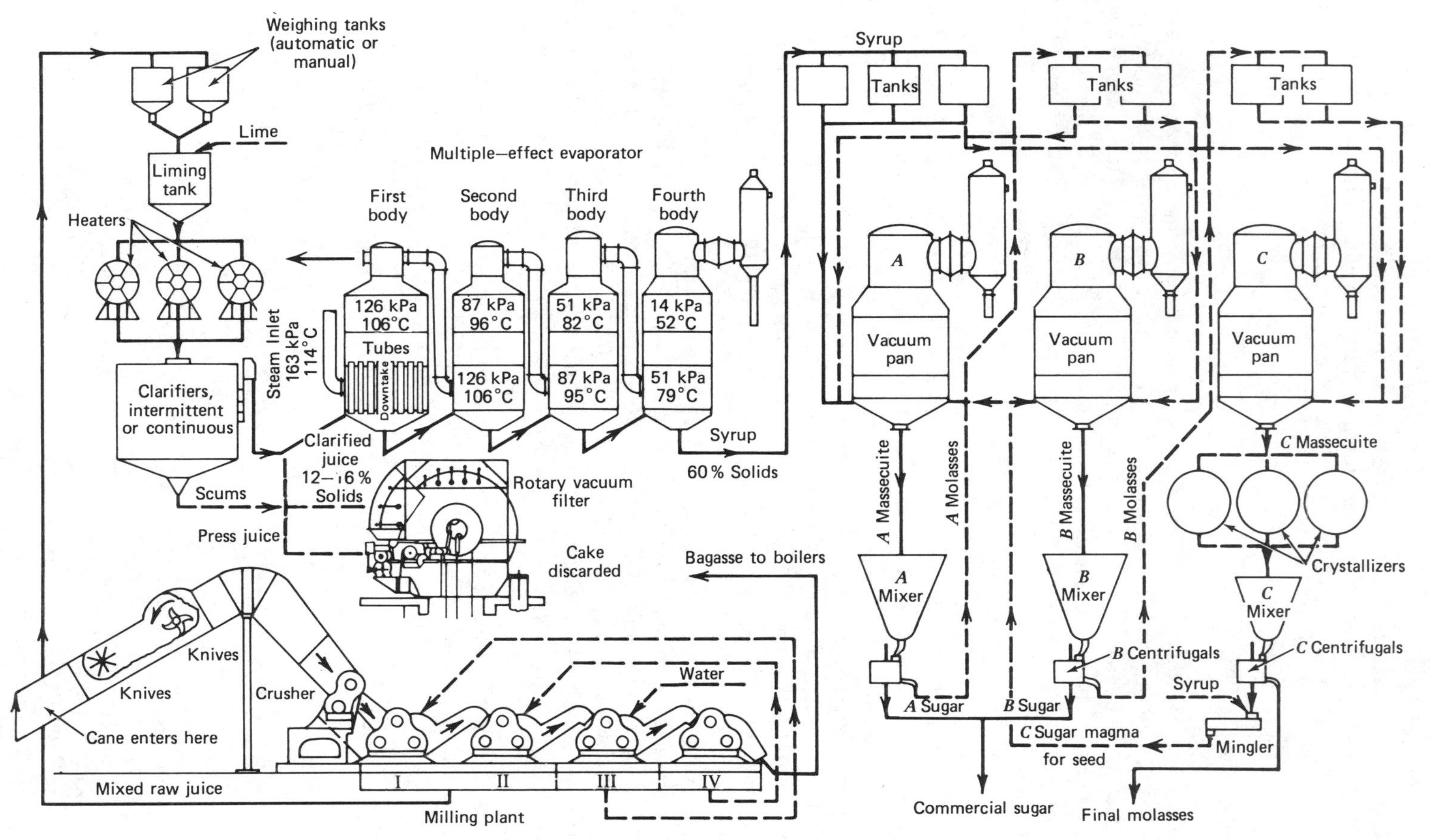

Figure 1. Flow diagram of a raw-sugar factory. To convert kPa to psia, multiply by 0.145.

turns at about 700 rpm, cuts the cane into pieces of 1–2 dm length, splits it up a bit, and also acts as a leveler to distribute the cane more evenly on the carrier. The second set, called shredder knives, turn faster and combine a cutting and a hammer action by having a closer clearance with the housing. These quite thoroughly cut up and shred the cane into a fluffy mat of pieces a few cm in the largest dimension. In preparing cane for diffusion, it is desirable to break every plant cell. Therefore, cane for diffusion is put through an even finer shredder called a buster or fiberizer. No juice is extracted in the shredders. In milling, the cane then goes to the crusher rolls which are similar to the mills, but have only two rolls, which have large teeth and are widely spaced. These complete the breaking up of the cane to pieces of the order of 1–3 cm. A large amount of juice is removed here (see also Fruit juice).

Milling. The prepared cane passes to a series of mills called a tandem or milling train (14). These mills are composed of massive horizontal cylinders or rolls in groups of three, one on the top and two on the bottom in a triangle formation. The rolls are 50–100 cm dia and 1–3 m long, and have grooves that are 2–5 cm wide and deep around them. There may be anywhere from 3–7 of these three-roll mills in tandem, hence the name. These mills, together with their associated drives and gearing, are among the most massive machinery used by industry. The bottom two rolls are fixed, and the top roll is free to move up and down. The top roll is hydraulically loaded with a force equivalent of about 500 t. The rolls turn at 2–5 rpm, and the velocity of the cane through them is 10–25 cm/s. After passing through the mill, the fibrous residue from the cane, called bagasse, is carried to the next mill by bagasse carriers and is directed from the first squeeze in a mill to the second by a turnplate. In order to achieve good extraction, a system of imbibition is used: bagasse going to the final mill is sprayed with water to extract whatever sucrose remains; the resultant juice from the last mill is then sprayed on the bagasse mat going to the next to last mill, and so on. The combination of all these juices is collected from the first mill and is mixed with the juice from the crusher. The result is called mixed juice and is the material that goes forward to make the sugar.

The mills are powered with individual steam turbines. The exhaust steam from the turbines is used to evaporate the water from the cane juice. The capacity of sugarcane mills is 30–300 t of cane per hour.

Diffusion. Diffusion is used universally with sugar beets but is little used with sugarcane (15). The process in cane is mostly lixiviation (washing) with only a little true diffusion from unbroken plant cells. Since the lixiviation is much faster, great effort is expended in preparing the cane by breaking it so thoroughly that nearly all of the plant cells are ruptured. In many instances, diffusers were added to an already existing mill, and, therefore, the diffuser unit was placed after the crusher rolls. In the diffusers, the shredded cane travels countercurrent to hot (75°C) water. In the ring diffuser, the cane moves around in an annular ring. In tower diffusers, the cane moves vertically, and in rotating drum diffusers, it travels in a spiral. For a more complete description of diffusers, see Sugars, beet sugar. Whatever the apparatus, the juice obtained is much like juice from mills.

Milling achieves ca 95% extraction of the sucrose in the cane, diffusion ca 97% extraction. Diffusion juice contains somewhat less suspended solids (dirt and fiber), and is of higher purity (sucrose as percent of solids). The diffusion plant costs much less and takes much less energy to run. The bagasse from diffusion contains much more water.

Bagasse. The bagasse from the last mill is about 50 wt % water and will burn directly (2,16). Diffusion bagasse is dripping wet and must be dried in a mill or some sort of bagasse press. Most bagasse is burned in the boilers that run the factories (see Bagasse).

Clarification. The juice from either milling or diffusion is about 12–18 brix (percent solids), 10–15 pol (polarization) (percent sucrose), and 70–85% purity (pol/brix) (17). These figures depend upon geographical location, age of cane, variety, climate, cultivation, condition of juice extraction system, and other factors. As dissolved material, it contains in addition to sucrose some invert sugar, salts, silicates, amino acids, proteins, enzymes, and organic acids; the pH is 5.5–6.5. It carries in suspension cane fiber, field soil, silica, bacteria, yeasts, molds, spores, insect parts, chlorophyll, starch, gums, waxes, and fats. It looks brown and muddy with a trace of green from the chlorophyll.

In the juice from the mill, the sucrose is inverting (hydrolyzing to glucose and fructose) under the influence of native invertase enzyme or an acid pH. The first step of processing is to stop the inversion by raising the pH to ca 7.5 and heating to nearly 100°C to inactivate the enzyme and stop microbiological action. At the same time, a large fraction of the suspended material is removed by settling. The cheapest source of hydroxide is lime, and this has the added advantage that calcium makes many insoluble salts. Clarification by heat and lime, a process called defecation, was practiced in Egypt many centuries ago and remains in many ways the most effective means of purifying the juice. Phosphate is added to juices deficient in phosphate to increase the amount of calcium phosphate precipitate which makes a floc that helps clarification (see Flocculating agents). When the mud settles poorly, polyelectrolyte flocculants such as polyacrylamides are sometimes used. The heat and high pH serve to coagulate proteins which are largely removed in clarification.

The equipment used for clarification (18) is of the Dorr clarifier type. It consists of a vertical cylindrical vessel composed of a number of trays with conical bottoms stacked one over the other. The limed raw juice enters near the center of each tray and flows toward the circumference. A sweep arm in each tray turns quite slowly and sweeps the settled mud toward a central mud outlet. The clear juice from the top circumference overflows into a header.

Diffusion juice contains less suspended solids than mill juice. In many diffusion operations, some or all of the clarification is carried out in the diffusers by adding lime.

The mud from clarification is filtered on Oliver rotary vacuum filters to recovery the juice. The mud mostly consists of field soil and very finely divided fiber. It also contains nearly all the protein (0.5 wt % of the juice solids) and cane wax. The mud is returned to the fields.

Although clarification removes most of the mud, the resulting juice is not necessarily clear. The equipment is often run at beyond its capacity and the control slips a little so that the clarity of the clarified juice is not optimum. Suspended solids that slip past the clarifiers will be in the sugar. Clarified juice is dark brown. The color is darker than raw juice because the initial heating causes significant darkening.

In an attempt to produce plantation white sugar, or at least a very good raw sugar, the carbonatation process was originated in Java in 1880. In this process, a large excess of lime is added to produce a juice pH of >10.5. Then, carbon dioxide from the lime kiln is bubbled through to form a great volume of precipitates which is removed by filtration (19). Juices from carbonatation are much better clarified than settled juices, and they are somewhat decolorized. The use of this process is small and decreasing.

Another scheme that makes a better clarification is the sulfitation process (20). In this method, lime is added as usual, but then sulfur dioxide from a sulfur burner is bubbled through the juice. The precipitate is settled as in the ordinary clarification. The bleaching effect of the sulfite makes a lighter-colored sugar. This method is little used in the West but is used extensively in the Orient.

There is another type of clarification that is not used at this point in the process. This is the Talodura process (21–22) for clarification of the thick syrup after the evaporator station in the raw-sugar process. This process involves lime and phosphate precipitation followed by aeration and dosage with a specific polyacrylamide flotation aid.

These supplemental clarification schemes essentially complete the clarification that was not done in the initial clarification. Juices so clarified can often be processed into direct consumption sugar.

Evaporation. Cane juice has a sucrose concentration of nominally 15 brix (percent). The solubility of sucrose in water is about 72 brix. The concentration of sucrose must reach the solubility point before crystals can start growing. This involves the removal by evaporation of 93% of the water in the cane juice. Since water has the largest of all latent heats of vaporization, this involves a very large amount of energy. In the energy crunch of the late 1970s, the DOE found that the sugar industry was one of the largest users of energy. The sugar industry already knew this very well and had been using multiple-effect evaporators for saving energy for more than a century (see Evaporation).

The working of a multiple-effect evaporator can be seen in Figure 1. In each succeeding effect, the vapors from the previous effect are condensed to supply the heat. This works only because each succeeding effect is operating at a lower pressure and hence boils at lower temperature. The result is that 1 kg of steam is used to evaporate 4 kg of water. The steam used is exhaust steam from the turbines in the mill or turbines driving electrical generators. The steam has therefore already been used once and here in the second use it is made to give fourfold duty.

The usual evaporator equipment is vertical-body juice-in-tube units. Several variations are in use, but the result is the same. The only auxiliary equipment is the vacuum pump. Today, steam-jet ejectors are general, although mechanial pumps were formerly used.

Since the cane juice contains significant amounts of inorganic ions, including calcium and sulfate, the heating surfaces are quick to scale and require frequent cleaning. In difficult cases, the heating surfaces must be cleaned every few days. This requires shutting down the whole mill or at least one heat-exchange unit while the cleaning is done. Inhibited hydrochloric acid or mechanical cleaners are usually employed. Magnesium oxide is sometimes used instead of lime as a source of hydroxide. Magnesium costs more, but it makes less boiler scale on the heaters. It is also easier to remove because it is more soluble; however, for the same reason, more gets into the sugar. Whether it is used or not depends upon the influence, standing, and persuasiveness of the chief engineer who must keep the plant running and the chief chemist who must make good sugar.

The evaporation is carried on to a final brix of ca 65–68. The juice, after evaporation, is called syrup and is very dark brown, almost black, and a little turbid. For further information, see ref. 23.

Crystallization. The crystallization (qv) of the sucrose from the concentrated syrup is traditionally a batch process (24). The solubility of sucrose changes rather little with temperature. It is 68 brix at room temperature and 74 brix at 60°C. For this reason, only a small amount of sugar can be crystallized out of a solution by cooling. The sugar must instead be crystallized by evaporating the water. Sucrose solutions up to a supersaturation of 1.3 are quite stable. Above this supersaturation, spontaneous nucleation occurs, and new crystals form. The sugar boiler therefore evaporates water until the supersaturation is ca 1.25 and then seeds the pan. The seeding consists of introducing just the right number of small sugar crystals (powdered sugar) so that, when all have grown to the desired size, the pan will be full. After seeding, the evaporation and feeding of syrup are balanced so that the supersaturation is as high as possible in order to achieve the fastest possible rate of crystal growth, without exceeding 1.3.

The boiling point of a saturated sugar solution at 101.3 kPa (1 atm) is 112°C. Sugar is heat-sensitive and, at this temperature, thermal degradation is too great. The boiling is therefore done under the highest practical vacuum at a boiling point of ca 65°C. The sugar boiler therefore must manipulate the vacuum along with the steam and feed. A proofstick on the vacuum pan allows the contents of the pan to be sampled while under vacuum. When the pan is full, the steam and feed are stopped, the vacuum is broken, and the batch, or strike, is dropped into a receiver below.

A strike is ca 50 metric tons of sugar and it is boiled in 90 min. At the end of this time, the mixture of crystals and syrup, called massecuite, must still be fluid enough to be stirred and discharged from the pan. In practice, about half of the sugar in the pan is in crystal form and half remains in the syrup. In this case, the pan yield is said to be 50%. Some very good sugar boilers are able to achieve as much as a 60% yield on first strikes.

In actual practice, there are many more fine details to sugar boiling than is briefly explained here. Because modern pans have been extensively instrumented, automatic boiling can be used.

The sugar-boiling process can be made continuous by continually introducing seed as well as syrup, and also continually removing some of the product sugar. If the continuous boiling is done in one vessel, then the product will always contain crystals of all sizes. The market demand for crystals of a narrow size range can be met by having several vessels in series. In one scheme, all are within one body boiling under the same vacuum. In another scheme, several batch pans are connected in a continuous fashion.

Vacuum Pans. Vacuum pans have a small heating element in comparison to the very large liquor and vapor space above it (25). The heating element was formerly steam coils but is now usually a chest of vertical tubes called a calandria. The sugar is inside the tubes. There is a large center opening (downcomer) for circulation. A typical pan is shown in Figure 2.

The vacuum pan has a very large discharge opening: typically 1 m dia. At the end of a strike, the massecuite contains more crystals than syrup and is therefore very viscous. This large opening is required to empty the pan in a reasonable time. At the top or dome of the pan, there are various entrainment separators. The pan may also be equipped with a mechanical stirrer. This is usually an impeller in or below the central downcomer, driven by a shaft coming down all the way from the top. The pan shown has no such stirrer but relies on natural circulation. The strike is started with

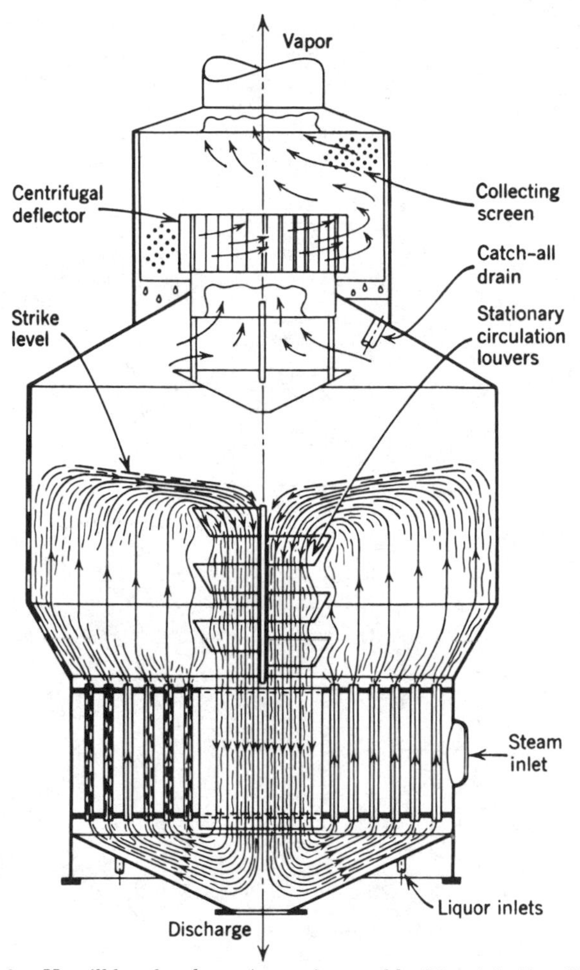

Figure 2. Hamill low-head pan (manufactured by Honolulu Iron Works).

the liquor just above the top of the calandria. The strike level can not be very near the top because vapor space must be allowed for separation of entrainment. In operation, the boiling is very vigorous with much splashing of liquid.

The vacuum is maintained mostly by condensing the vapors in a barometric condenser. In some cases, a surface condenser is used. This serves as a source of distilled water and recovers heat. More often, however, a jet condenser is used in which the cold condensing water is sprayed into the hot vapor and both condensate and condenser water are mixed. A supplementary vacuum pump is required to remove noncondensable gases.

Centrifuging. The massecuites from the vacuum pans enter a holding tank called a mixer that has a very slowly turning paddle to prevent the crystals from settling (26). The mixer is a feed for the centrifuges. In batch-type centrifuges, the mother liquor is separated from the crystals in batches of about 1 t at a time. A typical batch centrifuge is shown in Figure 3.

The basket is lined with a screen having openings that will retain crystals of the minimum size required. Very fine crystals will go mostly with the mother liquor. The basket is attached to the drive shaft only at the bottom so that discharge devices and

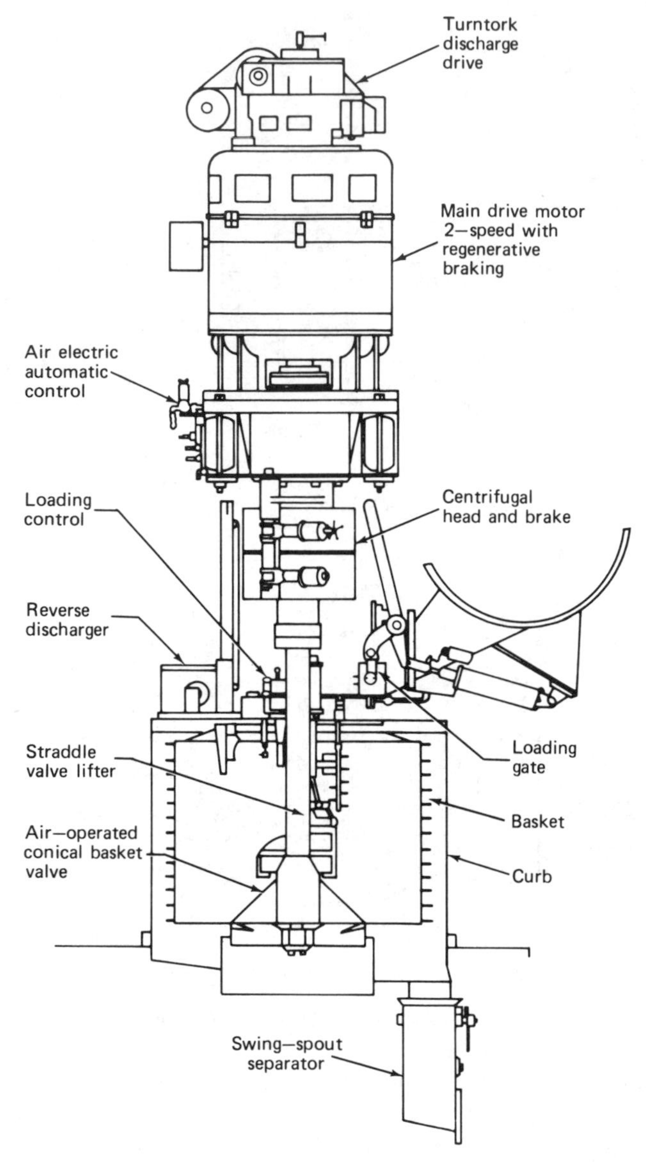

Figure 3. Suspended batch-type automatic centrifuge with reverse-discharge drive (27).

wash-water sprays can be mounted through the top. The essential feature of the suspended centrifuge is a universal joint in the centrifugal head, which is inside the brake. If, as is usually the case, the basket is not loaded evenly, this universal joint allows the basket-shaft-sugar mass to rotate around its own center of gravity instead of the center of the shaft. This greatly reduces vibration which would otherwise shake apart the equipment, and likely the whole building along with it. The diameter of the basket is ≥1 m and the top speed is such that a force of ca 1000 times G is exerted. This works out to 1000–2000 rpm, depending on the diameter. In operation, the loading gate opens, allowing massecuite to fill the basket. The thickness of the massecuite against the wall of the basket is 20 cm. After an initial spin, for most uses, the sugar is sprayed with wash water, sometimes several times. A final spin then allows the greatest possible separation. After this, the centrifuge must be stopped and the sugar plowed out while the centrifuge is turning very slowly. It is then washed, brought up to speed, and the cycle started over. The entire cycle takes as little as 3 min on refined sugars to as long as 2 h on very viscous final strikes. The apparatus has been completely automated, a masterpiece of ingenuity.

The starting and stopping of the centrifugal batch consumes a great amount of energy, much of which can be saved by the use of continuous centrifuges (Fig. 4). They consist of a conical basket with a screen as in the basket centrifuge. The point of the cone is down and the feed is at the point. The centrifugal force causes the sugar to crawl up the cone, while the mother liquor passes through. At the top, the sugar flies off and is caught in the outer housing. The sugar crystals are badly broken when they hit the curb, and this limits the use of the continuous centrifuge to sugar that is to be reprocessed. It has its greatest use in final strikes which are the most difficult of all to centrifuge.

Boiling Systems. In raw-sugar manufacture, the first strike of sugar is called the A strike, and the mother liquor obtained from this strike from the centrifuges is called A molasses (28). The pan yield in sugar boiling is about 50%. Because crystallization is an efficient purification process, the product sugar is much purer than the cane juice and the molasses much less pure. As an approximation, crystallization reduces the impurities by a factor of 10 or more in the product sugar. Therefore, almost all of the impurities remain in the molasses. Enough molasses accumulates from boiling two first strikes to boil a second strike. The B sugar from the second strike is only half as pure as that from the first strike, but the B molasses is twice as impure. This can go on to a third strike. At this point, 7/8 of the sugar from the cane juice is in the form of crystals and 1/8 in the C molasses. In practice, three strikes is about all that can be gotten from cane juice. The trick is to maneuver to obtain good sugar, but at the same time have the C or final molasses as impure as possible. The purity of the feed to the final strike is adjusted to obtain the lowest possible purity of final molasses. Some of the

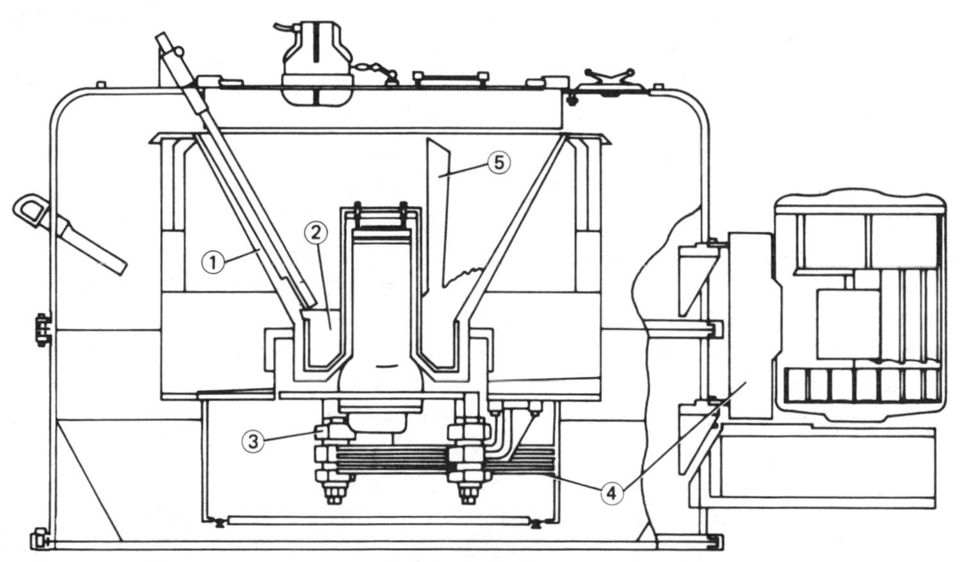

Figure 4. Continuous centrifuge. 1, Stainless steel basket; 2, loading bowl; 3, support; 4, drive; 5, massecuite "side-feed" (27).

C sugar is redissolved and started over, some is used as footing for A and B strikes. The C sugar is of very small crystal size so it is taken into the A or B pans as seed and grown to an acceptable size. This practice is actually a step backward because it hides impure C sugar in the center of better A and B sugars. The product raw sugar is a mixture of A and B sugars.

There are many variations in the boiling scheme, such as two and four boilings, blending molasses, and returning molasses to the same strike from which it came. All of these tricks are used, depending on cane purity and capabilities of the equipment available.

Crystallizers. When the steam is turned off at the end of a sugar boiling, evaporation ceases immediately and the mixture of crystals and supersaturated syrup in the pan starts toward equilibrium, which is the point of saturation (29). In relatively pure sugar solutions, this equilibrium is reached in a few minutes well before the syrup is removed from the crystals in the centrifuges. However, for low purity solutions, the crystallization is slower and reaching equilibrium can take a significant amount of time. In the final strike, the time can amount to days, so final strikes are not sent directly to the centrifuges, but instead to crystallizers, holding tanks in which the crystals grow as much as possible and the supersaturation in the molasses is reduced to 1.0. Since the intention in handling the final molasses is to remove as much sugar as possible, advantage is taken of the small temperature coefficient of solubility, and the massecuite is also cooled. The crystallizers are large tanks, some open-top, with a slow-moving stirrer that is sometimes also a cooling coil. At the end of the holding time, the massecuite is warmed slightly as it enters the centrifuge to lower the viscosity and achieve better separation. The limiting factor in exhaustion of molasses is the viscosity. A little more water can always be boiled out, but the molasses must remain fluid enough to run out of the pan, into the centrifuge, and to flow between the sugar crystals on the centrifuge screens.

Storage. The raw sugar, the A and B sugar from the centrifuges, may be slightly dried or cooled but usually is not. This sugar goes to the warehouse, eventually to be shipped to a refinery, where it will be warehoused again. Raw sugar can be stored for up to several years. Its chief enemies are heat and moisture. Moisture allows microbiological action to devour the sucrose. The index of moisture as it affects storage is the ratio of moisture to nonsucrose solids, called the safety factor. Sugar with a safety factor greater than 0.25 will not keep. High temperature causes an easily detected increase in color long before the sucrose loss can be measured. Sugars at 30°C can be stored for years; at 38°C, for months; and at 45°C, for weeks.

Very High Pol Raw. A recent development has been the production of very high pol raw. It was formerly believed that raw sugar must not be washed in the centrifuges because the adhering film of molasses would "protect" the sugar. However, if the raw sugar is washed slightly in the centrifuge, and then dried and cooled, the result is a raw sugar with the very high pol (% sucrose) of 99.4. This material is not sticky, is free flowing, and keeps very well. This sugar does not qualify as raw sugar in some tariff rules; therefore, very high pol raw that enters international trade must be "polluted" with something to reduce the pol to the raw-sugar class for that country. This is done by adding some high test molasses (invert syrup) as the sugar is loaded into the ship. This is, of course, a step backward to conform to regulations.

Direct-Consumption Sugar. Many raw-sugar mills produce direct-consumption sugar by performing a supplemental clarification on the cane juice, as mentioned under clarification, or by filtering it, followed by some decolorization such as with sulfite, hypochlorite, peroxide, or powdered carbon. From these improved syrups, one strike of direct-consumption sugar can be obtained; later strikes go into raw sugar.

Refining

Sugar refineries are located in large cities. They are near seacoasts with harbors and facilities for receiving raw sugar by ship. They thus can receive sugar from anywhere in the world, although each refiner has favorites that suit the refinery, market, or have been the traditional supplier. Refineries are open all year, although the busy season is in the summer.

Refineries are always large. Their capacity is expressed in terms of daily melt. Melt is the sugar term for dissolving, and means the amount of sugar melted or processed each day. The smallest refineries have a daily melt of 450 t, and large ones have as much as ten times that amount. The yield of refined sugar is nominally 93% of the raw-sugar input.

Raw sugar is light to dark brown in color and sticky. The size of the sucrose crystals is ca 1 mm. Refiners would like to have raw sugar that is high in sucrose and of uniform quality; however, they must be prepared to refine anything. Raw sugars are about 98 pol, although they are always described in terms of the equivalent "raw value" expressed as 96 pol, a base value from the 1920s when raw sugars were of this pol.

A simplified flow diagram for a typical bone-char refinery is given in Figure 5. Refineries tend to be more complex than this diagram. The refining process was established in the late 1800s and several refineries have occupied the same location since then.

Terminology changes slightly in the refinery. In raw sugar, a syrup is a concentrated solution going to the pans. After the boiling, the solution separated from the crystals is molasses. In the refinery, it is liquors that are fed to the pans, and syrups that are separated from the crystals. Local jargon adds to the confusion with such terms as greens and jets, meaning syrups, and barrel syrup, meaning final molasses.

Affination. The first step in refining is to remove the molasses film from the outside of the raw-sugar crystals. This is done by a washing process known as affination (30). A syrup that is not quite saturated with sucrose is mingled with the incoming raw sugar in a large trough containing a mixer paddle and scroll. This mixture is then centrifuged and washed in the centrifuge rather more than less. A uniform crystal size is important in raw sugars because a mixture of different sizes or broken crystals does not wash well in the affination centrifuge. The syrup formed is called affination syrup and is used for mingling. The sugar is called washed sugar and is ten shades lighter in color than the raw sugar. It is estimated that 90% of refining is done in this first step. About 10% of the sugar becomes part of the affination syrup, which thus keeps increasing in volume and is sent to the recovery house.

The recovery house is a route through a set of equipment in the same building. It uses the same processes that are used in the main refinery, but in a manner more like a raw-sugar operation. As the name suggests, sugar is recovered in the recovery house, but the main object is to transfer impurities into molasses that contains the least possible amount of sucrose. The recovered sugar is called remelt and is sent back to process.

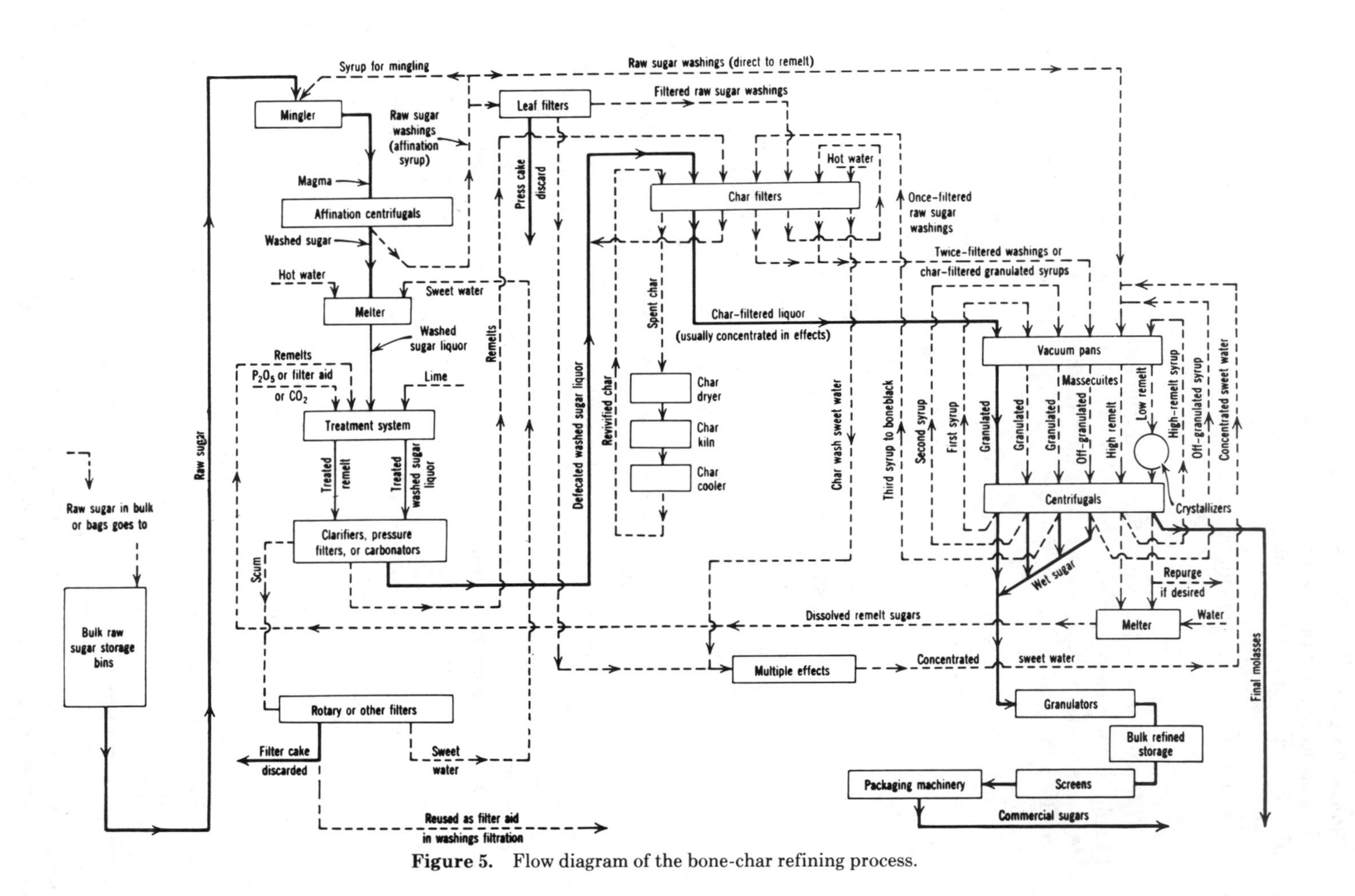

Figure 5. Flow diagram of the bone-char refining process.

Melting. The washed sugar is melted in hot water, and usually the pH is adjusted with lime. Water that contains a little sugar from anywhere in the refinery is called sweetwater, and if it does not contain much impurity, is used in the melter. The washed sugar liquor coming from the melter is adjusted to the operating concentration, usually about 65 brix (percent sucrose). The trend is to operate refineries at higher brix up to 68, because if water is not added, it does not have to be boiled away later. The washed sugar liquor is dark brown and quite turbid, and appears much darker than the sugar from which it came. The melter liquor is strained through a plain screen to catch debris in the raw sugar.

Clarification. The object of clarification is the complete removal of all particulate matter (31). The particles in the sugar come from all sources, eg, field soil and fiber (*bagacilio*) which escaped clarification in the raw-sugar factory; all microbiological life, including yeasts, molds, bacteria, and their spores; colloids and very high molecular weight polysaccharides; and foreign contaminants such as insect and rodent droppings. The very diversity of the nature of the particulate matter and the wide range of particle sizes makes clarification a difficult and critical step in the refining of sugar. One of three processes is used: filtration, carbonatation, or phosphatation.

Filtration. The most straightforward process is filtration (qv). This removes particles down to the submicrometer size, but not easily. In order to conserve energy in evaporating water, the sugar liquor is at as high a brix as possible, which means high viscosity. The temperature is raised to reduce the viscosity as much as possible. But, sugar solutions simply cannot be filtered without a filter aid. The first application of diatomaceous earth, or kieselguhr, was to sugar filtration and sugar remains the largest user of these materials. Without filter aid, the wide range of particle sizes quickly plugs the filtration medium, resulting in greatly reduced, almost zero, flow. With a tight filter aid that achieves good clarity, the flow is still very slow (a faster filter aid does not achieve good clarity). Filtration, as the sole means of clarification, is being replaced by other methods (see Diatomite).

Chemical Defecation. In order to make the liquor more suitable for clarification, it has long been the practice to add various substances that form precipitates in the sugar liquor and that help coagulate the impurities. Today, only two processes are used, carbonatation and phosphatation; both use lime. The addition of lime in the raw sugar process was the first use of defecation. It is applied to both the washed sugar liquor and also to the affination syrup, which carries a heavier load of particulate matter.

Carbonatation. The carbonatation process used in a cane-sugar refinery is similar to the carbonatation used in the raw-sugar factory, except for the higher concentration of sugar. Beet-sugar carbonatation is also similar, except it is operated at a higher pH. In the preparation of the lime, attention must be paid to the temperature of kilning the lime, its particle size, and its hydration. The lime dosage is 0.4–1.2% on melt. It is added as a slurry, and is mixed rapidly with the sugar liquor. The pH rises to about 12.5. To avoid destruction of sucrose and reducing sugars at this pH, the holding time in the small mixing tank is kept to one minute.

The flow then goes to the first saturation tank, where carbon dioxide gas from the flue is bubbled through. (Those kilning their own lime use the carbon dioxide from the lime kiln for this step.) The pH in the first saturator is 9.5 and should be lower for better filterability and higher for better absorption of gas. Several saturation tanks are used: two, three, or four. The pH of the last tank is 7.5, at which point the minimum lime-salts solubility is reached. The holding time is $\leq$1 h in each tank.

In the course of precipitation of the calcium carbonate as calcite, insoluble lime salts are coprecipitated and many other impurities are entangled in the forming precipitate. Considerably more precipitate is formed than is necessary for good clarity. The extra precipitate is formed in order to get good filterability. The final step is a filtration on any type filter.

Phosphatation. In the phosphate clarification scheme, lime and phosphoric acid (food grade) are added simultaneously with good mixing. The phosphoric acid is added in proportion to the melt at about 0.01–0.02%. The lime is added to bring the pH to 7.8. The calcium phosphate precipitate forms a floc of no particular crystal structure. It is even better at scavenging impurities by entrapment than the carbonate precipitate. It also has the useful property of attaching or entrapping air bubbles. Thus, at the same time that the floc is being formed, some air is injected, mixed, or pumped into the system. Raising the temperature a few degrees also helps tiny air bubbles materialize throughout the liquor. The precipitate then floats to the surface as a scum of 80% organic matter and is scraped off without any filtration. The mixing is very thorough just as the reagents are added, gentle in a floc-development section, and then minimal in the flotation zone.

The phosphate clarifiers are also called frothing clarifiers and have many sizes and shapes with scrapers going forward, backward, and around. Some are heated and some are deep. Sugar is recovered from the scums by clarifying again. The phosphate system uses only about $^1/_{10}$ as much reagents as the carbonate system, and so produces only $^1/_{10}$ the scum volume.

No matter what the method of clarification, the clarified liquor is brilliantly clear without any sign of turbidity. It is, however, dark, rather like a cup of weak coffee.

Decolorization. The key process in sugar refining is decolorization (32). Color is the principal control in every sugar refinery. It is the main property that distinguishes refined sugar from raw sugar. The word color is used loosely. It usually means visual appearance, but in technical sugar work it means colorant, the material causing the color. It can be classified in three groups: plant pigments; melanoidins resulting from the reaction of amino acids with reducing sugars; and caramels resulting from the destruction of sucrose. Many, but not all, compounds in each of these classes have been identified. In sugar work, color refers collectively to the optical sum of all the colorants.

Measurement of color is by light transmission at a wavelength of 420 nm. It is expressed as

$$A = -\log T/bc$$

where A is the attenuation index; T is the transmission; b is the cell depth, cm; and c is the concentration of sugar, g/mL. By dividing by the concentration of sugar, one obtains an index of the amount of colorant relative to the amount of sugar. Since many of the colorants are pH sensitive, ie, they show some indicator effect, the color is always measured at pH 7.00. Sucrose concentration must also be kept constant because it affects the refractive index, which in turn affects the light scattering by the residual particulate matter that is always present in spite of clarification. This is important in measuring the color of refined sugars that have very little color but still have scattering particles. The value of A from the equation can be multiplied by 1000 to give milli-absorbancy units or sugar-color units. Raw sugars have a color of 1000–2000. Refined sugars have a specification of less than 35.

Decolorization has traditionally been accomplished by carbon adsorbents, although other processes are available. Bleaching-type decolorizations with hypochlorite, sulfite, or peroxide have largely fallen into disuse because they do not remove the colorant, only camouflage it. Sugars so decolorized do not have good keeping qualities, and these processes are now used only for direct-consumption sugars. One common property of sugar colorant is that much of it is of an anionic nature and, therefore, it is susceptible to ion exchange. Some refineries now use only ion exchange for color removal. The anionic property can also be used to precipitate the color with a suitable cationic material. However, carbon adsorbents remain the principal method of decolorization. Powdered carbons can only be used once or twice and are too expensive. Bone char was the standard method, but requires a large investment in plant and uses considerable energy. Granular carbon has a very large capacity for color and is being used in all new installations (see Carbon and artificial graphite).

Carbon adsorbents are general adsorbents; they adsorb everything out of the sugar solution, including sugar, with little selectivity. Because the colorant is the combination of many substances, any theoretical treatment would involve the summation of many adsorption isotherms, most of which are unknown, and is thus impractical even with a computer. No theoretical treatment of sugar-color adsorption by carbon adsorbents has ever been successful. The best correlation of color removed is with contact time, measured in seconds for very finely powdered carbons, and hours for larger-particle granular carbons.

Bone char and granular carbon behave similarly in decolorization of sugar. For the contact with sugar liquor, both are contained in beds called cisterns (also misnamed filters) ca 3 m dia and 7 m tall and holding 30–40 t carbon. The liquor flows downward with a contact time of 2–4 h. The first liquors are water white with a very gradual yellowing. The cistern stays on stream until the color of the liquor becomes too great to be handled by the remainder of the refining process. The decolorization is always greater than 90%. The bone-char cycle is about 4 d; for granular carbon, it is 4 wk. The first liquors from bone-char treatment are lighter than the first liquors from carbon. However, from the adsorbent point of view, the two systems are different.

Bone char is made by heating degreased cattle bones to about 700°C in the absence of air. It is about 6–10% carbonaceous residue and 90% calcium phosphate from the bone with an open pore structure supplied by the bone. The surface area available to nitrogen is 100 m^2/g. The particle size is about 1 mm. Besides being a carbon adsorbent, it has ion-exchange properties that permit removal of considerable ash from the sugar. These same ion-exchange properties result in a buffering effect that keeps the pH of the sugar liquor from falling.

After the decolorization cycle, the sugar is washed out of the bed (sweetened off) and then the bed is washed with cool water to remove as much as possible of the adsorbed inorganic salts (mostly calcium sulfate). The sweet water is of low purity and cannot be recycled. The organic coloring matter is adsorbed so tightly that no amount of washing will remove it. The water is therefore drained from the char and the char moved from the cistern to a kiln where the organic matter is burned off at 500°C, with a little oxygen in the kiln atmosphere to burn away freshly deposited carbon and keep the pores open. The kilns used for regeneration of bone char are either pipe kilns in which the char is inside pipes and the flame outside, or Herreschoff-type kilns (Fig. 6), in which the char is in contact with the combustion gases. The Herreschoff kiln is preferred because it offers better control of kiln atmosphere, better heat transfer,

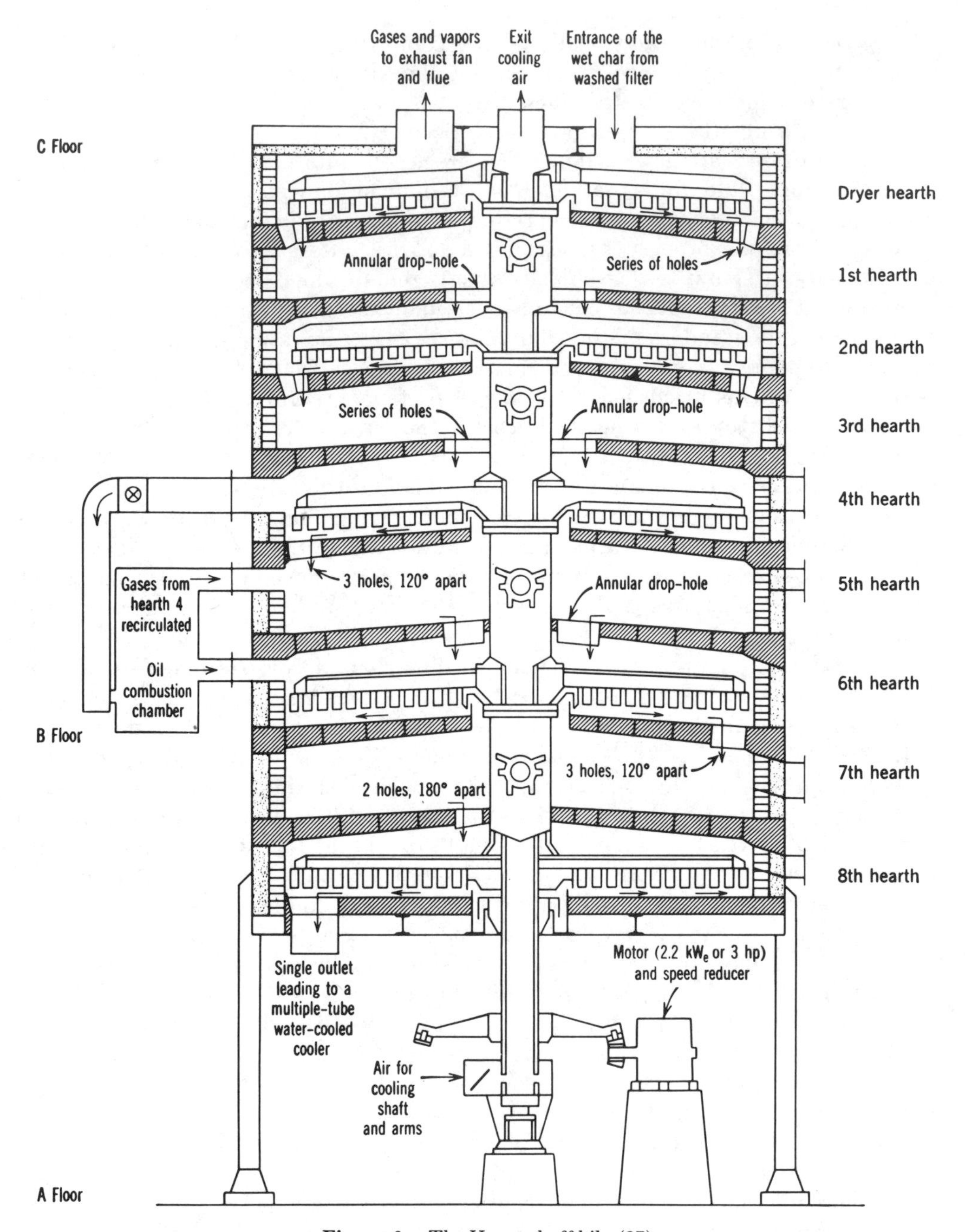

Figure 6. The Herreschoff kiln (27).

and smaller size, hence lower cost. The bone char is then cooled, returned to the cisterns, and settled with sugar liquor. The char loss is ca 0.5% per cycle. The amount of char used to decolorize sugar liquor is expressed as percent char burned (regenerated) on melt. For bone char, it is ca 10%, depending on the raw sugar being processed and the other refining processes.

On bone char, it has been observed ionic color removal process is affected by ionic constituents in the liquor (33). High calcium liquors decolorize well, but high sulfate liquors decolorize as much as ten times more poorly for an ion-concentration change of only 0.01 *N*.

Granular carbon, originally developed by Pittsburgh Coke and Chemical Co. for use in gas masks, is made from carefully selected mixtures of coal, heated to ca 1000°C, and steam-activated. It contains over 70% carbonaceous residue, the remainder being mostly siliceous ash. The surface area available to nitrogen is 1000 m^2/g, but more than half is in pores too small for the sugar colorant. The particle size is similar to bone char. Granular carbon has no ion-exchange properties, does not remove ash, and must contain dead-burned magnesite for pH control. The magnesite hydrolyzes very slowly and supplies hydroxide at just the right rate to keep the pH up.

After the sugar-decolorization cycle, the sugar is washed out. The sweet waters are high purity and are sent to the melter. No further washing is used because nothing can be washed out. Kilning is in Herreschoff kilns or rotary kilns at 700°C. The oxygen concentration in the kilns is higher than for bone char to better burn off the heavier loading. The granular carbon is quenched in water and hydraulically returned to the cisterns where it is sweetened on. The cycle is repeated. The carbon loss is ca 5% per cycle, making the life of the carbon ca 20 cycles. The carbon burned on melt is 1% or less.

Powdered carbon is made from a wide variety of starting materials, including wood, coal, agricultural wastes, and black ash from paper mills. All contain organic material that is charred by heat and activated by oxygen or steam. Powdered carbon is introduced into the sugar liquor in an agitated tank with a holding time of 20 min, although less than 1 s is long enough contact time. The mixture is then filtered with the help of filter aid. Powdered carbons are not regenerated. They do a very good job of decolorizing with a minimum of equipment. Use is 0.5% on melt.

Process variations include sending a sequence of liquors of increasing color to the adsorbent. Also, the sugar liquor may be sent through adsorbent in multiple passes, first through the older adsorbent and then through the adsorbent more recently regenerated. Many refineries use two passes. A variation on this scheme is the use of slugged beds of granular carbon. Every so often the flow is stopped, a slug of carbon is hydraulically removed from the bottom of the column, and a fresh slug added to the top. In this system, the liquor flow is upward and the velocity is such that the whole bed is lifted against the top of the column. The whole column may consist of 5–10 slugs, equivalent to 5–10 passes in one column. This approaches continuous adsorption. There is also a continuous adsorption process (CAP) in which the adsorbent is continuously removed from the bottom and added to the top. The flow in this system is such that the bed is expanded but not fluidized or raised.

Ion exchange (qv) is also used for decolorization. The chloride form of strong base quaternary ammonium anion exchangers very effectively exchange color for chloride. The chloride that goes into the sugar liquor is very soluble and forms a little more molasses. The ion exchanger is regenerated with brine which acquires a very high BOD (on the order of 10,000 ppm) and is removed with difficulty. The ion exchanger gradually becomes fouled with phenolic compounds and can only be cleaned with strong acids, bases, and hypochlorite, but the capacity and life of the ion exchangers makes the process very practical. The significant advantages of ion exchange are *in situ* regeneration without the use of heat, short contact time, and small size of equipment. Ion exchange is also used for ash removal or deionization of sugar.

Talofloc process takes advantage of anionic properties of the sugar colorant as in ion exchange (34). In this process, however, the colorants are precipitated by reacting with a quaternary ammonium compound. The compound chosen must be sufficiently soluble in sugar solution to dissolve, but it must also be of just the right size to make an insoluble precipitate with sugar colorant. The compound chosen was dioctadecyldimethylammonium chloride. The precipitated colorant is scavenged from the sugar solution by a phosphate clarification. The dose is 300–500 ppm of the flocculant. The disadvantage of the system is that the additive is used only once and cannot be regenerated.

Crystallization. The color of the washed, clarified, and decolorized liquor going into the crystallization process ranges from water white to slightly yellow (35). Many refiners polish-filter the sugar liquor at this stage to make sure that it is sparkling clear with no turbidity. Others rely on good operation upstream and do not polish-filter. In many cases, the brix has become too low, either on purpose or by error; these liquors first go to the evaporators to bring the brix to ≥68. The vacuum pans are the same as were described under Raw Sugar Manufacture, and their operation is the same. They are operated even more carefully to produce crystals of the desired size. Great care is taken to avoid conglomerates and fines. Boiling rate and throughput are important. A new strike of some 50 metric tons must be dropped every 90 min to keep up with the production schedule.

The boiling schemes used in the refinery are more extensive and variable than those used in the raw house. This is because the starting material is of much higher purity. Ordinarily, three, four, or five strikes of refined sugar are obtained. Table 2 shows a typical scheme involving four boilings. In developing this table, the sugar produced can be assumed to be 100% pure (eg, ≥99.9). All the impurities, therefore, are in the syrups, which about double in impurities with each strike.

The syrup from the fourth strike may be handled in different ways. It may be used

Table 2. Material Balance in Four-Boiling System

		Input: pan feed liquor, 200 t solids
		Eight strikes
first strikes	each	25 t feed solids, purity 99%
53% yield	strike	→ 13.25 t (53 wt %) sugar
		11.75 t (47 wt %) syrup solids, purity 98%
		combine 2 syrups
		Four strikes
second strikes	each	23.5 t feed solids, purity 98%
50% yield	strike	→ 11.75 t (50 wt %) sugar
		11.75 t (47 wt %) syrup solids, purity 96%
		combine 2 syrups
		Two strikes
third strikes	each	23.5 t feed solids, purity 96%
47% yield	strike	→ 11.05 t (47 wt %) sugar
		12.45 t (53 wt %) syrup solids, purity 92%
		combine 2 syrups
		One strike
fourth strikes		24.91 t feed solids, purity 92%
44% yield		→ 10.96 t (44 wt %) sugar
		13.95 t (56 wt %) syrup solids, purity 88%
		Output: 186.05 t sugar, 13.95 t syrup solids

in the recovery house, but is more likely used in making speciality syrups or brown sugars. It may also be sent back to decolorization or clarification, and recycled.

The refined sugar centrifuges are always batch type because they leave the crystals intact. The centrifuging is easy and the cycles are short.

The drying of the sugar from the centrifuges is done in a rotary dryer using hot air. This dryer is universally misnamed the granulator because by drying in motion, it keeps the sugar crystals from sticking together, or keeps them granular. The hot sugar from the granulator is cooled in an exactly similar rotary drum using cold air (see Drying).

Conditioning. The sugar from the coolers would appear to be finished, but after a few days storage it becomes wet with water trapped inside the grain because of the very high rate of crystallization and drying. After a few days, this moisture migrates outside the crystal and the sugar is wet again. The moisture is removed by a process known as conditioning, in which the sugar is stored for four days with a current of air passing through it to carry away the moisture. In one system, a single silo is used with sugar being continuously added to the top and removed from the bottom, and a current of dry air blowing upward. In another system, the sugar is stored in a number of small bins. It is continuously transferred from bin to bin with dry air blowing around the conveyors that move the sugar.

Packing, Storing and Shipping. Sugar is sometimes stored in bulk and then packaged as needed. Others package the sugar and then warehouse the packages. The present trend is away from consumer-sized (less than 50 kg) packages and toward bulk shipments.

Products

Sugar marketing is primarily local. Little refined sugar is shipped across national boundaries. Each refinery serves its own local marketing area. True competition starts only in overlap areas that are about equidistant from two refineries. Beet sugar competes with cane sugar on much the same terms. The great dumping ground for sugar in the United States is Chicago, where prices are lowest.

Refined granulated sugar is the usual output of a sugar refinery. By far the largest part of the production is in this one class. The particle size of the refined granulated sugar for table use varies from region to region. It is much finer in North America than in Europe. Different particle sizes have different names and are not standardized. If particle size is of consequence, it must be established by the local supplier.

There can be different grades of refined granulated sugar, but many refiners make only one high grade. The largest user of sugar is the soft-drink industry, which uses 25% of the sugar in the United States and 16% worldwide (see Carbonated beverages). The United States is beginning to use more HFCS (high fructose corn sweetener) because of price, but whatever the source, soft-drink manufacturers require the highest grade of sugar. Every refinery must make bottler's-grade sugar and most make little or nothing else. Different brands of sugars do not vary in quality. Canner's-grade sugar is low in flat sour bacteria which can cause trouble in sweetened nonacidic products such as canned corn. All sugar now meets canner's-grade requirements.

Large-grain speciality sugars are used for candy and cookies. White large-grain sugar can be made only from the very purest of liquors; therefore, those interested in the very best sugar specify coarse grain. The highest quality best sugar is made by re-refining large-grain sugar. It is available only from C and H (California and Hawaiian Sugar Co., Crockett, Calif.) as CON-AA.

Fine-grain sugar, used because it is quick dissolving, consists of small crystals obtained by screening.

Powdered sugar is made by grinding granulated sugar and adding 3% corn starch to help prevent caking. The fineness is designated by labels such as 4X, 6X, 10X. However, the label is misleading; 12X is not twice as fine as 6X.

Cubes are made by mixing a syrup with granulated sugar to just the right consistency to form cubes. These are then dried. The process is expensive and the price of cubes is high relative to ordinary granulated sugar. In Belgium, 30% of the sugar sold is cubes.

Reagent-grade sugar is sold by chemical supply houses at a very high price. However, better sugar can be obtained at every grocery store. The problem is the length of time the sugar is stored. Sugar is not completely stable, and fresher sugar is found anywhere but the chemical supply house. For critical applications, any refinery will supply first-strike sugar.

Brown sugar can be made in two ways. In one, white sugar is coated with a syrup that has been selected to give the right color and taste. The other method is more difficult but yields a superior brown sugar. A strike is boiled from a selected liquor, but a special boiling technique is used, one that is wrong from the point of view of refining white sugar; it purposely makes conglomerates and a range of crystal sizes and occludes mother liquor. The brown-sugar massecuites are not washed in the centrifuge. Brown sugars are known in the trade as soft sugars. Although there are 13 grades of brown sugars, most refiners make only one or two, and the grades are not reproducible from one area or refiner to another.

Brown sugar comprises only a small part of the output of most refiners, ranging from only 3% in warm climates to perhaps 10% in cold regions. The area of highest brown-sugar consumption in the world is Vancouver, Canada, where brown sugar accounts for 20% of total use. In this region, a favorite is a distinctly yellow sugar. Brown sugar is not raw sugar, but as its manufacture is described above, it is distinctly refined. The difference between raw sugar and brown sugar is not so much the sucrose content, the color, or taste, but rather the absence of field soil, cane fiber, bacteria, yeasts, molds, insect parts, and rat hairs which may be present in raw sugars. Neither contains a significant amount of protein, vitamins, or minerals.

Liquid sugar is a solution of sucrose and/or invert sugar. It is convenient for the user because no handling of sugar bags or bulk sugar is required. It can be held in tanks and pipes, and the sugar obtained with the turn of a valve. It is manufactured either by shipping what would otherwise be pan-feed liquor, or else by remelting granulated sugar. A polish filtration and a powdered-carbon treatment may also be used.

Syrups (qv) are made for local preference. Most have a high percentage of invert because a mixture of sucrose, glucose, and fructose can be raised to a much higher concentration without the formation of crystals. The higher concentration means a higher viscosity which is desirable in syrups. Blends are made with different sugars for table syrups. Corn syrups are often used in the blends. Maple syrup is often blended with cane sugar to extend the maple syrup. Some maple syrups contain as little as 5% maple syrup.

High test molasses is not molasses at all, but a partially inverted cane syrup that is boiled to high concentrations. It is light brown in color.

Molasses is an often misused name, as in high test molasses. True molasses is the syrup left over after the sugar has been crystallized. Edible molasses, often sold as blackstrap molasses or New Orleans molasses, are not really molasses, but edible syrups from some later strike. They still contain too much sugar to be thrown away as molasses. The real molasses is indeed black, has an extremely strong taste, and is used for its sugar content, either as cattle feed or for fermentation (see Syrups).

BIBLIOGRAPHY

"Sugar Manufacture" in *ECT* 1st ed., Vol. 13, pp. 203–227, by L. A. Wills, Consultant; "Sugar, Cane Sugar" in *ECT* 2nd ed., Vol. 19, pp. 166–203, by H. G. Gerstner, Colonial Sugars Company.

1. J. Bricout and J. C. Fontes, *Sugar J.* **39,** 31 (March 1977).
2. M. Patureau, *By-Products of the Cane Sugar Industry*, Elsevier Publishing Co., Amsterdam, 1969.
3. V. Kollonitsch, *Sucrose Chemicals*, International Sugar Research Foundation, Bethesda, Md., 1970.
4. J. L. Hickson, *Sucrochemistry*, American Chemical Society, Washington, D.C., 1977.
5. N. Deerr, *History of Sugar*, Chapman and Hall, London, 1949.
6. E. Artschwager and E. W. Brandes, *Sugarcane, Agricultural Handbook No. 122*, U.S. Department of Agriculture, 1958.
7. A. C. Barnes, *The Sugar Cane*, 2nd ed., John Wiley & Sons, Inc., New York, 1974.
8. A. Hugill, *Sugar and All That*, Gentry Books, London, 1978.
9. J. E. Irvine, *Sugar J.* **43,** 22 (Feb. 1981).
10. U.S. Pat. 4879 (1846), N. Rillieux.
11. L. G. Nickell, *Hawaii. Plant. Rec.* **58,** 293 (1973).
12. J. E. Irvine, *Sugar J.* **42,** 24 (Oct. 1979).
13. G. P. Meade and J. C. P. Chen, *Cane Sugar Handbook*, 10th ed., John Wiley & Sons, Inc., New York, 1977, Chapt. 5.
14. *Ibid.*, p. 53.
15. *Ibid.*, p. 81.
16. *Ibid.*, Chapt. 6.
17. *Ibid.*, Chapt. 7.
18. *Ibid.*, Chapt. 8.
19. *Ibid.*, p. 153.
20. *Ibid.*, p. 148.
21. *Ibid.*, p. 162.
22. Br. Pats. 1,397,297 (1975) and 1,428,790 (1976), J. T. Rundell and R. R. Potter (to Tate and Lyle).
23. T. W. Baker in ref. 13, Chapt. 11.
24. Ref. 13, p. 261.
25. Ref. 13, p. 235.
26. Ref. 13, p. 334.
27. Ref. 13, Chapt. 1.
28. Ref. 13, p. 268.
29. Ref. 13, Chapt. 13.
30. Ref. 13, p. 427.
31. Ref. 13, p. 430.
32. F. G. Carpenter and M. A. Clarke in ref. 13, Chapt. 19.
33. F. G. Carpenter, D. Larry, and V. R. Deitz, *Proceedings of the 1961 Technical Session on Bone Char*, Bone Char Project, Washington, D.C., 1962, pp. 259–293.
34. M. C. Bennett, F. G. Gardiner, J. C. Abram, and J. T. Rundell, *Proceedings of the International Society for Sugar Cane Technology* **14,** 1569 (1971); Br. Pat. 1,224,990 (March 10, 1977) and U.S. Pat. 3,698,951 (Oct. 17, 1972), (to Tate and Lyle).
35. H. G. Gerstner in ref. 13, Chapt. 20.

FRANK G. CARPENTER
United States Department of Agriculture

BEET SUGAR

Sugar beets thrive in temperate to relatively cold climates. *Beta vulgaris* is a biennial plant that accomplishes its vegetative growth in the first year and its seed production in the second. Thus, for sugar-production purposes, it is harvested at the end of the first year's growth. It may be noted that the sugar beet, var. *crassa*, is white, as opposed to the familiar red color of var. *cruenta*. Growth requires a frostfree period of five or six months, and the equivalent of an annual rainfall of ca 20 cm. The sum of the daily mean temperatures during the beet-growing period should average about 2500°C. Insufficient rainfall during the growing period can be supplemented by irrigation. The harvest season, normally in the fall of the year, should be dry to permit mechanical harvesting and transportation.

Sugar beets are often profitable crops in areas where few other crops, except potatoes, can be grown successfully. Sugar beets are an excellent crop to fit into a crop-rotation scheme, as their culture tends to improve the soil. Beets are deep-rooted, and consequently bring up plant food from a considerable depth, which is thus made available to other shallow-rooted crops. In general, sugar beets cannot be planted in the same field more often than once every three or four years, since they are attractive hosts to certain nematodes.

Because sugar beets are grown in temperature climates, it is possible to plant them in heavily populated areas and to manufacture the sugar relatively close to the areas of consumption. This fact, plus the relative simplicity of the process of manufacture, has resulted in single-factory production, ie, in plants where the product is made from the raw vegetable material in a single process.

Sugar-Beet Composition

The composition of sugar beets is given in Table 1. Of the nitrogenous compounds, ca 58% is protein which is removed by the factory process. Of the other nitrogen compounds, betaine constitutes 40% which is carried through to the molasses. The

Table 1. Composition of Sugar Beets[a]

Component	Percent
juice	ca 92
insoluble matter[b]	ca 5
water, chemically bound	ca 3
Soluble solids	11–25
sucrose in solids[c]	87.5
sucrose in beet	10–22
Nonsucrose substances soluble in juice	
organic nitrogen compounds	ca 44
nitrogen-free organic compounds	36
inorganic compounds	20

[a] From refs. 1–5.
[b] So-called marc.
[c] Expressed as purity of the juice.

remainder consists of amino acids and purines, and amides and ammonium salts; the latter cause difficulties in processing. The nitrogen-free organic substances other than sucrose are ca one half organic acids with varying amounts of other organic compounds including raffinose, invert sugars, and small amounts of fats and saponin. The principal acid is citric acid which, together with oxalates, is eliminated as the insoluble calcium salts in the purification stage. Most other acids have soluble lime salts and are not eliminated. The invert sugars are almost completely destroyed in liming. Purification removes 95% of the saponin, a glycoside chiefly composed of oleanolic and glucuronic acids which is present just under the skins of the beet roots. It protects the root against nematodes and other small organisms, which are destroyed by saponin. A small amount of saponin is frequently found as a trace impurity in the sugar product. The principal anions of the soluble organic salts are phosphate, sulfate, and chloride. Traces of iron, silicon, and aluminum are also present. Phosphates are eliminated in the purification step, whereas most other inorganic compounds pass through to the molasses.

The beet marc consists of the individual cell membranes, and of nearly equal concentrations of cellulose, pectic matter, and hemicellulose (pentosans), plus 4–5% lignin. The cell walls also contain small amounts of protein and inorganic matter.

Manufacture

The manufacture of beet sugar is a seasonal operation, since factory operations are confined to the time intervals during which the sugar beets are harvested, plus the additional time during which the beets can be stored in piles without undue loss of sucrose. In most areas of the United States and Canada, except for California and Arizona, the harvest period lasts from the beginning of October until the end of November, by which time the harvest must be completed to avoid the freezing of the beets in the ground. It is customary to build storage piles of beets and to continue factory operations for about two more months, resulting in an operating season of ca 120 days. In northern California, where there are separate fall and spring plantings and harvests, factory operations may total 300 days a year. In southern California and Arizona, factory operations start about May 1 and continue into July.

In order to extend their operating seasons, a few factories are built with extra capacity at the so-called beet end, in order to produce excess concentrated, purified thick juice which is stored in suitable tanks (see Fig. 1). The beet end includes the point of entry of the sugar beets through the extracted, purified, and concentrated thick juice; the sugar end comprises the remainder or crystallization part of the process. At factories with storage facilities for excess thick juice, the sugar end continues to operate after the fresh beet supplies are completely used. The sugar is crystallized from the supplies of stored thick juice.

Diffusion Process. The sugar in the sugarbeet is contained within the parenchyma cells, which are the most common type of cells in vegetable tissue. The parenchyma cells are enclosed by cell walls made of ca equal quantities of cellulose and protopectin. The interior of the cell, or vacuole, holds the cell sap, a water solution usually containing ca 17.5 wt % sucrose and 2.5 wt % other substances. The walls of the mature cell vacuole are lined with protoplasm composed of proteins, chiefly albumins (see Proteins). On heating to a sufficiently high temperature (60°C), the protoplasm is clotted or slowly coagulated. On heating to 70–80°C, the clotted protoplasm forms a stringy network through which diffusion of dissolved molecules can take place. Small molecules pre-

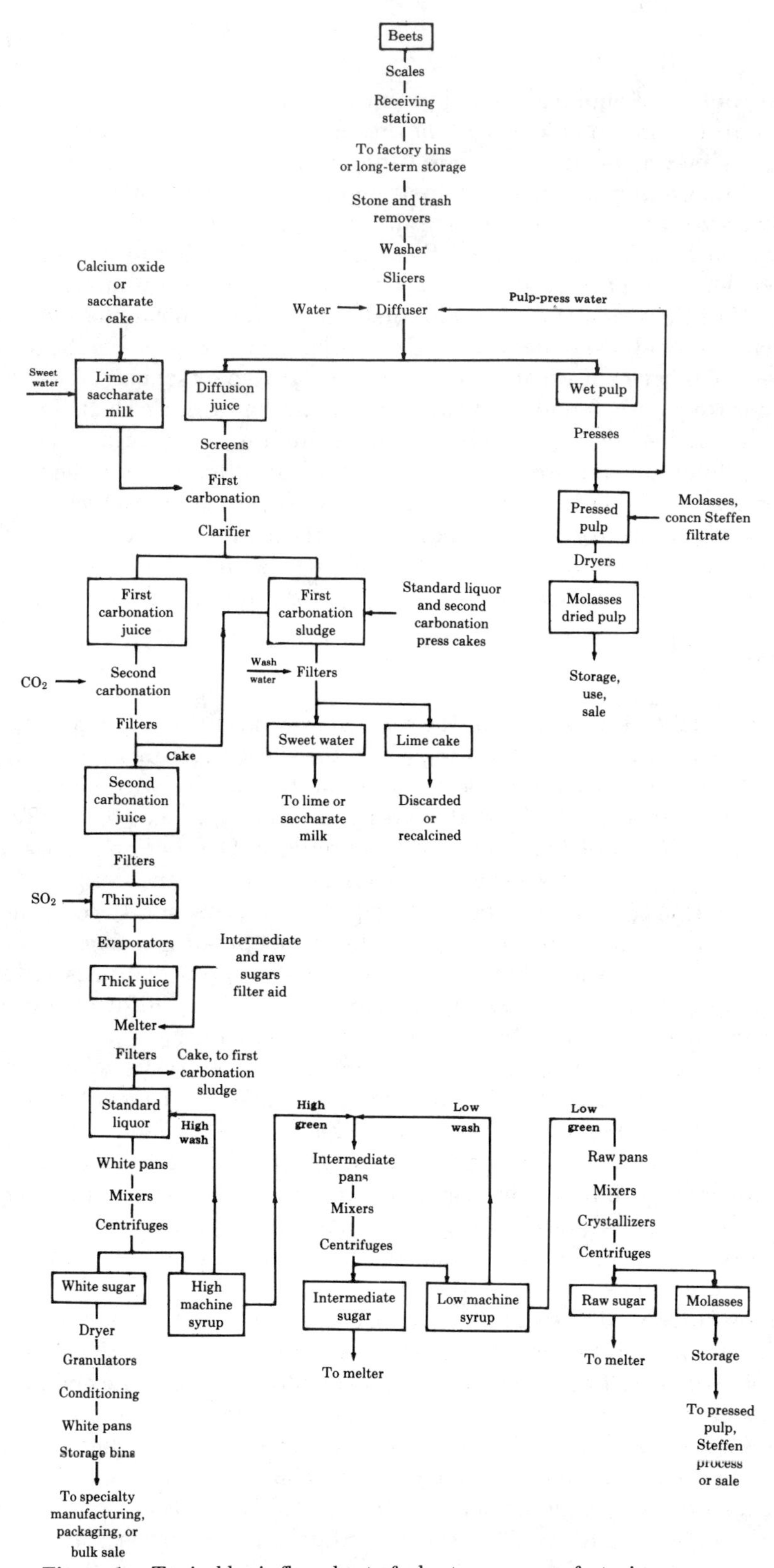

Figure 1. Typical basic flow sheet of a beet-sugar manufacturing process.

sumably pass through most readily, followed by sucrose, alkaline earth complexes, and colloids. Extraction rates are actually slightly different as shown in Table 2.

The unexpectedly slow extraction of the small sodium and potassium ions is due to the electrostatic retarding effect of the high molecular weight polymeric anions. The low rate of calcium extraction is due to the presence of calcium in the form of insoluble salts (see also Diffusion separation methods).

Preparation. Since sugar beets have about the same specific gravity as water, it is convenient to move them from factory bin or slab storage to the processing area in water flumes (6). These flumes are equipped with rock and stone removers, vegetation and trash catchers, and beet washers. At the factory, the beets are elevated to a higher point by a conveyor belt, elevator, or beet pump, where they are deposited in supply hoppers after they are passed over the beet-slicing mechanisms. In the United States, horizontal-axis, rotating-drum slicers are used. Serrated knives are set in the periphery of the drum and the beets, entering into the interior of the drum, are sliced into cossettes (long, thin strips, of either V-shaped or square cross-section). More and more, vertical-axis rotating-disk slicers are coming into use (see Fruit juices). In these, the knives are set along the diameter of a revolving disk, and the beets contact the knives from above, while the cossettes pass through the disk. Typical cossettes may be 2- or 3-mm thick and up to 15-cm long.

Diffusion. The sucrose is extracted from the cossettes by passing them through a diffuser, countercurrent to a flow of hot leaching water (7–9). At one end of the equipment, the cossettes enter and the raw juice leaves. At the other end, the exhausted cossette pulp leaves and hot water enters.

As a result of the countercurrent flow of leaching water, at any point in the diffuser the concentration of sucrose in the liquid outside the cossettes is lower than in the beet cells, thus providing the necessary gradient for sucrose extraction. From the point of maximum temperature at one end to the other end of the diffuser, the temperature is gradually lowered; the diffuser supply water enters at ca 60°C. The temperature pattern adopted is set relatively high as the cossettes enter the diffuser, mainly to kill the beet cells in order to start diffusion of sucrose into the surrounding liquid phase. The high temperature also kills most of the soil microorganisms entering with the cossettes. The temperature at the cossette inlet end of the diffuser is 75–80°C. Ideally, the temperature elsewhere in the diffuser is 70–73°C, dropping off at the other end to the temperature of the battery supply water. If the temperature pattern is too high, or if the supply water is too alkaline, pectic substances within the beet cells are extracted, causing difficulties later in the processing of the juice and in removing the

Table 2. Rates of Extraction of Various Substances from Killed[a] Beet Tissue

	Extraction, %		Extraction, %
sucrose	98	total ash	70
"harmful" nitrogen[b]	95	magnesium	60
chloride	90	sodium	60
potassium	80	total nitrogen	57
phosphate	75	proteins	30
sulfate	70	calcium	10

[a] Heated to render cells permeable for diffusion.
[b] Mostly amides.

water from the exhausted cossettes or pulp. On the other hand, if the temperature pattern is too low, microbial or enzymatic inversion of sucrose and fermentations by lactic acid producing and other microorganisms are increased, resulting in serious losses of sucrose. Thus, a median temperature pattern is followed with periodic addition of chemical sterilizing agents, eg, freshly prepared formaldehyde, to limit microbial action. In a typical single addition, 6.8 kg hydrolyzed paraformaldehyde is added in shots to the part of the diffuser where the infection appears to be centered. Continuous addition is not effective, because the microorganisms acquire a tolerance. The formaldehyde is destroyed in the liming and carbonation steps (10). The enriched raw juice leaving the diffuser is screened to remove small pieces of cossettes and, at that point, may contain ca 14.5 wt % total solids, which, with a sucrose content of 12 wt %, has a purity of ca 83 wt %.

Equipment. In North America, three types of diffusers are in general use (11): the slope diffuser, rotary diffuser, and tower diffuser. The slope diffuser is a covered, sloping trough, ca 4–7.3 m wide and 15.8–19 m long, depending on the capacity. The cossettes are carried up the trough by perforated scroll flights. Hot water enters at the upper end, flows down against the direction of cossette movement, and leaves the lower end of the diffuser as diffusion or raw juice. The diffuser is equipped with steam jackets for heating.

Today, two other types of diffusers are preferred. The RT-4, Figure 2, consists of a horizontal rotating cylinder built on the Archimedes screw principle (12). There are no internal moving parts, and the partitions are so arranged and screened as to propel the cossettes in one direction while the juice moves in the other. While the cossettes move one pitch forward for each revolution, the juice is screened out, and is conducted two pitches backward. In this fashion, a countercurrent extraction is achieved. The RT-4 has been built with capacities ranging from ca 2,100–10,000 metric tons per day, in 13 diameters ranging from 4.2–7.0 m.

The BMA tower diffuser consists of a vertical tank with a rotating vertical central shaft equipped with conveying arms that move the cossettes upward against a downward flow of juice. The cossettes are slurried with final juice in an external mixer and tower. A large mechanically wiped screen section is built into the bottom of the tower to separate the juice from the cossettes. Fresh water is admitted at the top, and the exhausted cossettes (pulp) are extracted from the top by means of a screw conveyor. The center shaft is driven by a variable-speed electric-drive and gear-reduction system. Between the rows of rotating arms, stationary radial arms keep the mass from turning. The capacities of this countercurrent extractor range from ca 1000–6000 t/d in 11 different diameters, ranging from 3.3–7.9 m.

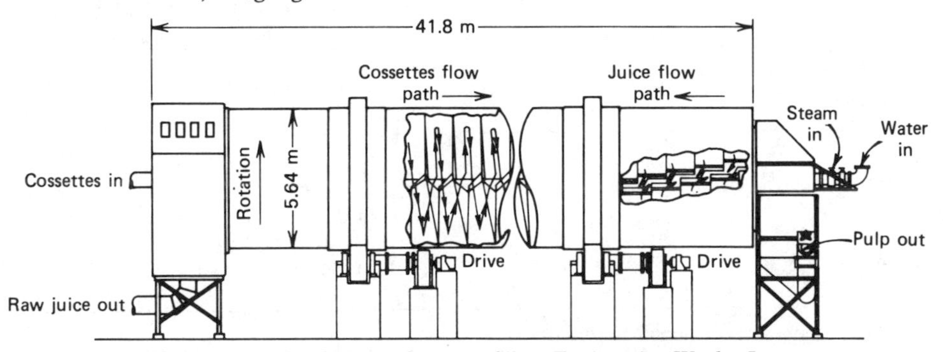

Figure 2. RT-4 diffuser. Courtesy Silver Engineering Works, Inc.

Dried Pulp. The pulp leaving the diffuser with a water content of ca 95% is pressed in screw-type presses to 75–80% moisture (13). The water, which still contains some sucrose, is returned to the diffuser at a point somewhat below the discharge end, usually after heat sterilization. The pressed pulp is enriched by the addition of molasses or concentrated Steffen filtrate (see below), or both, to a maximum of ca 35% of the final dried product. This enriched pulp is dried in rotary dryers, and is cooled and stored with a moisture content of 6–10%. The once common practice of storing and selling unpressed wet pulp is no longer followed because of the unpleasant fermentation odors and the losses of solids by liquefaction in storage. Molasses-dried beet pulp is excellent cattle feed and its sale contributes appreciably to the profits of the beet-sugar factory (see Pet and livestock feed). Dried pulp production amounts to 6–8 wt % of beets sliced, increasing with rising percentage of sucrose in the beets.

Purification with Lime and Carbon Dioxide. Although similar in principle, the method of purifying beet-sugar juice differs considerably from that employed in the raw cane-sugar factory because of the different amounts of invert sugars contained in the juices of the sugarcane stalk and the beet root (14). Cane juice contains ca 5–10 wt % invert sugar (based on sucrose), whereas in the healthy sugar beet it is <1%. Enough invert sugars are present in cane-sugar juice to make their destruction uneconomical. Furthermore, destruction of large amounts would result in the formation of color bodies that are expensive to remove, and would reduce the invert content of the final blackstrap molasses, and thus its value for many uses. Thus, only ca 10% of the amount of lime used in beet processing is used in cane processing. In a beet-sugar factory, ca 2 wt % of lime on beets is added in a straight house (ie, not using the Steffen process), and about 3% in a Steffen house.

In a raw-cane-sugar mill, a much smaller amount of lime is added. Throughout the cane mill and the cane refinery, the pH values of the process juices must be kept nearly neutral or slightly acidic to avoid the destruction of invert sugars. In a beet-sugar factory, the small amounts of invert sugar present are destroyed by massive liming, while the coloring matter is adsorbed on the surfaces of the fine calcium carbonate crystals formed in the first carbonation. The final molasses contains ca 50% total sugars in both beet and cane processing, but in cane blackstrap molasses as much as one third of the total sugar may be invert sugar, whereas in beet discard molasses the invert content seldom exceeds 4% of total sugars.

Chemistry. All North American beet-sugar factories use lime and carbon dioxide for juice purification (15). The effectiveness of this process is based on the direct action of the lime on impurities, and the special adsorptive properties of the calcium carbonate precipitate formed after carbonation.

Treatment with lime gives precipitates and soluble products; 60–70% of the nonsucrose products removed by precipitation are acids that form insoluble lime salts. Oxalates are almost completely removed by precipitation, as are hydroxycitrates, citrates, tartrates, phosphates, and some sulfates. The other nonsucrose products are removed through coagulation by calcium or hydroxyl ions. The substances coagulated are proteins, saponins, and various vegetable coloring matters. In addition, small amounts of iron, aluminum, and magnesium are precipitated as the hydroxides.

The reactions leading to soluble products are mostly caused by the hydroxyl ions introduced by the lime, whereas free acids are neutralized. Ammonium salts give off ammonia. Asparagine and glutamine are converted to the respective amides, which are hydrolyzed, also with evolution of ammonia, and accumulate as calcium salts in the juices. Allantoin decomposes slowly to ammonia, carbon dioxide, calcium glycolate, and a precipitate of calcium oxalate. Oxamic acid decomposes also rather slowly, similarly yielding free ammonia and calcium oxalate. Likewise, as a result of action in the hot, alkaline medium, the invert sugars dextrose and levulose are decomposed in complex fashion to give yellow or brown juices. The small amount of fats extracted from the beets is saponified, leaving glycerol in solution and precipitating the calcium salts of fatty acids. Pectin precipitates as calcium pectinate, which decomposes in the alkaline solution to methyl alcohol and pectic acid. The latter forms a gelatinous precipitate of calcium pectate, a substance that causes clogging when the juice is filtered later.

Process. On passing carbon dioxide through the limed mixture, either simultaneously with or after liming, calcium carbonate is precipitated. The second phase of the purification is the adsorption of nonsucrose materials, particularly colloids, on the calcium carbonate crystals. Thus, physical conditions in carbonation are adjusted to favor the production of a precipitate with a high surface area, which involves high pH (about 10.2) and supersaturation of calcium carbonate. In practice, the calcium carbonate precipitate formed in the first carbonation has a very large surface area, and adsorption on this positively charged surface is the essence of the purification taking place in carbonation. The nonsucrose substances adsorbed are negatively charged colloids, such as gums, calcium salts of various inorganic and organic compounds, especially amino acids, and the colored products that result from the decomposition of the invert sugars.

The fine calcium carbonate crystals, with their adsorbed impurities, tend to cluster around pieces of coagulated organic matter, which are so large that they settle readily in tray clarifiers, and can be removed satisfactorily from the underflow by vacuum filters. The clear overflow from the tray clarifier is carbonated a second time at ca pH 9. The calcium carbonate crystals thus formed are relatively coarse, and are easily removed in leaf-type pressure filters. Occasionally, small amounts of milk of lime are added to the second carbonation. This addition may act as a safety factor, partially rectifying errors in purification resulting from defects in the first carbonation. A flow sheet of first and second carbonation of a typical Dorr system is shown in Figure 3. This process was used in 1981 in 60% of North American beet-sugar factories.

Modifications. Many modifications of liming and carbonation procedures have been devised. Predefecation or preliming and main liming in advance of carbonation has been adopted in some form by most European and some North American factories. About one tenth of the total lime to be used is gradually added to the diffusion juice over a period of ca 20 min; many nitrogenous substances are precipitated at their isoelectric points and are not redispersed on main liming. This predefecation sludge is slimy and not easily removed. In order to give it a more granular character, some sludge from the first carbonation clarifier may be returned and mixed with the predefecation mixture.

In separate main liming, the remaining and largest part of the lime is added immediately after heating and predefecation. Precipitation and destruction reactions are completed in 10–20 min. Since lime is more soluble in cold than in hot juice, in a vigorous defecation the lime is added to cold juice, followed by heating to ca 80°C. In a mild defecation, the lime is added to the hot juice. In the Dorr-system defecation, which is very mild, the hot juice is simultaneously limed and carbonated with the aid of recirculation and high alkalinities are avoided. Mild defecation has the advantage

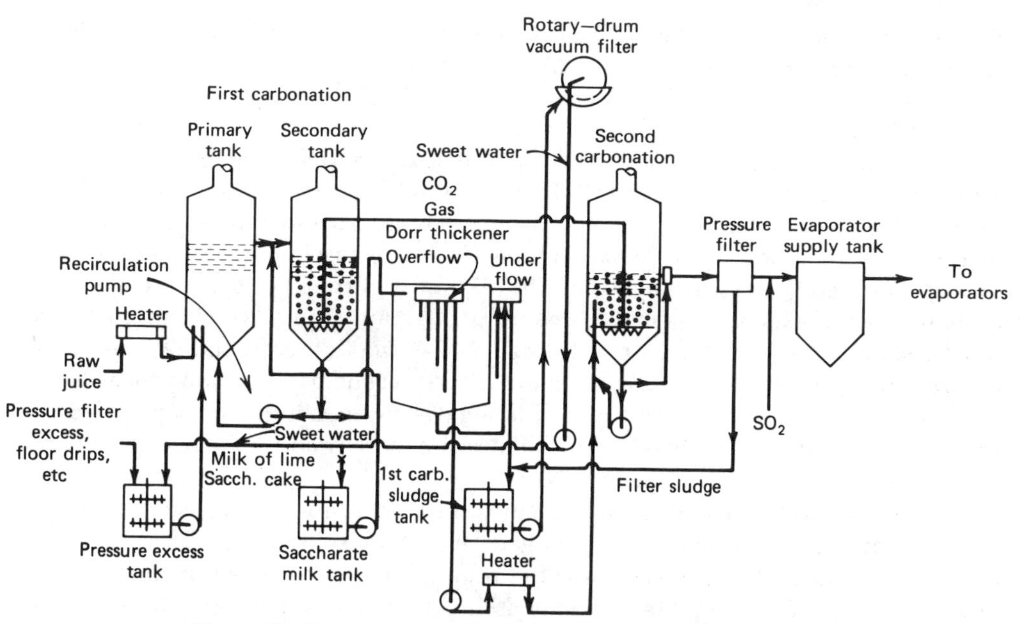

Figure 3. Dorr-system continuous first and second carbonation.

that the first carbonation sludge is less gummy, and easier to separate, and the sugar is easily removed by washing of the filters. However, neither defecation nor destruction reactions are completed. Therefore, the resultant thin juice may be thermolabile, and may increase excessively in color while in the concentration and crystallization stages.

Many factories using Dorr-system purification, have reduced the size of the first, or primary tank, changing it merely to a doughnut encircling the upper part of the secondary tank (see Fig. 4). This is known as the Benning carbonator. The juice from

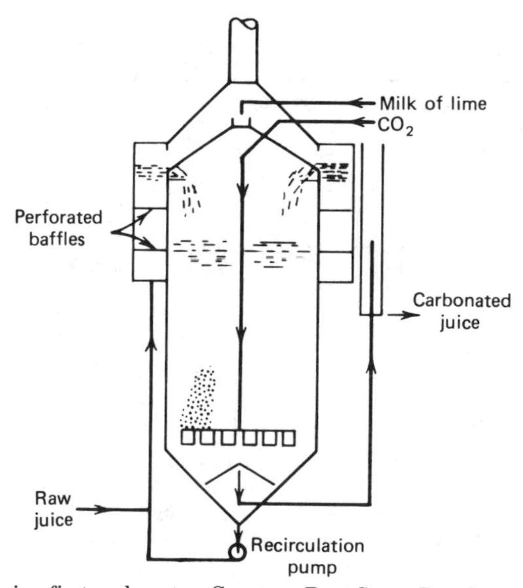

Figure 4. Benning first carbonator. Courtesy Beet Sugar Development Foundation.

the heater, which follows the diffuser or preliming, enters the lower part of the encircling primary tank, and overflows into the top of the enclosed secondary tank. The liming agent is added at the point of overflow. Purification results are the same as with the standard two-tank system.

A process coming into increasing use in North America is the BMA-65 (Braunschweigische Maschinenbau Anstalt-1965) (16). In this process, a mixture of the raw juice with ca 10% of the first carbonation sludge is heated and defecated–carbonated in a series of tanks, which involve a preliming (in effect), intermediate liming, and main liming, followed by first carbonation. The alkalinity of the juice being purified gradually increases in its passage up to a maximum in main liming, which results in good purification and stable flocs. After the removal of the first carbonation sludge with a sedimenting clarifier and rotary vacuum filters, second carbonation proceeds as in the Dorr system. Invert sugars are completely destroyed, amides are saponified, and the final juices are only lightly colored. There are more than 100 installations of this type throughout the world (see Fig. 5).

Other Purification Treatments. The most promising alternatives or supplements to liming and carbonation are treatments with activated carbon (17) or ion-exchange resins. In some beet-sugar factories, the defecation process is supplemented with powdered, activated vegetable carbon to assist in the removal of colloidal substances. The carbon is added to the thick juice after the evaporators, and removed in standard liquor filters. The filter cake might be reslurried in clear, first-carbonation juice, and is finally removed in the second-carbonation filtration.

With the development of efficient granular carbons, such as those made from bituminous coal with suitable binders, the exhausted carbon can be regenerated in kilns. With Herreshoff-type kilns, total carbon losses are 3–6% per cycle. Common dosages are on the order of 2 kg/t beets, or 680 g per 45-kg (100-lb) bag of sugar manufactured. In one case, the beet-sugar manufacturer uses granular carbon with a continuous system in which the treated juice flows from the third or fourth evaporator effects and contains ca 45% solids. The treated juice travels upward in vertical tanks,

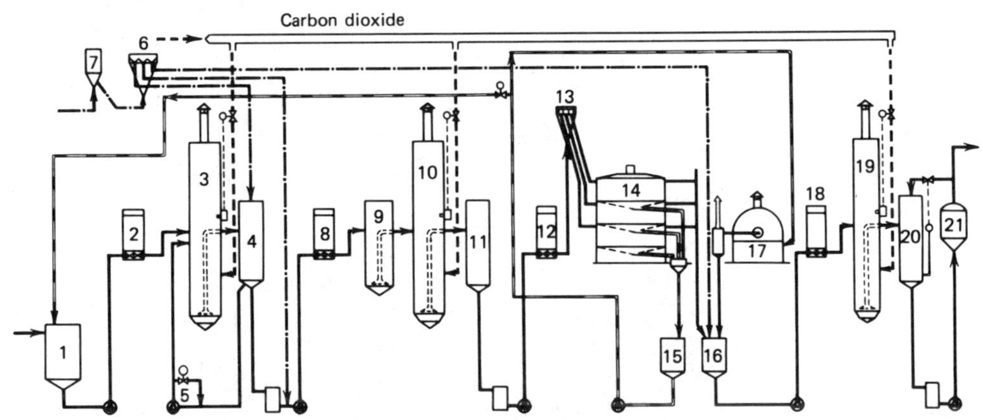

Figure 5. BMA-64 juice purification. 1, mixed juice tank; 2, mixed juice preheater; 3, precarbonation; 4, intermediate pump; 5, recirculation pump; 6, proportional milk-of-lime distributor; 7, milk-of-lime metering; 8, intermediate juice preheater; 9, main liming; 10, first carbonation; 11, pumping tank, first carbonation; 12, carbonation juice distribution; 14, clarifier; 15, thickened sludge tank; 16, clear juice tank; 17, rotary vacuum filter; 18, clear juice heater; 19, second carbonation; 20, post-reaction vessel, second carbonation; 21, pump head tank. Courtesy Silver Engineering Works, Inc.

and the vegetable carbon downward. The carbon is removed at the bottom of the tank, and fed continuously to a regenerating furnace. The carbon removes ca 40% of the color, most of the saponins, and virtually all the beet odor (see Absorptive separation liquids).

The use of ion-exchange resins has been investigated (18) (see Ion exchange). Deliming resins reduce the lime-salts content of the thin juice. These cation-exchange resins are used in the sodium-ion form and are regenerated with NaCl. High lime salts in the thin juices cause scaling of the evaporator heat-exchange surfaces and may lead to turbidity in aqueous solutions of the product. They also increase the amounts of molasses formed.

In one factory, a mixture of thin juice and machine syrup is treated with an ion-exchange resin as an effective supplement to lime and carbon dioxide purification. Machine syrup is obtained by centrifuging the massecuite. After addition of a small amount of MgO, the waste liquid from the ion-exchange regeneration is spray-dried and the reconstituted nonsucrose substances (RNS) are sold as fertilizer.

In the United States, several factories are successfully using the Quentin process (19), which is widely used in Europe. In this process, intermediate machine syrup is treated with cation-exchange resin in the magnesium form. Magnesium salts are far less melassigenic (molasses-forming) than sodium or potassium salts, and the purity of the final molasses can be lowered to 45–50%, thus reducing sugar losses. The economic utility of this process is partially dependent on the availability of cheap magnesium brine for resin regeneration.

Sulfitation. A small quantity of sulfur dioxide, obtained from burning elemental sulfur or purchased liquid sulfur, is added to the thin juice (20). The sulfur dioxide is not used for bleaching the juice, but to catalytically inhibit the Maillard or browning reaction between reducing sugars and amino acids. The browning reaction is the chief cause of increased coloration of the process juices during evaporation and crystallization. Only ca 25 ppm sulfur dioxide in the thin juice is required, although doses are commonly between 100 and 200 ppm since sulfur dioxide is gradually oxidized to sulfate as it passes through the sugar end. Occasionally, additional small amounts of sulfur dioxide are added to the thick juice to maintain a concentration of at least 25 ppm in the pan boilings. After sulfitation, the thin juice is sent to the evaporators, sometimes after a filtration.

Evaporation and Fuel Economy. For maximum fuel economy, all high pressure steam from the boilers is expanded to exhaust-steam pressure through turbines driving electric generators and engines; all exhaust steam is used in the first effect of the evaporator; all possible evaporation takes place in the evaporator to reduce the load on the vacuum pans; and for heating of vacuum pans, heaters, or melters, vapors are employed from one of the evaporator effects at the lowest possible pressure. It is customary to heat the beet-end juice gradually, using portions of the latter-effect vapors to start and finishing with first-effect vapors. Thus in the case of a quintuple-effect evaporator, diffusion juice may be heated with fourth-effect vapor initially and then with third-effect vapor; after carbonation, juice may be heated with second-effect vapor, and juice on its way to the evaporator first effect with second-effect vapor, and then with first-effect vapor. Similar heating arrangements are made in the sugar end. The use of vapor portions for other purposes than heating the next effect conserves heat and requires adjustment of the amount of heating surface for the individual effects. The heating surface areas decrease from the first to the last effect. The use of heat is illustrated in Table 3 for a five-effect evaporator with four Kestner, long-tube, single-pass effects, and a fifth calandria effect which normally has little function, but may be used when other bodies are temporarily out of service for descaling or other maintenance (see Evaporation).

Crystallization. In the thick juice from the evaporators (see Fig. 1), two remelted sugars are dissolved, filter aid is added, usually diatomaceous earth, and a tight filtration follows (21) (see Diatomite). The filtered liquor, or standard liquor, provides the feed material for the vacuum-pan boiling of the first product. Pan boiling is very similar to the same unit process in the sugarcane refinery, except that beet sugar crystallizes differently.

Sufficient feed liquor in the vacuum pan permits its concentration to a critical sucrose supersaturation, where the liquor still covers the heating element and crystallization may be induced by adding a small amount of powdered sugar. When a sufficient number of crystals have been formed, the supersaturation is dropped to the point where spontaneous nucleation no longer occurs. Alternatively, the charge of feed liquor may be concentrated to a supersaturation suitable for growing crystals, and a carefully measured and prepared mass of seed crystal is added; consequently, the fully grown crystals are as numerous as the seed crystals added. With either procedure, the crystals are grown to full size with additions of feed liquor. The content of a vacuum pan is known as a strike. This massecuite consists of a mass of crystals and mother liquor. It is discharged into mixer tanks for temporary storage to supply the centrifugal separators. In the centrifugal separators, the sugar is spun free of syrup and washed briefly with hot water. After drying and screening, the product is white granulated sugar.

The spun-off syrup is known as green syrup (although it is brown, not green). It provides the feed syrup for the vacuum pan for the second or intermediate boiling. The wash syrup is returned to the white or first-boiling feed. Centrifugation of the intermediate massecuite provides intermediate sugar, which becomes part of the standard liquor. The green syrup from the intermediate boiling (low machine syrup) provides the feed supply for the third, or raw boiling. The intermediate wash syrup is recycled to the intermediate pan-feed supply. The massecuite from the raw boiling is held for 16–74 h in crystallizers where the temperature is gradually lowered, allowing time for all the sugar to crystallize. The sugar from the raw boiling becomes a part of the standard liquor, whereas all the syrup spun from the sugar is called molasses (see also Crystallization).

Table 3. Distribution of Steam Between Evaporator and Process Heating[a]

Vapor	To next effect for evaporation, %	For process heating, %	Pressure, kPa[b]
exhaust steam	100		377
first	73	27	308
second	33	67	219
third	5	95	136
fourth	0	100	127

[a] In a typical run, thin juice entering the evaporator at 13.5% dry substance may emerge from the last effect at about 64%.

[b] To convert kPa to psi, multiply by 0.145.

Steffen Process for Desugaring Molasses. In North America, ca 25% of the factories are using the Steffen process for desugaring molasses (22). This process is highly empirical. Its mechanism is not fully understood, and it is not even known for certain if the calcium saccharate precipitates formed in the process are colloidal micelles of variable composition or true compounds. About 95% of the sucrose is recovered from the molasses. Normal molasses, containing ca 50% sucrose, is diluted with water to ca 6% sucrose, and cooled to ca 6°C, or in any case, well below 18°C. A little more than 100% dry lime, based on sucrose (wt/wt) is added; two competing reactions occur: the slaking of the quicklime and the formation of the calcium saccharate. The slight excess of lime represses the slaking reaction in favor of the calcium saccharate formation. The latter is a function of the physical surface area of the quicklime particles. The remainder of the lime is added as finely powdered quicklime, and quickly mixed with the cold, limed molasses solution. After formation of the saccharate precipitate, the cold saccharate cake is removed in rotary vacuum filters and the cold filtrate is heated by steam injection, resulting in the precipitation of more saccharates. The mixture is thickened in a small thickener and the underflow is filtered in rotary vacuum filters to remove the hot saccharate cake. The final, hot filtrate may be carbonated to reduce the lime content, filtered, and concentrated to ca 60% solids. In this form, it is added to the pressed pulp before drying, or it may be used for the production of monosodium glutamate (see Amino acids, monosodium glutamate). At some factories the hot Steffen filtrate is discarded. The cold and hot saccharate cakes are slurried in sweet water to form the saccharate milk, which serves as the liming agent for carbonation. On carbonation these saccharates are decomposed, yielding free sucrose and calcium hydroxide, which later is converted to calcium carbonate (see Fig. 6).

The trisaccharide, raffinose, is present in sugar beets in appreciable quantities and in even greater amounts in beets that have been subjected to cold temperatures. Calcium raffinates are precipitated along with the saccharates in the Steffen process and the raffinose is liberated in the first carbonation. In the factory, raffinose appears in the molasses, and is recycled to the Steffen process. Thus, raffinose tends to build up in the Steffen factory. Since its specific rotation at 20°C is +105.2°, considerably higher than that of sucrose (+65.5°), its presence leads to appreciable errors in the polarization analysis for sucrose. In addition, large amounts of raffinose cause the sucrose crystals to assume needlelike shapes to such an extent that 45 kg sugar may no longer fit in a normal 45-kg (100-lb) bag. Raffinose is adsorbed on certain specific crystal faces, and thus considerably reduces the crystal surface area available for crystallization. Since crystallization rates depend on surface area, the rates are reduced.

Because of the buildup of raffinose and other impurities, molasses is discarded in Steffen factories for a 24-h period every week or so. Sometimes a small amount of molasses is discarded continuously. The discarded molasses is either added to the pressed pulp before drying or sold for yeast production or similar uses. It has a somewhat higher content of nitrogenous substances than cane blackstrap molasses.

A new enzyme, galactosidase, hydrolyzes raffinose to 65.5% sucrose and 34.5% galactose (23). The latter, a reducing sugar, is destroyed in liming. The process is effective and economical if the percentage of raffinose in the molasses is >4–5%. The enzyme is secreted by a mold, *Mortierella vinacea*, and is manufactured by growing the mold, which is then made into pellets. The enzyme acts as a catalyst, and is not used up, but is gradually lost by the solution and attrition of the pellets. Its use elim-

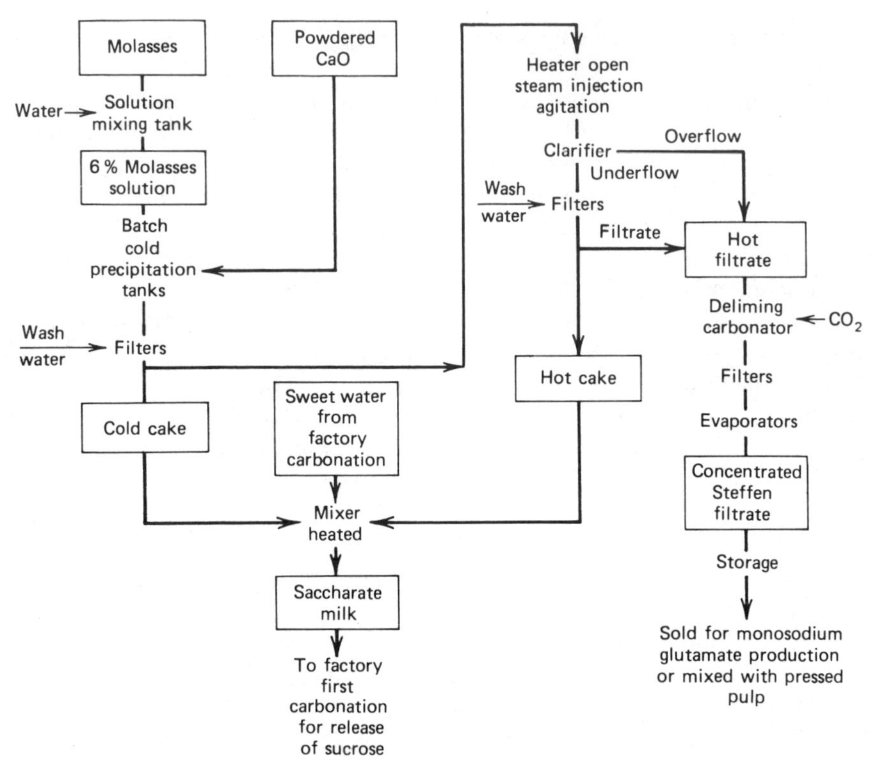

Figure 6. Typical flow sheet of the batch Steffen process for the desugarization of molasses.

inates the discarding of molasses. Molasses produced with galactosidase contains 3–3.8% raffinose. In the United States, two factories have used galactosidase for molasses with a high raffinose content.

The Granulated Product. The granulated sugar produced in beet-sugar factories is dried, screened, cooled, and bagged or stored in bulk (24). It is also used for the manufacture of secondary products, such as liquid and powdered sugars. More beet sugar than cane sugar is stored in large quantities because of seasonal production. Bulk-storage capacities as high as 34,000 metric tons of sugar are not uncommon.

Quickly dried granulated sugar frequently requires additional aeration or storage with occasional movement to permit the escape of traces of moisture, and cooling to prevent inversion or possible increase in coloration.

Beet-sugar processing equipment such as filters, vacuum pans, centrifuges, granulators, dryers, and screens, is similar to that employed in cane refineries. Beet-sugar products as manufactured in the United States are essentially identical with refined cane-sugar products. In the past, a rather widespread prejudice against beet sugar was caused by the fact that in the earlier days of the beet-sugar industry much of the product was not properly processed. However, this situation changed with the introduction of the three-stage-boiling system with two remelt sugars in the early 1930s, instead of the two-stage-boiling system used previously with only one remelt sugar.

In Europe, it was the custom to manufacture a raw beet sugar in factories near the beet fields. The raw sugar was shipped to refineries for processing into white sugar. The practice still persists in some localities, and a considerable amount of raw sugar is sold for direct consumption. Furthermore, some cane refineries in Europe process beet sugar which is sometimes mixed with raw cane sugar.

Automation. Many new beet-sugar factories are centrally controlled (25). Partial or complete-computer control is anticipated by the whole industry. Automation has markedly reduced labor costs, and should similarly reduce energy consumption.

Steps have been taken in Europe to computerize certain unit operations such as diffusion, crystallization, carbonation, evaporation, and boiler-house and pulp-dryer operations.

Material Balance and Quality Control

Yields and Losses. Sucrose is relatively fragile when in solution. Its ready inversion to dextrose and levulose, its easy fermentation, and its caramelization at only moderately high temperatures make strict chemical accounting necessary (26).

Sucrose in Sliced Beets. This is determined by weighing all the beet cossettes, usually in continuously integrating moving-belt scales or batch scales. Representative samples of cossettes are taken at hourly intervals and immediately analyzed for their sucrose content. The product of the weight of the cossettes and their percent sucrose content gives the weight of sucrose entering the factory in this form. The beet grower is paid on this basis.

Sucrose in Molasses. In measuring molasses produced, a representative sample is analyzed for sucrose. The weight of the molasses is determined with batch scales, or estimated with volume flowmeters. A material balance gives the amounts of molasses in storage tanks as determined manometrically, the amounts added to the pulp to be dried, and the amounts sold.

The molasses worked up in the Steffen process is usually weighed into the process on scales or measured with volumetric tanks accompanied by sample analyses.

Granulated Sugar Produced. The granulated sugar from the factory is accurately weighed in bucket scales, in which each load of ca 227 kg (500 lb) is determined within 0.5%.

Yields and Losses Statements. These statements are made for time intervals of one day to one month, depending on individual company practice. Losses of sucrose in the beets between the time of purchase and the time of slicing are called the outside-factory losses. A part of these losses derives from metabolic changes in the beets while in transit to the factory, in the factory bins awaiting immediate processing, or in long-term storage in piles. Direct physical losses of beet-root tails or broken pieces

Table 4. Per Capita U.S. Sweeteners Consumption, 1980[a]

Sweetener	kg	% of caloric sweetener
beet sugar	11.7	20.5
domestic-grown cane sugar	9.6	16.9
imported cane sugar	16.6	29.2
total cane and beet sugar	(37.9)	(66.6)
total corn sweeteners	18.4	32.3
miscellaneous (honey, etc)	0.64	1.1
total caloric sweeteners	56.94	*100.0*
total noncaloric sweeteners	3.2	

[a] Ref. 28.

Table 5. Canadian and U.S. Beet-Sugar Manufacturers, Locations, and Capacities

Company	Location	Capacity, t/d
The Amalgamated Sugar Company	Mini-Cassia, Rupert, Idaho	5850
	Twin Falls, Idaho	4172
	Nampa, Idaho	8163
	Nyssa, Ore.	5760
The American Crystal Sugar Company	Clarksburg, Calif.	2721
	Crookston, Minn.	3946
	Drayton, N.D.	4898
	East Grand Forks, Minn.	6530
	Moorhead, Minn.	4172
	Hillsboro, N.D.	4263
Amstar Corporation	Chandler, Ariz.	3401
Spreckels Sugar Division	Manteca, Calif.	3809
	Mendota, Calif.	3628
	Woodland, Calif.	2993
British Columbia Sugar Refining Co., Ltd.		
Alberta Sugar Company	Taber, Alberta	4400
Manitoba Sugar Company	Winnipeg (Fort Garry), Manitoba	3000
The Great Western Sugar Company	Bayard, Neb.	1995
	Billings, Mont.	3673
	Fort Morgan, Colo.	2948
	Gering, Neb.	1905
	Greeley, Colo.	1905
	Kemp (Goodland) Kan.	3084
	Loveland, Colo.	3084
	Lovell, Wyo.	1905
	Mitchell, Neb.	1950
	Ovid, Colo.	2404
	Scottsbluff, Neb.	2993
	Sterling, Colo.	2222
Northern Ohio Sugar Co.	Fremont, Ohio	2902
Holly Sugar Corporation	Brawley, Calif.	5896
	Sidney, Mont.	3628
	Worland, Wyo.	1995
	Torrington, Wyo.	2902
	Hereford, Texas	5896
	Hamilton City, Calif.	2540
	Tracy, Calif.	3900
Michigan Sugar Company	Caro, Mich.	2177
	Carrollton (Saginaw) Mich.	2177
	Croswell, Mich.	1723
	Sebewaing, Mich.	2358
Minn-Dak Farmers Cooperative	Wahpeton, N.D.	4082
Monitor Sugar Company	Bay City, Mich.	3809
Raffinerie de Sucre du Quebec	Mont-Saint-Hilaire, Quebec[a]	1596
Southern Minnesota Beet Sugar Cooperative	Renville, Minn.	5896
Union Sugar Company (Subs. Consolidated Foods Corp.)	Betteravia, Calif.	4535

[a] Raw sugar.

occur also and some losses occur from leaching out of sucrose while the beet juice is in the water flumes. Outside-factory losses may involve 1–5% of the sucrose purchased, depending on circumstances. The total outside loss or shrink is found by the difference between the sucrose in beets purchased and beets sliced.

Total inside-factory losses are found by subtracting the weight of granulated sugar produced from the amount of sucrose in the beets sliced, with correction for sucrose in the molasses worked in the Steffen house, if any. An effort is made to account for all losses. The readily accountable losses include the sucrose in molasses (which is a gain if more molasses is worked than produced), the exhausted beet pulp, the lime flume, the Steffen hot filtrate, and the factory wastewaters. Often estimates are made of inversion in the sugar end based on the amount of invert in the molasses produced, and of fermentation in the diffuser by measuring the increases of lactic acid and invert contents from the cossettes to the raw juice. The difference between the total inside-factory losses and the total accounted losses is the unaccounted loss. With good operations, the sucrose loss in molasses produced varies between 16 and 18% of the sucrose in the cossettes. The total of the other accounted losses is 2–5%, and the unaccounted losses usually amount to 1.5–3% of the sucrose in the cossettes.

Extraction denotes the percentage of granulated sucrose recovered of the total sucrose in the beets, when purchased or when entering the factory. For a so-called straight house, the extraction is 78–85%. The extraction of a Steffen house is, of course, enhanced by the sucrose recovered and may amount to 88–94% of the sucrose in the beets purchased, depending on the amount of molasses worked. However, in many Steffen factories the extractions are calculated only on the basis of the sucrose in beets purchased, and do not include the sucrose in the molasses worked.

Analysis. Analyses of the process materials are based on the use of the refractometer or Brix hydrometer for measuring the total solids in solution; of the polariscope for the determination of sucrose; and on copper-reduction methods for invert sugars (27). Purity and totally dissolved solids are determined in the raw juice, standard liquor, intermediate and raw massecuites, wash and green syrups, and molasses. Sucrose alone is determined in the cossettes, raw juice, lime flume, Steffen filtrates, saccharate cakes, and molasses. The pH values of process liquors are measured throughout the process since the liquors are kept basic to prevent sucrose inversion. First and second carbonations are controlled by titrated alkalinities (which are more closely correlated with pH). Calcium oxide is determined in carbonation juices, and total lime salts in thin juice. Sulfur dioxide is controlled by iodimetric analyses of the thin juice and the white massecuite. Detailed analyses are required of quicklime, coke, carbon dioxide gas, granular carbon, and saccharate cakes. Quality control follows color and turbidity, sediment, specks, odor, taste, ash, sulfite sulfur, particle size (by sieve test), saponins and solution-foam grade (a measure of the tendency of the dissolved sugar to foam when boiled). Numerous determinations must be made of both mesophilic and thermophilic microorganisms, as strict limits must be met for bottlers of carbonated beverages (qv), canners, and other food manufacturers.

Economic Aspects

The 1980-per-capita consumption of sweeteners in the United States is given in Table 4. Sugar economists predict that by 1985, corn-based high fructose syrup will have taken over a sufficiently large share of the market to raise the percentage of corn-based sweeteners to 48% of the caloric sweetener market (see Sweeteners).

The current beet-sugar manufacturing companies in the United States and Canada are given in Table 5 (29).

BIBLIOGRAPHY

"Beet Sugar" under "Sugar Manufacture" in *ECT* 1st ed., Vol. 13, pp. 217–227, by L. A. Willis, Consultant; "Beet Sugar" in *ECT* 2nd ed., Vol. 19, pp. 203–220, by R. A. McGinnis, Spreckels Sugar Co.

1. R. A. McGinnis, ed., *Beet-Sugar Technology*, 3rd ed., The Beet-Sugar Development Foundation, Fort Collins, Colo., 1982, pp. 25–63.
2. F. Schneider, ed., *Technologie des Zuckers*, 2nd ed., Verein der Zuckerindustrie, M. & H. Schaoer, Hannover, 1968, Chapt. 1.
3. P. M. Silin, *Technology of Beet-Sugar Production and Refining*, Pishchepromizdat, Moscow, 1958; *Israel Program for Scientific Translations*, Jerusalem, 1964, pp. 43–87.
4. K. Vukov, *Physics and Chemistry of Sugar Beet in Sugar Manufacture*, Elsevier Scientific Publishing Co., Amsterdam, 1977, pp. 63–97.
5. *Die Herstellung*, 2nd ed., VEB Fachbuchverlag, Leipzig, 1980, pp. 68–78.
6. Ref. 1, pp. 101–114; ref. 2, Chapts. 3–4; ref. 3, pp. 88–106; ref. 4, pp. 417–425; ref. 5, pp. 131–162.
7. Ref. 1, pp. 120–132; ref. 2, Chapt. 5; ref. 3, pp. 114–175; ref. 4, pp. 426–446; ref. 5, pp. 164–171.
8. H. Bruniche-Olsen, *Solid-Liquid Extraction*, Nyt Fordisk Forlag, Copenhagen, 1962.
9. T. Baloh, *Z. Zuckerind.* **27,** 363 (1977).
10. A. Carruthers and J. F. T. Oldfield, *J. Am. Soc. Sugar Beet Technol.* **13,** 105 (1964).
11. Ref. 1, pp. 132–151; ref. 2, pp. 196–219; ref. 3, pp. 154–175; ref. 5, pp. 171–199.
12. Ref. 1, pp. 135–136; ref. 2, p. 206; ref. 3, pp. 167–169; ref. 5, pp. 176–180.
13. Ref. 1, pp. 625–637; ref. 2, pp. 869–874.
14. Ref. 1, Chapts. 7–9; ref. 2, Chapt. 6; ref. 3, pp. 179–210; ref. 5, pp. 223–289.
15. Ref. 1, pp. 159–229; ref. 3, pp. 231; ref. 5, p. 223.
16. Ref. 1, pp. 214–215; ref. 2, pp. 303–306.
17. Ref. 1, pp. 300–332; ref. 3, pp. 258–270.
18. Ref. 2, Chapt. 10; ref. 5, pp. 280–286.
19. Ref. 1, pp. 326–330; ref. 2, pp. 616–618; ref. 5, p. 289.
20. Ref. 1, pp. 265–274; ref. 2, pp. 527–528; ref. 3, pp. 228–230; ref. 5, p. 249.
21. Ref. 1, Chapt. 12; ref. 2, Chapt. 8; ref. 3, pp. 297–373; ref. 5, pp. 370–425.
22. Ref. 1, Chapt. 18; ref. 2, pp. 986–987; ref. 3, pp. 389, 392.
23. Ref. 1, pp. 613–615; J. Obara, S. Hashimoto, and H. Suzuki, *Sugar Technol. Rev.* **4**(3), (1977).
24. Ref. 1, Chapt. 13; ref. 2, pp. 483–487; ref. 5, pp. 506–526.
25. Ref. 1, pp. 693–708.
26. Ref. 1, pp. 741–756; ref. 5, pp. 625–654.
27. Ref. 1, pp. 756–758; ref. 5, pp. 655–689.
28. *Outlook and Situation*, U.S. Department of Agriculture, Sept. 1981, Table 20, p. 44.
29. Ref. 1, pp. 798–799.

General Reference

P. Honig, ed., *Principles of Sugar Technology*, 3 vols., Elsevier Publishing Co., New York, 1953, 1959, 1963, 2045 pages. A comprehensive treatise, dealing primarily with cane sugar manufacture. There are sections of value to the beet-sugar industry, including those on the Properties of Sucrose, Sucrose Crystallography and Crystallization Fundamentals, Heat Transfer, Centrifugation, and Microbiology.

R. A. McGinnis
Consultant

SUGAR DERIVATIVES

Sucrose [*57-50-1*] is produced worldwide in far greater quantity and in higher purity than any other organic chemical. The annual production of raw cane and beet sugar exceeded 90×10^6 metric tons in 1977–1978. Despite this abundant availability of sucrose and its relatively low cost, probably no more than 0.1 wt % is consumed as a chemical feedstock. The potential value of sucrose as a raw material for the chemical industry has been recognized for many years and has been the subject of considerable research. The development of nonfood uses for sucrose and its derivatives has been actively pursued under the sponsorship of the sugar industry, initially through the Sugar Research Foundation, Inc. and subsequently through the International Sugar Research Foundation, Inc., both with offices in Bethesda, Maryland. Since March 1978, the World Sugar Research Organization Ltd., based in London, UK, has taken over the role of promoting sugar research internationally.

The focus of interest on sucrose as a chemical raw material has given rise to the name sucrochemistry for the field of carbohydrate chemistry concerned specifically with sucrose and its derivatives. Numerous and diverse uses for sucrose and its compounds have been proposed or patented, though in practice very few applications have been successfully developed. Nevertheless, as an industrial raw material for the chemical industry, sucrose is becoming increasingly competitive in cost with organic raw materials derived from petroleum. Apart from the production of fuel alcohol by the fermentation of sugarcane juice or molasses, as, for example, in Brazil (1), little use has so far been made of this renewable resource (see Chemurgy; Fuels from biomass).

Sucrose can be used as a starting material in degradative reactions, from which chemically simpler products are usually obtained, or in syntheses in which the carbon framework of the sucrose molecule is retained. The potential for sugar as a chemical feedstock follows the underlying chemistry of sucrose.

Sucrose

Sucrose is a nonreducing disaccharide in which the hexose sugars D-glucose [*50-99-7*] and D-fructose [*57-48-7*] are combined through their glycosidic hydroxyl groups. Glucose is in the α-pyranose ring form and fructose is in the β-furanose ring form. The numbering of the hydroxyl groups is shown in Figure 1. Sucrose is relatively stable, since it lacks the potential free carbonyl group of a reducing sugar, and has the chemical properties of an octahydric alcohol. It is, however, readily hydrolyzed to its component hexoses by the appropriate enzyme, eg, yeast invertase, or by acid catalysis. Thus, in degradative reactions, as in fermentation (qv), sucrose generally loses its unique identity and gives similar products as a mixture of glucose and fructose under equivalent conditions.

The synthetic chemistry of sucrose is that of a polyhydric alcohol, in which the relative reactivities of the hydroxyl groups differ according to whether they are primary or secondary and to their position in the molecule. This leads to the possibility of selective reaction at specific positions on the sucrose molecule and multiple substitution, indicating an almost infinite diversity of potential derivatives. Differences in reactivity

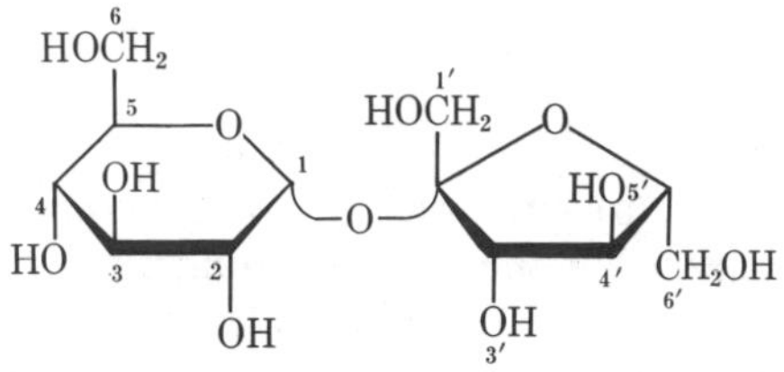

Figure 1. Sucrose.

depend on a number of factors, eg, the solvent used, type of reaction, steric interference, and temperature. Generally, the primary hydroxyl groups are the most reactive and the 2-, 3-, 3′-, and 4′-hydroxyl groups the least reactive.

Sucrose is insoluble in most organic solvents, so that reactions in solution require the use of polar or hydrogen-bonding solvents (see Table 1) (2). Most commonly, pyridine, dimethyl sulfoxide, *N,N*-dimethylformamide and hexamethylphosphoric triamide are used. Occasionally, a reaction takes place in an aqueous medium or in the absence of a solvent. The need for costly or polar solvents, which are difficult to recover, is frequently a determining factor in the economics of a synthesis starting from sucrose.

Degradative Reactions of Sucrose and Other Sugars. ***Thermal.*** Sucrose, when heated at temperatures above its melting point (180°C), rapidly decomposes with the formation of a complex mixture of volatile compounds, ie, 2-butanone; 2-methyl-2-cyclopentenone; 2-hydroxy-3-methyl-2-cyclopenten-1-one; 2-acetofuran; 5-methyl-2(5*H*)-furanone; furfural; 5-methylfurfural; 5-hydroxymethylfurfural; 2,5-dimethylfuran; γ-butyrolactone; 3-hydroxy-2-methyl-4*H*-pyran-4-one; and phenol; and an involatile residue containing glucose; levoglucosan [*498-07-7*] (**1**); levoglucosenone [*37112-31-5*] (**2**); 1,4:3,6-di-anhydroglucose [*4451-30-3*] (**3**); and polymers (3). The intensely dark-colored product, known as burnt sugar caramel, is widely used as a coloring agent in foods and drinks. In manufacturing, catalysts, eg, sodium hydroxide, ammonia, and ammonium bisulfite, are added to increase the tinctorial power and

Table 1. Solubility of Sucrose in Nonaqueous Solvents[a]

Solvent	Solubility (at 100°C), g/100 g
dimethyl sulfoxide	58.7
morpholine	45.1
dipropyl sulfoxide	42
N-methyl-2-pyrrolidinone	33.5
N,N-dimethylformamide	29.6
hexamethylphosphoric triamide	v. sol.
2-methylpiperazine	29.5
propane-1,2-diol	~11
pyridine	5.99
pyrazine	2.23
sulfolane	<1
dimethylsulfolane	<1
1,4-dioxane	<1

[a] Ref. 2.

to modify the properties of the caramel for particular applications (4). Sucrose can be replaced by glucose, high dextrose equivalent (DE) corn syrup, or invert syrup, which caramelize at lower temperatures, typically 130°C. The product is usually specified by rigorous analytical tests; however, its composition is undefined which leads to problems in food-additive regulatory control.

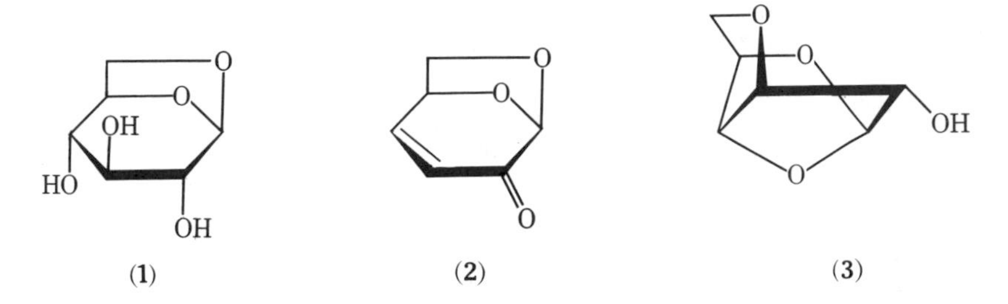

(1) (2) (3)

Hydrolysis. In acidic aqueous solution, sucrose is hydrolyzed at a rate that depends on temperature, pH, and solids concentration, to give an equimolar mixture of glucose and fructose (5). Under alkaline conditions, sucrose is more stable than in acid solution, but at high pH it is rapidly degraded on heating of the solution. Since glucose and fructose are extremely unstable under alkaline conditions, only their products of decomposition are obtained on alkaline hydrolysis of sucrose.

Acidic. In acid and alkaline solution, glucose and fructose are interconvertible through their common intermediate enediol form of the hexose (**4**) (Fig. 2). The equivalent enzyme-catalyzed interconversion in the presence of glucose isomerase, for example from *Actinoplanes missouriensis*, is used in the commercial production of high fructose corn syrups (HFCS) (6). The intermediate enediol (**4**) readily loses one molecule of water to give arabinohexos-2-ulose [*26345-59-5*] (**5**) (Fig. 2). The further decomposition of the latter explains the formation of the numerous degradation products identified as components of the mixture obtained on heating sucrose, glucose, or fructose in acid or alkaline solution.

Under mildly acidic conditions and at high temperature (typically 100–120°C), sugars undergo progressive loss of three molecules of water with the formation of 5-hydroxymethylfurfural (**6**) (7). Pentoses, derived from the hydrolysis of pentosans, the main components of hemicelluloses, undergo a similar reaction when heated in the presence of dilute sulfuric acid to form furfural. This process is used for the commercial production of furfural from corn cobs or sugarcane bagasse (see Furan derivatives).

At high temperatures, typically 180°C, and in the presence of strong acids, eg, HCl, hexose sugars give a mixture of levulinic acid (**7**) and formic acid as a result of the further breakdown of the intermediate (**6**) (8). Levulinic acid is the source of a wide range of products, eg, the esters, which have been proposed for use as solvents for nitrocellulose, α-angelicalactone, 2,6-dimethyl-3-pyridazinone, and 5-methyl-2-pyrrolidinone (9). None of these routes to these derivatives, however, is commercially significant.

Alkaline. Under strongly alkaline conditions, glucose and fructose decompose rapidly by the formation of the hexosenediol moiety. Elimination of a hydroxy group from the carbon atom β to the carbonyl group, a reaction catalyzed by calcium ion, yields the corresponding deoxyhexosulose which can then react in any of several di-

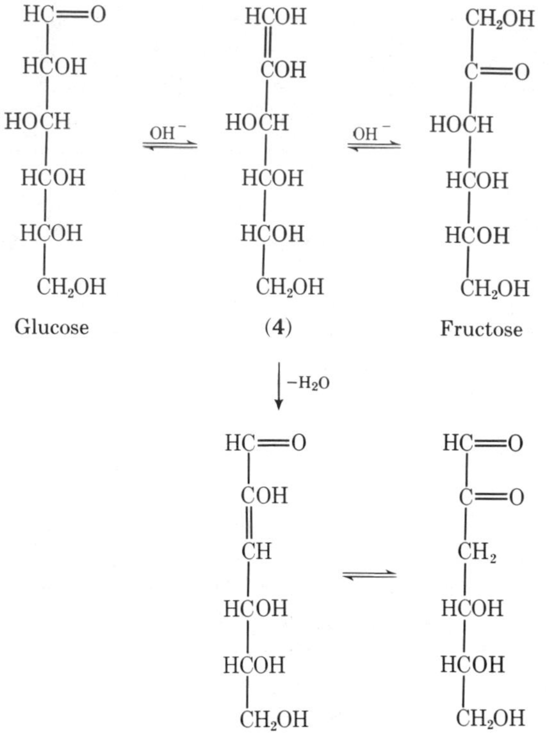

Glucose		(4)		Fructose
HC=O		HCOH		CH_2OH
HCOH		‖ COH		C=O
HOCH	$\overset{OH^-}{\rightleftharpoons}$	HOCH	$\overset{OH^-}{\rightleftharpoons}$	HOCH
HCOH		HCOH		HCOH
HCOH		HCOH		HCOH
CH_2OH		CH_2OH		CH_2OH

(4) $\downarrow -H_2O$

		(5)
HC=O		HC=O
COH		C=O
‖ CH	$\rightleftharpoons$	CH_2
HCOH		HCOH
HCOH		HCOH
CH_2OH		CH_2OH

Figure 2. Dehydration of glucose and/or fructose to give *arabino*-hexos-2-ulose.

rections according to the reaction conditions (10). Under alkaline conditions, the deoxyhexosulose can undergo a benzilic acid disproportionation-type reaction with the formation of saccharinic acids; the proportions in which the different isomers form depend on the conditions and on the nature of the alkali cation (11). Calcium ions tend to favor rearrangement without fission; sodium hydroxide, however, promotes fission of the hexose carbon chain either by splitting the dicarbonyl bond or by reverse aldol condensation. The carbonyl fragments either recombine, followed by an irreversible elimination of water, leading to dark unsaturated anionic polymers (caramel), or they are transformed into simple, stable, organic acids, eg, lactic, formic, acetic, oxalic, and erythronic acids, by internal rearrangement or atmospheric oxidation.

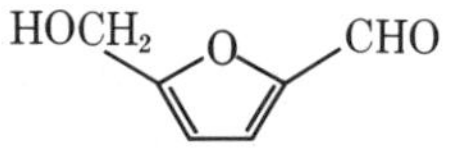

$HOCH_2$–(furan ring, O)–CHO

(**6**)

$CH_3C(=O)CH_2CH_2CO_2H$

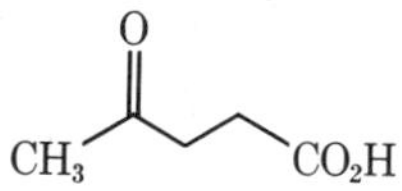

(**7**)

Thus, sugar heated in solution with excess lime gives predominantly lactic acid, which can be isolated in high yield (12) (see Hydroxycarboxylic acids). However, this is not a favorable production route compared with fermentation or the synthetic route from acrylonitrile (qv).

Oxidation. In the presence of oxygen, the double bond of the enediol form of reducing sugars undergoes fission with the formation of the corresponding lower aldonic acid. For example, glucose is oxidized to arabonic acid [*13752-83-5*] when the alkaline solution is shaken with oxygen under pressure at ambient temperature, in yields of up to 75% (13). However, direct oxidation of glucose in the presence of a palladium catalyst results in the formation of gluconic acid [*526-95-4*] (14). The direct oxidation of glucose to gluconic acid by oxygen also takes place in the presence of the enzyme glucose oxidase (15). By means of the immobilized enzyme, invert sugar obtained from sucrose can be converted into a mixture of gluconic acid and fructose (see Enzymes, immobilized). Oxidation of glucose to gluconic acid can also be carried out by hypobromite or by electrolytic oxidation in the presence of bromide (16).

Oxidation of sucrose with nitrogen tetroxide gives D-glucaric acid [*87-73-0*] or, in the presence of a vanadium pentoxide catalyst, oxalic acid [*144-62-7*] (17). However, the latter is not an economically favorable manufacturing process (see Oxalic acid).

Reduction. Sucrose is not reducible, but upon hydrolysis it gives a mixture of sorbitol [*50-70-4*] and mannitol [*69-65-8*] in the approximate molar ratio of 3:1 on electrolytic reduction or hydrogenation in the presence of a nickel or palladium catalyst. Mannitol can be separated from the resulting syrup by crystallization.

Hydrogenation of sucrose in alkaline solution at high temperatures gives a mixture of products, which include glycerol [*56-81-5*], propane-1,2-diol [*57-55-6*], and ethylene glycol [*107-21-1*] formed by the reduction of the carbonyl fragmentation products of the sugar (18). Glycerol (qv) was manufactured by this route until the process was discontinued in 1969 because it became uncompetitive with the synthetic route from propylene.

Ammoniation. Sugars on heating in solution with aqueous ammonia undergo a complex sequence of reactions in which a wide range of nitrogen heterocyclic compounds form, including derivatives of imidazole, pyrazine, piperazine, and pyridine (19). The reaction is rarely of value for the synthesis of specific products, owing to the low yields of individual compounds and the difficulty of their separation. Nevertheless, the preparation of, for example, 4-hydroxymethyl-1*H*-imidazole [*822-55-9*] involves this route (20). 2-Methylpiperazine [*109-07-9*] is produced in 27% yield by the high pressure hydrogenation of sugars in the presence of ammonia. The possibility of manufacturing 2-methylpiperazine by this route has been studied in detail, but the process is not economically favorable (21).

The reaction of reducing sugars with amino acids (qv), known as the Maillard reaction, is involved in the nonenzymic browning of foods during cooking, processing, and storage (22). It also contributes to color formation during the extraction of sugar from beet and cane and the processing of sugar liquors.

Synthetic Derivatives. Sucrose undergoes those reactions characteristic of alcohols, giving, for example, esters, ethers, acetals, and urethanes. The hydroxyl group can be replaced by hydrogen, halide, azide, amino, thiol, nitrile, thiocyanate, thioacetate, and other monovalent groups. The primary hydroxyl groups can be oxidized to aldehyde or carboxylic acid and secondary hydroxyl groups to ketone. Unsaturation can be introduced and internal anhydro rings created. The importance of these derivatives as synthetic intermediates has been limited by their availability. Because of the necessary costly solvents and the low yields of the pure derivative, uses have been developed only for certain esters and ethers in specialized fields of application.

Physical properties and CAS Registry Numbers of some sucrose derivatives are listed in Table 2.

Table 2. Physical Properties of Some Sucrose Derivatives

Derivative	CAS Registry No.	Melting point, °C	Specific rotation, degrees $[\alpha]_D$ (solvent)	Ref.
Esters				
sucrose octaacetate	*[126-14-7]*	86–87	+59.6 ($CHCl_3$)	
sucrose 2,3,6,3′,4′-pentaacetate (**8**)	*[35867-25-5]*	154–156	+22.0 ($CHCl_3$)	23
sucrose 2,3,6,1′,3′,4′-hexaacetate (**9**)	*[56038-06-3]*		+40 ($CHCl_3$)	24
sucrose 1′,3,3′,4′,6,6′-hexapivalate	*[69075-35-0]*	155	+54.5 (CH_3OH)	25
3′,4′-ditosylsucrose hexaacetate	*[67909-46-0]*	150–152	+54.1 ($CHCl_3$)	26
Ethers				
sucrose octabenzyl ether	*[18685-22-8]*		+38.6 ($CHCl_3$)	27
6,1′,6′-tritritylsucrose pentaacetate	*[35867-26-6]*	235–236	+68.9 ($CHCl_3$)	28
octakis(trimethylsilyl)sucrose	*[19159-25-2]*	63–65	+35.9 ($CHCl_3$)	29
6,6′-bis(*tert*-butyldimethylsilyl)-sucrose hexaacetate	*[63775-77-9]*		+58.0 ($CHCl_3$)	24
Acetals				
4,6-*O*-benzylidenesucrose hexaacetate	*[52706-46-4]*	160–161	+45 ($CHCl_3$)	30
4,6-*O*-isopropylidenesucrose hexaacetate	*[71196-27-5]*	168–170	+45.7 ($CHCl_3$)	30
1′,2:4,6-di-*O*-isopropylidenesucrose tetraacetate	*[67909-39-1]*	85–87	+12.8 ($CHCl_3$)	31
1′,2-*O*-(diphenylsilylene)sucrose hexaacetate	*[74645-80-0]*	142–144	+60.6 ($CHCl_3$)	32
1′,2:6,6′-di-*O*-(diphenylsilylene)sucrose tetraacetate	*[74638-39-4]*	234–236	+9.6 ($CHCl_3$)	32
Epoxides				
α-D-glucopyranosyl-3,4-anhydro-β-D-*ribo*-hexulofuranoside hexaacetate			+59.1 ($CHCl_3$)	26
α-D-glucopyranosyl-3,4-anhydro-β-D-*lyxo*-hexulofuranoside hexaacetate			+70.2 ($CHCl_3$)	26
Halides				
6,6′-dibromo-6,6′-dideoxysucrose hexamesylate	*[41671-36-7]*	171–173	+39.5 (acetone)	33
4,6,6′-tribromo-4,6,6′-trideoxy-*galacto*-sucrose pentamesylate	*[41671-38-9]*	140–141	+54.8 (acetone)	33
6,6′-dichloro-6,6′-dideoxysucrose	*[40984-16-5]*	85–88	+55 (H_2O)	34
6,6′-dichloro-6,6′-dideoxysucrose hexaacetate	*[40984-14-3]*	117–118	+55 ($CHCl_3$)	35
4,6,6′-trichloro-4,6,6′-trideoxy-*galacto*sucrose	*[57783-44-5]*	115–116	+88 (CH_3OH)	36
4,6,1′,6′-tetrachloro-4,6,1′,6′-tetradeoxy*galacto*sucrose tetramesylate	*[59343-74-7]*	120–121	+65.5 ($CHCl_3$)	36
6,6′-dideoxy-6,6′-diiodosucrose hexamesylate	*[35903-17-4]*	218–219	+41.8 (acetone)	37
Nitrogen derivatives				
4,6,6′-triazido-4,6,6′-trideoxy-*galacto*sucrose pentamesylate	*[41725-46-6]*	151–153	+45.2 (acetone)	33
4,6,1′,6′-tetra-azido-4,6,1′,6′-tetra-deoxy*galacto*sucrose tetramesylate	*[41787-81-9]*	171–173	+45.7 (acetone)	33
Deoxy derivatives				
6,6′-dideoxysucrose hexamesylate	*[35903-18-5]*	191–192	+44.4 ($CHCl_3$)	33
6,6′-dideoxy-5,5′-dienosucrose hexamesylate	*[35903-19-6]*	110–114	+13.7 (acetone)	38

Esters. Sucrose is readily esterified to the octaester in high yield by reaction with a molar excess of the anhydride or chloride of a sterically unhindered organic acid in the presence of pyridine or other suitable base. Thus, sucrose octabenzoate [*2425-84-5*] is prepared by treating an alkaline aqueous solution of sucrose with benzoyl chloride in the presence of an inert solvent, eg, tetrachloroethane, toluene, or dioxan (39–40). The octabenzoate passes into the organic phase as it is formed and can be recovered by evaporation of the solvent.

Sucrose octaacetate is prepared by the direct reaction of acetic anhydride with sucrose in the presence of a base, eg, sodium acetate or pyridine. The mixed esters of sucrose are obtained if a mixture of the anhydrides of two organic acids is used for the acylation. For example, in the reaction of acetic anhydride with sucrose in the presence of potassium carbonate at 110–130°C with isobutyric acid as the solvent, the mixed octaester is obtained in 94% yield (41). The ratio of acetate to isobutyrate groups is determined by the molar proportion of acetic anhydride to isobutyric acid used and by the relative reactivity of the respective anhydrides.

With sterically hindered acyl chlorides, eg, 2,4,6-trimethylbenzenesulfonyl chloride, reaction takes place almost exclusively with the primary hydroxyl groups of sucrose, even though normally the secondary hydroxyl group at position 4 is more reactive than the primary hydroxyl at position 1′ on the fructose moiety (43). Steric hindrance of the trimethylacetyl (pivaloyl) group is less marked. Nevertheless, in the reaction of pivaloyl chloride with sucrose in the presence of pyridine, conditions can be regulated so that the 1′,3,3′,4′,6,6′-hexapivalate can be isolated directly in 45% yield; steric factors reduce the rate of esterification of the more reactive 2- and 4-hydroxyl groups (25).

The sulfonate esters of sucrose, typically the octa-*p*-toluenesulfonate (tosylate) and octa(methanesulfonate) [*35781-94-3*] (octamesylate), have been widely studied and are of considerable synthetic utility in carbohydrate chemistry. These esters are readily prepared and purified and are frequently obtained as crystals (44). They are usually prepared by the action of the appropriate sulfonyl chloride on sucrose in pyridine solution (45). The pyridine acts as a basic catalyst which also neutralizes the hydrogen chloride formed; thus, there is no acid decomposition of unreacted sucrose.

An important alternative route to sucrose esters is by transesterification between sucrose and an ester, whereby an acyl group is transferred to sucrose in the presence of a basic catalyst, eg, potassium carbonate. The reaction is normally reversible, though the position of equilibrium can be shifted in either direction by continuously removing one component of the reaction. For example, sucrose heated in solution in dimethylformamide with the methyl ester of a fatty acid in the presence of anhydrous potassium carbonate forms predominantly the sucrose monoester (mainly the 6- and 6′-isomers) with the liberation of methanol. In the presence of an excess of the methyl ester, further hydroxyl groups react to give the di- and higher esters of sucrose. If a triglyceride rather than a methyl ester is used, then equilibrium is reached between sucrose and the mono- and diesters of sucrose and glycerol. If the glycerol is not removed, the reaction remains at its equilibrium composition. Since glycerol mono- and diesters of fatty acids are effective emulsifying agents and are particularly suited for food applications, the mixture of sucrose and glycerol esters resulting from the reaction of sucrose with natural triglycerides, eg, tallow, palm oil, or coconut oil, can be used without further separation (46). The solvent customarily used in this reaction, ie, dimethylformamide, is a costly component both for recovery and elimination from the food-grade product. A solventless process has been developed in which sucrose and a triglyceride react in the presence of a potassium carbonate catalyst at 125°C to give the mixture of sucrose monoesters and glyceryl esters directly (47). The mixed product, known as a sucroglyceride, can be separated into its component sucrose esters and mono- and diglycerides by means of food-grade solvents (48).

Ethers. An alcohol reacts directly with a reactive alkyl halide in the presence of a base to form an ether. Sucrose reacts, for example, with benzyl chloride in the presence of sodium hydroxide to give the octabenzyl ether in good yield (27). The reaction is neither readily controlled nor selective.

If the alkyl halide is strongly hindered and reactive, eg, triphenylmethyl chloride (trityl chloride), the reaction with sucrose takes place readily in the presence of pyridine with the almost exclusive formation of ethers of the primary hydroxyl groups (23). The triphenylmethyl ether bond is easily broken by hydrolysis under acidic conditions; thus, the trityl ethers offer a means of selectively protecting the primary hydroxyl groups of sucrose (49).

In the presence of hydrogen chloride in hot acetic acid, 6,1′,6′-tritritylsucrose pentaacetate is detritylated with internal migration of the acetyl group from *C*-4 to *C*-6 to give crystalline sucrose 2,3,6,3′,4′-pentaacetate (**8**), in which the free hydroxyl groups are present at positions 4, 1′, and 6′ (28).

The trimethylsilyl ethers of sucrose are readily formed by the reaction of sucrose with trimethylsilyl chloride in the presence of a base, eg, pyridine or hexamethyldisilazane. The fully substituted silyl ether, octakis(trimethylsilyl)sucrose, is sufficiently thermally stable and volatile to be separated quantitatively from other sugar silyl ethers by gas-liquid chromatography (29).

tert-Butyldimethylsilyl chloride is a selective silylating agent, which reacts preferentially with the primary hydroxyl groups of sucrose (see Silicon compounds, silylating agents) (50). The sucrose silyl ether bond is readily cleaved under acid conditions or in the presence of fluoride ion, which also promotes the simultaneous migration of an acetyl group from the 4- to the 6-hydroxyl during desilylation of the acetate ester (24). For example, 6,6′-bis-(*tert*-butyldimethylsilyl)sucrose hexaacetate on desilylation by fluoride gives sucrose 2,3,6,1′,3′,4′-hexaacetate (**9**), in which the secondary hydroxyl group at position 4 is free.

Benzyl ethers of sucrose form from the reaction of benzyl chloride in the presence of a strong base; sodium hydride in dimethyl sulfoxide is the preferred reagent (51). The benzyl group is cleaved by catalytic hydrogenation to regenerate the free hydroxyl (52). The easy removal of the benzyl ether blocking group under neutral conditions makes possible the selective protection of hydroxyl groups, for example, in the presence of ester substituents.

The methyl ethers of sucrose do not form selectively and the ether bonds are not readily cleaved. The methyl ethers are useful in structural analysis for identifying the positions of unsubstituted hydroxyl groups by exhaustive methylation. The usual method for preparing the methyl ethers involves reaction with methyl iodide in the presence of silver oxide (53). When acyl groups, which can migrate under such conditions, are present or the reaction is hindered, diazomethane–boron trifluoride in methylene chloride is the preferred methylating agent (54).

Ethers of sucrose are also formed by addition to a reactive ethylenic bond. For example, acrylonitrile reacts readily with sucrose in aqueous solution in the presence of a base to give sucrose 1′,6,6′-triscyanoethyl ether (**10**). Further reaction with acrylonitrile takes place in chloroform solution to give sucrose octakiscyanoethyl ether (55) (see also Cyanoethylation).

Reactive alkylene oxides add to sucrose in aqueous solution or in the absence of a solvent to give octakis(hydroxyalkyl) ethers, eg, (**11**). The reaction with ethylene oxide or propylene oxide is used to prepare stable polyols for use as cross-linking agents in semirigid polyurethane foams (56) (see Foamed plastics). Higher alkyloxiranes do not react easily and useful products have not been obtained by this route.

Acetals. Sucrose adds to the reactive double bond of vinylethers, for example, of 3,4-dihydro-2*H*-pyran, to give the hemiacetals having a degree of substitution up to 7 (57). It was not until 1974 that the preparation of a cyclic acetal of sucrose was first reported. Previously, it had been considered that steric factors precluded the formation of the simple cyclic acetals and ketals by sucrose. Benzylidene bromide in pyridine reacts with sucrose to give 4,6-*O*-benzylidenesucrose, which is isolated as the hexaacetate in 35% yield (58). Subsequently, the preparation of 4,6-*O*-isopropylidenesucrose hexaacetate and of 1′,2:4,6-di-*O*-isopropylidenesucrose (**12**) tetraacetates by the reaction of sucrose with 2,2-dimethoxypropane in dimethylformamide in the presence of *p*-toluenesulfonic acid as catalyst was reported (30). The diacetal is of interest in that the eight-membered 1′,2-isopropylidene ring bridges the two hexose rings of sucrose (31). Its formation is favored by the relatively rigid conformation of sucrose, which brings the appropriate hydroxyl groups into juxtaposition for the formation of the bond.

(**8**) R, R″ = H

(**9**) R = H, R′, R″ = CCH_3 (with C=O)

(**10**) R = CH_2CH_2CN, R′ = H

(**11**) R = R′ = $CH_2CH(OH)CH_3$

A similar eight-membered cyclic silicon acetal is formed in the reaction of sucrose with dimethoxydiphenylsilane; 1′,2-*O*-(diphenylsilylene)sucrose is isolated as the hexaacetate. The expected 4,6-*O*-silylene acetal was not formed, but the 1′,2:6,6′-di-*O*-(diphenylsilylene)sucrose was isolated as the tetraacetate in 4% yield. The latter has the 12-membered ring structure bridging the 6- and 6′-positions of sucrose (32).

Urethanes. Sucrose reacts readily with an isocyanate to give the corresponding urethane. For example, a diisocyanate, eg, toluene diisocyanate reacts with sucrose to give a polyurethane. In this reaction, the polyfunctionality of sucrose allows multiple cross-linking with the formation of rigid polymers (59). Sucrose tends to give brittle products when used in the production of polyurethane foams, and ethoxylated or propoxylated sucrose are the preferred cross-linking polyols.

Long-chain alkyl cyanates react with sucrose to give monoalkylurethanes, which have surface-active properties. The reaction of sucrose with sodium isocyanate and an alkylsulfonyl chloride gives the alkylsulfonylurethane of sucrose, which is also of interest for surfactant applications (60) (see Surfactants and detersive systems).

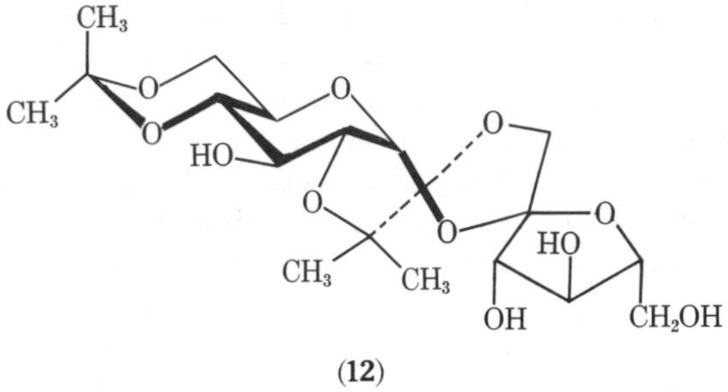

(12)

Anhydrosucrose. Sucrose can form internal ether bridges creating multiple ring systems. The epoxide (oxirane), unlike the stable anhydro rings of sucrose, is readily opened by nucleophilic displacement, which gives the epoxides considerable synthetic utility. Sucrose anhydrides form on treatment of the appropriate sulfonyl ester of sucrose with sodium methoxide in methanol. Anhydro-ring formation takes place between the primary hydroxyl groups, eg, the mesyl or tosyl ester, at positions 6-, 1′-, and 6′- with the secondary positions at 3-, 4′-, and 3′- of sucrose, respectively (62).

Epoxides. Sucrose epoxides form from suitably protected sucrose derivatives in which a hydroxyl group is in the *trans* configuration relative to a vicinal leaving group, usually a sulfonate ester, by the action of a base, eg, sodium methoxide, in methanol. For example, 3′,4′-ditosylsucrose hexaacetate yields a mixture of two epoxides; the 3,4-anhydro-β-D-lyxohexulofuranoside (**13**) (76%) and the corresponding *ribo* form (12%) on treatment with methanolic sodium methoxide at 70°C for 1 min (26). The direction of formation of the epoxide ring depends on the position of initial attack by the methoxide ion. The 3′,4′-riboepoxide is obtained in 96% yield on similar treatment of 3′-*p*-tosylsucrose heptaacetate. The 3′,4′-riboepoxide ring is opened by nucleophilic displacement; for example, by the azide ion on treatment with sodium azide in aqueous ethanol at 80°C, with the formation predominantly of the corresponding 4′-azido-4′-deoxy-xylofuranoside; the latter is isolated as the heptaacetate.

Halides. The sulfonyl esters of sucrose readily undergo nucleophilic displacement of the sulfonyl groups with the formation of the corresponding deoxysucrose derivatives on treatment with the appropriate anion in a suitable solvent. This reaction provides a valuable synthetic route to a wide range of deoxysucrose derivatives in which one or more hydroxyl groups have been replaced by halogen or an equivalent group.

In practice, the primary sulfonyl esters of sucrose are much more reactive than those at the secondary positions in nucleophilic displacement, with the 6- and 6′-positions reacting preferentially to the 1′-position. The relative reactivity in bimolecular nucleophilic-type (S_N2) reactions reflect the activation energy of the transition complex, which is determined by conformational forces and steric interference (62). The polarizability of the leaving group and of the nucleophile also determines the course of the reaction. For example, octa-*O*-mesylsucrose [*35781-94-3*] treated with a solution of sodium bromide in phosphoric hexamethyltriamide at 85°C for 2 h gives

6,6′-dibromo-6,6′-dideoxysucrose hexamesylate [*41671-36-7*] (36%) and 4,6,6′-tribromo-4,6,6′-trideoxy*galacto*sucrose pentamesylate [*41671-38-9*] (39%), suggesting that the 4-mesyloxy group is more reactive in this displacement than the primary 1′-ester (33). As would be expected for an S_N2 displacement, inversion of configuration at the secondary carbon occurs giving the galacto-configuration.

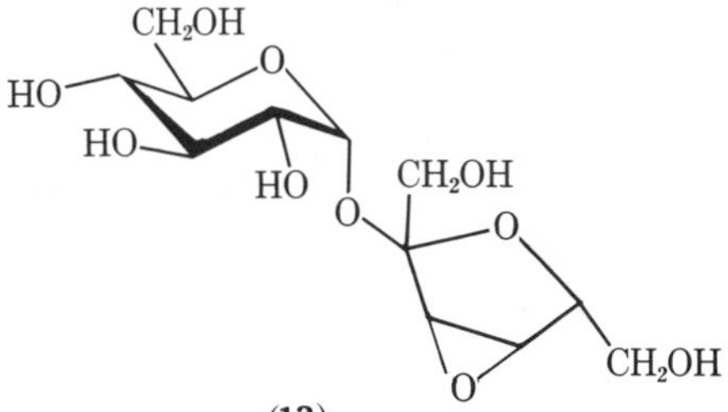

(13)

Although displacement takes place readily with chloride, bromide, and iodide nucleophiles, fluoride ion generally results in elimination with the formation of the unsaturated derivative rather than of the expected fluorodeoxysucrose. The preparation of 6-deoxy-6-fluorosucrose [*76410-48-5*], 6,1′- [*76410-45-2*] and 6,6′-dideoxy-6,6′-difluorosucrose [*76410-46-3*], and 6,1′,6′-trideoxy-6,1′,6′-trifluorosucrose [*76410-47-4*] (39) has been reported by the use of the strongly polar leaving group, *N,N*-diethylaminosulfur trifluoride (DAST) (63–64).

Sucrose can be chlorinated directly with an excess of sulfuryl chloride in pyridine, initially at −70°C and subsequently at room temperature. Under these conditions, 4,6,6′-trichloro-4,6,6′-trideoxy*galacto*sucrose [*57783-44-5*] can be isolated in 50% yield as the chlorosulfate ester (36). Under forcing conditions, a complex mixture of chlorosulfate esters of chlorodeoxysucrose is obtained, from which 4,6,1′,6′-tetrachloro-4,6,1′,6′-tetradeoxy*galacto*sucrose [*59343-74-7*] can be isolated in 40% yield as the tetra-*O*-mesyl ester.

Other chlorinating agents can be used for the selective chlorination of sucrose in high yield. 6,6′-Dichloro-6,6′-dideoxysucrose [*40984-16-5*] is obtained in 55% yield from the reaction of sucrose with methanesulfonyl chloride–*N,N*-dimethylformamide complex (Vilsmeier reagent) (34). Treatment of sucrose with a carbon tetrachloride–triphenylphosphine reagent in pyridine, which reacts selectively with primary hydroxyl groups, gives 6,6′-dichloro-6,6′-dideoxysucrose in 92% yield (35).

Sucrose–Nitrogen Compounds. Nucleophilic displacement of sulfonyl esters of sucrose by the strongly nucleophilic azide ion takes place readily at the primary positions and at position 4 with inversion of configuration on treatment of the ester with sodium azide in hexamethylphosphoric triamide at 90°C. After 48 h, octa-*O*-mesylsucrose [*35781-94-3*] gives 4,6,6′-triazido-4,6,6′-trideoxy*galacto*sucrose pentamesylate [*41725-46-6*] and 4,6,1′,6′-tetraazido-4,6,1′,6′-tetradeoxy*galacto*sucrose tetramesylate [*41787-81-9*] in 60% and 10% yield, respectively, which indicates the higher reactivity at the secondary 4-position compared with the 1′-primary ester (33). The azidosucrose derivatives are reduced to the corresponding aminosucrose on catalytic hydrogenation (42).

Deoxysucrose Derivatives. Deoxysucrose derivatives are usually obtained by the reductive dehalogenation of the corresponding halogen derivatives. For example, catalytic hydrogenation of 6,6′-dideoxy-6,6′-diiodosucrose hexamesylate [*35903-17-4*] gives the 6,6′-dideoxysucrose ester [*35903-18-5*] (37). The latter also forms from the catalytic hydrogenation of 6,6′-dideoxy-5,5′-dienosucrose hexamesylate [*35903-19-6*] (**14**) (65).

Elimination rather than nucleophilic displacement occurs upon treating a suitable sucrose derivative with fluoride ion. For example, 6,6′-dideoxy-6,6′-diiodosucrose hexamesylate on heating with silver fluoride in pyridine eliminates halogen with the formation of the corresponding 6,6′-dideoxy-5,5′-dienosucrose hexamesylate (**14**) (38).

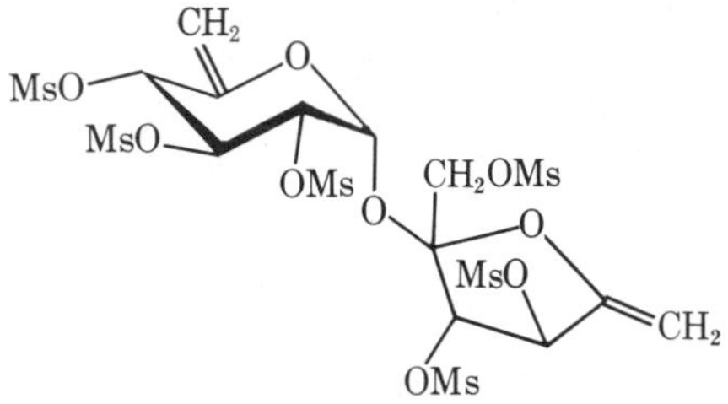

(**14**) Ms = SO_2CH_3

Enzymic Conversion Products. Under the action of certain enzymes, sugars undergo transglycosylation, a process by which a di- or higher saccharide is formed by the transfer of a sugar residue. For example, sucrose in the presence of yeast invertase gives, in addition to its hydrolysis products glucose and fructose, the trisaccharides 6-kestose [*562-68-5*] (**15**) and neokestose [*3688-75-3*] (**16**) (66). The predominant trisaccharide formed by mold invertase, eg, *Aspergillus oryziae*, is 1-kestose [*470-69-9*] (**17**). Raffinose [*512-69-6*] (**18**) forms in a similar reversion reaction by the transfer of a galactosyl residue from melibiose [*585-99-9*] (**19**) to sucrose in the presence of the enzyme α-D-galactoside hydrolase (67).

Simple rearrangement of hexose linkages also occurs in the presence of a suitable enzyme with the migration of the glucopyranosyl residue of sucrose from the 2′ position to the 5′ position to give the isomeric reducing disaccharide leucrose [*7158-70-5*] (5-*O*-(α-D-glucopyranosyl)-D-fructose), which was first observed in the products of fermentation of *Leuconostoc mesenteroides* (68). Other bacteria, eg, *Protaminobacter rubrum* and *Serratia plymuthica*, effect a similar rearrangement of sucrose with migration to position 6′ to give isomaltulose [*13718-94-0*] or palatinose [*13718-94-0*] (6-*O*-(α-D-glucopyranosyl)-D-fructose) (**20**) (69). In a recently described process, the enzyme system from *Erwinia rhapontici* is immobilized on an inert support or the cells are trapped in beads of calcium alginate to allow the conversion of sucrose to take place in a continuous flow system (70). A solution of sucrose at ca 55 wt % solids at 30°C and neutral pH is pumped through a column of immobilized enzyme to give 90–95% conversion of sucrose to isomaltulose, which is recovered by crystallization (mp 118–122°C) (see Enzymes, immobilized).

Palatinose can replace sucrose, which it closely resembles in physical properties, in the manufacture of food products for human or animal consumption (71). It has only 37% of the sweetness of sucrose at 7 wt % concentration (see Sweeteners). The reducing group of palatinose can be reduced to an alcohol giving a mixture of the two glucosylhexitols, 6-α-D-glucopyranosylsorbitol [*534-73-6*] (isomaltitol) and 1-α-D-glucopyranosylmannitol [*20942-99-8*] (Palatinit, trademark of Süddeutsche Zucker, A.G. Mannheim, FRG), mp 140–145°C. The sugar alcohols have ca 45% of the sweetness of sucrose and are claimed to be suitable replacements for sucrose in confectionery applications (72). The viscosity of isomaltitol solutions is similar to those of sucrose at equivalent concentration and temperature. The physical properties are compared in Table 3. Palatinit is incompletely digested or absorbed on ingestion leading to a food-calorie (kJ) equivalence below 50% of that of sucrose (73). It is also claimed to be less cariogenic than sucrose and is being developed as a low food-calorie, noncariogenic sweetener, though it has not yet been approved for food use (74).

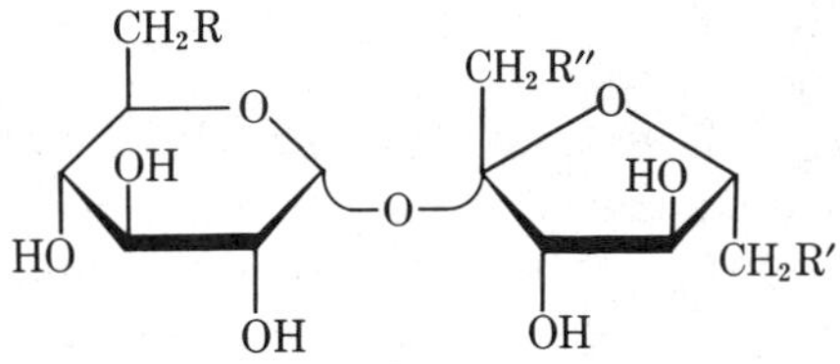

(**15**) R′ = R″ = OH, R =
(**16**) R = R″ = OH, R′ =
(**17**) R = R′ = OH, R″ =

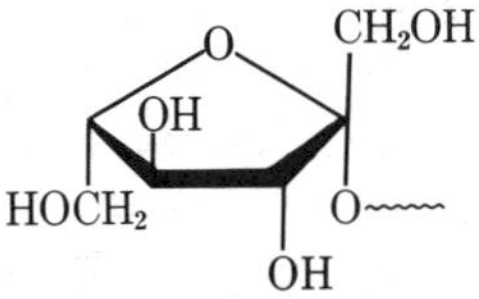

(**18**) R′ = R″ = OH, R =

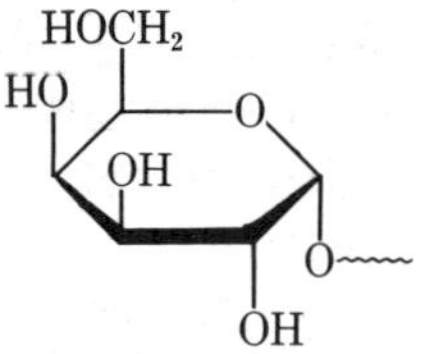

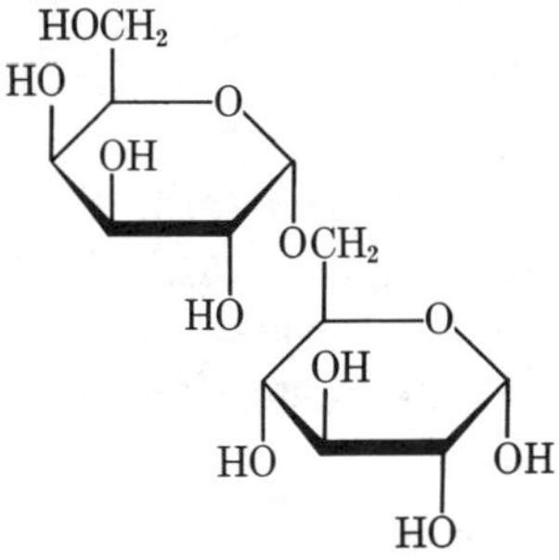

(**19**)

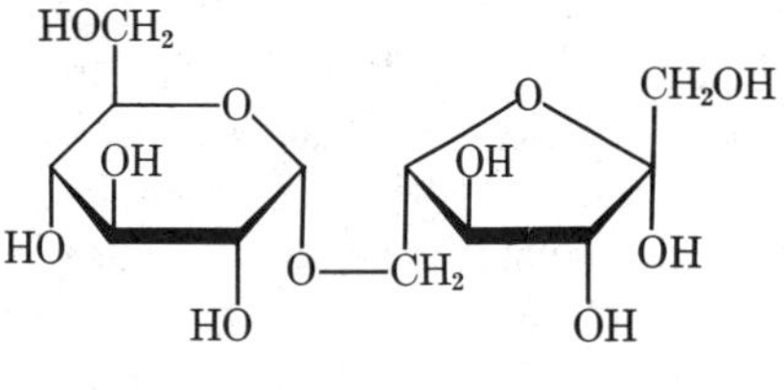

(**20**)

Table 3. Properties of Glycosylhexitols[a]

Chemical name	CAS Registry No.	Trivial name	Melting point, °C	Specific rotation, $[\alpha]_D$	Relative sweetness (sucrose = 1)
4α-D-glucopyranosyl-sorbitol	*[585-88-6]*	maltitol		+90°	0.75
6α-D-glucopyranosyl-sorbitol	*[534-73-6]*	isomaltitol	168	+90.5°	0.45
1α-D-glucopyranosyl-mannitol	*[20942-99-8]*		173.5	+90.5°	0.45
4β-D-galactopyranosyl-sorbitol	*[585-86-4]*	lactitol	146 from ethanol, 72 from aqueous ethanol	+14°	0.5

[a] Ref. 72.

Uses

Although numerous potential industrial applications of sucrose derivatives are described in the scientific and patent literature, they have not been widely adopted and only a few derivatives are commercially available. The production capacity for sucrose derivatives is seldom more than 1000 metric tons per year. There are no recent published figures for production and use of sucrose derivatives or for sucrose consumption as a chemical feedstock. The total amount of sucrose used as a chemical feedstock would not exceed 10^5 t/yr.

Sucrose octaacetate, an intensely bitter crystalline material, has limited use as a denaturant for alcohol and sugar. It is not generally permitted as a food additive, so that proposals to use it as a bittering agent or as an emulsifier cannot be adopted (see Food additives). As a plasticizer for cellulose acetate, it is not particularly suitable; the mixed acetate–isobutyrate ester is preferable.

Sucrose diacetate hexaisobutyrate [*27216-37-1*] (SAIB) is a viscous liquid of low volatility and is stable to hydrolysis and to discoloration. Since it is nontoxic, it is used as a clouding agent to disperse essential oils in soft drinks. It and sucrose octabenzoate [*2425-84-5*] are used in the manufacture of transparent paper, as they have refractive indexes close to that of cellulose (75).

The sucrose octabenzoate of commerce (see Table 4) has an average degree of esterification of 7.4. It is a noncrystalline, clear, colorless solid (mp 95–101°), is very soluble in nonhydroxylic organic solvents to give viscous solutions, and is insoluble in water. It is used as a plasticizer for cellulose nitrate and acetate–butyrate esters, acrylic and poly(vinyl chloride)-acetate films, and in nitrocellulose lacquers (39). It is recommended for use in admixture with other plasticizers (qv), eg, SAIB, to reduce migration and to improve pigment wetting, dispersion, and adhesion. It is also of value as a nonreactive diluent in uv-cured resin systems.

Sucrose monoesters of long-chain fatty acids are surface-active in aqueous systems. They have considerable potential as nonionic surfactants, because they are completely nontoxic and biodegradable. Sucrose monostearate [*25168-73-4*], monopalmitate [*26446-38-8*], monooleate [*25496-92-8*], and distearate [*27195-16-0*] are commercially available and have been manufactured in Japan since 1960. A new

Table 4. Properties of Sucrose Benzoate [*2425-84-5*][a]

Property	Value
molecular formula	$C_{12}H_{15}O_4(C_6H_5CO_2)_7$ approx
molecular weight	ca 1070
appearance	glassy clear solid
color (liquid), APHA	60
specific gravity, 25°/25°C	1.25
refractive index at 25°C	1.577
softening point, °C	98
saponification equivalent	153
hydroxyl content, meq OH/g	0.9
viscosity (molten), mPa·s (= cP)	
at 100°C	4000
at 150°C	590

[a] Ref. 76.

process not involving the toxic solvent dimethylformamide has since been developed by Ryoto Co., Ltd., of Tokyo. The plant has a production capacity of 3000 t/yr.

The esters are bland, nonallergenic, nontoxic, and nonirritating and are thus particularly suitable for formulation as detergents for domestic use and for use as dispersing, stabilizing, and emulsifying agents in the pharmaceutical, dairy, food, and agricultural industries (77). In several countries, eg, Japan, France, Belgium, and Switzerland, sucrose esters are permitted food additives. The most important limitation on their wider use is cost. In low volume, high value applications, cost is not limiting but the consumption of sucrose esters is small. For their use in domestic or industrial detergents, efficacy and low cost are important if they are to be competitive with conventional petroleum-derived anionic or nonionic surfactants.

In a process developed in the United Kingdom, the use of solvents in the transesterification reaction is avoided, except for the subsequent separation and purification of the food-grade sucrose esters (46). The sucroglyceride is suitable for direct use in many applications without further processing; for example, in the production of margarine, mayonnaise, ice cream, artificial cream, and dessert toppings. Sucrose esters in chocolate manufacture improve surface appearance, and in spray-dried and freeze-dried beverages they aid dispersion and dissolution. In baking, sucrose monoesters are effective as a dough-strength improver, where they increase loaf volume and improve the crumb structure and antistaling properties of bread. The incorporation of nonwheat protein into bread is made possible without loss of loaf quality by the addition of up to 0.4 wt % sucrose monopalmitate, monostearate, or monolaurate [*25339-99-5*] (78).

Sucrose ester surfactants, being nonallergenic and nonirritating, are particularly suited for use in cosmetics (qv) and shampoos. They are compatible with the skin, and preserve the protective oily mantle which is removed by ionic detergents (79).

Polyesters of sucrose with long-chain fatty acids are not absorbed on ingestion but reduce blood cholesterol levels. Sucrose octaoleate [*34816-23-4*], for example, has been evaluated for dietary use as a noncaloric fat substitute and for the prevention and treatment of hypercholesterolemia (80).

Calcium sucrose phosphate [*12676-30-1*] has been considered for use as a dietary additive for the control of dental caries. Its efficacy depends on buffering the dental plaque by reducing the pH fall which leads to the erosion of the dental hydroxyapatite matrix (81) (see Dentifrices). Several chlorodeoxysucrose derivatives, eg, 1′,4,6′-trichloro-1′,4,6′-trideoxy*galacto*sucrose [*56038-13-2*] (**21**), are anticariogenic as shown by their interference with the bacterial metabolism of *Streptococcus mutans.* The latter is the plaque-forming organism normally involved in tooth decay. The sucrose derivatives reduce bacterial adhesion and plaque formation and block sugar metabolism and thus the pH fall (82).

Inversion of configuration of the 4-hydroxyl group of sucrose gives β-D-fructofuranosyl-α-D-*galacto*pyranoside [*13322-96-8*] (**22**), known by the trivial name galactosucrose. This crystalline isomer of sucrose closely resembles sucrose in physical properties but is unexpectedly tasteless (83). By contrast, 4-chloro-4-deoxygalactosucrose [*56038-27-8*] (**23**) is five times sweeter than sucrose. Intensification of sweetness increases with further substitution of chlorine, eg, 1′,4,6′-trichloro-1′,4,6′-trideoxygalactosucrose is 650 times sweeter than sucrose (84). This derivative has a pure sweet taste that is indistinguishable from that of sucrose with no side flavors or aftertaste. On the other hand, 2,6,1′,6′-tetrachloro-2,6,1′,6′-tetradeoxymannosucrose [*73159-31-6*] (**24**) is intensely bitter, having twice the bitterness of sucrose octaacetate (85).

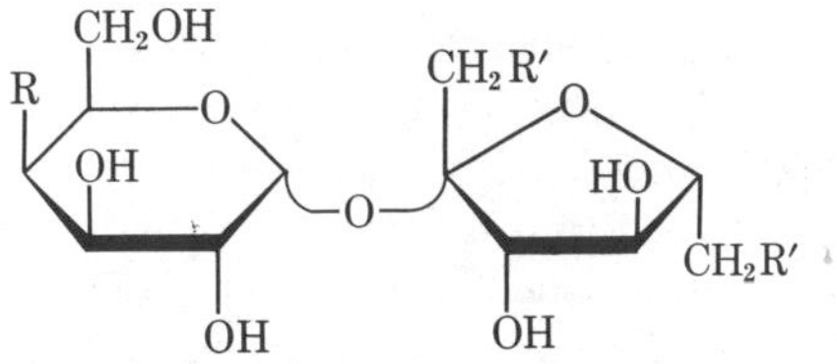

(21) R = R′ = Cl
(22) R = R′ = OH
(23) R = Cl, R′ = OH

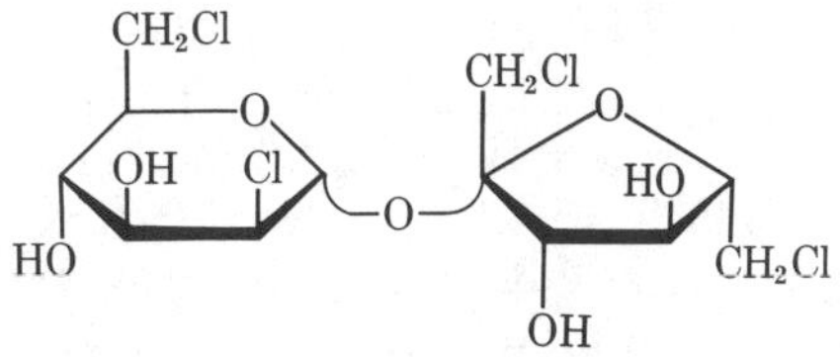

(24)

The levorotatory isomer of sucrose L-sucrose [*69257-56-3*] (α-L-glucopyranosyl-β-L-fructofuranoside) has identical sweetness to sucrose but is not metabolized on ingestion and is therefore noncaloric. Its high cost of production precludes its early development as a dietary sweetening agent for which it has been proposed (86).

Although 1′,4,6′-trichloro-1′,4,6′-trideoxygalactosucrose has no pharmacological properties, 6,6′-dichloro-6,6′-dideoxysucrose is effective as a male antifertility agent in rats and primates, allowing complete and reversible control of male reproductive fertility (87) (see Contraceptives).

An important and potentially large market for sucrose and its derivatives is in synthetic resins and plastics. Sucrose has been evaluated as a filler or substitute for phenol in melamine and novalak-type resins, though there is no evidence that unmodified sucrose participates in the cross-linking reaction with formaldehyde (88).

Octa-*O*-allylsucrose [*14699-90-2*] was investigated as a component of air-drying resin films, but the films proved too brittle for practical use (89). Sucrose polyesters of drying oil acids derived from linseed oil, tung oil, or soya oil, eg, sucrose heptalinoleate [*56449-51-5*], are superior to linseed oil as an air-drying paint vehicle in terms of film adhesion and alkali resistance (see Driers and metallic soaps). An ester with a lower degree of esterification is more readily prepared by ester interchange between sucrose and methyl linoleate in dimethyl sulfoxide (90). In the manufacture of alkyd-type surface coatings, sucrose esters of drying-oil acids do not give sufficiently marked advantages over conventional resins to offset their higher cost (see Alkyd resins).

Sucrose and its derivatives have been widely developed as components of polyurethane resins, in particular rigid polyurethane foams (60) (see Urethane polymers). The poly(hydroxypropyl)ether of sucrose is the preferred polyol because it has good miscibility with the diisocyanate cross-linking reagent and the fluorocarbon blowing agent used in the manufacture of rigid foams. Flame-retardant properties can be introduced into the resin by copolymerizing with halogen or phosphorus derivatives of sucrose (91).

BIBLIOGRAPHY

"Sugar Derivatives" in *ECT* 1st ed., Vol. 13, pp. 261–270, by J. L. Hickson, Sugar Research Foundation, Inc.; "Sugar Derivatives" in *ECT* 2nd ed., Vol. 19, pp. 221–233, by J. L. Hickson, International Sugar Research Foundation, Inc.

1. V. Yand and S. C. Trindade, *Chem. Eng. Prog.* **72,** 11 (1979); L. S. Pimental, *Biotech. Bioeng.* **22,** 1989 (1980).
2. O. K. Kononenko and K. M. Herstein, *Chem. Eng. Data Ser.* 1(1), 87 (1956).

3. R. R. Johnson, E. D. Alford, and G. W. Kinzer, *J. Agric. Food Chem.* **17**(1), 22 (1969); D. Gardiner, *J. Chem. Soc. C*, 1473 (1966).
4. R. N. Greenshields and A. W. MacGillivray, *Process Biochem.* **7**(12), 11, 16 (1972); R. N. Greenshields, *Process Biochem.* **8**(4), 17 (1973).
5. K. J. Parker, *Sucr. Belge* **89,** 119 (1970).
6. C. Bucke in G. G. Birch, N. Blakebrough, and K. J. Parker, eds., *Enzymes and Food Processing*, Applied Science Publishers, Ltd., London, 1981, pp. 51–63.
7. E. F. L. J. Anet, *Aust. J. Chem.* **18,** 240 (1965).
8. L. F. Wiggins, *Adv. Carbohydr. Chem.* **4,** 306 (1950).
9. R. H. Leonard, *Ind. Eng. Chem.* **48,** 1331 (1956).
10. A. Ishizu, B. Lindberg, and O. Theander, *Carbohydr. Res.* **5,** 329 (1967).
11. E. F. L. J. Anet, *Aust. J. Chem.* **14,** 293 (1961).
12. R. Montgomery and R. A. Ronca, *Ind. Eng. Chem.* **45,** 1136 (1953).
13. H. S. Isbell, *J. Res. Natl. Bur. Stand.* **29,** 227 (1942).
14. Br. Pat. 1,208,101 (Oct. 7, 1970), G. J. K. Acres and A. E. R. Budd (to Johnson Matthye and Co., Ltd.).
15. D. Scott in G. Reed, ed., *Enzymes in Food Processing*, Academic Press, Inc., New York, 1975, pp. 219–254, 519–547.
16. H. S. Isbell, H. L. Frush, and F. J. Bates, *Ind. Eng. Chem.* **24,** 375 (1932).
17. S. D. Deshpande and S. N. Vyas, *Ind. Eng. Chem. Prod. Res. Dev.* **18**(1), 69 (1979).
18. U.S. Pat. 3,030,429 (April 17, 1962), F. Conradin, Tamins, G. Bertossa, Ems, and J. Giesen (to Inventa, A.G.).
19. M. J. Kort, *Adv. Carbohydr. Chem.* **25,** 311 (1970).
20. J. R. Totter and W. J. Darby in E. C. Horning, ed., *Organic Syntheses*, Coll. Vol. 3, John Wiley & Sons, Inc., New York, 1955, pp. 460–462.
21. O. K. Kononenko, *Final Report of Project 155 to the Sugar Research Foundation*, Herstein Laboratories, Inc., New York, 1967.
22. G. P. Ellis, *Adv. Carbohydr. Chem.* **14,** 63 (1959).
23. B. Helferich, *Adv. Carbohydr. Chem.* **3,** 79 (1948).
24. F. Franke and R. D. Guthrie, *Aust. J. Chem.* **31,** 1285 (1978).
25. L. Hough, M. J. Chaudhary, and A. C. Richardson, *J. Chem. Soc. Chem. Comm.*, 664 (1978).
26. M. R. Jenner, Ph.D. thesis, University of London, England, 1976, pp. 123–127.
27. M. Gomberg and C. C. Buchler, *J. Am. Chem. Soc.* **43,** 1904 (1921).
28. G. G. McKeown, R. S. E. Serenius, and L. D. Hayward, *Can. J. Chem.* **35,** 28 (1957).
29. C. C. Sweeley, R. Bentley, M. Makita, and W. W. Wells, *J. Am. Chem. Soc.* **85,** 2497 (1963).
30. R. Khan, K. S. Mufti, and M. R. Jenner, *Carbohydr. Res.* **65,** 109 (1978).
31. R. Khan and K. S. Mufti, *Carbohydr. Res.* **43,** 247 (1975).
32. Ref. 26, pp. 70–74.
33. L. Hough and K. S. Mufti, *Carbohydr. Res.* **27,** 47 (1973).
34. R. Khan, M. R. Jenner, and K. S. Mufti, *Carbohydr. Res.* **39,** 253 (1975).
35. A. K. M. Annisuzzaman and R. L. Whistler, *Carbohydr. Res.* **61,** 511 (1978).
36. H. Parolis, *Carbohydr. Res.* **48,** 132 (1976).
37. C. H. Bolton, L. Hough, and R. Khan, *Carbohydr. Res.* **21,** 133 (1972).
38. L. Hough and K. S. Mufti, *Carbohydr. Res.* **25,** 497 (1972).
39. E. P. Lira and R. F. Anderson in J. L. Hickson, ed., *Sucrochemistry*, American Chemical Society, Washington, D.C., 1972, pp. 223–230.
40. Jpn. Pat. Appl. 52/95625 (Aug. 11, 1977), K. Hotta and F. Yamagishi (to Dai-Ichi Kogyo Seiyaku Co., Ltd.).
41. Jpn. Pat. Appl. 50/89314 (July 17, 1975), K. Kamemaru, N. Nishinohara, T. Niimura, and Y. Yabushita (to Unitika, Ltd.).
42. R. Khan, K. S. Mufti, and M. R. Jenner, *Carbohydr. Res.* **30,** 183 (1973).
43. L. Hough, S. P. Phadnis, and E. Tarelli, *Carbohydr. Res.* **44,** C12 (1975).
44. D. H. Ball and F. W. Parrish, *Adv. Carbohydr. Chem.* **24,** 139 (1969).
45. R. U. Lemieux and J. P. Barrette, *Can. J. Chem.* **37,** 1964 (1959).
46. K. J. Parker, K. James, and J. Hurfold in Ref. 40, pp. 97–114.
47. Br. Pat. 1,399,053 (June 25, 1975), K. J. Parker, R. A. Khan, and K. S. Mufti (to Tate & Lyle, Ltd.).
48. Br. Pat. 1,500,341 (Feb. 8, 1978), K. James (to Tate & Lyle, Ltd.).
49. L. Hough and K. Mufti, *Carbohydr. Res.* **21,** 144 (1972).

50. F. Franke and R. D. Guthrie, *Aust. J. Chem.* **30,** 639 (1977).
51. C. M. McCloskey, *Adv. Carbohydr. Chem.* **12,** 137 (1957); M. E. Tate and C. T. Bishop, *Can. J. Chem.* **41,** 1801 (1963).
52. K. Frendenberg, H. Toepfler, and C. Anderson, *Ber.* **61,** 1750 (1928).
53. E. L. Hirst and E. Percival in R. L. Whistler, M. L. Wolfrom, and J. N. Be Miller, eds., *Methods in Carbohydrate Chemistry*, Vol. 2, Academic Press, Inc., New York, 1963, pp. 145–150.
54. M. G. Lindley, G. G. Birch, and R. Khan, *Carbohydr. Res.* **43,** 360 (1975).
55. S. A. Barker, J. S. Brimacombe, M. R. Harnden, and J. A. Jarvis, *J. Chem. Soc.*, 3403 (1963).
56. J. W. Lemaistre and R. B. Seymour, *J. Org. Chem.* **13,** 782 (1948).
57. S. A. Barker, J. S. Brimacombe, J. A. Jarvis, and J. M. Williams, *J. Chem. Soc.* 3158 (1962).
58. R. Khan, *Carbohydr. Res.* **32,** 375 (1974).
59. K. C. Frisch and J. E. Kresta in Ref. 40, pp. 238–256.
60. W. Gerhardt, *Tenside* **5,** 10 (1968).
61. R. Khan, *Carbohydr. Res.* **22,** 441 (1972).
62. A. C. Richardson, *Carbohydr. Res.* **10,** 395 (1969).
63. T. J. Tewson and M. J. Welch, *J. Org. Chem.* **43,** 1090 (1978).
64. U.S. Pat. 4,228,150 (Oct. 14, 1980), J. F. Robyt and J. N. Zikopoulos (to Iowa State University Research Foundation, Inc.).
65. R. Khan and M. R. Jenner, *Carbohydr. Res.* **48,** 306 (1976).
66. D. Gross in R. L. Whistler and M. L. Wolfrom, eds., *Methods in Carbohydrate Chemistry*, Vol. 1, Academic Press, Inc., New York and London, 1962, pp. 360–366.
67. J. B. Pridham and M. W. Walter, *Biochem. J.* **92,** 20P (1964).
68. F. H. Stodola, E. S. Sharpe, and H. J. Koepsell, *J. Am. Chem. Soc.* **78,** 2514 (1956).
69. Eur. Pat. 1,099 (March 21, 1979), W. Crueger, L. Drath, and M. Munit (to Bayer A.G.).
70. Brit. Pat. Appl. 2,063,268 (Nov. 4, 1980), C. Bucke and P. S. J. Cheerham (to Tate & Lyle, Ltd.).
71. Brit. Pat. Appl. 2,066,639 (Nov. 7, 1979), C. Bucke and P. S. J. Cheerham (to Tate & Lyle, Ltd.).
72. H. Schiweck, *Proceedings of the ERGOB Conference*, European Research Group on Oral Biology, Geneva, S. Karger Publishers, Basel, Switzerland, 1978, pp. 138–144.
73. G. Siebert and U. Grupp, Ref. 73, pp. 109–113.
74. E. J. Karle and F. Gehring, *Dtsch. Zahnaerztl. Z.* **31,** 189 (1978).
75. C. H. Coney in Ref. 39, pp. 213–222.
76. Velsicol Chemical Corporation, 1982.
77. L. Bobichon in Ref. 39, pp. 115–120.
78. P. A. Seib, W. J. Hoover, and C. C. Tsen in Ref. 39, pp. 121–135.
79. L. Nobile, P. Rovesti and M. B. Sampa, *Am. Perfum.* **79**(7), 19 (1964).
80. R. W. Fallat, C. J. Glueck, R. Lutmer, and F. H. Mattson, *Am. J. Clin. Nutr.* **29,** 1204 (1976).
81. J. H. Curtin, J. Gagolski, and B. M. Smythe, *Food Technol. Aust.* **19,** 508, 511, 513 (1967).
82. D. B. Drucker and J. Verran, *Arch. Oral Biol.* **24,** 965 (1980).
83. R. Khan, *Carbohydr. Res.* **25,** 232 (1972).
84. L. Hough and R. Khan, *TIBS* **3,** 61 (1978).
85. Brit. Pat. Appl. 41115/78 (Oct. 18, 1978), R. Khan, M. R. Jenner, and H. Lindseth (to Tate & Lyle, Ltd.).
86. Belg. Pat. 866,171 (April 22, 1977), W. A. Szarek and J. K. N. Jones (to Queens University at Kingston).
87. W. C. L. Ford and G. M. H. Waites, *Int. J. Andrology Suppl.* **2,** 541 (1978).
88. W. Flavell and G. L. Redfearn in J. Yudkin, J. Edelman, and L. Hough, eds., *Sugar*, Butterworths, London, 1971, pp. 69–79.
89. M. Zief and E. Yanovsky, *Ind. Eng. Chem.* **41,** 1697 (1949).
90. R. N. Faulkner in Ref. 39, pp. 176–196.
91. U.S. Pat. 3,642,646 (Feb. 15, 1972), S. H. Marcus (to Standard Oil Co.); U.S. Pat. 3,694,430 (Sept. 26, 1973), J. S. Heckles and E. J. Quinn (to Armstrong Cork Co.).

General References

V. Kollonitsch, *Sucrose Chemicals*, The International Sugar Research Foundation, Inc., Bethesda, Md., 1970.

L. Hough, J. L. Hickson, W. Flavell, and G. L. Redfearn, J. G. Buchanan, and K. J. Parker in J. Yudkin, J. Edelman and L. Hough, eds., *Sugar*, Butterworths, London, 1971, pp. 49–92.

J. L. Hickson, ed., *Sucrochemistry*, A.C.S. Symposium Series 41, American Chemical Society, Washington, D.C., 1977.
J. L. Hickson and R. Khan in G. G. Birch and K. J. Parker, eds., *Sugar: Science and Technology*, Applied Science Publishers, Ltd., London, 1979, pp. 151–210.
R. Khan and A. J. Forage, *Sugar Technology Rev.* **7**, 175 (1979/80).
R. Khan, *Adv. Carbohydr. Chem. Biochem.* **33**, 235 (1976).
M. R. Jenner in C. K. Lee, ed., *Developments in Food Carbohydrate-2*, Applied Science Publishers, Ltd., London, 1980, pp. 91–143.
The properties of sucrose derivatives are comprehensively summarized in tabulated form in these reviews.

K. J. PARKER
Tate & Lyle, Ltd.

SUGAR ECONOMICS

Sugar (sucrose [*57-50-1*]) is produced primarily from sugarcane, which is grown in tropical and subtropical climates, and from sugar beets, which are grown in more temperate climates. Sugarcane originated in the islands of the South Pacific and subsequently spread to southeastern Asia, India, China, the Philippines, and Hawaii.

The beet-sugar industry was established and developed in Western Europe comparatively recently. In 1747, a German chemist, Andreas Marggraf, proved that sugar produced from cane and beets was identical. Despite his findings, it was not until 1799 that the first factory was established for the production of beet sugar. The production of beet sugar in Europe soon spread from France and Germany into other continental countries, including Italy, Holland, Austria, and Russia. Early attempts to establish a U.S. beet-sugar industry in the mid-1800s were short-lived. The first successful beet-sugar factory in the United States was not established until after the Civil War.

More than 80 countries produce ca 97×10^6 metric tons of sugar in a cane-to-beet ratio of ca 1.5:1 (see Tables 1 and 2). Sugar has a long history of production, import, export, and price restrictions. Efforts are made, as indicated in the International Sugar Agreement negotiated in 1977, to keep the world price of sugar at $0.286–0.506/kg through export quotas, buffer stocks, and various other mechanisms (2). Internal sugar prices and production in many countries are controlled. For example, in the United States when sugar prices are low, import duties, fees, and quotas are imposed to keep

Table 1. World Cane and Beet-Sugar Production, 10^3 t [a,b]

Year	Cane	Beet	Total production
1909–1910	8,042	6,648	14,690
1929–1930	17,381	9,359	26,740
1949–1950	18,307	10,695	29,002
1969–1970	43,265	29,727	72,992
1974–1975	50,138	33,931	84,069
1979–1980	54,493	32,703	87,196
1980–1981	58,639	38,146	96,785

[a] Ref. 1.
[b] Figures expressed in terms of raw value (raw sugar to white sugar in the proportion of 100:92 and on a crop-year basis ending Aug. 1981). These values are to be distinguished from those listed in Table 3, which are given on a statistical-year basis.

Table 2. 1980–1981 Worldwide Production, Imports, Exports, and Consumption of Sugar, 10^3 t[a,b]

Country	Production		Imports	Exports	Consumption[c]
	Beet	Cane			
West Europe					
Belgium-Luxembourg	870		35	610	390
Denmark	464		1	270	250
France	4253		365	2610	2135
FRG	2982		165	1000	2235
Greece	230		165	0	340
Ireland	161		50	45	170
Italy	1945		200	300	1773
Netherlands	951		35	290	740
UK	1202		1300	120	2480
Austria	456		0	95	375
Finland	123		170	70	225
Iceland	0		11	0	11
Malta, Gibraltar, Cyprus	0		42	0	42
Norway	0		175	0	175
Portugal	10		363	0	357
Spain	997	3	0	0	1130
Sweden	327		45	30	355
Switzerland	105		165	2	270
Turkey	935		340	4	1270
Yugoslavia	728		110	65	790
East Europe					
Albania	40		0	0	40
Bulgaria	230		205	0	455
Czechoslovakia	830		80	170	740
GDR	600		300	70	830
Hungary	480		100	20	550
Poland	1130		255	35	1380
Romania	570		110	0	680
USSR	7000		5550	90	12400
North and Central America					
Barbados		110	0	100	16
Belize		104	0	95	9
Canada	89		875	25	975
Costa Rica		194	0	95	139
Cuba		6500	0	5950	540
Dominican Republic		1100	0	950	220
El Salvador		175	0	30	140
Guadeloupe		95	0	85	10
Guatemala		437	0	245	235
Hawaii		940	0	950	38
Honduras		209	0	95	115
Jamaica		220	12	115	115
Mexico		2550	640	0	3225
Nicaragua		209	0	85	120
Panama		200	0	135	70
Puerto Rico		159	0	100	95
Trinidad		125	15	85	55
other United States	2643	1544	4700	400	8800
other countries		113	35	50	102
South America					
Argentina		1700	0	700	1070
Bolivia		275	0	100	148

Table 2 (*continued*)

Country	Production Beet	Production Cane	Imports	Exports	Consumption[c]
Brazil		8600	0	2400	6550
Chile	205		253	25	450
Colombia		1300	0	240	1025
Ecuador		360	0	60	305
Guyana		300	0	260	35
Paraguay		95	0	20	73
Peru		520	140	0	610
Surinam		8	3	1	10
Uruguay	37	43	30	0	105
Venezuela		300	380	0	700
Africa					
Algeria	10		485	0	500
Egypt		679	650	0	1330
Ethiopia		166	0	15	160
Kenya		453	0	130	320
Madagascar		116	0	25	100
Malawi		160	0	115	43
Mauritius		565	0	530	42
Morocco	378	42	280	0	680
Mozambique		225	0	65	160
Reunion		260	0	230	16
South Africa		1800	0	600	1320
Sudan		315	105	0	410
Swaziland		365	0	335	23
Tanzania		130	20	10	140
Tunisia	9		205	29	185
other countries		1042	1192	260	1972
Asia					
Bangladesh		155	0	0	173
The Peoples' Republic of China	680	2300	975	110	3970
Hong Kong		0	120	25	105
Indonesia		1430	350	0	1733
India		5650	53	3	5600
Iran	50	130	500	0	900
Iraq	10	15	330	0	440
Japan	576	227	1650	55	2700
Malaysia		65	470	20	535
Pakistan	38	890	14	0	880
Philippines		2435	0	1480	1260
Singapore		0	125	10	115
Republic of Korea		0	675	150	520
Taiwan		758	0	340	450
Thailand		1650	85	1025	650
other countries		185	1857	0	2032
Oceania					
Australia		3500	0	2765	800
Fiji		425	0	390	37
New Zealand		0	145	0	160
other countries		0	42	0	42
World	*32,344*	*50,996*	*27,748*	*28,009*	*89,191*

[a] Ref. 1.
[b] All figures given in terms of raw value and on a statistical-year basis ending August 31.
[c] Includes consumption of stock from previous year's production.

Table 3. World Cane and Beet-Sugar Production, Consumption, and Prices[a,b]

	(a) Initial stocks, 10^3 t	(b) Production, 10^3 t	(c) Consumption, 10^3 t	(d) Final stocks, 10^3 t	d/c, %	Price[c], $/kg
1967–1968	18,907	65,626	64,492	19,636	30.45	0.0570–0.0297
1968–1969	19,636	66,828	66,912	19,157	28.63	0.0609–0.0319
1969–1970	19,157	72,981	70,590	21,124	29.92	0.0869–0.0629
1970–1971	21,124	71,030	72,760	18,751	25.77	0.0920–0.0686
1971–1972	18,751	72,176	74,333	16,890	22.72	0.1309–0.0880
1972–1973	17,853	75,550	76,354	16,398	21.48	0.2009–0.1228
1973–1974	16,398	79,932	78,939	16,122	20.42	0.2603–0.1956
1974–1975	16,122	78,263	76,355	17,470	22.88	1.2459–0.3370
1975–1976	17,470	81,581	78,881	20,478	25.96	0.8433–0.2924
1976–1977	20,478	86,950	81,358	25,206	30.98	0.3282–0.1657
1977–1978	25,206	91,155	85,258	30,634	35.93	0.2209–0.1555
1978–1979	30,634	90,990	89,667	31,353	34.97	0.1971–0.1415
1979–1980	31,353	84,870	89,608	25,921	28.93	0.3285–0.1665
1980–1981[d]	25,921	88,157	88,697	24,908	28.08	0.9040–0.3791
1981–1982[e]	24,908	95,431	91,015	29,315	32.21	0.6162–0.2565

[a] Refs. 2–3.
[b] Figures expressed in terms of raw value and (except for prices) on a statistical-year basis (as compared to those given in Table 1, which are given on a crop-year basis) ending Aug. 1981. For simplicity, imports and exports, which are roughly equal each year, have been omitted. Thus, (a) + (b) − (c) approximates (d).
[c] Fob Caribbean ports, calendar-year basis; ranges based on monthly averages.
[d] Preliminary.
[e] Estimated.

the U.S. raw sugar price above $0.37/kg, thus helping domestic cane and beet growers remain in business.

Most sugar in the world is traded through private arrangements between sellers and buyers. Only ca 15–20% is sold on the free market. Free-market sugar prices exhibit a cyclical pattern, with the ratio of final stocks to consumption being the single best indicator of the future direction of prices (see Table 3). Though there is a strong inverse correlation between the ratio and prices, prediction of price levels from the ratio is difficult. Econometric models, no matter how complex, invariably fail to indicate future sugar prices because too many variables, eg, market psychology, political turmoil, and speculative activity, cannot be accurately quantified. The typical price cycle includes a period of a few years of low prices followed by a relatively short period of high prices and then a return to low prices.

Two factors recently have begun to affect the sugar market and are likely to play increasingly bigger roles. In some areas, notably the European Economic Community (EEC), Japan, Canada, and the United States, corn syrups containing high percentages of fructose are being used to replace sugar in many industrial uses. By 1985, high fructose corn syrups could replace ca 4% of the world's sugar consumption.

The recent global energy crisis has led many countries to consider producing ethanol from biomass, eg, sugarcane (see Fuels from biomass). Brazil is the leader in this area and processes cane into alcohol or sugar. The economics of the processes depend upon a complex formula, which includes the relationship of world sugar and oil prices. If the trend of producing alcohol from sugarcane continues, the price swings of the world sugar market could be dampened by a switch to alcohol production when sugar prices are low, thus removing some excess sugar from the market, and a swing back to more sugar production when sugar prices start to rise.

Qualitatively, crystalline fructose [*57-48-7*] could compete with sucrose; however, the costs of producing fructose are prohibitive. Thus, its competitiveness will be determined by associated technology and processing costs.

BIBLIOGRAPHY

"Sugar Economics" in *ECT* 2nd ed., Vol. 19, pp. 233–236, by J. L. Hickson, International Sugar Research Foundation, Inc.

1. F. O. Licht, *World Sugar Statistics, 1980–81*, Ratzeburg, FRG, Sept. 1981.
2. International Sugar Agreement, 1977, negotiated in Geneva in 1977 under the auspices of the United Nations Conference on Trade and Development, ratified by members of the International Sugar Organization, London, England, and entered into force Jan. 1978.
3. *World Sugar Statistics*, F. O. Licht, Ratzeburg, FRG, various issues.
4. *Sugar Year Book*, International Sugar Organization, London, various issues.

General References

J. L. Hickson, *Sucrochemistry*, American Chemical Society, Washington, D.C., 1977.

R. H. Cottrell, *Beet-Sugar Economics*, Caxton Printers, Ltd., Caldwell, Idaho, 1952.

G. P. Meade and J. C. P. Chen, *Cane Sugar Handbook*, John Wiley & Sons, Inc., New York, 1977.

V. Kollonitsch, *Sucrose Chemicals*, The International Sugar Research Foundation (now World Sugar Research Organization), W. S. Cowell, Ltd., Ipswich, UK, 1970.

Sweeteners: Issues and Uncertainties, National Academy of Sciences, Washington, D.C., 1975.

International Sugar Economic Year Book and Directory, F. O. Licht, Ratzeburg, FRG, published yearly.

D. Smith, *Cane Sugar World*, Palmer Publications, New York, 1978.

KIM BADENHOP
B. W. Dyer & Co.

SPECIAL SUGARS

Fructose

D-Fructose [*57-48-7; 30237-26-4*] (fructose, levulose, fruit sugar) is a monosaccharide constituting one half of the sucrose molecule. It was first isolated from hydrolyzed cane sugar (invert sugar) over 100 years ago (1–2). Fructose is one of the sweeteners in fruits and is primarily mixed with glucose (dextrose) and sucrose (3) (see Carbohydrates; Sweeteners). It also occurs in honey with glucose (4). Despite this ubiquity, fructose has until recently remained a noncommercial product because of the expense involved in its isolation. The development of technologies for preparing fructose from glucose in the isomerized mixture led to a greater availability of pure, crystalline fructose in the 1970s (5–7). The price for pure fructose is high enough so that the product was not competitive in 1981 with sucrose and the corn syrups as a commercial sweetener (see Syrups). However, because of certain recognized short-term physiological effects, fructose is used as a sweetener for special dietary purposes.

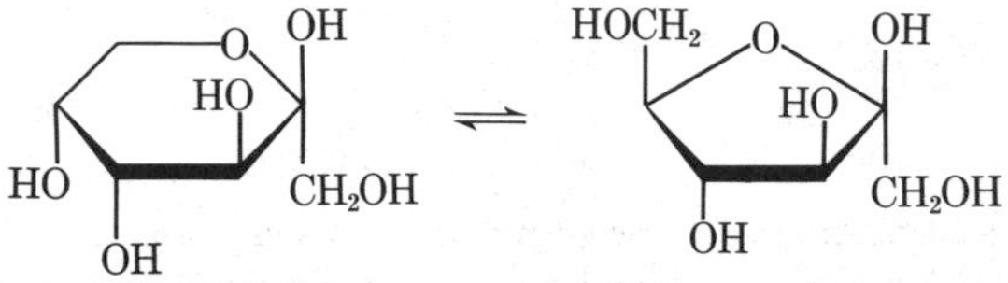

β-D-fructopyranose ⇌ β-D-fructofuranose

β-D-fructose

Pure D-fructose is a white, hygroscopic, crystalline substance and is not to be confused with the high fructose corn syrups which may contain 42–90 wt % fructose. The nonfructose part of these syrups is glucose plus small amounts of glucose polymers. Fructose is highly soluble in water. At 20°C, it is 79% soluble as compared with 47% for glucose and 67% for sucrose.

The sweetness of fructose is 1.3–1.8 times that of sucrose (8). This property makes fructose attractive as an alternative for sucrose and other commercially available sweeteners. As with other properties, the sweetness of a given substance depends on the conditions under which it is used, and it is unwise to assume that the sweetness determined for dilute solutions of a pure material is maintained when the sweetener is used as an ingredient in a food or diet. Fructose is probably sweetest when cold and freshly made up in low concentrations at a slightly acidic pH (5). Part of this state results from the difference in the fructose structure when cold (fructopyranose, sweet) as compared to that when the sugar is warm (fructofuranose, less sweet). Fructose, as well as other sweeteners, must be incorporated into a given formulation so that its contribution to the sweetness and other properties of the product can be evaluated.

Fructose is a highly reactive substance. When stored in solution at high temperatures, fructose browns rapidly as well as polymerizes to dianhydrides [*38837-99-9, 50692-21-2, 50692-22-3, 50692-23-4, 50692-24-5*]. Fructose also reacts very rapidly with amines and proteins in the Maillard or browning reaction (5). An appreciation of these properties allows the judicious choice of conditions under which fructose can be used successfully in food applications.

Because of its relatively high degree of sweetness, fructose has been the object of commercial production for decades. Isolation of fructose from either hydrolyzed sucrose or hydrolyzed fructose polymers, eg, inulin (Jerusalem artichoke), has not proved economically competitive with the very low price for sucrose processed from sugarcane or sugar beets (see Alcohols, polyhydric).

Commercial quantities of crystalline fructose became available when the Finnish Sugar Company developed ion-exchange methods first for hydrolyzing sucrose and then for separating the hydrolysate into the constituents, ie, glucose and fructose. The latter step involves the calcium form of a sulfonated-polystyrene exchange resin. Further economies in production were realized when the same company developed a method for crystallizing fructose from an aqueous rather than a water–alcohol solution (5).

More recent technologies also involve ion-exchange separation of fructose from glucose in a mixture obtained by the isomerization of glucose by means of immobilized glucose isomerase or microbial cells containing the enzyme (7). This technique is used by Hoffmann-La Roche at their plant in Thomson, Illinois. Another procedure for making crystalline fructose is detailed in ref. 9. Basically, glucose (dextrose) is oxidized by glucose-2-oxidase to glucosone which is then selectively hydrogenated to fructose. This procedure has the advantage of not requiring isomer separation in order to isolate the crystalline product. However, it will be a number of years before the process is used commercially. At such a time, the inventors claim that fructose will be available at prices competitive with sucrose from sugarcane or sugar beets (see Sugar, sugar economics).

Commercially available crystalline fructose in Europe cost ca $17.6/kg in 1968–1969 with ca 7500 metric tons marketed. Ten years later, the amount of product available was ca 20,000 t, and the cost had dropped to ca $8.80/kg. In the United States in 1981, the wholesale price is just under $2.20/kg. The lower price in the United States can be traced to the startup of the new Hoffmann-LaRoche fructose plant in Thomson, Illinois (10). At the retail level in 1981, crystalline fructose cost almost 10 times as much as sugar (sucrose). Thus, economically, crystalline fructose will not be competitive with other commercially available sweeteners unless there is a breakthrough in processing that allows a significant decrease in price.

Fructose has been successfully incorporated into formulas for the preparation of jams, jellies, preserves–marmalades, baked goods, cake mixes, puddings, dry beverage powders, gelatin desserts, tabletop sweeteners, ice cream, candies, and yogurt (see Food processing). In food formulations, by far the most prevalent use for fructose is in the preparation of foods for special dietary purposes (2,5).

Such specialty foods include dry beverage mixes, candies, and dry cake and cookie mixes. The special dietary purposes are realized because fructose has a high sweetening power that theoretically allows consumption of less sweetener, ie, fewer food calories, in a fructose-based foodstuff compared with the same food made with sucrose. However, substituting fructose for other sweeteners has not yet been shown to be an effective method of weight control or reduction.

A further advantage of fructose in foods is based on fructose not promoting as high a rise in blood insulin as sucrose. Fructose-based foods, then, may be helpful in the diets of diabetics and others who find it necessary to watch their blood-sugar levels carefully. Again, caution should be exerted in using fructose for this purpose, as long-range studies of the effects of fructose feeding in humans have not been conducted (11–13).

Maltose

Maltose [*69-79-4*] (malt sugar) occurs occasionally in plants and fruits (14–15). It is much more frequently recognized as a structural component of starch. Pure maltose is isolated with difficulty from a directed starch hydrolysate, ie, high maltose corn syrup, by precipitation with ethanol. Purification can be achieved by way of the β-maltose octaacetate. Removal of the acetate groups allows crystallization of the monohydrate of β-maltose. Commercial maltose typically contains 5–6 wt % of the trisaccharide maltotriose with traces of glucose (16). Maltose is about half as sweet as sucrose and, in the pure state, has been used in intravenous feeding in Europe and Japan. Energy from maltose used in this manner becomes accessible to the body at a slower rate than energy supplied by intravenous feeding of monosaccharides, eg, glucose (17).

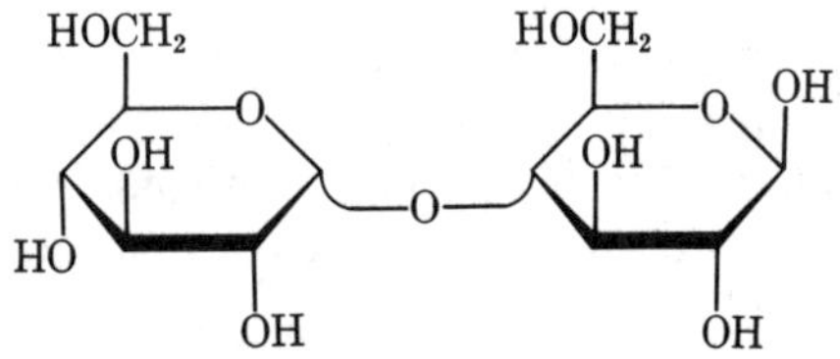

4-*O*-α-D-glucopyranosyl-D-glucose
maltose

The maltose in malt syrups is important in brewing (see Beer). High maltose syrups from starch typically contain ca 8–9 wt % glucose, 40–52 wt % maltose, and the remainder as higher saccharides (18–19). Such syrups are used in the preparation of confections, preserves, and other foodstuffs.

Hydrogenation of high maltose syrups gives a mixture of sugar alcohols, from which maltitol [*585-88-6*] can be isolated in crystalline form. Maltitol is almost as sweet as sucrose (0.9 times) and has been promoted as a sweetener in various food applications (20).

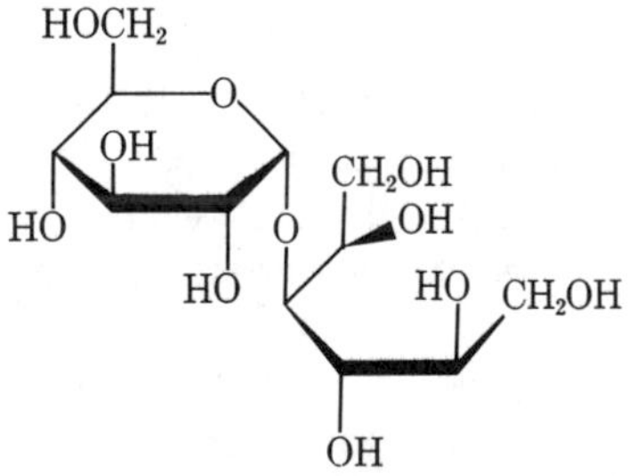

maltitol

Lactose

Lactose [*63-42-3*] (milk sugar) is the only sugar available commercially that is derived from animal rather than plant sources. The concentration of lactose in milk products ranges from 4.8 wt % in whole milk to 73.5 wt % in sweet dried whey, a by-product from the manufacture of cheese (21). There have been reports of the presence of lactose in plant materials, eg, sapote and acacia, but this has not been confirmed (22–23).

4-*O*-β-D-galactopyranosyl-D-glucose
β lactose

Lactose is prepared commercially from whey as the crystalline α-monohydrate. It is available in varying degrees of purity. The fermentation grade is 98 wt % pure, whereas USP lactose is refined to 99.8 wt % purity (24). Although the α-monohydrate is the commercially available form of lactose, the sugar can be crystallized at high temperature to give the β-anhydride [*56907-28-9*]. The sugar is not very soluble in water (ca 20 g/100 g water at 25°C), nor is it very sweet (less than one third the sweetening power of sucrose). Typically, lactose, which is a reducing sugar, reacts with amines and amino compounds with resultant browning.

The amount of lactose consumed in the U.S. diet is ca 25 g/d per capita (25). The commercial production of lactose, as distinguished from whey products, in 1980 was ca 64,000 metric tons. Approximately 16% of the product was used in animal feeds; the remainder was produced for human use (26). In 1979, ca 32,000 t of lactose were sold domestically. Of this total, 20,500 t were used in infant food formulations, 4500 t were used in pharmaceuticals, 3200 t were used in dairy products (see Milk and milk products), and the rest was used in dietetic foods and miscellaneous applications (26).

Food applications for lactose have been explored in great depth, both at the University of California and the Eastern Regional Laboratories of the USDA (27). Lactose has been used as a nutritional sugar, flavor enhancer, and texture controller and color retainer in dairy beverages, fruit and vegetable juices, confections, infant formulas, and bakery products, eg, doughnuts and cookies. It has also been used as a carrier for synthetic sweeteners (27). Lactose, and the lactose in substances such as whey, have been hydrolyzed commercially by enzymes to yield products that can be tolerated physiologically much more easily by people who have a lactose intolerance (28–30).

BIBLIOGRAPHY

"Commercial Sugars" in *ECT* 1st ed., Vol. 13, pp. 244–251, by J. L. Hickson, Sugar Research Foundation; "Special Sugars" in *ECT* 2nd ed., Vol. 19, pp. 237–242, by J. L. Hickson, International Sugar Research Foundation, Inc.

1. C. P. Barry and J. Honeyman, *Adv. Carbohydr. Chem.* **7,** 53 (1952).
2. T. E. Doty and E. Vanninen, *Food Technol. (Chicago)* **26,** 24 (1975).
3. R. E. Wrolstad and R. S. Shallenberger, *J. Assoc. Off. Anal. Chem.* **64,** 91 (1981).
4. J. W. White, Jr. in E. Crane, ed., *Honey, A Comprehensive Survey*, Crane, Russak & Co., Inc., New York, 1975, p. 157.
5. E. Vanninen and T. Doty in G. G. Birch and K. J. Parker, eds., *Sugar: Science and Technology*, Applied Science Publications Ltd., London, 1979, p. 311.
6. T. Doty in P. Koivistoinen and L. Hyvönen, eds., *Carbohydrate Sweetners in Foods and Nutrition*, Academic Press, New York, 1980, p. 259.
7. W. P. Chen, *Process Biochem.* **15,** 36 (1980).
8. R. S. Shallenberger, *J. Food Sci.* **28,** 584 (1963).
9. Eur. Pat. Application (Nov. 6, 1979), S. L. Neidelman and W. F. Amon (to Cetus Corporation).
10. *Chem. Week*, 49 (Oct. 7, 1981).
11. P. A. Crapo and J. M. Olefsky, *Nutr. Today* **15,** 10 (1980).
12. P. A. Crapo, O. G. Kolterman, and J. M. Olefsky, *Diabetes Care* **3,** 575 (1980).
13. K. K. Kimura and C. J. Carr, *Dietary Sugars in Health and Disease, I. Fructose, P.B. 262–764*, National Technical Information Service, U.S. Department of Commerce, 1976.
14. J. H. Pazur in W. Pigman and D. Horton, eds., *The Carbohydrates*, 2nd ed., Vol. IIA, Academic Press, New York, 1970, p. 107.
15. R. S. Shallenberger in H. L. Sipple and K. W. McNutt, eds., *Sugars in Nutrition*, Academic Press, New York, 1974, p. 74.
16. *Maltose Monohydrate*, ·NRC, Catalog No. M-105, Pfanstiehl Laboratories, Inc., Waukegan, Ill., 1981.
17. I. MacDonald in ref. 15, p. 310.
18. H. M. Pancoast and W. R. Junk, *Handbook of Sugars*, 2nd ed., Avi Publishing Co., Inc., Westport, Conn., 1980, p. 178.
19. S. M. Cantor in ref. 15, p. 116.
20. C. A. M. Hough in C. A. M. Hough, K. J. Parker, and A. J. Vlitos, eds., *Developments in Sweeteners—1*, Applied Science Publishers, Ltd., London, 1979, p. 75.
21. T. A. Nickerson, *Food Technol.* **32,** 40 (1978).
22. J. H. Pazur in ref. 14, p. 104.
23. R. S. Shallenberger in ref. 15, p. 71.
24. *Lactose*, Foremost Foods Co., Foremost-McKesson, Inc., San Francisco, Calif., 1970.
25. *National Food Situation*, U.S. Department of Agriculture, Economic Research Service, Washington, D.C., 1974, p. 32.
26. *Statistical Reports*, Whey Products Institute, Chicago, Ill., 1980.
27. *Food Eng.* **50,** 154 (1978).
28. N. S. Shah and T. A. Nickerson, *J. Food Sci.* **439,** 1575 (1978).
29. V. H. Holsinger, *Food Technol. (Chicago)* **32,** 35 (1978).
30. Ref. 18, p. 368.

General References

W. Pigman and D. Horton, eds., *The Carbohydrates*, Vol. IA, 2nd ed., Academic Press, New York, 1970.

H. L. Sipple and K. W. McNutt, eds., *Sugars in Nutrition*, Academic Press, New York, 1974.

G. G. Birch and K. J. Parker, eds., *Sugar: Science and Technology*, Applied Science Publishers, Ltd., London, 1979.

P. Koivistoninen and L. Hyvönen, eds., *Carbohydrate Sweeteners in Foods and Nutrition*, Academic Press, New York, 1980.

C. A. M. Hough, K. J. Parker, and A. J. Vlitos, eds., *Developments in Sweeteners—1*, Applied Science Publications, Ltd., London, 1979.

G. N. BOLLENBACK
The Sugar Association, Inc.

SULFA DRUGS. See Antibacterial agents, synthetic, sulfonamides.

SULFAMATES. See Sulfamic acid.

SULFAMIC ACID AND SULFAMATES

Sulfamic acid [*5329-14-6*] (HSO_3NH_2, amidosulfuric acid) is a monobasic, inorganic, dry acid and the monoamide of sulfuric acid. Sulfamic acid is produced and sold in the form of water-soluble crystals and granules. This acid was known and prepared in laboratories for nearly a hundred years before it became a commercially available product. It was first prepared from the reaction of lead imidosulfonate and hydrogen sulfide (1). Later work resulted in identification and preparation of sulfamic acid in its pure form (2). In 1936, a new and practical process was developed which shortly thereafter became the basis for its commercial preparation (3–4). This process involves the reaction of urea with sulfur trioxide and sulfuric acid; it continues to be the main method for production of sulfamic acid.

Sulfamic acid has a unique combination of properties which makes it particularly well-suited for scale removing and chemical cleaning operations. These are the main commercial applications, but sulfamic acid also is used in sulfation reactions, pH adjustment, and a variety of chemical processing applications. Salts of sulfamic acid are used in electroplating and electroforming operations as well as for manufacturing flame retardants and weed and brush killers.

Properties

Sulfamic Acid. ***Physical.*** Sulfamic acid is a dry acid formed in orthorhombic crystals. The acid is highly stable and may be kept for years without change in properties. The pure crystals are nonvolatile, nonhygroscopic, and odorless. Shipment of sulfamic acid does not require a DOT hazard label. Selected physical properties of sulfamic acid are listed in Table 1; other properties are given in refs. 5–8.

Chemical Properties. Selected chemical properties of sulfamic acid are listed in Table 2; other properties are listed in ref. 9.

Although the acid is relatively strong, corrosion rates are low in comparison to other acids, as shown in Table 3. The low corrosion rate can be reduced further by addition of corrosion inhibitors.

Inorganic Reactions. Sulfamic acid is highly stable up to its melting point. Thermal decomposition begins at 209°C; at 260°C, it produces sulfur dioxide, sulfur trioxide, nitrogen, water, and traces of other products, chiefly nitrogen compounds. Aqueous sulfamic acid solutions are quite stable at room temperature. At higher temperatures, however, acidic solutions and the ammonium salt hydrolyze to sulfates, with rates increasing rapidly with elevated temperatures, lower pHs, and increased concentrations. These hydrolysis reactions are exothermic. Concentrated solutions

Table 1. Selected Physical Properties of Sulfamic Acid

Property	Value
mol wt	97.09
mp, °C	205
decomposition temperature, °C	209
density (at 25°C), g/cm^3	2.126
refractive indexes, 25 ± 3°C	
α	1.553
β	1.563
γ	1.568
solubility, wt %	
in water, °C	
0	12.80
20	17.57
40	22.77
60	27.06
80	32.01
in formamide, °C	
25	0.1667
in methanol, °C	
25	0.0412
in ethanol (2% benzene), °C	
25	0.0167
in acetone, °C	
25	0.0040
in ether, °C	
25	0.0001

Table 2. Selected Chemical Properties of Sulfamic Acid

Property	Value
dissociation constant (at 25°C)	0.101
heat of formation, kJ/mol[a]	−685.9
heat of solution, kJ/mol[a]	19.0
pH of aqueous solutions (at 25°C)	
normality	
1.00	0.41
0.75	0.50
0.50	0.63
0.25	0.87
0.10	1.18
0.05	1.41
0.01	2.02

[a] To convert J to cal, divide by 4.184.

that are heated in closed containers or in vessels with insufficient venting can generate sufficient internal pressure to cause container rupture. An ammonium sulfamate, 60 wt % aqueous solution, exhibits runaway hydrolysis when heated to 200°C at pH 5 and to 130°C at pH 2. The danger is minimized in a well-vented container, however, since the 60 wt % solution boils at 107°C (8,10). The hydrolysis reactions are.

$$HSO_3NH_2 + H_2O \rightarrow NH_4HSO_4$$

$$NH_4SO_3NH_2 + H_2O \rightarrow (NH_4)_2SO_4$$

Table 3. Relative Corrosion Rates of 3 wt % Aqueous Solutions of Acids at 22 ± 2°C; Sulfamic Acid = 1.0

Metal	H_2SO_4	HCl
1010 steel	2.6	4.2
cast iron	3.2	3.2
galvanized iron	63.0	very rapid corrosion
tin plate	81.0	23.0
304 stainless	10.0	very rapid corrosion
zinc	2.2	very rapid corrosion
copper	1.5	6.7
brass	1.5	2.8
bronze	4.0	7.0
aluminum	0.6	5.3

Alkali metal sulfamates are stable in neutral or alkaline solutions even at boiling temperatures. Rates of hydrolysis for sulfamic acid in aqueous solutions have been measured at different conditions (8,10); see Table 4.

Sulfamic acid readily forms various metal sulfamates by reaction with the metal or the respective carbonates, oxides, or hydroxides. The ammonium salt is formed by neutralizing the acid with ammonium hydroxide:

$$Zn + 2\ HSO_3NH_2 \rightarrow Zn(SO_3NH_2)_2 + H_2$$

$$CaCO_3 + 2\ HSO_3NH_2 \rightarrow Ca(SO_3NH_2)_2 + H_2O + CO_2$$

$$FeO + 2\ HSO_3NH_2 \rightarrow Fe(SO_3NH_2)_2 + H_2O$$

$$Ni(OH)_2 + 2\ HSO_3NH_2 \rightarrow Ni(SO_3NH_2)_2 + H_2O$$

$$NH_4OH + HSO_3NH_2 \rightarrow NH_4SO_3NH_2 + H_2O$$

Nitrous acid reacts very rapidly and quantitatively with sulfamic acid:

$$HSO_3NH_2 + HNO_2 \rightarrow H_2SO_4 + H_2O + N_2$$

This reaction can be used for the quantitative analysis of nitrites (3,11). The reaction with concentrated nitric acid gives nitrous oxide (12–13):

$$HSO_3NH_2 + HNO_3 \rightarrow H_2SO_4 + H_2O + N_2O$$

Chlorine, bromine, and chlorates oxidize sulfamic acid to sulfuric acid and to nitrogen (1,14):

$$2\ HSO_3NH_2 + KClO_3 \rightarrow 2\ H_2SO_4 + N_2 + KCl + H_2O$$

Table 4. Hydrolysis of Sulfamic Acid Aqueous Solutions at 80°C

	Wt % of HSO_3NH_2 originally present that is lost to hydrolysis		
Time, h	1% solution	10% solution	30% solution
1	4.5	7.8	7.9
2	9.1	15.1	15.1
3.1	13.3		22.0
4.2		22.7	
5	16.9	28.3	27.5
6	20.6	34.3	32.8
7	24.2	39.5	37.5
8	27.3	43.7	

Chromic acid, permanganic acid, and ferric chloride do not oxidize sulfamic acid.

Hypochlorous acid reacts at low temperatures to form *N*-chlorosulfamic and *N,N*-dichlorosulfamic acids:

$$HSO_3NH_2 + HOCl \rightarrow \underset{[17172\text{-}27\text{-}9]}{HSO_3NHCl} + H_2O$$

$$HSO_3NH_2 + 2\,HOCl \rightarrow \underset{[17085\text{-}87\text{-}9]}{HSO_3NCl_2} + 2\,H_2O$$

A similar reaction occurs when sodium is present with the formation of sodium *N*-chlorosulfamate [*13637-90-6*] and sodium *N,N*-dichlorosulfamate [*13637-67-7*]:

$$NaSO_3NH_2 + 2\,NaOCl \rightarrow NaSO_3NCl_2 + 2\,NaOH$$

Sulfamoyl chloride [*7778-42-9*] forms from the reaction with thionyl chloride (15–16):

$$HSO_3NH_2 + SOCl_2 \rightarrow NH_2SO_2Cl + HCl + SO_2$$

Certain metal iodides, eg, of sodium, potassium, cesium, or rubidium, react with sulfamic acid to form the triodides (17):

$$2\,HSO_3NH_2 + 3\,CsI + H_2O \rightarrow CsI_3 + Cs_2SO_4 + (NH_4)_2SO_3$$

An exception exists to the monobasic nature of sulfamic acid when it dissolves in liquid ammonia. Sodium, potassium, etc add both to the amido and sulfonic portions of the molecule to give salts, eg, $NaSO_3NHNa$ [*83930-11-4*], etc.

Sodium sulfate and sulfamic acid form the complex:

$$6HSO_3NH_2.5Na_2SO_4.15H_2O$$

Sulfamic acid and its salts retard the precipitation of barium sulfate and prevent precipitation of silver and mercury salts by alkali. It has been suggested that salts of the type $AgNHSO_3K$ [*15293-60-4*] form with mercury, gold, and silver (18). Upon heating of such solutions, the metal deposits slowly in mirror form on the wall of a glass container.

Studies of chemical and electrochemical behavior of various metals in sulfamic acid solutions are described in ref. 19.

Organic Reactions. Primary alcohols react with sulfamic acid to form alkyl ammonium sulfate salts (20–22):

$$ROH + HSO_3NH_2 \rightarrow ROSO_2ONH_4$$

Sulfation by sulfamic acid has been used, especially in the FRG, in the preparation of detergents from dodecyl, oleyl, and other higher alcohols. It is also used in sulfating phenols and phenol–ethylene oxide condensation products. Secondary alcohols react in the presence of an amide catalyst, eg, acetamide or urea (23). Pyridine has also been used. Tertiary alcohols do not react. Phenols are considered to react, yielding phenyl ammonium sulfates. These reactions include those of naphthols, cresol, anisole, anethole, pyrocatechol, resorcinol, and hydroquinone. Ammonium aryl sulfates form as intermediates and sulfonates form by subsequent rearrangement (24–25). Studies of sulfoesterification of higher alcohols with sulfamic acid are described in ref. 25.

Amides react in certain cases to form ammonium salts of sulfonated amides (21). For example, treatment with benzamide yields ammonium *N*-benzoylsulfamate [*83930-12-5*], $C_6H_5CONHSO_3^-NH_4^+$, and treatment with ammonium sulfamate yields diammonium imidodisulfonate [*13597-84-1*], $HN(SO_2ONH_4)_2$. Ammonium sulfamate or sulfamic acid and ammonium carbonate dehydrate liquid or solid amides to nitriles (qv) (27).

Primary, secondary, and tertiary amines react with sulfamic acid to form amine sulfamates (28):

$$HSO_3NH_2 + RNH_2 \rightarrow RNH_2.HOSO_2NH_2$$

Guanidine salts can be prepared by reaction of thiocyanates and sulfamates (21).

Aldehydes form addition products with sulfamic acid salts. These are stable in neutral or slightly alkaline solutions but are hydrolyzed in acid and strongly alkaline solutions. With formaldehyde, the calcium salt of the methylol (hydroxymethyl) derivative [*82770-57-8*], $Ca(O_3SNHCH_2OH)_2$, is obtained as a crystalline solid.

Cadmium, cobalt, copper, and nickel sulfamates react with lower aliphatic aldehydes. These stable compositions are suitable for use in electroplating solutions for deposition of the respective metal (see Electroplating).

Fatty acid acyl halides react with sulfamic acid (30). Unsaturated compounds, in general, do not react readily and yields are low.

The *N*-alkyl and *N*-cyclohexyl derivatives of sulfamic acid are comparatively stable; the *N*-aryl derivatives are very unstable and can only be isolated in salt form. A series of thiazolylsulfamic acids has been prepared.

Cellulose sulfated with sulfamic acid degrades less than if sulfated with sulfuric acid (22). Cellulose esters of sulfamic acids are formed by the reaction of sulfamyl halides in the presence of tertiary organic bases (see Cellulose derivatives).

Other organic reactions of sulfamic acid are described in refs. 31 and 32.

Sulfamates. Sulfamates are formed readily by the reaction of sulfamic acid with the appropriate metal or its oxide, hydroxide, or carbonates. Approximations of heats of neutralization are −54.61 kJ/mol (−13.05 kcal/mol) for the NaOH reaction and −47.83 kJ/mol (−11.43 kcal/mol) for NH_4OH at 26–30°C. Sulfamates prepared from weak bases form acidic solutions, whereas those prepared from strong bases produce neutral solutions. The pH of 5 wt % solution of ammonium sulfamate is 5.2. Crystals of ammonium sulfamate become deliquescent at relative humidities of 70% and higher. Both the ammonium and potassium sulfamates [*7773-06-0*, *13823-50-2*] liberate ammonia at elevated temperatures and form the corresponding imidodisulfonates (12). The inorganic sulfamates are quite water-soluble except for the basic mercury salt. Some relative solubilities of sulfamates at 25°C in water are: ammonium sulfamate, 103 g/100 g; sodium sulfamate, 106 g/100 g; magnesium sulfamate, 119 g/100 g; calcium sulfamate, 67 g/100 g; barium sulfamate, 34.2 g/100 g; zinc sulfamate, 115 g/100 g; and lead sulfamate, 218 g/100 g. Literature references for properties of a number of sulfamates are listed in Table 5.

Manufacture

Sulfamic acid is manufactured by the reaction of urea, sulfur trioxide, and sulfuric acid (3–4,45). The exothermic reaction is considered to take place in two steps:

$$NH_2CONH_2 + SO_3 \rightarrow [HSO_3NHCONH_2]$$

(aminocarbonylsulfamic acid intermediate)

$$[HSO_3NHCONH_2] + H_2SO_4 \rightarrow 2\ HSO_3NH_2 + CO_2$$

Table 5. Literature References for Properties of Sulfamates

Compound	CAS Registry No.	Refs.
ammonium	[7773-06-0]	33
aluminum	[10101-13-0]	2, 33
barium	[13770-86-0]	27, 33–34
cadmium	[14017-36-8]	33
calcium	[13770-92-8]	33
cobalt	[14017-41-5]	33, 35
copper	[14017-38-0]	27, 33, 36–38
iron	[14017-39-1]	12, 33
lead	[32849-69-7]	27, 33, 36
lithium	[21856-68-8]	33
magnesium	[13770-91-7]	33, 43
manganese	[83929-95-7]	33
nickel	[13770-89-3]	33, 37
potassium	[13823-50-2]	33
silver	[14325-99-6]	12, 15, 18, 27, 33–34
sodium	[13845-18-6]	33
strontium	[83929-96-8]	33
thallium	[21856-70-2]	33
uranyl	[82783-83-3]	33
zinc	[13770-90-6]	33
1:1 anilinium	[10310-62-0]	21, 36, 39–41
1:1 amylaminium	[82323-98-6]	40
1:1 benzylaminium	[82323-99-7]	42
1:1 hydroxylaminium	[82324-00-3]	12, 21
1:1 trimethylaminium	[6427-17-4]	43
1:1 hydrazinium	[39935-03-0]	2, 12, 36, 40
α-naphthylaminium	[83929-97-9]	39–40
β-naphthylaminium	[83929-98-0]	
1:1 piperidinium	[82324-01-4]	44
1:1 *p*-toluidinium	[68734-85-0]	39–40

or overall:

$$NH_2CONH_2 + SO_3 + H_2SO_4 \rightarrow 2\ HSO_3NH_2 + CO_2$$

A liquid mixture of equimolar quantities of urea and sulfuric acid is added to a large excess of liquid sulfur trioxide while the mixture is agitated and cooled to prevent the formation of carbon dioxide. After completion of the initial reaction, the mass is heated further to produce sulfamic acid. Excess sulfur trioxide is removed by distillation or by other methods. Other production processes have been reported, including the use of urea and fuming sulfuric acid (46–47).

Sulfamates are produced by reaction of the acid with the appropriate hydroxide or carbonate. The most important commercial salt is ammonium sulfamate. It is made by adding anhydrous ammonia and the acid to ammonium sulfamate mother liquor from a preceding crystallization operation to form a hot concentrated solution. This solution is passed to crystallizers from which ammonium sulfamate crystals are recovered; the mother liquor is recycled (48). The manufacture of ammonium sulfamate also is described in FRG and Japanese patents (49–50).

Shipments and Economic Aspects

Sulfamic acid is produced commercially by a number of companies, including DuPont in the United States, Hoechst in the FRG, Nissan in Japan, and smaller companies in Taiwan and India. The 1981 world production capacity was ca 52,000 metric tons. Although not supplied in bulk, truckloads and carloads of packaged material are available from producers. Less than truckload quantities are readily supplied by a network of chemical distributors. In the United States, sulfamic acid is produced and supplied in crystal and in granular form and is packaged in fiber drums and in bags. In mid-1981, the truckload price for the crystal grade (fob USA) was $0.84/kg and, for the granular grade, it was $0.66/kg.

Ammonium sulfamate is also commercially produced and available in the United States. It is supplied in truckloads or carloads of 45 kg net weight paper bags. Less-than-truckload quantities are supplied through chemical distributors. The commercial grade contains >99 wt % ammonium sulfamate, and the mid-1981 truckload price was $1.10/kg. Other sulfamates are commercially available, primarily through plating specialty sources (51).

Specifications

Sulfamic Acid. Specifications and typical analyses of the commercial technical grades of sulfamic acid available in 1981 are listed in Table 6.

Table 6. Specifications and Analyses of Commercially Available Sulfamic Acid [a]

	Specifications		Typical analyses	
Property	Crystal	Granular	Crystal	Granular
sulfamic acid, wt %	99.5 min	91.0 min	99.7	92.2
sulfates (as SO_4), wt %	0.42 max		0.20	
iron (as Fe), wt %	0.01 max		0.001	
insoluble matter, wt %	0.02 max		0.001	
moisture, wt %	0.10 max		0.04	
ammonium bisulfate, wt %				3.5
sulfuric acid, wt %				3.5
urea, wt %				0.1
magnesium, wt %				0.15
sulfur trioxide, wt %				0.2
color			white	off-white
U.S. sieve series analysis				
particle size, wt %				
2 mm (10 mesh)		1.0 max		trace
1.2–2 mm (10–16 mesh)	3.0 max		0.1	
0.15 mm (100 mesh)		15.0 max		4.2
bulk density, kg/m^3			ca 1200	ca 1120

[a] Values for product manufactured by E. I. du Pont de Nemours & Co. Inc. Typical properties are not guaranteed and all numbers are subject to updating.

Analytical and Test Methods

The determination of sulfamic acid is made by titration of an accurately prepared sulfamic acid solution with sodium nitrite solution. It is based on the reaction

$$HSO_3NH_2 + NaNO_2 \rightarrow NaHSO_4 + H_2O + N_2$$

An automatic titrator is used with a silver-to-platinum polarized electrode system with potassium bromide as the electrometric indicator. Potassium bromide depolarizes the silver anode. At the end point, the first excess of nitrite depolarizes the platinum cathode. The resulting potential surge actuates the delivery valve and stops the titration. Standard 1 *N* nitrite is used to titrate a solution prepared by dissolving ca 22 g of sulfamic acid in 1 L of distilled water. The temperature of the sulfamic acid solution should not be allowed to exceed 25°C and preferably should be lower.

$$\text{wt \% sulfamic acid} = \frac{(\text{mL } N/1 \text{ NaNO}_2 \times 9.710)}{\text{g sample in final aliquot}}$$

This method is considerably more accurate than either the "total acid less impurities" method or the nitrite titration with an outside starch-paste indicator. Both of these methods have been used.

For ammonium sulfamate assay determination, the same procedure is used as for sulfamic acid after the initial addition of excess sulfuric acid to convert the salt to sulfamic acid.

Health and Safety Factors

Sulfamic acid and its solutions cause eye burns and irritate the nose, throat, and skin. Workers should wear cup-type, rubber, or soft plastic-framed goggles, equipped with approved impact-resistant glass or plastic lenses. Goggles should be carefully fitted to ensure maximum protection and comfort. Exposure to the skin can be minimized by wearing rubber gloves when handling sulfamic acid and its solutions, and hands should be washed thoroughly after handling. Breathing of the dust should be avoided and adequate ventilation should be provided.

In case of eye contact, the eyes should be flushed immediately with plenty of water for at least 15 min and a physician should be consulted. Exposed skin should also be flushed with plenty of water. Anyone who has ingested the acid should immediately drink large amounts of water and a physician should be called. Vomiting should not be induced.

Toxicity data for the acid are as follows: oral LD_{50} (rats), 1600 mg/kg and oral LD_{50} (mice), 3100 mg/kg (52). There are a number of regulations pertaining to the use of sulfamic acid in the cleaning of food-processing equipment and in the manufacture of food packages and packaging materials (53). However, because of possible changes of or additions to regulations, procedures or up-to-date requirements should always be checked, including hazard classification, labeling, food-use clearances, worker-exposure limitations, and waste-disposal limitations.

Sulfamic acid is approved for use in all departments under the Meat, Poultry, Rabbit, and Egg Products Inspection Program of the USDA. Prior to the use of sulfamic acid as a cleaner, food products and packaging materials should be removed from the room or carefully protected; after use, all surfaces must be rinsed with potable water. Sulfamic acid migration to food from paper and paperboard products used in food packaging is generally recognized as safe when used in approved applications as described (see Packaging, industrial). This applies to both aqueous and fatty foods.

Uses

Sulfamic Acid. ***Removal of Residues from Industrial Processing Equipment.*** Properties of sulfamic acid that make it particularly well-suited for scale-removal operations and chemical cleaning include the following: the acid is available in dry form, permitting convenient transportation, storage, handling, and packaging operations; it is a strong acid in aqueous solutions and is very effective in solubilizing hard-water scales; sulfamic acid is nonvolatile and chloride-free; aqueous solutions do not emit objectionable or corrosive fumes and can be used on stainless steel with no problem of chloride stress corrosion; it is readily reactive with most deposits to form highly water-soluble compounds; this property promotes good rinsibility and minimizes redeposition of dissolved solids; it is less corrosive than many other strong acids on most materials of construction; it is also compatible with corrosion inhibitors, wetting agents, pH indicators, and other components of dry-cleaning formulations; it does not have a high toxicity rating. Even so, all suppliers recommend caution in handling and personal protection to avoid any injury (see Water, industrial water treatment).

Acid cleaners based on sulfamic acid are used in a large variety of applications, eg, air-conditioning (qv) systems; marine equipment, including salt water stills; wells (water, oil, and gas); household equipment, eg, copperware, steam irons, humidifiers, dishwashers, toilet bowls, and brick and other masonry; dairy equipment, eg, pasteurizers, evaporators, preheaters, and storage tanks; industrial boilers, condensers, heat exchangers, and preheaters; food-processing (qv) equipment; brewery equipment (see Beer); sugar evaporators; and paper-mill equipment (see also Metal surface treatments; Evaporation; Pulp).

Manufacture of Dyes and Pigments and pH Adjustment. Use of sulfamic acid in the manufacture of dyes and pigments involves removal of excess nitrite from diazotization reactions and is based on the reaction

$$NO_2^- + HSO_3NH_2 \rightarrow HSO_4^- + N_2 + H_2O$$

(see Azo dyes). Sulfamic acid is also used in some dyeing operations for pH adjustment; however, it is useful in lowering pH levels in a variety of other systems. The low pH persists at elevated temperatures and there are no objectionable fumes.

Paper-Pulp Bleaching. Sulfamic acid additions to chlorination bleaching stages are effective in reducing pulp-strength degradation associated with high temperatures (54) (see Bleaching agents). Other benefits are noted when sulfamic acid is added to the hypochlorite bleaching stage (55), including: reduction of pulp-strength losses as a result of high temperature or low pH; increased production by means of higher temperatures and lower pHs at the same pulp-strength level; savings in chemical costs, eg, lower consumption of buffer, caustic soda, and higher priced bleaching agents; and improved efficiency through reducing effects of variation in temperature and pH.

Chlorine Vehicle and Stabilizer. Sulfamic acid reacts with hypochlorous acid to produce *N*-chlorosulfamic acids, compounds in which the chlorine is still active but more stable than in hypochlorite form. The commercial interest in this area is for chlorinated water systems in paper mills, ie, for slimicides, cooling towers, and similar applications (56) (see Industrial antimicrobial agents).

Manufacture of Sulfamates. Chief salts manufactured from sulfamic acid in the United States are the ammonium and nickel sulfamates.

Analytical and Laboratory Operations. Sulfamic acid has been recommended as a reference standard in acidimetry (57). It can be purified by recrystallization to give a stable product that is 99.95 wt % pure. The reaction with nitrite as used in the sulfamic acid analytical method has also been adapted for determination of nitrites with the acid as the reagent. This reaction is used commercially in other systems for removal of nitrous acid impurities, eg, in sulfuric and hydrochloric acid purification operations.

Sulfation and Sulfamation. Sulfamic acid can be regarded as an ammonia-SO_3 complex and has been used commercially and always in anhydrous systems. Sulfation of mono, ie, primary and secondary, alcohols; polyhydric alcohols; unsaturated alcohols; phenols; and phenolethylene oxide condensation products has been performed with sulfamic acid (see Sulfonation and sulfation). The best known application of sulfamic acid for sulfamation is the preparation of sodium cyclohexylsulfamate [*139-05-9*], which is a synthetic sweetener (see Sweeteners).

Sulfamates. ***Ammonium Sulfamate.*** A number of flame retardants (qv) used for cellulosic materials, including fabrics and paper products, are based on ammonium sulfamate (58). These products are water-soluble and, therefore, nondurable if treated fabrics are washed or exposed to weathering conditions. For most fabric and paper constructions, efficient flame retardancy can be provided with no apparent effect on color or appearance and without stiffening or adverse effects on the hand of fabrics. A wide variety of materials are treated, including hazardous work area clothing, drapes, curtains, decorative materials, blankets, sheets, and specialty industrial papers.

Ammonium sulfamate is highly effective in nonselective products to control weeds, brush, stumps, and trees (59) (see Herbicides).

Other. Nickel sulfamate is made by the combination of nickel carbonate and sulfamic acid. It is almost exclusively used in the plating industry, with its solutions used for both plating and electroforming. The principal value of this system is low internal stress in deposits and high plating rates (51). Other sulfamates used in plating solutions include the respective salts of cobalt, cadmium, ferrous iron, and lead (see Electroplating). Ferrous sulfamate is used in nuclear fuel processing solutions (60). Certain amine sulfamates impart softening properties to papers and textiles (61). Such materials exhibit long effective service, particularly at low humidity (see Quaternary ammonium compounds).

BIBLIOGRAPHY

"Sulfamic Acid" in *ECT* 1st ed., Vol. 13, pp. 285–294, by Gilberta G. Torrey, E. I. du Pont de Nemours & Co., Inc., Grasselli Chemicals Department; "Sulfamic Acid and Sulfamates" in *ECT* 2nd ed., Vol. 19, pp. 242–249, by D. Santmyers and R. Aarons, E. I. du Pont de Nemours & Co., Inc.

1. H. Rose, *Pogg. Anal.* **33,** 235 (1834); **42,** 415 (1837); **61,** 397 (1844).
2. E. Berglund, *Ber.* **9,** 1896 (1876).
3. P. Baumgarten and I. Marggraff, *Ber.* **63,** 1019 (1930).
4. P. Baumgarten, *Ber.* **69,** 1929 (1936); U.S. Pat. 2,102,350 (Dec. 14, 1937), P. Baumgarten (to Du-Pont).
5. J. Donnay and H. Ondik, *Crystal Data Determinative Tables*, Vol. 2, 3rd ed., U.S. Dept. of Commerce, National Bureau of Standards, Joint Committee on Powder Standards, Washington, D.C., 1975, pp. O–202.

6. A. Cameron and F. Duncanson, *Acta Crystallogr.* **1332,** 1563 (1976).
7. P. G. Sears and co-workers, *J. Inorg. Nucl. Chem.* **35,** 2087 (1973).
8. M. E. Cupery, *Ind. Eng. Chem.* **30,** 627 (June 1938).
9. J. Kurtz and J. Farrar, *J. Am. Chem. Soc.* **91,** 6057 (1969); G. Nash, *J. Chem. Eng. Data,* **13,** 271 (1968).
10. J. K. Hunt, *Chem. Eng. News* **30,** 707 (1952); *Chem. Week,* 69 (1951); J. Notley, *J. Appl. Chem. Biotechnol.* **23,** 717 (Oct. 10, 1973); A. Tsypin and E. Fomenko, *Tr. Gos. Nauchno-Issled. Proektn. Inst. Azotn. Promsti. Prod. Org. Sint.* **27,** 21 (1974).
11. F. L. Hahn and P. Baumgarten, *Ber.* **63,** 3028 (1930).
12. E. Divers and T. Haga, *J. Chem. Soc.* **69,** 1634 (1896).
13. F. Ephraim and E. Lasocky, *Ber.* **44,** 395 (1911); P. Baumgarten, *Ber.* **71,** 80 (1938).
14. W. Traube and E. von Drathen, *Ber.* **51,** 111 (1918).
15. F. Ephraim and M. Gurewitsch, *Ber.* **43,** 138 (1910).
16. H. Denivelle, *Bull. Soc. Chim.* **3,** 2150 (1936).
17. P. Sakellaridis, *Bull. Soc. Chem. Biol.* **9–10,** 610 (Sept./Oct. 1951).
18. K. Hofmann and co-workers, *Ber.* **45,** 1731 (1912).
19. O. Tubertini and co-workers, *Ann. Chim. (Rome)* **57,** 555 (1967); R. Piontelli and co-workers, *Electrochim. Met.* **3**(1); **42**(4), (1968); **2**(2), 141 (1967); A. LaVecchia, *Electrochim. Met.* **3**(1), 71 (1968); *Symposium on Sulfamic Acid and Its Electrometallurgic Applications (Proceedings),* Polytechnic School of Milan, Milan, Italy, May 25, 1966.
20. U.S. Pat. 1,931,962 (Oct. 24, 1933) and Ger. Pat. 558,296 (Aug. 22, 1930), K. Marx, K. Brodersen, and M. Quaedvlieg (to I. G. Farben).
21. Ger. Pat. 565,040 (Nov. 25, 1932), K. Brodersen and M. Quaedvlieg (to I. G. Farben).
22. W. N. Carton, *Anal. Chem.* **23,** 1016 (1951); H. Dietz and co-workers, *Agric. News Lett. E. I. du Pont de Nemours & Co., Inc.* **9,** 35 (1941).
23. U.S. Pat. 2,452,943 (Nov. 2, 1948), J. Malkemus and co-workers (to Colgate Palmolive Peet); Ger. Pat. 3,372,170 (Oct. 26, 1963) H. Remy (to Hoechst); U.S. Pat. 3,395,170 (June 28, 1966), J. Walts and L. Schenck (to General Aniline & Film).
24. K. Hofmann and E. Biesalski, *Ber.* **45,** 1394 (1912).
25. A. Quilico, *Gazz. Chim. Ital.* **37,** 793 (1927); *Atti. Accad. Lincei* **141,** 512 (1927).
26. S. Loktev and co-workers, *Abh. Dtsch. Akad. Wiss. Berlin Kl. Chem. Geol. Biol.* **6,** 107 (1967).
27. J. Boivin, *Can. J. Res. Sect. B* **28,** 671 (1950).
28. L. Goodson, *J. Am. Chem. Soc.* **69,** 1230 (1946).
29. U.S. Pat. 2,259,563 (Oct. 21, 1941), W. Hill (to American Cyanamid).
30. Brit. Pat. 372,389 (April 28, 1932), (to I. G. Farben).
31. L. Audrieth, M. Sveda, H. Sisler, and M. Butler, *Chem. Rev.* **26,** 49 (1940).
32. K. Andersen, *Comprehensive Organic Chemistry,* Vol. 3, Pergamon Press, Oxford, UK, 1979, p. 363.
33. E. Berglund, *Bull. Soc. Chim.* **29,** 422 (1878); *Lunds Univ. Acta* **13,** 4 (1875).
34. P. Eitner, *Ber.* **26,** 2836 (1892).
35. F. Ephraim and W. Flugel, *Helv. Chim. Acta* **7,** 724 (1924).
36. C. Paal and F. Kretschmer, *Ber.* **27,** 1241 (1894).
37. A. Callegari, *Gazzetta* **36**(2), 63 (1906).
38. M. Delepine and R. Demars, *Bull. Sci. Pharmacol.* **29,** 14 (1922).
39. A. Quilico, *Gazz. Chim. Ital.* **56,** 620 (1926).
40. C. Paal and H. Janicke, *Ber.* **28,** 3160 (1895); C. Paal and S. Daybeck, *Ber.* **30,** 880 (1897).
41. C. Paal, *Ber.* **34,** 2748 (1901).
42. C. Paal and L. Lowitsch, *Ber.* **30,** 869 (1897).
43. G. Thies, dissertation, Frederick-Wilhelms University, Berlin, 1935.
44. C. Paal and M. Hubaleck, *Ber.* **34,** 2757 (1901).
45. U.S. Pat. 2,880,064 (March 31, 1959), M. Harbaugh and G. Pierce (to E. I. du Pont de Nemours & Co., Inc.); U.S. Pat. 2,191,754 (March 27, 1940), M. Cupery (to E. I. du Pont de Nemours & Co., Inc.).
46. K. Toyokura and co-workers, *J. Chem. Eng. Jpn.* (Feb. 1979); Jpn. Pat. 70 24,649 (Nov. 27, 1967), S. Ito (to Bur. Ind. Tech.).
47. Ger. Pat. 2,637,948 (Aug. 24, 1976), R. Graeser and co-workers (to Hoechst); Ger. Pat. 2,106,019 (Feb. 9, 1971), R. Graeser and co-workers (to Hoechst); Brit. Pat. 1,068,942 (Aug. 9, 1962), W. Morris (to Seery, Defense); Brit. Pat. 1,062,329 (Dec. 22, 1964), A. Sowerby (to Marchon Products).
48. U.S. Pat. 2,487,480 (Nov. 8, 1949), G. Rohrmann (to E. I. du Pont de Nemours & Co., Inc.).

49. Ger. Pat. 1,915,723 (March 27, 1969), H. Hofmeister (to Hoechst); Ger. Pat. 2,850,903 (Nov. 24, 1978), G. Muenster (to Hoechst).
50. Jpn. Pat. 71 00,816 (Jan. 21, 1964), N. Sasaki and co-workers (to Mitsui Toatsu); Jpn. Pat. 69 28,374 (Aug. 28, 1964), Yamaguchi and co-workers (to Nitto Chem.); Jpn. Pat. 34 84,193 (April 10, 1964), Azakmi and co-workers (to Toyo Koatsu).
51. *Metal Finishing, 1981 Guidebook and Directory*, Metal and Plastics Publications, Inc., Hackensack, N.J., p. 286.
52. *NIOSH 1979 Registry of Toxic Effects of Chemical Substances*, Vol. 2, U.S. Department of Health, and Human Services, Washington, D.C., p. 286.
53. List of chemical compounds authorized for use in the *Meat and Poultry Inspection Program MPI-8*, U.S. Department of Agriculture, Washington, D.C., March 1977; Code of Federal Regulations (FDA) 21 CFR182.90; 21 CFR 176.170(a)(2); 38 CFR 11077.
54. R. Tobar, *TAPPI* **47,** 688 (1964); U.S. Pat. 3,308,012 (March 7, 1967), R. Tobar (to E. I. du Pont de Nemours & Co., Inc.).
55. U.S. Pat. 3,177,111 (April 6, 1965), L. Larson (to Weyerhaeuser).
56. U.S. Pat. 3,328,294 (June 27, 1967), R. Self and co-workers (to Mead); U.S. Pat. 3,749,672 (July 31, 1973), W. Golton and A. Rutkiewic (to E. I. du Pont de Nemours & Co., Inc.); U.S. Pat. 3,767,586 (Oct. 23, 1973), A. Rutkiewic (to E. I. du Pont de Nemours & Co., Inc.); U.S. Pat. 3,170,833 (Feb. 23, 1965), R. Owen and S. Thomas (to Cortez Chem.).
57. M. Caso and M. Cefola, *Anal. Chim. Acta* **21,** 205 (1959).
58. U.S. Pat. 2,723,212 (Nov. 8, 1955), R. Aarons and P. Wilson (to E. I. du Pont de Nemours & Co., Inc.).
59. U.S. Pat. 2,277,744 (March 3, 1942), M. Cupery and A. Tanberg (to E. I. du Pont de Nemours & Co., Inc.); U.S. Pat. 2,368,274–276 (Jan. 30, 1945), R. Torley (to American Cyanamid); U.S. Pat. 2,709,648 (May 31, 1955), M. Ryker and P. Wolf (to E. I. du Pont de Nemours & Co., Inc.).
60. N. Bibler, *Nucl. Technol.* **34** (Aug. 1977); L. Gray, *Nucl. Technol.* **40** (Sept. 1978); R. Walser, *U.S. Atomic Energy Commission Report*, ARH-SA-69, 1970.
61. F. Blakemore, *Am. Paper Converter*, **23**(6), 29 (1949); U.S. Pat. 2,526,462 (Oct. 17, 1950), O. Edelstein (to Pond Lily).

ELMER B. BELL
E. I. du Pont de Nemours & Co., Inc.

SULFANILAMIDE. See Antibacterial agents, synthetic, sulfonamides.

SULFANILIC ACID, p-$H_2NC_6H_4SO_3H$. See Sulfonation and sulfation; Amines, aromatic, aniline and its derivatives.

SULFATED ACIDS, ALCOHOLS, OILS, ETC. See Sulfonation and sulfation; Surfactants and detersive systems.

SULFATION. See Sulfonation and sulfation.

SULFIDES. See Sulfur compounds.

SULFITE PROCESS. See Pulp.

SULFITES. See Barium compounds; Sulfur compounds, sulfuric and sulfurous esters.

SULFOALKYLATION. See Sulfonation and sulfation.

SULFOCHLORINATION. See Sulfonation and sulfation; Sulfurization and sulfur chlorination.

SULFOLANES AND SULFONES

Physical Properties

Sulfolane [*126-33-0*], $C_4H_8SO_2$, is the most common of the few commercially available sulfones. Also known as tetrahydrothiophene-1,1-dioxide and tetramethylene sulfone, sulfolane is a colorless, highly polar, water-soluble compound and has the physical properties given in Table 1.

sulfolane

Table 1. Physical Properties of Sulfolane

Property	Value
molecular weight	120.17
boiling point, °C	287.3
melting point, °C	28.5
specific gravity, 30/30°C	1.266
100/4°C	1.201
density (at 15°C), g/cm^3	1.276
flash point, °C	165–178
viscosity, mPa·s (= cP)	
at 30°C	10.3
at 100°C	1.4
at 200°C	1.0
refractive index, n_D (at 30°C)	1.48
heat of fusion, kJ/kg[a]	11.44
dielectric constant	43.3
surface tension (at 30°C), mN/m (= dyn/cm)	35.5

[a] To convert J to cal, divide by 4.184.

Since sulfolane was first described in the chemical literature in 1916, it has been noted for its exceptional chemical and thermal stability and unusual solvent properties. Research for a commercial synthetic route to sulfolane began ca 1940, and market development quantities were available in 1959. Since then, the use of and applications for sulfolane have risen dramatically.

Chemical Properties

The thermal stability of sulfolane in the presence of various chemical substances is given in Table 2. With the exception of sulfur and aluminum chloride, sulfolane is relatively unreactive chemically. Despite its relative chemical inertness, sulfolane undergoes reactions, eg, halogenations, ring cleavage by alkali metals, ring additions catalyzed by alkali metals, reaction with Grignard reagents, and formation of weak chemical complexes.

Halogenation. Chlorine can be added to sulfolane by a uv-initiated process to give 3-chloro- [*3844-04-0*]; 3,4-dichloro- [*3001-57-8*]; and 3,3,4-trichlorosulfolane [*42829-14-1*] (2–3).

Bromination of sulfolane by BrCl under uv irradiation gives 2-bromosulfolane [*29325-66-4*] which reacts further to give *cis*-2,5-dibromosulfolane [*30186-52-8*] (4). Continued irradiation converts the cis-isomer to *trans*-2,5-dibromosulfolane [*30186-54-0*] which yields first the trans-2,4 isomer [*30186-53-9*] and then the trans-3,4-isomer [*15091-30-2*] upon further irradiation.

Ring Cleavage by Alkali Metals. The sulfolane ring can be cleaved by sodium or potassium metals in xylene at 66–140°C, by sodium amide in liquid ammonia at −33°C, and by sodium ethoxide at 240–250°C (5). The reaction products of the alkali-metal ring cleavage are dimeric bis-1,8-octanesulfinate salts under static reaction conditions and butanesulfinate salts under stirred reaction conditions. Ring cleavage by sodium amide in liquid ammonia gives sodium 3-butenesulfinate, and ring cleavage by sodium ethoxide gives sodium butadienesulfinate.

Ring Additions Catalyzed by Alkali Metals. The addition of tributyltin chloride and olefins, eg, styrene, isoprene, and butadiene, to sulfolane is catalyzed by alkali metals, eg, sodium, lithium, and sodium amide (6–9). The addition of tributyltin

Table 2. Thermal Stability of Sulfolane in the Presence of Various Substances[a]

Substance	Wt % based on sulfolane	Reflux time, h	Sulfolane recovery, wt %	Remarks
none		5	>98	darkened after 30 min
aluminum chloride	40	5	56	blackened with evolution of heat and HCl
potassium carbonate	40	5	>98	darkened after 30 min
sodium acetate	40	5	>98	darkened after 30 min
sodium hydroxide (25 wt %)	100	6	>98	darkened after 2 h
sulfur	10	7	25	H_2S evolved, black sludge as residue
sulfuric acid (93 wt %)	25	6[b]	89	darkened after 30 min

[a] Ref. 1.

[b] At 140–150°C.

chloride to sulfolane in the presence of sodium amide results in the formation of 2,5-bis(tributyltin)sulfolane [*41392-14-7*]. The addition of styrene to sulfolane in the presence of sodium yields 67% monostyrenated and 17% distyrenated sulfolane. Under similar conditions, isoprene gives 63% mono- and 7% disubstituted product.

Grignard Reactions. Sulfolane and its alkyl homologues react with ethylmagnesium bromide in ether, benzene, and tetrahydrofuran (10–11). The reaction products are sulfolanyl 2-mono- [*82770-58-9*] and 2,5-dimagnesium bromides [*82770-59-0*].

Lewis Acid Complexes. Sulfolane complexes with Lewis acids, eg, boron trifluoride and phosphorus pentafluoride (12). For example, at room temperature, sulfolane and boron trifluoride combine in a 1:1 mole ratio with the evolution of heat to give a white, hygroscopic solid which melts at 37°C.

Production and Economic Aspects

In early 1981, sulfolane was produced domestically by the Phillips Chemical Company at Borger, Texas, and by the Shell Chemical Company at Norco, Louisiana. Industrially, sulfolane is synthesized by hydrogenating 3-sulfolene [*77-79-2*] (2,5-dihydrothiophene-1,1-dioxide), which is the reaction product of butadiene and sulfur dioxide as shown below:

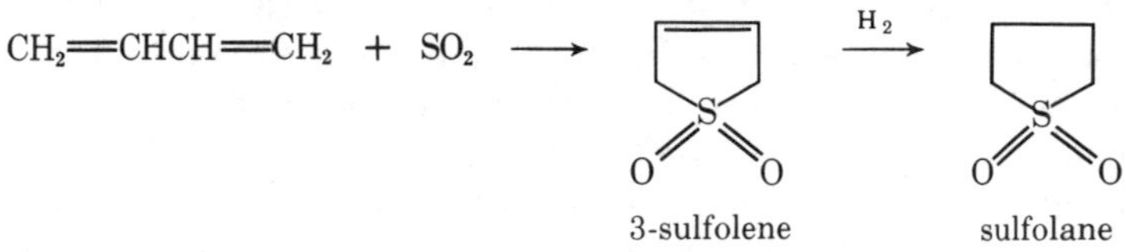

$$CH_2{=}CHCH{=}CH_2 + SO_2 \longrightarrow \text{3-sulfolene} \xrightarrow{H_2} \text{sulfolane}$$

3-sulfolene sulfolane

Commercially, sulfolane is available as anhydrous sulfolane and as sulfolane containing 3 wt % deionized water. The estimated domestic production capacity for sulfolane is ca 7300 metric tons per year. The price history of Phillips sulfolane (contains 3 wt % deionized water) from 1973–1980 was: 1973, $1.43/kg; 1974, $1.54/kg; 1975, $1.70/kg; 1976, $1.98/kg; 1977, $2.18/kg; 1978, $2.54/kg; 1979, $2.76/kg; and 1980, $3.48/kg.

Toxicity

No mortality in laboratory animals was induced in percutaneous doses up to 3.8 g/kg body weight (13–14). Subcutaneous administration of sulfolane gives LD_{50} values for rats, mice, and rabbits of 1.606, 1.360, and 1.900–3.500 g/kg body weight, respectively (15). LD_{50} values for sulfolane by oral administration to laboratory animals were 1.9–5.0 g/kg body weight (13–16). In most cases, the cause of death was believed to result from convulsions which lead to anoxia. When administered intraperitoneally, sulfolane is excreted both unchanged and as 3-hydroxysulfolane [*13031-76-0*] (17).

Sulfolane causes minimal and transient eye and skin irritation (13–14). Inhalation of sulfolane vapors in a saturated atmosphere is not considered biologically significant. However, when aerosol dispersions have been used to elevate atmospheric concentration, blood changes and convulsions have been observed in laboratory animals (15,18).

Uses

Extractive Solvent. ***Aromatic Hydrocarbons.*** Sulfolane is used principally as a solvent for extraction of benzene, toluene, and xylene from mixtures with aliphatic hydrocarbons (19–23) (see BTX processing). The sulfolane process was introduced in 1959 by Shell Development Company, and that process is licensed by Universal Oil Products. A sulfolane extraction process is also licensed by the Atlantic Richfield Company. In 1980, 55 sulfolane units operated worldwide and these accounted for ca 1600 t/yr of sulfolane consumption (see Extraction).

In general, the sulfolane extraction unit consists of four basic parts: extractor, extractive stripper, extract recovery column, and water-wash tower. The hydrocarbon feed is first contacted with sulfolane in the extractor where the aromatics and some light nonaromatics dissolve in the sulfolane. The rich solvent then passes to the extractive stripper where the light nonaromatics are stripped. The bottom stream, which consists of sulfolane and aromatic components, which are essentially free of nonaromatics, enters the recovery column where the aromatics are removed. The sulfolane is returned to the extractor. The nonaromatic raffinate obtained initially from the extractor is contacted with water in the wash tower to remove dissolved sulfolane, which is subsequently recovered in the extract recovery column. Benzene and toluene recoveries in the process are routinely over 99%, and xylene recoveries exceed 95%.

Normal and Branched Aliphatic Hydrocarbons. The urea-adduction method for separating normal and branched aliphatic hydrocarbons can be carried out in sulfolane (24–25). The process obviates the necessity of handling and washing the solid urea–normal-paraffin adduct formed when a solution of urea in sulfolane is contacted with the hydrocarbon mixture. Overall recovery by this process is 85% with a normal-paraffin purity of 98% (see Clathration).

Fatty Acids and Fatty Acid Esters. Sulfolane exhibits selective solvency for fatty acids and fatty acid esters depending on the molecular weight and degree of unsaturation (26–27). Applications for this process are enriching the unsaturation level in animal and vegetable fatty oils to provide products with better properties for use in paints, synthetic resins, food products, plastics, and soaps.

Wood Delignification. The production of wood pulp for the paper industry consists of removing lignin (qv) from wood chips, thus freeing the cellulose fibers. An aqueous solution containing 30–70 wt % sulfolane efficiently extracts the lignin from aspen, western hemlock, and southern pine wood chips with pulp yields of 50–75% (28–29) (see Pulp).

Extractive Distillation Solvent. Extractive distillation is a technique for separating components in narrow boiling range mixtures which are difficult to separate by ordinary fractionation. The process consists of allowing a higher boiling liquid that has a special affinity for one or more of the components in the mixture to flow downward in a distillation column countercurrent to the ascending vapors and thereby to enhance volatility differences of the mixture components. Sulfolane is a suitable extractive-distillation solvent for carrying out the separation of close boiling alcohols; mono- and diolefins, eg, isoprene and butadiene; electrochemical fluorination products; and aromatic hydrocarbons (30–38) (see Azeotropic and extractive distillation).

Gas Treating. The second largest commercial use for sulfolane is the removal of acidic components, eg, H_2S, CO_2, COS, CS_2, and mercaptans, from sour gas streams.

The process, known as the Sulfinol process, was introduced in 1963 and consists of contacting the gas stream with a mixture of sulfolane, an alkanolamine (usually diisopropylamine), and water (39–43) (see Alkanolamines; Carbon dioxide). The acid gases are absorbed chemically by the amine and absorbed physically by the sulfolane, which results in the advantages of high acid-gas loading and ease of solvent regeneration.

Other gas-treating processes involving sulfolane are atmospheric CO_2 removal in nuclear submarines, ammonia and H_2S removal from waste streams, and H_2S and SO_2 removal from gas mixtures (44–46). The latter process differs from the Sulfinol process in that the H_2S and SO_2 are converted to high purity, elemental sulfur (see Gas cleaning).

Polymer Solvent. Sulfolane is a solvent for a variety of polymers, eg, polyacrylonitrile, poly(vinylidene cyanide), poly(vinyl chloride), poly(vinyl fluoride), and polysulfones (47–51). Sulfolane solutions of polyacrylonitrile, poly(vinylidene cyanide), and poly(vinyl chloride) have been patented for fiber-spinning processes where the relatively low solution viscosity, good thermal stability, and comparatively low solvent toxicity of sulfolane are advantageous.

Polymer Plasticizer. Nylon, cellulose, and cellulose esters can be plasticized with sulfolane to improve the flexibility and to increase the elongation of the polymer (52–53). More importantly, sulfolane is a preferred plasticizer for the synthesis of cellulose hollow fibers, which are used as permeability membranes in reverse-osmosis (qv) cells (53–55). In the preparation of the hollow fibers, a molten mixture of sulfolane and cellulose triacetate are extruded through a die to form the hollow fiber. The sulfolane is subsequently extracted from the fiber with water to give a permeable, plasticizer-free, hollow fiber (see Hollow-fiber membranes).

Polymerization Solvent. Sulfolane can be used alone or in combination with a cosolvent as a polymerization solvent for polyureas, polysulfones, polysiloxanes, polyether polyols, polybenzimidazoles, polyphenylene ethers, and poly(1,4-benzamide) (poly(imino-1,4-phenylenecarbonyl)) (56–62). Advantages of using sulfolane as a polymerization solvent include increased polymerization rate, ease of polymer purification, better solubilizing characteristics, and improved thermal stability. The increased polymerization rate has been attributed not only to an increase in the reaction temperature because of the higher boiling point of sulfolane, but also to a decrease in the activation energy of polymerization as a result of the contribution from the sulfonic group of the solvent.

Miscellaneous. Because of its high dielectric constant, low volatility, and solubilizing characteristics, sulfolane has been patented for use in a wide variety of electronic and electrical applications, eg, as a coil-insulating component, battery solute, solvent in electronic display devices, capacitor impregnants, and solvent in electroplating baths (65–67). Textile applications for sulfolane include concentrated, storage-stable basic dyes; fabric treating prior to dyeing to improve the dye adsorption; and fiber treating to improve the tensile strength, pilling resistance; and drawing properties (68–70). The curing time of polysulfide-based sealants and fluoropolymer rubbers decrease significantly upon the incorporation of small amounts of sulfolane into the formulation (71–72). Sulfolane also exhibits catalytic activity for some reactions increasing both the conversion and the selectivity. Examples of systems where sulfolane functions catalytically are the synthesis of 1,4-dicyanobutene and substituted ketones (73–74).

Table 3. Physical Properties of 3-Sulfolene

Property	Value
molecular weight	118.154
specific gravity (at 70°C)	1.314
melting point, °C	65
boiling point	decomposes
flash point, °C	113

Sulfones

3-Sulfolene (2,5-dihydrothiophene-1,1-dioxide) is the next commercially important sulfone after sulfolane. Besides its precursor role in sulfolane manufacture, 3-sulfolene is an intermediate in the synthesis of sulfolanyl ethers, which are used as hydraulic-fluid additives (see Hydraulic fluids). 3-Sulfolene or its derivatives also have been used in cosmetics (qv) and slimicides. Selected physical properties of 3-sulfolene are listed in Table 3.

Other sulfones of commercial potential include dimethyl sulfone [*67-71-0*] (sulfonylbismethane), diiodomethyl *p*-tolyl sulfone [*20018-09-1*], and 4,4′-dihydroxydiphenyl sulfone [*80-09-1*] (Bisphenol S). Dimethyl sulfone has been patented as an extractive distillation solvent, an electroplating-bath solvent, and an ink and adhesive solvent (33,75–77). Diiodomethyl *p*-tolyl sulfone has been patented for use as an antifungal preservative (78–80); and 4,4′-dihydroxydiphenyl sulfone has been patented for a variety of applications, eg, an electroplating solvent, a washfastening agent, and a component in a phenolic resin (81–82) (see also Polymers containing sulfur, polysulfone resins).

$CH_3S(=O)_2CH_3$

dimethyl sulfone

$CH_3-C_6H_4-S(=O)_2-CHI_2$

diiodomethyl *p*-tolyl sulfone

$HO-C_6H_4-S(=O)_2-C_6H_4-OH$

Bisphenol S

BIBLIOGRAPHY

"Sulfolane" in *ECT* 2nd ed., Vol. 19, pp. 250–254, by G. S. Morrow, Shell Chemical Company.

1. T. E. Jordon and F. Kipnis, *Ind. Eng. Chem.* **41,** 2635 (1949).
2. V. I. Dornov and co-workers, *Khim. Seraorg. Soedin. Soderzh. Naftyakh Nefteprod.* **8,** 133 (1968).
3. V. I. Dornov and V. A. Snegotskaya, *Khim. Geterotsikl. Soedin. Sb.* **3,** 5 (1971).
4. V. I. Dornov and co-workers, *Zh. Org. Khim.* **6,** 2029 (1970).
5. V. I. Dornov and co-workers, *Khim. Seraorg. Soedin. Soderzh. Neftyakh Nefterprod.* **8,** 144 (1968).
6. Ger. Offen. 2,246,939 (April 5, 1973), D. J. Peterson, J. F. Ward, and R. A. D'Amico (to Procter and Gamble Co.).
7. U.S. Pat. 3,988,144 (Oct. 26, 1976), D. J. Peterson, J. F. Ward, and R. A. D'Amico (to Procter and Gamble Co.).
8. E. M. Asatryan and co-workers, *Arm. Khim. Zh.* **29,** 553 (1976); *Chem. Abstr.* **86,** 5246v (1977).

9. E. M. Asatryan and co-workers, *Tezisy Dokl. Molodezhnaya Konf. Org. Sint. Bioorg. Khim.* **44,** (1976).
10. T. E. Bezmenova and co-workers, *Khim. Seraorg. Soedin. Soderzh. Neftyakh Nefteprod.* **8,** 140 (1968).
11. U.S. Pat. 2,656,362 (Oct. 26, 1953), H. E. Faith (to Allied Laboratories, Inc.).
12. J. J. Jones, *Inorg. Chem.* **5,** 1229 (1966).
13. V. K. H. Brown, L. W. Ferrigan, and D. E. Stevenson, *Br. J. Industr. Med.* **23,** 302 (1966).
14. H. F. Smyth and co-workers, *Am. Ind. Hyg. Assoc. J.* **30,** 470 (1969).
15. M. E. Andersen and co-workers, *Toxicol. Appl. Pharmacol.* **40,** 463 (1977).
16. J. J. Roberts and G. P. Warwick, *Biochem. Pharmacol.* **6,** 217 (1961).
17. G. R. Middleton, R. W. Young, and L. J. Jenkins, Jr., *Annual Research Report Armed Forces Radiobiology Research Institute*, National Technical Information Service, Springfield, Va., 1974, pp. 77–79.
18. Z. K. Filippova, *Khim. Seraorg. Soedin. Soderzh. Neftyakh. Nefteprod.* **8,** 701 (1968).
19. D. B. Broughton and G. F. Asselin, *Seventh World Petroleum Congress Proceedings* **4,** 65 (1968).
20. F. S. Beardmore and W. C. G. Kosters, *J. Inst. Pet. London* **49**(469), 1 (1963).
21. M. A. Plummer, *Hydrocarbon Process.* **52**(6), 91 (1973).
22. M. A. Plummer, *Hydrocarbon Process.* **59**(9), 203 (1980).
23. *Aromatics Extraction Process*, EM-9964, Atlantic Richfield Company.
24. U.S. Pat. 3,617,499 (Nov. 2, 1971), D. M. Little (to Phillips Petroleum Co.).
25. U.S. Pat. 3,645,889 (Feb. 29, 1972), D. O. Hanson (to Phillips Petroleum Co.).
26. U.S. Pat. 2,360,860 (Oct. 24, 1944), R. C. Morris and E. C. Shokal (to Shell Development Co.).
27. J. Wisniak, *Br. Chem. Eng.* **15**(1), 76 (1970).
28. L. P. Clermont, *TAPPI* **53,** 2243 (1970).
29. L. D. Starr and co-workers, *TAPPI Alkaline Pulping Conference Preprints* **1975,** 195.
30. U.S. Pat. 2,570,205 (Oct. 9, 1951), C. S. Carlson, E. Smith, and P. V. Smith, Jr. (to Standard Oil Development Co.).
31. B. S. Rawat, K. L. Mallik, and I. B. Gulati, *J. Appl. Chem. Biotechnol.* **22,** 1001 (1972).
32. U.S. Pat. 3,024,028 (May 17, 1977), D. M. Haskell (to Phillips Petroleum Co.).
33. U.S. Pat. 4,054,613 (Oct. 18, 1977), D. M. Haskell, E. E. Hopper, and B. L. Munro (to Phillips Petroleum Co.).
34. U.S. Pat. 4,090,923 (May 23, 1978), D. M. Haskell and C. O. Carter (to Phillips Petroleum Co.).
35. U.S. Pat. 3,689,373 (May 5, 1972), W. M. Hutchinson (to Phillips Petroleum Co.).
36. Brit. Pat. 1,333,039 (Oct. 10, 1973), M. F. Kelly and K. D. Uitta (to Universal Oil Products).
37. Can. Pat. 962,212 (Feb. 4, 1975), H. L. Thompson (to Universal Oil Products).
38. Brit. Pat. 1,392,735 (May 26, 1975), H. L. Thompson (to Universal Oil Products).
39. C. L. Dunn, *Hydrocarbon Process. Pet. Refiner* **43,** 150 (1964).
40. U.S. Pat. 3,630,666 (Dec. 28, 1971), L. V. Kunkel (to Amoco Production Co.).
41. U.S. Pat. 4,025,322 (May 24, 1977), E. J. Fisch (to Shell Oil Co.).
42. U.S. Pat. 4,100,257 (July 11, 1978), G. Sartori and D. W. Savage (to Exxon Research and Engineering Co.).
43. U.S. Pat. 3,965,244 (June 22, 1976), J. A. Sykes, Jr. (to Shell Oil Co.).
44. P. R. Gustafson and R. R. Miller, *Naval Research Laboratories Report 6926*, July 1969.
45. U.S. Pat. 3,551,102 (Dec. 29, 1970), G. R. Hettick and D. M. Little (Phillips Petroleum Co.).
46. Ger. Pat. 2,108,284 (Aug. 31, 1972), G. Schulze (to Badische Anilin-und-Soda-Fabrik A.-G.).
47. U.S. Pat. 2,706,674 (April 19, 1955), G. M. Rothrock (to E. I. du Pont de Nemours & Co., Inc.).
48. U.S. Pat. 2,548,169 (April 10, 1951), F. F. Miller (to B. F. Goodrich Co.).
49. U.S. Pat. 2,617,777 (Nov. 11, 1952), E. Heisenberg and J. Kleine (to Vereinigte Glanzstoff-Fabriken).
50. U.S. Pat. 2,953,818 (Sept. 27, 1960), L. R. Barton (to E. I. du Pont de Nemours & Co., Inc.).
51. U.S. Pat. 3,474,030 (Oct. 21, 1969), D. M. Little (to Phillips Petroleum Co.).
52. U.S. Pat. 3,361,697 (Jan. 2, 1968), W. E. Garrison and T. J. Hyde (E. I. du Pont de Nemours & Co., Inc.).
53. U.S. Pat. 2,471,272 (May 24, 1949), G. W. Hooker and N. R. Peterson (to The Dow Chemical Co.).
54. U.S. Pat. 3,619,459 (Nov. 9, 1971), P. G. Schrader (to The Dow Chemical Co.).
55. T. E. Davis and G. W. Skiens, *Polym. Prepr. Am. Chem. Soc. Div. Polym. Chem.* **12,** 378 (1971).
56. U.S. Pat. 3,476,709 (Nov. 4, 1969), F. B. Jones (to Phillips Petroleum Co.).
57. U.S. Pat. 2,703,793 (March 8, 1955), M. A. Naylor, Jr. (to E. I. du Pont de Nemours & Co., Inc.).

58. U.S. Pat. 3,175,994 (March 30, 1965), A. Katchman and G. D. Cooper (to General Electric Co.).
59. Jpn. Pat. 70 06,533 (March 5, 1970), M. Ikeda and co-workers (to Takeda Chemical).
60. U.S. Pat. 3,784,517 (Jan. 8, 1974), F. L. Hedberg and C. S. Marvel (to United States Department of the Air Force).
61. Jpn. Pat. 74 02,359 (Jan. 19, 1974), S. Izawa and K. Mizushiro (to Asahi Dow, Ltd.).
62. U.S. Pat. 3,951,914 (April 20, 1976), S. L. Kwolek (to E. I. du Pont de Nemours & Co., Inc.).
63. Ger. Pat. 2,739,571 (March 23, 1978), L. G. Virsberg (to ASEA AB).
64. U.S. Pat. 3,891,458 (June 24, 1975), M. Eisenberg (to Electrochimica Corp.).
65. U.S. Pat. 4,110,015 (Aug. 29, 1978), T. B. Reddy (to American Cyanamid Co.).
66. R. Tobazeon and E. Gartner, *IEE Conf. Publ. London* **129,** 225 (1975).
67. Ger. Pat. 2,207,703 (Aug. 31, 1972), F. Huba and J. E. Bride (to Diamond Shamrock Corp.).
68. Brit. Pat. 1,349,511 (April 3, 1974), J. F. Dawson and J. Schofield (to Yorkshire Chemicals, Ltd.).
69. Jpn. Pat. 77 25,173 (Feb. 24, 1977), M. Ono and co-workers (to Mitsubishi Rayon Co., Ltd.).
70. Brit. Pat. 1,263,082 (Feb. 9, 1972), R. G. Roberts (to Courtaulds, Ltd.).
71. U.S. Pat. Off. T 974,003 (Sept. 1978), E. G. Miller, Def. Pub.
72. Ger. Pat. 2,449,095 (April 17, 1975), R. E. Kolb (to Minnesota Mining and Manufacturing Co.).
73. Ger. Pat. 2,256,039 (May 17, 1973), O. T. Onsager (to Halcon International Inc.).
74. Jpn. Pat. 74 25,925 (July 4, 1974), T. Kawaguchi and co-workers (to Kurary Co., Ltd.).
75. U.S. Pat. 4,046,647 (Sept. 6, 1977), E. P. Harbulak (to M and T Chemicals, Inc.).
76. Jpn. Pat. 80 50,072 (April 11, 1980), K. Hirano (to Pilot Ink Co., Ltd.).
77. Jpn. Pat. 80 66,980 (May 20, 1980), (to Toa Gosei Chemical Industry Co., Ltd.).
78. U.S. Pat. 4,078,888 (March 4, 1978), F. W. Arbir and F. C. Becker (to Abbott Laboratories).
79. *Fed. Regist.* **44,** 52189 (Sept. 1979).
80. Jpn. Pat. 78 73,434 (June 29, 1978), K. Kariyone and co-workers (to Fujisawa Pharmaceutical Co., Ltd.).
81. Brit. Pat. Appl. 2,001,679 (Feb. 7, 1979), S. A. Lipowski (to Diamond Shamrock Corp.).
82. Jpn. Pat. 79 28,357 (March 2, 1979), J. Hiroshima and I. Kai (to Asahi Organic Chemicals Industry Co., Ltd.).

General References

J. A. Reddick and W. E. Bunger, *Organic Solvents*, Vol. 2, 3rd ed., Wiley Interscience, a division of John Wiley & Sons, New York, 1970, pp. 467–468.
Technical Information on Sulfolane, Bulletin 524, Phillips Chemical Company.

MERLIN LINDSTROM
RALPH WILLIAMS
Phillips Research Center

SULFONAMIDES. See Antibacterial agents, synthetic, sulfonamides.